SYMBOLS

Symbol	Meaning
n	An integer (1, 2, 3, . . .)
n	Principal quantum number
n	Turns per unit length
N	Number of turns
$\mathbf{N}$	Normal (perpendicular) force
p	Object distance
p	Pressure
P	Power
PE	Potential Energy
q	Image distance
Q	Electric charge
Q	Heat
r	Angle of reflection, refraction
r	Internal resistance
r, R	Radius
R	Range
R	Rate of flow
R	Electrical resistance
s	Displacement from equilibrium position
s	Distance; displacement
S	Shear modulus
t	Time
T	Period
T	Temperature
T	Tension
v	Speed
$\mathbf{v}$	Velocity
V	Potential difference
V	Volume
w	Energy density
w	Weight
W	Work
X_C	Capacitive reactance
X_L	Inductive reactance
Y	Young's Modulus
Z	Atomic number
Z	Impedance
α (alpha)	Angular acceleration
α (alpha)	Temperature coefficient of resistivity
β (beta)	Sound intensity level
γ (gamma)	Electromagnetic photon
γ (gamma)	Surface tension
Δ (delta)	"Change in"
η (eta)	Viscosity
θ (theta)	An angle
λ (lambda)	Wavelength
μ (mu)	Coefficient of friction
μ (mu)	Magnetic permeability
ρ (rho)	Density
ρ (rho)	Resistivity
σ (sigma)	Stefan-Boltzmann constant
Σ (sigma)	"Sum of"
τ (tau)	Torque
ϕ (phi)	Phase angle
Φ (phi)	Magnetic flux
ω (omega)	Angular velocity

POWERS OF TEN

10^{-10}	= 0.000,000,000,1	10^{0}	= 1
10^{-9}	= 0.000,000,001	10^{1}	= 10
10^{-8}	= 0.000,000,01	10^{2}	= 100
10^{-7}	= 0.000,000,1	10^{3}	= 1000
10^{-6}	= 0.000,001	10^{4}	= 10,000
10^{-5}	= 0.000,01	10^{5}	= 100,000
10^{-4}	= 0.000,1	10^{6}	= 1,000,000
10^{-3}	= 0.001	10^{7}	= 10,000,000
10^{-2}	= 0.01	10^{8}	= 100,000,000
10^{-1}	= 0.1	10^{9}	= 1,000,000,000
10^{0}	= 1	10^{10}	= 10,000,000,000

MULTIPLIERS FOR SI UNITS

a	atto-	10^{-18}	da	deka-	10^{1}
f	femto-	10^{-15}	h	hecto-	10^{2}
p	pico-	10^{-12}	k	kilo-	10^{3}
n	nano-	10^{-9}	M	mega-	10^{6}
μ	micro-	10^{-6}	G	giga-	10^{9}
m	milli-	10^{-3}	T	tera-	10^{12}
c	centi-	10^{-2}	P	peta-	10^{15}
d	deci-	10^{-1}	E	exa-	10^{18}

FIFTH EDITION

PHYSICS

F I F T H E D I T I O N

PHYSICS

ARTHUR BEISER

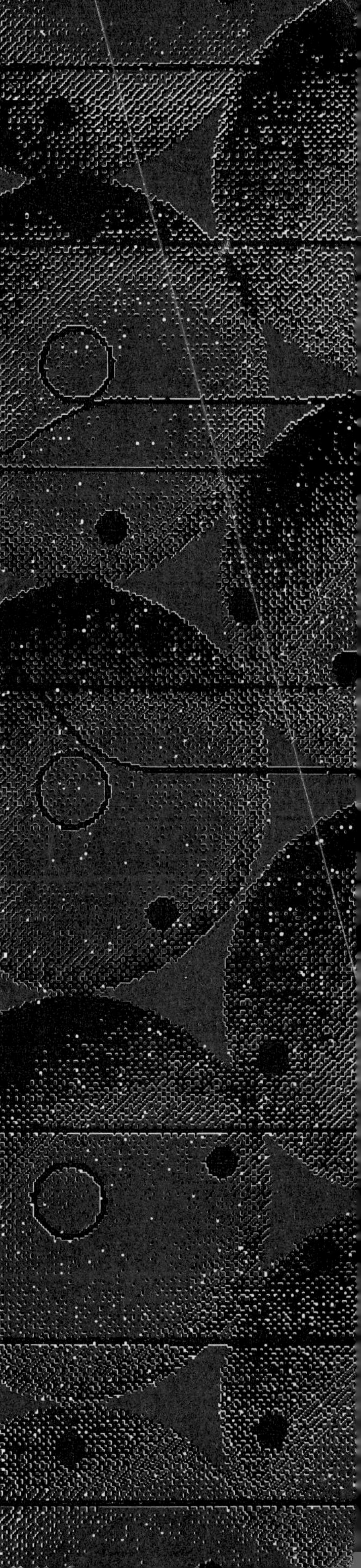

ADDISON-WESLEY PUBLISHING COMPANY

Reading, Massachusetts • Menlo Park, California
New York • Don Mills, Ontario • Wokingham, England
Amsterdam • Bonn • Sydney • Singapore • Tokyo • Madrid • San Juan

Sponsoring Editor	Stuart Johnson
Production Supervisor	Peggy J. Flanagan
Cover and Text Designer	Marshall Henrichs
Technical Art Consultant	Joe Vetere
Photo Research	Laurie Sciuto
Copy Editor	Jacqueline Dormitzer
Manufacturing Supervisor	Roy Logan
Compositor	Ruttle, Shaw & Wetherill, Inc.
Printer	Rand McNally & Company

Cover photograph: This photo shows the effects of a vibrating tuning fork in a bowl of water. The strobe exposure, made by Ed Braverman of Boston, was approximately 1/3000 of a second.

Reprinted with corrections April, 1992.

Library of Congress Cataloging-in-Publication Data

Beiser, Arthur.
Physics / Arthur Beiser. — 5th ed.
p. cm.
Includes index.
ISBN 0-201-16867-7
1. Physics. I. Title.
QC23.B4144 1991
530—dc20 90-953
CIP

11 12 13 14 15 16 17 18 19 20-RNT-0099

PREFACE

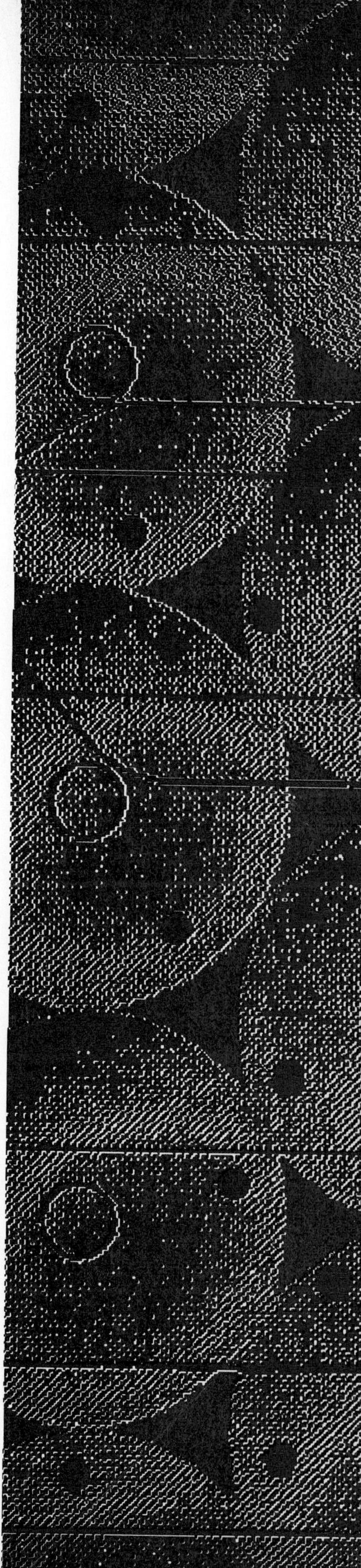

This book is intended for use with one-year, non-calculus introductory courses in physics. The emphasis is on basic physical ideas and how they are manifested in the world around us, both in nature and in technology. Physics is the master science whose scope ranges from elementary particles to the universe as a whole, and all engineering—from the design of the simplest building to that of the most advanced computer—rests on physical principles. Because physics is so rich in insights, only a sampling of them can be given here, but enough are presented to reward students for their attention to the fundamentals of the subject.

The mathematical level of *Physics* has been kept low so as not to exclude less-prepared students. Although problem-solving is important, it is not the only reason to study physics and has not been allowed to dominate the book. The majority of the exercises involve straightforward applications of the text.

As many formulas as possible are derived rather than merely stated. To provide motivation for following complex derivations, they are often given in separate sections after the significance of their results has been made clear. When a derivation would be too lengthy or would need mathematics beyond the level of the book, the train of thought involved is described or the result is otherwise justified. As little as possible is pulled out of a hat.

Physics is a development of an earlier textbook of mine, *The Mainstream of Physics,* which was published nearly thirty years ago. The wide use of *Mainstream* and of the four previous editions of *Physics* testify to the merits of the approach to physics teaching they embody.

New Features

In revising *Physics* for its fifth edition, one of the chief aims was to make the book more attractive to its users. Much of the text was completely rewritten in a less formal style for easier reading; few paragraphs escaped at least some change. Many arguments were recast for greater clarity and interest.

A new feature is a set of 17 brief biographies of major contributors to physics. These sketches are meant not only to provide a historical perspective but also to emphasize that physics is an ever-evolving body of knowledge brought into being by men and women seeking to understand how the physical universe works.

Changes in Organization

The sequence of topics in *Physics* has been altered at the request of users of the previous edition. The book now starts with straightline motion before going on to vectors and motion in two dimensions. The mass-energy relation is given only brief mention in the chapter on energy, pending a fuller discussion in a later chapter on relativity. Circular motion has been shifted to follow energy and momentum. This permits using energy considerations when analyzing circular motion and brings the related subject of rotational motion directly afterward. Equilibrium is next, in an appropriate place as a special case of linear and rotational motion. Capacitance and inductance are now included respectively in the chapters on electric energy and on electromagnetic induction. Finally, the two previous chapters on the atom were combined into a single one.

Changes in Content

A number of topics are either new or expanded from earlier treatments. Some examples are relative velocity, conservative and nonconservative forces, sound waves in pipes, the greenhouse effect, work done by and on a gas, entropy, energy and momentum transport by electromagnetic waves, X-ray diffraction, single-slit diffraction, polarization by reflection, general relativity, and the history of the universe in the light of elementary-particle theories. To make room, some topics were either condensed or eliminated, in particular those that overlap chemistry.

Illustrations

The illustrations are full partners to the text, and much attention was given to them in preparing this edition. There are about a hundred more figures than in the previous edition, and many of the existing figures were revised or entirely redrawn for greater effectiveness. Sequences of figures accompany complicated derivations, a figure for each step. In this way students unaccustomed to abstract arguments have a visual pathway besides a formal pathway to appreciate how an important formula is obtained. In addition, figures are used liberally with key concepts and equations to show as well as tell the point being made. A selection of photographs is included as part of the effort to bring alive the ideas of physics.

Student Aids

Several aids to the student are incorporated in the book:

EXAMPLES Nearly three hundred worked examples show how physical principles are put into practice.

IMPORTANT TERMS AND FORMULAS A glossary of important terms is given at the end of each chapter together with the formulas needed to solve problems based on the chapter material. These lists act as chapter summaries.

MATH APPENDIXES Only basic algebra and simple trigonometry are used in the book, and these are reviewed in two appendixes to the extent required. Powers-of-ten notation for small and large numbers is carefully explained and the use of calculators for finding powers and roots is described. The appendixes are self-contained and include exercises to sharpen rusty skills (or to practice new ones).

EXERCISES Each chapter ends with exercises of four kinds. The total is 2655, twenty percent more than in the previous edition.

Multiple Choice. These exercises, about thirty per chapter, serve as a quick, painless check on the student's grasp of the chapter contents. They act as a kind of warm-up for the regular exercises and have proved to be an effective learning tool. Correct answers provide reinforcement and encouragement; incorrect ones identify areas of weakness.

Questions. A typical chapter has over a dozen questions that test the student's understanding of the ideas in the chapter.

Exercises. A typical chapter has about forty exercises arranged in groups that correspond to sections in the text. They range from quite easy to moderately challenging. The simpler exercises are meant to provide practice in manipulating symbols and numbers, to build confidence, and to promote an intuitive feeling for the magnitudes of the quantities involved. Other exercises are more complex, with the strategy of attack not always obvious, but all of them should be within the reach of most students.

Problems. The problems, up to twenty per chapter, are more difficult or elaborate than the exercises. Some require serious thought and are correspondingly more fun for the abler student.

Answers to the odd-numbered questions, exercises, and problems are at the back of the book.

Supplements

Although the text contains all the information needed to cope with the end-of-chapter exercises, many students welcome additional assistance. Two supplements are available for them.

The *Study Guide,* prepared by Thomas O'Kuma of Lee College, lists important terms, comments on various physical concepts and applications, and provides step-by-step solutions to model problems.

The *Student Solutions Manual,* prepared by Craig Watkins of Massachusetts Institute of Technology, contains worked solutions to all the odd-numbered exercises and problems. Understanding these solutions should bring the unsolved even-numbered exercises and problems within the reach of the student. Together with the examples in the text, almost a thousand solutions are presented that span all levels of difficulty.

The instructor has not been forgotten. The *Instructor's Manual* includes notes on central ideas in the text, sample exams, lists of appropriate films and videos, and my answers to the even-numbered questions, exercises, and problems.

OmniTest, prepared by Craig Watkins and ips Publishing, enables instructors to choose problems from a computerized test bank and to create their own using the test-edit program. *OmniTest* runs on the IBM PC and compatibles.

Another important supplement is the *Laboratory Manual with Computer Activities,* prepared by Paul Robinson of Bullard High School, Fresno, California. This manual has experiments and computer activities that emphasize understanding and applying concepts rather than cookbook exercises. There is an Instructor's Manual to accompany this lab manual.

Acknowledgments

In revising *Physics* for this edition I have had the benefit of constructive criticism from the following reviewers, whose generous help was of great value:

William Melton	University of North Carolina at Charlotte
F. M. Phelps	Central Michigan University
John Ritter	Richland College
William DeBuvitz	Middlesex County College
Russell Poch	Howard Community College
Richard Delaney	College of Aeronautics
Glenn Sowell	University of Nebraska—Lincoln
David Mills	College of the Redwoods
Barry Gilbert	Rhode Island College
John Garlow	Tarrant County Community College
Robert Kernell	Old Dominion University
David Markowitz	University of Connecticut
Carl Nave	Georgia State University
Jerry Reid	Central Piedmont Community College
Chang Shih	College of Du Page
William Cochran	Youngstown State University
John Shelton	College of Lake County
Anthony Donfor	University of the District of Columbia
Paul Varlashkin	East Carolina University
Joseph Priest	Miami University
Robert Rasera	University of Maryland—Baltimore County
Paul Chow	California State University—Northridge
Frederick Glaser	Pan American University
Thomas O'Kuma	Lee College
William Deutschman	
Jerome Raskin	

A major contribution was made by Craig Watkins of Massachusetts Institute of Technology, who read both the manuscript and the proofs and checked the answers to all the exercises. Users of the book have reason to be grateful for his eagle eye.

Finally, I want to thank my friends at Addison-Wesley, especially Stuart Johnson, Peggy J. Flanagan, and Laurie Sciuto, who did so splendid a job of turning the manuscript into a book.

ARTHUR BEISER

CONTENTS

FIFTH EDITION

PHYSICS

CHAPTER 1

DESCRIBING MOTION

Everything in the universe is in motion, from the electrons in atoms to the stars in the sky. We see people and cars moving on the ground, birds and airplanes in the sky, fish and ships in the sea. The most solid of buildings sway in the wind, and, though they do so at the speed of a growing fingernail, even the continents shift across the face of the earth. Understanding motion is thus a natural start toward understanding how the world around us works, which is the goal of physics. Historically, Galileo's experiments with moving bodies began the era of modern science. The subject of this chapter is motion in a straight line, the simplest motion of all.

1.1 Units

A few basic units are enough for all physical quantities.

The raw material of physics consists of measurements. Every measurement is a comparison. When we say a boat is 18 meters long, we mean that its length is 18 times a certain unit of length called the meter. The result of every measurement has two parts. One is a number (18 for the boat) to answer the question "How many?" The other is a unit (here the meter) to answer the question "Of what?"

All quantities in the physical world can be expressed in terms of only six basic units. For the first third of this book we will need only the units of length (the **meter,** m), time (the **second,** s), and mass (the **kilogram,** kg). Every other mechanical unit is some combination of two or all three of these. The unit of force, for instance, is the kg·m/s^2, which is called the **newton** (N) for convenience. The other basic units are the **kelvin** (K) for temperature, the **ampere** (A) for electric current, and the **candela** (cd) for luminous intensity (Table 1.1).

These units are part of the Système International, or SI, which is the current version of the metric system introduced in France two centuries ago to replace the hodgepodge of traditional units that were then making commerce and industry difficult. Today SI units are used by all scientists and in most of the world in everyday life as well. Although SI units have been legal in the United States for many years, they have not yet displaced British units, such as the foot and the pound, outside the laboratory.

Whenever possible, basic units are defined in terms of quantities in nature that do not change. The second, for instance, is now defined in terms of the microwave radiation given off under certain circumstances by the ^{133}Cs atom: 1 s is equal to the time needed for 9,192,631,770 cycles of this radiation to be emitted. The meter in turn is defined in terms of the second and the speed of light: 1 m is the distance traveled in 1/299,792,458 s by light waves in a vacuum.

Because a basic unit in a system may not be convenient in size for a given measurement, other units have come into use within each system. Thus long distances are usually given in miles rather than in feet, or in kilometers rather than in meters. The great advantage of the SI system is that it is a decimal system, which makes calculations easy, whereas the British system is quite irregular in this respect. (1 km = 1000 m, for example, but 1 mi = 5280 ft.) The chief units in each system, together with their equivalents in the other system, are given inside the back cover of the book. The prefixes used with SI units of all kinds are listed in Table 1.2; powers-of-ten notation is reviewed in Appendix B.

Table 1.1 Basic SI units

Quantity	Unit
Length	meter (m)
Time	second (s)
Mass	kilogram (kg)
Temperature	kelvin (K)
Electric current	ampere (A)
Luminous intensity	candela (cd)

Table 1.2 Subdivisions and multiples of SI units are widely used. Each is designated by a prefix according to the corresponding power of ten.

Prefix	Power of Ten	Abbreviation	Pronunciation	Example
atto-	10^{-18}	a	at′ toe	1 aC = 1 attocoulomb = 10^{-18} C
femto-	10^{-15}	f	fem′ toe	1 fm = 1 femtometer = 10^{-15} m
pico-	10^{-12}	p	pee′ koe	1 pf = 1 picofarad = 10^{-12} f
nano-	10^{-9}	n	nan′ oe	1 ns = 1 nanosecond = 10^{-9} s
micro-	10^{-6}	μ	my′ kroe	1 μA = 1 microampere = 10^{-6} A
milli-	10^{-3}	m	mil′ i	1 mg = 1 milligram = 10^{-3} g
centi-	10^{-2}	c	sen′ ti	1 cL = 1 centiliter = 10^{-2} L
kilo-	10^{3}	k	kil′ oe	1 kN = 1 kilonewton = 10^{3} N
mega-	10^{6}	M	meg′ a	1 MW = 1 megawatt = 10^{6} W
giga-	10^{9}	G	ji′ ga	1 GeV = 1 gigaelectronvolt = 10^{9} eV
tera-	10^{12}	T	ter′ a	1 Tm = 1 terameter = 10^{12} m
peta-	10^{15}	P	pe′ ta	1 Ps = 1 petasecond = 10^{15} s
exa-	10^{18}	E	ex′ a	1 EJ = 1 exajoule = 10^{18} J

Converting Units

Suppose we find from a European map that Amsterdam is 648 km from Berlin and want to know what this distance is in miles. Two rules apply to such a conversion:

1. **Units are treated in an equation in exactly the same way as algebraic quantities, and may be multiplied and divided by one another;**
2. **Multiplying or dividing a quantity by 1 does not affect its value.**

To convert 648 km to its equivalent in miles, we note from the table inside the back cover that

$$1 \text{ km} = 0.621 \text{ mi}$$

Therefore

$$0.621 \frac{\text{mi}}{\text{km}} = 1$$

and multiplying or dividing any quantity by 0.621 mi/km does not affect its value but only changes the units in which it is given. Hence we have

$$d = (648 \cancel{\text{km}})\left(0.621 \frac{\text{mi}}{\cancel{\text{km}}}\right) = 402 \text{ mi}$$

because km/km = 1 and so drops out. (When multiplied out completely, [648] [0.621] = 402.408. As discussed in Appendix B–4, we can keep only as many significant figures as there are in the least accurately known quantity used in a calculation. Since this means three significant figures here, the result must be expressed as 402 mi.)

EXAMPLE 1.1

The piston displacement of the Volvo MD21A diesel engine is 2.11 liters, where 1 liter = 1000 cm^3. Express this volume in cubic inches.

SOLUTION The displacement of the engine is V = 2110 cm^3. Since 1 cm = 0.394 in, 1 cm^3 = (0.394 in.)^3 = 0.0612 in^3, and

$$V = (2110\,\cancel{\text{cm}^3})\left(0.0612\frac{\text{in}^3}{\cancel{\text{cm}^3}}\right) = 129\ \text{in}^3$$ ◆

EXAMPLE 1.2

Express a speed of 80 km/h in meters per second.

SOLUTION Here two units are to be converted, which we can do in a single step:

$$v = \left(80\frac{\cancel{\text{km}}}{\cancel{\text{h}}}\right)\left(1000\frac{\text{m}}{\cancel{\text{km}}}\right)\left(\frac{1}{3600\text{s}/\cancel{\text{h}}}\right) = 22\ \text{m/s}$$ ◆

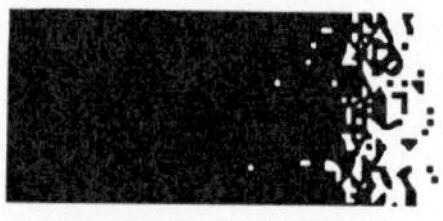

1.2 Frame of Reference

All motion is relative to a frame of reference.

When something changes its position with respect to its surroundings, we say it *moves*. There are two separate ideas here. One is that of *change*: When something has moved, the world is not exactly the same as it was before. The other idea is that of **frame of reference:** If we are to notice that something is moving, we must be able to check its position relative to something else. The choice of a frame of reference for reckoning motion depends on the situation. For a car, the most convenient frame of reference is the earth's surface; for a sailor, it is the deck of the ship; for a planet, it is the sun; for an electron in an atom, it is the atom's nucleus.

The key to solving many problems in physics is the proper choice of a frame of reference. Newton was able to interpret the motions of the earth and planets in terms of the gravitational pull of the sun only because he used the Copernican model of the solar system, which has the sun at its center (see Section 6.9). If instead he had used the older, Ptolemaic model, which has the sun and the other planets moving around the earth, the paths of the planets would have appeared to be very irregular. Newton would not have been able to discover the law of gravity by analyzing these motions.

1.3 Speed

Average and instantaneous speeds may differ.

As we all know, the speed of a moving object is the rate at which it covers distance. But there is more to the idea of speed than this simple statement.

The **average speed** $\bar{v}$ of an object that travels the distance x in the time interval t is

$$\bar{v} = \frac{x}{t} \qquad \textit{Average speed} \quad (1.1)$$

$$\text{Average speed} = \frac{\text{distance traveled}}{\text{time interval}}$$

Thus a car that went 180 km in 3.0 h had an average speed of

$$\bar{v} = \frac{x}{t} = \frac{180\text{ km}}{3.0\text{ h}} = 60\text{ km/h}$$

The average speed of the car is only part of the story of its trip, however. Knowing $\bar{v}$ does not tell us whether the car had the same speed for the entire 3 h or sometimes went faster than 60 km/h and sometimes slower.

Suppose we are in a car starting from rest and note the time at which its odometer shows it has covered 100 m, 200 m, 300 m, and so on. The data might appear as in Table 1.3. When we plot these data on a graph, as in Fig. 1.1, we find that the line joining the various points is not a straight line but has an upward curve.

Table 1.3

Total distance, m	0	100	200	300	400	500
Elapsed time, s	0	28	40	49	57	63

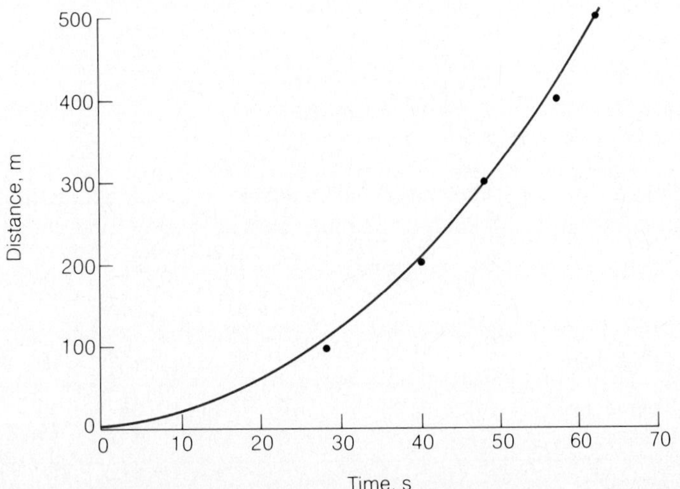

Figure 1.1

A graph of the data in Table 1.3. Each point represents the result of one measurement. The smooth curve is a reasonable fit to the data even though it does not quite go through every point.

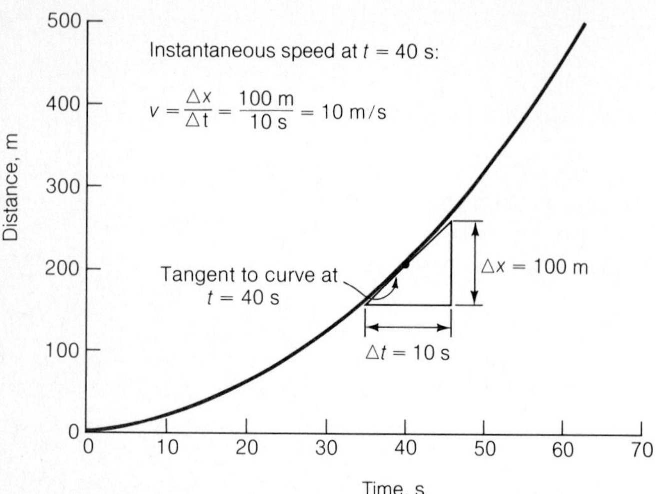

Figure 1.2

The procedure for finding the instantaneous speed at $t = 40$ s from the data in Table 1.3. Because the graph of distance versus time is not a straight line, the instantaneous speed changes continuously.

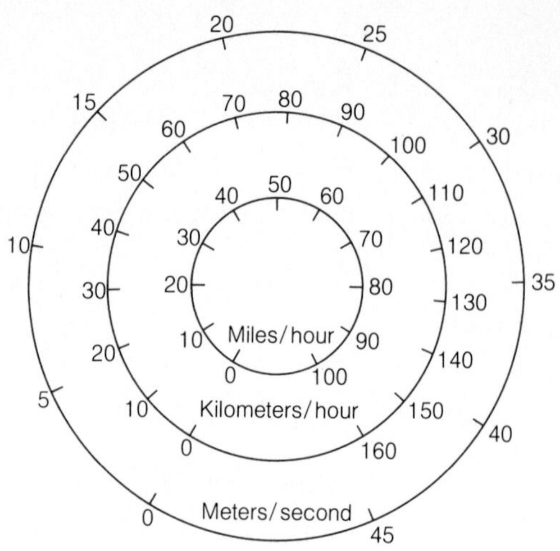

Figure 1.3

A speedometer calibrated in m/s, km/h, and mi/h. The speedometer of a vehicle indicates its instantaneous speed.

In each successive equal time interval (as marked off at the bottom of the graph), the car covers a greater distance than before—it is going faster and faster.

Even though the car's speed is changing, at every moment it has a certain definite value (indicated by its speedometer). To find this **instantaneous speed** v at a particular time t, we draw a straight line tangent to the distance-time curve at that value of t. The length of the line does not matter. Then we find v from the tangent line by using the formula

$$v = \frac{\Delta x}{\Delta t} \qquad \textit{Instantaneous speed} \quad (1.2)$$

where Δx is the distance interval between the ends of the tangent and Δt is the time interval between them. (Δ is the Greek capital letter *delta* and customarily means "change in.") From Fig. 1.2 the instantaneous speed of the car at $t = 40$ s is

$$v = \frac{\Delta x}{\Delta t} = \frac{100 \text{ m}}{10 \text{ s}} = 10 \text{ m/s}$$

Figure 1.3 is a speedometer dial calibrated in m/s, km/h, and mi/h. Evidently 10 m/s corresponds to 36 km/h or 22 mi/h. Table 1.4 shows the instantaneous speeds of the car at 10-s intervals as found from the graph of Fig. 1.2.

Table 1.4

Elapsed time, s	0	10	20	30	40	50	60
Instantaneous speed, m/s	0	2.5	5.0	7.5	10	12.5	15

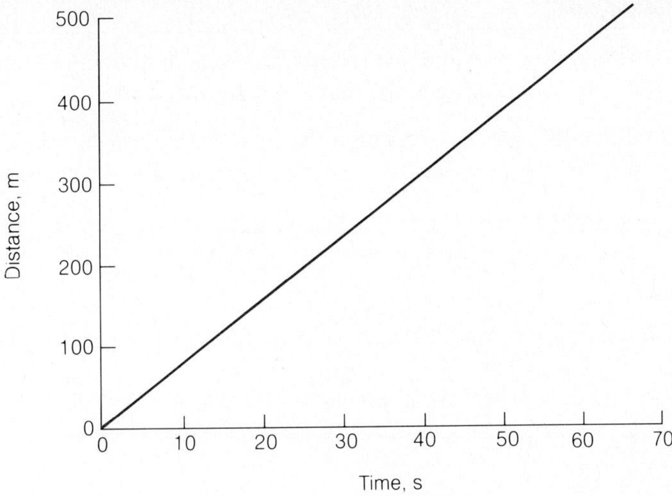

Figure 1.4

The distance-time graph of a car traveling at a constant speed of 7.5 m/s is a straight line.

When the instantaneous speed of an object does not change, it is moving at **constant speed.** Figure 1.4 is a distance-time graph of a car that has a constant speed of 7.5 m/s: The curve is, of course, a straight line. For the case of constant speed, Eq. (1.2) can be rewritten to provide two useful formulas. The first gives the distance covered in a given period of time:

$$x = vt \qquad v = \textit{constant} \qquad (1.3)$$

$$\text{Distance} = (\text{speed})(\text{time})$$

The second formula gives the time needed to cover a given distance:

$$t = \frac{x}{v} \qquad v = \textit{constant} \qquad (1.4)$$

$$\text{Time} = \frac{\text{distance}}{\text{speed}}$$

If the speed of an object is constant, we can calculate how far it will go in a given period of time. Or, given the distance, we can calculate the time required.

EXAMPLE 1.3

A woman standing in front of a cliff claps her hands, and 2.50 s later she hears an echo. How far away is the cliff? Assume that the speed of sound is 343 m/s.

SOLUTION The total distance a sound travels in 2.50 s is

$$x = vt = (343 \text{ m/s})(2.50 \text{ s}) = 858 \text{ m}$$

The sound must travel to the cliff and back, so the cliff is half this distance away, which is $\frac{1}{2}(858 \text{ m}) = 429 \text{ m}$. ◆

EXAMPLE 1.4

If the highway speed limit is raised from 55 mi/h to 65 mi/h, how much time is saved on a 100-mi trip made at the maximum permitted speed?

SOLUTION At $v_1 = 55$ mi/h the time needed for the trip is $t_1 = x/v_1$, and at $v_2 = 65$ mi/h the time needed is $t_2 = x/v_2$. Hence the time saved is

$$\Delta t = t_1 - t_2 = \frac{x}{v_1} - \frac{x}{v_2} = \frac{100\text{ mi}}{55\text{ mi/h}} - \frac{100\text{ mi}}{65\text{ mi/h}} = 0.28\text{ h}$$

Since 1 h = 60 min, the time saved in minutes is

$$\Delta t = (0.28\text{ h})(60\text{ min/h}) = 17\text{ min}$$

◆

1.4 Velocity

How fast in what direction.

The speed of a moving object specifies only how fast the object is going, regardless of its direction. If we are told that a car has a constant speed of 40 km/h, we do not know where it is headed even though it may be moving along a straight road. A more complete description that includes both speed and direction is called **velocity.** If we are told that a car has a constant velocity of 40 km/h toward the west, we know all there is to know about its motion and can easily figure out where it will be in an hour, in two hours, or at any other time.

Because the rate of change of distance tells us only "how fast," the definition of velocity involves a different quantity, **displacement.** The displacement of an object that has moved is its change in position. Suppose an airplane leaves New York City and flies a distance of 640 km in an hour and then lands. From this information all we know is that the airplane must have landed somewhere inside a circle of radius 640 km whose center is New York. The airplane could have turned around and come back to New York and still flown 640 km. If the airplane was displaced by 640 km to the west, however, we know that it landed in Cleveland (Fig. 1.5).

An object in straight-line motion that is at the point x_1 at the time t_1 and at the point x_2 at the time t_2 has been displaced by $x_2 - x_1$. The **average velocity** of the object in the time interval $t_2 - t_1$ is

$$\bar{v}_{vel} = \frac{x_2 - x_1}{t_2 - t_1} \qquad \textit{Average velocity} \quad (1.5)$$

$$\text{Average velocity} = \frac{\text{displacement}}{\text{time interval}}$$

The average velocity and average speed of a moving object may not be numerically the same. The airplane just described had an average speed of 640 km/h. If it landed in Cleveland, its average velocity would be 640 km/h to the west.

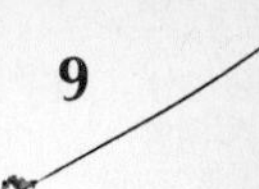

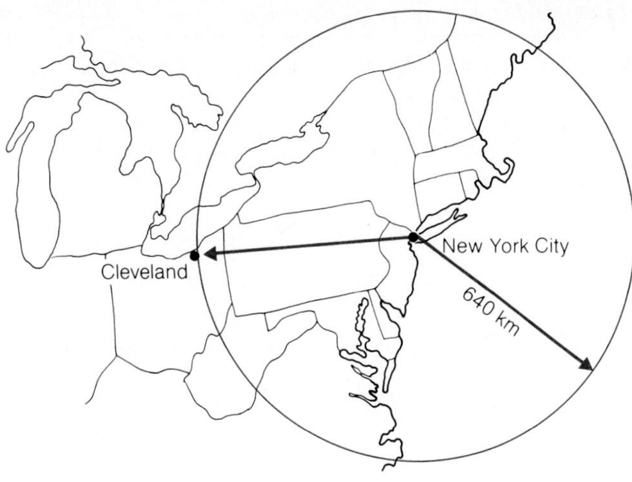

Figure 1.5

An airplane that leaves New York and flies for a distance of 640 km may land anywhere within the circle shown. If the airplane undergoes a displacement of 640 km to the west, it will land in Cleveland.

If instead the airplane returned to New York, its displacement would be 0 and its average velocity for the flight would also be 0.

On the other hand, the instantaneous speed of an object is always equal to the *magnitude* of its **instantaneous velocity.** An airplane whose instantaneous speed is 640 km/h might have an instantaneous velocity of 640 km/h to the west or 640 km/h to the north, or it might be flying in a circle at 640 km/h, but the magnitude of the airplane's instantaneous velocity is 640 km/h in each case.

1.5 Acceleration

Vroom!

In the real world few objects move at constant velocity for very long. An object whose velocity is increasing or decreasing, or whose direction is changed, is said to be **accelerated** (Fig. 1.6). In this chapter we are concerned only with straight-line motion; accelerations that involve changes in direction are discussed later.

Just as velocity is the rate of change of displacement with time, acceleration is the rate of change of velocity with time. If an object's velocity is v_0 to begin with

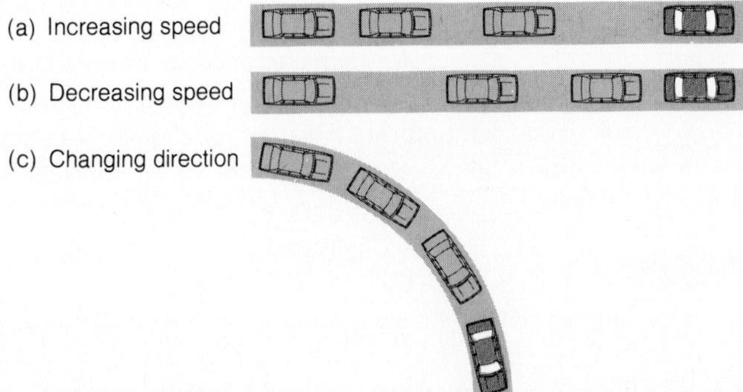

Figure 1.6

The successive positions of three accelerated cars after equal periods of time. (a) The car is going faster and faster, so it travels a longer distance in each period of time. (b) The car is going slower and slower, so it travels a shorter distance in each period of time. (c) The car's speed is constant, but its direction changes.

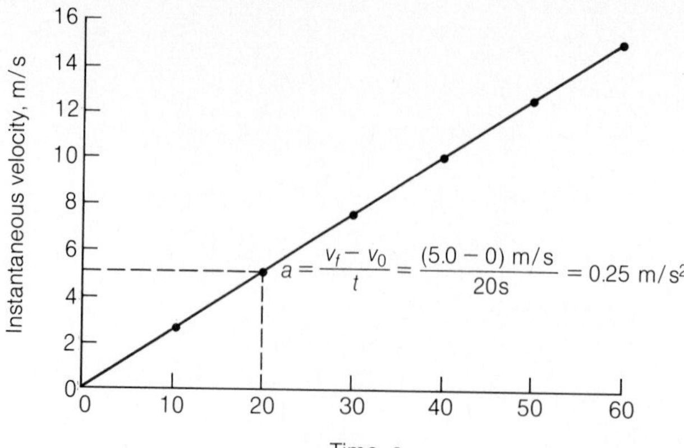

Figure 1.7

A graph of instantaneous velocity versus time for the data in Table 1.4. Although the car's velocity is not constant, it varies in a uniform way with time. This is an example of constant acceleration.

and changes to v_f during a time interval t, its acceleration a is given by

$$a = \frac{v_f - v_0}{t} \qquad \textit{Acceleration} \quad (1.6)$$

$$\text{Acceleration} = \frac{\text{change in velocity}}{\text{time}}$$

Let us return to the data on the car in Tables 1.3 and 1.4 and plot a graph of its instantaneous velocity v versus time, as in Fig. 1.7. All the points lie on a straight line, so v is directly proportional to t. Although the car's velocity is not constant, it varies in a uniform way with time; as time goes on, the velocity increases exactly in proportion. What is constant here is the car's acceleration.

The car whose motion we have been considering started out at $v_0 = 0$, and after $t = 20$ s its velocity was $v_f = 5.0$ m/s. Hence the car's acceleration is

$$a = \frac{v_f - v_0}{t} = \frac{(5.0 - 0)\text{ m/s}}{20\text{ s}} = 0.25\text{ m/s}^2$$

If we make the same calculation for the later time $t = 40$ s when v_f is 10 m/s, we also get

$$a = \frac{(10 - 0)\text{ m/s}}{40\text{ s}} = 0.25\text{ m/s}^2$$

Because the value of a is the same, the acceleration is constant. If the acceleration varied, we would get different values of a at different times.

Not all accelerations are constant, of course, but a great many are very nearly so. In what follows, all accelerations are assumed constant unless stated otherwise.

EXAMPLE 1.5

Discuss the motion of the car whose velocity-time graph is shown in Fig. 1.8.

SOLUTION A horizontal line on a v-t graph means that v does not change, so the acceleration is 0 at first. A line sloping upward means a positive acceleration (v in-

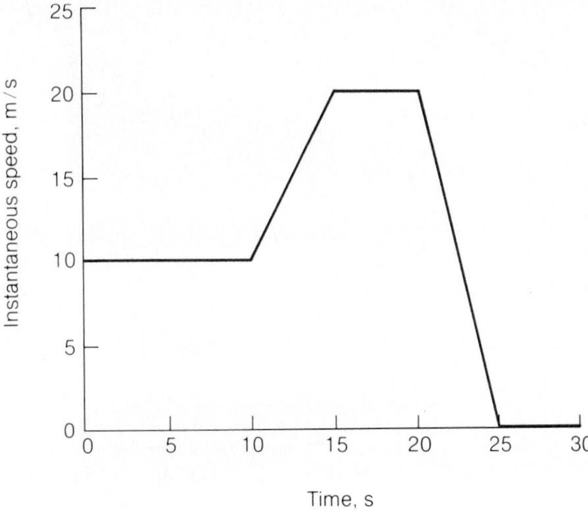

Figure 1.8
Instantaneous velocity versus time for a certain car.

creasing), and a line sloping downward means a negative acceleration (v decreasing). Hence the car begins to move at the constant velocity of 10 m/s, then is accelerated at 2 m/s^2 to a velocity of 20 m/s and travels at 20 m/s for 5 s. The brakes are now applied and the car undergoes a negative acceleration of -4 m/s^2 until it comes to a stop at $t = 25$ s. ◆

1.6 Velocity and Acceleration

How fast after an acceleration?

Equation (1.6) can be solved for the final velocity v_f of an object whose initial velocity is v_0. The result is

$$v_f = v_0 + at \qquad \textit{Final velocity} \quad (1.7)$$

Final velocity = initial velocity + velocity change

See Fig. 1.9.

In accelerated motion we have three quantities—displacement, velocity, and acceleration—that can be either positive or negative. In the case of a car, a negative velocity implies motion opposite to a specified direction, which we can call backward motion (though it does not matter which way the car actually faces). A negative displacement means the car ends up behind its starting point. A negative acceleration means the car is slowing down if v_0 is positive or that it is increasing in backward velocity if v_0 is negative. When v_0 is negative and a is positive, the car is going backward at a decreasing velocity.

All this may seem confusing, but if we are careful to put the right signs on the known quantities, our calculations will always give the right answers. The examples that follow in this section and in Sections 1.7 and 1.8 concern a car that starts out at 20 m/s (about 45 mi/h) with an acceleration of -1 m/s^2 (about -2.2 mi/h per

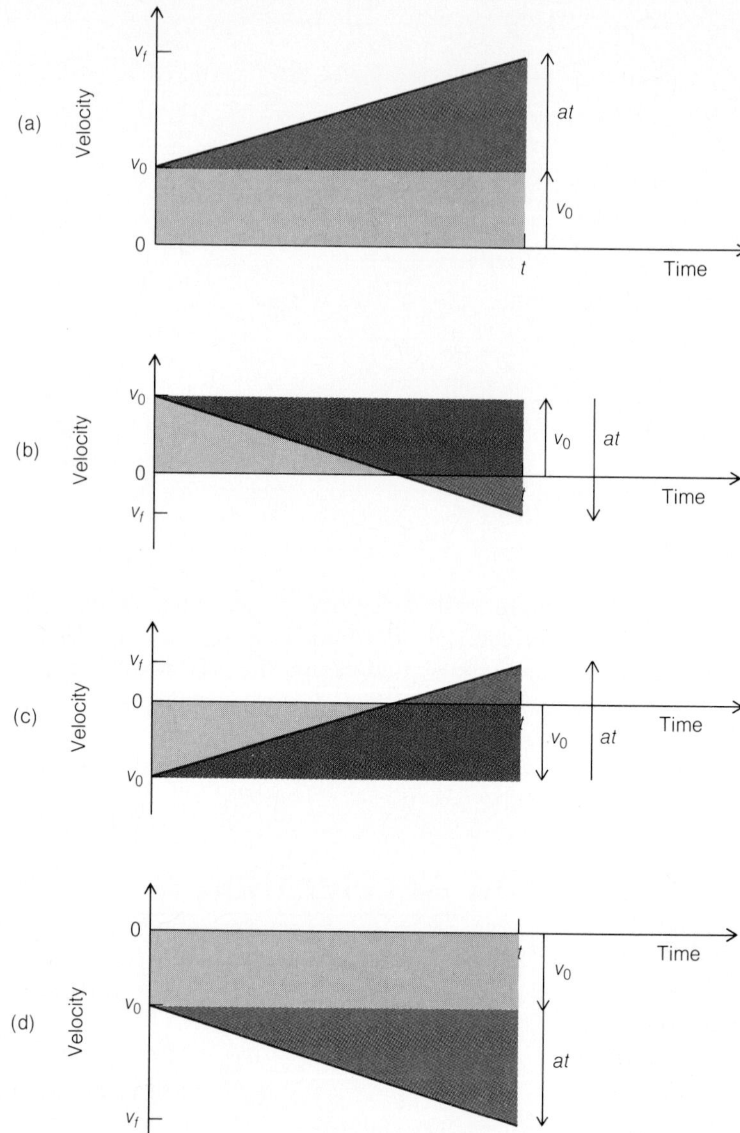

Figure 1.9

The final velocity v_f of an accelerated object is equal to its initial velocity v_0 plus the change in velocity at that occurred during the time t. (a) v_0 positive, a positive; (b) v_0 positive, a negative; (c) v_0 negative, a positive; (d) v_0 negative, a negative.

second), which corresponds to a light foot on the brake pedal. Then, when the car has come to a stop, its driver takes her foot off the brake and instantly shifts into reverse to give the car the same backward acceleration of -1 m/s^2 it had before. This is not a very realistic situation, but it is an instructive one.

EXAMPLE 1.6

A car has an initial velocity of 20 m/s and an acceleration of -1.0 m/s^2. Find its velocity after 10 s and after 50 s.

SOLUTION The velocity after 10 s is

$$v_f = v_0 + at = 20 \text{ m/s} - (1.0 \text{ m/s}^2)(10 \text{ s}) = (20 - 10) \text{ m/s} = 10 \text{ m/s}$$

which is half of what it was initially. After 50 s the velocity of the car, if a stays the same, is

$$v_f = v_0 + at = 20 \text{ m/s} - (1.0 \text{ m/s}^2)(50 \text{ s}) = (20 - 50) \text{ m/s} = -30 \text{ m/s}$$

which is in the opposite direction and greater than the original velocity v_0. ◆

1.7 Displacement, Time, and Acceleration

How far an accelerated object goes in a given time.

As we have seen, the final velocity of an object that has been accelerated for the time t is given by $v_f = v_0 + at$. The next question to ask is, how far does the object go during the time interval t?

If we know the average velocity $\bar{v}$ during the time interval t, we can find $x = \bar{v}t$, the displacement of the object. Because a is constant, v is changing at a uniform rate, and the average velocity can be found in the same way as the average of two numbers:

$$\bar{v} = \frac{v_0 + v_f}{2} \qquad \textit{Average velocity under constant acceleration} \qquad (1.8)$$

If a is not constant, this formula does not hold.

Here the initial velocity is v_0 and the final velocity is $v_0 + at$. Hence

$$\bar{v} = \frac{v_0 + v_0 + at}{2} = v_0 + \tfrac{1}{2}at$$

The displacement is accordingly

$$x = \bar{v}t = (v_0 + \tfrac{1}{2}at)\,t$$

which yields the very useful formula

$$x = v_0t + \tfrac{1}{2}at^2 \qquad \textit{Distance under constant acceleration} \qquad (1.9)$$

Figure 1.10 illustrates this result. When the initial velocity is $v_0 = 0$, we have simply

$$x = \tfrac{1}{2}at^2 \qquad v_0 = 0 \qquad (1.10)$$

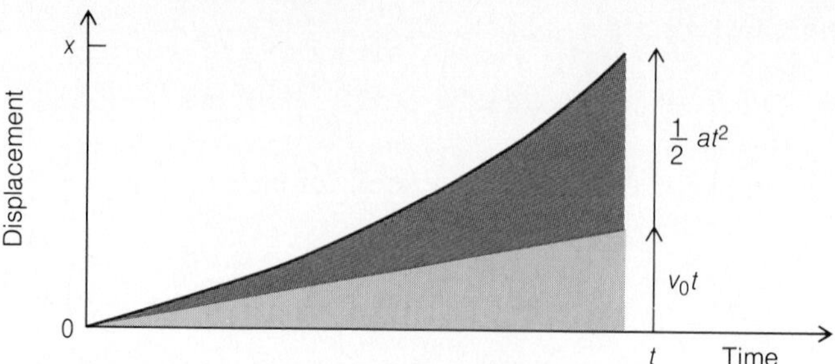

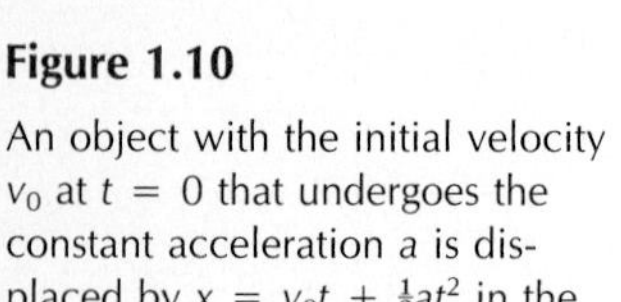

Figure 1.10

An object with the initial velocity v_0 at $t = 0$ that undergoes the constant acceleration a is displaced by $x = v_0t + \frac{1}{2}at^2$ in the time t.

EXAMPLE 1.7

We return to the car with the initial velocity of 20 m/s and the acceleration of -1.0 m/s^2. Find its displacement after the first 10 s and after the first 50 s from the moment the acceleration beings.

SOLUTION In the first 10 s the car travels

$$x = v_0t + \tfrac{1}{2}at^2 = (20 \text{ m/s})(10\text{s}) + \tfrac{1}{2}(-1.0 \text{ m/s}^2)(10 \text{ s})^2$$
$$= (200 - 50) \text{ m} = 150 \text{ m} = 0.15 \text{ km}$$

and its displacement after 50 s is

$$x = v_0t + \tfrac{1}{2}at^2 = (20 \text{ m/s})(50 \text{ s}) + \tfrac{1}{2}(-1.0 \text{ m/s}^2)(50 \text{ s})^2$$
$$= (1000 - 1250) \text{ m} = -250 \text{ m} = -0.25 \text{ km}$$

This result means that the car is 0.25 km *behind* its starting point after 50 s have elapsed (Fig. 1.11). ◆

1.8 Displacement, Velocity, and Acceleration

Additional formulas for accelerated motion

It is not hard to find relationships among x, v_0, v_f, and a that do not directly involve the time t. The first step is to rewrite Eq. (1.6) for acceleration as

$$t = \frac{v_f - v_0}{a}$$

Now we substitute this expression for t into the formula for the displacement:

$$x = v_0t + \tfrac{1}{2}at^2$$

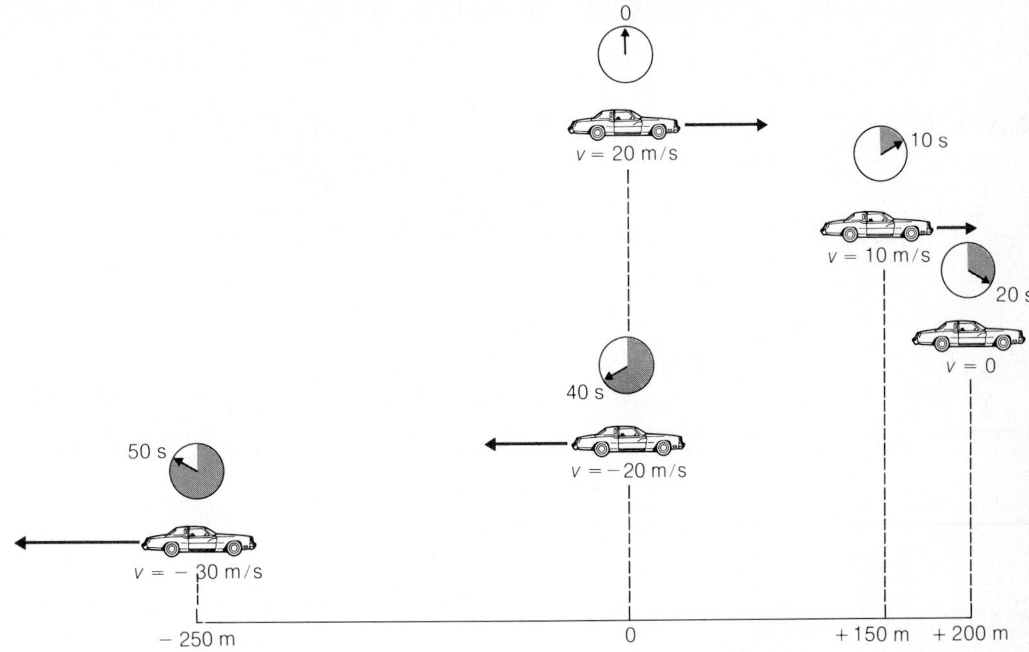

Figure 1.11

Position and velocity, at various times, of a car that has a constant acceleration of -1.0 m/s^2 and whose initial velocity is 20 m/s. At $t = 20$ s the car has come to a stop, and then it begins to move in the negative (backward) direction. At $t = 40$ s the car is back where it started, but moving in the direction opposite to its initial one. At $t = 50$ s the car is 250 m behind its starting point (Example 1.7).

The result is

$$x = v_0 \frac{(v_f - v_0)}{a} + \tfrac{1}{2} a \frac{(v_f - v_0)^2}{a^2} = \frac{v_0 v_f}{a} - \frac{v_0^2}{a} + \frac{v_f^2}{2a} - \frac{v_0 v_f}{a} + \frac{v_0^2}{2a}$$

$$= \frac{v_f^2 - v_0^2}{2a} \qquad \textit{Displacement under constant acceleration} \quad (1.11)$$

Another, often useful, way to write this result is:

$$v_f^2 = v_0^2 + 2ax \qquad \textit{Velocity under constant acceleration} \quad (1.12)$$

EXAMPLE 1.8

How far will the car in Examples 1.6 and 1.7 have gone when it comes to a stop?

SOLUTION When the car is at rest, $v_f = 0$. Since $v_0 = 20$ m/s and $a = -1.0$ m/s^2, we have from Eq. (1.11)

$$x = \frac{v_f^2 - v_0^2}{2a} = \frac{0 - (20\ \text{m/s})^2}{(2)(-1.0\ \text{m/s}^2)} = \frac{400\ \text{m}^2/\text{s}^2}{2.0\ \text{m/s}^2} = 200\ \text{m} = 0.20\ \text{km}$$

The car comes to a stop after it has gone 0.20 km. Then, since the acceleration is negative and is assumed here to continue, it begins to move in the negative (backward) direction, as in Fig. 1.11. ♦

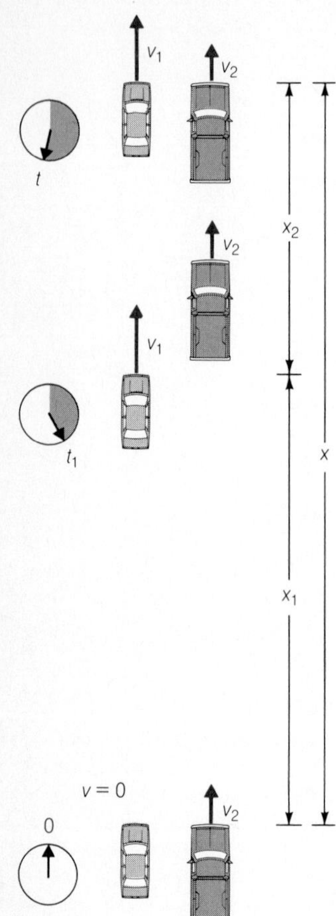

Figure 1.12

EXAMPLE 1.9

A car is stationary in front of a red traffic light. As the light turns green, a truck goes past at a constant velocity of 15 m/s. At the same moment, the car begins to accelerate at 1.25 m/s^2. When it reaches 25 m/s, the car continues at this velocity. When does the car pass the truck? How far have they gone from the traffic light at that time?

SOLUTION (a) The car's final velocity is $v_1 = 25$ m/s, and the truck's constant velocity is $v_2 = 15$ m/s (Fig. 1.12). The car needs the time

$$t_1 = \frac{v_1}{a} = \frac{25 \text{ m/s}}{1.25 \text{ m/s}^2} = 20 \text{ s}$$

to reach the velocity v_1. To find the distance the car covers during its acceleration, we use Eq. (1.11) with $v_0 = 0$ and $v_f = v_1$. The result is

$$x_1 = \frac{v_1^2}{2a} = \frac{(25 \text{ m/s})^2}{(2)(1.25 \text{ m/s}^2)} = 250 \text{ m} = 0.25 \text{ km}$$

If t is the total time the car needs to catch up with the truck, the car travels the further distance

$$x_2 = v_1 (t - t_1)$$

for a total distance of

$$x = x_1 + x_2 = x_1 + v_1 (t - t_1)$$

The truck meanwhile covers the same distance in the same time at the constant velocity v_2, so

$$x = v_2 t$$

Setting equal these two formulas for x gives

$$v_2 t = x_1 + v_1(t - t_1)$$

$$t = \frac{x_1 - v_1 t_1}{v_2 - v_1} = \frac{250 \text{ m} - (25 \text{ m/s})(20 \text{ s})}{(15 \text{ m/s}) - (25 \text{ m/s})} = 25 \text{ s}$$

The car passes the truck after 25 s.

(b) We can use either formula for x. The second one gives

$$x = v_2 t = (15 \text{ m/s})(25 \text{ s}) = 375 \text{ m} = 0.38 \text{ km}$$ ◆

Another useful formula that involves displacement and velocity follows from Eq. (1.8). Multiplying both sides by the time t gives

$$\bar{v}t = \left(\frac{v_0 + v_f}{2}\right) t$$

Since $\bar{v}t$ is the displacement x when the acceleration is constant,

$$x = \left(\frac{v_0 + v_f}{2}\right) t \qquad \textit{Displacement under constant acceleration} \qquad (1.13)$$

This formula differs from Eq. (1.11) in that it contains the time t but not the acceleration a. Thus Eq. (1.13) is handy when the value of a neither is known nor has to be determined.

EXAMPLE 1.10

The driver of a train traveling at 30 m/s applies the brakes when he passes an amber signal. The next signal is 1.5 km down the track, and the train reaches it 75 s later. Find the speed of the train at the second signal and the train's acceleration.

SOLUTION (a) Since we are given v_0, x, and t and are to find v_f, Eq. (1.13) is the one to use here. First we solve it for v_f, which gives

$$v_f = \frac{2x}{t} - v_0$$

Now we substitute the known quantities:

$$v_f = \frac{(2)(1500\ \text{m})}{75\ \text{s}} - 30\ \text{m/s} = 10\ \text{m/s}$$

(b) From Eq. (1.7) we have

$$v_f = v_0 + at$$

$$a = \frac{v_f - v_0}{t} = \frac{(10 - 30)\ \text{m/s}}{75\ \text{s}} = -0.27\ \text{m/s}^2$$

◆

All problems that involve motion under constant acceleration can be solved by using the formulas developed thus far in this chapter, which are listed in Table 1.5 for convenience. The way to pick the most appropriate formula is to list the known and unknown quantities and then see which formula relates them and no

Table 1.5 Formulas for motion under constant acceleration

Formula	Eq.
$x = \left(\frac{v_0 + v_f}{2}\right) t$	(1.13)
$x = v_0 t + \frac{1}{2}at^2$	(1.9)
$v_f = v_0 + at$	(1.7)
$v_f^2 = v_0^2 + 2ax$	(1.12)

others. For instance, in the previous example only the first equation in Table 1.5 contains v_0, x, and t, which are given, and v_f, which we were to find. Once we had v_f, we could have used any of the other formulas to obtain a; the third one in Table 1.5 was chosen because it was the easiest to use here.

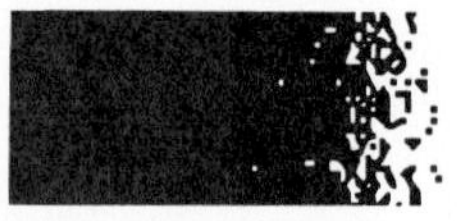

1.9 Acceleration of Gravity

Experiment and observation are the foundations of science.

Drop a stone, and it falls. Does the stone fall at constant speed, or is it accelerated? Does the motion of the stone depend on its shape, its size, or its color?

Long ago such questions were answered by Greek philosophers, notably Aristotle, by reasoning based on what to them were self-evident principles. One such principle was that each kind of material had a ''natural'' place where it belonged and toward which it tried to go. Thus fire rose naturally upward toward the sun and stars. Stones were ''earthy'' and so fell down toward their home in the earth. A big stone was more earthy than a small one and so should fall faster.

For over two thousand years almost nobody felt it necessary to perform experiments to seek information on the physical universe. Then Galileo (1564–1642) revolutionized science by doing just that: performing experiments. Galileo's universe turned out to be perhaps less charming than that of Aristotle, but it had the merit of fitting the facts. Modern science owes its success in understanding and utilizing natural phenomena to its reliance on experiment and observation.

Figure 1.13

If air resistance is neglected, all freely falling objects near the earth's surface have the same acceleration of $g = 9.8\ \text{m/s}^2$. The longer an object falls, the faster it falls.

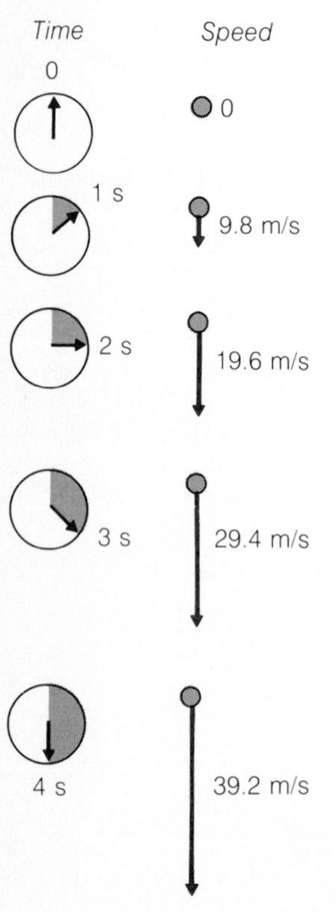

What Galileo found, as the result of careful measurements, was that *all freely falling objects have the same acceleration* at the same place near the earth's surface. This acceleration, which is called the acceleration of gravity (symbol g), does not have the same value everywhere, for reasons given in Chapter 6. The worldwide average of g at sea level is 9.81 m/s^2, but g is 9.83 m/s^2 at the poles, 9.78 m/s^2 at the equator, and still less atop Mount Everest. In this book we use

$$g = 9.80\ \text{m/s}^2 \qquad \textit{Acceleration of gravity}$$

which is the average for the United States. In British units $g = 32.2\ \text{ft/s}^2$.

An object falling freely from rest thus has an instantaneous speed of 9.8 m/s (32 ft/s) after the first second, a speed of 19.6 m/s (64 ft/s) after the next second, and so on. The longer the time during which a stone falls after being dropped, the greater its speed when it hits the ground (Fig. 1.13). But the stone's *acceleration* is always the same.

Another aspect of Galileo's work deserves comment. His conclusion that all things fall with the same constant acceleration is an *idealization* of reality. The actual accelerations with which objects fall depend on many factors besides the location on the earth—for instance, the size and shape of the object and the density and state of the atmosphere. A stone, for example, falls faster than a feather does in air because of the effects of buoyancy and air resistance. Galileo saw that the basic phenomenon was a constant acceleration downward, with other factors acting merely to cause deviations from the constant value. In a vacuum the stone and feather fall with exactly the same acceleration (Fig. 1.14).

According to legend, Galileo dropped a bullet and a cannonball from the Leaning Tower of Pisa to show that all falling objects have the same acceleration.

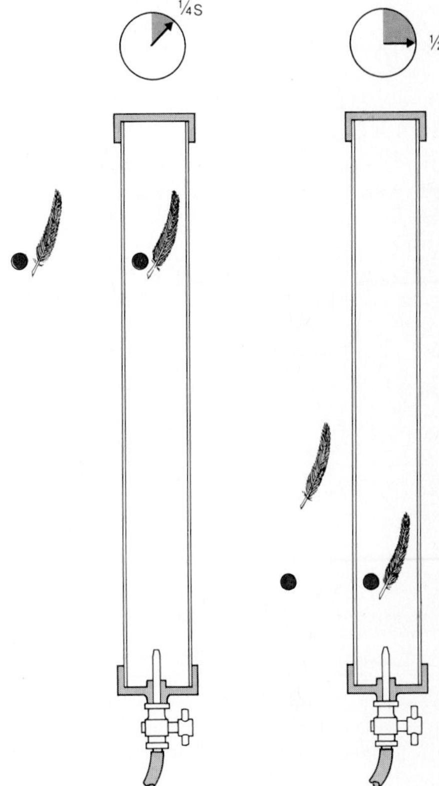

Figure 1.14

All objects that fall in a vacuum near the earth's surface have the same downward acceleration. In air, however, a stone falls faster than a feather because air resistance affects it less.

Why Raindrops Don't Kill

The drag force due to air resistance on an object of a given mass, size, and shape depends on the speed of the object—the faster it goes, the more the drag. In the case of a falling object, the drag force increases as the speed increases until finally it cannot go any faster. The object then continues to fall at a constant **terminal speed** (Fig. 1.15 and Table 1.6). The terminal speed of a person in free fall is about 54 m/s (120 mi/h), whereas it is only about 6.3 m/s (14 mi/h) with an open parachute. In the absence of air resistance, raindrops would reach the ground at speeds high enough to be dangerous.

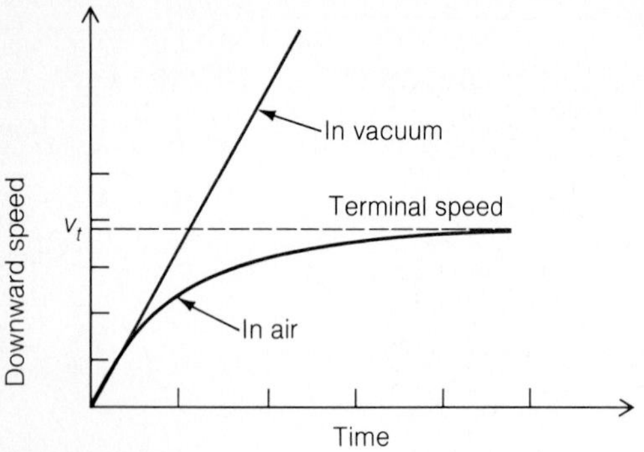

Figure 1.15

This graph compares how the speed of a falling body varies with time when it is in a vacuum and when it is in air. In a vacuum the speed is given by $v = gt$ and increases without limit until the body strikes the ground. In air the terminal speed of v_t is eventually reached, and the body continues to fall at this speed.

The terminal speed of a person falling from an airplane is considerably reduced by a parachute, which allows a safe landing.

A cloud consists of tiny water droplets or ice crystals whose terminal speeds are so small, often 1 cm/s or less, that very little updraft is needed to keep them suspended indefinitely. When a cloud is rapidly cooled, however, the water droplets or ice crystals grow in size and weight until their terminal speeds become too great for updrafts to keep them aloft. The result is a fall of raindrops or snowflakes from the cloud.

Table 1.6 Some approximate terminal speeds (1 m/s = 2.2 mi/h)

Ball	Terminal Speed	Ball	Terminal Speed
16-lb shot	145 m/s	Tennis ball	30
Baseball	42	Basketball	20
Golf ball	40	Ping-Pong ball	9

BIOGRAPHICAL NOTE

Galileo Galilei

Galileo Galilei (1564–1642) was born in Pisa; his given name, following a custom of that part of Italy, is a variation of his family name. When he was 17, Galileo entered the University of Pisa to study medicine but soon decided instead to make physics and mathematics his life's work. Galileo's father at first opposed this change in plan, arguing that medicine was a more practical career choice, but finally gave his permission. At about this time Galileo made his first scientific discovery, that the time needed by a pendulum for each back-and-forth cycle is the same regardless of how wide the swings are.

After leaving the university in 1585, Galileo spent a few years in Florence, where his family then lived, studying various subjects and writing papers on hydrostatics and the centers of mass of solid objects. These papers led to a three-year appointment as professor of mathematics at Pisa. Galileo now knew exactly what he wanted to do: to investigate natural phenomena by observation and experiment using mathematical reasoning to interpret the results. Modern science began with Galileo.

From Pisa, Galileo went on to the University of Padua, where he continued work he had begun at Pisa on accelerated motion, falling bodies, and the paths taken by projectiles. He invented the thermometer at Padua. Using a telescope he had built, Galileo was the first person to see the four largest satellites of Jupiter, the phases of Venus, sunspots, and the mountains of the moon. Turning his telescope to the Milky Way, he found that it consisted of individual stars. Galileo regarded his astronomical discoveries as "infinitely stupendous," and his contemporaries agreed.

In 1610 the now-famous Galileo returned to Florence as court mathematician to Cosimo, Duke of Tuscany. All went well for some years, but then Galileo's support of the Copernican model of the universe led to serious trouble. Copernicus (1473–1543) held that the sun and stars were fixed in space, that the planets (the earth among them) revolved around the sun, that the earth rotated daily on its axis, and that the moon revolved around the earth. Galileo pointed out that his astronomical findings fit this model well. However, the earlier Ptolemaic model, in which the earth was the stationary center of the universe with all the heavenly bodies circling it, was a firm part of church doctrine. The Holy Office told Galileo to stop advocating the Copernican system, and Galileo obeyed.

When a friend of his was elected pope in 1631, Galileo resumed teaching the merits of the Copernican system. The pope turned against him, and in 1633 Galileo was convicted of heresy in Rome. Galileo, then 70, escaped being burned at the stake, the fate of earlier heretics, but was forced to publicly deny that the earth moved. (Legend has it that Galileo then muttered, "Yet it does move.") Galileo's condemnation, a tragedy for him, was also a tragedy on a larger scale, for the church was now locked into a permanent conflict with science.

Under house arrest for the rest of his life, Galileo summed up his earlier work on the behavior of moving bodies in *Two New Sciences*. The manuscript of this book had to be smuggled out of Italy for publication in the freer atmosphere of Holland. Galileo became blind soon afterward, which did not stop him from inventing a pendulum clock and a method of navigation based on observations of the satellites of Jupiter. Still under confinement, Galileo died in 1642 at 78.

1.10 Free Fall

What goes up must come down.

We can apply the formulas for motion under constant acceleration to objects in free fall. It is important to keep in mind that the direction of the acceleration of gravity g is always downward, no matter whether we are dealing with a dropped object or with one that is initially thrown upward.

EXAMPLE 1.11

A stone is dropped from the top of New York's Empire State Building, which is 450 m high. How long does it take the stone to reach the ground? (Neglect air resistance.) What is its velocity when it strikes the ground? (See Fig. 1.16.)

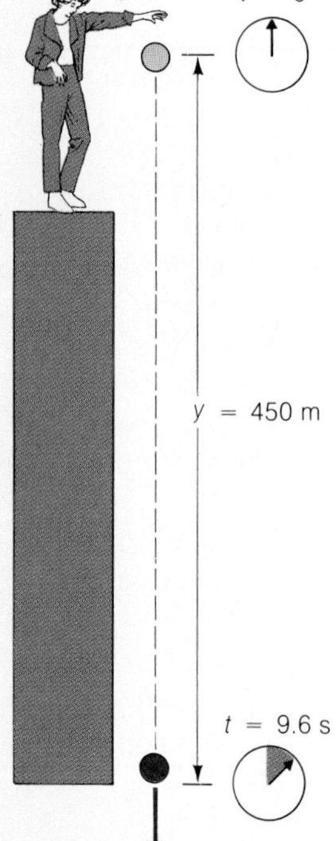

Figure 1.16

SOLUTION In a problem like this one, where the motion is entirely downward, it is easiest to consider down as the positive direction, so that $y = 450$ m and $g = 9.8$ m/s^2. The general formula for the distance traveled in the time t by an accelerated object is

$$y = v_0 t + \tfrac{1}{2}at^2$$

Here $v_0 = 0$, since the stone is simply dropped from rest, and the acceleration is

$$a = g$$

Hence we have

$$y = \tfrac{1}{2}gt^2$$

where y represents vertical distance from the starting point. Solving for t gives

$$t = \sqrt{\frac{2y}{g}} = \sqrt{\frac{(2)(450\text{ m})}{9.8\text{ m/s}^2}} = \sqrt{92}\text{ s} = 9.6\text{ s}$$

Knowing the time of fall makes it simple to find the stone's final velocity:

$$v_f = v_0 + at = 0 + gt = (9.8\text{ m/s}^2)(9.6\text{ s}) = 94\text{ m/s}$$ ◆

The speed (that is, magnitude of velocity) that a dropped object has when it reaches the ground is the same as the speed with which it must be thrown upward from the ground to rise to the same height (Fig. 1.17). To prove this statement, we refer to the formula $v_f^2 = v_0^2 + 2ax$ and replace the x with y to give

$$v_f^2 = v_0^2 + 2ay$$

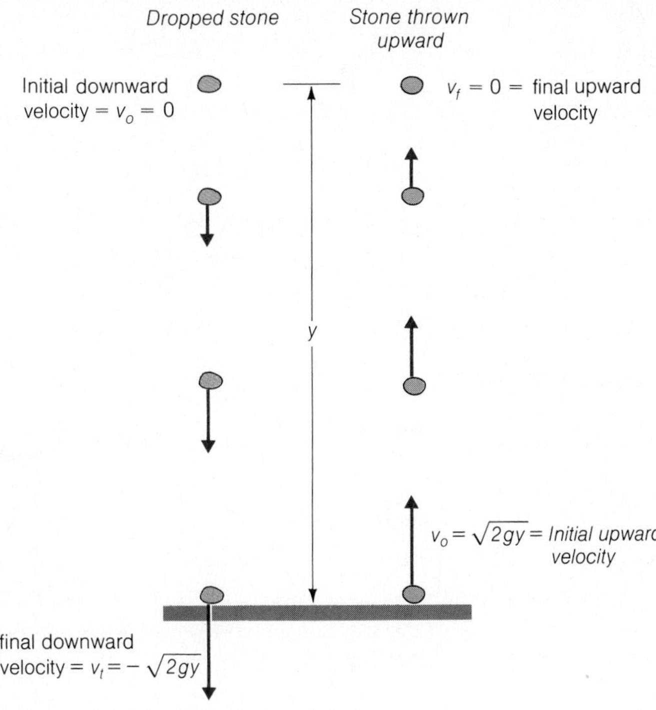

Figure 1.17

A stone dropped from a height y reaches the ground with the speed $\sqrt{2gy}$. In order to reach the height y, a stone thrown upward from the ground must have the minimum speed $\sqrt{2gy}$.

When a stone is dropped, $a = g$ and $v_0 = 0$, so

$$v_f^2 = 0 + 2gy \qquad v_f = \sqrt{2gy}$$

When the stone is thrown upward, on the other hand, $a = -g$ (since the downward acceleration is opposite in direction to the upward initial speed), and at the top of its path $v = 0$. Hence

$$0 = v_0{}^2 - 2gy \qquad v_0 = \sqrt{2gy}$$

The speed is the same in both cases.

EXAMPLE 1.12

A stone is thrown upward with an initial speed of 16 m/s (Fig. 1.18). (a) What will its maximum height be? (b) When will it return to the ground? (c) Where will it be in 0.8 s? (d) Where will it be in 2.4 s?

SOLUTION (a) We make use of Eq. (1.11) to find the highest point the stone will reach, with up as positive (+) and down as negative (−). Here $v_0 = 16$ m/s and $a = -9.8$ m/s^2, and at the top of the path $x = y$ and $v_f = 0$. Hence

$$y = \frac{v_f^2 - v_0^2}{2a} = \frac{0 - (16\text{ m/s})^2}{2(-9.8\text{ m/s}^2)} = \frac{256\text{ (m/s)}^2}{19.6\text{ m/s}^2} = 13\text{ m}$$

(b) When will the stone strike the ground? An object takes precisely as long to fall from a certain height y as it does to rise that high (provided that y is its maximum

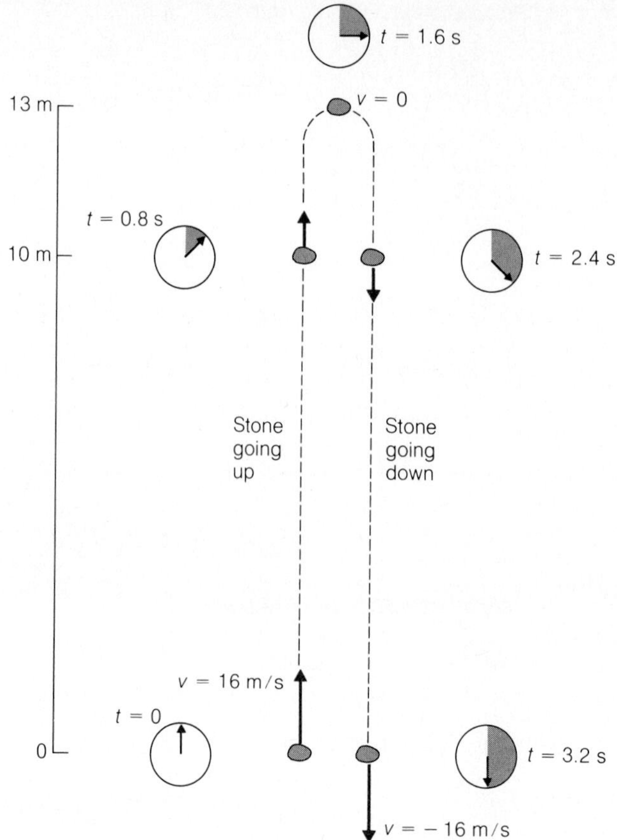

Figure 1.18

The path of a stone thrown upward with an initial speed of 16 m/s. Air resistance is neglected (Example 1.12).

height, as it is here), just as an object's final speed when dropped from a height y is the same as the initial speed required for it to get that high. From Eq. (1.10)

$$y = \tfrac{1}{2}gt^2$$

from which we find that

$$t = \sqrt{\frac{2y}{g}} = \sqrt{\frac{(2)(13\text{ m})}{9.8\text{ m/s}^2}} = 1.6\text{ s}$$

Because the stone takes as long to rise as to fall, the total time it is in the air is twice 1.6 s, or 3.2 s. (We could just as well have started from $v_f = v_0 + at$ with $v_f = 0$ and $a = -g$ at the top of the path.)

(c) To find the height of the stone a given time after it was thrown upward, we make use of Eq. (1.9), $x = v_0t + \tfrac{1}{2}at^2$, with $x = y$, $v_0 = 16$ m/s, and $a = -9.8$ m/s^2. For $t = 0.8$ s,

$$y = (16\text{ m/s})(0.8\text{s}) - \tfrac{1}{2}(9.8\text{ m/s}^2)(0.8\text{ s})^2 = 10\text{ m}$$

(d) When we substitute $t = 2.4$ s in the formula given above, the result is again

$$y = (16 \text{ m/s})(2.4 \text{ s}) - \tfrac{1}{2}(9.8 \text{ m/s}^2)(2.4 \text{ s})^2 = 10 \text{ m}$$

What this result means is that at 0.8 s the stone is at a height of 10 m on its way up, then it goes on further to its maximum height of 13 m, and at 2.4 s it is once more at a height of 10 m but now on the way down. ♦

The examples of this chapter have been worked out not because they are in themselves especially significant, but because they illustrate the power of the mathematical approach to physical phenomena. By defining certain quantities and relating them to each other and to events that actually occur in the real world, a structure of equations may be built up. This structure is an instrument enabling us to solve problems that otherwise would each require a separate, perhaps difficult or impossible, experiment. We must remember that the validity of the theoretical structure depends on its experimental basis; but once this is established, we may proceed to work out its consequences with pencil and paper.

Important Terms

To measure a quantity means to compare it with a standard quantity of the same kind called a **unit**. In a **system of units,** a set of basic units is specified from which all other units in the system are derived. **SI** units are used in everyday life in much of the world and universally by scientists.

The **average speed** of a moving object is the distance it covers in a time interval divided by the time interval. The object's **instantaneous speed** at a certain moment is the rate at which it is covering distance at that moment.

The **velocity** of an object describes both its speed and its direction of motion.

The **acceleration** of an object is the rate at which its velocity changes with time. Changes in direction as well as of speed are considered accelerations.

The **acceleration of gravity** is the acceleration of a freely falling body near the earth's surface. The symbol of the acceleration of gravity is g, and its value is 9.8 m/s^2 (32 ft/s^2).

Important Formulas

Speed: $v = \dfrac{x}{t}$

Constant speed: $x = vt$

$t = \dfrac{x}{v}$

Acceleration: $a = \dfrac{v_f - v_0}{t}$

Motion under constant acceleration:

$$x = \frac{v_0 + v_f}{2}t$$

$$x = v_0t + \tfrac{1}{2}at^2$$

$$v_f = v_0 + at$$

$$v_f^2 = v_0^2 + 2ax$$

Free fall from rest:

$$y = \tfrac{1}{2}gt^2$$

$$v_f = \sqrt{2gy}$$

(Air resistance is assumed negligible in the following exercises and problems.)

MULTIPLE CHOICE

1. The prefix micro represents
a. 1/10 b. 1/100
c. 1/1000 d. 1/1,000,000

2. A centimeter is
a. 0.001 m b. 0.01 m
c. 0.1 m d. 10 m

3. Of the following, the shortest is
a. 1 mm b. 0.01 in.
c. 0.001 ft d. 0.00001 km

4. Of the following, the longest is
a. 10^4 in. b. 10^4 m
c. 10^3 ft d. 0.1 mi

5. A height of 5 ft 8 in. is equivalent to
a. 173 cm b. 177 cm
c. 207 cm d. 223 cm

6. The number of cubic centimeters in a cubic foot is approximately
a. 1.7×10^3 b. 1.7×10^4
c. 2.8×10^4 d. 1.7×10^5

7. The number of seconds in a month is approximately
a. 2.6×10^6 b. 2.6×10^7
c. 2.6×10^8 d. 2.6×10^9

8. On a distance-time graph, a horizontal straight line corresponds to motion at
a. zero speed
b. constant speed
c. increasing speed
d. decreasing speed

9. On a distance-time graph, a straight line sloping upward to the right corresponds to motion at
a. zero speed
b. constant speed
c. increasing speed
d. decreasing speed

10. On a velocity-time graph, the motion of a car traveling along a straight road with the uniform acceleration of 2 m/s^2 would appear as a
a. horizontal straight line
b. straight line sloping upward to the right
c. straight line sloping downward to the right
d. curved line whose downward slope to the right increases with time

11. The acceleration of a stone thrown upward is
a. greater than that of a stone thrown downward
b. the same as that of a stone thrown downward
c. smaller than that of a stone thrown downward
d. zero until it reaches the highest point in its motion

12. A stone is thrown upward from a roof at the same time as an identical stone is dropped from there. The two stones
a. reach the ground at the same time
b. have the same velocity when they reach the ground
c. have the same acceleration when they reach the ground
d. None of the above

13. A bicycle travels 12 km in 40 min. Its average speed is
a. 0.3 km/h b. 8 km/h
c. 18 km/h d. 48 km/h

14. A car that travels at 40 km/h for 2.0 h, at 50 km/h for 1.0 h, and at 20 km/h for 0.50 h has an average speed of
a. 31 km/h b. 40 km/h
c. 45 km/h d. 55 km/h

15. A pitcher takes 0.10 s to throw a baseball, which leaves his hand with a velocity of 30 m/s. The ball's acceleration was
a. 3 m/s^2 b. 30 m/s^2
c. 300 m/s^2 d. 3000 m/s^2

16. How long does a car with an acceleration of 2.0 m/s^2 take to go from 10 m/s to 30 m/s?
a. 10 s b. 20 s
c. 40 s d. 400 s

17. A car undergoes a constant acceleration of 6.0 m/s^2 starting from rest. In the first second it travels
a. 3 m b. 6 m
c. 18 m d. 36 m

18. An airplane requires 20 s and 400 m of runway to become airborne, starting from rest. Its velocity when it leaves the ground is
a. 20 m/s b. 32 m/s
c. 40 m/s d. 80 m/s

19. A car has an initial velocity of 15 m/s and an acceleration of 1.0 m/s^2. In the first 10 s after the acceleration begins, the car travels
a. 50 m b. 150 m
c. 155 m d. 200 m

20. A car has an initial velocity of 15 m/s and an acceleration of -1.0 m/s^2. In the first 10 s after the acceleration begins, the car travels
a. 25 m b. 50 m
c. 100 m d. 145 m

21. How far does the car in the previous question go before coming to a stop?
a. 113 m b. 150 m
c. 225 m d. 450 m

22. A wheel falls from an airplane flying horizontally at an altitude of 490 m. The wheel strikes the ground in

a. 10 s b. 50 s
c. 80 s d. 100 s

23. A stone is dropped from a cliff. After it has fallen 30 m, its velocity is

a. 17 m/s b. 24 m/s
c. 44 m/s d. 588 m/s

24. Two balls are thrown vertically upward, one with an initial velocity twice that of the other. The ball with the greater initial velocity will reach a height

a. $\sqrt{2}$ that of the other
b. twice that of the other
c. 4 times that of the other
d. 8 times that of the other

25. A ball thrown vertically upward at 25 m/s continues to rise for approximately

a. 2.5 s b. 5 s
c. 7.5 s d. 10 s

26. In the previous question, how much time will elapse before the ball strikes the ground?

a. 2.5 s b. 5 s
c. 7.5 s d. 10 s

27. A ball is thrown downward at 5 m/s from a roof 10 m high. Its velocity when it reaches the ground is

a. 13 m/s b. 14 m/s
c. 15 m/s d. 20 m/s

28. Another ball is thrown upward at 5 m/s from the same roof. Its velocity when it reaches the ground is

a. 13 m/s b. 14 m/s
c. 15 m/s d. 20 m/s

QUESTIONS

1. In the following pairs of length units, which is the shortest? Inch, centimeter; yard, meter; mile, kilometer.

2. Figure 1.19 is a graph that shows the displacement of an object along a straight line plotted against time. Describe the motion of the object.

3. Can a rapidly moving object have the same acceleration as a slowly moving one? If so, give an example.

4. The acceleration of a certain moving object is constant in magnitude and direction. Must the path of the object be a straight line? If not, give an example.

5. Figure 1.20 shows how the velocity of a car varied during a period of time. (a) In what time interval was the car's acceleration highest? (b) In what time interval did the car cover the most distance?

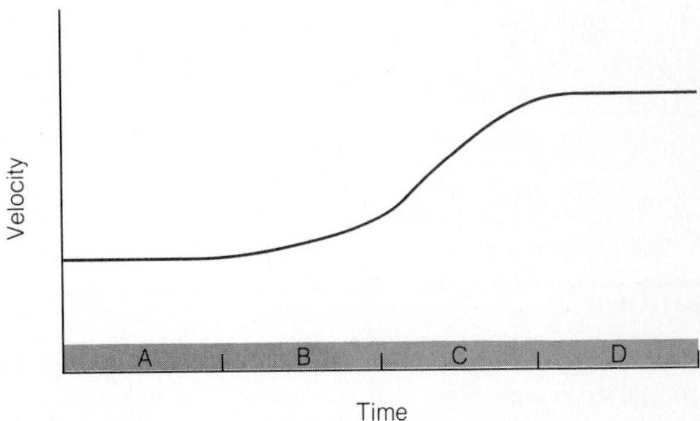

Figure 1.20

Question 5.

Figure 1.19

Question 2.

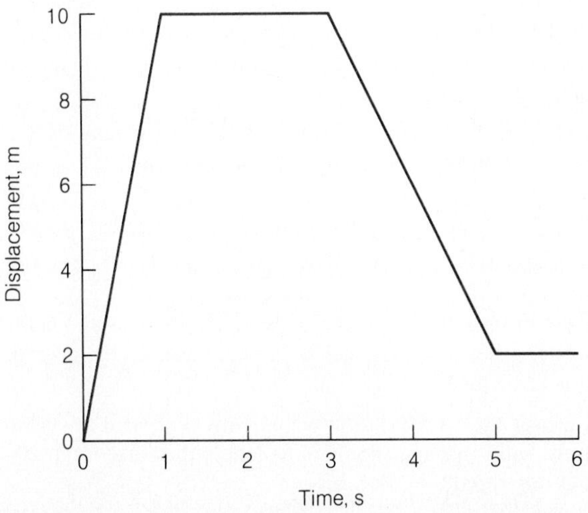

6. Figure 1.21 shows displacement-time graphs for nine cars.

a. Which cars are or have been moving in the forward direction?
b. Which cars are or have been moving in the backward direction?
c. Which car has the highest constant velocity?
d. Which car has the highest constant velocity in the forward direction?
e. Which car has the highest constant velocity in the backward direction?
f. Which cars have the same velocity?
g. Which car has not moved at all?
h. Which car has accelerated from rest to a constant velocity?

i. Which car has been brought to a stop from an initial velocity in the forward direction?

j. Which car has been brought to a stop from an initial velocity in the backward direction?

7. A movie is shown that appears to be of a ball falling through the air. From what appears on the screen, is there any way to determine whether the movie is actually of a ball being thrown upward but the film is being run backward in the projector?

8. The acceleration of gravity at the surface of Venus is 8.9 m/s^2. Would a ball thrown upward on Venus return to the ground sooner or later than a ball thrown upward with the same velocity on the earth?

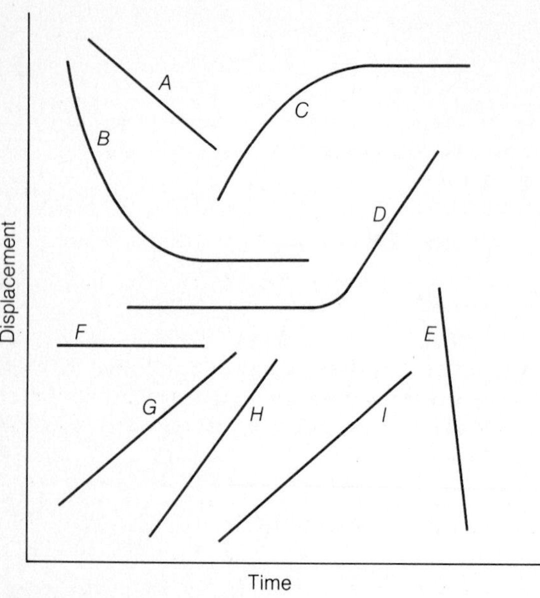

Figure 1.21
Question 6.

EXERCISES

1.1 Units

1. The tallest tree in the world is a Sequoia in California that is 368 ft high. How high is this in meters? In kilometers?

2. A moderately large tree gives off 300 gal of water daily to the atmosphere. How many cubic meters is this?

3. The speedometer of a European car is calibrated in kilometers per hour. What is the car's speed in miles per hour when the speedometer reads 50?

4. In 1968 the horse Dr. Fager ran a mile in 1 min 32.2 s. Find his average speed in meters per second.

5. An acre contains 4840 yd^2, where 1 yd = 3 ft. How many square meters is this? How many acres are there in a square kilometer?

6. Find the volume in cubic meters of a swimming pool whose dimensions are 50 ft × 20 ft × 6 ft.

7. A "board foot" is a unit of lumber measure that corresponds to the volume of a piece of wood 1 ft square and 1 in. thick. How many cubic inches are there in a board foot? How many cubic feet? How many cubic centimeters?

8. Water emerges from the nozzle of a fountain in Arizona at 75 km/h and reaches a height of 170 m. Express the water speed in meters per second and in miles per hour and the height in feet.

1.3 Speed

9. A pitcher throws a baseball at 90 mi/h. How much time does it take the ball to reach the batter 60 ft away?

10. In 1977 Steve Weldon ate 91.44 m of spaghetti in 28.73 s. At the same speed, how long would it take him to eat 15.00 m of spaghetti?

11. The starter of a race stands at one end of a line of runners. What is the difference between the arrival of the sound of his pistol at the nearest runner and at the most distant runner 10 m farther away? The speed of sound in air is 343 m/s. (In a sprint, 0.01 s can mean the difference between winning and coming in second.)

12. The speed of light is 3.0×10^8 m/s. How long does it take light to reach the earth from the sun, which is 1.5×10^{11} m away?

13. The average lifetime of many unstable elementary particles is about 10^{-23} s, and they move at nearly the speed of light (3.0×10^8 m/s) after being created in high-energy collisions between other elementary particles. As measured in the laboratory, how far does such a particle travel on the average before decaying?

14. A car travels 540 km in 4.5 h. How far will it go in 8.0 h at the same average speed? How long will it take to go 200 km at this speed?

15. An airplane takes off at 9:00 A.M. and flies at 300 km/h until 1:00 P.M. At 1:00 P.M. its speed is increased to 400 km/h, and it maintains this speed until it lands at 3:30 P.M. What is the airplane's average speed for the entire flight?

16. A car travels at 100 km/h for 2.0 h, at 60 km/h for the next 2.0 h, and finally at 80 km/h for 1.0 h. Find its average speed for the entire 5.0 h.

17. A police car moving at 140 km/h chases a speeding car moving at 110 km/h. How much closer to the speeding car does the police car get each minute?

18. A police car leaves in pursuit of a holdup car 0.50 h after the latter has left the scene of the crime at 120 km/h. How fast must the police car go if it is to catch up with the holdup car in 1.00 h?

19. A Honda moving at 100 km/h is 80 km behind a Ford moving in the same direction at 60 km/h. How far does the Honda travel before it catches up with the Ford?

1.6 Velocity and Acceleration

20. A baseball moving at 20 m/s is struck by a bat and moves off in the opposite direction at 30 m/s. If the impact lasted 0.010 s, what was the baseball's acceleration?

21. (a) A car's velocity increases from 8 m/s to 20 m/s in 10.0 s. Find its acceleration. (b) The car's velocity then decreases from 20 m/s to 10 m/s in 5.0 s. Find its acceleration in this time interval.

22. The tires of a certain car begin to lose their grip on the pavement at an acceleration of 5.0 m/s^2. If the car has this acceleration, how many seconds does it require to reach a speed of 25 m/s starting from 10 m/s?

23. A car starts from a velocity of 10 m/s with an acceleration of 2.0 m/s^2. Find the time needed for the car to reach 20 m/s.

24. An airplane reaches its takeoff velocity of 60 m/s in 30 s starting from rest. Find the time it spends in going from 40 m/s to 60 m/s.

25. The velocity-time graph in Fig. 1.22 represents the motion of a golf cart. Find its acceleration at t = 25 s and at t = 35 s.

Figure 1.22
Exercise 25.

26. The brakes of a car moving at 14 m/s are suddenly applied and the car comes to a stop in 4.0 s. (a) What was its acceleration? (b) How long would the car take to come to a stop starting from 20 m/s with the same acceleration? (c) How long would the car take to slow down from 20 m/s to 10 m/s with the same acceleration?

27. A passenger in an airplane flying from New York to Los Angeles notes the time at which he passes over various cities and towns. With the help of a map he determines the distances between these landmarks, and compiles the table shown below. Plot the displacement of the airplane versus time from these data, and describe the airplane's motion with the help of the graph.

Time (P.M.):	4:00	5:12	5:41	6:14
Distance (km):	0	660	926	1267
Time (P.M.):	6:39	7:54	9:18	10:00
Distance (km):	1525	2300	3028	3392

28. The velocity-time graph in Fig. 1.23 represents the motion of a certain car. (a) Find the car's acceleration at t = 15, 25, 35, and 45 s. (b) Find the distance covered by the car from t = 0 to t = 30 s and from t = 30 s to t = 50 s.

1.7 Displacement, Time, and Acceleration

29. A Porsche reaches a velocity of 42.0 km/h from a standing start in 15.5 s. What distance does it cover while doing so?

30. A Ferrari covers 100 m from a standing start in 6.0 s at constant acceleration. Find its final velocity.

31. A DC-8 airplane has a takeoff velocity of 80 m/s, which it reaches 35 s after starting from rest. (a) How much time does the airplane spend in going from 0 to 20 m/s? What

Figure 1.23
Exercise 28.

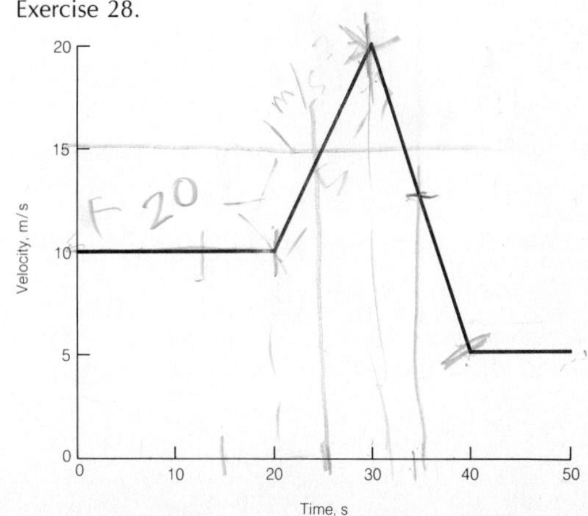

distance does it cover in doing so? (b) How much time does the airplane spend in going from 60 to 80 m/s? What distance does it cover in doing so? (c) What is the minimum length of the runway?

32. A car starts from rest with an acceleration of 0.40 m/s^2. One minute later the acceleration stops, and the car continues at constant velocity for 5.0 min. The car's brakes are then applied, and the car comes to rest in 2.0 min with a uniform deceleration. (a) Show the car's motion on a velocity-time graph. (b) Find the total displacement of the car.

1.8 Displacement, Velocity, and Acceleration

33. A spacecraft takes 200 km after being launched to reach the escape speed from the earth of 11.2 km/s. Find its average acceleration in terms of g.

34. A golf cart has an acceleration of 0.4 m/s^2. What is its velocity after it has covered 10 m starting from rest?

35. A spacecraft has an acceleration of magnitude 5.0 g. What distance is needed for it to attain a velocity of 10 km/s?

36. A car moving at 20 m/s slows down at -1.5 m/s^2 to a velocity of 10 m/s. How far did the car go during the slowdown?

37. An express train passes a certain station at 20 m/s. The next station is 2 km away and the train reaches it 1.0 min later. (a) Did the train's velocity change? (b) If it did, what was its velocity at the second station? (Assume a constant acceleration.)

38. The time interval between a driver noticing a hazard ahead and her application of the brakes is typically 0.68 s. The maximum acceleration produced by the brakes of an average car might be 6.5 m/s^2. Find the distance required to stop such a car from the moment a hazard is seen when the initial velocity is (a) 25 km/h, (b) 50 km/h, and (c) 100 km/h.

1.10 Free Fall

39. Is it true that an object dropped from rest falls three times farther in the second second after being released than it does in the first second?

40. How fast must a ball be thrown upward to reach a height of 12 m?

41. Divers in Acapulco, Mexico, leap from a point 36 m above the sea. What is their velocity when they enter the water?

42. A stone is dropped from a cliff 490 m above its base. How long does the stone take to fall?

43. A ball dropped from the roof of a building takes 4.0 s to reach the street. How high is the building?

44. A bullet is fired vertically upward and returns to the ground in 20 s. Find the height it reaches.

45. Find the initial and final velocities of a ball thrown vertically upward that returns to the thrower 3.0 s later.

46. A stone is thrown vertically upward at 9.8 m/s. When will it reach the ground?

47. A ball is thrown vertically downward at 10 m/s. What is its velocity 1.0 s later? 2.0 s later?

48. A ball is thrown vertically upward at 10 m/s. What is its velocity 1.0 s later? 2.0 s later?

49. A lead pellet is propelled vertically upward by an air rifle with an initial velocity of 16 m/s. Find its maximum height.

50. At what velocity should a stone be thrown downward from a cliff 120 m high to reach the ground in 4.0 s?

51. An apple is thrown vertically downward from a cliff 48 m high and reaches the ground 2.0 s later. What was the apple's initial velocity?

52. From Fig. 1.15 we can see that the air resistance experienced by a falling object is not an important factor until a speed of about half its terminal speed is reached. The terminal speed of a golfball is 40 m/s. How much time is needed for a dropped golfball to reach a speed of half this? How far does it fall in this time?

53. The acceleration of gravity at the surface of Mars is 3.7 m/s^2. A stone thrown upward on Mars reaches a height of 15 m. (a) Find its initial velocity. (b) What is the total time of flight?

54. A movie scene in which a car is supposed to fall from a cliff 36 m high is filmed using a model car that falls from a model cliff 1.0 m high. The film will be shown at 24 frames/s. How fast should the camera be run for the result to be realistic?

55. In a novel a pirate forces a captured sailor to "walk the plank" (something unknown in the history of piracy), from which he falls into the sea 7.0 m below. The author claims the sailor's entire life flashes before him during his fall. (a) How much time is available for this? (b) With what velocity will the sailor enter the water?

56. A bagel is thrown vertically upward with a velocity of 20 m/s from a bridge. On the way down the bagel just misses the bridge and falls into the water 5.0 s after having been thrown. Find the height of the bridge above the water.

57. A British parachutist bails out at an altitude of 150 m and accidentally drops his monocle. If he descends at the constant velocity of 6.0 m/s, how much time separates the arrival of the monocle on the ground from the arrival of the parachutist himself?

58. A Russian balloonist floating at an altitude of 150 m accidentally drops his samovar and starts to ascend at the constant velocity of 1.2 m/s. How high will the balloon be when the samovar reaches the ground?

59. An orangutan throws a coconut vertically upward at the foot of a cliff 40 m high while his mate simultaneously drops another coconut from the top of the cliff. The two coconuts collide at an altitude of 20 m. What was the initial velocity of the coconut that was thrown upward?

60. When a flea jumps, it accelerates through about 0.80 mm (a little less than the length of its legs) and is able to reach a height of as much as 10 cm. (a) Find the flea's acceleration (assumed constant) and its velocity at takeoff when this occurs. (b) When a person jumps, the acceleration distance is about 50 cm. If the acceleration of a person were the same as that of a flea, find the velocity at takeoff and the height that would be reached.

61. A helicopter is climbing at 8.0 m/s when it drops a pump near a leaking boat. The pump reaches the water 4.0 s afterward. How high was the helicopter when the pump was dropped? When the pump reached the water?

PROBLEMS

62. Figure 1.24 is a plot of displacement versus time for a beetle moving along a straight stick. (a) Plot the beetle's velocity versus time. (b) What is the beetle's average velocity during the period shown? (c) What is its average speed during the same period?

63. A snake is slithering toward you at 1.5 m/s. If you start walking when it is 5.0 m away, how fast must you go in order that the snake not overtake you when you have gone 40.0 m?

64. A car covers one-quarter of the distance to the next town at 40 km/h, another quarter at 50 km/h, and the rest at 90 km/h. Find the car's average speed for the entire distance.

65. A man drives one-third of the distance home at 25 mi/h. How fast must he drive the rest of the way in order to average 35 mi/h for the entire trip?

66. A woman on a bicycle climbs a hill at 4.0 m/s and then returns down the same hill at 8.0 m/s. What is her average speed for the entire trip?

67. The odometer of a car is checked at 1.0-min intervals, and the readings below are obtained. Calculate the speed of the car in each time interval and plot the results on a graph. Describe the motion of the car with the help of this graph.

Time (min):	0	1.0	2.0	3.0
Distance (km):	42.20	42.74	43.64	44.90
Time (min):	4.0	5.0	6.0	7.0
Distance (km):	46.34	47.78	49.22	50.66
Time (min):	8.0	9.0	10.0	
Distance (km):	51.74	52.10	52.10	

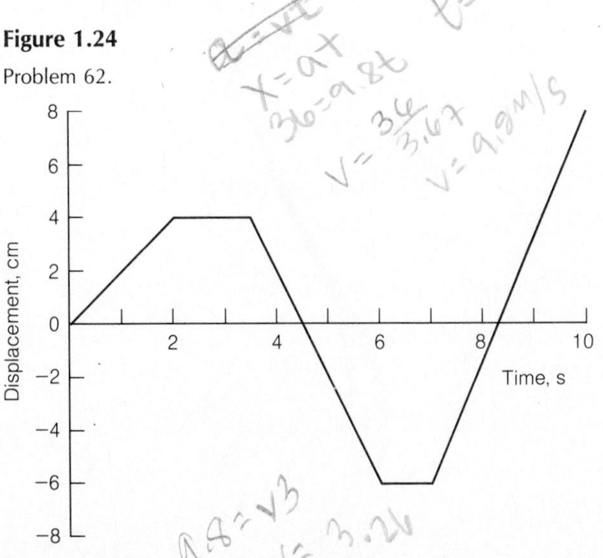

Figure 1.24
Problem 62.

68. A European train passes successive kilometer posts at the times given below. Plot the data on a graph and determine whether the train's speed is constant over the entire distance. If it not constant, plot the train's speed in each time interval versus time and find the acceleration. What are the train's initial and final speeds?

Distance (km):	0	1.0	2.0	3.0	4.0
Time (s):	0	64	114	156	193
Distance (km):		5.0	6.0	7.0	8.0
Time (s):		227	259	292	324

69. A car that is leaking oil at a constant rate is uniformly accelerated from rest starting out just when a drop falls. The first drop down the road is 0.80 m from the car's starting point. How far from the starting point is the fourth drop?

70. A sprinter accelerates from rest until he reaches a velocity of 12.0 m/s and then continues running at this velocity. If he takes 11.0 s to cover 100 m, what was his acceleration and how long did it last?

71. A bus travels 400 m between two stops. It starts from rest and accelerates at 1.5 m/s^2 until it reaches a velocity of 9.0 m/s. The bus continues at this velocity and then decelerates at 2.0 m/s^2 until it comes to a halt. Find the total time required for the journey.

72. A rocket is launched upward with an acceleration of 100 m/s^2. Eight seconds later the fuel is exhausted. Find (a) the highest velocity the rocket attains; (b) the maximum altitude; (c) the total time of flight; and (d) the velocity with which the rocket strikes the ground.

73. A girl throws a ball vertically upward at 10 m/s from the roof of a building 20 m high. (a) How long will it take the ball to reach the ground? (b) What will its velocity be when it strikes the ground?

74. A girl throws a ball vertically downward at 10 m/s from the roof of a building 20 m high. (a) How long will it take the ball to reach the ground? (b) What will its velocity be when it strikes the ground?

75. An elevator has a maximum acceleration of ± 1.5 m/s^2 and a maximum velocity of 6.0 m/s. Find the shortest period of time required for it to take a passenger to the tenth floor of a building from street level, a height of 50 m, with the elevator coming to a stop at this floor.

76. A person in an elevator drops an apple from a height 2.0 m above the elevator's floor and, with a stopwatch, times the fall of the apple to the floor. What is found (a) when the elevator is ascending with an acceleration of 1.0 m/s^2; (b) when it is descending with an acceleration of 1.0 m/s^2; (c) when it is ascending at a constant velocity of 3.0 m/s; (d) when it is descending at a constant veloicity of 3.0 m/s; and (e) when the cable has broken and it is descending in free fall?

77. When a certain ball is dropped to the ground from a height of 2.0 m, it bounces back up to its original height. (a) Plot a graph of the height of the ball versus time for the first 3.0 s of its motion. (b) Plot a graph of the ball's velocity during the same period.

78. A champagne bottle is held upright 1.2 m above the floor as the wire around its cork is removed. The cork then pops out, rises vertically, and falls to the floor 1.4 s later. (a) What height above the bottle did the cork reach? (b) What was the cork's initial velocity? (c) Its final velocity?

Answers to Multiple Choice

1. d	**8.** a	**15.** c	**22.** a
2. b	**9.** b	**16.** a	**23.** b
3. b	**10.** b	**17.** a	**24.** c
4. b	**11.** b	**18.** c	**25.** a
5. a	**12.** c	**19.** d	**26.** b
6. c	**13.** c	**20.** c	**27.** c
7. a	**14.** b	**21.** a	**28.** c

C H A P T E R

2

VECTORS

Some physical quantities are completely specified by just a number and a unit. If we say that a person's mass is 70 kg, that the area of a farm is 160 acres, or that the frequency of a sound wave is 660 cycles/s, there is nothing more to add. These are examples of **scalar quantities.**

Other quantities have directions as well as magnitudes associated with them. Your mass is the same whether you stand on your feet or on your head. However, if you drive north from Denver at 80 km/h you will end up in Canada, whereas if you drive south from Denver you will end up in Mexico. To say that your car's speed is 80 km/h, a scalar statement, is evidently not enough to describe its motion completely. A quantity whose direction is significant is called a **vector quantity.** Thus velocity, which incorporates direction as well as speed, is a vector quantity. Vector quantities occur often in physics, and their arithmetic is different from the ordinary arithmetic of scalar quantities.

2.1 Vector Quantities

Which way as well as how much.

Displacement, velocity, and acceleration, which were considered in Chapter 1, are vector quantities. Another vector quantity is **force.** A force is often spoken of as a "push" or a "pull." There is more to the concept of force than this, but it is all we need for the time being. Forces are responsible for all changes in motion: A force is needed to start a stationary object moving, to change its direction of motion, and to stop it. We have to know the direction as well as the magnitude of a force to know what its effects will be. Push this book down on the table and nothing happens. Push the book up and it will rise into the air. The SI unit of force (and of weight, since weight is a force) is the **newton** (N), which is equivalent to a little less than 0.225 lb. The weight of an apple is about 1 N.

We can represent a vector quantity by a straight line with an arrowhead at one end to show the direction of the quantity. The length of the line is proportional to the magnitude of the quantity. Such a line is called a **vector.**

Figure 2.1 shows how a 640-km westward displacement can be represented by a vector. The compass rose establishes the orientation of the displacement, and the distance scale establishes the relationship between length in the diagram and the actual length.

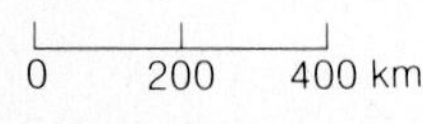

Figure 2.1

The arrowed line is a vector that represents a 640-km westward displacement.

Other vector quantities besides displacements can also be represented by vectors. The length of the vector in each case is proportional to the magnitude of the quantity it represents, and its direction is the direction of the quantity. Thus a velocity whose magnitude is 25 m/s is represented on a scale of 1 mm = 1 m/s by an arrow 25 mm long, and a force whose magnitude is 1500 N is represented on a scale of 1 mm = 100 N by an arrow 15 mm long (Fig. 2.2). The directions of the arrows indicate the directions of the quantities.

The symbols of vector quantities are customarily printed in boldface type (**F** for force), whereas italics are used both for scalar quantities (*m* for mass) and for the magnitudes of vectors. Thus if we denote the 640-km westward displacement by the symbol **A,** its magnitude is $A = 640$ km. In handwriting, vector quantities are indicated by placing arrows over their symbols—for instance, $\vec{A}$.

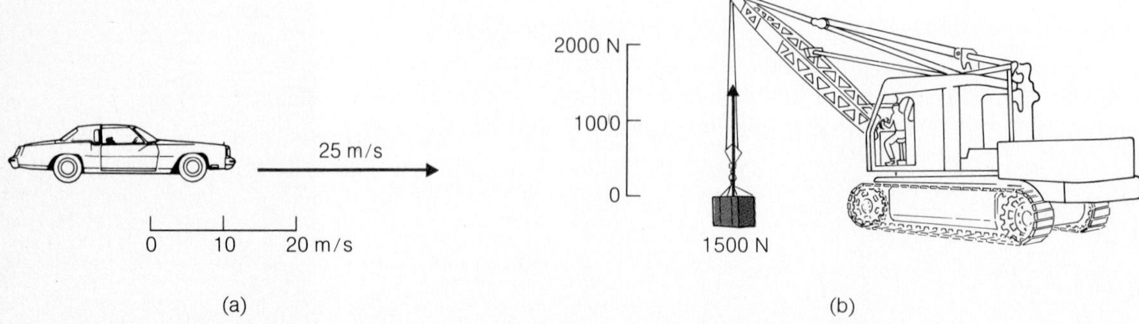

Figure 2.2

(a) A vector that represents a velocity of 25 m/s. (b) A vector that represents a force of 1500 N.

2.2 Vector Addition

The sum of two or more vectors is called their resultant.

Ordinary arithmetic is used to add two or more scalar quantities of the same kind together. Ten kg of potatoes plus 4 kg of potatoes equals 14 kg of potatoes. The same arithmetic is used for vector quantities of the same kind when the directions are the same. A total upward force of 100 N is required to lift a 100-N weight, regardless of whether the force is applied by one, two, or ten people. It is the total force on an object, not the individual forces, that affects the motion of the object.

What do we do when the directions are different? A person who walks 10.0 km north and then 4.0 km west does *not* undergo a displacement of 14.0 km from his starting point, even though he has walked a total distance of 14.0 km. A vector diagram provides a convenient method of determining his displacement. The procedure is to make a scale drawing of the successive displacements **A** and **B,** as in Fig. 2.3, and to join the starting point and the end point with a single vector **R.** The net displacement is **R,** whose length corresponds to 10.8 km and whose direction corresponds to 22° west of north. These values can be found by using a ruler and protractor.

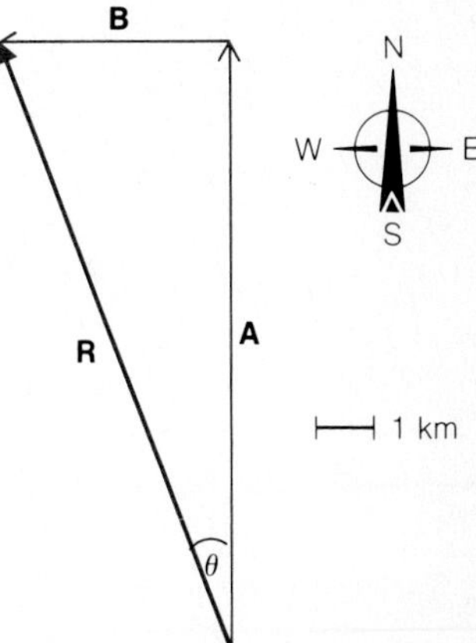

Figure 2.3

The displacements of 10.0 km north, **A**, and 4.0 km west, **B**, add to give a resultant displacement **R** of 10.8 km in a direction 22° west of north.

The general rule for adding vectors is illustrated in Fig. 2.4. To add the vector **B** to the vector **A,** we follow these two steps:

---

1. **Shift B parallel to itself until its tail is at the head of A. In its new position B must still have its original length and direction.**
2. **Draw a new vector R (called the *resultant*) from the tail of A to the head of B. The vector R is equal to A + B.**

To find the magnitude R of the resultant **R,** we measure the length of **R** on the diagram and compare this length with the scale. The direction of **R** with respect to **A** or **B** may be found with a protractor. The order in which vectors are added does not matter, so that

$$\mathbf{A} + \mathbf{B} = \mathbf{B} + \mathbf{A} \qquad \textit{Vector addition} \quad (2.1)$$

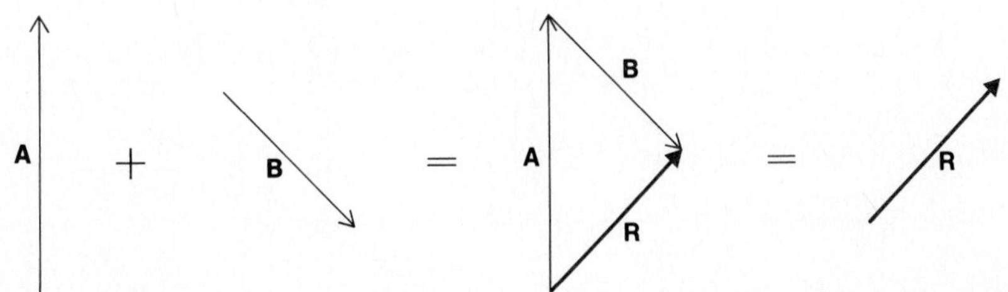

Figure 2.4

The addition of two vectors to obtain the resultant **R**.

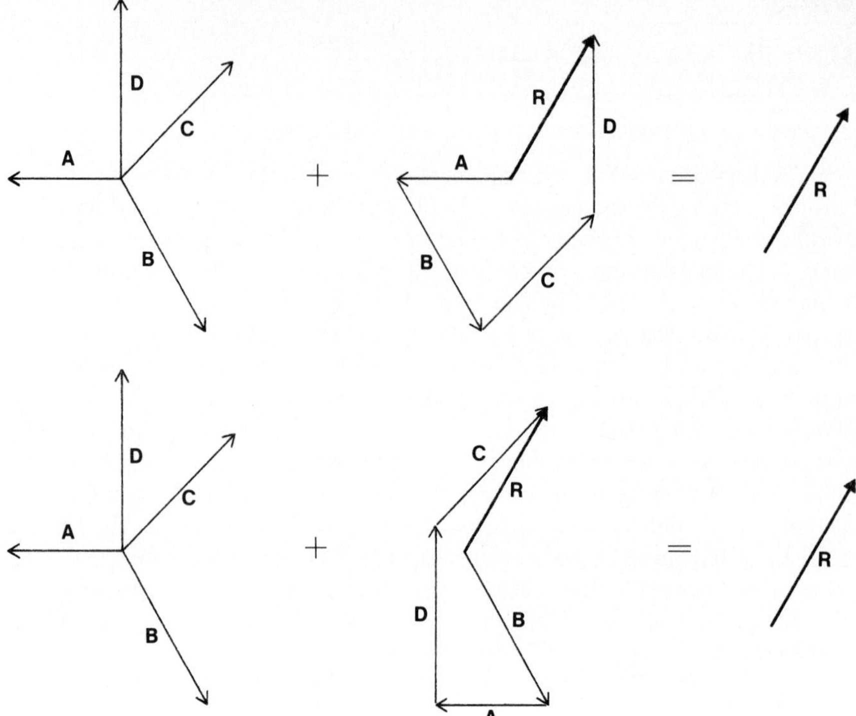

Figure 2.5

The addition of four vectors to obtain the resultant **R**. The order in which the vectors are added does not affect the magnitude or direction of **R**.

Any number of vectors of the same kind can be added in this way. First place the tail of each vector at the head of the previous one, being sure to keep their lengths and original directions unchanged. Then draw a vector **R** from the tail of the first vector to the head of the last one. Figure 2.5 shows how four vectors are added together in two different ways to get the same resultant **R.**

Trigonometric Method

For an approximate result a ruler and protractor are sufficient. Trigonometry provides more accuracy and is especially easy to apply when the vectors to be added are perpendicular to each other. Basic trigonometry is reviewed in Appendixes A.7 and A.8 and is summarized in Fig. 2.6.

Figure 2.6

Basic trigonometric formulas.

$$\sin\theta = \frac{\text{opposite side}}{\text{hypotenuse}} = \frac{A}{C}$$

$$\cos\theta = \frac{\text{adjacent side}}{\text{hypotenuse}} = \frac{B}{C}$$

$$\tan\theta = \frac{\text{opposite side}}{\text{adjacent side}} = \frac{A}{B}$$

Pythagorean theorem: $A^2 + B^2 = C^2$

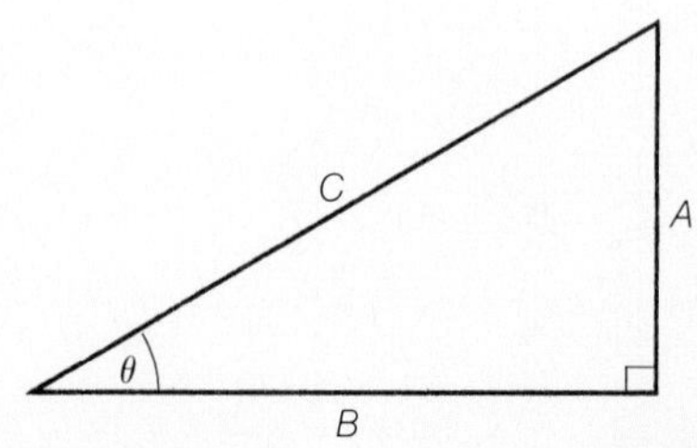

EXAMPLE 2.1

A person walks 10.0 km north and then 4.0 west, as in Fig. 2.3. Use trigonometry to find the person's displacement.

SOLUTION Because **A** is perpendicular to **B,** we can use the Pythagorean theorem to find the magnitude R of the resulting displacement **R.** According to this theorem, the square of the hypotenuse (the long side) of a right triangle equals the sum of the squares of the other two sides. Hence

$$R^2 = A^2 + B^2$$
$$R = \sqrt{A^2 + B^2} = \sqrt{(10.0\,\text{km})^2 + (4.0\,\text{km})^2} = 10.8\,\text{km}$$

The angle θ between **R** and **A** is specified by

$$\tan\theta = \frac{B}{A} = \frac{4.0\,\text{km}}{10.0\,\text{km}} = 0.40$$

When we look at a table of trigonometric functions, we find that the angle whose tangent is closest to 0.40 is 22°, since tan 22° = 0.404. To the nearest degree,

$$\theta = 22°$$

Alternatively, we can use an electronic calculator to find θ by entering 0.40 and pressing the $\tan^{-1}$ (sometimes arctan or inv tan) key. ◆

2.3 Vector Subtraction

The negative of a vector points in the opposite direction.

It is sometimes necessary to subtract one vector from another. To subtract **A** from **B,** for example, we first form the negative of **A,** denoted −**A,** a vector that has the same length as **A** but points in the opposite direction (Fig. 2.7). We then add −**A** to **B** in the usual manner. This procedure may be summarized as

$$\mathbf{B} - \mathbf{A} = \mathbf{B} + (-\mathbf{A}) \qquad \textit{Vector subtraction} \quad (2.2)$$

Vector subtraction will be used in Chapter 6 to find the acceleration of an object that moves in a curved path.

Figure 2.7

Vector subtraction.

2.4 Relative Velocity

Changing the frame of reference changes the velocity.

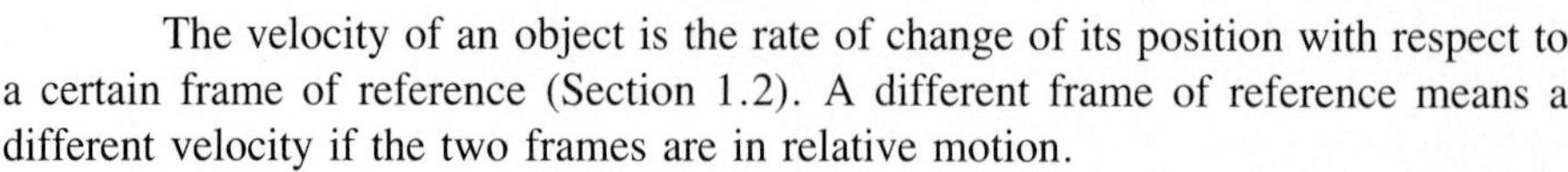

The velocity of an object is the rate of change of its position with respect to a certain frame of reference (Section 1.2). A different frame of reference means a different velocity if the two frames are in relative motion.

All motion is relative to a chosen frame of reference. Here the photographer has turned the camera to keep pace with one of the cyclists. Relative to this cyclist, the road and the other cyclists are all moving.

Suppose we are in a boat cruising on a river that is itself flowing past the shore. The water in the river is one frame of reference and the shore is another frame of reference. We are interested in three velocities:

$$\mathbf{v}_{BW} = \text{velocity of the boat relative to the water}$$
$$\mathbf{v}_{WS} = \text{velocity of the water relative to the shore}$$
$$\mathbf{v}_{BS} = \text{velocity of the boat relative to the shore}$$

The boat's velocity relative to the shore is the vector sum of its velocity relative to the water and the velocity of the water relative to the shore:

$$\mathbf{v}_{BS} = \mathbf{v}_{BW} + \mathbf{v}_{WS} \tag{2.3}$$

The subscript letters that identify each velocity are chosen so that the first letter refers to the object whose motion is being described. The second letter refers to the frame of reference for that velocity. In Eq. (2.3) the first subscript letter on the left-hand side corresponds to the first subscript letter on the right-hand side, and the last subscript letter on the left-hand side corresponds to the last subscript letter on the right-hand side. This scheme avoids the confusion that can arise in situations less straightforward than one of a boat in a river.

Figure 2.8

A boat crossing a river must head upstream (that is, against the current) to reach a point on the other shore opposite its starting point.

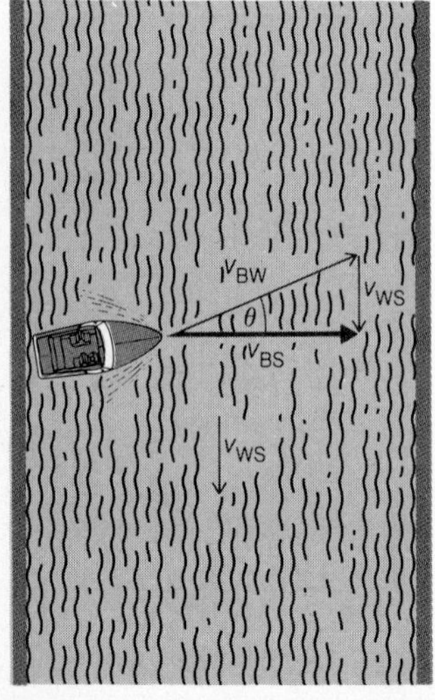

EXAMPLE 2.2

A boat moving at 5.0 m/s relative to the water is crossing a river 300 m wide whose current is 2.0 m/s relative to the shore, as in Fig. 2.8. (a) In what direction should the boat head if it is to reach a point on the other shore directly opposite its starting point? (b) How long does the boat take to cross the river if it heads in this direction?

SOLUTION (a) According to Eq. (2.3)

$$\mathbf{v}_{BS} = \mathbf{v}_{BW} + \mathbf{v}_{WS}$$

Here the boat's velocity $\mathbf{v}_{BS}$ relative to the shore is perpendicular to the river current $\mathbf{v}_{WS}$. Hence

$$\sin\theta = \frac{v_{WS}}{v_{BW}} = \frac{2.0\ \text{m/s}}{5.0\ \text{m/s}} = 0.40$$
$$\theta = 24^\circ$$

The boat must head upstream (against the current) at 24° relative to a direct course across the river.

(b) Using the Pythagorean theorem, we find the magnitude v_{BS} of the boat's velocity relative to the shore as follows:

$$v_{BW}^2 = v_{BS}^2 + v_{WS}^2$$
$$v_{BS} = \sqrt{v_{BW}^2 - v_{WS}^2} = \sqrt{(5.0\text{ m/s})^2 - (2.0\text{ m/s})^2} = 4.6\text{ m/s}$$

The time needed for the boat to cross the river is

$$t = \frac{x}{v_{BS}} = \frac{300\text{ m}}{4.6\text{ m/s}} = 65\text{ s}$$

♦

EXAMPLE 2.3

A car is moving north at 80 km/h, and a bus is moving east at 50 km/h; both of these velocities are relative to the earth's surface. (a) Find the velocity of the car relative to the bus. (b) Find the velocity of the bus relative to the car.

SOLUTION (a) The three velocities here are

$\mathbf{v}_{CE}$ = velocity of the car relative to the earth

$\mathbf{v}_{BE}$ = velocity of the bus relative to the earth

$\mathbf{v}_{CB}$ = velocity of the car relative to the bus

Using the same reasoning that led to Eq. (2.3) gives

$$\mathbf{v}_{CE} = \mathbf{v}_{CB} + \mathbf{v}_{BE}$$

The order of the subscript letters follows the rule given earlier. Solving for $\mathbf{v}_{CB}$ gives

$$\mathbf{v}_{CB} = \mathbf{v}_{CE} - \mathbf{v}_{BE}$$

Thus the velocity of the car from the frame of reference of the bus is the difference $\mathbf{v}_{CE} - \mathbf{v}_{BE}$ between their velocities relative to the earth (Fig. 2.9).

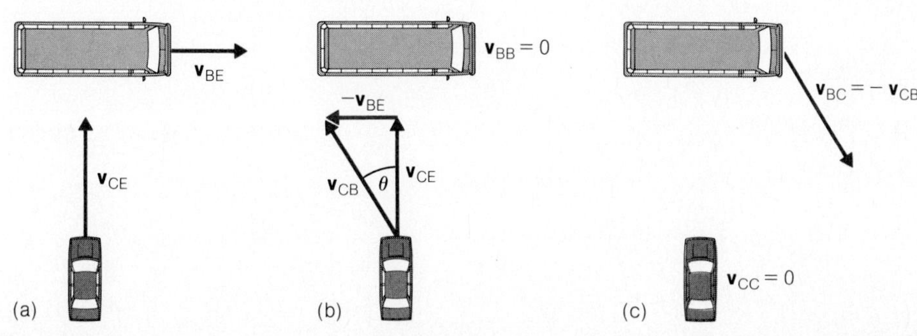

Figure 2.9

(a) The velocities of the car and bus in Example 2.3 from the frame of reference of the earth. (b) From the frame of reference of the bus, the bus is at rest ($\mathbf{v}_{BB} = 0$) and the earth is moving relative to it at the velocity $-\mathbf{v}_{BE}$. Hence the car is moving relative to the bus at the velocity $\mathbf{v}_{CB} = \mathbf{v}_{CE} + (-\mathbf{v}_{BE})$. (c) The velocity of the bus relative to the car is equal in magnitude and opposite in direction to the velocity of the car relative to the bus. Relative to the car, the car's own velocity is $\mathbf{v}_{CC} = 0$.

This result can also be obtained by noting that, to someone in the bus, the bus is at rest and the earth is moving relative to it at the velocity $-\mathbf{v}_{BE}$. To someone on the bus, the car is moving at $\mathbf{v}_{CE}$ relative to the earth, while the earth is moving at $-\mathbf{v}_{BE}$ relative to the bus. The sum of these velocities is $\mathbf{v}_{CB} = \mathbf{v}_{CE} + (-\mathbf{v}_{BE}) = \mathbf{v}_{CE} - \mathbf{v}_{BE}$ as before.

Since $\mathbf{v}_{CE}$ is perpendicular to $\mathbf{v}_{BE}$, the magnitude of $\mathbf{v}_{CB}$ is

$$\begin{aligned} v_{CB} &= \sqrt{v_{CE}{}^2 + (-v_{BE})^2} \\ &= \sqrt{v_{CE}{}^2 + v_{BE}{}^2} = \sqrt{(80\ \text{km/h})^2 + (50\ \text{km/h})^2} = 94\ \text{km/h} \end{aligned}$$

The angle θ between $\mathbf{v}_{CB}$ and $\mathbf{v}_{CE}$ is found from the given values of v_{CE} and v_{BE}:

$$\begin{aligned} \tan\theta &= \frac{v_{BE}}{v_{CE}} = \frac{50\ \text{km/h}}{80\ \text{km/h}} = 0.625 \\ \theta &= 32^\circ \end{aligned}$$

The direction of $\mathbf{v}_{CB}$ is 32° west of north.

(b) The velocity $\mathbf{v}_{BC}$ of the bus relative to the car has the same magnitude as the velocity $\mathbf{v}_{CB}$ of the car relative to the bus but is in the opposite direction:

$$\mathbf{v}_{BC} = -\mathbf{v}_{CB}$$

Hence v_{BC} is 94 km/h, and the direction of $\mathbf{v}_{BC}$ is 32° east of south, as in Fig. 2.9(c). ♦

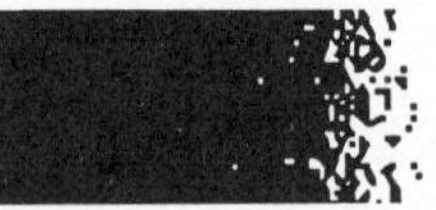

2.5 Resolving a Vector

How to replace a vector with two or more other vectors called its components.

Just as we can add two or more vectors to give a single resultant vector, so we can break a vector up into two or more other vectors. The latter process is called **resolving** the vector. The new vectors are called the **components** of the original one (Fig. 2.10). As we will often find, the best way to analyze a physical situation may be to resolve a vector into components. Almost invariably the components are chosen

Figure 2.10

(a) In vector addition, two or more vectors are combined to form a single vector. (b) In vector resolution, a single vector is replaced by two or more other vectors whose sum is the same as the original vector. Here **A** and **B** are the components of the vector **C**.

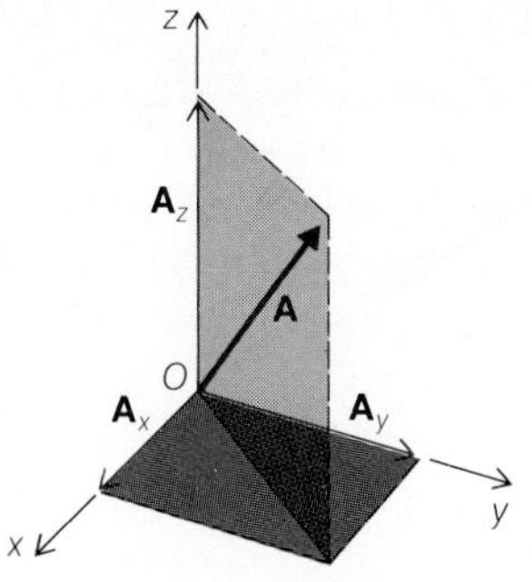

Figure 2.11

How a vector **A** can be resolved into three mutually perpendicular components $\mathbf{A}_x$, $\mathbf{A}_y$, and $\mathbf{A}_z$.

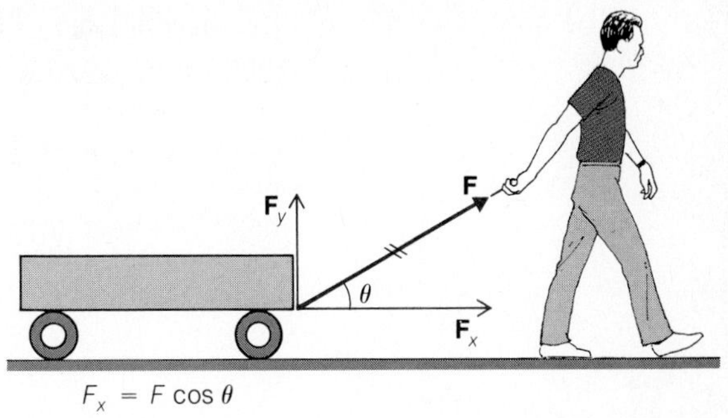

Figure 2.12

The resolution of a force vector into horizontal and vertical components. The original vector is indicated by two short lines across it.

to be perpendicular to one another. Usually two components, in directions labeled x and y, are enough, but sometimes three are needed, in x-, y-, and z-directions (Fig. 2.11).

Figure 2.12 shows a boy pulling a wagon with a rope at the angle θ above the ground. Only part of the force he exerts affects the horizontal motion, since the wagon moves horizontally while the force **F** is not a horizontal one. We can resolve **F** into two components, $\mathbf{F}_x$ and $\mathbf{F}_y$, where

$$\mathbf{F}_x = \text{horizontal component of } \mathbf{F}$$
$$\mathbf{F}_y = \text{vertical component of } \mathbf{F}$$

It is a good habit to distinguish between the original vector and its components when all are on the same diagram by drawing two short lines across the original vector, as in Fig. 2.12. This reminds us that the original vector can now be ignored since it has been replaced by its components.

The magnitudes of $\mathbf{F}_x$ and $\mathbf{F}_y$ are

$$F_x = F \cos\theta \qquad (2.4)$$
$$F_y = F \sin\theta \qquad (2.5)$$

The horizontal component $\mathbf{F}_x$ is responsible for the wagon's motion, while the vertical component $\mathbf{F}_y$ pulls upward on it. $\mathbf{F}_x$ is the projection of **F** in the horizontal direction, and $\mathbf{F}_y$ is the projection of **F** in the vertical direction.

EXAMPLE 2.4

If the force the boy exerts on the wagon in Fig. 2.12 is 60.0 N and $\theta = 24°$, find F_x and F_y. Does their sum equal 60.0 N? Should it?

SOLUTION We have (Fig. 2.13)

$$F_x = F \cos\theta = (60.0\text{ N})(\cos 24°) = 54.8\text{ N}$$
$$F_y = F \sin\theta = (60.0\text{ N})(\sin 24°) = 24.4\text{ N}$$

Figure 2.13

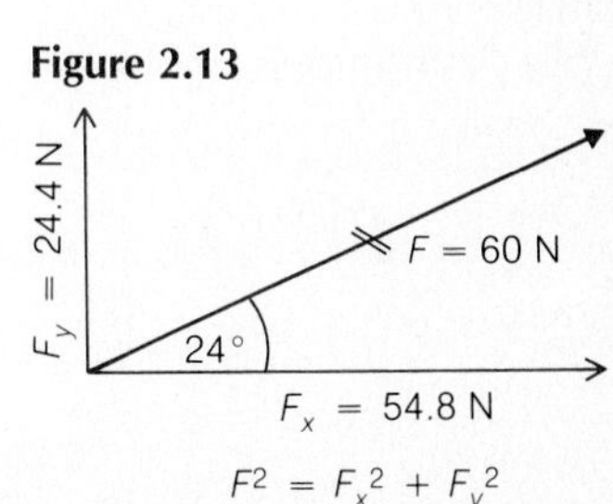

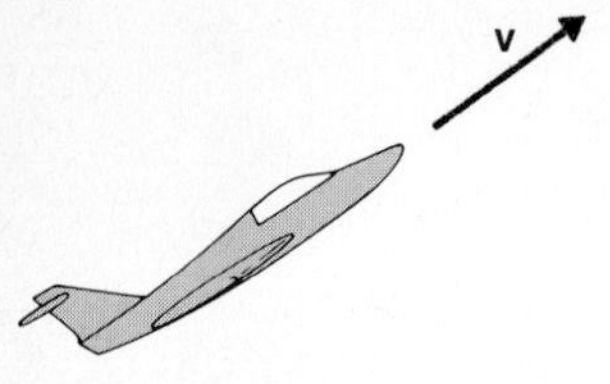

The algebraic sum of the magnitudes F_x and F_y is 54.8 N + 24.4 N = 79.2 N, although the boy has exerted a force of only 60.0 N. Where is the mistake?

The answer is that there is no mistake. $\mathbf{F}_x$ and $\mathbf{F}_y$ are vectors whose directions are different, so they can *only* be added vectorially. The algebraic sum of the magnitudes F_x and F_y has no meaning. If we add $\mathbf{F}_x$ and $\mathbf{F}_y$ vectorially, with the help of the Pythagorean theorem we find that

$$F^2 = F_x^2 + F_y^2$$
$$F = \sqrt{F_x^2 + F_y^2} = \sqrt{(54.8\ \text{N})^2 + (24.4\ \text{N})^2} = 60.0\ \text{N}$$

which is equal to the force the boy exerts. ◆

EXAMPLE 2.5

A woman on the ground sees an airplane climbing at an angle of 35° above the horizontal. She gets into her car and by driving at 120 km/h is able to stay directly below the airplane. What is the airplane's speed?

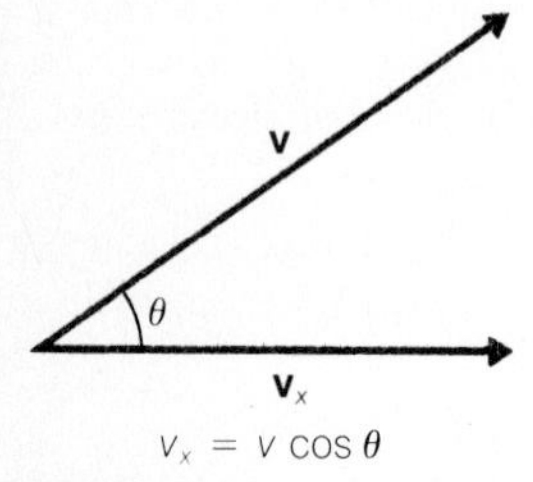

$v_x = v \cos \theta$

Figure 2.14

SOLUTION The car's velocity is equal to the horizontal component $\mathbf{v}_x$ of the airplane's velocity **v** (Fig. 2.14). Since

$$v_x = v \cos \theta$$

we have for the airplane's speed

$$v = \frac{v_x}{\cos \theta} = \frac{120\ \text{km/h}}{\cos 35^\circ} = 146\ \text{km/h}$$

◆

The components of a vector do not have to be horizontal and vertical—they can have whatever orientation is convenient for the situation we are considering. A car on an inclined driveway is an example (Fig. 2.15). If its brakes are released, the force that causes the car to roll down the driveway is the component $\mathbf{F}_\parallel$ of the car's weight **w** along the driveway. The steeper the driveway, the greater the force $\mathbf{F}_\parallel$.

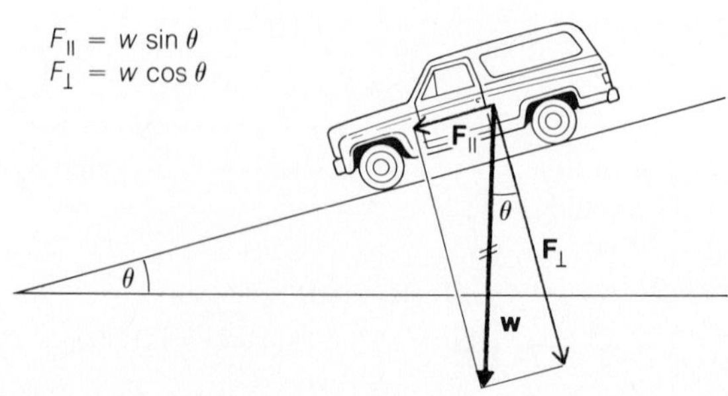

Figure 2.15

Weight is a force that acts vertically downward, but it can be resolved into components parallel and perpendicular to a surface of arbitrary orientation.

EXAMPLE 2.6

A car weighing 12.0 kN (2700 lb) is parked on a driveway that is at a 15° angle with the horizontal. Find the components of the car's weight parallel and perpendicular to the driveway.

SOLUTION The weight **w** of anything is the gravitational force the earth exerts on it, a force that acts vertically downward. Because **w** is vertical and $\mathbf{F}_\perp$ is perpendicular to the road, the angle θ between **w** and $\mathbf{F}_\perp$ is equal to the angle $\theta = 15°$ between the road and the horizontal. Hence

$$F_\parallel = w \sin \theta = (12.0 \text{ kN})(\sin 15°) = 3.1 \text{ kN}$$

$$F_\perp = w \cos \theta = (12.0 \text{ kN})(\cos 15°) = 11.6 \text{ kN}$$

◆

2.6 Vector Addition by Components

Divide and conquer.

As we have seen, we can add any number of vectors graphically just by drawing them head-to-tail. A line from the tail of the first vector to the head of the last is the resultant **R**. This is easy to do but not very accurate.

A problem of this kind can be solved by trigonometry in two ways. One is to begin by adding the first two vectors together, then adding a third to the resultant of the first two, a fourth to the resultant of the first three, and so on. This method usually means a lot of work with many chances for error.

A better way to add several vectors is to work in terms of their components. The procedure is as follows for vectors that lie in the same plane:

1. Resolve the initial vectors into their components in the x- and y-directions.
2. Add the components in the x-direction to give $\mathbf{R}_x$ and add the components in the y-direction to give $\mathbf{R}_y$. That is,

$$\begin{aligned} \mathbf{R}_x &= x\text{-component of } \mathbf{R} \\ &= \mathbf{A}_x + \mathbf{B}_x + \mathbf{C}_x + \cdots = \text{sum of } x\text{-components} \end{aligned} \tag{2.6}$$

$$\begin{aligned} \mathbf{R}_y &= y\text{-component of } \mathbf{R} \\ &= \mathbf{A}_y + \mathbf{B}_y + \mathbf{C}_y + \cdots = \text{sum of } y\text{-components} \end{aligned} \tag{2.7}$$

3. Find the magnitude and direction of the resultant **R** from the components $\mathbf{R}_x$ and $\mathbf{R}_y$. From the Pythagorean theorem,

$$R = \sqrt{R_x^2 + R_y^2} \tag{2.8}$$

The direction of **R** can be found from the values of the components by trigonometry. The best way to do this depends on the problem at hand.

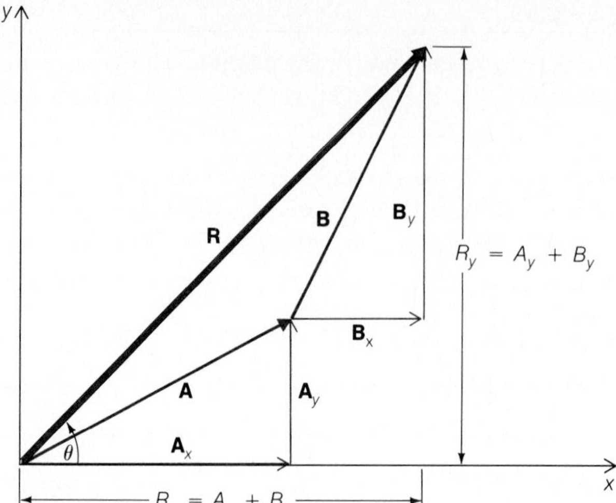

Figure 2.16
Vector addition by components.

Figure 2.16 shows the connection between head-to-tail vector addition and the component method for the case of two vectors.

EXAMPLE 2.7

The sailboat *Ardent Spirit* is headed due north at a forward speed of 6.0 knots (kn). The pressure of the wind on its sails causes the boat to move sideways to the east at 0.5 kn. A tidal current is flowing to the southwest at 3.0 kn. What is the velocity of *Ardent Spirit* relative to the earth's surface?

(A *knot* is a unit of speed equal to one nautical mile per hour. The nautical mile is widely used in air and sea navigation because it is the same in length as one minute [1′] of latitude, where 60′ = 1°. Since 1 nautical mile = 1.852 km = 6076 ft, 1 kn = 1.852 km/h = 1.151 mi/h.)

SOLUTION For convenience, we shall call north the $+y$-direction, south the $-y$-direction, east the $+x$-direction, and west the $-x$-direction. With the help of Fig. 2.17 we see that the magnitudes of the components of the three velocity vectors are

$$A_x = 0 \qquad C_x = -(3.0\ \text{kn})(\cos 45°)$$
$$A_y = 6.0\ \text{kn} \qquad = -2.1\ \text{kn}$$
$$B_x = 0.5\ \text{kn} \qquad C_y = -(3.0\ \text{kn})(\sin 45°)$$
$$B_y = 0 \qquad = -2.1\ \text{kn}$$

We then add together the components in each direction:

$$R_x = A_x + B_x + C_x \qquad R_y = A_y + B_y + C_y$$
$$= 0 + 0.5\ \text{kn} - 2.1\ \text{kn} \qquad = 6.0\ \text{kn} + 0 - 2.1\ \text{kn}$$
$$= -1.6\ \text{kn} \qquad = 3.9\ \text{kn}$$

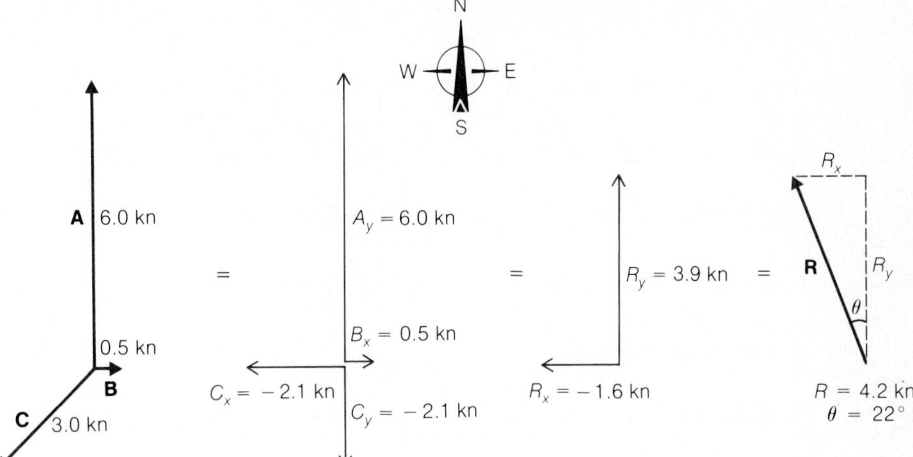

Figure 2.17

A sailboat headed north at 6.0 kn is acted upon by a sideways force due to the wind that sets it to the east at 0.5 kn. A tidal current flows southeast at 3.0 kn. The resultant velocity of the boat relative to the earth's surface has a magnitude of 4.2 kn and is directed 22° west of north (Example 2.7).

The magnitude of the resultant velocity is

$$R = \sqrt{R_x^2 + R_y^2} = \sqrt{1.6^2 + 3.9^2}\ \text{kn} = 4.2\ \text{kn}$$

Thus the speed of the boat relative to the earth is 4.2 kn. Its direction, as shown in the last part of the diagram, is west of north at an angle θ with north. We can find the value of θ by first finding the value of tan θ:

$$\tan\theta = \frac{R_x}{R_y} = \frac{1.6}{3.9} = 0.410 \quad \text{so} \quad \theta = 22^\circ$$

◆

2.7 Motion in a Vertical Plane

Only the vertical component of an object's motion is affected by gravity.

In Chapter 1 we considered two idealized kinds of straightline motion. One was the motion of something that could travel only horizontally—a car on a level road, for instance. Gravity could be ignored here. The other was vertical motion, and we had to take into account the downward acceleration of gravity g.

Now we come to the general case of an object moving freely near the earth's surface. If the object's initial direction is not exactly up or down, gravity will cause its path to be curved rather than straight.

Our strategy in attacking problems of this kind is to resolve the object's acceleration $\mathbf{a}$ (assumed constant) and initial velocity $\mathbf{v}_0$ into their horizontal components $\mathbf{a}_x$ and $\mathbf{v}_{0x}$ and vertical components $\mathbf{a}_y$, $\mathbf{v}_{0y}$. With the help of the formulas of Chapter 1 we then examine separately the object's motion in each of these directions. Finally we combine $\mathbf{v}_x$ and $\mathbf{v}_y$ to find $\mathbf{v}$, and x and y to find the displacement $\mathbf{s}$, at any time t after the start of the motion by vector addition.

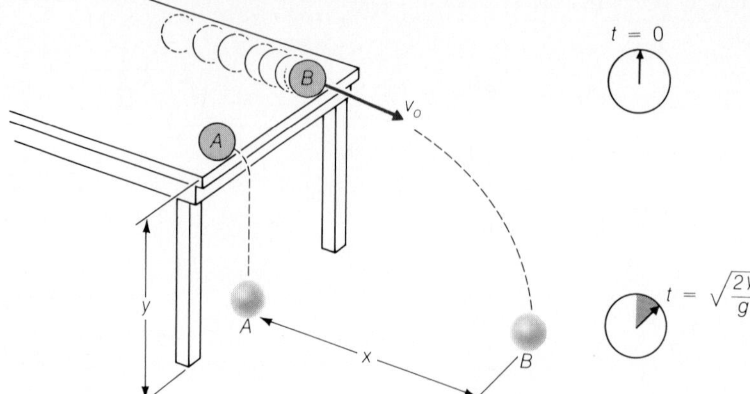

Figure 2.18

Ball A is dropped from the edge of a table while ball B is simultaneously rolled off the edge with the initial horizontal speed v_0.

Suppose that we drop a ball A from the edge of a table while rolling an identical ball B off to the side (Fig. 2.18). At the moment the balls leave the table, A has zero velocity while B has the horizontal velocity $\mathbf{v}_0$. The velocity components of the balls therefore have the magnitudes

$$v_{Ax} = 0 \qquad v_{Bx} = v_0 \qquad v_{Ay} = 0 \qquad v_{By} = 0$$

Both balls reach the floor at the same time, even though B has traveled some distance x away from the table. The reason for this behavior is that the acceleration of gravity is the same for all bodies near the earth regardless of their state of motion. Both A and B started out with no vertical velocity, both underwent the same downward acceleration, and so both took the same period of time to fall.

If we consider vertically downward to be the $+y$-direction, an object in free fall is displaced by

$$y = \tfrac{1}{2} gt^2$$

in the time t when it starts with no vertical component of velocity. Hence both balls require the time

$$t = \sqrt{\frac{2y}{g}}$$

to reach the floor. If the table is 1.0 m high, then

$$t = \sqrt{\frac{(2)(1.0 \text{ m})}{9.8 \text{ m/s}^2}} = 0.45 \text{ s}$$

While it is falling, ball B is also moving horizontally with the speed v_0. When it strikes the floor, it will have been displaced horizontally by $x = v_0 t$. Let us say that $v_0 = 5.0$ m/s. Therefore

$$x = v_0 t = (5.0 \text{ m/s})(0.45 \text{ s}) = 2.3 \text{ m}$$

EXAMPLE 2.8

Find the speeds with which balls A and B strike the floor.

SOLUTION The final velocity of A has only the single component

$$v_{Ay} = gt$$

and so

$$v_A = v_{Ay} = (9.8\ \mathrm{m/s^2})(0.45\ \mathrm{s}) = 4.4\ \mathrm{m/s}$$

The final velocity of B, however, has both horizontal and vertical components, namely

$$v_{Bx} = v_0 \qquad v_{By} = gt$$

From Fig. 2.19 we see that since v_{Bx} is perpendicular to v_{By}, their vector sum $\mathbf{v}_B$ has the magnitude

$$v_B = \sqrt{v_{Bx}^2 + v_{By}^2} = \sqrt{v_0^2 + (gt)^2}$$

Since $v_0 = 5.0$ m/s and $t = 0.45$ s,

$$v_B = \sqrt{(5.0\ \mathrm{m/s})^2 + [(9.8\ \mathrm{m/s^2})(0.45\ \mathrm{s})]^2} = 6.7\ \mathrm{m/s}$$

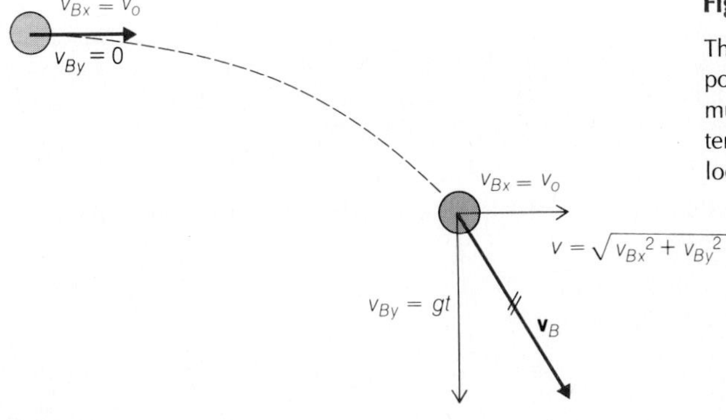

Figure 2.19

The horizontal and vertical components of the velocity of ball B must be added vectorially to determine the magnitude of its velocity.

◆

2.8 Projectile Flight

How to find the initial angle for maximum range.

In considering the flight of a projectile, let us ignore the curvature of the earth, the variation of g with altitude, and (for the moment) air resistance. If the initial velocity $\mathbf{v}_0$ of the projectile makes an angle of θ with level ground, its components have the magnitudes

$$v_{0x} = v_0 \cos\theta \qquad \textit{Initial velocity components} \qquad (2.9)$$

$$v_{0y} = v_0 \sin\theta \qquad (2.10)$$

We can once again use the formulas for straight-line motion to examine the horizontal and vertical aspects of the projectile's flight separately, since these are independent of each other. Under the assumptions above, the horizontal velocity component v_x remains constant during the flight. The vertical component v_y, however, gradually drops to zero owing to the downward acceleration of gravity and then becomes more and more negative (meaning the projectile falls faster and faster) until the ground is reached. If we consider vertically upward to be the $+y$-direction, at the time t the projectile's velocity components become

$$v_x = v_{0x} \qquad \textit{Velocity components at later time} \qquad (2.11)$$

$$v_y = v_{0y} - gt \qquad (2.12)$$

If the projectile starts from the point x_0, y_0, then at the time t its horizontal and vertical coordinates of position will be

$$x = x_0 + v_{0x}t \qquad \textit{Position of projectile} \qquad (2.13)$$

$$y = y_0 + v_{0y}t - \tfrac{1}{2}gt^2 \qquad (2.14)$$

EXAMPLE 2.9

A ball is thrown at 20.0 m/s at an angle of 65° above the horizontal. The ball leaves the thrower's hand at a height of 1.80 m (Fig. 2.20). At what height will it strike a wall 10.0 m away? What would the height be if v_0 were 30.0 m/s?

SOLUTION (a) First we find the time of flight t from the horizontal component of the ball's initial velocity, which is

$$v_{0x} = v_0 \cos\theta = (20.0 \text{ m/s})\cos 65° = 8.45 \text{ m/s}$$

Since $x = 10.0$ m is the horizontal distance to the wall,

$$t = \frac{x}{v_{0x}} = \frac{10.0 \text{ m}}{8.45 \text{ m/s}} = 1.18 \text{ s}$$

The vertical component of the ball's initial velocity is

$$v_{0y} = v_0 \sin\theta = (20.0 \text{ m/s})\sin 65° = 18.13 \text{ m/s}$$

The height at which the ball strikes the wall is therefore

$$\begin{aligned} y &= y_0 + v_{0y}t - \tfrac{1}{2}gt^2 \\ &= 1.80 \text{ m} + (18.13 \text{ m/s})(1.18 \text{ s}) - \tfrac{1}{2}(9.80 \text{ m/s}^2)(1.18 \text{ s})^2 \\ &= 16.4 \text{ m} \end{aligned}$$

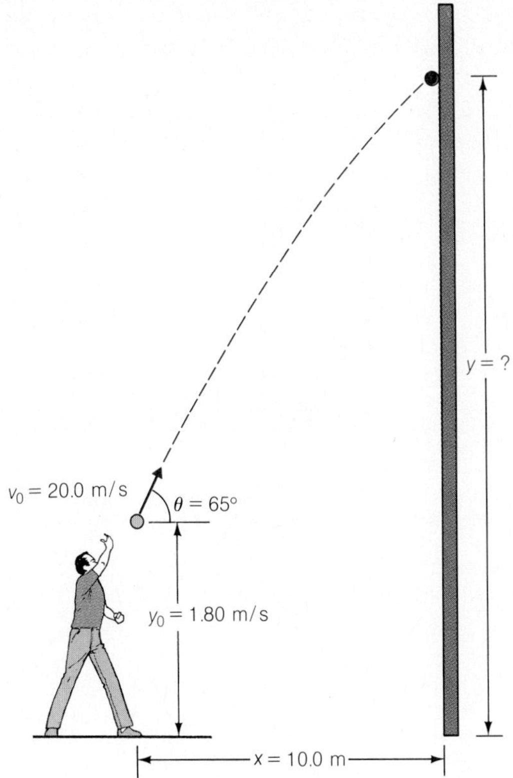

Figure 2.20

(b) A similar calculation gives $y = 20.2$ m when $v_0 = 30.0$ m/s. The greater the initial velocity, the shorter the time of flight and the less the drop due to gravity during the flight. ◆

We can use the formulas derived above to find the range of a projectile launched from the ground, that is, how far from its starting point it will return to the ground. We first calculate the time of flight T (Fig. 2.21). The projectile will continue to rise until the vertical component of its velocity, given by

$$v_y = v_0 \sin\theta - gt$$

is zero. At this time t,

$$0 = v_0 \sin\theta - gt \quad \text{and} \quad t = \frac{v_0 \sin\theta}{g}$$

The projectile needs the same amount of time to return to the ground, and so its total time of flight is

$$T = 2t = \frac{2v_0 \sin\theta}{g} \qquad \textit{Time of flight} \quad (2.15)$$

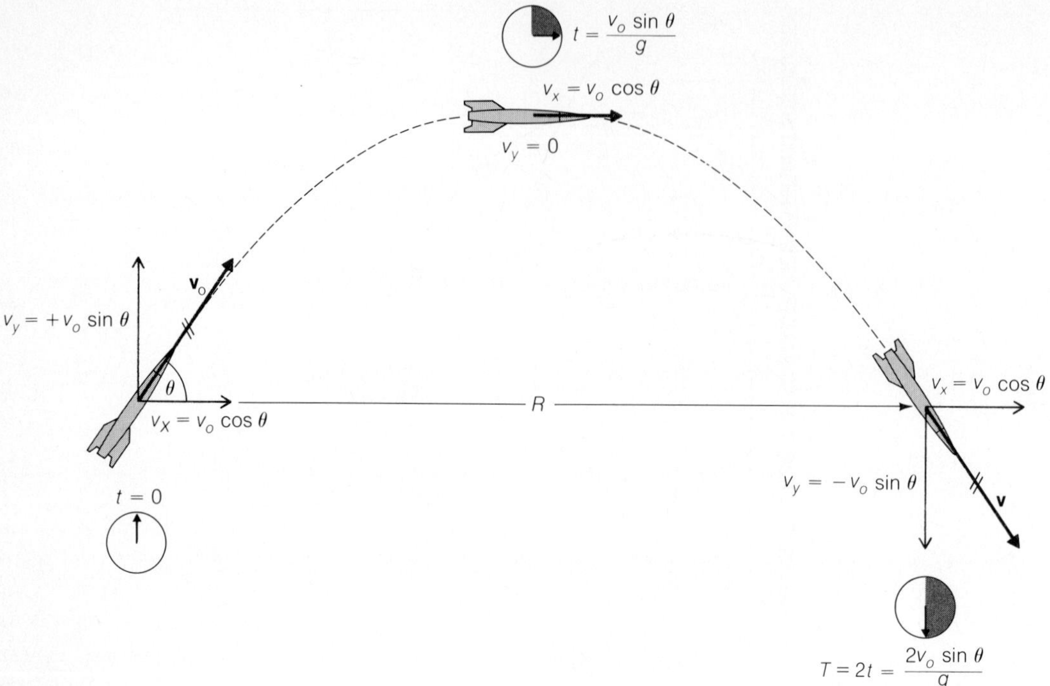

Figure 2.21

In projectile flight, the horizontal component of velocity is constant in the absence of air resistance.

Since v_x is constant, the projectile's range R is

$$R = v_x T = (v_0 \cos\theta)\left(\frac{2v_0 \sin\theta}{g}\right) = \frac{2v_0^2}{g}\sin\theta\cos\theta$$

This formula can be simplified by making use of the trigonometric identity

$$\sin\theta\cos\theta = \tfrac{1}{2}\sin 2\theta$$

The range may therefore be written

$$R = \frac{v_0^2}{g}\sin 2\theta \qquad \textit{Range of projectile} \quad (2.16)$$

We note that R is a maximum when $\sin 2\theta = 1$, because 1 is the highest value the sine function can have. Since $\sin 90° = 1$, the maximum range R_{max} occurs when the initial angle θ is 45°:

$$R_{max} = \frac{v_0^2}{g}$$

Any other angle, greater or smaller, will result in a shorter range (Fig. 2.22).

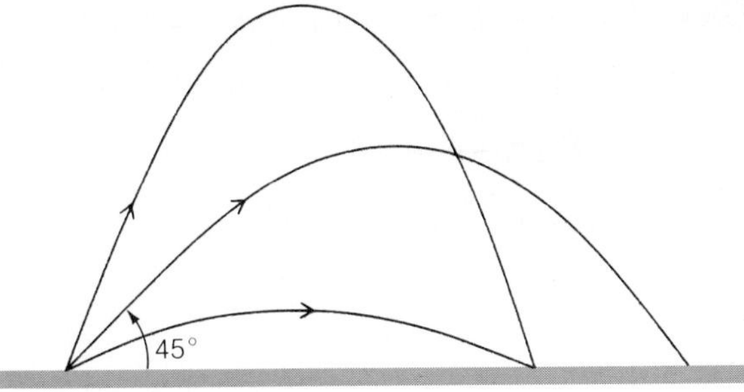

Figure 2.22

In the absence of air resistance, the maximum range of a projectile occurs when it is sent off at an angle of 45°.

Suppose θ_1 is an angle that leads to a certain range and we want to find the other angle θ_2 that will give the same range. To find θ_2 we use the trigonometric identity

$$\sin \phi = \sin (180° - \phi)$$

If we let $\phi = 2\theta_1$, we have

$$\sin 2\theta_1 = \sin (180° - 2\theta_1)$$

From this we conclude that the two possible angles are θ_1 and

$$\theta_2 = 90° - \theta_1 \tag{2.17}$$

Javelin about to be thrown at an angle of 45°.

EXAMPLE 2.10

An arrow leaves a bow at 30 m/s. (a) What is its maximum range? (b) At what two angles could the archer point the arrow if it is to reach a target 70 m away?

SOLUTION (a) The maximum range is

$$R_{max} = \frac{v_0^2}{g} = \frac{(30 \text{ m/s})^2}{9.8 \text{ m/s}^2} = 92 \text{ m}$$

(b) From Eq. (2.16) we obtain

$$\sin 2\theta_1 = \frac{Rg}{v_0^2} = \frac{(70 \text{ m})(9.8 \text{ m/s}^2)}{(30 \text{ m/s})^2} = 0.762$$

Thus

$$2\theta_1 = \sin^{-1} 0.762 = 50° \quad \text{and} \quad \theta_1 = 25°$$

The other angle for the same range is

$$\theta_2 = 90° - \theta_1 = 90° - 25° = 65°$$

♦

Effect of Air Resistance

Air resistance, of course, reduces the range of a projectile below the value given by Eq. (2.16). At a given speed the retardation is many times less for a ball heavy for its size, such as a baseball, than for a ball light for its size, such as a Ping-Pong ball. This is evident from the list of terminal speeds in Table 1.6. For a given projectile, the drag force increases with speed. A baseball for which $v_0 = 35$ m/s (about 80 mi/h) has a maximum range in air about 70% of R_{max} in vacuum, whereas when $v_0 = 50$ m/s (about 110 mi/h), the comparable ratio is down to 55%.

Because the horizontal component of a projectile's velocity decreases steadily as it moves through the air, its path is not symmetric about its highest point, as Fig. 2.23 shows. In this situation—that is, in real life—the maximum range occurs for an initial angle of less than 45°. The greater the drag force, the lower the angle for R_{max}. For a baseball struck hard by a bat, the optimum angle is about 40°.

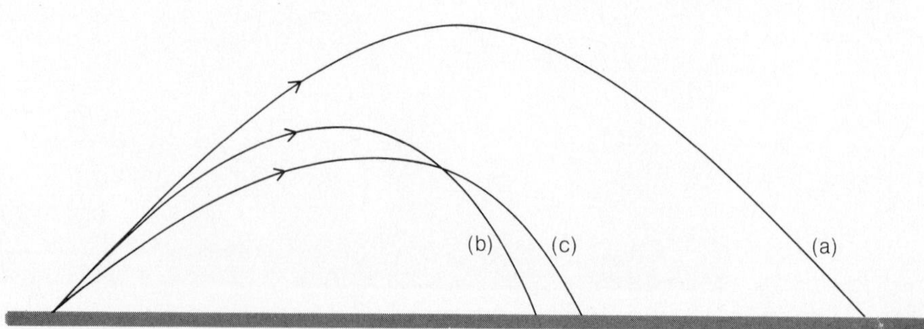

Figure 2.23

Paths of a baseball with the same initial speed after being struck by a bat. (a) Initial angle of 45° in vacuum. (b) Initial angle of 45° in air. (c) Initial angle of 40° in air.

Important Terms

A **scalar quantity** has magnitude only. A **vector quantity** has both magnitude and direction. Thus time is a scalar quantity, and force is a vector quantity.

A **vector** is an arrowed line whose length is proportional to the magnitude of some vector quantity and whose direction is that of the quantity. A **vector diagram** is a scale drawing of the various forces, velocities, or other vector quantities involved in the motion of a body.

In the graphical method of **vector addition,** the tail of each successive vector is placed at the head of the previous one, with their lengths and original directions kept unchanged. The **resultant** is a vector drawn from the tail of the first vector to the head of the last.

A vector can be **resolved** into two or more other vectors, called the **components** of the original vector. Usually the components of a vector are chosen to be in mutually perpendicular directions.

Important Formulas

Vector addition: $\mathbf{A} + \mathbf{B} = \mathbf{B} + \mathbf{A}$

Vector subtraction: $\mathbf{B} - \mathbf{A} = \mathbf{B} + (-\mathbf{A})$

Components of a vector:

$$\mathbf{A} = \mathbf{A}_x + \mathbf{A}_y$$
$$A_x = A \cos \theta$$
$$A_y = A \sin \theta$$

Vector addition by components:

$$R = \sqrt{R_x^2 + R_y^2}$$

where $R_x = A_x + B_x + C_x + \cdots$

$R_y = A_y + B_y + C_y + \cdots$

Angle θ between **R** *and* $\mathbf{R}_x$: $\tan \theta = \dfrac{R_y}{R_x}$

Projectile motion:

$$v_x = v_{0x}$$
$$v_y = v_{0y} - gt$$
$$x = x_0 + v_{0x}t$$
$$y = y_0 + v_{0y}t - \tfrac{1}{2}gt^2$$

Range of projectile: $R = \dfrac{v_0^2}{g}\sin 2\theta$

Angles for same range: $\theta_2 = 90° - \theta_1$

(Air resistance is assumed negligible in the following exercises and problems.)

MULTIPLE CHOICE

1. Which of the following units could be associated with a vector quantity?
 a. km/s^2 b. kg/s
 c. hours d. m^3

2. The minimum number of unequal forces whose vector sum can equal zero is
 a. 1 b. 2
 c. 3 d. 4

3. The magnitude of the resultant of two forces is a minimum when the angle between them is
 a. 0 b. 45°
 c. 90° d. 180°

4. An engine block is supported by a rope hoist attached to an overhead beam. When the block is pulled to one side by a horizontal force exerted by another rope, the tension in the rope hoist
 a. is less than before
 b. is unchanged
 c. is greater than before
 d. may be any of the above, depending on the magnitude of the horizontal force

5. Which of the following pairs of displacements cannot be added to give a resultant displacement of 2 m?
 a. 1 m and 1 m b. 1 m and 2 m
 c. 1 m and 3 m d. 1 m and 4 m

6. Which of the following sets of forces cannot have a vector sum of zero?
 a. 10, 10, and 10 N b. 10, 10, and 20 N
 c. 10, 20, and 20 N d. 10, 20, and 40 N

7. Which of the following sets of displacements might be able to return a car to its starting point?
   a. 2, 8, 10, and 25 km
   b. 5, 20, 35, and 65 km
   c. 60, 120, 180, and 240 km
   d. 100, 100, 100, and 400 km

8. An airplane whose airspeed is 200 km/s is flying in a wind of 80 km/h. The speed of the airplane relative to the ground must be between
   a. 80 and 200 km/h b. 80 and 280 km/h
   c. 120 and 200 km/h d. 120 and 280 km/h

9. Fred walks 8 km north and then 5 km in a direction 60° east of north. His resultant displacement from his starting point is
   a. 11 km b. 12 km
   c. 13 km d. 14 km

10. The resultant of a 4-N force acting upward and a 3-N force acting horizontally is
   a. 1 N b. 5 N
   c. 7 N d. 12 N

11. The angle between the resultant in Multiple Choice 10 and the vertical is approximately
   a. 37° b. 45°
   c. 53° d. 60°

12. An airplane travels 100 km to the north and then 200 km to the east. The displacement of the airplane from its starting point is approximately
   a. 100 km b. 200 km
   c. 220 km d. 300 km

13. At what angle east of north should the airplane in Multiple Choice 12 have headed in order to reach its destination in a straight flight?
   a. 22° b. 45°
   c. 50° d. 63°

14. Two forces of 10 N each act on an object. The angle between the forces is 120°. The magnitude of their resultant is
   a. 10 N b. 14 N
   c. 17 N d. 20 N

15. A vector **A** lies in a plane and has the components $\mathbf{A}_x$ and $\mathbf{A}_y$. The magnitude A_x of $\mathbf{A}_x$ is equal to
   a. $A - A_y$ b. $\sqrt{A} - \sqrt{A_y}$
   c. $\sqrt{A - A_y}$ d. $\sqrt{A^2 - A_y^2}$

16. An escalator has a velocity of 3.0 m/s at an angle of 60° above the horizontal. The vertical component of its velocity is
   a. 1.5 m/s b. 1.8 m/s
   c. 2.6 m/s d. 3.5 m/s

17. The following forces act on an object: 10 N to the north, 20 N to the southeast, and 5 N to the west. The magnitude of their resultant is
   a. 5 N b. 10 N
   c. 13 N d. 23 N

18. The direction of the resultant of the forces in Multiple Choice 17 is
   a. east b. 24° south of east
   c. 45° south of east d. 63° south of east

19. Ball *A* is thrown horizontally, and ball *B* is dropped from the same height at the same moment.
   a. Ball *A* reaches the ground first.
   b. Ball *B* reaches the ground first.
   c. Ball *A* has the greater speed when it reaches the ground.
   d. Ball *B* has the greater speed when it reaches the ground.

20. A ball rolls off the edge of a horizontal roof at 10.0 m/s. Two seconds later the speed of the ball will be
   a. 10.0 m/s b. 19.6 m/s
   c. 29.6 m/s d. 22.0 m/s

21. A ball is thrown at a 30° angle above the horizontal with a speed of 3.0 m/s. After 0.50 s the horizontal component of its velocity will be
   a. 1.5 m/s b. 2.6 m/s
   c. 4.9 m/s d. 5.5 m/s

22. A ball is thrown at a 40° angle below the horizontal at 8.0 m/s. After 0.40 s the horizontal component of its velocity will be
   a. 5.1 m/s b. 6.1 m/s
   c. 6.7 m/s d. 10.4 m/s

23. After 0.40 s the magnitude of the vertical component of the ball of Multiple Choice 22 is
   a. 3.9 m/s b. 5.1 m/s
   c. 9.1 m/s d. 10.0 m/s

24. Two identical projectiles are fired at the same angle. The initial speed of *B* is twice that of *A*. The range of *B* is
   a. $R_B = R_A$ b. $R_B = \sqrt{2}\, R_A$
   c. $R_B = 2R_A$ d. $R_B = 4R_A$

QUESTIONS

1. Is it correct to say that scalar quantities are abstract, idealized quantities with no precise counterparts in the physical world, whereas vector quantities can be said to properly represent reality?

2. A bird sits on an overhead power cable, which sags slightly under its weight. Is it possible to prevent any such sagging by applying enough tension to the cable?

3. The resultant of three vectors is zero. Must they all lie in a plane?

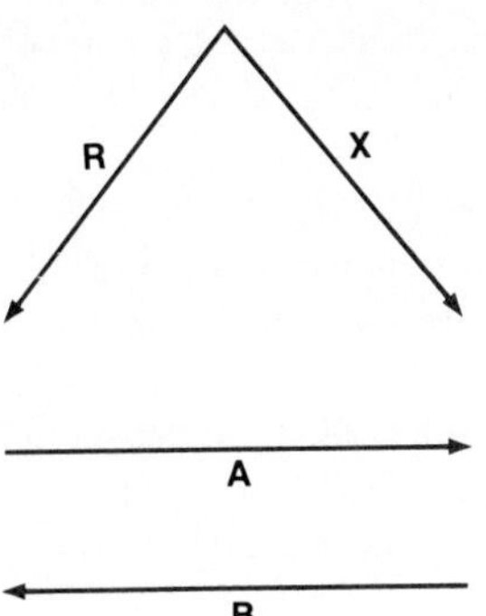

Figure 2.24
Question 4.

4. The resultant of two forces **X** and **Y** is **R.** **X** and **R** are shown in Fig. 2.24. Is **A** or **B** equal to **Y**?

5. Three forces, each of 10 N, act on the same object. What is the maximum total force they can exert on the object? The minimum total force?

6. Two boats with the same speed relative to the water set out to cross a river. Boat *A* heads directly across the river, and boat *B* heads upstream at an angle such that it arrives at a point on the other side opposite its starting point. (a) Which boat arrives at the other side first? (b) Which boat travels the greatest distance relative to the water? (c) Relative to the earth?

7. Three balls are thrown simultaneously at the same speed from a roof. Ball *A* is thrown vertically upward, ball *B* is thrown horizontally, and ball *C* is thrown vertically downward. (a) Do the balls reach the ground at the same time? If not, in which order do they reach the ground? (b) Do the balls have the same speeds when they reach the ground? If not, what is the order of their speeds?

8. A person at the masthead of a sailboat moving at constant velocity drops a wrench. The person is 20 m above the boat's deck at the time, and the stern of the boat is 20 m aft of the mast. Is there a minimum speed the sailboat can have such that the wrench does not land on the deck?

9. A hunter aims a rifle directly at a squirrel on a branch of a tree. The squirrel sees the flash of the rifle's firing. Should the squirrel stay where it is or drop from the branch in free fall at the instant the rifle is fired?

10. Does the speed of a projectile sent off at a 45° angle of elevation vary in its path? If so, where is the speed greatest and where is it least?

EXERCISES

2.2 Vector Addition

1. A 100-N sack of potatoes is suspended by a rope. A person pushes sideways on the sack with a force of 40 N. What is the tension in the rope?

2. A driver becomes lost and travels 12.0 km west, 5.0 km south, and 8.0 km east. Find the magnitude and direction of the car's displacement from the starting place.

3. A person walks 70 m to an elevator and then ascends 40 m. Find the magnitude and direction of the person's displacement from the starting place.

4. A woman jogs 4.2 km east and the 2.9 km south at a speed of 8.0 km/h. How much time would she have saved if she had jogged directly to her destination? In what direction should she have headed for this?

5. Two cars leave a crossroads at the same time, one headed north at 50 km/h and the other headed east at 70 km/h. How far apart are they after 0.5 h? After 2.0 h?

6. The resultant of two perpendicular forces has a magnitude of 40 N. If the magnitude of one of the forces is 25 N, what is the magnitude of the other force?

7. The ketch *Minots Light* is heading northwest at 7.0 kn through a tidal stream that is flowing southwest at 3.0 kn. Find the magnitude and direction of its velocity relative to the earth's surface.

8. A car traveled 16 km north and then 20 km east. The total time of travel was 0.75 h. Find the average speed and average velocity of the car for this trip.

9. A city block is bounded by north-south and east-west streets. The north-south sides of the block are 100 m long, and the east-west sides are 160 m long. Isabel starts from the southwest corner of the block and walks north, east, and south around the block to end up at the southeast corner 3.5 min later. Find her average speed and average velocity for the trip.

2.4 Relative Velocity

10. A woman is rowing at 8.0 km/h relative to the water in a river 1.5 km wide in which the current is 5.0 km/h. In what direction should she head in order to get across the river in the shortest possible time? How much time will she take?

11. A boat moving at 15 km/h relative to the water is crossing a river 2.0 km wide in which the current is flowing at 5.0 km/h. (a) If the boat heads directly for the other shore, how long will the trip take? (b) In what direction should the boat head if it is to reach a point on the other shore directly opposite the starting point? (c) How long will the crossing take in case (b)?

12. A car is moving at 50 km/h toward the northeast. (a) What is the velocity of the car as seen from a bus moving southwest at 40 km/h? (b) What is the velocity of the bus as seen from the car?

13. A car is moving at 50 km/h toward the south. (a) What is the speed of this car as seen from a truck moving east at 80 km/h? (b) What is the speed of the truck as seen from the car?

2.5 Resolving a Vector

14. A horizontal and a vertical force combine to give a resultant force of 10 N that acts in a direction 40° above the horizontal. Find the magnitudes of the horizontal and vertical forces.

15. A woman pushes a 50-N lawn mower with a force of 25 N. If the handle of the lawn mower is 45° above the horizontal, how much downward force is the lawn mower exerting on the ground?

16. A sailboat cannot sail directly toward the wind but must "tack" back and forth at a certain angle with respect to the direction from which the wind is blowing. Which sailboat has the greater component of velocity to windward, the *Alpha*, whose velocity is 5.0 km/h at an angle of 40° away from the wind, or the *Beta*, whose velocity is 6.0 km/h at an angle of 50° away from the wind?

17. A horse is towing a barge along a canal with a rope that makes an angle of 25° with the canal. The horse exerts a force of 500 N on the rope. How much force is exerted on the barge in the direction of its motion?

18. Two tugboats are towing a ship. Each exerts a horizontal force of 5.0 tons, and the angle between the two ropes is 30°. What is the resultant force exerted on the ship?

19. On a windless day, raindrops that fall on the side windows of a car moving at 10 m/s are found to make an angle of 50° with the vertical. Find the speed of the raindrops relative to the ground.

20. The shadow of an airplane taking off moves along the runway at 170 km/h. The sun is directly overhead and the airplane's air speed is 200 km/h. Find the angle at which the airplane is climbing.

21. An airplane is heading southeast when it takes off at an angle of 25° above the horizontal at 200 km/h. (a) What is the vertical component of its velocity? (b) What is the horizontal component of its velocity? (c) What is the component of the velocity of the plane toward the south?

22. An airplane whose speed is 150 km/h climbs from a runway at an angle of 20° above the horizontal. What is its altitude 1.00 min after takeoff? How many kilometers does it travel in a horizontal direction in this period of time?

2.6 Vector Addition by Components

23. Two billiard balls are rolling on a flat table. One has the velocity components $v_x = 1.0$ m/s, $v_y = 2.0$ m/s. The other has the velocity components $v_x = 2.0$ m/s, $v_y = 3.0$ m/s. If both balls started from the same point, what is the angle between their paths?

24. In going from one city to another, a car whose driver tends to get lost goes 30 km north, 50 km west, and 20 km in a direction 30° south of east. How far apart are the cities? In what direction should the car have headed to travel directly from the first city to the second?

25. An airplane flies 200 km east from city *A* to city *B*, then 200 km south from city *B* to city *C*, and finally 100 km northwest to city *D*. How far is it from city *A* to city *D*? In what direction must the airplane head to return directly to city *A* from city *D*?

26. The yacht *Quicksilver* is headed north at 7.0 kn and the yacht *Lianda* is headed southeast at 8.0 kn. (a) Find the velocity of *Quicksilver* relative to *Lianda*. (b) Find the velocity of *Lianda* relative to *Quicksilver*.

27. A car is headed east at 50 km/h, and a truck is headed northwest at 70 km/h. (a) Find the velocity of the car relative to the truck. (b) Find the velocity of the truck relative to the car.

2.7 Motion in a Vertical Plane

28. An airplane is flying at an altitude of 8000 m at a speed of 900 km/h. At what distance ahead of a target must it drop a bomb?

29. A rifle is aimed directly at the bull's-eye of a target 50 m away. If the bullet's speed is 350 m/s, how far below the bull's-eye does the bullet strike the target?

30. An arrow shot horizontally from a cliff at 15 m/s lands 30 m away. How high is the cliff?

31. An airplane is in level flight at a speed of 400 km/h and an altitude of 1000 m when it drops a bomb. Find the bomb's speed when it strikes the ground.

32. A ball is rolled off the edge of a swimming pool and strikes the water 2.0 m away 0.30 s later. What was the initial speed of the ball? Its speed when it struck the water?

33. A rescue line is to be thrown horizontally from the bridge of a ship 30 m above sea level to a lifeboat 30 m away. What speed should the line have?

34. A horizontal stream of water comes out of a nozzle 1.2 m above the ground and strikes the ground 3.0 m away. At what speed does the water leave the nozzle? At what speed does it strike the ground?

35. An airplane is flying at an altitude of 500 m over the ocean at a speed of 60 m/s. A ship is moving along a path directly under that of the airplane but in the opposite direction at 10 m/s. At what horizontal distance from the ship should the pilot of the airplane drop a parcel meant to land on the ship's deck?

36. A ball is rolled off the edge of a table with a horizontal velocity of 1.0 m/s. What will be the ball's velocity 0.10 s later?

37. A ball is thrown horizontally from the roof of a building 20 m high at 30 m/s. What will be the ball's velocity when it strikes the ground?

2.8 Projectile Flight

38. In April 1959 Miss Victoria Zacchini was fired 47 m from a cannon in Madison Square Garden, New York City. What was the minimum muzzle speed of the cannon?

39. Find the minimum initial speed of a champagne cork that travels a horizontal distance of 11 m.

40. Find the range of an arrow that leaves a bow at 50 m/s at an angle of 50° above the horizontal.

41. A football leaves the toe of a punter at an angle of 40° above the horizontal. What is its minimum initial speed if it travels 40 m?

42. A ball is thrown at 20.0 m/s at an angle of 60° above the horizontal. A wind blowing in the opposite direction reduces the ball's horizontal component of velocity by 5.0 m/s. How far away does the ball land?

43. A blunderbuss can fire a slug 100 m vertically upward. (a) What is its maximum horizontal range? (b) With what speeds will the slug strike the ground when fired upward and when fired so as to have maximum range?

44. A person can throw a ball a maximum distance of L. If the ball is thrown upward with the same initial speed, how high will it go?

PROBLEMS

45. A boat moving downstream in a river at constant velocity relative to the water takes 160 s to cover a measured kilometer between marks on the shore. The boat takes 270 s to cover the same distance when headed upstream at the same velocity relative to the water. (a) Find the boat's velocity relative to the water. (b) Find the velocity of the current in the river relative to the shore.

46. A sphere that weighs 100 N rests in the angle formed by two boards 90° apart, as shown in Fig. 2.25. The boards are very smooth, so the only forces they exert on the sphere are perpendicular to their surfaces. Find the magnitude of each of these forces.

47. The following forces act on an object resting on a level, frictionless surface: 10 N to the north, 20 N to the east, 10 N at an angle 40° south of east, and 20 N at an angle 50° west of south. Find the magnitude and direction of the resultant force acting on the object.

48. Find the magnitude and direction of the resultant of a 40-N force that acts at an angle of 63° clockwise from the $+y$-axis, a 15-N force that acts at an angle of 120° clockwise from the $+y$-axis, and a 30-N force that acts at an angle of 310° clockwise from the $+y$-axis.

49. The following horizontal forces act on an object: **A** has a magnitude of 6.0 N and is in the $+y$ direction; **B** has a magnitude of 10.0 N and is in the $-x$ direction; **C** has a magnitude of 8.0 N and is at an angle of 45° clockwise from the $+x$ direction. Find the mgnitude and direction of **A** + **B** − **C**.

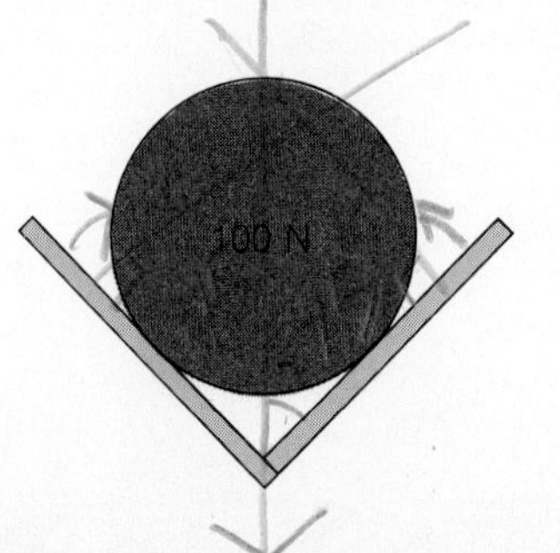

Figure 2.25 Problem 46.

50. The vector **A** has a magnitude of 20 cm and points 20° clockwise from the $+y$-direction. The vector **B** has a magnitude of 10 cm and points 60° clockwise from the $+y$-direction. Find the magnitude and direction **A** + **B**, **A** − **B**, and **B** − **A**.

51. The vector **A** has a magnitude of 10 cm and points 37° clockwise from the $+y$-direction. The vector **B** has a magnitude of 10 cm and points 37° clockwise from the $+x$-direction. Find the magnitude and direction of **A** + **B**, **A** − **B**, and **B** − **A**.

52. George walks 4.0 km in one direction and 2.0 km in another direction to end up 5.0 km from where he started. What is the angle between the two directions?

53. Nadia also walks 4.0 km in one direction and 2.0 km in another direction. However, she ends up only 3.0 km from where she started. What is the angle between the two directions?

54. A ball is thrown at 20 m/s from the roof of a building 25 m high at an angle of 30° below the horizontal. At what height above the ground will the ball strike the side of another building 20 m away from the first?

55. A ball is thrown horizontally toward the north from a rooftop at 8.0 m/s. A wind blowing from the east gives the ball a westward horizontal component of velocity of 10 m/s. (a) What is the speed of the ball relative to the ground after 2.0 s? (b) What angle does its velocity make relative to the vertical at this time? (c) What angle does its velocity make relative to due north at this time?

56. What percentage increase in initial speed is required to increase the range of a javelin by 20%?

57. A baseball is struck by a batter and caught 100 m away 5.0 s later. (a) What was the baseball's initial speed? (b) What maximum height did it reach? (c) What was its speed 2.0 s after being struck? (d) What was its height at that time?

58. A ball is thrown at 12 m/s at an angle of 35° above the horizontal. (a) Find its velocity 1.0 s later. (b) At what time after it was thrown will the ball be headed at an angle of 20° above the horizontal? (c) At 20° below the horizontal?

59. (a) Derive a formula that gives the maximum height H reached by a projectile whose initial speed is v_0 that leaves the ground at an angle of θ above the horizontal. (b) Derive a formula for the ratio R/H between the projectile's range R and its maximum height H. What is the value of R/H when R is a maximum? (c) The angle ϕ is defined in Fig. 2.26. Show that $\tan \theta = 2 \tan \phi$.

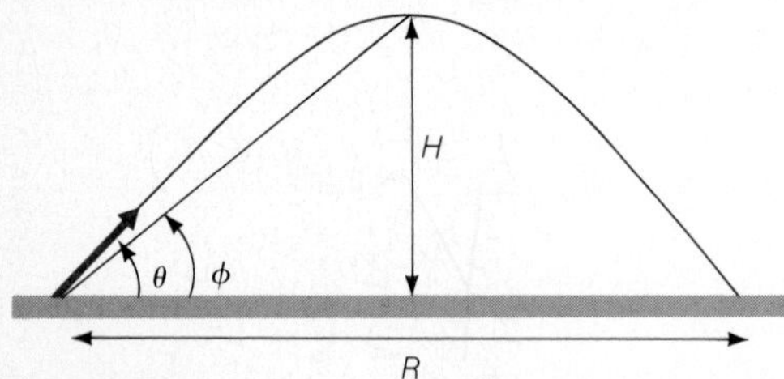

Figure 2.26

Problem 59.

60. A golf ball leaves a tee at 60 m/s and strikes the ground 200 m away. At what two angles with the horizontal could it have begun its flight? Find the time of flight and maximum altitude in each case.

61. A shell is fired at a velocity of 300 m/s at an angle of 30° above the horizontal. (a) How far does it go? What are its time of flight and maximum altitude? (b) At what other angle could the shell have been fired to have the same range? What would its time of flight and maximum altitude have been in this case?

Answers to Multiple Choice

1. a	7. c	13. d	19. c
2. b	8. d	14. a	20. d
3. d	9. a	15. d	21. b
4. c	10. b	16. c	22. b
5. d	11. a	17. b	23. c
6. d	12. c	18. b	24. d

CHAPTER

FORCE AND MOTION

Thus far we have managed to learn a good deal about motion while avoiding some very basic questions. To begin with, what makes anything move in the first place? Once moving, why do some things change direction, why do others speed up or slow down, and why do still others seem able to keep moving indefinitely at constant velocity? Questions such as these led Isaac Newton (1642–1727) to formulate three principles, based on observations he and others had made, that summarize so much of the behavior of moving bodies that they have become known as the *laws of motion*. These laws are the subject of this chapter.

3.1 First Law of Motion

Constant velocity is as natural as being at rest.

Imagine a ball on a level floor. Left alone, it stays where it is. If we give it a push, the ball rolls for a while and then comes to a stop. The smoother the ball and the floor, the farther the ball rolls before stopping. With a perfectly round ball and a perfectly smooth and level floor, and with no air to impede the motion, would the ball ever stop?

There will never be a perfect ball and a perfect floor, of course. But we can come close. The result is that, as the resistance to its motion decreases, the ball does indeed go farther and farther for the same push. We can reasonably conclude that, under ideal conditions, the ball would keep rolling forever.

This conclusion was first reached by Galileo. Later it was stated by Newton, who was born in the year of Galileo's death, as his **first law of motion:**

(a) Sudden start

(b) Sudden stop

Figure 3.1

(a) When a car suddenly starts to move, the inertia of the passengers tends to keep them at rest relative to the earth, and so their heads move backward relative to the car. (b) When the car suddenly stops, inertia tends to keep the passengers moving, and so their heads move forward relative to the car. Both effects illustrate the first law of motion.

An object at rest will remain at rest and an object in motion will continue in motion at constant velocity (constant speed in a straight line) in the absence of any interaction with something else.

The reluctance of an object to change its state of rest or of uniform motion is a property of matter known as **inertia.** When a car suddenly starts to move, its passengers feel themselves pushed backward (Fig. 3.1). What is actually happening is that inertia tends to keep their bodies where they initially were while the car begins to move. When the car suddenly stops, on the other hand, the passengers feel themselves pushed forward. What is actually happening is that inertia tends to keep their bodies moving while the car comes to a halt.

3.2 Mass

A measure of inertia.

A measure of the inertia of an object at rest is its **mass.** The more something resists being set in motion (assuming it is able to move freely), the greater its mass. The inertia of a bowling ball exceeds that of a basketball, as we can tell by kicking them in turn (Fig. 3.2), so the mass of the bowling ball exceeds that of the basketball.

We can compare the inertias—and hence the masses—of two objects, A and B, in a straightforward way. What we do in essence is put a small spring between them, push them together so the spring is compressed, and tie a string between them to hold the assembly in place (Fig. 3.3). Now we cut the string. The compressed spring pushes the objects apart, and A flies off to the left at the speed v_A while B flies off to the right at the speed v_B. Object A has a lower speed than object B, and we interpret this difference to mean that A exhibits more inertia than B.

We repeat the experiment a number of times using springs of different stiffness, so that the recoil speeds are different in each case. What we find each time

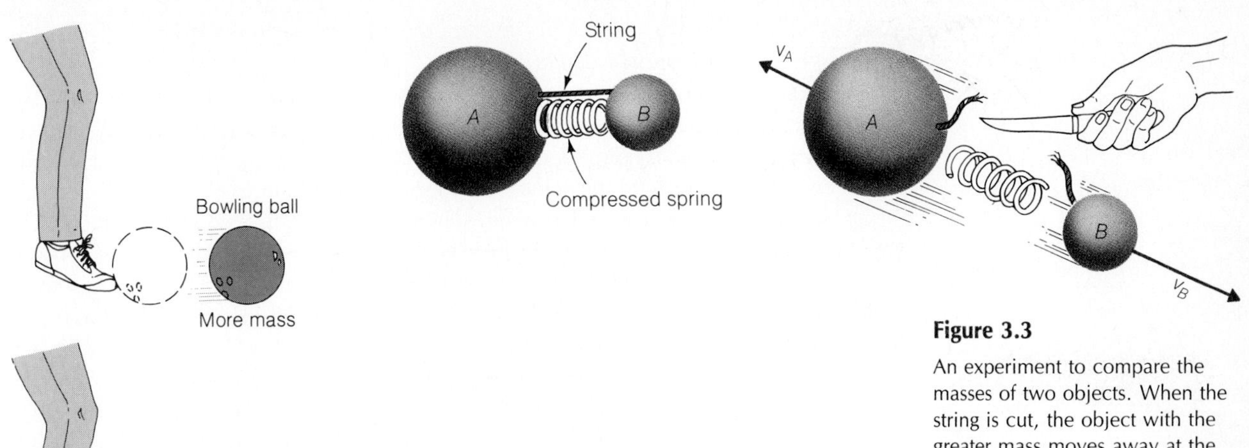

Figure 3.2

The mass of an object determines its inertia.

Figure 3.3

An experiment to compare the masses of two objects. When the string is cut, the object with the greater mass moves away at the lower speed.

is that A moves more slowly than B, and that, regardless of the exact values of v_A and v_B, their *speed ratio* v_B/v_A is always the same.

The fact that the speed ratio v_B/v_A is constant gives us a way to specify exactly what we mean by mass. If we denote the masses of A and B by the symbols m_A and m_B, respectively, we define the ratio of these masses to be

$$\frac{m_A}{m_B} = \frac{v_B}{v_A} \tag{3.1}$$

The object with the greater mass has the lower speed, and vice versa. This procedure

The greater the mass of an object, the greater its resistance to a change in its state of rest or of motion, as this shot-putter can testify.

gives us an experimental method for finding the ratio between the masses of any two objects—it is an *operational definition*.

The next step is to choose an object to be the standard unit of mass. By international agreement this object is a platinum cylinder at Sèvres, France, called the *standard kilogram*. The mass of any other object in the world can be determined by a recoil experiment with the standard kilogram. (There are easier ways to measure mass, needless to say, but we are interested in basic principles for the present.) A liter of water, which is slightly more than a quart, has a mass of 1 kg.

Why bother to define mass in this roundabout way? After all, everybody knows that the mass of an object refers to the amount of matter it contains. The trouble is that ''amount of matter'' is a vague concept: It could refer to an object's volume, to the number of atoms it contains, or to yet other properties. It has proved most fruitful to choose the inertia of an object as a measure of the quantity of matter it contains and to define the object's mass in terms of this inertia as shown in an appropriate experiment.

3.3 Force

Any influence that can change the velocity of an object.

According to the first law of motion, the velocity of an object (which may be 0) remains constant as long as it does not interact with something else. When the object interacts with something else, its velocity changes.

We can interact with a football by kicking it, and the result is a change in the football's velocity from $\mathbf{v}_1 = 0$ to some value $\mathbf{v}_2$ (Fig. 3.4). Or the interaction

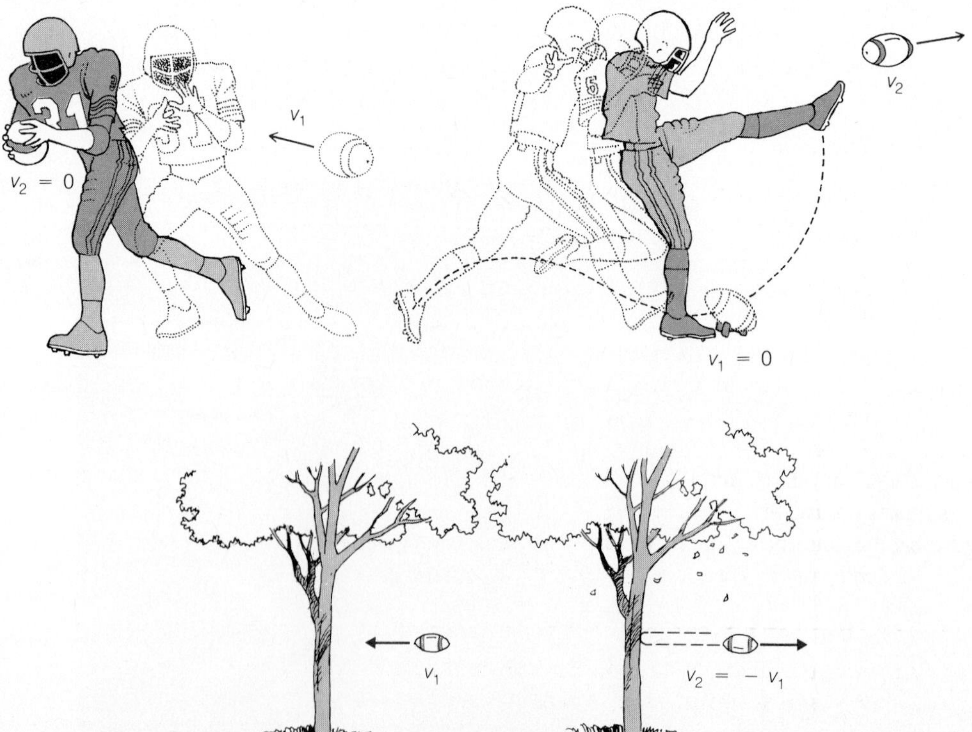

Figure 3.4

Three examples of how an interaction may give rise to a change in the velocity of an object.

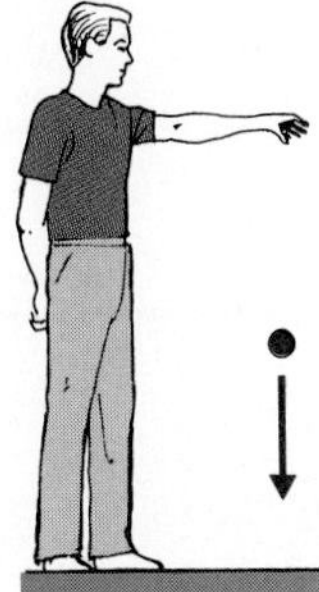

Figure 3.5

A downward force exerted by the earth causes dropped objects to fall.

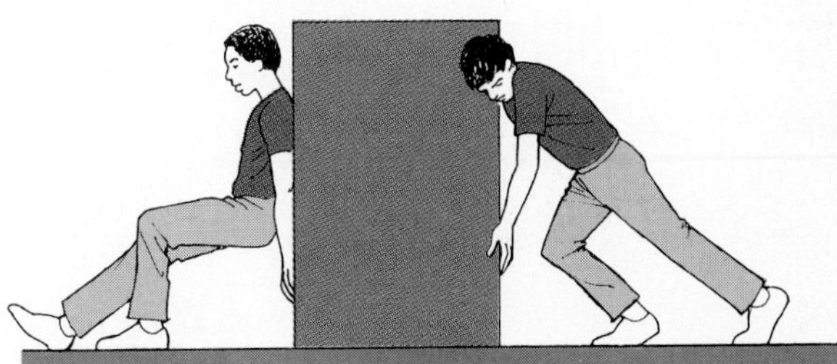

Figure 3.6

When several forces act on an object, they may cancel one another out to leave no net force.

can take the form of catching a moving football, in which case again the football's velocity changes. An interaction can lead to a change in the *direction* of **v** as well as to a change in its magnitude, as, for instance, when a football bounces off a tree. The concept of **force** can be used to put all this on a precise basis. In general:

A force is any influence that can change the velocity of an object.

This definition is in accord with the notion of a force as a "push" or a "pull," but it goes further since no direct contact is implied. No hand reaches up from the earth to pull a dropped stone downward, yet the increasing downward velocity of the falling stone testifies to the action of a force on it (Fig. 3.5).

Two or more forces may act on an object without affecting its state of motion if the forces cancel one another. What is needed for a velocity change is a **net force,** also called an **unbalanced force.** When an object is acted on by a set of forces whose resultant is zero, the forces cancel, and the object is then in **equilibrium** (Fig. 3.6). But each of the forces acting by itself would accelerate the object.

In terms of net force the first law of motion becomes:

If no net force acts on it, an object at rest will remain at rest and an object in motion will remain in motion at constant velocity.

3.4 Second Law of Motion

How an object's acceleration is related to its mass and to the net force acting on it.

Newton's **second law of motion** gives us a quantitative definition of force:

The net force acting on an object equals the product of the mass and the acceleration of the object. The direction of the force is the same as that of the acceleration.

In equation form,

$$\mathbf{F} = m\mathbf{a} \qquad \textit{Second law of motion} \quad (3.2)$$

$$\text{Net force} = (\text{mass})(\text{acceleration})$$

According to the second law, a net force that gives an object twice the acceleration another force does must be twice as great as the other one (Fig. 3.7). An object moving to the right but going slower and slower is accelerated to the left. Hence there must be a force toward the left acting on it (Fig. 3.8). The first law of motion is a special case of the second: When the net force on something is zero, its acceleration is also zero.

The definition of mass given earlier also fits nicely into the second law. The greater the mass of an object acted on by a given force, the smaller its acceleration and hence the smaller its final speed starting from rest (Fig. 3.9).

Figure 3.7

The acceleration of an object is proportional to the net force applied to it. Successive positions of a block are shown at 1-s intervals while forces of F, $2F$, and $3F$ are applied.

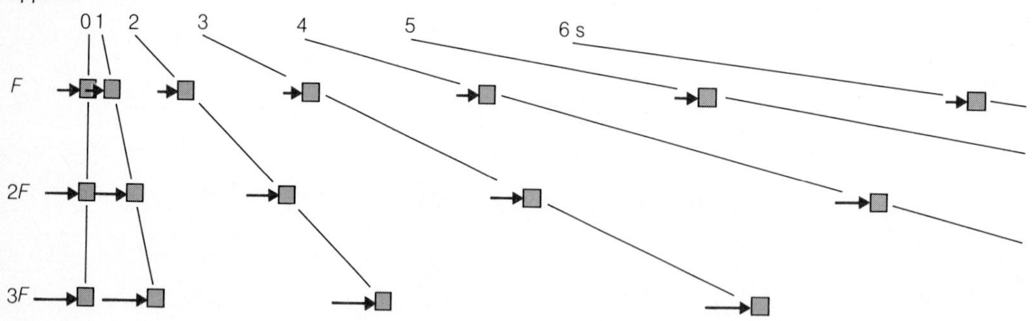

Figure 3.8

A force and the acceleration it produces are always in the same direction.

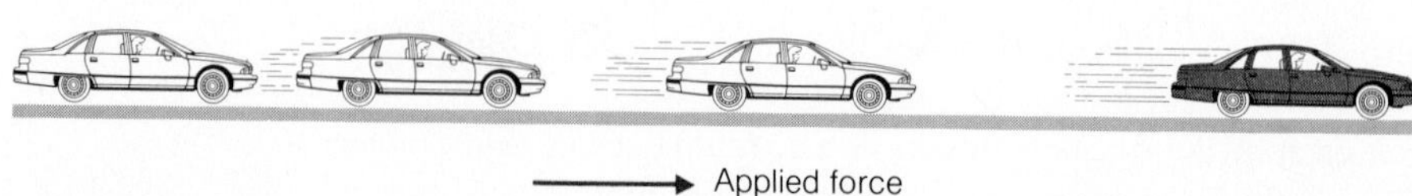

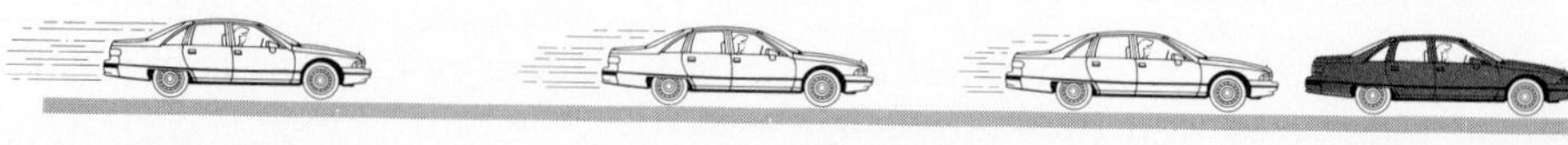

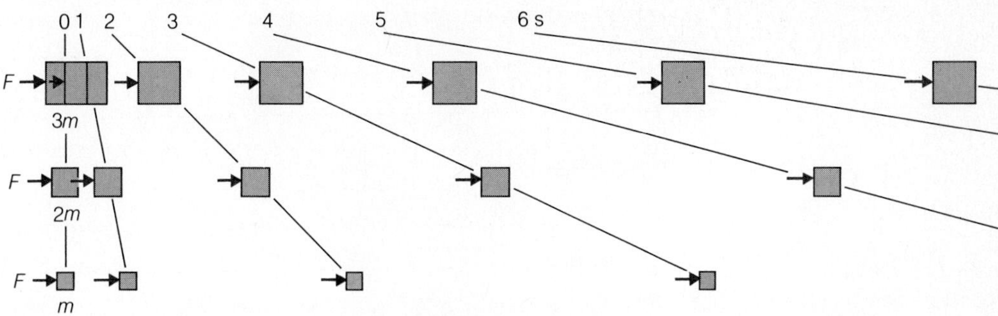

Figure 3.9

When the same force is applied to objects of different masses, the resulting accelerations are inversely proportional to the masses. Successive positions of blocks of mass m, $2m$, and $3m$ are shown at 1-s intervals while identical forces of F are applied.

The relationship between force and acceleration means that the less acceleration an object has, the smaller the net force on it. If you drop to the ground from a height, you can reduce the force of the impact by bending your knees as you strike the ground so you come to a stop gradually instead of suddenly. The same reasoning can be applied to make cars safer. If a car's body is built to crumple progressively in a crash, the forces acting on the passengers will be smaller than if the car were rigid (Fig. 3.10).

In the SI system the unit of force is the **newton** (N):

A newton is that net force which, when applied to a 1-kg mass, gives it an acceleration of 1 m/s².

The newton is not a basic unit and in calculations may have to be replaced by its equivalent in terms of the meter, the second, and the kilogram. This equivalent is easy to find:

$$F = ma$$

$$1\ \text{N} = (1\ \text{kg})(1\ \text{m/s}^2) = 1\ \text{kg} \cdot \text{m/s}^2$$

A newton is about 0.225 lb, a little less than ¼ lb. A pound is nearly 4½ N.

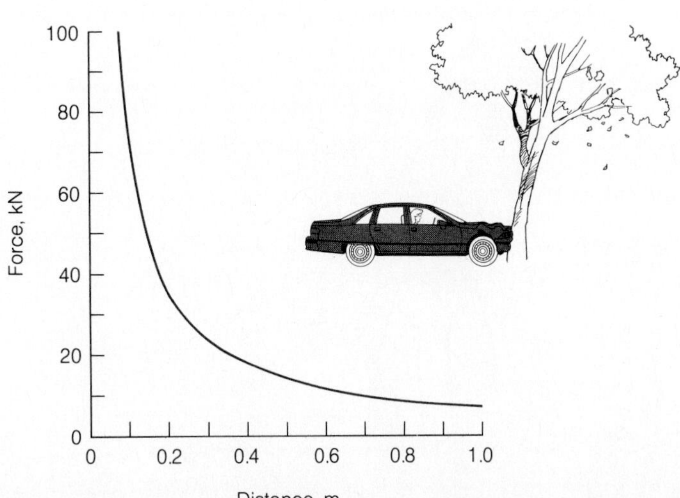

Figure 3.10

The farther something moves while being stopped, the less the force it experiences. This graph shows how the force on a 70-kg person in a car that crashes at 50 km/h (31 mi/h) varies with the distance the person moves during the impact. (10 kN = 10^4 N = 2248 lb)

A gradual landing in sand reduces the acceleration of this long jumper when she strikes the ground and thereby reduces the force on her body.

The second law of motion is a powerful tool for analyzing problems that involve force and motion. Let us examine a few such problems to become familiar with how the second law is applied.

EXAMPLE 3.1

A 60-g tennis ball approaches a racket at 30 m/s, is in contact with the racket's strings for 5.0 ms (1 ms = 1 millisecond = 10^{-3} s), and then rebounds at 30 m/s (Fig. 3.11). What was the average force the racket exerted on the ball?

SOLUTION The tennis ball experienced a change in velocity of

$$\Delta v = v_f - v_0 = (-30 \text{ m/s}) - (30 \text{ m/s}) = -60 \text{ m/s}$$

so its acceleration was

$$a = \frac{\Delta v}{\Delta t} = \frac{-60 \text{ m/s}}{5.0 \times 10^{-3} \text{ s}} = -1.2 \times 10^4 \text{ m/s}^2$$

The corresponding force is, since 60 g = 0.060 kg,

$$F = ma = (0.060 \text{ kg})(-1.2 \times 10^4 \text{ m/s}^2) = -720 \text{ N} = -0.72 \text{ kN}$$

Figure 3.11

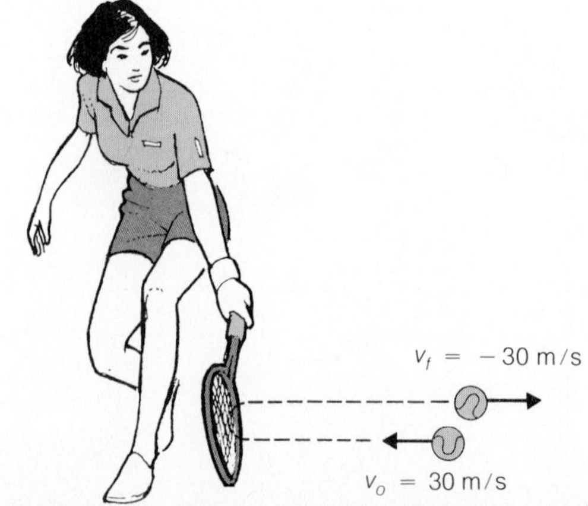

The minus sign means that the force was in the opposite direction to that of the ball when it approached the racket. The British equivalent of 0.72 kN is 162 lb. This force may seem high, but not if we remember that it was exerted for only 0.0050 s. ◆

EXAMPLE 3.2

During performances of the Bouglione Circus in 1976, John Tailor was fired from a compressed-air cannon whose barrel was 20 m long. Tailor emerged from the cannon (twice daily, three times on Saturdays and Sundays) at 40 m/s. If Tailor's mass was 70 kg, find the average force on him during the firing of the cannon (Fig. 3.12).

SOLUTION We start by finding Tailor's acceleration with the help of Eq. (1.12),

$$v_f^2 = v_0^2 + 2ax$$

Here $v_0 = 0$, $v_f = 40$ m/s, and $x = 20$ m, so

$$v_f^2 = 0 + 2ax$$

$$a = \frac{v_f^2}{2x} = \frac{(40 \text{ m/s})^2}{(2)(20 \text{ m})} = 40 \text{ m/s}^2$$

The corresponding average force is

$$F = ma = (70 \text{ kg})(40 \text{ m/s}^2) = 2800 \text{ N} = 2.8 \text{ kN}$$

which is about 630 lb.

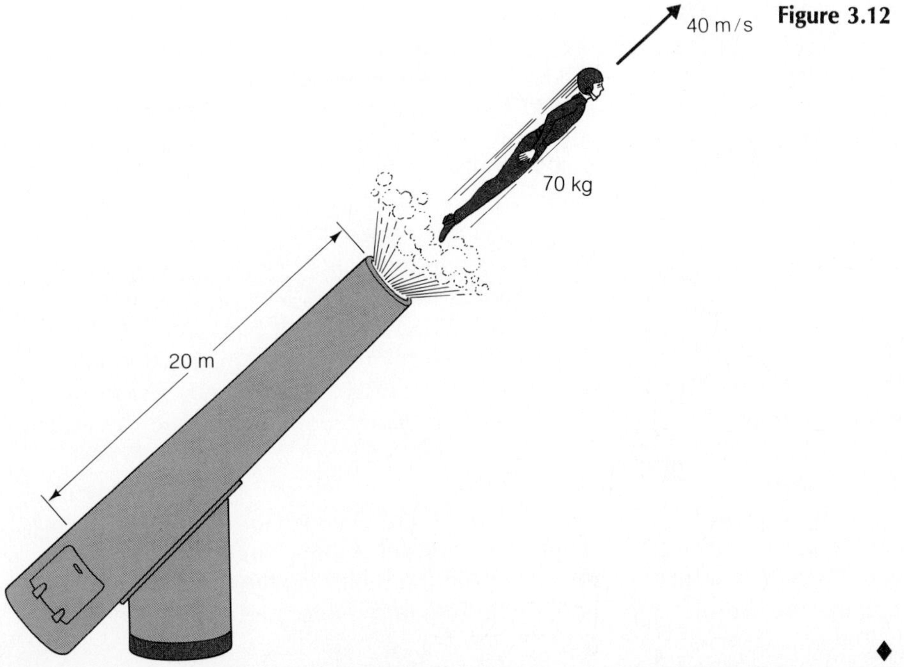

Figure 3.12

◆

BIOGRAPHICAL NOTE

SIR ISAAC NEWTON

(See Appendix, Note 1, page 627)

Isaac Newton

Issac Newton (1642–1727) was born to a farming family in Woolsthorpe, England, in the year of Galileo's death. The young Newton showed an aptitude for science, and although his widowed mother would have preferred that he become a farmer, he went to Cambridge University in 1661. An outbreak of plague led the university to close in 1665, the year Newton graduated, and he returned to Woolsthorpe for 18 months.

As Newton later wrote, "In those days I was in the prime of my age for invention, and minded mathematics and philosophy more than at any time since." In those 18 months Newton developed the binomial theorem, invented differential and integral calculus, discovered the law of gravitation, and proved that white light is a composite of light of all colors—an astonishing catalog.

Newton went back to Cambridge when the university reopened in 1667 and two years later became Lucasian Professor of Mathematics when his teacher, Isaac Barrow, resigned to make way for him. At this time Newton revealed little of the work he had done at Woolsthorpe, the start of a habit of secrecy that was to continue for the rest of his life. Many of his results were made known only after others had rediscovered them, which naturally led to quarrels over priority. In fact, what brought Newton his first acclaim was none of the Woolsthorpe achievements but a reflecting telescope he had devised and built with his own hands.

At Cambridge, Newton carried out both experimental and theoretical research in a number of fields of physics. He also spent much time on chemistry, though without the same kind of success. Despite his reticence, word of Newton's accomplishments got around, and he was persuaded to collect his ideas on mechanics into a book. The *Principia,* which appeared in 1687, begins with what today are known as Newton's laws of motion and goes on to apply them to a variety of situations. Next, the book shows how the inverse-square law of gravity can be derived from Kepler's laws of planetary motion, and also proves that a uniform sphere can be considered gravitationally as a point mass located at its center. Then more complex phenomena are analyzed, among them hydrodynamics, the motion of objects through resistive media, wave motion, some aspects of orbital motion, and the origin of the earth's equatorial bulge. Although Newton had originally obtained many of these results using calculus, he employed only geometrical reasoning in the *Principia,* at least in part to avoid disputes about the validity of this new form of mathematics.

After writing the *Principia,* Newton began to turn away from science. He became a member of Parliament in 1689 and later an official, eventually the Master, of the British Mint. At the Mint Newton helped reform the currency and fought counterfeiters. In 1704 he published *Opticks,* a summary of his earlier work in this field; he waited this long to avoid criticism from a rival, Robert Hooke, who had died the year before. Newton's free time in the last 30 years of his life was mainly spent in theological studies, notably in trying to date events in the Bible. He died at 85, a figure of honor whose stature remains great to this day.

EXAMPLE 3.3

A horizontal force of 10 N is applied to a 4.0-kg block that is at rest on a perfectly smooth, level surface. Find the speed of the block and how far it has gone after 6.0 s.

SOLUTION We begin by finding the block's acceleration, which is

$$a = \frac{F}{m} = \frac{10\ \text{N}}{4.0\ \text{kg}} = \frac{10\ \text{kg}\cdot\text{m/s}^2}{4.0\ \text{kg}} = 2.5\ \text{m/s}^2$$

The speed of the block after 6.0 s is

$$v = at = (2.5\ \text{m/s}^2)(6.0\ \text{s}) = 15\ \text{m/s}$$

For the distance the block travels in $t = 6.0$ s at an acceleration of $a = 2.5\ \text{m/s}^2$, we require Eq. (1.9), $x = v_0t + \frac{1}{2}at^2$. The block started from rest, so $v_0 = 0$ and

$$x = \tfrac{1}{2}at^2 = \tfrac{1}{2}(2.5\ \text{m/s}^2)(6.0\ \text{s})^2 = 45\ \text{m}$$

After 6.0 s a 4.0-kg mass acted on by a net force of 10 N will have gone 45 m and have a speed of 15 m/s (Fig. 3.13).

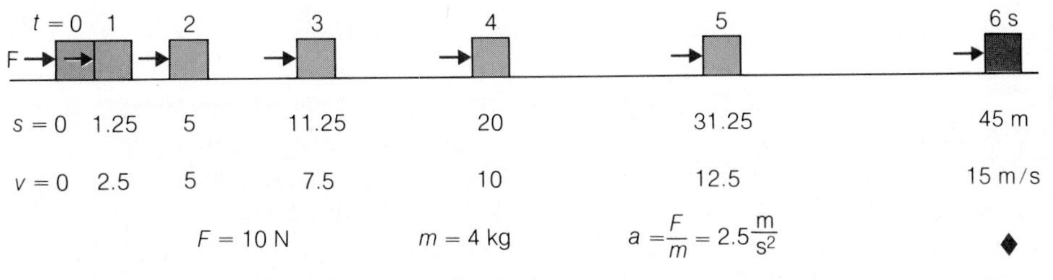

Figure 3.13 Successive distances and speeds of a 4-kg mass acted on by a 10-N force.

◆

Muscular Forces

The forces an animal exerts are produced by contractions of its skeletal muscles. A muscle is a bundle of parallel fibers that tapers at each end into a tendon, which provides the connection to a bone. In some cases a muscle end forks into two or even three tendons. The bones linked by a muscle are hinged together at a joint. The motion of the bones relative to the joint is controlled by the muscle, which is usually teamed with another muscle on the opposite side.

A muscle fiber contracts when it is given an electrical stimulus by a nerve ending. The force of the contraction is the same for each fiber. The greater the required total force, the greater the number of fibers that are stimulated. The maximum force a muscle can exert thus depends on the number of fibers it contains, which is proportional to its cross-sectional area. Maximum forces of up to 70 N/cm^2 (100 lb/in^2) have been reported. An athlete might have a biceps muscle in his arm 8 cm (3 in.) in diameter, which means it could produce forces up to 3500 N (790 lb). As we will see in Chapter 8, the geometries of animal skeletons and muscles favor range of movement over force, so the actual forces a person's hands and feet can exert are much smaller than those produced by the muscles themselves.

An animal of a certain type whose length is L has, in general, muscles whose cross-sectional areas and hence strengths are roughly proportional to L^2. Hence another animal of the same type but twice as long has muscles that are $(2L)^2/L^2 = 4$ times stronger than the corresponding ones in the first animal.

To be sure, the mass of an animal depends on its volume and so on L^3. This means that the larger the animal is, the stronger its muscles have to be to carry out the same tasks. Because mass varies as L^3 whereas strength varies as L^2, large animals are weaker in relation to their masses than smaller ones. This is obvious in nature. Many insects, for instance, can carry objects several times their own weights, but animals the size of humans are limited to loads comparable to their own weights.

Whether the muscles of a certain kind of animal are intrinsically stronger or weaker than those of an animal of a different kind is another matter. For instance, human muscles are considerably stronger than those of insects, figured on the basis of force exerted per unit cross-sectional area. Apart from the structural problems a human-sized insect would have, it would be a rather feeble creature.

3.5 Weight

Weight is a force.

The force with which the earth attracts an object is called its **weight.** If you weigh 700 N (157 lb), that means the earth is pulling you down with a force of 700 N. Weight (a vector quantity) is different from mass (a scalar quantity), which is a measure of inertia. Weight and mass, however, are closely related.

The weight of a stone is the force that gives it the acceleration of gravity g when it is dropped. If the stone's mass is m, then the downward force on it, which is its weight w, can be found from the second law of motion, $F = ma$. Since $F = w$ and $a = g$,

$$w = mg \qquad \textit{Weight} \quad (3.3)$$

$$\text{Weight} = (\text{mass})(\text{acceleration of gravity})$$

The mass of an object is a more basic property than its weight because the pull of gravity on it is not the same everywhere. This pull is less on a mountaintop than at sea level, and less at the equator than near the poles because the earth bulges slightly at the equator. A person who weighs 700 N in Lima, Peru, would weigh about 703 N in Oslo, Norway. On the surface of Mars the same person would weigh only 266 N, so he or she would be able to jump much higher there than on the earth's surface. However, a ball could not be thrown any faster on Mars because $F = ma$ and the ball's mass m is the same.

In working out problems that concern the second law of motion, it is usually best to begin by sketching the forces that act on the object involved. This is called a **free-body diagram.** Forces that the object exerts on anything else should not be shown, because such forces do not affect the object's motion.

EXAMPLE 3.4

A loaded elevator whose total mass is 800 kg is suspended by a cable whose maximum permissible tension is 20,000 N. What is the greatest upward acceleration possible for the elevator under these circumstances? What is the maximum possible downward acceleration? (Tension can be a confusing concept. See Fig. 3.14.)

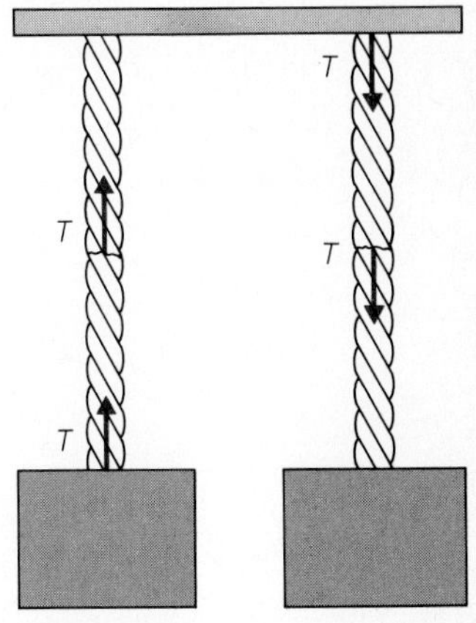

Figure 3.14

The tension T in a cable is the magnitude of the force any part of the cable exerts on the adjoining part. The tension is the same in both directions in the cable, and, if the cable has no mass, it is the same along the entire cable. Since only cables of negligible mass will be considered here, we can think of T as the magnitude of the force either end of the cable exerts on whatever it is attached to.

SOLUTION When the elevator is at rest or moving at constant velocity, the tension in the cable is just its weight of

$$w = mg = (800 \text{ kg})(9.8 \text{ m/s}^2) = 7840 \text{ N}$$

To accelerate the elevator upward, an additional tension F is required in order to provide a net upward force (Fig. 3.15). Since the total tension T cannot exceed 20,000 N, the greatest accelerating force available is

$$F = T - w = 20{,}000 \text{ N} - 7840 \text{ N} = 12{,}160 \text{ N}$$

The elevator's acceleration when this net force is applied is

$$a = \frac{F}{m} = \frac{12{,}160 \text{ N}}{800 \text{ kg}} = 15.2 \text{ m/s}^2$$

For the elevator to exceed the downward acceleration of gravity $g = 9.8$ m/s^2, a downward force besides its own weight is needed. Since this cannot be provided by a supporting cable, the maximum acceleration is 9.8 m/s^2.

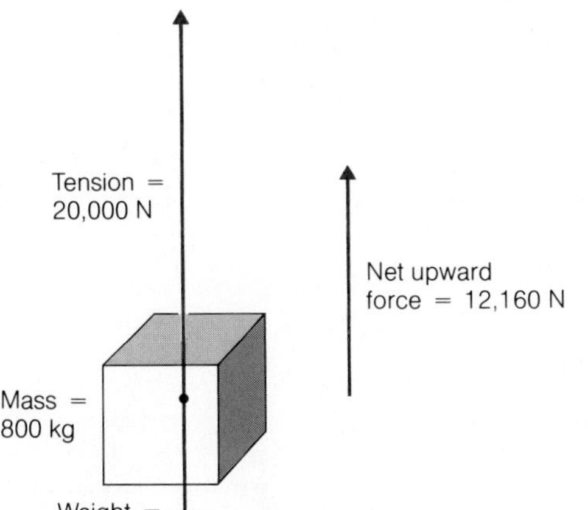

Figure 3.15

A free-body diagram showing that the net upward force on an elevator of mass 800 kg is 12,160 N when the tension in its supporting cable is 20,000 N.

◆

EXAMPLE 3.5

Figure 3.16 shows a 12-kg block, A, which hangs from a string that passes over a pulley and is connected at its other end to a 30-kg block, B, which rests on a frictionless table. Find the accelerations of the two blocks under the assumption that the string is massless and the pulley is massless and frictionless. What is the tension in the string?

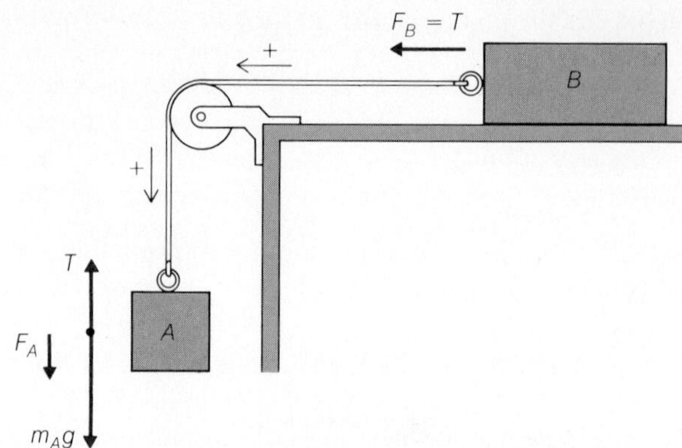

Figure 3.16

Both blocks have the same acceleration.

SOLUTION Because the blocks are joined by the string, their accelerations have the same magnitude a even though they are different in direction. The net force on B is equal to the tension T in the string. From the second law of motion, taking the left as the + direction so that a will come out positive,

$$F_B = T = m_B a$$

In the case of A, the net force is the difference between its weight $m_A g$, which acts downward, and the tension T in the string, which acts upward. If we consider the downward direction as positive, so that the two accelerations will have the same sign,

$$F_A = m_A g - T = m_A a$$

We can eliminate T, whose value we do not know at this point, by substituting $T = m_B a$ from the first equation in the second equation. This step permits us to find a:

$$m_A g - m_B a = m_A a$$

$$m_A g = (m_A + m_B)a$$

$$a = \frac{m_A g}{m_A + m_B} = \frac{(12\text{ kg})(9.8\text{ m/s}^2)}{12\text{ kg} + 30\text{ kg}} = 2.8\text{ m/s}^2$$

The tension in the string is

$$T = m_B a = (30\text{ kg})(2.8\text{ m/s}^2) = 84\text{ N}$$

If B were fixed in place, the tension would equal the weight of A, which is $m_A g =$ 118 N. Here the tension is less because B moves in response to the pull of A's weight, but it is not zero because of B's inertia. ♦

EXAMPLE 3.6

Figure 3.17 shows the same two blocks, A and B, suspended by a string on either side of a massless, frictionless pulley. Find the accelerations of the two blocks and the tension in the string.

SOLUTION Here A moves upward and B moves downward, both with accelerations having the same magnitude a. Applying the second law of motion to the two blocks gives

$$F_A = T - m_A g = m_A a$$
$$F_B = m_B g - T = m_B a$$

where we have considered up as + for A and down as + for B. From the first equation $T = m_A a + m_A g$. Using this value of T in the second equation permits us to find a:

$$m_B g - m_A a - m_A g = m_B a$$
$$(m_B - m_A)g = (m_A + m_B)a$$
$$a = \frac{(m_B - m_A)g}{m_A + m_B} = \frac{(30\text{ kg} - 12\text{ kg})(9.8\text{ m/s}^2)}{12\text{ kg} + 30\text{ kg}}$$
$$= 4.2\text{ m/s}^2$$

The tension in the string may be found from the first equation above:

$$T = m_A a + m_A g = m_A(a + g) = (12\text{ kg})(4.2 + 9.8)\text{m/s}^2 = 168\text{ N}$$

In Section 7.7 we will learn how to take into account the effect of the mass of an actual pulley. ◆

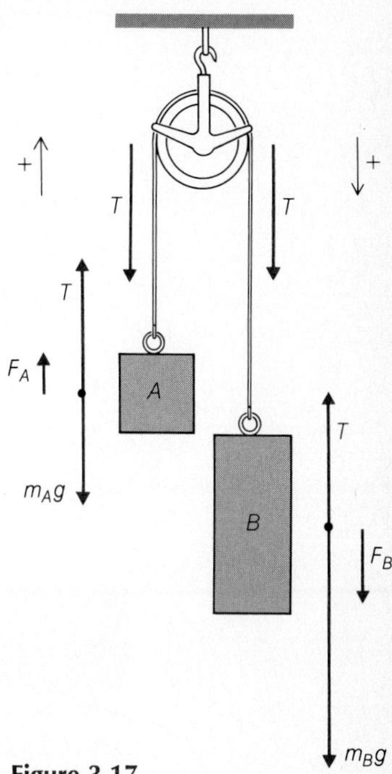

Figure 3.17
The force on each block is the difference between its weight and the tension in the string.

EXAMPLE 3.7

A rope whose working strength is 2000 N is used to tow a 1000-kg car up a 10° incline, as in Fig. 3.18. Find the maximum acceleration that can be given to the car.

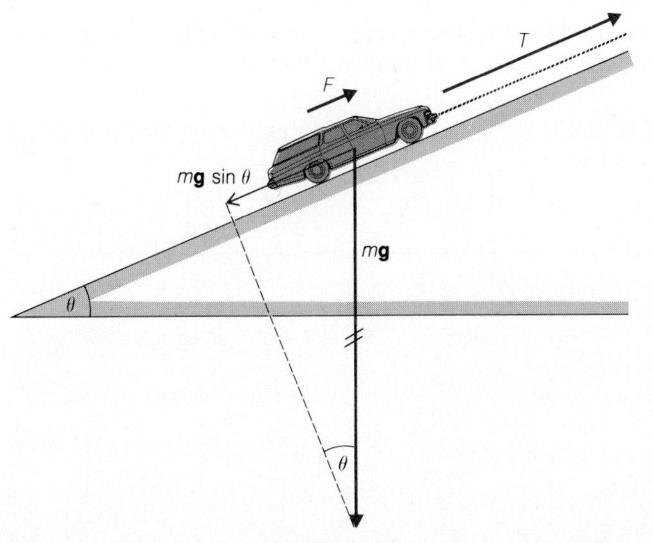

Figure 3.18

SOLUTION The component of the weight $m\mathbf{g}$ of the car that is parallel to the incline has the magnitude $mg \sin \theta$. If T is the maximum tension in the rope, the maximum net force along the incline that can be applied to the car is

$$F = T - mg \sin \theta$$

From the second law of motion,

$$T - mg \sin \theta = ma$$

$$a = \frac{T}{m} - g \sin \theta = \frac{2000 \text{ N}}{1000 \text{ kg}} - (9.8 \text{ m/s}^2)(\sin 10°) = 0.30 \text{ m/s}^2$$

On a level road, $\theta = 0$ and $\sin \theta = 0$, and the maximum acceleration would be 2.0 m/s^2. ◆

3.6 British Units of Mass and Force

The slug and the pound.

In the British system the unit of mass is the **slug** and the unit of force is the **pound** (lb). (These units will not be used in this book but may be of interest.) An object whose mass is 1 slug experiences an acceleration of 1 ft/s^2 when a net force of 1 lb acts on it. Table 3.1 compares units of mass and weight in the SI and British systems.

The slug is an unfamiliar unit because in everyday life weights rather than masses are specified in the British system. We go shopping for 10 lb of apples, not 1/3 slug of apples. In the SI system, on the other hand, masses are normally specified. European grocery scales are calibrated in kilograms, not in newtons.

As a practical matter, especially when traveling outside the United States, it can be handy to know how to find the weight in pounds of a mass in kilograms, or the mass in kilograms of a weight in pounds. The relations between kilograms and pounds (for $g = 9.80$ m/s^2 $= 32.2$ ft/s^2 only!) are as follows:

> One kilogram corresponds to 2.21 pounds in the sense that the weight of 1 kilogram is 2.21 pounds.

Table 3.1 Units of mass and weight

System of units	Unit of mass	Unit of weight	Acceleration of gravity g	To find mass m given weight w	To find weight w given mass m
SI	Kilogram (kg)	Newton (N)	9.80 m/s^2	$m \text{ (kg)} = \dfrac{w\text{(N)}}{9.80 \text{ m/s}^2}$	$w \text{ (N)} = m \text{ (kg)} \times 9.80 \text{ m/s}^2$
British	Slug	Pound (lb)	32.2 ft/s^2	$m \text{ (slugs)} = \dfrac{w\text{(lb)}}{32.2 \text{ ft/s}^2}$	$w \text{ (lb)} = m \text{ (slugs)} \times 32.2 \text{ ft/s}^2$

Conversion of units: 1 slug = 14.6 kg; 1 kg = 0.0685 slug; 1 N = 0.225 lb; 1 lb = 4.45 N

One pound corresponds to 0.454 kilogram in the sense that the mass of 1 pound is 0.454 kilogram.

Thus a 60.0-kg woman weighs (60.0 kg)(2.21 lb/kg) = 133 lb, and a 170-lb man has a mass of (170 lb)(0.454 kg/lb) = 77.2 kg.

3.7 Third Law of Motion

Action and reaction.

We push against a heavy table and it does not move. Evidently the table is pushing back just as hard as we push against it. The table stays in place because our force on it is matched by the opposing force of friction between its base and the floor. We ourselves do not move because the force of the table on us is matched by a similar opposing force between our shoes and the floor.

Suppose now that we and the table are on a frozen lake whose surface is so slippery on a warm day that there is no friction. Again we push on the table, which this time moves away from us as a result (Fig. 3.19). But we can stick to the ice no better than the table can, and we find ourselves sliding backward. No matter what we do, pushing on the table always means that the table pushes back on us.

Considerations of this kind led Newton to his **third law of motion:**

When an object exerts a force on another object, the second object exerts on the first a force of the same magnitude but in the opposite direction.

If we call one of the interacting objects A and the other B, then according to the third law of motion

$$\mathbf{F}_{AB} = -\mathbf{F}_{BA} \qquad \textit{Third law of motion} \quad (3.4)$$

Here $\mathbf{F}_{AB}$ is the force A exerts on B and $\mathbf{F}_{BA}$ is the force B exerts on A (Fig. 3.20).

The third law of motion always applies to two different forces on two different bodies—the **action force** that one body exerts on another, and the equal but opposite **reaction force** that the second exerts on the first.

Figure 3.20

According to the third law of motion, $\mathbf{F}_{AB} = -\mathbf{F}_{BA}$.

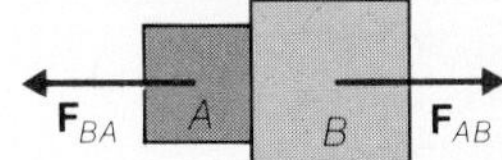

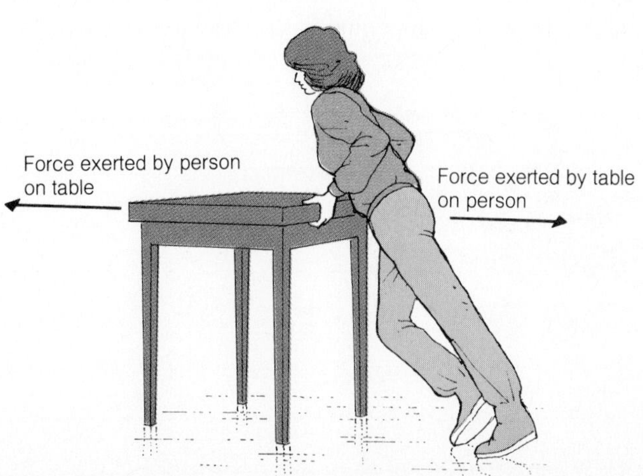

Figure 3.19

Pushing a table on a frozen lake results in person and table moving apart in opposite directions. There is no way to exert a force without an equal and opposite force coming into being. The two forces act on different objects.

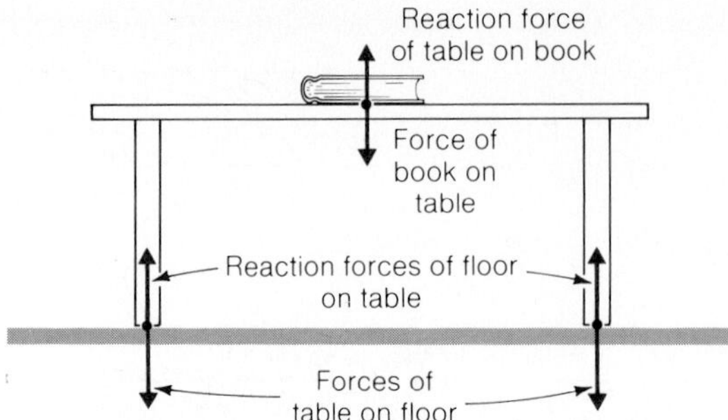

Figure 3.21

The action-reaction forces between a book and a table and between the table and the floor.

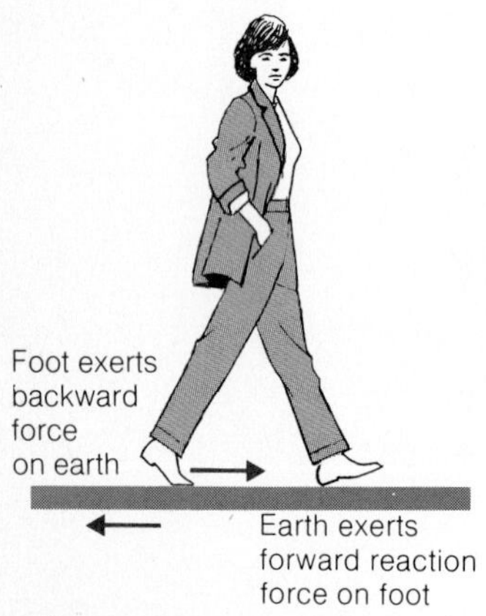

Figure 3.22

When we push backward on the earth with one foot, the opposite reaction force of the earth pushes forward on us. The latter force causes us to move forward.

This book weighs about 20 N. When it rests on a table, it presses down on the table with a force of 20 N (Fig. 3.21). The table pushes upward on the book with a reaction force of 20 N. Why doesn't the book fly upward into the air? The answer is that the upward force of 20 N on the book merely balances its weight of 20 N, which acts downward. If the table were not there to cancel out the latter 20-N force, the book would, of course, be accelerated downward.

The process of walking is another example of the third law at work. We push backward with one foot, and the ground pushes forward on us. The forward reaction force exerted by the ground causes us to move forward. At the same time, the backward force of our foot causes the earth to move backward (Fig. 3.22). Owing to the earth's enormously greater mass, its motion is too small to be detected, but it does move.

Suppose we find ourselves on a frozen lake with a melted surface. Now we cannot walk because the lack of friction prevents us from exerting a backward force on the ice that would produce a forward force on us. But what we can do is exert a force on some object we may have with us, say a snowball. We throw the snowball forward by applying a force to it. At the same time, the snowball is pressing back on us with the identical force but in the opposite direction, and in consequence we find ourselves moving backward (Fig. 3.23).

It is sometimes arbitrary which force of an action-reaction pair to consider action and which reaction. For instance, we cannot really say that the gravitational pull of the earth on an apple is the action force and that the pull of the apple on the earth is the reaction force, or the other way around. When we push on the ground while walking, however, it is legitimate to call this force the action force and the force with which the ground pushes back on us the reaction force.

Figure 3.23

If we throw a snowball while standing on a frozen lake, the reaction force pushes us backward.

EXAMPLE 3.8

A 1.0-kg block A and a 3.0-kg block B are in contact on a frictionless horizontal surface, as in Fig. 3.24. (a) A horizontal force of $F = 10$ N is applied to block A. Find the force this block exerts on block B. (b) The same force is now applied to block B. Find the force block A exerts on block B in this case.

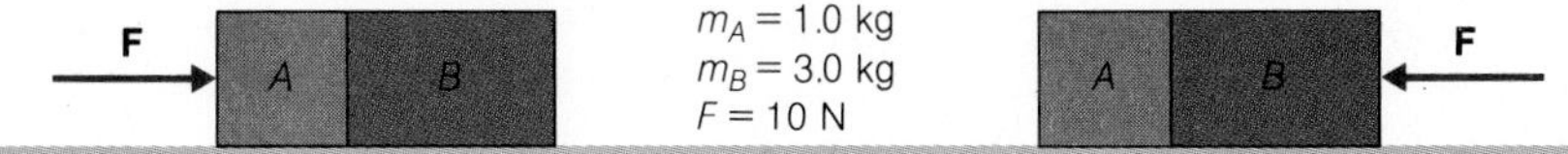

Figure 3.24

SOLUTION (a) Because the blocks stick together, they have the same acceleration, which is

$$a = \frac{F}{m_A + m_B} = \frac{10\text{ N}}{4.0\text{ kg}} = 2.5\text{ m/s}^2$$

To give block B this acceleration, the force block A exerts on it must be

$$F_{AB} = m_B a = (3.0\text{ kg})(2.5\text{ m/s}^2) = 7.5\text{ N}$$

(b) The force block B exerts on block A has the magnitude

$$F_{BA} = m_A a = (1.0\text{ kg})(2.5\text{ m/s}^2) = 2.5\text{ N}$$

The reaction force of block A on block B must have the same magnitude but the opposite direction. As a check, we note that the net force to the left on block B is 10 N − 2.5 N = 7.5 N, which is what is needed to give this block the acceleration a. ◆

3.8 Friction

Sometimes useful, sometimes not.

Frictional forces act to impede relative motion between two surfaces in contact. Friction is quite different from inertia. Inertia refers to the tendency of an object to maintain its state of rest or motion at constant velocity when no net force acts on it. Even the smallest net force, however, can accelerate an object despite its inertia. Friction, on the other hand, refers to an actual force that arises to oppose relative motion between contacting surfaces.

Often friction is desirable. The fastening action of nails, screws, and bolts and the resistive action of brakes depend on it. Walking would be impossible without friction. In many situations, however, friction merely reduces efficiency, and great efforts are made in industry to minimize it through the use of lubricants—notably grease and oil—and special devices—notably the wheel. About half the power of a typical automobile engine is wasted in overcoming friction in the engine itself and in its drive train.

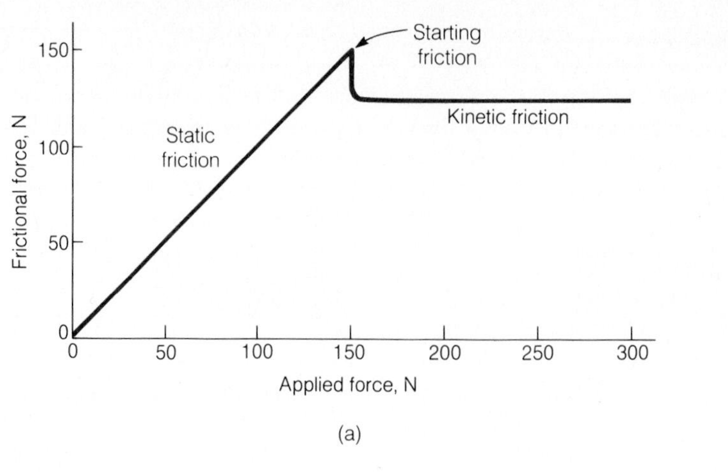

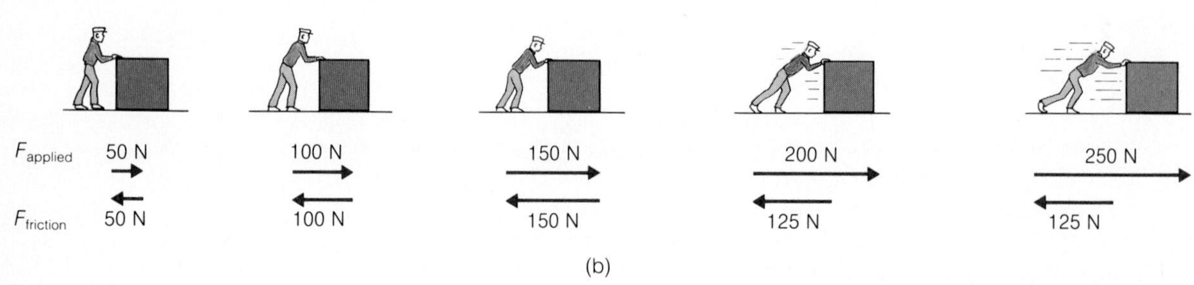

Figure 3.25

(a) As force is applied to a box on a level floor, the frictional resistive force increases to a certain maximum, decreases somewhat as the box begins to move, and then remains constant. (b) The net force on the box equals the applied force minus the frictional force. The frictional force is always equal to or less than the applied force.

Let us consider what happens when we try to move a box across a level floor, as in Fig. 3.25. As we begin to push, the box stays in place because the floor exerts a force on the bottom of the box, which opposes the force we apply. This opposition force is friction, and it arises from the nature of the contact between the floor and the box. As we push harder, the frictional force also increases to match our efforts, until finally we are able to exceed the frictional force and begin to move the box. The frictional force is parallel to the contacting surfaces of the floor and the box.

As indicated in Fig. 3.25(a), the force between two stationary surfaces in contact that prevents motion between them is called **static friction.** In a given case, static friction has a certain maximum value called **starting friction.** When the applied force exceeds starting friction, one surface breaks away and begins to move relative to the other. The **kinetic friction** (or **sliding friction**) that occurs after that is less than the starting friction, so less force is needed to keep the box of Fig. 3.25 in motion than the force needed to set it moving in the first place.

Figure 3.26

A lubricant reduces friction by providing a soft material able to flow readily to separate surfaces in contact.

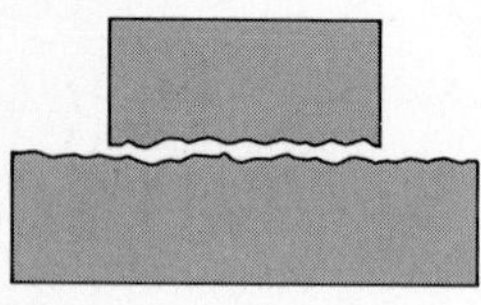

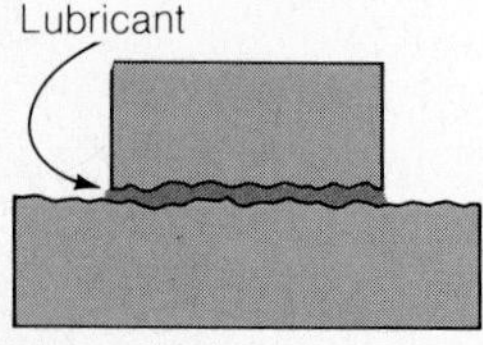

Because starting friction is always greater than kinetic friction, a car's tires provide more traction when they are not slipping. This holds true whether the intention is to accelerate the car forward (spinning the wheels on snow or ice is not very effective) or to slow it down (locking the wheels by applying too much force to the brakes reduces the retarding force on the car).

Friction has two chief causes. One is the mechanical interlocking of irregularities in the two surfaces, as in Fig. 3.26, which prevents one surface from sliding evenly past the other. The second cause is the tendency for materials in very close contact to stick together because of attractive forces between their respective atoms; the origin of such forces is discussed later. The tiny "welds" that occur where projections on the surfaces are pressed together must be broken in order to move one surface over the other. As the motion continues, new welds form and must be broken in turn.

Lubricants reduce friction by separating two contacting surfaces with a layer of softer material. Instead of rubbing against each other, the surfaces rub against the lubricant (Fig. 3.26). Depending on the specific application, the best lubricant may be a gas, a liquid, or a solid. Most lubricants are oils derived from petroleum. Grease consists of oil to which a thickening agent has been added to prevent the oil from running out from between the surfaces involved. The joints of the limbs of the human body are lubricated by a substance called *synovial fluid,* which resembles blood plasma.

Figure 3.27

Both the flattening of a wheel and the indentation of the surface it presses on contribute to rolling friction.

There is no relative motion between the rim of a wheel and a smooth surface over which it rolls if wheel and surface are both rigid and thus no frictional resistance must be overcome. By reducing friction so drastically, the wheel makes it possible to carry loads from one place to another without the enormous forces that dragging them would require. Most aspects of our technological civilization rely on the wheel in one way or another.

Neither wheels nor the surfaces on which they travel can ever be perfectly rigid. Figure 3.27 shows the flattening of the wheel and the indentation of the surface, which both contribute to **rolling friction.** Because the wheel and the surface must be constantly deformed as the wheel rolls, a force is needed to keep the wheel rolling. However, this force is usually many times smaller than that needed to overcome sliding friction. Balls and rollers are widely used to reduce the friction on a rotating shaft by replacing sliding friction with rolling friction. Tire deformation accounts for most of the frictional force on a car at speeds of up to 50 to 80 km/h, when air resistance starts to dominate.

3.9 Coefficient of Friction

The greater the normal force, the greater the friction, regardless of the area in contact.

It is a matter of experience that the frictional force exerted by one surface on another depends on two factors:

1. The kinds of surfaces that are in contact;
2. The perpendicular force with which either surface is pressed against the other. This perpendicular force is usually referred to as the **normal force,** symbol F_N.

The more tightly two objects are pressed together, the greater the friction between them. For this reason an empty box is easier to push across a floor than a similar box loaded with something heavy (Fig. 3.28). Equally familiar is the effect of the nature of the contacting surfaces. For instance, we need more than three times as much force to push a wooden box across a wooden floor as we do to push a steel box of the same weight across a steel floor. Interestingly enough, the area in contact between the two surfaces is not important. It is just as hard to push a small 50-kg box over a given floor as it is to push a large 50-kg box of the same material over the same floor.

Figure 3.28

The greater the normal force with which one surface is pressed against the other, the greater the force of friction between them.

Table 3.2 Approximate coefficients of static and kinetic friction for various materials in contact

Materials in contact	Coefficient of static friction, μ_s	Coefficient of kinetic friction, μ_k
Wood on wood	0.5	0.3
Wood on stone	0.5	0.4
Steel on steel (smooth)	0.15	0.09
Metal on metal (lubricated)	0.03	0.03
Leather on wood	0.5	0.4
Rubber tire on dry concrete	1.0	0.7
Rubber tire on wet concrete	0.7	0.5
Glass on glass	0.94	0.4
Steel on Teflon	0.04	0.04
Bone on bone (dry)		0.3
Bone on bone (lubricated)		0.003

To a good degree of approximation, the following formulas relate the normal force F_N pressing one surface against another to the frictional force F_f that results:

$$F_f \leq \mu_s F_N \qquad \textit{Static friction} \quad (3.5)$$

$$F_f = \mu_k F_N \qquad \textit{Kinetic friction} \quad (3.6)$$

The quantity μ (Greek letter *mu*) is called the **coefficient of friction** and is a constant for a given pair of surfaces. Table 3.2 lists the coefficients of static and kinetic friction for a number of sets of surfaces.

When an object resting on a surface is pushed, the force of static friction increases with the applied force until the limiting value of $\mu_s F_N$ is reached. That is, when no motion occurs, $\mu_s F_N$ gives the starting frictional force, not the actual frictional force: The actual frictional force has the same magnitude as the applied force, but is in the opposite direction. When the applied force is greater than the starting frictional force of $\mu_s F_N$, motion begins and the coefficient of kinetic friction must then be used. In this case Eq. (3.6) gives the frictional force, which no longer depends on the applied force and is the same over a fairly wide range of speeds.

As mentioned earlier, the normal force between an object and a surface with which it is in contact can be taken either as the force the object exerts on the surface or as the reaction force the surface exerts on the object, since both have the same magnitude. If we use a free-body diagram to help solve problems, however, we are limited to forces that act *on* the object. When friction is involved, this means we must use the reaction force on the object for the normal force F_N. In the examples that follow, this convention will be used.

EXAMPLE 3.9

A 100-kg wooden crate is at rest on a level stone floor (Fig. 3.29). (a) What is the minimum horizontal force needed to start the crate moving? (b) What is the minimum horizontal force needed to keep the crate in motion at constant speed? (c) What will happen if a horizontal force of 500 N is applied to the crate?

SOLUTION (a) The starting frictional force is given by Eq. (3.5) using the coefficient of static friction for these surfaces from Table 3.2:

$$F_f(\text{starting}) = \mu_s F_N = \mu_s mg = (0.5)(100 \text{ kg})(9.8 \text{ m/s}^2) = 490 \text{ N}$$

The crate will begin to move only if a force of more than 490 N is applied to it.

(b) When the crate is moving, the coefficient of kinetic friction must be used to find the frictional force, which is

$$F_f(\text{kinetic}) = \mu_k mg = (0.4)(100 \text{ kg})(9.8 \text{ m/s}^2) = 392 \text{ N}$$

Although at least 490 N of force must be exerted to set the crate in motion, only 392 N of force is needed to continue the motion at constant speed.

(c) Since the applied force of $F_A = 500$ N exceeds the starting force of 490 N, the crate will begin to move. Since the applied force also exceeds the force of sliding friction $F_f = 392$ N, the crate will be accelerated. The net force on the moving crate is

$$F = F_A - F_f = 500 \text{ N} - 392 \text{ N} = 108 \text{ N}$$

The crate's acceleration is, from the second law of motion $F = ma$,

$$a = \frac{F}{m} = \frac{108 \text{ N}}{100 \text{ kg}} = 1.08 \text{ m/s}^2$$

◆

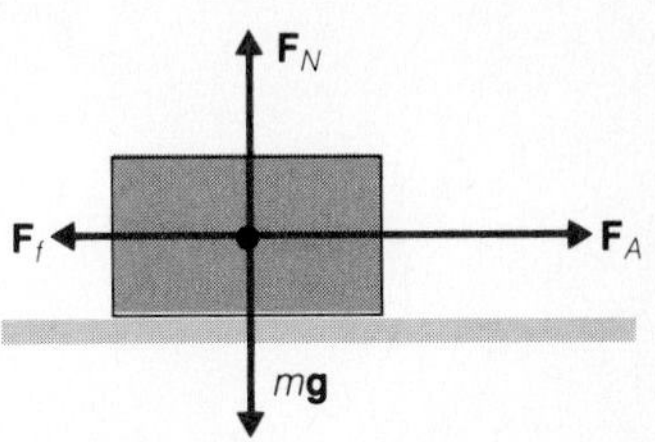

Figure 3.29

Free-body diagram of a box on a level floor. The normal force $\mathbf{F}_N$ has the magnitude mg here. When the applied force $\mathbf{F}_A$ exceeds the frictional force $\mathbf{F}_f$, the net force $\mathbf{F}_A - \mathbf{F}_f$ acts to accelerate the box.

EXAMPLE 3.10

A person whose shoes have leather heels is walking on a wooden floor. (a) Find the maximum angle the forward leg may make with the vertical in order that the heel not slip on the floor. (b) How is this angle affected if the floor is wet, which reduces the coefficient of static friction?

SOLUTION (a) The geometry of the situation is shown in Fig. 3.30. The force **F** that the foot exerts on the floor, if there is no slip, has the reaction force $\mathbf{F}_R$, whose components parallel and perpendicular to the floor are, respectively,

$$F_x = F \sin \theta \qquad F_y = F \cos \theta$$

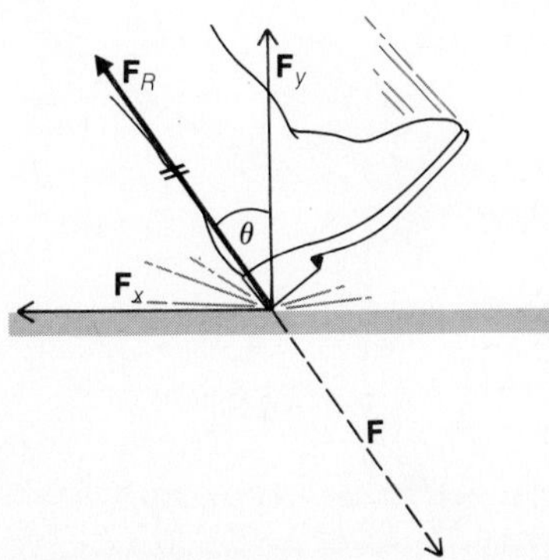

Figure 3.30

If slipping is not to occur, the angle θ must be smaller than a certain critical angle that depends on the coefficient of static friction between the heel and the floor.

The condition for the heel not to slip is that the frictional force $\mathbf{F}_f$ have the same magnitude as $\mathbf{F}_x$. Since

$$F_f = \mu_s F_N = \mu_s F \cos \theta$$

we have for the limiting angle

$$F_f = F_x \qquad \mu_s F \cos \theta = F \sin \theta \quad \text{and} \quad \tan \theta = \mu_s$$

From Table 3.2 the coefficient of static friction for leather on wood is 0.5, so

$$\tan \theta = 0.5 \quad \text{and} \quad \theta = 27°$$

If θ is equal to or less than 27°, the heel will not slip.

(b) Reducing μ_s reduces θ, so smaller steps have to be taken on a slippery surface. Slipping is a sudden process because, as it starts, the smaller coefficient of kinetic friction takes over, and the frictional force drops sharply. This is why it is hard to keep from falling once slipping begins. ♦

EXAMPLE 3.11

A wooden chute is being built, along which wooden crates of merchandise are to be slid down into the basement of a store. (a) What angle with the horizontal should the chute make if the crates are to slide down at constant speed? (b) With what force must a 100-kg crate be pushed in order to start it sliding down the chute if the angle of the chute is that found in (a)?

SOLUTION (a) Figure 3.31 shows the various forces that act on the crate. The component $\mathbf{F}$ of the crate's weight $m\mathbf{g}$ parallel to the chute has the magnitude

$$F = mg \sin \theta$$

and the reaction force $\mathbf{F}_N$ of the chute on the crate has the magnitude

$$F_N = mg \cos \theta$$

Figure 3.31

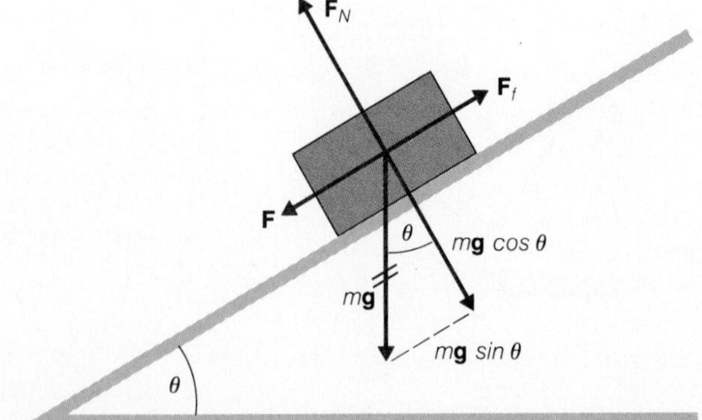

When the crate slides down at constant speed, there is no net force acting on it, according to Newton's first law of motion. Hence the downward force along the chute must exactly balance the force of kinetic friction, which means that

$$F = \mu_k F_N$$

$$mg \sin \theta = \mu_k mg \cos \theta$$

$$\mu_k = \frac{\sin \theta}{\cos \theta} = \tan \theta$$

From Table 3.2 the value of μ_k for wood on wood is 0.3, and so $\theta = 17°$.

(b) We note that the coefficient of static friction here is $\mu_s = 0.5$. Hence the force of static friction to be overcome is

$$F_f = \mu_s F_N = \mu_s mg \cos \theta$$

This is greater than the force of kinetic friction, and so a crate will not begin to move without a push. The force component along the plane due to the crate's own weight is $F = mg \sin \theta$. Here F is less than F_f, and so, if we call F_A the applied force parallel to the plane required to move the crate,

$$\text{Applied force} + \text{forward component of weight} = \text{backward frictional force}$$

$$F_A + mg \sin \theta = \mu_s mg \cos \theta$$

$$F_A = mg(\mu_s \cos \theta - \sin \theta)$$

$$= (100\text{ kg})(9.8\text{ m/s}^2)(0.5 \cos 17° - \sin 17°) = 182\text{ N}$$ ◆

Important Terms

The **inertia** of an object refers to the apparent resistance it offers to changes in its state of motion. The property of matter that manifests itself as inertia is called **mass.** The unit of mass in the SI system is the **kilogram.**

A **force** is any influence that can cause an object to be accelerated. The unit of force in the SI system is the **newton.**

The **weight** of an object is the gravitational force exerted on it by the earth. The weight of an object is proportional to its mass.

Newton's **first law of motion** states that, in the absence of a net force acting on it, an object at rest will remain at rest and an object in motion will continue in motion at constant velocity. The **second law of motion** states that a net force acting on an object causes it to have an acceleration proportional to the magnitude of the force and inversely proportional to the object's mass. The acceleration is in the same direction as the force. The **third law of motion** states that when an object exerts a force on another object, the second object exerts on the first a force of the same magnitude but in the opposite direction.

The **free-body** diagram of an object shows only the forces that act *on* the object. Such a diagram is helpful in working out problems that involve the second law of motion.

The term **friction** refers to the resistive forces that arise to oppose the motion of an object past another with which it is in contact. **Kinetic friction** is the frictional resistance an object in motion experiences. **Static friction** is the frictional resistance that prevents a stationary object from being set in motion.

The **coefficient of static friction** is the constant of proportionality for a given pair of contacting surfaces not in relative motion that relates the maximum frictional force between them to the normal force with which one presses against the other. The **coefficient of kinetic friction** is this constant for surfaces in relative motion.

Important Formulas

Second law of motion: $\mathbf{F} = m\mathbf{a}$

Weight: $w = mg$

Third law of motion: $\mathbf{F}_{AB} = -\mathbf{F}_{BA}$

Frictional force: $F_f \leq \mu_s F_N$ (Static friction)

$F_f = \mu_k F_N$ (Kinetic friction)

MULTIPLE CHOICE

1. As an airplane climbs,
- **a.** its mass increases
- **b.** its mass decreases
- **c.** its weight increases
- **d.** its weight decreases

2. The weight of an object
- **a.** is the quantity of matter it contains
- **b.** refers to its inertia
- **c.** is basically the same quantity as its mass but expressed in different units
- **d.** is the force with which it is attracted to the earth

3. The acceleration of gravity on Mars is 3.7 m/s^2. Compared with her mass and weight on the earth, an astronaut on Mars has
- **a.** less mass and less weight
- **b.** less mass and more weight
- **c.** the same mass and less weight
- **d.** less mass and the same weight

4. An automobile that is towing a trailer is accelerating on a level road. The force that the automobile exerts on the trailer is
- **a.** equal to the force the trailer exerts on the automobile
- **b.** greater than the force the trailer exerts on the automobile
- **c.** equal to the force the trailer exerts on the road
- **d.** equal to the force the road exerts on the trailer

5. When a horse pulls a wagon, the force that causes the horse to move forward is the force
- **a.** he exerts on the wagon
- **b.** the wagon exerts on him
- **c.** he exerts on the ground
- **d.** the ground exerts on him

6. The action and reaction forces referred to in Newton's third law of motion
- **a.** act on the same object
- **b.** act on different objects
- **c.** need not be equal in magnitude and need not have the same line of action
- **d.** must be equal in magnitude but need not have the same line of action

7. A woman whose weight is 500 N is standing on the ground. The force the ground exerts on her is
- **a.** less than 500 N
- **b.** equal to 500 N
- **c.** more than 500 N
- **d.** any of the above, depending on her location on the earth's surface

8. A jumper of weight w presses down on the ground with a force of magnitude F and becomes airborne as a result. The magnitude of the force the ground exerted on the jumper must have been
- **a.** equal to w and less than F
- **b.** equal to w and equal to F
- **c.** more than w and equal to F
- **d.** more than w and more than F

9. When an object exerts a normal force on a surface, the force is
- **a.** equal to the object's weight
- **b.** frictional in origin
- **c.** parallel to the surface
- **d.** perpendicular to the surface

10. To set an object in motion on a surface usually requires
- **a.** less force than to keep it in motion
- **b.** the same force as that needed to keep it in motion
- **c.** more force than to keep it in motion
- **d.** only as much force as is needed to overcome inertia

11. The frictional force between two surfaces in contact does not depend on
- **a.** the normal force pressing one against the other
- **b.** the areas of the surfaces
- **c.** whether the surfaces are stationary or in relative motion
- **d.** whether a lubricant is used or not

12. When a 1.0-N force acts on a 1.0-kg object that is able to move freely, the object receives
- **a.** a speed of 1.0 m/s
- **b.** an acceleration of 0.102 m/s^2
- **c.** an acceleration of 1.0 m/s^2
- **d.** an acceleration of 9.8 m/s^2

13. When a 1.0-N force acts on a 1.0-N object that is able to move freely, the object receives
- **a.** a speed of 1.0 m/s
- **b.** an acceleration of 0.102 m/s^2
- **c.** an acceleration of 1.0 m/s^2
- **d.** an acceleration of 9.8 m/s^2

14. The weight of 600 g of salami is

a. 0.061 N b. 5.9 N
c. 61 N d. 5.9 kN

15. A certain force gives a 5.0-kg object an acceleration of 2.0 m/s^2. The same force would give a 20-kg object an acceleration of

a. 0.5 m/s^2 b. 2.0 m/s^2
c. 4.9 m/s^2 d. 8.0 m/s^2

16. A force of 10 N gives an object an acceleration of 5.0 m/s^2. What force would be needed to give it an acceleration of 1.0 m/s^2?

a. 1 N b. 2 N
c. 5 N d. 50 N

17. An upward force of 600 N acts on a 50-kg dumbwaiter. The dumbwaiter's acceleration is

a. 0.82 m/s^2 b. 2.2 m/s^2
c. 11 m/s^2 d. 12 m/s^2

18. A net force of 1.0 N acts on a 2.0-kg object, initially at rest, for 2.0 s. The distance the object moves during that time is

a. 0.5 m b. 1.0 m
c. 2.0 m d. 4.0 m

19. A 500-g ball lying on the ground is kicked with a force of 250 N. If the kick lasted 0.020 s, the ball flew off with a speed of

a. 0.01 m/s b. 0.1 m/s
c. 2.5 m/s d. 10 m/s

20. An 80-kg firefighter slides down a brass pole 8.0 m high in a fire station while exerting a frictional force of 600 N on the pole. He reaches the bottom of the pole in the time of

a. 1.5 s b. 1.9 s
c. 2.6 s d. 7.0 s

21. A brick has the dimensions 8 cm × 16 cm × 32 cm. The force of starting friction between the brick and a wooden floor is

a. greatest when the brick rests on the 8 cm × 16 cm face
b. greatest when the brick rests on the 8 cm × 32 cm face
c. greatest when the brick rests on the 16 cm × 32 cm face
d. the same regardless of which face it rests on

22. The coefficient of static friction between two wooden surfaces is

a. 0.5 b. 0.5 N
c. 0.5 kg/N d. 0.5 N/kg

23. A force of 40 N is needed to set a 10-kg steel box moving across a wooden floor. The coefficient of static friction is

a. 0.08 b. 0.25
c. 0.4 d. 2.5

24. The coefficient of static friction for steel on ice is 0.1. The force needed to set a 70-kg skater in motion is about

a. 0.1 N b. 0.7 N
c. 7 N d. 70 N

25. A horizontal force of 150 N is applied to a 51-kg carton on a level floor. The coefficient of static friction is 0.5 and that of kinetic friction is 0.4. The frictional force acting on the carton is

a. 150 N b. 200 N
c. 250 N d. 500 N

26. A car is moving at 20 m/s on a road such that the coefficient of static friction between its tires and the road is 0.80. The minimum distance in which the car can be stopped is

a. 26 m
b. 33 m
c. 51 m
d. undeterminable unless the car's mass is known

27. The coefficients of static and kinetic friction for wood on wood are, respectively, 0.5 and 0.3. If a 100-N wooden box is pushed across a horizontal wooden floor with just enough force to overcome the force of static friction, its acceleration is

a. 0.2 m/s^2 b. 0.5 m/s^2
c. 2.0 m/s^2 d. 5.0 m/s^2

28. A toboggan reaches the foot of a hill at a speed of 4 m/s and coasts on level snow for 15 m before coming to a stop. The coefficient of kinetic friction is

a. 0.004 b. 0.05
c. 0.16 d. 0.27

QUESTIONS

1. When a body is accelerated, a force is invariably acting on it. Does this mean that, when a force is applied to a body, it is invariably accelerated?

2. Compare the tension in the coupling between the first two cars in a train being pulled by a locomotive with the tension in the coupling between the last two cars when the train's speed is increasing.

3. It is less dangerous to jump from a high wall onto loose earth than onto a concrete pavement. Why?

4. Measurements are made of distance versus time for three moving objects. The distances are found to be directly proportional to t, t^2, and t^3, respectively. What can you say about the net force acting on each of the objects?

5. When a force equal to its weight is applied to a body free to move, what is its acceleration?

6. A person in an elevator suspends a 1.0-kg mass from a spring balance. What is the nature of the elevator's motion when the balance reads 9.0 N? 9.8 N? 10.0 N?

7. Can anything ever have a downward acceleration greater than g? If so, how can this be accomplished?

8. When you "weigh" an object on one pan of an equal-arm balance by putting standard masses on the other pan until the beam is horizontal, are you actually determining the object's weight or its mass? What if you use a spring balance?

9. Can we conclude from the third law of motion that a single force cannot act on a body?

10. Since the opposite forces of the third law of motion are equal in magnitude, how can anything ever be accelerated?

11. An engineer designs a propeller-driven spacecraft. Because there is no air in space, he incorporates a supply of oxygen as well as a supply of fuel for the motor. What do you think of the idea?

12. A car with its engine running and in gear goes up a hill and then down on the other side. What forces cause it to move upward? Are they the same as the forces that then cause it to move downward?

13. A bird is in a closed box that rests on a scale. Does the reading on the scale depend on whether the bird is standing on the bottom of the box or is flying around inside it? What would happen if the bird were in a cage instead?

14. Ships are often built on ways that slope down to a nearby body of water. Normally a ship is launched before most of its interior and superstructure have been installed, and is completed when afloat. Is this done because the additional weight would cause the ship to slide down the ways prematurely?

15. A person is pushing a box across a floor. Can the frictional force on the box ever exceed its own weight? Can the frictional force ever exceed the applied force?

EXERCISES

3.4 Second Law of Motion

1. A 25-g snail goes from rest to a speed of 3.0 mm/s in 5.0 s. How much horizontal force does the ground exert on it?

2. A 430-g soccer ball lying on the ground is kicked and flies off at 25 m/s. If the duration of the impact was 0.010 s, what was the average force on the ball?

3. An empty truck whose mass is 2000 kg has a maximum acceleration of 1.0 m/s^2. What is its maximum acceleration when it is carrying a 1000-kg load?

4. A car whose mass (including the driver) is 1600 kg has a maximum acceleration of 1.2 m/s^2. If three 80-kg passengers are also in the car, find its maximum acceleration then.

5. A net horizontal force of 4000 N is applied to a 1400-kg car. What will the car's speed be after 10 s if it started from rest?

6. A bicycle and its rider together have a mass of 80 kg. If the bicycle's speed is 6.0 m/s, how much force is needed to bring it to a stop in 4.0 s?

7. A 12,000-kg airplane launched by a catapult from an aircraft carrier is accelerated from 0 to 200 km/h in 3.0 s. (a) How many times the acceleration of gravity is the airplane's acceleration? (b) What is the average force the catapult exerts on the airplane?

8. A force of 20 N gives an object an acceleration of 5.0 m/s^2. (a) What force would be needed to give the same object an acceleration of 1.0 m/s^2? (b) What force would be needed to give an acceleration of 10 m/s^2?

9. A 2000-kg truck is braked to a stop in 15 m from an initial speed of 12 m/s. How much force was required?

10. A 100-kg motorcycle carrying a 70-kg rider comes to a stop in 40 m from a speed of 50 km/h when its brakes are applied. Find the force exerted by the brakes and the force experienced by the rider.

11. A driver shifts into neutral when her 1200-kg car is moving at 80 km/h and finds that its speed has dropped to 65 km/h 10 s later. What was the average drag force acting on the car?

12. The driver of the car in Exercise 11 presses down on the gas pedal and thereby applies a force of 1300 N to the car. How long does it take the car to go from 65 km/h back to 80 km/h?

13. A billy goat running at 5.0 m/s butts his head against a tree. If his head has a mass of 4.0 kg and it comes to a stop in 5.0 mm, find the average force of the impact. What do you think will happen to the goat?

14. A car moving at 10 m/s (22.4 mi/h) strikes a stone wall. (a) The car is very rigid and the 80-kg driver comes to a stop in a distance of 20 cm. What is his average acceleration and how does it compare with the acceleration of gravity g? How much force acted on him? Express this force in both newtons and pounds. (b) The car is so constructed that its front end gradually collapses on impact, and the driver comes to a stop in a distance of 100 cm. Answer the same questions for this situation.

3.5 Weight

15. Two objects whose masses are 5.0 kg and 10.0 kg are suspended by strings as in Fig. 3.32. Find the tension in each string.

16. A 100-kg man slides down a rope at constant speed. (a) What is the minimum breaking strength the rope must have? (b) If the rope has precisely this strength, will it support the man if he tries to climb back up?

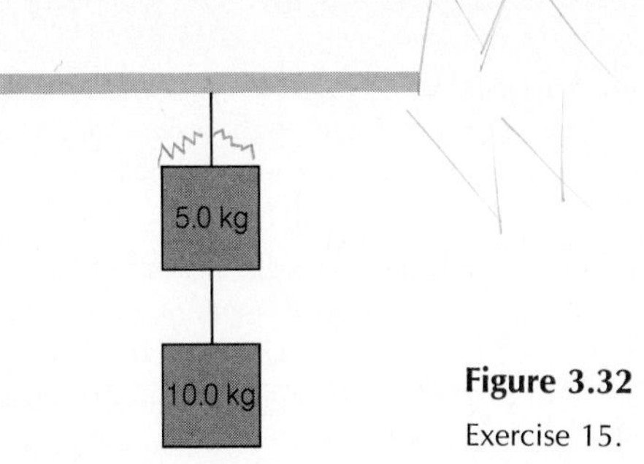

Figure 3.32
Exercise 15.

17. A force of 20 N acts on a body whose weight is 8.0 N. (a) What is the mass of the body? (b) What is its acceleration?

18. A force of 20 N acts on a body whose mass is 4.0 kg. (a) What is the weight of the body? (b) What is its acceleration?

19. A mass of 8.0 kg and another of 12.0 kg are suspended by a string on either side of a frictionless pulley. Find the acceleration of each mass.

20. How much force must you supply to give a 1.0-kg object an upward acceleration of $2g$? A downward acceleration of $2g$?

21. (a) A 1000-kg elevator has a downward acceleration of 1.0 m/s^2. What is the tension in its supporting cable? (b) The same elevator has an upward acceleration of 1.0 m/s^2. What is the tension in its cable now?

22. (a) A 400-kg load of bricks on a 50-kg pallet is raised with an acceleration of 0.40 m/s^2. What is the tension in the supporting rope? (b) The empty pallet is lowered with a downward acceleration of 0.60 m/s^2. What is the tension in the rope now?

23. A parachutist whose total mass is 100 kg is falling at 50 m/s when her parachute opens. Her speed drops to 7.0 m/s in a vertical distance of 40 m. What total force did her harness have to withstand? How many times her weight is this force?

24. A balloon whose total mass is M is falling with the constant acceleration a. How much ballast should be thrown overboard to make the balloon float at constant altitude?

25. A person stands on a scale in an elevator. When the elevator is at rest, the scale reads 700 N. When the elevator starts to move, the scale reads 600 N. (a) Is the elevator going up or down? (b) Does it have a constant speed? If so, what is this speed? (c) Does it have a constant acceleration? If so, what is this acceleration?

26. A 60-kg person stands on a scale in an elevator. How many newtons does the scale read (a) when the elevator is ascending with an acceleration of 1.0 m/s^2; (b) when it is descending with an acceleration of 1.0 m/s^2; (c) when it is ascending at the constant speed of 3.0 m/s; (d) when it is descending at the constant speed of 3.0 m/s; (e) when the cable has broken and the elevator is descending in free fall?

27. A sprinter presses on the ground with a force equal to three times his own weight at a 50° angle with the horizontal at the start of a race. What is his forward acceleration?

28. A 10-kg crate and a 100-kg crate are both sliding without friction down a plane inclined at 20° with the horizontal. What is the acceleration of each crate?

3.9 Coefficient of Friction

29. A 100-kg wooden crate rests on a level wooden floor. What is the minimum force required to move it at constant speed across the floor?

30. A woman prevents a 2.0-kg brick from falling by pressing it against a vertical wall. The coefficient of static friction is 0.60. What force must she use? Is this more or less than the weight of the brick?

31. An eraser is pressed against a vertical blackboard with a horizontal force of 10 N. The coefficient of static friction between eraser and blackboard is 0.20. Find the force parallel to the blackboard required to move the eraser.

32. A very strong 100-kg man is having a tug-of-war with a 1500-kg elephant. The coefficient of static friction for both on the ground is 1.0 (a) Find the force needed to move the elephant. (b) What would happen if the man could actually exert such a force?

33. A tennis ball whose initial speed is 2.0 m/s rolls along a floor for 5.0 m before coming to a stop. Find the coefficient of rolling friction.

34. The coefficient of rolling friction of a baseball rolling on the ground is 0.080. If the ball comes to a stop in 5.0 s, what was its initial speed?

35. In what distance will a car skid to a stop on a dry concrete road if its brakes are locked when it is moving at 80 km/h?

36. A driver sees a cow in the road ahead and applies the brakes so hard that they lock and the car skids to a stop in 24 m. If the road is concrete and it is a dry day, what was the car's speed in kilometers per hour when the brakes were applied?

37. Find the maximum possible accelerations the engines of the following cars can provide on a level road when the coefficient of static friction is μ. In each case the front wheels of the car support 60% of its weight.

a. a car with rear-wheel drive
b. a car with front-wheel drive
c. a car with four-wheel drive

38. A 1250-kg car reaches a speed of 2.0 m/s after being pushed by two people for 12 m starting from rest. If the coefficient of rolling friction is 0.0070, what force did the people exert on the car?

39. A 2.0-kg wooden block whose initial speed is 3.0 m/s slides on a smooth floor for 2.0 m before it comes to a stop. (a) Find the coefficient of kinetic friction. (b) How much force would be needed to keep the block moving at constant speed across the floor?

40. A truck moving at 15 m/s is carrying a 2000-kg steel girder that rests on the bed of the truck without any fastenings. The coefficient of static friction between the girder and the truck bed is 0.50. (a) Find the minimum distance in which the truck can come to a stop without having the girder move forward. (b) If the girder had a mass of 3000 kg, would this distance be different?

41. The coefficient of static friction between a tire and a certain road is 0.70. (a) What is the maximum possible acceleration of a car on this road? (b) What is the minimum distance in which the car can be stopped when it is moving at 100 km/h?

42. A railway boxcar is set in motion along a track at the same initial velocity as a truck on a parallel road. If the only horizontal forces acting on both vehicles are due to rolling friction with the respective coefficients of 0.0045 and 0.040, which vehicle will come to a stop first? How many times farther will the other vehicle travel?

PROBLEMS

46. A box slides down a frictionless plane that makes an angle of 35° with the horizontal. A horizontal force of half the box's weight presses the box against the plane. What is the box's acceleration?

47. A 40-kg kangaroo exerts a constant force on the ground in the first 60 cm of her jump, and rises 2.0 m higher. When she carries a baby kangaroo in her pouch, she can rise only 1.8 m higher. What is the mass of the baby kangaroo?

48. In the arrangement shown in Fig. 3.33 find the mass M that will result in the 2-kg block remaining at rest. Assume that the pulleys are massless and frictionless.

49. Three objects whose masses are 2.0 kg, 3.0 kg, and 4.0 kg are joined by strings, as in Fig. 3.34. The 2.0-kg object is pulled upward with a force of 120 N. Find the tensions in the strings between the objects.

Figure 3.33
Problem 48.

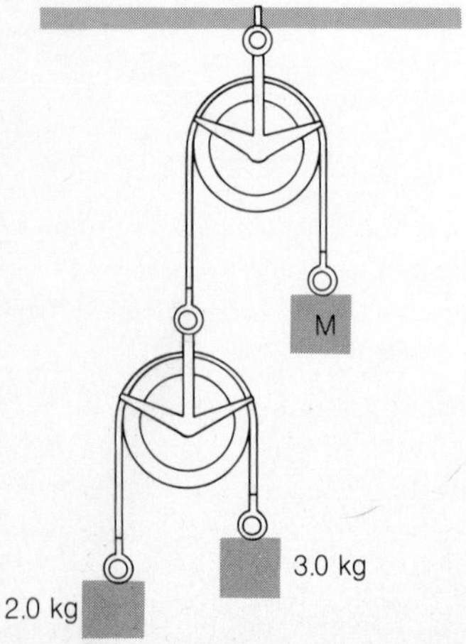

Figure 3.34
Problem 49.

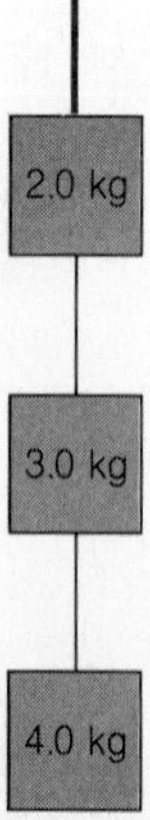

43. A sled slides down a snow-covered hill at constant speed. If the hillside is 10° above the horizontal, what is the coefficient of kinetic friction between the runners of the sled and the snow?

44. A cyclist finds that she is able to coast at constant speed along a road that slopes downward at an angle of 1° with the horizontal. If she and her bicycle together have a mass of 70 kg, what is the force required to propel them at the same constant speed along a level road?

45. A 200-kg crate is being slid down a ramp that makes an angle of 20° with the horizontal. The coefficient of kinetic friction is 0.30. How much force parallel to the plane must be applied to the crate if it is to slide down at constant speed? In which direction must the force be applied?

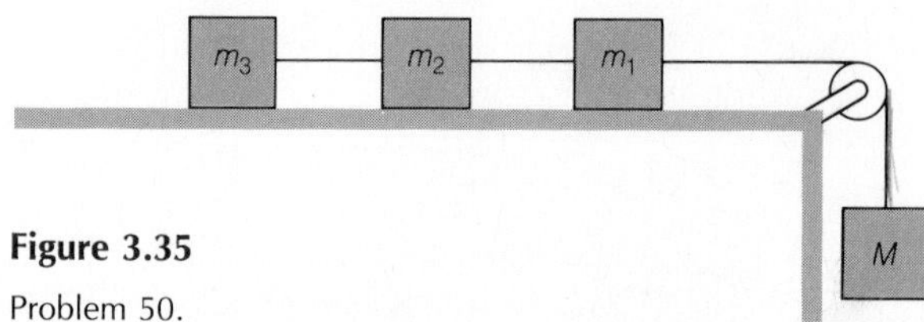

Figure 3.35
Problem 50.

50. Find the accelerations of the blocks shown in Fig. 3.35 and the tensions in the strings that join them. Neglect friction and the mass of the pulley.

51. The drag force R on a falling object of mass m due to air resistance is proportional to the square of the object's speed v, so that $R = kv^2$ where k depends on the properties of the object. Find the terminal speed of the object in terms of m, k, and g.

52. If μ is the coefficient of static friction between the cart A and block B in Fig. 3.36 what acceleration must the cart have if B is not to fall?

53. A 5.0-kg block resting on a horizontal surface is attached to a 5.0-kg block that hangs freely by a string passing over a pulley, an arrangement similar to that shown in Fig. 3.16. (a) If there is no friction between the first block and the surface,

Figure 3.36
Problem 52.

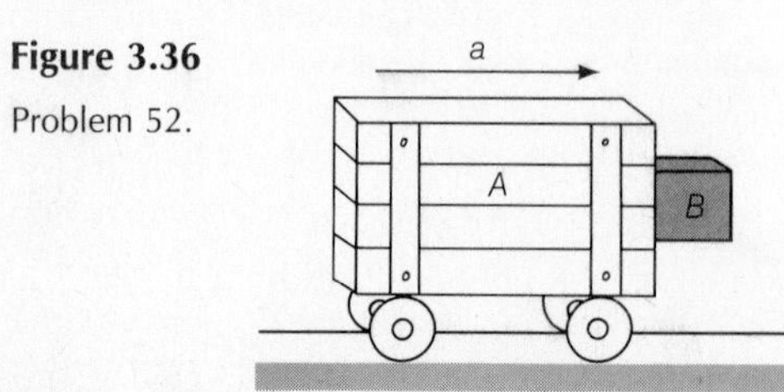

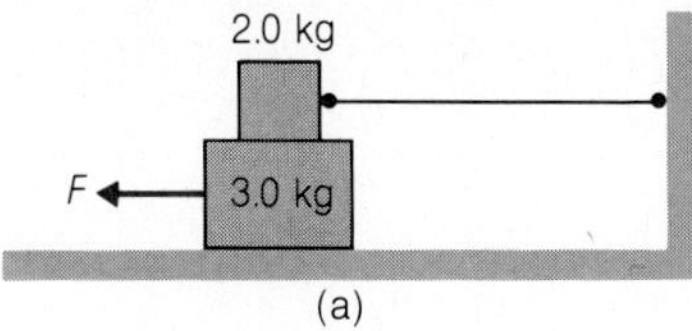

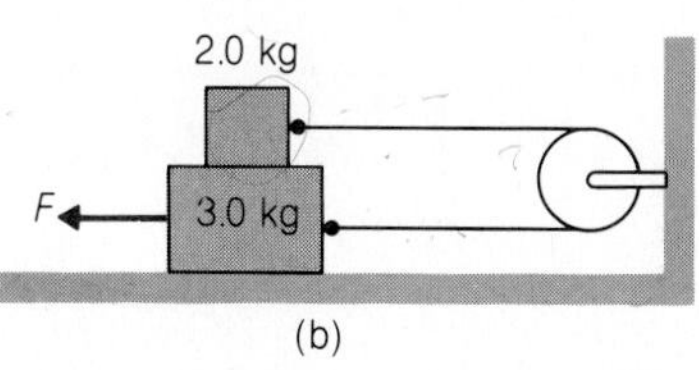

Figure 3.37
Problem 55.

what is the block's acceleration? (b) If the coefficient of kinetic friction between the first block and the surface is 0.20, what is the block's acceleration?

54. A 10.0-kg block resting on a horizontal surface is attached to a 5.0-kg block that hangs freely by a string passing over a pulley, an arrangement similar to that shown in Fig. 3.16. (a) If there is no friction between the first block and the surface, what is the block's acceleration? What is the block's acceleration if the coefficient of kinetic friction between the first block and the surface is (b) 0.40? (c) 0.50? (d) 0.60?

55. Two wooden blocks rest on a wooden table, as in Fig. 3.37. The coefficient of kinetic friction is 0.30. Find the minimum force needed to move the 3.0-kg block at constant speed in each case.

56. A horse pulls a 300-kg sled at constant speed over level snow by a rope that is 35° above the horizontal. If the coefficient of kinetic friction is 0.10, find the force the horse exerts on the rope.

57. A ship whose mass is 2.0×10^7 kg rests on launching ways that slope down to the water at an angle of 6°. The ways are greased to reduce the coefficient of kinetic friction to 0.11. Will the ship slide down the ways into the water by itself? If not, find the force needed to winch the ship into the water.

58. Two blocks joined by a string are sliding down an inclined plane, as in Fig. 3.38. The coefficient of kinetic friction between the 5.0-kg block and the plane is 0.40 and that between the 1-kg block and the plane is 0.10. Find the tension in the string.

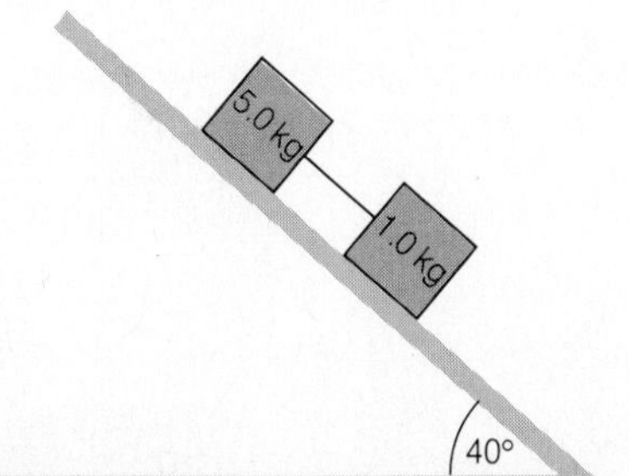

Figure 3.38
Problem 58.

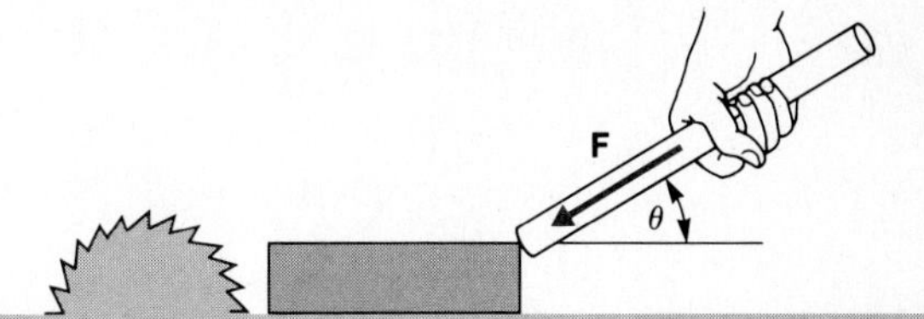

Figure 3.39
Problem 63.

59. A block slides down an inclined plane 9.0 m long that makes an angle of 38° with the horizontal. The coefficient of kinetic friction is 0.25. If the block starts from rest, find the time required for it to reach the foot of the plane.

60. If the block in Problem 59 has a mass of 50 kg, find the force parallel to the plane required to move it upward at constant speed.

61. A skier starts from rest and slides 50 m down a slope that makes an angle of 40° with the horizontal. She then continues sliding on level snow. (a) If the coefficient of kinetic friction between skis and snow is 0.10 and air resistance is neglected, what is the speed of the skier at the foot of the slope? (b) How far away from the foot of the slope does she come to a stop?

62. A block takes twice as long to slide down an inclined plane that makes an angle of 35° with the horizontal as it does to fall freely through the same vertical distance. What is the coefficient of kinetic friction?

63. A stick is used to push a block of wood toward the blade of a circular saw as in Fig. 3.39. (a) If the weight of the block of wood is w, the angle between the stick and table is θ, and the coefficient of kinetic friction between block and table is μ, verify that the force that must be applied to the stick to move the block at constant speed is $\mu w/(\cos\theta - \mu\sin\theta)$. (b) Show that if the stick is held at too steep an angle, the block cannot be moved, no matter how much force is applied. (c) Find the value of the critical angle for $\mu = 0.25$.

64. A workman is walking on the outside of an oil storage tank in the form of a horizontal cylinder of radius 4.0 m. When the workman is 0.6 m below the top of the tank (measured vertically), he slips. What is the coefficient of static friction between his shoes and the tank? (*Hint:* At any point on it, the circumference of a circle is perpendicular to a radius to that point from the center of the circle.)

Answers to Multiple Choice

1. d	**8.** c	**15.** a	**22.** a
2. d	**9.** d	**16.** b	**23.** c
3. c	**10.** c	**17.** b	**24.** d
4. a	**11.** b	**18.** b	**25.** a
5. d	**12.** c	**19.** d	**26.** a
6. b	**13.** d	**20.** c	**27.** c
7. b	**14.** b	**21.** d	**28.** b

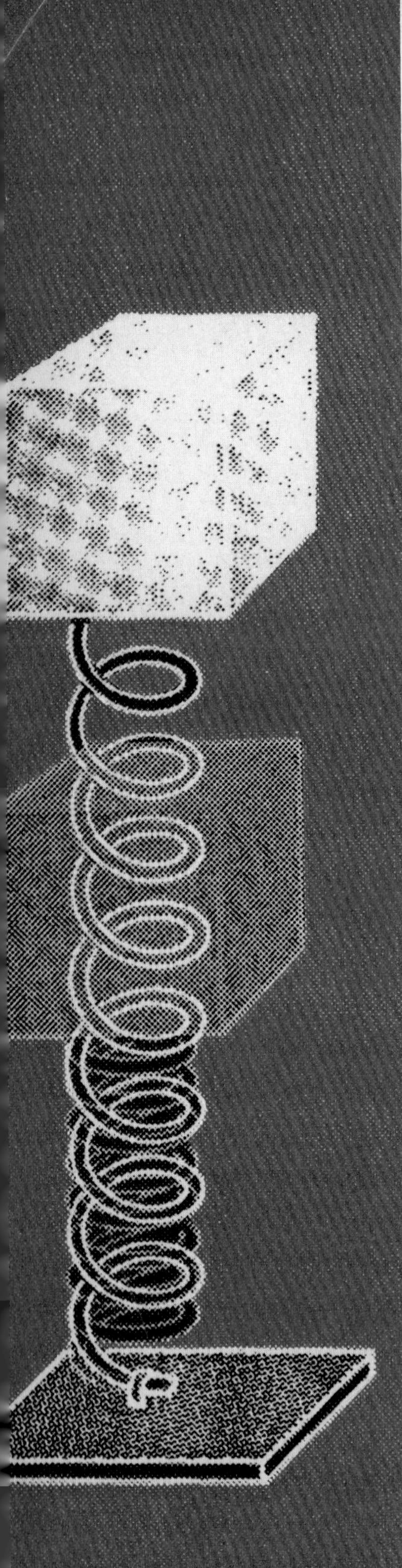

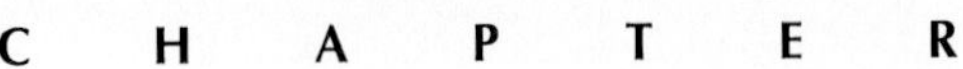

CHAPTER 4

ENERGY

The word **energy** is part of everyday life. We say an active person is energetic. We hear a candy bar described as being full of energy. We complain about the cost of the electric energy that lights our lamps and turns our motors. We worry about someday running out of the energy stored in coal and oil. We argue about whether nuclear energy is a blessing or a curse. Exactly what is meant by energy?

In general, energy refers to an ability to accomplish change. All changes in the physical world involve energy in some way. But "change" is not a very precise concept, and we must be sure of exactly what we are talking about before going further. The best way to introduce the complicated and many-sided idea of energy is to begin with the simpler idea of work.

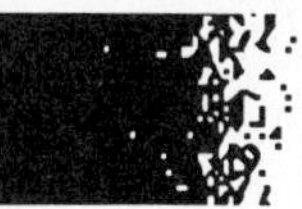

4.1 Work

A measure of the change a force produces.

As we know, forces can produce changes. Forces set things in motion, change their paths, and bring them to a stop. Forces push things together and pull them apart. The physical quantity called **work** is a measure of the amount of change (in a broad sense) to which a force gives rise when it acts on something.

Suppose we push against a wall. When we stop, nothing has happened even though we exerted a force on the wall. But if we apply the same force to a ball, the ball flies through the air when we let it go (Fig. 4.1). The difference is that the ball moved during our push, whereas the wall did not.

Or we might try to lift a heavy barbell. If we fail, the world is exactly the same afterward. If we succeed, though, the barbell is now up in the air, which represents a change. As before, the difference is that in the second case an object moved while we exerted a force on it.

To make these ideas precise, work is defined as follows:

The work done by a constant force F acting on an object while it undergoes a displacement x is equal to the magnitude of the force component F_x in the direction of the displacement multiplied by the magnitude x of the displacement.

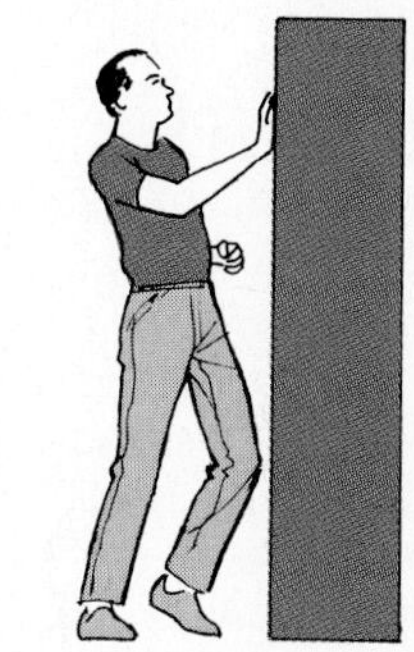

Figure 4.1

Work is done by a force when the object on which it acts is displaced while the force is applied.

If the angle between the force **F** and the displacement **x** is θ, as in Fig. 4.2 (a), then

$$F_x = F \cos \theta$$

and so the work done is

$$W = F_x x = (F \cos \theta)\, x$$

We can therefore define work by the equation

$$W = Fx \cos \theta \qquad \textit{Work} \quad (4.1)$$

Work is a scalar quantity. No direction is associated with work, even though it depends on two vector quantities, force and displacement. Figure 4.2 summarizes

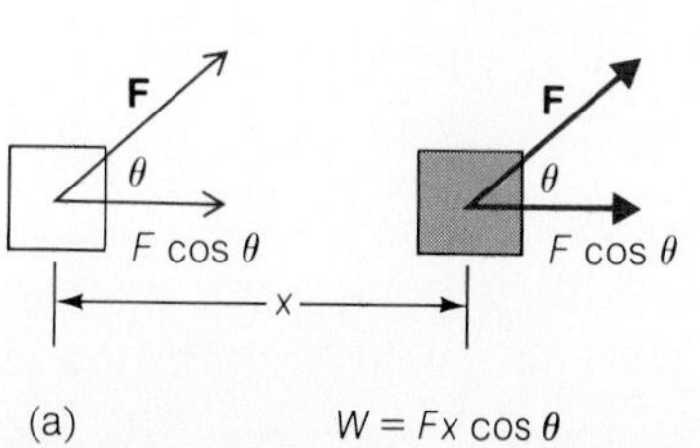

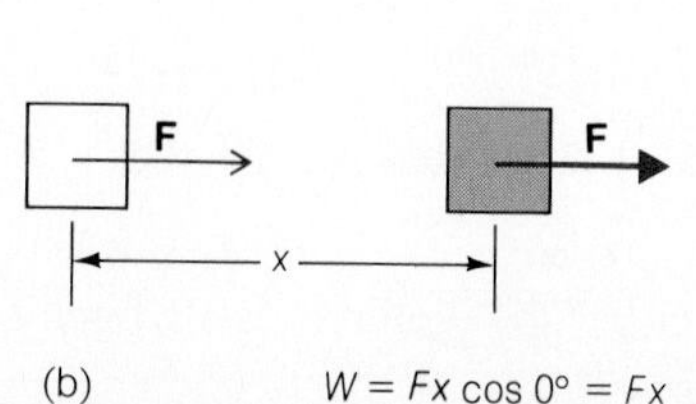

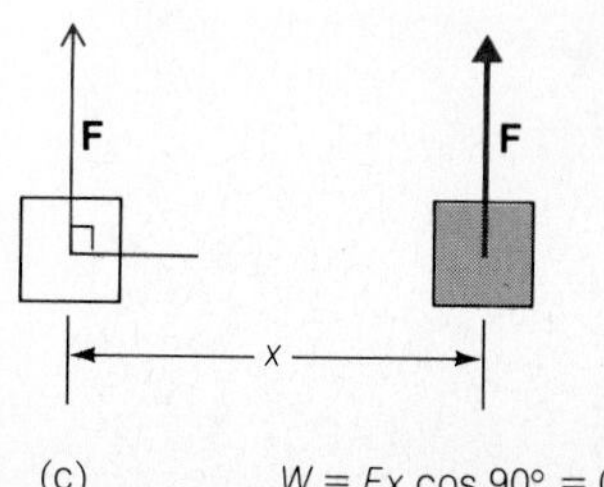

Figure 4.2

The work done by a force depends on the angle θ between the force and its displacement. (a) In general, $W = Fx \cos \theta$. (b) When **F** is parallel to **x**, $W = Fx$. (c) When **F** is perpendicular to **x**, $W = 0$ since **F** then has no component in the direction of **x** and therefore does not affect its motion.

Work is done in lifting a barbell, but no work is done while it is being held in the air even though this may be tiring.

the meaning of Eq. (4.1). When **F** is in the same direction as **x**, $\theta = 0$ and $\cos\theta = 1$, so

$$W = Fx \qquad (\mathbf{F} \text{ parallel to } \mathbf{x}) \tag{4.2}$$

But if **F** is perpendicular to **x**, $\theta = 90°$, and $\cos\theta = 0$, so in this case $W = 0$.

The preceding definition is a great help in clarifying the effects of forces. Unless a force acts through a displacement, no work is done no matter how great the force. (Although we may get tired after pushing on a wall for a long time, we still have done no work on the wall if the wall remains in place. We get tired because muscle fibers constantly contract and release when exerting a steady force even if no motion occurs.) Even if an object moves, no work is done on it unless a force acting on it has a component in its direction of motion.

If the direction of $\mathbf{F}_x$ is opposite to that of the displacement **x**, then $\cos\theta = \cos 180° = -1$. The work done by the force on the object in this case is negative. An example is the work done by a frictional force, since $\mathbf{F}_f$ is always opposite to **x**. In such a situation we can regard the work as being done *by* the moving object instead of *on* it, as when $\mathbf{F}_x$ is in the same direction as **x**.

The SI unit of work is the **joule**, abbreviated J. One joule is the work done by a force of 1 N acting through a distance of 1 m. That is,

$$1 \text{ joule} = 1 \text{ J} = 1 \text{ N·m}$$

The joule is named after the English scientist James Joule (1818–1889) and is pronounced ''jool.'' To raise an apple from your waist to your mouth takes about 1 J of work.

In the British system the unit of work is the *foot-pound*, abbreviated ft·lb. One ft·lb is equal to the work done by a force of 1 lb acting through a distance of 1 ft. A ft·lb is a little more than $1\frac{1}{3}$ J.

EXAMPLE 4.1

A person pulls an 80-kg crate 20 m across a level floor using a rope that is 30° above the horizontal. The person exerts a force of 150 N on the rope (Fig. 4.3). How much work is done?

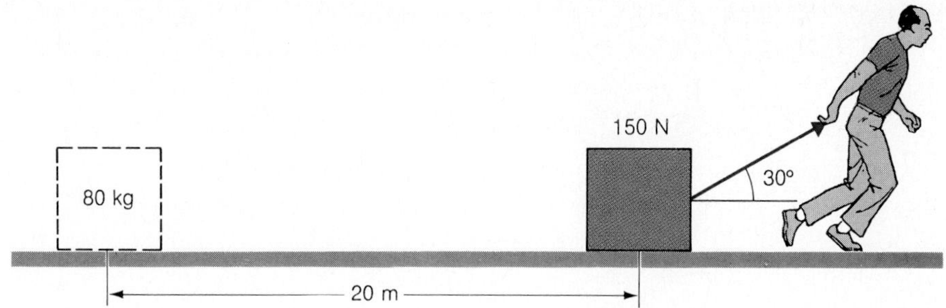

Figure 4.3

SOLUTION The work done is

$$W = Fx \cos \theta = (150 \text{ N})(20 \text{ m})(\cos 30°)$$
$$= (150)(20)(0.866) \text{ J} = 2.6 \times 10^3 \text{ J} = 2.6 \text{ kJ}$$

Evidently the 80-kg mass of the crate has no significance here—the force exerted on the crate determines how much work is done. ♦

4.2 Work Done Against Gravity

This work depends only on the height, not on the path.

The work needed to lift an object of mass m against gravity is easy to figure out. The force of gravity on the object is simply its weight $w = mg$. Therefore the force that must be exerted to lift the object has the same magnitude of mg and is in the same direction as the height h to which the object is raised. Since $F_x = mg$, $x = h$, and $\cos \theta = \cos 0 = 1$, the work done is

$$W = mgh \qquad \textit{Work done against gravity} \quad (4.3)$$

To lift an object of mass m to the height h requires the work mgh (Fig. 4.4).

It is important to note that only the height h is involved in work done against the force of gravity. The particular route taken by an object being raised is not significant (Fig. 4.5). Excluding any frictional effects, exactly as much work must

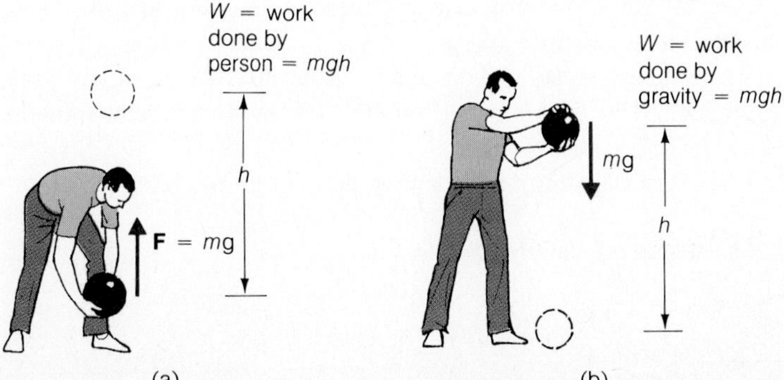

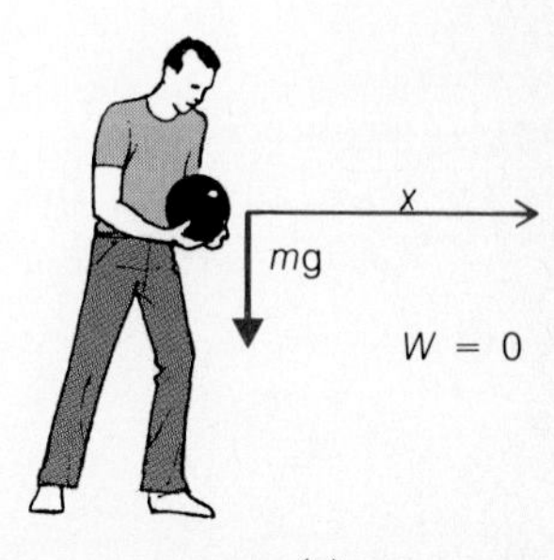

Figure 4.4

(a) The work mgh must be done to lift an object of mass m to a height h. (b) When an object of mass m falls from a height h, the force of gravity does the work mgh on it. (c) The force of gravity does no work on objects that move parallel to the earth's surface.

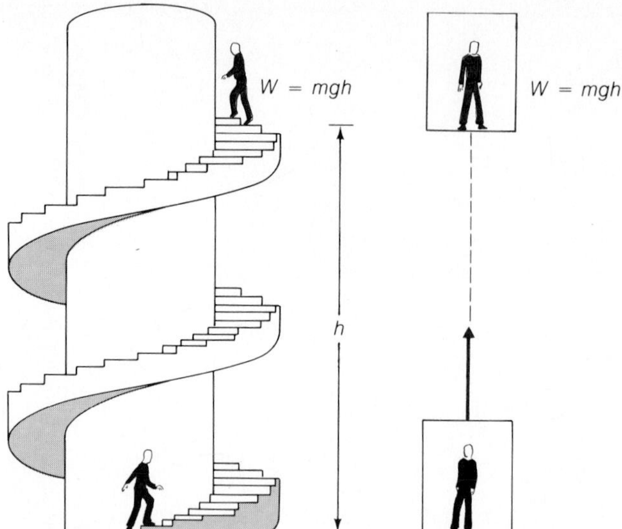

Figure 4.5

In the absence of friction, the work done in lifting a mass m to a height h is mgh regardless of the exact path taken.

be done to climb a flight of stairs as to go up in an elevator to the same floor (though not by the person involved!). Moving an object parallel to the earth's surface requires no work to be done against gravity because $h = 0$.

EXAMPLE 4.2

Eating a banana enables a person to perform about 4.0×10^4 J of work. To what height does eating a banana enable a 60-kg woman to climb?

SOLUTION From $W = mgh$ we have

$$h = \frac{W}{mg} = \frac{4.0 \times 10^4 \text{ J}}{(60 \text{ kg})(9.8 \text{ m/s}^2)} = 68 \text{ m}$$

◆

4.3 Power

The rate of doing work.

Often the time needed to carry out a task is as important as the actual work needed. If we are willing to wait, even a tiny motor can raise an elevator as high as we like. If we are in a hurry, however, we want a motor whose work output is rapid in terms of the total work needed. Thus the rate at which work is being done is significant. This rate is called **power**. The more powerful something is, the faster it can do work.

If an amount of work W is done in a time interval t, the power involved is

$$P = \frac{W}{t} \qquad \textit{Power} \quad (4.4)$$

$$\text{Power} = \frac{\text{work done}}{\text{time interval}}$$

In the SI system, where work is measured in joules and time in seconds, the unit of power is the **watt**:

$$1 \text{ watt} = 1 \text{ W} = 1 \text{ J/s}$$

The **kilowatt-hour** (kWh) is often used as a unit of work. Since 1 kW = 1000 W = 1000 J/s and 1 h = 3600 s,

$$1 \text{ kilowatt-hour} = 1 \text{ kWh} = 3.60 \times 10^6 \text{ J} = 3.60 \text{ MJ}$$

The **horsepower** (hp) is the traditional unit of power in engineering. Two centuries ago, in order to sell the steam engines he had perfected, James Watt had to compare their power outputs with that of a power source more familiar to his customers. After various tests he found that a typical horse could perform work at a rate of 497 W for as long as ten hours per day. To avoid disputes, Watt multiplied this figure by 1.5 to establish the unit he called the horsepower. Watt's horsepower therefore represents a rate of doing work of (1.5)(497W) = 746 W:

$$1 \text{ hp} = 746 \text{ W} = 0.746 \text{ kW} \qquad 1 \text{ kW} = 1.34 \text{ hp}$$

Few actual horses can develop this much power for very long. The early steam engines ranged from 4 to 100 hp, with the 20-hp model the most popular.

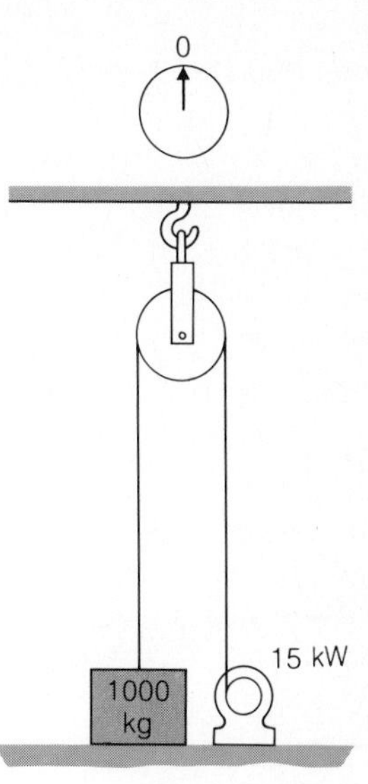

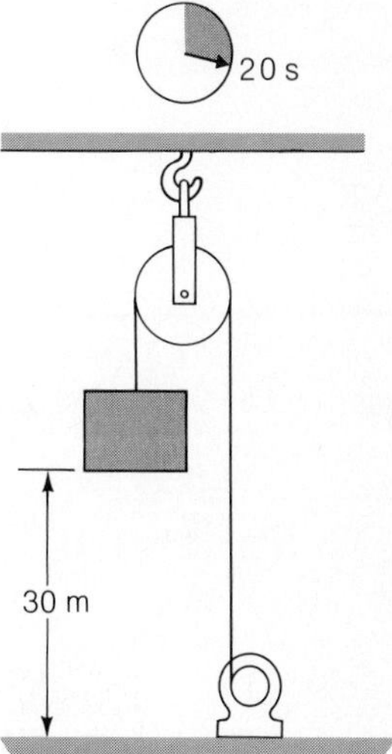

Figure 4.6

EXAMPLE 4.3

An electric motor with an output of 15 kW provides power for the elevator of a six-story building. If the total mass of the loaded elevator is 1000 kg, what is the minimum time needed for it to rise the 30 m from the ground floor to the top floor (Fig. 4.6)?

SOLUTION The work done in raising the elevator through a height h is

$$W = mgh$$

Since $P = W/t$, the time needed for the motor to raise the elevator by 30 m is

$$t = \frac{W}{P} = \frac{mgh}{P} = \frac{(1000 \text{ kg})(9.8 \text{ m/s}^2)(30 \text{ m})}{15 \times 10^3 \text{ W}} = 20 \text{ s}$$

♦

Force, Speed, and Power

How much power is delivered when a constant force **F** does work on an object moving at the constant velocity **v**? If θ is the angle between **F** and **v**, then

$$P = \frac{W}{t} = \frac{Fx\cos\theta}{t}$$

Since $x/t = v$,

$$P = Fv\cos\theta \qquad \textit{Power} \quad (4.5)$$

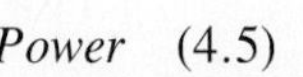

When **F** is in the same directon as **v**,

$$P = Fv \qquad (\mathbf{F} \text{ parallel to } \mathbf{v}) \tag{4.6}$$

$$\text{Power} = (\text{force})(\text{speed})$$

EXAMPLE 4.4

A boat is powered by three 2600-kW diesel engines. At the top speed of 27.4 m/s, the boat's propellers produce a total force of 250 kN. (This force equals in magnitude the resistance of the water to the boat's motion at that speed; hence the force does not accelerate the boat.) Find the efficiency of the boat's propulsion system.

SOLUTION The efficiency is $P_{\text{output}}/P_{\text{input}}$. Here

$$P_{\text{output}} = Fv = (250 \text{ kN})(27.4 \text{ m/s}) = 6850 \text{ kW}$$

$$P_{\text{input}} = (3)(2600 \text{ kW}) = 7800 \text{ kW}$$

and so

$$\text{Efficiency} = \frac{P_{\text{output}}}{P_{\text{input}}} = \frac{6850 \text{ kW}}{7800 \text{ kW}} = 0.88 = 88\%$$

◆

A person in good physical condition is usually capable of a continuous power output of about 75 W, which is 0.1 hp. An athlete such as a runner or a swimmer may have a power output several times greater during a distance event. What limits the power output of a trained athlete is not muscular development but the supply of oxygen via the bloodstream from the lungs to the muscles, where it is needed for the metabolic processes that extract work from nutrients. However, for a momentary effort such as that of a weight lifter or a jumper, an athlete's power output may exceed 5 kW.

EXAMPLE 4.5

A 40-kg woman uses 100 W of power to ride a 10-kg bicycle on a level road at 6.0 m/s. If she can develop 250 W, what is the angle of the steepest hill she can climb at this speed?

SOLUTION The friction and air resistance the woman must overcome at 6.0 m/s are the same when she goes uphill at the same speed, so she has 250 W − 100 W = 150 W available for the climb. The additional force she needs for a constant vertical speed of v_h is the total weight mg of her and her bicycle, so from Eq. (4.6)

$$v_h = \frac{P}{F} = \frac{P}{mg} = \frac{150 \text{ W}}{(50 \text{ kg})(9.8 \text{ m/s}^2)} = 0.306 \text{ m/s}$$

With the help of Fig. 4.7 we find that

$$\sin\theta = \frac{v_h}{v} = \frac{0.306 \text{ m/s}}{6.0 \text{ m/s}} = 0.051 \qquad \theta = 3°$$

Figure 4.7

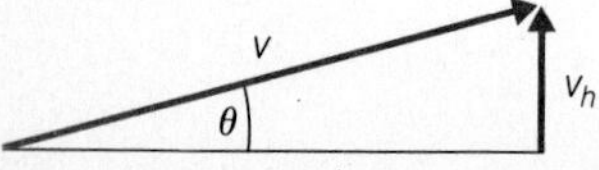

The steepest hill she can climb at 6.0 m/s has a slope of 3°. On a steeper hill she would have to go more slowly. ◆

4.4 Energy

Three broad categories.

Now that we know what work is, we can define energy:

Energy is the property that gives something the capacity to do work.

When we say that something has energy, we mean it is able (directly or indirectly) to exert a force on something else and do work on it. On the other hand, when we do work on something, we add to it an amount of energy equal to the work done. The unit of energy is the same as that of work, the joule.

There are three broad categories of energy:

1. **Kinetic energy**, the energy something possesses by virtue of its motion;
2. **Potential energy**, the energy something possesses by virtue of its position;
3. **Rest energy**, the energy something possesses by virtue of its mass.

In these descriptions the word *something* was used instead of *object* because, as we will see later, such nonmaterial entities as force fields and massless particles may also possess energy.

All modes of energy possession fit into one or another of these three categories. For instance, it is convenient for many purposes to think of heat as a separate form of energy, but what this term actually refers to is the sum of the kinetic energies of the randomly moving atoms and molecules in a body of matter. "Chemical energy" is actually the electric potential energy of electrons in the atoms and molecules of which matter is composed; this energy is given off in chemical reactions, such as the burning of gasoline in a car engine. Wave motion involves the interchange of kinetic and potential energies. Rest energy—the $E_0 = m_0c^2$ that most of us have heard of, if perhaps not fully understood—will be considered in Chapter 25.

4.5 Kinetic Energy

The energy of motion.

When we do work on a ball by throwing it, what becomes of this work?

Let us suppose we apply the constant force **F** to the ball for a distance x before it leaves our hand, as in Fig. 4.8(a). The work done on the ball is therefore, since $\cos\theta = 1$,

$$W = Fx \qquad (4.7)$$

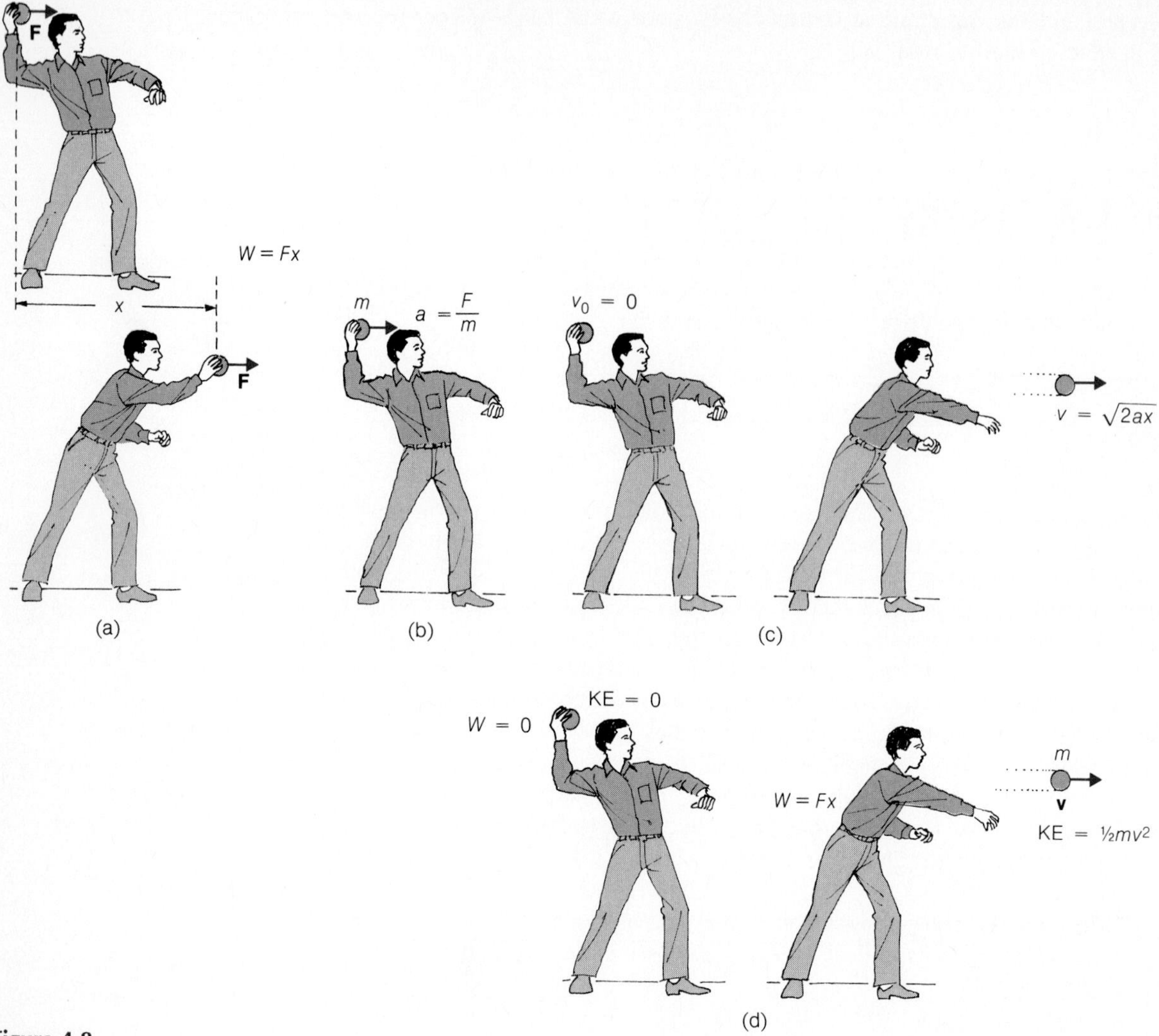

Figure 4.8

Successive steps in deriving the formula KE $= \frac{1}{2}mv^2$ for the kinetic energy of a moving object.

The mass of the ball is m. As we throw it, its acceleration has the magnitude

$$a = \frac{F}{m}$$

according to the second law of motion, $\mathbf{F} = m\mathbf{a}$ (Fig. 4.8b).

We know from the formula $v_f^2 = v_0^2 + 2ax$ that when an object starting from rest ($v_0 = 0$) undergoes an acceleration of magnitude a through a distance x, its final speed v is related to a and x by

$$v_f^2 = 2ax \tag{4.8}$$

This relationship is pictured in Fig. 4.8(c).

If we now substitute F/m for a in Eq. (4.8), we find that

$$v_f^2 = 2ax = 2\left(\frac{F}{m}\right)x$$

which we can rewrite as

$$Fx = \tfrac{1}{2}mv_f^2 \tag{4.9}$$

The quantity on the left-hand side, Fx, is the work our hand has done in throwing the ball, as in Eq. (4.7). The quantity on the right-hand side, $\frac{1}{2}mv_f^2$, must therefore be the energy acquired by the ball as a result of the work we did on it. This energy is **kinetic energy**, energy of motion. That is, we interpret the preceding equation as follows (Fig. 4.8d):

$$\begin{aligned} Fx &= \tfrac{1}{2}mv_f^2 \\ \text{Work done on ball} &= \text{kinetic energy of ball} \end{aligned}$$

The symbol for kinetic energy is KE. The kinetic energy of an object of mass m and velocity v is therefore

$$\text{KE} = \tfrac{1}{2}mv^2 \qquad \textit{Kinetic energy} \tag{4.10}$$

A moving object is able to perform an amount of work equal to $\frac{1}{2}mv^2$ in the course of being stopped. Although Eq. (4.10) was derived for a particular situation, it is a perfectly general result.

EXAMPLE 4.6

A 600-g hammer head strikes a nail at a speed of 4.0 m/s and drives it 5.0 mm into a wooden board (Fig. 4.9). What is the average force on the nail?

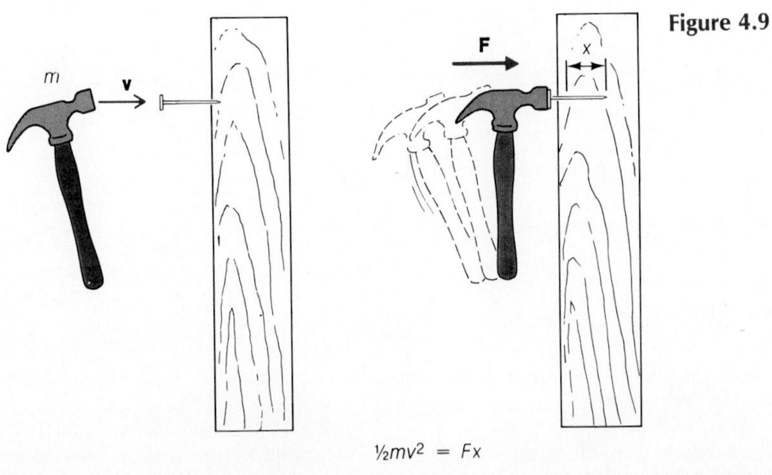

Figure 4.9

SOLUTION The initial kinetic energy of the hammer head is $\frac{1}{2}mv^2$, and the work done on the nail is Fx. Hence $Fx = \frac{1}{2}mv^2$ and

$$F = \frac{mv^2}{2x} = \frac{(0.60\ \text{kg})(4.0\ \text{m/s})^2}{2(0.0050\ \text{m})} = 960\ \text{N} = 0.96\ \text{kN}$$

which is 216 lb. ♦

The v^2 factor means that KE goes up very rapidly with increasing speed. Consider a 1200-kg car at 25 km/h (16 mi/h) and at 100 km/h (62 mi/h). Since 1 km/h = 0.278 m/s, the respective speeds are 7.0 and 27.8 m/s, and so the corresponding KE values are (Fig. 4.10):

$$\text{KE}_1 = \tfrac{1}{2}mv_1^2 = \tfrac{1}{2}(1200\ \text{kg})(7.0\ \text{m/s})^2 = 2.9 \times 10^4\ \text{J}$$
$$\text{KE}_2 = \tfrac{1}{2}mv_2^2 = \tfrac{1}{2}(1200\ \text{kg})(27.8\ \text{m/s})^2 = 46.4 \times 10^4\ \text{J}$$

At 100 km/h the car has 16 times as much kinetic energy as it does at 25 km/h. The fact that kinetic energy, and hence ability to do work (that is, damage), is proportional to the square of the speed is responsible for the severity of automobile accidents at high speeds.

EXAMPLE 4.7

What is the power output of the engine of a 1200-kg car if the car can go from 25 km/h to 100 km/h in 12 s?

SOLUTION The work needed to accelerate the car is

$$W = \text{KE}_2 - \text{KE}_1 = 46.4 \times 10^4\ \text{J} - 2.9 \times 10^4\ \text{J} = 43.5 \times 10^4\ \text{J}$$

The power needed to provide this amount of work in 12 s is

$$P = \frac{W}{t} = \frac{43.5 \times 10^4\ \text{J}}{12\ \text{s}} = 3.63 \times 10^4\ \text{W} = 36.3\ \text{kW}$$

which is equivalent to

$$P = (36.3\ \text{kW})\left(1.34 \frac{\text{hp}}{\text{kW}}\right) = 49\ \text{hp}$$ ♦

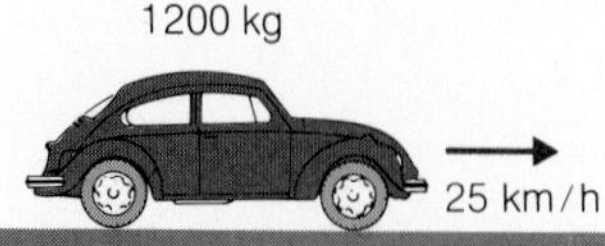

KE$_1$ = 2.9 × 10^4 J

100 km/h

KE$_2$ = 46.4 × 10^4 J

Figure 4.10

Kinetic energy is proportional to v^2. A car going 4 times as fast as another car of the same mass has 16 times as much KE.

Running Speeds

We can use the relationship $Fx = \frac{1}{2}mv^2$ between work done and the resulting kinetic energy to arrive at an interesting conclusion about animal running speeds. According to this formula,

$$v = \sqrt{\frac{2Fx}{m}}$$

Let us interpret v as an animal's speed, F as the force its leg muscles exert over the distance x, and m as its mass. As we saw in Section 3.4, the mass of an animal is approximately proportional to L^3, where L is a representative linear dimension such as its length, and the forces its muscles can exert are approximately proportional to L^2. The distance through which a muscle acts is proportional to L. Hence the quantity Fx/m depends on L as $(L^2)(L)/L^3 = 1$, which means that Fx/m, and hence v, should not vary with L at all! In fact, although different animals have different running abilities, there is indeed little correlation with size over a wide span. A hare can run about as fast as a horse.

An animal's highest aerobic speed, which is the fastest it can run while supplying its muscles with the oxygen they require on a continuous basis, tends to be about half its maximum speed over short distances in which an oxygen debt can be incurred. This agrees with human experience. Distance races, such as marathons, are typically run at average speeds about half those of sprints.

4.6 Potential Energy

The energy of position.

When we drop a stone from a height h, it falls faster and faster until it strikes the ground. If we lift the stone afterward, we see that it has done work by making a shallow hole in the ground (Fig. 4.11). At its original height the stone must have had the capacity to do work even though it was not moving at the time. The work the stone could do by falling to the ground is called its **potential energy,** symbol PE.

As we know, the work needed to raise a stone of mass m to a height h is

$$W = mgh \qquad (4.3)$$

The same amount of work can be done *by* the stone after dropping through the same height h. Accordingly the potential energy of the stone is

$$PE = mgh \qquad \textit{Gravitational potential energy} \qquad (4.11)$$

The gravitational potential energy of an object depends on the reference level from which its height h is measured. For example, the potential energy of a 1.0-kg book held 10 cm above a desk is

$$PE = mgh = (1.0\text{ kg})(9.8\text{ m/s}^2)(0.10\text{ m}) = 0.98\text{ J}$$

with respect to the desk (Fig. 4.12). However, if the book is 1.0 m above the floor

Figure 4.11

A raised stone has potential energy with respect to the ground.

Raised stone has PE = *mgh*

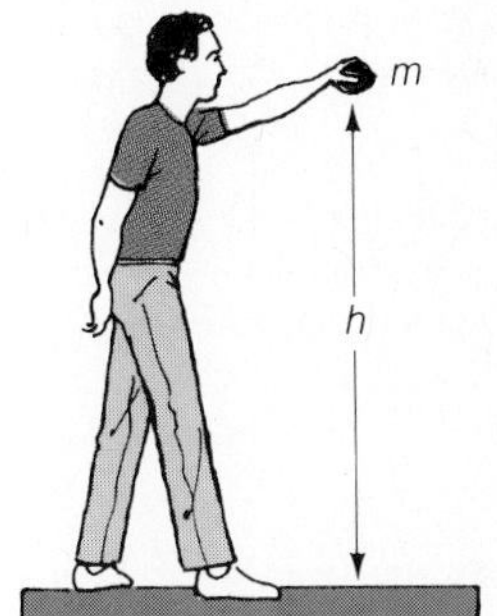

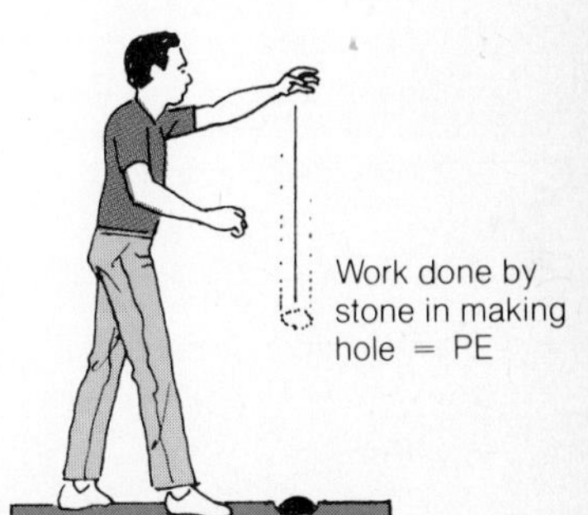

of the room, its potential energy is

$$PE = mgh = (1.0\ \text{kg})(9.8\ \text{m/s}^2)(1.0\ \text{m}) = 9.8\ \text{J}$$

with respect to the floor. And the book may conceivably be 100 m above the ground, so its potential energy is

$$PE = mgh = (1.0\ \text{kg})(9.8\ \text{m/s}^2)(100\ \text{m}) = 980\ \text{J}$$

with respect to the ground. The height h in the formula PE $= mgh$ means nothing unless the base height $h = 0$ is specified.

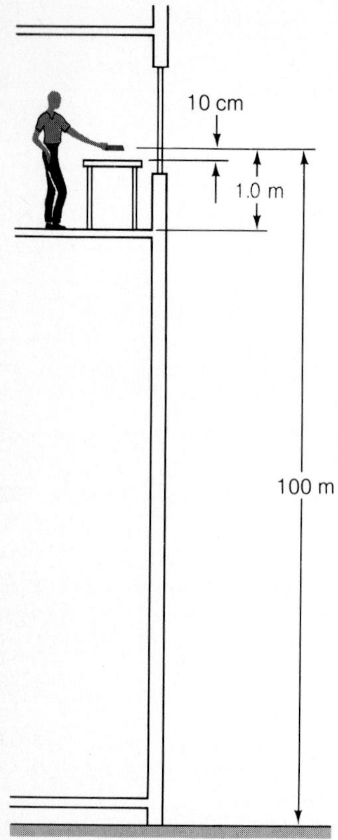

Figure 4.12

The potential energy of an object depends on the reference level from which its height h is measured. The object here is a 1.0-kg book.

EXAMPLE 4.8

Compare the potential energy of a 1200-kg car at the top of a hill 30 m high (Fig. 4.13) with its kinetic energy when moving at 100 km/h.

SOLUTION The car's potential energy relative to the bottom of the hill is

$$PE = mgh = (1200\ \text{kg})(9.8\ \text{m/s}^2)(30\ \text{m}) = 3.5 \times 10^5\ \text{J}$$

Earlier we found that the kinetic energy of the car at 100 km/h is 4.6×10^5 J, which is greater than its potential energy at the top of the hill. This means that a crash at 100 km/h (62 mi/h) into a stationary obstacle will yield more work—that is, do more damage—than dropping the car 30 m (98 ft).

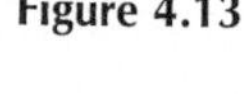

Figure 4.13

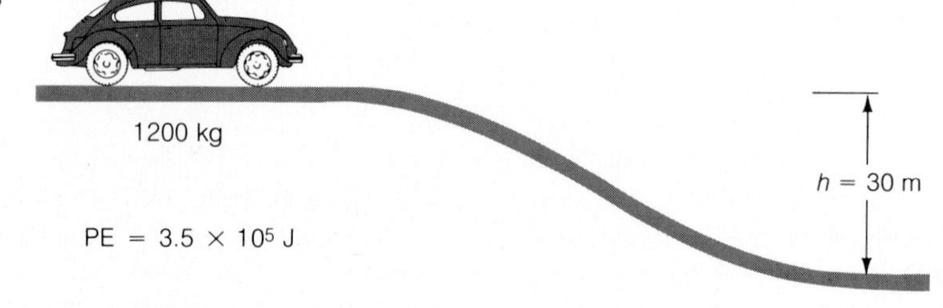

◆

EXAMPLE 4.9

Electricity demand varies a great deal throughout the day, and to provide generating capacity for peak periods that lies idle at other times is very expensive. It is more efficient to even out the variations by always running a generating plant at full capacity and in some way storing the surplus energy of off-peak periods to be tapped when needed. But how can this be done on the scale required? In one approach, at Dinorwig, in Wales, turbines connected to electric motor-generators pump water 568 m up to a mountain reservoir at night and at other times of low energy demand. When more electric power is needed, the water is allowed to fall back through the turbines to run the motor-generators in their generator mode. If 85% of the potential energy of the stored water is converted into electricity, how much water must pass through the turbines per second to produce the system's maximum power output of 1680 MW?

SOLUTION The PE of a mass m of stored water is mgh. If 85% of this PE can be recovered as work W done to produce electricity in the generators,

$$W = 0.85\ \text{PE} = 0.85mgh$$

The corresponding power output P is

$$P = \frac{W}{t} = 0.85gh\ \frac{m}{t}$$

Since $P = 1680\ \text{MW} = 1680 \times 10^6\ \text{W}$, the rate of water flow is

$$\frac{m}{t} = \frac{P}{0.85gh} = \frac{1680 \times 10^6\ \text{W}}{(0.85)(9.8\ \text{m/s}^2)(568\ \text{m})} = 3.55 \times 10^5\ \text{kg/s}$$

This is 355 tonnes per second, where 1 tonne = 1 "metric ton" = 10^3 kg. A tonne of water occupies 1 m^3, so 355 tonnes occupies 355 m^3, the volume of a small house. ◆

What we have been speaking of as the potential energy of an object is actually not a property of the object itself. It is a property of the *system* of the object and whatever exerts on the object a force that depends on the object's position. Thus the PE of a stone held above the earth's surface is shared by both the stone and the earth. When the stone is let go, both it and the earth move toward each other. But the earth's motion is too small to detect because of its immense mass relative to that of the stone. It is therefore appropriate for us to assign the entire PE of the system to the stone. In the case of two interacting objects more nearly equal in mass, however, we must keep in mind that the PE belongs to the system, not to just one of the objects.

The elastic potential energy of this bow becomes kinetic energy of the arrow when the bowstring is released.

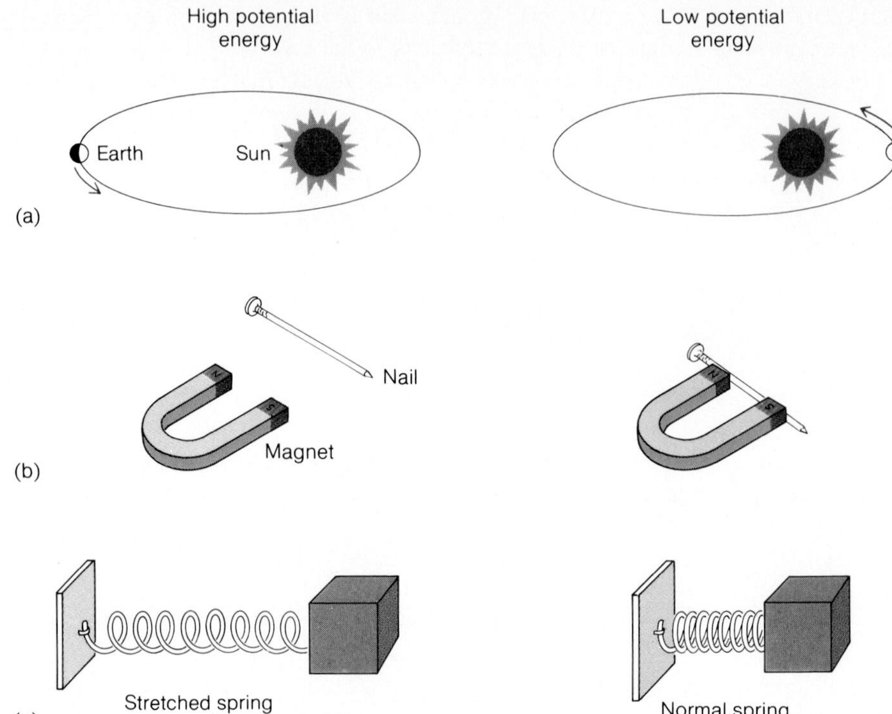

Figure 4.14

Three examples of potential energy. In each case the PE is a property of the entire system.

Until now we have considered only one kind of potential energy, namely, the gravitational PE of an object raised above some reference level near the earth's surface. The concept of potential energy is a much more general one, however, for it refers to the energy something has as a consequence of its position regardless of the nature of the force acting on it. The earth itself, for instance, has gravitational potential energy with respect to the sun, since if its orbital motion were to cease, it would fall toward the sun (Fig. 4.14). An iron nail has magnetic potential energy with respect to a nearby magnet, since it will fly to the magnet if released. An object at the end of a stretched spring has elastic potential energy with respect to its position when the spring has its normal extension, since if let go the object will move as the spring contracts. In each of these cases the object in question has the potentiality of doing work in its original position.

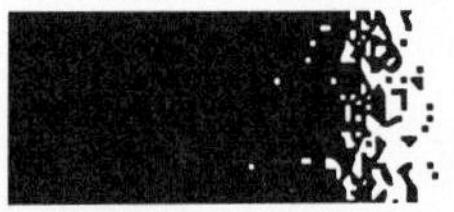

4.7 Conservation of Energy

A fundamental law of nature.

A **conservation principle** states that, in a system of any kind isolated from the rest of the universe, a certain quantity keeps the same value it originally had no matter what changes the system undergoes.

As an example, the law of conservation of mass revolutionized chemistry by holding that the total mass of the products of a chemical reaction is the same as the total mass of the original substances. Thus the increase in mass of a piece of iron when it rusts indicates that the iron has combined with some other material, rather than having decomposed, as the early chemists believed. In fact, the gas oxygen was discovered in the course of seeking this other material.

Given one or more conservation principles that apply to a given system, we know at once which classes of events can take place in the system and which cannot. Thus when iron rusts, the gain in mass means that it has combined chemically with something else. In physics we can often draw conclusions about the behavior of the particles that make up a system without a detailed study by basing our analysis on the conservation of some particular quantities. The power of this approach is shown by the great success of physics in understanding natural phenomena, a success largely due to the variety of conservation principles that have been discovered.

The first conservation principle we will study is that of **conservation of energy:**

The total amount of energy in a system isolated from the rest of the universe always remains constant, although energy transformations from one form to another may occur within the system.

This principle is perhaps the most fundamental generalization in all of science; no violation of it has ever been found.

A falling ball provides a simple example of conservation of energy. As it falls, its initial potential energy is converted into kinetic energy, so that the total energy of the stone remains the same. The potential energy of a 1-kg ball 50 m above the ground is $mgh = 490$ J, and its total energy is 490 J until it interacts with the ground and transfers energy to it (Fig. 4.15).

Another example is the motion of a planet about the sun. Planetary orbits are elliptical, so that at different points in its orbit the planet is at different distances from the sun. When the planet is close to the sun, it has a low PE, just as a stone near the ground has a low PE; when the planet is far from the sun, it has a high PE (Fig. 4.14a). Since the sum of the planet's PE and KE must be constant, we conclude that the kinetic energy of the planet is at a maximum when it is nearest the sun and

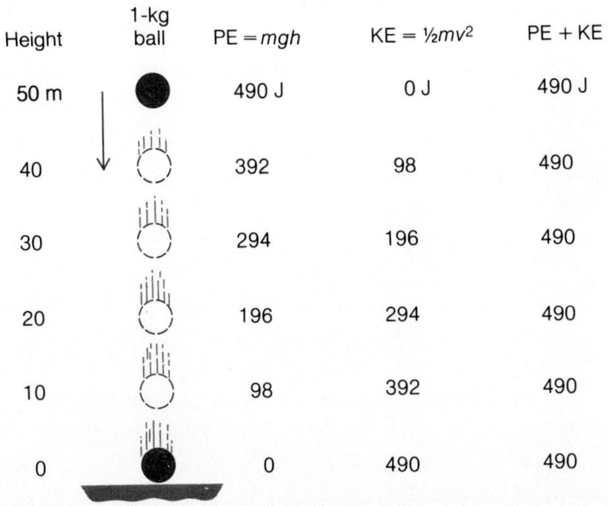

Height	1-kg ball	PE = mgh	KE = $\frac{1}{2}mv^2$	PE + KE
50 m		490 J	0 J	490 J
40		392	98	490
30		294	196	490
20		196	294	490
10		98	392	490
0		0	490	490

Figure 4.15
The total energy of a falling ball remains constant as its potential energy is transformed into kinetic energy.

at a minimum when it is farthest from the sun. The earth's orbital speed varies between 29 and 30 km/s for this reason.

Newton's laws of motion enable us—in principle—to solve all problems that involve forces and moving objects. However, these laws are easy to use only in the simplest cases because in order to apply them we must take into account all the various forces acting on each object at every point in its path. This is usually a difficult and complicated procedure. The advantage of the principle of conservation of energy is that it tells us about the relationship between the initial and final states of motion of some object or system of objects without our having to investigate what happens in between.

EXAMPLE 4.10

A skier is sliding downhill at a constant speed of 8.0 m/s when she reaches an icy patch on which her skis move with negligible friction (Fig. 4.16). If the icy patch is 10 m high, what is the skier's speed at its bottom?

SOLUTION Because the path of the skier on the icy patch is complicated, to apply **F** = *m***a** here would be very difficult. Using conservation of energy, however, makes the problem quite easy. At the top of the icy patch, the skier's initial kinetic energy is

$$\mathrm{KE}_1 = \tfrac{1}{2}mv_1^2$$

where $v_1 = 8.0$ m/s, and her potential energy relative to the bottom of the patch is

$$\mathrm{PE} = mgh$$

where $h = 10$ m. At the bottom the skier's kinetic energy is $\mathrm{KE}_2 = \mathrm{KE}_1 + \mathrm{PE}$, and so

$$\tfrac{1}{2}mv_2^2 = \tfrac{1}{2}mv_1^2 + mgh$$

$$v_2^2 = v_1^2 + 2gh$$

$$v_2 = \sqrt{v_1^2 + 2gh} = \sqrt{(8.0\ \mathrm{m/s})^2 + (2)(9.8\ \mathrm{m/s^2})(10\ \mathrm{m})} = 16\ \mathrm{m/s}$$

We did not have to know the skier's mass to find her final speed.

Figure 4.16

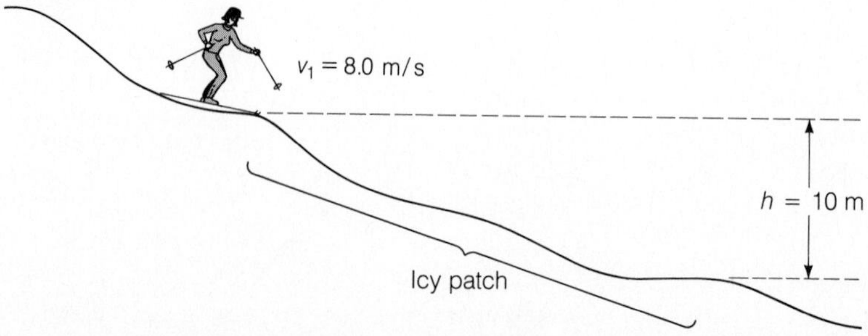

◆

Center of Mass

Sometimes we must take into account the size and shape of the object whose motion we are examining, which we have not had to do thus far. The first step is to identify the object's **center of mass** (CM), which is the point at which we can consider all its mass concentrated. Often the location of the CM is obvious; in the case of a uniform sphere, for instance, it is the geometric center. (In Chapter 8 we will learn how to find the CMs of more complicated bodies.) We now treat the object as a particle located at the CM. This procedure works provided the various forces on the object all act through its CM. If they don't, the object tends to turn under their influence, and what happens then is the subject of Chapter 7.

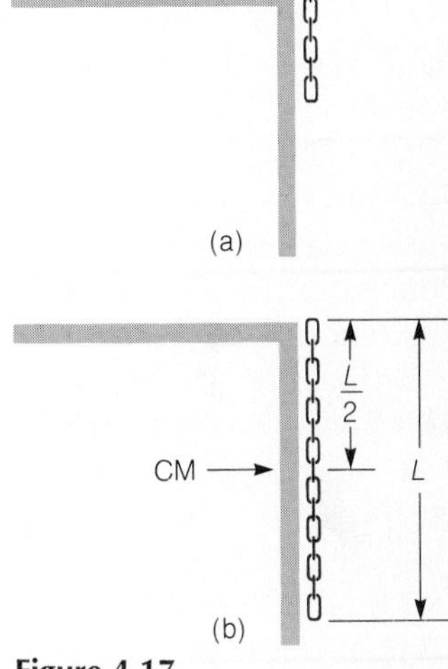

Figure 4.17

EXAMPLE 4.11

A chain L long lying on a frictionless table starts from rest to slide over the edge of the table, as in Fig. 4.17(a). What will the speed of the chain be as its last link passes over the edge?

SOLUTION The center of mass is at the chain's midpoint. When the last link of the chain has just passed over the edge of the table, as in Fig. 4.17(b), its CM is $L/2$ below the edge, and the chain as a whole has lost $mgh = mg(L/2)$ of potential energy. The lost PE has become KE, and so

$$\tfrac{1}{2}mgL = \tfrac{1}{2}mv^2$$
$$v = \sqrt{gL}$$

◆

4.8 Conservative and Nonconservative Forces

Only a conservative force gives rise to potential energy.

As we have seen, we can do work on something and thereby give it kinetic or potential energy, which in turn can reappear as work. When we throw a stone upward into the air, the work we do appears first as kinetic energy. As the stone rises, the kinetic energy gradually becomes potential energy. At its highest point the stone has only potential energy, which, as the stone begins to fall, is converted back into kinetic energy. Finally, when the stone strikes the ground and makes a hole in it, the kinetic energy turns to work.

But what happens to the energy used in opposing frictional forces? To find the answer, all we need do is rub one piece of wood against another. After a short time, it is obvious that the rubbed surfaces of the wood are warmer than before. This is a quite general observation: Work done against frictional forces always produces a rise in the temperature of the objects involved. What is happening is that the energy that disappears from the motion of each object as a whole reappears as additional KE of random motion of the molecules of which the objects are composed. This additional molecular energy is manifested as a rise in temperature, as will be explained in Chapter 13.

BIOGRAPHICAL NOTE

James Joule

James Joule (1818–1889), the son of an English brewer, was a sickly child and was educated largely at home. For a while he was taught by John Dalton, the founder of the modern atomic theory of matter, who inspired him to follow a scientific career.

Joule's research began at 18, when he tried to find out why an electric motor he had built was inefficient. This led to a study of the heat produced by an electric current, no easy task in those early days of electricity. His careful measurements, a feature of all his work, showed that this heat is produced at a rate proportional to the square of the current, the I^2R law discussed in Chapter 18. Joule went on from there to examine the heat given off in a wide variety of other processes.

Half a century earlier Count Rumford had realized that there was a connection of some kind between heat and what we today call energy. Rumford, born Benjamin Thompson in Massachusetts, supported the British during the Revolutionary War and thought it wise to flee to Europe afterward. Before Rumford, heat was regarded as an invisible, weightless fluid called *caloric*. Absorbing caloric caused an object to grow warmer; losing caloric caused it to grow cooler. In 1798 Rumford was in charge of boring cannon for a German prince and was impressed by the large amount of heat given off. If heat were a fluid, cutting into solid metal should allow it to escape. However, even a dull drill that cut no metal could produce unlimited heat, which implied that the metal contained an infinite amount of caloric. This made no sense to Rumford, who concluded that heat and "motion" were related in some way. Nevertheless, many respected scientists continued to believe in caloric for many years.

Joule spent a decade in measuring the heat produced when energy of some kind disappears. In the best-known procedure, a falling weight on a string caused a set of paddle wheels to rotate in an insulated container of water. The temperature of the water rose, and Joule found that there was always a fixed ratio between the work done by the weight as it dropped and the heat given to the water. Determinations made in other ways—for instance, by compressing air—gave exactly the same ratio between work and heat, a ratio that was called the mechanical equivalent of heat.

Joule's chemical and electrical experiments agreed with the mechanical ones, and the result was his announcement of the law of conservation of energy in 1847, when he was 29. The German physicist Julius Mayer had stated this law earlier than Joule, but on the basis of less extensive and less precise data, and Hermann von Helmholtz, another German physicist, considered the implications of energy conservation in greater detail than Joule did. Joule's work, however, was conclusive in establishing the law, and for this reason the SI unit of energy was named for him.

Joule did other useful work after 1847, but his productivity soon declined. Although he was a modest man—"I have done two or three little things, but nothing to make a fuss about," he once wrote—many honors came his way, and when his funds eventually ran out, Queen Victoria awarded him a pension. He died at 71.

This discussion is summarized in the **work-energy theorem:**

Work done *on* object = change in object's KE
+ change in object's PE
+ work done *by* object

The kinetic energy of the object may either increase or decrease; the same is true of its potential energy. It is the work done *by* the object that may be converted into heat.

EXAMPLE 4.12

A 25-kg box is pulled up a ramp 20 m long and 3.0 m high by a constant force of 120 N (Fig. 4.18). If the box starts from rest and has a speed of 2.0 m/s at the top, what is the force of friction between box and ramp?

SOLUTION According to the work–energy theorem, the total work W done by the applied force of 120 N must equal the work W_f done against friction (which becomes heat) plus the change ΔKE in the box's kinetic energy and the change ΔPE in its potential energy. Hence

$$W = W_f + \Delta\text{KE} + \Delta\text{PE} \quad \text{and} \quad W_f = W - \Delta\text{KE} - \Delta\text{PE}$$

The total work done is

$$W = Fs = (120\ \text{N})(20\ \text{m}) = 2400\ \text{J}$$

Since the box starts from rest, its change in kinetic energy is equal to its kinetic energy at the top of the ramp, and so

$$\Delta\text{KE} = \tfrac{1}{2}mv^2 = \tfrac{1}{2}(25\ \text{kg})(2.0\ \text{m/s})^2 = 50\ \text{J}$$

The change in the box's potential energy is

$$\Delta\text{PE} = mgh = (25\ \text{kg})(9.8\ \text{m/s}^2)(3.0\ \text{m}) = 735\ \text{J}$$

Hence the work done against friction is

$$W_f = W - \Delta\text{KE} - \Delta\text{PE} = (2400 - 50 - 735)\ \text{J} = 1615\ \text{J}$$

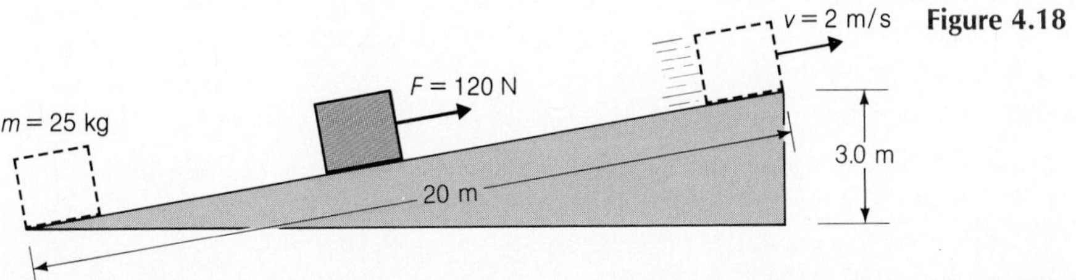

Figure 4.18

Because $W_f = F_f s$, the frictional force is

$$F_f = \frac{W_f}{s} = \frac{1615\text{ J}}{20\text{ m}} = 81\text{ N}$$ ◆

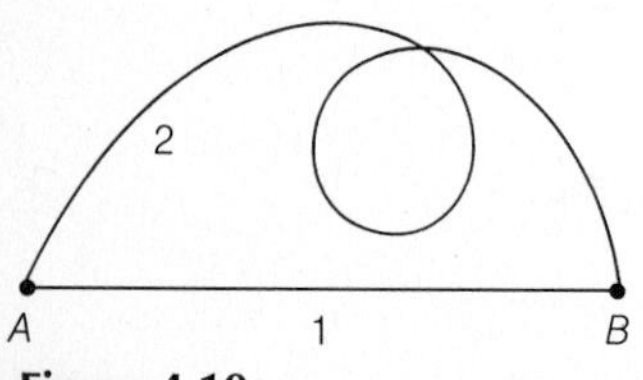

Figure 4.19

The work done by a conservative force (such as gravity) that acts on an object going from A to B is the same for either path. The work done by a nonconservative force (such as friction) will be different if the object takes path 2 rather than path 1.

Clearly we can divide forces into two categories. **Conservative forces** include the gravitational, magnetic, and elastic forces shown in Fig. 4.14. The work done by (or against) a force of this kind depends only on the endpoints of the motion of an object it acts on. The path taken between the endpoints does not matter. Such a force can give rise to a potential energy: If the path is reversed, the work done can reappear as kinetic energy.

The work done by (or against) a **nonconservative force,** on the other hand, depends on the path taken. Friction is an example. In Fig. 4.19 the work done to move something from A to B against a frictional force will be greater if path 2 is taken instead of path 1. If a conservative force were acting, the work would be the same. A nonconservative force does not give rise to potential energy: The work done cannot be returned by reversing the path. Nonconservative forces are also called **dissipative forces.**

Of course, the work done against a nonconservative force in going from A to B is not really lost. This work merely is changed into a form of energy—heat, light, sound, and so forth—that is not recoverable by bringing the object back to A. The term nonconservative forces is somewhat misleading because the forces actually do conserve energy. What they do not conserve is the KE + PE of an object they act on.

Important Terms

Work is a measure of the change (in a general sense) a force gives rise to when it acts on something. When an object is displaced while a force acts on it, the work done by the force is equal to the product of the displacement and the component of the force in the direction of the displacement. The SI unit of work is the **joule.**

The rate at which work is done is called **power.** The SI unit of power is the **watt,** which is equal to 1 J/s.

Energy is that which may be converted into work. When something has energy, it is able to perform work or, in a general sense, to change some aspect of the physical world. The unit of energy is also the joule.

The three broad categories of energy are **kinetic energy,** which is the energy something has by virtue of its motion; **potential energy,** which is the energy something has by virtue of its position; and **rest energy,** which is the energy something has by virtue of its mass.

The principle of **conservation of energy** states that the total amount of energy in a system isolated from the rest of the universe always remains constant, although energy transformations from one form to another may occur within the system.

Work done by a **conservative force** (such as gravity) is independent of the path taken; such a force can give rise to a potential energy. Work done by a **nonconservative force** (such as friction) varies with the path taken and is dissipated; such a force cannot give rise to a potential energy.

Important Formulas

Work: $W = Fx \cos \theta$

Work in lifting object: $W = wh = mgh$

Power: $P = \frac{W}{t} = Fv \cos \theta$

Kinetic energy: $\text{KE} = \frac{1}{2}mv^2$

Gravitational potential energy: $\text{PE} = wh = mgh$

MULTIPLE CHOICE

1. Which of the following is not a unit of power?
- a. joule-second
- b. watt
- c. newton-meter per second
- d. horsepower

2. To keep a vehicle moving at the speed v requires a force F. The power needed is

a. Fv b. $\frac{1}{2}Fv^2$
c. F/v d. F/v^2

3. A dropped ball loses PE as it falls. Gravity is an example of
- a. conservative force
- b. nonconservative force
- c. dissipative force
- d. any of the above, depending on the reference level

4. The work done in moving an object from A to B against a nonconservative force
- a. cannot be recovered by moving the object from B to A
- b. does not depend on the path taken between A and B
- c. is always entirely converted into heat
- d. disappears forever

5. A 2.00-kg book is held 1.00 m above the floor for 50 s. The work done is

a. 0 b. 10.2 J
c. 100 J d. 980 J

6. The work done in lifting 30 kg of bricks to a height of 20 m on a building under construction is

a. 61 J b. 600 J
c. 2940 J d. 5880 J

7. The amount of work a 0.60-kW electric drill can do in 1.0 min is

a. 0.6 J b. 36 J
c. 0.6 kJ d. 36 kJ

8. A sedentary person requires about 6 million J of energy per day. This rate of energy consumption is equivalent to about

a. 70 W b. 335 W
c. 600 W d. 250 kW

9. A 40-kg boy runs up a staircase to a floor 5.0 m higher in 7.0 s. His power output is

a. 29 W b. 0.28 kW
c. 1.4 kW d. 14 kW

10. A 60-kg woman whose average power is 30 W can climb a mountain 2.0 km high in

a. 1.0 h b. 1.1 h
c. 5.4 h d. 11 h

11. A 200-kg bucket of cement is accelerated upward by a force of 2200 N. The acceleration continues for 5.00 s. The power needed at the end of this period is

a. 1.44 kW b. 11.8 kW
c. 13.2 kW d. 121 kW

12. An 800-kg white horse has a power output of 1.00 hp. The maximum force it can exert at a speed of 3.0 m/s is

a. 0.25 kN b. 2.2 kN
c. 2.6 kN d. 3.6 kN

13. A road slopes upward so that it climbs 1.0 m for each 12 m of distance along the road. A car whose weight is 10 kN moves up the road at a constant speed of 24 m/s. The minimum power the car's engine develops is

a. 1.7 kW b. 2.0 kW
c. 20 kW d. 0.24 MW

14. A 1.0-kg mass has a potential energy of 1.0 J relative to the ground when it is at a height of

a. 0.12 m b. 1.0 m
c. 9.8 m d. 32 m

15. A 1.0-N weight has a potential energy of 1.0 J relative to the ground when it is at a height of

a. 0.10 m b. 1.0 m
c. 9.8 m d. 32 m

16. A total of 4900 J is expended in lifting a 50-kg mass. The mass was raised to a height of

a. 10 m b. 98 m
c. 960 m d. 245 km

17. A 1.0-kg mass has a kinetic energy of 1.0 J when its speed is

a. 0.45 m/s b. 1.0 m/s
c. 1.4 m/s d. 4.4 m/s

18. A 1.0-N weight has a kinetic energy of 1.0 J when its speed is

a. 0.45 m/s b. 1.0 m/s
c. 1.4 m/s d. 4.4 m/s

19. Car A has a mass of 1000 kg and a speed of 60 km/h, and car B has a mass of 2000 kg and a speed of 30 km/h. The kinetic energy of car A is
- a. half that of car B
- b. equal to that of car B
- c. twice that of car B
- d. four times that of car B

20. A 1000-kg car whose speed is 80 km/h has a kinetic energy of

a. 2.52×10^4 J b. 2.47×10^5 J
c. 2.42×10^6 J d. 3.20×10^6 J

21. A 1000-kg car has a KE of 450 kJ. The car's speed is

a. 21 m/s b. 30 m/s
c. 45 m/s d. 900 m/s

22. The height above the ground of a child on a swing varies from 50 cm at the lowest point to 200 cm at the highest point.

The maximum speed of the child is

a. about 5.4 m/s
b. about 7.7 m/s
c. about 29 m/s
d. dependent on the child's mass

23. The 2.0-kg head of an ax is moving at 60 m/s when it strikes a log. If the blade of the ax penetrates 20 mm into the log, the average force it exerts is

a. 3 kN b. 90 kN
c. 72 kN d. 180 kN

24. The car of a roller coaster starts from rest at *A* in Fig. 4.20. If there is no friction or air resistance, the car's speed at *B* is

a. 4.5 m/s b. 14 m/s
c. 20 m/s d. 0.39 km/s

25. The car's speed at *C* is

a. 13 m/s b. 15 m/s
c. 20 m/s d. 0.16 km/s

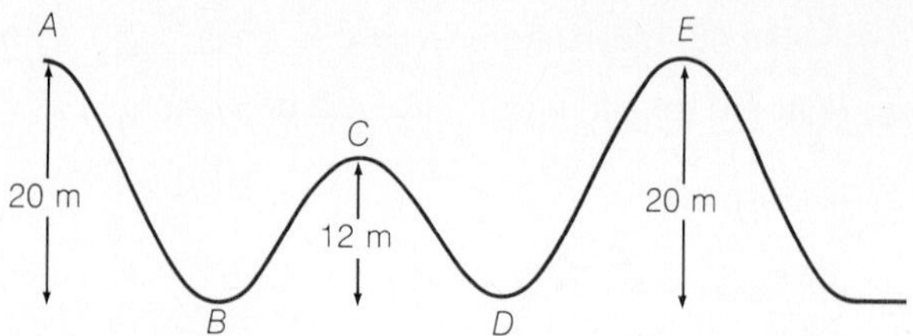

Figure 4.20
Multiple Choice 24–27.

26. The car's speed at *D* is

a. 13 m/s b. 15 m/s
c. 20 m/s d. 0.24 km/s

27. The car's speed at *E* is

a. 0 b. 4.5 m/s
c. 13 m/s d. 15 m/s

QUESTIONS

1. Under what circumstances (if any) is no work done on a moving object even though a net force acts on it?

2. The potential energy of a golf ball in a hole is negative with respect to the ground. Under what circumstances (if any) is its kinetic energy negative?

3. Does every moving body possess kinetic energy? Does every stationary body possess potential energy?

4. A golf ball and a Ping-Pong ball are dropped in a vacuum chamber. When they have fallen halfway to the bottom, how do their speeds compare? Their kinetic energies? Their potential energies?

5. Identical twins Ann and Betty set out to climb a mountain. Ann chooses a slope that averages 30° above the horizontal, and Betty chooses a slope that averages 40°. Compare the amounts of work the twins do.

6. (a) Where is the KE of the pendulum bob of Fig. 4.21 a minimum? A maximum? (b) Where is the PE of the bob a minimum? A maximum?

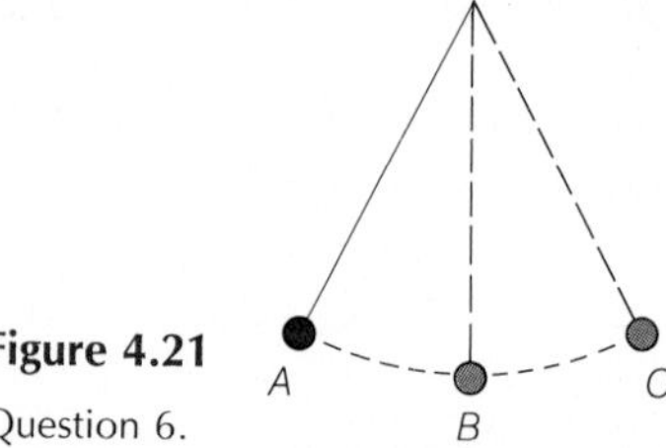

Figure 4.21
Question 6.

7. When the valve connecting the two tanks shown in Fig. 4.22 is opened, water from *A* flows into *B* until their levels are the same. What percentage of the original PE of the water in tank *A* is changed into KE and eventually into heat in this process?

8. How can the existence of nonconservative forces be reconciled with the principle of conservation of energy?

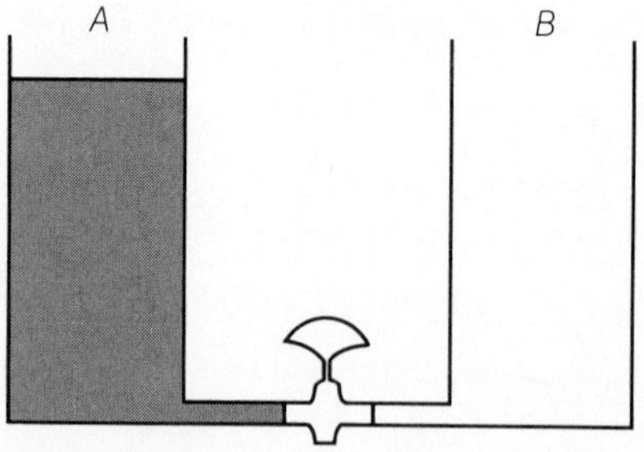

Figure 4.22
Question 7.

EXERCISES

4.1 Work

1. The sun exerts a force of 4.0×10^{28} N on the earth, and the earth travels 9.4×10^{11} m in its annual orbit around the sun. How much work is done by the sun on the earth in the course of a year?

2. A horse is towing a barge with a rope that makes an angle of 20° with the canal. If the horse exerts a force of 400 N, how much work does it do in moving the barge 1.0 km?

3. A 500-N box is pushed across a horizontal floor with a force of 250 N. The coefficient of kinetic friction is 0.40. (a) Find the work done in pushing the box 20 m. (b) How much work went into overcoming friction and how much into accelerating the box?

4. A 50-kg box is pushed across a horizontal floor by a horizontal force of 180 N. The coefficient of kinetic friction is 0.30. (a) Find the work done in pushing the box 20 m. (b) How much work went into overcoming friction and how much into accelerating the box?

5. A 20-kg wooden box is pushed a distance of 15 m on a horizontal stone floor by a force just sufficient to overcome the friction between box and floor. The coefficient of kinetic friction is 0.40. (a) What is the required force? (b) How much work does the force do?

4.2 Work Done Against Gravity

6. The work done to lift a 30-kg mass is 4.0 kJ. If the mass is at rest before and after its elevation, how high does it go?

7. A uniform 200-kg lamppost 9.0 m long is lying on the ground. How much work is needed to raise it to a vertical position?

8. A woman drinks a bottle of beer and proposes to work off its 460-kJ energy content by exercising with a 20-kg barbell. If each lift of the barbell from chest height to over her head is through 60 cm and the efficiency of her body is 10% under these circumstances, how many times must she lift the barbell?

9. (a) A force of 130 N is used to lift a 12-kg mass to a height of 8.0 m. How much work is done by the force? (b) A force of 130 N is used to push a 12-kg mass on a horizontal, frictionless surface for a distance of 8.0 m. How much work is done by the force?

10. An 80-kg person carrying a 10-kg backpack climbs to the summit of a 3000-m mountain starting from sea level along a route that averages 30° above the horizontal. Find the work the person does.

11. A horizontal force of 5.0 N is used to push a box up a ramp 5.0 m long that is at an angle of 15° above the horizontal. How much work is done by the force?

4.3 Power

12. A 15-kW motor is used to hoist an 800-kg bucket of concrete to the twentieth floor of a building under construction, a height of 90 m. If no energy is lost, how much time is required?

13. A weightlifter raises a 150-kg barbell from the floor to a height of 2.2 m in 0.80 s. What is his average power output during the lift?

14. In 1932 five members of the Polish Olympic ski team climbed from the 5th to the 102nd floor of the Empire State Building, a distance of approximately 350 m, in 21 min. If one of these men had a mass of 70 kg, how much power did he develop during the ascent?

15. The anchor windlass of a boat must be able to raise a total load (anchor plus chain) of 800 kg at a speed of 0.50 m/s. What should the minimum rating of the motor be, in kilowatts?

16. A car needs 13 kW to move along a level road at 60 km/h. What is the sum of the air resistance and rolling friction on the car at that speed?

17. Each of the four engines of a DC-8 airplane develops 7500 hp when the cruising speed is 240 m/s. How much thrust does each engine produce under these circumstances?

18. A motorboat requires 160 hp to move at the constant speed of 8.0 m/s. How much resistive force does the water exert on it at that speed?

19. A hoist whose efficiency is 80% is powered by an 8.0-kW motor. At what speed can the hoist raise an 800-N load?

20. A 75-kg man carrying 10-kg pack climbs a mountain 2800 m high in 10 h. (a) What is his average power output? (b) The efficiency with which his body utilizes food energy for climbing is 15%. How many joules of food energy were needed for the climb?

21. A crane whose motor has a power input of 5.0 kW raises a 1200-kg beam through a height of 30 m in 90 s. Find the efficiency.

22. A trash compactor with a $\frac{1}{4}$-hp motor can apply a crushing force of 8000 N. How fast does its ram move? Assume 80% efficiency.

23. The bilge pump of a boat is able to raise 200 L of water per minute through a height of 1.2 m. If the pump is 60% efficient, how much power must be supplied to the pump? The mass of 1 liter of water is 1 kg.

24. In an effort to lose weight, a person runs 5.0 km per day at an average speed of 4.0 m/s. While running, the person's body processes consume energy at a rate of 1.0 kW. Fat has an energy content of about 40 kJ/g. How many grams of fat are metabolized during each run?

25. A person's metabolic processes can usually operate at a power of 6.0 W/kg of body mass for several hours at a time. If a 60-kg woman carrying a 12-kg pack is walking uphill with an energy-conversion efficiency of 20%, at what rate, in meters per hour, does she ascend?

4.5 Kinetic Energy

26. Find the kinetic energy of a 115-kg ostrich running at 15 m/s.

27. Find the speed of a 2.0-g insect whose kinetic energy is 0.010 J.

28. A raindrop of mass 1.0 mg falls at 3.0 m/s when there is no wind. Find its KE when a wind of 5.0 m/s is blowing.

29. Find the kinetic energy of a 70-kg runner who covers 400 m in 45 s at constant speed.

30. Is the work needed to bring a car's speed from 10 to 20 km/h less than, equal to, or more than the work needed to bring its speed from 90 to 100 km/h? If the amounts of work are different, what is the ratio between them?

31. A 2.0-kg ball is at rest when a horizontal force of 5.0 N is applied. In the absence of friction, what is the speed of the ball after it has gone 10 m?

32. A 1.0-kg trout is hooked by a fisherman and swims off at 2.5 m/s. The fisherman stops the trout in 50 cm by braking his reel. How much tension is exerted on the line?

33. A certain 800-kg car has a motor whose power output is 30 kW. If the car is carrying two passengers whose total mass is 150 kg, how long will it need to accelerate from 70 to 110 km/h?

34. A 70-kg sprinter pushes on the starting blocks for 0.20 s and leaves them with a speed of 5.0 m/s. He then continues to accelerate for a further 5.0 s until his speed is 12 m/s. Find the reaction force exerted by the starting blocks on the sprinter and his average power output in each of the accelerations.

35. What is the power output of a 1200-kg car that can go from 25 km/h to 100 km/h in 12 s? Neglect air resistance and rolling friction.

4.6 Potential Energy

36. A 3.0-kg stone is dropped from a height of 100 m. Find its kinetic and potential energies when it is 50 m from the ground.

37. A ball is dropped from a height of 60 m. How high above the ground will it be when one-third of its total energy is KE?

38. A 60-kg woman jumps off a wall 1.2 m high and lands with her knees stiff. If her body is compressed by 25 mm on impact, what is the average force exerted on her by the ground? What would the force be if she had bent her knees on impact, so that she came to a stop in 25 cm?

39. In the operation of a certain pile driver, a 500-kg hammer is dropped from a height of 5.0 m above the head of a pile. If the pile is driven 20 cm into the ground with each impact of the hammer, what is the average force on the pile when struck?

40. A bullet moving at 500 m/s has a kinetic energy of 2500 J and a potential energy of 0.50 J at a certain moment. What is the bullet's mass and how high above the ground is it at that moment?

41. A ball is dropped from a height of 1.0 m and loses 10% of its kinetic energy when it bounces on the ground. To what height does it rise?

42. A certain frog's hind legs produce a force of 2.5 times its weight through a vertical distance of 90 mm, at which point the frog becomes airborne. (a) What is the frog's speed at takeoff? (b) What height above the ground does the frog reach? Assume a vertical jump.

43. A waterfall is 30.0 m high, and 1.00×10^4 kg of water flows over it per second. (a) How much power does this flow represent? (b) If all this power could be converted to electricity, how many 100-W light bulbs could be supplied?

4.7 Conservation of Energy

44. A force of 600 N is used to lift a 500-N statue, initially at rest, to a height of 5.0 m. (a) What is the change in the statue's PE? (b) In its KE?

45. A woman skis down a slope 100 m high. Her speed at the foot of the slope is 20 m/s. What percentage of her initial potential energy was dissipated?

46. At her highest point a 40-kg girl on a swing is 2.0 m from the ground, while at her lowest point she is 0.80 m from the ground. What is her maximum speed? On another swing a 50-kg boy undergoes exactly the same motion. What is his maximum speed?

47. Find the height of the bar a pole-vaulter can clear on the basis of the following assumptions: His running speed is 8.0 m/s; his center of mass is initially 1.1 m above the ground, and he pulls himself upward along the pole 0.60 m as the pole swings into a vertical position; and all his initial KE is converted into work done to raise his center of mass sufficiently to clear the bar.

48. A man uses a rope and system of pulleys to lift an 80-kg object to a height of 2.0 m. He exerts a force of 220 N on the rope and pulls a total of 8.0 m of rope through the pulleys in the course of raising the object, which is at rest afterward. (a) How much work does the man do? (b) What is the change in the object's PE? (c) If the answers to (a) and (b) are different, explain.

49. (a) A force of 8.0 N is used to push a 0.50-kg ball over a horizontal, frictionless table a distance of 3.0 m. If the ball starts from rest, what is its final KE? (b) The same force is used to lift the same ball a height of 3.0 m. If the ball starts from rest, what is its final KE?

50. A force of 200 N is used to lift a 15-kg object, initially at rest, to a height of 8.0 m. No friction is present. What is the speed of the object at this height?

51. An 800-kg car coasts down a hill 40 m high with its engine off and its driver's foot pressing on the brake pedal. At the top of the hill the car's speed is 6.0 m/s, and at the bottom it is 20.0 m/s. How much energy was converted to heat on the way down?

PROBLEMS

52. A person pulls a 100-kg crate at constant speed 20 m across a level floor using a rope that is 30° above the horizontal. If the coefficient of kinetic friction between crate and floor is 0.30, how much work is done by the person?

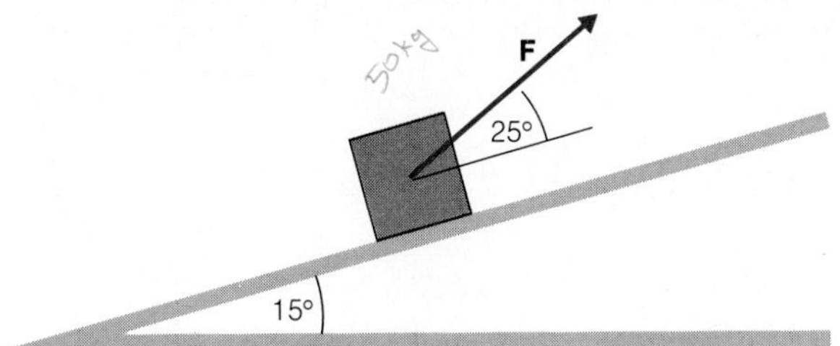

Figure 4.23

Problem 54.

53. A 10-kg block is pulled up a plane 20° above the horizontal at constant speed by a force parallel to the plane. The coefficient of kinetic friction between the block and the plane is 0.40. How much work is done by the force in moving the block 2.0 m along the plane?

54. A 50-kg box is pulled 10 m up a plane 15° above the horizontal at constant speed by a force acting 25° above the plane, as in Fig. 4.23. The coefficient of kinetic friction between the box and the plane is 0.30. How much work is done by the applied force?

55. A 70-kg block of balsa wood is 1.5 m high and has a base 60 cm × 60 cm. The block is tilted about one edge until it falls over on its side. How much work had to be done to tilt the block to the point where it began to fall over?

56. A well 50 cm in diameter is to be dug to a depth of 20 m. The earth taken out must be lifted 1.2 m above the ground into a truck that carries it away. If each m^3 of the earth has a mass of 1300 kg, what is the minimum work needed to remove the earth from the well?

57. A 7.0-kg shot is thrown 18 m. What was its minimum initial KE?

58. An object of mass m is accelerated uniformly from rest for a time t during which it covers the distance x. Find the work done on the object.

59. A 2.0-kg block is pulled up a ramp by a constant force parallel to the ramp of 15 N. The ramp is 2.0 m long and inclined at 30° above the horizontal. If the coefficient of kinetic friction between the block and the ramp is 0.30, what is the speed of the block at the top of the ramp?

60. A brick slides down a wooden plank 2.0 m long tilted so that one end is at a height of 1.0 m. The brick's speed at the bottom is 2.5 m/s. The plank is then sanded smooth and waxed so that the coefficient of friction is half what it was before, and the brick is slid down it again. What is the new speed of the brick at the bottom?

61. The bob of a pendulum 1.2 m long is pulled aside so that the string is 40° from the vertical. When the bob is released, with what speed will it pass through the bottom of its path?

62. Steam enters a 10-MW-output turbine at 800 m/s and emerges at 100 m/s. Assuming 90% mechanical efficiency, determine what mass of steam passes through the turbine per second.

63. An escalator 14 m long is carrying a 70-kg person from one floor to another floor 8.0 m higher. The linear speed of the escalator is 1.0 m/s. (a) How much work does the escalator do in carrying the person to the top? What is its power output while doing so? (b) The person is walking up the escalator at 0.8 m/s. Answer the same questions for this situation. (c) The person is walking down the escalator at 0.8 m/s. Answer the same questions for this situation.

64. A 20-kg crate is dropped from a height of 15 m and reaches the ground in 1.9 s. If the force of air resistance on the crate is constant, what is its magnitude? What percentage of the original potential energy of the crate is lost?

65. The power required to propel a 1200-kg car at 40 km/h on a level road is 30 kW. (a) How much resistance must the car overcome at this speed? (b) How much power is needed for the car to ascend an 8° hill at the same speed?

66. A 60-kg woman is riding a 10-kg bicycle at a constant speed of 20 km/h. Friction and air resistance total 25 N. Find her power output (a) when the road is horizontal and (b) when it goes up a 5° hill.

67. The motion of a bicyclist is opposed by the rolling resistance F_{roll} of the tires on the road, which is independent of speed v, and by the air resistance of bicycle and rider, which is proportional to v^2. Thus the total resistive force can be expressed by the formula $F = F_{roll} + kv^2$, where F_{roll} and k are constants for any particular combination of bicycle, rider, and rider's posture. If a certain bicyclist develops 22.8, 122.4, and 375.6 W of power at the respective speeds 4.0, 8.0, and 12.0 m/s on a level road, what are the corresponding values of F_{roll} and k?

68. A wooden block whose initial speed is 6.0 m/s starts to slide up a plane inclined at 25° above the horizontal. (a) If the coefficient of kinetic friction is 0.30, how far up the plane does the block go? (b) What will the block's speed be after it has slid back down the plane to its starting point?

Answers to Multiple Choice

1. a	**8.** a	**15.** b	**22.** a
2. a	**9.** b	**16.** a	**23.** d
3. a	**10.** d	**17.** c	**24.** c
4. a	**11.** c	**18.** d	**25.** a
5. a	**12.** a	**19.** c	**26.** c
6. d	**13.** c	**20.** b	**27.** a
7. d	**14.** a	**21.** b	

CHAPTER

5

MOMENTUM

Because the universe is so complex, we need the help of many different quantities to describe its various aspects economically. Besides length, time, and mass we have already found velocity, acceleration, force, work, and energy to be useful, and more are to come. None of these quantities is sacred—we can certainly give up any of them, but only at the expense of making it harder to understand what is going on around us. The idea behind defining each physical quantity is to single out something that unifies a wide range of observations. Then we can boil down a large number of separate discoveries about nature into a brief, clear statement, for instance, the second law of motion or the law of conservation of energy. In this chapter we learn how the concepts of linear momentum and impulse can help us to analyze the behavior of moving bodies.

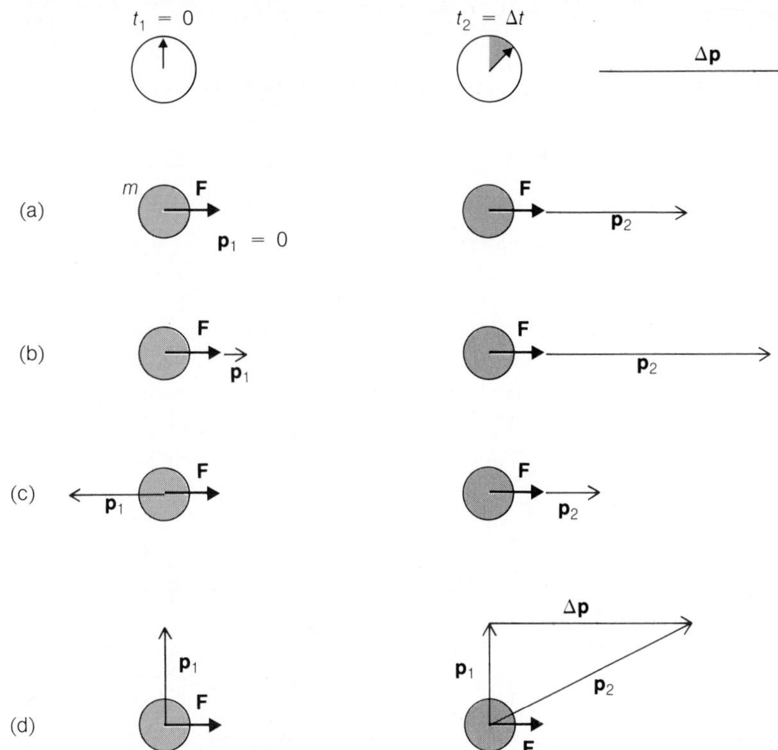

Figure 5.3

Applying a constant force **F** to a mass m for a time Δt changes its momentum by $\Delta \mathbf{p} = \mathbf{F}\,\Delta t$. At (a) the mass is initially at rest; at (b) its initial momentum is in the same direction as **F**; at (c) its initial momentum is in the opposite direction to **F**; and at (d) its initial momentum is perpendicular to **F**. Since momentum is a vector quantity, the momentum change $\Delta \mathbf{p}$ must be added to the initial momentum $\mathbf{p}_1$ by the process of vector addition.

In the SI the unit of impulse is the **newton-second** (N·s), and the unit of momentum is the kg·m/s; they are actually the same, of course, but it is often convenient to distinguish between them in this way.

EXAMPLE 5.2

The head of a golf club is in contact with a 46-g golf ball for 0.50 ms (1 ms = 1 millisecond = 10^{-3} s), and as a result the ball flies off at 70 m/s. Find the average force that was acting on the ball during the impact.

SOLUTION The ball starts from rest, hence its momentum change is

$$\Delta mv = (0.046 \text{ kg})(70 \text{ m/s}) = 3.22 \text{ kg·m/s}$$

From Eq. (5.3) we therefore have

$$F = \frac{\Delta mv}{\Delta t} = \frac{3.22 \text{ kg·m/s}}{5.0 \times 10^{-4} \text{ s}} = 6.4 \times 10^{3} \text{ N}$$

whose British equivalent is 1450 lb. A force of the same magnitude but acting in the opposite direction (the *recoil force*) acts on the club's head during the impact, in accord with the third law of motion. No golf club could withstand such a static load, but the impact is so brief that its only effect on the shaft is to bend it temporarily by a few centimeters. ◆

5.3 Conservation of Momentum

A vector principle.

Energy and work are scalar quantities that have magnitude only. Despite the fundamental and all-inclusive character of the law of conservation of energy, it cannot by itself enable us to solve most problems that involve interacting bodies. A simple example is the firing of a rifle. The requirement that energy be conserved means that the kinetic energies of the bullet and the recoiling rifle, plus the heat and sound energy that are liberated, must equal the chemical energy of the detonated explosive. However, this does not tell us how the total energy is divided among the rifle, the bullet, and the atmosphere. Indeed, because energy is a scalar quantity, conserving it does not even imply that the bullet and rifle must move in opposite directions. To solve many problems in dynamics when we have no detailed knowledge of the active forces, an additional principle of a vector nature is required.

Let us consider two particles A and B that collide and thereby exert forces on each other. At every instant during their interaction these forces, $\mathbf{F}_{AB}$ acting on B and $\mathbf{F}_{BA}$ acting on A, obey Newton's third law of motion,

$$\mathbf{F}_{AB} = -\mathbf{F}_{BA}$$

This is shown in Fig. 5.4. Hence the impulses exchanged must be equal and opposite,

$$\mathbf{F}_{AB}\,\Delta t = -\mathbf{F}_{BA}\,\Delta t$$

and the *total* momentum of the system of A and B together is the same after the collision as it was before.

This result can be generalized to any system of two or more particles. If no net force from outside the system acts on it, the total linear momentum of the system, which is the sum

$$\Sigma\mathbf{p} = \mathbf{p}_1 + \mathbf{p}_2 + \mathbf{p}_3 + \cdots \tag{5.4}$$

of the individual momenta of its particles, cannot change. (The symbol Σ is the Greek capital letter sigma and stands for "sum of.") Thus we have the principle of **conservation of linear momentum:**

When the vector sum of the external forces acting on a system of particles equals zero, the total linear momentum of the system remains constant.

Of course, as we saw above, interactions within the system may change the *distribution* of the total momentum $\Sigma\mathbf{p}$ among the various particles in the system, but $\Sigma\mathbf{p}$ stays the same.

The total of the kinetic energies of the particles need not be constant, however. The KE involved in an explosion is zero to start with, because it all comes from the chemical energy stored in the explosive material, but it is not zero afterward.

Figure 5.4

Particles A and B collide and exert the forces $\mathbf{F}_{AB}$ and $\mathbf{F}_{BA}$ on each other. As shown in the graph, these forces are equal in magnitude and opposite in direction at all times. Their impulses are also equal and opposite, so the total momentum of A and B is left unchanged by the collision, although it may be distributed differently between them.

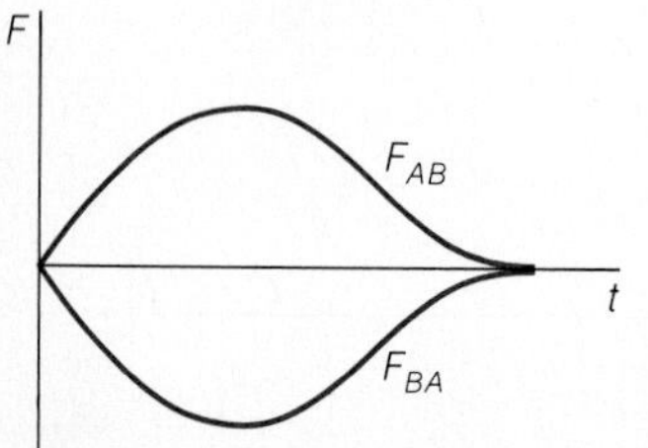

In a collision some or even all of the KE of relative motion of the colliding objects may disappear into heat and sound energy. In both cases the total momentum does not change.

EXAMPLE 5.3

An astronaut in orbit outside an orbiting space station throws his 800-g camera away in disgust when it jams (Fig. 5.5). If he and his space suit together have a mass of 100 kg and the speed of the camera is 12 m/s, how far away from the space station will he be in 1 h?

SOLUTION The initial momentum relative to the space station of the system of camera + astronaut is 0. If we call the camera 1 and the astronaut 2, then

$$\text{Final momentum} = \text{initial momentum}$$
$$m_1\mathbf{v}_1 + m_2\mathbf{v}_2 = 0$$
$$\mathbf{v}_2 = -\frac{m_1}{m_2}\mathbf{v}_1$$

where $\mathbf{v}_1$ and $\mathbf{v}_2$ are the final velocities of the camera and astronaut. The minus sign means that these velocities are in opposite directions along the same line. If we take the camera's direction as positive, the speed of the astronaut is

$$v_2 = -\frac{m_1}{m_2}v_1 = -\left(\frac{0.800\text{ kg}}{100\text{ kg}}\right)(12\text{ m/s}) = -0.096\text{ m/s}$$

After $t = 1\text{ h} = 3600\text{ s}$ the astronaut will be

$$s = v_2t = (0.096\text{ m/s})(3600\text{ s}) = 346\text{ m}$$

away from the space station. The total momentum of the system of astronaut plus camera remains zero until other forces act on them.

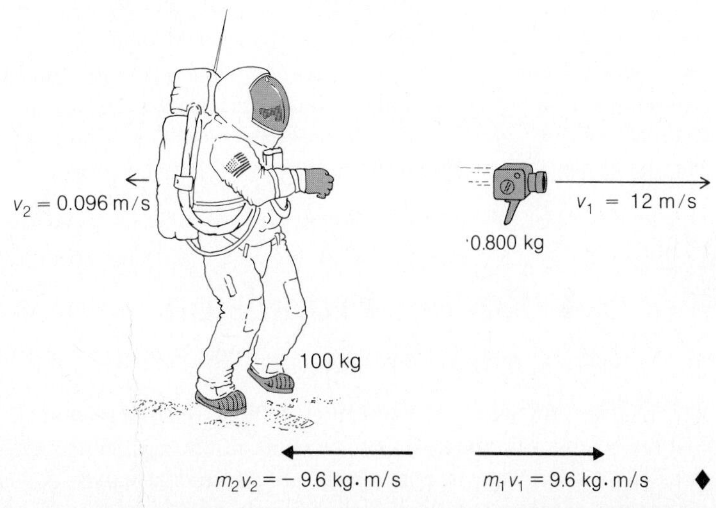

Figure 5.5

The total momentum of a system of particles remains constant if no net external force acts on the system. The initial momentum of this system was 0; hence the forward momentum of the thrown camera is equal in magnitude to the backward momentum of the astronaut who threw it.

EXAMPLE 5.4

A certain cannon has a range of 2 km. One day a shell it fires explodes at the top of its path into two equal fragments. One fragment falls vertically downward with no initial speed. How far away from the cannon does the other fragment land?

SOLUTION The situation is shown in Fig. 5.6. At the top of the shell's path, it has the momentum mv_x in the horizontal direction. The fragment that falls straight down has no horizontal momentum after the explosion. The other fragment has the mass $m/2$ and the horizontal speed v_x', so

$$\text{Final momentum} = \text{initial momentum}$$

$$\frac{m}{2}v_x' = mv_x$$

$$v_x' = 2v_x$$

The second fragment has twice the speed of the shell when it breaks up. Because the range from here is directly proportional to the horizontal speed (Section 2.7), this fragment goes twice as far from the top of the path as the intact shell would have gone, which is an additional 1 km. Hence the fragment reaches the ground 3 km from the cannon.

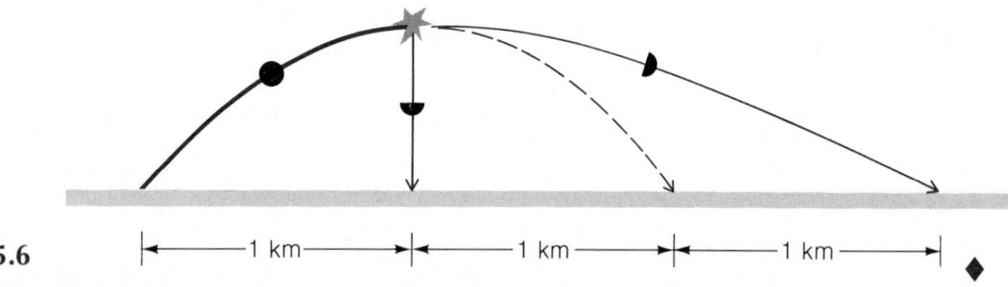

Figure 5.6

With the help of advanced mathematics, it is possible to show that if the laws of nature are the same at every point in space, then the principle of conservation of linear momentum must follow as an inevitable consequence. It is also possible to show that if the laws of nature do not change with time, so that they have always been the same as they are now and always will remain the same, then energy must be conserved in all interactions. Thus these principles, as well as being useful relationships for solving practical problems, give us a hint of a profound order underlying the physical universe.

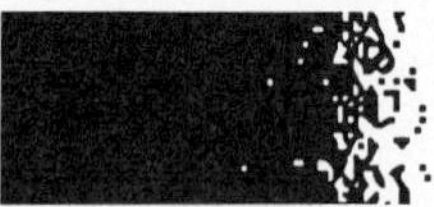

5.4 Rocket Propulsion

Momentum conservation is the basis.

The principle underlying rocket flight is conservation of momentum. The total momentum of a rocket on its launching pad is zero. When it is fired, the exhaust gases shoot downward at high speed, and the rocket moves upward to balance the momentum of the gases (Fig. 5.7). Rockets do not operate by "pushing" against

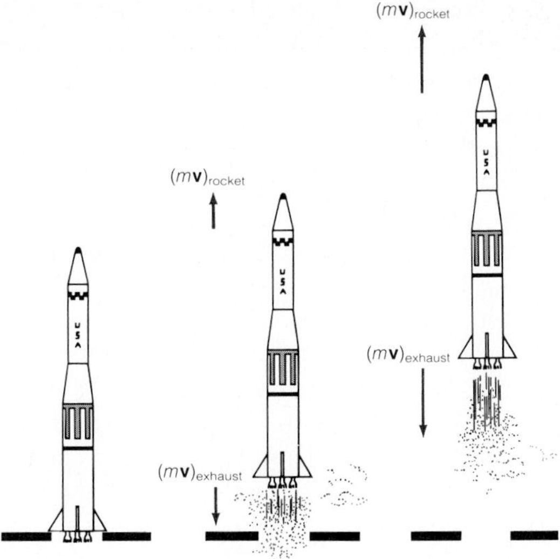

Figure 5.7

Conservation of momentum in rocket flight. The downward momentum of the exhaust gases is exactly balanced by the upward momentum of the rocket itself.

their launching pads, the air, or anything else; in fact, they perform best in space, where there is no atmosphere to impede their motion. The energy of the rocket and its exhaust comes from chemical energy stored in the fuel (Fig. 5.8). The total momentum of the system of rocket plus exhaust, which is initially zero, does not remain constant after a launch from the earth, because of the impulses provided by air resistance and the earth's gravitational pull.

The upward force exerted on a rocket by the expulsion of exhaust gases is called **thrust.** If a mass Δm of exhaust gas comes out at the velocity $\mathbf{v}$ relative to the rocket in the time interval Δt, the momentum change of the gas relative to the rocket is $\Delta\,(m\mathbf{v})$. From Eq. (5.3) $\mathbf{F}\,\Delta t = \Delta\,(m\mathbf{v})$, so the associated force on the rocket is

$$\mathbf{F} = \frac{\Delta(m\mathbf{v})}{\Delta t} = \mathbf{v}\,\frac{\Delta m}{\Delta t} \qquad \textit{Thrust} \quad (5.5)$$

The thrust of a rocket is the product of the exhaust speed and the rate at which fuel is consumed. In the case of a rocket that burns 100 kg of fuel per second with an exhaust speed of 3 km/s, the thrust is

$$F = v\,\frac{\Delta m}{\Delta t} = (3000\text{ m/s})(100\text{ kg/s}) = 3 \times 10^5\text{ N}$$

which is about 33 tons.

When the spacecraft is launched from the earth's surface, the thrust of its motor must exceed its initial weight $m_0 g$ for it to rise from the ground. If the thrust remains constant, the spacecraft acceleration increases as its fuel is burnt. A maximum acceleration of several times g in magnitude is usual.

Rocket propulsion is a gradual rather than an instantaneous process, with the fuel burned and ejected as exhaust gases at a certain rate instead of in one lump. As a result part of the momentum of the exhaust is wasted in pushing forward unburned fuel. When this factor is taken into account, the ultimate speed of a rocket (neglecting air resistance and gravity) turns out to be directly proportional to the speed of the exhaust gases and to the logarithm of the ratio of the rocket's initial mass to its final

Figure 5.8

In a liquid-fueled rocket, the reacting substances are mixed and ignited in a combustion chamber and the resulting exhaust gases escape at high speed through a steerable nozzle. The liberated chemical energy becomes kinetic energy of the rocket and its exhaust. Liquid hydrogen and liquid oxygen are used to power many rockets, and their product is water vapor, H_2O. The combination of 1 kg of hydrogen and 8 kg of oxygen releases 245 MJ.

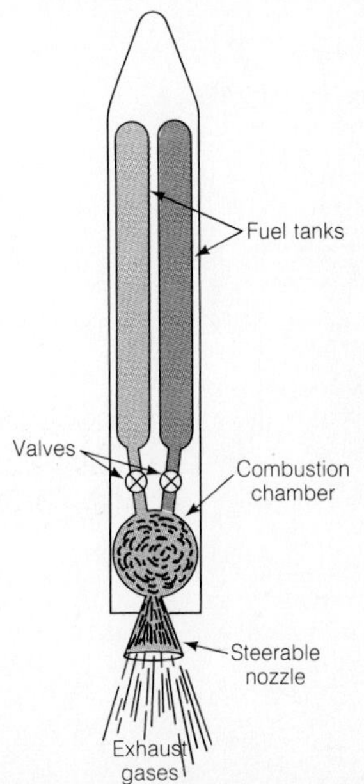

mass after the fuel has been consumed. A typical modern rocket might have an exhaust speed of 3 km/s with 75% of its initial mass consisting of fuel, which gives a final speed of 4.2 km/s. This is over 9000 mi/h, which is very fast by ordinary standards but still only about half the minimum speed needed to put a satellite in orbit around the earth (see Section 6.7).

To attain higher speeds than a single rocket is capable of, two or more rocket stages can be used. The first stage is a large rocket whose payload is another, smaller rocket. When the fuel of the first stage has been consumed, its fuel tanks and engine are cast loose. Then the second stage is fired starting from a high initial speed instead of from rest and without the burden of the fuel tanks and engine of the first stage. This process can be repeated a number of times, depending on the final speed required. The *Saturn V* launch vehicle that propelled the *Apollo 11* spacecraft to the moon in July 1969 used three stages, as shown in Fig. 5.9. At original ignition the entire assembly was 111 m long and had a mass of 2.9×10^6 kg (3240 tons).

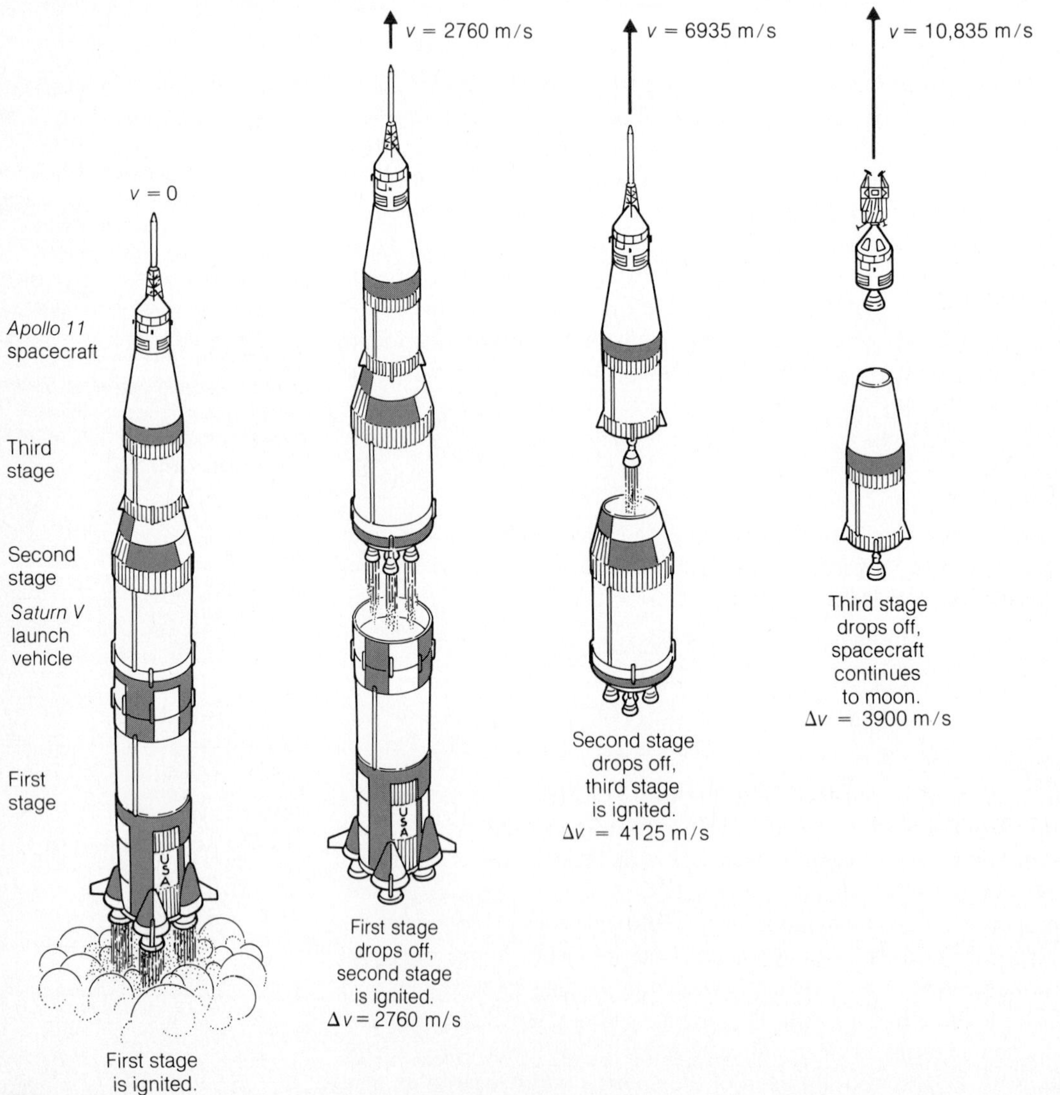

Figure 5.9

The *Saturn V* launch vehicle that propelled the *Apollo 11* spacecraft to the moon for the first manned landing used three rocket stages.

Apollo II lifts off its pad on July 16, 1969 to begin the first human visit to the moon.

5.5 Inelastic Collisions

KE is not conserved in such collisions.

Momentum is conserved in all collisions as well as in all explosions. When two or more objects collide, their final total momentum equals their initial total momentum. *The essential effect of a collision is to redistribute the total momentum of the colliding objects.*

All collisions conserve momentum, but not all of them conserve kinetic energy as well. Collisions fall into three categories (Fig. 5.10):

1. *Elastic collisions,* which conserve kinetic energy.
2. *Inelastic collisions,* which do not conserve kinetic energy. Some KE is lost to heat, sound energy, and so forth.
3. *Completely inelastic collisions,* in which the objects stick together afterward. In such collisions the KE loss is the maximum possible.

Given the initial masses and velocities of the objects involved, both elastic and completely inelastic collisions can be analyzed to give the final velocities. Although few actual collisions fit exactly into these extreme categories, many come close, so it is worth examining them.

Figure 5.10

The three possible outcomes of a ball striking the ground illustrate the three categories of collisions.

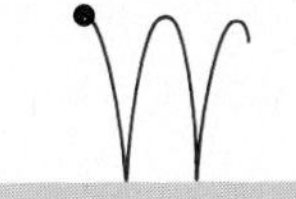

Perfectly elastic; KE is conserved.

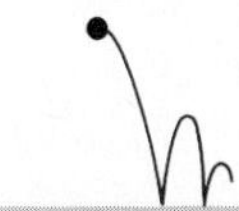

Inelastic; some KE is lost.

Perfectly inelastic; maximum KE is lost.

EXAMPLE 5.5

A 5-kg lump of clay that is moving at 10 m/s to the left strikes a 6-kg lump of clay moving at 12 m/s to the right. The two lumps stick together after they collide. Find the final speed of the composite object and the kinetic energy dissipated in the collision.

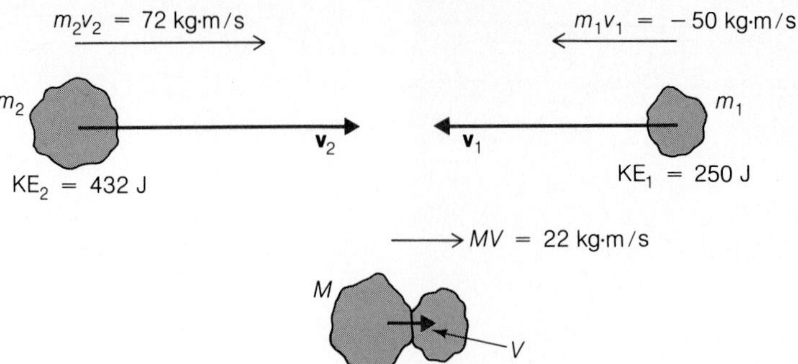

Figure 5.11

In a completely inelastic collision, the colliding bodies stick together. Kinetic energy is not conserved in such an event. Linear momentum is conserved in all collisions.

SOLUTION This is an example of a completely inelastic collision. If we call the mass of the final object M and its velocity V, conservation of linear momentum requires that

$$\text{Final momentum} = \text{initial momentum}$$
$$MV = m_1 v_1 + m_2 v_2$$

Adopting the convention that motion to the right is + and to the left is −, we have

$$m_1 = 5\text{ kg} \qquad m_2 = 6\text{ kg} \qquad M = m_1 + m_2 = 11\text{ kg}$$
$$v_1 = -10\text{ m/s} \qquad v_2 = +12\text{ m/s} \qquad V = ?$$

Solving for V yields

$$V = \frac{m_1 v_1 + m_2 v_2}{M} = \frac{(5\text{ kg})(-10\text{ m/s}) + (6\text{ kg})(12\text{ m/s})}{11\text{ kg}} = 2\text{ m/s}$$

Since V is positive, the composite body moves off to the right (Fig. 5.11).

Energy and momentum are distinct concepts. The lumps of clay before the collision have the kinetic energies

$$\text{KE}_1 = \tfrac{1}{2}m_1 v_1^2 = \tfrac{1}{2}(5\text{ kg})(-10\text{ m/s})^2 = 250\text{ J}$$
$$\text{KE}_2 = \tfrac{1}{2}m_2 v_2^2 = \tfrac{1}{2}(6\text{ kg})(12\text{ m/s})^2 = 432\text{ J}$$

After the collision the new lump of clay has the kinetic energy

$$\text{KE}_3 = \tfrac{1}{2}MV^2 = \tfrac{1}{2}(11\text{ kg})(2\text{ m/s})^2 = 22\text{ J}$$

The total kinetic energy prior to the collision was 432 + 250 or 682 J, whereas afterward it is only 22 J. The difference of 660 J was dissipated largely into heat in the collision, with some probably being lost to sound energy as well. ◆

It is important to keep in mind the directional character of linear momentum. Sometimes a problem involves bodies that move along the same line, as in the preceding example, but in general they may move in two or three dimensions, and we must then take this into account by a vector calculation.

EXAMPLE 5.6

A 60-kg man is sliding east on the frictionless surface of a frozen pond at a velocity of 0.50 m/s. He is struck by a 1.00-kg snowball whose velocity is 20 m/s toward the north. If the snowball sticks to the man, what is his final velocity?

SOLUTION Here linear momentum must be conserved separately in both the east-west and the north-south directions, which we will call the x- and y-axes, respectively. Since

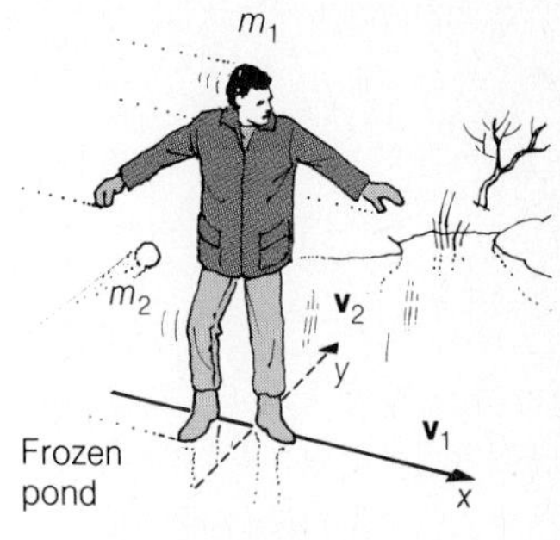

Before collision

$$m_1 = 60\ \text{kg} \qquad m_2 = 1.00\ \text{kg} \qquad M = m_1 + m_2 = 61\ \text{kg}$$
$$v_{1x} = 0.50\ \text{m/s} \qquad v_{2x} = 0 \qquad V_x = ?$$
$$v_{1y} = 0 \qquad v_{2y} = 20\ \text{m/s} \qquad V_y = ?$$

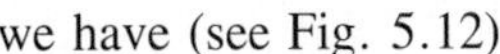

we have (see Fig. 5.12)

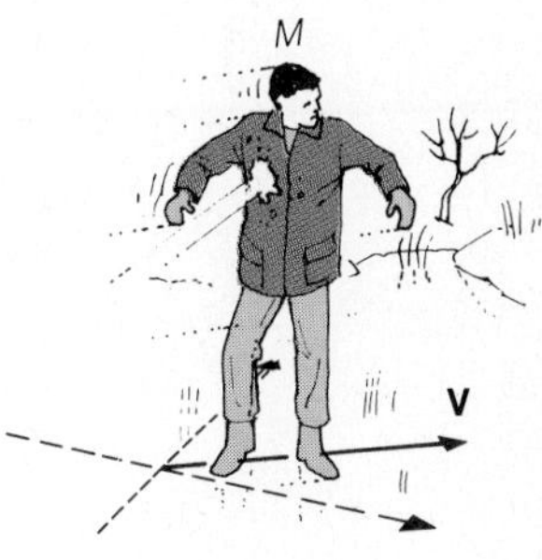

After collision

Figure 5.12

$$\text{Final momentum} = \text{initial momentum}$$
$$x\text{-direction:} \quad MV_x = m_1v_{1x} + m_2v_{2x} = (60\ \text{kg})(0.50\ \text{m/s}) = 30\ \text{kg·m/s}$$
$$y\text{-direction:} \quad MV_y = m_1v_{1y} + m_2v_{2y} = (1.00\ \text{kg})(20\ \text{m/s}) = 20\ \text{kg·m/s}$$

Hence the magnitude MV of the momentum of man + snowball after the collision is

$$MV = \sqrt{(MV_x)^2 + (MV_y)^2} = \sqrt{(30\ \text{kg·m/s})^2 + (20\ \text{kg·m/s})^2} = 36\ \text{kg·m/s}$$

and the corresponding final speed is

$$V = \frac{MV}{M} = \frac{36\ \text{kg·m/s}}{61\ \text{kg}} = 0.59\ \text{m/s}$$

We can specify the direction in which the man + snowball combination moves after the collision in terms of the angle θ between the $+y$-direction (which is north) and MV:

$$\tan\theta = \frac{MV_x}{MV_y} = \frac{30\ \text{kg·m/s}}{20\ \text{kg·m/s}} = 1.5$$
$$\theta = 56°$$

Thus man + snowball move in a direction 56° to the east of north (Fig. 5.13). ◆

Figure 5.13

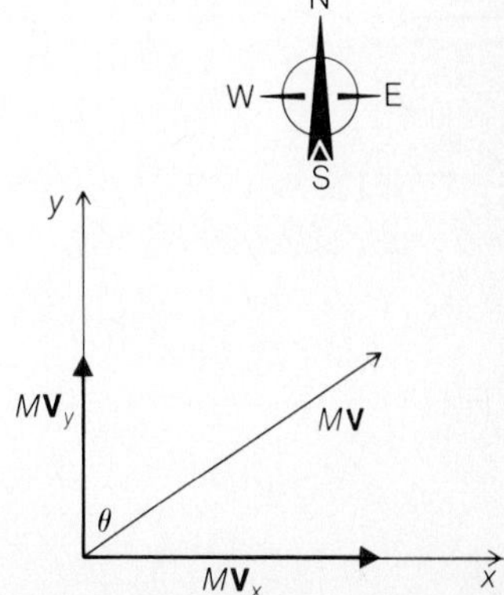

5.6 Elastic Collisions

They conserve kinetic energy.

In an elastic collision no kinetic energy is lost. Suppose we have an object of initial speed v_1 that collides elastically with another object of initial speed v_2. Both objects move along the same straight line before and after the collision. If the speeds of the objects after the collision are, respectively, v_1' and v_2', then

$$v_1 - v_2 = -(v_1' - v_2') \tag{5.6}$$
$$\text{Initial relative velocity} = -(\text{final relative velocity})$$

This equation follows from the conservation of both KE and momentum (see Problem 49). What it means is that the effect of the collision is to reverse the direction of the relative velocity without changing its magnitude. Thus *the relative speed with which the objects approach each other equals the relative speed with which they move apart afterward.*

EXAMPLE 5.7

A ball rolling on a level table strikes head-on an identical ball that is at rest, as in Fig. 5.14. What is the result of the collision?

SOLUTION Here $v_2 = 0$, so from Eq. (5.6)

$$v_1 = v_2' - v_1'$$

From conservation of momentum, since the balls have the same mass m,

$$m_1v_1 + m_2v_2 = m_1v_1' + m_2v_2'$$
$$mv_1 + 0 = mv_1' + mv_2'$$
$$v_1 = v_1' + v_2'$$

Setting equal these two formulas for v_1 gives

$$v_2' - v_1' = v_1' + v_2'$$
$$-2v_1' = 0$$
$$v_1' = 0$$

The first ball has no speed after the collision, which means it comes to a stop. For the final speed of the second ball, which was originally at rest, we can use either of the formulas for v_1 with $v_1' = 0$. Thus

$$v_1 = v_2' - v_1' = v_2' - 0 = v_2'$$

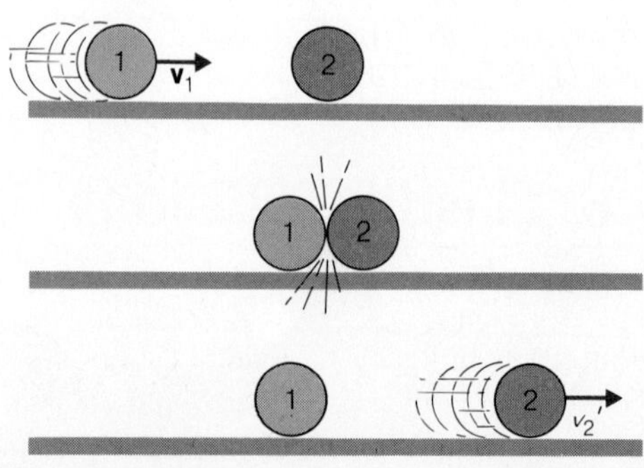

Figure 5.14
When a rolling ball makes a head-on collision with an identical stationary ball, the first ball stops and the second begins moving with the first's initial velocity.

The second ball begins to move with the speed v_1. The two balls have exchanged their initial velocities. ♦

The results of the preceding problem help us to understand the operation of the toy shown in Fig. 5.15, which consists of a number of identical steel balls suspended by strings. When one ball at the left is pulled out and released, it swings to strike the row of stationary balls, and one ball at the right swings out in response. Why not two balls, or indeed all the others?

The answer is that both momentum and KE can be conserved *only* if a single ball swings out on the right. If two balls were to swing out, their joint mass would be twice that of the ball on the left, and to conserve momentum their initial speeds would have to be half that of the ball on the left. But the combined kinetic energies of the two balls would then be half that of the ball on the left:

	One ball on left		Two balls on right
Momentum:	mv	$=$	$m\left(\frac{v}{2}\right) + m\left(\frac{v}{2}\right) = mv$
Kinetic energy:	$\frac{1}{2}mv^2$	$\neq$	$\frac{1}{2}m\left(\frac{v}{2}\right)^2 + \frac{1}{2}m\left(\frac{v}{2}\right)^2 = \frac{1}{4}mv^2$

Only if half the initial KE is lost can two balls swing out. The same reasoning shows why, if two balls are pulled out at the left and released, as in Fig. 5.15(b), two balls at the right will swing out after the collision, and so on.

What about the case when more than half the balls are pulled out to the left? In that event the balls at the left cannot give up all their momentum and KE to the balls at the right, and so several of the pulled-out balls continue to move to the right after the collision.

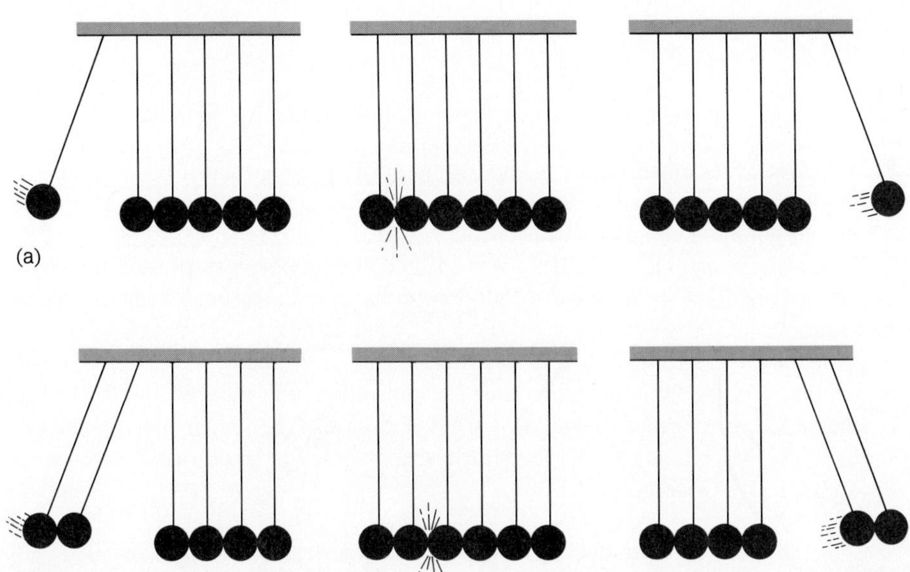

Figure 5.15

The behavior of this arrangement of suspended steel balls is determined by the conservation of both momentum and kinetic energy.

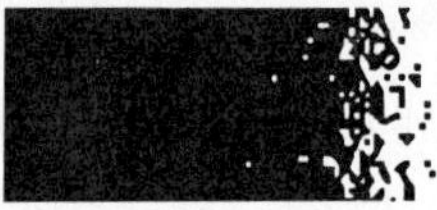

5.7 Energy Transfer

It is a maximum when the colliding objects have the same mass.

A moving object strikes a stationary one, and as a result the second object is set in motion. What is the mass ratio between the two that will cause the struck object to have the highest possible speed after the impact? What mass ratio will lead to the greatest transfer of energy to the struck object? These are not questions of abstract interest only but are closely connected with a variety of actual problems that range from the design of golf clubs to the design of nuclear reactors.

For simplicity we will confine ourselves to elastic head-on collisions in which both objects move along the same straight line. Let us consider an object of mass m_1 and initial speed v_1 that strikes a stationary object of mass m_2, after which their respective speeds are v_1' and v_2'. From conservation of momentum,

$$m_1 v_1 = m_1 v_1' + m_2 v_2'$$

$$m_1(v_1 - v_1') = m_2 v_2' \tag{5.7}$$

From Eq. (5.6), since $v_2 = 0$,

$$v_1 = v_2' - v_1' \tag{5.8}$$

Combining Eqs. (5.7) and (5.8) gives for the final speed of m_1

$$v_1' = \frac{m_1 - m_2}{m_1 + m_2} v_1 \tag{5.9}$$

and for the final speed v_2' of m_2

$$v_2' = \frac{2m_1}{m_1 + m_2} v_1 \tag{5.10}$$

Some interesting general conclusions follow from these formulas (Fig. 5.16):

1. When m_1 is less than m_2, v_1' is opposite to v_1: The lighter object bounces off the heavier one. A ball striking a wall is an extreme example. Here, since m_2 is virtually infinite compared with m_1, $v_1' = -v_1$.
2. When $m_1 = m_2$, $v_1' = 0$ and $v_2 = v_1$. The colliding object stops, and the target one moves off with the same speed. This is the case considered in the preceding section.
3. When m_1 is greater than m_2, as in the case of a tennis serve, the colliding object continues in the same direction after the impact but with reduced speed while the target object moves ahead of it at a faster pace. When m_1 is much greater than m_2, the colliding object loses little speed while the target one is given a speed nearly twice v_1.

The ratio between the kinetic energy KE_2' transferred to the initially stationary object and the kinetic energy KE_1 of the colliding object can be found with the help

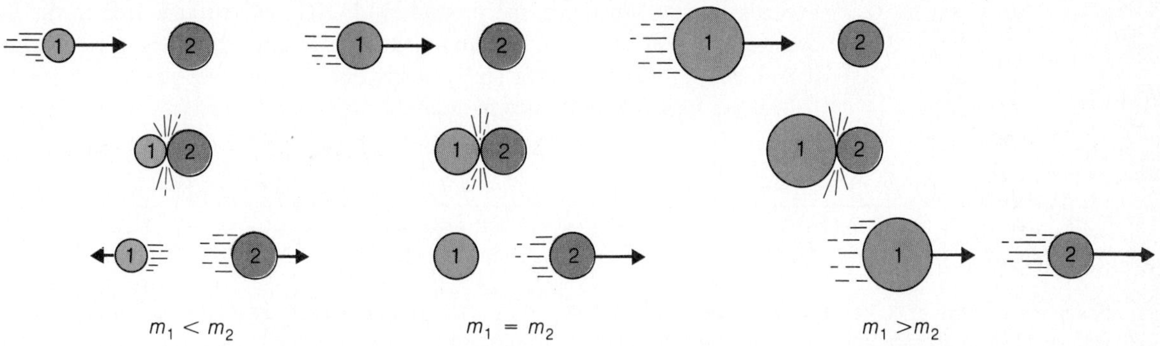

$m_1 < m_2$ (a) $m_1 = m_2$ (b) $m_1 > m_2$ (c)

Figure 5.16

How the effects of an elastic head-on collision with a stationary target depend on the relative masses of the two objects.

of Eq. (5.10):

$$\frac{\text{Transferred KE}}{\text{Original KE}} = \frac{\text{KE}_2'}{\text{KE}_1} = \frac{\frac{1}{2}m_2v_2'^2}{\frac{1}{2}m_1v_1^2} = \frac{4m_1m_2}{(m_1 + m_2)^2} = \frac{4(m_2/m_1)}{(1 + m_2/m_1)^2} \tag{5.11}$$

This formula is plotted in Fig. 5.17. Evidently the transfer of energy is a maximum for $m_1 = m_2$, when *all* the energy of m_1 is given to m_2. This is the situation illustrated in Fig. 5.16(b). We note that, for a given mass ratio m_2/m_1, it does not matter which object was initially moving and which object was at rest.

The fact that the energy transfer in a collision of the kind described here is a maximum when the objects involved have the same mass is an example of **impedance matching.** *Impedance* is a general term for opposition to energy transfer. Here it corresponds to inertia, but the concept of impedance has a broader significance. We will encounter impedance matching again later in this book when we look at energy transfer in wave motion and in direct-current and alternating-current electric circuits.

From Eq. (5.11) it would seem that the best mass for the head of a golf club would be the same as that of a golf ball, in order that all the energy of the club be given to the ball (assuming an elastic collision). In this case $v_2' = v_1$. However, according to Eq. (5.10) the greater the mass m_1 of the clubhead, the more the ball's velocity v_2' will exceed v_1, up to a limit of $2v_1$. The trouble with a heavy clubhead is twofold: It is hard to swing a heavy golf club as fast as a light one, and the more m_1

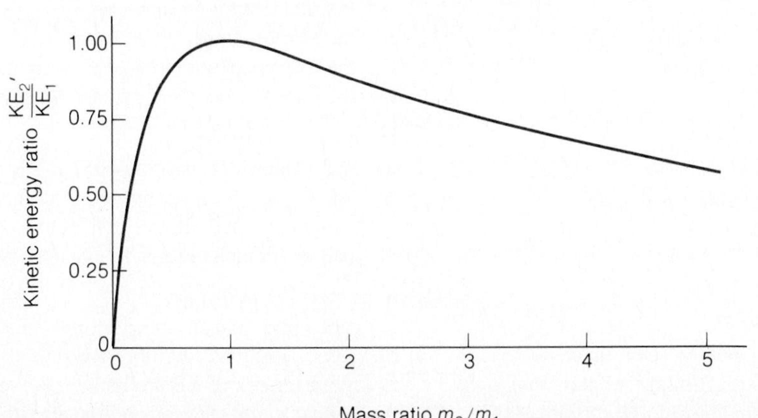

Figure 5.17

Energy transfer in an elastic head-on collision between a moving object and a stationary one.

exceeds m_2, the smaller is the proportion of KE that is transferred to the ball—the extra effort does not provide a commensurate return. Experience has led golfers to use clubheads whose masses are typically about four times the 46-g mass of a golf ball where maximum distance is required.

Important Terms

The **linear momentum** of an object is the product of its mass and velocity. Linear momentum is a vector quantity having the direction of the object's velocity.

The **impulse** of a force is the product of the force and the time during which it acts. Impulse is a vector quantity having the direction of the force. When a force acts on an object that is free to move, its change in momentum equals the impulse given it by the force.

The law of **conservation of momentum** states that when the vector sum of the external forces acting on a system of particles equals zero, the total linear momentum of the system remains constant.

The **thrust** of a rocket is the force that results from the expulsion of exhaust gases.

In an **elastic collision** kinetic energy is conserved. In an **inelastic collision** KE is not conserved. A **completely inelastic collision** is one in which the objects stick together on impact, which results in the maximum possible KE loss. Linear momentum is conserved in all collisions.

Important Formulas

Linear momentum: $\mathbf{p} = m\mathbf{v}$

Impulse and momentum change: $\mathbf{F}\,\Delta t = \Delta(m\mathbf{v})$

Thrust: $\mathbf{F} = \mathbf{v}\dfrac{\Delta m}{\Delta t}$

MULTIPLE CHOICE

1. An object at rest may have
- **a.** velocity **b.** momentum
- **c.** kinetic energy **d.** potential energy

2. An object in motion need not have
- **a.** velocity **b.** momentum
- **c.** kinetic energy **d.** potential energy

3. An object that has momentum must also have
- **a.** acceleration **b.** impulse
- **c.** kinetic energy **d.** potential energy

4. Momentum is most closely related to
- **a.** kinetic energy **b.** potential energy
- **c.** impulse **d.** power

5. Momentum can be expressed in
- **a.** N/s **b.** N·s
- **c.** N·m **d.** N·m/s

6. When the velocity of a moving object is doubled,
- **a.** its acceleration is doubled
- **b.** its momentum is doubled
- **c.** its kinetic energy is doubled
- **d.** its potential energy is doubled

7. An iron sphere of mass 30 kg has the same diameter as an aluminum sphere of mass 10.5 kg. The spheres are simultaneously dropped from a cliff. When they are 10 m from the ground, they have identical
- **a.** accelerations **b.** momenta
- **c.** potential energies **d.** kinetic energies

8. If a shell fired from a cannon explodes in midair,
- **a.** its total momentum increases
- **b.** its total momentum decreases
- **c.** its total kinetic energy increases
- **d.** its total kinetic energy decreases

9. What never changes when two or more objects collide is
- **a.** the momentum of each one
- **b.** the kinetic energy of each one
- **c.** the total momentum of all the objects
- **d.** the total kinetic energy of all the objects

10. In an elastic collision
- **a.** momentum is conserved but not KE
- **b.** KE is conserved but not momentum
- **c.** momentum and KE are both conserved
- **d.** neither momentum nor KE is conserved

11. In an inelastic collision
- **a.** momentum is conserved but not KE
- **b.** KE is conserved but not momentum
- **c.** momentum and KE are both conserved
- **d.** neither momentum nor KE is conserved

12. In a perfectly elastic collision between two objects, their relative speed after the collision is
   a. 0
   b. less than their relative speed before the collision
   c. equal to their relative speed before the collision
   d. more than their relative speed before the collision

13. A ball whose momentum is **p** strikes a wall and bounces off. The change in the ball's momentum is
   a. 0 b. **p**
   c. 2**p** d. −2**p**

14. Object A strikes object B, which is initially at rest. The maximum transfer of energy from A to B occurs when m_A is
   a. less than m_B b. equal to m_B
   c. greater than m_B
   d. any of the above, depending on the speed of A

15. In the collision in Multiple Choice 14, object B will acquire the most speed when m_A is
   a. less than m_B b. equal to m_B
   c. greater than m_B
   d. any of the above, depending on the speed of A

16. If a rocket of initial mass m is to rise from its launching pad, its initial thrust must exceed
   a. $\frac{1}{2}mg$ b. mg
   c. $2\,mg$ d. $\frac{1}{2}mg^2$

17. The average momentum of a 70-kg runner who covers 400 m in 50 s is
   a. 8.75 kg·m/s b. 57 kg·m/s
   c. 560 kg·m/s d. 5488 kg·m/s

18. A 30-kg girl and a 25-kg boy face each other on frictionless roller skates. The girl pushes the boy, who moves away at a speed of 1.0 m/s. The girl's speed is
   a. 0.45 m/s b. 0.55 m/s
   c. 0.83 m/s d. 1.2 m/s

19. An astronaut whose total mass is 100 kg ejects 1.0 g of gas from his propulsion pistol at a speed of 50 m/s. His recoil speed is
   a. 0.5 mm/s b. 5 mm/s
   c. 5 cm/s d. 50 cm/s

20. A 200-g ball moving at 5.0 m/s strikes a wall perpendicularly and rebounds elastically at the same speed. The impulse given to the wall is
   a. 1 N·s b. 2 N·s
   c. 2.5 N·s d. 10 N·s

21. A stationary 60-g tennis ball is struck by a racket that exerts an average force of 300 N on it for a time of 0.0050 s. The KE of the ball after the impact is
   a. 0.019 J b. 0.068 J
   c. 1.5 J d. 19 J

22. An 800-kg car moving at 80 km/h overtakes a 1200-kg car moving at 40 km/h in the same direction. If the two cars stick together, the wreckage has an initial speed of
   a. 8 km/h b. 40 km/h
   c. 56 km/h d. 60 km/h

23. The cars in Multiple Choice 22 are moving in opposite directions and collide head-on. If they stick together, the wreckage now has an initial speed of
   a. 8 km/h b. 40 km/h
   c. 56 km/h d. 60 km/h

24. The total KE lost by the cars in Multiple Choice 23 is
   a. 0
   b. less than that lost by the cars in Multiple Choice 22
   c. the same as that lost by the cars in Multiple Choice 22
   d. more than that lost by the cars in Multiple Choice 22

25. A 1200-kg car moving north at 50 km/h collides with a 1600-kg car moving east at 30 km/h. The cars stick together, and the wreckage moves off at an initial speed of
   a. 27 km/h b. 38 km/h
   c. 39 km/h d. 58 km/h

26. The direction in which the wreckage of the preceding collision moves off is
   a. 39° E of N b. 45° E of N
   c. 51° E of N d. 54° E of N

27. The additional force needed to maintain the speed of a pickup truck at a constant 20 m/s when rain begins to fall and accumulates in its back at a rate of 3.0 kg/min is
   a. 0 b. 1 N
   c. 10 N d. 60 N

QUESTIONS

1. Is it possible for an object to have more kinetic energy but less momentum than another object? Less kinetic energy but more momentum?

2. When the momentum of an object is doubled in magnitude, what happens to its kinetic energy?

3. When the kinetic energy of an object is halved in magnitude, what happens to its momentum?

4. When a rocket explodes in midair, how are its total momentum and total kinetic energy affected?

5. How is the principle of conservation of linear momentum related to the definition of mass given in Chapter 3 and to Newton's first law of motion? In what way does this principle go beyond the definition of mass and the first law of motion?

6. (a) When an object at rest breaks up into two parts that fly

off, must they move in exactly opposite directions? (b) When a moving object strikes a stationary one and the two do not stick together, must they move off in exactly opposite directions?

7. A railway car is at rest on a frictionless track. A man at one end of the car walks to the other end. (a) Does the car move while he is walking? (b) If so, in which direction? (c) What happens when the man comes to a stop?

8. An empty coal car coasts at a certain speed along a level railroad track without friction. (a) It begins to rain. What happens to the speed of the car? (b) The rain stops, and the collected water gradually leaks out. What happens to the speed of the car now?

9. Give an example of a collision in which all the KE of the participating objects is lost. What happens to their total momentum?

EXERCISES

5.1 Linear Momentum

1. Find the momentum of a 50-g bullet whose KE is 250 J.

2. Find the momentum of a giraffe galloping at 20 km/h whose KE is 6.0 kJ.

3. Car *A* has a mass of 1000 kg and is moving at 60 km/h. Car *B* has a mass of 2000 kg and is moving at 30 km/h. Compare the kinetic energies and momenta of the two cars.

5.2 Impulse

4. A 1000-kg car strikes a tree at 30 km/h and comes to a stop in 0.15 s. Find its initial momentum and the average force on the car while it is being stopped.

5. A 160,000-kg DC-8 airplane is flying at 870 km/h. (a) Find its momentum. (b) If the thrust its engines develop is 340 kN, how much time is needed for the airplane to reach this speed starting from rest? Neglect air resistance, changes in altitude, and the fuel consumed by the engines.

6. A 150-g baseball reaches a batter with a speed of 25 m/s. After it has been struck, it leaves the bat at 35 m/s in the opposite direction. If the ball was in contact with the bat for 1.0 ms, find the average force exerted on it during this period.

7. A 4.0-kg seagull flies at 15 m/s into the windshield of an airplane flying in the opposite direction at 180 m/s. If the impact lasts 1.0 ms, find the average force on the windshield.

8. Water emerging from a hose at a rate of 2.0 L/s and a speed of 8.0 m/s strikes a person. If the water loses all its momentum on impact, find the force on the person. (The mass of 1 L of water is 1 kg.)

9. A 2.0-kg pail rests on a scale. Water is poured into the pail from a height of 1.5 m at a rate of 100 g/s. (a) What is the reading on the scale (in kilograms) 1.0 min after the pouring began? (b) The pouring stops at 1.0 min. What is the reading on the scale afterward?

10. Wheat falls vertically at 50 kg/s on a horizontal conveyor belt whose speed is 4.0 m/s. How much force is needed to keep the speed of the belt constant? How much power is needed to operate the belt?

5.3 Conservation of Momentum

11. A hunter has a rifle that can fire 60-g bullets with a speed of 900 m/s. A 40-kg leopard springs at him at 10 m/s. How many bullets must the hunter fire into the leopard in order to stop him in his tracks?

12. A hunter in a rowboat loses the oars and decides to set the boat in motion by firing his rifle astern five times. The total mass of the boat and its contents is 150 kg, the mass of each bullet is 20 g, and the speed of the bullets is 600 m/s. Assume water resistance is negligible. How far does the boat go in 10 min?

13. A 50-kg girl throws a 5.0-kg pumpkin at 10 m/s to a 50-kg boy, who catches it. If both are on a frictionless frozen lake, how fast does each of them move backward?

14. A 70-kg astronaut is drifting forward in an orbiting space shuttle at 20 mm/s. A fellow astronaut throws her a 250-g orange, which she catches. If she then starts to move backward at 10 mm/s, what was the speed of the orange?

15. An astronaut outside an orbiting spacecraft uses a pistol that ejects a gas in order to maneuver in space. Suppose the astronaut and her space suit have a total mass of 100 kg and the pistol ejects 12 g of gas per second at a speed of 650 m/s. How long should the astronaut operate the pistol in order to have a speed of 1.0 m/s?

16. A spacecraft moving at 10.0 km/s breaks apart into two pieces of equal mass that continue moving in the original direction. If the speed of one of the pieces is 4.0 km/s, what is the speed of the other one?

17. A spacecraft moving at 10.0 km/s breaks apart into two pieces of equal mass, one of which moves off at 4.0 km/s in a direction opposite to the original direction. Find the speed and direction of the other piece.

18. An object at rest breaks up into two fragments that fly apart. Express the ratio of their kinetic energies in terms of their masses.

19. A 70-kg man and a 50-kg woman are in a 60-kg rubber dinghy. The man dives into the water with a horizontal speed of 3.0 m/s. If his swimming speed is 1.0 m/s, can he return to the dinghy? If not, can the woman change the dinghy's motion enough by diving off it at 3.0 m/s in the opposite direction? Could she then return to the dinghy herself if her swimming speed were also 1.0 m/s?

20. The elastic potential energy of the compressed spring of

Fig. 3.3 is 12 J, and the masses of the two objects are, respectively, $m_A = 2.0$ kg and $m_B = 1.0$ kg. What speed does each object have when the string is cut?

21. A certain cannon has a range of 2.0 km. One day a shell that it fires explodes at the top of its path into two equal fragments that move horizontally immediately afterward. One fragment lands next to the cannon. How far away from the cannon does the other fragment land?

5.4 Rocket Propulsion

22. The motor of a 1000-kg rocket standing vertically on the ground is being tested. Fuel is being burned at the rate of 5.0 kg/s with an exhaust speed of 2.8 km/s. Find the force needed to hold the rocket down.

23. The motors of a spacecraft provide a total thrust of 1.8 MN at takeoff. If the exhaust speed is 2.5 km/s, at what rate is fuel being consumed?

24. A rocket fired from the earth's surface ejects 2% of its mass at a speed of 2.0 km/s in the first second of its flight. Find the initial acceleration of the rocket.

25. A certain rocket consumes 25 kg of fuel per second, and its exhaust has a speed of 3000 m/s. If the initial mass of the rocket is 5000 kg of which 3500 kg is fuel, what are its initial and final accelerations? Ignore the decrease in g with altitude.

26. A rocket launched vertically from the earth has a mass of 2500 kg at takeoff. (a) If the rocket's initial acceleration is 0.50 g and it is using fuel at the rate of 15 kg/s, what is the speed of the exhaust gases? (b) What is the acceleration of the rocket 1 min later? Ignore the decrease in g with altitude.

5.5 Inelastic Collisions

27. A neutron of mass 1.67×10^{-27} kg and speed 1.00×10^5 km/s collides with a stationary deuteron of mass 3.34×10^{-27} kg. The two particles stick together. What is the speed of the composite particle (called a triton)?

28. A 30-kg girl who is running at 3.0 m/s jumps on a stationary 10-kg sled on a frozen lake. How fast does the sled then move?

29. A 3.0-g bullet moving at 4.0 km/s strikes an 8.0-kg wooden block resting on a frictionless table. The bullet passes through the block and comes out with a speed of 0.20 km/s. What is the speed of the block?

30. A moving stone strikes a stationary lump of clay whose mass is three times that of the stone and becomes embedded in it. What proportion of the stone's initial KE is lost?

31. An 0.50-kg stone moving at 4.0 m/s overtakes a 4.0-kg lump of clay moving at 1.0 m/s. The stone becomes embedded in the clay. (a) What is the speed of the composite body after the collision? (b) How much KE is lost in the collision?

32. A 1000-kg car moving east at 80 km/h collides head-on with a 1500-kg car moving west at 40 km/h, and the two cars stick together. (a) Which way does the wreckage move and with what initial speed? (b) How much KE is lost in the collision?

33. A 1000-kg car moving east at 80 km/h overtakes a 1500-kg car moving east at 40 km/h and collides with it. The two cars stick together. (a) What is the initial speed of the wreckage? (b) How much KE is lost in the collision?

34. A 1200-kg car traveling east at 30 km/h collides with an 1800-kg car traveling north at 20 km/h. The cars stick together after the collision. What is the velocity (magnitude and direction) of the wreckage?

35. An 0.50-kg stone moving north at 4.0 m/s collides with a 4.0-kg lump of clay moving west at 1.0 m/s. The stone becomes embedded in the clay. What is the velocity (magnitude and direction) of the composite body after the collision?

5.6 Elastic Collisions

36. A neutron of mass 1.67×10^{-27} kg and speed 1.00×10^5 m/s collides head-on with a stationary deuteron of mass 3.34×10^{-27} kg. The particles do not stick together, and the deuteron moves off at 6.67×10^4 m/s. What is the speed of the neutron? Is the collision elastic?

37. The 176-g head of a golf club is moving at 45 m/s when it strikes a 46-g golf ball and sends it off at 65 m/s. Find the final speed of the clubhead after the impact, assuming that the mass of the club's shaft can be neglected.

38. A billiard ball moving at 4.00 m/s strikes another billiard ball at rest and moves off at 3.46 m/s in a direction 30° from its original line of motion. What is the velocity of the target ball?

5.7 Energy Transfer

39. A 10-kg iron ball rolling at 2.0 m/s on a horizontal surface strikes a 1.0-kg wooden ball of the same size that is at rest. What is the speed of each ball after the collision? What proportion of the iron ball's original KE was transferred to the wooden ball?

40. A 46-g golf ball flies off at 1.5 times the speed of the clubhead that struck it. What was the mass of the clubhead? What proportion of the clubhead's initial KE was transferred to the ball?

PROBLEMS

41. Some years ago the record distance for the shotput was 22 m and that for the discus throw was 68 m. The mass of the shot was 7.25 kg and that of the discus was 2.00 kg. Neglecting air resistance and assuming that each athlete hurled his projectile at the optimal angle, compare the initial KEs and momenta of the projectiles. What does this comparison suggest?

42. A stream of n particles per m^3, each with velocity $\mathbf{v}$ and mass m, strikes a wall of area A perpendicular to $\mathbf{v}$. The particles stick to the wall. Find the average force on the wall. (*Hint:* Begin by showing that the number of particles that strike the wall in the time Δt is $nAv\Delta t$.)

43. A certain howitzer has a barrel 7.6 m long and fires a 65-kg projectile 20 cm in diameter at a muzzle speed of 400 m/s. Use three different ways, each based on an important physical principle, to find the average force on the projectile while it is in the howitzer barrel.

44. Coal falls vertically at 100 kg/s on a horizontal conveyor belt whose speed is 4 m/s. Compare the power needed to keep the speed of the belt constant with the rate at which the coal gains KE. If there is any difference, account for it.

45. An object at rest explodes into three fragments of equal mass. Two of the fragments fly off in directions 60° apart with equal speeds of 10 m/s. What are the speed and direction of the third fragment?

46. A 500-kg cannon on a wheeled carriage is at the foot of a 25° slope with its barrel horizontal. The cannon fires a 30-kg shell at 300 m/s in a direction opposite to the slope. (a) What is the cannon's initial speed up the slope? (b) How far along the slope does the cannon travel before it comes to a stop and starts to roll back down? Disregard friction.

47. A ballistic pendulum consists of a wooden block of mass M suspended by long strings from the ceiling. A bullet of mass m and speed v is fired horizontally into the block, which swings away until its height is the amount h above its original height. (a) Find a formula that gives v in terms of g and the readily measurable quantities m, M, and h. (b) A 5.0-g bullet is fired into the 2.0-kg block of a ballistic pendulum, which rises by 10 cm. Find the bullet's speed.

48. As shown in Fig. 5.18, a magnet A attached to a string R long is pulled out to one side so the string is horizontal. When the magnet is let go, it swings downward and strikes an iron cube B of the same mass that is resting on a frictionless surface. The two stick together and swing upward on the other side. What is the maximum value of θ, the angle between the string and the vertical?

49. Verify Eq. (5.6) in the following way. Set equal the initial and final KEs of the two objects and use the algebraic relationship $a^2 - b^2 = (a + b)(a - b)$ to obtain $m_1(v_1 + v_1')(v_1 - v_1') = m_2(v_2' + v_2)(v_2' - v_2)$. Then set equal the initial and final momenta of the objects to obtain $m_1(v_1 - v_1') = m_2(v_2' - v_2)$. From these equations obtain Eq. (5.6).

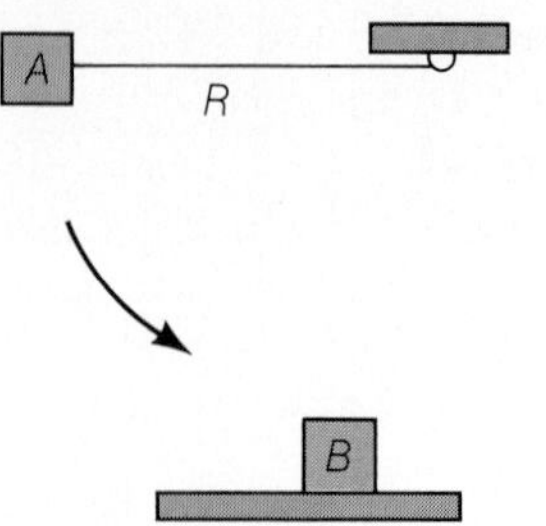

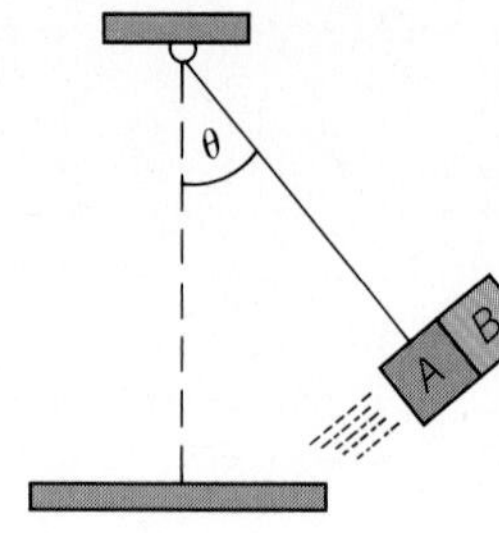

Figure 5.18

Problem 48.

50. A billiard ball moving at 4.0 m/s strikes a pair of billiard balls at rest. One of the latter moves off at 2.0 m/s in a direction 60° from the line of motion of the first ball, and the other moves off at 3.0 m/s in a direction 30° from the line of motion of the first ball on the other side. What is the velocity of the first ball after the collision?

51. A billiard ball at rest is struck by another ball of the same mass whose speed is 5.0 m/s. After an elastic collision the striking ball goes off at an angle of 40° with respect to its original direction of motion and the struck ball goes off at an angle of 50° with respect to this direction. Find the final speeds of both balls.

52. A billiard ball at rest is struck by another billiard ball of the same mass whose speed is 4.0 m/s. After an elastic collision the striking ball goes off at an angle of 25° with respect to its original direction of motion. Find the angle the struck ball makes with this direction and the final speeds of both balls.

Answers to Multiple Choice

1. d	8. c	15. c	22. c
2. d	9. c	16. b	23. a
3. c	10. c	17. c	24. d
4. c	11. a	18. c	25. a
5. b	12. c	19. a	26. a
6. b	13. d	20. b	27. b
7. a	14. b	21. d	

CHAPTER

6

CIRCULAR MOTION AND GRAVITATION

Almost everything in the natural world travels in a curved path. Often these paths are circles or very nearly circles. For instance, the orbits of the earth and the other planets about the sun are almost circular, as is the orbit of the moon around the earth. On a smaller scale a handy way to visualize an atom is to imagine its electrons circling around its central nucleus. Of course, circular motion is no novelty in our own experience—it is hard to think of any machine in which circular motion of some kind is not involved.

6.1 Centripetal Force

A curved path requires an inward pull.

Tie a ball to the end of a string and whirl it around your head, as in Fig. 6.1. What you will find is that your hand must pull on the string to keep the ball going in a circle. Let go of the string, and the ball flies off tangentially to the circle.

The force needed to make an object follow a curved path is called **centripetal force,** which means "center-seeking force." Without it, circular motion cannot occur. In general:

Centripetal force is the force perpendicular to the velocity of an object moving along a curved path. The centripetal force is directed toward the center of curvature of the path.

An object that moves in a circle at constant speed is said to undergo **uniform circular motion.** As shown in Fig. 6.2, the centripetal force $\mathbf{F}_c$ on such an object always points to the center of the circle. This means that $\mathbf{F}_c$ is perpendicular to the direction of $\mathbf{v}$ at all times, so the centripetal force does no work on the object, and the object's speed v remains constant.

A centripetal force acts whenever rotational motion occurs, since such a force is needed to change the path of a particle from the straight line that it would normally follow to a curved path. Gravitation provides the centripetal forces that keep the planets moving around the sun and the moon around the earth. Friction between tires and road provides the centripetal force needed by a car in rounding a curve in a level road (Fig. 6.3). If the tires are worn and the road is wet or icy, the frictional force may be too small to permit the car to turn.

Figure 6.1

When a ball is whirled at the end of a string, the tension in the string provides the centripetal force that keeps the ball moving in a circle.

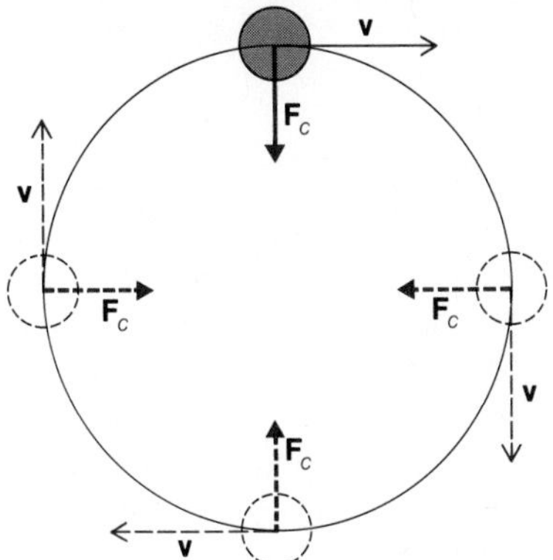

Figure 6.2

Even though the velocity **v** of an object traveling in a circle at constant speed has the same magnitude along its path, the direction of **v** changes constantly. The inward force that causes this change in direction is called centripetal force, $\mathbf{F}_c$.

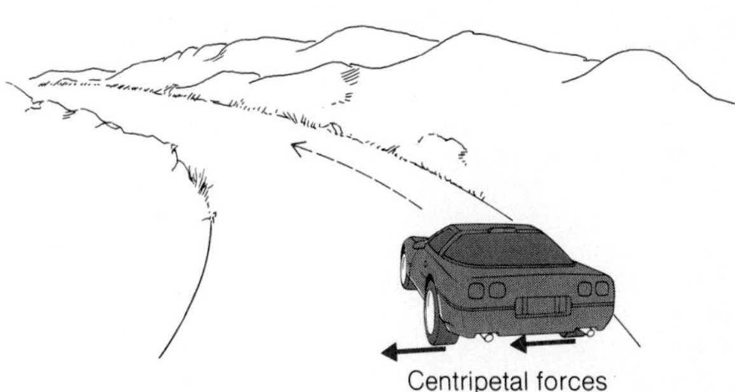

Figure 6.3

The centripetal force exerted when a car rounds a curve on a level road is provided by friction between its tires and the road.

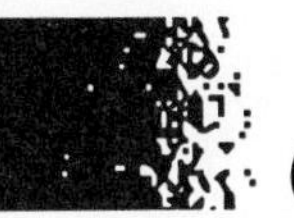

6.2 Centripetal Acceleration

It produces a change in direction without a change in speed.

How much centripetal force do we need to keep a given object moving in a circle at a certain speed? To find out, we begin by asking what the object's acceleration is. In the following derivation the object will be a particle, but the same arguments and conclusion hold for the circular motion of the center of mass of an object of any size.

In Fig. 6.4 (a) a particle is shown traveling along a circular path of radius r at the constant speed v. At $t = 0$ the particle is at the point A, where its velocity is $\mathbf{v}_A$, and at $t = \Delta t$ the particle is at the point B, where its velocity is $\mathbf{v}_B$. The change $\Delta\mathbf{v}$ in the particle's velocity in the time interval Δt is $\Delta\mathbf{v} = \mathbf{v}_B - \mathbf{v}_A$, and its acceleration is $\mathbf{a} = \Delta\mathbf{v}/\Delta t$.

The vector triangle whose sides are $-\mathbf{v}_A$, $\mathbf{v}_B$, and $\Delta\mathbf{v}$ is similar to the space triangle whose sides are OA, OB, and s, as we can see from Fig. 6.4(c). Since v is the constant speed of the particle, the magnitudes of $-\mathbf{v}_A$ and $\mathbf{v}_B$ are both v. Also, OA and OB are radii of the circle, so their lengths are both r. Corresponding sides of similar triangles are proportional, hence

$$\frac{\Delta v}{v} = \frac{s}{r} \quad \text{and} \quad \Delta v = \frac{vs}{r}$$

The distance the particle actually covers in going from A to B is the arc joining these points, the length of which is $v\,\Delta t$. The distance s, however, is the chord joining A and B, as in Fig. 6.4(d). We are finding the *instantaneous* acceleration of the particle, which means that A and B are very close together. In this case the

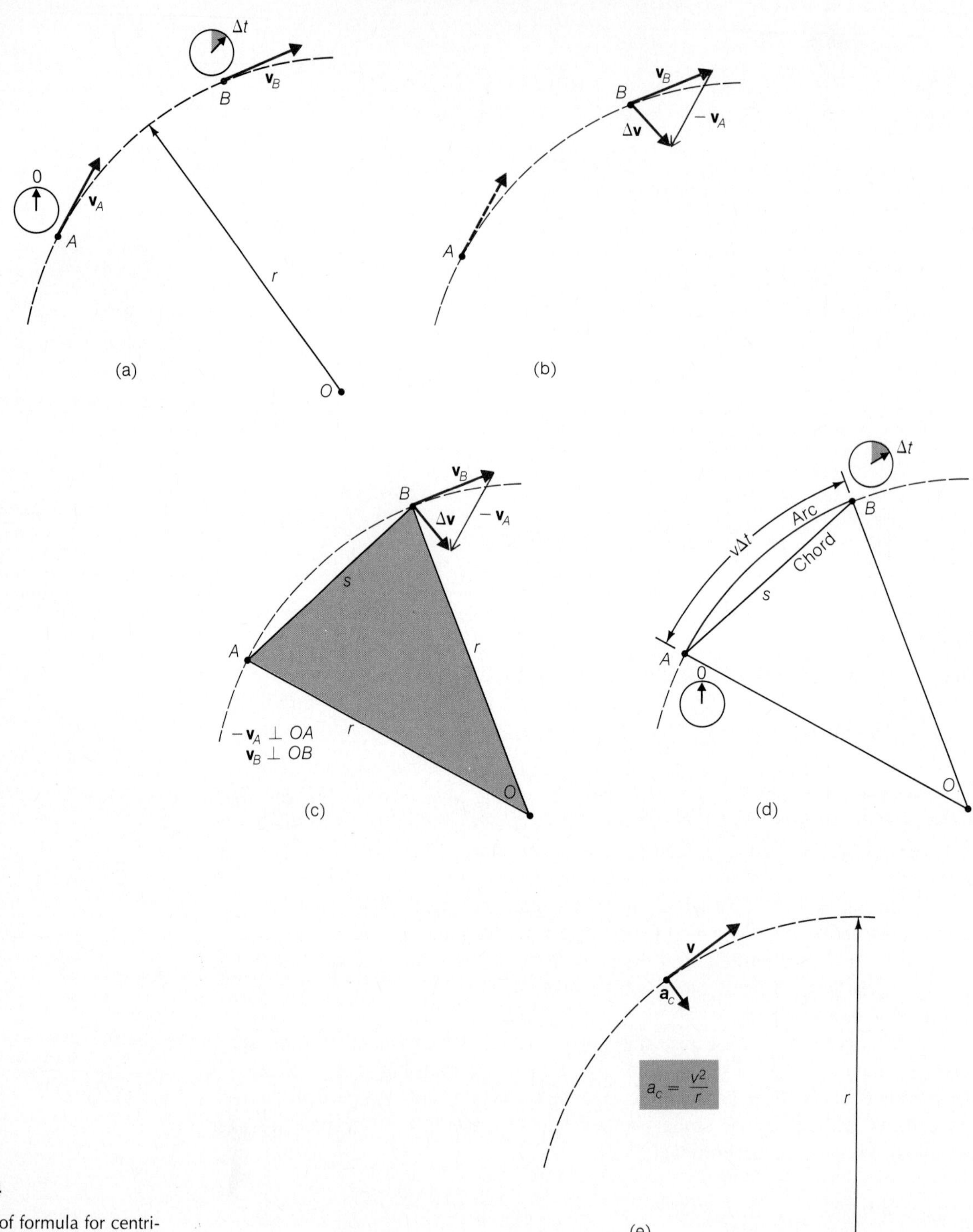

Figure 6.4

Derivation of formula for centripetal acceleration.

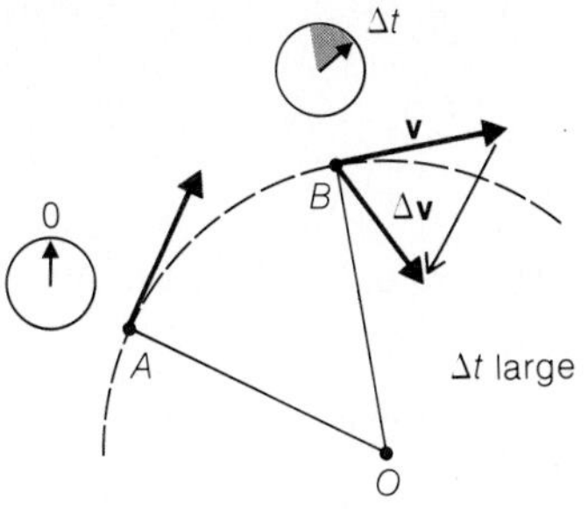

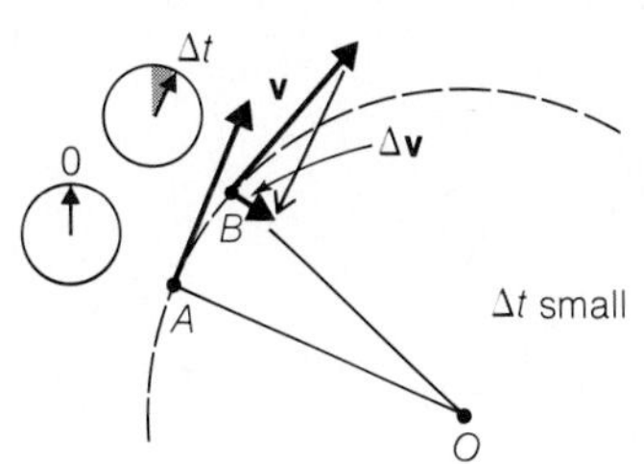

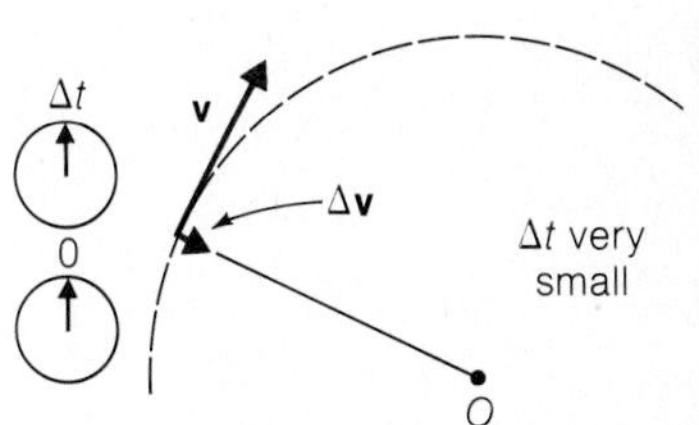

Figure 6.5

The centripetal acceleration of a particle in uniform circular motion points toward the center of the circle.

chord and arc are equal. Hence

$$s = v\,\Delta t$$

and we have

$$\Delta v = \frac{v^2 \Delta t}{r}$$

Since the magnitude of the particle's acceleration is $a_c = \Delta v/\Delta t$,

$$a_c = \frac{v^2}{r} \qquad \textit{Centripetal acceleration} \quad (6.1)$$

The inward **centripetal acceleration** of a particle in uniform circular motion is proportional to the square of its speed and inversely proportional to the radius of its path. See Fig. 6.4(e).

In Fig. 6.4(c) $\Delta\mathbf{v}$ does not quite point toward O, the center of the particle's circular path. When A and B are very close together, however, $\Delta\mathbf{v}$ *does* point toward O (Fig. 6.5). Because the acceleration $\mathbf{a}_c$ is an instantaneous acceleration, we only care about the case when Δt and hence s are extremely small. The direction of $\mathbf{a}_c$ is accordingly radially inward.

Period of Circular Motion

The time T needed by an object in uniform circular motion to complete an orbit is called its **period.** If r is the radius of the circle, its circumference is $2\pi r$ (Fig. 6.6). The speed of the object is the distance it covers (here $2\pi r$) divided by the time

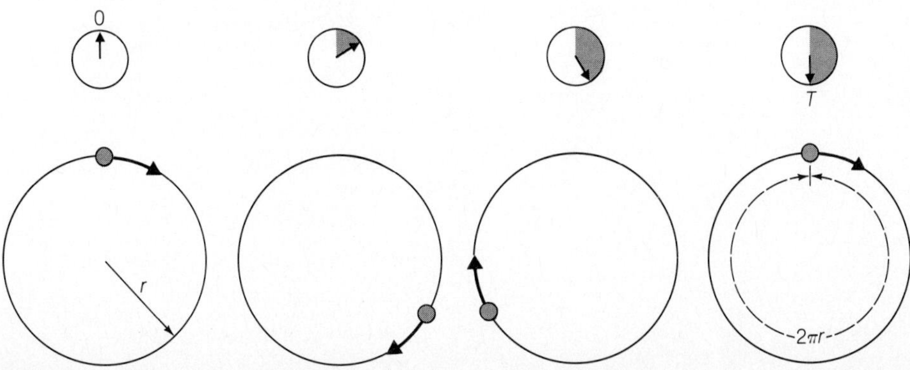

Figure 6.6

An object in uniform circular motion whose period is T has a speed of $2\pi r/T$.

needed (here T), so

$$v = \frac{2\pi r}{T} \qquad \textit{Speed in orbit of period T} \quad (6.2)$$

This formula is often useful in working out problems in uniform circular motion.

EXAMPLE 6.1

The ball in Fig. 6.1 is whirled in a horizontal circle 80 cm in radius. If the ball makes two revolutions per second, what is its centripetal acceleration?

SOLUTION The ball takes $T = 0.5$ s per orbit. Using Eq. (6.2) for the ball's speed gives, from Eq. (6.1), the centripetal acceleration

$$a_c = \frac{v^2}{r} = \left(\frac{2\pi r}{T}\right)^2 \left(\frac{1}{r}\right) = \frac{4\pi^2 r}{T^2} = \frac{(4\pi^2)\,(0.8\text{ m})}{(0.5\text{ s})^2} = 126\text{ m/s}^2$$

which is nearly 13 times the acceleration of gravity g. ◆

6.3 Magnitude of Centripetal Force

How much of an inward pull.

From the second law of motion we can see that the centripetal force $\mathbf{F}_c$ on an object of mass m in uniform circular motion is $\mathbf{F}_c = m\mathbf{a}_c$, which has the magnitude

$$F_c = \frac{mv^2}{r} \qquad \textit{Centripetal force} \quad (6.3)$$

Figure 6.7 illustrates how F_c varies with m, v, and r. Now we know why a car rounding a curve is hard to control when it is heavy (large m), when it is going fast (large v means very large v^2), and when the curve is sharp (small r).

Figure 6.7

Centripetal force.

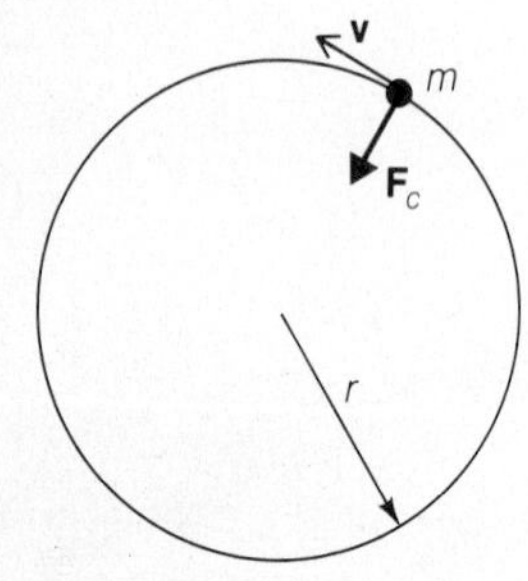

The centripetal force on an object in uniform circular motion is equal in magnitude to mv^2/r.

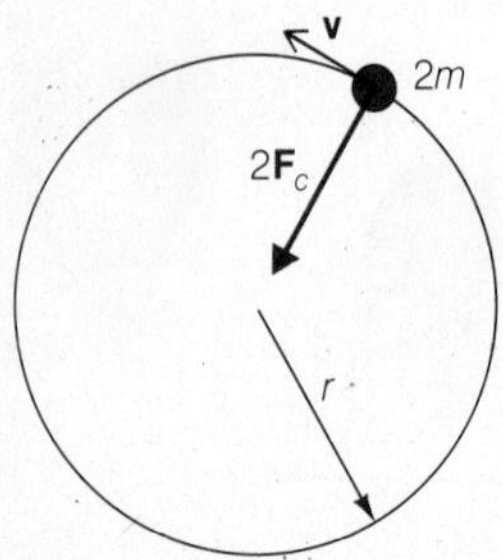

Doubling the mass doubles the required centripetal force.

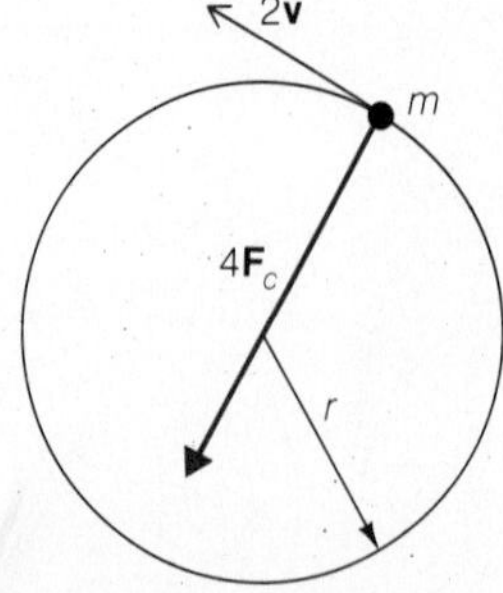

Doubling the speed, however, quadruples the required centripetal force.

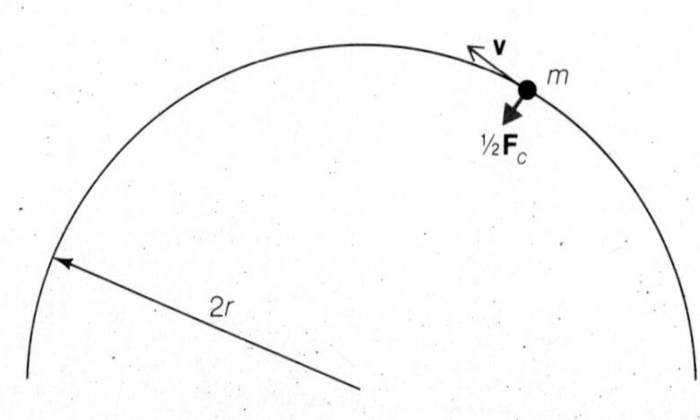

Doubling the radius of the circle halves the required centripetal force.

EXAMPLE 6.2

(a) Find the centripetal force needed by a 1200-kg car to make a turn of radius 40 m at a speed of 25 km/h (16 mi/h). (b) Assuming the road is level, find the minimum coefficient of static friction between the car's tires and the road that will permit the turn to be made (Fig. 6.8).

SOLUTION (a) The car's speed is

$$v = (25 \text{ km/h}) \left(0.278 \frac{\text{m/s}}{\text{km/h}}\right) = 7.0 \text{ m/s}$$

Accordingly the centripetal force is

$$F_c = \frac{mv^2}{r} = \frac{(1200 \text{ kg})(7.0 \text{ m/s})^2}{40 \text{ m}} = 1470 \text{ N} = 1.5 \text{ kN}$$

(b) The frictional force is given by

$$F_f = \mu F_N$$

To find the coefficient of friction μ_s, we substitute 1470 N for F_f and the car's weight of $w = mg$ for the normal force F_N. Thus

$$\mu_s = \frac{F_f}{F_N} = \frac{F_c}{mg} = \frac{1470 \text{ N}}{(1200 \text{ kg})(9.8 \text{ m/s}^2)} = 0.125$$

which is available under most driving conditions.

It is worth noting that the coefficient of friction required here does not depend on the car's mass, since

$$\mu_s = \frac{F_c}{mg} = \frac{mv^2/r}{mg} = \frac{v^2}{gr}$$

Because the required value of μ depends on v^2, high-speed turns on a level road can be dangerous. In this example, increasing the speed to 60 km/h (37 mi/h) increases the needed μ_s to 0.72, which is too much for a wet road, and the car would skid.

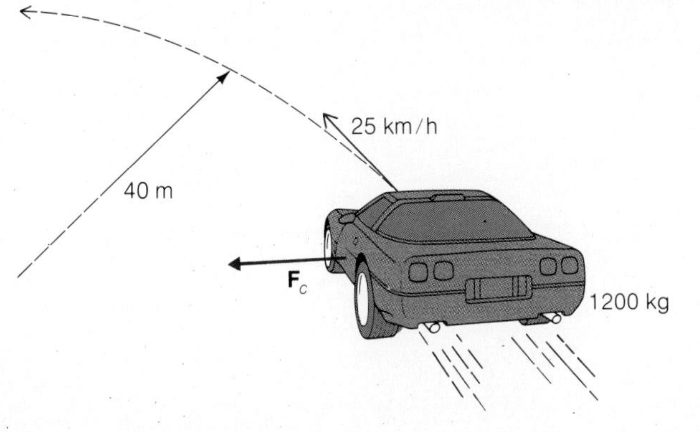

Figure 6.8

A centripetal force of 1470 N (330 lb) is needed by this car to make the turn shown. The corresponding coefficient of static friction between its tires and the road is $\mu_s = 0.125$.

◆

Braking during a fast turn may be risky. Suppose a car is rounding a curve on a flat road at the highest speed possible without the tires slipping on the road. This means that the centripetal force on the car is the most that can be transferred to its tires from the road by static friction. If the driver now applies the brakes, an additional force is exerted on the tires at right angles to the centripetal force on them. The resultant of these two forces is now more than static friction between the tires and the road can provide. The tires lose their grip on the road, the lower coefficient of kinetic friction takes over, and the car skids toward the outside of the curve. As Table 3.2 shows, the coefficient of static friction between a rubber tire and a concrete road is about 50% greater than the corresponding coefficient of sliding friction.

The likelihood of a loss of control increases as the braking force is increased. If a skid does occur, pressing harder on the brake pedal will not help, because the tires are already slipping over the road. The only way to regain control is to ease up on the brakes and to straighten the wheels momentarily, which will allow the tires to recapture their grip on the road.

EXAMPLE 6.3

Usually the friction between its tires and the road is enough to provide a car with the centripetal force it needs to make a turn. However, if the car's speed is high or the road surface is slippery, the available frictional force may not be enough and the car will skid. To reduce the chance of skids, highway curves are often **banked** so that the roadbed tilts inward. The horizontal component of the reaction force of the road on the car (the action force is the car pressing on the road) then furnishes the required centripetal force. Find the proper banking angle for a car making a turn of radius r at the speed v.

SOLUTION The reaction force **F** of the road on the car is perpendicular to the roadbed, as in Fig. 6.9, since friction is not involved here. This force can be resolved

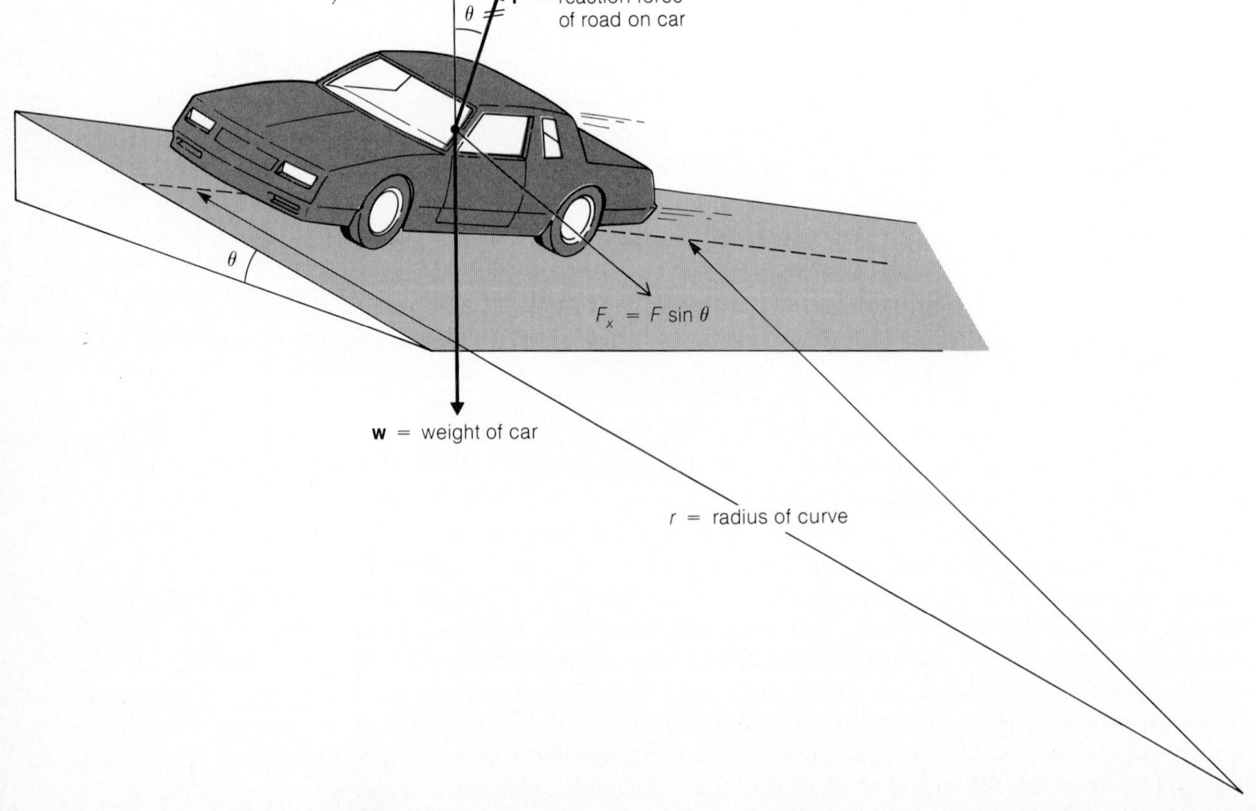

Figure 6.9

When a car rounds a banked curve, the horizontal component $\mathbf{F}_x$ of the reaction force **F** of the road on the car provides it with the required centripetal force.

At high speeds friction between its tires and the road may not provide enough centripetal force for a vehicle to round a turn on a level road. If the turn is banked, additional centripetal force is provided by the reaction force of the road on the vehicle.

into two components, $\mathbf{F}_y$, which supports the weight $\mathbf{w}$ of the car, and $\mathbf{F}_x$, which is available to provide centripetal force. From the diagram,

$$F_x = F \sin \theta = \text{horizontal component of reaction force}$$

$$F_y = F \cos \theta = \text{vertical component of reaction force}$$

where θ is the angle between the roadbed and the horizontal. Since F_x furnishes the centripetal force F_c,

$$F \sin \theta = \frac{mv^2}{r}$$

The vertical component F_y of the reaction force equals the car's weight mg, and so

$$F \cos \theta = mg$$

We divide the first of these equations by the second to obtain

$$\frac{F \sin \theta}{F \cos \theta} = \frac{mv^2}{mgr}$$

$$\tan \theta = \frac{v^2}{gr} \qquad \textit{Banking angle} \quad (6.4)$$

The tangent of the proper banking angle θ varies directly with the square of the car's speed and inversely with the radius of the curve. The mass of the car does not matter. When a car goes around a curve at precisely the design speed, the reaction force of the road provides the centripetal force. If the car goes more slowly than this, friction tends to keep it from sliding down the inclined roadway; if the car goes faster, friction tends to keep it from skidding outward.

The same considerations apply to an airplane making a turn, in which case Eq. (6.4) specifies the angle its wings should make with the horizontal (Fig. 6.10). ◆

Figure 6.10

(a) In level flight at constant velocity, the lift produced by an airplane's wings equals its weight to give no resultant force perpendicular to the airplane's direction of motion. (b) When the airplane banks, a resultant force comes into being whose horizontal component provides the centripetal force needed for a turn. The vertical component of the resultant force is downward, so the airplane loses altitude as it turns. (c) If the airplane's speed is increased as it banks, the greater lift can balance its weight to permit a horizontal turn.

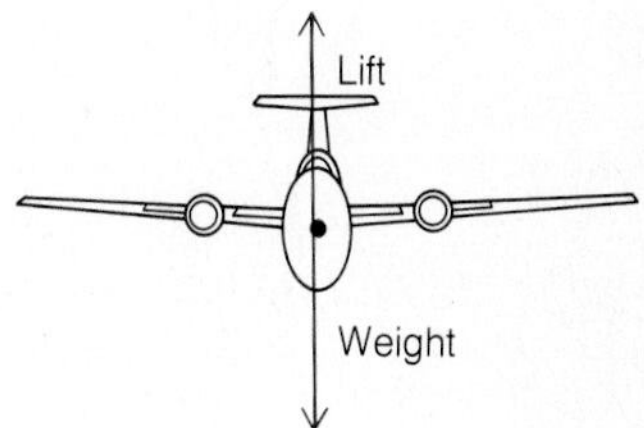

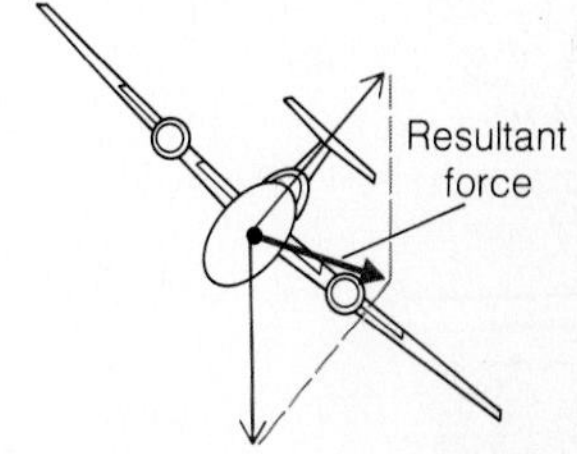

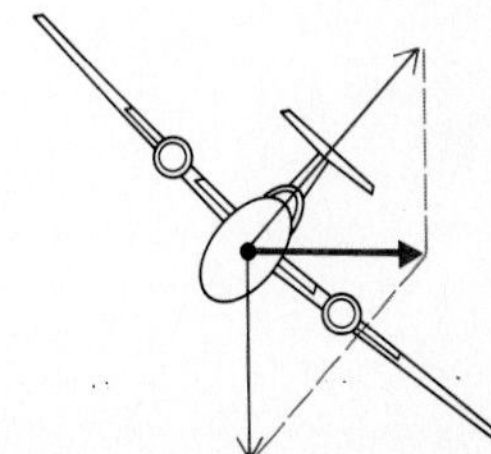

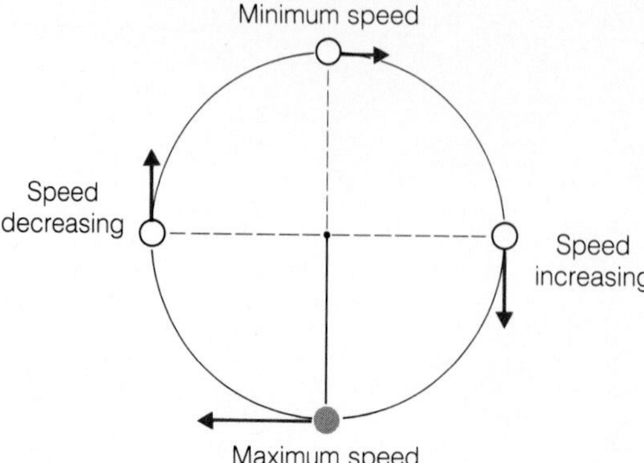

Figure 6.11

The speed of an object moving freely in a vertical circle is not constant. Shown are velocity vectors at four different positions on the circle.

6.4 Motion in a Vertical Circle

Total energy is constant but speed is not.

Let us now look at what happens when an object moves in a vertical rather than a horizontal circle. The object's total energy remains constant, and all of it is kinetic energy at the bottom of the circle (Fig. 6.11). At the top of the circle, however, the total energy is divided between KE and the maximum potential energy relative to the bottom, so the object's speed is least there. The motion is circular but not uniform. The object goes faster as it goes downward and slower as it goes upward.

EXAMPLE 6.4

A ball is being swung at constant energy at the end of an 80-cm string, as in Fig. 6.12. If its speed at the top of the circle is 3.5 m/s, what is its speed at the bottom?

SOLUTION At the top of the circle the ball has the potential energy

$$PE_1 = mgh = mg(2r) = 2mgr$$

and at the bottom $PE_2 = 0$. Since the total energy is constant,

$$KE_2 + PE_2 = KE_1 + PE_1$$

$$\tfrac{1}{2}mv_2^2 + 0 = \tfrac{1}{2}mv_1^2 + 2mgr$$

$$v_2^2 = v_1^2 + 4gr$$

$$v_2 = \sqrt{v_1^2 + 4gr}$$

Here $v_1 = 3.5$ m/s and $r = 0.80$ m, so the speed of the ball at the bottom of its path is

$$v_2 = \sqrt{(3.5\text{ m/s})^2 + (4)(9.8\text{ m/s}^2)(0.80\text{ m})} = 6.6\text{ m/s}$$ ◆

Figure 6.12

$KE_1 = \frac{1}{2}mv_1^2$

$PE_1 = mgh = 2\,mgr$

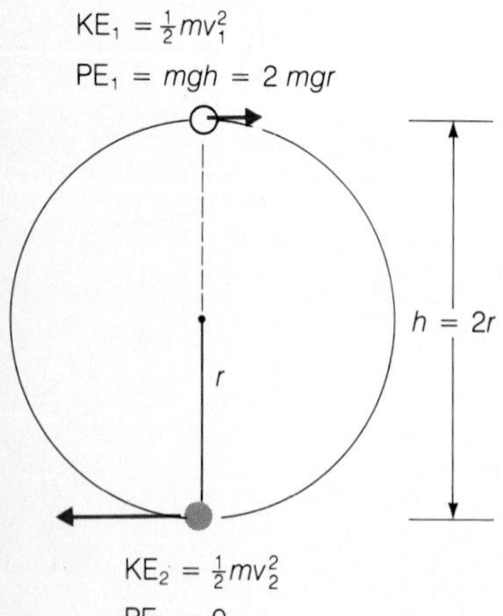

$KE_2 = \frac{1}{2}mv_2^2$

$PE_2 = 0$

When an object is whirled in a vertical circle at the end of a string, how does the tension in the string vary with the object's position?

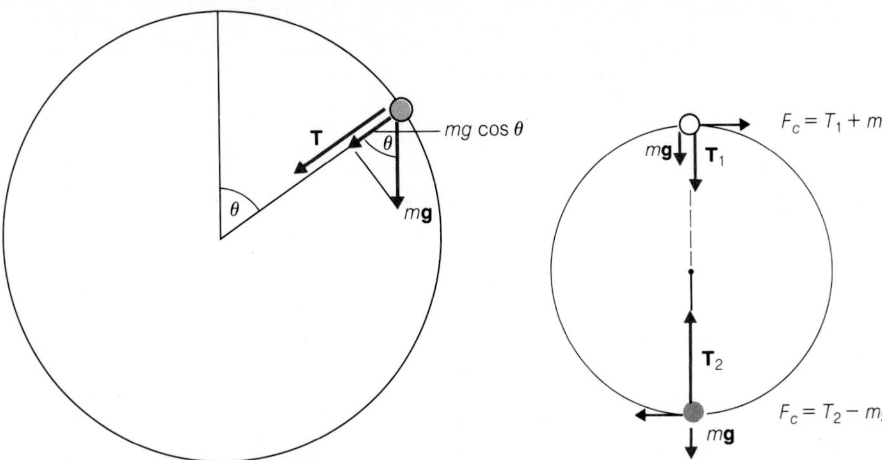

Figure 6.13

The centripetal force on an object moving in a vertical circle is the sum of the tension T in the string and the component $mg \cos \theta$ of the object's weight toward the center of the circle.

Figure 6.14

For a given speed v, the tension in the string is least at the top of the circle, most at the bottom.

The centripetal force $\mathbf{F}_c$ on the object at any point is the vector sum of the tension $\mathbf{T}$ in the string and the component of the object's weight $m\mathbf{g}$ toward the center of the circle. As we can see from Fig. 6.13, when the string is at the angle θ with the vertical,

$$F_c = T + mg \cos \theta$$

and so

$$T = F_c - mg \cos \theta \qquad \textit{String tension} \quad (6.5)$$

At the top of the circle, $\theta = 0$ and $\cos \theta = 1$, so

$$T_1 = \frac{mv_1^2}{r} - mg \qquad \textit{Top of circle} \quad (6.6)$$

At the bottom of the circle, $\theta = 180°$ and $\cos \theta = -1$, so

$$T_2 = \frac{mv_2^2}{r} + mg \qquad \textit{Bottom of circle} \quad (6.7)$$

For a constant speed v, the tension in the string is least at the top of the circle and most at the bottom (Fig. 6.14).

EXAMPLE 6.5

An airplane pulls out of a dive in a circular arc whose radius is 1000 m. The speed of the airplane is a constant 200 m/s. Find the force with which the 80-kg pilot presses down on his seat at the bottom of the arc.

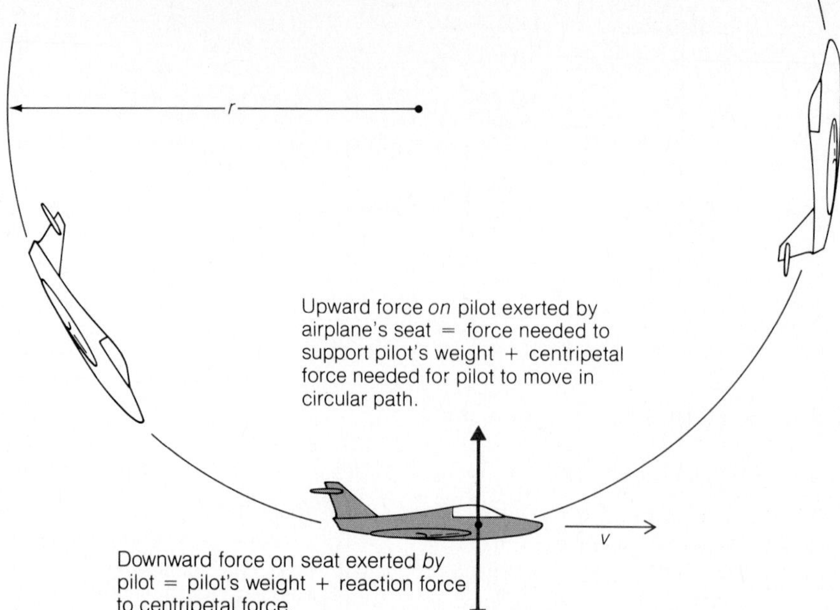

Figure 6.15

Forces exerted by and on the pilot of an airplane making a vertical circle at constant speed.

SOLUTION The downward force **F** the pilot exerts on his seat is the reaction to the upward force of the seat on him that both supports his weight and keeps him in his circular path (Fig. 6.15). The force of the seat on the pilot plays the same role the tension in the string does for an object being whirled in a vertical circle. Hence Eq. (6.7) applies here with T_2 corresponding to F, and we have

$$F = \frac{mv^2}{r} + mg = \frac{(80\text{ kg})(200\text{ m/s})^2}{1000\text{ m}} + (80\text{ kg})(9.8\text{ m/s}^2) = 3984\text{ N} = 4.0\text{ kN}$$

The pilot presses down on his seat with a force of 3984 N, more than five times his weight.

Because there has been no compensating increase in his muscular strength, the pilot may be unable to move his arms and legs in order to control the airplane. A further complication is the tendency of the pilot's blood to leave his head because of inertia, leaving him with impaired vision ("blacked out") and perhaps unconscious. Special pressure suits have been devised that prevent disturbances in blood supply and in the positions of internal organs during severe accelerations in flight.

The forces exerted by and on a pilot or astronaut vary with the individual's mass. Hence it is often convenient to speak instead of the acceleration, by custom in units of g. Here the pilot's acceleration is

$$a = \frac{v^2}{r} = \frac{(200\text{ m/s})^2}{1000\text{ m}} = 40\text{ m/s}^2 = 4.1\,g$$

◆

Critical Speed

Anyone who has whirled a ball on a string knows that, if the ball's speed at the top is less than a certain critical value v_0, the string goes slack and the upper part of the ball's path flattens out. To find this critical speed we set the string tension $T_1 = 0$ in Eq. (6.6), with the result

$$T_1 = \frac{mv_0^2}{r} - mg = 0$$

$$v_0 = \sqrt{rg} \qquad \textit{Critical speed} \quad (6.8)$$

For a ball with an 80-cm string, as in the first example in this section,

$$v_0 = \sqrt{(0.8.0 \text{ m})(9.8 \text{ m/s}^2)} = 2.8 \text{ m/s}$$

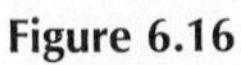

EXAMPLE 6.6

A sled starts from rest and slides down the frictionless track shown in Fig. 6.16 to loop the loop without falling off. If $r = 10$ m, what is the minimum value of h?

SOLUTION In the loop the force exerted on the sled by the track takes the place of tension in a string. Equation (6.8) therefore applies here for the minimum speed at the top of the loop, so $v_0 = rg$, and the sled's kinetic energy there is

$$\text{KE} = \tfrac{1}{2}mv_0^2 = \tfrac{1}{2}mrg$$

At the top of the track the sled is $h - 2r$ above the top of the loop, and its potential energy there relative to the top of the loop is

$$\text{PE} = mg(h - 2r)$$

Because there is no friction,

$$\text{PE} = \text{KE}$$

$$mg(h - 2r) = \tfrac{1}{2}mrg$$

$$h = \frac{r}{2} + 2r = \frac{5}{2}r$$

With $r = 10$ m, $h = 25$ m.

Figure 6.16

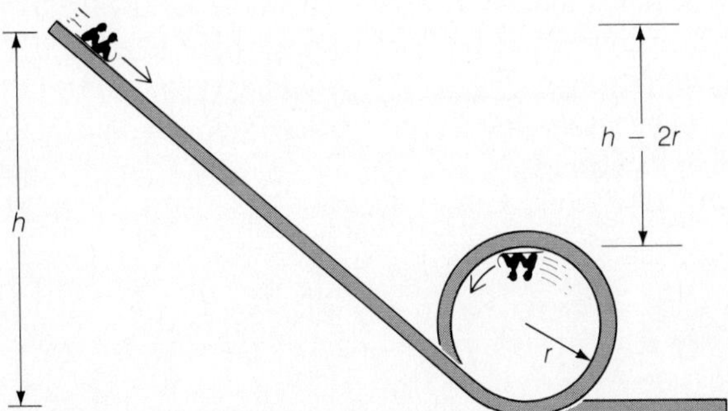

◆

If there is no friction or air resistance, a sled that starts from a height 2.5 times the radius of a vertical loop above the bottom of the loop will not leave the track at the top of the loop.

6.5 Gravitation

A fundamental force.

The earth and the other planets have approximately circular orbits around the sun. We conclude that the planets are being acted on by centripetal forces that act toward the sun. This much was generally understood by the middle of the seventeenth century, when Newton turned his mind to the question of exactly what the nature of the centripetal forces was.

Newton proposed that the inward force exerted by the sun that is responsible for the planetary orbits is one example of a universal interaction, called **gravitation,** that occurs between all objects in the universe by virtue of their possession of mass. Another example of gravitation, according to Newton, is the attraction of the earth for nearby objects. Thus the centripetal acceleration of the moon and the downward acceleration g of objects dropped near the earth's surface are both due to the gravitational pull of the earth.

Newton was able to arrive at the form of the **law of universal gravitation** from an analysis of the motions of the planets about the sun (Section 6.9):

Every object in the universe attracts every other object with a force directly proportional to the product of their masses and inversely proportional to the square of the distance separating them.

The law of gravitation in equation form is

$$F_{\text{grav}} = G\frac{m_A m_B}{r^2} \qquad \textit{Gravitational force} \quad (6.9)$$

where m_A and m_B are the masses of any two objects and r is the distance between them. The quantity G is a universal constant whose value is

$$G = 6.67 \times 10^{-11} \frac{\text{N}\cdot\text{m}^2}{\text{kg}^2} \qquad \textit{Gravitational constant}$$

The direction of the gravitational force is always along a line joining the two objects A and B. The force on A exerted by B is equal in magnitude to that on B exerted by A but is in the opposite direction (Fig. 6.17). A homogeneous spherical object, or one composed of homogeneous spherical shells, behaves gravitationally as if all its mass were concentrated at its center.

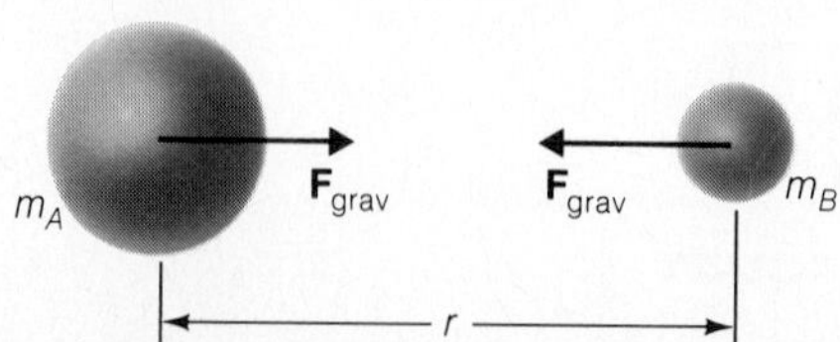

Figure 6.17
The gravitational forces between two spherical objects.

EXAMPLE 6.7

A grocer puts a 100-kg lead block under the pan of his scale. By how much does this increase the reading of the scale when 1 kg of onions are on the pan, if the centers of mass of the lead and of the onions are 0.3 m apart?

SOLUTION The gravitational force of the lead on the onions is

$$F = G\frac{m_A m_B}{r^2} = \left(6.67 \times 10^{-11} \frac{\text{N}\cdot\text{m}^2}{\text{kg}^2}\right) \frac{(100\text{ kg})(1\text{ kg})}{(0.3\text{ m})^2} = 7.4 \times 10^{-8}\text{ N}$$

The increase in the scale reading is therefore

$$m = \frac{F}{g} = \frac{7.4 \times 10^{-8}\text{ N}}{9.8\text{ m/s}^2} = 7.6 \times 10^{-9}\text{ kg} = 0.0000076\text{ g}$$

so it is hardly worth the effort. Blowing gently on the onions will increase the reading over a million times more. ◆

It is impossible to determine the gravitational constant G from astronomical data alone, as Newton realized. A direct measurement of the gravitational force between two known masses a known distance apart is needed. The difficulty here, as the foregoing example illustrates, is that gravitational forces are extremely small between objects of laboratory size. The value of G was finally established in 1798, more than a century after Newton's work, by Henry Cavendish. He used an instrument called a *torsion balance* (Fig. 6.18), which is the rotational analog of an ordinary spring balance. The forces exerted on the small spheres in Cavendish's experiment could be found from the resulting twist in the fine suspending thread.

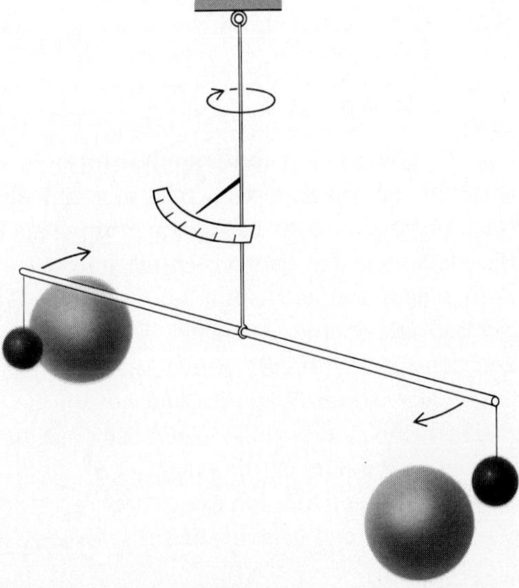

Figure 6.18

Torsion balance for measuring gravitational forces.

BIOGRAPHICAL NOTE

Henry Cavendish

Henry Cavendish (1731–1810) was born to an aristocratic English family—both his grandfathers were dukes—and inherited a large fortune. Educated at Cambridge, he thereafter avoided contact with people, especially women (to whom he would never speak). When he died at 79, he was alone, at his insistence.

Cavendish cared only for science and carried out a huge amount of both experimental and theoretical research at his home in London. Curiosity alone motivated Cavendish, and he was not concerned about whether others learned about his results. Many of Cavendish's findings came to light only long after his death. This was especially true of his work in electricity, which included deriving the law of electrical force from the observation that there is no electric field inside a charged conductor. This law was independently established and made known some years later by the French physicist Charles Coulomb, and it has been called Coulomb's law ever since.

Cavendish did publish a few papers. In 1776 he announced the discovery of hydrogen and listed its chief properties, and in 1784 he reported that burning hydrogen produced water. Other important research in chemistry remained buried in his notebooks. Cavendish is best known for the 1798 paper on his measurement of the gravitational constant G using the torsion balance invented by his contemporary John Mitchell. So careful was Cavendish that his figure was not improved upon for a century. Knowing G enabled Cavendish to calculate the mass of the earth, the first time this had ever been done.

What Is Gravity?

Gravity is a **fundamental force** in the sense that it cannot be explained in terms of any other force. Only four fundamental forces are known: gravitational, electromagnetic, weak nuclear, and strong nuclear. These forces seem to be responsible for everything that happens in the universe. The weak and strong forces have very short ranges and act within atomic nuclei. The weak force is closely related to the electromagnetic force, and the strong force seems to be related to both of them. Electromagnetic forces, which (like gravity) are unlimited in range, act between electrically charged particles and govern the structures and behavior of atoms, molecules, solids, and liquids. When one object touches another, an electromagnetic force is what each exerts on the other.

Gravitational forces act between all bodies in the universe and hold together planets, stars, and galaxies of stars. Exactly how gravity, the weakest of the four, is related to the other fundamental forces is not clear at present, although some hints are beginning to emerge.

How can we be so sure that the law of gravitation, which was obtained from data on the solar system, holds everywhere else in the universe as well? There is no simple answer to this legitimate query. What we must do is look at the large body of knowledge that bears on the subject. For instance, we observe that all the matter near the earth's surface has the same acceleration in free fall, which suggests the same gravitational behavior. Spectroscopic analysis (see Chapter 27) of the light and radio waves that reach us from space shows that the matter of the rest of the universe acts the same as the matter found on the earth. Cosmic ray particles from far away in our galaxy and perhaps beyond seem identical with the matter around us. And so on. Nowhere do we find reason to suspect that any objects in the universe do not obey the law of gravitation, and it would not make sense to propose that such objects exist with no reason to do so.

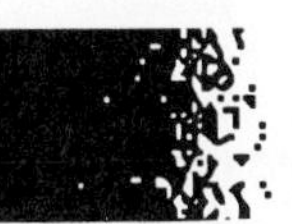

6.6 Gravity and the Earth

Why the earth is round.

The earth is round because gravitational forces squeeze it into this shape. If a part of the earth were to stick out very far, the pull of the rest would lead to strong pressures on the material underneath. This material would then flow out sideways until the bump became level or nearly so. The downward forces around the rim of a large hole would similarly cause the underlying material to flow into it. The earth's mountains and ocean basins are actually very small-scale irregularities—the total range from the Pacific depths to the summit of Mount Everest is less than 20 km, not much compared with the earth's radius of 6400 km.

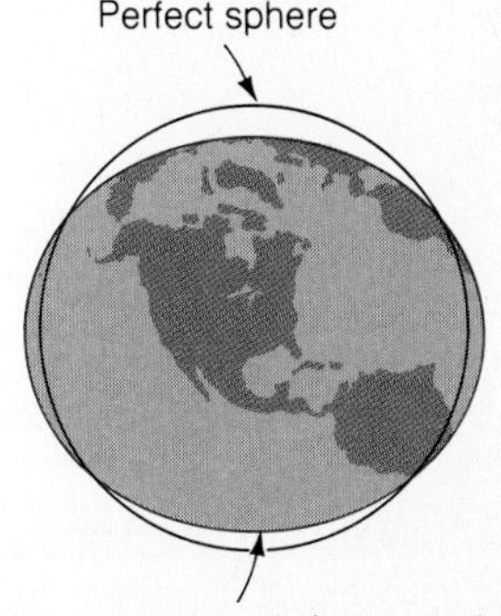

Figure 6.19

The earth is an oblate spheroid in shape because of its rotation, with a difference of 43 km between its polar and equatorial diameters.

The smaller an object is, the more likely its rigidity will be able to withstand the tendency of gravity to impose a spherical form. Thus the two satellites of Mars have been able to stay oblong in shape because their longest dimensions are, respectively, only 23 and 11 km.

Because the earth is rotating, its equatorial region bulges outward, just as a ball on a string swings outward when whirled around (Fig. 6.19). The earth is about 0.34% away from being a perfect sphere (apart from surface irregularities). Venus, whose "day" is 243 of our days, turns so slowly that its distortion is negligible. Saturn, at the other extreme, spins so rapidly that it is out of round by 9.6%.

Given the values of G, g, and the earth's radius r_e, we can calculate the earth's mass. Let us consider an object of mass m on the earth's surface, say an apple. The gravitational pull of the earth on the apple is the apple's weight of $w = mg$. As mentioned earlier, a spherical object behaves gravitationally as though its mass were concentrated at its center. Thus the earth-apple system can be represented by two particles of masses M and m a distance r_e apart, where M is the earth's mass and r_e is its radius (Fig. 6.20). According to Newton's law of gravitation, the force the earth exerts on the apple is

$$F = G\frac{Mm}{r_e^2}$$

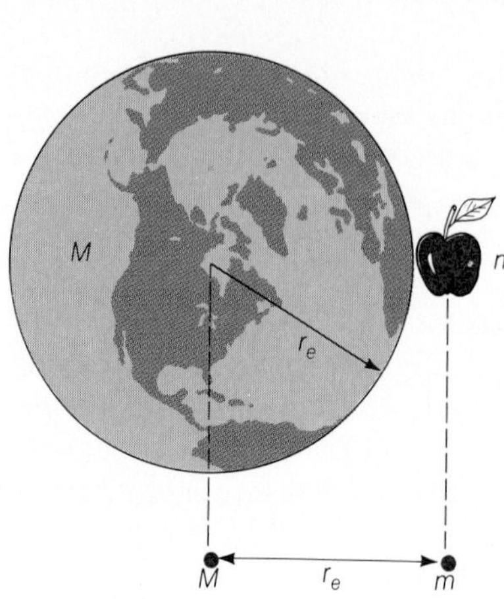

Figure 6.20

The mass M of the earth can be calculated from a knowledge of G, g, and r_e.

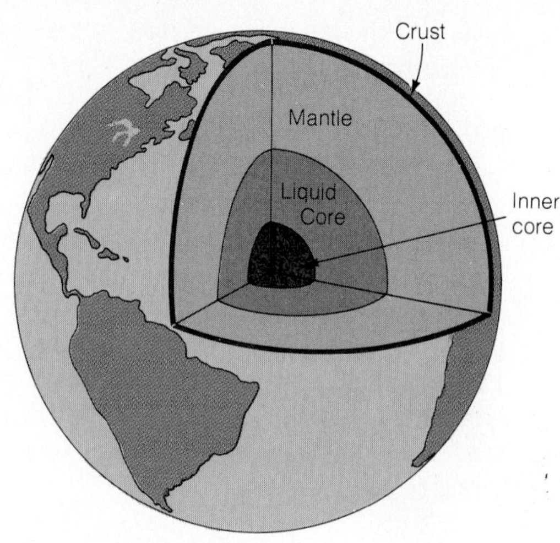

Figure 6.21

Structure of the earth. The mantle is composed of dense rock, the core probably of molten iron with a solid inner core.

This force must equal the apple's weight $w = mg$, and so

$$G\frac{Mm}{r_e^2} = mg$$

$$M = \frac{gr_e^2}{G} \tag{6.10}$$

The apple's mass has dropped out of the calculation, as of course it must. The mass of the earth is

$$M = \frac{gr_e^2}{G} = \frac{(9.8\ \text{m/s}^2)(6.4 \times 10^6\ \text{m})^2}{6.7 \times 10^{-11}\ \text{N}\cdot\text{m}^2/\text{kg}^2} = 6.0 \times 10^{24}\ \text{kg}$$

Astronauts in the *Apollo 11* spacecraft saw this view of the earth as they orbited the moon, part of whose bleak landscape appears in the foreground. Gravity is responsible for the roundness of the earth and other large astronomical bodies.

Figure 6.21 shows the structure of the earth on the basis of indirect, but persuasive, evidence from several lines of inquiry. The outer skin is a relatively thin **crust** of rock about 5 km thick under the oceans and an average of 35 km thick under the continents. The **mantle**, about 2900 km thick, consists of dense rock probably similar in composition to certain surface rocks and to stony meteorites. Because the materials of the mantle and crust are relatively light, they provide only 67% of the earth's mass although they constitute 80% of its volume, and so the **core** must be very dense to make up the rest of the mass. As will be discussed in Section 12.4, earthquake wave studies show that the core must be a liquid with a solid inner core, and the existence of the earth's magnetic field means that the liquid must be a metal in order to support the required electric currents. (The earth's interior is too hot for permanent magnetism to occur there.) Iron, which is abundant in the universe generally, seems to meet all the requirements, and geologists agree that the earth's core consists chiefly of molten iron.

EXAMPLE 6.8

The moon is 3.84×10^8 m from the earth and circles the earth once every 27.3 days (Fig. 6.22). Compare the moon's centripetal acceleration with the acceleration it would experience on the basis of Newton's law of gravitation.

SOLUTION The moon's orbital period is $T = (27.3 \text{ d})(86{,}400 \text{ s/d}) = 2.36 \times 10^6$ s. From Eq. (6.2) the corresponding orbital speed is

$$v = \frac{2\pi r}{T} = \frac{(2\pi)(3.84 \times 10^8 \text{ m})}{2.36 \times 10^6 \text{ s}} = 1.02 \times 10^3 \text{ m/s}$$

The centripetal acceleration of the moon is accordingly

$$a_c = \frac{v^2}{r} = \frac{(1.02 \times 10^3 \text{ m/s})^2}{3.84 \times 10^8 \text{ m}} = 2.7 \times 10^{-3} \text{ m/s}^2$$

and is directed toward the center of the earth.

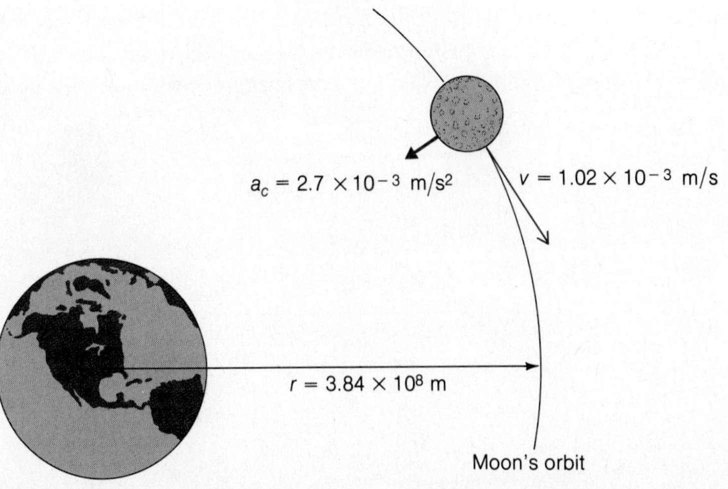

Figure 6.22
The moon is accelerated toward the center of the earth.

The ratio between the moon's orbital radius and the earth's radius is

$$\frac{r}{r_e} = \frac{3.84 \times 10^8 \text{ m}}{6.4 \times 10^6 \text{ m}} = 60$$

From Eq. (6.9) the gravitational force the earth exerts on the moon ought to be $(1/60)^2 = 1/3600$ as strong as the force the earth would exert on it if the moon were at the earth's surface. The acceleration a of the moon toward the earth, in turn, ought to be 1/3600 of the acceleration of an object at the earth's surface. Since the latter acceleration is g,

$$a = \frac{g}{3600} = \frac{9.8 \text{ m/s}^2}{3600} = 2.7 \times 10^{-3} \text{ m/s}^2$$

In fact, this is the same as a_c, the moon's observed centripetal acceleration. Newton used the agreement between the two figures as evidence for the universal validity of his law of gravitation. We note that a knowledge of G, which Newton did not have, was not needed. ◆

Acceleration of Gravity

According to Eq. (6.10) the acceleration of gravity is given by

$$g = \frac{GM}{r_e^2} \qquad \textit{Acceleration of gravity} \quad (6.11)$$

Figure 6.23 shows how the weight mg of a 100-kg astronaut decreases with distance from the earth's center. At the surface he weighs 980 N. When he is 100 times farther from the center, his weight is 10^{-4} as great, only 0.098 N—the weight of a cigar at the earth's surface. The astronaut's mass of 100 kg, of course, is the same everywhere in the universe.

Chapter 1 mentioned that g varies around the earth, from 9.78 m/s^2 at the equator to 9.83 m/s^2 at the poles. Because the earth bulges at the equator (see Fig. 6.19), the closer we get to it, the farther we are from the earth's center. This accounts for 0.02 m/s^2 of the maximum difference of 0.05 m/s^2. The rest follows from the earth's turning around its axis. The centripetal acceleration of a point on the earth's

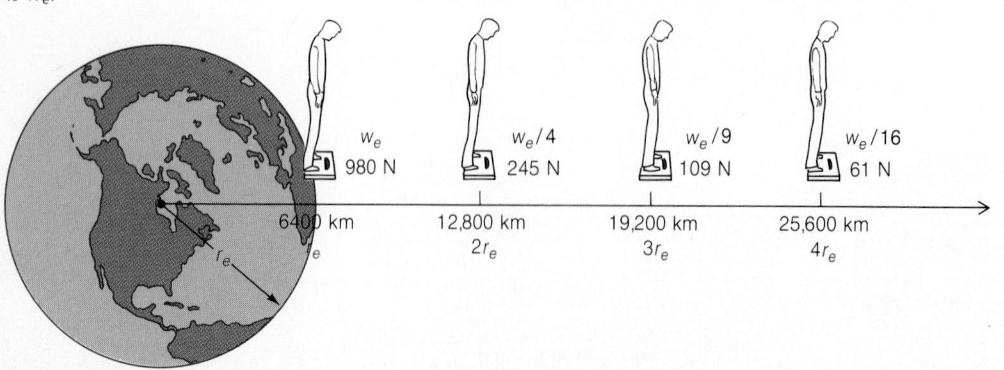

Figure 6.23

The gravitational force of the earth on an object varies inversely with the square of the object's distance from the center of the earth. Shown here is how the weight of a 100-kg person decreases with distance; the person's weight at the earth's surface is w_e.

surface increases from 0 at the poles to 0.03 m/s^2 at the equator. The earth's gravitational pull provides this acceleration, and the acceleration of a falling body relative to the surface is reduced accordingly.

6.7 Earth Satellites

Why they don't fall down.

The first artificial satellite, *Sputnik I*, was launched in 1957 by the Soviet Union. Since then thousands of others have been put into orbits around the earth, most of them by the United States and the Soviet Union. Men and women have been in orbit regularly since 1961, when a Russian cosmonaut circled the globe at an altitude of about 160 km. The first American went into orbit the following year.

What keeps an artificial earth satellite from falling down? The answer, of course, is that it *is* falling down, but, like the moon, at just such a rate as to circle the earth in a stable orbit. Let us use what we know about gravitation and circular motion to investigate the orbits of earth satellites. In the following discussion we will neglect the frictional resistance of the atmosphere, which ultimately brings down all artificial satellites.

Near the earth the gravitational force on an object is its weight mg. For uniform circular motion around the earth, this force must provide the object with the centripetal force mv^2/r. The condition for a stable orbit is therefore $mg = mv^2/r$, so that

$$v = \sqrt{rg} \qquad \textit{Satellite speed} \qquad (6.12)$$

If we let g_0 be the value of g at the earth's surface, where $r = r_e$, then from Eq. (6.11) the value of g at any radius r is

$$g = \left(\frac{r_e}{r}\right)^2 g_0 \qquad \textit{Acceleration of gravity at radius } r \qquad (6.13)$$

From Eqs. (6.12) and (6.13) the satellite speed at the radius r is

$$v = \sqrt{rg} = \sqrt{\frac{r_e^2 g_0}{r}} \qquad \textit{Orbit of earth satellite} \qquad (6.14)$$

For a circular orbit just above the earth's surface,

$$v_0 = \sqrt{r_e g_0} = \sqrt{(6.4 \times 10^6 \text{ m})(9.8 \text{ m/s}^2)} = 7.9 \times 10^3 \text{ m/s}$$

which is about 28,400 km/h. Anything sent off tangentially to the earth's surface at this speed will become a satellite of the earth. If it is sent off at a higher speed, its orbit will be elliptical rather than circular (Fig. 6.24).

If the object's speed is great enough, it can escape permanently from the earth. Readers of *Through the Looking Glass* may recall the Red Queen's remark, "Now, here, you see, it takes all the running you can do to stay in the same place.

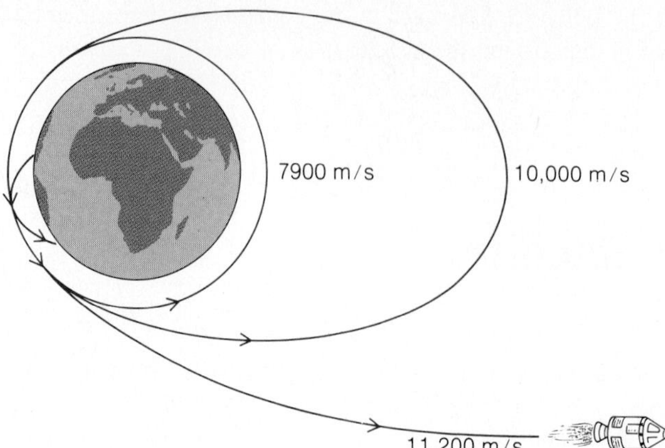

Figure 6.24

Depending on its speed, an object projected horizontally above the earth's surface may fall back to the earth, revolve around the earth in a circular orbit, revolve around the earth in an elliptical orbit, or escape permanently from the earth into space.

If you want to get somewhere else, you must run at least twice as fast as that!'' The correct ratio is actually $\sqrt{2}$, so that the **escape speed** is $\sqrt{2}v_0 = 11.2 \times 10^3$ m/s in the case of the earth.

A satellite can be placed in a circular orbit of any radius if it is provided with a small rocket motor to give it the required impulse at the right altitude. More than 3000 satellites are now in orbits that range from about 160 km to 35,900 km above the earth. The closer satellites are used to keep watch on the earth's surface, both for military purposes and for such scientific ones as keeping track of the weather and studying earth resources. Five satellites at an altitude of 1230 km are used in the highly accurate Transit system of marine navigation.

Some satellites circle the equator with a period of exactly 1 day so that they remain indefinitely over a particular location. A satellite in such a **geostationary orbit** can ''see'' about a third of the earth's surface. Nearly 200 of the satellites now in geostationary orbits are used to relay radio communications from one point to another, notably for intercontinental telephone calls and television broadcasts. Such satellites, as well as most of the others, obtain their electrical power from sunlight with the help of photoelectric cells.

EXAMPLE 6.9

Find the altitude of a geostationary orbit.

SOLUTION According to Eq. (6.2) the speed of an object in a circular orbit of period T is

$$v = \frac{2\pi r}{T}$$

where here $T = 1$ day $= 86{,}400$ s. Another formula for the satellite speed is given by Eq. (6.14),

$$v = \sqrt{\frac{r_e^2 g_0}{r}}$$

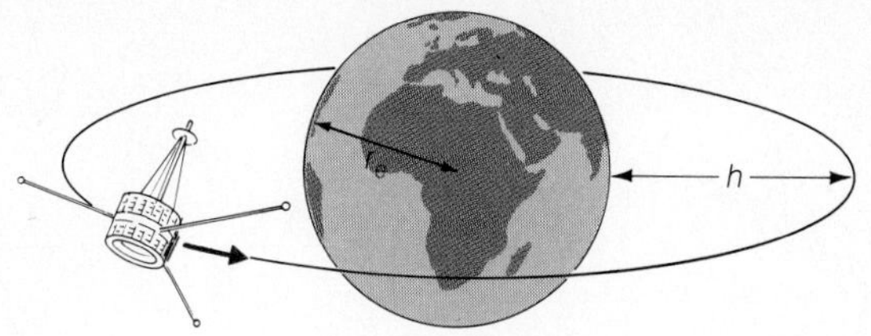

Figure 6.25
(Not to scale.)

Setting the two formulas equal enables us to eliminate v:

$$\frac{2\pi r}{T} = \sqrt{\frac{r_e^2 g_0}{r}}$$

$$r = \sqrt[3]{\frac{r_e^2 g_0 T^2}{4\pi^2}} = \sqrt[3]{\frac{(6.4 \times 10^6 \text{ m})^2(9.8 \text{ m/s}^2)(8.64 \times 10^4 \text{s})^2}{4\pi^2}}$$

$$= \sqrt[3]{759 \times 10^{20}} \text{ m} = \sqrt[3]{75.9 \times 10^{21}} \text{ m} = 4.23 \times 10^7 \text{ m}$$

The corresponding altitude h (Fig. 6.25) above the earth's surface is

$$h = r - r_e = 42.3 \times 10^6 \text{ m} - 6.4 \times 10^6 \text{ m} = 35.9 \times 10^6 \text{ m}$$

which is about 22,000 miles. ♦

6.8 Weightlessness

How to lose weight without dieting.

Because an earth satellite is always falling toward the earth, an astronaut inside one feels "weightless." In reality, there *is* a gravitational force acting on him. What is missing to his senses is the upward reaction force provided by a stationary platform underneath him—the seat of a chair, the floor of a room, the ground itself. Instead of pushing back, the floor of the satellite falls just as fast as he does toward the earth.

It is useful to distinguish between the *actual weight* of an object, which is the gravitational force acting on it, and its *apparent weight,* which is the force it exerts on whatever it rests upon. We can think of the apparent weight of a person as the reading on a bathroom spring scale the person is standing on. Astronauts in an earth satellite have no apparent weight because they do not press down on the floor of the satellite.

A person who jumps off a diving board is just as "weightless" in her fall as an astronaut, since nothing restricts her acceleration toward the earth either. But a person standing on the ground is acted on by *both* the downward force of gravity and the upward reaction force of the ground. The latter force is what prevents her from simply dropping all the way down to the center of the earth. The human body (indeed, all living things on the earth) evolved in the presence of both these forces, and various

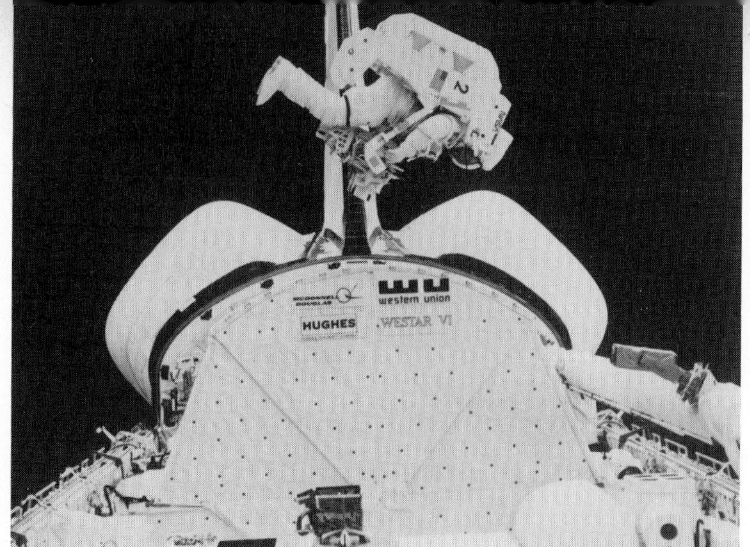

The earth exerts a gravitational force on an astronaut in orbit. The astronaut feels "weightless" because no reaction force acts on him or her.

body functions, such as blood circulation, do not seem to take place efficiently in a "weightless" state. Future spacecraft designed for long journeys may be set in rotation so that inertia will cause astronauts to press against the cabin sides, which will then press back (action-reaction again) and so bring about a situation corresponding to that on the earth's surface.

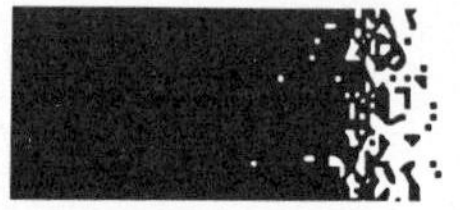

6.9 Kepler's Laws and Gravitation

How the planets move.

Until the sixteenth century the earth was generally believed to be the center of the universe. According to the **Ptolemaic system,** which was a detailed picture based on this idea, the sun and moon revolve around the earth in circular orbits, while each planet moves in a small circle whose center follows a larger circle around the earth. The stars were supposed to be fixed in a crystal sphere that turns once a day.

Another view of the heavens was put forward by Nicolaus Copernicus (1473–1543), a Polish scientist, who proposed instead that the earth and the other planets revolve around the sun, while the moon revolves around the earth. The stars are far away in space, and the earth rotates daily on its axis. Copernicus supported his hypothesis with detailed calculations based on circular orbits in which the planets and moon move at constant speeds.

On the basis of measurements he and others (notably the Dane Tycho Brahe) had made, the German astronomer Johannes Kepler (1571–1630) further developed the Copernican system. He showed that the planetary motions have these regularities, which have become known as Kepler's laws:

1. Each planet has an elliptical orbit with the sun at one focus (Fig. 6.26).
2. Each planet moves so that a radius vector from the sun to it sweeps out equal areas in equal times (Fig. 6.27).
3. The ratio between the square of a planet's period of revolution and the cube of its average distance from the sun has the same value for all the planets (Fig. 6.28).

Newton used these laws to arrive at his law of universal gravitation. Let us see how the formula $F_{\text{grav}} = Gm_Am_B/r^2$ follows from Kepler's third law.

Figure 6.26

How to draw an ellipse. The longer the string relative to the distance between the tacks, the more nearly circular is the ellipse. The points in an ellipse that correspond to the positions of the tacks are called foci. The orbits of the planets are ellipses with the sun at one focus.

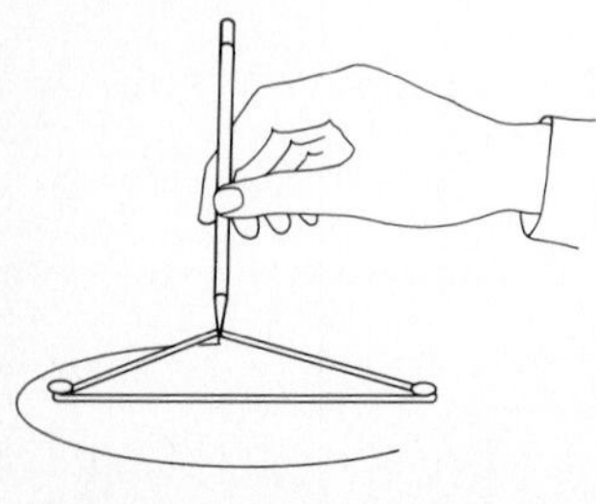

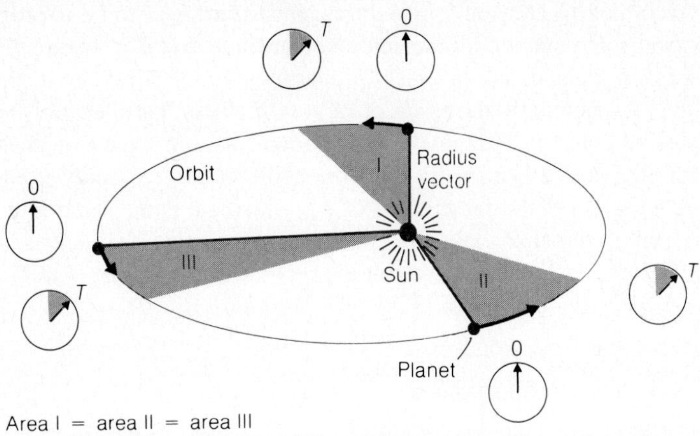

Figure 6.27

Kepler's second law. As a planet goes around the sun in its elliptical orbit, its radius vector from the sun sweeps out equal areas in equal times.

For simplicity we assume that the planets move in circular orbits around a stationary sun. A planet of mass m_A, orbital radius r, and speed v must be acted on by the centripetal force

$$F_c = \frac{m_A v^2}{r}$$

If the period of the orbit is T, then, as we know, $v = 2\pi r/T$ and

$$F_c = \frac{4\pi^2 m_A r}{T^2}$$

Kepler's third law states that

$$\frac{T^2}{r^3} = K$$

where K has the same value for all the planets. Hence $T^2 = Kr^3$ and

$$F_c = \frac{4\pi^2 m_A}{Kr^2}$$

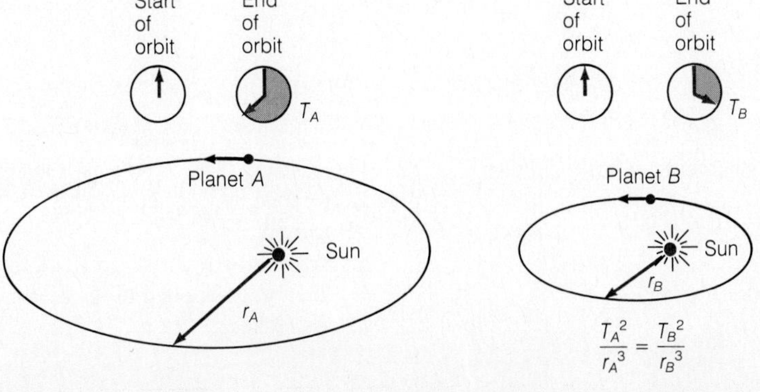

Figure 6.28

Kepler's third law. The ratio T^2/r^3 is the same for all the planets, where T is the period of a planet's revolution around the sun and r is the average radius of its orbit.

According to Newton's hypothesis, this centripetal force is provided by the gravitational force exerted by the sun. We conclude that F_{grav} is directly proportional to the mass of a planet and inversely proportional to the square of its distance from the sun.

Newton's third law of motion requires that the force a planet exerts on the sun be equal in magnitude to the force the sun exerts on the same planet. If the formula above is correct, then we should be able to apply it either way for a given planet and get the same value of F_{grav}. Since r is the same in both cases, F_{grav} must be proportional to *both* the planet's mass m_A and the sun's mass m_B. Because the sun's mass is constant, we can express the quantity $4\pi^2/K$ as Gm_B, so that

$$F_{grav} = G\frac{m_A m_B}{r^2}$$

where G is a universal constant. Extending this formula to *any* two bodies in the universe gives Newton's law of gravitation.

Important Terms

An object traveling in a circle at constant speed is said to be undergoing **uniform circular motion.**

The velocity of an object in circular motion continually changes in direction, although its magnitude may remain constant. The acceleration that causes the object's velocity to change is called **centripetal acceleration,** and it points toward the center of the object's circular path.

The inward force that provides an object in circular motion with its centripetal acceleration is called **centripetal force.**

Newton's **law of universal gravitation** states that every object in the universe attracts every other object with a force directly proportional to both their masses and inversely proportional to the square of the distance separating them.

Kepler's laws describe how the planets move around the sun.

Important Formulas

Period of orbit: $T = \frac{2\pi r}{v}$

Centripetal acceleration: $a_c = \frac{v^2}{r}$

Centripetal force: $F_c = \frac{mv^2}{r}$

Law of gravitation: $F_{grav} = G\frac{m_A m_B}{r^2}$

Acceleration of gravity: $g = \left(\frac{r_e}{r}\right)^2 g_0$

Satellite orbit: $v = \sqrt{rg} = \sqrt{\frac{r^2{}_e g_0}{r}}$

MULTIPLE CHOICE

1. The acceleration of an object undergoing uniform circular motion is constant in

- **a.** magnitude only
- **b.** direction only
- **c.** both magnitude and direction
- **d.** neither magnitude nor direction

2. The centripetal force on a car rounding a curve on a level road is provided by

- **a.** gravity
- **b.** friction between its tires and the road
- **c.** the torque applied to its steering wheel
- **d.** its brakes

3. The radius of the path of an object in uniform circular motion is doubled. The centripetal force needed if its speed remains the same is
   a. half as great as before
   b. the same as before
   c. twice as great as before
   d. four times as great as before

4. The centripetal force needed to keep the earth in orbit is provided by
   a. inertia
   b. its rotation on its axis
   c. the gravitational pull of the sun
   d. the gravitational pull of the moon

5. Newton's law of gravitation cannot explain
   a. Kepler's laws
   b. the size of the earth
   c. the shape of the earth
   d. the motion of the moon

6. The shape of the earth is closest to that of
   a. a perfect sphere b. an egg
   c. a football d. a grapefruit

7. The reasons why the earth's core is believed to be molten iron do not include iron's
   a. ability to conduct electric current
   b. ability to be permanently magnetized
   c. high density
   d. relative abundance in the universe

8. The gravitational acceleration of an object
   a. has the same value everywhere in space
   b. has the same value everywhere on the earth's surface
   c. varies somewhat over the earth's surface
   d. is greater on the moon because of its smaller diameter

9. A hole is drilled to the center of the earth and a stone is dropped into it. When the stone is at the earth's center, compared with the values at the earth's surface
   a. its mass and weight are both unchanged
   b. its mass and weight are both zero
   c. its mass is unchanged and its weight is zero
   d. its mass is zero and its weight is unchanged

10. If the earth were three times farther from the sun than it is now, the gravitational force exerted on it by the sun would be
   a. three times as large as it is now
   b. nine times as large as it is now
   c. one-third as large as it is now
   d. one-ninth as large as it is now

11. The moon's mass is 1.2% of the earth's mass. Relative to the gravitational force the earth exerts on the moon, the gravitational force the moon exerts on the earth
   a. is smaller
   b. is the same
   c. is greater
   d. depends on the phase of the moon

12. The speed needed to put a satellite in orbit does not depend on
   a. the radius of the orbit
   b. the shape of the orbit
   c. the value of g at the orbit
   d. the mass of the satellite

13. An astronaut is "weightless" in a spacecraft
   a. as it takes off from the earth
   b. in certain orbits only
   c. in all orbits
   d. only when the escape speed is exceeded

14. According to Kepler's third law, the time needed for a planet to go around the sun
   a. depends on its mass
   b. depends on the average radius of its orbit
   c. depends on its speed of rotation
   d. is the same for all the planets

15. The speed of a planet in its elliptical orbit around the sun
   a. is constant
   b. is highest when it is closest to the sun
   c. is lowest when it is closest to the sun
   d. varies, but not with respect to its distance from the sun

16. A 0.50-kg ball moves in a circle 40 cm in radius at a speed of 4.0 m/s. Its centripetal acceleration is
   a. 10 m/s^2 b. 20 m/s^2
   c. 40 m/s^2 d. 80 m/s^2

17. The centripetal force on the ball in Multiple Choice 16 is
   a. 10 N b. 20 N
   c. 40 N d. 80 N

18. A toy cart at the end of a string 0.70 m long moves in a circle on a table. The cart has a mass of 2.0 kg, and the string has a breaking strength of 40 N. The maximum speed of the cart is approximately
   a. 1.9 m/s b. 3.7 m/s
   c. 11.7 m/s d. 16.7 m/s

19. On a rainy day the coefficient of friction between a car's tires and a certain level road surface is reduced to half its usual value. The maximum safe speed for rounding the curve is
   a. unchanged
   b. reduced to 25% of its usual value
   c. reduced to 50% of its usual value
   d. reduced to 71% of its usual value

20. A car is traveling at 50 km/h on a road such that the coefficient of friction between its tires and the road is 0.50. The minimum turning radius of the car is
   a. 9.9 m b. 39 m
   c. 0.95 km d. 5.1 km

21. A 2.0-kg stone at the end of a string 1.0 m long is whirled in a vertical circle. The tension in the string is 52 N when the stone is at the bottom of the circle. The stone's speed then is

a. 4 m/s b. 5 m/s
c. 6 m/s d. 7 m/s

22. A car traveling at 20 m/s just leaves the road at the top of a hump. The radius of curvature of the hump is

a. 20 m b. 41 m
c. 3.9 km
d. impossible to determine without knowing the car's mass

23. A woman whose mass is 60 kg on the earth's surface is in a spacecraft at a height of twice the earth's radius (that is, 2 earth radii) above the earth's surface. Her mass there is

a. 6.7 kg b. 15 kg
c. 20 kg d. 60 kg

24. A man whose mass is 80 kg on the earth's surface is also in the spacecraft of Multiple Choice 23. His weight there is

a. 87 N b. 196 N
c. 261 N d. 784 N

25. What would a woman who weighs 600 N on the earth weigh on a planet that has the same mass as the earth but half its radius?

a. 150 N b. 300 N
c. 1200 N d. 2400 N

26. A 200-kg satellite circles the earth in an orbit 7.0×10^6 m in radius. At this altitude $g = 8.2$ m/s^2. The speed of the satellite is

a. 38 m/s b. 0.85 km/s
c. 7.6 km/s d. 7.9 km/s

27. Earth satellite *A* has an orbit four times greater in radius than the orbit of satellite *B*. The orbital speed of *A* is

a. $v_B/4$ b. $v_B/2$
c. $2v_B$ d. $4v_B$

QUESTIONS

1. Under what circumstances, if any, can an object move in a circular path without being accelerated?

2. A car makes a clockwise turn on a level road at too high a speed and overturns. Do its left wheels leave the ground first, or its right wheels?

3. A person swings an iron ball in a vertical circle at the end of a string. At what point in the circle is the string most likely to break? Why?

4. Where should you stand on the earth's surface to experience the most centripetal acceleration? The least?

5. A track team on the moon could set new records for the high jump (if they did not need space suits, of course) because of the smaller gravitational pull there. Could sprinters also improve their times for the 100-m dash?

6. Compare the meanings of g and G.

7. Why is Eq. (6.12), which gives the speed a satellite must have for a stable orbit around the earth, the same as Eq. (6.8), which gives the critical speed of a ball being whirled in a vertical circle at the end of a string?

8. For the moon to have the same orbit it has now, what would its speed have to be if (a) the moon's mass were double its present mass, and (b) the earth's mass were double its present mass?

9. An earth satellite is placed in an orbit whose radius is half that of the moon's orbit. Is its time of revolution longer or shorter than that of the moon?

10. Two satellites are launched from a certain station with the same initial speeds relative to the earth's surface. One is launched toward the west, the other toward the east. Will there by any difference in their orbits? If so, what will the difference be and why?

11. Must a space vehicle in continuously powered flight reach the escape speed in order to leave the earth permanently?

12. An airplane makes a vertical circle in which it is upside down at the top of the loop. Will the pilot fall out of his seat if he has no belt to hold him in place?

EXERCISES

6.2 Centripetal Acceleration

1. A phonograph record 30 cm in diameter rotates $33\frac{1}{3}$ times per minute. (a) What is the linear speed of a point on its rim? (b) What is the centripetal acceleration of a point on its rim?

2. The minute hand of a large clock is 0.50 m long. (a) What is the linear speed of its tip in meters per second? (b) What is the centripetal acceleration of the tip of the hand?

3. What is the minimum radius at which an airplane flying at 300 m/s can make a U-turn if its centripetal acceleration is not to exceed $4.0g$?

4. If the earth were to spin so fast that a person at the equator were weightless, what would the length of the day be?

5. An astronaut in training is seated at the end of a horizontal arm 7.0 m long. How many revolutions per second must the arm make for the astronaut to experience a horizontal acceleration of $4.0g$?

6. A car starts from rest to go around a circular track 200 m in radius. If the car's speed increases at a rate of 4.0 m/s^2, how long will it take for the car's centripetal acceleration to equal its linear acceleration?

6.3 Magnitude of Centripetal Force

7. What is the centripetal force needed to keep a 3.0-kg mass moving in a circle of radius 0.50 m at a speed of 8.0 m/s?

8. A string 1.0 m long breaks when its tension is 100 N. What is the greatest speed at which it can be used to whirl a 1.0-kg stone? (Neglect the gravitational pull of the earth on the stone.)

9. A 10-kg counterweight is bolted to the rim of a flywheel 1.8 m in diameter that is rotating 300 times per minute. How much force must the bolts withstand?

10. A car is traveling at 50 km/h on a level road where the coefficient of static friction between its tires and the road is 0.70. Find the minimum turning radius of the car.

11. A 2000-kg car is rounding a curve of radius 200 m on a level road. The maximum frictional force the road can exert on the tires of the car is 4000 N. What is the highest speed at which the car can round the curve?

12. A box is resting on the flat floor in the rear of a station wagon moving at 15 m/s. What is the minimum radius of a turn the station wagon can make if the box is not to slip? Assume that $\mu_s = 0.40$.

13. A dime is placed 10 cm from the center of a record. The coefficient of static friction between coin and record is 0.30. Will the coin remain where it is or will it fly off when the record turns at $33\frac{1}{3}$ rev/min? At 78 rev/min?

14. British railway engineers refer to the centripetal acceleration of a train rounding a curve in terms of "cant deficiency," the banking angle that would be needed for a passenger to feel no sideways force relative to the train. What is the centripetal acceleration that corresponds to the maximum permitted cant deficiency of 4.25°?

15. An airplane traveling at 500 km/h banks at an angle of 45° as it makes a turn. What is the radius of the turn in kilometers? Assume that the rudder is not used in making the turn.

16. A highway curve has a radius of 300 m. (a) At what angle should it be banked for a traffic speed of 100 km/h? (b) If the curve is not banked, what is the minumum coefficient of friction required between tires and road?

17. A curve in a road 8.0 m wide has a radius of 60 m. How much higher than its inner edge should the outer edge of the road be if it is to be banked properly for cars traveling at 30 km/h?

6.4 Motion in a Vertical Circle

18. A string 1.0 m long is used to whirl a 0.50-kg stone in a vertical circle. What is the tension in the string when the stone is at the top of the circle moving at 5.0 m/s?

19. A string 0.80 m long is used to whirl a 2.0-kg stone in a vertical circle. What must be the speed of the stone at the top of the circle if the string is to be just taut? How does this speed compare with that required for a 1.0-kg stone in the same situation?

20. The 200-g head of a golf club moves at 45 m/s in a circular arc of 1.0 m radius. How much force must the player exert on the handle of the club to prevent it from flying out of his hands at the bottom of the swing? Assume that the shaft of the club has negligible mass.

21. A road has a hump 12 m in radius. What is the minimum speed at which a car will leave the road at the top of the hump?

22. A pendulum has a string 1.0 m long. If the tension in the string is twice the bob's weight when the string is vertical, what is the speed of the bob at that point in its path?

23. A physics teacher swings a pail of water in a vertical circle 1.0 m in radius at constant speed. What is the maximum time per revolution if the water is not to spill?

6.5 Gravitation

6.6 Gravity and the Earth

24. Two identical lead spheres whose centers are 2.0 m apart attract each other with a force of 1.0×10^{-5} N. Find the mass of each sphere.

25. A 2.0-kg mass is 1.0 m away from a 5.0-kg mass. What is the gravitational force (a) that the 5.0-kg mass exerts on the 2.0-kg mass; (b) that the 2.0-kg mass exerts on the 5.0-kg mass? (c) If both masses are free to move, what are their respective accelerations in the absence of other forces?

26. A bull and a cow elephant, each of mass 2000 kg, attract each other gravitationally with a force of 1.0×10^{-5} N. How far apart are they?

27. The earth's average orbital radius is 1.5×10^{11} m and its average orbital speed is 3.0×10^4 m/s. From these figures, together with the value of G, find the mass of the sun.

28. The mass of the planet Saturn is 5.7×10^{26} kg and that of the sun is 2.0×10^{30} kg. The average distance between them is 1.4×10^{12} m. (a) What is the gravitational force the sun exerts on Saturn? (b) Assuming that Saturn has a circular orbit, find its orbital speed.

29. An object dropped near the earth's surface falls 4.9 m in the first second. How far does the moon fall toward the earth in each second? Why doesn't the moon ever reach the earth?

30. If a planet existed whose mass and radius were both half those of the earth, what would the acceleration of gravity at its surface be in terms of g?

31. If a planet existed whose mass and radius were both twice those of the earth, what would the acceleration of gravity at its surface be in terms of g?

32. The moon's radius is 27% of the earth's radius, and its mass is 1.2% of the earth's mass. (a) What is the acceleration of gravity on the surface of the moon? (b) How much would a 60-kg person weigh there?

6.7 Earth Satellites

33. In terms of the earth's radius, how far is it from the center of the earth to a point outside the earth where $g = 0.98$ m/s^2?

34. What is the acceleration of a meteor when it is 1 earth's radius above the surface of the earth? Two earth's radii?

35. The radius of the earth is 6.4×10^6 m. What is the acceleration of a meteor when it is 8.0×10^6 m from the center of the earth?

36. A satellite is to be put into orbit around the moon just above its surface. What should its speed be? Assume that the moon's radius is half that of the earth and that the acceleration of gravity at its surface is $g/6$.

37. Find the speed of an earth satellite whose orbit is 400 km above the earth's surface. What is the period of the orbit?

38. Find the altitude of an earth satellite whose period is 40 h.

39. Saturn's mass is 5.7×10^{26} kg, and its radius is 6.0×10^7 m. (a) What is the acceleration of gravity on Saturn's surface? (b) What is the minimum speed a satellite of Saturn must have?

6.8 Weightlessness

40. If the apparent weight of a person in an airplane is not to exceed 2.0 mg, what is the maximum radius of a circular arc in which the airplane can pull out of a dive at a constant speed of 300 km/h?

41. If the apparent weight of a person in an airplane is not to exceed 2.0 mg, what is the maximum horizontal acceleration the airplane can have?

42. A 50-kg skier goes over a bump 10 m in radius at a speed of 9.0 m/s. What is her apparent weight at the top of the bump? How fast would she have to go to fly off the top?

6.9 Kepler's Laws and Gravitation

43. Venus is 0.723 as far from the sun as the earth is. How many earth days are there in a year on Venus?

44. Phobos and Deimos, the two satellites of Mars, have orbits with average radii of 9380 km and 23,500 km, respectively. The period of Phobos is 0.319 earth days. What is the period of Deimos?

PROBLEMS

45. A 500-g model airplane flies in a horizontal circle while attached to the ground with a wire 10 m long that is at an angle of 40° above the ground. If the airplane takes 5.0 s to complete each circle, what is the tension in the wire?

46. A girl on a merry-go-round who is sitting 5.0 m from its axis is holding a string with a ball at its end. If the string makes an angle of 3° with the vertical, how long does it take the merry-go-round to make a complete rotation?

47. Derive a formula that gives the centripetal acceleration of a point P on the earth's surface in terms of the earth's radius r_e, its period of revolution T, and the latitude θ of the point (see Fig. 6.29).

48. Figure 6.30 shows a *conical pendulum,* which consists of a bob of mass m that moves in a horizontal circle while suspended by a string that traces out a cone in space. (a) Use physical reasoning to explain why the angle θ increases as the period T of the pendulum decreases. (b) Show that $\theta = \cos^{-1}(gT^2/4\pi^2L)$.

49. A bicycle and rider of total mass 80 kg go around a banked circular track 40 m in radius at a speed of 16 m/s. If the reaction force of the track on the bicycle is perpendicular to the track, find its magnitude.

50. A car whose speed is 90 km/h rounds a curve 180 m in radius that is properly banked for a speed of 45 km/h. Find the minimum coefficient of friction between tires and road that will permit the car to make the turn.

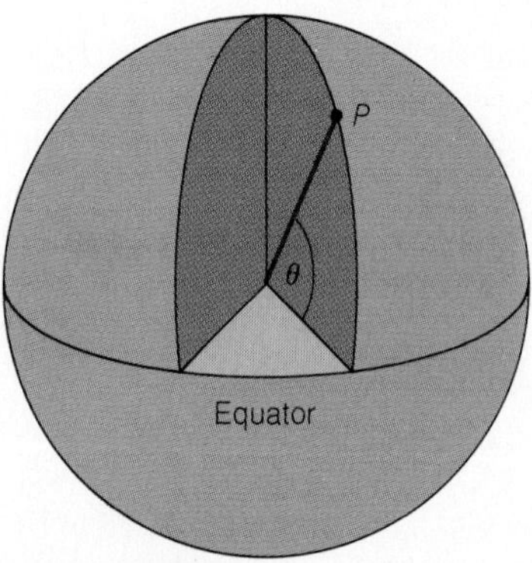

Figure 6.29
The latitude of the location P on the earth's surface is θ (Problem 47).

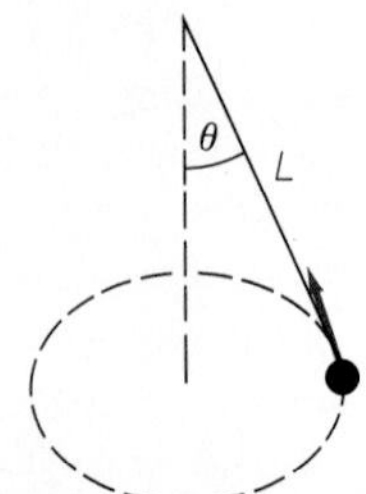

Figure 6.30
Conical pendulum (Problem 48).

51. The moon's mass is 7.3×10^{22} kg, and the average radius of its orbit is 3.8×10^8 m. At what point could an object be placed between the earth and the moon where it would experience no resultant force? (Neglect the gravitational attractions of the sun and the other planets.)

52. Most of the stars in the galaxy of which the sun is a member (the "Milky Way") are concentrated in an assembly

This spiral galaxy of stars in the constellation Andromeda closely resembles the Milky Way galaxy of which the sun is a member.

about 100,000 light-years across whose shape is roughly that of a fried egg. The sun is about 30,000 light-years from the center of the galaxy and revolves around it with a period of about 2.0×10^8 years. A reasonable estimate for the mass of the galaxy may be obtained by considering this mass to be concentrated at the galactic center with the sun revolving around it like a planet around the sun. On this basis, calculate the mass of the galaxy. How many stars having the mass of the sun is this equivalent to? (1 year = 3.16×10^7 s, 1 light year = 9.46×10^{15} m, and $m_{sun} = 1.99 \times 10^{30}$ kg.)

53. The moon circles the earth every 27.3 days in an orbit of radius 3.8×10^8 m, and the earth circles the sun every 365 days in an orbit of radius 1.5×10^{11} m. From these figures, without knowing the value of G, Newton was able to find the ratio between the masses of the sun and the earth. Do the same.

54. Show that the period of a satellite orbiting a planet just above its surface depends only on G and the planet's density ρ (Greek letter *rho*). (Density is mass per unit volume; the volume of a sphere of radius r is $4\pi r^3/3$.)

55. A simplified explanation for the tides is as follows: The moon attracts water at A in Fig. 6.31 more strongly than water at B. Because of these unequal forces, the moon pulls water at A away from the rest of the earth and also pulls the earth as a whole away from water at B, which tends to be left behind. As the earth rotates, the resulting bulges are held in place by the moon, so a given place on the earth experiences two tidal cycles per day. The gravitational force the sun exerts on the earth is nearly 180 times greater than the force the moon exerts on the earth, yet the moon is more effective in producing the tides than the sun. To see why, perform the following calculations. First find the difference between the force the sun exerts on 1 kg of water at a point on the equator nearest the sun and the force the sun would exert on 1 kg of water at the earth's center. Then make the same calculation for the forces the moon exerts on 1 kg of water at these locations, and compare the results. The moon's mass is 7.3×10^{22} kg, the sun's mass is 2.0×10^{30} kg, the earth's radius is 6.4×10^6 m, the earth's orbital radius is 1.5×10^{11} m, and the moon's orbital radius is 3.8×10^8 m.

Note: There is an easy way to make these calculations. When $x << 1$, $1/(1 - x)^2 \approx 1 + 2x$. Hence if R is the distance from the sun (or moon) to the earth's center and r is the earth's radius, then

$$\frac{1}{(R-r)^2} = \frac{1}{R^2\left(1 - \frac{r}{R}\right)^2} \approx \frac{1}{R^2}\left(1 + 2\frac{r}{R}\right)$$

Figure 6.31

Problem 55.

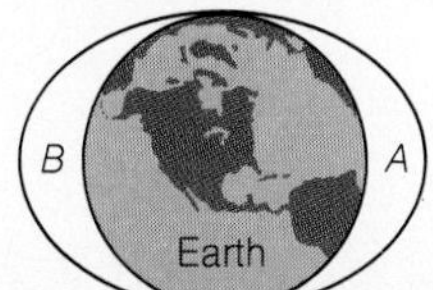

56. Because the earth is an oblate spheroid (that is, shaped like a grapefruit), a person at the equator is about 22 km farther from the earth's center than a person at the North or South Pole. In addition, part of the gravitational pull on a person at the equator goes into providing the centripetal force required there. Both effects decrease the weight of a person at the equator as registered on a scale as compared with its value at either pole. Find the percentage difference in weight due to each source.

Note: When $x << 1$, $1/(1 + x)^2 \approx 1 - 2x$, so if $r << R$, $1/(R + r)^2 \approx (1/R^2)(1 - 2r/R.)$

57. The French astronomer Cassini discovered 4 of Saturn's satellites in the seventeenth century (another 19 have been discovered since). The orbital radii and periods of these satellites are as follows:

Satellite	Orbit	Period
Tethys	2.95×10^8 m	1.89 days
Dione	3.77×10^8 m	2.74 days
Rhea	5.27×10^8 m	4.52 days
Iapetus	35.60×10^8 m	79.30 days

(a) Do these satellites obey Kepler's third law? (b) Find the mass of Saturn from the data above and the value of G.

Answers to Multiple Choice

1. a	**8.** c	**15.** b	**22.** b
2. b	**9.** c	**16.** c	**23.** d
3. a	**10.** d	**17.** b	**24.** a
4. c	**11.** b	**18.** b	**25.** d
5. b	**12.** d	**19.** d	**26.** c
6. d	**13.** c	**20.** b	**27.** b
7. b	**14.** b	**21.** a	

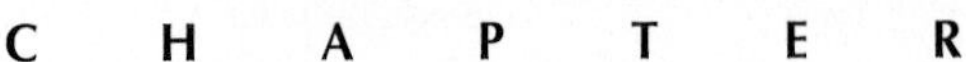

CHAPTER 7

ROTATIONAL MOTION

Until now we have been looking at translational motion, motion in which something shifts in position from one moment to the next. But rotational motion is just as common. The sun, the moon, planets, stars, and galaxies of stars all rotate as they sweep through space. Down on the earth, wheels, pulleys, propellers, drills, and audio discs rotate in order to do their jobs. In the atomic world, protons, neutrons, and electrons all rotate, and their rotations affect how they form atoms and how atoms form molecules, liquids, and solids. Here our concern is mainly with the rotational motion of a rigid body, one whose shape does not change as its spins. Strictly speaking, there is no such thing, but in practice many rotating bodies distort too little to matter. As we will find, all the formulas that describe rotational motion are exact analogs of the formulas we have been using to describe translational motion.

7.1 Angular Measure

The radian is the most convenient unit.

We are accustomed to measuring angles in degrees, where 1° is 1/360 of a full rotation. That is, a complete turn represents 360°.

A better unit for our present purposes is the **radian** (rad). The radian is defined with the help of a circle drawn with its center at the vertex of the angle in question. If the circle's radius is r and the arc cut by the angle is s as in Fig. 7.1, then the angle in radians is given by

$$\theta = \frac{s}{r} = \frac{\text{arc length}}{\text{radius}} \qquad \textit{Radian measure} \quad (7.1)$$

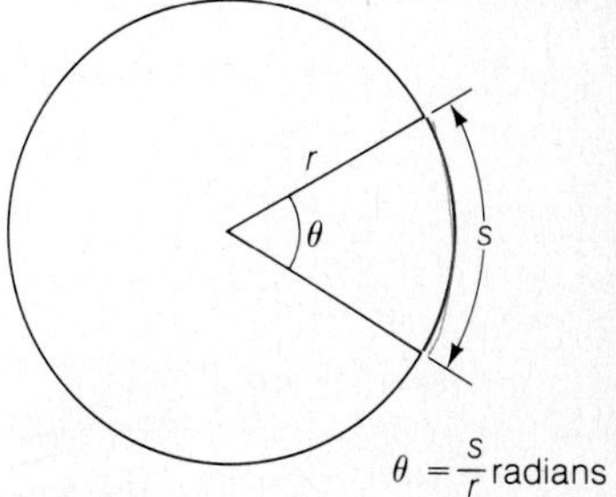

Figure 7.1

The ratio of arc to radius gives the magnitude of an angle in radians.

The angle θ between two radii of a circle, in radian measure, is the ratio of the arc s to the radius r. An angle of 1 rad has an arc length that is the same as the radius.

It is easy to find the conversion factor between degrees and radians, and vice versa. There are 360° in a complete circle, and the number of radians in a complete circle is

$$\theta = \frac{s}{r} = \frac{2\pi r}{r} = 2\pi$$

because the circumference of a circle of radius r is $2\pi r$ (Fig. 7.2). Hence

$$360° = 2\pi \text{ rad}$$

from which we find that

$$1° = 0.01745 \text{ rad} \quad \text{and} \quad 1 \text{ rad} = 57.30° \qquad \textit{Radians and degrees}$$

Figure 7.2

(a) 360° = 2π rad, so 1 rad = 57.30°. (b) A circle contains 2π rad, which is about 6.28 rad.

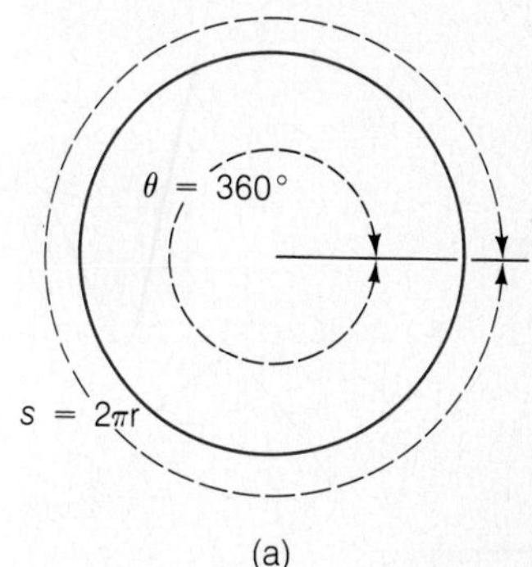

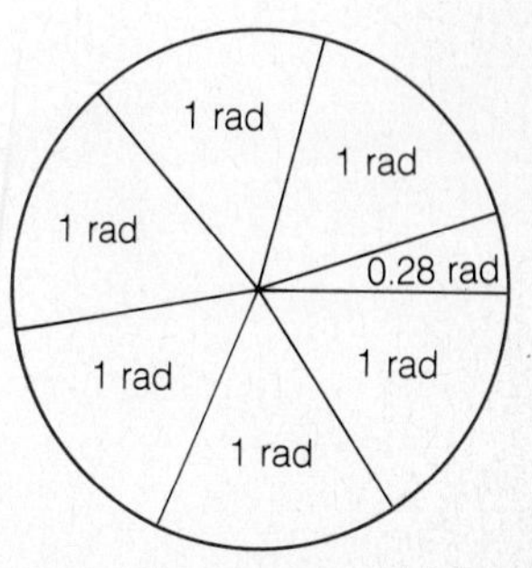

The conversions (2π rad/360°) and (360°/2π rad) are easier to remember.

The radian is an odd kind of unit because it has no dimensions—an angle in radians is a ratio of lengths, so it is really a pure number. In calculations the unit "rad" is always dropped at the end unless the result is an angular quantity.

EXAMPLE 7.1

A phonograph record 30.0 cm in diameter turns through an angle of 120°. How far does a point on its rim travel?

SOLUTION First the angle is converted from degrees to radians:

$$\theta = (120°)\left(\frac{2\pi \text{ rad}}{360°}\right) = 2.09 \text{ rad}$$

The radius of the record is 15.0 cm, and so, from Eq. (7.1),

$$s = r\theta = (15.0 \text{ cm})(2.09 \text{ rad}) = 31.4 \text{ cm}$$

♦

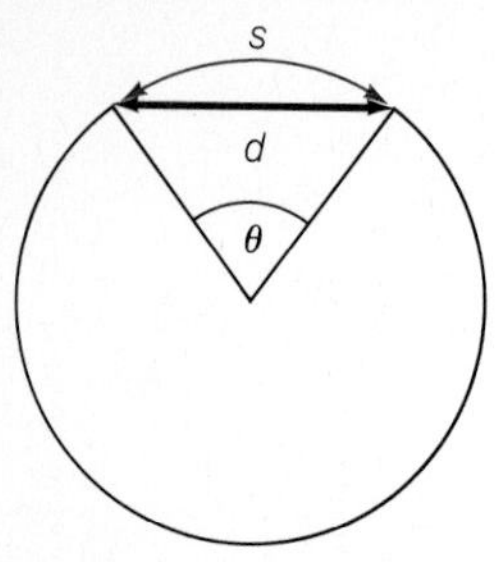

Figure 7.3

For small angles the difference between the arc s and the chord d is usually negligible.

EXAMPLE 7.2

A television image with 525 horizontal lines is being shown on a picture tube whose screen is 50 cm high. If a viewer's eyes can resolve detail to 0.0003 rad—about 1′ (one minute of arc), where 60′ = 1°—how far away should the eyes be from the screen in order to just be able to see the separate lines?

SOLUTION The distance between adjacent lines is $s = 0.50$ m/525, hence

$$r = \frac{s}{\theta} = \frac{0.50\text{ m}}{(525)(0.0003\text{ rad})} = 3.2\text{ m}$$

Because θ is so small here, we can ignore the difference between the arc s and the chord d (Fig. 7.3). Even when $\theta = 10°$, this difference is only 0.11%. At large angles, however, it is appreciable: When $\theta = 90°$, for instance, the difference between s and d is 10%. ◆

Table 7.1 Degrees to radians

0° =	0 rad	
30° =	$\pi/6$ rad =	0.524 rad
45° =	$\pi/4$ rad =	0.785 rad
60° =	$\pi/3$ rad =	1.047 rad
90° =	$\pi/2$ rad =	1.571 rad
180° =	π rad =	3.142 rad
270° =	$3\pi/2$ rad =	4.712 rad
360° =	2π rad =	6.283 rad
720° =	4π rad =	12.566 rad

Sometimes it is useful to express angles in radian measure in terms of π itself. For example, an angle of 90° is $\frac{1}{4}$ of a complete circle, and so

$$90° = (\tfrac{1}{4}\text{ circle})\left(2\pi\frac{\text{rad}}{\text{circle}}\right) = \frac{\pi}{2}\text{ rad}$$

Of course, this has the same numerical value as

$$(90°)\left(\frac{2\pi\text{ rad}}{360°}\right) = 1.571\text{ rad}$$

because $\pi/2 = 1.571$. Table 7.1 gives some other examples of radian measure in terms of π and as decimals.

7.2 Angular Speed

Round and round it goes.

If a rotating body turns through the angle θ in the time t, its average **angular speed** ω (Greek letter *omega*) is

$$\omega = \frac{\theta}{t} \qquad \textit{Angular speed} \quad (7.2)$$

When θ is in radians and t in seconds, which are the usual units for these quantities, the unit of ω is the rad/s.

Two other common units of angular speed are the revolution per second (rps) and the revolution per minute (rpm), where

$$1\frac{\text{rev}}{\text{s}} = \left(1\frac{\cancel{\text{rev}}}{\text{s}}\right)\left(2\pi\frac{\text{rad}}{\cancel{\text{rev}}}\right) = 2\pi\text{ rad/s} = 6.28\text{ rad/s}$$

$$1\frac{\text{rev}}{\text{min}} = \left(1\frac{\cancel{\text{rev}}}{\cancel{\text{min}}}\right)\left(2\pi\frac{\text{rad}}{\cancel{\text{rev}}}\right)\left(\frac{1\,\cancel{\text{min}}}{60\text{ s}}\right) = \frac{2\pi}{60}\text{ rad/s} = 0.105\text{ rad/s}$$

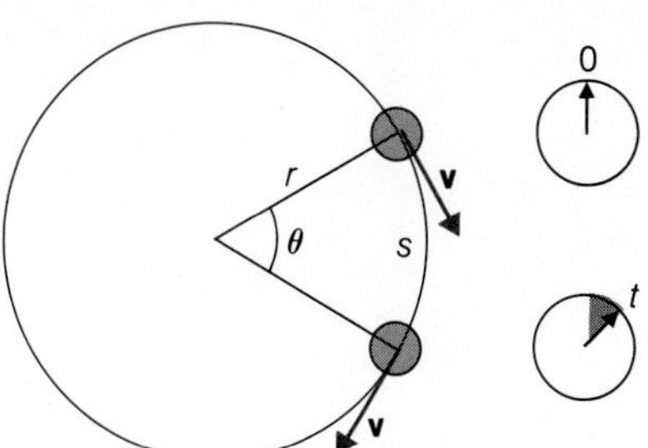

Figure 7.4

The angular speed of a particle in uniform circular motion is $\omega = v/r$.

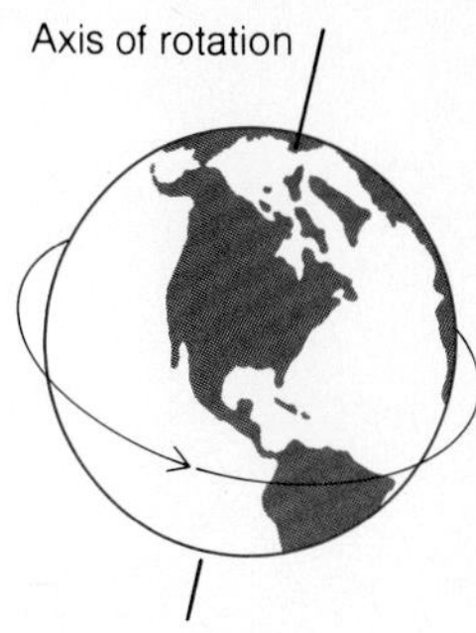

Figure 7.5

The axis of rotation is that line of particles which does not move in a rotating body. (It may be a line in space.)

Suppose we have a particle moving with the uniform speed v in a circle of radius r, as in Fig. 7.4. This particle travels the distance $s = vt$ in the time t. The angular distance through which it moves in that time is

$$\theta = \frac{s}{r} = \frac{vt}{r}$$

so that its angular speed is

$$\omega = \frac{\theta}{t} = \frac{vt}{rt}$$

or

$$\omega = \frac{v}{r} \qquad \textit{Angular speed} \quad (7.3)$$

$$\text{Angular speed} = \frac{\text{linear speed}}{\text{path radius}}$$

This relationship can be written another way:

$$v = \omega r \qquad (7.4)$$

Linear speed = (angular speed)(path radius)

The formulas of this section are valid only when ω is expressed in radian measure.

The **axis of rotation** of a rigid body turning in place is that line of particles which does not move (Fig. 7.5). Sometimes the axis of rotation is a line in space. All other particles of the rotating body move in circles about the axis. Since $v = \omega r$, the farther a particle is from the axis, the greater its linear speed, although all the particles of the body (except those on the axis) have the same angular speed.

A drill bit must have a particular linear speed for cutting into a given material without damage. The larger the radius of the drill bit, the slower it needs to turn to achieve the correct linear speed.

EXAMPLE 7.3

Find the linear speeds of points 2.0 cm and 15.0 cm from the axis of a phonograph record rotating at $33\frac{1}{3}$ rpm (Fig. 7.6).

SOLUTION The angular speed of the record is

$$\omega = (33\tfrac{1}{3}\text{ rpm})\left(\frac{2\pi\text{ rad/s}}{60\text{ rpm}}\right) = 3.5\text{ rad/s}$$

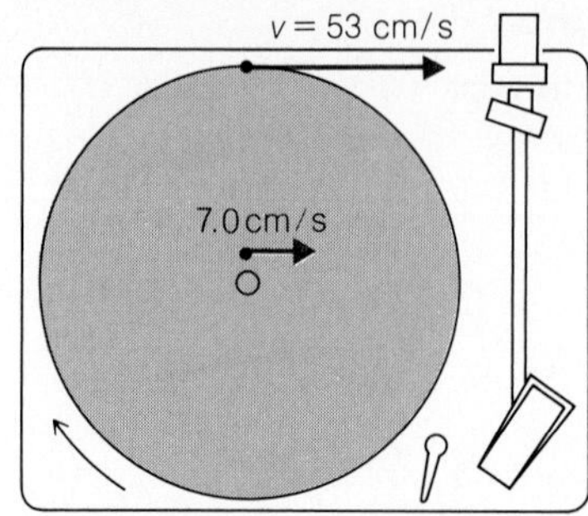

Figure 7.6 $\omega = 33\frac{1}{3}$ rpm

Hence a point 2.0 cm from the axis has a linear speed of

$$v = \omega r = \left(3.5\,\frac{\text{rad}}{\text{s}}\right)(2.0\text{ cm}) = 7.0\text{ cm/s}$$

and a point on the record's rim, where $r = 15.0$ cm, has a linear speed of

$$v = \left(3.5\,\frac{\text{rad}}{\text{s}}\right)(15.0\text{ cm}) = 53\text{ cm/s}$$ ◆

7.3 Rotational Kinetic Energy

A spinning body has KE even though it may be going nowhere.

The correspondences between the linear and angular versions of displacement and speed are rather obvious. Less obvious—in fact, not obvious at all—is the rotational equivalent of mass.

Mass is a measure of the reluctance of a body at rest to be set in translational motion or, once in motion, to have its velocity change. Certainly the same body is also reluctant to be set in rotation, and if already rotating, it tends to continue to do so. We need a measure of this rotational inertia. Perhaps the easiest way to find a suitable one is by considering the kinetic energy of a rotating body.

A rotating body has KE because its particles are moving even though the body as a whole remains in place. The speed of a particle that is the distance r from the axis of a rigid body rotating with the angular velocity ω is, as we know, $v = \omega r$ (Fig. 7.7). If the particle's mass is m, its kinetic energy is therefore

$$\text{KE} = \tfrac{1}{2}mv^2 = \tfrac{1}{2}m\omega^2r^2 \tag{7.5}$$

The body consists of many particles that need not have the same mass or be the same distance from the axis. However, all the particles have the common angular speed ω. This means that the total kinetic energy of all the particles may be written

$$\text{KE} = \Sigma\tfrac{1}{2}mv^2 = \tfrac{1}{2}(\Sigma mr^2)\omega^2 \tag{7.6}$$

Figure 7.7

The kinetic energy of each particle of a rotating body depends on the square of its distance r from the axis of rotation.

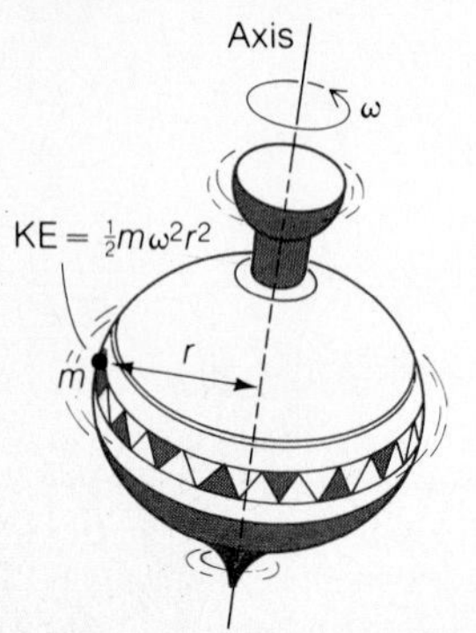

where the symbol Σ means, as before, "sum of." Equation (7.6) states that the kinetic energy of a rotating rigid body is equal to half the sum of the mr^2 values of all its particles multiplied by the square of its angular speed ω.

The quantity

$$I = \Sigma mr^2 \qquad \textit{Moment of inertia} \quad (7.7)$$

is called the **moment of inertia** of the body. It has the same value regardless of the body's state of motion. The farther a given particle is from the axis of rotation, the faster it moves and the greater is its contribution to the KE of the body. The moment of inertia of a body depends on the way in which its mass is distributed relative to its axis of rotation. It is perfectly possible for one body to have a greater moment of inertia than another even though its mass may be much the smaller of the two.

The KE of a body of moment of inertia I rotating with the angular velocity ω is therefore

$$\text{KE} = \tfrac{1}{2}I\omega^2 \qquad \textit{Rotational kinetic energy} \quad (7.8)$$

Evidently the rotational analog of mass is moment of inertia, just as the rotational analog of linear speed is angular speed. We will find further support for the correspondence of mass and moment of inertia in the rest of this chapter.

7.4 Moment of Inertia

The rotational analog of mass.

A rigid body is made up of a large number of separate particles whose masses are m_1, m_2, m_3, and so on (Fig. 7.8). To find the moment of inertia of such a body about a specified axis, we first multiply the mass of each of these particles by the square of its distance from the axis ($r_1{}^2$, $r_2{}^2$, $r_3{}^2$, and so on). Then we add all the mr^2 values. That is,

$$I = \Sigma mr^2 = m_1r_1{}^2 + m_2r_2{}^2 + m_3r_3{}^2 + \cdots \quad (7.9)$$

The unit of I is the kg·m^2.

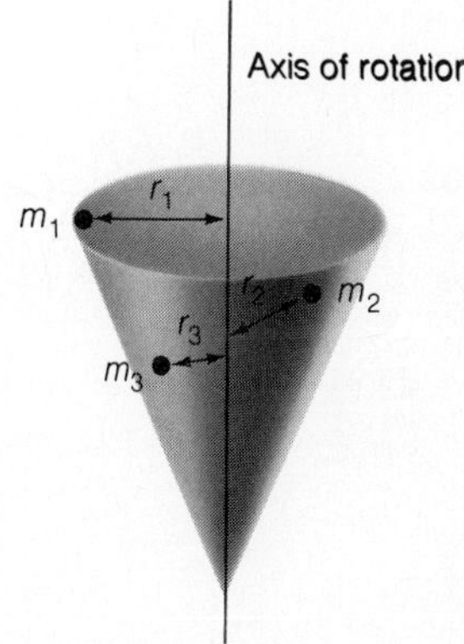

Figure 7.8

A rigid body consists of a large number of particles each of which has a certain mass m and distance r from the axis.

EXAMPLE 7.4

Find the moment of inertia of a thin ring of mass M and average radius R about an axis passing through its center and perpendicular to the plane in which it lies.

SOLUTION As in Fig. 7.9, we proceed by subdividing the ring into n segments, each of which is at a distance R from the axis. Hence

$$I = m_1R^2 + m_2R^2 + m_3R^2 + \cdots + m_nR^2$$
$$= (m_1 + m_2 + m_3 + \cdots + m_n)R^2$$

Figure 7.9

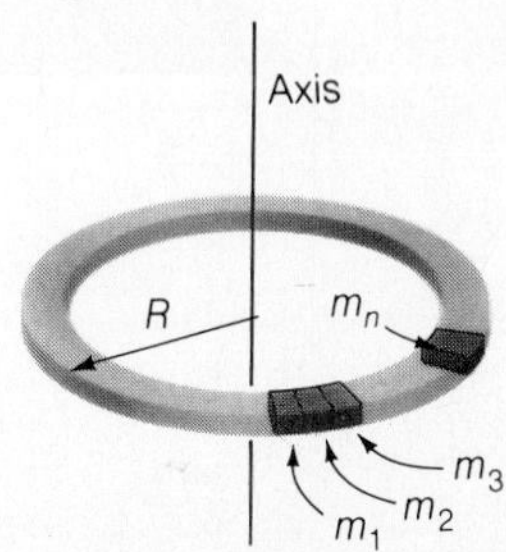

But the sum of the masses of the segments is the same as the total mass M of the ring, and so

$$I = MR^2$$ ♦

When a body consists of a continuous distribution of matter, the more particles we imagine it to contain, the more accurate will be our value of its moment of inertia. While I can be calculated for a few simple bodies without difficulty by Eq. (7.9), in general either considerable labor or the use of integral calculus is required. Figure 7.10 gives the moments of inertia of several regularly shaped bodies in terms of the total mass M and dimensions of each.

EXAMPLE 7.5

Rotating flywheels have been proposed for energy storage in electric power plants. The flywheels would be set in motion during off-peak periods by electric motors, which would act as generators to return the stored energy during times of heavy demand. (a) Find the kinetic energy of a 1.0×10^5-kg cylindrical flywheel whose radius is 2.0 m that rotates at 400 rad/s. (b) For how many hours could the flywheel supply energy at the rate of 1.0 MW?

SOLUTION (a) From Fig. 7.10 the moment of inertia of a solid cylinder is $I = \frac{1}{2}MR^2$. The kinetic energy of the cylinder is therefore

$$\begin{aligned} \text{KE} &= \tfrac{1}{2}I\omega^2 = \tfrac{1}{2}(\tfrac{1}{2}MR^2)\omega^2 = \tfrac{1}{4}MR^2\omega^2 \\ &= (\tfrac{1}{4})(1.0 \times 10^5\,\text{kg})(2.0\,\text{m})^2(400\,\text{rad/s})^2 = 1.6 \times 10^{10}\,\text{J} \end{aligned}$$

(b) Because power = energy/time, 1 MW = 10^6 W, and 1 h = 3600 s, we have

$$t = \frac{\text{KE}}{P} = \frac{1.6 \times 10^{10}\,\text{J}}{1.0 \times 10^6\,\text{W}} = 1.6 \times 10^4\,\text{s} = 4.4\,\text{h}$$ ♦

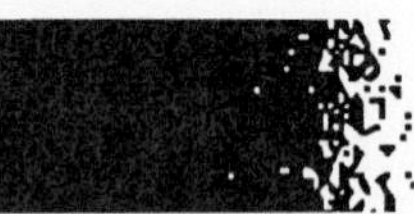

7.5 Combined Translation and Rotation

The total KE is the sum of translational KE and rotational KE.

When a rigid body is both moving through space and undergoing rotation, its total kinetic energy is the sum of its translational and rotational kinetic energies. The translational KE is calculated on the basis that the body is a particle whose linear speed is the same as that of the body's center of mass. The rotational KE is calculated on the basis that the body is rotating about an axis that passes through the center of

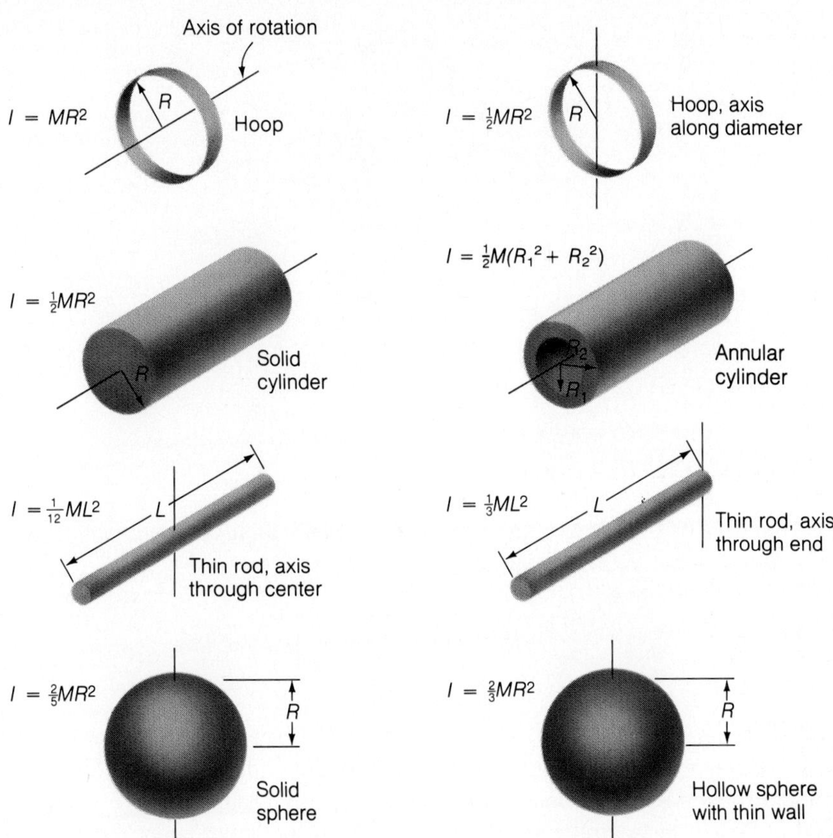

Figure 7.10
Moments of inertia of various bodies each of mass M, about indicated axes.

mass. Thus

$$\text{KE} = \text{KE}_{\text{translation}} + \text{KE}_{\text{rotation}} = \tfrac{1}{2}mv^2 + \tfrac{1}{2}I\omega^2 \tag{7.10}$$

where m is the body's mass, v is the velocity of its center of mass, I is its moment of inertia about an axis through the center of mass, and ω is its angular speed about that axis.

EXAMPLE 7.6

Consider a uniform cylinder of radius R and mass m that is poised at the top of an inclined plane (Fig. 7.11). Will it have a greater speed at the bottom if it slides down without friction or if it rolls down without sliding?

SOLUTION In the case of sliding, we set the cylinder's initial PE of mgh equal to its final KE of $\frac{1}{2}mv^2$, and find as usual that

$$mgh = \tfrac{1}{2}mv^2$$
$$v = \sqrt{2gh}$$

In the case of rolling, the cylinder has both translational and rotational KE at the

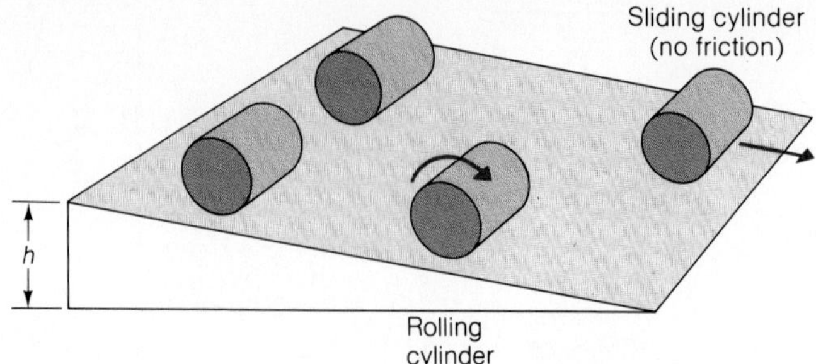

Figure 7.11

bottom, so that

$$mgh = \tfrac{1}{2}mv^2 + \tfrac{1}{2}I\omega^2$$

The moment of inertia of a cylinder is $I = \frac{1}{2}mR^2$. If it rolls without slipping, its linear and angular velocities are related by the formula $\omega = v/R$, so

$$mgh = \tfrac{1}{2}mv^2 + \tfrac{1}{2}(\tfrac{1}{2}m\,R^2)\left(\frac{v^2}{R^2}\right) = \tfrac{1}{2}mv^2 + \tfrac{1}{4}mv^2 = \tfrac{3}{4}mv^2$$

$$v = \sqrt{\tfrac{4}{3}gh}$$

The cylinder moves more slowly when it rolls down the plane than when it slides without friction because some of the available energy is absorbed by its rotation. The total energy of the cylinder at the bottom of the plane is the same in both cases. ♦

EXAMPLE 7.7

A solid ball starts rolling from rest at the top of an inclined plane 10.0 m long and reaches the bottom 7.0 s later. What angle does the plane make with the horizontal?

SOLUTION Our strategy here is to find h, since $\sin\theta = h/L$ (Fig. 7.12) and we know that $L = 10.0$ m. First we determine the linear speed v_f of the ball at the bottom of the plane from L and t, the time needed for it to roll down. From Eq. (1.13), $x = (v_0 + v_f)t/2$, we have here with $x = L$ and $v_0 = 0$

$$v_f = \frac{2L}{t} = \frac{(2)\,(10.0\text{ m})}{7.0\text{ s}} = 2.86\text{ m/s}$$

Because $I = (2/5)mR^2$ for a sphere and $\omega = v/R$ (Eq. 7.3), conservation of energy

Figure 7.12

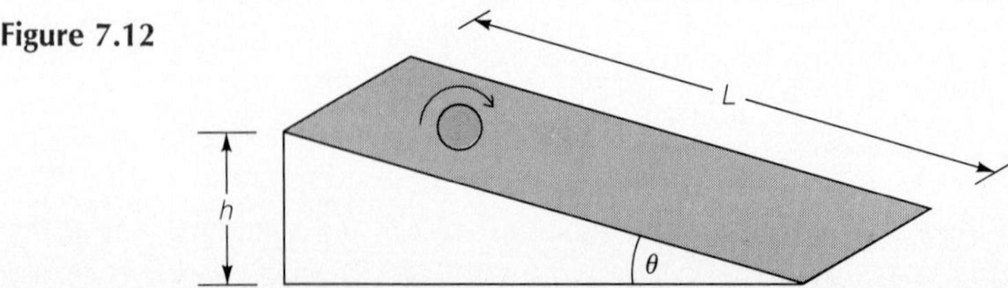

gives

$$mgh = \tfrac{1}{2}mv^2 + \tfrac{1}{2}I\omega^2$$

$$= \tfrac{1}{2}mv^2 + (\tfrac{1}{2})(\tfrac{2}{5})(mR^2)\left(\frac{v^2}{R^2}\right) = \tfrac{1}{2}mv^2 + \tfrac{1}{5}mv^2 = 0.7\,mv^2$$

$$h = \frac{0.7\,v^2}{g} = \frac{(0.7)(2.86\text{ m/s})^2}{9.8\text{ m/s}^2} = 0.584\text{ m}$$

The angle θ of the plane is therefore

$$\theta = \sin^{-1}\frac{h}{L} = \sin^{-1}\frac{0.584\text{ m}}{10\text{ m}} = \sin^{-1}0.0584 = 3°$$

◆

7.6 Angular Acceleration

The formulas for accelerated rotation have the same forms as those for accelerated linear motion.

A rotating body need not have a uniform angular speed ω, just as a moving particle need not have a uniform linear speed v. If the angular speed of a body changes by an amount $\Delta\omega$ in the time interval Δt, its average **angular acceleration** α (Greek letter *alpha*) is

$$\alpha = \frac{\Delta\omega}{\Delta t} \qquad \textit{Angular acceleration} \quad (7.11)$$

The unit of angular acceleration is the rad/s^2.

A particle moving in a circle of radius r that has an angular acceleration α also has a linear acceleration a_T tangential to its path, because its orbital speed is changing (Fig. 7.13). From the definition $\omega = v/r$, if r is constant

$$\Delta\omega = \frac{\Delta v}{r} \quad \text{and} \quad \alpha = \frac{\Delta\omega}{\Delta t} = \frac{\Delta v}{r\Delta t}$$

Because the component of the particle's linear acceleration along the direction of its velocity **v** has the magnitude $a_T = \Delta v/\Delta t$, we see that

$$\alpha = \frac{a_T}{r} \qquad (7.12)$$

$$\text{Angular acceleration} = \frac{\text{tangential acceleration}}{\text{path radius}}$$

Conversely,

$$a_T = \alpha r \qquad (7.13)$$

Tangential acceleration = (angular acceleration)(path radius)

Figure 7.13

The angular and tangential accelerations of a particle moving in a circle are proportional to each other.

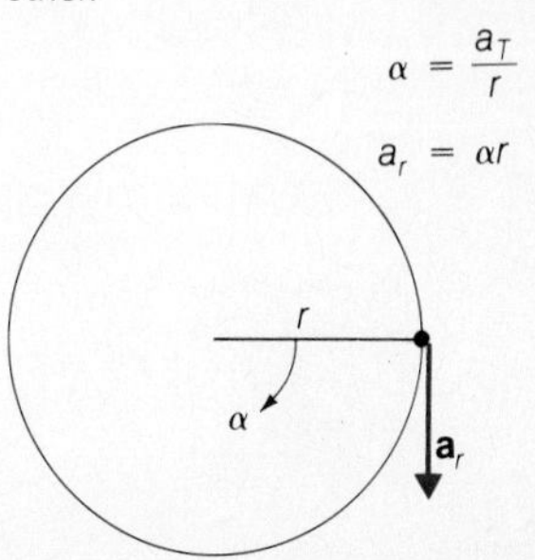

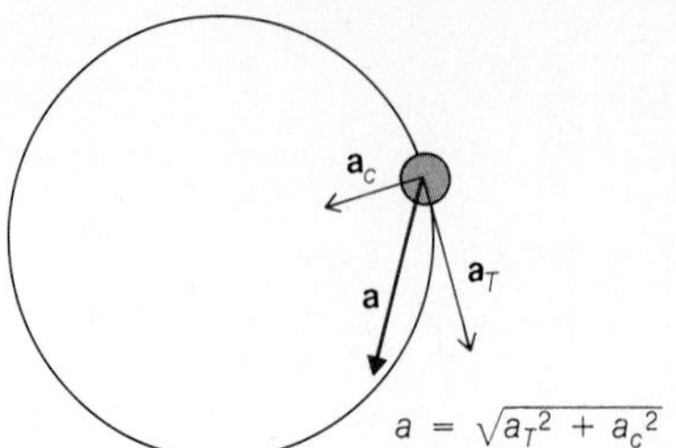

Figure 7.14

The tangential and centripetal accelerations of a particle in circular motion are always perpendicular.

We must be careful to distinguish between the **tangential acceleration** a_T of a particle, which represents a change in its speed, and its **centripetal acceleration** a_c, which represents a change in its direction of motion. A particle in circular motion with the speed v has, as we know, the centripetal acceleration

$$a_c = \frac{v^2}{r}$$

directed toward the center of its circular path. The accelerations $\mathbf{a}_T$ and $\mathbf{a}_c$ are therefore always perpendicular (Fig. 7.14). *All* particles in circular motion have centripetal accelerations. Only those particles whose speed changes, however, have tangential accelerations. The centripetal acceleration of a particle moving in a circle of radius r can be expressed in terms of its angular velocity ω as

$$a_c = \omega^2 r \qquad \textit{Centripetal acceleration} \qquad (7.14)$$

because $v = \omega r$.

EXAMPLE 7.8

Find the magnitude of the total linear acceleration of a particle moving in a circle of radius 0.40 m at the instant when the angular speed is 2.0 rad/s and the angular acceleration is 5.0 rad/s^2.

SOLUTION From Eqs. (7.13) and (7.14), for the particle's tangential and centripetal accelerations, we have

$$a_T = \alpha r = \left(5.0 \frac{\text{rad}}{\text{s}^2}\right)(0.40 \text{ m}) = 2.0 \text{ m/s}^2$$

$$a_c = \omega^2 r = \left(2.0 \frac{\text{rad}}{\text{s}}\right)^2 (0.40 \text{ m}) = 1.6 \text{ m/s}^2$$

We must add $\mathbf{a}_T$ and $\mathbf{a}_c$ vectorially, because they are vector quantities and are in different directions. As shown in Fig. 7.14, the magnitude a of the vector sum $\mathbf{a}$ of $\mathbf{a}_T$ and $\mathbf{a}_c$ is

$$a = \sqrt{a_T^2 + a_c^2} = \sqrt{2.0^2 + 1.6^2} \text{ m/s}^2 = 2.6 \text{ m/s}^2$$

◆

Table 7.2

Linear motion	Angular motion	
$x = v_0t + \frac{1}{2}at^2$	$\theta = \omega_0 t + \frac{1}{2}\alpha t^2$	(7.15)
$x = \left(\frac{v_0 + v_f}{2}\right)t$	$\theta = \left(\frac{\omega_0 + \omega_f}{2}\right)t$	(7.16)
$v_f = v_0 + at$	$\omega_f = \omega_0 + \alpha t$	(7.17)
$v_f^2 = v_0^2 + 2ax$	$\omega_f^2 = \omega_0^2 + 2\alpha\theta$	(7.18)

The formulas we obtained in Chapter 1 for the linear motion of a particle under constant acceleration all have counterparts in angular motion. Because the derivations are the same, they are simply listed in Table 7.2.

EXAMPLE 7.9

A motor starts rotating from rest with an angular acceleration of 12.0 rad/s^2. (a) What is the motor's angular speed 4.0 s later? (b) How many revolutions does it make in this period of time?

SOLUTION (a) From Eq. (7.17) we find

$$\omega_f = \omega_0 + \alpha t = 0 + (12.0\ \text{rad/s}^2)(4.0\ \text{s}) = 48\ \text{rad/s}$$

(b) Because we know ω_0, ω_f, α, and t, we can use Eq. (7.15), (7.16), or (7.18) to find θ. Eq. (7.15) has the advantage that it involves only the quantities we are given, so any errors made or rounding done in finding ω_f cannot affect the value of θ. From Eq. (7.15)

$$\theta = \omega_0 t + \tfrac{1}{2}\alpha t^2 = 0 + \tfrac{1}{2}(12.0\ \text{rad/s}^2)(4.0\ \text{s})^2 = 96\ \text{rad}$$

Because 1 rev $= 2\pi$ rad,

$$\theta = \frac{96\ \text{rad}}{2\pi\ \text{rad/rev}} = 15.3\ \text{rev}$$

◆

7.7 Torque

The rotational analog of force.

According to Newton's second law of motion, a net force applied to a body gives it a linear acceleration. What can give a body an angular acceleration?

Everyday experience can guide us here. When we want to turn a bolt, we exert a force **F** on a wrench whose jaws are on the bolt's head (Fig. 7.15). The stronger the force, the greater the turning effect. Also, the longer the perpendicular

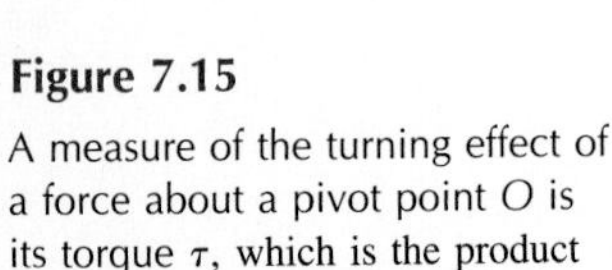

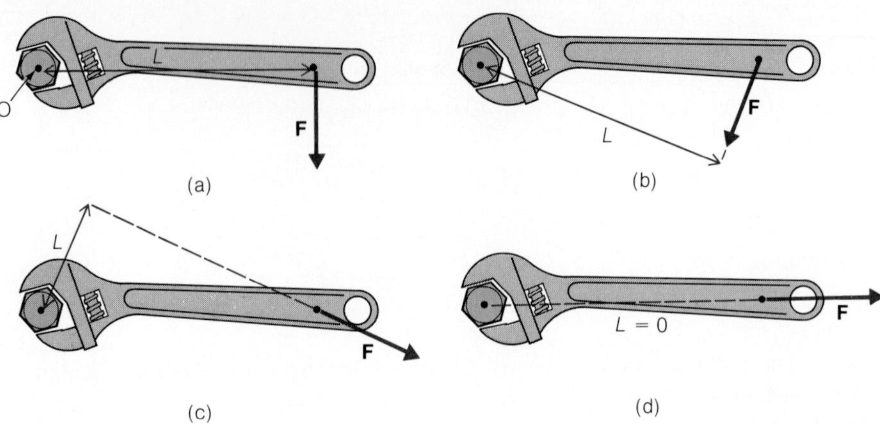

Figure 7.15

A measure of the turning effect of a force about a pivot point O is its torque τ, which is the product FL of the magnitude F of the force and the moment arm L. In (a) the moment arm L is longest, and so τ is a maximum. In (d) the line of action of the force passes through O, and so $L=0$ and $\tau = 0$.

distance L from the line of action of the force to the pivot point O, which is the bolt, the greater the turning effect. The distance L is called the **moment arm** of the force.

The product of the magnitude F of the force and its moment arm L is called the **torque** of the force about O. To twist something is to apply a torque to it. The symbol for torque is τ (Greek letter *tau*), so that

$$\tau = FL \qquad \textit{Torque} \quad (7.19)$$

Torque = (force)(moment arm)

Torque is expressed in newton·meters (N·m).

Let us now see how torque is related to angular acceleration. Figure 7.16 shows a particle of mass m that moves in a circle of radius r. A force **F** that acts on the particle tangentially to the particle's path gives it the acceleration a_T according to the formula

$$F = ma_T$$

Because the tangential acceleration a_T here is equal to αr,

$$F = mr\alpha$$

Multiplying both sides of this equation by r gives

$$Fr = mr^2\alpha$$

We recognize Fr as the torque τ of the force F about the axis of rotation of the particle and mr^2 as the particle's moment of inertia I. The latter equation therefore states that

$$\tau = I\alpha \qquad (7.20)$$

Torque = (moment of inertia)(angular acceleration)

Although Eq. (7.20) was derived for the case of a single particle, it is also valid for any rotating body, provided that the torque and moment of inertia are both calculated about the same axis.

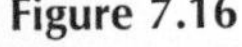

Figure 7.16

A force **F** that acts on a particle moving in a circle of radius r exerts a torque of magnitude $\tau = Fr$ when **F** is perpendicular to the particle's path.

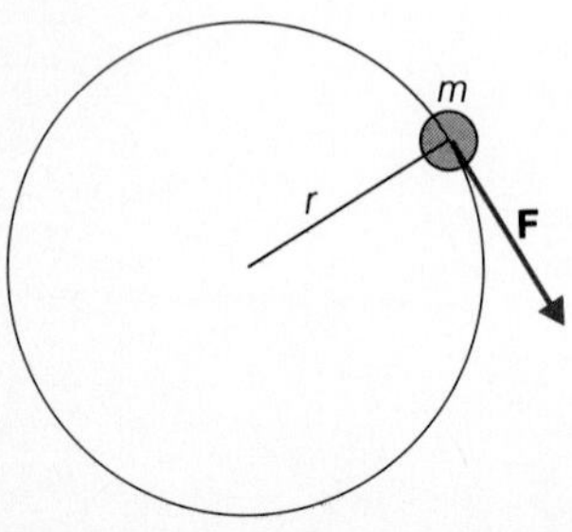

The formula $\tau = I\alpha$ is the fundamental law of motion for rotating bodies in the same sense that $F = ma$ is the fundamental law of motion for bodies moving through space. In rotational motion, torque plays the same role that force does in translational motion. The angular acceleration experienced by a body when the torque τ acts on it is

$$\alpha = \frac{\tau}{I}$$

Thus α is directly proportional to the torque and inversely proportional to the body's moment of inertia.

EXAMPLE 7.10

A 2-kg grindstone 10 cm in radius is turning at 120 rad/s. The motor is switched off, and a chisel is pressed against the grindstone with a force whose tangential component is 2 N (Fig. 7.17). How long will it take the grindstone to come to a stop?

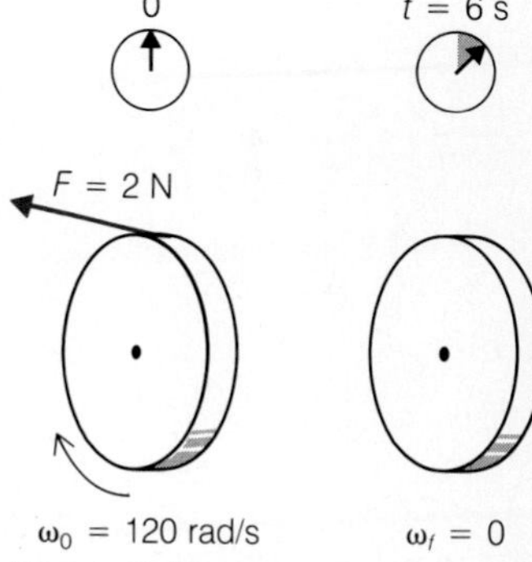

Figure 7.17

SOLUTION The grindstone is a solid cylinder and so its moment of inertia is

$$I = \tfrac{1}{2}MR^2 = \tfrac{1}{2}(2\ \text{kg})(0.1\ \text{m})^2 = 0.01\ \text{kg}\cdot\text{m}^2$$

The torque the chisel exerts on the grindstone is

$$\tau = -Fr = -(2\ \text{N})(0.1\ \text{m}) = -0.2\ \text{N}\cdot\text{m}$$

where the minus sign means that the torque acts to slow down the grindstone. The angular acceleration of the grindstone is

$$\alpha = \frac{\tau}{I} = \frac{-0.2\ \text{N}\cdot\text{m}}{0.01\ \text{kg}\cdot\text{m}^2} = -20\ \text{rad/s}^2$$

We now call on Eq. (7.17), $\omega_f = \omega_0 + \alpha t$, and set $\omega_f = 0$ to correspond to the grindstone coming to a stop. The result is

$$\alpha t = -\omega_0$$

$$t = \frac{-\omega_0}{\alpha} = \frac{-120\ \text{rad/s}}{-20\ \text{rad/s}^2} = 6\ \text{s}$$

◆

EXAMPLE 7.11

Figure 7.18 shows a 12-kg block, A, and a 30-kg block, B, connected by a string that passes over a pulley, just as in Example 3.6 at the end of Section 3.5 (Fig. 3.17). Here, however, the pulley is not massless but is a uniform disk whose mass is 10 kg. Find the accelerations of the blocks.

SOLUTION In the earlier problem, where the pulley was massless, the string had the same tension on both sides of the pulley. Here, owing to the inertia of the pulley,

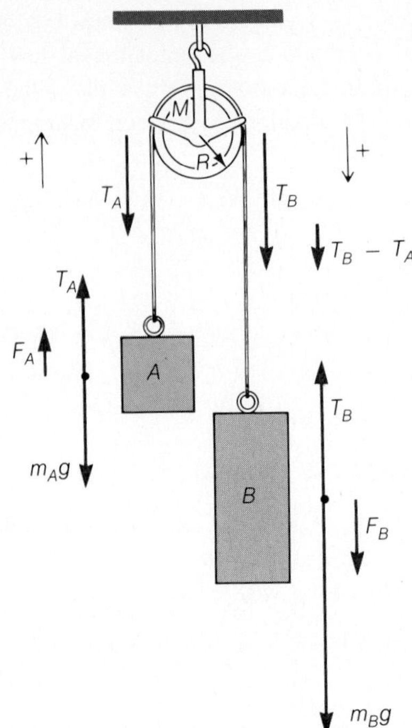

Figure 7.18

The inertia of the pulley reduces the accelerations of the suspended blocks.

the tensions are different. The net force F_A on block A is the difference between the tension T_A in the string supporting it and its weight of $m_A g$. If the acceleration of A is a (considered positive as in the figure),

$$F_A = T_A - m_A g = m_A a$$
$$T_A = m_A(g + a)$$

The net force $F_B = m_B g - T_B$ on block B is directed downward (considered positive). Its acceleration is also a, and

$$F_B = m_B g - T_B = m_B a$$
$$T_B = m_B(g - a)$$

The difference $T_B - T_A$ between the tensions equals the net force on the pulley's rim. This means that the torque on the pulley is

$$\tau = (T_B - T_A)R = m_B(g - a)R - m_A(g + a)R$$
$$= (m_B - m_A)gR - (m_A + m_B)aR$$

From Fig. 7.10 the moment of inertia of the pulley is $I = \frac{1}{2}MR^2$. Because $\tau = I\alpha$ and $\alpha = a/R$, we have

$$(m_B - m_A)gR - (m_A + m_B)aR = I\alpha = \tfrac{1}{2}MR^2\left(\frac{a}{R}\right) = \tfrac{1}{2}MRa$$

from which we obtain

$$a = \frac{(m_B - m_A)g}{M/2 + (m_A + m_B)} = \frac{(18\text{ kg})(9.8\text{ m/s}^2)}{(10\text{ kg})/2 + 42\text{ kg}} = 3.75\text{ m/s}^2$$

As we would expect, this is less than the 4.2 m/s^2 acceleration found earlier when the pulley was considered massless. It is interesting to note that we did not have to know the pulley's radius. ◆

When a net force acts on a body able to move freely, the body will have both linear and angular accelerations unless the line of action of the force passes through the body's center of mass (Fig. 7.19). In the latter case, if the body was not rotating initially, it will keep its original orientation during its motion.

7.8 Work and Power

No surprises here.

Mechanical energy is usually transmitted by rotary motion. The power output of almost every modern engine emerges via a rotating shaft, and more often than not this power is expended in some form of rotation as well. The tires of a car, the propeller of a ship, the bit of a drill, the vanes of a centrifugal pump, and the rotor of a dynamo all function by turning.

Let us consider a shaft of radius r on which a tangential force **F** acts, as in Fig. 7.20. After a time t the shaft has turned through the angle θ, the point at which the force is applied has moved through the distance s, and the force has done the amount of work

$$W = Fs = Fr\theta$$

Figure 7.19

A force whose line of action passes through the center of gravity of a body cannot cause it to rotate.

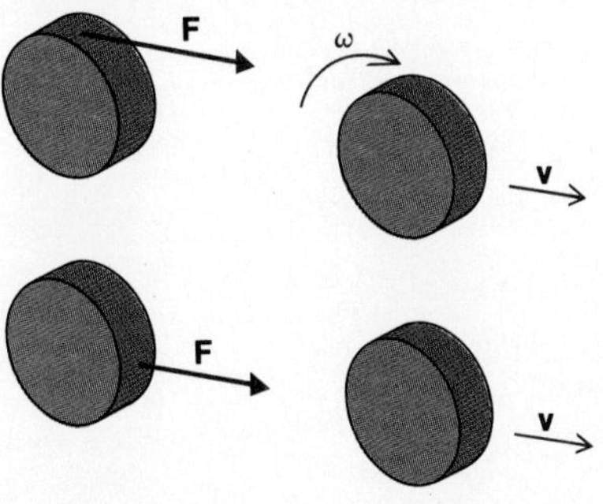

Figure 7.20

The power transmitted by a rotating shaft is the product of the torque τ on the shaft and its angular speed ω, hence $P = \tau\omega$.

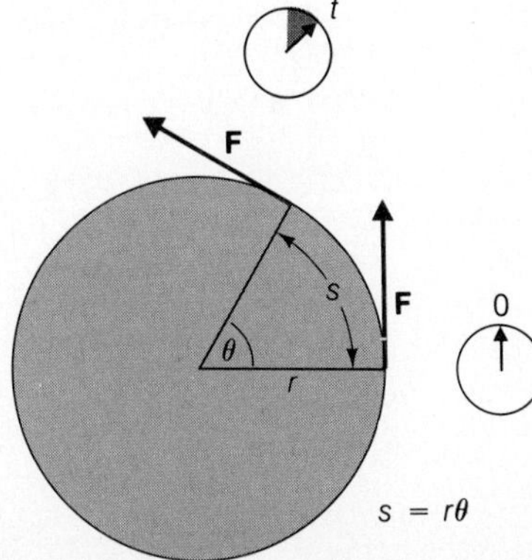

We recognize that Fr is the torque τ applied to the shaft, so

$$W = \tau\theta \qquad \textit{Rotational work} \qquad (7.21)$$

$$\text{Work} = (\text{torque})(\text{angular displacement})$$

Power is the rate at which work is done, which gives us here

$$P = \frac{W}{t} = \frac{Fr\theta}{t} = Fr\frac{\theta}{t}$$

Because $Fr = \tau$ and $\theta/t = \omega$, the angular speed of the shaft, we have

$$P = \tau\omega \qquad \textit{Rotational power} \qquad (7.22)$$

$$\text{Power} = (\text{torque})(\text{angular speed})$$

When energy is to be transmitted at a certain rate P by means of a rotating shaft, the higher the angular speed, the lower the torque needed, and vice versa. Equation (7.22) is the rotational analog of the formula

$$P = Fv$$

$$\text{Power} = (\text{force})(\text{speed})$$

EXAMPLE 7.12

The alternator on a truck engine produces 250 W of electric power when it turns at 3600 rpm, which is 377 rad/s. The alternator's pulley has a radius of 40.0 mm. If the alternator is 95% efficient, find the difference between the tensions in the taut and slack parts of the V-belt connecting it to the engine.

SOLUTION The efficiency of the alternator is Eff $= P_{out}/P_{in}$, and so the input power to its pulley is

$$P_{in} = \frac{P_{out}}{\text{Eff}} = \frac{250\text{ W}}{0.95} = 263\text{ W}$$

The torque on the alternator pulley is

$$\tau = \frac{P_{in}}{\omega}$$

Therefore the net force on the rim of the pulley must be

$$F = \frac{\tau}{r} = \frac{P_{in}/\omega}{r} = \frac{P_{in}}{\omega r} = \frac{263\text{ W}}{(377\text{ rad/s})(0.0400\text{ m})} = 17.4\text{ N}$$

which is about 4 lb. This is the difference between the tensions in the two parts of the V-belt (Fig. 7.21).

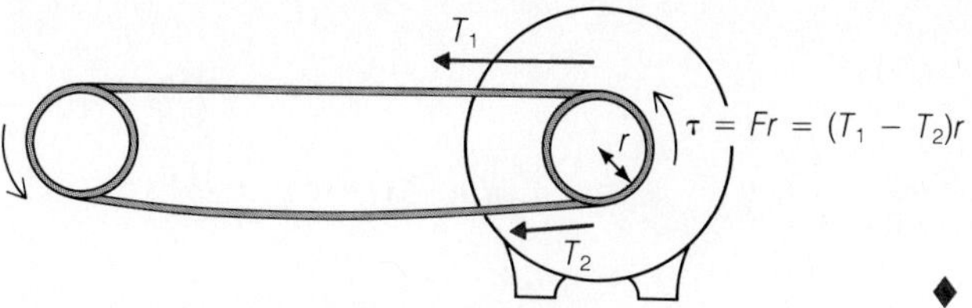

Figure 7.21

The torque on a V-belt pulley is due to the difference in the tensions in the taut and slack parts of the V-belt.

7.9 Angular Momentum

Its conservation is an important physical principle.

We have all noticed that rotating objects tend to keep spinning. A top would spin forever if friction did not slow it down. The rotational equivalent of linear momentum is called **angular momentum,** and **conservation of angular momentum** is the formal way to describe the tendency of a spinning object to continue to spin.

The magnitude L of the angular momentum of a body depends on its moment of inertia I and angular speed ω in the same way its linear momentum depends on its mass m and linear speed v:

$$L = I\omega \qquad \textit{Angular momentum} \quad (7.23)$$

Angular momentum = (moment of inertia) (angular speed)

Like linear momentum, angular momentum is a vector quantity. The direction associated with the angular momentum **L** of a spinning object is given by the right-hand rule shown in Fig. 7.22.

We recall from Chapter 5 that, if no net force acts on a body, its linear momentum is constant. Similarly, if no net torque acts on a rigid body, its angular momentum is constant. In fact, the angular momentum of a body does not change when $\Sigma\tau = 0$ even if it is *not* a rigid body. If a body is so altered during its motion that its moment of inertia changes, ω also changes to keep **L** the same. Thus we have the useful principle of **conservation of angular momentum**:

When the sum of the external torques acting on a system of particles is zero, the total angular momentum of the system remains constant.

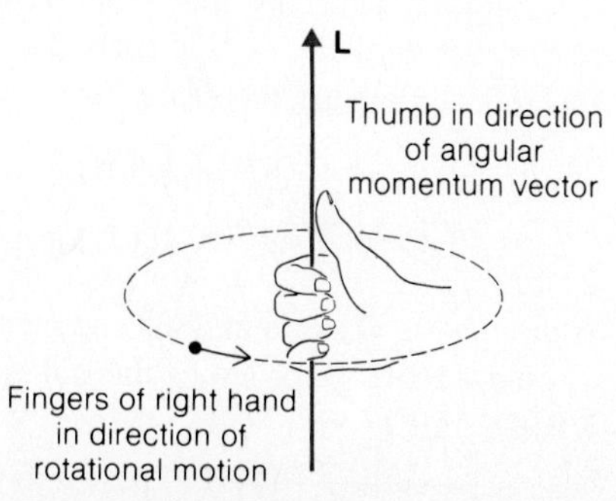

Figure 7.22

Angular momentum is a vector quantity whose direction is given by the right-hand rule shown here.

Figure 7.23

Angular momentum is conserved when a skater executes a spin.

When a skater brings her arms and legs close to her axis of rotation after starting a spin, her moment of inertia is decreased and she spins faster as a result.

In equation form, if $I_1\omega_1$ is the initial angular momentum of a system of particles and $I_2\omega_2$ is its final angular momentum, then if no torque has acted on the system

$$I_1\omega_1 = I_2\omega_2 \qquad \textit{Conservation of angular momentum} \qquad (7.24)$$

A skater or ballet dancer doing a spin capitalizes on conservation of angular momentum. In Fig. 7.23 a skater is shown starting her spin with her arms and one leg outstretched. By bringing her arms and extended leg inward, she reduces her moment of inertia considerably and consequently spins faster.

EXAMPLE 7.13

A skater has the moment of inertia 3.0 kg·m^2 when her arms are outstretched and 1.0 kg·m^2 when her arms are brought to her sides. She starts to spin at the rate of 1.0 rev/s when her arms are outstretched, and then pulls her arms to her sides. (a) What is her final angular speed? (b) How much (if any) work did she have to do?

SOLUTION (a) Because the conversion factors would only cancel anyway, we can use the rev/s as the unit of angular speed. Here $I_1 = 3.0$ kg·m^2, $I_2 = 1.0$ kg·m^2, and $\omega_1 = 1.0$ rev/s. From conservation of angular momentum

$$I_1\omega_1 = I_2\omega_2$$

$$\omega_2 = \frac{I_1}{I_2}\omega_1 = \left(\frac{3.0\ \text{kg}\cdot\text{m}^2}{1.0\ \text{kg}\cdot\text{m}^2}\right)(1.0\ \text{rev/s}) = 3.0\ \text{rev/s}$$

The skater spins three times faster than she did with her arms outstretched.

(b) The work the skater does must equal the change (if any) in her kinetic energy. To calculate KE here, we must use the rad/s as the unit of angular velocity. Because 1 rev $= 2\pi$ rad, $\omega_1 = 2\pi$ rad/s and $\omega_2 = 6\pi$ rad/s. The skater's initial and final kinetic energies are therefore

$$\text{KE}_1 = \tfrac{1}{2}I_1\omega_1^2 = \tfrac{1}{2}(3.0\ \text{kg}\cdot\text{m}^2)(2\pi\ \text{rad/s})^2 = 59\ \text{J}$$

$$\text{KE}_2 = \tfrac{1}{2}I_2\omega_2^2 = \tfrac{1}{2}(1.0\ \text{kg}\cdot\text{m}^2)(6\pi\ \text{rad/s})^2 = 178\ \text{J}$$

Although the skater's angular momentum stayed the same, her kinetic energy did not. The new KE is three times the old KE, and the work she had to do is $W = \text{KE}_1 - \text{KE}_2 = 119$ J. ♦

Table 7.3 Comparison of linear and angular quantities

Linear quantity		Angular quantity	
Distance	$x = v_0t + \frac{1}{2}at^2$	Angle	$\theta = \omega_0t + \frac{1}{2}\alpha t^2$
	$x = \left(\frac{v_0 + v_f}{2}\right)t$		$\theta = \left(\frac{\omega_0 + \omega_f}{2}\right)t$
Speed	$v = v_0 + at$	Angular speed	$\omega = \omega_0 + \alpha t$
	$v^2 = v_0^2 + 2ax$		$\omega^2 = \omega_0^2 + 2\alpha\theta$
Acceleration	$a = \Delta v/\Delta t$	Angular acceleration	$\alpha = \Delta\omega/\Delta t$
Mass	m	Moment of inertia	I
Force	$F = ma$	Torque	$\tau = I\alpha$
Momentum	$p = mv$	Angular momentum	$L = I\omega$
Work	$W = Fx$	Work	$W = \tau\theta$
Power	$P = Fv$	Power	$P = \tau\omega$
Kinetic energy	$KE = \frac{1}{2}mv^2$	Kinetic energy	$KE = \frac{1}{2}I\omega^2$

Like the conservation principles of energy and of linear momentum, the conservation of angular momentum turns out to be a consequence of a symmetry property of the universe. If the laws of nature are independent of direction in space—that is, if the laws of nature are the same regardless of how an observer is oriented with respect to an event of some kind—then angular momentum must be conserved in all interactions, as observed.

As we have seen, there are many points of correspondence between angular and linear motion at constant speed and under constant acceleration. Table 7.3 lists the principal ones. Although the symbols are different, the formulas are the same because the basic concepts involved are so closely related.

Spin Stabilization

Because angular momentum is a vector quantity, a torque must be applied to change the orientation of the axis of rotation of a spinning body as well as to change the magnitude of its angular velocity. The greater the magnitude of **L**, the more torque is needed for it to deviate from its original direction. This is the principle behind the spin stabilization of projectiles such as footballs and rockets. Such projectiles are set spinning about axes in their directions of motion so that they do not tumble and thereby increase air resistance. A top is another illustration of the vector nature of angular momentum. A stationary top set on its tip falls over at once, but a rotating top stays upright until most of its angular momentum is lost by friction between its tip and the ground (Fig. 7.24).

Figure 7.24

The faster a top spins, the more stable it is. When most of its angular momentum has been lost through friction, the top falls over.

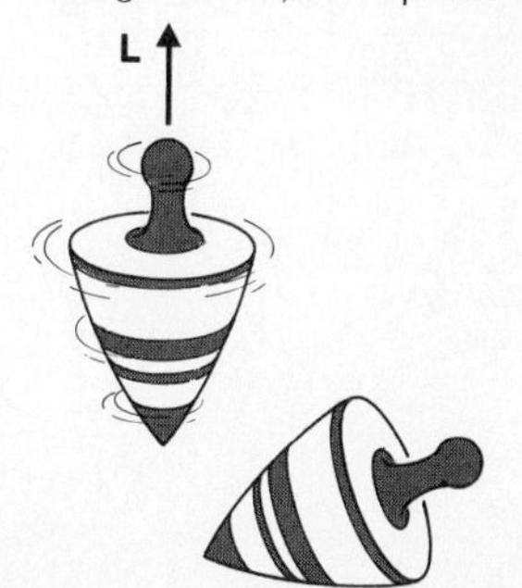

A **gyroscope** is a disk with a high moment of inertia whose axis is mounted on gimbals, as in Fig. 7.25. When it is set in motion, conservation of angular momentum keeps the disk in a fixed orientation in space regardless of how its support is moved. Thus a gyroscope can be used on an airplane as an artificial horizon when the true horizon is not visible and as the basis of a compass that operates independently of the earth's magnetic field.

The spinning earth behaves like a gyroscope as it revolves around the sun. The earth's axis is tilted by 23.5° with respect to the axis of its orbit, and the direction of the tilt is very nearly constant relative to the orbit axis (Fig. 7.26). As a result, for half of each year one hemisphere of the earth receives more direct sunlight than

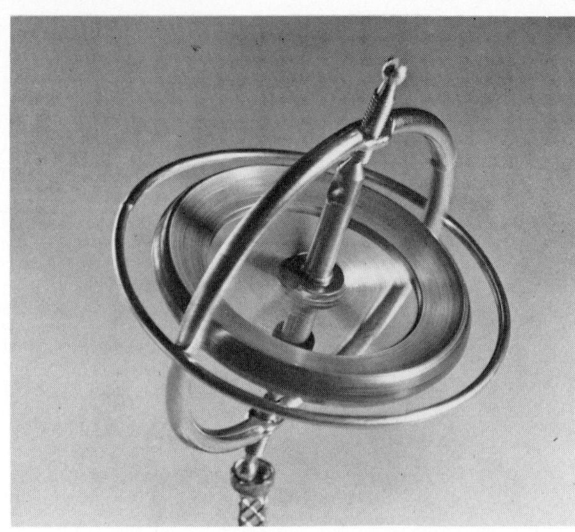

Figure 7.25

A gyroscope.

the other hemisphere, and in the other half it receives less. (A beam of light that arrives at an angle to a surface delivers less energy per square meter than does a similar perpendicular beam, as we can see in Fig. 7.27.) It is tempting to think that the seasons occur because the sun–earth distance changes during the year. This cannot be the case, however, because the seasons are reversed in the two hemispheres: Summer in the United States is winter in Australia.

As mentioned above, the direction of the earth's axis is actually not quite constant. Because of gravitational torques exerted by the sun and moon on the tilted earth's equatorial bulge, this direction traces out a cone every 26,000 years. The effect is called **precession.** The changing direction of the axis of a spinning top is another example of precession.

Conservation of angular momentum provides an elegant way to reduce the angular speed of a spin-stabilized object in space where friction is not available. The spacecraft that carried the European satellite *ESRO 1* into orbit was set rotating at about 100 rpm to stabilize it before the last rocket stage was separated. Afterward

Figure 7.26

The origin of the seasons. As the earth moves around the sun, the daylight side of the northern hemisphere is tilted away from the sun in January and toward the sun in June. As a result January is a winter month and June a summer month in this hemisphere. The seasons are reversed in the southern hemisphere.

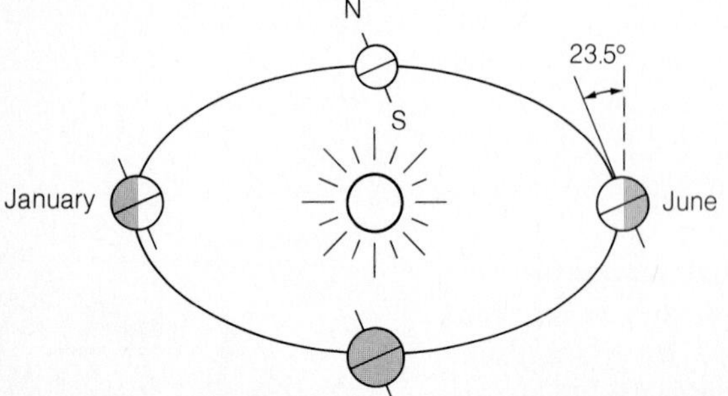

Figure 7.27

Sunlight that reaches a surface at an angle has its energy spread over a larger area than sunlight that reaches the surface perpendicularly.

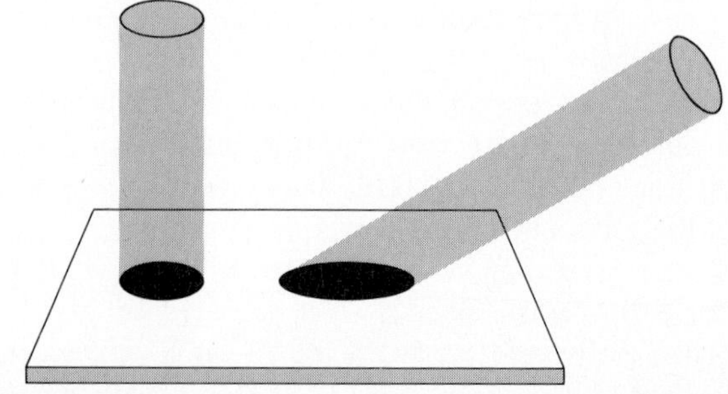

the rotation rate was slowed to about 20 rpm by allowing two strings with masses at their ends to unwind from the spinning satellite, thereby increasing I and lowering ω. The strings were then let go, and the masses flew off into space together with the unwanted angular momentum.

7.10 Kepler's Second Law

It follows from conservation of angular momentum.

Kepler's laws of planetary motion were mentioned in Section 6.9. The second of these laws states that the speed of a planet in its elliptical orbit around the sun varies so that a radius vector from the sun to it sweeps out equal areas in equal times (Fig. 6.27). Let us see how this law follows from the conservation of angular momentum. Angular momentum must be conserved by a planet because the gravitational force the sun exerts on it acts along the line joining the centers of the two bodies and so results in no torque on the planet.

Figure 7.28 shows the position of a planet at two different times the brief interval Δt apart. In this interval the planet's radius vector moves through the angle $\Delta\theta = \omega\,\Delta t$, where ω is the planet's angular velocity in that part of its orbit. In the same time interval, the radius vector increases from r to $r + \Delta r$, so that it sweeps out what is very close to a triangle whose base is $r + \Delta r$ and whose altitude is $r\Delta\theta$. The area ΔA of this triangle is

$$\text{Area} = \tfrac{1}{2}(\text{base})(\text{altitude})$$
$$\Delta A = \tfrac{1}{2}(r + \Delta r)(r\Delta\theta)$$

Because Δt is small, so is Δr (both are exaggerated in the figure for clarity), and $r + \Delta r$ is very close to r. We may therefore neglect Δr, which gives

$$\Delta A = \tfrac{1}{2}r^2\Delta\theta$$

The rate at which the radius vector sweeps out area is therefore

$$\frac{\Delta A}{\Delta t} = \tfrac{1}{2}r^2\frac{\Delta\theta}{\Delta t} = \tfrac{1}{2}r^2\omega$$

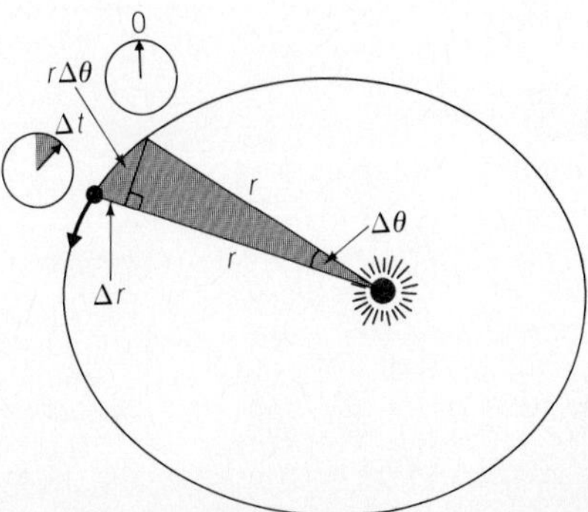

Figure 7.28

Kepler's second law of planetary motion is a consequence of the conservation of angular momentum.

The moment of inertia of the planet about an axis through the sun is $I = mr^2$, where m is the planet's mass. The planet's angular momentum is

$$L = I\omega = mr^2\omega$$

The law of areas thus becomes

$$\frac{\Delta A}{\Delta t} = \frac{L}{2m}$$

Because L must be constant, $\Delta A/\Delta t$ must also be constant, and planets move faster when they are near the sun than they do when they are far from it. In the case of the earth, the difference in speed is about 1 km/s, somewhat over 3%.

Important Terms

The **radian** is a unit of angular measure equal to 57.30°. If a circle is drawn whose center is at the vertex of an angle, the angle in radian measure is equal to the ratio between the arc of the circle cut by the angle and the radius of the circle. A full circle contains 2π radians.

The **angular speed** ω of a rotating body is the angle through which it turns per unit time. The **angular acceleration** α of a rotating body is the rate of change of its angular speed with respect to time.

The **axis of rotation** of a rigid body turning in place is that line of particles which does not move.

All particles in circular motion experience centripetal accelerations, but only those particles whose angular speed changes have **tangential accelerations.**

The **moment of inertia** I of a body about a given axis is the rotational analog of mass in linear motion. Its value depends on the way in which the mass of the body is distributed about the axis.

The **torque** of a force about a particular axis is the product of the magnitude of the force and the perpendicular distance from the line of action of the force to the axis. The latter distance is called the **moment arm** of the force.

The **angular momentum** L of a rotating body is the product $I\omega$ of its moment of inertia and angular speed. The principle of **conservation of angular momentum** states that the total angular momentum of a system of particles remains constant when no net external torque acts on it.

Important Formulas

Angle in radian measure: $\theta = \frac{s}{r}$

$$\theta = \left(\frac{\omega_0 + \omega_f}{2}\right)t$$

Angular speed: $\omega = \frac{\theta}{t}$

Linear speed: $v = \omega r$

Moment of inertia: $I = \Sigma mr^2$

Rotational kinetic energy: $\text{KE} = \frac{1}{2}I\omega^2$

Motion under constant angular acceleration:

$$\theta = \omega_0 t + \tfrac{1}{2}\alpha t^2$$

$$\omega_f = \omega_0 + \alpha t$$

$$\omega_f^2 = \omega_0^2 + 2\alpha\theta$$

Angular acceleration: $\alpha = \frac{\Delta\omega}{\Delta t} = \frac{a_T}{r}$

Torque: $\tau = FL = I\alpha$

Work and power: $W = \tau\theta$

$P = \tau\omega$

Angular momentum: $L = I\omega$

MULTIPLE CHOICE

1. In a rigid object undergoing uniform circular motion, a particle that is a distance R from the axis of rotation

a. has an angular speed proportional to R
b. has an angular speed inversely proportional to R
c. has a linear speed proportional to R
d. has a linear speed inversely proportional to R

2. The centripetal acceleration of a particle in circular motion

a. is less than its tangential acceleration
b. is equal to its tangential acceleration
c. is more than its tangential acceleration
d. may be more or less than its tangential acceleration

3. The rotational analog of force in linear motion is

a. moment of inertia
b. angular momentum
c. torque
d. weight

4. The rotational analog of mass in linear motion is

a. moment of inertia b. angular momentum
c. torque d. angular speed

5. A quantity not directly involved in the rotational motion of an object is

a. mass b. moment of inertia
c. torque d. angular speed

6. All rotating objects at sea level that have the same mass and angular velocity also have the same

a. angular momentum
b. moment of inertia
c. kinetic energy
d. gravitational potential energy

7. The moment of inertia of an object does not depend on

a. its mass
b. its size and shape
c. its angular speed
d. the location of the axis of rotation

8. Of the following properties of a yo-yo moving in a circle, the one that does not depend on the radius of the circle is the yo-yo's

a. angular speed
b. angular momentum
c. linear speed
d. centripetal acceleration

9. A yo-yo being swung in a circle need not possess

a. angular speed
b. angular momentum
c. angular acceleration
d. centripetal acceleration

10. The total angular momentum of a system of particles

a. remains constant under all circumstances
b. changes when a net external force acts on the system
c. changes when a net external torque acts on the system
d. may or may not change under the influence of a net external torque, depending on the direction of the torque

11. A hoop and a disk of the same mass and radius roll down an inclined plane. At the bottom they have the same

a. angular speed b. angular momentum
c. KE of rotation d. gravitational PE

12. A solid disk rolls from rest down an inclined plane. A hole is then cut in the center of the disk, and the resulting doughnut-shaped object is allowed to roll down the same inclined plane. Its speed at the bottom of the plane is

a. less than that of the solid disk
b. the same as that of the solid disk
c. greater than that of the solid disk
d. any of the above, depending on the size of the hole

13. A solid iron sphere A rolls down an inclined plane, while an identical sphere B slides down the plane in a frictionless manner.

a. Sphere A reaches the bottom first.
b. Sphere B reaches the bottom first.
c. They reach the bottom together.
d. Which one reaches the bottom first depends on the angle of the plane.

14. At the bottom of the plane, the total KE of sphere A in Multiple Choice 13 is

a. less than that of sphere B
b. equal to that of sphere B
c. more than that of sphere B
d. more or less than that of sphere B, depending on the angle of the plane

15. The number of radians in half a circle is

a. $\pi/2$ b. π
c. 2π d. 4π

16. An angle of 20° is equivalent to

a. 0.17 rad b. 0.20 rad
c. 0.35 rad d. 3.18 rad

17. An angle of 540° is equivalent to

a. $2\pi/3$ rad b. π rad
c. $3\pi/2$ rad d. 3π rad

18. An angle of $\pi/6$ rad is equivalent to

a. 15° b. 30°
c. 60° d. 90°

19. A wheel is 1 m in diameter. When it makes 30 rev/min, the linear speed of a point on its circumference is

a. $\pi/2$ m/s b. π m/s
c. 30π m/s d. 60π m/s

20. The shaft of a motor turns at a constant 3000 rpm. In 40 s it turns through

a. 2000 rad b. 6283 rad
c. 12,000 rad d. 12,566 rad

21. An object undergoes a uniform angular acceleration. In the time t since the object started rotating from rest, the number of turns it makes is proportional to

a. $\sqrt{t}$ b. t
c. t^2 d. t^3

22. A wheel that starts from rest has an angular speed of 20 rad/s after being uniformly accelerated for 10 s. The total angle through which it has turned in these 10 s is

a. 2π rad b. 40π rad
c. 100 rad d. 200 rad

23. A motor takes 8.00 s to go from 60 rad/s to 140 rad/s at constant angular acceleration. The total angle through which its shaft turns during this time is

a. 320 rad b. 640 rad
c. 800 rad d. 1600 rad

24. A flywheel rotating at 10 rev/s is brought to rest by a constant torque in 15 s. In coming to a stop, the flywheel makes

a. 75 rev b. 150 rev
c. 472 rev d. 600 rev

25. A constant torque of 60 N·m acts on a 20-kg flywheel whose moment of inertia is 12 kg·m^2. The time needed for the flywheel to reach 75 rad/s starting from rest is

a. 15 s b. 25 s
c. 30 s d. 180 s

26. A torque of 30 N·m acts on a 5.0-kg wheel of moment of inertia 2.0 kg·m^2. If the wheel starts from rest, after 10 s it will have turned through

a. 750 rad b. 1500 rad
c. 3000 rad d. 6000 rad

27. The final KE of the wheel in Multiple Choice 26 will be

a. 9 kJ b. 23 kJ
c. 45 kJ d. 56 kJ

28. A tangential force of 10 N acts on the rim of a wheel 80 cm in diameter. The wheel takes 1.5 s to complete its first revolution starting from rest. The wheel's moment of inertia is

a. 0.72 kg·m^2 b. 0.96 kg·m^2
c. 1.8 kg·m^2 d. 4.5 kg·m^2

29. A motor develops 56 kW when its shaft turns at 300 rad/s. The torque on the shaft

a. is 30 N·m
b. is 187 N·m
c. is 17 kN·m
d. depends on the shaft diameter

QUESTIONS

1. Many flywheels have most of their mass concentrated around their rims. What is the advantage of this?

2. The density of an object is its mass divided by its volume. The density of iron is about three times the density of aluminum. If an iron cylinder and an aluminum cylinder have the same length and the same mass, which has the greater moment of inertia?

3. A square and a rectangle of the same mass are cut from a sheet of metal. Which has the greater moment of inertia about a perpendicular axis through its center?

4. Will a car coast downhill faster if it has heavy tires or light tires?

5. A hollow cylinder and a solid cylinder having the same mass and diameter are released from rest simultaneously at the top of an inclined plane. Which reaches the bottom first?

6. Which will roll to the bottom of an inclined plane first, a can that contains ice or an identical can that contains water?

7. An aluminum cylinder of radius R, a lead cylinder of radius R, and a lead cylinder of radius $2R$ all roll from rest down an inclined plane. In what order will they reach the bottom?

8. A person on a bicycle presses down on each pedal through the front half of its circle. In which position is the torque a maximum? In which positions is it zero?

9. What determines the moment arm of a force?

10. The lower the center of gravity of an object, the more stable it is. However, it is easier to balance a billiard cue vertically when its tip is on one's finger with its heavy handle up in the air rather than with its handle on the finger. Why?

11. Why does a diver tuck her knees to her chest when turning somersaults during a dive?

12. All helicopters have two propellers. Some have both propellers on vertical axes but rotating in opposite directions, and others have one on a vertical axis and one on a horizontal axis perpendicular to the helicopter body at the tail. Why is a single propeller never used?

13. If the polar ice caps melt, how will the length of the day be affected?

14. The dimensions of all mechanical quantities can be expressed in terms of the fundamental dimensions of length L, time T, and mass M. For example, the dimensions of acceleration are LT^{-2}. Find the dimensions in terms of L, T, and M of force, torque, energy, power, linear momentum, angular momentum, and impulse.

EXERCISES

7.1 Angular Measure

1. An apple pie is cut into nine equal pieces. What angle (in radians) is included between the sides of each piece?

2. How many radians does the second hand of a clock turn through in 30 s? In 90 s? In 105 s? Express the answers in terms of π.

3. The resolution of the human eye is about $1'$, where $60' = 1°$. (a) Express an angle of $1'$ in radians. (b) What is the length of the smallest detail that can be discerned when an object 25 cm away is being examined? (This is the distance of most distinct vision.)

4. In a good light a golden eagle can detect a 50-cm hare at a distance of 2 km. What angle does this represent, in radians?

5. A phonograph record 30 cm in diameter turns through an angle of 130°. Through what linear distance does a point on its rim travel?

6. Show that the ratio between the arc and the chord in Fig. 7.3 is given by $\theta/[2 \sin(\theta/2)]$, where θ is expressed in radians.

7.2 Angular Speed

7. What is the angular speed in radians per second of the hour, minute, and second hands of a clock?

8. A car makes a U-turn in 5.0 s. What is its average angular speed?

9. In 1976 Kazuya Shiozaki made 49,299 turns in 5 h 37 min of skipping rope. Find the average angular speed of the rope.

10. The shaft of a motor rotates at the constant angular speed of 3000 rpm. How many radians will it have turned through in 10 s?

11. The blades of a rotary lawnmower are 30 cm long and rotate at 315 rad/s. Find the linear speed of the blade tips and their angular speed in rpm.

12. A drill bit 6.0 mm in diameter is turning at 1200 rpm. Find the linear speed of a point on its circumference.

13. In 1941 a wind-driven electric power station was set up at Grandpa's Knob, Vermont. The propeller was 40 m in diameter. When the propeller was turning at 30 rev/min, what was the linear speed of one of the blade tips in kilometers per hour?

14. A steel cylinder 40 mm in radius is to be machined in a lathe. At how many revolutions per second should it rotate in order that the linear speed of the cylinder's surface be 70 cm/s?

15. A rotating platform is to be used to test aircraft equipment under accelerations of $6g$. If the equipment is 50 cm from the axis, what angular speed is needed?

16. The propeller of a boat rotates at 100 rad/s when the speed of the boat is 6.0 m/s. The diameter of the propeller is 40 cm. What is the speed of the tip of the propeller?

17. A barrel 80 cm in diameter is rolling with an angular speed of 5.0 rad/s. What is the instantaneous velocity of (a) its top, (b) its center, and (c) its bottom, all with respect to the ground?

7.3 Rotational Kinetic Energy

7.4 Moment of Inertia

18. A 7.0-kg bowling ball 30 cm in diameter rolls at a speed of 5.0 m/s. What is its total kinetic energy?

19. The baton of a drum majorette is a 300-g uniform rod 80 cm long. What is its kinetic energy when it is twirled about its center at an angular velocity of 10 rad/s?

20. The wheel shown in Fig. 7.29 has a diameter of 1.2 m. Its rim has a mass of 12 kg, and each spoke has a mass of 1.0 kg. Find the moment of inertia of the wheel.

21. The **radius of gyration** of an object is the distance from a specified axis of rotation to a point at which the object's entire mass may be considered to be concentrated from the point of view of rotational motion about that axis. Thus the moment of inertia of an object of mass M and radius of gyration k is $I = Mk^2$. (a) The radius of gyration about its center of a hollow sphere of radius R and mass M is $k = \sqrt{\frac{2}{3}}R$. Find its moment of inertia. (b) Find the radius of gyration of a solid sphere about its center.

22. An 80-kg flywheel 35 cm in radius has a radius of gyration of 30 cm. (a) Find its moment of inertia. (b) Find the radius of a solid disk with the same mass and moment of inertia. What does this suggest about the cross-sectional form of the flywheel?

23. The earth's mass is 6.0×10^{24} kg, and its radius is 6.4×10^6 m. (a) Find its rotational kinetic energy under the assumption that it is a uniform sphere. (b) The earth's core is actually much denser than the surrounding mantle and crust, as shown in Fig. 6.21. How does this affect its rotational kinetic energy?

24. The seat of an adjustable piano stool has the mass m and the moment of inertia I. If the seat is given an initial angular speed of ω in such a direction that the seat rises and there is no friction, how much higher is the seat when it stops turning?

Figure 7.29
Exercise 20.

7.5 Combined Translation and Rotation

25. A hoop rolls down an inclined plane. What fraction of its total KE is associated with its rotation?

26. A thin-walled hollow sphere rolls down an inclined plane at the same speed as a hollow cylinder whose outer diameter is the same. What is the ratio between the inner and outer diameters of the cylinder?

27. A solid sphere of mass M_1 and radius R_1 and a hollow sphere of mass M_2 and radius R_2 start rolling down an inclined plane at the same moment. Find the ratio of their linear speeds at the bottom of the plane.

28. A hollow cylinder of mass M and radius R rolls down an inclined plane of height h and reaches the bottom with a linear speed of $v = \sqrt{8gh/7}$. Find I for the cylinder.

29. A 3.0-kg hoop 1.0 m in diameter rolls down an inclined plane 10 m long that is at an angle of 20° with the horizontal. (a) What is the angular speed of the hoop at the bottom of the plane? (b) What is its linear speed? (c) What is its rotational kinetic energy? (d) What is its total kinetic energy?

7.6 Angular Acceleration

30. A wheel starts from rest at a constant angular acceleration and turns through 500 radians in 10 s. (a) What is this angular acceleration? (b) What is the final angular speed?

31. A truck undergoes an acceleration of 0.25 m/s^2. If its wheels are 1.0 m in diameter, what is their angular acceleration?

32. A circular saw blade rotating at 15 rev/s is brought to a stop in 125 revolutions. How much time did this take?

33. An engine idling at 10 rev/s is accelerated at 2.5 rev/s^2 to 20 rev/s. How many revolutions does it make during this acceleration?

34. A phonograph turntable slows down to a stop in 20 s from an initial angular speed of 3.5 rad/s. (a) What is its angular acceleration? (b) How many turns does it make while slowing down?

35. A record is dropped on a turntable rotating at $33\frac{1}{3}$ rpm. The record slips briefly before reaching the same speed as the turntable. If the turntable rotates through 20° while the record is slipping, find the record's angular acceleration.

7.7 Torque

36. A **couple** consists of two forces of equal magnitude that act in opposite directions along parallel lines of action a distance d apart. If the magnitude of each force is F, what is the torque exerted by a couple?

37. The torque required to fracture a person's tibia (the large bone in the lower leg) is about 100 N·m. If the distance between the pivoting heel of the safety binding on a ski and the toe release is 35 cm, what is the maximum horizontal force at which the release should be set to open?

38. A rope 1.0 m long is wound around the rim of a drum of radius 12 cm and moment of inertia 0.020 kg·m^2. The rope is pulled with a force of 2.5 N. (a) Assuming that the drum is free to rotate without friction, find its final angular speed. (b) What is its final kinetic energy? (c) How much work is done by the force?

39. A torque of 500 N·m is applied to a turbine rotating at 200 rad/s. After 40 s its speed has doubled. What is the turbine's moment of inertia?

40. A cylinder whose axis is fixed has a string wrapped around it that is pulled with a force equal to the cylinder's weight. Show that the acceleration of the string is equal to $2g$.

41. A 200-kg cylindrical flywheel 30 cm in radius is acted on by a torque of 20 N·m. (a) If it starts from rest, how much time is required to accelerate it to an angular speed of 10 rad/s? (b) What is its kinetic energy at this angular speed?

7.8 Work and Power

42. The cutting tool of a lathe exerts a torque of 20 N·m on a steel shaft. Find the power needed to turn the shaft at 8.0 rev/s.

43. A 1.0-kW motor rotates at 125 rad/s. How much torque can it exert?

44. A diesel engine can exert 140 N·m of torque when it is turning at 250 rad/s. Find the corresponding power output.

45. A high-speed elevator is being planned to lift a total load of 4000 kg at 6.0 m/s. The winding drum is to be 1.8 m in diameter. Neglecting losses, how much power is required from the driving motor? At what angular speed should the power be developed?

46. A lathe whose motor has a power output of 2.0 kW is being used to cut a round bar 12 cm in diameter that is turning at 200 rpm. If 10% of the motor's power is lost in friction, what is the maximum force the cutting tool can exert on the bar?

47. An electric motor develops 5.0 kW at 2000 rpm. The motor delivers its power through a spur gear 20 cm in diameter. If two teeth of the gear transmit torque to another gear at the same instant, find the force exerted by each gear tooth.

48. A V-8 engine that develops 200 kW at 90 rev/s is coupled through a frictionless 5:1 reduction gear to a windlass drum 80 cm in diameter. (a) What is the heaviest mass the windlass can raise? (b) At what linear speed will it raise such a load?

49. A 30-rev/s motor operates a pump through a V-belt drive. The motor pulley is 20 cm in diameter, and the tension in the belt is 180 N on one side and 70 N on the other. What is the power output of the motor?

7.9 Angular Momentum

50. (a) What is the angular momentum of a particle of mass m that moves in a circle of radius r at the speed v? (b) An earth satellite follows an elliptical orbit in which its maximum altitude above the earth's surface is 2000 km and its minimum altitude is 400 km. Use the result of (a) to find the ratio between the maximum and minimum speeds of the satellite.

51. Two objects, each of mass m, are moving at the same speed v in opposite directions along parallel straight lines s apart. What is the angular momentum of the system about a point equidistant from the objects? If the point were elsewhere, would the angular momentum be different?

52. A phonograph record of mass m and radius R is dropped on a turntable with the same radius and the mass M whose initial angular speed is ω. The turntable is rotating freely with no torque acting on it. What is the final angular speed of the system? Assume that the turntable is a uniform disk.

53. A model train of mass m runs at the speed v (relative to the track) around a circular track mounted on the edge of a disk of mass M and radius R. The disk can turn freely about a vertical axis through its center. If the disk was at rest when the train started to move, what is the angular speed of the disk?

PROBLEMS

54. The tires of a car on a muddy road are 35 cm in radius and are rotating at 40 rad/s. What is the maximum distance a lump of mud can be thrown behind a tire?

55. A neutron speed selector consists of two disks of a neutron-absorbing material mounted 1.0 m apart on a shaft that turns at 1000 rpm, as in Fig. 7.30. Each disk has a radial slot, and the angular separation of the slots is 3″ (1° = 60′ and 1′ = 60″). What is the speed of the neutrons that can pass through this device?

56. Two women are throwing a ball at the speed v back and forth on a merry-go-round of radius R and angular speed ω. Carla is at the center and Rose is at the rim. (a) In what direction relative to the line between them should Carla throw the ball so it reaches Rose? (b) How long does the ball take to reach Rose? (c) and (d) Answer the same questions for Rose's throw to Carla.

57. A uniform thin rod of length L is pivoted about a horizontal axis at one end. If it is released from a horizontal position, what will its maximum angular velocity be?

58. A uniform knitting needle 20 cm long is balanced on its point on a table. After a moment it falls over. Find the speed with which the upper end of the needle strikes the table.

59. A bottle of mass m is tied to the free end of a string wrapped around a uniform cylinder of mass M and radius R whose axis is horizontal. If the cylinder can turn freely, what is the linear speed of the bottle after it has fallen through a height h?

Figure 7.30

Problem 55 (3″ = 3 sec of arc).

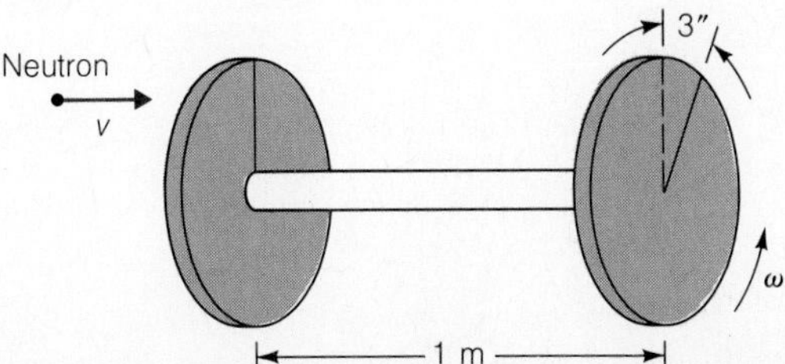

60. A uniform cylinder of radius R and mass m rolls down a plane inclined at an angle of θ with the horizontal. What is the cylinder's linear acceleration down the plane? Does this acceleration depend on whether the cylinder starts from rest or has an initial speed?

61. A string is wrapped around a thin-walled hollow cylindrical spool of mass M and radius R. The outer end of the string is held in place, and the spool is allowed to fall with the string unwinding as it descends. (a) Show that the linear acceleration a of the spool is $g/2$ by starting from the fact that the tension in the string is $M(g - a)$. (b) Verify this result by using conservation of energy to find the linear speed v of the spool after it has fallen a distance h and then obtaining a from v and h.

62. Find the linear acceleration of the spool in Problem 61 if it is a solid cylinder.

63. The pulley in the arrangement described in Example 3.5 (Fig. 3.16) is a uniform disk whose mass is 10 kg. Find the acceleration of the blocks.

64. Blocks whose masses are 200 g and 250 g are suspended from the ends of a string that passes over a pulley 30 cm in diameter, as in Fig. 7.18. The pulley's moment of inertia is 0.12 kg·m^2. If the blocks start from rest at the same level, what are their linear speeds when they are 30 cm apart?

65. The uniform door to a bank vault is 2.0 m high and 1.0 m wide and is supported by frictionless hinges. To determine the door's mass, a person pushes against its free end with a constant force of 250 N always directed perpendicularly to its plane and finds that 1.5 s is needed to turn the door through 90°. What is the door's mass? (*Note:* The formula for the moment of inertia of a uniform rectangular plate about one of its edges is the same as one of the formulas given in Fig. 7.10.)

66. An object of mass m on a frictionless table is attached to the end of a string that passes through a frictionless hole in the table, as shown in Fig. 7.31. The object is moving in a circle of radius R with the angular speed ω when the string is pulled through the hole until the object's orbital radius is $R/2$. (a) What is the new angular speed of the object? (b) How much work had to be done to pull in the string?

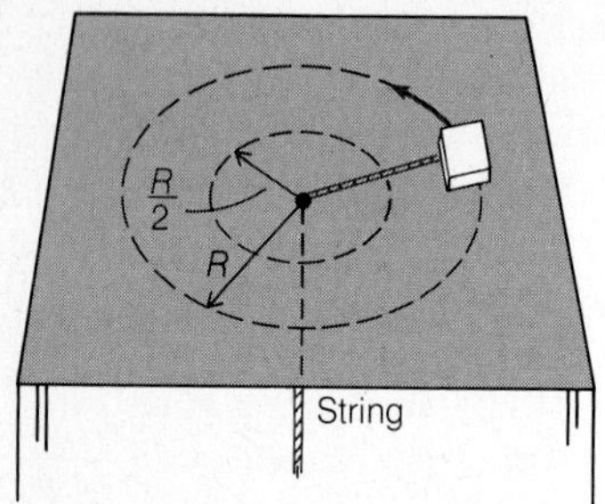

Figure 7.31

Problem 66.

67. Two 0.40-kg balls are joined by a 1.0-m string and set whirling through the air at 5.0 rev/s about a vertical axis through the center of the string. After a while the string stretches to 1.2 m. (a) What is the new angular speed? (b) What are the initial and final kinetic energies of each ball? If these are different, account for the difference.

68. A woman is sitting on a piano stool with 3.0 kg dumbbells in her outstretched arms 60 cm from the axis of the stool. She is given a push and starts to turn at 1.0 rev/s. Her arms become tired, and she performs 70 J of work to pull the dumbbells to her lap, where they are 10 cm from the axis. Find the woman's moment of inertia about the axis of the stool. Neglect the contribution of her arms to the moment of inertia of the system.

69. A disk of moment of inertia 1.0 kg·m^2 that is rotating at 100 rad/s is pressed against a similar disk that is initially at rest but is able to rotate freely. The two disks stick together and rotate as a unit. (a) What is the final angular speed of the combination? (b) If any kinetic energy was lost, where did it go? (c) The two disks are now separated. What are their new angular speeds? (d) If any kinetic energy was lost in the separation, where did it go?

Answers to Multiple Choice

1. c	**9.** c	**17.** d	**25.** a
2. d	**10.** c	**18.** b	**26.** a
3. c	**11.** d	**19.** a	**27.** b
4. a	**12.** a	**20.** d	**28.** a
5. a	**13.** b	**21.** c	**29.** b
6. d	**14.** b	**22.** c	
7. c	**15.** b	**23.** c	
8. a	**16.** c	**24.** a	

CHAPTER

8

EQUILIBRIUM

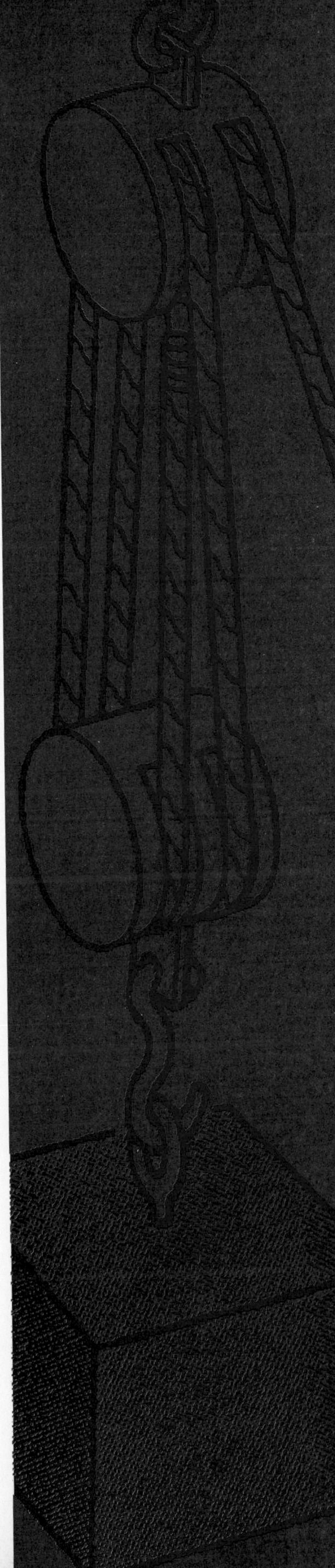

Thus far we have been concerned with moving objects, the study of which is called dynamics. Now, in a change of pace, we look at statics, the study of objects at rest. Our interest centers on the conditions under which something acted on by two or more forces is nevertheless not accelerated. It is not enough that such forces should balance out to give no net force: If the lines of action of the forces do not meet at a common point, they will produce a torque that starts the object spinning. Evidently an object at rest cannot have either a net force or a net torque acting on it. This is a simple enough idea, but one that is not always so simple to apply.

8.1 Translational Equilibrium

It occurs when force components in each direction add up to zero.

We begin by considering only forces that act through the same point on an object. This allows us to ignore torques. Later we consider situations in which torques are important.

When the forces that act on an object have a vector sum of zero, the object is said to be in **translational equilibrium.** With no net force there is no linear acceleration. Of course, such an object need not be at rest but may be moving along a straight path at constant speed.

The condition for translational equilibrium is easy to express:

$$\Sigma \mathbf{F} = 0 \qquad \textit{Translational equilibrium} \quad (8.1)$$

As before, Σ means "sum of," and here **F** refers to the various forces that act *on* the object under study.

In many equilibrium situations all the forces lie in the same plane. When this is the case, we can establish a set of x-y coordinate axes wherever convenient in the plane and then resolve each force **F** into the components $\mathbf{F}_x$ and $\mathbf{F}_y$. Thus we can replace Eq. (8.1), which is a vector equation, with the two scalar equations

$$\text{Sum of } x \text{ force components} = \Sigma F_x = 0 \qquad (8.2)$$

$$\text{Sum of } y \text{ force components} = \Sigma F_y = 0 \qquad (8.3)$$

Is is usually much easier to calculate the components of each force present and to make use of Eqs. (8.2) and (8.3) than it is to work with the forces themselves in Eq. (8.1).

There are five steps to follow in working out problems concerning the translational equilibrium of an object:

1. **Draw a sketch of the forces that act on the object. (As we know, this is called a *free-body* diagram.) Do not show the forces that the object exerts on anything else, because such forces do not affect the equilibrium of the object itself.**
2. **Choose a convenient set of coordinate axes and resolve the various forces acting on the object into components along these axes. Be sure to use + and − signs consistently. If the *y*-axis is vertical and the *x*-axis is horizontal, for instance, an upward force would be considered as +, a downward force as −, a force to the right as +, and a force to the left as − (Fig. 8.1).**
3. **Set the sum of the force components along each axis equal to 0, as specified in Eqs. (8.2) and (8.3). This is the condition for equilibrium.**
4. **Solve the resulting equations for the unknown quantity or quantities.**
5. **Substitute numerical values of the known quantities to find the answer.**

Figure 8.1

Sign conventions for the x- and y-components of a force.

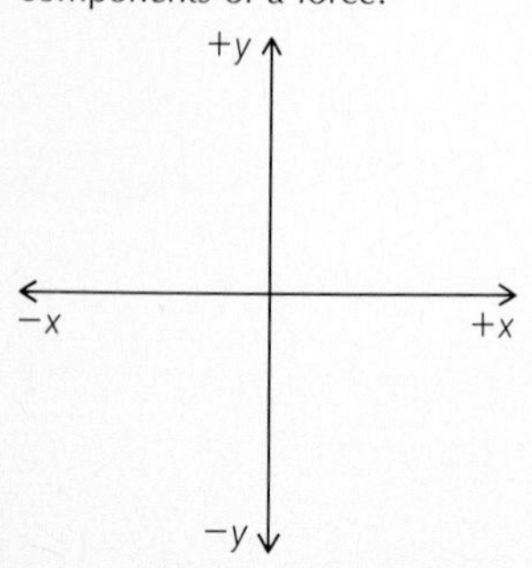

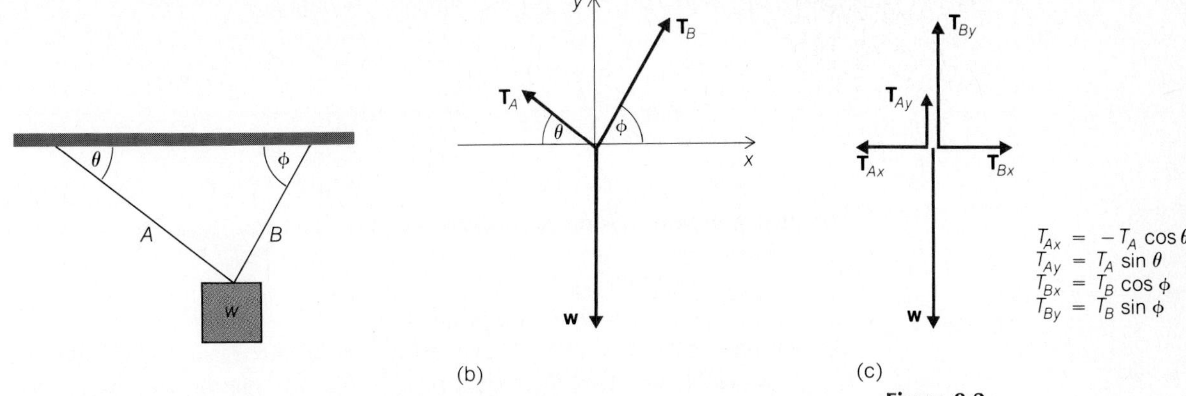

Figure 8.2

(a) A box of weight w is suspended by two ropes. (b) A free-body diagram of the forces acting on the box. (c) At equilibrium the sum ΣF_x of the horizontal force components and the sum ΣF_y of the vertical force components each equals zero.

An example will make clear this procedure. Figure 8.2 shows a box of weight w that is supported by two weightless ropes, A and B. The ropes are at the respective angles θ and ϕ with the horizontal. We note that the tension T in a rope is a force along its length in a direction away from its point of attachment at either end, as we saw in Fig. 3.14.

We begin by resolving the tension in each rope into components in the x- and y-directions, so that we have T_{Ay} and T_{By} upward, T_{Ax} to the left, and T_{Bx} to the right. From Fig. 8.2(b) we have

$$T_{Ax} = -T_A \cos\theta \qquad T_{Bx} = T_B \cos\phi \qquad \textit{Horizontal components}$$

$$T_{Ay} = T_A \sin\theta \qquad T_{By} = T_B \sin\phi \qquad \textit{Vertical components}$$

We may now replace the actual tensions T_A and T_B by their components, as in Fig. 8.2(c). For the forces in the vertical and horizontal directions to cancel separately,

$$\Sigma F_x = T_{Bx} + T_{Ax} = 0 \qquad \textit{Horizontal direction} \quad (8.4)$$

$$\Sigma F_y = T_{Ay} + T_{By} - w = 0 \qquad \textit{Vertical direction} \quad (8.5)$$

EXAMPLE 8.1

In the preceding situation $w = 100$ N, $\theta = 37°$, and $\phi = 60°$. Find the tension in each rope.

SOLUTION From Eq. (8.4) we find that

$$T_{Bx} + T_{Ax} = 0$$

$$T_B \cos\phi - T_A \cos\theta = 0$$

$$T_B = T_A \frac{\cos\theta}{\cos\phi} = T_A \frac{\cos 37°}{\cos 60°} = 1.6T_A$$

From Eq. (8.5) we find that

$$T_{Ay} + T_{By} - w = 0$$

$$T_A \sin\theta + T_B \sin\phi - 100\ \text{N} = 0$$

Substituting $1.6T_A$ for T_B in the last equation, we obtain

$$T_A \sin\theta + 1.6T_A \sin\phi - 100\ \text{N} = 0$$
$$T_A(\sin 37^\circ + 1.6 \sin 60^\circ) = 100\ \text{N}$$
$$T_A = 50\ \text{N}$$

The tension in rope A is 50 N. Because $T_B = 1.6T_A$ here,

$$T_B = (1.6)(50\ \text{N}) = 80\ \text{N}$$

The tension in rope B is 80 N. The algebraic sum of the tensions in the two ropes is 130 N, which is more than the weight being supported, but this sum means nothing. The vector sum of $\mathbf{T}_A$ and $\mathbf{T}_B$ is 100 N acting upward, and this cancels the downward force $\mathbf{w} = 100$ N. ◆

EXAMPLE 8.2

A 100-N box is suspended from the end of a horizontal strut, as in Fig. 8.3. Find the tension in the cable supporting the strut under the assumption that the strut's weight is negligible.

SOLUTION It is easiest to consider the equilibrium of the end of the strut. The three forces that act on the end of the strut are the tension $\mathbf{T}$ in the cable, the outward force $\mathbf{F}_s$ exerted by the strut itself, and the weight $\mathbf{w}$ of the box. The horizontal and vertical components of the tension T are, from the diagram,

$$T_x = -T\cos 30^\circ \qquad T_y = T\sin 30^\circ$$

The end of the strut is in equilibrium when

$$\Sigma F_x = T_x + F_s = 0 \quad \text{and} \quad \Sigma F_y = T_y - w = 0$$

All we need is the second of these equations to find T:

$$T_y - w = 0$$
$$T\sin 30^\circ = w$$
$$T = \frac{w}{\sin 30^\circ} = \frac{100\ \text{N}}{0.500} = 200\ \text{N}$$

◆

Figure 8.3

(a) A box of weight w is suspended from the end of a horizontal strut held in place by a cable attached to the wall. (b) A free-body diagram of the forces acting on the end of the strut. (c) At equilibrium $\Sigma F_x = 0$ and $\Sigma F_y = 0$.

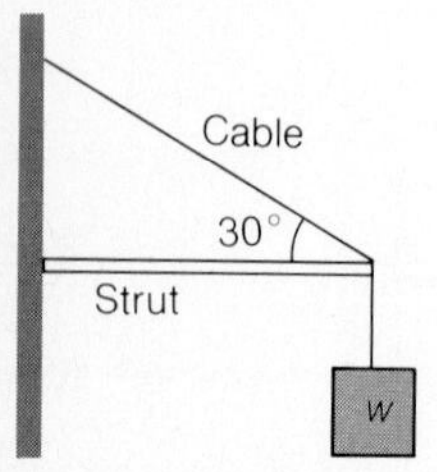

(a)

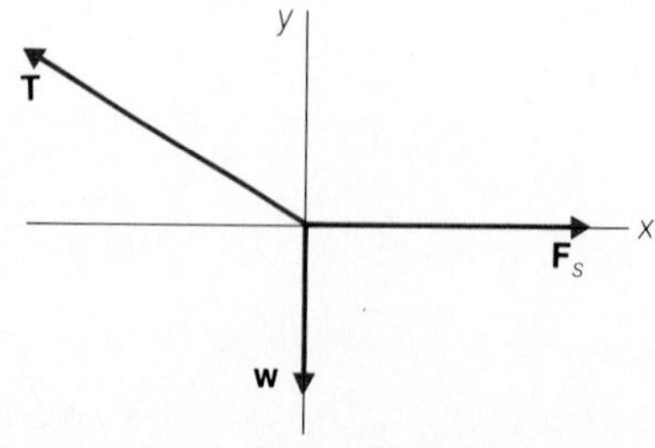

(b)

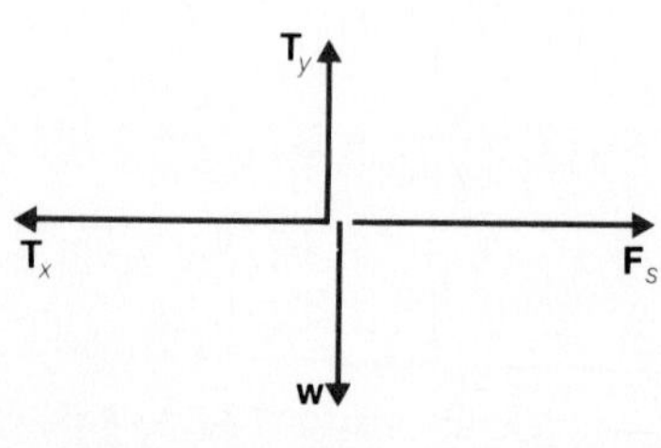

(c)

EXAMPLE 8.3

Find the inward force the strut of the previous example exerts on the wall.

SOLUTION Because this force is horizontal, we need the equation that expresses the equilibrium of the strut in the x-direction, which is

$$\Sigma F_x = T_x + F_s = 0$$

With $T = 200$ N and $T_x = -T\cos 30°$,

$$\Sigma F_x = -T\cos 30° + F_s = 0$$

$$F_s = T\cos 30° = (200\text{ N})(\cos 30°) = 173\text{ N}$$

◆

EXAMPLE 8.4

A 200-kg motor is suspended from the ceiling by a cable 6.0 m long, as in Fig. 8.4(a). If the coefficient of static friction between his shoes and the floor is $\mu_s = 0.6$, the maximum horizontal force a 70-kg person can exert on the motor before he starts to slip is $F = \mu_s mg = 412$ N. Find the horizontal displacement of the motor when a horizontal force of 412 N is applied to it.

Figure 8.4

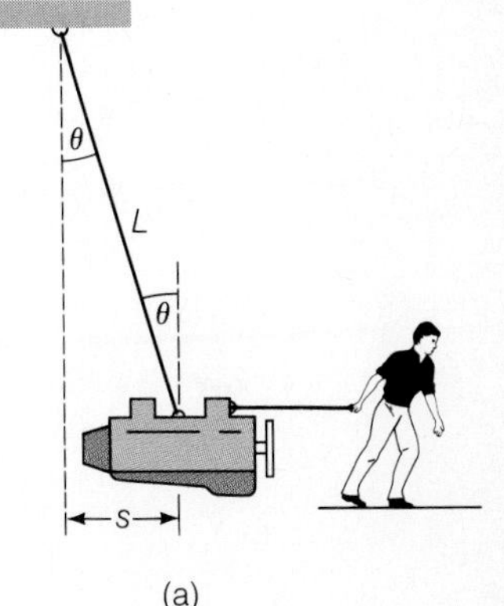

(a)

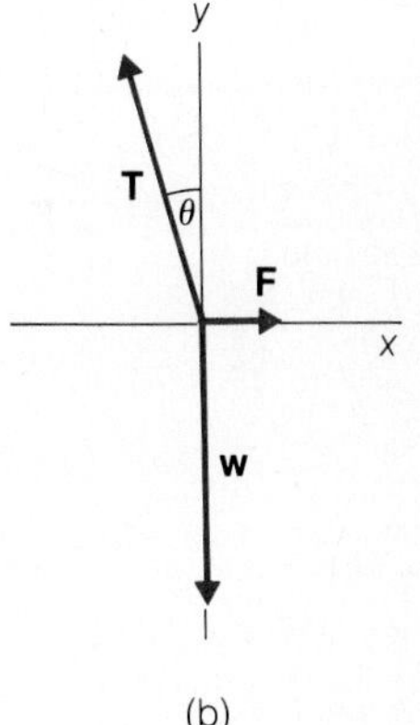

(b)

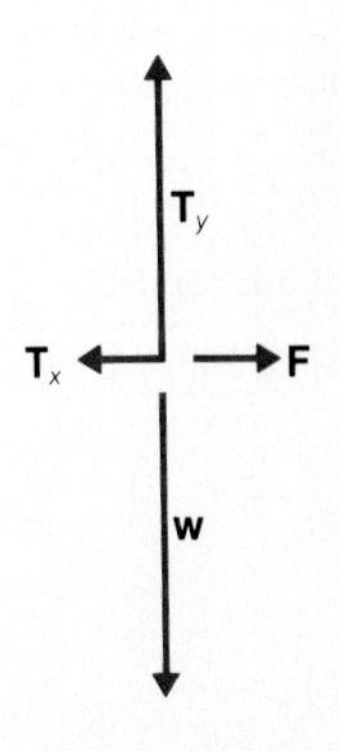

(c)

SOLUTION Since we know that L in Fig. 8.4(a) is 6.0 m, we must find the angle θ in order to determine the displacement s. We begin by resolving the tension **T** in the cable into its horizontal and vertical components. These are

$$T_x = -T\sin\theta \qquad T_y = T\cos\theta$$

Next we substitute into Eqs. (8.2) and (8.3) to obtain

$$\Sigma F_x = T_x + F = -T\sin\theta + F = 0$$

$$\Sigma F_y = T_y - w = T\cos\theta - w = 0$$

We now solve the first equation for $\sin\theta$ and the second for $\cos\theta$:

$$T\sin\theta = F \quad \text{or} \quad \sin\theta = \frac{F}{T} \qquad T\cos\theta = w \quad \text{or} \quad \cos\theta = \frac{w}{T}$$

Because $\sin\theta/\cos\theta = \tan\theta$,

$$\tan\theta = \frac{\sin\theta}{\cos\theta} = \frac{F/T}{w/T} = \frac{F}{w} = \frac{F}{mg} = \frac{412\text{ N}}{(200\text{ kg})(9.8\text{ m/s}^2)} = 0.210$$

and

$$\theta = \tan^{-1} 0.210 = 12°$$

It is not necessary to know the tension T in order to find θ. From Fig. 8.4(a) we have for the horizontal displacement s of the motor

$$s = L\sin\theta = (6.0\text{ m})(\sin 12°) = 1.2\text{ m}$$

◆

8.2 Rotational Equilibrium

The sum of the torques about any point must be zero.

When the lines of action of the various forces that act on an object intersect at a common point, they do not tend to set the object in rotation. Such forces are said to be **concurrent.**

If the lines of action of the various forces do *not* intersect, the forces are **nonconcurrent,** and the object may be set into rotation even though the vector sum of the forces may equal zero. In Fig. 8.5(a) the three applied forces are concurrent and, if their vector sum is zero, the object is in equilibrium. In Fig. 8.5(b) the same forces are nonconcurrent, and they combine to produce a counterclockwise torque.

By convention a torque that tends to produce a counterclockwise rotation is considered positive and a torque that tends to produce a clockwise rotation is considered negative (Fig. 8.6). Thus the condition for an object to be in **rotational equilibrium** is that the sum of the torques acting on it about any point, using the convention given above for plus and minus signs, be zero:

$$\Sigma\tau = 0 \qquad \textit{Rotational equilibrium} \qquad (8.6)$$

Of course, if the various forces that act do not all lie in the same plane, the sum of the torques in each of three mutually perpendicular planes must be zero.

It is possible to prove that if the sum of the torques on an object in translational equilibrium is zero about any point, it is also zero about all other points. Hence the location of the point about which torques are calculated in an equilibrium problem

Figure 8.5

(a) The lines of action of concurrent forces intersect at a common point. (b) When the lines of action do not intersect at a common point, the forces are nonconcurrent and the object cannot be in rotational equilibrium.

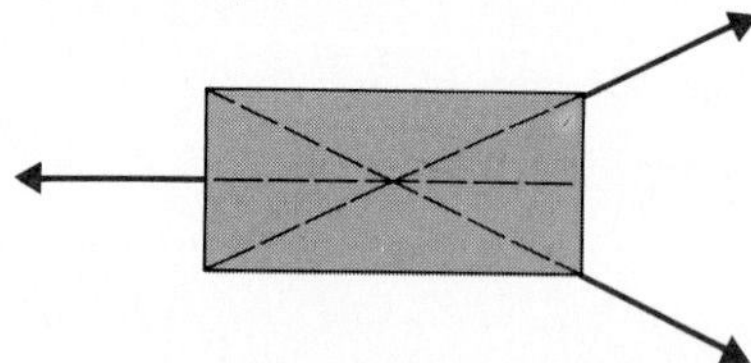

(a) Concurrent forces

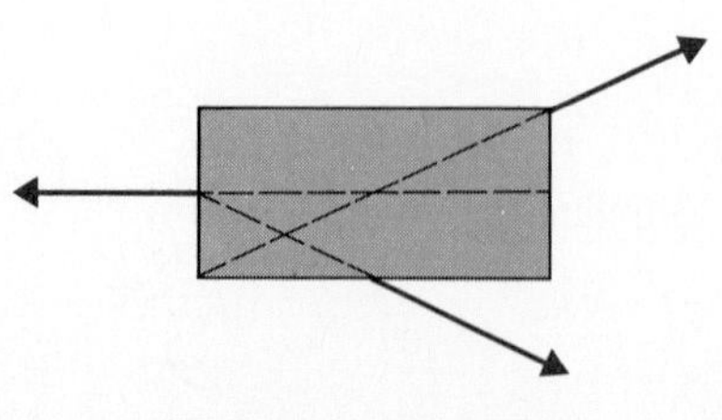

(b) Nonconcurrent forces

Figure 8.6

(a) A torque that tends to produce a counterclockwise rotation is considered positive. (b) A torque that tends to produce a clockwise rotation is considered negative.

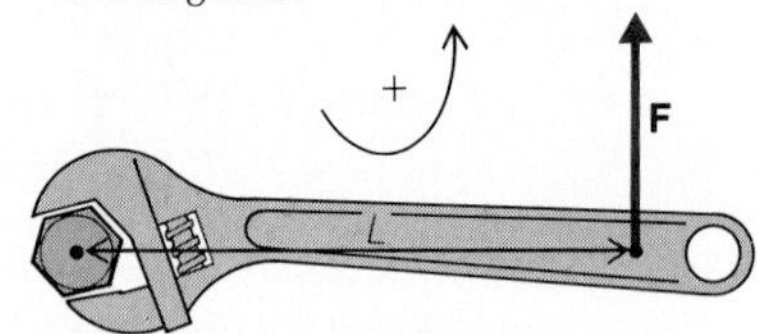

(a) $\tau = FL$

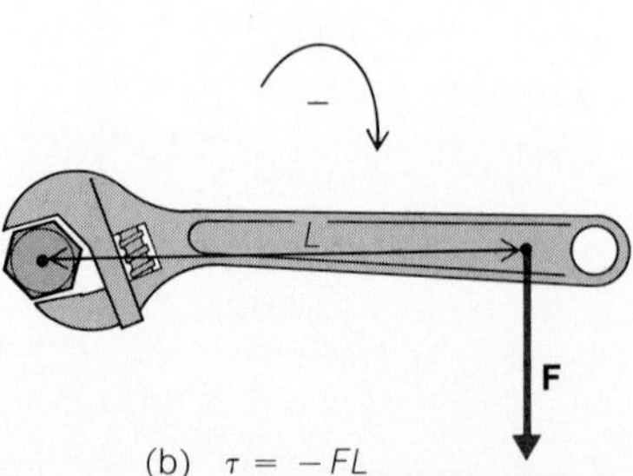

(b) $\tau = -FL$

is completely arbitrary; *any* point will do. (Of course, *all* torques must be calculated about this point.) Let us verify this statement with an example.

The complex structure of this crane keeps its own mass low while enabling it to shift heavy loads. The design of the crane is based on the conditions for translational and rotational equilibrium.

EXAMPLE 8.5

Figure 8.7 shows a rod 2.0 m long that has weights of 10 N and 30 N at its ends. We assume that the weight of the rod is negligible. At what point should the rod be picked up if it is to have no tendency to rotate? In other words, where is the balance point of the rod?

SOLUTION 1 We first compute torques about the unknown balance point. If x is the distance of the 30-N weight from this point, the 10-N weight is $(2.0\text{ m} - x)$ from it on the other side. The torques these weights exert are

$$\tau_1 = w_1 L_1 = +30x\text{ N}$$

$$\tau_2 = -w_2 L_2 = -10(2.0\text{ m} - x)\text{ N}$$

The torque τ_2 is negative because it acts clockwise. Equilibrium will result when

$$\Sigma\tau = \tau_1 + \tau_2 = 30x\text{ N} - 10(2.0\text{ m} - x)\text{ N} = 0$$

$$30x = 10(2.0\text{ m} - x)$$

$$40x = 20\text{ m}$$

$$x = 0.5\text{ m}$$

When the rod is picked up 0.5 m from the 30-N weight, the two weights exert opposite torques of the same magnitude (15 N·m) about this point, so the rod is in balance. ♦

SOLUTION 2 Let us solve the same problem by calculating torques about the middle of the rod, as shown in Fig. 8.8. Here y represents the distance between the balance point and the middle of the rod. We now have three torques to consider:

$$\tau_1 = w_1 L_1 = +(30\text{ N})(1.0\text{ m}) = +30\text{ N}\cdot\text{m}$$

$$\tau_2 = F L_2 = -40y\text{ N}$$

$$\tau_3 = w_2 L_3 = -(10\text{ N})(1.0\text{ m}) = -10\text{ N}\cdot\text{m}$$

Figure 8.7

Torques are computed about the unknown balance point of the rod in Solution 1.

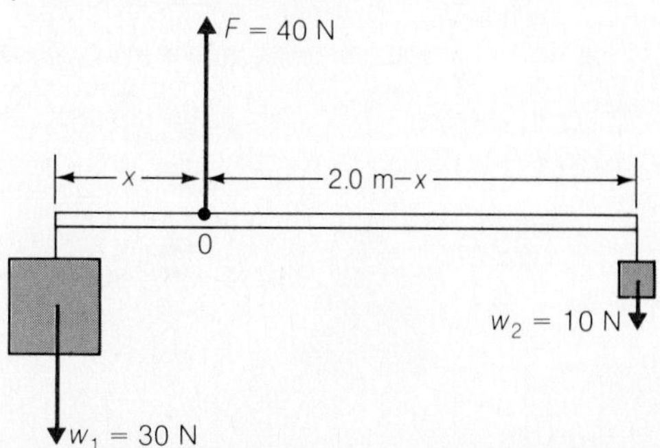

Figure 8.8

Torques are computed about the center of the rod in Solution 2.

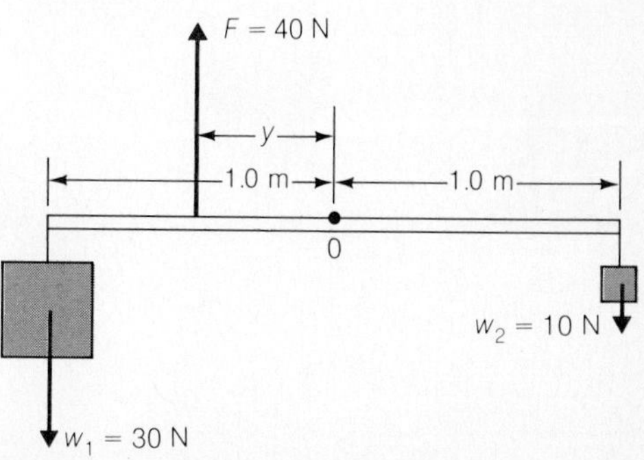

The condition for equilibrium is

$$\Sigma\tau = \tau_1 + \tau_2 + \tau_3 = 30 \text{ N·m} - 40y \text{ N} - 10 \text{ N·m} = 0$$

from which we find

$$y = 0.5 \text{ m}$$

The location of the balance point is the same regardless of the point about which we calculate torques. It is usually wise to calculate torques about the point of application of one of the forces that act on an object, because this makes it unnecessary to consider the torque produced by that force. ♦

Not all equilibrium situations are necessarily stable. For instance, a cone balanced on its apex is in equilibrium, but it will fall over when disturbed even slightly (Fig. 8.9). This is an example of an **unstable equilibrium.** The same cone on its base will return to its original position if tipped over a little; hence it is in **stable equilibrium** on its base. There is a third possibility as well, illustrated by a cone lying on its side. If such a cone is displaced, it stays in equilibrium in its new position without tending either to move further or to return to where it was before. A cone on its side is said to be in **neutral equilibrium.**

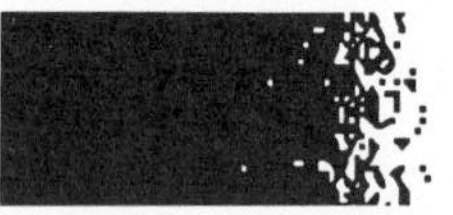

8.3 Center of Gravity

The balance point of an object.

The **center of gravity** (CG) of an object is that point from which it can be suspended without tending to rotate (Fig. 8.10). Each particle in an object has a certain weight and therefore exerts a torque about whatever point the object is suspended from. There is only one point for an object about which all these torques cancel out no matter how it is oriented; this is its center of gravity. For equilibrium purposes we can therefore regard the entire weight of an object as a downward force acting from its CG.

Figure 8.9

These cones are all in equilibrium, but only one of them is in a stable position.

Figure 8.10

A body suspended from its center of gravity is in equilibrium in any orientation.

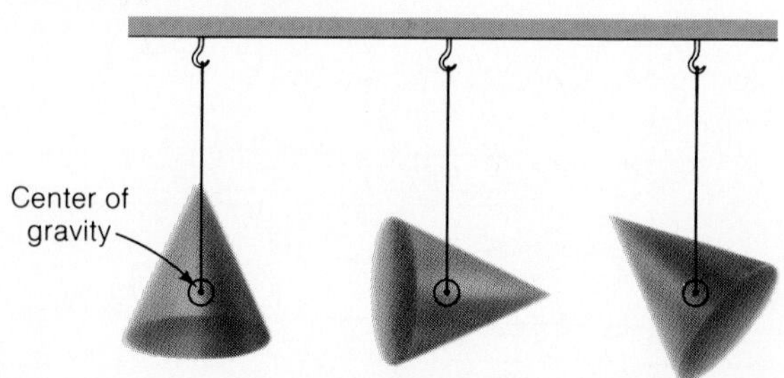

We can now recognize the distinctions between the different kinds of equilibrium shown in Fig. 8.9. The left-hand cone is in stable equilibrium because its CG has to be raised to tip it over; the center cone is in unstable equilibrium because any change in its orientation lowers its CG; and the right-hand cone is in neutral equilibrium because the height of its CG does not change when it is rolled along on its side.

If the rod in Figs. 8.7 and 8.8 had the weight w instead of being weightless, we could take into account its effect on the location of the balance point by including the torque due to a force of magnitude w acting downward at the center of the rod. The CG of a uniform object of regular shape is located at its geometrical center. The CG of an irregular object need not even be located within the object itself. The CG of a seated person, for example, is a few inches in front of the person's abdomen.

EXAMPLE 8.6

A beam that projects beyond its supports is called a **cantilever.** A diving board is an example. Find the forces exerted by the two supports of the 4.0-m, 50-kg uniform diving board shown in Fig. 8.11 when a 60-kg woman stands at its end.

SOLUTION The weight of the diving board is $w_1 = m_1 g_1 = 490$ N, and that of the woman is $w_2 = m_2 g = 588$ N. It is obvious that the force F_1 exerted by the left-hand support must be downward. (If we nevertheless were to consider F_1 as upward, the result would be a negative value for F_1, signifying the opposite direction.) We can calculate F_1 without knowing F_2 by computing torques about the point where F_2 is applied. The CG of the board is at its middle, which is 1.2 m from the pivot point. Hence the three torques that act about this point are

$$\tau_1 = +F_1 x_1 = +(F_1)(0.8\text{ m}) = 0.8 F_1\text{ m}$$
$$\tau_2 = -w_1 x_2 = -(490\text{ N})(1.2\text{ m}) = -588\text{ N·m}$$
$$\tau_3 = -w_2 x_3 = -(588\text{ N})(3.2\text{ m}) = -1882\text{ N·m}$$

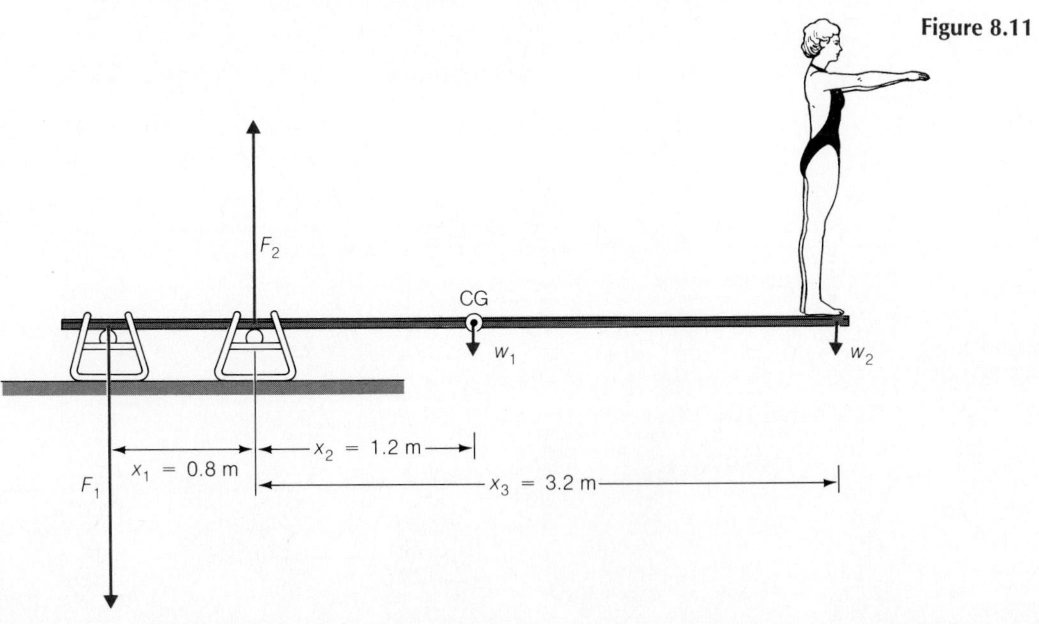

Figure 8.11

Adding the torques and setting their sum equal to zero gives

$$\Sigma\tau = \tau_1 + \tau_2 + \tau_3 = 0.8F_1 \text{ m} - 588 \text{ N·m} - 1882 \text{ N·m} = 0$$
$$F_1 = 3088 \text{ N}$$

We can find F_2 by considering the translational equilibrium in the y-direction of the loaded board:

$$F_2 = F_1 + w_1 + w_2 = 3088 \text{ N} + 490 \text{ N} + 588 \text{ N} = 4166 \text{ N}$$ ◆

EXAMPLE 8.7

A person holds a 10-kg pail of water with his upper arm at his side and his forearm outstretched, as in Fig. 8.12. The palm of his hand is 35 cm from his elbow, his upper arm is 30 cm long, and his biceps muscle is attached to his forearm 5 cm from his elbow. The person's forearm (including his hand) has a mass of 3 kg, and its CG is 16 cm from the elbow. Find the force the biceps muscle exerts to support the forearm and pail.

SOLUTION We will calculate torques about the elbow, which simplifies the calculation because then we need to consider only the vertical component F_y of the muscular force **F**, the weight $w_1 = m_1g = 29.4$ N of the forearm, and the weight $w_2 = m_2g = 98$ N of the pail. Because

$$\tan\theta = \frac{5 \text{ cm}}{30 \text{ cm}} = 0.167 \quad \text{and} \quad \theta = 9.5°$$

we have

$$F_y = F\cos\theta = 0.986\,F$$

The torques about the elbow are τ_1 exerted by the muscle, τ_2 by the forearm's weight,

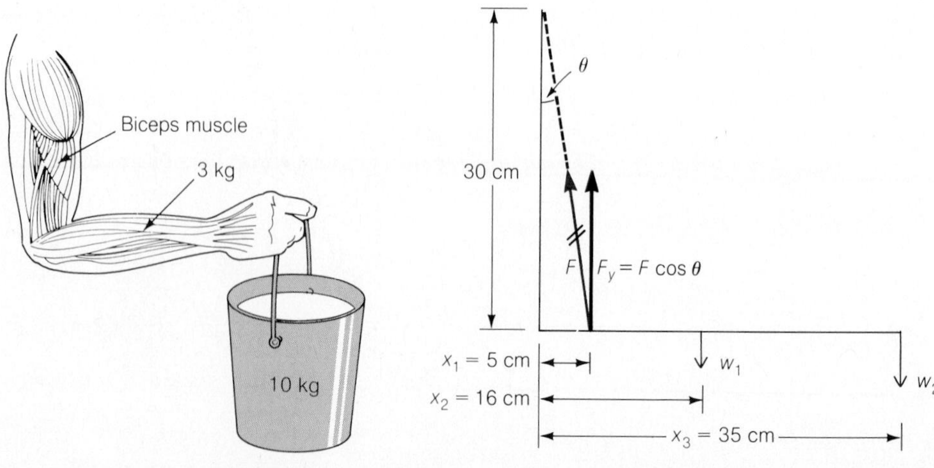

Figure 8.12
The force exerted by the elbow on the forearm can be disregarded if torques are calculated about the elbow. The horizontal component of the tension in the biceps muscle has no moment arm about the elbow and so is also disregarded.

and τ_3 by the pail of water. These torques are, respectively,

$$\tau_1 = F_y x_1 = (0.986F)(0.05 \text{ m}) = 0.0493\, F \text{ m}$$
$$\tau_2 = -w_1 x_2 = -(29.4 \text{ N})(0.16 \text{ m}) = -4.7 \text{ N·m}$$
$$\tau_3 = -w_2 x_3 = -(98 \text{ N})(0.35 \text{ m}) = -34.3 \text{ N·m}$$

The sum of the torques about the elbow must be 0 for equilibrium, hence

$$\Sigma\tau = \tau_1 + \tau_2 + \tau_3 = 0.0493\, F \text{ m} - 4.7 \text{ N·m} - 34.3 \text{ N·m} = 0$$

from which we find

$$F = \frac{(4.7 + 34.3) \text{ N·m}}{0.0493 \text{ m}} = 791 \text{ N}$$

This force, which is equivalent to 178 lb, is more than six times the combined weights of the forearm and the pail of water. We might regard the body as being inefficiently designed, since such large muscular forces are required for ordinary tasks. However, when we reflect on the large span this arrangement enables the hand to move through and the speed at which it can do so, it is clear that inefficiency in one sense has been traded for efficiency in another. ◆

EXAMPLE 8.8

A ladder 4.0 m long is leaning against a frictionless wall with its lower end 1.6 m away from the wall, as in Fig. 8.13. If the ladder weighs 150 N, what forces does it exert on the wall and on the ground?

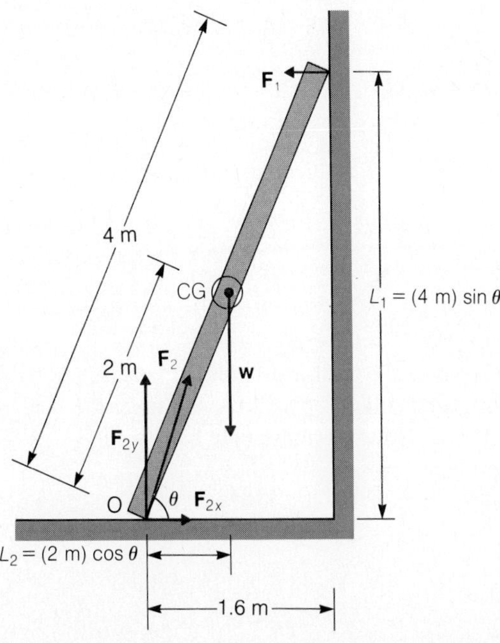

Figure 8.13

To simplify the calculation, the pivot point about which torques are calculated is chosen as the bottom of the ladder.

SOLUTION Because the wall is frictionless, the only force it can exert on the ladder is the horizontal reaction force $\mathbf{F}_1$. The other forces on the ladder are its weight **w**, which acts downward from its midpoint, and the reaction force $\mathbf{F}_2$ exerted by the ground. It is worth mentioning that the direction of $\mathbf{F}_2$ need not lie along the line of the ladder.

Because $\mathbf{F}_{2y}$, the upward component of $\mathbf{F}_2$, and **w** are the only vertical forces and **w** is in the $-$ y-direction,

$$\Sigma F_y = F_{2y} - w = 0$$
$$F_{2y} = w = 150\text{ N}$$

Similarly the horizontal force $\mathbf{F}_1$ acting to the left must have the same magnitude as $\mathbf{F}_{2x}$ acting to the right:

$$\Sigma F_x = F_{2x} - F_1 = 0$$
$$F_{2x} = F_1$$

Before going further we need the value of the angle θ. From Fig. 8.13

$$\cos\theta = \frac{1.6\text{ m}}{4.0\text{ m}} = 0.400$$
$$\theta = \cos^{-1} 0.400 = 66°$$

Let us find F_1 first. The easiest way is to calculate torques about the lower end of the ladder, because then we will have only the two torques produced by **w** and by $\mathbf{F}_1$ to consider. We have

$$\tau_1 = F_1 L_1 = (F_1)(4.0\text{ m})\sin 66° = 3.65 F_1\text{ m}$$
$$\tau_2 = -wL_2 = -(150\text{ N})(2.0\text{ m})\cos 66° = -122\text{ N·m}$$

The sum of these torques must be zero for the ladder to be in equilibrium, so

$$\tau = \tau_1 + \tau_2 = 0$$
$$\tau_1 = -\tau_2$$
$$3.65 F_1\text{ m} = -(-122\text{ N·m}) = 122\text{ N·m}$$
$$F_1 = \frac{122\text{ N·m}}{3.65\text{ m}} = 33.4\text{ N}$$

The force the wall exerts on the ladder is horizontal with the magnitude 33.4 N, and this is equal and opposite to the force the ladder exerts on the wall.

Now we must find $\mathbf{F}_2$. Since $F_{2x} = F_1$ and $F_{2y} = w$, the total force the ground exerts on the ladder is

$$F_2 = \sqrt{F_{2x}^2 + F_{2y}^2} = \sqrt{F_1^2 + w^2}$$
$$= \sqrt{(33.4\text{ N})^2 + (150\text{ N})^2} = 154\text{ N}$$

The force the ladder exerts on the ground has the same magnitude. The angle θ between $\mathbf{F}_2$ and the vertical is

$$\theta = \tan^{-1}\frac{F_{2x}}{F_{2y}} = \tan^{-1}\frac{33.4\text{ N}}{150\text{ N}} = \tan^{-1}0.223 = 13^\circ$$ ◆

8.4 Finding a Center of Gravity

A straightforward calculation.

An experimental way to find the CG of an irregular object is shown in Fig. 8.14. An analytical procedure follows from the definition of CG. What we do is imagine the object divided into two or more separate elements whose CGs are known, then consider these elements as particles joined by weightless rods, and finally calculate the balance point of the assembly. If the object is complex—for instance, the hull of a ship—a great many elements must be considered for an accurate result, but the method is still straightforward.

To see how such a calculation can be made, let us consider a system of three particles whose masses are m_1, m_2, and m_3. They are located x_1, x_2, and x_3, respectively, from one end of a weightless rod, as in Fig. 8.15. The CG of the system is located at some distance X from the end of the rod. This means that the torque exerted about this end by a single particle of mass $M = m_1 + m_2 + m_3$ located at X will be exactly the same as the torque exerted by the actual system of the three particles.

Figure 8.14

To find the center of gravity of a flat body, suspend it and a plumb bob successively from two different points on its edge. The center of gravity is located at the intersection of the two lines of action of the plumb bob.

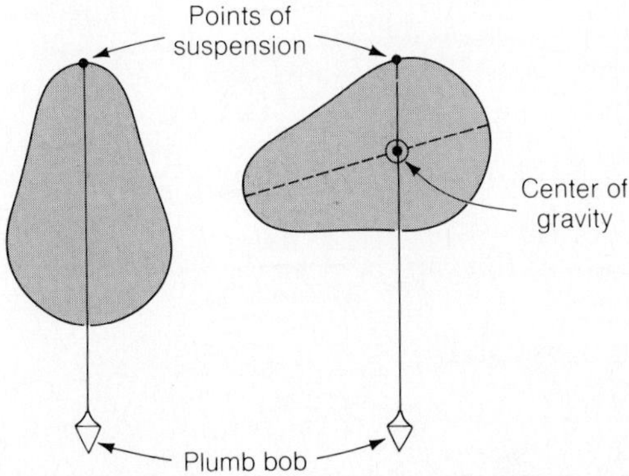

Figure 8.15

The center of gravity of a complex object can be found by considering it as a system of particles and applying Eq. (8.7).

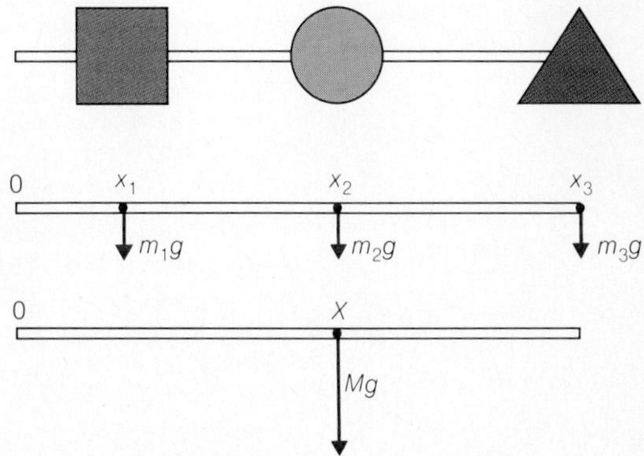

Hence

$$\Sigma\tau = m_1gx_1 + m_2gx_2 + m_3gx_3 = MgX = (m_1 + m_2 + m_3)gX$$

so that

$$X = \frac{m_1x_1 + m_2x_2 + m_3x_3}{m_1 + m_2 + m_3}$$

This formula can be generalized to a system of any number of particles (or objects that can be represented by particles located at their CGs) in a straight line, giving

$$X = \frac{m_1x_1 + m_2x_2 + m_3x_3 + \cdots}{m_1 + m_2 + m_3 + \cdots} \qquad (8.7)$$

Center of gravity of composite object

EXAMPLE 8.9

The hand, forearm, and upper arm of a woman have the respective masses 0.4 kg, 1.2 kg, and 1.9 kg, and their CGs are, respectively, 0.60 m, 0.40 m, and 0.15 m from her shoulder joint (Fig. 8.16). Find the distance of the CG of her entire unbent arm from the shoulder joint.

SOLUTION From Eq. (8.7)

$$X = \frac{(0.4\text{ kg})(0.60\text{ m}) + (1.2\text{ kg})(0.40\text{ m}) + (1.9\text{ kg})(0.15\text{ m})}{(0.4 + 1.2 + 1.9)\text{ kg}}$$
$$= 0.29\text{ m}$$

The CG of the entire arm is 0.29 m from the shoulder joint.

Figure 8.16

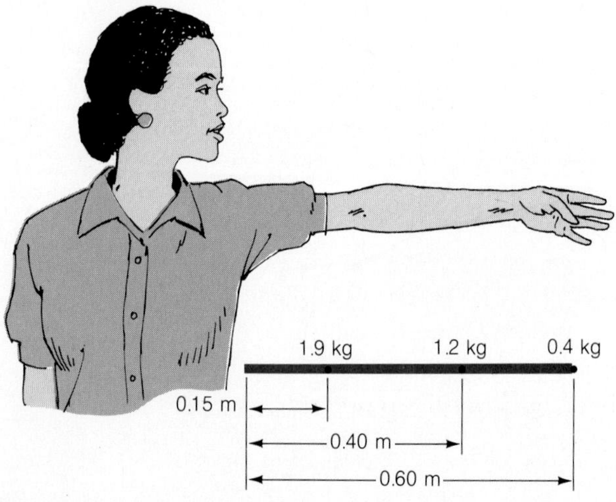

♦

If the irregular object whose CG is to be found lies in a plane rather than along a straight line (or is three-dimensional), the same procedure is applied along two (or three) coordinate axes. An example will make this clear.

EXAMPLE 8.10

Find the CG of the L-shaped steel plate shown in Fig. 8.17.

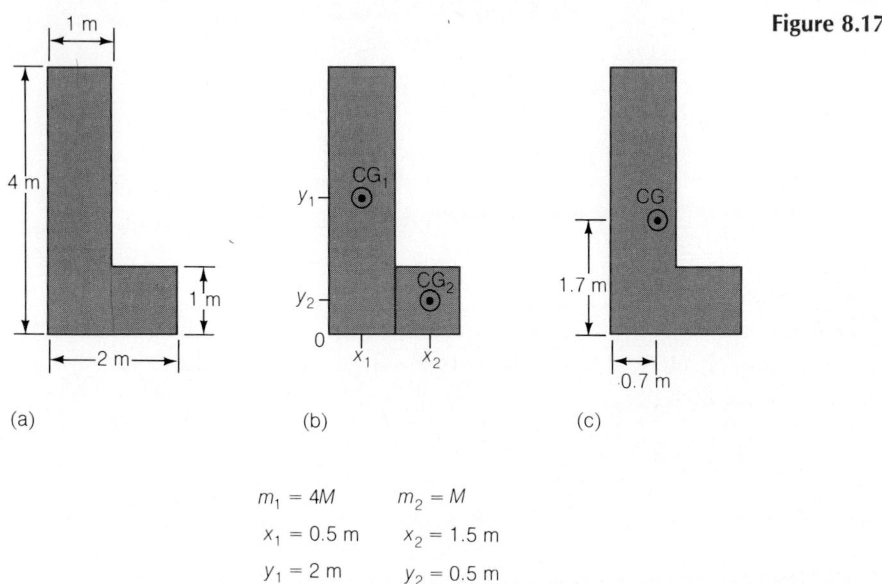

Figure 8.17

$m_1 = 4M$ $m_2 = M$

$x_1 = 0.5$ m $x_2 = 1.5$ m

$y_1 = 2$ m $y_2 = 0.5$ m

SOLUTION We imagine the plate to consist of two sections, one a rectangle 4 m long by 1 m wide and the other a square 1 m on a side. The CGs of these sections are at their geometric centers, as shown in Fig. 8.17(b). Now we replace each section by a particle at its CG. If the steel plate has a mass of M per square meter, particle 1 has a mass of $4M$ and particle 2 has a mass of M, because their areas are, respectively, 4 m^2 and 1 m^2. The x- and y-coordinates of the CG of the entire plate are as follows:

$$X = \frac{m_1x_1 + m_2x_2}{m_1 + m_2} = \frac{(4M)(0.5 \text{ m}) + (M)(1.5 \text{ m})}{4M + M} = 0.7 \text{ m}$$

$$Y = \frac{m_1y_1 + m_2y_2}{m_1 + m_2} = \frac{(4M)(2 \text{ m}) + (M)(0.5 \text{ m})}{4M + M} = 1.7 \text{ m}$$

The location of this point is shown in Fig. 8.17(c). We did not need to know the value of M in this case, because the steel plate is uniform. ♦

Net Force and Center of Gravity

The response of an object to a net force acting on it depends on whether the line of action of the force passes through the object's CG or not. In the case where the line of action passes through the CG, the force has no effect on the object's

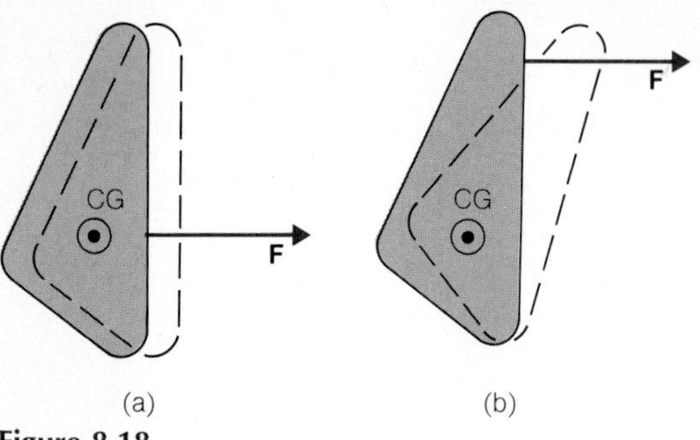

Figure 8.18

(a) A body remains in rotational equilibrium when the line of action of an applied force passes through its CG. (b) When the line of action of an applied force does not pass through its CG, the body is given a rotational as well as a translational acceleration.

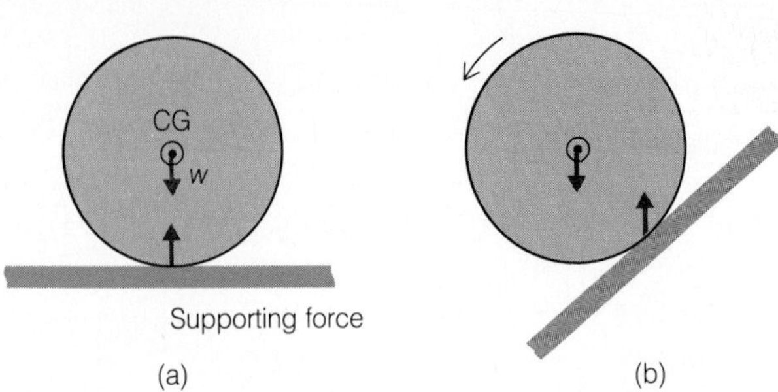

Figure 8.19

A wheel rolls downhill on an inclined surface because the line of action of the supporting force does not pass through its CG. This leads to a net torque on the wheel.

rotational motion (Fig. 8.18). If the object is not rotating to begin with, such a force will simply accelerate it in the direction of the force. However, if the line of action of the force does not pass through the CG, the object will start to spin as well as to move.

A wheel on a level surface has its CG directly above its point of contact with the surface, as in Fig. 8.19(a). The line of action of the reaction force of the surface on the wheel, which is what supports the wheel, passes through the wheel's CG. Unless it is pushed, the wheel has no acceleration. If the surface is tilted, though, the line of action of the supporting force does not intersect the wheel's CG. The result is a torque on the wheel that starts it rolling (Fig. 8.19b).

We have all noticed that a car tilts backward when it is accelerated, and that it tilts forward when the brakes are applied. These effects occur because the forces applied to the car in each case are horizontal ones that act where the tires touch the road, and are not in line with the car's CG. The car rotates when the resulting torques act on it until the springs of the suspension system restore rotational equilibrium (Fig. 8.20).

Figure 8.20

A car tilts backward when accelerated (a) and forward when braked (b) because the lines of action of the applied forces in each case do not pass through the car's CG. In (a) the car is assumed to have rear-wheel drive. If the car has front-wheel drive, the reaction forces act on its front tires, but the situation is otherwise the same.

8.5 Mechanical Advantage

Three basic machines.

A machine is a device that transmits force or torque for a definite purpose. All machines, however complicated, are actually combinations of only three basic machines: the lever, the inclined plane, and the hydraulic press. Thus the train of gears that carries power from the engine of a car to its wheels is a development of the lever; the screw jack that can raise one end of the car from the ground is a development of the inclined plane; and the brake system that permits a touch of the foot to stop the car is a development of the hydraulic press.

The **mechanical advantage** (MA) of a machine is the ratio between the output force F_{out} it exerts and the input force F_{in} that is furnished to it:

$$\text{MA} = \frac{F_{out}}{F_{in}} \qquad \textit{Mechanical advantage} \quad (8.8)$$

A mechanical advantage greater than 1 means that the output force exceeds the input force, whereas a mechanical advantage less than 1 means that the output force is smaller than the input force. Usually the MA is greater than 1, which makes it possible for a relatively small applied force to accomplish a task ordinarily beyond its capacity. However, sometimes an MA less than 1 is useful. An example is a pair of scissors, where the range of motion is increased at the expense of a reduced force.

The ideal mechanical advantage of a machine is its MA under perfect conditions. The actual MA, on the other hand, takes into account friction and any other dissipative factors. The efficiency of a machine is equal to the ratio between its actual and ideal MAs. In some machines, such as the simple lever, the efficiency may be close to 100%, whereas in others, such as the screw, it may be less than 10%. In the latter case the low efficiency is actually an advantage, because it prevents the screw from backing out by itself. We will consider only ideal mechanical advantages here.

The **principle of equilibrium** is the basis for calculating mechanical advantage: When it is provided with the input force F_{in}, a machine will exactly balance a load equal to F_{out}. In the case of the lever shown in Fig. 8.21, the condition for equilibrium is that the torque produced by F_{in} about the fulcrum be the same in magnitude as that produced by F_{out}. If we call the lever arms of the respective forces L_{in} and L_{out}, we have

$$F_{in}L_{in} = F_{out}L_{out}$$

$$\text{MA} = \frac{F_{out}}{F_{in}} = \frac{L_{in}}{L_{out}} \qquad \textit{The lever} \quad (8.9)$$

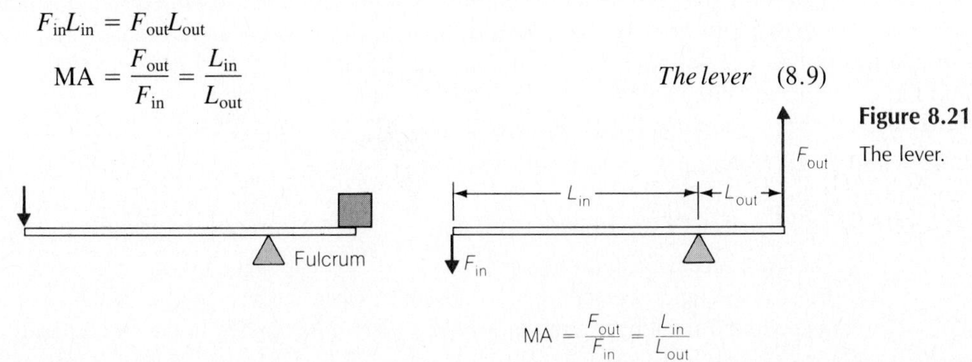

Figure 8.21 The lever.

BIOGRAPHICAL NOTE

Archimedes

Archimedes (287?–212 B.C.) was born in Syracuse, Sicily, at that time a Greek colony. He studied in Alexandria, Egypt, a center of learning in the ancient world, under a teacher who had been a pupil of the great geometer Euclid. Archimedes then returned to Syracuse, where family wealth allowed him to devote his life to science and mathematics.

The principle of the lever—that the ratio between the input and output forces equals the inverse ratio of the respective lever arms—was first proved by Archimedes. He claimed that, given a long enough lever and a place to put the fulcrum, he could move the earth. Challenged by King Heiron of Syracuse to move something really large, even if not the earth, Archimedes used a pulley system (which is a development of the lever) to drag a fully laden ship across dry land.

Another discovery of Archimedes was the basic law of hydrostatics that bears his name. According to Archimedes' principle, the buoyant force on an object immersed in a fluid equals the weight of the fluid the object displaces. Legend has it that this idea came to Archimedes in his bath while he was thinking of a way to test whether King Heiron's new crown was pure gold or not. He then ran naked to the palace crying, "Eureka!"—"I've got it!" The crown turned out to be only partly gold, and its maker was executed.

Archimedes did considerable work in mathematics, both in algebra and in geometry. Among his accomplishments was a calculation of π that put it between 223/71 and 220/70, an excellent approximation for those days. He was proudest of having found the formula for the ratio of the volumes of a sphere and the cylinder that encloses it, and asked that the formula be engraved on his tombstone.

Archimedes also invented a number of ingenious mechanical devices, including a planetarium that could show the motions of the sun, the moon, and the planets across the sky. Ancient historians gave Archimedes credit for a number of weapons that kept the Romans from landing during a siege of Syracuse by their fleet from 215 to 212 B.C. Among these weapons supposedly were huge lenses that focused the sun's rays on the Roman ships and set them afire. The Romans eventually conquered Syracuse, and Archimedes was killed by one of their soldiers when he refused to stop working out a geometrical problem he had scratched in the sand of the marketplace.

The ideal mechanical advantage of the lever is equal to the inverse ratio of the lever arms. A 4:1 ratio of lever arms, for instance, means an MA of 4 when the input force is applied to the longer arm, an MA of $\frac{1}{4}$ when it is applied to the shorter arm.

EXAMPLE 8.11

A steel rod 2.00 m long is to be used to pry up one end of a 500-kg crate. If a workman can exert a downward force of 350 N on one end of the rod, where should he place a block of wood to act as the fulcrum?

SOLUTION The weight of the crate is $w = mg = 4900$ N. A force of half this is needed to lift one end, so the MA must be

$$\text{MA} = \frac{F_{\text{out}}}{F_{\text{in}}} = \frac{w/2}{F_{\text{in}}} = \frac{2450\ \text{N}}{350\ \text{N}} = 7$$

Hence the ratio of the lever arms should be 7:

$$L_{\text{in}} = 7L_{\text{out}}$$

Since the lever is 2.00 m long,

$$L_{\text{in}} + L_{\text{out}} = 7L_{\text{out}} + L_{\text{out}} = 2.00\ \text{m}$$

$$L_{\text{out}} = \frac{2.00\ \text{m}}{8} = 0.25\ \text{m} = 25\ \text{cm}$$

The fulcrum should be placed 25 cm from the crate. ♦

The lever of Fig. 8.21, where the fulcrum is between the load and the applied force, is called a Class I lever. In a Class II lever, of which the wheelbarrow is an example, the load is between the fulcrum and the applied force, as in Fig. 8.22. In

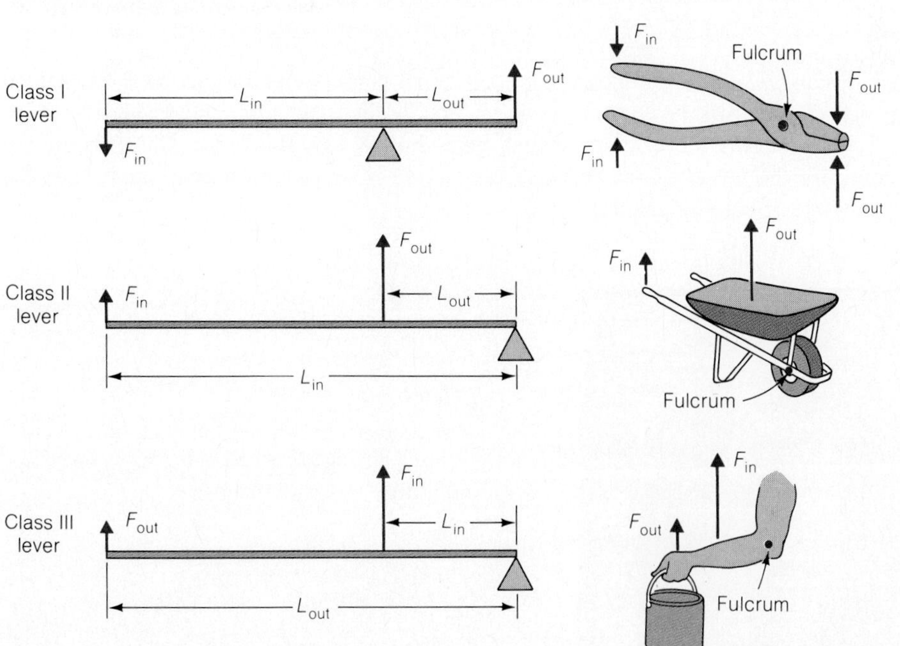

Figure 8.22

The three classes of lever.

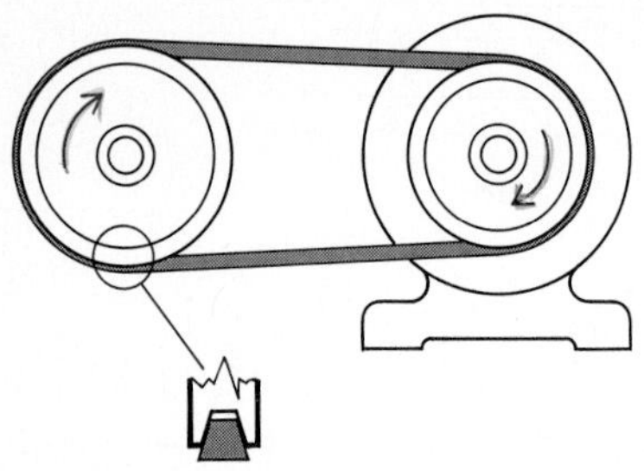

(a) V belt drive

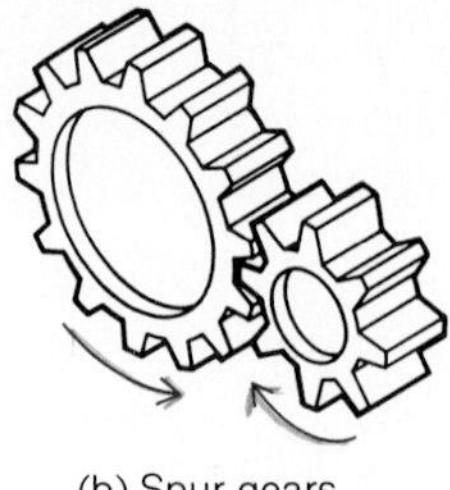

(b) Spur gears

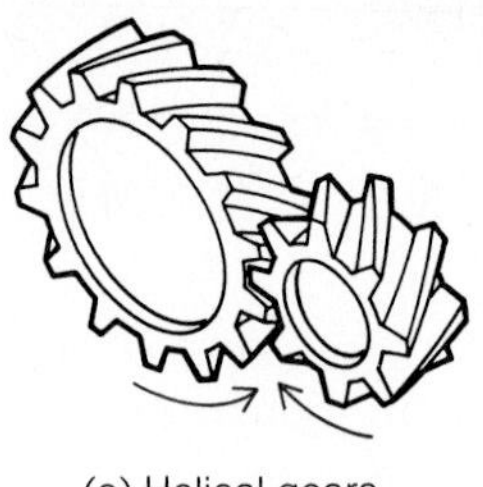

(c) Helical gears

Figure 8.23

(a) A V belt has the advantage over a flat belt that it cannot slide sideways on the pulleys; the large area in contact with the sides of the pulleys helps prevent slipping. (b) Spur gears have their teeth cut parallel to the axes of rotation. (c) Helical gears have curved teeth cut in a spiral pattern at an angle to their axes; several teeth of meshing helical gears are in contact at all times, which makes a set of these gears smoother in operation and able to bear greater torque.

a Class III lever, of which the human forearm is an example, the load and fulcrum are at the ends with the applied force between them.

The various kinds of elementary lever are all handicapped by the limited angle through which they can operate. Certain developments of the lever, however, can be used on a continuous basis, and one or another of them is an important element in nearly every motor-driven machine. In V belt and gear drive systems, the MA is given by the ratio of the diameters of the driven and driving pulleys or gears (Fig. 8.23).

The ranges of motion of the ends of a lever are in proportion to their lengths: If the left-hand end of the lever in Fig. 8.24 is moved through an arc of s_{in}, the right-hand end will be moved through an arc of s_{out}, where

$$\frac{s_{in}}{s_{out}} = \frac{L_{in}}{L_{out}}$$

Hence

$$\text{MA} = \frac{F_{out}}{F_{in}} = \frac{s_{in}}{s_{out}} \qquad \textit{Mechanical advantage} \quad (8.10)$$

The same relationship follows from the law of conservation of energy. If there is no friction, the work $F_{in}s_{in}$ done *on* a machine by the input force must equal the work $F_{out}s_{out}$ done *by* the machine, which gives Eq. (8.10). Hence Eq. (8.10) holds not only for the lever but for all machines and in fact is often easier to apply than the principle of equilibrium, Eq. (8.8).

A chain drive combines features of both belt and gear drives. Its MA equals the ratio of the diameters of the driven and driving sprocket wheels; the greater the MA, the greater the multiplication of the driving torque.

Figure 8.24

Each end of a lever moves through a different distance if its arms are not equal in length. The ratio of distances equals the ratio of lever arms.

$$\text{MA} = \frac{s_{in}}{s_{out}} = \frac{L_{in}}{L_{out}}$$

s_{in} L_{out} s_{out} L_{in}

EXAMPLE 8.12

Find the mechanical advantage of the block and tackle shown in Fig. 8.25.

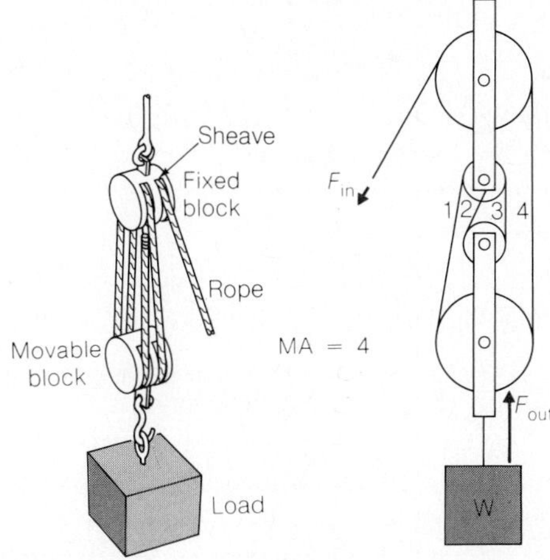

Figure 8.25

The ideal mechanical advantage of a block and tackle is equal to the number of strands of rope that support the movable block, in this case 4.

SOLUTION When the free end of the rope is pulled with the force F_{in} through a distance d, the movable block is raised through a height of $\frac{1}{4}d$, since there are four strands that must be shortened. Hence

$$\text{MA} = \frac{s_{in}}{s_{out}} = \frac{d}{\frac{1}{4}d} = 4$$

The mechanical advantage of this block and tackle is 4.

In general, the MA of a block and tackle is equal to the number of strands of rope that support the movable block and thereby the load. The strand to which the tension F_{in} is applied in Fig. 8.25 does not contribute to supporting the movable block; the upper pulley merely serves to change the direction of the applied force. ◆

The Inclined Plane

The inclined plane is the second of the three basic machines. The third, the hydraulic press, will be considered in Chapter 10.

We are so accustomed to using the inclined plane in everyday life that it may be hard to think of it as a "machine." So instinctive an act as choosing a gradual slope of a hill to walk up instead of a steep slope is based on the principle of the inclined plane. Figure 8.26 shows an inclined plane along which a crate of weight w

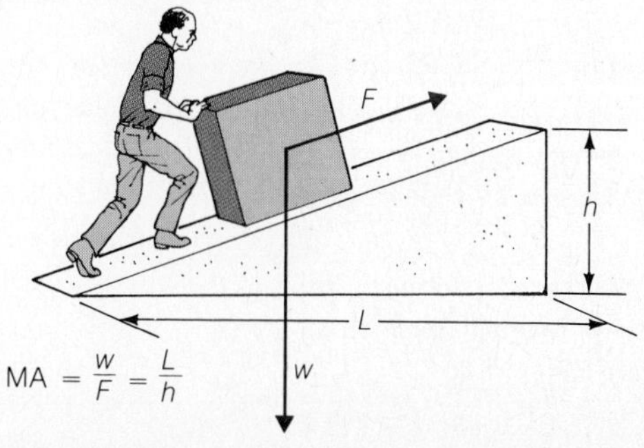

Figure 8.26

The inclined plane.

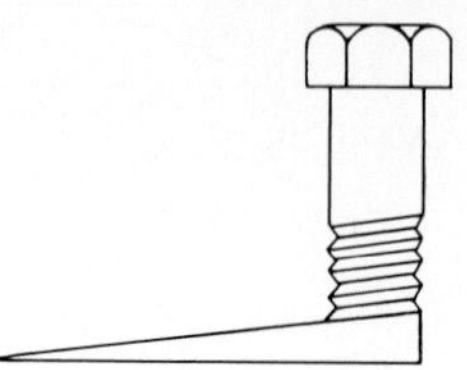

Figure 8.27

A screw is an inclined plane wrapped around a cylinder.

Figure 8.28

The wedge.

is being pushed. The plane is L long and h high. From Eq. (8.10) we have

$$\text{MA} = \frac{s_{in}}{s_{out}} = \frac{L}{h} \qquad \textit{Inclined plane} \quad (8.11)$$

EXAMPLE 8.13

A 150-kg safe on frictionless casters is to be raised 1.2 m off the ground to the bed of a truck. Planks 4.0 m long are available for the safe to be rolled along. How much force is needed to push the safe up to the truck?

SOLUTION Since MA $= F_{out}/F_{in}$, from Eq. (8.11) we have

$$F_{in} = \frac{F_{out}}{\text{MA}} = \frac{mg}{L/h} = \frac{mgh}{L} = \frac{(150\ \text{kg})(9.8\ \text{m/s}^2)(1.2\ \text{m})}{4.0\ \text{m}} = 441\ \text{N}$$

◆

The **screw** and the **wedge** are both developments of the inclined plane. As we can see from Fig. 8.27, a screw is an inclined plane wrapped around a cylinder to form a continuous helix. Aside from its use in splitting logs, leveling objects, and holding doors open, the wedge (Fig. 8.28) provides the operating principle of all cutting tools. A knife or a chisel is obviously a wedge, but so are the teeth of a saw and the abrasive chips of a grindstone.

Important Terms

When the net force acting on an object is zero, the object is in **translational equilibrium.** When the net torque acting on an object is zero, the object is in **rotational equilibrium.**

The **center of gravity** of an object is that point from which it can be suspended in any orientation without tending to rotate. The weight of an object can be considered as a downward force acting on its center of gravity.

A **machine** is a device that transmits force or torque. The three basic machines are the **lever,** the **inclined plane,** and the **hydraulic press.**

The **mechanical advantage** of a machine is the ratio between the output force (or torque) it exerts and the input force (or torque) that is furnished to it. The **ideal mechanical advantage** is its value under ideal circumstances, whereas the **actual mechanical advantage** is its value when friction is taken into account.

The **efficiency** of a machine is the ratio between its actual and ideal mechanical advantages; it is always less than 100%.

Important Formulas

Translational equilibrium: $\Sigma F_x = 0$

$\Sigma F_y = 0$

$\Sigma F_z = 0$

Rotational equilibrium: $\Sigma \tau = 0$ about any point

Mechanical advantage: $\text{MA} = \dfrac{F_{\text{out}}}{F_{\text{in}}} = \dfrac{s_{\text{in}}}{s_{\text{out}}}$

MULTIPLE CHOICE

1. Which of the following sets of horizontal forces could leave an object in equilibrium?
- **a.** 25, 50, and 100 N
- **b.** 5, 10, 20, and 50 N
- **c.** 8, 16, and 32 N
- **d.** 20, 20, and 20 N

2. Which of the following sets of horizontal forces could not leave an object in equilibrium?
- **a.** 6, 8, and 10 N
- **b.** 10, 10, and 10 N
- **c.** 10, 20, and 30 N
- **d.** 20, 40, and 80 N

3. If an object is free to move in a plane, the number of scalar equations that must be satisfied for it to be in equilibrium is
- **a.** 2 **b.** 3
- **c.** 4 **d.** 6

4. If an object is free to move in three dimensions, the number of scalar equations that must be satisfied for it to be in equilibrium is
- **a.** 2 **b.** 3
- **c.** 4 **d.** 6

5. An object in equilibrium may not have
- **a.** any forces acting on it
- **b.** any torques acting on it
- **c.** velocity
- **d.** acceleration

6. Two ropes are used to support a stationary weight W. The tensions in the ropes must
- **a.** each be $W/2$
- **b.** each be W
- **c.** have a vector sum of magnitude W
- **d.** have a vector sum of magnitude greater than W

7. A weight is suspended from the middle of a rope whose ends are at the same level. In order for the rope to be perfectly horizontal, the forces applied to the ends of the rope
- **a.** must be equal to the weight
- **b.** must be greater than the weight
- **c.** might be so great as to break the rope
- **d.** must be infinite in magnitude

8. If the sum of the torques on an object in equilibrium is zero about a certain point, it is
- **a.** zero about no other point
- **b.** zero about some other points
- **c.** zero about all other points
- **d.** any of the above, depending on the situation

9. In an equilibrium problem the point about which torques are computed
- **a.** must pass through one end of the object
- **b.** must pass through the object's center of gravity
- **c.** must intersect the line of action of at least one force acting on the object
- **d.** may be located anywhere

10. The center of gravity of an object
- **a.** is always at its geometrical center
- **b.** is always in the interior of the object
- **c.** may be outside the object
- **d.** is sometimes arbitrary

11. An object of mass m is moving at the speed v toward another object, also of mass m, that is at rest. The speed of the center of gravity of the system is
- **a.** $v/2$ **b.** v
- **c.** $2v$ **d.** v^2

12. Which of the following cannot be increased by using a machine of some kind?
- **a.** Force **b.** Torque
- **c.** Speed **d.** Work

13. The minimum number of pulleys in a block and tackle needed to achieve an MA of 6 is
- **a.** 3 **b.** 4
- **c.** 5 **d.** 6

14. The highest MA that can be obtained with a system of two pulleys is
- **a.** 1 **b.** 2
- **c.** 3 **d.** 4

15. A 100-N box is suspended by a rope from an overhead support. If a horizontal force of 58 N is applied to the box, the rope will make an angle with the vertical of
- **a.** 30° **b.** 45°
- **c.** 60° **d.** 75°

16. A 5.0-N picture is supported by two strings that run from its upper corners to a nail on the wall. If each string makes a 40° angle with the vertical, the tension in each is

a. 3.3 N b. 3.9 N
c. 5.0 N d. 10.0 N

17. The system shown in Fig. 8.29 is in equilibrium. The mass of *A* is

a. 0.5 kg b. 1 kg
c. $\sqrt{2}$ kg d. 2 kg

18. A weight of 1.0 kN is suspended from the end of a horizontal strut 1.5 m long that projects from a wall. A supporting cable 2.5 m long goes from the end of the strut to a point on the wall 2.0 m above the strut. The tension in the cable is

a. 0.8 kN b. 1.0 kN
c. 1.25 kN d. 1.67 kN

19. The inward force the strut in Multiple Choice 18 exerts on the wall is

a. 0.75 kN b. 0.94 kN
c. 1.25 kN d. 1.34 kN

20. A 60-kg object is attached to one end of a 40-kg steel tube 2.4 m long. The distance from the loaded end to the balance point is

a. 48 cm b. 60 cm
c. 80 cm d. 160 cm

21. A uniform wooden plank 2.00 m long that weighs 200 N is supported by a sawhorse 400 mm from one end. What downward force on the end nearest the sawhorse is needed to keep the plank level?

a. 200 N b. 300 N
c. 500 N d. 800 N

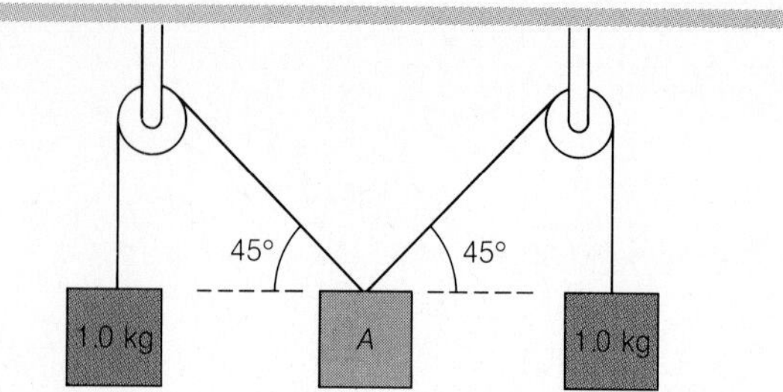

Figure 8.29

Multiple Choice 17.

22. A person pries up one end of a 200-kg crate with a steel pipe 2.00 m long. If the force exerted is 350 N, the distance of the fulcrum from the crate is

a. 30 cm b. 53 cm
c. 71 cm d. 86 cm

23. One end of a plank 4.0 m long is supported 1.0 m above the ground. A 10-kg box on frictionless wheels is placed on the plank. The force parallel to the plank needed to keep the box stationary is

a. 25 N b. 95 N
c. 98 N d. 0.39 kN

24. A machine with an ideal MA of 5 and an actual MA of 4 is used to raise a 10-kg load by 4.0 m. The work input to the machine is

a. 40 J b. 0.16 kJ
c. 0.39 kJ d. 0.49 kJ

QUESTIONS

1. What forces should be included in a free-body diagram of an object? What forces should not be included? Should the object's weight be included? If the object rests on a surface, should the reaction force of the surface be included?

2. If the earth followed a circular orbit around the sun, its speed would be constant. In this case would the earth be in equilibrium?

3. A ladder rests against a frictionless wall. (a) In what direction must be the force the ladder exerts on the wall? (b) How is the force the ladder exerts on the ground related to the weight of the ladder? Why?

4. Would you expect any difference in how soon you get tired between standing with your feet together and standing with them some distance apart? Why?

5. The sign convention for angular momentum is the same as that for torque. Is the angular momentum of the minute hand of a clock positive or negative when you look at the clock's face? When you turn the clock over?

6. In an equilibrium problem, under what circumstances is it necessary to consider the torques exerted by the various forces?

7. About which point should the torques exerted by the various forces that act on a body be calculated when this is necessary?

8. Why is a person with his feet apart more stable than if his feet were together?

9. What must be true of the MA of a machine meant to increase force? Of a machine meant to increase speed?

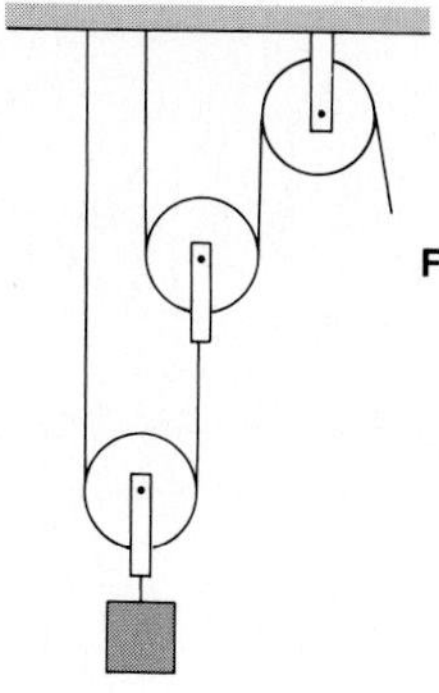

Figure 8.30

Questions 11 and 12.

10. What effect does lubrication have on the ideal MA of a machine? On its actual MA?

11. What is the MA of the pulley system shown in Fig. 8.30?

12. If the three pulleys in Fig. 8.30 were arranged instead as an ordinary block and tackle, what would be the maximum MA?

EXERCISES

8.1 Translational Equilibrium

1. A horizontal beam 6.0 m long projects from the wall of a building. A guy wire that makes a 40° angle with the horizontal is attached to the outer end of the beam. When a weight of 100 N is attached to the end of the beam, what is the tension in the guy wire? (Neglect the weight of the beam.)

2. A 40,000-kg airplane is provided with 130 kN of thrust by its engines as it takes off at an angle of 25° above the horizontal. If the velocity of the airplane is constant, how much upward force on it is provided by the vertical component of the lift produced by the flow of air over its wings?

3. A 50-kg object is suspended from two ropes that each make an angle of 30° with the vertical. What is the tension in each rope?

4. The arrangement shown in Fig. 8.31 is in equilibrium. Find the tensions in the strings *A* and *B*.

5. The arrangement shown in Fig. 8.32 is in equilibrium. Find the mass of *A*.

6. A 60-kg girl sitting in a canvas sling is being hoisted up a mast by a rope that passes through a pulley at the masthead and then down to a winch on the deck. She presses her feet against the mast to hold herself clear. How much force must her legs exert if her CG is 60 cm from the mast when she is 4.0 m below the masthead?

7. A horizontal force of 80 N acts on a 5.0-kg object suspended by a string. What are the direction and magnitude of the force the string must provide to keep the object at rest?

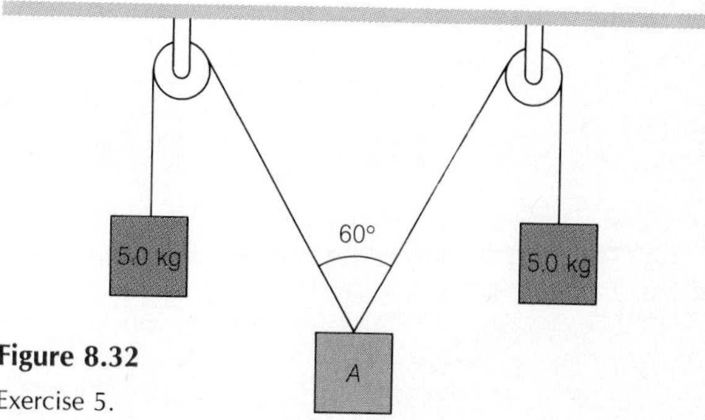

Figure 8.32

Exercise 5.

8. A 500-kg load of bricks is lifted by a crane alongside a building under construction. A horizontal rope is used to pull the load to where it is required, and the supporting cable is then at an angle of 10° from the vertical. What is the tension in the rope? In the supporting cable?

9. A 600-N crate is suspended by two ropes, one that makes an angle of 20° with the vertical and the other that makes an angle of 30° with the vertical. Find the tension in each rope.

10. A gong is suspended by two ropes. One makes an angle of 25° with the vertical and is under a tension of 70 N. The other rope makes an angle of 40° with the vertical. Find the mass of the gong and the tension in the second rope.

11. Find the maximum load that can be supported by the arrangement shown in Fig. 8.33 if the force on the strut is not to exceed 3000 N.

12. The femur is the large bone in the thigh. When a fractured femur is set, the thigh muscles tend to pull the segments together so tightly that the healed bone would be shorter than its original length. To prevent this, a traction apparatus like

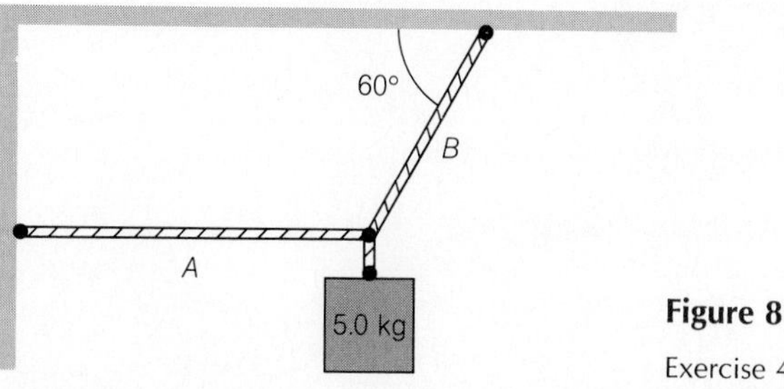

Figure 8.31

Exercise 4.

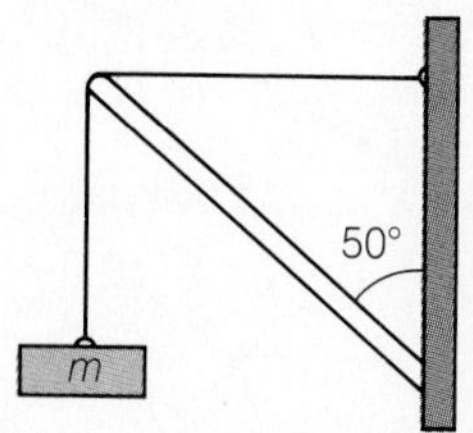

Figure 8.33

Exercise 11.

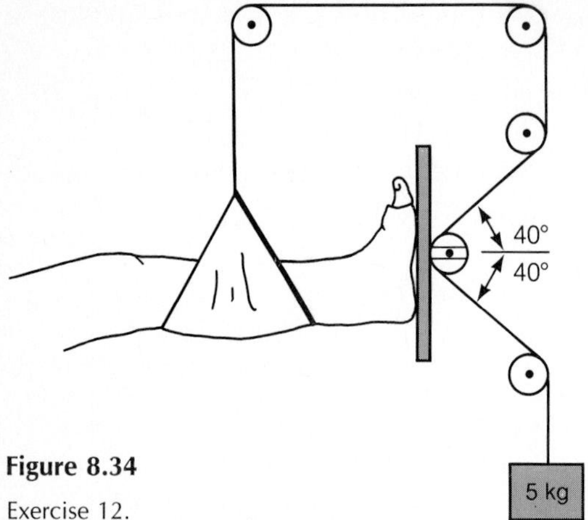

Figure 8.34

Exercise 12.

the one shown in Fig. 8.34 can be attached to the lower leg to exert a force that opposes the muscular contraction. The advantage of this particular arrangement is that the injured leg can be moved around a fair amount without much effect on the applied force. Find the magnitude of the resultant force on the femur produced by the apparatus and its direction relative to the horizontal for the situation shown. Assume a mass of 3.5 kg for the lower leg.

13. A bird sits on a telephone wire midway between two poles 20 m apart. The wire, assumed weightless, sags by 50 cm. If the tension in the wire is 90 N, what is the mass of the bird?

14. A car is stuck in the mud. To get it out, the driver ties one end of a rope to the car and the other to a tree 30 m away. He then pulls sideways on the rope at its midpoint. If he exerts a force of 500 N, how much force is applied to the car when he has pulled the rope 1.5 m to one side?

15. Find the compressive force on the boom of the crane shown in Fig. 8.35 and the tension in the horizontal supporting cable. Assume both cables are fixed at the end of the boom.

8.2 Rotational Equilibrium

16. When a person stands on one foot with his heel raised, the entire reaction force of the floor, which is equal to his weight w, acts upward on the ball of his foot, as in Fig. 8.36. In order to raise his heel, he must apply the upward force F_1 via his Achilles tendon, so the downward force F_2 on his ankle is greater than his weight w. Find the values of F_1 and F_2 for a 75-kg man for whom $L_1 = 50$ mm and $L_2 = 150$ mm. If the distance L_1 were greater, would this mean an increase or a decrease in the muscular force F_1?

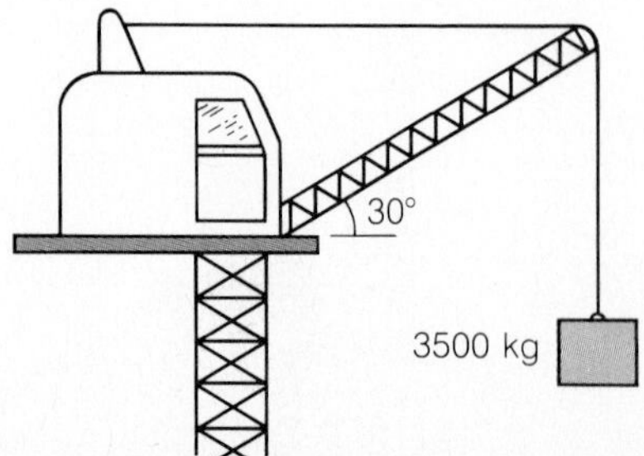

Figure 8.35

Exercise 15.

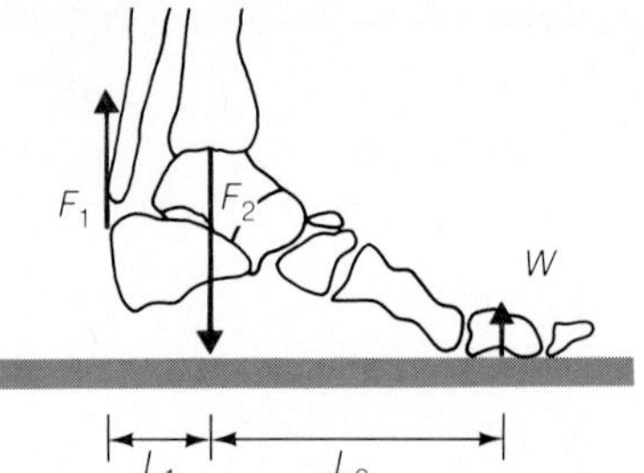

Figure 8.36

Exercise 16.

17. A 15-kg child and a 25-kg child sit at opposite ends of a 4.0-m seesaw pivoted at its center. Where should a third child whose mass is 20 kg sit in order to balance the seesaw?

18. The driver of a car applies a force of 200 N to the brake pedal shown in Fig. 8.37. What is the resulting tension T in the brake cable?

19. The upper end of a weightless ladder rests against a smooth, frictionless wall at a point h above the ground. The coefficient of static friction between the lower end of the ladder and the floor is μ. What horizontal distance can a person move in climbing up the ladder before the ladder starts to slip?

20. A weightless ladder 6.0 m long rests against a frictionless wall at an angle of 60° above the horizontal. A 75-kg person is 1.2 m from the top of the ladder. What horizontal force at the bottom of the ladder is needed to keep it from slipping?

21. A fishing rod 3.0 m long is attached to a pivoting holder at its lower end. The rod is 50° above the horizontal. A fisherman grasps the rod 1.0 m from its lower end. What horizontal force must the fisherman exert to keep the rod at this angle when reeling in a 10-kg fish from directly below the rod's upper end?

22. The lower end of a weightless rod 2.00 m long is hinged to a wall, and a load of 200 N is suspended from its upper end, as in Fig. 8.38. A horizontal rope 60 cm long joins the middle of the rod to the wall. Find the tension in the rope.

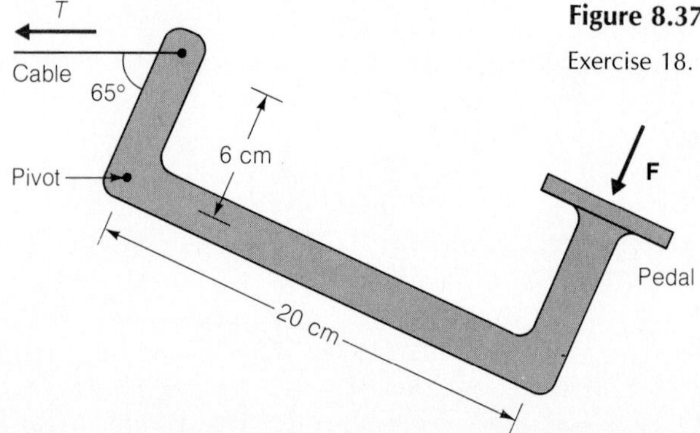

Figure 8.37

Exercise 18.

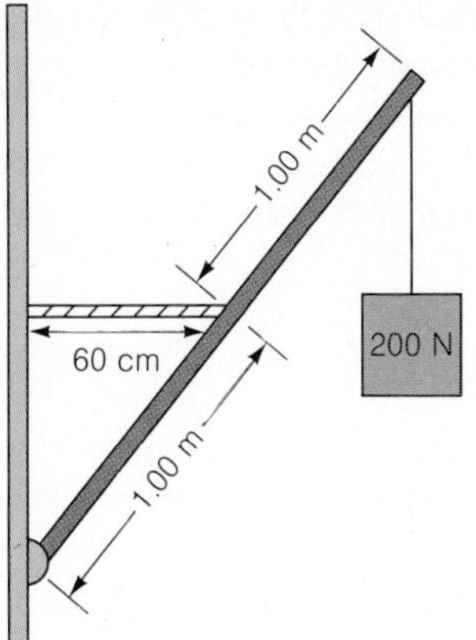

Figure 8.38

Exercise 22.

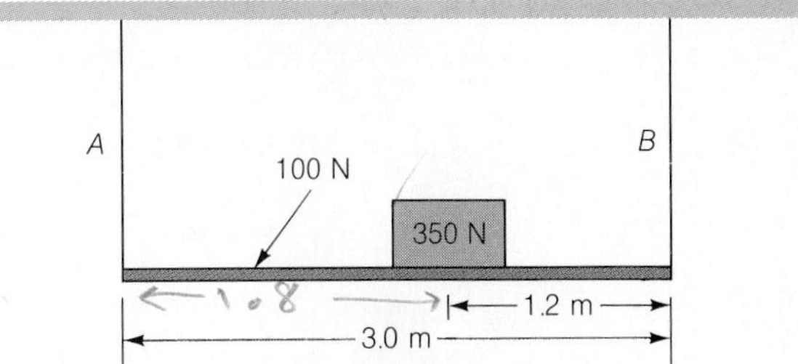

Figure 8.39

Exercise 28.

8.3 Center of Gravity

23. A uniform plank 5.0 m long that weighs 300 N rests on two sawhorses 1.0 m from each end. How near either end of the plank can a 700-N painter stand without tipping the plank?

24. A square table has legs at three of its corners. Will the table fall over? If not, what is the maximum mass that can be placed on the unsupported corner before it falls over?

25. The wheels of a certain truck are 1.8 m apart, and the truck falls over when tilted sideways at 30° from the horizontal. How high above the road is its CG?

26. A 60-kg woman stands upright with her feet 20 cm apart and her arms outstretched on both sides. In her right hand she holds a dumbbell 70 cm from the centerline of her body. Find the maximum mass the dumbbell can have if she is not to topple over.

27. A crate whose total mass is 80 kg rests on a floor such that the coefficient of static friction is 0.40. The crate is 1.5 m high and 1.0 m square, and its CG is at its geometric center. (a) Find the minimum horizontal force needed to start the crate sliding across the floor. (b) What is the maximum height above the floor at which this force can be applied without tipping the crate over? (*Hint:* Assume the crate has just begun to tip over so only the far edge of its base is touching the floor, and calculate moments about this edge.)

28. A uniform beam that weighs 100 N is supported by two vertical wires *A* and *B*. A box that weighs 350 N is placed on the beam, as shown in Fig. 8.39. Find the tensions in the two wires.

29. Three people are carrying a horizontal ladder 4.00 m long. One of them holds the front end of the ladder, and the other two hold opposite sides of the ladder the same distance from its far end. What is the distance of the latter two people from the far end of the ladder if each person supports one-third of the ladder's weight?

30. A person holds a 10-kg board that is 2.00 m long in a horizontal position by supporting it with one hand underneath it 40 cm from one end while the other hand presses down at that end, as shown in Fig. 8.40. Find the force each hand exerts. What percentage of the board's weight is each force?

31. The front and rear axles of a 24-kN truck are 4.0 m apart. The CG of the truck is located 2.5 m behind its front axle. Find the weight supported by the front wheels of the truck.

32. A 250-N bag of cement is placed on a 4.0-m-long plank 1.8 m from one end. The plank itself weighs 80 N. Two men pick up the plank, one at each end. How much weight must each support?

33. The CG of a 150-kg polar bear standing on all fours is 100 cm from her hind paws and 80 cm from her fore paws. Find the force the ground exerts on each of her paws.

34. A butcher with his upper arm vertically at his side and his forearm horizontal presses down on a scale with his hand, which is 40 cm from his elbow joint. His forearm and hand together weigh 30 N, and their CG is 20 cm from his elbow joint. The triceps muscle on the back of his upper arm is attached 2.5 cm behind the center of the elbow joint and it is exerting an upward force of 1.0 kN. What is the reading on the scale in kilograms?

35. A 40-kg boom 3.0 m long is hinged to a vertical mast and held in position by a rope at its end that is attached to the mast 1.0 m above the hinge pin. If the boom is horizontal and uniform, find the tension in the rope.

36. A 20-kg object is suspended from one end of a 30-kg wooden beam 3.0 m long. Where should the beam be picked up so that it remains horizontal?

37. The front wheels of a certain car are found to support 600 kg and the rear wheels 400 kg. The car's wheelbase (distance between axles) is 2.2 m. How far from the forward axle is the CG of the car?

Figure 8.40

Exercise 30.

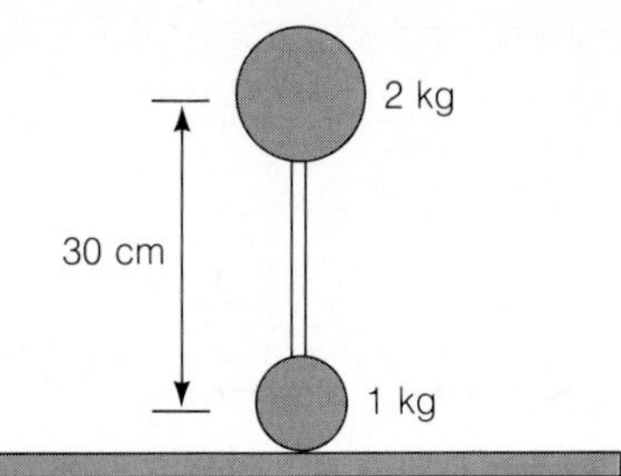

Figure 8.41
Exercise 39.

8.4 Finding a Center of Gravity

38. The moon does not actually revolve around the earth: Both revolve around the center of mass of the earth-moon system. Find the location of this center of mass in terms of r_e, the earth's radius, given that the earth-moon distance is $60r_e$ and that the masses of the two bodies are 6.0×10^{24} kg and 7.3×10^{22} kg, respectively.

39. The asymmetric dumbbell shown in Fig. 8.41 is balanced on a frictionless surface. Eventually the dumbbell falls over with the 2-kg ball moving to the right. What is the final position of the 1-kg ball? (*Hint:* Consider the motion of the CG of the dumbbell.)

40. A 60.0-kg woman is sitting with her upper body and lower legs vertical and her thighs horizontal. The mass of her upper body is 40.0 kg and its CG is 32 cm above her hips; the mass of her thighs is 13.0 kg and their CG is 16 cm in front of her hips; and the mass of her lower legs and feet is 7.0 kg and their CG is 23 cm below her knees and 37 cm in front of her hips. Find the location of the CG of the seated woman.

41. Find the location of the CG of the flat object of uniform thickness shown in Fig. 8.42.

8.5 Mechanical Advantage

42. A woman uses a chain hoist whose efficiency is 95% to lift a 100-kg motor. She exerts a force of 120 N on the chain and pulls a total of 8.0 m of chain through the pulleys of the hoist. Through what height is the motor raised?

43. A vertical-axis capstan drum 60 cm in diameter was used to pull in the anchor chain of a whaler a century ago. Six horizontal arms 1.2 m long protruded from the top of the drum, and a man on each arm exerted a force of 300 N on it. The efficiency of the system was 85%. (a) Find the force applied to the anchor chain. (b) How far must each man have walked to raise 100 m of chain?

44. In the frictionless pulley system shown in Fig. 8.43, a force of 40 N is needed to support the 10-kg object. What is the mass of the lower pulley?

45. A block and tackle is used to pull a car out of the mud. A block with two pulleys is fastened to the car's bumper, and another block with three pulleys is tied to a tree. One end of a rope is made fast to the bumper and threaded through the various pulleys. Neglecting friction, determine how much force is applied to the car when two people exert a total force of 900 N on the free end of the rope.

46. A steel rod 2.00 m long is to be used to pry up one end of a 400-kg crate. If a workman can exert a downward force of 400 N on one end of the rod, where should he place a block of wood to act as the fulcrum?

47. A ramp 25 m long slopes down 1.2 m to the edge of a lake. What force is needed to pull a 300-kg boat on an 80-kg trailer up along the ramp if friction is negligible?

48. The pitch p of a screw is the distance between adjacent threads; when the screw is turned through a complete rotation, it moves the distance p. Find the ideal MA of a screw that is turned by a screwdriver whose handle has the radius R.

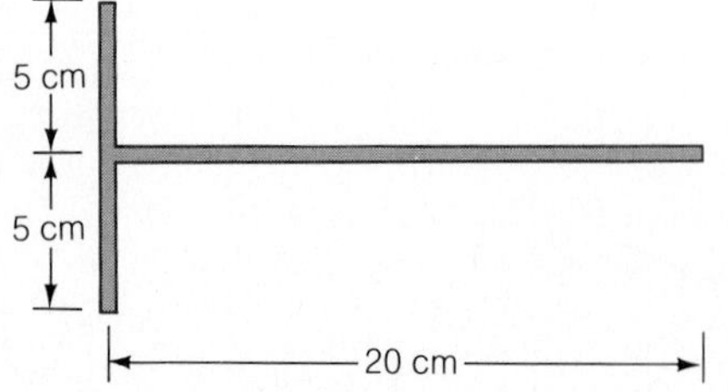

Figure 8.42
Exercise 41.

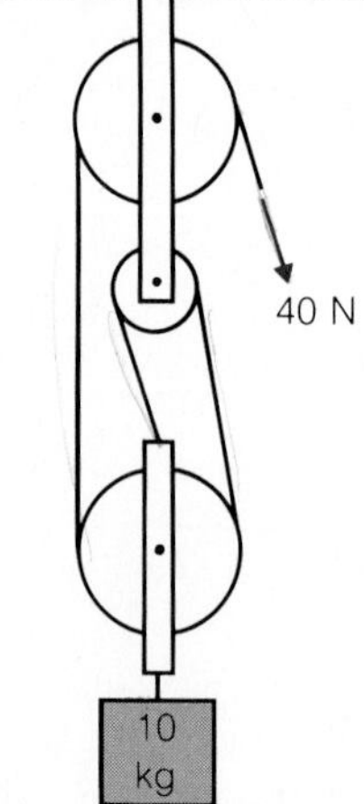

Figure 8.43
Exercise 46.

PROBLEMS

49. Two uniform rods pivoted together at one end stand like an inverted V on a table. If the angle between each rod and the table is θ, find the minimum coefficient of static friction between each rod and the table.

50. A 500-kg load is supported by a pair of hinged legs, as shown in Fig. 8.44. There is no friction between the legs and the floor. Find the tension in the rope that joins the midpoints of the legs.

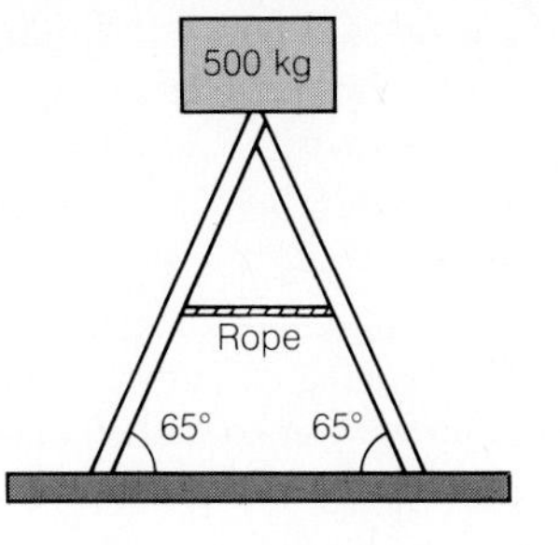

Figure 8.44
Problem 50.

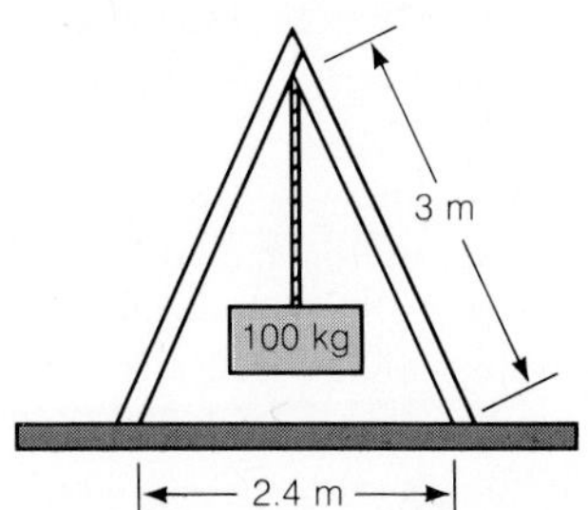

Figure 8.45
Problem 51.

51. A 100-kg load is supported by two hinged legs, as shown in Fig. 8.45. Find the force each leg exerts on the ground.

52. A 50-kg load is supported by two hinged legs, as shown in Fig. 8.46. Find the force each leg exerts on the ground.

53. A boom hinged at the base of a vertical mast is used to lift a 1500-kg load, as shown in Fig. 8.47. Find the tension in the cable from the top of the mast to the end of the boom. The weight of the boom itself can be neglected.

54. (a) Find the tension T in the wire and the force F the horizontal beam exerts when the structure shown in Fig. 8.48 is at rest. What is the direction of **F**? (b) Find T and F when the structure is rotating twice per second. What is the direction of **F** now? Neglect the masses of the beam and the wire.

55. An object is attached to the midpoint of a string of length L whose ends are tied to rings that fit loosely on a horizontal rod, as shown in Fig. 8.49. If the coefficient of static friction between the rings and the rod is μ, find the maximum distance x between the rings before they begin to slip in terms of L and μ.

Figure 8.46
Problem 52.

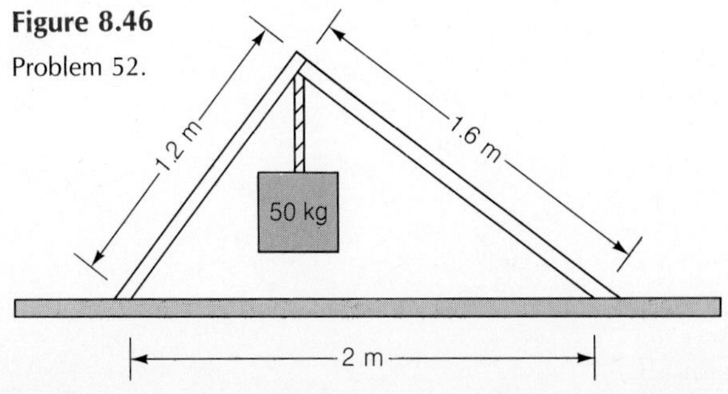

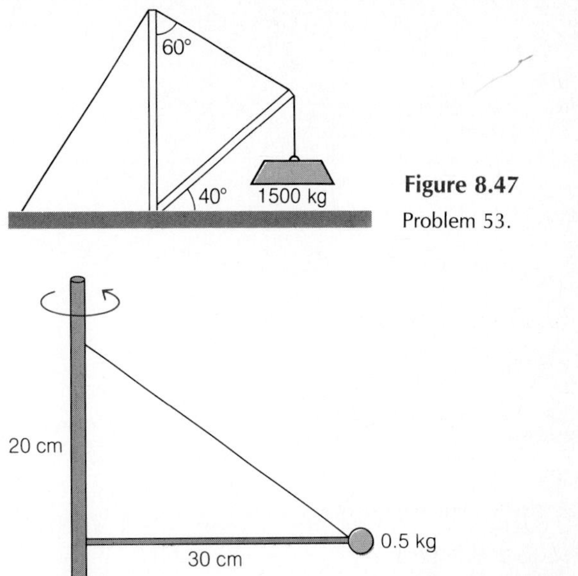

Figure 8.47
Problem 53.

Figure 8.48
Problem 54.

56. The flat, uniform metal plate shown in Fig. 8.50 has a circular hole 10 cm in radius. Find the location of the CG of the plate.

57. A 200-kg uniform horizontal beam 4.0 m long is supported at one end by a rigid post and at the other by a rope

Figure 8.49
Problem 55.

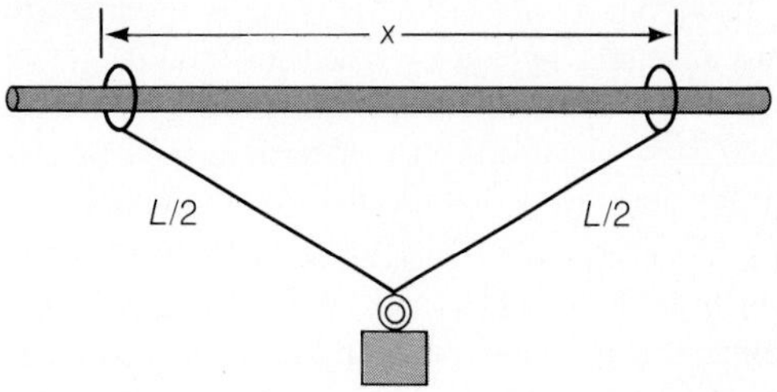

Figure 8.50
Problem 56.

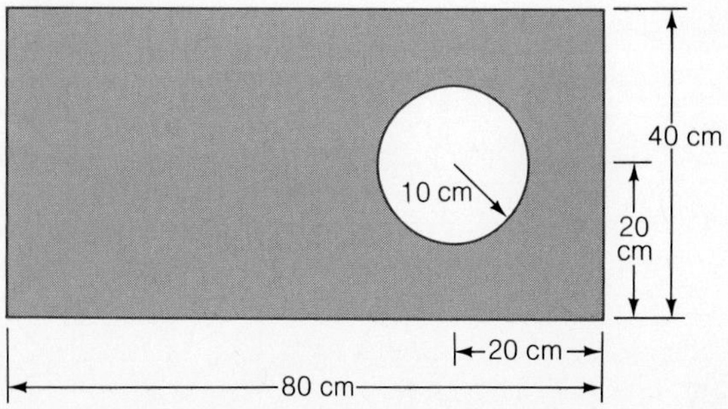

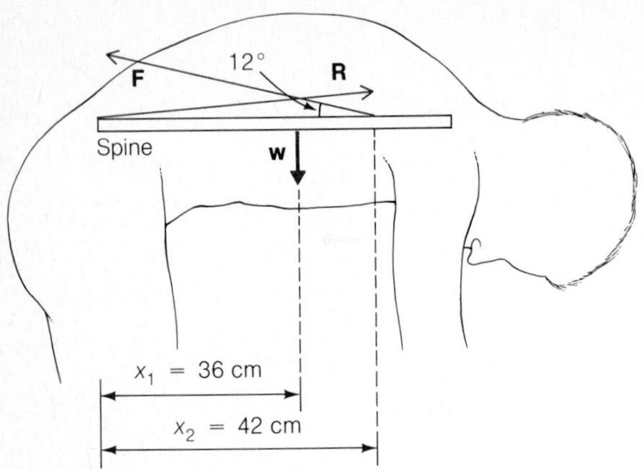

Figure 8.51

Problem 58.

that makes an angle of 40° with the beam. A load of 1000 kg is suspended from the outer end of the beam. Find the tension in the rope.

58. Figure 8.51 is a simplified diagram of a person bending over with the back horizontal. Typically a person's upper body (head, torso, and arms) comprises 65% of the total weight and has its CG 72% of the height above the feet. In the case of an 80-kg person 1.8 m tall, the upper body weight is 510 N and its CG is 36 cm above the base of the spine. This weight is supported by the back muscles, whose effect is equivalent to that of a single force **F** acting 42 cm above the base of the spine at an angle of 12°. (a) Find the magnitude of **F**. (b) Find the magnitude of the reaction force **R** that acts on the base of the spine. How many times greater than the weight of the upper body is this force? (Because bending over, particularly when picking something up, puts so much stress on the lower spine, this posture is best avoided. When an object is to be picked up, the back should be kept as vertical as possible and the actual lifting done by the leg muscles.)

59. The arm of a certain person has a mass of 3.0 kg and its CG is 28 cm from the shoulder joint. When the arm is outstretched so that it is horizontal, the shoulder muscle supporting it has a line of action 15° above the horizontal. (a) If this muscle is attached to the upper arm 13 cm from the shoulder joint, what is the force developed by the muscle? (b) Find the force when the outstretched hand 60 cm from the shoulder joint is used to support a 1.0-kg load.

60. A uniform 10-kg ladder 2.5 m long is placed against a frictionless wall with its base on the ground 80 cm from the wall. Find the magnitudes of the forces exerted on the wall and on the ground.

61. A uniform 15-kg ladder 3.0 m long rests against a frictionless wall at a point 2.4 m off the floor. Find the vertical and horizontal components of the force exerted by the ladder on the floor.

62. A uniform 12-kg ladder 2.8 m long rests against a vertical frictionless wall with its lower end 1.1 m from the wall. If the ladder is to stay in place, what must be the minimum coefficient of friction between the bottom of the ladder and the ground?

63. A door 2.40 m high and 0.80 m wide has hinges at the top and bottom of one edge. The entire 200-N weight of the door is supported by the upper hinge. Find the magnitude and direction of (a) the force the door exerts on the upper hinge; (b) the force the lower hinge exerts on the door.

64. A door 3.0 m high and 1.2 m wide has hinges on one edge that are 30 cm above the bottom and 30 cm below the top. The entire 500-N weight of the door is supported by the lower hinge. Find the magnitude and direction of (a) the force the door exerts on the lower hinge; (b) the force the upper hinge exerts on the door.

Answers to Multiple Choice

1. d	**7.** d	**13.** c	**19.** a
2. d	**8.** c	**14.** d	**20.** a
3. b	**9.** d	**15.** a	**21.** b
4. d	**10.** c	**16.** a	**22.** b
5. d	**11.** a	**17.** c	**23.** a
6. c	**12.** d	**18.** c	**24.** d

CHAPTER

MECHANICAL PROPERTIES OF MATTER

The usual way a physicist approaches a complex situation is first to set up a simple model that represents the main features of the situation. Then, if predictions based on this model agree reasonably well with observation and experiment, the model is further refined until agreement is even better. Our work in physics thus far is an illustration of this process. We began by treating objects as though they were just particles, and later we went on to consider them as rigid bodies. Actually, of course, there is no such thing as a rigid body—we can stretch, compress, or twist the strongest piece of steel if we apply suitable forces. In this chapter we come even closer to reality by looking at some of the mechanical properties of materials.

9.1 Density

Mass per unit volume.

A characteristic property of every substance is its **density,** which is its mass per unit volume. When we speak of lead as a "heavy" metal and of aluminum as a "light" one, what we really mean is that lead has a greater density than aluminum: A cubic meter of lead has a mass of 11,300 kg, whereas a cubic meter of aluminum has a mass of only 2700 kg.

Although in the SI system the proper unit of density is the kg/m^3, densities are frequently given in g/cm^3. Because there are 10^3 g in a kilogram and 100 cm in a meter, hence 10^6 cm^3 in a cubic meter,

$$1 \text{ g/cm}^3 = 10^3 \text{ kg/m}^3$$

The densities of various common substances are given in Table 9.1. The symbol for

Table 9.1 Densities of various substances at atmospheric pressure and room temperature

	Density	
Substance	**kg/m^3**	**g/cm^3**
Air	1.3	1.3×10^{-3}
Alcohol (ethyl)	7.9×10^2	0.79
Aluminum	2.7×10^3	2.7
Balsa wood	1.3×10^2	0.13
Blood (37°C)	1.06×10^3	1.06
Bone	1.6×10^3	1.6
Carbon dioxide	2.0	2.0×10^{-3}
Concrete	2.3×10^3	2.3
Copper	8.9×10^3	8.9
Gasoline	6.8×10^2	0.68
Gold	1.9×10^4	19
Helium	0.18	1.8×10^{-4}
Hydrogen	0.09	9×10^{-5}
Ice	9.2×10^2	0.92
Iron and steel	7.8×10^3	7.8
Lead	1.1×10^4	11
Mercury	1.4×10^4	14
Nickel	8.9×10^3	8.9
Nitrogen	1.3	1.3×10^{-3}
Oak	7.2×10^2	0.72
Oxygen	1.4	1.4×10^{-3}
Water, pure	1.00×10^3	1.00
Water, sea	1.03×10^3	1.03

density is ρ (Greek letter *rho*). If the volume V of a substance has the mass m,

$$\rho = \frac{m}{V} \qquad \textit{Density} \quad (9.1)$$

$$\text{Density} = \frac{\text{mass}}{\text{volume}}$$

EXAMPLE 9.1

A 50-g bracelet is suspected of being gold-plated lead instead of pure gold. When it is dropped into a full glass of water, 4.0 cm^3 of water overflows (Fig. 9.1). Is the bracelet pure gold? If not, what proportion of its mass is gold?

SOLUTION The density of the bracelet is

$$\rho = \frac{m}{V} = \frac{50\text{ g}}{4.0\text{ cm}^3} = 12.5\text{ g/cm}^3$$

This is less than the density of gold, which is 19 g/cm^3, so the bracelet is not pure gold. To find the proportion of gold it contains, we note that

$$m_{\text{gold}} + m_{\text{lead}} = 50\text{ g}$$

$$V_{\text{gold}} + V_{\text{lead}} = \frac{m_{\text{gold}}}{\rho_{\text{gold}}} + \frac{m_{\text{lead}}}{\rho_{\text{lead}}} = 4.0\text{ cm}^3$$

From the first equation, $m_{\text{lead}} = 50\text{ g} - m_{\text{gold}}$. Substituting this expression for m_{lead} and the values $\rho_{\text{gold}} = 19\text{ g/cm}^3$ and $\rho_{\text{lead}} = 11\text{ g/cm}^3$ in the second equation yields

$$\frac{m_{\text{gold}}}{19\text{ g/cm}^3} + \frac{50\text{ g}}{11\text{ g/cm}^3} - \frac{m_{\text{gold}}}{11\text{ g/cm}^3} = 4.0\text{ cm}^3$$

$$m_{\text{gold}} = 14\text{ g}$$

Hence the proportion of gold in the bracelet is 14 g/50 g = 0.28 = 28%. ◆

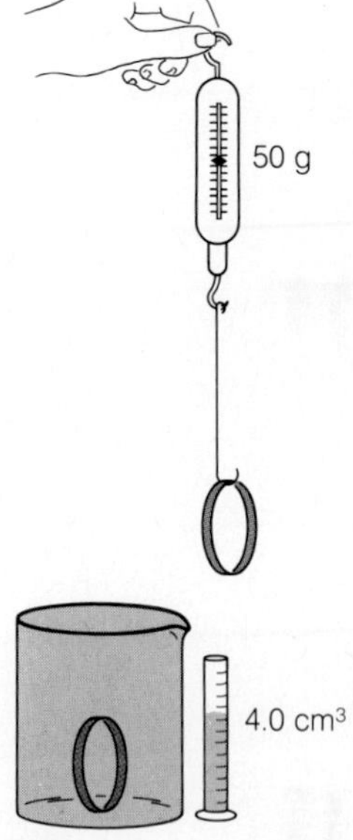

Figure 9.1

The density of this bracelet is $\rho = m/V = (50\text{ g})/(4.0\text{ cm}^3) = 12.5\text{ g/cm}^3$.

EXAMPLE 9.2

We want to pump water from a well 20 m deep at the rate of 500 liters/min. Assuming 60% efficiency, how powerful should the pump motor be?

SOLUTION The required power output of the pump is

$$P_{\text{out}} = \frac{\text{work}}{\text{time}} = \frac{W}{t} = \frac{mgh}{t} = \frac{\rho Vgh}{t}$$

Because 1 liter = 1 L = 10^{-3} m^3 and 1 min = 60 s, the flow rate of 500 liters/min is equivalent to

$$\frac{V}{t} = \left(500\,\frac{\text{L}}{\text{min}}\right)\left(10^{-3}\frac{\text{m}^3}{\text{L}}\right)\left(\frac{1}{60\ \text{s/min}}\right) = 8.33 \times 10^{-3}\ \text{m}^3/\text{s}$$

The power output is therefore

$$P_{\text{out}} = \rho g h\left(\frac{V}{t}\right) = \left(1.00 \times 10^3\,\frac{\text{kg}}{\text{m}^3}\right)\left(9.8\,\frac{\text{m}}{\text{s}^2}\right)(20\ \text{m})\left(8.33 \times 10^{-3}\,\frac{\text{m}^3}{\text{s}}\right)$$
$$= 1633\ \text{W}$$

In general, efficiency = power output/power input, so the pump motor must have a power of

$$P_{\text{in}} = \frac{P_{\text{out}}}{\text{efficiency}} = \frac{1633\ \text{W}}{0.60} = 2722\ \text{W} = 2.7\ \text{kW}$$ ◆

The **specific gravity** of a substance is its density relative to that of water and is a pure number. Because the density of water is almost exactly 1 g/cm^3, the specific gravity of a substance is very nearly equal to the numerical value of its density when expressed in g/cm^3. Because the density of aluminum is 2.7 g/cm^3, its specific gravity is 2.7. Specific gravity is also called **relative density.**

9.2 Elasticity

Tension, compression, and shear.

Although most solid bodies seem to be solid and unyielding, we can nevertheless deform them either temporarily or permanently by applying **stresses.** (When equal and opposite forces act on a body, the stress on the body is the magnitude of either force divided by the area over which it is exerted; see Section 9.4.) Stresses fall into three categories: tensions, compressions, and shears. These are illustrated in Fig. 9.2.

A **tensile stress** is applied to an object when equal and opposite forces act away from each other on its ends. Such a stress tends to lengthen the object. A **compressive stress** is applied when such forces act toward each other. Such a stress tends to shorten the object. A **shear stress** is applied when the equal and opposite

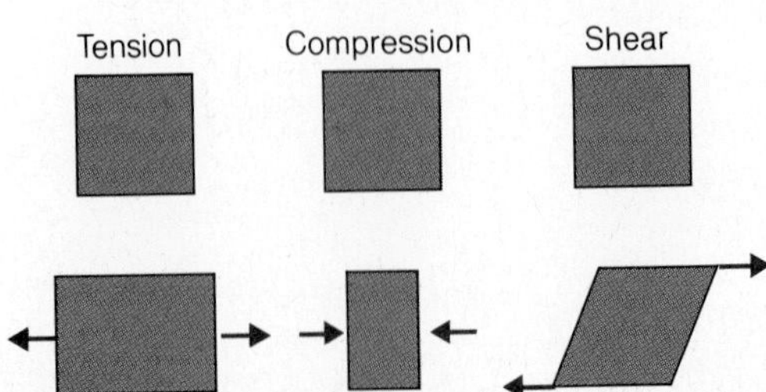

Figure 9.2
The three types of stress.

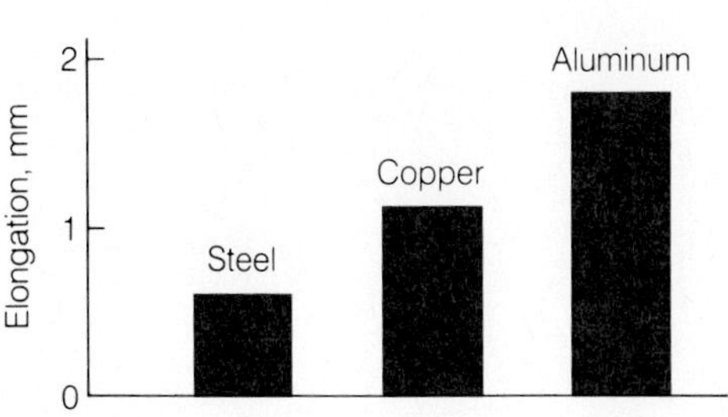

Figure 9.3

The increases in length of three wires 1 m long and 1 mm in diameter when a tensile force of 100 N is applied.

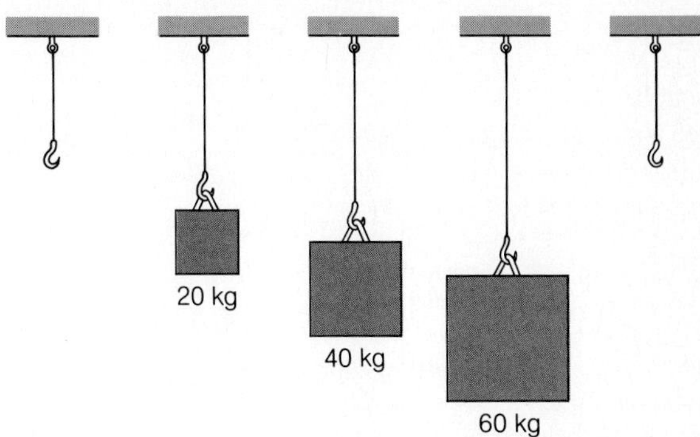

Figure 9.4

The elongation of a wire is proportional to the stress applied to it, provided the elastic limit is not exceeded. When the stress is removed, the wire returns to its original length.

forces have different lines of action. Such a stress tends to alter the shape of the object without changing its volume.

Thus tension stretches an object, compression shrinks it, and shear twists it.

The response of an object to a given stress depends on its composition, shape, temperature, and so on. Figure 9.3 shows the increases in length of identical steel, copper, and aluminum wires fixed in place at one end when a force of 100 N is applied to the other end.

The amount by which a particular object is deformed is directly proportional to the applied stress, provided the stress is not too much. In the case of tension, for example, we might find that supporting a 20-kg mass with a certain thin wire stretches the wire by 1 mm (Fig. 9.4). Doubling the mass to 40 kg will stretch it by 2 mm, tripling the mass to 60 kg will stretch it by 3 mm, and so on. When the mass is removed, the wire returns to its original length.

This proportionality is called **Hooke's law** and may be written

$$F = kx \qquad \textit{Hooke's law} \quad (9.2)$$

where F is the applied tension, x the resulting elongation, and k a constant whose value depends on the nature and dimensions of the object under stress. The force

Forces in the earth's crust acting over thousands or millions of years can deform even the strongest rock.

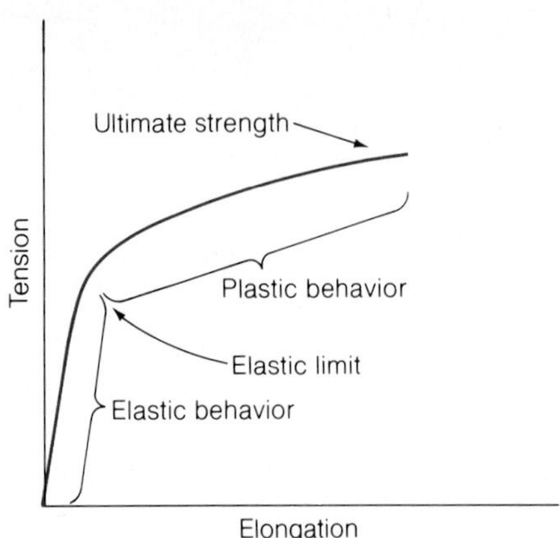

Figure 9.5

Graph of the elongation of an iron rod as more and more tension is applied to it. When a tension below the elastic limit is applied and then removed, the rod returns to its original length, which is elastic behavior. When a tension exceeding the elastic limit is applied and removed, the rod does not contract fully but remains permanently longer, which is plastic behavior.

constant k is higher for materials such as steel than it is for materials such as lead. It is proportional to the cross-sectional area of the object, so a thick wire of a given material has a higher value of k than a thin wire of the same material. Relationships like Eq. (9.2) also apply to solids under shear and to all states of matter under compression.

The term **elastic limit** refers to the maximum stress that can be applied to an object without its being permanently deformed as a result. When its elastic limit is exceeded, the object may or may not be far from breaking. Brittle substances like glass or cast iron break at or near their elastic limits. Bone is another material that remains elastic until its breaking point is reached, and so it cannot be permanently changed in size or shape by applying a force. The elastic limit of many materials decreases with temperature, which is why metals are often heated before being formed into shape by bending or hammering.

Most metals (cast iron is one exception) can be deformed a great deal beyond their elastic limits, a property known as **ductility.** Copper is a very ductile metal; although a copper wire whose cross-sectional area is 1 mm^2 reaches its elastic limit when a force of about 150 N is applied, it will not rupture until the force has more than doubled.

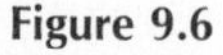

Figure 9.6

A force applied to an object with a crack causes a concentration of stress at the end of the crack. The smaller the diameter of the end, the greater the stress concentration and the more likely the crack is to spread.

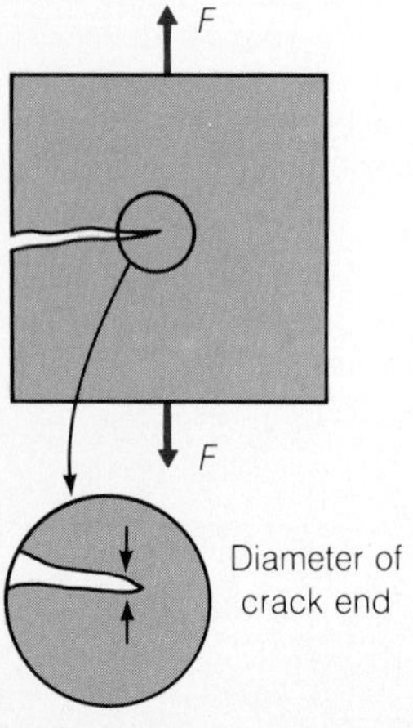

A graph of the elongation of an iron rod versus the tension applied to it is shown in Fig. 9.5. At first the graph is a straight line, which corresponds to Hooke's law. Past the elastic limit the graph flattens out, which means that each increase in tension by a given amount produces a greater increase in length than it did below the elastic limit: The rod stretches more readily. If the tension is removed after having exceeded the elastic limit, the rod remains longer than it was originally; it has undergone **plastic deformation.** The **ultimate strength** of the rod is the greatest tension it can withstand, and it corresponds to the highest point on the curve.

An object may fail through **fatigue** after repeated applications of stresses that are well under its original breaking strength. Tiny defects in the internal structure of the material grow a little each time a stress acts, and eventually cracks appear that lead to rupture (Fig. 9.6). The chief factors are the level of stress and the total number of stress cycles. The number of cycles needed for failure decreases rapidly with increasing stress. A metal bar that can tolerate a million applications of a force equal to 20% of its ultimate strength may break after only 10,000 or 20,000 applications of a force twice as great. The problem is clearly most severe in machinery—an engine operating at 3000 rpm undergoes 180,000 stress cycles per hour. Most failures in aircraft components are due to fatigue.

9.3 Structure of Solids

How elasticity and plasticity originate.

Most solids are **crystalline,** with their atoms arranged in regular patterns. Usually a sample of such a solid does not consist of a single perfect crystal but is an assembly of a great many tiny crystals. A few solids, such as pitch, wax, and glass, have irregular structures much like those of liquids. They are said to be **amorphous** (''without form'') solids.

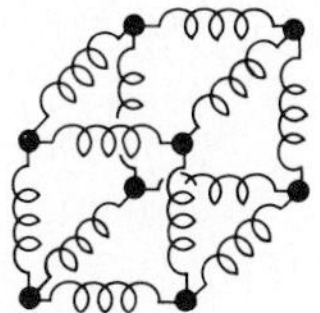

Figure 9.7

A simple model of a crystalline solid. Each spring represents the bond between adjacent atoms, which resist being pulled farther apart or pressed closer together than their normal spacing. No such springs actually exist, but they help us to visualize the origins of elastic behavior in solids.

The atoms in a crystal have a certain normal spacing and resist being pulled farther apart or being squeezed closer together. For a simple mental picture of a crystal, we can imagine little springs joining the atoms, as in Fig. 9.7. Each spring represents a **bond** between a pair of atoms. (Such bonds are electrical in nature, as we will learn in Chapter 28.) In terms of this model, it is easy to see why even the smallest stress can deform a crystal and why it returns to its original form when a stress below the elastic limit is removed.

Few crystals have completely regular structures. A common crystal defect is a missing line of atoms, which is called a **dislocation**. Dislocations make it possible to see how a solid can be permanently deformed without breaking. We might think the sliding of layers of atoms over one another is responsible. However, calculations show that sliding of this kind—which involves the simultaneous rupture of millions of bonds between atoms—would need forces about a thousand times stronger than those actually required.

Figure 9.8(a) shows a crystal with a missing row of atoms. When equal and opposite forces are applied to the crystal along different lines of action, as in Fig. 9.8(b), the dislocation shifts to the right as the atoms in the layer with the missing row shift their bonds, one row at a time, with the atoms in the layer above. In Fig. 9.8(c) the dislocation has reached the end of the crystal, which is now permanently deformed. Because the bonds were shifted one row at a time instead of all at once, much less force was needed than would have been required to deform a perfect solid.

To increase the strength of a solid, it is necessary to impede the motion of dislocations in its structure. There are two chief ways to do this in a metal. One is to increase the number of dislocations by hammering the metal or squeezing it between rollers. The dislocations then become so numerous and tangled together that they interfere with each other's motion. This effect is called **work hardening.**

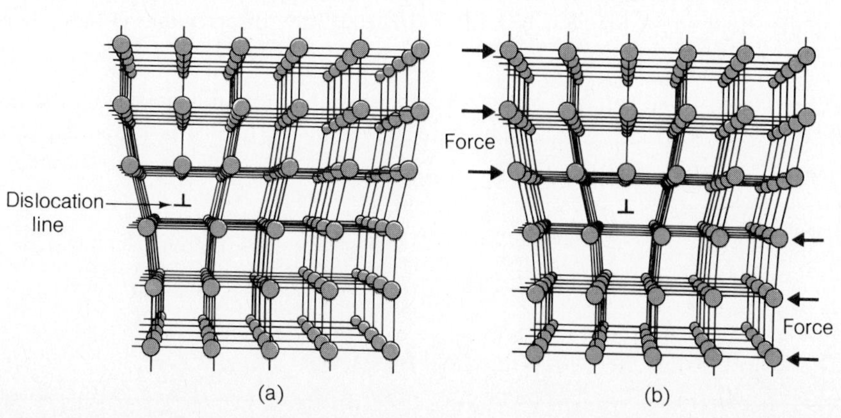

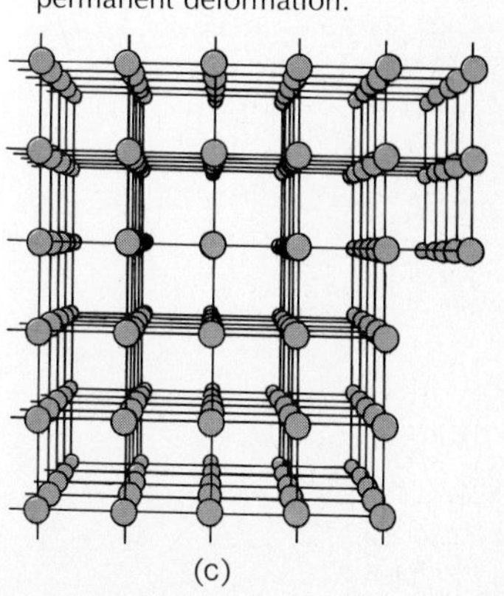

Figure 9.8

The motion of a dislocation in crystal under stress results in a permanent deformation.

Another approach is to add foreign atoms to the metal that act as roadblocks to the progress of dislocations. Thus adding small amounts of carbon, chromium, manganese, and other elements to iron turns it into the much stronger steel by hindering the ability of dislocations to move through the metal. The most suitable composition for steel depends on how it is going to be used. For example, a carbon content of more than about 0.75% plus appropriate treatment gives a very hard steel (the steel used in files and razor blades typically contains 1.25% carbon), but such steels are difficult to weld. When welding is required, low-carbon steels are preferred.

9.4 Young's Modulus

Strain is proportional to stress below the elastic limit.

Suppose we apply equal and opposite forces F to the ends of a bar so that it is under tension or compression, as in Fig. 9.9. If the cross-sectional area of the bar is A, the **stress** on it is defined as the ratio between F and A:

$$\text{Stress} = \frac{\text{force}}{\text{area}} = \frac{F}{A} \tag{9.3}$$

The SI unit of stress is the newton per square meter, which is called the **pascal** (Pa) in honor of the French scientist and philosopher Blaise Pascal (1623–1662). Thus

$$1 \text{ pascal} = 1 \text{ Pa} = 1 \text{ N/m}^2$$

The pascal is a very small unit: The stress you exert by pressing hard with your thumb on a table is about a million pascals. In the British system the unit of stress is the pound per square inch, or **psi**:

$$1 \text{ psi} = 1 \text{ lb/in}^2 = 6.89 \times 10^3 \text{ Pa}$$

EXAMPLE 9.3

A 70-kg swami lies on a bed of nails 1.0 cm apart whose points each have an area of 1.0 mm^2. If the area of the swami's body in contact with the bed is 0.50 m^2 and the threshold stress for pain is 1 MPa (10^6 Pa), how disagreeable is the experience for the swami?

SOLUTION The bed has 1 nail/cm^2. Because the area in contact with the nails is 0.50 m^2 = 5000 cm^2, the swami's weight of mg = 686 N is supported by 5000 nails. Each nail's area is 1.0 mm^2 = 1.0×10^{-6} m^2, so the stress each nail exerts on the swami is

$$p = \frac{F}{A} = \frac{686 \text{ N}}{(5000)(1.0 \times 10^{-6} \text{ m}^2)} = 1.37 \times 10^5 \text{ Pa} = 0.137 \text{ MPa}$$

This is well below the 1-MPa threshold for pain. ◆

EXAMPLE 9.4

A copper wire 1.00 mm in diameter and 2.0 m long is used to support a mass of 5.0 kg. (a) By how much does the wire stretch under this load? (b) What is the minimum diameter the wire can have if its elastic limit is not to be exceeded?

SOLUTION (a) From Eq. (9.6)

$$\Delta L = \frac{L_0}{Y}\frac{F}{A}$$

Here, because the radius of a wire 1.00 mm in diameter is 5.0×10^{-4} m, we have

$$\begin{aligned} L_0 &= 2.0\,\text{m} \\ F &= mg = (5.0\,\text{kg})(9.8\,\text{m/s}^2) = 49\,\text{N} \\ Y &= 1.1 \times 10^{11}\,\text{N/m}^2 \\ A &= \pi r^2 = \pi(5.0 \times 10^{-4}\,\text{m})^2 = 7.85 \times 10^{-7}\,\text{m}^2 \end{aligned}$$

and so

$$\Delta L = \frac{(2.0\,\text{m})(49\,\text{N})}{(1.1 \times 10^{11}\,\text{N/m}^2)(7.85 \times 10^{-7}\,\text{m}^2)} = 1.1 \times 10^{-3}\,\text{m} = 1.1\,\text{mm}$$

(b) We note from Table 9.3 that the elastic limit of copper is 1.5×18^8 Pa. Hence

$$\frac{F}{A} = 1.5 \times 10^8\,\text{Pa} = 1.5 \times 10^8\,\text{N/m}^2$$

$$A = \frac{49\,\text{N}}{1.5 \times 10^8\,\text{N/m}^2} = 3.27 \times 10^{-7}\,\text{m}^2$$

Because the cross-sectional area of a wire of radius r is $A = \pi r^2$,

$$\begin{aligned} r &= \sqrt{\frac{A}{\pi}} = \sqrt{\frac{3.27 \times 10^{-7}\,\text{m}^2}{\pi}} \\ &= \sqrt{1.04 \times 10^{-7}\,\text{m}^2} = \sqrt{10.4 \times 10^{-8}\,\text{m}^2} = 3.2 \times 10^{-4}\,\text{m} \end{aligned}$$

and the corresponding diameter is 6.4×10^{-4} m, which is 0.64 mm. ◆

Bone

It is clear from Table 9.3 that bone is an excellent structural material. Bone is a nonuniform substance in which flexible fibers of a protein called collagen provide most of the tensile strength and crystals of the calcium compound hydroxyapatite provide most of the compressive strength. The different properties of these components lead to different values of Young's modulus for bone in tension and in compression as well as to different ultimate strengths. Because a baby's bones are largely collagen, they tend to bend rather than break under stress.

The compressive load on the leg bones of an animal depends on its weight, which in turn varies as the cube L^3 of a representative linear dimension L such as its length or height. An animal 3 times as long as another of the same form will weigh about 27 times as much. The strength of a bone, however, depends on its cross-sectional area, which for similar animals varies as L^2. Thus animals widely different in size cannot resemble one another: A large animal must have relatively thicker leg bones than a small one because L^3 increases faster than L^2. It is no accident that a hippopotamus has thicker legs for its size than a mouse does, and that the largest animals of all, the whales, live in the oceans, where their immense body weights are supported by buoyancy rather than by legs.

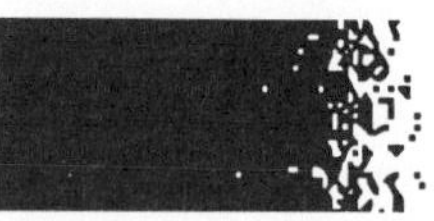

9.5 Shear

The shear modulus of a material is a measure of its rigidity.

Shear stresses change the shape of an object on which they act but not its volume. The situation is much like that of a book whose covers are pushed out of alignment, as in Fig. 9.11. The layers of atoms, which are analogous to the pages of the book, are displaced sideways, but the spacing of the layers, which corresponds to the thickness of the pages, remains the same.

Let us consider a block of thickness d whose lower face is fixed in place and on whose upper face the force F acts (Fig. 9.12). A measure of the relative distortion of the block caused by the shear stress is the angle ϕ, called the **angle of shear.** Because this angle is always small, its value in radians is equal to the ratio s/d between the displacement s of the block's faces and the distance d between them. The greater the area A of these faces, the less they will be displaced by the shear force F.

Figure 9.11

In shear there is a change in shape without a change in volume. The angle ϕ is the angle of shear.

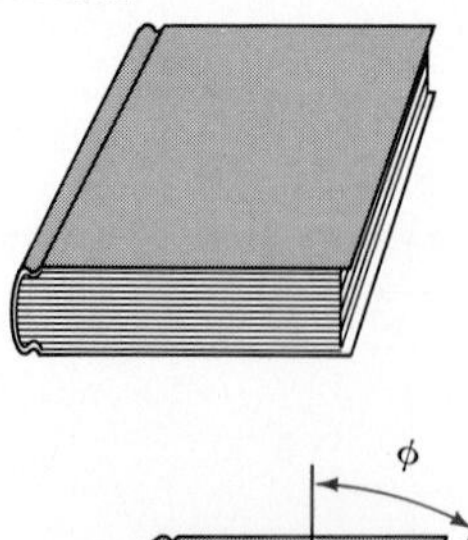

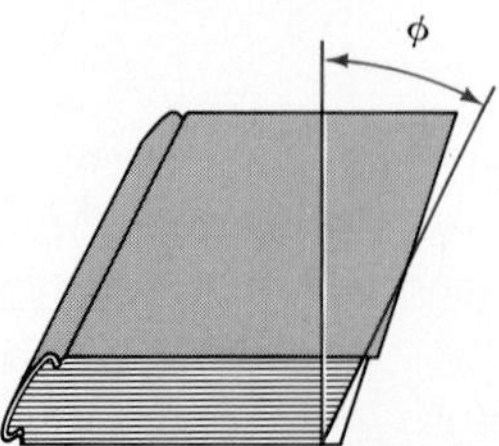

Figure 9.12

The angle of shear ϕ (in radians) is equal to s/d. The greater the shear modulus S, the more rigid the material.

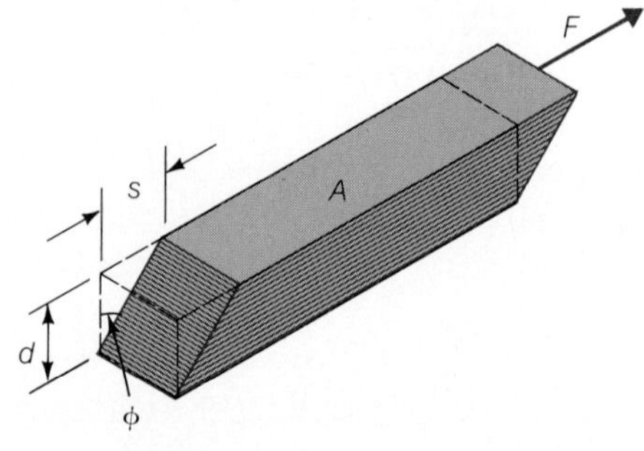

Therefore

$$\text{Shear stress} = \frac{\text{force}}{\text{area}} = \frac{F}{A}$$

$$\text{Shear strain} = \text{angle of shear} = \phi = \frac{s}{d}$$

and the stress-strain equation for shear is

$$\text{Shear modulus} = \frac{\text{shear stress}}{\text{shear strain}}$$

$$S = \frac{F/A}{\phi} = \frac{F/A}{s/d}$$

Hence we have

$$\phi = \frac{s}{d} = \frac{1}{S}\frac{F}{A} \qquad \textit{Shear} \quad (9.7)$$

Here the applied forces are *parallel* to the faces on which they act and not perpendicular as in the case of tension and compression.

The quality S is called the **shear modulus,** and the values of S for various substances are given in Table 9.2. The higher the value of S, the more rigid the material; S is sometimes referred to as the **modulus of rigidity** for this reason. It is interesting to note that the shear modulus of any material is usually a good deal smaller than its Young's modulus. This means that it is easier to slide the atoms of a solid past one another than it is to pull them apart or squeeze them together.

The shear strength of a material is the greatest shear stress an object of that material can withstand before breaking. The greater the shear strength of a material, the more force must be applied to cut a sheet of it with a pair of scissors (or their industrial equivalent) or to punch a hole in the sheet.

EXAMPLE 9.5

Ordinary mild steel ruptures when a shear stress of about 3.5×10^8 Pa is applied. Find the force needed to punch a 10-mm diameter hole in a steel sheet 3.0 mm thick.

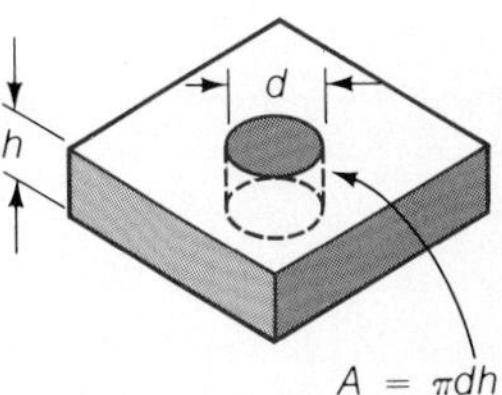

Figure 9.13

SOLUTION The area across which the shear stress is exerted here is the cylindrical inner surface of the hole (Fig. 9.13), so that

$$A = \pi dh = \pi(1.0 \times 10^{-2}\ \text{m})(3.0 \times 10^{-3}\ \text{m}) = 9.4 \times 10^{-5}\ \text{m}^2$$

Since we are given that $(F/A)_{\text{max}} = 3.5 \times 10^8$ Pa, we have

$$F_{\text{max}} = (3.5 \times 10^8\ \text{Pa})(9.4 \times 10^{-5}\ \text{m}^2) = 3.3 \times 10^4\ \text{N}$$

which is nearly 4 tons. ♦

9.6 Bulk Modulus

What happens when an object is squeezed uniformly.

If we squeeze something on all sides, it shrinks in volume by some amount ΔV from its original volume of V_0. Only force components perpendicular to an object's surface act to compress it; parallel components lead to shearing strains. If the compressive force per unit area F/A is the same over the entire surface of the object, as in Fig. 9.14, then

$$\text{Volume stress} = \frac{\text{force}}{\text{area}} = \frac{F}{A}$$

$$\text{Volume strain} = \frac{\text{change in volume}}{\text{original volume}} = \frac{\Delta V}{V_0}$$

and

$$\text{Bulk modulus} = -\frac{\text{volume stress}}{\text{volume strain}}$$

$$B = -\frac{F/A}{\Delta V/V_0}$$

Thus we have for the relative change in volume

$$\frac{\Delta V}{V_0} = -\frac{1}{B}\frac{F}{A} \qquad \textit{Uniform compression} \quad (9.8)$$

The minus sign corresponds to the fact that an increase in force leads to a decrease in volume. The quantity B is called the **bulk modulus.** Typical values of B are given in Table 9.2.

The perpendicular stress F/A is usually called **pressure,** symbol p, as will be discussed in Chapter 10. Hence we can also write

$$\frac{\Delta V}{V_0} = -\frac{p}{B} \qquad (9.9)$$

Figure 9.14
Bulk compression.

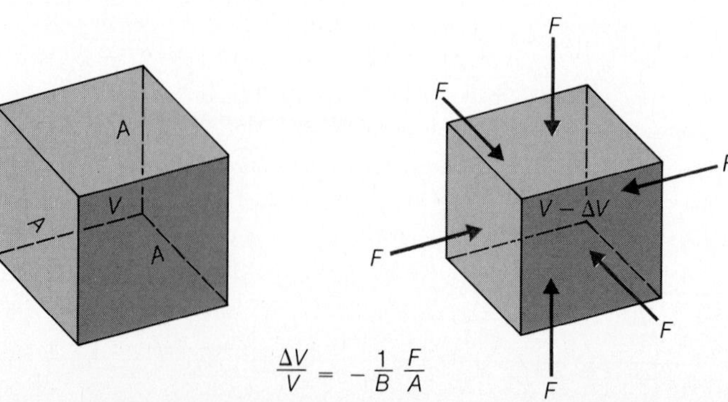

Table 9.4 Bulk moduli of liquids at room temperature

Liquid	Bulk modulus, B ($\times 10^9$ Pa)
Alcohol, ethyl	0.90
Benzene	1.05
Kerosene	1.3
Mercury	26
Oil, lubricating	1.7
Water	2.3

Liquids can support neither tensions nor shears, but they do tend to resist compression. Bulk moduli for several liquids are given in Table 9.4. Interatomic forces within a liquid are weaker than those within a solid, which is why bulk moduli of liquids are smaller than those of solids. A lot of pressure is nevertheless needed to compress a liquid by more than a slight amount. For instance, to compress a volume of water by 1% requires an inward force per unit area of 2.3×10^7 Pa, which is equivalent to 3300 lb/in^2.

EXAMPLE 9.6

Verify the preceding statement.

SOLUTION The bulk modulus of water is 2.3×10^9 Pa and, by hypothesis, $\Delta V/V_0 = -0.01$. Hence

$$\frac{F}{A} = -B\frac{\Delta V}{V_0} = -(2.3 \times 10^9\ \text{Pa})(-0.01) = 2.3 \times 10^7\ \text{Pa}$$ ♦

9.7 Building over Space

How materials govern design.

Three successive inventions for building over space have shaped the course of architecture. The earliest, and still the most widely used, is the **post-and-beam** arrangement shown in Fig. 9.15(a), in which two vertical posts hold up a horizontal beam. Before steel came into general use in the last century, the width that could be spanned by a beam was severely limited by the nature of the materials available, which were mainly wood and stone. Wooden beams cannot support really large loads unless the span is narrow, and obtaining timber of adequate size and quality has always been a problem in much of the world. Stone can be quite strong in compression but is very much less so in tension, and the lower part of a beam is under tension. Narrow doorways and numerous interior supports were accordingly needed in buildings of even modest size that used post-and-beam construction until steel beams came into use in the nineteenth century.

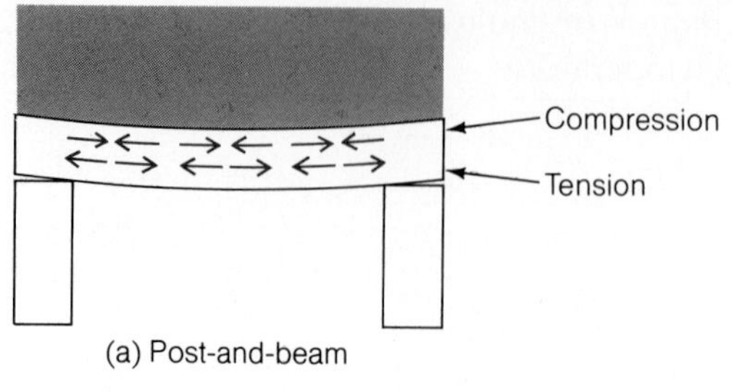

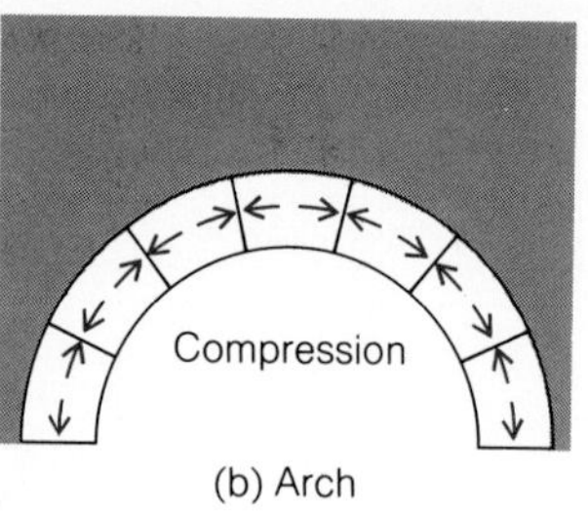

Tension

Compression

(c) Cantilever

Figure 9.15

Three ways to build over space. (a) A horizontal beam experiences tensile as well as compressive stresses. (b) The principal stresses in an arch are compressive. An arch made of stones or bricks, which are weak in tension but strong in compression, can support a considerable load. (c) Modern structural materials such as steel and reinforced concrete are strong in tension as well as compression and can be used as cantilevers supported only at one end.

Figure 9.16

The members of a truss are arranged in triangles.

Figure 9.17

The structural design of a proposed 135-story building uses trusses to cope with wind loads.

The great advantage of the **arch** is that the principal stresses imposed on it by a load are compressive (Fig. 9.15(b)). Stone and brick are sufficiently strong in compression for quite large arches, and also for domes, which are the three-dimensional equivalent of arches. The arch first came into wide use in Roman times, although it had been known thousands of years earlier. Innumerable ancient structures based on the arch and the dome still stand, testimony to the sound engineering principle behind them.

The third, and most recent, development in building over space is the **cantilever,** a beam held in place at one end only (Fig. 9.15(c)). Reinforced concrete as well as steel can be used in cantilever construction. The tall buildings of today do not have the load-bearing walls of the past but have internal skeletons instead. The outer parts of their floors and their thin "curtain" walls, often almost entirely glass, are supported by cantilever beams that extend outward from central frames.

Solid beams are often replaced by **trusses.** A truss is a structure made up of straight members of wood or metal arranged in triangles (Fig. 9.16). A truss using the same amount of material as a solid beam is much more rigid; or, for the same rigidity, a truss is much lighter than the corresponding solid beam. The members of a truss are in tension or in compression, not in shear, and hence can be relatively thin. The longest truss in the world is a 375-m bridge span.

Tall buildings are themselves vertical cantilevers because of the enormous forces winds can exert on them. The higher a skyscraper, the more severe the bending loads it must be able to withstand. For this reason a 135-story building proposed for New York City's Columbus Circle was designed as a trussed tube to maximize stiffness (Fig. 9.17).

Truss bridge over the Amoskeag River in New Hampshire.

Important Terms

The **density** of a substance is its mass per unit volume.

The **specific gravity** of a substance is its density relative to that of water. Specific gravity is also called **relative density.**

The three categories of stress forces are **tension,** in which equal and opposite forces that act away from each other are applied to a body; **compression,** in which equal and opposite forces that act toward each other are applied to a body; and **shear,** in which equal and opposite forces that do not act along the same line of action are applied to a body. A tensile stress tends to elongate a body, a compressive stress to shorten it, and a shearing stress to change its shape without changing its volume.

Hooke's law states that the amount of deformation experienced by a body under stress is proportional to the magnitude of the stress. Thus the elongation of a wire is proportional to the tension applied to it.

The **elastic limit** is the maximum stress to which a solid can be subjected without being permanently altered. Hooke's law is valid only when the elastic limit is not exceeded.

The **stress** on an object is the applied force per unit area; its unit is the **pascal,** equal to 1 N/m^2. **Strain** is the resulting change in a dimension of the object relative to its original value. A **modulus of elasticity** of a material is the ratio between a particular kind of applied stress and the resulting strain, provided the elastic limit is not exceeded. **Young's modulus** applies to tension and compression; the **shear modulus** applies to shear; and the **bulk modulus** applies to uniform volume compression.

Important Formulas

Density: $\rho = \frac{m}{V}$

Hooke's law: $F = kx$

Tension or linear compression: $\frac{\Delta L}{L_0} = \frac{1}{Y}\frac{F}{A}$

Shear: $\phi = \frac{s}{d} = \frac{1}{S}\frac{F}{A}$

Uniform compression: $\frac{\Delta V}{V_0} = -\frac{1}{B}\frac{F}{A} = -\frac{p}{B}$

MULTIPLE CHOICE

1. Which of the following quantities is independent of the size and shape of an object composed of a given material?
a. Volume b. Mass
c. Weight d. Density

2. When equal and opposite forces are exerted on an object along different lines of action, the object is said to be under
a. tension b. compression
c. shear d. elasticity

3. Ductility refers to the ability of a metal to
a. be deformed temporarily
b. be deformed permanently
c. shrink under compression
d. break under tension

4. When a force acts on an object, the stress on it is equal to
a. the relative change in its dimensions
b. the applied force per unit area
c. Young's modulus
d. the elastic limit

5. A shear stress that acts on an object affects its
a. length b. width
c. volume d. shape

6. Another name for the shear modulus of a material is
a. Young's modulus
b. modulus of rigidity
c. bulk modulus
d. ductility

7. The only elastic modulus that applies to liquids is
a. Young's modulus
b. shear modulus
c. modulus of rigidity
d. bulk modulus

8. According to Hooke's law, the force needed to elongate an elastic object by an amount x is proportional to
a. x b. $1/x$
c. x^2 d. $1/x^2$

9. The stress on a wire supporting a load does not depend on
a. the wire's length
b. the wire's diameter
c. the mass of the load
d. the acceleration of gravity

10. The density of brass is 8.4 g/cm^3. The volume of a 200-g brass monkey is approximately
a. 0.042 cm^3 b. 0.41 cm^3
c. 24 cm^3 d. 1.7 L

11. Gasoline has a specific gravity of 0.68. One liter of gasoline weighs
a. 0.68 N b. 1.4 N
c. 6.7 N d. 6.8 N

12. A cable stretches by the amount a under a certain load. If it is replaced by a cable of the same material but half as long and half the diameter, the same load will stretch it by
a. $a/4$ b. $a/2$
c. a d. $2a$

13. The original cable in Multiple Choice 12 could support a maximum load of w without exceeding the elastic limit. The elastic limit will not be exceeded by the new cable up to a load of
a. $w/4$ b. $w/2$
c. w d. $2w$

14. An iron wire 1.0 m long with a square cross section 2.0 mm on a side is used to support a 100-kg load. Its elongation is
a. 0.0027 mm b. 0.27 mm
c. 1.3 mm d. 3.7 mm

15. Young's modulus for aluminum is 7.00×10^{10} Pa. The force needed to stretch by 1.00 mm an aluminum wire 2.00 mm in diameter and 800 mm long is
a. 2.75 N b. 275 N
c. 1.10 kN d. 2.75 kN

16. The elastic limit of aluminum is 1.80×10^8 Pa. The maximum mass the wire in Multiple Choice 15 can support without exceeding the elastic limit is
a. 58 kg b. 116 kg
c. 232 kg d. 565 kg

17. A wire 10 m long with a cross-sectional area of 0.10 cm^2 stretches by 13 mm when a load of 100 kg is suspended from it. The Young's modulus for this wire is
a. 0.77×10^{10} Pa
b. 7.5×10^{10} Pa
c. 7.7×10^{10} Pa
d. 9.3×10^{10} Pa

18. The ultimate strength in compression of aluminum is 2×10^8 Pa. The maximum mass an aluminum cube 1 cm on each edge can support is about
a. 2×10^2 kg b. 2×10^3 kg
c. 2×10^4 kg d. 2×10^5 kg

19. The force needed to punch a hole 8.0 mm square in a steel sheet 3.0 mm thick whose shear strength is 2.5×10^8 Pa is
a. 2.7 kN b. 6 kN
c. 24 kN d. 48 kN

20. When a pressure of 2.0 MPa is applied to a sample of kerosene, it contracts by 0.15%. The bulk modulus of kerosene is
a. 1.7 MPa b. 1.997 MPa
c. 2.003 MPa d. 1.3 GPa

QUESTIONS

1. Two boys wish to break a string. Are they more likely to do this if each takes one end of the string and they pull against each other, or if they tie one end of the string to a tree and both pull on the free end? Why?

2. A rubber band is easy to stretch at first, but after it has been extended by a certain amount, additional stretching becomes more difficult at a greater rate. Make a rough stress-strain graph for rubber in tension.

3. A three-legged stool has one leg of aluminum, one of copper, and one of steel. The legs have the same dimensions. If the load on the stool is on its exact center, which leg is under the greatest stress and which under the least stress? Which leg experiences the greatest strain and which the least strain?

4. A cable is replaced by another one of the same length and same material but of twice the diameter. How does this affect the maximum load that can be supported by the cable? What would be the result of using a cable of the same diameter and material but twice the length of the original one?

5. Two wires are made of the same material, but wire A is twice as long and has twice the diameter of wire B. Find the elongation of wire B relative to that of wire A when both are subjected to the same load.

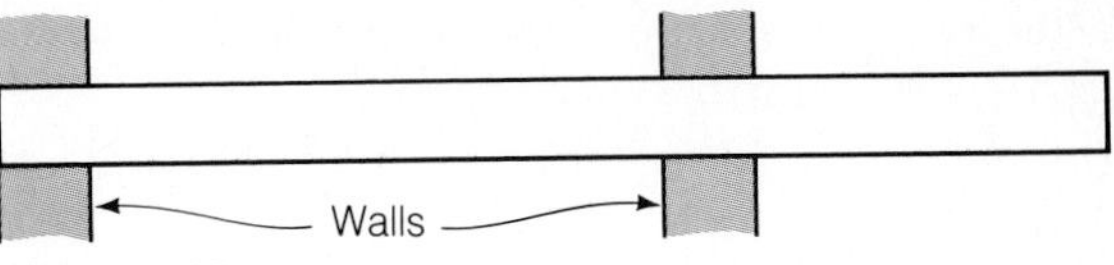

Figure 9.18

Question 7.

6. The elastic moduli of a particular material are related to one another, and each of them can be expressed in terms of the other two. For instance, the theory of elasticity shows that Young's modulus Y can be expressed in terms of the shear modulus S and the bulk modulus B by the formula

$$Y = \frac{9SB}{(3B + S)}$$

Check this formula for three of the materials listed in Table 9.2. Give several reasons why you would not expect perfect agreement.

7. Figure 9.18 shows the concrete floor of a building that is continued outside to form a cantilevered balcony. Because of their expense, steel rods are only used to reinforce concrete where tensile stresses occur. Show where reinforcing rods are most needed in the floor and balcony of the diagram.

EXERCISES

9.1 Density

1. A room is 5.0 m long, 4.0 m wide, and 3.0 m high. What is the mass of the air it contains?

2. A coil of sheet steel 0.80 mm thick and 500 mm wide has a mass of 156 kg. Find the length of the steel in the coil.

3. Mammals have approximately the same density as fresh water. Find the volume in liters of a 55-kg woman and the volume in cubic meters of a 140,000-kg blue whale.

4. A 1200-kg concrete slab that measures 2.0 m $\times$ 1.0 m $\times$ 20 cm is delivered to a building under construction. Does it contain steel reinforcing rods or is it plain concrete?

5. The radius of the earth is 6.37×10^6 m, and its mass is 5.98×10^{24} kg. (a) Find the average density of the earth. (b) The average density of rocks at the earth's surface is 2.7×10^3 kg/m^3. What must be true of the matter of which the earth's interior is composed? Is it likely that the earth is hollow and peopled by another species, as the ancients believed?

6. If gold costs \$300/oz, how many millimeters on a side does a \$10,000 cube of gold measure?

7. Water and oil are poured into the ends of a glass tube bent into a U-shape, as in Fig. 9.19. Find the density of the oil in grams per cubic centimeter.

8. A 200-g bottle has a mass of 340 g when filled with water and 344 g when filled with blood plasma. What is the density of the plasma?

9. A 1.0-kW motor drives a pump that raises gasoline from an underground tank through a height of 3.2 m. If the pump is 65% efficient, find the rate of flow of gasoline in L/min.

9.2 Elasticity

10. When a coil spring is used to support a 12-kg object, the spring stretches by 40 mm. What is the force constant of the spring?

11. A coil spring has a force constant of 1000 N/m. How much will it stretch when it is used to support an object whose mass is 8.0 kg?

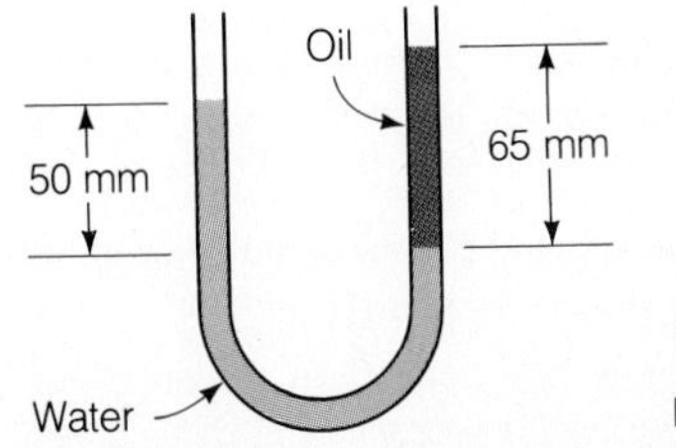

Figure 9.19

Exercise 7.

12. The length of a spring is measured when a series of forces is applied to it. The results are as follows:

Applied force, N	0	5	10	15	20	25	30
Spring length, cm	10.0	10.8	11.7	12.5	13.3	14.2	16.4

From a graph of applied force versus elongation for the spring, estimate its elastic limit.

9.4 Young's Modulus

13. A nylon rope 10.0 mm in diameter breaks when a load of 25 kN is applied to it. What would you estimate for the breaking strength of a nylon rope 6.0 mm in diameter? 14.0 mm in diameter?

14. The actual dimensions of a "2 × 4" timber are 38 × 90 mm. Find the maximum mass a pine 2 × 4 standing on end can support.

15. How safe is it for a 75-kg circus acrobat to balance on the index finger of his right hand, whose bones have a minimum cross-sectional area of 0.50 cm^2?

16. Find the maximum force a hammer can apply to a steel nail 2.5 mm in diameter if the nail's elastic limit is not to be exceeded.

17. A human hair 60 μm in diameter can just support a mass of 55 g (equivalent to about 2 oz of weight). How does the ultimate strength in tension of hair compare with that of aluminum?

18. In a test firing, the three main engines of the space shuttle *Columbia* developed a total thrust of 4.5×10^6 N. Eight steel bolts 90 mm in diameter were used to hold the *Columbia* to its launch pad during the firing. What was the ratio between the untimate strength of the bolts and the applied stress?

19. The table below shows by how much a metal rod 120 cm long whose cross-sectional area is 16 mm^2 stretches when various loads are applied. From a stress-strain graph for the metal, find its Young's modulus.

Load, kN	0	1	2	3	4	5
ΔL, mm	0	0.63	1.25	1.88	2.50	3.13

20. A cube of pine wood 50 mm on an edge is held in the jaws of a vise by a force of 2000 N. By how much is the wood compressed?

21. A steel wire 1.0 m long and 1.0 mm square in cross section supports a mass of 6.0 kg. By how much does it stretch?

22. A solid steel post 4.0 m long and 30 mm in radius supports a load of 3000 kg. By how much is it shortened?

23. An 80-kg man has femurs (the femur is the bone of the thigh) 42 cm long and 11 cm^2 in average cross-sectional area. Find the approximate change in length of each femur as the man walks.

24. A sagging floor is jacked up and a steel girder 3.0 m long whose cross-sectional area is 40 cm^2 is put in place underneath. When the jack is removed, a sensitive strain gauge shows that the girder has been compressed by 0.20 mm. Find the load the girder is supporting.

25. A brass wire 2.0 m long whose cross-sectional area is 5.0 mm^2 stretches by 2.6 mm when a force of 600 N is applied. Find the value of Young's modulus for the wire.

26. By how much can a copper wire 2.0 m long be stretched before its elastic limit is exceeded?

27. An iron pipe 3.0 m long is used to support a sagging floor. The inside diameter of the pipe is 10 cm and its outside diameter is 11 cm. When the force on it is 15 kN, by how much is it compressed?

28. A certain kind of brick has a density of 2400 kg/m^3 and an ultimate strength in compression of 9.6×10^7 Pa. Find the maximum height to which a wall of this brick can be built before the lowest brick disintegrates.

29. A wall of lead bricks exactly 1 m high is used to shield a sample of radium. Each brick was originally a cube exactly 10 cm on an edge. What is the change in height of the lowest brick when the wall has been erected?

30. Suppose you have cylinders of the same length and diameter of bone, aluminum, and steel. For each cylinder find the ratio between the maximum tensile and compressive forces it can withstand and its mass. How does bone compare with aluminum and steel as a structural material?

9.5 Shear

31. A 30-mm cube of raspberry gelatin on a table is subjected to a shearing force of 0.50 N. The upper surface is displaced by 5.0 mm. What is the shear modulus of the gelatin?

32. Two steel plates are riveted together with 10 rivets each 5.0 mm in diameter. If the maximum shear stress the rivets can withstand is 3.5×10^8 Pa, how much force applied parallel to the plates is needed to shear off the rivets?

33. A punch press that exerts a force of 20 kN is employed to punch 1.0-cm-square holes in sheet aluminum. If the shear strength of aluminum is 70 MPa, what is the maximum thickness of aluminum sheet that can be used?

34. A slot 200 mm long and 13 mm wide is to be punched in a steel sheet 1.5 mm thick whose shear strength is 3.4×10^8 Pa. Find the force needed.

9.6 Bulk Modulus

35. The pressure at a depth of 1.0 km in the ocean exceeds sea-level atmospheric pressure by about 10 MPa. If an iron

anchor whose volume at the surface is 400 cm^3 is lowered to a depth of 1.0 km, by how much does its volume decrease?

36. The deepest known point in the oceans, 11 km below the surface, is found in the Marianas Trench southwest of Guam in the Pacific. At that depth the water pressure exceeds sea-level pressure by 0.11 GPa. The density of sea water at sea level is 1.03×10^3 kg/m^3 and its bulk modulus is 2.3 GPa. What is the density of sea water at a depth of 11 km?

37. If an aluminum object is placed in a vacuum, by what percentage will its volume increase? Atmospheric pressure is 1.013×10^5 Pa.

PROBLEMS

38. One gram of gold can be beaten out into a foil 1.00 m^2 in area. (Thus an ounce of gold can yield 300 ft^2 of foil.) How many atoms thick is such a foil? The mass of a gold atom is 3.27×10^{-25} kg.

39. A helicopter whose mass is 900 kg has rotor blades 5.0 m in radius. Find the speed of the air forced downward by its blades when the helicopter hovers at rest above the ground.

40. The parachute of a 60-kg woman fails to open, and she falls into a snowbank at a terminal speed of 60 m/s, coming to a stop 1.5 m below the surface of the snow. If the area of the woman's body that strikes the snow is 0.20 m^2 and the stress required for serious injury to body tissues is 5.0×10^5 Pa, is she likely to survive the impact?

41. A steel cable whose cross-sectional area is 2.5 cm^2 supports a 1000-kg elevator. The elastic limit of the cable is 3.0×10^8 Pa. What is the maximum upward acceleration that can be given the elevator if the tension in the cable is to be no more than 20% of the elastic limit?

42. A certain material has a density of ρ and a tensile strength of U. How long a rod of this material can be suspended from one end without breaking under its own weight? What is this length in the case of aluminum? Steel? (Assume the density of steel to be the same as that of iron.)

43. A copper wire and a steel wire are being used side by side to support a load. (a) If they have the same diameter and each stretches by the same amount, what proportion of the load does each one support? (b) What should the ratio of their diameters be if they are each to support half the load?

44. A bar 2.0 m long is suspended horizontally from its ends by steel wires 1.0 m long. The wires have radii of 1.0 mm. A mass of 20 kg is then attached 40 cm from one end of the bar. What is the difference in length between the two wires?

45. A wire 1.0 m long of negligible mass with a cross-sectional area of 1.0 mm^2 is supported at both ends so that it is very nearly horizontal. When a mass of 1.0 kg is attached to the midpoint of the wire, the wire stretches so that the midpoint sags by 8.0 mm. Find the approximate value of Young's modulus for the wire's material.

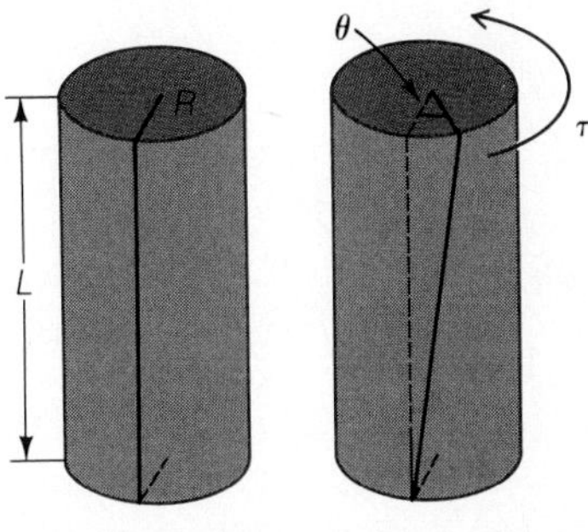

Figure 9.20
Problem 46.

46. When a torque τ is applied to a cylinder, the angle θ (in radians) through which it twists depends on the cylinder's length L, radius r, and shear modulus S according to the formula

$$\theta = \frac{2\tau L}{\pi S r^4}$$

The geometry of the situation is shown in Fig. 9.20. If the angle of twist of a solid steel drive shaft 50 mm in diameter and 2.5 m long is not to exceed 2°, find the maximum power the shaft can transmit at 900 rpm.

Answers to Multiple Choice

1. d	**6.** b	**11.** c	**16.** a
2. c	**7.** d	**12.** d	**17.** b
3. b	**8.** a	**13.** a	**18.** b
4. b	**9.** a	**14.** c	**19.** c
5. d	**10.** c	**15.** b	**20.** d

C H A P T E R

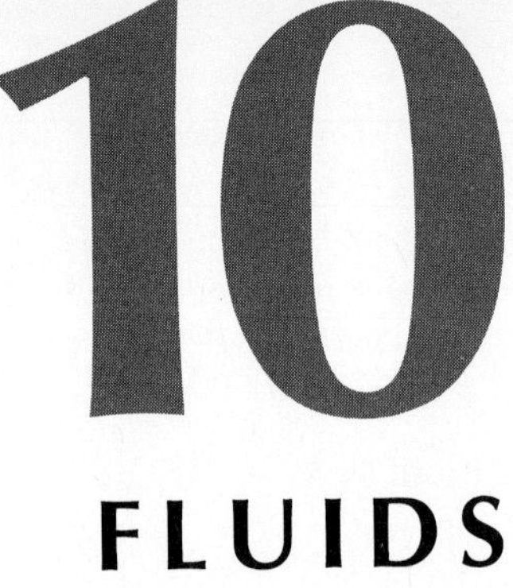

10 FLUIDS

As the name implies, a fluid is a substance that flows readily. Gases and liquids are fluids. Its ability to flow enables a fluid to exert an upward force on an immersed object such as a ship, to multiply an applied force in a hydraulic press, and to provide "lift" to an airplane wing. In what follows we will find that there is nothing mysterious about these properties of fluids, properties that follow from the laws of physics we already know.

10.1 Pressure

Normal force per unit area.

When a force **F** acts perpendicular to a surface of area A, the **pressure** p exerted on the surface is the ratio between the magnitude F of the force and the area. Because a perpendicular force is also described as a **normal force,** we can say that pressure is the magnitude of normal force per unit area:

$$p = \frac{F}{A} \qquad \text{Pressure} \quad (10.1)$$

$$\text{Pressure} = \frac{\text{normal force}}{\text{area}}$$

As we learned in Chapter 9, the SI unit of pressure is the **pascal** (Pa), equal to 1 N/m^2. Because the pascal is very small, the **bar** is widely used instead, where

$$1 \text{ bar} = 10^5 \text{ Pa}$$

The average pressure of the earth's atmosphere at sea level is 1.013 bar. Meteorologists commonly use the **millibar** (mb) for convenience, where 1 mb $= 10^{-3}$ bar, so that 1013 mb corresponds to average atmospheric pressure. Actual atmospheric pressures usually range between 970 and 1040 mb.

Three ways to measure pressure are shown in Fig. 10.1. As a rule what is found is the difference between the unknown pressure and atmospheric pressure. This difference is **gauge pressure,** whereas the true pressure, called **absolute pressure,** includes atmospheric pressure. That is,

$$p = p_{\text{gauge}} + p_{\text{atm}} \qquad (10.2)$$

$$\text{Absolute pressure} = \text{gauge pressure} + \text{atmospheric pressure}$$

A tire inflated to a gauge pressure of 2 bar contains air at an absolute pressure of about 3 bar because sea-level atmospheric pressure is 1.013 bar (Fig. 10.2).

EXAMPLE 10.1

The flat roof of a house is 10.0 m long and 8.0 m wide and has a mass of 7500 kg. Before a severe storm the people in the house closed its doors and windows so tightly that the air pressure inside remained at 1013 mb even when the outside pressure fell to 980 mb. (These are absolute pressures, of course.) Compare the upward force on the roof with its weight.

SOLUTION The area of the roof is (10.0 m) (8.0 m) = 80 m^2. The difference Δp between the pressures on the inside and outside of the roof is

$$\Delta p = (1013 \text{ mb} - 980 \text{ mb})(100 \text{ Pa/mb}) = 3.3 \text{ kPa}$$

Because the pressure on the inside of the roof is greater, the net force on it is upward with the magnitude

$$F = A\Delta p = (80 \text{ m}^2)(3.3 \text{ kPa}) = 264 \text{ kN}$$

This is 59,400 lb! The roof's weight is

$$w = mg = (7500 \text{ kg})(9.8 \text{ m/s}^2) = 74 \text{ kN}$$

Thus the upward force on the roof is nearly four times its weight. If the roof is not well attached to the walls of the house, and if the windows do not break first, the roof will be lifted off during the storm. Clearly it is unwise to seal a building when a large drop in air pressure is coming. ◆

Figure 10.1

Three types of pressure gauge. (a) An aneroid measures pressure in terms of the amount by which the thin, flexible ends of an evacuated metal chamber are pushed in or out by the external pressure. (b) A manometer measures pressure in terms of the difference in height h of two mercury columns, one open to the atmosphere and the other connected to the source of the unknown pressure. (c) A Bourdon tube straightens out when the internal pressure exceeds the external pressure.

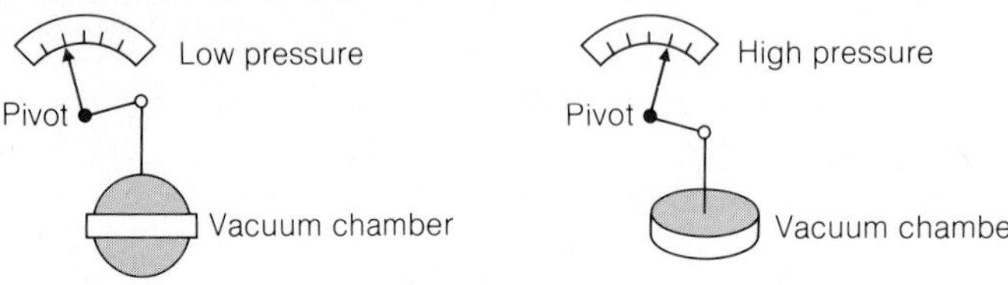

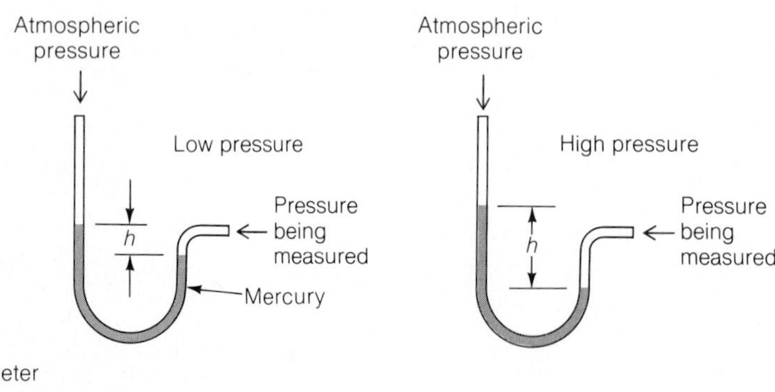

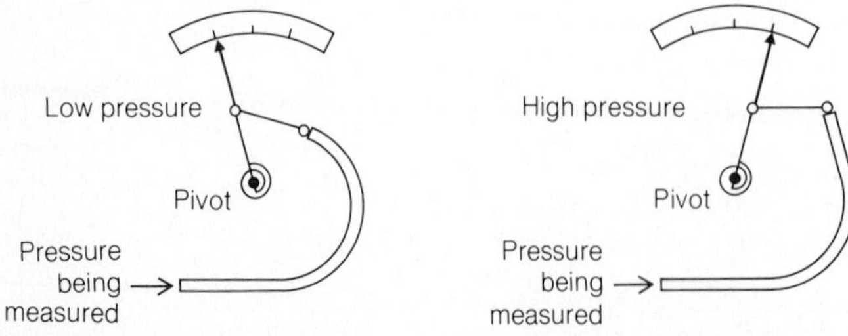

Figure 10.2

On an absolute pressure scale, 0 corresponds to a perfect vacuum. On a gauge pressure scale, 0 corresponds to atmospheric pressure, which averages 1.013 bars (1.013×10^5 Pa). Shown is an absolute pressure of 3 bar, which is equivalent to a gauge pressure of 2 bar.

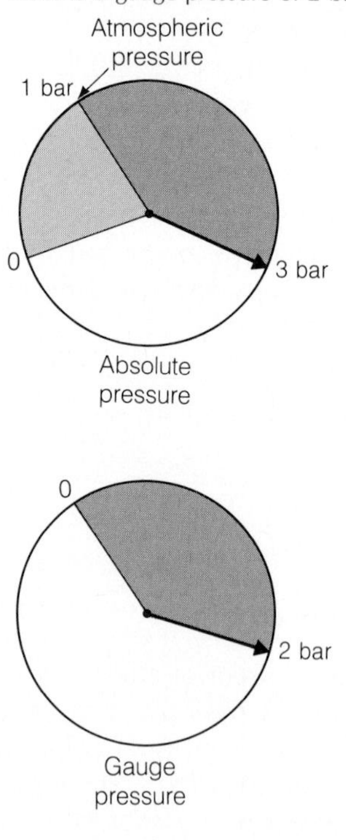

Three Properties of Pressure in a Fluid

Pressure is a useful quantity because fluids flow under stress instead of being deformed elastically as solids are. The absence of rigidity in fluids has three significant consequences:

---

1. The forces a fluid at rest exerts on the walls of its container, and vice versa, always act perpendicular to the walls.

If this were not so, any sideways force by a fluid on a wall would be accompanied, according to the third law of motion, by a sideways force back on the fluid, which would cause the fluid to move parallel to the wall. But the fluid is at rest, so the force must be perpendicular to the container walls. A *moving* fluid is another matter. As we will learn in Section 10.8, frictional forces act between a moving fluid and, for instance, the walls of a pipe or the bank of a river. The difference between a body of liquid in contact with a solid and two solids in contact is that in the former case there is no static friction between them.

---

2. An external pressure exerted on a fluid is transmitted uniformly throughout the volume of the fluid.

If this were not so, the fluid would flow from a region of high pressure to one of low pressure, which would equalize the pressure. We must keep in mind, however, that this statement, which is known as **Pascal's principle,** refers to a pressure imposed from outside the fluid. The fluid at the bottom of a container is always under more pressure than that at the top owing to the weight of the overlying fluid. A notable example is the earth's atmosphere (Fig. 10.3), although such pressure differences are ordinarily significant only for liquids.

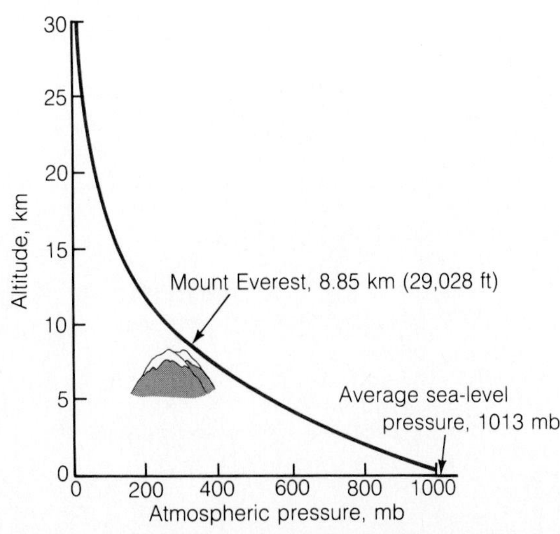

Figure 10.3

Variation of atmospheric pressure with height above sea level. At an altitude of 50 km, the pressure is down to about 1/1000 of its sea-level value. (1 mb = 100 Pa.)

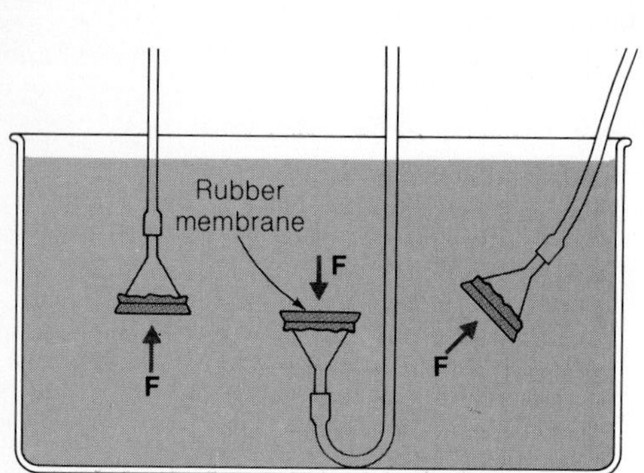

Figure 10.4

At a given depth, the force exerted by the pressure in a fluid is the same in all directions.

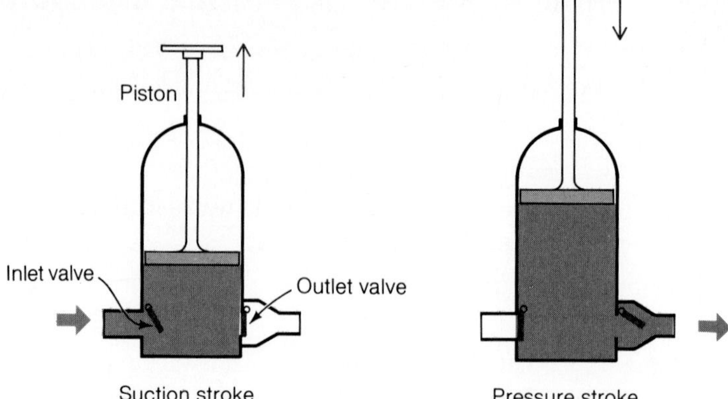

Figure 10.5

A piston pump. During the suction stroke, the inlet valve opens and the fluid is drawn into the cylinder. When the piston is pushed down, the inlet valve closes, the outlet valve opens, and the fluid is forced out.

3. The pressure on a small surface in a fluid is the same regardless of the orientation of the surface.

If this were not so, again, the fluid would flow in such a way as to equalize the pressure (Fig. 10.4).

A *pump* is a device for creating a pressure difference. The operation of the familiar piston pump is illustrated in Fig. 10.5. A pump of this kind can be used to provide high pressure (to inflate a tire or a football, for instance) or low pressure (to pump water from a well or the bilge of a boat, for instance). The heart circulates blood through the lungs and through the body by means of pumps of this kind, with the contraction and relaxation of the muscular walls of the heart chambers taking the place of piston strokes. Rotary pumps are more suited to motor drive; two examples are shown in Fig. 10.6.

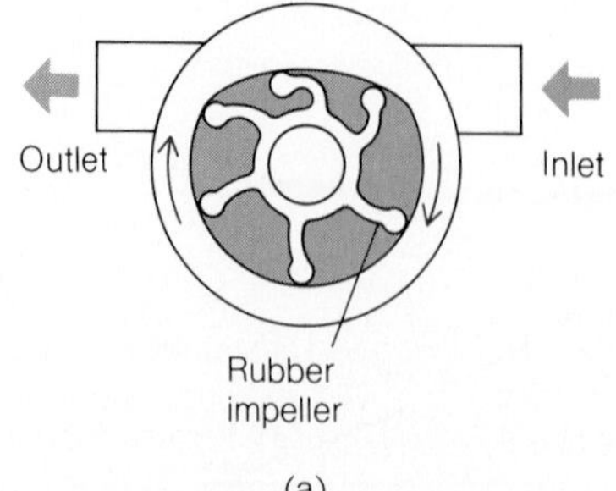

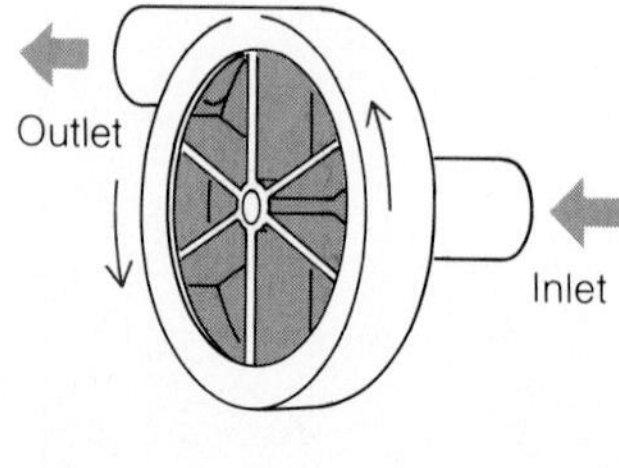

Figure 10.6

Two types of rotary pump. (a) Impeller pump. The fluid is compressed when the rubber fins of the impeller are bent over at the flattened top of the chamber. (b) Centrifugal pump. The rotating blades set the fluid in motion.

10.2 Hydraulic Press

The pressure is the same on both pistons, but the force is not.

Two of the three basic machines, the lever and the inclined plane, were considered in Chapter 8. The third, the **hydraulic press,** is based on Pascal's principle, which is property 2 given in the previous section: An external pressure on a fluid is transmitted uniformly throughout the volume of the fluid.

In the hydraulic press of Fig. 10.7, a force F_{in} acts on a piston of area A_{in} to produce the pressure $p = F_{in}/A_{in}$ on the fluid inside. This pressure is transmitted by the fluid to the output piston, where $p = F_{out}/A_{out}$. Because the pressure is the same on both pistons,

$$p = \frac{F_{out}}{A_{out}} = \frac{F_{in}}{A_{in}}$$

The mechanical advantage of the hydraulic press is therefore

$$\text{MA} = \frac{F_{out}}{F_{in}} = \frac{A_{out}}{A_{in}} = \frac{L_{in}}{L_{out}} \qquad \textit{Hydraulic press} \quad (10.3)$$

A small input force can be considerably increased merely by having the output piston much larger in area than the input piston.

The fluid is assumed incompressible, and so the volume of the fluid transferred from one cylinder to the other as the pistons move must be the same. Hence the piston displacements are inversely proportional to the forces on them: The input piston must move through a large range in order to move the output piston through a small one. If the piston areas are in the ratio 100:1, a force of 1 N can lift a weight

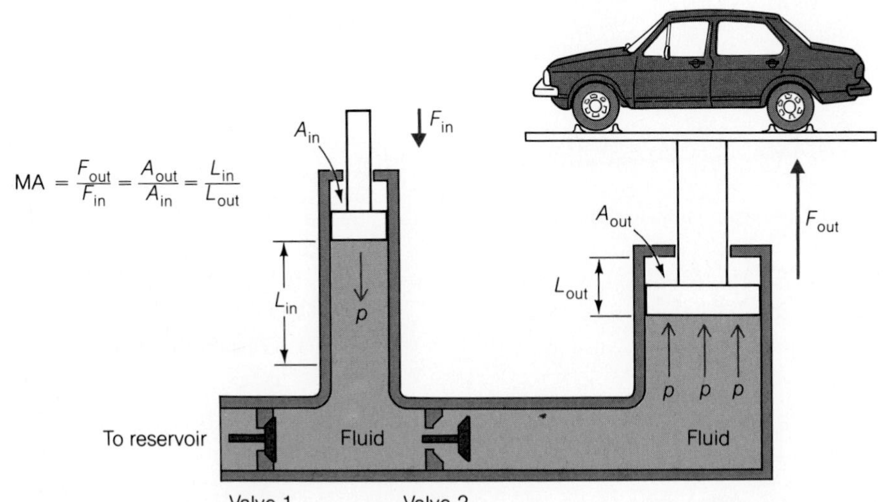

Figure 10.7

The hydraulic press makes use of the fact that pressure exerted on a fluid is transmitted equally throughout the fluid. Valve 1 is closed and valve 2 open on the downstroke of the input piston. Valve 1 is open and valve 2 closed on the upstroke, when additional fluid is drawn into the input cylinder to enable its piston to make another stroke.

The hydraulic rams on this backhoe convert liquid pressure into applied forces. The pressure is provided by an engine-driven pump.

of 100 N, but for every centimeter the weight is raised, the input piston must move down 100 cm. The purpose of the valves indicated in Fig. 10.7 is to permit the output piston to be raised by a series of short strokes of the input piston.

The principle of the hydraulic press can be applied in a variety of ways. Common industrial uses include presses of various kinds, garage lifts, control systems in airplanes, and vehicle brakes.

EXAMPLE 10.2

The input and output pistons of a hydraulic jack are, respectively, 10 mm and 40 mm in diameter. A lever with a mechanical advantage of 6.0 is used to apply force to the input piston (Fig. 10.8). How much mass can the jack lift if a force of 180 N (about 40 lb) is applied to the lever and friction is negligible in the system?

SOLUTION The mechanical advantage of the jack is

$$\text{MA} = \frac{A_{\text{out}}}{A_{\text{in}}} = \frac{(\pi r_{\text{out}})^2}{(\pi r_{\text{in}})^2} = \left(\frac{d_{\text{out}}}{d_{\text{in}}}\right)^2 = \left(\frac{40 \text{ mm}}{10 \text{ mm}}\right)^2 = 16$$

The total mechanical advantage of the lever + jack system is

$$\text{MA}_{\text{total}} = (\text{MA}_{\text{lever}})(\text{MA}_{\text{jack}}) = (6.0)(16) = 96$$

Hence the output force of the system when the input force is 180 N is

$$F_{\text{out}} = (\text{MA})(F_{\text{in}}) = (96)(180 \text{ N}) = 1.73 \times 10^4 \text{ N}$$

If the force equals a weight that the jack is lifting, the corresponding mass is

$$m = \frac{w}{g} = \frac{1.73 \times 10^4 \text{ N}}{9.8 \text{ m/s}^2} = 1.8 \times 10^3 \text{ kg}$$

which is nearly 2 tons. ♦

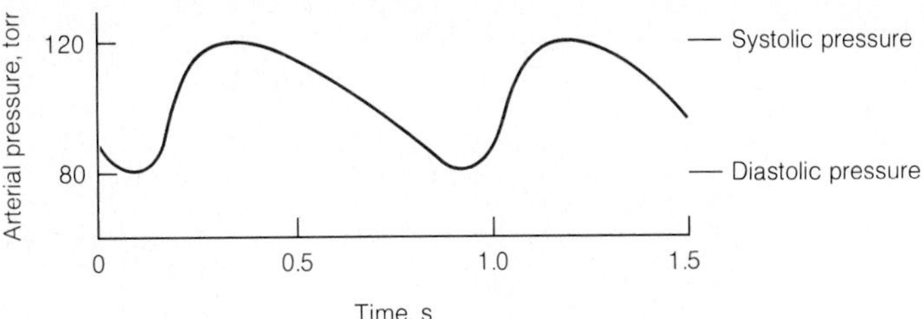

Figure 10.12

Arterial blood pressure rises and falls as the heart contracts and relaxes. The pressures shown are typical for a healthy person. The time scale corresponds to 70 heartbeats per minute.

cuff is then further reduced until the gurgling stops, which corresponds to the restoration of normal blood flow. The pressure at this time, called **diastolic,** represents the pressure in the artery between the contractions of the heart.

Physicians express blood pressures in **torr,** where 1 torr is the pressure exerted by a column of mercury 1 mm high; it is equivalent to 133 Pa. The torr was formerly referred to as the "millimeter of mercury," abbreviated mm Hg. The unit is named after Evangelista Torricelli (1608–1647), the Italian physicist who invented the barometer, which measures atmospheric pressure. Average atmospheric pressure is 760 torr. In a healthy person the systolic and diastolic blood pressures are, respectively, about 120 and 80 torr (Fig. 10.12).

EXAMPLE 10.4

The average pressure of the blood in a person's arteries is 100 torr at the same elevation as the heart. Find the average pressure in an artery in the head of a standing person, say 40 cm above the heart, and in an artery in the foot, say 120 cm below the heart. The density of blood is 1.06×10^3 kg/m^3.

SOLUTION The difference in pressure in each case is $\Delta p = \rho g h$. Here

$$\Delta p_1 = \rho g h_1 = (1.06 \times 10^3\ \mathrm{kg/m^3})(9.8\ \mathrm{m/s^2})(0.4\ \mathrm{m}) = 4.16 \times 10^3\ \mathrm{Pa}$$
$$\Delta p_2 = \rho g h_2 = (1.06 \times 10^3\ \mathrm{kg/m^3})(9.8\ \mathrm{m/s^2})(1.2\ \mathrm{m}) = 12.5 \times 10^3\ \mathrm{Pa}$$

Because 1 torr = 133 Pa, $\Delta p_1 = 31$ torr and $\Delta p_2 = 94$ torr. Hence the pressure in the artery in the head is (100 − 31) torr = 69 torr and that in the artery in the foot is (100 + 94) torr = 194 torr. The arteries that lead to the head expand and contract as needed to keep the flow of blood to the brain constant despite changes in the elevation of the head relative to the heart. Such expansions and contractions require a few seconds to be completed, which explains why sitting up suddenly from a horizontal position may lead to a momentary dizzy sensation. ◆

10.4 Buoyancy

Sink or swim.

When an object is placed in water (or any other fluid), an upward force acts on it. This effect, called **buoyancy,** makes it possible for balloons to float in the air and ships to float in the sea.

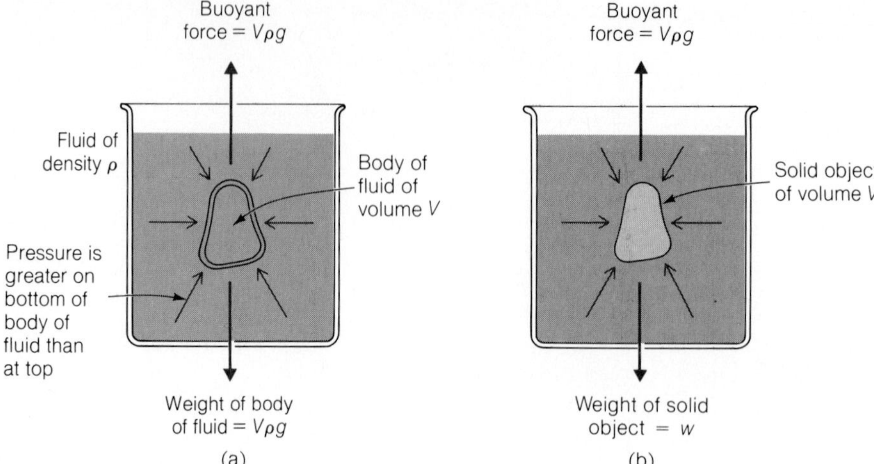

Figure 10.13

The buoyant force on a submerged object is equal to the weight of the fluid it displaces.

Imagine a solid object of volume V submerged in a fluid of density ρ. A body of fluid of the same size and shape in the same place is acted on by a buoyant force that is the vector sum of all the forces the rest of the fluid exerts on it, as in Fig. 10.13(a). The buoyant force must be upward because the pressure (and hence the upward force) on the bottom of the body is greater than the pressure (and hence the downward force) on its top. The horizontal forces on the sides of the body cancel one another out. Because the body of fluid is in equilibrium, the buoyant force on it equals its weight, which is $V\rho g$.

Now we replace the body of fluid by the solid object, as in Fig. 10.13(b). The various pressures remain the same, so the buoyant force of $V\rho g$ also remains the same. We conclude that

$$F_{\text{buoyant}} = V\rho g \qquad \textit{Archimedes' principle} \quad (10.6)$$

$$\text{Buoyant force} = \text{weight of displaced fluid}$$

This result is known as **Archimedes' principle,** after its discoverer.

The buoyant force on an object in a fluid is equal to the weight of the fluid the object displaces.

Archimedes' principle holds whether the object floats or sinks. If the object's weight is greater than the buoyant force, it sinks. If its weight is less than the buoyant force, it floats, in which case the volume V refers only to the submerged part.

EXAMPLE 10.5

A bracelet that appears to be gold is suspended from a spring scale that reads 50 g in air and 46 g when the bracelet is immersed in a glass of water. Is the bracelet pure gold?

SOLUTION First we find the volume of the bracelet. The buoyant force on it in water is

$$F_b = (\Delta m)g = V\rho_{water}g$$

where Δm is the difference between the scale readings. Because the density of water is 1 g/cm^3,

$$V = \frac{\Delta m}{\rho_{water}} = \frac{4 \text{ g}}{1 \text{ g/cm}^3} = 4 \text{ cm}^3$$

If m is the scale reading in air, the density of the bracelet is

$$\rho = \frac{m}{V} = \frac{50 \text{ g}}{4 \text{ cm}^3} = 12.5 \text{ g/cm}^3$$

The density of gold is 19 g/cm^3, and so the bracelet cannot be pure gold. ◆

EXAMPLE 10.6

An iceberg is a chunk of freshwater ice that has broken off from an ice cap, such as those that cover Greenland and Antarctica, or from a glacier at the edge of the sea. (More than three-quarters of the world's fresh water is in the form of ice.) Find the proportion of the volume of an iceberg that is submerged.

SOLUTION If V_{ice} is the iceberg's total volume and V_{sub} is its submerged volume, then

$$\text{Weight of iceberg} = \text{weight of displaced water}$$

$$V_{ice}\rho_{ice}g = V_{sub}\rho_{seawater}g$$

$$\frac{V_{sub}}{V_{ice}} = \frac{\rho_{ice}}{\rho_{seawater}} = \frac{9.2 \times 10^2 \text{ kg/m}^3}{1.03 \times 10^3 \text{ kg/m}^3} = 0.89$$

Thus 89% of the volume of an iceberg is below sea level (Fig. 10.14).

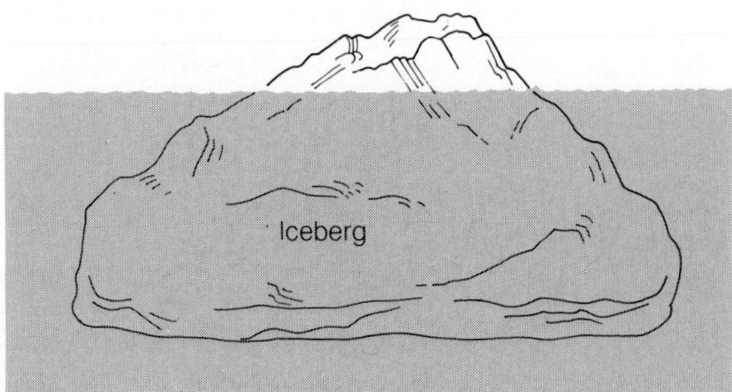

Figure 10.14
Only 11% of the volume of an iceberg is above sea level.

◆

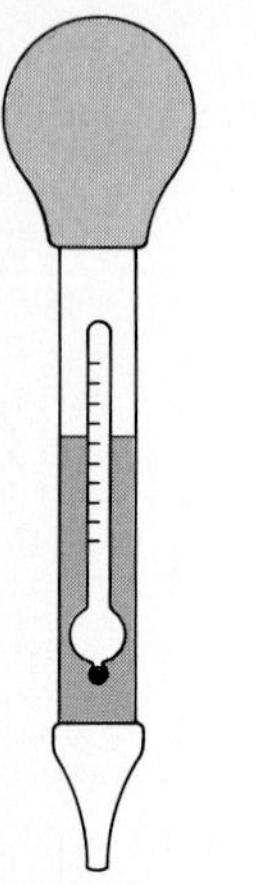

Figure 10.15

A hydrometer.

The **hydrometer** is a simple instrument based on Archimedes' principle that is used to measure the density of a liquid (Fig. 10.15). By squeezing and then releasing the rubber bulb, the liquid is drawn into a glass tube that contains a smaller sealed tube with a weight at its bottom. The higher the sealed tube floats, the denser the liquid.

The density of the acid in a car's storage battery depends on its state of charge, which means that a hydrometer can be used to establish the condition of the battery. A fully charged battery typically has a specific gravity of about 1.28 (that is, its density is 1.28 g/cm^3), which drops to about 1.15 when the battery is completely discharged. A hydrometer can also be used to determine the freezing point of the ethylene glycol antifreeze solution in the cooling system of a car, because the freezing point depends on the proportion of ethylene glycol it contains. Ethylene glycol has a specific gravity of 1.11, and so the greater the specific gravity of the solution, the higher the proportion of antifreeze and the lower the freezing point. For example, a specific gravity of 1.04 corresponds to a 28% solution and to a freezing point of −14°C.

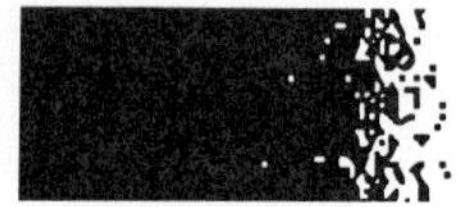

10.5 Fluid Flow

A simple model for a complex subject.

Fluids in motion often behave in complex and unpredictable ways. However, we can understand many aspects of fluid flow on the basis of a simple model that in many cases is reasonably realistic. The liquids in this model are supposed to be incompressible and to have no viscosity. (*Viscosity* is the term used to describe internal frictional resistances in a fluid; see Section 10.8.) With no viscosity, layers of fluid slide freely past one another and past other surfaces, so that we can apply the model to such liquids as water but not to such liquids as honey.

Another approximation we will make is that the fluid undergoes **laminar** (or **streamline) flow** only. In streamline flow, which is illustrated in Fig. 10.16, every particle of liquid passing a particular point follows the same path (called a *streamline*) as the particles that passed that point before. Furthermore, the direction in which the individual fluid particles move is always the same as the direction in which the fluid as a whole moves.

At the other extreme is **turbulent flow,** in which whirls and eddies occur, such as those in a cloud of cigarette smoke or at the foot of a waterfall. Turbulence generally develops at high speeds and when there are obstructions or sharp bends in the path of the fluid.

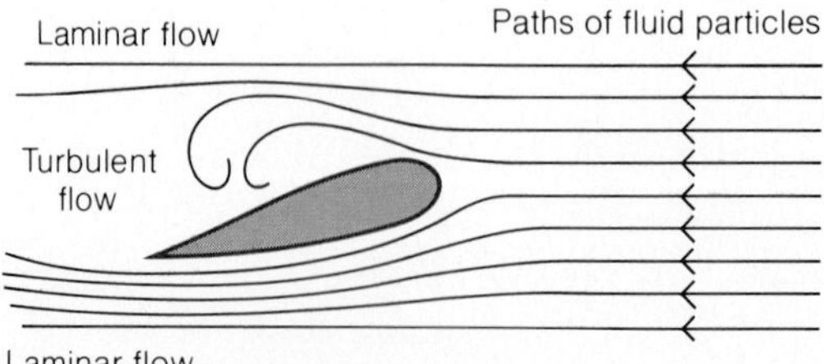

Figure 10.16

Laminar and turbulent flows around an obstacle.

BIOGRAPHICAL NOTE

Daniel Bernoulli

Daniel Bernoulli (1700–1782) was born to a remarkable Swiss family of mathematicians. His father and uncle were among the early developers of calculus, and two of Daniel's brothers, a cousin, and two nephews also did distinguished work in mathematics. Daniel himself studied medicine at the University of Basle. But mathematics was in his blood, and he spent the years from 1725 to 1733 in Russia as professor of mathematics in Saint Petersburg (now Leningrad).

In Saint Petersburg, Bernoulli began to apply mathematics to various aspects of physics, especially hydrodynamics (the study of fluid motion). He returned to Basle in 1733 and later became professor of physics there. In 1738 Bernoulli published *Hydrodynamica,* a summary of his research in this field. In it appears what is now called Bernoulli's equation, which he derived from an early version of the law of conservation of energy. Also notable in *Hydrodynamica* was the first attempt at a quantitative kinetic theory of gases. Here Bernoulli showed that assuming a gas to consist of a great many randomly moving particles can account for the relationship between the pressure and the volume of a gas sample. Bernoulli taught at Basle until he was 76, and died six years later.

---

energy principle to a parcel of the liquid of volume v_1tA_1 as it enters at the left in the time t and to the same parcel as it leaves at the right. The mass of the parcel is

$$m = \rho V = \rho v_1 t A_1 = \rho v_2 t A_2$$

because vA has the same value at 1 and 2. The net amount of work ΔW done on the liquid parcel as it passes from 1 to 2 must equal the net change in its potential energy ΔPE as its height goes from h_1 to h_2, plus the net change in its kinetic energy ΔKE as its speed goes from v_1 to v_2. That is,

$$\Delta W = \Delta\text{PE} + \Delta\text{KE}$$

The work done *on* the parcel at 1 is the force p_1A_1 acting on it multiplied by the distance v_1t through which the force acts. The work done *by* the parcel at 2 is the force p_2A_2 multiplied by the distance v_2t through which the force acts. The *net* work done on the parcel is therefore

$$\Delta W = p_1A_1v_1t - p_2A_2v_2t = \frac{p_1m}{\rho} - \frac{p_2m}{\rho}$$

The change in the potential energy of the parcel in going from 1 to 2 is

$$\Delta\text{PE} = mgh_2 - mgh_1$$

and the change in its kinetic energy is

$$\Delta KE = \tfrac{1}{2}mv_2^2 - \tfrac{1}{2}mv_1^2$$

Therefore

$$\Delta W = \Delta PE + \Delta KE$$

$$\frac{p_1 m}{\rho} - \frac{p_2 m}{\rho} = mgh_2 - mgh_1 + \tfrac{1}{2}mv_2^2 - \tfrac{1}{2}mv_1^2$$

When we divide through by the common factor m, multiply by ρ, and rearrange terms, the result is Bernoulli's equation,

$$p_1 + \rho gh_1 + \tfrac{1}{2}\rho v_1^2 = p_2 + \rho gh_2 + \tfrac{1}{2}\rho v_2^2$$

EXAMPLE 10.9

Water flows upward through the pipe shown in Fig. 10.22 at 15 L/s. If water enters the lower end of the pipe at 3.0 m/s, what is the difference in pressure between the two ends?

SOLUTION The first step is to find the water speed v_2 at the upper end of the pipe. From Eq. (10.7), $R = vA$, this is

$$v_2 = \frac{R}{A_2} = \frac{(15\text{ L/s})(10^{-3}\text{ m}^3/\text{L})}{(20\text{ cm}^2)(10^{-4}\text{ m}^2/\text{cm}^2)} = 7.5\text{ m/s}$$

We now substitute the known quantities v_1, v_2, and $h_2 - h_1$ into Bernoulli's equation. The density of water is $\rho = 1.0 \times 10^3$ kg/m^3, hence

$$\begin{aligned}\Delta p = p_1 - p_2 &= \rho g(h_2 - h_1) + \tfrac{1}{2}\rho(v_2^2 - v_1^2)\\ &= \left(1.0 \times 10^3 \frac{\text{kg}}{\text{m}^3}\right)\left(9.8 \frac{\text{m}}{\text{s}^2}\right)(1.5\text{ m})\\ &\quad + \tfrac{1}{2}\left(1.0 \times 10^3 \frac{\text{kg}}{\text{m}^3}\right)\left[\left(7.5 \frac{\text{m}}{\text{s}}\right)^2 - \left(3.0 \frac{\text{m}}{\text{s}}\right)^2\right]\\ &= 38\text{ kPa}\end{aligned}$$

Figure 10.22

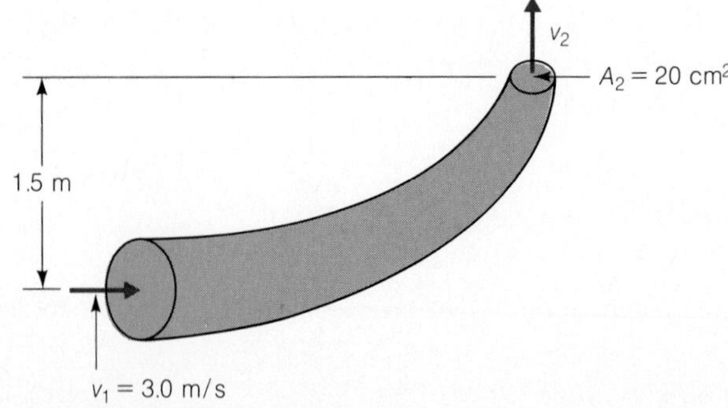

◆

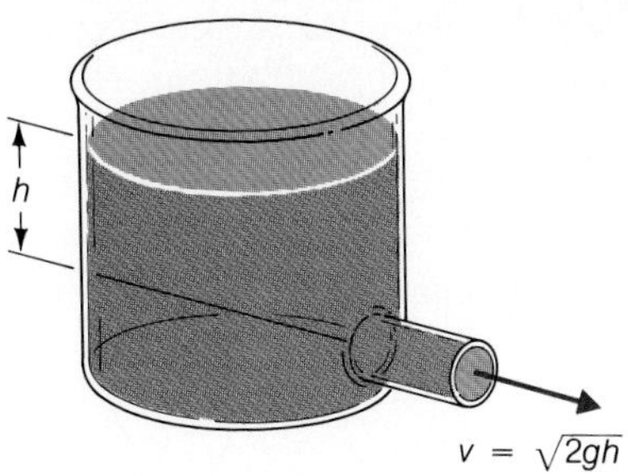

Figure 10.23
Torricelli's theorem.

10.7 Applications of Bernoulli's Equation

Heavier-than-air flight is one of them.

In many situations the speed, pressure, or height of a liquid is constant, and simplified forms of Bernoulli's equation hold. Thus when a liquid column is stationary, we see that the pressure difference between two depths in it is

$$p_2 - p_1 = \rho g(h_1 - h_2) \qquad \textit{Liquid at rest} \quad (10.11)$$

which is just what Eq. (10.4) states. Evidently the latter formula is included in Bernoulli's equation.

Another straightforward result occurs when $p_1 = p_2$. As an example, Fig. 10.23 shows a liquid emerging from an opening in a tank. The liquid pressure equals atmospheric pressure both at the top of the tank and at the opening. If the opening is small compared with the cross section of the tank, the liquid level in the tank will fall slowly enough for the liquid speed at the top of the tank to be assumed zero. If the speed of the liquid as it leaves the opening is v and the difference in height between the top of the liquid and the opening is h, Bernoulli's equation reduces to

$$\tfrac{1}{2}\rho v^2 = \rho g h$$
$$v = \sqrt{2gh} \qquad \textit{Torricelli's theorem} \quad (10.12)$$

The speed with which the liquid comes out is the same as the speed of a body falling from rest from the height h. This result is called **Torricelli's theorem,** and, like the relationship between pressure and depth, it is a special case of Bernoulli's equation. The rate at which liquid flows through the opening may be found from Eq. (10.7) if its area A is known. The volume of liquid per unit time is

$$R = vA = A\sqrt{2gh} \qquad \textit{Flow from orifice} \quad (10.13)$$

EXAMPLE 10.10

A boat strikes an underwater rock and opens a pencil-sized crack 7.0 mm wide and 150 mm long in its hull 65 cm below the waterline. The crew takes 5.00 min to locate the crack and plug it up. How much water entered the boat in the meantime?

SOLUTION The area of the crack is (7.0 mm)(150 mm) = 1050 mm^2 = 1.05 × 10^{-3} m^2, and so, from Eq. (10.13),

$$R = (1.05 \times 10^{-3}\ \text{m}^2)\sqrt{(2)(9.8\ \text{m/s}^2)(0.65\ \text{m})} = 3.75 \times 10^{-3}\ \text{m}^3/\text{s}$$

In 5.00 min = 300 s the volume of water that came in was

$$V = Rt = (3.75 \times 10^{-3}\ \text{m}^3/\text{s})(300\ \text{s}) = 1.1\ \text{m}^3$$

This is over a ton, a substantial amount. ◆

Here is an illustration of a further special case of Bernoulli's equation.

EXAMPLE 10.11

Geysers in Yellowstone National Park and elsewhere shoot columns of water as high as 50 m into the air at more or less regular intervals. If the water at the mouth of a geyser is assumed to be initially at rest, what subsurface pressure is needed for the water to reach such a height?

SOLUTION If we let 1 refer to the mouth of the geyser and 2 to the top of the water column, $h_1 = 0$, $h_2 = 50$ m, $p_2 = p_{\text{atm}} = 1.0 \times 10^5$ Pa, and $v_1 = v_2 = 0$. From Bernoulli's equation

$$\begin{aligned} p_1 - p_2 &= \rho g(h_2 - h_1) \\ p_1 &= p_{\text{atm}} + \rho g h_2 \\ &= 1.0 \times 10^5\ \text{Pa} + (1.0 \times 10^3\ \text{kg/m}^3)(9.8\ \text{m/s}^2)(50\ \text{m}) \\ &= 1.0 \times 10^5\ \text{Pa} + 4.9 \times 10^5\ \text{Pa} = 5.9 \times 10^5\ \text{Pa} \end{aligned}$$

The absolute pressure at the mouth of the geyser is 5.9 bar, and the gauge pressure there is 4.9 bar. ◆

Lift

The most interesting special case of Bernoulli's equation occurs when there is no change in height during the motion of the liquid (Fig. 10.24). Here

$$p_1 + \tfrac{1}{2}\rho v_1^2 = p_2 + \tfrac{1}{2}\rho v_2^2 \qquad \textit{Pressure and speed} \quad (10.14)$$

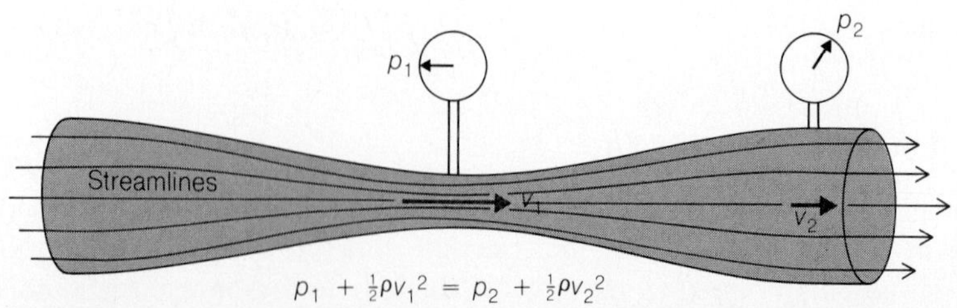

Figure 10.24

In a horizontal pipe the pressure is greatest when the velocity is least, and vice versa.

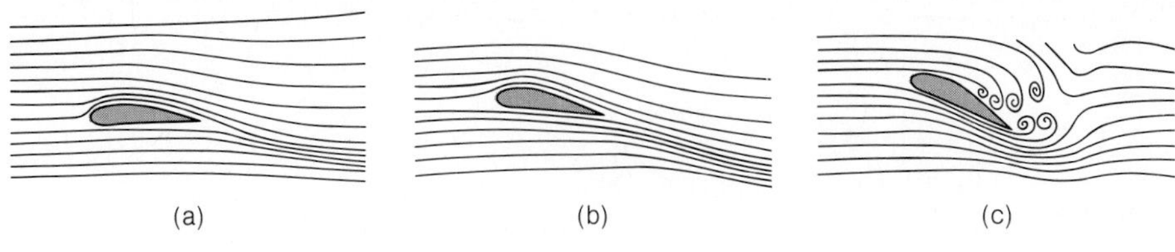

Figure 10.25

Air flow past a wing. (a) The air speed is greater over the upper surface, as indicated by the closer streamlines, and so the pressure is lower there and a net upward force called lift acts on the wing. (b) When the wing is angled upward, air deflected from its lower surface provides additional lift. (c) If the angle of attack is too great, turbulence reduces lift and increases drag.

which means that the pressure in the liquid is least where the speed is greatest, and vice versa.

A familiar application of Eq. (10.14) is the lifting force produced by the flow of air past the wing of an airplane, as in Fig. 10.25. (Air is, of course, a compressible fluid and so does not fit our model exactly, but the behavior predicted by Bernoulli's equation is not a bad approximation for gases at moderate speeds.) Air moving past the upper surface of the wing must travel faster than air moving past the lower surface. This is indicated by the closeness of the streamlines near the upper surface. The difference in speed gives a reduced pressure over the top of the wing, which leads to an upward force on the wing. The greater the difference in air speeds around the upper and lower surfaces, the greater the lift that is produced, provided the wing shape is not so extreme that turbulence results (Fig. 10.25c). An additional source of lift is the deflection of air from the angled bottom of the wing, which leads to an upward force on it.

Air flows around the tip of each wing from the high-pressure region under it to the low-pressure region above it (Fig. 10.26). The result is a trail of whirling air (called a **vortex**) behind each wingtip that can be dangerous for small aircraft flying too closely behind large ones.

The lift force F_{lift} on a wing of area A due to the difference in air speeds between its surfaces is, from Eq. (10.14),

$$F_{\text{lift}} = (p_1 - p_2)A = \tfrac{1}{2}\rho A(v_2^2 - v_1^2)$$

Here v_1 is the air speed below the wing and v_2 is its speed above the wing. How these speeds are related to the speed v of the airplane depends on the shape of the wing and on the angle the wing makes with the direction of the airflow ahead of it. In general $(v_2^2 - v_1^2)$ is proportional to v^2, so F_{lift} is proportional to Av^2. Thus the faster an airplane moves, the smaller the wing area needed to provide it with lift. Because a small wing also exhibits less resistance to motion through the air (less

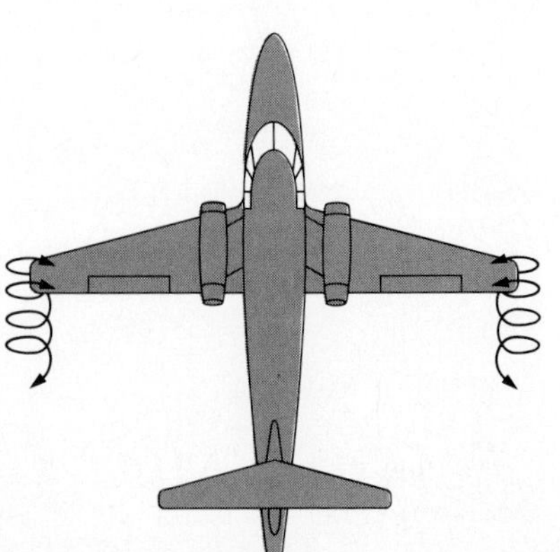

Figure 10.26

Wingtip vortexes occur because air flows from the high-pressure region under a wing to the low-pressure region above it. They may extend for long distances behind a large aircraft.

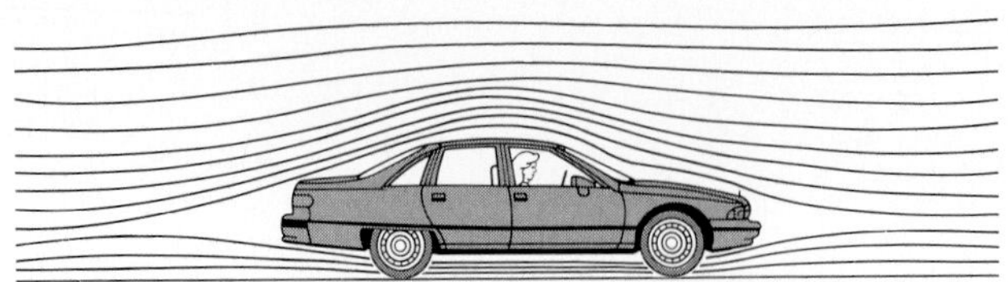

Figure 10.27

A car streamlined to reduce drag experiences aerodynamic lift and may be hard to control at high speeds unless its shape is modified with "spoilers" that interfere with the laminar flow over the car.

"drag") than a large one, it is clear why high-speed airplanes have relatively small wings. The penalty such airplanes must pay is that their takeoff and landing speeds need to be correspondingly high.

Similar considerations apply to the wings of birds and other flying animals such as bats, even though their wings provide propulsion as well as lift. A bird's weight w is proportional to its volume and hence to L^3, where L is a representative linear dimension such as its length. The area of the bird's wings is proportional to L^2, so in order that F_{lift} be equal to w—the condition for level flight—v^2L^2 must be proportional to L^3. Hence v^2 must be proportional to L, and v proportional to $\sqrt{L}$: The larger the bird, the greater its minimum flying speed.

Of course, different birds have different aerodynamic properties, so we would not expect strict accordance to this relationship, but it does seem to be followed at least in an approximate way. Anyone who has watched birds has observed that small ones can take off quickly with only a few flaps of their wings, whereas large ones, such as ducks, geese, and swans, which require more initial speed, have more trouble in becoming airborne. The huge flying reptiles that lived in the time of the dinosaurs probably had to launch themselves from cliffs and, like present-day sailplanes, could stay aloft for long periods only with the help of rising air currents.

At low speeds most of the resistance a car offers to motion comes from the flexing of its tires. As the speed increases, aerodynamic drag becomes more and more important, until finally the car's engine cannot provide enough power to overcome it and the car can go no faster. Drag depends on the shape of a car as well as on its speed, so much attention is paid nowadays to streamlining. A streamlined car, however, is much like a wing section (Fig. 10.27), and the resulting lift at high speeds reduces the grip of the tires on the road. Because both steering and braking depend on this grip, the result can be dangerous. Ridges ("spoilers") are often placed on the rear ends of fast cars to reduce lift by interfering with laminar flow, but an increase in drag is the inevitable result.

Air flow over a car being checked in a wind tunnel. Laminar flow gives least resistance.

10.8 Viscosity

Internal friction in a fluid.

The viscosity of a fluid is a kind of internal friction that prevents neighboring layers of the fluid from sliding freely past one another. Figure 10.28 shows how the speed of a fluid in a pipe varies with distance from the axis. The fluid in contact with the pipe's wall is stationary and, as we would expect, the speed increases to a maximum along the pipe's axis. The smaller the viscosity of the fluid, the greater the various speeds will be, but the characteristic parabolic shape of the speed profile will persist as long as the flow remains laminar.

Figure 10.28

Because of viscosity, the velocity of fluid in a pipe varies from zero at the pipe wall to a maximum along the axis.

Viscosity is given a quantitative definition in terms of the experiment shown in Fig. 10.29, in which a plate of area A is being pulled across a layer of fluid of thickness s. A hollow cylinder having a smaller rotating cylinder inside it, with a fluid layer between them, is a better arrangement for actual measurements.

For most fluids the force F required to pull the plate at the constant speed v turns out to be proportional to A and v and inversely proportional to s. The faster the motion and the thinner the layer of fluid, the more force is needed for a given plate area. Different fluids offer different degrees of resistance to the motion, but the force needed varies as Av/s for most of them provided the speed is not so great that turbulence occurs. Thus we can write

$$F = \frac{\eta A v}{s} \tag{10.15}$$

where η, the constant of proportionality, is called the **viscosity** of the fluid (η is the Greek letter *eta*). The SI unit of viscosity is the N·s/m^2, which is known as the **poiseuille** (Pl). An older unit, the *poise*, remains in common use, where 10 poise = 1 N·s/m^2; the *centipoise*, 0.01 poise, is equal to 10^{-3} N·s/m^2.

The viscosities of some common fluids are listed in Table 10.1. The viscosity of a liquid decreases with temperature as its molecules becomes less and less tightly bound to one another (see Chapter 13), a phenomenon familiar to anyone who has heated honey or molasses. One of the reasons a drop in body temperature is so dangerous is that the viscosity of the blood then increases, which impedes its flow. The viscosity of a gas, in contrast to that of a liquid, increases with temperature

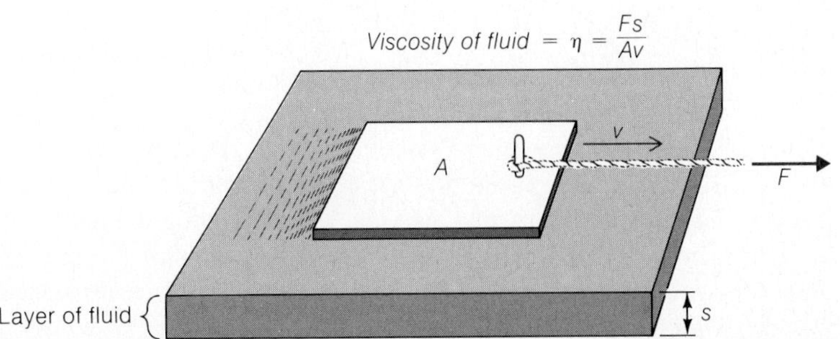

Figure 10.29

The viscosity of a fluid is defined in terms of the force needed to pull a flat plate at constant speed across a layer of the fluid.

Table 10.1 Viscosities of various fluids at atmospheric pressure and the indicated temperature

Substance	Temperature (°C)	Viscosity (Pl)
Gases		
Air	0	1.7×10^{-5}
	20	1.8×10^{-5}
	100	2.2×10^{-5}
Water vapor	100	1.3×10^{-5}
Liquids		
Alcohol (ethyl)	20	1.2×10^{-3}
Blood plasma	37	1.3×10^{-3}
Blood, whole (varies with speed)	37	$\sim 2 \times 10^{-3}$
Glycerin	20	0.83
Water	0	1.8×10^{-3}
	20	1.0×10^{-3}
	40	0.66×10^{-3}
	60	0.47×10^{-3}
	80	0.36×10^{-3}
	100	0.28×10^{-3}

because the higher the temperature, the faster the gas molecules move and the more often they collide with one another (again, see Chapter 13).

The viscosity of motor oil is usually given in terms of SAE (Society of Automotive Engineers) index numbers. The higher the number, the greater the viscosity, so that SAE 30 oil is ''thicker'' than SAE 10 oil at a given temperature. Because viscosities of liquids decrease with temperature, SAE 30 oil has about the same viscosity at 20°C (68°F) as SAE 10 oil at 0°C (32°F). If SAE 30 oil were used in very cold weather, its resistance to flow would make an engine hard to start and it might not circulate adequately afterward; SAE 10 oil is better in such conditions. On the other hand, SAE 10 oil would be too thin on a hot summer day to provide proper lubrication, and SAE 30 oil would be appropriate then. ''Multigrade'' oils are formulated to be usable over a wide range of temperatures.

Let us return to a fluid moving through a pipe. If the pipe is cylindrical with the length L and inside radius r, and a fluid of viscosity η is in laminar flow through it under the influence of a pressure difference $\Delta p = p_1 - p_2$, as in Fig. 10.30, the rate of flow is

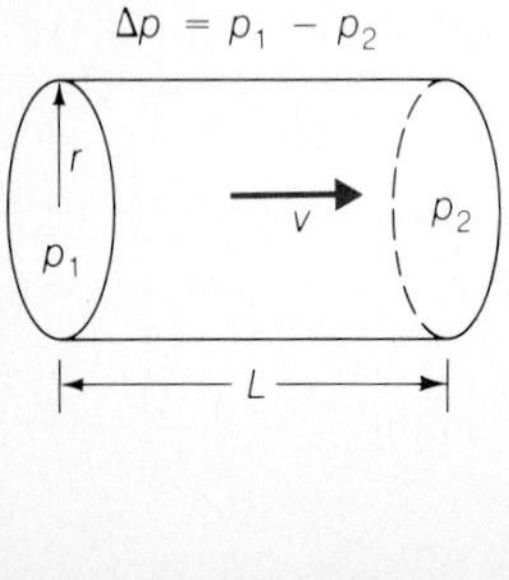

$$R = \frac{\pi r^4 \, \Delta p}{8 \eta L} \qquad \textit{Poiseuille's law} \qquad (10.16)$$

Figure 10.30

Poiseuille's law describes fluid flow through a pipe.

Equation (10.16) is known as **Poiseuille's law**. The dependence of the rate of flow on the viscosity η and on the pressure gradient $\Delta p/L$ are both about what we might expect, but the variation with r^4 is remarkable: Halving the radius of a pipe reduces R by a factor of 16 if Δp stays the same. The radius of the pipe plays a far more important role than its length does with respect to viscous resistance. (See Problem 71 for a way to justify the form of Eq. 10.16.)

EXAMPLE 10.12

Atherosclerosis is the medical term for a common condition in which arteries are narrowed by deposits of tissue called plaque. If the flow of blood is to continue at its usual rate, a higher pressure is required. A consequence of atherosclerosis is therefore an elevated blood pressure, which has many undesirable effects on the body, one of them being that the heart must work harder to circulate the blood. (a) Find the percentage increase in pressure needed to maintain R constant when the radius of an artery is decreased by 10%. (b) What is the corresponding increase in the power needed to maintain the flow of blood through that artery?

SOLUTION (a) For a constant rate of flow R, the product $r^4 \Delta p$ must be constant. Here $r_2 = 0.9r_1$, so

$$r_1^4 \, \Delta p_1 = r_2^4 \, \Delta p_2$$

$$\frac{\Delta p_2}{\Delta p_1} = \left(\frac{r_1}{r_2}\right)^4 = \left(\frac{r_1}{0.9r_1}\right)^4 = 1.52$$

The blood pressure must increase by 52% if R is to be unchanged.

(b) From Eq. (10.9) $P = \Delta pR$. Here R is the same, so

$$\frac{P_2}{P_1} = \frac{\Delta p_2}{\Delta p_1} = 1.52$$

The power output of the heart must increase by 52% also. Narrowing of the arteries usually means a reduced flow of blood as well as an increase in blood pressure and in the work done by the heart. ◆

10.9 Reynolds Number

Fluid-flow systems are dynamically similar when their Reynolds numbers are the same.

Laminar flow is regular and straightforward to analyze. Turbulent flow is irregular and largely unpredictable in its details. Because actual fluid flows often involve some degree of turbulence, calculations based on laminar flow are seldom able to establish exactly what will happen in a new situation. At the same time, full-size experiments are too expensive and time consuming when designing, say, a new airplane or ship where many different shapes must be studied to find the most efficient one. The obvious solution is to work with small-scale models. But under what circumstances is the fluid flow around its model the same as the flow around an actual airplane or ship?

Work done by the British physicist Osborne Reynolds a century ago showed that fluid flows in systems of the same kind will be **dynamically similar**—follow similar patterns—if a quantity now called the **Reynolds number** N_R is the same in them. The Reynolds number for a given system depends on four factors: the density

ρ and viscosity η of the fluid, the average speed v, and a characteristic length L of the system. The value of N_R is given by

$$N_R = \frac{\rho v L}{\eta} \qquad \textit{Reynolds number} \quad (10.17)$$

The Reynolds number is a dimensionless quantity. That is, N_R is a pure number with no units associated with it.

EXAMPLE 10.13

Wind-tunnel data for lift and drag gathered on symmetrical wing sections 30 cm from leading to trailing edge are being used to help design the 3.0-m-long keel of a racing sailboat. What wind-tunnel air speed would give a flow pattern that corresponds to that of fresh water around the keel at a boat speed of 4 m/s (about 8 knots)?

SOLUTION From Eq. (10.17), for the same Reynolds number at 20°C,

$$\begin{aligned} v_{\text{wing}} &= \left(\frac{\rho_{\text{water}}}{\rho_{\text{air}}}\right)\left(\frac{L_{\text{keel}}}{L_{\text{wing}}}\right)\left(\frac{\eta_{\text{air}}}{\eta_{\text{water}}}\right)(v_{\text{keel}}) \\ &= \left(\frac{1.0 \times 10^3 \text{ kg/m}^3}{1.3 \text{ kg/m}^3}\right)\left(\frac{3.0 \text{ m}}{0.30 \text{ m}}\right)\left(\frac{1.8 \times 10^{-5} \text{ Pl}}{0.66 \times 10^{-3} \text{ Pl}}\right)(4 \text{ m/s}) = 84 \text{ m/s} \end{aligned}$$

which is 163 knots. ♦

Poiseuille's law holds only for laminar flow. When the flow in a pipe is turbulent, more energy is lost to friction, and the flow rate R for the same Δp is reduced. The Reynolds number can be used as a guide to the nature of the flow in a pipe. Experiments show that in this situation a Reynolds number of less than about 2000 corresponds to laminar flow and a Reynolds number of more than about 3000 corresponds to turbulent flow. Between 2000 and 3000 the flow may be of either kind and may even shift back and forth between them. In other situations—for instance, fluid flow past an obstacle—marked changes in patterns of flow also occur in the vicinity of particular N_R values, from one type of turbulence to another as well as from laminar to turbulent flow.

EXAMPLE 10.14

In Section 10.5 we considered a garden hose 6 mm in inside radius through which water passed at 2.5 m/s. (a) Assuming the water temperature is 20°C, establish the nature of the flow in the hose from its Reynolds number. (b) What is the maximum speed for laminar flow?

SOLUTION (a) From Table 10.1 the viscosity of water at 20°C is 1.0×10^{-3} Pl. With $L = 2r = 12$ mm $= 0.012$ m, the Reynolds number of the flow is

$$N_R = \frac{\rho v L}{\eta} = \frac{(1.0 \times 10^3 \text{ kg/m}^3)(2.5 \text{ m/s})(0.012 \text{ m})}{1.0 \times 10^{-3} \text{ Pl}} = 30{,}000$$

Because 1 Pl = 1 N·s/m^2 = 1 kg/m·s, the units cancel. Here N_R is greater than 3000, so the motion of the water in the hose is turbulent.

(b) To find the highest water speed for laminar flow, we solve Eq. (10.17) for v with $N_R = 2000$:

$$v = \frac{\eta N_R}{\rho L} = \frac{(1.0 \times 10^{-3}\,\text{Pl})(2000)}{(1.0 \times 10^3\,\text{kg/m}^3)(0.012\,\text{m})} = 0.17\ \text{m/s}$$

This is not very fast. ♦

10.10 Surface Tension

A liquid surface is like a membrane under tension.

The surface of a liquid behaves remarkably like a membrane under tension. Thus a steel sewing needle placed horizontally on the surface of some water in a dish does not sink even though its density is nearly eight times that of water (Fig. 10.31). The needle rests in a depression in the water surface just as if that surface were a sheet of rubber stretched across the dish. The term **surface tension** for this effect is quite appropriate. Another example of surface tension is the tendency of a liquid drop to assume a spherical shape, just as an inflated balloon does.

The origin of surface tension lies in the fact that molecules on the surface of a body of liquid are acted on by a net inward force, whereas those in the interior are acted on by equal forces in all directions (Fig. 10.32). Because the surface molecules are all being pulled inward, the surface tends to contract to the minimum possible area. A sphere has the least area relative to its volume of any object, and so a liquid sample not acted on by any external forces (such as gravitation or air resistance if it is moving) will take on a spherical form. A liquid surface can support a small object such as a needle because the weight of the object is not enough to rupture the surface, which would involve first stretching it to a greater degree than that shown in Fig. 10.31.

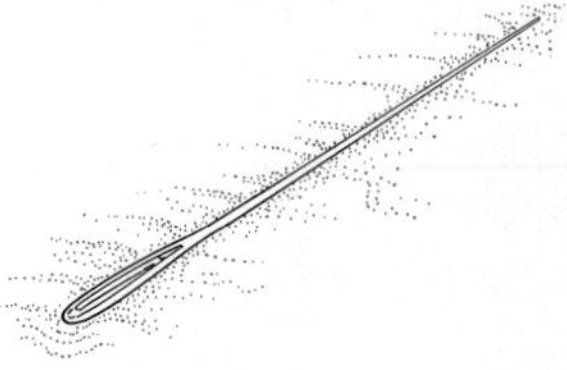

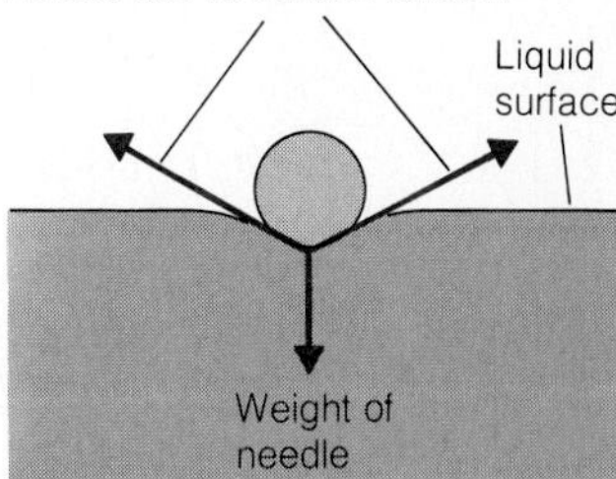

Figure 10.31

The surface tension of a water surface supplements buoyancy in supporting a steel needle. The elastic character of a liquid surface resembles that of a membrane under tension.

Suppose we dip the wire frame of Fig. 10.33 into a glass of soapy water. The result is a thin film of liquid across the opening of the frame. Each surface of the film is under tension, and a force of $2F$ is needed to hold the sliding wire in place against this tension, where F is the force needed for each surface. The length of the sliding wire is L, and the ratio between F and L is defined as the *surface tension* γ (Greek letter *gamma*) of the liquid:

$$\gamma = \frac{F}{L} \qquad \textit{Surface tension} \quad (10.18)$$

The unit of γ is the N/m.

In the case of water, $\gamma = 0.073$ N/m at 20°C. Surface tensions generally decrease with temperature, so that $\gamma = 0.059$ N/m for water at 100°C. Among common liquids only mercury has a higher surface tension than water, with $\gamma = 0.44$ N/m. At 20°C the surface tension of ethyl alcohol is a third that of water. The high surface tension of water is useful for many small creatures, such as the insects that

Figure 10.32

Because a molecule on the surface of a body of liquid has no molecules above it to interact with, the net force on it is toward the body of the liquid.

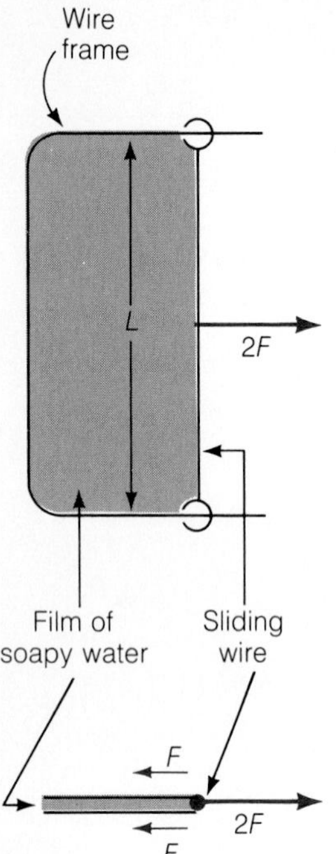

Figure 10.33

Each surface of the liquid film exerts a force of F on the sliding wire, so an external force of $2F$ is needed to hold it in place. The ratio $\gamma = F/L$ is the surface tension of the liquid.

walk on the surface of a pond and the larvae that are suspended from the surface while they develop.

Elsewhere the high γ of water can be a handicap. An example is the alveoli of the lungs, which are tiny sacs at the end of the bronchial tubes in whose walls oxygen from the air enters the bloodstream while carbon dioxide leaves. The alveoli are lined with mucus, and if this mucus had the normal surface tension of other tissue fluids, normal breathing would not provide the pressure needed to expand the alveoli and fill them with air. However, the alveoli secrete a substance that lowers the surface tension of the mucus enough to enable the expansion to occur readily. This substance is an example of a **surfactant**, whose addition to a liquid reduces its surface tension. Detergents are efficient surfactants, and even an extremely dilute solution of detergent in water can penetrate small crevices and be absorbed by tightly woven textiles. Surfactants are also called "wetting agents" because they allow a water solution to spread uniformly over a surface instead of forming into separate droplets.

Capillarity

A familiar phenomenon is the rise of most liquids in a capillary tube. Capillarity is responsible for many familiar effects, such as the ability of paper and cloth fibers to absorb water. Two factors are involved: the **cohesion** of the liquid, which refers to the attractive forces its molecules exert on one another, and the **adhesion** of the liquid to the surface of a solid, which refers to the attractive forces the solid exerts on the liquid molecules.

If the adhesive forces exceed the cohesive ones, as they do in the case of water and glass, then the liquid tends to stick to the solid. It will rise in a capillary tube of that material since the attraction of a liquid molecule to the wall of the tube is greater than the attraction of this molecule to the others. In such an event the liquid surface is concave, as in Fig. 10.34(a), with an angle of contact θ that is characteristic of the liquid-solid combination. For water and clean glass, $\theta \approx 0$, as it is for any liquid that "wets" a particular solid, because adhesion must be much greater than cohesion for this to occur. For kerosene and glass, $\theta = 26°$, which reflects the smaller amount by which adhesion exceeds cohesion for these substances.

Figure 10.34

When the angle of contact θ is less than 90°, the liquid in a capillary tube rises; when θ is more than 90°, the liquid falls.

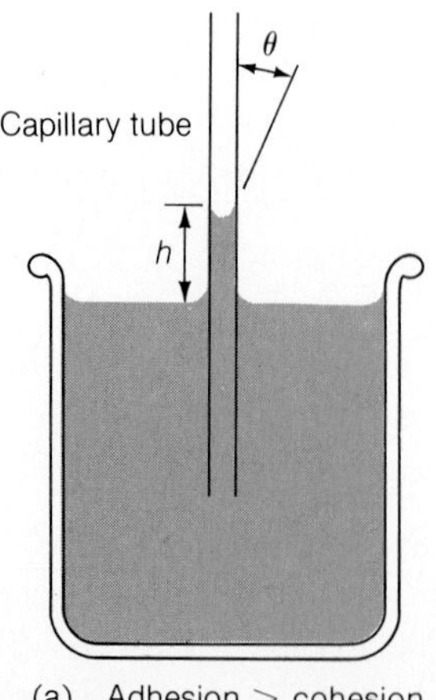

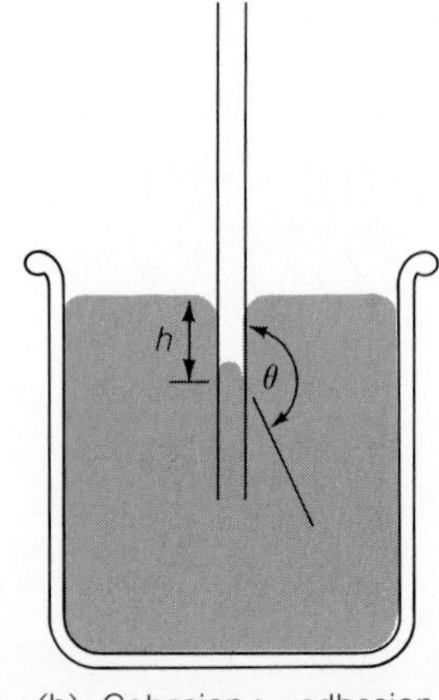

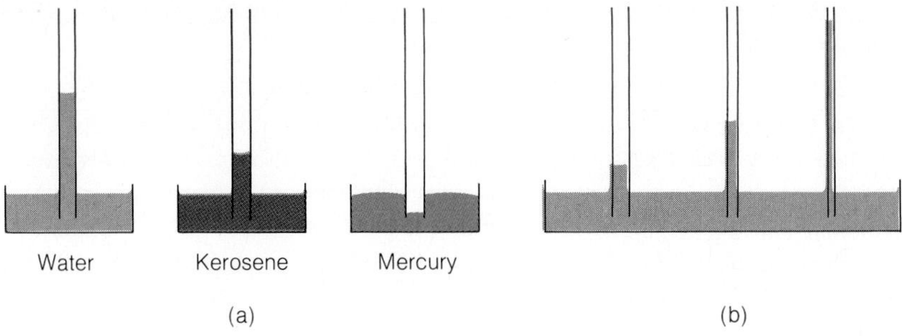

Figure 10.35

(a) Different liquids behave differently in tubes of the same size. (b) The narrower a tube, the higher a liquid rises in it (or falls if $\theta > 90°$).

When cohesion is greater than adhesion, as it is for mercury and glass, the liquid level in a capillary tube is lower than the level of the surrounding liquid. Now the liquid surface in the tube is convex, with an angle of contact greater than 90°. For mercury and glass, $\theta = 140°$. Water spreads out into a thin film on a clean horizontal glass surface, but mercury pulls itself into tiny droplets.

It is straightforward to calculate the height h to which a liquid rises (or falls) in a capillary tube. Let us consider the situation shown in Fig. 10.34(a). The upward force on the liquid column is exerted by the vertical component $F \cos \theta$ of the surface tension of the ring of liquid that is adhering to the tube at the top of the column. If the tube has an inner radius of r, the ring is $2\pi r$ in length, and the surface tension force is $F = \gamma L = 2\pi\gamma r$. The downward force is the weight of the elevated liquid, which is $w = \rho g V = \pi \rho g r^2 h$ where ρ is the density of the liquid and $\pi r^2 h$ its volume. Hence

$$F \cos \theta = w$$

$$2\pi\gamma r \cos \theta = \pi \rho g r^2 h$$

$$h = \frac{2\gamma \cos \theta}{r \rho g} \qquad \textit{Capillarity} \quad (10.19)$$

This formula agrees with experience (Fig. 10.35). The greater the surface tension γ of the liquid, the greater its adhesion to the tube (which means a small θ and a large $\cos \theta$); the narrower the tube and the less the liquid density, the higher the liquid column. If cohesion exceeds adhesion, as in Fig. 10.34(b), the same formula applies. Now $\theta > 90°$, so $\cos \theta$ is negative, and h is negative also.

EXAMPLE 10.15

The water channels in a certain plant are 0.0050 mm in radius. If the contact angle is zero, what is the maximum height to which water can rise in the plant under the influence of surface tension?

SOLUTION At 20°C, $\gamma = 0.073$ N/m. With $\theta = 0$ here, $\cos \theta = 1$ and, from Eq. (10.19),

$$h = \frac{2\gamma \cos \theta}{r \rho g} = \frac{(2)(0.073 \text{ N/m})}{(5.0 \times 10^{-6} \text{ m})(1.0 \times 10^{3} \text{ kg/m}^3)(9.8 \text{ m/s}^2)} = 3.0 \text{ m}$$

♦

Important Terms

The **pressure** on a surface is the perpendicular force per unit area that acts on it. **Gauge pressure** is the difference between absolute pressure and atmospheric pressure.

Pascal's principle states that an external pressure exerted on a confined fluid is transmitted uniformly throughout its volume.

The **hydraulic press** is a machine consisting of two fluid-filled cylinders of different diameters connected by a tube. The input force is applied to a piston in one of the cylinders, and the output force is exerted by a piston in the other cylinder. The MA of a hydraulic press is equal to the inverse ratio of the cylinder areas.

Archimedes' principle states that the buoyant force on a submerged object is equal to the weight of fluid it displaces.

In **laminar** (or **streamline**) **flow** every particle of fluid passing a particular point follows the same path, whereas in **turbulent flow** irregular whirls and eddies occur. The greater the speed of a fluid in streamline flow, the lower its pressure.

The **viscosity** of a fluid is a measure of its internal friction.

The **Reynolds number** of a system that involves fluid flow determines the nature of the flow. Systems with the same Reynolds number are **dynamically similar,** which means their patterns of fluid flow are the same.

The **surface tension** of a liquid refers to the tendency of its surface to contract to the minimum possible area in any situation.

Important Formulas

Pressure: $p = F/A = p_{\text{gauge}} + p_{\text{atm}}$

Pressure at depth h in a fluid: $p = p_{\text{external}} + \rho g h$

Archimedes' principle: $F_{\text{buoyant}} = V\rho g$

Equation of continuity: $R = v_1 A_1 = v_2 A_2$

Bernoulli's equation: $p + \rho g h + \frac{1}{2}\rho v^2 = \text{constant}$

Poiseuille's law: $R = \dfrac{\pi r^4 \,\Delta p}{8\eta L}$

Reynolds number: $N_R = \dfrac{\rho v L}{\eta}$

Capillarity: $h = \dfrac{2\gamma \cos\theta}{r\rho g}$

MULTIPLE CHOICE

1. Atmospheric pressure does not correspond to approximately

a. 14.7 lb/in^2 b. 98 N/m^2
c. 1013 mb d. 1.013×10^5 Pa

2. The hydraulic press is able to produce a mechanical advantage because

a. the force a fluid exerts on a piston is always parallel to its surface
b. an external pressure exerted on a fluid is transmitted uniformly throughout its volume
c. at any depth in a fluid the pressure is the same in all directions
d. the pressure in a fluid varies with its speed

3. In the operation of a hydraulic press, it is impossible for the output piston to exceed the input piston's

a. displacement b. speed
c. force d. work

4. Fluid at rest at the bottom of a container is

a. under less pressure than the fluid at the top
b. under the same pressure as the fluid at the top
c. under more pressure than the fluid at the top
d. any of the above, depending on the circumstances

5. The pressure at the bottom of a vessel filled with liquid does not depend on the

a. acceleration of gravity
b. liquid density
c. height of the liquid
d. area of the liquid surface

6. A person stands on a very sensitive scale and inhales deeply. The reading on the scale

a. does not change
b. increases
c. decreases
d. depends on the expansion of the person's chest relative to the volume of air inhaled

7. Buoyancy occurs because, with increasing depth in a fluid,

a. the pressure increases
b. the pressure decreases
c. the density increases
d. the density decreases

8. In order for an object to sink when placed in water, its average specific gravity (relative density) must be

a. less than 1

b. equal to 1
c. more than 1
d. any of the above, depending on its shape

9. A cake of soap placed in a bathtub of water sinks. The buoyant force on the soap is

a. zero
b. less than its weight
c. equal to its weight
d. more than its weight

10. The density of fresh water is 1.00 g/cm^3 and that of seawater is 1.03 g/cm^3. A ship will float

a. higher in fresh water than in seawater
b. at the same level in fresh water and in seawater
c. lower in fresh water than in seawater
d. any of the above, depending on the shape of its hull

11. Bernoulli's equation is based on

a. the second law of motion
b. the third law of motion
c. the conservation of momentum
d. the conservation of energy

12. An express train goes past a station platform at high speed. A person standing at the edge of the platform tends to be

a. attracted to the train
b. repelled from the train
c. attracted or repelled, depending on the ratio between the speed of the train and the speed of sound
d. unaffected by the train's passage

13. The volume of liquid flowing per second out of an orifice at the bottom of a tank does not depend on

a. the area of the orifice
b. the height of liquid above the orifice
c. the density of the liquid
d. the value of the acceleration of gravity

14. The Reynolds number for fluid flow in a pipe does not depend on

a. the length of the pipe
b. the diameter of the pipe
c. the viscosity of the fluid
d. the speed of the fluid

15. Of the following liquids, the one with the highest surface tension is

a. cold water b. hot water
c. soapy water d. alcohol

16. Water neither rises nor falls in a silver capillary. This suggests that the contact angle between water and silver is

a. 0 b. 45°
c. 90° d. 180°

17. A 20-kg piston rests on a sample of gas in a cylinder 10 cm in radius. The absolute pressure on the gas is

a. 0.0064 bar b. 0.062 bar
c. 1.019 bars d. 1.075 bars

18. The input piston of a hydraulic press is 20 mm in diameter, and the output piston is 10 mm in diameter. An input force of 1.0 N will produce an output force of

a. 0.25 N b. 0.50 N
c. 2.0 N d. 4.0 N

19. A manometer (Fig. 10.1b) whose upper end is evacuated and sealed can be used to measure atmospheric pressure. Such an instrument is called a mercury barometer, and the average height of the mercury column in it is 760 mm. If water were used in a barometer instead of mercury, the height of the column of water would be about

a. 56 mm b. 760 mm
c. 10 m d. 20 m

20. A viewing window 30 cm in diameter is installed with its center 3.0 m below the surface of an aquarium tank filled with seawater. The force the window must withstand is approximately

a. 22 N b. 0.22 kN
c. 2.1 kN d. 8.6 kN

21. An aquarium tank 25 cm long and 15 cm wide contains water 10 cm deep but no fish. The gauge pressure on its bottom is

a. 0.98 Pa b. 98 Pa
c. 0.98 kPa d. 9.8 kPa

22. If the tank in Multiple Choice 21 contained fish as well as water, with the water depth the same 10 cm, the pressure at the bottom would be

a. greater
b. the same
c. smaller
d. any of the above, depending on the total mass of the fish

23. An object suspended from a spring scale is lowered into a pail filled to the brim with a liquid, and 4 N of the liquid overflows. The scale indicates that the object weighs 6 N in the liquid. The weight in air of the object is

a. 2 N b. 4 N
c. 6 N d. 10 N

24. A wooden board 2.0 m long, 30 cm wide, and 40 mm thick floats in water with 10 mm of its thickness above the surface. The mass of the board is

a. 1.8 kg b. 18 kg
c. 24 kg d. 176 kg

25. A force of 1000 N is required to raise a concrete block to the surface of a freshwater lake. The force required to lift it out of the water is approximately

a. 700 N b. 1062 N
c. 1140 N d. 1800 N

26. The depth in fresh water at which the water density is 1% greater than its value at the surface is approximately

a. 2.3×10^2 m b. 2.3×10^3 m
c. 2.3×10^4 m d. 2.3×10^5 m

27. The total cross-sectional area of all the capillaries of a certain person's circulatory system is 0.25 m². If blood flows through the system at the rate of 100 cm³/s, the average speed of blood in the capillaries is

a. 0.4 mm/s b. 4 mm/s
c. 25 mm/s d. 400 mm/s

28. A certain person's heart beats 1.2 times per second and pumps 1.0×10^{-4} m³ of blood per beat against an average pressure of 14 kPa. The power output of the heart is

a. 1.2 W b. 1.4 W
c. 1.7 W d. 12 W

29. Water leaves the safety valve of a boiler at a speed of 30 m/s. The gauge pressure inside the boiler is

a. 4.5 millibars b. 1.5 bars
c. 4.5 bars d. 450 bars

30. A horizontal pipe of 10 cm² cross-sectional area is joined to another horizontal pipe of 50 cm² cross-sectional area. The speed of water in the small pipe is 6.0 m/s, and the pressure there is 200 kPa. The speed of water in the large pipe is

a. 0.24 m/s b. 1.2 m/s
c. 3 m/s d. 13 m/s

31. The pressure in the large pipe of Multiple Choice 30 is

a. 183 kPa b. 202 kPa
c. 217 kPa d. 235 kPa

32. The rate of flow of water through the large pipe in Multiple Choice 30 is

a. 0.6 L/s b. 1.2 L/s
c. 3 L/s d. 6 L/s

QUESTIONS

1. A little water is boiled for a few minutes in a tin can, and the can is sealed while it is still hot. Why does the can collapse as it cools?

2. When a person drinks a soda through a straw, where does the force come from that causes the soda to move upward?

3. What is the absolute pressure of the air in a flat tire?

4. A jar is filled to the top with water, and a piece of cardboard is slid over the opening so that there is only water in the jar. If the jar is now turned over, will the cardboard fall off? What will happen if there is any air in the jar?

5. Why are the windows on commercial airliners much smaller than those on light single-engine aircraft?

6. The rolling resistance of a tire is due to its distortion as it rolls. Can you think of two ways to reduce rolling resistance?

7. The three containers shown in Fig. 10.36 are filled with water to the same height. Compare the pressures at the bottoms of the containers.

8. An ice cube floats in a glass of water filled to the brim. What will happen when the ice melts?

9. A wooden block is in such perfect contact with the bottom of a water tank that there is no water beneath it. Is there a buoyant force on the block?

10. Two spheres of the same diameter but of different mass are dropped from a tower. If air resistance is the same for both, which will reach the ground first? Why?

Figure 10.36

Question 7.

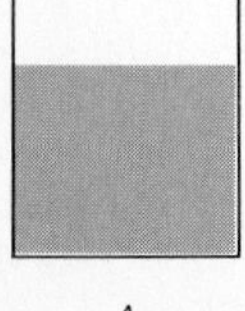

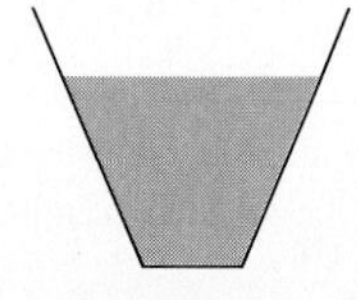

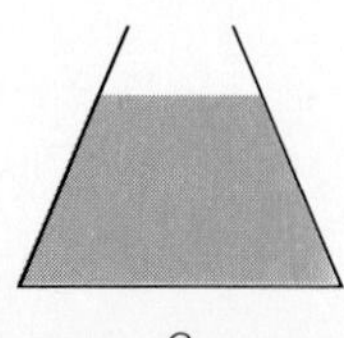

11. In central Sweden a bridge carries the Göta Canal over a highway. What, if anything, happens to the load on the bridge when a boat passes across it?

12. A ship catches fire, and its steel hull expands as a result of the increase in temperature. What happens to the volume of water the ship displaces? What happens to the ship's freeboard, which is the height of its deck above sea level?

13. An aluminum canoe is floating in a swimming pool. After a while it begins to leak and sinks to the bottom of the pool. What happens to the water level in the pool?

14. A helium-filled balloon rises to a certain altitude in the atmosphere and floats there instead of rising indefinitely. Why?

15. Why does a steel ship float when the density of steel is nearly eight times that of water? Why does it sink if it springs a leak and water enters its hull?

16. Will there be any difference between the rates of flow from pipes *A* and *B* in Fig. 10.37? If so, from which pipe will the liquid flow faster?

17. Why is it more dangerous when ice forms on an airplane's wings than when it forms on its fuselage?

Figure 10.37

Question 16.

18. Because its surface is not perfectly smooth, a spinning baseball drags nearby air around with it. Use this fact to explain why a baseball thrown with a spin moves in a curved path.

19. Does a fluid whose density is greater than that of another fluid necessarily have a greater viscosity as well?

EXERCISES

10.1 Pressure

1. A 2.0-kg brick has the dimensions 7.5 cm × 15 cm × 30 cm. Find the pressures exerted by the brick on a table when it is resting on its various faces.

2. The two upper front teeth of a person exert a total force of 30 N on a piece of steak. If the biting edge of each tooth measures 10 mm × 1.0 mm, what is the pressure on the steak?

3. A tire inflated to a gauge pressure of 2.0 bars has a nail 3.0 mm in diameter embedded in it. What is the force tending to push the nail out of the tire?

4. The weight of a car is equally supported by its four tires, each of which has an area of 150 cm^2 in contact with the ground. If the gauge pressure of the air in the tires is 1.8 bars, find the mass of the car.

5. A Super Constellation airplane whose mass is 50,000 kg is in level flight. The area of its wings is 153 m^2. What is the average difference in pressure between the upper and lower surfaces of its wings?

6. A plumber's helper is a rubber cup with a wooden handle. If the diameter of the cup is 125 mm, what is the maximum suction it can exert on a clogged pipe? How much force is needed on the handle to produce this suction?

7. The force on a phonograph needle whose point is 0.10 mm in radius is 0.20 N. What is the pressure in atmospheres that it exerts on the record?

8. A piston weighing 12 N rests on a sample of gas in a cylinder 5.0 cm in diameter. (a) What is the gauge pressure in the gas? (b) What is the absolute pressure in the gas?

9. A hypodermic syringe whose cylinder is 10 mm in diameter and whose needle is 1.0 mm in diameter is used to inject a liquid into a vein in which the blood pressure is 2.0 kPa. Find the minimum force needed on the plunger of the syringe.

10. A 7.0-g bullet leaves the 60-cm barrel of a 0.30-caliber rifle (whose bore diameter is 0.30 in. = 7.62 mm) at 600 m/s. Find the average pressure of the gases in the rifle barrel while the bullet is being accelerated.

10.2 Hydraulic Press

11. A force of 50 N is applied to the input piston of a hydraulic system. The area of the input piston is 60 cm^2, and that of the output piston is 15 cm^2. (a) How much force does the output piston exert? (b) If the input piston moves 2.0 cm per stroke, how far does the output piston move?

12. A lever with a mechanical advantage of 5.0 is attached to the pump piston of a hydraulic press. The area of the pump piston is 10 cm^2, and that of the output piston is 125 cm^2. (a) Find the force the output piston exerts when a force of 120 N is applied to the pump lever. (b) If each stroke of the lever moves the pump piston 3.0 cm, how many strokes are needed to move the output piston 30 cm?

10.3 Pressure and Depth

13. The height of water at two identical dams is the same, but dam *A* holds back a lake containing 2 km^3 of water, whereas dam *B* holds back a lake containing 1 km^3 of water. What is the ratio of the total force exerted on dam *A* to that exerted on dam *B*?

14. The average pressure on a rectangular dam is equal to the pressure at half the height H of the water. Show that the total force on the dam is proportional to H^2.

15. In 1960 the U.S. Navy bathyscaphe *Trieste* descended to a depth of 10,920 m in the Pacific Ocean near Guam. Neglecting the increase in water density with depth, find the pressure on the *Trieste* at the bottom of its dive.

16. In a barometer on a particular day, a mercury column 766 mm high balanced the pressure of the atmosphere. What was that pressure in pascals?

17. A person sucking hard on a thin tube can reduce the pressure in it to 90% of atmospheric pressure. How high can water rise up the tube?

18. The pressure of the blood in a patient's vein is 12 torr. Find the minimum height above the vein that a container of plasma must be held for the plasma to enter the vein. (The actual height is always much greater than this figure in order to achieve a reasonable flow.)

19. A standing woman whose head is 35 cm above her heart bends over so that her head is 35 cm below her heart. What is the change (in torr) in the blood pressure in her head?

20. If the density of the atmosphere were uniform at sea-level density instead of decreasing with altitude, how thick a layer would it form?

21. A submarine is at a depth of 30 m in seawater. The interior of the submarine is maintained at 1 atm pressure. Find the force that must be withstood by a square hatch 80 cm on a side.

22. Water and olive oil rise to the indicated heights in the device shown in Fig. 10.38. Find the density of olive oil.

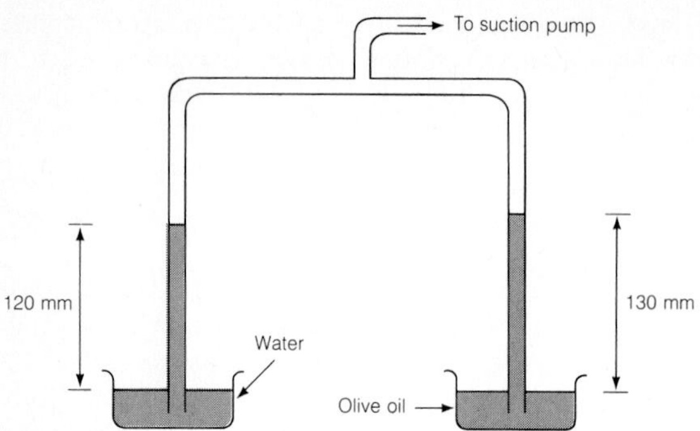

Figure 10.38
Exercise 22.

10.4 Buoyancy

23. A wooden board 10.0 cm thick floats in water with its upper surface parallel to and 3.0 cm above the water surface. What is the density of the wood?

24. A cube of oak (relative density 0.72) 10 cm on each edge is floating in fresh water with one face parallel to the water surface. Find the height of this face above the water.

25. The density of a person is slightly less than that of water. Assuming that these densities are the same, find the buoyant force of the atmosphere on a 70-kg person.

26. A barge 35 m long and 6.0 m wide has a mass of 2.0×10^5 kg. What depth of seawater is needed to float it?

27. A woman whose mass is 60 kg stands on a rectangular swimming raft 3.0 m long and 2.0 m wide that is floating in fresh water. By how much does the raft rise after she dives off?

28. Six empty 200-L drums whose mass is 20 kg each are used as floats for a boat dock in a freshwater lake. The dock platform is 1.8×9.0 m and is made of pine planks 40 mm thick. What is the maximum number of 75-kg people who can stand on the platform without getting their feet wet? The density of pine is 370 kg/m^3.

29. When a 500-g statue of a falcon suspended from a spring scale is immersed in water, the scale reads 400 g, and when it is suspended in benzene, the scale reads 412 g. What is the density of benzene?

30. How much force is needed to raise a 200-kg iron anchor when it is submerged in seawater?

31. A 30-kg balloon is filled with 100 m^3 of hydrogen. How much force is needed to hold it down?

32. A design has been proposed for a modern helium-filled airship that is to have a useful lift of 5.0×10^5 kg at sea level. (a) How many m^3 of helium must the airship contain? (b) The length of the airship would be 400 m. If it were a cylinder, what would its diameter be?

33. The largest rigid airship ever built was the German *Hindenburg*, which was 245 m long and had a capacity of 2.0×10^5 m^3 of hydrogen. It was destroyed in a fire at Lakehurst, N.J., in 1937 after crossing the Atlantic. What was the total payload of the *Hindenburg* including its structure but not including the hydrogen it contained?

34. A steel tank whose capacity is 200 L has a mass of 36 kg. (a) Will it float in seawater when empty? (b) When filled with fresh water? (c) When filled with gasoline?

35. A paperweight has an apparent mass of 70 g when immersed in water and 76 g when immersed in oil of density 0.70 g/cm^3. Find the mass and volume of the paperweight.

10.5 Fluid Flow

36. A pipe whose inside diameter is 30 mm is connected to three smaller pipes whose inside diameters are 15 mm each. If the liquid speed in the larger pipe is 1.0 m/s, what is its speed in the smaller pipes?

37. Water flows at a speed of 1.0 m/s through a hose whose internal diameter is 10 mm. What should the diameter of the nozzle be if the water is to emerge at 5.0 m/s?

38. Half a liter of gasoline emerges per second from a fuel hose at a speed of 50 cm/s. What is the diameter of the nozzle?

39. The left ventricle of a certain running man pumps 20 L of blood per minute into his aorta at an average pressure of 140 torr. Find the total power output of his heart under the assumption that his right ventricle has an output 20% as great as that of his left ventricle.

40. The power output of an athlete's heart during strenuous activity might be 8.0 W. If the blood pressure increases by 150 torr when it passes through the athlete's heart, find the corresponding rate of flow of blood in liters per minute.

10.6 Bernoulli's Equation

10.7 Applications of Bernoulli's Equation

41. A water tank has a hole 3.0 cm^2 in area located 2.0 m below the water surface. How many kilograms of water emerge from the hole per second?

42. A tank contains gasoline under a gauge pressure of 3.0 bars. Neglecting the difference in height between the liquid surface and the valve, determine at what speed the gasoline emerges when the valve is opened.

43. During the pumping phase of the heart's action, the gauge pressure of blood in the major arteries of a normal person at the level of the heart is about 120 torr. If one of these arteries is cut and blood spurts out vertically, how high will it go?

44. Figure 10.39 shows a hose whose internal cross-sectional area is 2.0 cm^2 being used to siphon water out of a tank. Find the speed with which the water emerges and the rate of flow in L/s.

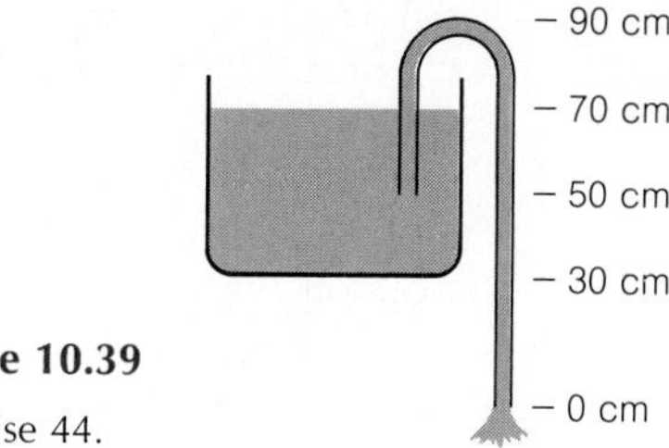

Figure 10.39

Exercise 44.

45. Kerosene leaks out of a hole 12 mm in diameter at the bottom of a tank at a rate of 40 L/min. How high is the kerosene in the tank?

10.8 Viscosity

46. A grease nipple on a car has a hole 0.50 mm in diameter and 5.0 mm long. If the viscosity of the grease used is 80 $N \cdot s/m^2$, what is the pressure needed to force 1.0 cm^3 of grease into the nipple in 10 s? How many times atmospheric pressure is this?

47. Find the pressure gradient needed to pump 16 L of water at 20°C per minute through a pipe 6.0 mm in radius. Take the viscosity of the water into account.

48. A hypodermic syringe has a cylinder 8.0 mm in diameter and a needle 0.40 mm in diameter and 30 mm long. How much force is needed on the plunger to send water at 20°C through the needle at 0.10 cm^3/s?

49. A typical capillary blood vessel is 1.0 mm long and has a radius of 2.0 μm. (a) If the pressure difference between its ends is 20 torr, what is the average speed of the blood that flows through such a capillary? (b) If the heart pumps 80 cm^3 of blood through a person's body per second, how many capillaries does the body contain?

10.9 Reynolds Number

50. Water at 40°C flows at 10 cm/s through a tube 10 mm in diameter. What is the Reynolds number of the system? Is the flow laminar or turbulent?

51. At what speed should air at 20°C flow through a tube for the Reynolds number to be the same as that of water at 20°C flowing through the tube at 20 cm/s? Assume that the viscosity of air at 20°C is 1.8×10^{-5} Pl.

10.10 Surface Tension

52. Ethyl alcohol rises 14.2 mm in a glass capillary tube of radius 0.40 mm. The contact angle is 0. Find the surface tension of ethyl alcohol.

53. Find the radius of a glass capillary tube in which water at 20°C rises 20 mm.

54. A glass capillary tube whose radius is 0.50 mm is dipped in a dish of mercury, whose surface tension is 0.44 N/m. The contact angle is 140°. Find the height of the mercury in the capillary relative to the mercury surface in the dish.

PROBLEMS

55. Obtain Archimedes' principle by setting the work involved in raising an object of mass m and volume V through the height h in a fluid equal to the change in the potential energy of the system of object + fluid.

56. Obtain Archimedes' principle by showing that the vector sum of the forces acting on the faces of a cube submerged in a fluid is upward in direction and has a magnitude equal to the weight of the fluid displaced. For simplicity assume that a face of the cube is parallel to the fluid surface.

57. Calculate the density of seawater at a depth of exactly 5 km. Use the bulk modulus of water given in Table 9.4.

58. A person's head is 40 cm above his heart. The average arterial blood pressure at the level of the heart is 100 torr. Find the average arterial pressure in the person's head when standing and when lying down in an elevator accelerated upward at 3.0 m/s^2.

59. A lead fishing sinker suspended from a spring scale is immersed in oil whose density is 820 kg/m^3. The reading on the scale is 0.2 N. Find the mass of the sinker.

60. A helium-filled weather balloon 2.0 m in radius whose total mass including instruments is 20 kg is tethered to the ground by a rope. A wind causes the rope to make an angle of 25° with the vertical. Find the tension in the rope.

61. A sailboat has 8000 kg of lead ballast attached to the bottom of its keel, the center of gravity of which is 1.5 m below the water surface when the boat is level. When the boat is tilted by 20° from the vertical, what is the torque exerted by this ballast about an axis along the length of the boat in the plane of the water surface? Assume seawater.

62. Water emerges from a fire hose at a speed of 20 m/s and a rate of flow of 50 L/s. (a) Find the force with which the nozzle must be held. (b) Find the required power of the pump motor, assuming 50% overall efficiency.

63. A tank truck filled with gasoline springs a leak 1.5 m below the liquid surface. If the hole has an area of 100 cm^2 and is at the back of the tank where its wall is vertical, what is the force acting on the truck?

64. A 1360-hp pump throws a jet of water 130 m into the air in Geneva, Switzerland. (a) With what speed does the water leave the mouth of the fountain? (b) If the overall efficiency is 60%, how many kilograms of water per minute are thrown into the air?

65. A horizontal stream of water leaves an orifice 1.0 m above the gound and strikes the ground 2.0 m away. What is the gauge pressure behind the orifice?

66. An open tank of height H that is filled with water has a hole the distance y below the top. (a) How far from the tank does the water strike the ground? (b) At what value of y is the distance a maximum?

67. A horizontal pipe 2.0 mm in radius at one end gradually increases in size so that it is 5.0 mm in radius at the other end. The pipe is 4.0 m long. Water is pumped into the small end of the pipe at a speed of 8.0 m/s and a pressure of 2.0 bars. (a) Find the speed and pressure of the water at the pipe's large end. (b) The pipe is turned so as to be vertical with the small end underneath, so the water flows upward. Find the speed and pressure of the water at the large end of the pipe now.

68. A horizontal pipe of 45 cm^2 cross-sectional area is joined to another horizontal pipe of 15 cm^2 cross-sectional area; their total length is 2.0 m. Water at 20°C flows through the pipe at 18 L/s, and the pressure in the large pipe is 200 kPa. (a) What is the speed of the water in each pipe? (b) What is the nature of the flow in each pipe? (c) What is the pressure in the small pipe? (d) The pipes are turned so as to be vertical with the large end down, and the water flows upward at the same rate. What is the pressure at the upper end of the small pipe now?

69. Water flows upward through the pipe shown in Fig. 10.40 at 40 L/s. If the pressure at the lower end is 80 kPa, find the water speed at both ends and the pressure at the upper end.

70. Gasoline enters the pipe shown in Fig. 10.41 from the left at 3 m/s and 150 kPa. Find the speed and pressure at which gasoline leaves the pipe and the rate of flow in kilograms per second.

71. The essential features of Poiseuille's law can be obtained by considering the dimensions of the quantities involved. It is reasonable to expect that the rate of flow R depends in some way on the pipe radius r, the viscosity η, and the pressure gradient $\Delta p/L$. We can try to express this by writing $R = K\,\eta^a r^b (\Delta p/L)^c$, where a, b, and c are unknown exponents and K is a dimensionless constant. (a) Express each side of this equation in terms of the dimensions of each quantity (for instance, the dimensions of R are L^3T^{-1}). (b) Set equal the exponents of L, T, and M on both sides of the result. (c) Finally, solve for a, b, and c. The value of K cannot be found in this way but requires a proper derivation.

Figure 10.40
Problem 69.

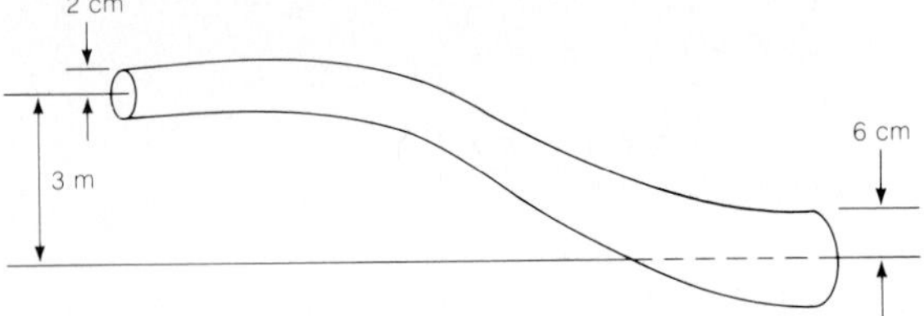

Figure 10.41
Problem 70.

72. The drag force on an object moving through a fluid at a speed v low enough for turbulence not to occur is proportional to both v and the viscosity η of the fluid, so that $F_{drag} = K\eta v$. The constant K depends on the size and shape of the object, but not on its composition, because a thin layer of fluid sticks to its surface so that the friction that occurs is between successive layers of fluid. In the case of a spherical object of radius r, $K = 6\pi r$. The resulting formula for F_{drag} is called **Stokes's Law.** (a) Verify the plausibility of Stokes's law by the method of dimensional analysis described in Problem 71. (b) Verify that a sphere of density ρ and radius r falling freely in a fluid of viscosity η and density ρ' has a terminal speed of $2r^2g(\rho - \rho')/9\eta$. (c) An iron ball is dropped into a tank of water at 20°C, and another identical ball is dropped into a tank of ethyl alcohol at the same temperature. Which ball will have the greater terminal speed?

73. The height h by which a liquid rises (or falls) between two vertical parallel plates the distance s apart when placed in a pan of the liquid is given by $h = 2\gamma \cos\theta/s\rho g$. (a) Obtain this result in the same way Eq. (10.19) was derived by considering a parcel of elevated liquid h high, s wide, and x long. (b) Two vertical glass plates 1 mm apart are dipped into a pan of water at 20°C. Find the height to which the water rises.

74. What is the minimum inside diameter the tube of a mercury barometer should have if the error due to capillarity is not to exceed 0.25%? Will the error be in the direction of lower or higher pressure than the true value? Mercury has a surface tension of 0.44 N/m, and the contact angle between glass and mercury is 140°.

Answers to Multiple Choice

1. b	**9.** b	**17.** d	**25.** d
2. b	**10.** c	**18.** a	**26.** b
3. d	**11.** d	**19.** c	**27.** a
4. c	**12.** a	**20.** c	**28.** c
5. d	**13.** c	**21.** c	**29.** c
6. d	**14.** a	**22.** b	**30.** b
7. a	**15.** a	**23.** d	**31.** c
8. c	**16.** c	**24.** b	**32.** d

CHAPTER

HARMONIC MOTION

Many events in both nature and technology are periodic, with a certain motion repeating itself over and over again. As its name suggests, **simple harmonic motion** is the most basic kind of oscillation. A tree swaying in the wind, a child on a swing, the pistons of a car engine, the vibrating atoms in a solid—all are undergoing simple harmonic motion, or very nearly so. In fact, *every* periodic event is either simple harmonic or else the result of several such motions mixed together.

In harmonic motion the energy of the vibrations goes back and forth between kinetic and potential forms. The potential energy may be elastic rather than gravitational; we will find later that the notion of potential energy has even wider scope. The electrical equivalent of potential energy, in particular, can lead to electrical oscillations that closely resemble harmonic motion. It is these oscillations that make possible the electronic recording and transmission of sight and sound.

11.1 Elastic Potential Energy

The work done to deform an elastic object.

Open the lid of a jack-in-the-box and out pops Jack. How fast does he move? How high does he go? Not a very dignified problem for a serious science like physics, one might think, but the principles behind a jack-in-the-box apply equally well to any other system that stores elastic potential energy.

When we pull out a spring, it resists being stretched, and if we then let go, the spring returns to its original length. As we know, this is typical of elastic behavior. On the other hand, when we pull out a piece of taffy, it also resists being stretched, but if we then let go, nothing happens: The deformation is permanent. This is typical of plastic behavior.

In the case of the stretched spring, a **restoring force** comes into being that tries to return the spring to its normal length. The more we stretch the spring, the more the restoring force we must overcome (Fig. 11.1). Exactly the same thing happens when we compress the spring: It resists being shortened, and if we let go, the spring goes back to its normal length. Again a restoring force arises, and again the more the compression, the more the restoring force to be overcome.

The amount x by which an elastic solid is stretched or compressed by a force is directly proportional to the magnitude F of the force, provided the elastic limit is not exceeded. This proportionality is called Hooke's law, as mentioned in Chapter 9. Thus we can write.

$$F = kx \qquad \textit{Hooke's law} \quad (11.1)$$

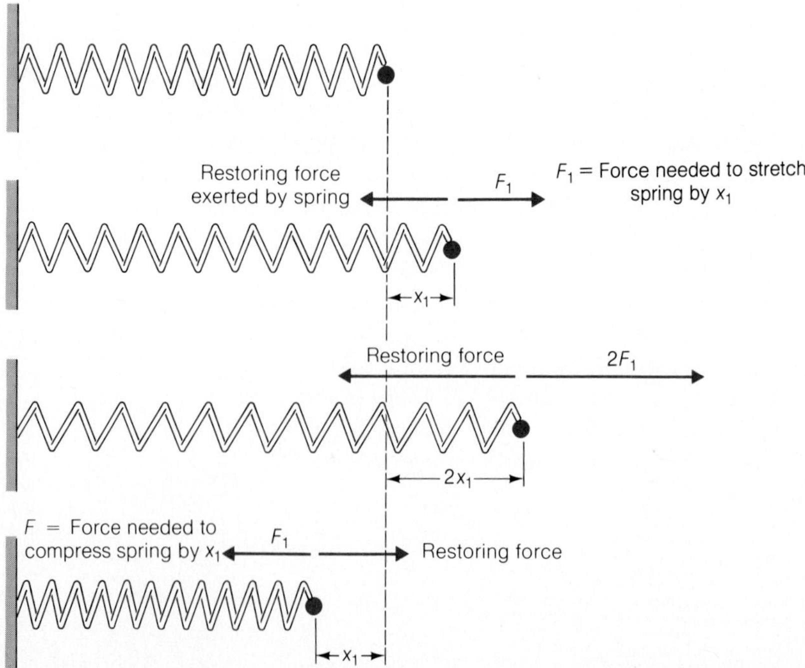

Figure 11.1

When a spring (or other elastic body) is stretched or compressed, a restoring force comes into being that tries to return the spring to its normal length.

where k is a constant whose value depends on the nature and dimensions of the object. A stiff spring has a higher value of k than a weak one.

EXAMPLE 11.1

A spring of force constant k_1 is connected end-to-end to a spring of force constant k_2, as in Fig. 11.2. (a) What is the force constant k of the set of springs? (b) If $k_1 =$ 10 N/m and $k_2 =$ 30 N/m, what is k?

SOLUTION (a) When a force F is applied to the set, each spring is acted on by F. Hence the elongations of the springs are, respectively,

$$x_1 = \frac{F}{k_1} \qquad x_2 = \frac{F}{k_2}$$

and the total elongation of the set is

$$x = x_1 + x_2 = \frac{F}{k_1} + \frac{F}{k_2} = F\frac{(k_1 + k_2)}{k_1 k_2}$$

Because $F = kx$ for the set, its force constant is

$$k = \frac{F}{x} = \frac{k_1 k_2}{k_1 + k_2}$$

(b) $$k = \frac{(10\ \text{N/m})(30\ \text{N/m})}{10\ \text{N/m} + 30\ \text{N/m}} = 7.5\ \text{N/m}$$

The force constant of the set is smaller than either k_1 or k_2.

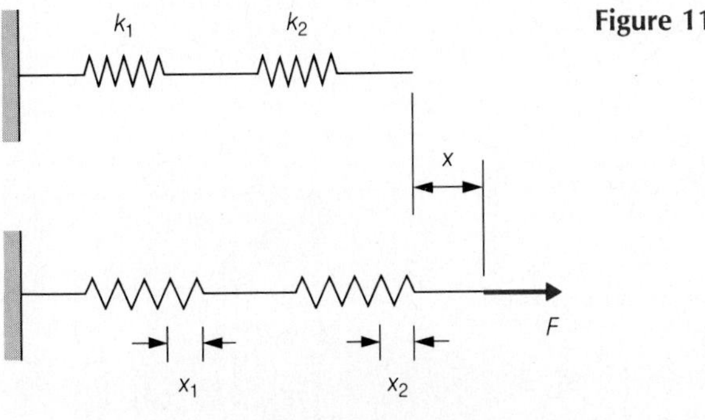

Figure 11.2

◆

The work needed to stretch (or compress) an object that obeys Hooke's law is easy to figure out. The work done by a force is the product of the magnitude of the force and the distance through which it acts. Here the force used in stretching the object is not constant but is proportional to the elongation x at each point in the stretching process. Because F is proportional to x, the *average* force $\overline{F}$ applied while

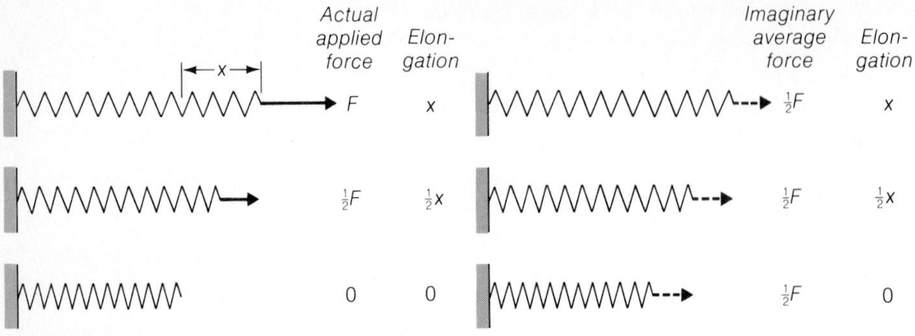

Figure 11.3

To compute the work done in stretching a body that obeys Hooke's law, the varying force that actually acts during the expansion may be replaced by the average force.

the body is stretched from its normal length by an amount x to its final length is

$$\overline{F} = \frac{F_{\text{initial}} + F_{\text{final}}}{2} = \frac{0 + kx}{2} = \tfrac{1}{2}kx$$

since the initial force is 0 and final force is kx (Fig. 11.3). The work done in stretching the spring is the product of the average force $\overline{F} = \frac{1}{2}kx$ and the total elongation x, so that

$$W = \text{PE} = \tfrac{1}{2}kx^2 \qquad \textit{Elastic potential energy} \quad (11.2)$$

This formula is most often used in connection with springs: To stretch (or compress) a spring whose force constant is k by an amount x from its normal length requires $\frac{1}{2}kx^2$ of work to be done. This work goes into **elastic potential energy.** When the spring is released, its potential energy of $\frac{1}{2}kx^2$ is transformed into kinetic energy or into work done on something else (Fig. 11.4). Work done against frictional forces within the spring itself always absorbs some fraction of the available potential energy.

Figure 11.4

Some devices that make use of elastic potential energy in their operation.

EXAMPLE 11.2

The horizontal spring shown in Fig. 11.5 has a force constant k of 90 N/m. Attached to the free end of the spring is a 1.4-kg block. If the spring is pulled out 50 cm from its equilibrium position and then released, what will the block's speed be when it returns to the equilibrium position?

SOLUTION When the spring is released, its elastic potential energy starts to be converted into kinetic energy of the block. We will assume that the spring's mass is small compared with that of the block and that its internal friction may be neglected. At the equilibrium position of the spring, $x = 0$, and all the initial potential energy of $\frac{1}{2}kx^2$ is now kinetic energy $\frac{1}{2}mv^2$. Hence

$$\frac{1}{2}mv^2 = \frac{1}{2}kx^2$$

$$v = \sqrt{\frac{k}{m}}x$$

$$= \sqrt{\frac{90 \text{ N/m}}{1.4 \text{ kg}}}(0.50 \text{ m}) = 4.0 \text{ m/s}$$

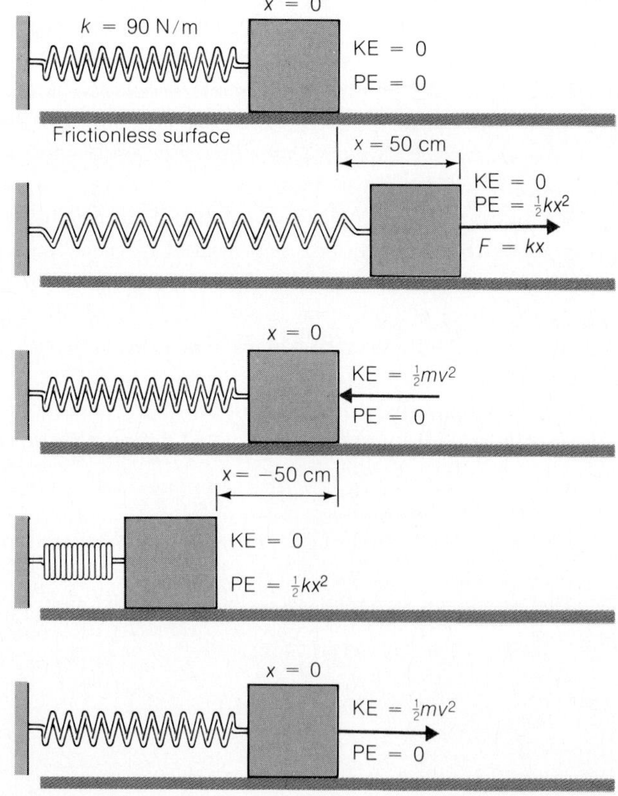

Figure 11.5

A 1.4-kg block attached to a spring whose force constant is 90 N/m is pulled 50 cm from its equilibrium position. When the spring is released, its elastic potential energy of $\frac{1}{2}kx^2$ is converted into kinetic energy $\frac{1}{2}mv^2$. As the block's momentum compresses the spring on the other side of the equilibrium position, the kinetic energy is converted back into elastic potential energy.

◆

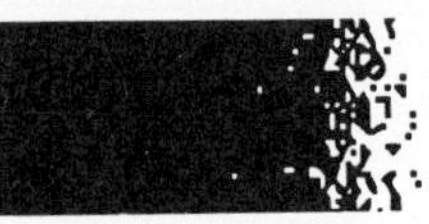

11.2 Simple Harmonic Motion

The energy of an oscillator shifts back and forth between PE and KE.

When a spring with something attached to it is pulled out and released, it does not merely go back to its equilibrium position and stop there. What happens is that the elastic potential energy $\frac{1}{2}kx^2$ of the spring is first turned into kinetic energy $\frac{1}{2}mv^2$ of the moving object. As the object's momentum compresses the spring on the other side of the equilibrium position, this KE is turned back into elastic PE (see Fig. 11.5). The compression $-x$ will have the same magnitude as the original extension x because the PE in both cases is the same and

$$\tfrac{1}{2}kx^2 = \tfrac{1}{2}k(-x)^2$$

Left to itself without friction, the spring-object system will keep oscillating back and forth indefinitely. The behavior of such a system is called **simple harmonic motion** (SHM).

Simple harmonic motion occurs whenever a force proportional to the displacement acts on a body in the opposite direction to its displacement from its normal position. The elastic restoring force of a stretched or compressed spring always tends to return the spring to its normal length, but the momentum of the moving mass compels it to overshoot in one direction and then in the other and thus to oscillate.

The **period** T of a system undergoing SHM is the time required for it to make one complete oscillation. (A complete oscillation is often called a **cycle**.) In the case of a spring, the period is the time the spring spends in going from its maximum extension, say, through its maximum compression and back to its maximum extension once more, as in Fig. 11.6.

To find out how the position of an oscillating object varies with time, we can attach a pen to the object and pull a sheet of paper past it at a constant speed (Fig. 11.7). The graph has the same form as that of sin θ plotted against θ, and accordingly the curve is said to be *sinusoidal*. Sinusoidal variations are found elsewhere in physics also—for example, in wave motion and in connection with alternating electric currents.

The maximum displacement A of an object undergoing harmonic motion on either side of its equilibrium position is called the **amplitude** of the motion:

$$A = x_{\max}$$

Amplitude = maximum displacement from equilibrium position

The period of all types of simple harmonic motion is given by the same formula,

$$T = 2\pi\sqrt{-\frac{x}{a}} \qquad \textit{Simple harmonic motion} \quad (11.3)$$

$$\text{Period} = 2\pi\sqrt{-\frac{\text{displacement}}{\text{acceleration}}}$$

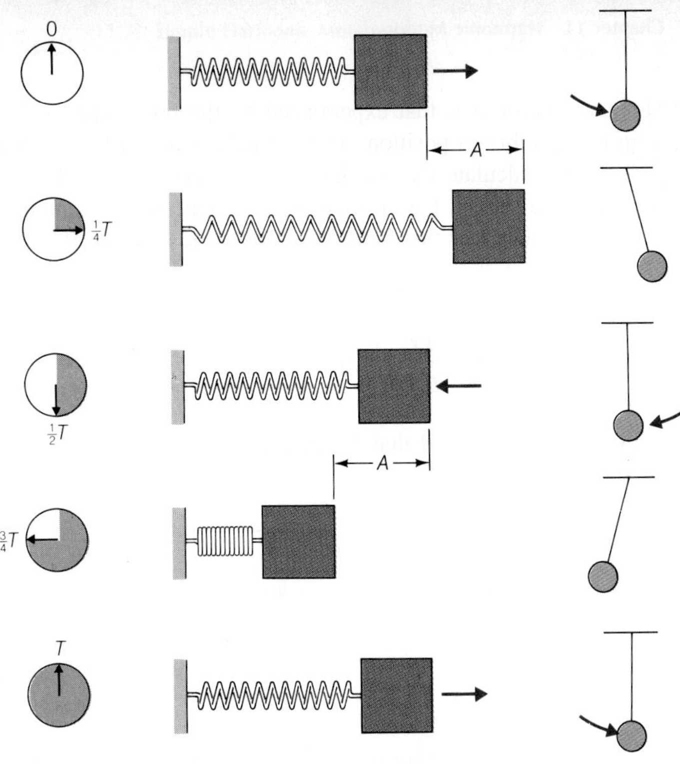

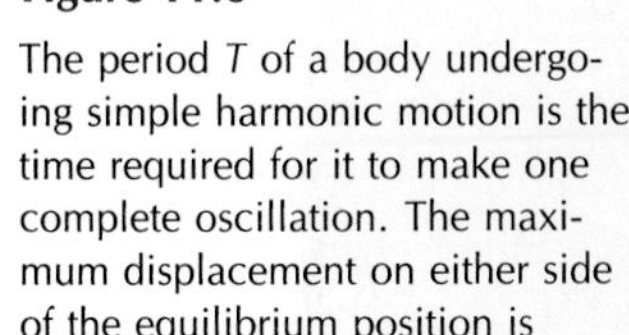

Figure 11.6

The period T of a body undergoing simple harmonic motion is the time required for it to make one complete oscillation. The maximum displacement on either side of the equilibrium position is called the *amplitude A* of the motion.

Figure 11.7

A pen attached to an oscillating object can be used to trace out a graph of the position of the object versus time on a sheet of paper pulled past the pen at a constant speed. The resulting graph has the same form as that of sin θ plotted against θ, and accordingly the curve is said to be *sinusoidal*.

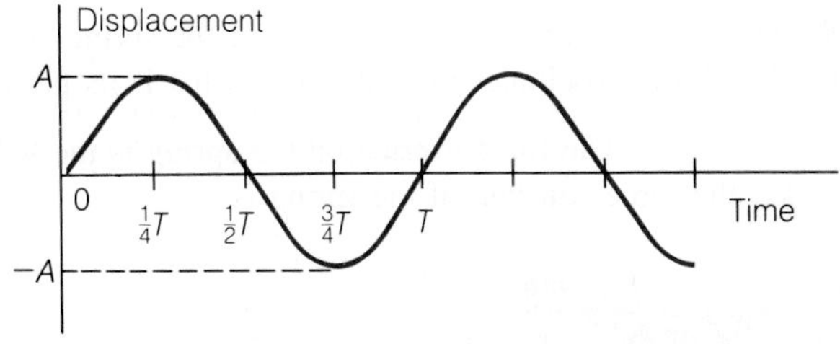

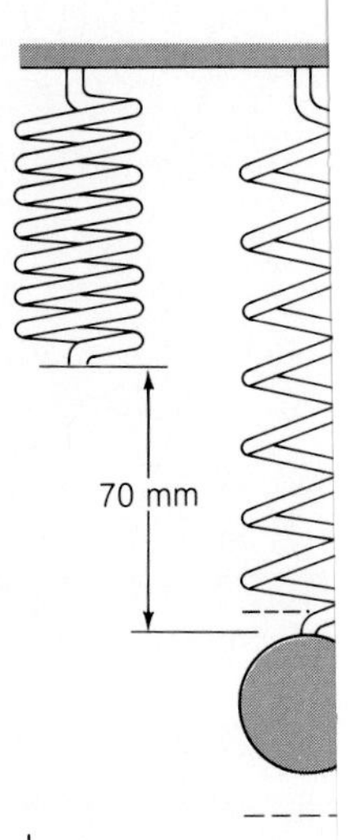

The period of the

$T = 2\pi$

We did not have t
$x/g = m/k$ here. T

$f = \frac{1}{T} =$

EX

A U-shaped glass
surface tension are
One side of the co
forth motion of the

SOLUTION The
sectional area of t
distance x, the othe
the gauge pressure
restoring force is
x is small compare

$a = \frac{F}{m} =$

Because F is proportional to $-x$, the motion is simple harmonic, and from Eq. (11.3) its period is

$$T = 2\pi\sqrt{-\frac{x}{a}} = 2\pi\sqrt{\frac{L}{2g}}$$ ◆

11.3 A Model of Simple Harmonic Motion

How to derive the formula for the period of SHM.

To go directly from the formula for restoring force $F_r = -kx$ to Eq. (11.3) for the period T of SHM requires the use of calculus. However, simpler math can make the same link if we are clever enough to find a suitable model. Such a model is shown in Fig. 11.10.

The model makes use of a particle moving in a vertical circle at constant speed. The particle is lit from above, and it casts a shadow on a horizontal screen below its orbit. As the particle goes around the circle, its shadow oscillates back and forth. The shadow moves fastest at the center, slows down as it approaches each end of the path, comes to a stop, and then reverses its direction. We might suspect that

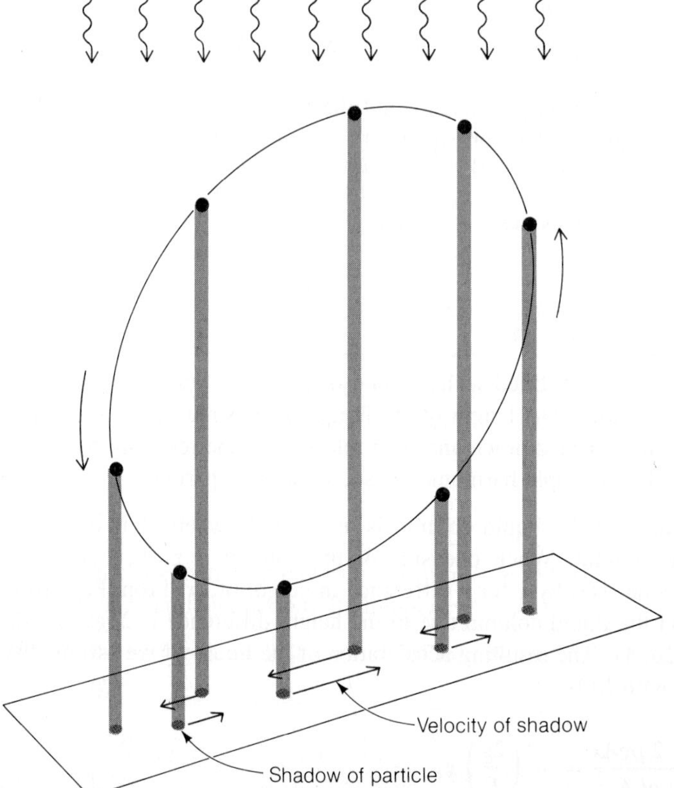

Figure 11.10
The shadow of a particle undergoing uniform circular motion executes simple harmonic motion.

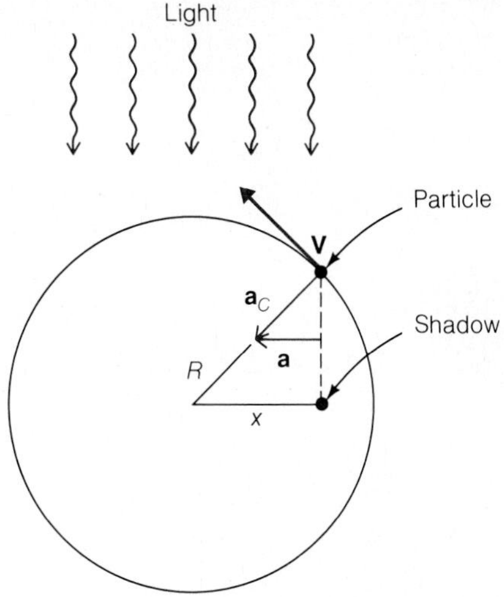

Figure 11.11

The acceleration of the shadow is the horizontal component of the particle's centripetal acceleration. The shadow's acceleration is proportional to its displacement x and is in the opposite direction.

the shadow is executing SHM. To verify this suspicion, we must show that the acceleration a of the shadow at any time is proportional to its displacement x from the center of its path and opposite in direction.

The acceleration of the shadow at any point is the horizontal component of the particle's acceleration a_c. As we know, a particle in uniform circular motion at the speed V in a circle of radius R experiences the centripetal acceleration

$$a_c = -\frac{V^2}{R}$$

where the minus sign indicates that the acceleration is inward toward the center of the circle. The horizontal component of this acceleration is the acceleration a of the shadow. Because corresponding sides of similar triangles are proportional, we see from Fig. 11.11 that

$$\frac{a}{a_c} = \frac{x}{R} \quad \text{and so} \quad a = \frac{x}{R}a_c$$

Because $a_c = -V^2/R$, the acceleration of the shadow is

$$a = -\frac{x}{R}\frac{V^2}{R} = -\frac{V^2}{R^2}x \tag{11.7}$$

Thus the shadow's acceleration is proportional to its displacement x and in the opposite direction. This means that the shadow is indeed executing SHM.

Now let us use this model to find the period of an object in SHM. The circumference of a circle of radius R is $2\pi R$, and a particle moving around the circle with the constant speed V covers this distance in the time

$$T = \frac{2\pi R}{V}$$

From Eq. (11.7) we find that

$$\frac{R}{V} = \sqrt{-\frac{x}{a}}$$

and so

$$T = 2\pi\sqrt{-\frac{x}{a}} \qquad (11.3)$$

This is the general formula for the period of SHM given earlier.

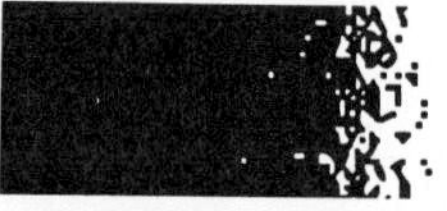

11.4 Position, Speed, and Acceleration

Formulas that apply to any harmonic oscillator.

The principle of conservation of energy gives us a way to express the speed of an object in SHM in terms of its frequency f, amplitude A, and displacement x. The total energy of the oscillator (object plus spring) is the sum of its kinetic and potential energies at any time, which are, respectively, $\frac{1}{2}mv^2$ and $\frac{1}{2}kx^2$. At either end of the motion, when $x = +A$ or $x = -A$, the object is stationary and has only the potential energy $\frac{1}{2}kA^2$. Hence

$$\begin{aligned} \text{Total energy} &= \text{KE} + \text{PE} \\ \tfrac{1}{2}kA^2 &= \tfrac{1}{2}mv^2 + \tfrac{1}{2}kx^2 \\ mv^2 &= k(A^2 - x^2) \\ v &= \sqrt{k/m}\sqrt{A^2 - x^2} \end{aligned}$$

From Eqs. (11.5) and (11.6) we know that

$$f = \frac{1}{T} = \frac{1}{2\pi}\sqrt{\frac{k}{m}}$$

which can be rewritten as

$$\sqrt{k/m} = 2\pi f \qquad (11.8)$$

The speed of the object when it has the displacement x is accordingly

$$v = 2\pi f\sqrt{A^2 - x^2} \qquad \textit{Speed at given displacement} \quad (11.9)$$

This formula gives only the absolute value of v; whether the sign of v is + or − depends on whether the body is at $+x$ or $-x$ and on whether it is on its way toward or away from the equilibrium position from there.

From Eq. (11.9) we see that the maximum speed v_{max} of the object, which occurs at the equilibrium position when $x = 0$, is

$$v_{max} = 2\pi fA \qquad \textit{Maximum speed} \quad (11.10)$$

The maximum speed is proportional to both the frequency and the amplitude of the motion.

The energy of an oscillating object shifts back and forth between kinetic and potential forms. To find the total energy, we can calculate either KE_{max} or PE_{max}, with the help, respectively, of Eq. (11.10) or (11.8):

$$KE_{max} = \tfrac{1}{2}mv^2_{max} = 2\pi^2 mf^2A^2 \qquad \textit{Total energy} \quad (11.11)$$

$$PE_{max} = \tfrac{1}{2}kA^2 = 2\pi^2 mf^2A^2 \qquad (11.12)$$

The total energy depends on the square of the frequency and the square of the amplitude. Note that the total energy is not the sum of Eqs. (11.11) and (11.12), because when KE or PE reaches its maximum, the other is zero.

To find the acceleration of an object in SHM, we refer back to Eq. (11.4), which states that $a = -kx/m$. On the basis of Eq. (11.8) this formula becomes

$$a = -4\pi^2 f^2 x \qquad \textit{Acceleration at given displacement} \quad (11.13)$$

The acceleration is always opposite in direction to the displacement, which, of course, is one of the conditions for SHM to occur. The maximum acceleration occurs at either end, when $x = \pm A$, and has the magnitude

$$a_{max} = 4\pi^2 f^2 A \qquad \textit{Maximum acceleration} \quad (11.14)$$

The maximum acceleration is proportional to the square of the frequency and to the amplitude.

Although these formulas were derived for the case of a vibrating spring, their validity is perfectly general and they apply to any type of harmonic oscillator, from the bob of a pendulum to an atom in a molecule.

Figure 11.12

When the downward acceleration of the piston exceeds g, the coin will be left behind by the piston on its downstroke.

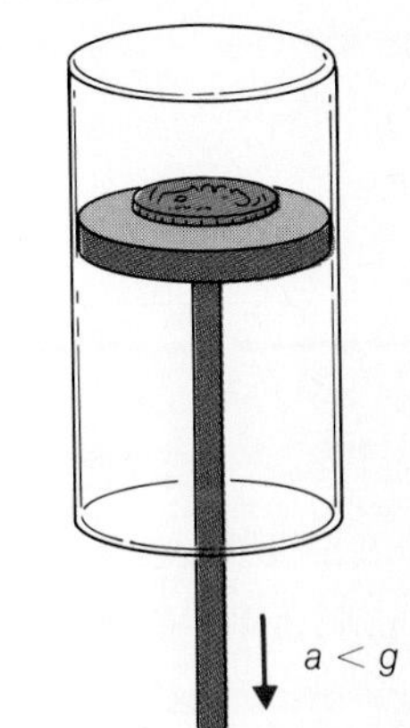

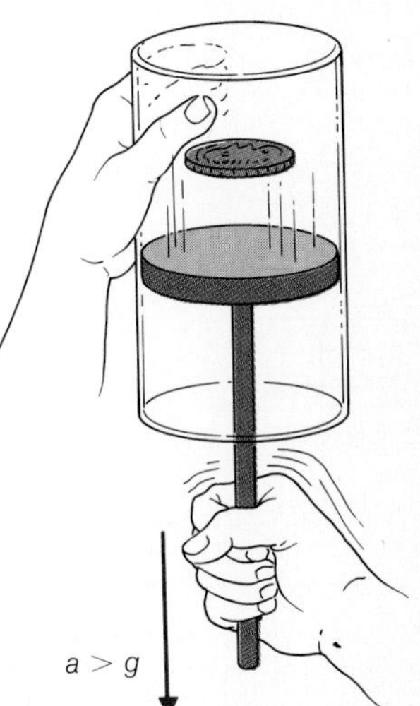

EXAMPLE 11.5

In a car engine, each piston moves up and down in an approximation of SHM. Given that a certain piston has a mass of 0.50 kg, a total travel (''stroke'') of 120 mm, and a frequency of oscillation of 60 Hz (corresponding to 3600 rpm), find the maximum force it experiences, assuming SHM.

SOLUTION The maximum force is given by $F_{max} = ma_{max}$. Here, because the amplitude is half the total travel, $A = 60 \text{ mm} = 0.060 \text{ m}$, and

$$F_{max} = ma_{max} = 4\pi^2 mf^2A = 4\pi^2(0.50 \text{ kg})(60 \text{ Hz})^2(0.060 \text{ m}) = 4264 \text{ N}$$
$$= 4.3 \text{ kN}$$

This is nearly 1000 lb. ◆

EXAMPLE 11.6

A piston undergoes SHM in a vertical direction with an amplitude of 60 mm. A coin is placed on top of the piston (Fig. 11.12). What is the lowest frequency at which the coin will be left behind by the piston on its downstroke?

SOLUTION The coin will leave the piston when the piston's downward acceleration exceeds the acceleration of gravity g. The maximum downward acceleration of the piston occurs at the highest point of its motion, when $a_{max} = 4\pi^2 f^2 A$ according to Eq. (11.14). Hence we set a_{max} equal to g and solve for the frequency f:

$$a_{max} = g = 4\pi^2 f^2 A$$

$$f = \frac{1}{2\pi}\sqrt{\frac{g}{A}}$$

$$= \frac{1}{2\pi}\sqrt{\frac{9.8 \text{ m/s}^2}{0.060 \text{ m}}} = 2.0 \text{ Hz}$$

◆

The variations of the displacement, velocity, and acceleration of a body undergoing SHM are plotted versus time in Fig. 11.13. The graphs are plotted on the assumption that the body is at $x = +A$ when $t = 0$. This corresponds to pulling out the body and letting it go at $t = 0$. At this instant the body's acceleration is a maximum and is opposite in direction to x, while the velocity is zero because the particle has not yet started to move.

When the particle is at the origin, $x = 0$ and the spring is at its normal length. Because the spring exerts no force on the body at this time, its acceleration is zero. The speed of the body is now a maximum because the system has no PE at this point.

When $x = -A$, the body is at the other extreme of its range, and its acceleration, again a maximum, is positive. The acceleration is once more in the direction of the origin, though now from the other side. All the energy of oscillation is potential, and so the body is stationary at this instant.

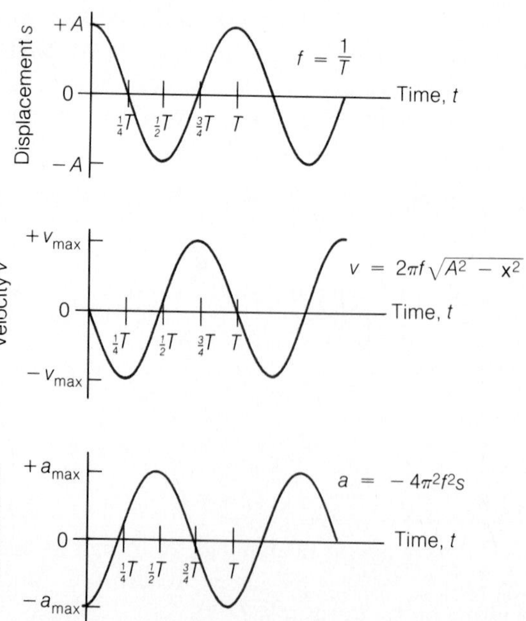

Figure 11.13

Displacement, velocity, and acceleration in simple harmonic motion.

11.5 Trigonometric Notation

How to find x, v, and a at any time.

There are situations in which we need to know the position, the velocity, and the acceleration of a harmonic oscillator as functions of time. Figure 11.14 shows the same model used in Fig. 11.10, with a particle moving in a circle having a shadow that undergoes SHM. The particle has an angular speed of $\omega = 2\pi f$, and ω is called the **angular frequency** of the harmonic motion of its shadow. The unit of ω is, as usual, the radian/s.

The angular displacement of the particle at any time t is $\theta = \omega t$, its linear speed in the circle is ωA, and its centripetal acceleration is $\omega^2 A$, where A is the radius

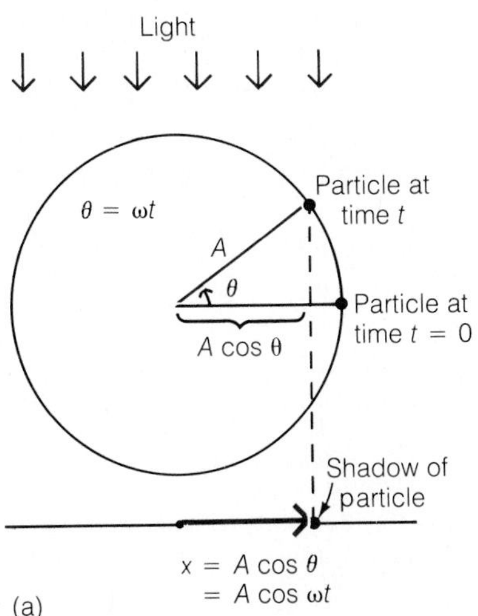

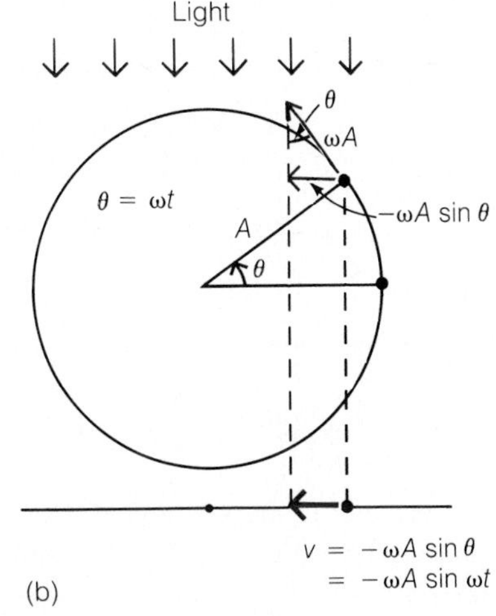

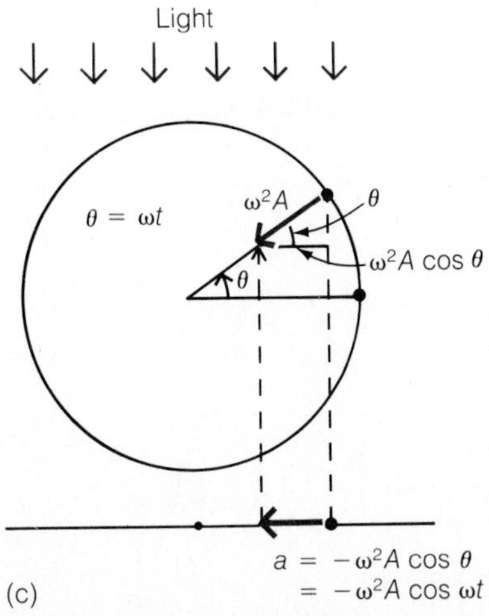

Figure 11.14

(a)–(c). The displacement, velocity, and acceleration of the shadow of the particle of Fig. 11.10 can be expressed in terms of the time t as shown here. The angular speed of the particle is ω, the magnitude of its linear velocity is ωA, and the magnitude of its centripetal acceleration is $\omega^2 A$.

of the circle. From Fig. 11.14 we see that

$$x = A \cos \theta = A \cos \omega t \tag{11.15}$$

$$v = -\omega A \sin \theta = -\omega A \sin \omega t \tag{11.16}$$

$$a = -\omega^2 A \cos \theta = -\omega^2 A \cos \omega t \tag{11.17}$$

The sign convention used is that a quantity directed to the right is considered positive and one directed to the left is considered negative. Graphs of x, v, and a versus time were given in Fig. 11.13.

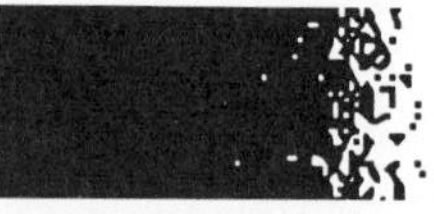

11.6 The Simple Pendulum

Its period is independent of its mass.

A pendulum undergoes SHM as it swings back and forth, provided the arc through which it moves is not too large. We can see why this limitation arises by using Eq. (11.3) to find the period of a pendulum. We assume that the entire mass of the pendulum is concentrated in its bob and that the bob's dimensions are small compared with the pendulum's length. In the next section we will look into more realistic pendulums.

Figure 11.15 shows a pendulum of length L whose bob has the mass m. The weight $m\mathbf{g}$ of the bob, which acts vertically downward, may be resolved into two forces, $\mathbf{T}$ and $\mathbf{F}$, which act, respectively, parallel to and perpendicular to the supporting string L. That is,

$$\mathbf{T} + \mathbf{F} = m\mathbf{g}$$

The force $\mathbf{F}$ is the restoring force that acts to return the bob to the midpoint of its motion. The space triangle hLx and the vector triangle $\mathbf{T}m\mathbf{g}\mathbf{F}$ are similar, with each having a right angle and two sides of one being parallel to the two corresponding sides of the other. Hence

$$\frac{F}{x} = \frac{mg}{L}$$

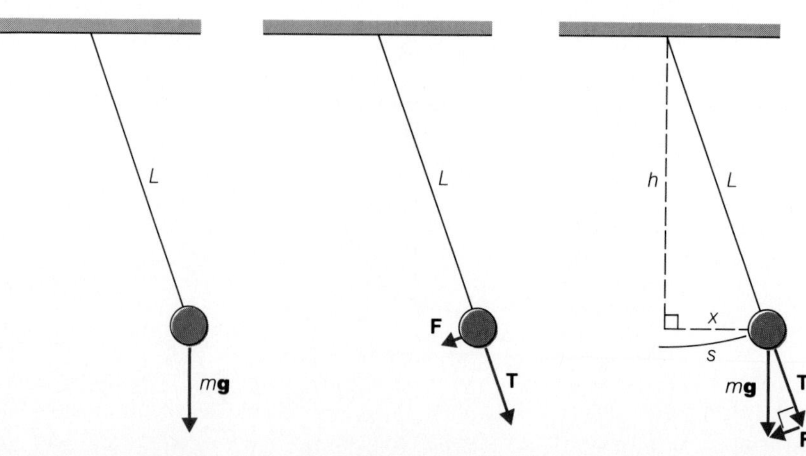

Figure 11.15

A pendulum executes simple harmonic motion when its oscillations are so small in amplitude that the chord x is very nearly equal in length to the arc s.

BIOGRAPHICAL NOTE

Christian Huygens

Christian Huygens (1629–1695) was born at The Hague into a Dutch family both wealthy and cultivated. Family friends included the artists Rembrandt and Frans Hals and the philosophers Spinoza and Descartes, and Christian's father had corresponded with Galileo. Christian was educated at home and later at the University of Leyden, where he studied mathematics and law. At Leyden he built several telescopes whose quality was high enough for him to discover the rings of Saturn and its largest satellite, which he named Titan. When his schooling was over, Huygens returned home, where his father's support allowed him to devote all his time to physics, astronomy, and mathematics.

Huygens' work covered a wide range. His collected papers fill 22 volumes and include many findings that now appear in every basic physics text. For instance, Huygens was the first to show that momentum is conserved in collisions and to derive the formula for the centripetal force on an object moving in a curved path. He was especially interested in measuring time, and he designed a practical pendulum clock that was soon installed in clock towers all over the Netherlands. The formula relating the period of a simple pendulum to its length was developed by Huygens, as was a method for constructing a pendulum whose period is independent of its amplitude for large as well as small amplitudes.

Today Huygens is best known for having founded the wave theory of light, which he used to explain reflection and refraction. Newton, however, thought that light consisted of a stream of particles, and his prestige kept the wave theory in the shade for a century.

In 1666 Huygens, by now famous, was invited to Paris, where an apartment, a laboratory, and an income were provided as part of a policy to bring glory to King Louis XIV by surrounding him with noted scholars. Huygens stayed in Paris for 15 years, still a productive scientist. During this period he visited Newton in London. Then Louis XIV began to turn against Protestants, and Huygens went back to The Hague. He spent most of his time there preparing his many findings for publication. His health declined in his later years, and he died at the age of 66.

The restoring force acting on the bob is therefore

$$F = -\frac{mgx}{L}$$

where the minus sign indicates that **F** points in the direction of decreasing x.

If the string is not far from the vertical, the horizontal distance x is nearly the same as the actual path length s. If they are equal, then F is proportional to $-s$

and the motion is simple harmonic. In this case the acceleration of the bob is

$$a = \frac{F}{m} = -\frac{g}{L}x$$

Substituting in Eq. (11.3), $T = 2\pi\sqrt{-x/a}$, gives

$$T = 2\pi\sqrt{\frac{L}{g}} \qquad \textit{Simple pendulum} \quad (11.18)$$

Provided that x is close to s, the period of a pendulum is proportional to the square root of its length and does not depend on the mass of its bob. If the arc through which the pendulum swings on either side of the vertical is 5°, its actual period exceeds that given by Eq. (11.18) by only 0.05%. Even if the arc is 20°, the discrepancy is less than 1%.

EXAMPLE 11.7

How long should a pendulum be for it to have a period of 1.00 s?

SOLUTION We first solve Eq. (11.18) for L:

$$T = 2\pi\sqrt{\frac{L}{g}} \qquad T^2 = \frac{4\pi^2 L}{g} \qquad L = \frac{gT^2}{4\pi^2}$$

Inserting the values $g = 9.8\ \text{m/s}^2$ and $T = 1.00$ s gives

$$L = \frac{(9.8\ \text{m/s}^2)(1.00\text{s})^2}{4\pi^2} = 0.25\ \text{m}$$

◆

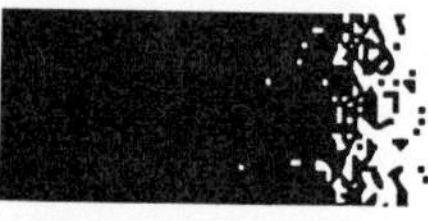

11.7 The Physical Pendulum

How to find the period of any swinging object.

An object of any shape will oscillate back and forth when it is pivoted at some point other than its center of gravity and given an initial push. Such an object is called a **physical pendulum.** We will find that the formula for the period of a simple pendulum also holds for a physical pendulum provided the length L is properly interpreted.

Figure 11.16 shows a physical pendulum pivoted at the point O. The pendulum has been pushed to one side so that the line from O to its center of gravity is at the angle θ from the vertical. The pendulum's weight mg acts from the center of gravity and produces the restoring torque

$$\tau = -mgx$$

where x is the horizontal distance between O and the center of gravity. The minus sign reflects the fact that the restoring torque is opposite in direction to the angular displacement θ of the pendulum.

When θ is small, the chord x is very nearly equal to the arc s. Assuming them to be equal is the same approximation made in analyzing the simple pendulum (see Fig. 11.15). Because h is the distance between the pivot point O and the center of gravity CG in Fig. 11.16, what we have for small θ is

$$x = s = h\theta$$

This means that

$$\tau = -mgh\theta$$

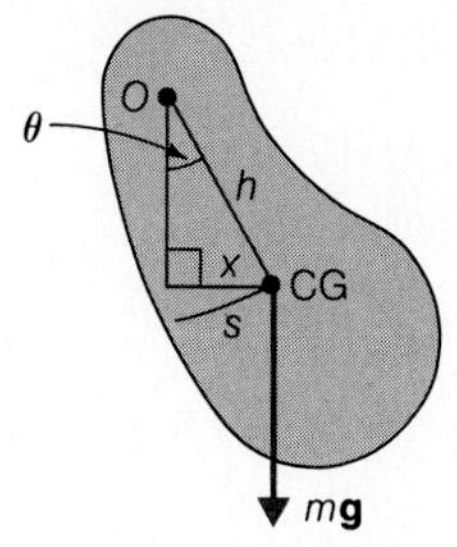

Figure 11.16

A physical pendulum pivoted at O. The center of gravity is marked CG.

We know that a restoring force of the form $-kx$ leads to SHM, so we might suspect that a restoring torque of the form $-K\theta$ leads to a rotational version of SHM. If I is the moment of inertia of the pendulum about O and α is the angular acceleration produced by the restoring torque, then

$$\alpha = \frac{\tau}{I} = -\frac{mhg}{I}\theta \tag{11.19}$$

Comparing this formula with the equivalent one for a harmonic oscillator,

$$a = -\frac{k}{m}x \tag{11.4}$$

suggests that the period of a physical pendulum can be given by the general formula for the period of a harmonic oscillator,

$$T = 2\pi\sqrt{-\frac{x}{a}} \tag{11.5}$$

with θ/α replacing x/a. This idea turns out to be correct, and we have

$$T = 2\pi\sqrt{-\frac{\theta}{\alpha}} = 2\pi\sqrt{\frac{I}{mgh}} \qquad \textit{Physical pendulum} \tag{11.20}$$

EXAMPLE 11.8

Analyze the process of walking by considering the leg as a physical pendulum.

SOLUTION A leg of length L may be crudely approximated by a thin rod hinged at one end. From Fig. 7.10 the moment of inertia of such a rod about one end is

$$I = \tfrac{1}{3}mL^2$$

where m is its mass. If the center of mass of the leg is at its middle, $h = L/2$, and the natural period of oscillation of the leg is

$$T = 2\pi\sqrt{\frac{I}{mgh}} = 2\pi\sqrt{\frac{mL^2/3}{mgL/2}} = 2\pi\sqrt{\frac{2L}{3g}}$$

The period of a leg 1.0 m long is

$$T = 2\pi\sqrt{\frac{2(1.0\text{ m})}{3(9.8\text{ m/s}^2)}} = 1.6\text{ s}$$

Each step represents only half a complete cycle, so if the leg is swinging freely, it takes 0.8 s and the rate of walking is 75 steps/min. Walking at a slower or a faster rate than this involves more effort. With a stride 80 cm long, 75 steps/min means a speed of 60 m/min, which is 3.6 km/h (2.24 mi/h).

The longer the legs of an animal, the longer its stride, but T is increased as well. The natural speed of walking varies as L/T, and because T is proportional to $\sqrt{L}$, this speed is proportional to $L/\sqrt{L} = \sqrt{L}$. The larger an animal, then, the faster it walks, although it takes an increase by a factor of 4 in leg length to double the natural speed. Running is quite a different matter because the muscular strength and the mass of an animal are involved as well as its size. As we saw in Section 4.5, all animals have roughly similar running speeds. When running, an animal's legs are bent while moving forward, which reduces their lengths and hence their natural periods of oscillation. This reduction makes it easier for the legs to be moved rapidly. ♦

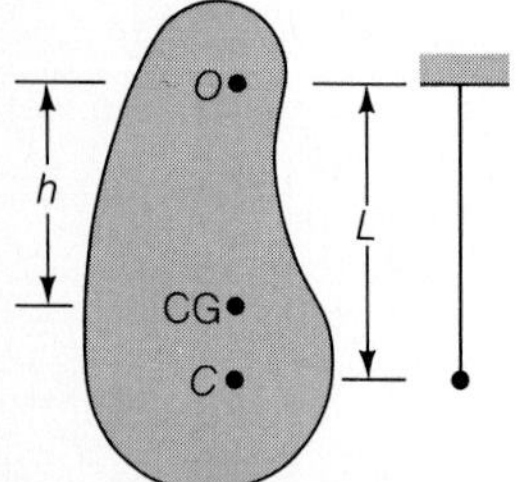

Figure 11.17

A simple pendulum of length $L = I/mh$ has the same period as that of a physical pendulum. The point C is the center of oscillation.

Center of Oscillation

The simple pendulum whose period is the same as that of a given physical pendulum may be found by setting equal the formulas for their respective periods of oscillation:

$$2\pi\sqrt{\frac{L}{g}} = 2\pi\sqrt{\frac{I}{mgh}}$$

$$L = \frac{I}{mh} \qquad (11.21)$$

Thus the mass of a physical pendulum can be regarded as concentrated at a point C the distance $L = I/mh$ from the point O. This point is called the **center of oscillation** (Fig. 11.17) and has two interesting properties:

1. If the pendulum is pivoted at C instead of at O, it will oscillate with the same period as before and O will be the new center of oscillation.
2. If the pendulum is struck along a line of action through C, there will be no reaction force on the pivot at O. A baseball that strikes a bat at the latter's center of oscillation does not produce a sting in the batter's hands, for example (Fig. 11.18). The center of oscillation is often called the **center of percussion** for this reason. The concept of center of percussion plays an important part in the design of many mechanical devices.

Figure 11.18

When a physical pendulum is struck at its center of percussion, its pivot experiences no reaction force.

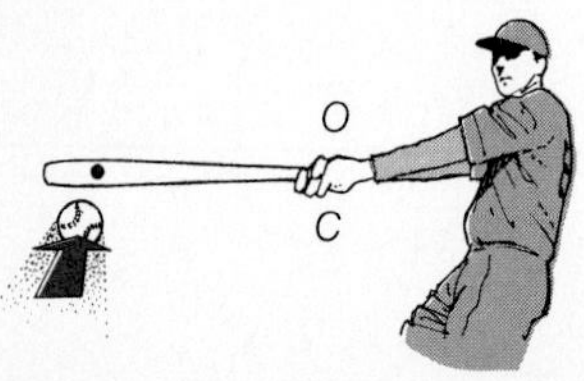

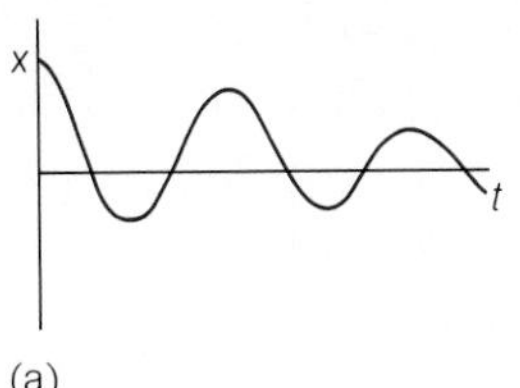

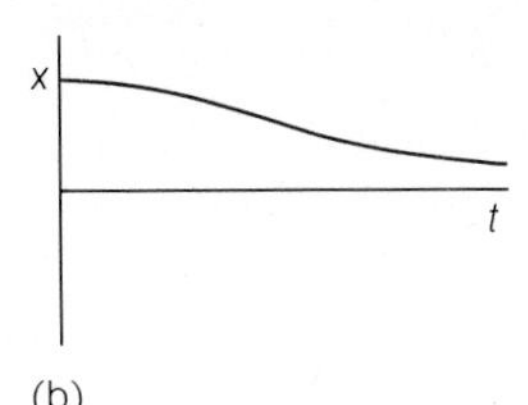

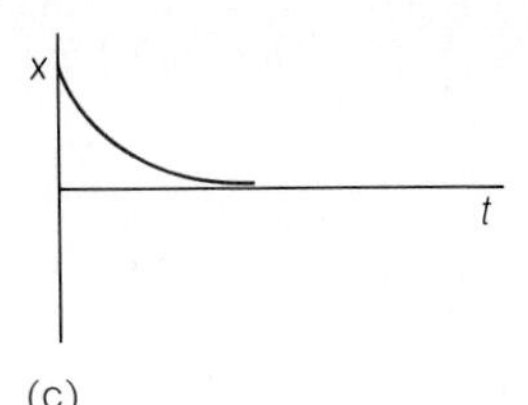

Figure 11.19

(a) Damped harmonic oscillator. (b) An overdamped oscillator returns to its equilibrium position very slowly. (c) A critically damped oscillator returns to its equilibrium position rapidly, but not so rapidly that it overshoots and begins to oscillate.

11.8 Damped Harmonic Oscillator

The greater the damping, the slower the oscillations and the sooner they die out.

Once set in motion, an ideal harmonic oscillator should continue oscillating forever. Actual oscillators do not behave this way. Although in some cases they may continue for a long time, if no further energy is fed into the oscillations their amplitude will decrease until they come to a stop. This phenomenon is called **damping.** If the amount of damping is very large, as it is for a car whose springs are damped with good shock absorbers, no oscillation at all occurs. After being displaced, the object gradually returns to its equilibrium position and goes no further.

In general, damping is caused by frictional forces. The potential energy of a stretched spring is not completely converted into kinetic energy at $x = 0$; some of it goes into work done against the frictional forces that are acting. At the end of each vibration the loss in energy leads to a shorter extension of the spring, until finally the amplitude becomes 0. Damping also acts to reduce the frequency of the motion—the greater the damping, the slower the oscillations.

Figure 11.19 shows the effects of different amounts of damping on a harmonic oscillator. In (a) the damping is small, and the oscillations steadily decrease in amplitude. In (b) the damping is so great that the displaced object never oscillates but returns to its equilibrium position very slowly; such an oscillator is said to be **overdamped**. A **critically damped** oscillator, as in (c), falls right on the line between the other two situations; it returns to its equilibrium position as rapidly as possible without overshooting, which would mean oscillating. Critical damping is desirable in many situations, such as that of the suspension of a car.

Important Terms

A body under stress has **elastic potential energy,** which is equal to the work done in deforming it.

Simple harmonic motion (SHM) is an oscillatory motion that occurs whenever a restoring force acts on a body in the opposite direction to its displacement from its equilibrium position, with the magnitude of the restoring force proportional to the magnitude of the displacement.

The **period** T of a system undergoing SHM is the time needed for it to make one complete oscillation. The **frequency** f of such a system is the number of complete oscillations it makes per unit time.

The **amplitude** of a body undergoing SHM is its maximum displacement on either side of its equilibrium position. The period of the motion is independent of the amplitude.

The **center of oscillation** of a pivoted object is that point at which it can be struck without producing a reaction force on its pivot.

In a **damped harmonic oscillator,** friction progressively reduces the amplitude of the vibrations.

Important Formulas

Elastic potential energy: $PE = \frac{1}{2}kx^2$

Harmonic oscillator: $T = \frac{1}{f} = 2\pi\sqrt{-\frac{x}{a}} = 2\pi\sqrt{\frac{m}{k}}$

Simple pendulum: $T = 2\pi\sqrt{\frac{L}{g}}$

Physical pendulum: $T = 2\pi\sqrt{\frac{I}{mgh}}$

MULTIPLE CHOICE

1. The product of the period and the frequency of a harmonic oscillator is always equal to
- **a.** 1
- **b.** π
- **c.** 2π
- **d.** the amplitude of the motion

2. The period of a simple harmonic oscillator is independent of its
- **a.** frequency
- **b.** amplitude
- **c.** force constant
- **d.** mass

3. The amplitude of the motion of an object undergoing SHM is
- **a.** its total range of motion
- **b.** its maximum displacement on either side of the equilibrium position
- **c.** its minimum displacement on either side of the equilibrium position
- **d.** the number of cycles per second it undergoes

4. The amplitude of a simple harmonic oscillator does not affect its
- **a.** frequency
- **b.** maximum speed
- **c.** maximum acceleration
- **d.** maximum KE

5. The amplitude of a simple harmonic oscillator is doubled. Which one or more of the following is also doubled?
- **a.** its frequency
- **b.** its period
- **c.** its maximum speed
- **d.** its total energy

6. Which one or more of the following statements about an object in SHM in a straight line is not true?
- **a.** Its acceleration can be 0 at times.
- **b.** Its acceleration can be greater than g at times.
- **c.** Its amplitude must be small.
- **d.** Its KE is constant.

7. An object undergoes simple harmonic motion. Its maximum speed occurs when its displacement from its equilibrium position is
- **a.** zero
- **b.** a maximum
- **c.** half its maximum value
- **d.** none of the above

8. In simple harmonic motion there is always a constant ratio between the displacement of the mass and its
- **a.** speed
- **b.** acceleration
- **c.** period
- **d.** mass

9. An object attached to a horizontal spring undergoes SHM on a frictionless surface. Its PE
- **a.** is 0
- **b.** equals its KE
- **c.** does not equal its KE
- **d.** may be any of the above, depending on its displacement

10. A pendulum comes very close to executing simple harmonic motion provided that
- **a.** its bob is not too heavy
- **b.** the supporting string is not too long
- **c.** the arc through which it swings is not too small
- **d.** the arc through which it swings is not too large

11. The period of a simple pendulum depends on its
- **a.** mass
- **b.** length
- **c.** total energy
- **d.** maximum speed

12. A pendulum clock is in an elevator. The clock will run fast when the elevator is
- **a.** rising at constant speed
- **b.** falling at constant speed
- **c.** accelerating upward
- **d.** accelerating downward

13. An object pivoted at an arbitrary point swings back and forth with the period T. The number of other points in the object at which it can be pivoted and have the same period is
- **a.** 0
- **b.** at least 1
- **c.** at least 2
- **d.** unlimited

14. An object pivoted at an arbitrary point is struck at its center of oscillation.

a. The object oscillates with a shorter period than if struck anywhere else.
b. The object oscillates with a longer period than if struck anywhere else.
c. There is no reaction force on the pivot.
d. The reaction force on the pivot is a maximum.

15. A spring whose force constant is k is cut in half. Each of the new springs has a force constant of

a. $\frac{1}{2}k$ b. k
c. $2k$ d. $4k$

16. A spring of force constant 1.0 N/m is joined end-to-end to a spring of force constant 2.0 N/m. The force constant of the combination is

a. 0.67 N/m b. 1.0 N/m
c. 1.5 N/m d. 3.0 N/m

17. A force of 0.2 N is needed to compress a certain spring by 2 cm. Its potential energy when compressed is

a. 2×10^{-3} J b. 2×10^{-5} J
c. 4×10^{-5} J d. 8×10^{-5} J

18. When a 1.00-kg mass is suspended from a spring, the spring stretches by 50.0 mm. The force constant of the spring is

a. 0.20 N/m b. 1.96 N/m
c. 49 N/m d. 196 N/m

19. If the suspended mass of Multiple Choice 18 oscillates up and down, its period will be approximately

a. 0.032 s b. 0.071 s
c. 0.45 s d. 4.5 s

20. The maximum speed of a particle that undergoes simple harmonic motion with a period of 0.50 s and an amplitude of 20 mm is

a. π cm/s b. 2π cm/s
c. 4π cm/s d. 8π cm/s

21. In order to oscillate at 10 Hz, a mass of 20 g should be suspended from a spring whose force constant is

a. 2.5 N/m b. 8.9 N/m
c. 12.6 N/m d. 79 N/m

22. A woman walks to the end of a diving board, whose end moves down by 35 cm as a result. If the woman bounces on the end of the board, the period of oscillation will be

a. 0.19 s b. 1.2 s
c. 1.4 s d. dependent on her mass

23. A harmonic oscillator of mass 40 g has a period of 10 s. To reduce the period to 5 s, the mass should be changed to

a. 10 g b. 20 g
c. 80 g d. 160 g

24. A 1.0-kg piston in a compressor undergoes 20 cycles/s in which its total travel is 14 cm. The maximum force on the piston

a. is 1.1 kN
b. is 1.5 kN
c. is 2.2 kN
d. cannot be calculated from the given data

25. A boy swings from a rope 4.9 m long. His approximate period of oscillation is

a. 0.5 s b. 3.1 s
c. 4.4 s d. 12 s

26. A pendulum whose period on the earth's surface is 4.0 s is installed in a satellite that circles the earth in an orbit of radius $1.5R_{earth}$. The pendulum's frequency in the satellite is

a. 0 b. 0.20 Hz
c. 0.25 Hz d. 0.31 Hz

QUESTIONS

1. At what point or points in its motion is the energy of a harmonic oscillator entirely potential? At what point or points is its energy entirely kinetic?

2. Must a spring obey Hooke's law in order to oscillate?

3. A wooden object is floating in a bathtub. It is pressed down and then released. Under what circumstances will its oscillations be simple harmonic in nature?

4. The total energy of a harmonic oscillator is doubled. How does this affect the frequency of the oscillation? Its amplitude? The maximum speed?

5. On what, if anything, does the ratio between the maximum kinetic energy and maximum potential energy of a harmonic oscillator depend?

6. The Empire State Building in New York City oscillates in the wind with a period of about 8 s. Do you think the period is different on weekdays, when the building is full of people, from what it is on weekends, when the building is empty?

7. In what part of a pendulum's swing is the tension in its supporting string a maximum?

8. A body pivoted at some point is given an initial displacement and then released. Under what circumstances will it oscillate back and forth? Under what circumstances will the oscillations be simple harmonic in character? Under what circumstances will it behave like a simple pendulum?

9. What is the frequency of a pendulum whose normal period is T when it is in an elevator in free fall? When it is in an elevator descending at constant velocity? When it is in an elevator ascending at constant velocity?

10. Verify that T is expressed in units of time in Eqs. (11.5), (11.18), and (11.20).

11. Where should the center of percussion of an ax be located?

12. A loudspeaker generates sound waves that correspond in frequency and amplitude to the electrical oscillations that reach it from the output of an audio amplifier. Why is it desirable for the cone of a loudspeaker to be damped?

EXERCISES

11.1 Elastic Potential Energy

1. The work needed to compress a spring by 40 mm is 0.80 J. What is the force constant of the spring?

2. A 5.0-kg object is dropped on a vertical spring from a height of 2.0 m. If the maximum compression of the spring is 40 cm, what is its force constant?

3. A force of 2.0 N is needed to push a 100-g jack-in-the-box into its box, an operation in which the spring is compressed 10 cm. What will be the maximum speed of the jack-in-the-box when it pops out?

4. A toy rifle employs a spring whose force constant is 200 N/m. In use, the spring is compressed 50 mm, and when released, it propels a 5.0-g rubber ball. What is the ball's speed when it leaves the rifle?

5. A rubber band 30 cm in circumference has a cross section of 2.0 mm × 2.0 mm and a Young's modulus of 7 MN/m^2. Find the energy stored in the rubber band when it is stretched to twice its normal length, assuming that Y is constant over this extension.

6. You have two springs whose force constants are, respectively, 50 N/m and 100 N/m. (a) The springs are connected end-to-end, and the combination is pulled out by a force of 10 N. What is the PE of the stretched springs? (b) The springs are then connected side-by-side and again pulled out by the same force. What is the PE of the stretched springs now?

11.2 Simple Harmonic Motion

7. A 100-g mass is suspended from a spring whose force constant is 50 N/m. The mass is then pulled down 10 mm and released. What is the amplitude of the resulting oscillations? What is their frequency?

8. A 25-kg portable gasoline-powered generating set is mounted on four springs, which are depressed by 4.0 mm when the set is put down. Find the natural frequency of vibration of the system.

9. A 1000-kg car has a period of vertical oscillation of 2.0 s. Find the effective force constant of its springs.

10. Two springs with the same force constant k are connected to an object of mass m, as in Fig. 11.20. Find the period of oscillation in each case in the absence of friction.

11. A 50-g teacup suspended from a spring oscillates with a period of 1.5 s. When a 120-g mug is suspended from the same spring, what is its period of oscillation?

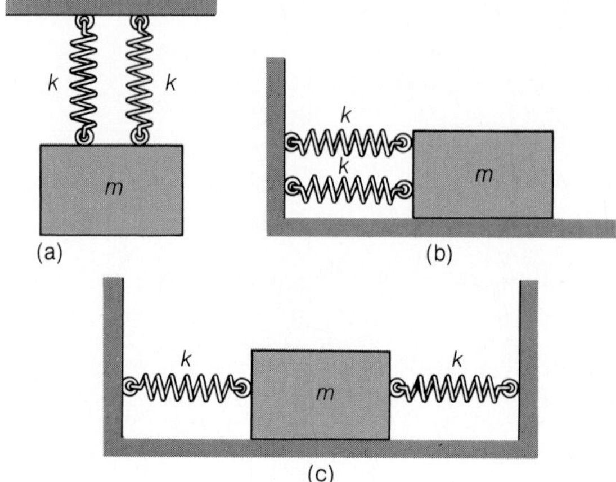

Figure 11.20

Exercise 10.

12. A spring has a 1.0-s period of oscillation when a 20-N weight is suspended from it. Find the elongation of the spring when a 50-N weight is suspended from it.

13. A 70-kg gymnast drops on a trampoline from a height of 60 cm. The trampoline sags 40 cm when the gymnast strikes it, and he then bounces up and down. If the motion is simple harmonic, find its period.

14. An object of unknown mass is suspended from a spring, and the system is set in oscillation. If the frequency of the oscillations is 2.0 Hz, by how much did the spring stretch when the object was suspended from it?

11.4 Position, Speed, and Acceleration

15. A harmonic oscillator has a period of 0.20 s and an amplitude of 10 cm. Find the speed of the moving object when it passes through the equilibrium position.

16. Atoms in a crystalline solid are in constant vibration at room temperature, where their motion has amplitudes in the neighborhood of 10^{-11} m. If the frequency of oscillation of one of the atoms in an iron bar is 2.5×10^{12} Hz, what are its maximum speed and acceleration?

17. A rack that undergoes SHM with an amplitude of 20 mm is used to test the ability of electronic equipment to survive accelerations. What frequency is needed to produce accelerations of $10g$?

18. The blade of a portable saber saw makes 3000 strokes per minute. Each stroke involves a total blade travel of 20 mm. Find the maximum speed of the blade.

19. The prongs of a tuning fork vibrate in SHM at a frequency of 660 Hz and with an amplitude of 1.0 mm at their tips. Find the maximum speed and acceleration of the prong tips. Express the acceleration in terms of g.

20. A 3.0-kg object is suspended from a spring and oscillates at 5.0 Hz with an amplitude of 15 cm. (a) What is the total energy of the motion? (b) What is the object's maximum acceleration?

21. At what displacement relative to the amplitude is the kinetic energy of a harmonic oscillator three times the potential energy?

22. An object is in SHM with an amplitude of 20 mm and a period of 0.20 s. (a) What is its speed when its displacement from its equilibrium position is 5.0 mm? (b) What is its displacement when its speed is 0.50 m/s?

23. An object whose mass is 1.0 kg hangs from a spring. When the object is pulled down 50 mm from its equilibrium position and released, it oscillates once per second. (a) What is the force constant of the spring? (b) What is the object's speed when it passes through its equilibrium position? (c) What is the maximum acceleration of the object?

24. An object whose mass is 0.0050 kg is in simple harmonic motion with a period of 0.040 s and an amplitude of 10.0 mm. (a) What is its maximum acceleration? (b) What is the maximum force on the object? (c) What is its acceleration when it is 5.0 mm from its equilibrium position? (d) What is the force on it at that point?

11.6 The Simple Pendulum

25. A chandelier is suspended from a high ceiling with a cable 6.0 m long. What is its period of oscillation?

26. A pendulum whose length is 1.53 m oscillates 24.0 times per minute in a particular location. What is the acceleration of gravity there?

27. Find the length of a pendulum whose period is 1.0 s on the surface of Mars, where the acceleration of gravity is 38% of its value on the earth's surface.

28. Find the period of a pendulum 50 cm long when it is suspended in (a) a stationary elevator; (b) an elevator falling at the constant speed of 5.0 m/s; (c) an elevator falling at the constant acceleration of 2.0 m/s^2; (d) an elevator rising at the constant speed of 5.0 m/s; (e) an elevator rising at the constant acceleration of 2.0 m/s^2.

29. The bob of a simple pendulum of length L is pulled to one side through an angle θ with the vertical and then let go. What is the maximum speed of the bob?

30. When a mass m is suspended from a massless spring whose normal length is d, the spring stretches to a new length of $2d$. Will the period of oscillation be greater if the system is set into vertical oscillations or if it swings back and forth like a pendulum? Assume that the spring remains $2d$ in length during the pendulum swings.

11.7 The Physical Pendulum

31. Show that the formula for the period of a physical pendulum reduces to that for the period of a simple pendulum for a bob suspended by a massless string.

32. A 200-g brass hoop 400 mm in diameter is suspended on a knife edge on which it rocks back and forth. The period of the oscillations is 1.27 s. Find the moment of inertia of the hoop about an axis through its circumference perpendicular to its plane.

33. An iron bar 80 cm long is suspended from one end. What is the period of its oscillations? What would be the length of a simple pendulum with the same period?

34. A broomstick suspended from one end oscillates with the same period as a simple pendulum 80 cm long. How long is the broomstick?

PROBLEMS

35. A steel wire 5.2 m long whose mass is 4.8 g hangs from an overhead beam. A 12-kg brass monkey is then attached to the lower end of the wire. How much additional energy is stored in the wire?

36. (a) Show that the PE per unit volume of an object under tension or compression is given by (stress)(strain)/2. (b) A 1.00-kg block of steel is placed in a vise whose jaws are tightened until the elastic limit of the steel is reached. How much elastic PE is stored in the block?

37. A spring 10 cm long whose force constant is 150 N/m is connected to a spring 20 cm long whose force constant is 100 N/m. The combination is then pulled out so that its ends are 50 cm apart. (a) Find the amount by which each spring is stretched. (b) A 400-g block on a frictionless surface is attached between the springs. Find its period of oscillation.

38. One end of a spring is attached to a car restricted to a horizontal track, and its other end is attached to an overhead support, as shown in Fig. 11.21. The car is pulled the distance x to one side and then released. Is its motion simple harmonic

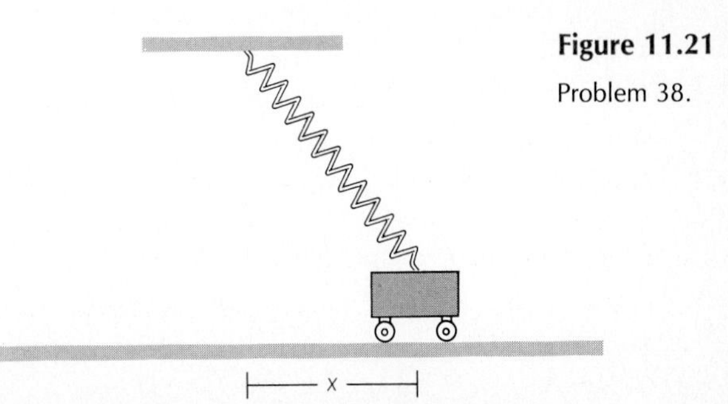

Figure 11.21
Problem 38.

if x is comparable to the height of the support? If so, what is its period?

39. A hole is bored through the earth along a diameter. Inside the hole the acceleration of gravity varies as rg/R, where r is the distance from the center of the earth, R is the earth's radius of 6.4×10^6 m, and g is the acceleration of gravity at the earth's surface. A stone is dropped into the hole and executes simple harmonic motion about the center of the earth. Why? Find the period of this motion.

40. A wooden cube of density ρ with the length L on each edge floats in a liquid of density ρ' so that its upper and lower faces are horizontal. The cube is pushed down and released. Verify that the cube then oscillates up and down in simple harmonic motion with a period of $2\pi\sqrt{L\rho/g\rho'}$.

41. A simple pendulum has a period of 2.56 s at a certain place. When the length of the pendulum is increased by 800 mm, its period increases to 3.14 s. Find the original length of the pendulum and the value of g at that place.

42. A **torsion pendulum** consists of an object of moment of inertia I suspended by a wire or thin rod, as in Fig. 11.22.

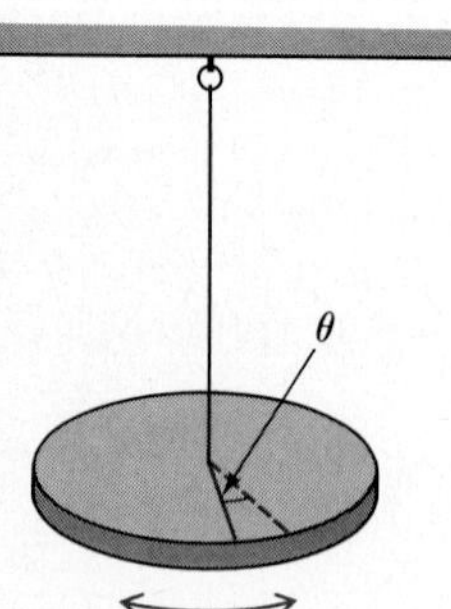

Figure 11.22

Problem 42.

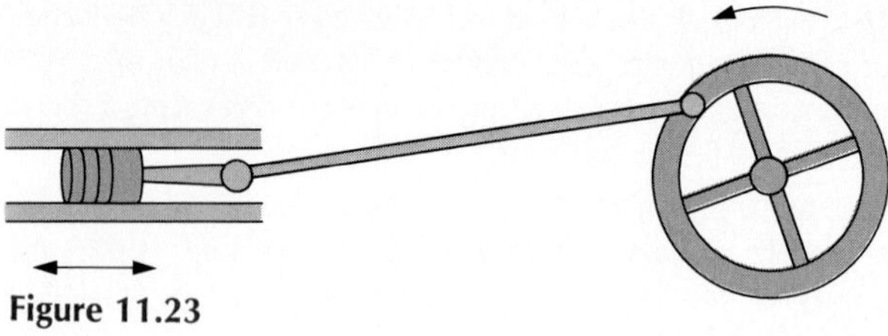

Figure 11.23

Problem 43.

When the object is turned through an angle θ, a restoring torque of $\tau = -K\theta$ arises. If the object is then released, it will oscillate back and forth in SHM. Find the period of the oscillations in terms of I and K.

43. Figure 11.23 shows a rotating wheel that drives a piston by means of a long connecting rod pivoted at both ends. The wheel's radius is 20 cm, and it is turning at 4.0 rev/s. (a) Where must the rod's attachment to the wheel be located at $t = 0$ in order that the equation $x = A \cos \omega t$ describe the piston's position? (b) What are the values of A and ω? (c) What are the maximum values of the piston's speed and acceleration?

Answers to Multiple Choice

1. a	**8.** b	**15.** c	**22.** b
2. b	**9.** d	**16.** a	**23.** a
3. b	**10.** d	**17.** a	**24.** a
4. a	**11.** b	**18.** d	**25.** c
5. c	**12.** c	**19.** c	**26.** a
6. c, d	**13.** b	**20.** d	
7. a	**14.** c	**21.** d	

CHAPTER

12

WAVES

Throw a stone into a lake, and water waves spread out from the splash. Clap your hands, and sound waves carry the noise all around you. Switch on a lamp, and light waves flood the room. Water, sound, and light waves differ in important ways, but all share the basic properties of wave motion. A wave is a periodic disturbance—a back-and-forth change of water height in water waves, of air pressure in sound waves, of electric and magnetic field strength in light waves—that moves away from a source and carries energy with it. Information, too, can be carried by waves, which is the case for everything we see and hear.

A vibrating system and a wave going through a medium have much in common. In both, a certain characteristic motion recurs at regular intervals; in both, potential energy is continuously changed into kinetic energy and back; in both, the properties of matter provide restoring forces. On the other hand, the energy in a vibrating system is confined to the system, whereas waves bring energy from one place to another.

12.1 Pulses in a String

An introduction to some important aspects of wave behavior.

If we give one end of a stretched string a quick shake, a kink or **pulse** travels down the string at some speed v (Fig. 12.1). If the string is uniform and completely flexible, the pulse keeps the same shape as it moves. It is worth examining the behavior of pulses in a string both because a wave can be regarded as a series of pulses and because it is easy to visualize what is going on.

The speed v of a pulse depends on the properties of the string—how heavy it is and how tightly it is stretched—and not on the shape of the pulse or on exactly how it is produced. Pulses move slowly down a slack, heavy rope, rapidly down a taut, light string. (''Heavy'' and ''light'' here refer to the mass per unit length of a string, not its total weight.) When the mass per unit length of a string is large, the pulse speed is low because the inertia of each segment of the string is large and it therefore responds slowly to the forces acting on it. When the string is tightly stretched, the pulse speed is high because the tendency of the string to straighten out is greater.

These observations are reflected in the formula

$$v = \sqrt{\frac{T}{m/L}} \qquad \textit{Waves in a string} \qquad (12.1)$$

for the speed of waves in a stretched string of mass m and length L that is under the tension T.

EXAMPLE 12.1

The stainless steel forestay of a racing sailboat is 20 m long, and its mass is 12 kg. To find its tension, the stay is struck by a hammer at the lower end and the return of the pulse is timed. If the time interval is 0.20 s, what is the tension in the stay?

SOLUTION If L is the length of the stay, the pulse travels the distance $2L$ in the time t, so its speed is $v = 2L/t$. From Eq. (12.1), with $L = 20$ m, $t = 0.20$ s, and $m = 12$ kg, the tension in the stay is

$$T = \frac{mv^2}{L} = \frac{m}{L}\left(\frac{2L}{t}\right)^2 = \frac{4mL}{t^2} = \frac{(4)(12\text{ kg})(20\text{ m})}{(0.20\text{ s})^2} = 2.4 \times 10^4\text{ N} = 24\text{ kN}$$

which is 5400 lb. ◆

The energy content of a moving pulse is partly kinetic and partly potential. As the pulse travels, its forward part is moving upward and its rear part is moving downward. Because the string has mass, there is a certain amount of KE associated with these up-and-down motions (Fig. 12.2). The PE is due to the tension in the string. Work had to be done in order to produce the pulse by pulling against the tension, and the deformed string accordingly possesses elastic PE.

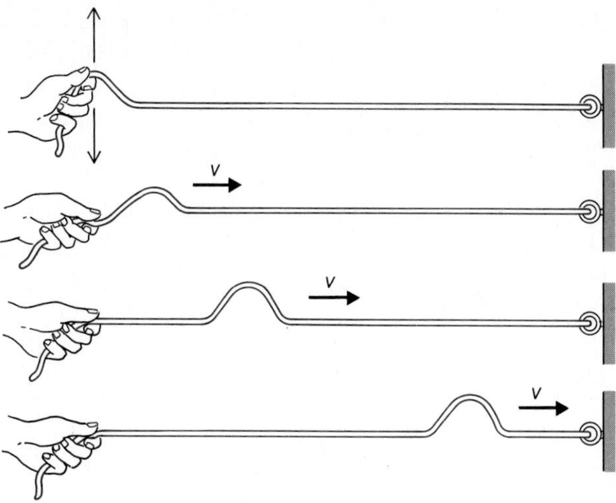

Figure 12.1

A pulse moves along a stretched string with a constant speed v.

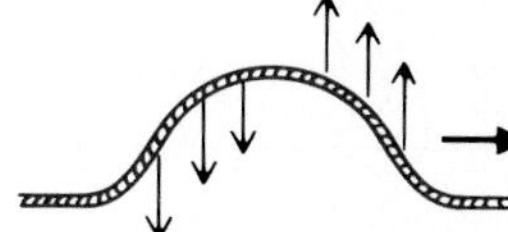

Figure 12.2

The forward part of this traveling pulse is moving upward, and the rear part is moving downward.

Reflection at a Boundary

When a pulse reaches the end of a string, it may be reflected back toward its starting point. Depending on how the end of the string is held in place, the reflected pulse may be inverted (upside down) or erect (right side up). Under just the right conditions, of course, the energy of the pulse may all be absorbed by the support, and the pulse will then disappear.

Suppose the end of the string is held firmly in place. When the pulse arrives there, the string exerts an upward force on the support (assuming the pulse is upward, as in Fig. 12.3). By the third law of motion, the support then exerts an equal and opposite reaction force on the string. This reaction force produces a pulse whose displacement is opposite to that of the original pulse but which otherwise has the same shape. The new inverted pulse then goes back along the string. Thus an upward pulse becomes a downward pulse upon reflection, and vice versa.

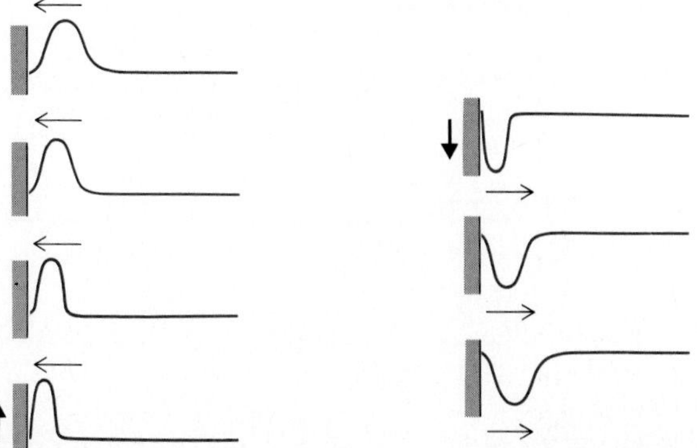

Figure 12.3

A pulse reaching a fixed end of the string is inverted on reflection.

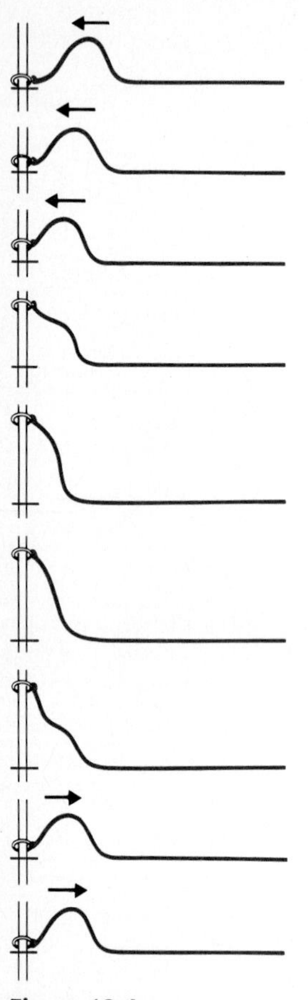

Figure 12.4

A pulse reaching a free end of the string is not inverted on reflection.

If the end of the string is not held firmly in place, however, the reflected pulse is not inverted. Figure 12.4 shows the end of a string attached to a ring free to move up and down a frictionless rod. When the pulse arrives at this end, the string moves upward until its KE is completely converted into elastic PE, whereupon the end of the string moves downward again to send out a pulse reversed in direction but otherwise the same as the original one.

If the end of the string is held in just the right manner between complete rigidity and complete freedom, the pulse will not be reflected at all but will disappear when it reaches the end. A little experimentation with pulses in an actual string will show that, whereas it is very easy to hold the far end of the string so that the reflected pulse is smaller than the original one, it is not easy at all to keep some reflection from taking place. It is equally hard to make perfect absorbers of sound, light, and water waves.

Reflection and Transmission at a Junction

So far we have been looking at pulses in a uniform string. Now let us connect two different strings together, one of them light (that is, with a low mass per unit length) and the other heavy (high mass per unit length). One end of the combination is fastened to something, and the other is given a shake to produce a pulse. Not surprisingly, the pulse passes from the first string to the second at the junction between them: But the transmission is not complete, since a reflected pulse also appears at the junction and proceeds in the opposite direction.

If the first string is the lighter one, the reflected pulse is inverted (Fig. 12.5). The greater inertia of the heavy string does not let it respond to the pulse as rapidly as the light string does, and an opposite reaction force occurs that causes the reflected pulse to be inverted, as though the junction were a rigid support. The energy of the original pulse is then split between the reflected and transmitted pulses. Since both strings have the same tension, the pulse travels more slowly in the heavy string. The length of the reflected pulse is the same as that of the original one, though its height is smaller since it has less energy. The length of the transmitted pulse, however, is shorter, because it moves more slowly than the original pulse while the time interval in which it comes into being is the same as that of the original pulse.

Figure 12.5

When a pulse passes from a light string to a heavy string, reflection occurs with the reflected pulse being inverted. The transmitted pulse is right side up. Because both strings have the same tension, the pulse travels more slowly in the heavy string.

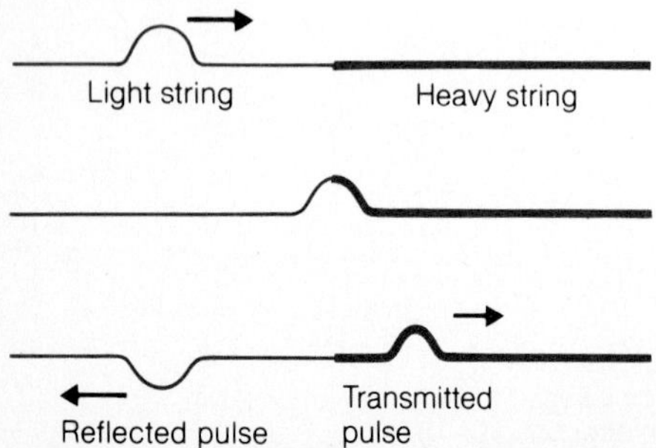

Figure 12.6

When a pulse passes from a heavy string to a light string, the reflected pulse stays right side up. The pulse speed is again less in the heavy string.

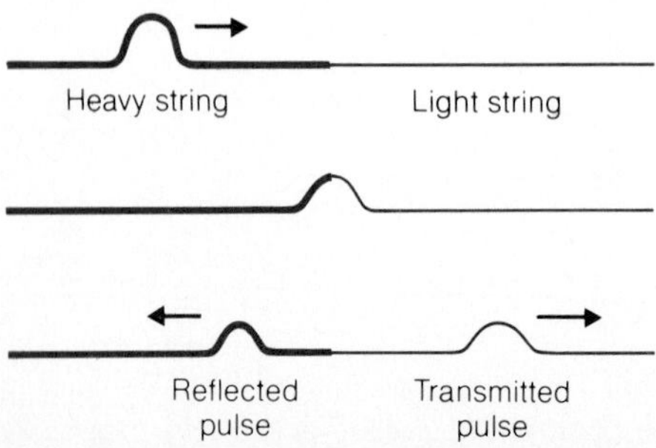

On the other hand, if the first string is the heavy one, the reflected pulse is erect (Fig. 12.6). The smaller inertia of the light string lets it follow the movements of the heavy one readily, and the situation is like that of a string whose end is able to move up and down freely. However, the light string does have some inertia, and so a reflected pulse again comes into being as well as a transmitted pulse. The pulse speed is higher in the light string, hence the pulse length is longer there than in the heavy one.

All types of waves, not just pulses in a stretched string, exhibit reflection and transmission at junctions between different media. For example, light waves are partially reflected and partially transmitted when they pass from air to glass. This is why we can see our images in a clear pane of glass (such as a shop window) even though the glass is transparent to light. The inversion of a pulse when it is reflected at a junction with a medium in which its speed is smaller also has a counterpart in the behavior of light waves, as we will find in Chapter 24.

12.2 Principle of Superposition

What happens when two pulses meet.

Opposite ends of a stretched string are given upward shakes, and the pulses thus produced move along the string toward each other. What happens when they meet? The result is a larger pulse at the moment the pulses come together, and then the separate pulses reappear and continue unchanged in their original directions of motion. Each pulse proceeds as though the other did not exist (Fig. 12.7).

The **principle of superposition** is a statement of this behavior:

When two pulses travel past a point in a string at the same time, the displacement of the string at that point is the sum of the displacements each pulse would produce there by itself.

What if one of the pulses is inverted relative to the other? According to the superposition principle, if the pulses have the same sizes and shapes, their displacements ought to cancel out when they meet, only to reappear later on after they have passed the crossing point. Such behavior is indeed observed in practice. At the instant of complete cancellation, the total energy of both pulses resides in the KE of the string segment where the cancellation occurs (Fig. 12.8).

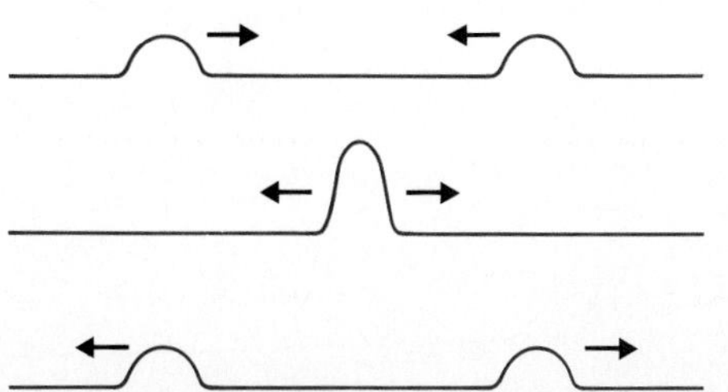

Figure 12.7

Two pulses moving in opposite directions along a stretched string. The pulses are unaffected by their crossing.

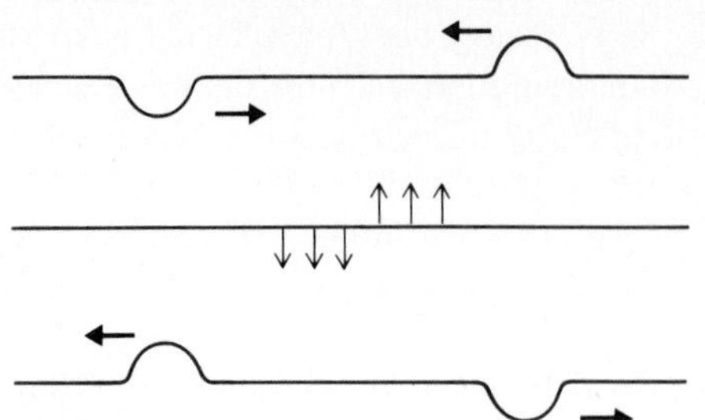

Figure 12.8

Complete cancellation occurs when two identical pulses with opposite displacements meet. At the instant of complete cancellation, the total energy of both pulses resides in the KE of the string segment where the cancellation occurs.

12.3 Periodic Waves

Wave speed equals the product of frequency and wavelength.

In a periodic wave, one pulse follows another in regular succession. Sound waves, water waves, and light waves are almost always periodic, although in each case a different quantity varies as the wave passes.

In periodic waves a certain waveform—the shape of the individual waves—is repeated at regular intervals. Periodic waves of all kinds usually have **sinusoidal** waveforms. A stretched string down which such waves move has the same appearance as a graph of sin x (or cos x) versus x that is moved along the x-axis with the wave speed v (Fig. 12.9).

Sinusoidal waves are common because the particles of matter in a medium that waves can travel through undergo simple harmonic motion when momentarily displaced from their equilibrium positions. The passage of a wave sets up coupled harmonic oscillations in the medium, with each particle behaving like a harmonic oscillator that has begun its cycle just a trifle later than the particle behind it (Fig. 12.10). The result in the case of waves in a stretched string is a waveform that is in essence a graph of how the position of a harmonic oscillator varies with time. As we know, this graph is a sine curve. (Light waves are also sinusoidal in character even though their existence does not involve the motion of material particles and they can travel through empty space.)

Three related quantities are useful in describing periodic waves:

1. The **wave speed** v, which is the distance through which each wave moves per second;
2. The **wavelength** λ (Greek letter *lambda*), which is the distance between adjacent crests or troughs;
3. The **frequency** f, which is the number of waves that pass a given point per second.

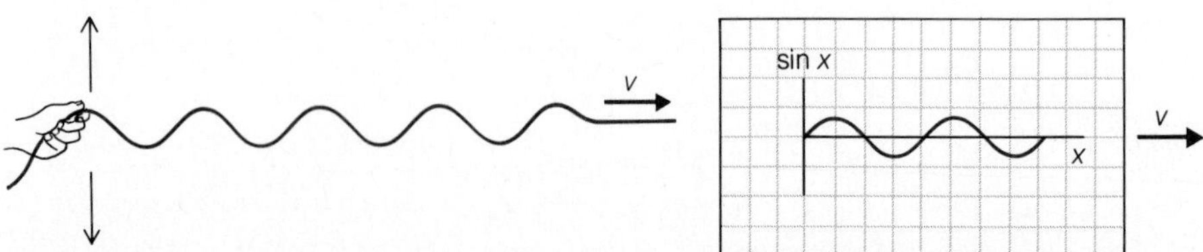

Figure 12.9

Most periodic waves have sinusoidal waveforms.

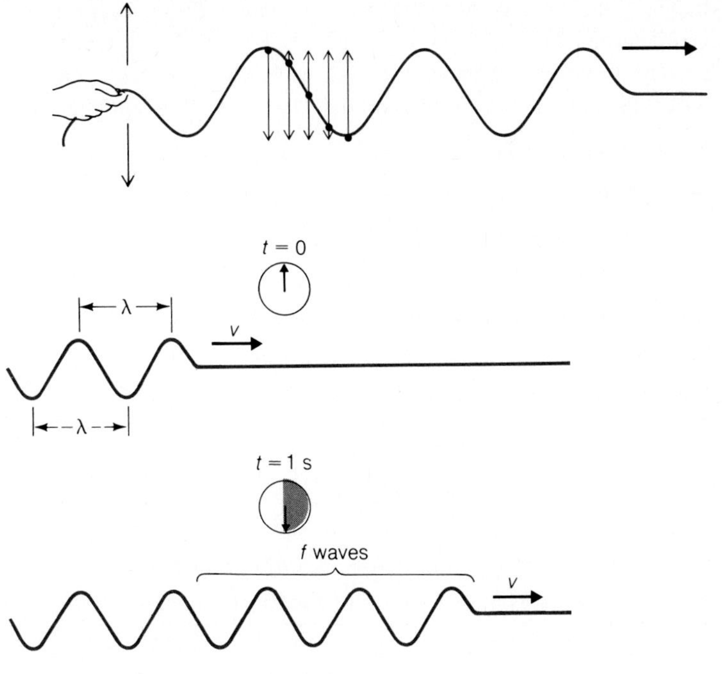

Figure 12.10

Each particle in the path of a sinusoidal wave executes SHM perpendicular to the wave direction.

v = wave speed
= distance traveled per second
= (number of waves passing a point per second) × (length of each wave)
= $f\lambda$

Figure 12.11

The speed of a wave is equal to the product of its frequency and wavelength.

The unit of frequency is the **hertz** (Hz). As in the case of oscillations, 1 Hz = 1 cycle/s. Multiples of the hertz are used for high frequencies:

1 kilohertz = 1 kHz = 10^3 Hz
1 megahertz = 1 MHz = 10^6 Hz
1 gigahertz = 1 GHz = 10^9 Hz

The wave speed, wavelength, and frequency of a train of waves are not independent of one another. In every second, f waves (by definition) go past a particular point, with each wave occupying a distance of λ (Fig. 12.11). Therefore a wave travels a total distance of $f\lambda$ per second, which is the wave speed v. Thus

$$v = f\lambda \qquad \textit{Wave speed} \quad (12.2)$$

Wave speed = (frequency)(wavelength)

which is a basic formula that applies to all periodic waves, sinusoidal or not.

EXAMPLE 12.2

A marine radar operating at a frequency of 9400 MHz emits groups of radio waves 0.0800 μs in duration. (The time needed for reflections of these groups to return indicates the distance of the target; see Section 22.2.) Radio waves, like light waves, are electromagnetic in nature and travel at 3.00×10^8 m/s. Find (a) the wavelength of these waves; (b) the length of each wave group, which is indicative of the precision with which the radar can measure distance; and (c) the number of waves in the group.

SOLUTION (a) From Eq. (12.2) the wavelength is

$$\lambda = \frac{v}{f} = \frac{3.00 \times 10^8 \text{ m/s}}{9.40 \times 10^9 \text{ Hz}} = 3.19 \times 10^{-2} \text{ m} = 3.19 \text{ cm}$$

(b) The length x of each wave group is simply

$$x = vt = (3.00 \times 10^8 \text{ m/s})(8.00 \times 10^{-8} \text{ s}) = 24.0 \text{ m}$$

(c) We can find the number n of waves in each group in either of these ways:

$$n = ft = (9.40 \times 10^9 \text{ Hz})(8.00 \times 10^{-8} \text{ s}) = 752 \text{ cycles} = 752 \text{ waves}$$

$$n = \frac{x}{\lambda} = \frac{24.0 \text{ m}}{3.19 \times 10^{-2} \text{ m}} = 752 \text{ wavelengths} = 752 \text{ waves}$$

◆

Sometimes it is more useful to consider the **period** T of a wave, which is the time required for one complete wave to pass a given point (Fig. 12.12). Because f waves pass by per second, the period of each wave is

$$T = \frac{1}{f} \qquad \textit{Wave period} \quad (12.3)$$

If five waves per second are passing by, for example, each wave has a period of $\frac{1}{5}$ s. In terms of period T, the formula for wave speed is

$$v = \frac{\lambda}{T} \qquad (12.4)$$

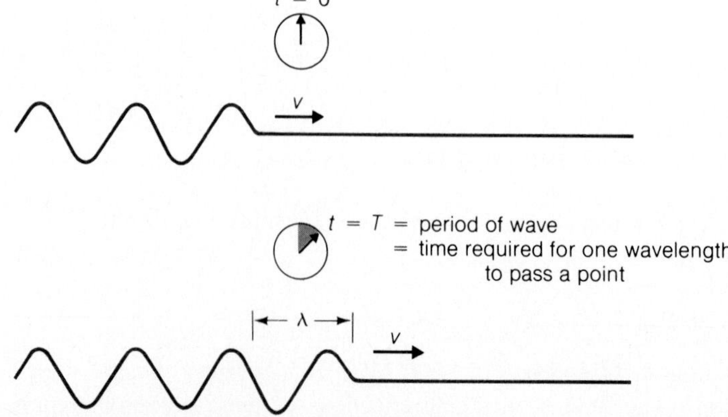

Figure 12.12

The period of a wave is the time required for one complete wave to pass by a given point.

EXAMPLE 12.3

An anchored boat is observed to rise and fall once every 4.0 s as waves whose crests are 25 m apart pass by it. Find the frequency and speed of the waves.

SOLUTION The frequency of the waves is

$$f = \frac{1}{T} = \frac{1}{4.0\text{ s}} = 0.25\text{ Hz}$$

and their speed is

$$v = \frac{\lambda}{T} = \frac{25\text{ m}}{4.0\text{ s}} = 6.25\text{ m/s}$$

♦

The **amplitude** A of a wave refers to the maximum displacement from their normal positions of the particles that oscillate back and forth as the wave travels by (Fig. 12.13). The amplitude of a wave in a stretched string is the height of the crests above the original line of the string (or the depth of the troughs below the original line). The speed, frequency, and wavelength of a wave are independent of its amplitude, just as the period of a harmonic oscillator is independent of its amplitude.

The correspondence between harmonic and wave motion leads to an interesting result. As we found in Section 11.4, a particle of mass m that undergoes simple harmonic motion of frequency f and amplitude A has a total energy of

$$E = 2\pi^2 m f^2 A^2$$

This dependence of energy on f^2 and on A^2 is also true for mechanical waves of all kinds. (A mechanical wave is one that involves moving matter, in contrast to, say, an electromagnetic wave.) Waves in a string are an example: Their energy per unit length is $2\pi^2\,(m/L)f^2A^2$, where m/L is the mass per unit length of the string.

Fourier's Theorem

We have been considering sinusoidal waves thus far. However, in a medium whose properties do not vary with wave frequency, everything that is true for a sinusoidal wave is also true for all other periodic waves. An interesting and important theorem by Jean Fourier (1768–1830) shows why this should be so.

What Fourier proved is that *any* periodic wave of frequency f, regardless of its waveform, can be thought of as a superposition of sinusoidal waves whose frequencies are f, $2f$, $3f$, and so on. The amplitudes of the various component waves

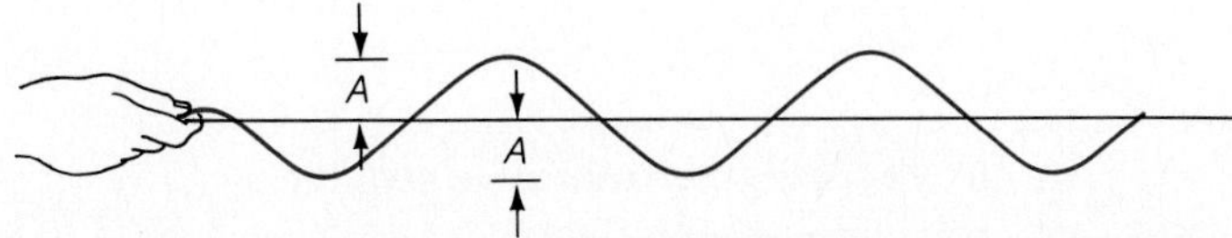

Figure 12.13

The quantity A is the amplitude of the wave.

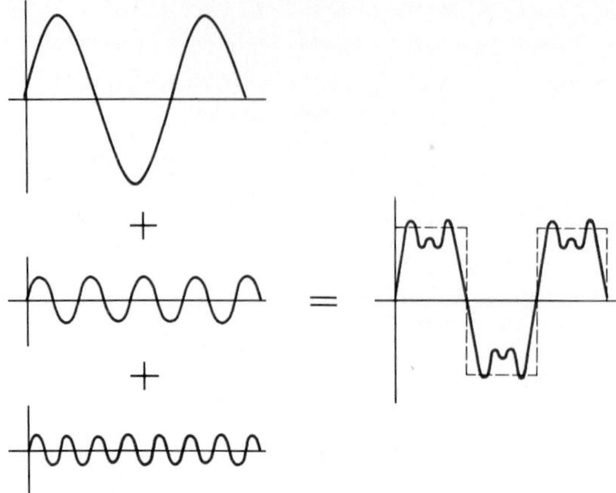

Figure 12.14

Fourier synthesis of a square wave.

depend on the precise character of the composite wave, and a mathematical procedure exists for finding these amplitudes. Figure 12.14 shows how just three waves of appropriate frequency and amplitude add up to give an approximation of a square wave. The more the waves that are included, the better the approximation becomes. Even an isolated pulse can be represented by a superposition of sinusoidal waves. Here, however, the frequencies must be very close to each other instead of being multiples of f, and a great many waves are needed.

12.4 Types of Waves

Waves may be transverse, longitudinal, or a combination of both.

Waves in a stretched string are **transverse waves** because the individual segments of the string vibrate perpendicular to the direction in which the waves travel. **Longitudinal waves** occur when the individual particles of a medium vibrate back and forth parallel to the direction in which the waves travel (Fig. 12.15). Longitudinal waves are easy to produce in a long coil spring; each portion of the spring is alternately compressed and extended as the waves pass by. Longitudinal waves, then, are essentially density fluctuations. Sound waves are longitudinal.

Waves on the surface of a body of water (or other liquid) are a combination of longitudinal and transverse waves. If we were somehow to tag individual water molecules and follow them when a train of waves passes by, we would find that their paths are like those shown in Fig. 12.16. Each molecule moves in a circle with a period equal to the period of the wave, and does not undergo a permanent shift. Because successive molecules reach the tops of their orbits at slightly different times, the water surface takes the form of a series of crests and troughs. At the crest of a wave the molecules move in the direction the wave is traveling, whereas in a trough the molecules move in the opposite direction.

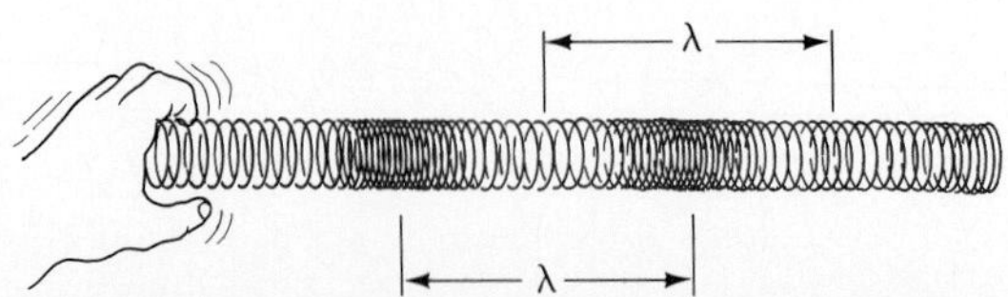

Figure 12.15

Longitudinal waves in a coil spring.

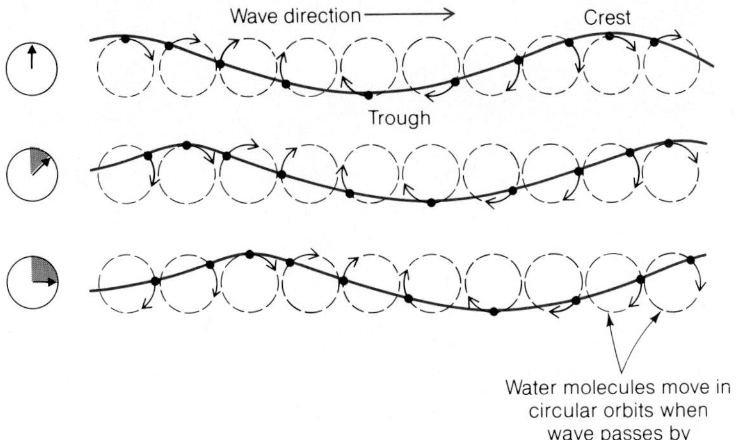

Figure 12.16

Water molecules move in circular orbits about their original positions when a typical deep-water wave passes by. At the crest of a wave the molecules are moving in the direction the wave is traveling, whereas in the trough the molecules are moving in the opposite direction. There is no net motion of water involved in the motion of such a wave.

The passage of a wave across the surface of a body of water, like the passage of a wave through any medium, involves the motion of a pattern. Energy is transported by virtue of the moving pattern, but there is no net transport of matter.

EXAMPLE 12.4

The water waves in Example 12.3 have an amplitude of 60 cm. Find the speed of an individual molecule of water on the surface.

SOLUTION The water molecules on the surface are moving in circles of radius 60 cm, so that as each wave passes by, the molecules travel a distance s equal to the circumference $2\pi r$ of the circle (Fig. 12.17). Hence

$$s = 2\pi r = 2\pi(0.60 \text{ m}) = 3.8 \text{ m}$$

Each wave takes $T = 4.0$ s to go past a given point, which means that the molecules must cover the 3.8-m circumference of their orbits in 4.0 s. The speed of each molecule is therefore

$$V = \frac{s}{T} = \frac{3.8 \text{ m}}{4 \text{ s}} = 0.95 \text{ m/s}$$

The speed of the *wave*, however, is 6.25 m/s, more than six times greater. Thus the motion of the pattern that constitutes a wave in a medium can be much more rapid than the motions of the individual particles of the medium, and energy can be transported by wave motion faster than might be possible through the net transport of matter.

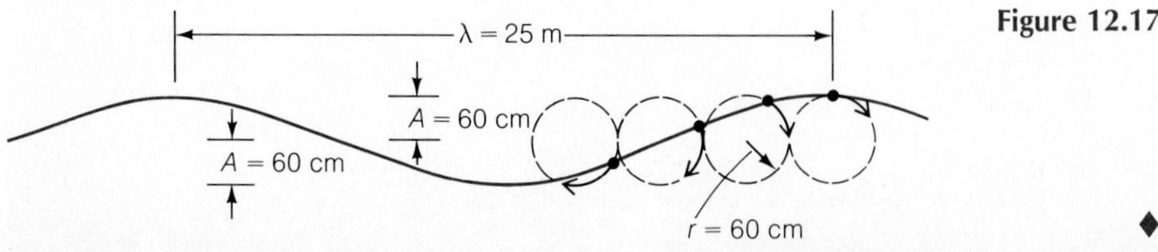

Figure 12.17

◆

When a wind blows over a body of water the height of the resulting waves depends on three factors: the speed of the wind, the length of time the wind blows, and the distance over which the wind is in contact with the water. These factors govern the rate at which energy is transferred to the water and thus govern the violence of the resulting disturbance.

The speed of water waves in deep water depends on their wavelength: The longer the waves, the faster they move. Thus waves 15 m long travel at 5 m/s, but waves 150 m long travel at 15 m/s. In shallow water the orbits of the water molecules touch the bottom, which slows down the progress of the waves: The shallower the water, the more slowly they move, regardless of wavelength. The wave speed in water 1 m deep is only half the wave speed in water 4 m deep. (The slowing of the lower parts of the orbits is what causes wave crests to fall forward, or ''break,'' where the bottom of a body of water rises near a shore.)

Earthquake Waves

Most earthquakes are caused by the sudden movement of rock along a fracture surface. When stresses developed within the earth's crust in a region become too great for the rock to support, a slippage occurs that sends out waves that shake the ground. The event responsible for an earthquake typically involves an area in the crust some tens of kilometers across located a few kilometers below the surface, although some earthquakes occur as much as several hundred kilometers down. Of the million or so earthquakes per year strong enough to be noticed, only a few liberate enough energy to do serious damage.

The upper deck of this highway in Oakland, California collapsed in the magnitude 7.1 earthquake whose waves shook San Francisco and its vicinity on October 17, 1989. Each step of 1 on the Richter scale of earthquake magnitude represents a change in vibration amplitude of a factor of 10. Quakes of magnitude 6 or more cause significant damage. The energy given off in a magnitude 8.7 quake, the strongest yet observed, is about double the energy content of the world's yearly consumption of coal and oil.

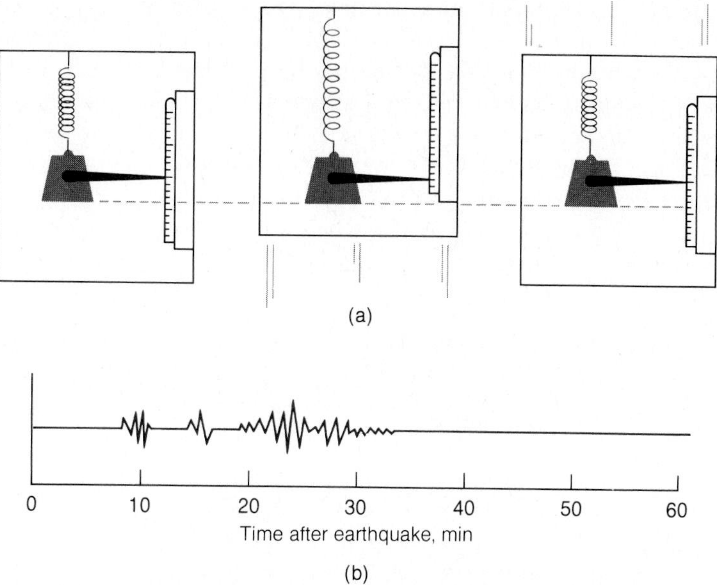

Figure 12.18

(a) A seismograph records the vibrations of earthquake waves. The suspended mass has a long period of oscillation, hence it remains very nearly stationary in space as the box and scale move up and down. A somewhat different seismograph based on the same principle responds to horizontal vibrations. (b) A seismogram of waves from an earthquake that occurred about 5000 km away. The first to arrive are longitudinal, the next are transverse, and the last are surface waves.

All three kinds of waves—transverse, longitudinal, and surface—are sent out by an earthquake and can be detected many thousands of kilometers away if the quake is a major one. The longitudinal waves are the fastest and usually have periods of 2 to 3 s; the slower transverse waves have periods of 10 to 15 s; and the still slower surface waves have periods of 10 to 60 s. The surface waves are usually the strongest (Fig. 12.18).

The longitudinal and transverse earthquake waves travel through the earth's interior, and by analyzing the waves that arrive at the various observatories around the world after an earthquake, it is possible to determine the structure of the earth's interior (see Fig. 6.21). In particular, the inability of transverse earthquake waves to go through the central part of the earth, whereas longitudinal waves can do so, indicates that this region must be liquid, because longitudinal waves can occur in a liquid whereas transverse waves cannot.

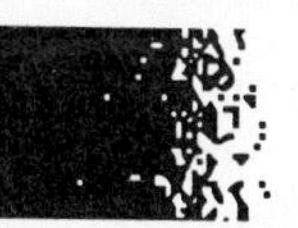

12.5 Standing Waves

In a given system only certain wavelengths are possible.

When we pluck a string whose ends are fixed, the string starts to vibrate in one or more loops (Fig. 12.19). These **standing waves** may be thought of as the result of waves that travel down the string in both directions, are reflected at the ends, proceed across to the opposite ends and are again reflected, and so on.

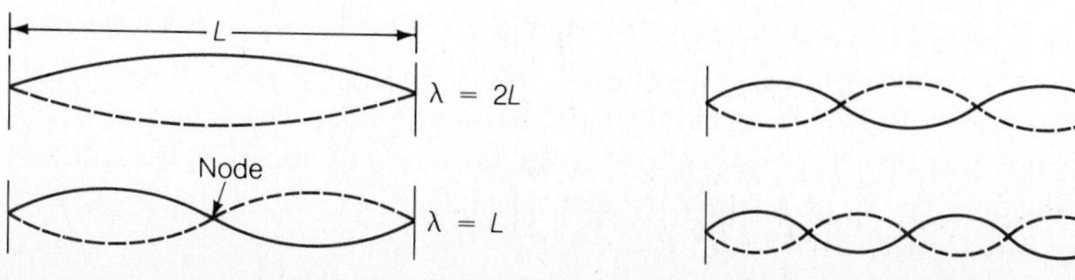

Figure 12.19

Standing waves in a stretched string.

To appreciate how standing waves come into being, we draw upon our knowledge of how pulses in a string are reflected and of what happens when two pulses traveling in opposite directions meet. When a pulse in a string is reflected at a rigid support, the reflected pulse is inverted. A similar inversion occurs for periodic waves. Hence, although waveform and wavelength stay the same, the wave train is in effect shifted by $\frac{1}{2}\lambda$ so that a crest arriving at the end of the string is reflected as a trough, and vice versa.

Now let us look into how two waves in the same string interact. Applied to waves, the principle of superposition states:

When two or more waves of the same nature travel past a point at the same time, the displacement at that point is the sum of the instantaneous displacements of the individual waves.

The principle of superposition holds for all types of waves, including waves in a stretched string, sound waves, water waves, and light waves.

Thus every wave train proceeds independently of any others that may also be present. Should two waves with the same wavelength come together in such a way that crest meets crest and trough meets trough, the resulting composite wave will have an amplitude greater than that of either of the original waves. In this case the waves are said to **interfere constructively** with each other. Should the waves come together in such a way that crest meets trough and trough meets crest, the composite wave will have an amplitude less than that of the larger of the original waves. In this case the waves are said to **interfere destructively** with each other (Fig. 12.20).

Figure 12.20

(a) Constructive interference. (b) Destructive interference.

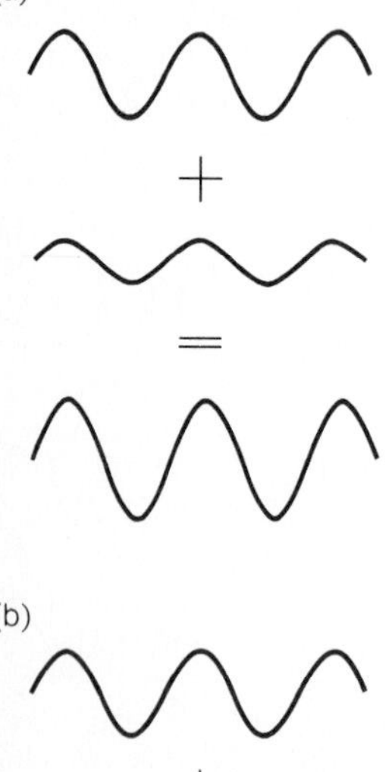

Let us apply these ideas to a stretched string whose ends are fixed. When the string is plucked, waves move back and forth between its ends, inverting each time they are reflected. These waves interfere with each other in such a way that destructive interference always occurs at the ends. If the string has length L, destructive interference will occur at the ends when the waves have wavelengths of $\lambda = 2L, L, 2L/3, L/2$, and so on (see Fig. 12.19). In the case of the shorter wavelengths, other points, called **nodes**, are present at intermediate positions where no motion of the string takes place.

Figure 12.21 shows how a standing wave pattern comes into being. The dotted curve represents a wave moving to the right, and the dashed curve a wave of the same wavelength and amplitude moving to the left. The sum of these waves is found by adding together their displacements at each point on the string, and is shown as a solid line. The addition is performed in Fig. 12.21 for four successive instants one-eighth of a period apart.

We note that the dotted and dashed curves do not always have displacements equal to zero at the ends of the string. There is no contradiction here because it is their resultant, the solid curve, that corresponds to reality, and this curve behaves as it should.

Any type of wave can occur as a standing wave between suitable reflectors. The vibrating air columns in wind instruments and organ pipes—and in the throat, mouth, and nose of a person speaking—are standing sound waves, for instance. Standing light waves play an important role in the operation of lasers, as we will find later. And standing waves of a rather remarkable kind are the key to understanding the structure of the atom.

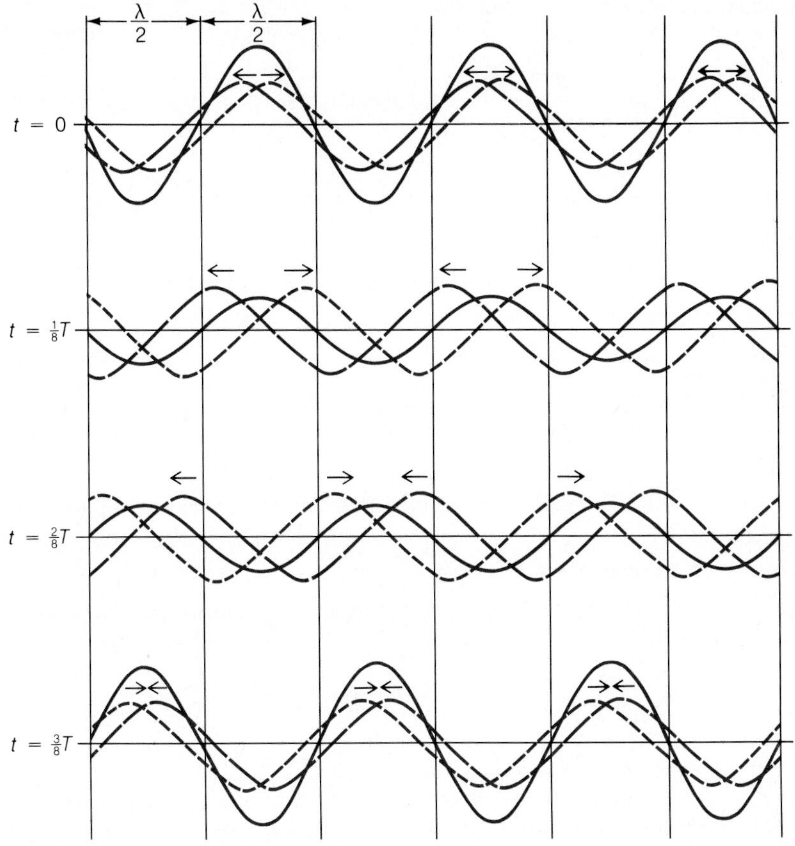

Figure 12.21

The origin of a standing wave. Waves (broken lines) that move to the left and to the right add up to give a stationary wave pattern (solid line).

The condition that nodes occur at each end of the string restricts the possible wavelengths of standing waves to

$$\lambda = \frac{2L}{n} \qquad n = 1, 2, 3, \ldots \qquad \textit{Standing waves} \quad (12.5)$$

The lowest possible frequency of oscillation f_1 of a stretched string corresponds to the longest wavelength, $\lambda = 2L$. Thus

$$f_1 = \frac{v}{\lambda} = \frac{v}{2L} \qquad (12.6)$$

Higher frequencies correspond to shorter wavelengths. From Eq. (12.5) we see that these higher frequencies can be represented in terms of f_1 by

$$f_n = nf_1 \qquad n = 2, 3, 4, \ldots \qquad (12.7)$$

The frequency f_1 is called the **fundamental frequency** of the string, and the higher frequencies f_2, f_3, and so on, are called **overtones**. Sometimes f_1 is called the first **harmonic,** f_2 the second harmonic, f_3 the third harmonic, and so on.

The wave velocity v is $\sqrt{T/(m/L)}$ according to Eq. (12.1). We can therefore express the fundamental frequency of a stretched string in terms of its length, linear density, and tension by the formula

$$f_1 = \frac{1}{2L}\sqrt{\frac{T}{m/L}} \qquad \textit{Fundamental frequency of stretched string} \quad (12.8)$$

This formula is the basis for the design of stringed musical instruments such as pianos and violins. A short, light, taut string means a high fundamental frequency of vibration; a long, heavy, slack string means a low fundamental frequency. To ''tune'' a stringed instrument, the tensions in the various strings are adjusted until their fundamental frequencies are correct.

EXAMPLE 12.5

The A string of a violin has a linear density of 0.60 g/m and an effective length of 330 mm. (a) Find the tension required for its fundamental frequency to be 440 Hz. (b) If the string is under this tension, how far from one end should it be pressed against the fingerboard in order to have it vibrate at a fundamental frequency of 495 Hz, which corresponds to the note B?

SOLUTION (a) First we square both sides of Eq. (12.8) and then solve for T:

$$f_1^2 = \frac{T}{4L^2(m/L)}$$

$$T = 4L^2f_1^2(m/L) = 4(0.330\text{ m})^2(440\text{ Hz})^2(6.0 \times 10^{-4}\text{ kg/m}) = 51\text{ N}$$

This is about 11 lb.

(b) According to Eq. (12.8), $f_A/f_B = L_B/L_A$, so here

$$L_B = L_A\left(\frac{f_A}{f_B}\right) = (330\text{ mm})\left(\frac{440\text{ Hz}}{495\text{ Hz}}\right) = 293\text{ mm}$$

Hence the string should be pressed (330 − 293) mm = 37 mm from one end. ♦

The 88 ''strings'' of a piano are metal wires whose fundamental frequencies range from 27 Hz for the lowest bass note to 4186 Hz for the highest treble note. This span is so great that the density as well as the length and tension of the wires must be varied in order that the instrument be of reasonable size: The treble strings are thin steel wires, and the bass strings are wound with copper wire to increase their linear density m/L. The total tension of all the strings in a piano is about 20 tons and is borne by a heavy cast iron frame.

Resonance

The fundamental frequency and overtones of a stretched string are its natural frequencies of oscillation: If the string is plucked, one or more of these frequencies will be excited. Eventually internal friction in the string will cause the various vibrations to die out. However, if we apply a periodic force to the string whose frequency is the same as one of the string's natural frequencies, the standing waves

Strong winds set up standing waves in the Tacoma Narrows Bridge in Washington State soon after its completion in 1940. The bridge collapsed as a result. Today bridges are stiffened to prevent such disaster.

continue as long as the periodic force supplies energy to the string. If the energy provided exceeds that lost to internal friction, the amplitude of the standing wave will increase until the string ruptures. (A column of soldiers can destroy a flimsy bridge by marching across it in step with one of its natural frequencies, although the bridge may be able to support the static load of the soldiers.)

The addition of energy to a system of some kind by a periodic force that varies with a frequency equal to one of the system's natural frequencies of vibration is called **resonance**. All rigid structures have natural frequencies of oscillation, even though their vibrations may be more complicated than those of a stretched string. These vibrations can be excited by a periodic force of the right frequency. The rattling of a car at certain speeds is a familiar example of resonance. The Mexican earthquake of 1985 was especially destructive in Mexico City because the frequency of the seismic waves happened to match the natural frequencies of many buildings there, notably slender ones of 8 to 15 stories. The acceleration of the ground was about $0.2g$, but resonance amplified the building vibrations to produce accelerations of over $1g$, well above what they could withstand.

12.6 Sound

Sound waves are pressure fluctuations in a solid, a liquid, or a gas.

Sound waves are longitudinal and consist of pressure fluctuations. The air (or other medium) in the path of a sound wave becomes alternately denser and rarer, and the resulting changes in pressure cause our eardrums to vibrate with the same frequency. This produces the physiological sensation of sound.

Most sounds are produced by vibrating objects. An example is the diaphragm of a loudspeaker (Fig. 12.22). When it moves outward, it pushes the air molecules in front of it closer together to form a region of high pressure that spreads out in front of the loudspeaker. The diaphragm then moves backward, expanding the volume available to nearby air molecules. Air molecules now flow toward the diaphragm, and a region of low pressure spreads out behind the high-pressure region. The continued vibrations of the diaphragm thus send out successive layers of condensation and rarefaction.

Figure 12.22

Sound consists of longitudinal waves, representing condensations and rarefactions in the air in its path.

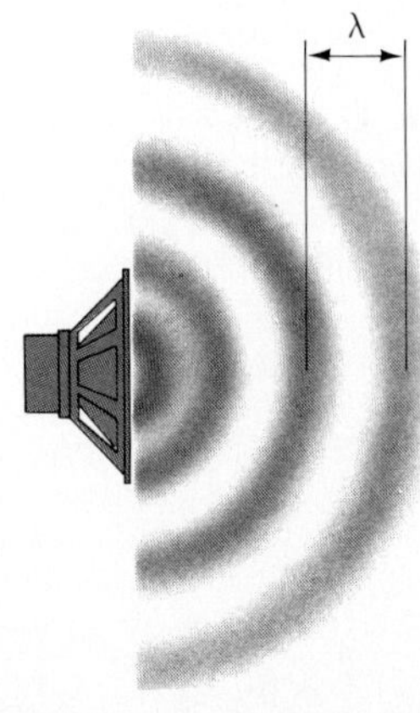

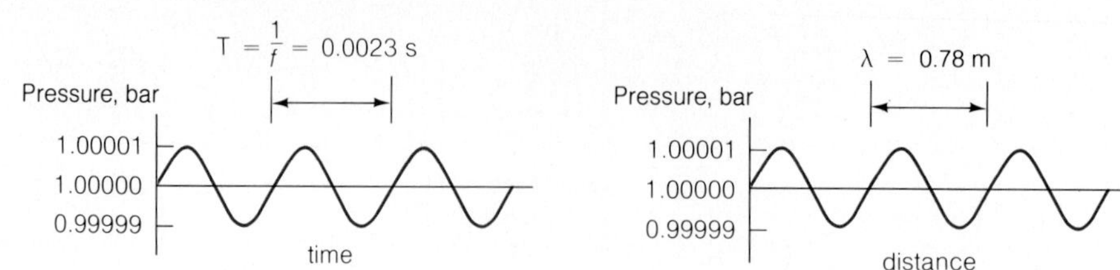

Figure 12.23

The pressure variations that constitute the musical note A in air at an undisturbed pressure of 1 bar. The amplitude is approximately that which a singer would produce.

The speed of sound in air at sea level and at 20°C is 343 m/s, which is 1126 ft/s. This speed increases at about 0.6 m/s per °C because the random speeds of air molecules increase with temperature and so make the passage of pressure fluctuations more rapid.

The musical note A, the frequency of which is 440 Hz, represents a wavelength λ in air of

$$\lambda = \frac{v}{f} = \frac{343 \text{ m/s}}{440 \text{ Hz}} = 0.78 \text{ m}$$

This is about 31 in. A person singing this note produces pressure waves whose amplitude might be 1 Pa, about 0.001% of sea-level atmospheric pressure (Fig. 12.23).

A normal ear responds to sound waves with frequencies from about 20 Hz to about 20,000 Hz, which correspond to wavelengths in air from 17 m (54 ft) to 17 mm (0.65 in). The fundamental frequency of the speaking voice averages about 145 Hz in men and about 230 Hz in women. Even considering the overtones that are present, the frequencies contained in ordinary speech are, for the most part, below 1000 Hz. Sound waves whose frequencies are about 20,000 Hz are called **ultrasonic** and can be detected by appropriate electromechanical devices. Ultrasonic waves are audible to many animals whose hearing organs are smaller than those of human beings and so are better able to resonate at higher frequencies. Figure 12.24 shows how pulses of ultrasound are used to determine the depth of the water under a boat or ship.

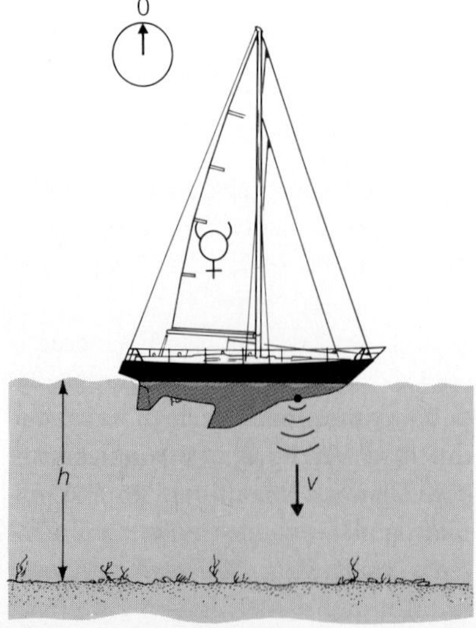

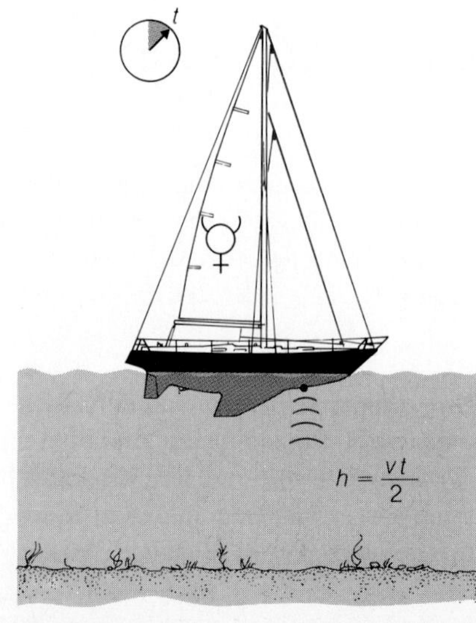

Figure 12.24

Echo sounding permits water depth to be rapidly and continuously determined from a boat. (a) A pulse of high-frequency sound waves is sent out by a suitable device on the boat's hull. (b) The time t needed for the pulse to return after being reflected by the sea floor is a measure of the water depth h.

Sound waves are transmitted by solids, liquids, and gases. In general, the stiffer the material, the faster the waves travel. This is reasonable when we reflect that stiffness implies particles that are tightly coupled together and so are more immediately responsive to one another's motions. As in the case of waves in a stretched string, the less dense the material, the less the inertia and the higher the speed of the waves. The speed of sound in a fluid medium is given by

$$v = \sqrt{\frac{B}{\rho}}$$

where B is the bulk modulus of the fluid and ρ its density. In a solid

$$v = \sqrt{\frac{Y}{\rho}}$$

where Y is the Young's modulus of the material. Table 12.1 lists the speed of sound in various materials.

The wavelength of the musical note A in seawater is

$$\lambda = \frac{v}{f} = \frac{1531 \text{ m/s}}{440 \text{ Hz}} = 3.5 \text{ m}$$

This is nearly five times the wavelength of the same note in air. When a sound wave produced in one medium enters another in which its speed is different, the frequency of the wave remains the same while the wavelength changes. This is in accord with the behavior of pulses that pass from one stretched string to another, as discussed earlier.

Table 12.1 Speed of sound

Medium	Speed, m/s
Gases (0°C)	
Carbon dioxide	259
Air	331
Air, 20°C	343
Helium	965
Liquids (25°C)	
Ethyl alcohol	1207
Water, pure	1498
Water, sea	1531
Solids	
Lead	1200
Wood, mahogany	~4300
Iron and steel	~5000
Aluminum	5100
Glass, pyrex	5170
Granite	~6000

EXAMPLE 12.6

Find the tension needed in a wire of cross-sectional area A and Young's modulus Y for the speeds of transverse and longitudinal (that is, sound) waves in it to be equal.

SOLUTION These speeds are given, respectively, by

$$v_t = \sqrt{\frac{T}{m/L}} \qquad v_s = \sqrt{\frac{Y}{\rho}}$$

and they will be equal when $T/(m/L) = Y/\rho$, or

$$T = \frac{m}{\rho L} Y$$

The wire's density ρ is $m/V = m/LA$, so the required tension is

$$T = YA$$

This is an extremely high tension—for a steel wire 1 mm in diameter; $T = 157$ kN, over 35,000 lb, well beyond the breaking strength of the wire, not to mention its elastic limit. ♦

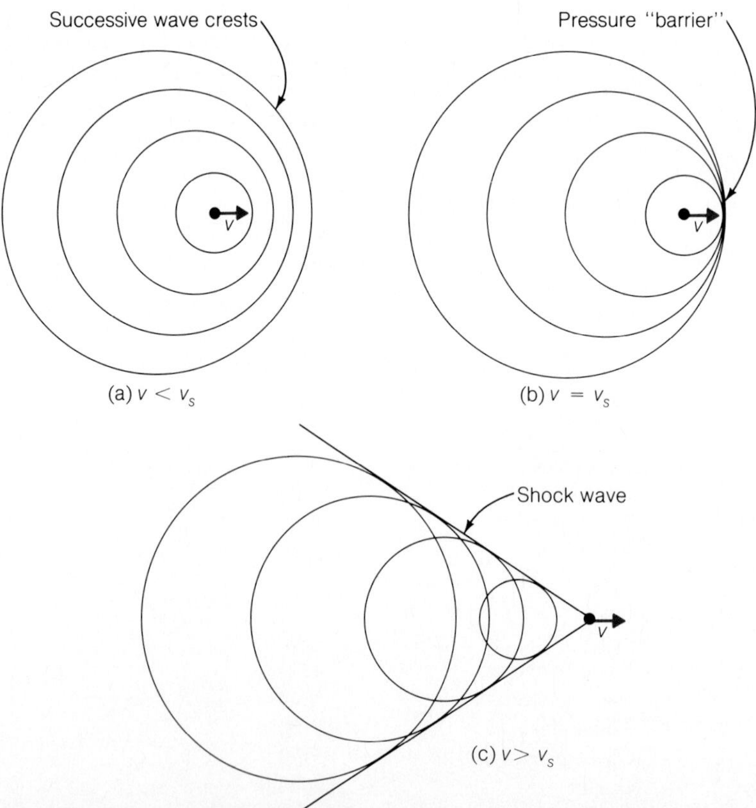

Figure 12.25
Waves produced by an object moving at (a) subsonic, (b) sonic, and (c) supersonic speeds.

As an object moves through a fluid—for instance, an airplane through the atmosphere—it disturbs the fluid, and the resulting pressure waves spread out in spherical shells at the speed of sound. If the airplane itself is moving at the speed of sound, the waves pile up in front of it, as in Fig. 12.25(b), to create a wall of high pressure—the "sound barrier." For the airplane to go faster than sound, a great deal of force is needed to push it through this barrier, more than the force needed to give a comparable acceleration at speeds either less than (subsonic) or more than (supersonic) that of sound.

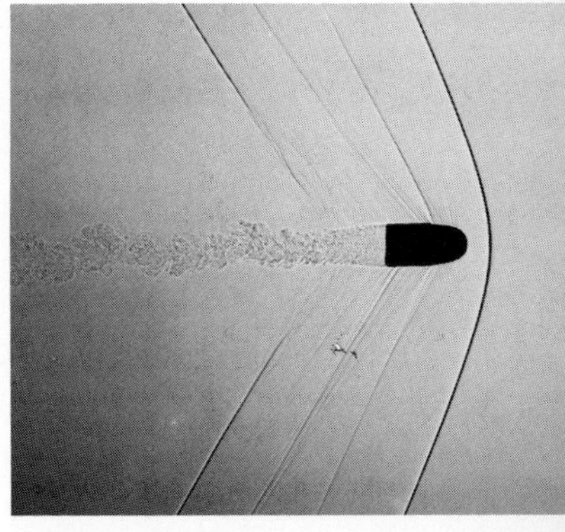

Shock waves produced by a bullet moving faster than the speed of sound.

At supersonic speeds, an airplane outdistances the waves it produces, which gives rise to the pattern of crests shown in Fig. 12.25(c). Where successive crests overlap, constructive interference takes place, and the result is a conical shell of high pressure with the airplane at its apex. This shell, which moves with the speed of sound, is called a **shock wave** because the pressure increases sharply when it passes. The shock wave is responsible for the sonic boom heard after a supersonic airplane has passed overhead. Because the rise in pressure is so sudden, such a shock wave can do physical damage to structures in its path even though the pressure change itself may not be very great. On a smaller scale, the crack of a whip is a sonic boom that occurs when the tip of the whip moves faster than sound.

EXAMPLE 12.7

The ratio between the speed of an airplane (or any other moving object) and the speed of sound is called the **Mach number** after the Austrian physicist Ernst Mach. Find the angle between the shock wave created by an airplane traveling at Mach 1.3 and the airplane's direction.

SOLUTION As shown in Fig. 12.26, in the time t the airplane travels the distance vt and the shock wave travels the distance $v_s t$. Hence the angle between the directions

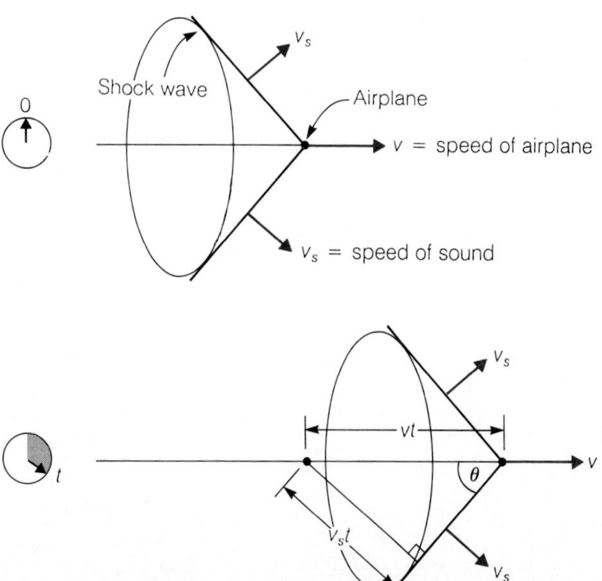

Figure 12.26

Geometry of a shock wave produced by a supersonic airplane.

of the airplane and of the shock wave is specified by

$$\sin\theta = \frac{v_s t}{vt} = \frac{v_s}{v}$$

Since a speed of Mach 1.3 means that $v = 1.3\ v_s$,

$$\sin\theta = \frac{v_s}{1.3\ v_s} = \frac{1}{1.3} = 0.769$$

$$\theta = \sin^{-1} 0.769 = 50°$$

◆

12.7 Interference of Sound Waves

Musical sounds have complex waveforms.

Longitudinal as well as transverse waves exhibit interference. The standing waves that occur in an air column illustrate this, because they are produced by the interference of sound waves going back and forth between its ends. The analysis is the same as in the case of standing waves in a stretched string, except that nodes (places where there is no oscillation) occur only at closed ends of the tube. At an open end, air molecules can move in and out freely, so open ends are **antinodes** where the maximum amplitude is always found.

Figure 12.27 shows the possible modes of vibration in tubes open at one end and open at both ends. The curves in each drawing indicate how the displacement extremes vary along the length of each tube. In practice, the vibrations at the open end of a tube extend a little beyond it, so the effective length of the tube is slightly more than its actual length.

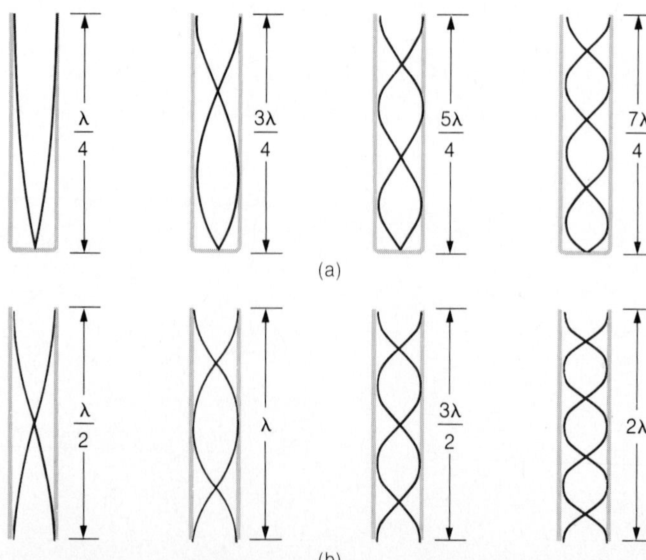

Figure 12.27

(a) Standing sound waves in a tube open at one end. The displacement maxima of the fundamental and the first three overtones are shown. A node occurs at the closed end and an antinode at the open end. (b) Standing sound waves in a tube open at both ends. An antinode occurs at both ends.

Sound waves can be produced in a variety of ways. Shown here are vibrating strings (guitar and bass), vibrating air columns (saxophone and trumpet), and vibrating membranes (cymbal and drum).

Wind instruments produce musical sounds by means of vibrating air columns. An organ has a separate pipe for each note. A woodwind, such as a flute or a clarinet, uses a single tube with holes whose opening or closing controls the effective length of the air column. Most brass instruments, such as the trumpet and the French horn, have valves connected to extra loops of tubing. Opening a valve adds to the length of the air column and thus produces a note of lower pitch. In a slide trombone the length of the air column is varied by sliding in or out a telescoping U-shaped tube. Vibrations of the players' lips as air is blown past them set up the standing waves in flutes, piccolos, and brass instruments. Vibrating reeds do this in clarinets, oboes, and saxophones.

The human voice is produced when the vocal cords in a person's throat set in vibration an air column that extends from the throat to the mouth and the nasal cavity above it. The shape of this column, which we adjust while speaking or singing by manipulating the mouth and tongue, determines the different vowel sounds by emphasizing some overtones and suppressing others. This shape also gives rise to the subtle differences that distinguish one person's voice from another's.

Natural sounds are never as regular as that shown in Fig. 12.23 but consist of mixtures of different frequencies that combine to give complex waveforms. The waveforms of sounds can be observed with the help of an oscilloscope, a device that displays electrical signals, such as those produced by a microphone in response to a sound, on the screen of a tube like a television picture tube. Figure 12.28 shows some waveforms of musical notes, which are mixtures of a fundamental frequency

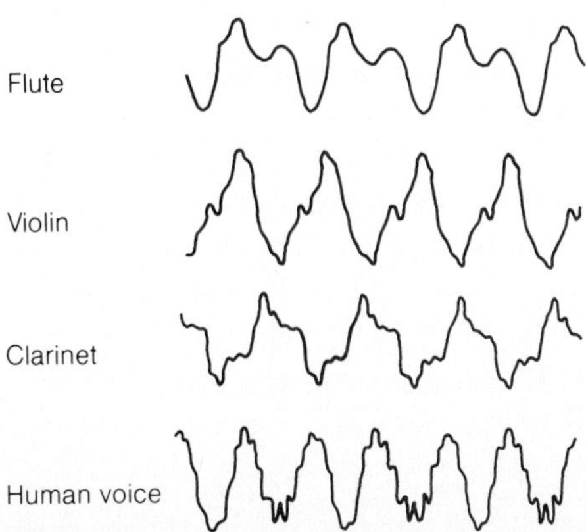

Figure 12.28

Waveforms of some musical sounds.

and some of its overtones. The precise mixture determines the quality, or timbre, of the note, which is different for each instrument and for different human voices. In the case of a flute, only the fundamental and first overtone are strong, giving a relatively simple waveform. In a clarinet, another woodwind, other overtones are also major contributors, leading to a "richer" sound.

Certain mixtures of frequencies are pleasing to the ear, and sounds that incorporate them are considered "musical." For instance, a tone combined with its first overtone, whose frequency is twice as great, appears harmonious to a listener. In music such an interval is called an **octave** because it includes eight notes. Another harmonious combination is a tone together with another whose frequency is 50% higher, so the frequencies are in the ratio 2:3. An example is C (262 Hz) plus G (392 Hz). Such an interval includes five notes (here C, D, E, F, G) and so is called a **fifth**. Somewhat less agreeable are two notes whose frequencies are in the ratio 4:5, for instance C and E (330 Hz), whose interval spans three notes and is called a **third**. The larger the numbers needed to express the ratio of frequencies, the less attractive the resulting sound is. Thus C and D are in the ratio 8:9 and seem discordant, and E and F, whose ratio is 15:16, are more discordant still. Ordinary sounds are mixtures of frequencies that have no special relationships with one another, and if the mixture seems particularly harsh, it qualifies as noise.

Beats

Let us look into what happens when two sounds of different frequency are combined. If two tuning forks (or other sources of single-frequency sound waves) whose frequencies are slightly different are struck at the same time, the sound that we hear fluctuates in intensity. At one instant we hear a loud tone, then virtual silence, then the loud tone again, then virtual silence, and so on. Why this happens is shown schematically in Fig. 12.29. The loud periods occur when the waves from the two forks interfere constructively, thus reinforcing one another, and the quiet periods occur when the waves interfere destructively, thus partially or wholly canceling one another out. These regular loudness pulsations are called **beats**.

A piano tuner can tell whether a piano string is in tune by listening for beats between the sound of that string and the sound of a tuning fork. If beats occur, the tension of the string is adjusted until they disappear. If you have access to a piano, you can exhibit beats by attaching a paper clip to one of the pair of strings used for each of the low notes. This changes the frequency of that string slightly. Striking the key for that note will set both strings vibrating, and you will hear beats.

Let us consider two waves with the frequencies f_1 and f_2, where $f_2 > f_1$. When the maxima of both waves occur together, there will be a loudness maximum, which is a beat. After a time T, their maxima will again coincide, and another beat will occur. In this time T, one of the waves has undergone one cycle more than the other, so that

$$f_2T - f_1T = 1 \quad \text{and} \quad f_2 - f_1 = \frac{1}{T}$$

The frequency of the beats is therefore

$$f_{\text{beat}} = \frac{1}{T} = f_2 - f_1 \qquad \textit{Beat frequency} \quad (12.9)$$

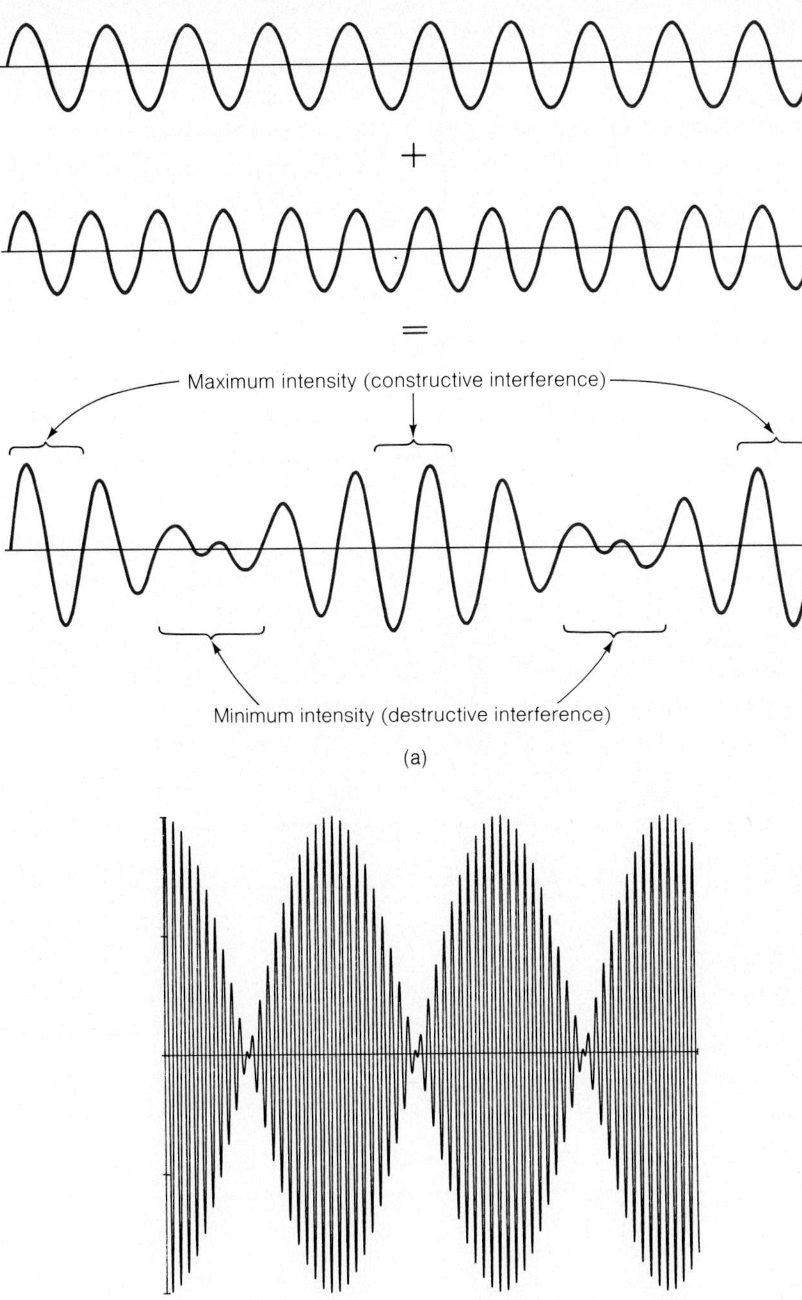

Figure 12.29

(a) The origin of beats in sound waves. (b) Beats produced by waves whose frequencies differ by 5%.

The sound we hear when beats are produced has a frequency that is the average of the two original frequencies, and the number of beats per second equals the difference between the two original frequencies. Thus the simultaneous vibrations of a 440-Hz tuning fork and an out-of-tune A string that vibrates at 444 Hz will produce a 442-Hz tone whose amplitude rises and falls four times per second.

Beats are hard to distinguish as such when the two frequencies differ by more than perhaps 10 Hz. When the frequencies are far enough apart, however, they give rise to an audible ''difference tone'' responsible for the lack of harmony of certain combinations of musical notes that was mentioned above. For instance, the notes E and F have the respective frequencies 330 Hz and 349 Hz, and when they are emitted together, the result is an additional 19-Hz difference tone that makes the combination seem dissonant.

12.8 Doppler Effect

Relative motion between an observer and a wave source changes the perceived frequency.

If we stand beside a road when a police car goes by with its siren blowing, we cannot help but notice that the pitch of the siren drops suddenly as the car passes by. What happens is that the pitch of the siren as the car approaches is *higher* than the pitch when the car is stationary, and the pitch when the car recedes is *lower* than its normal one. We also find a change of pitch if the siren is at rest and we go past it in a rapidly moving car: The pitch is higher than usual as we approach the siren, and lower as we recede. The change in frequency of a sound brought about by relative motion between source and listener is called the **Doppler effect.**

The origin of the Doppler effect is straightforward. As a moving source emits sound waves, it tends to overtake those traveling in the same direction (Fig. 12.30a). Hence the distance between successive waves is smaller than usual. This

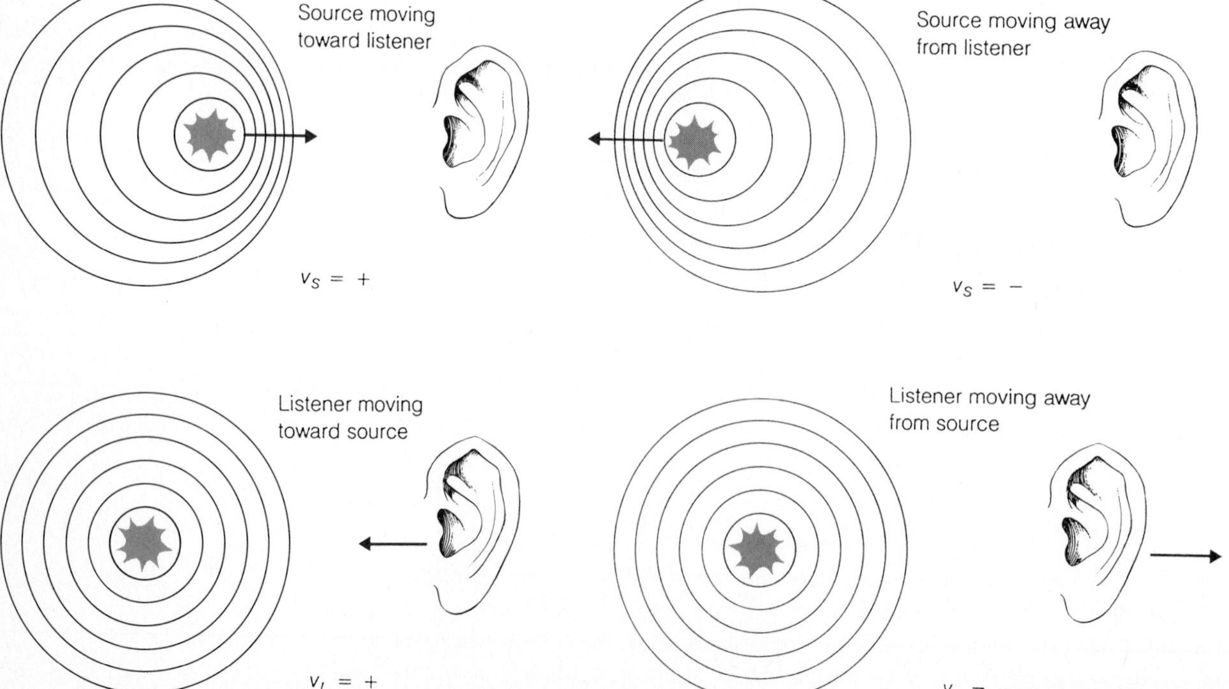

Figure 12.30

The Doppler effect occurs when there is relative motion between a source of sound and a listener. The sign convention for Eq. 12.10 is shown.

distance is the wavelength of the sound, so the corresponding frequency is higher. At the same time the source is moving away from those of its waves that travel in the opposite direction. This increases the distance between successive waves behind it and thereby reduces their frequency.

If the source is stationary, a listener moving toward it intercepts more sound waves per unit time than if she were at rest, and accordingly she hears a higher frequency (Fig. 12.30b). When the observer moves away from the source, fewer of the waves catch up with her per unit time, and she hears a lower frequency.

The relationship between the frequency f_L the listener hears and the frequency f_S produced by the source is

$$f_L = f_S\left(\frac{v + v_L}{v - v_S}\right) \qquad \textit{Doppler effect in sound} \quad (12.10)$$

Here v is the speed of sound in air, v_L is the speed of the listener (reckoned as + for motion toward the source, as − for motion away from the source), and v_s is the speed of the source (reckoned as + for motion toward the listener, as − for motion away from the listener). If the listener is stationary, $v_L = 0$, whereas if the source is stationary, $v_S = 0$.

EXAMPLE 12.8

The frequency of a train's whistle is 1000 Hz. (a) The train is approaching a stationary man at 40 m/s. What frequency does the man hear? (b) The train is stationary, and the man is driving toward it in a car whose speed is 40 m/s. What frequency does the man hear now? (Unless otherwise specified, a temperature of 20°C is to be assumed.)

SOLUTION (a) Here $f_s = 1000$ Hz, $v = 343$ m/s, $v_s = 40$ m/s, and $v_L = 0$. Hence the apparent frequency of the whistle is

$$f_L = f_S\left(\frac{v + v_L}{v - v_S}\right) = (1000\text{ Hz})\frac{343\text{ m/s}}{(343 - 40)\text{ m/s}} = 1132\text{ Hz}$$

(b) Again $f_S = 1000$ Hz and $v = 343$ m/s, but now $v_S = 0$ and $v_L = 40$ m/s. The apparent frequency of the whistle is therefore

$$f_L = f_S\left(\frac{v + v_L}{v - v_S}\right) = (1000\text{ Hz})\frac{(343 + 40)\text{ m/s}}{343\text{ m/s}} = 1117\text{ Hz}$$

We note that the answers to (a) and (b) are different. ♦

It is not difficult to derive Eq. (12.10). Let us consider first the situation shown in Fig. 12.30(a) of a sound source approaching a stationary listener with the speed v_S. The wavelength λ_S of the sound waves produced by the source when it is at rest is $\lambda_S = vT_S$, where $T_S = 1/f_S$ is the period of the waves. In the time T_S, the source moves toward the listener the distance $\Delta = v_S T_S$, so that the wavelength λ_L

in the direction of motion is

$$\lambda_L = \lambda_S - \Delta = (v - v_S)T_S = \frac{v - v_S}{f_S}$$

Since the speed of sound is a property of the medium in which the sound travels and does not depend on the speed of the source, the frequency f_L the listener perceives is related to λ_L by $f_L\lambda_L = v$. Hence

$$f_L = \frac{v}{\lambda_L} = \frac{vf_S}{v - v_S} = f_S\left(\frac{v}{v - v_S}\right) \qquad \textit{Moving source} \quad (12.10a)$$

If the source is moving away from the listener, the same reasoning gives $\lambda_L = \lambda_S + \Delta$ and $f_L = f_S v/(v + v_S)$. We can use Eq. (12.10a) for this case also simply by reckoning v_S as a negative quantity.

Next we consider a listener approaching a stationary sound source, as in Fig. 12.30(b). The listener's speed relative to the air is v_L, which means that it is $v + v_L$ relative to the waves. Thus the period of the waves appears to the listener to be $T_L = \lambda_S/(v + v_L)$ and their frequency to be

$$f_L = \frac{1}{T_L} = \frac{v + v_L}{\lambda_S} = f_S\left(\frac{v + v_L}{v}\right) \qquad \textit{Moving listener} \quad (12.10b)$$

If the listener is moving away from the sound source, the listener's speed relative to the waves is $v - v_L$. We can include this result in Eq. (12.10b) by reckoning v_L as negative in such a case. If both source and listener are in motion toward or away from each other, the source frequency f_S to be used in Eq. (12.10b) is the frequency f_L of the moving source from Eq. (12.10a), which gives Eq. (12.10):

$$f_L = f_S\left(\frac{v}{v - v_S}\right)\left(\frac{v + v_L}{v}\right) = f_S\left(\frac{v + v_L}{v - v_S}\right)$$

When $v_L = 0$, this formula becomes Eq. (12.10a), and when $v_S = 0$, it becomes Eq. (12.10b).

Frequency measurements can be made so accurately that relatively small speeds can be detected by means of the Doppler effect. For example, the speed of blood in an artery does not exceed about 0.4 m/s even in the aorta. When an ultrasound beam is directed at an artery, the waves reflected from the moving blood cells exhibit a Doppler shift because the cells then act as moving wave sources. From this shift the speed of the blood can be found (see Problem 53).

Doppler Effect in Light

The Doppler effect is an important tool of the astronomer, who is able to determine the speed of approach or recession of stars and galaxies from shifts in the characteristic frequencies of the light they emit. (These characteristic frequencies are discussed in Chapter 27.) This is the method by which the expansion of the universe was detected. Throughout the sky, characteristic frequencies in the light from distant galaxies are lower than normal, a phenomenon called the "red shift" because red light is at the low-frequency end of the visible spectrum. The magnitude of the frequency change increases with distance from the earth, which suggests that the entire universe is expanding so that all the objects in it recede from one another (Fig. 12.31).

The expansion apparently began about 15 billion years ago in the explosion of a condensed mass of primeval matter, an event usually called the **big bang**. The matter soon turned into the electrons, protons, and neutrons of which the present universe is composed, and as it expanded, individual aggregates formed that became the galaxies. Gravitational forces are slowing the expansion down, and it is possible (present data are insufficient to decide) that the expansion will eventually stop. If this happens, the universe must then contract, and its collapse (the **big crunch**) may be followed by another big bang. Otherwise the current expansion will continue forever.

The Doppler effect in light differs from that in sound because light, which does not depend on a material medium for its transmission, has the same relative speed c to all observers regardless of their state of motion (Chapter 25). The Doppler effect in light is represented by the formula

$$f = f_S \sqrt{\frac{1 + v/c}{1 - v/c}} \qquad \textit{Doppler effect in light} \qquad (12.11)$$

where f is the observed frequency, f_S is the frequency of the source, and v is the relative speed between source and observer. If source and observer are approaching each other, v is reckoned as $+$, and if they are receding from each other, v is reckoned as $-$. In a vacuum the speed of light is $c = 3.00 \times 10^8$ m/s; its value in air is very close to this. Unlike the case of sound, the Doppler effect in light does not distinguish between motion of a source and motion of an observer.

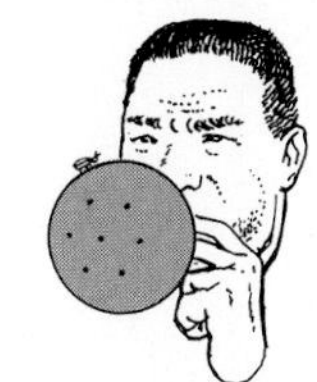

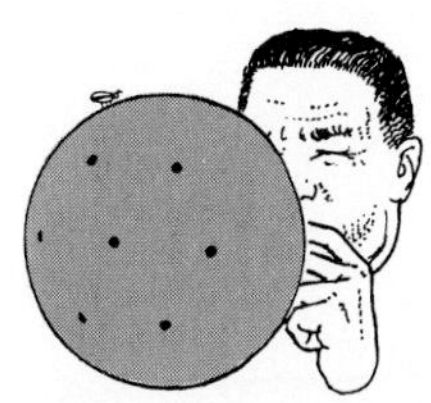

Figure 12.31

Two-dimensional analogy of the expanding universe. As the balloon is inflated, the spots on it become increasingly distant from one another. A bug on the balloon would find that the farther away a spot is from it, the faster the spot seems to be receding; this is true regardless of where the bug is located. In the case of the universe, the more distant a galaxy is from us, the faster it is moving away (as revealed by the Doppler effect), which means that the universe is expanding uniformly.

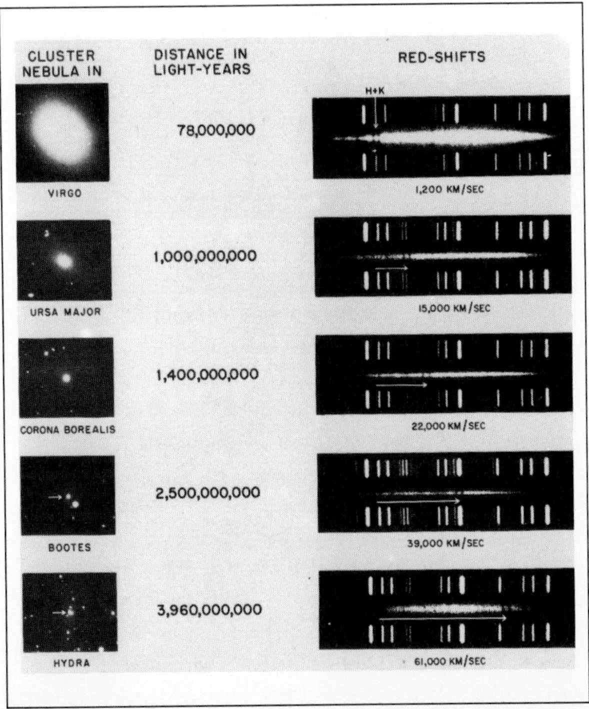

The red shift in the spectral lines of galaxies of stars increases with increasing distance from the earth. The lines marked H and K occur in the atomic spectrum of calcium. Reference spectra are shown above and below each galactic spectrum. Wavelength increases to the right. Interpreting such red shifts as Doppler shifts suggests that the entire universe has been expanding since an origin (the big bang) about 15 billion years ago.

Visible light consists of electromagnetic waves in a certain frequency interval to which the eye responds. Other electromagnetic waves, such as those used in radar and in radio communication, also exhibit the Doppler effect in accord with Eq. (12.11). Doppler shifts in radar waves are widely used by police to determine vehicle speeds, and such shifts in radio waves emitted by a set of earth satellites form the basis of the highly accurate Transit system of marine navigation.

EXAMPLE 12.9

A driver goes through a red light and, when he is arrested, claims the color he actually saw was green ($\lambda = 5.40 \times 10^{-7}$ m) and not red ($\lambda_S = 6.20 \times 10^{-7}$ m) because of the Doppler effect. The judge accepts this explanation and instead fines him for speeding at the rate of $1 for each km/h he exceeded the speed limit of 80 km/h. What was the fine?

SOLUTION To solve Eq. (12.11) for v, we first square both sides and then use routine algebra (see Appendix A) to obtain

$$v = c\left(\frac{f^2 - f_S^2}{f^2 + f_S^2}\right)$$

Here f, the observed frequency, is $c/\lambda = 5.56 \times 10^{14}$ Hz, and f_S, the source frequency, is $c/\lambda_S = 4.84 \times 10^{14}$ Hz. Hence

$$\begin{aligned} v &= (3.00 \times 10^8 \text{ m/s})\left[\frac{(5.56)^2 - (4.84)^2}{(5.56)^2 + (4.84)^2}\right] \\ &= 4.14 \times 10^7 \text{ m/s} = 1.49 \times 10^8 \text{ km/h} \end{aligned}$$

The fine is therefore $\$(1.49 \times 10^8 - 80) = \$148,999,920$. ◆

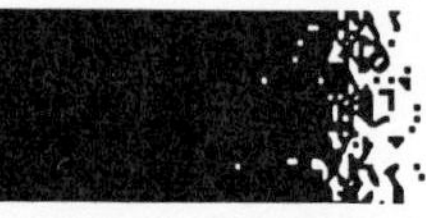

12.9 Sound Intensity

The decibel is the unit of sound intensity level.

The rate at which a wave of any kind carries energy per unit cross-sectional area is called its **intensity** I. The unit of intensity is the watt/m^2. One joule of energy per second flows through a 1-m^2 surface perpendicular to the path of a wave whose intensity is 1 W/m^2 (Fig. 12.32). The minimum intensity a 1000-Hz sound wave must have in order to be audible is about 10^{-12} W/m^2. This corresponds to pressure variations of less than a billionth of sea-level atmospheric pressure—the human ear is a very sensitive instrument. At the other extreme, sound waves with intensities over about 1 W/m^2 damage the ear.

The human ear does not respond linearly to sound intensity. Doubling the intensity of a particular sound produces the sensation of a somewhat louder sound, but one that seems far less than twice as loud. For this reason the scale customarily used to measure the intensity level of a sound is logarithmic, which is a reasonable approximation of the actual response of the human ear. The unit is the **decibel** (dB),

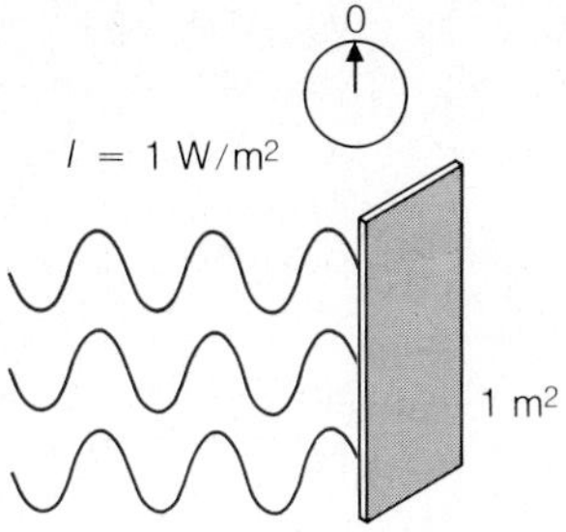

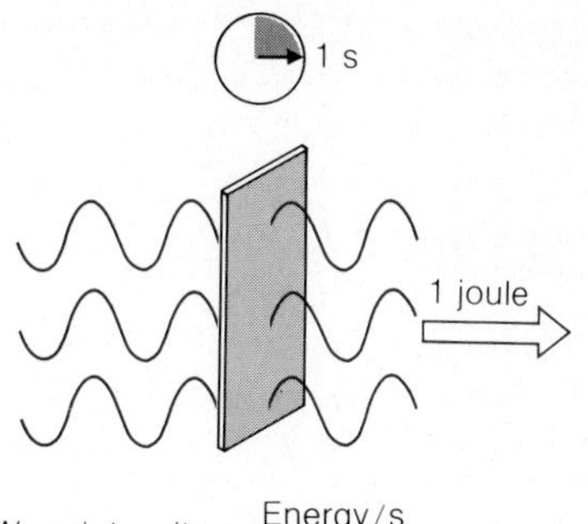

Figure 12.32

The intensity of a wave is a measure of the rate at which it transports energy.

named after Alexander Graham Bell (1847–1922), who invented the telephone. A sound that is barely audible ($I_0 = 10^{-12}$ W/m²) is given the value 0 dB. A 10-dB sound is, by definition, 10 times more intense than a 0-dB sound; a 20-dB sound is 10^2 (or 100) times more intense; a 30-dB sound is 10^3 (or 1000) times more intense; and so on. Formally, the **sound intensity level** β, in decibels, of a sound wave whose intensity in W/m² is I is given by

$$\beta = 10 \log \frac{I}{I_0} \qquad \textit{Sound intensity level} \quad (12.12)$$

(If $a = 10^b$, then b is the logarithm of a. Thus 3 is the logarithm of 1000 to the base 10 because $1000 = 10^3$. Logarithms are discussed in Appendix B.5.)

Ordinary conversation is usually about 60 dB, a sound intensity a million times greater than the faintest sound that can be heard. City traffic noise can be 80 dB, a rock band using amplifiers may reach 125 dB, and the noise of a jet airplane is about 140 dB at a distance of 30 m (Fig. 12.33). An extended exposure to sound intensity levels of over 90 dB usually leads to permanent hearing damage.

Figure 12.33

Decibel scale.

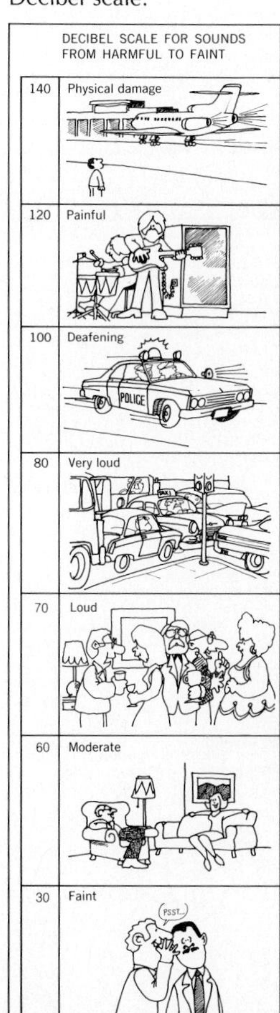

EXAMPLE 12.10

Five trumpets are being played, each at an average sound intensity level of 70 dB. What is the resulting sound intensity level?

SOLUTION The intensity of one trumpet is I_1, for a sound intensity level of

$$\beta_1 = 10 \log \frac{I_1}{I_0} = 70 \text{ dB}$$

The intensity of five trumpets is $5I_1$. Because $\log xy = \log x + \log y$ and $\log 5 = 0.70$, the sound intensity level of the trumpets is

$$\beta_5 = 10 \log \frac{5I_1}{I_0} = 10\left(\log 5 + \log \frac{I_1}{I_0}\right) = 10 \log 5 + 10 \log \frac{I_1}{I_0}$$
$$= 10(0.70)\text{dB} + \beta_1 = 7 \text{ db} + 70 \text{ dB} = 77 \text{ dB}$$

Hence increasing the intensity fivefold leads to only a 10% increase in the sound intensity level here. This is the reason why a solo instrument can be heard in a concerto even though a full orchestra is playing at the same time, and why one can carry on a conversation at a party even though many others are talking at the same time. ◆

EXAMPLE 12.11

A change in sound intensity level of 1 dB is about the minimum that can be detected by a person with good hearing; usually the change must be 2 or 3 dB to be readily apparent. What is the actual intensity ratio between two sounds that are 3 dB apart?

SOLUTION If the respective intensities and intensity levels of the two sounds are I_1, β_1 and I_2, β_2, then

$$\beta_2 - \beta_1 = 10\left(\log\frac{I_2}{I_0} - \log\frac{I_1}{I_0}\right)$$

Because $\log x - \log y = \log x/y$,

$$\log\frac{I_2}{I_0} - \log\frac{I_1}{I_0} = \log\left(\frac{I_2/I_0}{I_1/I_0}\right) = \log\frac{I_2}{I_1}$$

and so

$$\beta_2 - \beta_1 = 10\log\frac{I_2}{I_1}$$

Here $\beta_2 - \beta_1 = 3$, hence

$$3 = 10\log\frac{I_2}{I_1} \quad \text{or} \quad 0.3 = \log\frac{I_2}{I_1}$$

Now we take the antilogarithm of both sides:

$$\log^{-1} 0.3 = 2 = \log^{-1}\left(\log\frac{I_2}{I_1}\right) = \frac{I_2}{I_1}$$

To find $\log^{-1} 0.3$ with a calculator, enter 0.3 and press the 10^x key (Inv Log on some calculators). A 3-dB difference in intensity level corresponds to a factor of 2 in sound intensity. ◆

Because sound waves spread out as they move away from their source, their intensity decreases with distance. Let us consider the sound waves from a source whose power output is P; that is, P joules of energy flow from the source per second. At the distance r from the source, the total power P is distributed over the $4\pi r^2$ area of a sphere of radius r. Hence the intensity I of the sound at this distance is

$$I = \frac{P}{4\pi r^2} \qquad \textit{Intensity and distance} \quad (12.13)$$

The sound intensity is inversely proportional to the square of the distance from the source. If we go from 5 m away from a source to 10 m away, the intensity drops to $\frac{1}{4}$ its former value.

An inverse-square law holds for the intensity of all waves that spread out freely in three dimensions—the intensity of the light from a lamp also varies as $1/r^2$, for example. The purpose of the concave mirror in a searchlight is to avoid the $1/r^2$ decrease in intensity by concentrating the light waves from a lamp into as nearly parallel a beam as possible. Cupping one's hands around one's mouth similarly helps one's voice to carry farther.

EXAMPLE 12.12

The minimum sound intensity level needed for reasonable audibility is 20 dB. If a certain person speaking normally produces an intensity level of 40 dB at a distance of 1 m, what is the maximum distance at which the sound can be heard?

SOLUTION A difference of 20 dB is equivalent to an intensity ratio of $10^2 = 100$. Because $I_2/I_1 = r_1^2/r_2^2$,

$$r_2 = r_1\sqrt{I_1/I_2} = (1\text{ m})\sqrt{100} = 10\text{ m}$$ ◆

The ear is not equally sensitive to sounds of different frequencies. Maximum response occurs for sounds between 3000 and 4000 Hz, when the threshold for hearing is somewhat less than 0 dB. A 100-Hz sound, however, must have an intensity level of at least 40 dB to be heard. Sounds whose frequencies are below about 20 Hz **(infrasound)** and above about 20,000 Hz **(ultrasound)** are inaudible to almost everybody regardless of intensity. The intensity level at which a sound produces a feeling of discomfort in the ear is relatively constant at approximately 120 dB for all frequencies. Figure 12.34 shows how the thresholds of hearing and of discomfort vary

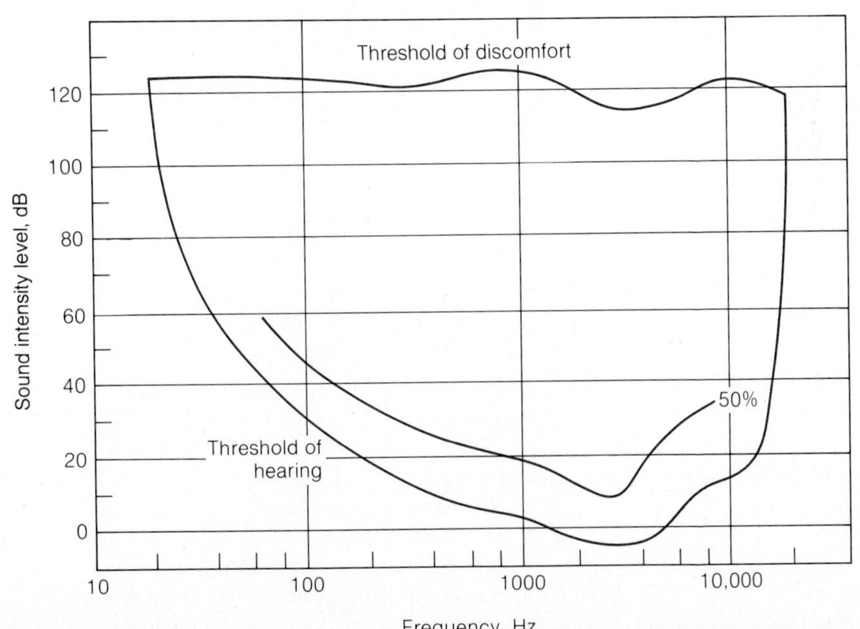

Figure 12.34

The response of the ear to sound varies with frequency. Only 1% of the U.S. population can hear sounds with intensity levels that fall below the lower curve; 50% can hear sound with intensity levels that fall below the curve above it. Hearing acuity decreases with age.

with frequency. Hearing deteriorates with age, most notably at the high-frequency end of the spectrum. A typical person 60 years old has a threshold of hearing about 10 dB higher than the lowest curve of Fig. 12.34 for 2000-Hz tones, nearly 30 dB higher for 8000-Hz tones, and nearly 70 dB higher for 12,000-Hz tones.

Important Terms

Wave motion involves the propagation of a change in a medium. Waves transport energy from one place to another.

The **frequency** of a series of periodic waves is the number of waves that pass a particular point per unit time, while their **wavelength** is the distance between adjacent crests or troughs. The **period** is the time required for one complete wave to pass a particular point. The **amplitude** of a wave is the maximum displacement of a particle of the medium on either side of its normal position when the wave passes.

Longitudinal waves occur when the individual particles of a medium vibrate back and forth in the direction in which the waves travel. **Transverse waves** occur when the individual particles of a medium vibrate from side to side perpendicular to the direction in which the waves travel. The vibrations of a stretched string are transverse waves.

The **principle of superposition** states that when two or more waves of the same nature travel past a given point at the same time, the amplitude at the point is the sum of the amplitudes of the individual waves. The interaction of different wave trains is called **interference. Constructive interference** occurs when the resulting composite wave has an amplitude greater than that of either of the original waves, and **destructive interference** occurs when the resulting composite wave has an amplitude less than that of either of the original waves.

Resonance occurs when periodic impulses are applied to a system at a frequency equal to one of its natural frequencies of oscillation.

Sound is a longitudinal wave phenomenon that results in periodic pressure variations. Sound intensity level is measured in **decibels.**

A **shock wave** is a shell of high pressure produced by the motion of an object whose speed exceeds that of sound.

The **Doppler effect** refers to the change in frequency of a wave when there is relative motion between its source and an observer.

Important Formulas

Waves in a string: $v = \sqrt{\dfrac{T}{m/L}}$

Wave motion: $v = f\lambda = \dfrac{\lambda}{T}$

Doppler effect in sound: $f_L = f_S\left(\dfrac{v + v_L}{v - v_S}\right)$

Doppler effect in light: $f = f_S\sqrt{\dfrac{1 + v/c}{1 - v/c}}$

Sound intensity level: $\beta(\text{dB}) = 10 \log \dfrac{I}{I_0}$

MULTIPLE CHOICE

1. The speed of waves in a stretched string depends on
- **a.** the tension in the string
- **b.** the amplitude of the waves
- **c.** the wavelength of the waves
- **d.** the acceleration of gravity

2. The higher the frequency of a wave,
- **a.** the lower its speed
- **b.** the shorter its wavelength
- **c.** the greater its amplitude
- **d.** the longer its period

3. Of the following properties of a wave, the one that is independent of the others is its
- **a.** amplitude
- **b.** speed
- **c.** wavelength
- **d.** frequency

4. In a transverse wave the individual particles of the medium
- **a.** move in circles
- **b.** move in ellipses
- **c.** move parallel to the direction of travel
- **d.** move perpendicular to the direction of travel

5. An example of a purely longitudinal wave is
   a. a sound wave
   b. an electromagnetic wave
   c. a water wave
   d. a wave in a stretched string

6. An example of a wave that has both longitudinal and transverse characteristics is
   a. a sound wave
   b. an electromagnetic wave
   c. a water wave
   d. a wave in a stretched string

7. Two waves meet at a time when one has the instantaneous amplitude A and the other has the instantaneous amplitude B. Their combined amplitude at this time is
   a. $A + B$
   b. $A - B$
   c. between $A + B$ and $A - B$
   d. indeterminate

8. A wave of amplitude A interferes with another wave of the same kind whose frequency is different but whose amplitude is also A. The resulting wave
   a. has an amplitude of $2A$
   b. varies in amplitude between 0 and A
   c. varies in amplitude between A and $2A$
   d. varies in amplitude between 0 and $2A$

9. A string 180 cm long has a fundamental frequency of vibration of 300 Hz. The length of a similar string under the same tension whose fundamental frequency is 200 Hz is
   a. 120 cm b. 147 cm
   c. 220 cm d. 270 cm

10. Of the following procedures, which one or ones will lower the pitch of a guitar string?
   a. Lengthen the string.
   b. Use a lighter string.
   c. Increase the tension in the string.
   d. Decrease the tension in the string.

11. Sound travels fastest in
   a. air b. water
   c. iron d. a vacuum

12. Sound waves do not travel through
   a. solids b. liquids
   c. gases d. a vacuum

13. The amplitude of a sound wave determines its
   a. pitch b. loudness
   c. overtones d. resonance

14. What we recognize as the quality, or timbre, of a musical sound is due to
   a. its fundamental frequency
   b. its overtone structure
   c. its amplitude
   d. the presence of beats

15. A pure musical tone causes a thin wooden panel to vibrate. This is an example of
   a. an overtone b. harmonics
   c. resonance d. interference

16. When a sound wave goes from air into water, the quantity that remains unchanged is its
   a. speed b. amplitude
   c. frequency d. wavelength

17. Beats are the result of
   a. diffraction
   b. constructive interference
   c. destructive interference
   d. both constructive and destructive interference

18. A sonic boom is heard after an airplane has passed overhead. This means that the airplane
   a. is accelerating
   b. is climbing
   c. was just then passing through the sound barrier
   d. is traveling faster than sound

19. Relative to the radio signals sent out by a spacecraft headed away from the earth, the signals that are received on the earth
   a. have a lower speed
   b. have a lower frequency
   c. have a shorter wavelength
   d. have all of the above characteristics

20. Sound waves whose frequency is 300 Hz have a wavelength relative to sound waves in the same medium whose frequency is 600 Hz that is
   a. half as great
   b. the same
   c. twice as great
   d. four times as great

21. Radio amateurs are permitted to communicate on the "10-meter band." What frequency of radio waves corresponds to a wavelength of 10 m? The speed of radio waves is 3.0×10^8 m/s.
   a. 3.3×10^{-8} Hz b. 3.0×10^7 Hz
   c. 3.3×10^7 Hz d. 3.0×10^9 Hz

22. A radio station broadcasts at a frequency of 660 kHz. The wavelength of these waves is
   a. 2.2 mm b. 455 m
   c. 4.55 km d. 1.98×10^{14} m

23. Waves in a lake are 5.00 m in length and pass an anchored boat 1.25 s apart. The speed of the waves is
   a. 0.25 m/s
   b. 4.00 m/s
   c. 6.25 m/s
   d. impossible to find from the information given

24. A boat at anchor is rocked by waves whose crests are 40 m apart and whose speed is 10 m/s. These waves reach the

boat once every

a. 400 s b. 30 s
c. 4 s d. 0.25 s

25. Two tuning forks of frequencies 310 and 316 Hz vibrate simultaneously. The number of times the resulting sound pulsates per second is

a. 0 b. 6
c. 313 d. 626

26. A steel wire has a fundamental frequency of vibration of 400 Hz. The fundamental frequency of another steel wire of the same length and under the same tension but whose diameter is twice as great is

a. 100 Hz b. 200 Hz
c. 400 Hz d. 800 Hz

27. In order to double the fundamental frequency of a wire under the tension T, the tension must be changed to

a. $0.5T$ b. $\sqrt{2T}$
c. $2T$ d. $4T$

28. How many times more intense is a 90-dB sound than a 40-dB sound?

a. 5 b. 50
c. 500 d. 10^5

29. Ten oboes produce a sound intensity level of 50 dB in a concert hall. The number of oboes needed to produce a level of 60 dB there is

a. 15 b. 20
c. 100 d. 200

QUESTIONS

1. A pulse sent down a long string eventually dies away and disappears. What happens to its energy?

2. When a wave passes a point in a stretched string, the energy of the string there changes back and forth between KE and elastic PE. What are the corresponding forms of energy for a water wave?

3. The amplitude of a wave is doubled. If nothing else is changed, how is the flow of energy affected?

4. A wave of frequency f and wavelength λ passes from a medium in which its speed is v to another medium in which its speed is $2v$. What are the frequency and wavelength of the wave in the second medium?

5. In general, in what state of matter does sound travel fastest? Why?

6. As a clarinet is played, the temperature of the air column inside it rises. What effect does this have on the frequencies of the notes it produces?

7. Verify that v has the dimensions of length/time in the formulas $v = \sqrt{T/(m/L)}$, $v = \sqrt{B/\rho}$, and $v = \sqrt{Y/\rho}$, which give the speed of waves, respectively, in a stretched string, in a fluid, and in a solid.

8. What is the relationship, if any, between the speed of sound in a given medium and the speed of a shock wave in that medium?

9. What property of a sound wave governs its pitch? Its loudness? Its quality?

10. How can constructive and destructive interference be reconciled with the principle of energy conservation?

11. A person has two tuning forks, one marked "256 Hz" and the other of unknown frequency. She strikes them simultaneously and hears 10 beats per second. "Aha," she says, "the other tuning fork has a frequency of 266 Hz." What is wrong with her conclusion?

12. A "double star" consists of two stars that revolve around their mutual center of mass. Telescopes cannot resolve a double star into its members, but an astronomer can recognize a double star from the characteristic frequencies in the light that reaches us from it. What property of these frequencies do you think enables the astronomer to identify their source as a double star?

13. The amplitude and wavelength of the wiggles in the grooves of a phonograph record correspond to the same quantities in the sound waves they represent. Inspection of a certain record reveals grooves whose wiggles have the same amplitude, but one has a wavelength three times shorter than the other. (a) If the audio system is linear, what will be the difference in the sound intensity the two sets of wiggles produce? (b) Will the difference in apparent loudness correspond to the difference in intensity?

EXERCISES

12.1 Pulses in a String

12.3 Periodic Waves

12.4 Types of Waves

1. A heavy rope hanging from the ceiling is given a transverse shake at its lower end. Show that the speed with which the resulting pulse travels up the rope increases as the square root of the height.

2. A certain groove in a phonograph record moves past the needle at a speed of 40 cm/s. The sound produced has a frequency of 3000 Hz. What is the wavelength of the wiggles in the groove?

3. A certain groove in a phonograph record moves past the needle at a speed of 30 cm/s. If the wiggles in the groove are 0.10 mm apart, what is the frequency of the sound that is produced?

4. Radio waves of very long wavelength can penetrate farther into seawater than those of shorter wavelength. The U.S. Navy communicates with submerged submarines by using radio waves whose frequency is only 76 Hz. What is the wavelength in air of those waves, in kilometers?

5. Water waves are observed approaching a lighthouse at a speed of 6.0 m/s. There is a distance of 7.0 m between adjacent crests. (a) What is the frequency of the waves? (b) What is their period?

6. In a ripple tank 90 cm across, a 6.0-Hz oscillator produces waves whose wavelength is 50 mm. Find the time the waves need to cross the tank.

7. An anchored boat rises by 1.2 m from its normal level and sinks by the same amount every 5.7 s as waves whose crests are 50 m apart pass it. Find the speed of the waves and the speed of the water molecules on the surface.

8. One end of a horizontal rope 12.0 m long is moved up and down seven times per second with a maximum displacement of 60 mm on each side of its equilibrium position. The rope has a mass of 1.00 kg and is under a tension of 40 N. Find the amplitude, speed, frequency, and wavelength of the resulting waves.

12.5 Standing Waves

9. The G (196 Hz) string of a guitar is 650 mm long. What is the speed of sound in the string?

10. The vibrating part of a violin string whose linear density is 4.7 g/m is 30 cm long. What tension should the string be under if its fundamental frequency is to be 440 Hz, the musical note A?

11. A stretched wire 1.00 m long has a fundamental frequency of 300 Hz. (a) What is the speed of the waves in the wire? (b) What are the frequencies of the first three overtones?

12. A steel wire 1.00 m long whose mass is 10.0 g is under a tension of 400 N. (a) What is the wavelength of its fundamental mode of vibration? (b) What is the frequency of this mode? (c) What is the wavelength of the sound waves produced when the string vibrates in this mode?

13. The G and A strings of a violin have the respective fundamental frequencies of 196 Hz and 440 Hz. If the G string has a linear density of 3.0 g/m, what should the linear density of the A string be if both are to be under the same tension?

14. The vibrating part of the G string of a certain violin is 330 mm long and has a fundamental frequency of 196 Hz when under a tension of 50 N. (a) Find the linear density of the string. (b) Where should the string be pressed in order for it to vibrate at 220 Hz?

15. The vibrating part of the E string of a violin is 330 mm long and has a fundamental frequency of 659 Hz. What is its fundamental frequency when the string is pressed against the fingerboard at a point 60 mm from its end? What are the first and second overtones of the string under these circumstances?

12.6 Sound

(Assume that the speed of sound in sea-level air is 343 m/s, which corresponds to a temperature of 20°C.)

16. The speed of sound is nearly the same in aluminum and steel, but the density of steel is almost three times that of aluminum. From this information find the approximate ratio of Young's moduli for the two metals.

17. A workman strikes a steel rail with a hammer, and the sound reaches an observer 500 m away both through the air and through the rail. How much time separates the two sounds?

18. An ultrasonic beam used in scanning body tissues has a frequency of 1.2 MHz. If the speed of sound in a particular tissue is 1540 m/s and the limit of resolution is equal to one wavelength, what is the size of the smallest detail that can be resolved?

19. There are two mechanisms by which a person determines the direction from which a sound comes. One compares the loudness of the sound at the left ear with that at the right ear, which is most effective at low frequencies. The other compares the phases of the waves that arrive at the two ears, which is most effective at high frequencies.(The **phase** of a wave is the part of its cycle it is in at a particular time and place.) The crossover point of equal effectiveness occurs at about 1200 Hz, and as a result sound with frequencies in the neighborhood of 1200 Hz are difficult to locate. How does the wavelength of a 1200-Hz sound compare with the distance between your ears?

20. A mine explodes at sea, and there is an interval of 5.0 s between the arrival of the sound through the water and its arrival through the air at a nearby ship. How far away is the mine from the ship?

21. An airplane is flying at 800 km/h at an altitude of 2.5 km. When the sound of the airplane's engines appears to a person on the ground to come from directly overhead, how far away is a point directly under the airplane?

22. A violin string is set in vibration at a frequency of 440 Hz. How many vibrations does it make while its sound travels 200 m in air?

23. A tuning fork vibrating at 440 Hz is placed in distilled water. (a) What are the frequency and wavelength of the waves produced within the water? (b) What are the frequency and wavelength of the waves produced in the surrounding air when the water waves reach the surface?

24. Find the angle between the shock wave created by an

airplane traveling at 500 m/s just above sea level and the direction of the airplane.

25. The angle between the shock wave created by an airplane and the direction of the airplane is 60°. Find the Mach number corresponding to the speed of the airplane.

26. The speed of surface waves in shallow water of depth h is $\sqrt{gh}$. How fast is a boat moving in water 1.5 m deep if the total angular width of its wake is 50°?

12.7 Interference of Sound Waves

27. When a clarinet produces the note E (330 Hz), how far is its reed from the first open hole? (The reed acts as a closed end to the air column.)

28. A 440-Hz tuning fork is held over the open top of a glass test tube held vertically. Water is added to the tube until the air column inside it is in resonance with the vibrations of the tuning fork. When this occurs, the air column is 190 mm long. What is the speed of sound in the laboratory?

29. A standing sound wave of 50 Hz is set up between the walls of a room. How far apart are the walls?

30. (a) Express the frequencies of the standing waves in a tube open at one end and closed at the other in terms of the tube's length and the speed of sound. (See Eqs. 12.6 and 12.7 for how this is done for standing waves in a stretched string.) (b) Do the same for the standing waves in a tube open at both ends. (c) The greater the number of overtones a musical note has, the "richer" it sounds. Which would have a richer tone, a tube open at one end or a tube open at both ends?

31. Determine whether the difference tone produced by the frequencies 264 Hz and 396 Hz will be consonant (pleasing to the ear) or dissonant.

32. Two identical steel wires have fundamental frequencies of vibration of 400 Hz. The tension in one of the wires is increased by 2%, and both wires are plucked. How many beats per second occur?

33. A wire under a tension of 50 N is vibrating at 200 Hz. If an identical wire under a tension of 55 N is also set vibrating, how many beats per second will be heard?

34. When two organ pipes open at both ends are sounded, 4 beats/s are heard. If the longer pipe is 1000 mm long, how long is the other pipe?

35. A whistle consists of two tubes, each closed at one end, that are 30 and 35 mm long. The chief frequency in the sound that is heard when the whistle is blown is that of the beats between the fundamental frequencies of the tubes. What is this frequency?

12.8 Doppler Effect

36. A person in a car is driving at 60 km/h toward a ferry whose whistle is blowing at 400 Hz. (a) What frequency does she hear? (b) The ferry leaves the dock and heads directly away from the driver at 15 km/h, still blowing its whistle. What frequency does she hear now?

37. A fire engine has a siren whose frequency is 500 Hz. What frequency is heard by a stationary observer when the engine moves toward him at 12 m/s? When it moves away from him at 12 m/s?

38. A police car moving at 130 km/h is chasing a truck moving in the same direction at 100 km/h. The police car has its siren on. If the truck driver finds the siren's frequency to be 600 Hz, what frequency does the police-car driver hear?

39. A latecomer to a concert hurries down the aisle toward his seat so fast that the note middle C (262 Hz) appears 1 Hz higher in frequency. How fast is he going?

40. A spacecraft moving away from the earth at 97% of the speed of light transmits data at the rate of 1.00×10^4 pulses/s. At what rate are they received?

41. A galaxy in the constellation Ursa Major is receding from the earth at 15,000 km/s. If one of the characteristic wavelengths of the light the galaxy emits is 5.50×10^{-7} m, what is the corresponding wavelength measured by astronomers on the earth?

42. The characteristic frequencies in the light from a distant galaxy of stars are found to be two-thirds as great as similar frequencies in the light from nearby stars. Find the recession speed of the distant galaxy.

12.9 Sound Intensity

43. How many times more intense is a 60-dB sound than a 50-dB sound? Than a 40-dB sound? Than a 20-dB sound?

44. The acoustic power output of a certain purring cat is 4.0×10^{-10} W. How far away can a person of normal hearing just detect the purr?

45. What is the intensity level in decibels of a sound whose intensity is 5.0×10^{-6} W/m^2?

46. The sound intensity level near a power lawn mower is 95 dB. What is the corresponding sound intensity?

47. A siren produces a 120-dB sound. What is its intensity in watts per square meter?

48. The sound intensity level in a busy street is 70 dB. At what rate does sound energy enter a room through an open window 80 cm square?

PROBLEMS

49. A steel wire originally 1.00 m long is stretched so that the resulting strain is 1.00%. Find the fundamental frequency of vibration of the wire.

50. The smallest frequency change a person with normal hearing can detect is about 1% for frequencies over 500 Hz. What is the speed of a car with a siren if a person at rest can just detect a change in the siren's pitch as it passes by? (*Hint:* v_S is much smaller than v, so $v^2 - v_S^2 \approx v^2$.)

51. A moving reflector approaches a stationary source of sound of frequency f with the speed u. A listener nearby hears both the original sound waves and the reflected sound waves. Obtain a formula for the number of beats per second the listener hears.

52. Train A is heading east at the speed v_A and train B is heading west at the speed v_B on an adjacent track. The locomotive on train A is blowing its whistle continuously. A passenger on train B observes the frequency of the sound from the whistle to be 307 Hz when the trains approach each other, 256 Hz when they are abreast, and 213 Hz when they recede from each other. From these figures find the speed of each train relative to the ground.

53. The speed with which blood flows through an artery can be determined from the Doppler shift in high-frequency sound waves sent into the artery and detected after reflection from the moving blood cells. What is done is to aim the source and receiver of the sound so that the reflections occur from cells moving away from them. Thus the sound waves striking a cell are Doppler shifted because of its motion away from the source, and the waves that reach the receiver are further shifted because they come from a reflector moving away from it. (a) Derive a formula for the frequency f_r of the waves that reach the receiver in terms of the source frequency f_S, the speed v of sound in blood (which is 1570 m/s), and the speed v_b of the blood. (b) Solve this formula for v_b. (c) Find the blood speed when the source frequency is 10^6 Hz and the Doppler shift is 40 Hz.

54. A riveting gun in a shipyard is producing 95 dB of noise. What is the sound intensity level when a second riveting gun begins to operate?

55. A chorus of 20 voices is singing at a sound intensity level of 70 dB. If the voices all have the same sound intensity level, what is it?

56. At a party the sound intensity level of the conversation is 65 dB when a record player is switched on and set to an intensity level of 70 dB. What is the sound intensity level in the room now?

57. According to government regulations, the maximum permitted daily exposure time in a workplace to noise of 90 dB is 8 h; to noise of 95 dB, 4 h; to noise of 100 dB, 2 h; and so on. Thus each increase of 5 dB means halving the permitted exposure time. What intensity ratio corresponds to a 5-dB change in sound intensity level?

58. The sound intensity level of a violin being played pianissimo is 50 dB at a distance of 8 m. Find the power output of the violin.

59. The average power output of a piano when it is playing a certain piece of music is 0.4 W. If the piano is assumed to radiate sound equally in all directions, what is the sound intensity level 10 m away?

60. A woman 50 m from a brass band in a park walks toward the band until the music she hears is 10 dB louder. How far from the band is she now?

Answers to Multiple Choice

1. a	**9.** d	**17.** d	**25.** b
2. b	**10.** a, d	**18.** d	**26.** b
3. a	**11.** c	**19.** b	**27.** d
4. d	**12.** d	**20.** c	**28.** d
5. a	**13.** b	**21.** b	**29.** c
6. c	**14.** b	**22.** b	
7. a	**15.** c	**23.** b	
8. d	**16.** c	**24.** c	

CHAPTER 13

THERMAL PROPERTIES OF MATTER

Suppose we had a microscope with unlimited power. What would we find if we were to look at a drop of water under higher and higher magnifications? Would we continue to see a clear, structureless liquid?

No such microscope exists, but there is plenty of evidence obtained in other ways to show that, on a very small scale, water consists of individual particles. In fact, so does all other matter, whether solid, liquid, or gas. The ancient Greeks anticipated this finding over two thousand years ago. What they did not even suspect, however, is that these particles are in constant, random motion, and that what we perceive as temperature is closely related to the average kinetic energy of the particles. This relationship is the key to understanding a great many physical phenomena, such as why gases become hot when compressed and why evaporation cools liquids.

13.1 Temperature

Assigning numbers to hot and cold.

All of us know what temperature signifies in terms of our sense impressions, and we do not have to understand what temperature means in terms of the structure of matter to make use of the concept of temperature in a variety of ways. For the moment it is sufficient for us simply to regard temperature as that which gives rise to sensations of hot and cold.

A **thermometer** is a device used to measure temperature. Most substances expand when heated and contract when cooled, and the thermometers of everyday life are based on this property of matter. More precisely, they are based on the different rates of expansion of different materials (Fig. 13.1). Thus the length of the

Figure 13.1

Three types of thermometer based on thermal expansion.

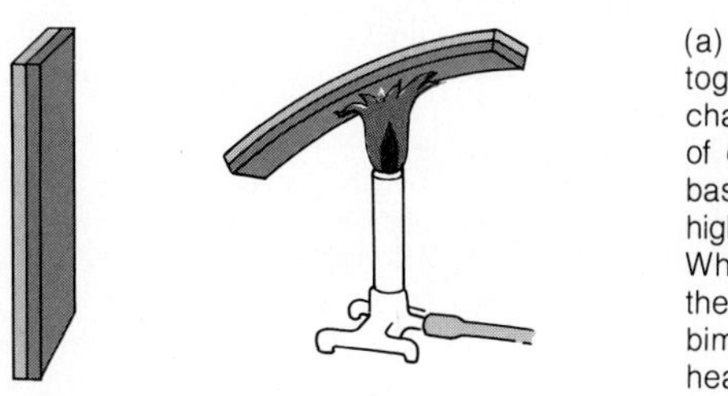

(a) Two strips of different metals that are joined together bend to one side or the other with a change in temperature owing to different rates of expansion in the two metals, which is the basis of household oven thermometers. The higher the temperature, the greater the deflection. When cooled, such a bimetallic strip bends in the opposite direction. A thermostat uses a bimetallic strip to operate a switch that turns a heating or cooling system on and off at preset temperatures.

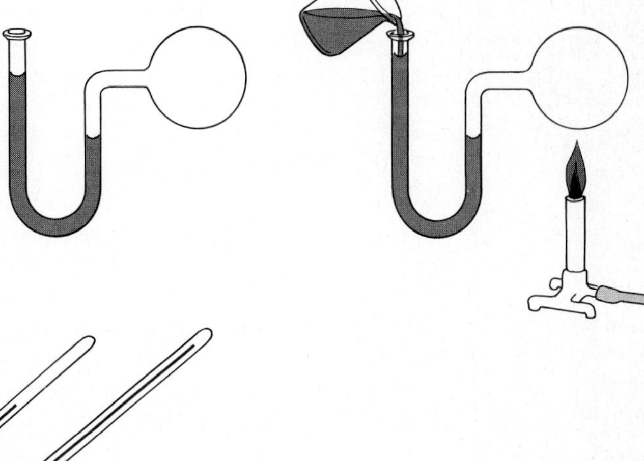

(b) In a constant-volume gas thermometer, which is a very sensitive laboratory instrument, the height of the mercury column at the left is adjusted until the mercury column at the right is at a fixed level. The difference in heights of the two mercury columns is a measure of the pressure needed to maintain the gas in a fixed volume, and hence a measure of the temperature.

(c) Mercury (or colored alcohol) expands more when heated than glass does, and so the length of the liquid column in a liquid-in-glass thermometer is a measure of the temperature of the thermometer bulb.

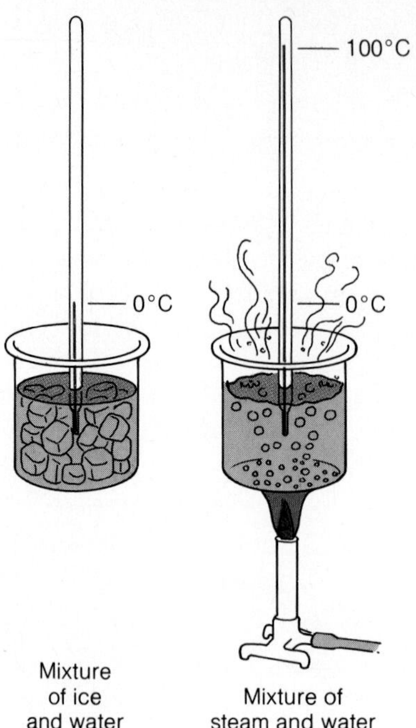

Figure 13.2

Calibrating a thermometer on the Celsius scale. A mixture of ice and water at atmospheric pressure is, by definition, at 0°C, and a mixture of steam and water at atmospheric pressure is, again by definition, at 100°C.

mercury column in the familiar mercury-in-glass thermometer indicates the temperature of the bulb because mercury expands more than glass when heated and contracts more than glass when cooled.

Thermal expansion is not the only property of matter that can be used to make a thermometer. The electrical resistance of most materials, for instance, varies with temperature. Another example is the color of an object heated until it glows. A poker thrust in a fire at first becomes a dull red, then bright red, orange, and yellow. At a high enough temperature it becomes "white hot." The color of a really hot object is thus a measure of its temperature, a fact used by astronomers when they study stars.

Before we can make a thermometer of any kind, we need a temperature scale. One way to establish such a scale is to note that water freezes into a solid (ice) and vaporizes into a gas (steam), both at definite temperatures at a given pressure. In the **Celsius scale** the freezing point of water at 1 atm pressure (or, more exactly, the point at which a mixture of ice and water is in equilibrium, with as much ice melting as water freezing) is called 0°, and the boiling point of water at 1 atm pressure (or, more exactly, the point at which a mixture of steam and water is in equilibrium) is called 100° (Fig. 13.2). Temperatures in the Celsius scale are written, for example, "40°C." In the United States the Celsius scale is sometimes called the **centigrade scale.**

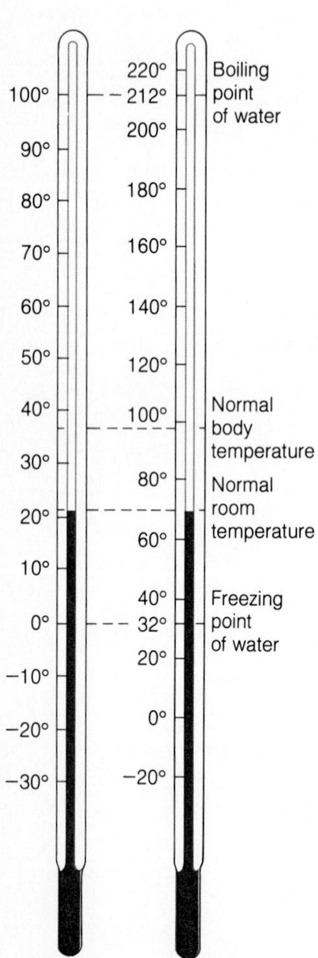

Figure 13.3

The Fahrenheit and Celsius temperature scales.

Although the Celsius scale is used in most of the world, a different temperature scale called the **Fahrenheit scale** is commonly used for nonscientific purposes in some English-speaking countries. In the Fahrenheit scale the freezing point of water is 32°F and the boiling point of water is 212°F (Fig. 13.3). This means that 180°F separates the freezing and boiling points of water, whereas 100°C separates them in the Celsius scale. Therefore Fahrenheit degrees are 100/180, or 5/9, as large as Celsius degrees. We can convert temperatures from one scale to the other with the help of the formulas

$$T_F = \tfrac{9}{5}T_C + 32° \qquad \textit{Celsius to Fahrenheit} \quad (13.1a)$$

$$T_C = \tfrac{5}{9}(T_F - 32°) \qquad \textit{Fahrenheit to Celsius} \quad (13.1b)$$

For instance, the Celsius equivalent of 70°F is $\tfrac{5}{9}(70° - 32°) = 21°C$.

13.2 Thermal Expansion

It is proportional to temperature change in nearly all cases.

A change in temperature causes most solids to change in length by an amount proportional to their original lengths and to the temperature change. A long steel bridge may vary in length by over a meter between summer and winter, which must be allowed for in its design.

If the original length of a rod of a certain material is L_0, its change in length ΔL after its temperature changes by ΔT is

$$\Delta L = aL_0\,\Delta T \qquad \textit{Thermal expansion} \quad (13.2)$$

Change in length = (a)(original length) (temperature change)

The quantity a, called the **coefficient of linear expansion,** is a constant whose value depends on the nature of the material. Different substances expand (and contract) to different extents. A lead rod, for example, changes in length by 60 times as much as a quartz rod of the same initial length when both are heated or cooled through the same temperature interval. Table 13.1 lists coefficients of linear expansion for various substances.

Table 13.1 Coefficients of linear expansion

Substance	Coefficient
Aluminum	2.4×10^{-5}/°C
Brass	1.8
Concrete	0.7–1.2
Copper	1.7
Iron	1.2
Lead	3.0
Quartz	0.05
Silver	2.0
Steel	1.2

Thermal expansion and contraction change the length of the Verrazano-Narrows Bridge across New York Bay by over a meter between summer and winter.

EXAMPLE 13.1

What is the increase in length of a steel girder that is 10 m long at 5°C when its temperature rises to 30°C?

SOLUTION The coefficient of linear expansion of steel is $1.2 \times 10^{-5}/°\text{C}$. From Eq. (13.2)

$$\Delta L = aL_0\,\Delta T = (1.2 \times 10^{-5}/°\text{C})(10\ \text{m})(25°\text{C}) = 3.0 \times 10^{-3}\ \text{m}$$
$$= 3.0\ \text{mm}$$

◆

EXAMPLE 13.2

How much force would be needed to stretch the girder by the same amount at constant temperature if the girder's cross sectional area were 200 cm^2?

SOLUTION Because the girder increases in length by 3.0 mm, the force is the same as that required to stretch it by 3.0 mm (Fig. 13.4). Equation (9.5) gives the change in length ΔL of a rod that is subjected to a tension or compression force F as

$$\Delta L = \frac{L_0}{Y}\frac{F}{A}$$

where A is the cross-sectional area of the rod and Y is Young's modulus for the material of the rod. The girder has $A = 200\ \text{cm}^2 = 0.020\ \text{m}^2$ and, from Table 9.2, Young's modulus for steel is $2.0 \times 10^{11}\ \text{N/m}^2$. Hence

$$F = \frac{YA\,\Delta L}{L_0} = \frac{(2.0 \times 10^{11}\ \text{N/m}^2)(0.020\ \text{m}^2)(3.0 \times 10^{-3}\ \text{m})}{10\ \text{m}}$$
$$= 1.2 \times 10^6\ \text{N}$$

The required force is 1.2 million newtons, which is equivalent to 135 tons! Clearly, thermal expansion can involve quite considerable forces.

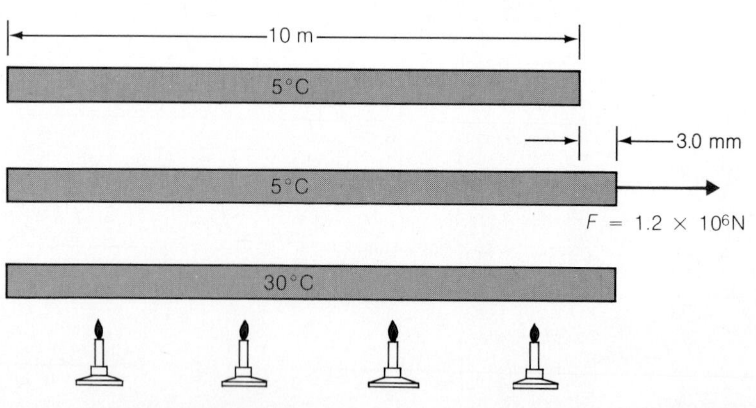

Figure 13.4

A 10-m steel girder expands 3.0 mm when its temperature is increased by 25°C. At constant temperature, a force of 1.2×10^6 N would be required to produce the same increase in length if the cross-sectional area of the girder were 200 cm^2.

◆

Table 13.2 Coefficients of volume expansion

Substance	Coefficient
Ethyl alcohol	7.5×10^{-4}/°C
Glass (average)	0.2
Glycerin	5.1
Ice	0.5
Mercury	1.8
Pyrex glass	0.09
Water	2.1

A formula similar to Eq. (13.2) holds for the changes in volume, ΔV, of a solid or liquid whose temperature changes by an amount ΔT. Here we have

$$\Delta V = bV_0 \, \Delta T \qquad \textit{Volume expansion} \quad (13.3)$$

where V_0 is the original volume and b is the **coefficient of volume expansion.** Table 13.2 lists coefficients of volume expansion for various substances. In general, the coefficients of linear and volume expansion are related by

$$b = 3a$$

so that we can readily determine the values of b for the materials in Table 13.1.

EXAMPLE 13.3

Calculate the volume of water that overflows when a Pyrex beaker filled to the brim with 250 cm^3 of water at 20°C is heated to 60°C.

SOLUTION First we note that a hole in a body expands or contracts by just as much as a solid object whose composition is that of the body and whose volume is that of the hole (Fig. 13.5). The reason is that the material around the hole is not affected by whether the hole is there or not. The boundary of the hole, and therefore the size and shape of the hole, thus behaves as it would if the hole were filled. This

Figure 13.5

A hole in a body expands or contracts with a change in temperature precisely as a solid object of the same size, shape, and composition would.

means that we can write for the increase in capacity of the beaker

$$\Delta V_b = b_P V_b \, \Delta T = (0.09 \times 10^{-4}/°C)(250 \text{ cm}^3)(40°C) = 0.09 \text{ cm}^3$$

The increase in the volume of the water is

$$\Delta V_w = b_w V_w \, \Delta T = (2.1 \times 10^{-4}/°C)(250 \text{ cm}^3)(40°C) = 2.1 \text{ cm}^3$$

and so the volume of water that overflows is

$$\Delta V_w - \Delta V_b = 2.0 \text{ cm}^3$$

◆

Water

The thermal expansion of water is unusual. Above 4°C water expands when heated, just as most other substances do. From 0°C to 4°C, however, the volume of a water sample *decreases* with increasing temperature (Fig. 13.6). Thus water has its maximum density at 4°C. Equally unusual is the expansion of water when it freezes, so that ice floats and exposed water pipes may burst in a severe winter. (The reason for this behavior will be discussed in Section 28.3.)

Because ice floats, a body of water freezes in winter from the top down, not from the bottom up as it would if ice were denser than water and cold water denser than warm water at all temperatures. Ice is a much poorer conductor of heat than water (the difference in heat conductivity is nearly a factor of 4), which means that the layer of ice that forms initially on the surface of a body of water impedes further freezing. Ice forming at the bottom would not have this effect. Many lakes, rivers, and arms of the sea are therefore able to escape total freezing in winter, and as a result their plant and animal life is able to survive until spring.

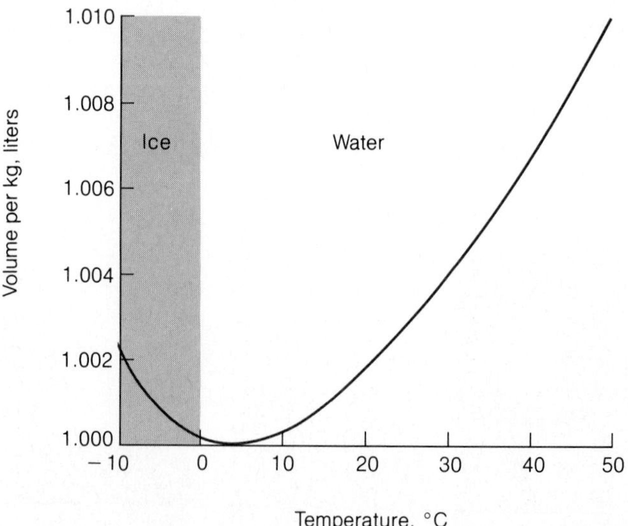

Figure 13.6

Water has its greatest density at 4°C. Because water expands when it freezes, ice floats.

13.3 Boyle's Law

At constant temperature, the volume of a gas is inversely proportional to its pressure.

The thermal behavior of a gas is rather different from the behaviors of solids and liquids because, unlike them, a gas does not have a specific volume at a given temperature but expands to fill its container. The only way to change the volume of a gas is to change the capacity of its container. Even though its volume may remain the same, however, another property of a confined gas does vary with its temperature, namely, the pressure it exerts on the container walls. The air pressure in the tires of a car drops in cold weather and increases in warm weather, an illustration of this property.

When the temperature of a sample of gas is held constant, the absolute pressure it exerts on its container is very nearly inversely proportional to the volume of the container. Expanding the container lowers the pressure; shrinking the container raises the pressure. Conversely, increasing the pressure on a gas sample reduces its volume; decreasing the pressure increases its volume (Fig. 13.7). This relationship is called **Boyle's law** after its discoverer, Robert Boyle (1627–1691). Though not exact, Boyle's law is an excellent approximation over a wide range of temperatures and pressures.

Boyle's law can be written as

$$p_1V_1 = p_2V_2 \qquad (T = \text{constant}) \qquad \textit{Boyle's law} \quad (13.4)$$

where p_1 is the absolute gas pressure when the volume of the gas is V_1, and p_2 is the absolute gas pressure when the volume is V_2.

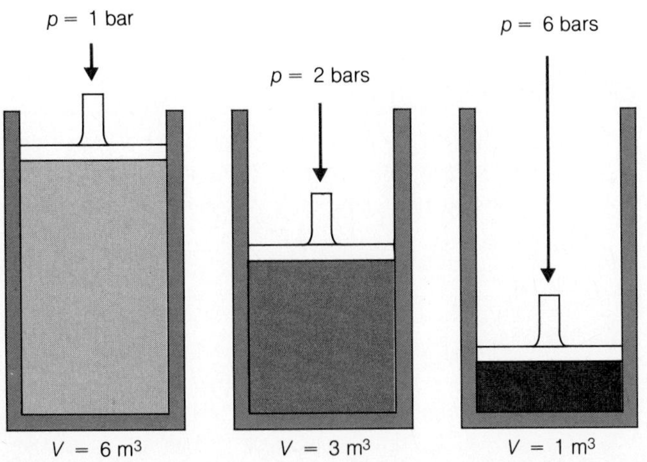

Figure 13.7

Boyle's law states that the volume of a gas sample is inversely proportional to its pressure at constant temperature. Thus $p_1V_1 = p_2V_2 = p_3V_3$, as shown.

BIOGRAPHICAL NOTE

Robert Boyle

Robert Boyle (1627–1691) was born in Ireland, the fourteenth child of the Earl of Cork. He was educated at home and at Eton, and as a youth traveled widely in Europe. In 1654 he settled in Oxford, England, where he devoted himself to experiments in physics and chemistry. Boyle hired as his assistant Robert Hooke, who later did noteworthy research on his own; Hooke's law, the proportionality between stress and strain below the elastic limit, was one of his discoveries. Boyle and Hooke built the earliest efficient vacuum pumps, and Boyle used an evacuated glass cylinder to show that objects of all kinds, from feathers to bullets, have exactly the same acceleration when dropped in a vacuum, as Galileo had predicted. Boyle also showed that electrical forces act in a vacuum, but sound does not pass through one.

Boyle conducted many experiments with gases, one result of which became known as Boyle's law: $pV =$ constant for a gas sample whose temperature is fixed. He referred to this inverse relationship as the "spring of air" by analogy with a coil spring's resistance to compression.

With the publication of *The Skeptical Chemist* in 1661, Boyle helped establish chemistry as a science by clearly distinguishing between elements and compounds. He was also the first to distinguish between acids and bases. In chemistry as in physics, Boyle was completely open about what he found and how he found it. This enabled others to repeat his experiments and check his observations and conclusions. Such openness was unusual at the time, when most scientists were secretive about their work; thus Boyle was a more significant figure in the development of science than his discoveries alone would suggest.

EXAMPLE 13.4

A scuba diver's 12-L tank is filled with air at a gauge pressure of 150 bars. If the diver uses 30 L of air per minute at the same pressure as the water pressure at her depth below the surface, how long can she remain at a depth of 15 m in seawater?

SOLUTION Since atmospheric pressure is ~1.0 bar, the absolute pressure of the air in the tank is

$$p_1 = p_{\text{atm}} + p_{\text{gauge}} = 1 \text{ bar} + 150 \text{ bars} = 151 \text{ bars}$$

The absolute pressure at a depth of 15 m is, according to Example 10.3,

$$p_2 = p_{\text{atm}} + \rho g h = 1.0 \text{ bar} + 1.5 \text{ bars} = 2.5 \text{ bars}$$

From Eq. (13.4) we have for the volume of air available at a pressure of 2.5 bars

$$V_2 = \frac{p_1V_1}{p_2} = \frac{(151 \text{ bars})(12 \text{ L})}{2.5 \text{ bars}} = 725 \text{ L}$$

However, 12 L of air remain in the tank, so only 713 L are usable and

$$t = \frac{713 \text{ L}}{30 \text{ L/min}} = 24 \text{ min}$$

◆

13.4 Charles's Law

At constant pressure, the volume of a gas is proportional to its absolute temperature.

Now let us see what happens to a gas when its temperature is changed. As mentioned earlier, if the volume of a gas is held constant, the pressure it exerts on its container depends on its temperature. According to Boyle's law, then, if we hold the gas pressure constant, its volume should vary with temperature. When this prediction was tested over 150 years ago by Jacques Charles and Joseph Gay-Lussac in France, they found that the change in volume ΔV of a gas sample is in fact related to a change ΔT in its temperature by the same formula, Eq. (13.3), that holds for solids and liquids:

$$\Delta V = bV_0 \, \Delta T$$

The significant thing about gases at constant pressure is that they *all* have very nearly the same coefficient of volume expansion b. By contrast, as Tables 13.1 and 13.2 indicate, the thermal coefficients for solids and liquids may have markedly different values for different substances. At 0°C the coefficient of volume expansion b_0 for all gases is very close to

$$b_0 = \frac{0.0037}{°\text{C}} = \frac{1/273}{°\text{C}} \qquad \text{(all gases)}$$

If we vary the temperature of a gas sample while holding its pressure constant, its volume changes by 1/273 of its volume at 0°C for each 1°C temperature change. A child's large balloon filled with air whose volume at 0°C is 1.000 m^3 has a volume of 1.037 m^3 at 10°C and 0.963 m^3 at −10°C (Fig. 13.8).

What happens when the balloon is cooled to −273°C? At that temperature the air in the balloon should have lost 273/273 of its volume at 0°C, and therefore should have vanished entirely!

Actually, all gases condense into liquids at temperatures above −273°C, so the question has no physical meaning. But −273°C is still an important temperature. Let us set up a new temperature scale, the **absolute temperature scale**, and designate −273°C as the zero point (Fig. 13.9). Temperatures in the absolute scale are expressed

Figure 13.8

Charles's law states that the volume of a gas sample is directly proportional to its absolute temperature at constant pressure. Thus $V_1/T_1 = V_2/T_2 = V_3/T_3$ as shown. (The changes in size are greatly exaggerated in the drawing.)

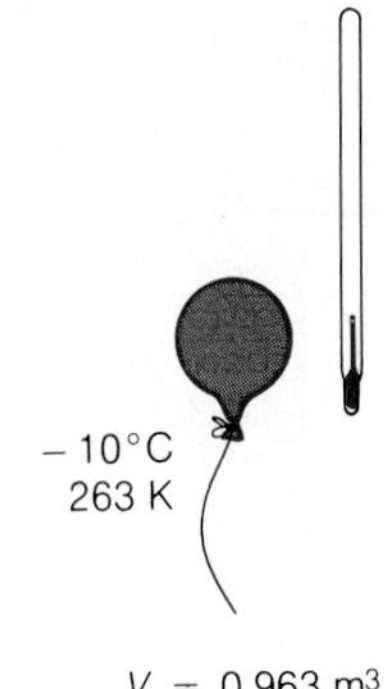

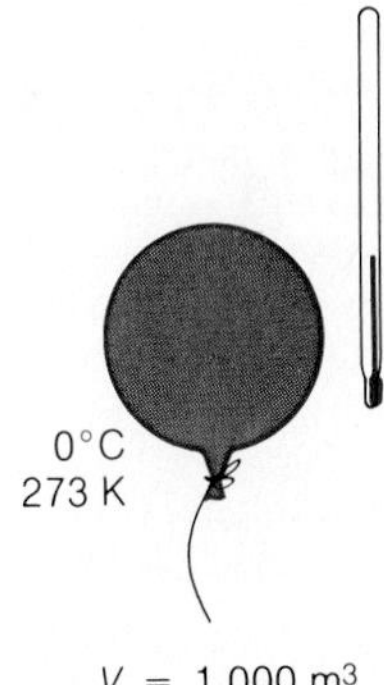

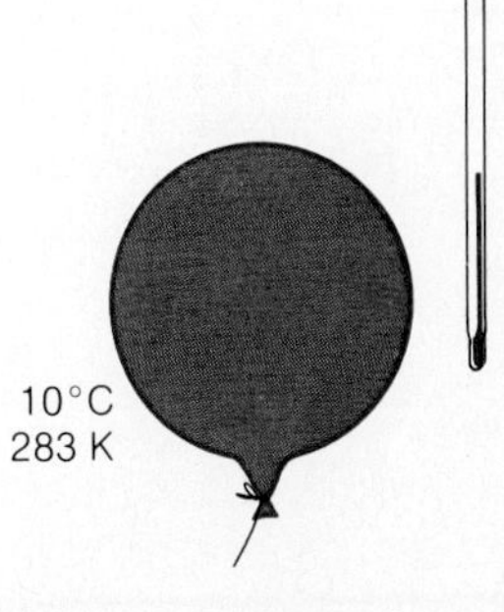

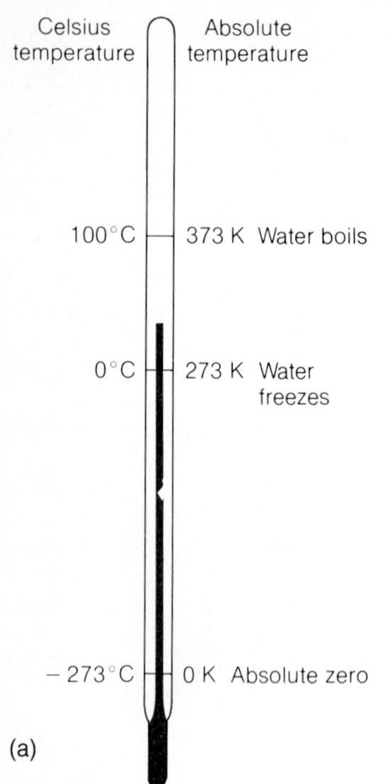

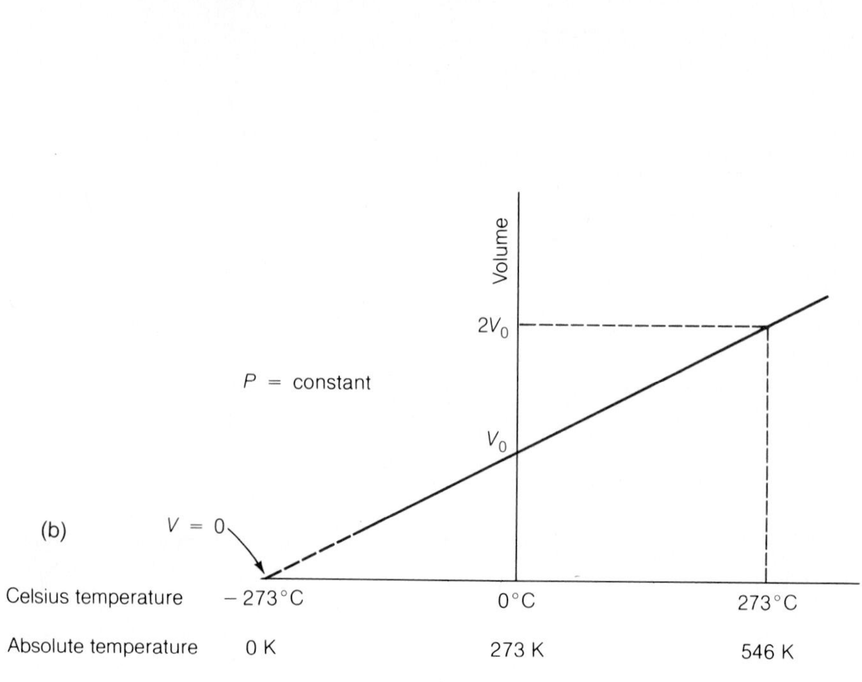

Figure 13.9

(a) The absolute temperature scale. (b) The volume of a gas at constant pressure is directly proportional to its absolute temperature.

In the operation of a snow-making machine, a mixture of compressed air and water is blown through a nozzle or set of nozzles. The expansion of the air cools the mixture sufficiently to freeze the water into the ice crystals of snow.

in **kelvins**, denoted K, after Lord Kelvin (1824–1907), a noted British physicist. To convert temperatures from one scale to the other we note that

$$T_K = T_C + 273 \qquad \textit{Celsius to absolute} \qquad (13.5a)$$

$$T_C = T_K - 273 \qquad \textit{Absolute to Celsius} \qquad (13.5b)$$

Thus the equivalent on the absolute scale of room temperature, 20°C, is

$$T_K = T_C + 273 = (20 + 273)\text{K} = 293 \text{ K}$$

The reason for setting up the absolute temperature scale is that, provided the pressure is constant, *the volume of a gas sample is directly proportional to its absolute temperature* (Fig. 13.9b). This relationship is called **Charles's law**. Like Boyle's law, Charles's law is not a basic physical principle, but deviations from it are usually quite small.

We can express Charles's law in the form

$$\frac{V_1}{T_1} = \frac{V_2}{T_2} \qquad (p = \text{constant}) \qquad \textit{Charles's law} \qquad (13.6)$$

where V_1 is the volume of a gas sample at the absolute temperature T_1 and V_2 is its volume at the absolute temperature T_2.

If there were a gas that did not liquify before reaching 0 K, then at 0 K its volume would shrink to zero. Because a negative volume has no meaning, it is natural to think of 0 K as **absolute zero**. Indeed 0 K is the lower limit to the temperature anything can have, but on the basis of a stronger argument than one based on imaginary gases. This argument is discussed later in this chapter. To five significant figures the Celsius equivalent of absolute zero is $-273.16°C$.

EXAMPLE 13.5

To what Celsius temperature must a gas sample initially at 20°C be heated if its volume is to double while its pressure remains the same?

SOLUTION The absolute equivalent of 20°C is

$$T_K = T_C + 273 = (20 + 273)\text{K} = 293 \text{ K}$$

Because $V_2 = 2V_1$, we have from Eq. (13.6)

$$T_2 = \frac{V_2T_1}{V_1} = \frac{(2V_1)T_1}{V_1} = 2T_1 = 2(293 \text{ K}) = 586 \text{ K}$$

The Celsius equivalent of 586 K is

$$T_C = T_K - 273 = (586 - 273)°\text{C} = 313°\text{C}$$ ♦

Ideal Gas Law

Boyle's law and Charles's law can be combined in a single formula called the **ideal gas law:**

$$\frac{p_1V_1}{T_1} = \frac{p_2V_2}{T_2} \qquad \textit{Ideal gas law} \quad (13.7)$$

When $T_1 = T_2$, the ideal gas law becomes Boyle's law, and when $p_1 = p_2$, it becomes Charles's law.

EXAMPLE 13.6

(a) A tank with a capacity of 1.0 m^3 contains helium gas at 27°C under a pressure of 20 atm. The helium is used to fill a balloon. When the balloon is filled, the gas pressure inside it is 1.0 atm, and its temperature has dropped to $-33°C$. (The gas has done work in expanding at the expense of its internal energy, and the cooling reflects this loss of internal energy.) What is the volume of the balloon at this time? (b) After a while the helium in the balloon absorbs heat from the atmosphere and returns to its original temperature of 27°C, and it expands further to maintain its pressure at 1.0 atm. What is the final volume of the balloon? (The gas pressure in the balloon is actually slightly greater than 1.0 atm to balance the tendency of the rubber to contract, but this is ignored here for convenience.)

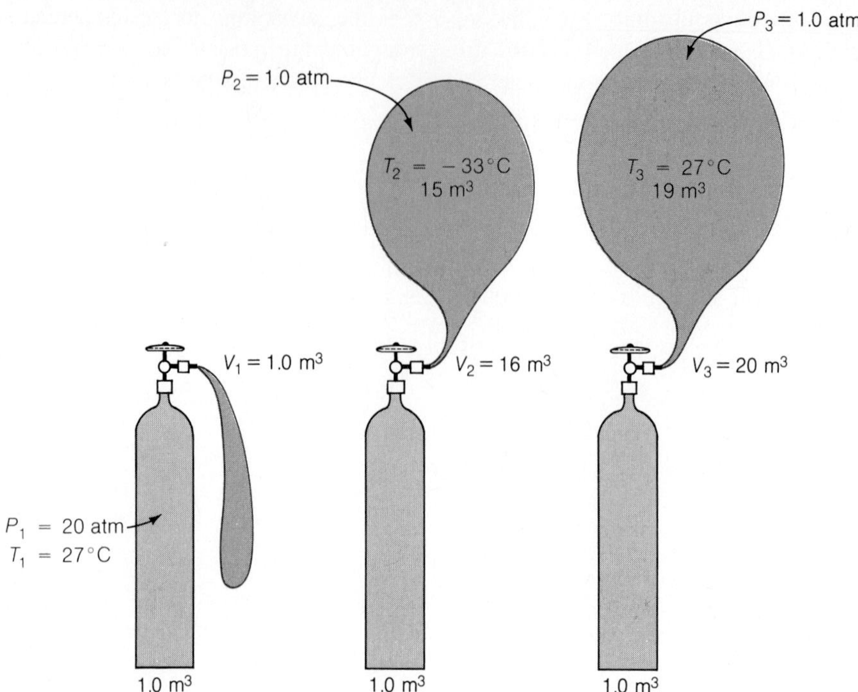

Figure 13.10

SOLUTION (a) The equivalents of 27°C and −33°C on the absolute scale are 300 K and 240 K, respectively. Applying the ideal gas law to the initial expansion gives

$$V_2 = \frac{T_2 p_1}{T_1 p_2} V_1 = \left(\frac{240\ \text{K}}{300\ \text{K}}\right)\left(\frac{20\ \text{atm}}{1.0\ \text{atm}}\right)(1.0\ \text{m}^3) = 16\ \text{m}^3$$

Because the tank's capacity is 1.0 m^3, the balloon's volume after the initial expansion is 15 m^3 (Fig. 13.10).

(b) When the helium has reached the outside air temperature of 27°C, which we will call state 3, then $T_1 = T_3$. Hence we need only apply Boyle's law to states 1 and 3 to obtain the eventual volume of the helium:

$$V_3 = \frac{p_1}{p_3} V_1 = \left(\frac{20\ \text{atm}}{1.0\ \text{atm}}\right)(1.0\ \text{m}^3) = 20\ \text{m}^3$$

Again we subtract the 1.0 m^3 volume of the tank to find the volume of the balloon itself, which is 19 m^3. ◆

The ideal gas law is obeyed approximately by all gases. The significant thing is not that the agreement with experiment is never perfect, but that *all* gases, no matter what kind, behave almost identically. An **ideal gas** is defined as one that obeys Eq. (13.7) exactly. Although no ideal gases actually exist, they do provide a target for theories of the gaseous state to aim at. It is reasonable to suppose that the ideal gas law is based on the essential nature of gases. Hence the next step is to account for this law and only afterward to seek reasons for its failure to be completely correct.

13.5 Atoms and Molecules

The ultimate particles of gases.

Elements are the simplest substances of which matter in bulk is composed. Over 100 are known, 92 found in nature and the rest artificially prepared. The ultimate particles of an element are its **atoms.**

Two or more elements may combine chemically to form a **compound**, a substance whose properties are different from those of the elements that compose it. The ultimate particles of a compound in gaseous form are **molecules**, which consist of the atoms of its constituent elements joined together in a definite way. Thus each molecule of water contains two hydrogen atoms and one oxygen atom. While the ultimate particles of elements are atoms, many elemental gases consist of molecules rather than atoms. Oxygen molecules, for instance, contain two oxygen atoms each. The molecules of other gases, such as helium and argon, are single atoms. Figure 13.11 shows schematically the composition of some common molecules.

Atoms and molecules are very small, and even a tiny bit of matter contains huge numbers of them. If each atom in a penny were worth one cent, all the money in the world would not be enough to pay for the penny. Elements in liquid and solid form are usually assemblies of individual atoms. Some compounds in these forms are also assemblies of individual molecules, but more often the situation is more complicated, as we will see in Chapter 28.

The masses of atoms and molecules are usually expressed in **atomic mass units** (u), where

$$1 \text{ atomic mass unit} = 1 \text{ u} = 1.66 \times 10^{-27} \text{ kg}$$

A list of the atomic masses of the elements is given in Appendix C. If we know the composition of a compound, we can calculate the corresponding molecular mass.

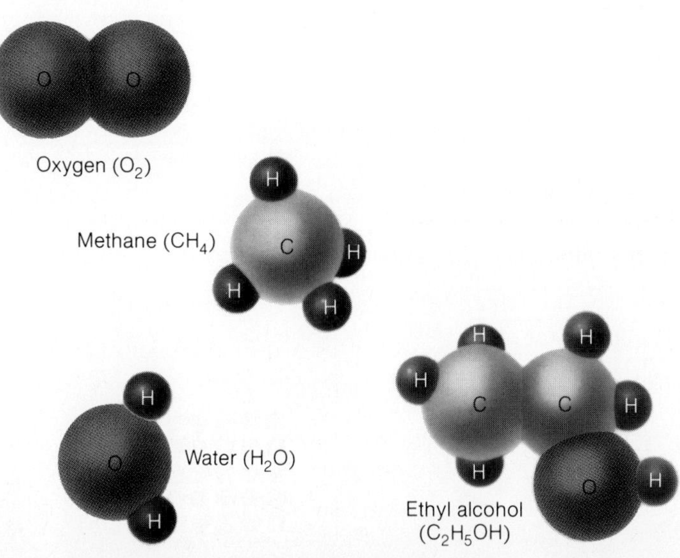

Figure 13.11

Molecular structures of several common substances.

EXAMPLE 13.7

How many H_2O molecules are present in 1.00 g of water? The atomic mass of hydrogen is 1.008 u and that of oxygen is 16.00 u.

SOLUTION We begin by finding the mass of the H_2O molecules in u:

$$\begin{aligned} 2\text{H} &= 2 \times 1.008\ \text{u} = \ \ 2.02\ \text{u} \\ \text{O} &= 1 \times 16.00\ \text{u} = \underline{16.00\ \text{u}} \\ &\qquad\qquad\qquad\quad 18.02\ \text{u} \end{aligned}$$

The mass of the H_2O molecule in kg is therefore

$$m = (18.02\ \text{u})(1.66 \times 10^{-27}\ \text{kg/u}) = 2.99 \times 10^{-26}\ \text{kg}$$

and so the number of H_2O molecules in 1.00 g $= 1.00 \times 10^{-3}$ kg of water is

$$\begin{aligned} \text{Molecules of H}_2\text{O} &= \frac{\text{mass of H}_2\text{O}}{\text{mass of H}_2\text{O molecule}} = \frac{1.00 \times 10^{-3}\ \text{kg}}{2.99 \times 10^{-26}\ \text{kg}} \\ &= 3.34 \times 10^{22}\ \text{molecules} \end{aligned}$$

◆

A considerable amount of experimentation and ingenious reasoning had to be carried out before the reality of atoms and molecules became definitely established. Although the full story of the kinetic-molecular theory of matter, a large part of which involves chemistry, will not be gone into here, it is easy to show that it can account for the ideal gas law. We will also see how the physical properties of solids and liquids, and the deviations of actual gases from an ideal gas, fit into the kinetic-molecular theory.

13.6 Kinetic Theory of Gases

Why gases behave as they do.

According to the **kinetic theory of gases**, a gas consists of a great many tiny individual molecules that do not interact with one another except in collisions. The molecules are supposed to be far apart compared with their dimensions and to be in constant motion, incessantly hurtling to and fro, as in Fig. 13.12. They are kept from escaping into space only by the solid walls of a container (or, in the case of the earth's atmosphere, by gravity). A natural consequence of the random motion and

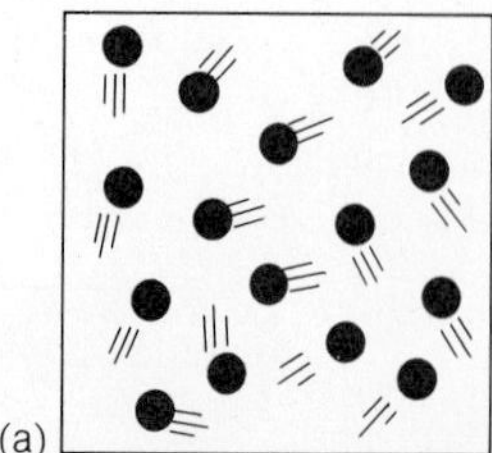

(a)

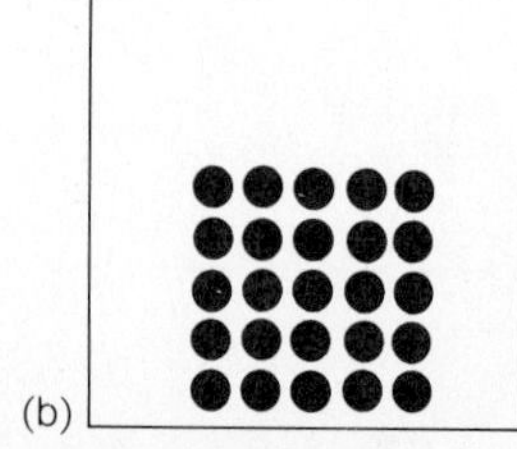

(b)

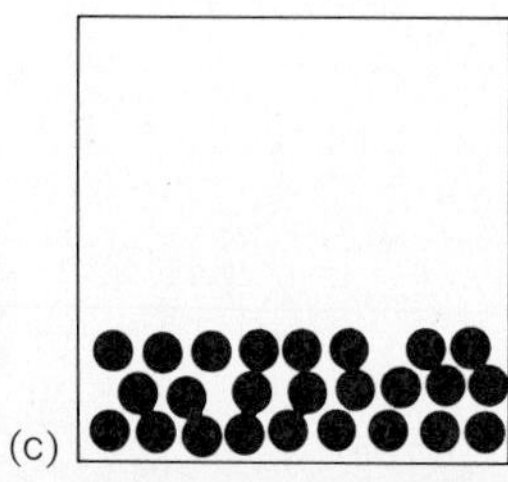

(c)

Figure 13.12

(a) The molecules of a gas are in constant, random motion. (b) The constituent particles of a solid are also in motion but oscillate about definite equilibrium positions. (c) The molecules of a liquid keep a more or less constant distance apart but move about freely.

large molecular separation is the tendency of a gas to completely fill its container and to be readily compressed or expanded.

The particles of a solid, on the other hand, are close together, and mutual attractive and repulsive forces hold them in place to provide the solid with its characteristic rigidity. In a liquid the intermolecular forces are also sufficient to keep the volume of a sample constant. However, they are not strong enough to prevent nearby molecules from sliding past one another, which allows liquids to flow.

Boyle's law follows directly from the picture of a gas as a group of randomly moving molecules. The pressure the gas exerts comes from the impacts of its molecules. The vast number of molecules in even a tiny gas sample means that their separate blows appear as a continuous force to our senses and measuring instruments. Figure 13.13 shows a simplified model of a gas confined to a box. Although the molecules are actually traveling about in all directions, the effects of their collisions with the walls of the box are the same as if one-third of them were moving back and forth between each pair of opposite walls.

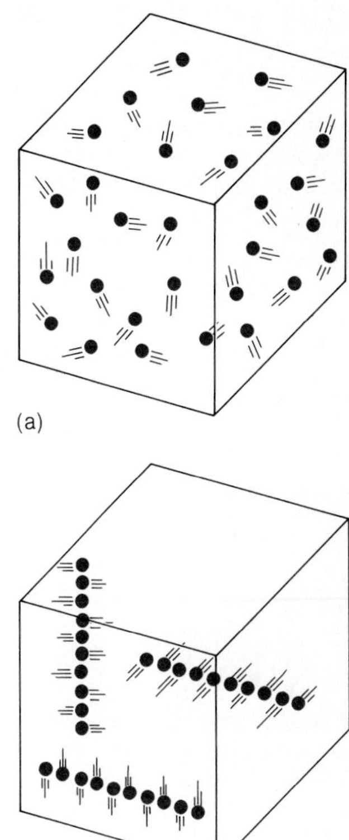

Figure 13.13

(a) An actual gas. (b) Simplified model of the gas.

When a cylinder containing a gas is doubled in volume, as in Fig. 13.14, those molecules moving up and down have twice as far to go between impacts. Since their speed is unchanged, the time between impacts is also doubled, and the pressure they exert on the top and bottom of the cylinder falls to half its original value. The expansion of the cylinder also means that the molecules moving horizontally are now spread over twice their former area, and the pressure on the sides of the cylinder accordingly falls to half its original value as well. Thus doubling the volume means halving the pressure, which is Boyle's law. Similar reasoning accounts for a rise in pressure when the volume is reduced.

Charles's law follows from the kinetic theory of gases provided that

---

The average kinetic energy of the random translational motions of the molecules of a gas is proportional to the absolute temperature of the gas.

This relationship is reasonable: We observe that compressing a gas quickly (so no heat can enter or leave the container) raises its temperature, and such a

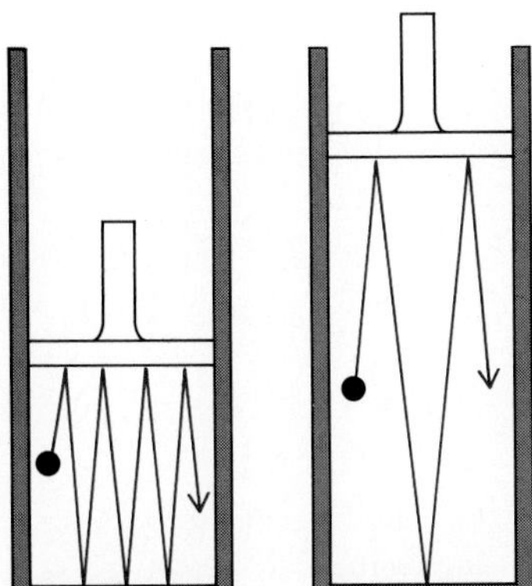

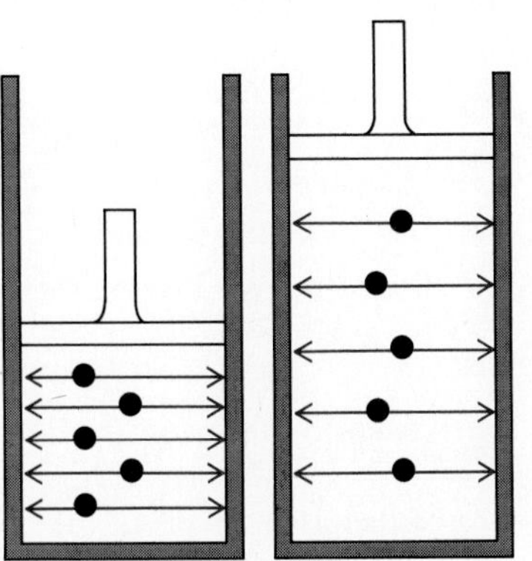

Figure 13.14

The origin of Boyle's law according to the kinetic theory of gases.

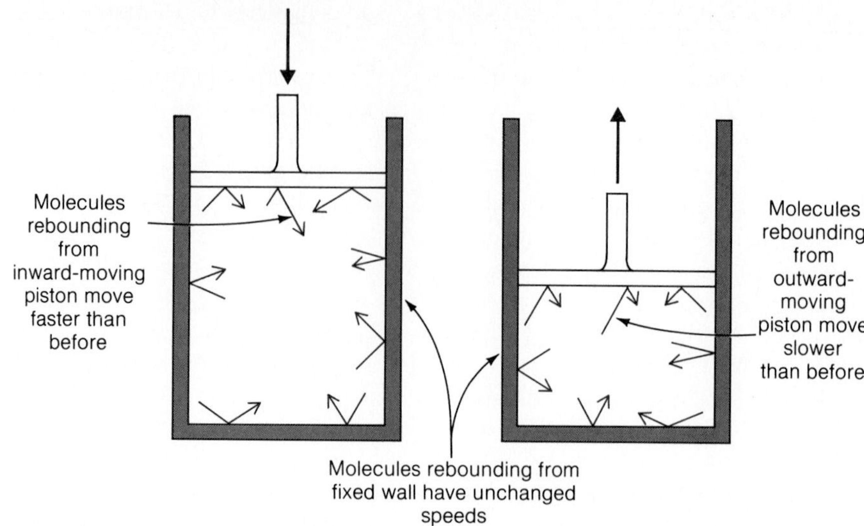

Figure 13.15

The temperature of a gas increases when it is compressed because the average energy of its molecules increases. The temperature of a gas decreases when it is expanded because the average energy of its molecules decreases.

compression must increase the average energy of the molecules because they bounce off the inward-moving piston more rapidly than they approach it (Fig. 13.15). A familiar example of the latter effect is a baseball rebounding with greater speed when struck by a bat. On the other hand, expanding a gas lowers its temperature, and such an expansion reduces molecular energies because molecules lose speed in bouncing off an outward-moving piston.

The significance of absolute zero in the kinetic theory of gases is straightforward: It is that temperature at which all molecular translational movement stops (Fig. 13.16). A more advanced analysis (see Section 26.7) shows that the molecules cannot be totally motionless, but the difference is small and not important here.

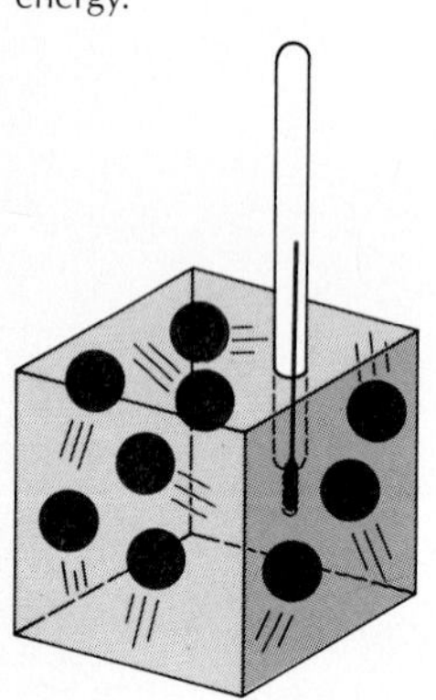

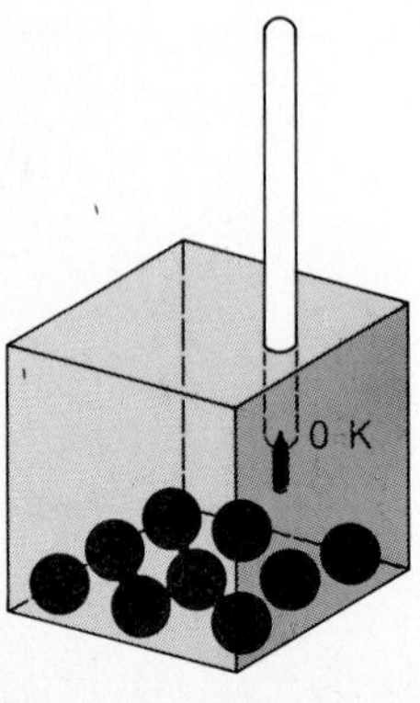

Figure 13.16

At absolute zero, the kinetic theory of gases predicts that molecular translational motion in a gas will cease. In reality, at absolute zero the molecules would retain a small minimum amount of kinetic energy.

The precise relationship between the average molecular kinetic energy KE_{av} and absolute temperature T is found to be

$$KE_{av} = \tfrac{3}{2}kT \qquad \textit{Molecular energy} \quad (13.8)$$

where k, known as **Boltzmann's constant**, has the value

$$k = 1.38 \times 10^{-23}\ \text{J/K}$$

Equation (13.8) holds for the molecules of all gases regardless of the masses of their molecules. It has been verified by direct measurements of molecular velocities and is derived in Section 13.10.

Thus we have an interpretation of temperature in terms of molecular motion that is much more precise and definite than simply describing temperature as that which is responsible for sensations of hot and cold.

EXAMPLE 13.8

One of the assumptions of the kinetic theory is that the average distance between the molecules in a gas is greater than the dimensions of the molecules themselves. Oxygen and nitrogen molecules are roughly 2×10^{-10} m in diameter, and 2.7×10^{25} of them are present in a cubic meter of air at room temperature and atmospheric pressure. What is the average separation of these molecules?

SOLUTION Because there are 2.7×10^{25} molecules/m^3, each molecule has on the average a volume to itself of

$$V_0 = \frac{1.0 \text{ m}^3}{2.7 \times 10^{25} \text{ molecules}} = 3.7 \times 10^{-26} \text{ m}^3$$

If we imagine this volume to be a cube of length L on each edge, $L^3 = V_0$ and

$$L = \sqrt[3]{V_0} = \sqrt[3]{37 \times 10^{-27} \text{ m}^3} = 3.3 \times 10^{-9} \text{ m}$$

This is the average separation of the molecules and is 17 times their diameters. ◆

13.7 Molecular Speeds

The higher the temperature, the greater the average speed.

We can use Eq. (13.8) to find the average speed of gas molecules whose mass is m:

$$\overline{\tfrac{1}{2}mv^2} = \tfrac{3}{2}kT$$

$$v_{\text{rms}} = \sqrt{\overline{v^2}} = \sqrt{\frac{3kT}{m}} \qquad \textit{Rms molecular speed} \quad (13.9)$$

This speed is denoted v_{rms} because it is the square root of the mean of the squared molecular speeds—the "root-mean-square" speed—and therefore different from the simple arithmetic average speed $\bar{v}$.

To emphasize their difference with a simple example, we can evaluate both kinds of average for an assembly of two molecules, one with a speed of 1.0 m/s and the other with a speed of 3.0 m/s. We find that

$$\bar{v} = \frac{v_1 + v_2}{2} = \frac{(1.0 + 3.0) \text{ m/s}}{2} = 2.0 \text{ m/s}$$

whereas

$$v_{\text{rms}} = \sqrt{\frac{v_1^2 + v_2^2}{2}} = \sqrt{\frac{(1.0)^2 + (3.0)^2}{2}} \text{ m/s} = \sqrt{5.0} \text{ m/s} = 2.24 \text{ m/s}$$

Clearly v_{rms} and $\bar{v}$ are not at all the same. The relationship between v_{rms} and $\bar{v}$ depends on the specific variation in molecular speeds being considered. For the distribution of molecular speeds found in a gas,

$$v_{\text{rms}} \approx 1.09\bar{v}$$

so the root-mean-square speed of Eq. (13.9) is about 9 percent greater than the arithmetic average $\bar{v}$.

EXAMPLE 13.9

Find the rms speed of oxygen molecules at 0°C.

SOLUTION Oxygen molecules have two oxygen atoms each. Since the atomic mass of oxygen is 16.00 u, the molecular mass is 32.00 u, and an O_2 molecule has a mass in kg of

$$m = (32.00\ \text{u})(1.66 \times 10^{-27}\ \text{kg/u}) = 5.31 \times 10^{-26}\ \text{kg}$$

At an absolute temperature of 273 K (which corresponds to 0°C), the rms speed of oxygen molecules is therefore

$$v_{\text{rms}} = \sqrt{\frac{3kT}{m}} = \sqrt{\frac{3(1.38 \times 10^{-23}\ \text{J/K})(273\ \text{K})}{5.31 \times 10^{-26}\ \text{kg}}} = 461\ \text{m/s}$$

This is a little over 1000 mi/h! Evidently molecular speeds are very large compared with those of the objects familiar to us. ◆

It is important to keep in mind that actual molecular speeds vary considerably on either side of v_{rms}. The graph in Fig. 13.17 shows the distribution of molecular speeds in oxygen at 273 K and in hydrogen at 273 K. The mass of an O_2 molecule is 16 times that of an H_2 molecule. Rms molecular speed decreases with molecular mass, hence at the same temperature molecular speeds in hydrogen are on the average greater than in oxygen. At the same temperature, however, the average molecular *energy* is the same for all gases.

In Fig. 13.18 we see the distributions of molecular speeds in oxygen at 73 K and at 273 K. The average molecular speed indeed increases with temperature. The

Figure 13.17

Molecular speeds in oxygen and hydrogen at 273 K (0°C). The smaller masses of H_2 molecules means that they have higher average speeds than O_2 molecules at the same temperature, because the average kinetic energy depends only on temperature.

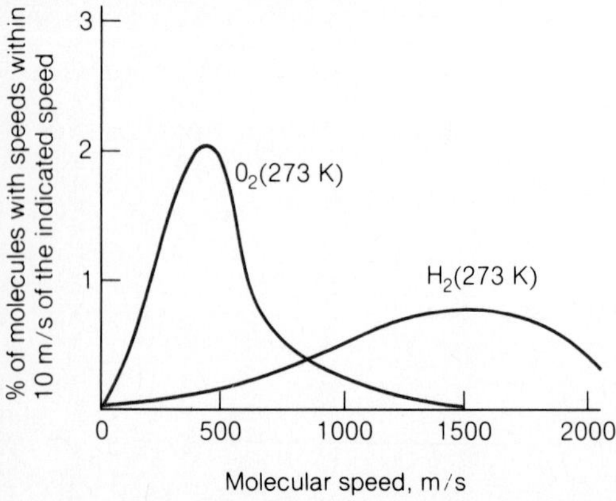

Figure 13.18

Molecular speeds in oxygen at 73 K (−200°C) and 273 K (0°C). The higher the temperature, the greater the average kinetic energy.

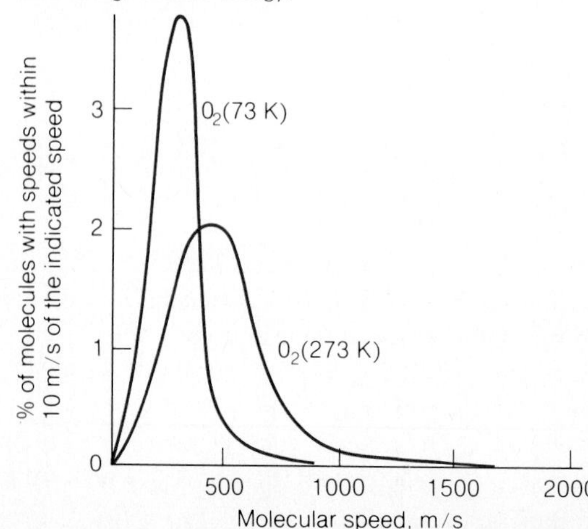

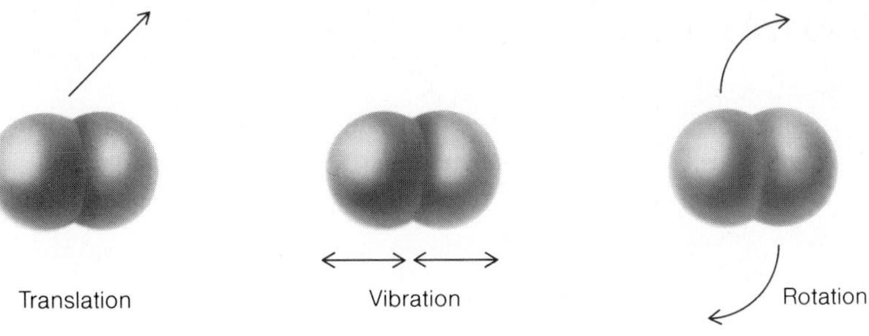

Figure 13.19

The molecules of a gas may have energies of vibration and rotation as well as of translation, but only the kinetic energy of their translational motions affects the temperature of the gas.

curves of Figs. 13.17 and 13.18 are not symmetrical, because the lower limit to v is fixed at $v = 0$ whereas there is, in principle, no upper limit. Actually, as the curves show, the likelihood of speeds many times greater than v_{rms} is small.

The distribution of molecular speeds in a gas has an interesting astronomical consequence. The higher the surface temperature of a planet, the faster the molecules of its atmosphere move, and the greater the chance they may exceed the escape speed of that planet and disappear into space. The smaller a planet, the lower its escape speed, and the closer a planet is to the sun, the warmer it is. Thus it is not surprising that Mercury, small and hot, has no atmosphere, whereas the giant outer planets of Jupiter, Saturn, Uranus, and Neptune have extremely dense atmospheres.

The kinetic theory of gases leads directly to the ideal gas law. As mentioned earlier, however, the ideal gas law is only a good approximation of reality. If we examine the initial assumptions that are made, it is easy to see why we should expect discrepancies between theory and experiment. For instance, it is assumed that gas molecules have negligible volumes, that they exert no forces on one another except in actual collisions, and that these collisions conserve kinetic energy of translational motion. (The last assumption means that the molecules are supposed to have no internal energy of their own, such as energy of rotation or vibration, whereas in fact they often do; see Fig. 13.19.) When the kinetic theory is worked out from more realistic assumptions, the results are in excellent agreement with observation.

13.8 Molecular Motion in Liquids

The kinetic theory of evaporation and boiling.

Liquids as well as gases consist of molecules in constant random motion. In 1827 the British botanist Robert Brown noticed that pollen grains in water are in continual, agitated movement. Similar **Brownian motion** is apparent whenever very small particles are suspended in a fluid medium, for example, smoke particles in air (Fig. 13.20). Brownian motion originates in the bombardment of the particles by molecules of the fluid, with successive molecular impacts coming from different directions and contributing different impulses to the particles. Albert Einstein, in 1905, found that he could account for Brownian motion quantitatively by assuming that, as a result of continual collisions with fluid molecules, the particles themselves have the same average kinetic energy as the molecules. Surprising as it may seem, this was the first direct verification of the reality of molecules, and it convinced a number of distinguished scientists who had previously been reluctant to believe that such things actually exist.

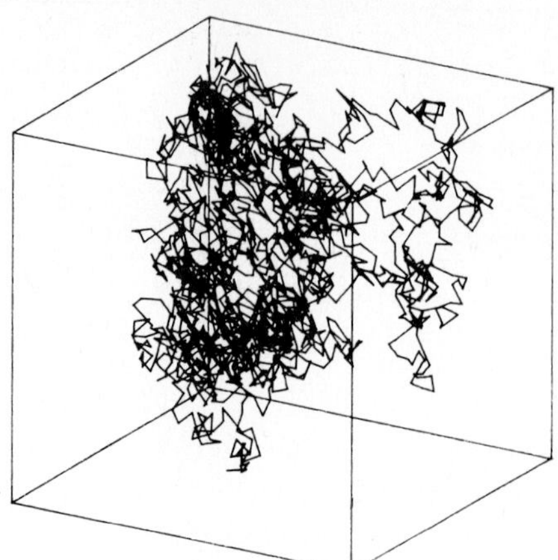

Figure 13.20

Computer-generated representation of Brownian motion. Successive line segments correspond to the movement of the particle in successive equal time intervals.

The kinetic-molecular theory can also explain evaporation. At its boiling point the molecules of a liquid have enough energy to break loose from the forces that hold them together, and the liquid becomes a gas. Evaporation, on the other hand, refers to the change of a liquid to a vapor *below* its boiling point. (A **vapor** is a gas whose temperature is below the boiling point of the substance.) Evaporation occurs only at the surface of a liquid, unlike boiling, in which bubbles of gas form throughout the interior of the liquid.

A hint as to what is going on when a liquid evaporates is that it grows cooler as it does so. The faster the evaporation, the more pronounced the cooling effect. Alcohol is effective at chilling the skin because it evaporates rapidly.

Evaporation can be understood on the basis of the distribution of molecular speeds in a liquid. Although not exactly the same as that found in a gas, this distribution resembles those shown in Figs. 13.17 and 13.18 in that a certain fraction of the molecules in any sample have much greater and much smaller speeds than the average. The fastest molecules have enough energy to escape through the liquid surface despite the attractive forces of the other molecules. The molecules left behind redistribute the available energy in collisions among themselves. Because the most energetic molecules have escaped, the average energy now is less than before and the liquid is at a lower temperature (Fig. 13.21).

When molecules from the vapor above a liquid surface happen to strike the surface, they may be trapped there, so that a constant two-way traffic of molecules to and from the liquid occurs. If the density of the vapor above the liquid is sufficiently great, just as many molecules return as leave it at any time. This situation is described

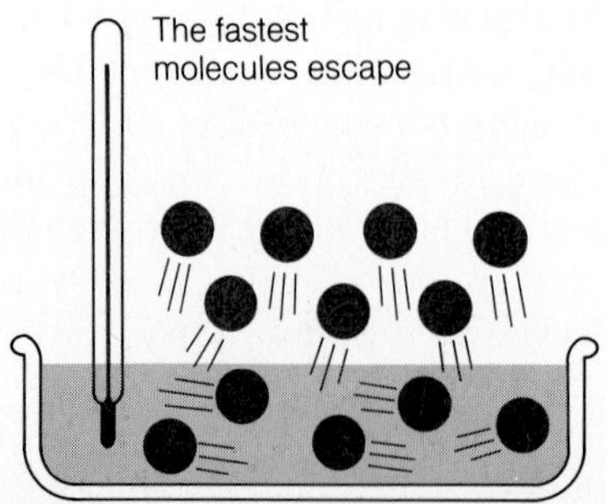

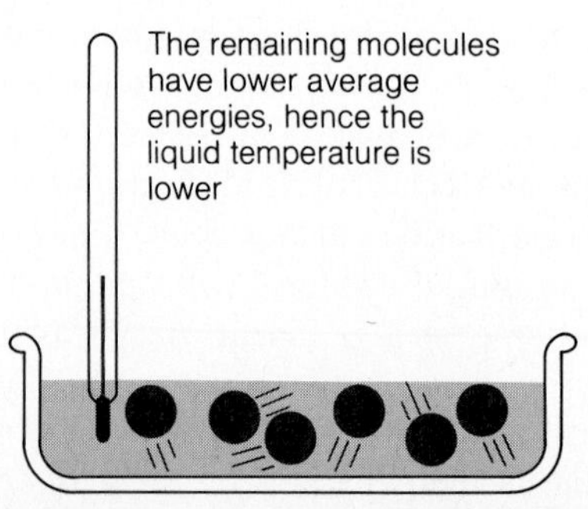

Figure 13.21

After evaporation, the remaining liquid is cooler than before.

Atmospheric pressure is $p = 1.013 \times 10^5$ Pa. Hence the work done during the expansion of 1.00 kg of water to steam is, from Eq. (15.3),

$$\begin{aligned} W &= p(V_2 - V_1) = (1.013 \times 10^5 \text{ Pa})(1.67 \text{ m}^3 - 0.00100 \text{ m}^3) \\ &= 1.69 \times 10^5 \text{ J} \\ &= 169 \text{ kJ} \end{aligned}$$

This is

$$\frac{W}{L_v} = \frac{169 \text{ kJ}}{2260 \text{ kJ}} = 0.075 = 7.5\%$$

of the heat of vaporization.

(b) The rest of the heat of vaporization becomes internal energy of the steam. From the first law of thermodynamics the internal energy ΔU added to the water as it becomes steam is

$$\Delta U = Q - W = 2260 \text{ kJ} - 169 \text{ kJ} = 2091 \text{ kJ}$$ ◆

The difference between a heat engine that runs continuously and a one-time-only event such as a dynamite blast is that the engine is not permanently changed as it turns heat into work. A sequence of processes brings the engine back to its starting point after each cycle. The p-V graph of a heat engine is therefore a closed curve that the engine traces over and over again in its operation, as in Fig. 15.6 (a).

The cycle shown in Fig. 15.6 (a) has four parts, each representing a different process. In expansions ab and bc the gas in the engine expands, thereby doing the work W_{ab} and W_{bc} on whatever the engine is connected to. This work is equal to the

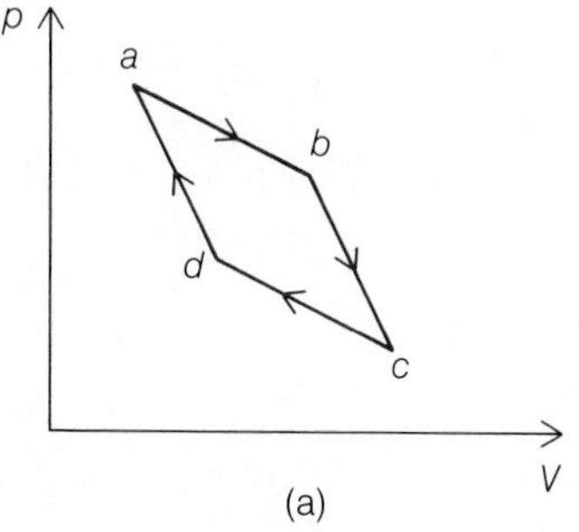

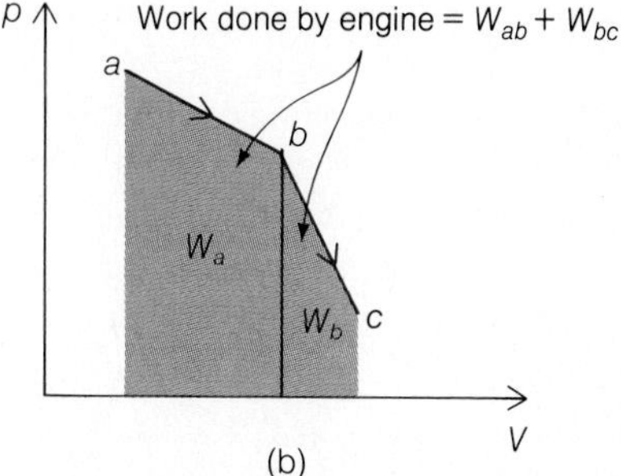

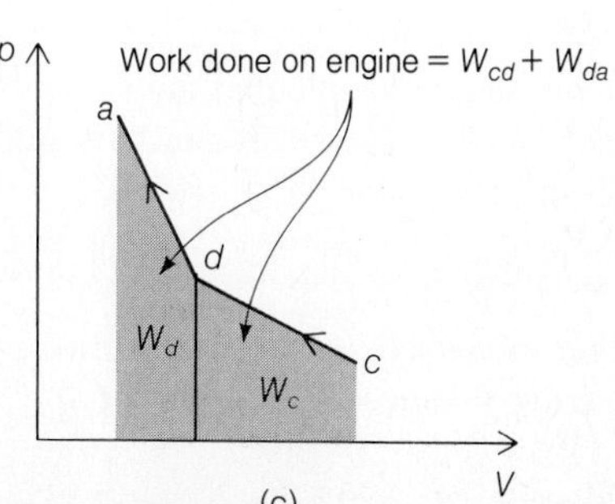

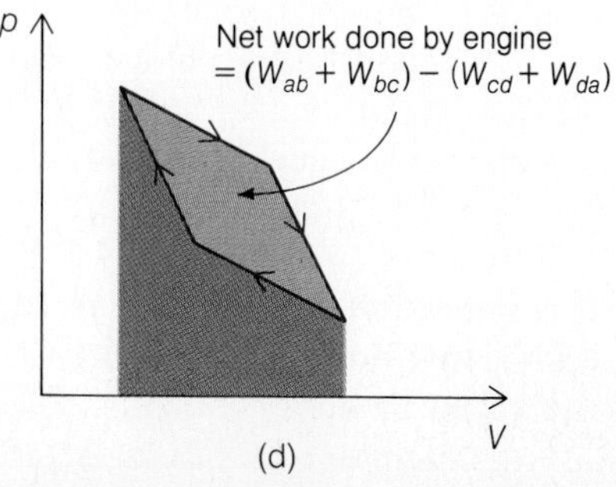

Figure 15.6

Indicator diagram of the operating cycle of an imaginary heat engine. The work done in an entire cycle is equal to the area enclosed by the curve *abcd* of the cycle.

area under the curves *ab* and *bc* on the graph (Fig. 15.6b). The compressions *cd* and *da* then bring the engine back to its original state. In these compressions the work W_{cd} and W_{da} is done on the engine itself (Fig. 15.6c). Thus the net work done in each cycle is

$$\text{Net work} = \text{work done by engine} - \text{work done on engine}$$
$$W = (W_{ab} + W_{bc}) - (W_{cd} + W_{da})$$

As we see in Fig. 15.6(d), W is equal to the area enclosed by the cycle *abcd*. The larger this area, the more work is done. A graph of this kind for a heat engine is called its **indicator diagram.**

15.3 Second Law of Thermodynamics

No engine can be perfectly efficient.

Heat is the easiest and cheapest form of energy to obtain, because all we need do to liberate it is to burn a fuel such as wood, coal, or oil. The real problem is to turn heat into mechanical energy so it can power cars, ships, airplanes, electric generators, and machines of all kinds.

To appreciate the problem, we recall that heat consists of the kinetic energies of moving atoms and molecules. In order to change heat into a more usable form, we must extract some of the energy of the random motions of atoms and molecules and convert it into the regular motions of a piston or a wheel. Such conversions cannot take place efficiently, for the same reason that it is easier to shatter a wineglass than to reassemble the fragments: The natural tendency of all physical systems is toward increasing disorder. The **second law of thermodynamics** is an expression of this tendency, whose role in the evolution of the universe is quite as central as are those of the various conservation principles.

According to the first law of thermodynamics, an engine cannot operate without a source of energy, but this law tells us nothing about the character of possible sources of energy. For instance, there is an immense amount of internal energy in the atmosphere, yet we cannot run a car by just taking in air, extracting some of its internal energy, and then exhausting liquid air. Or, to give an even more extreme case, it is energetically possible for a puddle of water to rise by itself into the air, cooling and freezing from water to ice as its internal energy changes into potential energy (Fig. 15.7). After all, a block of ice dropped from high enough melts when it hits the ground, its initial PE first being converted to KE and then into heat. Needless to say, water does not rise upward of its own accord, and we need a way to express this conclusion.

The second law of thermodynamics is the physical principle that supplements the first law in limiting our choice of heat sources for our engines. It can be stated in a number of equivalent ways, a common one being as follows:

It is impossible to construct an engine, operating in a cycle (that is, continuously), that does nothing other than take heat from a source and perform an equivalent amount of work.

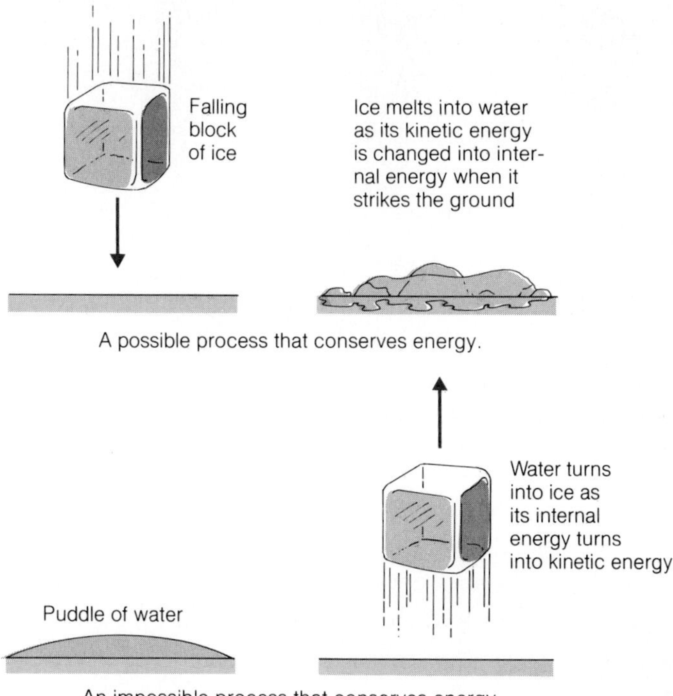

Figure 15.7

The second law of thermodynamics provides a way to identify impossible processes that are in accord with all other physical principles.

According to the second law of thermodynamics, then, no engine can be completely efficient—some of its heat input *must* be ejected. As we will see, the greatest efficiency any heat engine is capable of depends on the temperatures of its heat source and of the reservoir to which it exhausts heat. The greater the difference between these temperatures, the more efficient the engine. The second law is a consequence of the fact that

---

The natural direction of heat flow is from a reservoir of internal energy at a high temperature to a reservoir of internal energy at a low temperature, regardless of the total energy content of each reservoir.

This statement, in fact, may be regarded as an alternative expression of the second law (Fig. 15.8).

If we are to utilize the internal energy content of the atmosphere or the oceans, we must first provide a reservoir at a lower temperature than theirs in order to extract heat from them. There is no nearby reservoir in nature suitable for this purpose, for if there were, heat would flow into it until its temperature reached that of its surroundings. To establish a low-temperature reservoir, we must employ a refrigerator (which is a heat engine running in reverse by using up energy to extract heat), and in so doing we will perform more work than we can successfully obtain from the heat of the atmosphere or the oceans.

To sum up, the first law of thermodynamics tells us that we cannot get something for nothing. The second law singles out heat from other kinds of energy and recognizes that all conversions of heat into any other form must be inefficient.

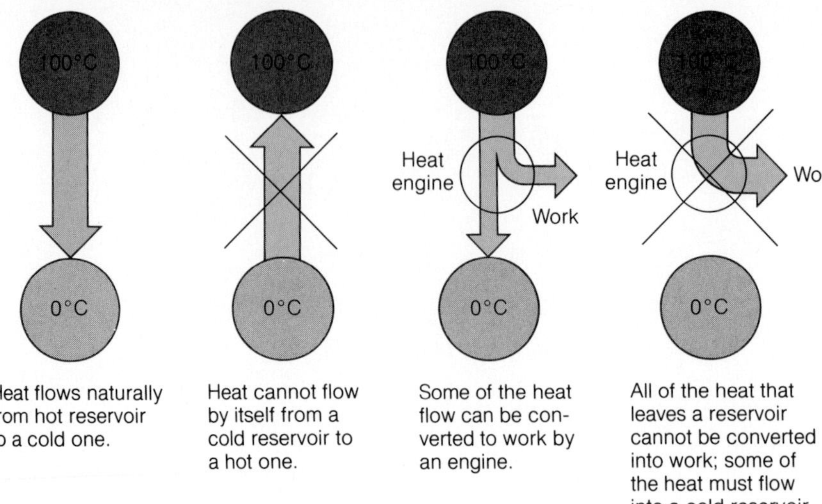

Figure 15.8

The second law of thermodynamics.

We cannot break even, either. In Section 15.8 we will see why heat is different in this respect from other kinds of energy.

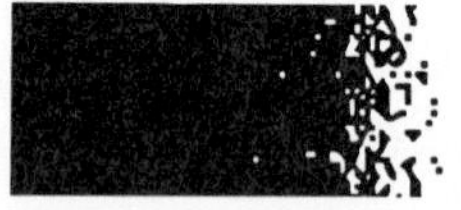

15.4 Carnot Engine

Because it is reversible without losing energy, no other engine can be more efficient.

Every heat engine behaves in the same general way: It absorbs heat at a certain temperature, converts some of the heat into work, and exhausts the rest at a lower temperature. The second law of thermodynamics expresses the fact that a heat engine cannot be 100% efficient in turning heat into work. But suppose we have an engine that is not subject to friction or to the loss of stored heat by conduction or radiation. What is the maximum efficiency of such an ideal engine?

Maximum efficiency can be reached only if all the processes that occur in the engine's operation do not involve changes in the engine that need extra work to reverse, because any such changes obviously mean the waste of energy. That is, *every process in the engine must be reversible without losing any energy.*

The flow of heat from a hot reservoir to a cooler one is not reversible in this way, because work would have to be done to ''push'' the heat back the other way. However, heat flow at constant temperature *is* reversible without losing energy. An **isothermal** process is one in which the temperature of the substance that undergoes a change of some kind stays constant. If the substance is gas in a container, we can imagine that the container is a good conductor of heat and is surrounded by a constant-temperature reservoir of heat. No process that occurs in the real world is ever wholly isothermal, just as no actual gas exactly resembles an ideal gas, but many processes are quite close to being isothermal.

In an isothermal process in a system of some kind, all the heat absorbed by the system is turned into work. If the system has work done on it, all the work is given off as heat.

At the other extreme from being in such close contact with a heat reservoir that its temperature never changes, a system might be so completely isolated from its

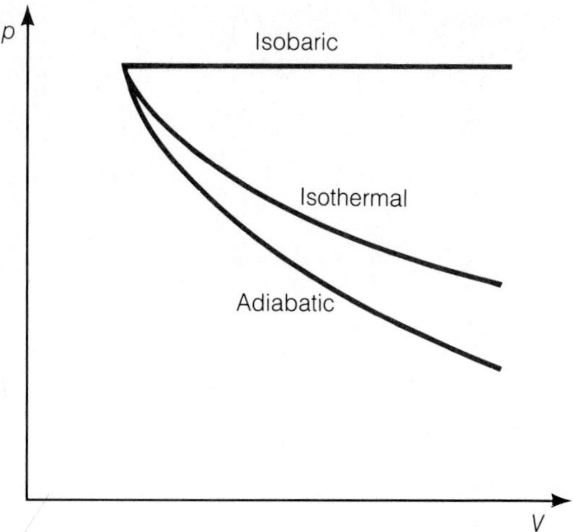

Figure 15.9

Isobaric (p = constant), isothermal (T = constant), and adiabatic ($Q = 0$) processes.

surroundings that heat can neither enter nor leave it. Any process undergone by a system in this situation is called **adiabatic.** An adiabatic process can involve a temperature change and still be reversible without losing energy because no heat transfer is involved that would be subject to the second law of thermodynamics. Most rapid thermodynamic processes are approximately adiabatic because heat transfer takes time and the process may be over before much heat has passed through the walls of the system.

In an adiabatic process any work done by the system comes from the internal energy of the system. Usually this means that the temperature of the system decreases. If the system has work done on it, its internal energy increases by the same amount. Figure 15.9 compares isobaric, isothermal, and adiabatic processes on a p-V diagram.

An imaginary heat engine that uses only isothermal and adiabatic processes was devised in 1824 by the French engineer Sadi Carnot. A **Carnot engine** consists of a cylinder that is filled with an ideal gas and has a movable piston at one end. The four stages in its operating cycle are shown in Fig. 15.10, together with a graph of each stage on a p-V diagram. These stages are as follows:

1. An amount of heat Q_1 is added to the gas, which expands isothermally at its initial temperature T_1. The heat added equals exactly the work done by the gas, which is why its temperature does not change.
2. The heat source is removed, and the expansion is allowed to continue. The second expansion is adiabatic and takes place at the expense of the energy stored in the gas, and so the gas temperature falls from T_1 to T_2. During expansions 1 and 2 the piston exerts a force on whatever it is attached to and thereby performs work.
3. Having done work in pushing the piston outward, the engine must now be returned to its initial state in order for it to be able to do further work. The third stage involves an isothermal compression of the gas at the constant temperature T_2 during which an amount of heat Q_2 is given off. The heat given off exactly equals the work done on the gas by the piston, and so its temperature does not change.
4. The gas is returned to its initial temperature, pressure, and volume by an adiabatic compression in which heat is neither added to it nor removed from it. Work is done on the gas in this compression, which is why its temperature rises.

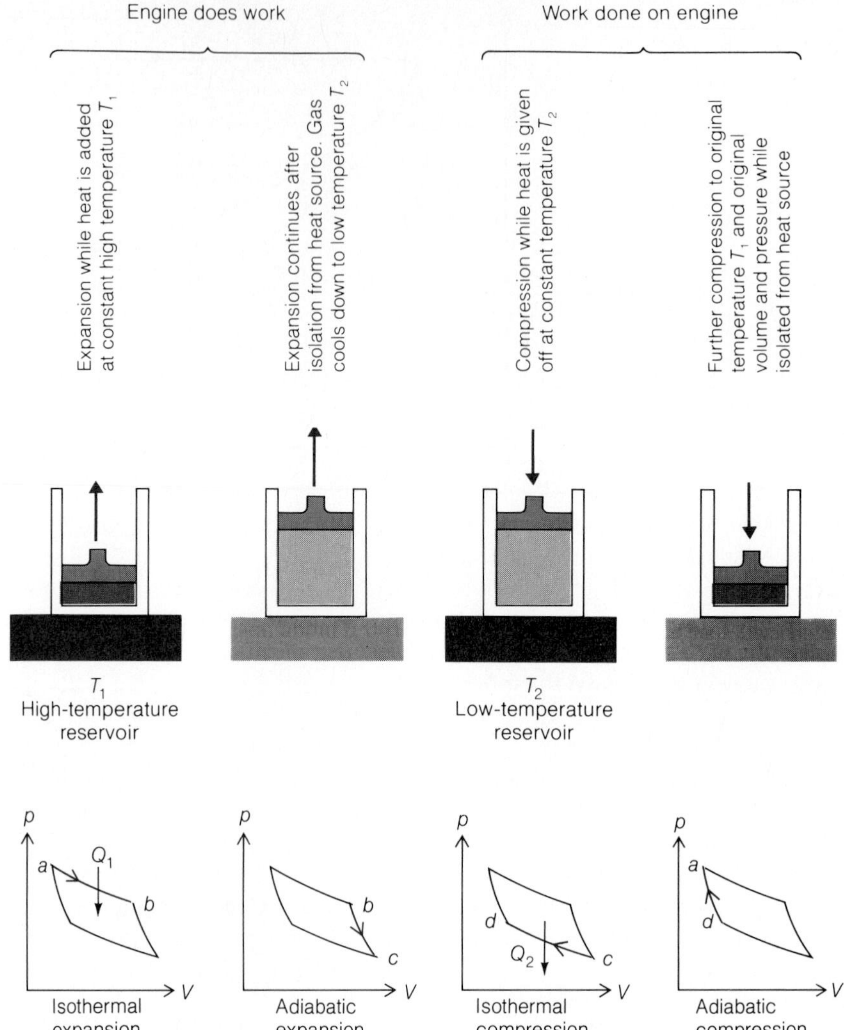

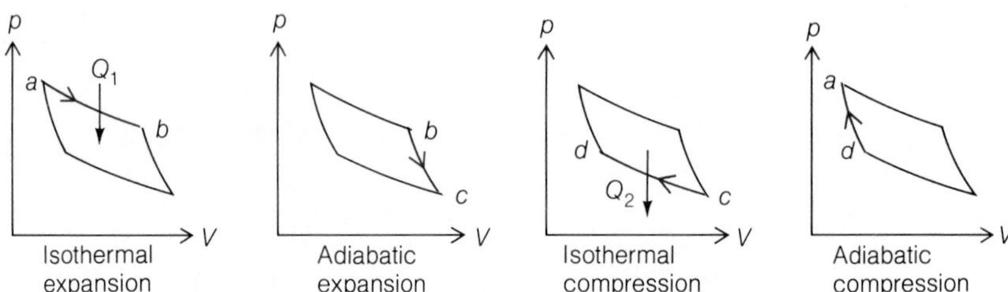

Figure 15.10

The Carnot cycle. This imaginary engine, which uses an ideal gas as its working substance, is the most efficient possible operating between reservoirs at the temperatures T_1 and T_2.

15.5 Engine Efficiency

It depends on the ratio between the temperatures at which the engine absorbs and rejects heat.

In each cycle a Carnot engine performs some net amount of work W, which is the difference between the work it does during the two expansions and the work done on it during the two compressions. It has taken in the heat Q_1 and ejected the heat Q_2 (Fig. 15.11). We note from Fig. 15.10 that the heat Q_2 *must* be ejected in order that the engine return to its initial state, from which it can begin another cycle. According to the first law of thermodynamics,

$$W = Q_1 - Q_2 \tag{15.4}$$

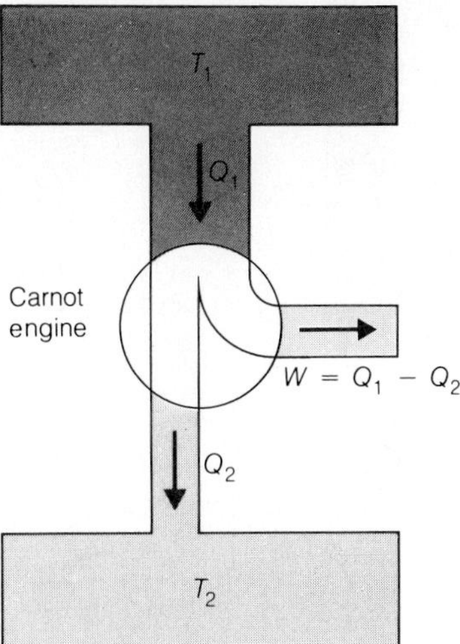

Figure 15.11

The efficiency of a Carnot engine depends on the ratio between Q_2 and Q_1.

because there is no internal energy change per cycle. The efficiency of the engine is the ratio between its work output W and its heat input Q_1, so that

$$\text{Eff} = \frac{W}{Q_1} = \frac{Q_1 - Q_2}{Q_1} = 1 - \frac{Q_2}{Q_1} \tag{15.5}$$

The smaller the ratio of the ejected heat Q_2 to the absorbed heat Q_1, the more efficient the engine.

As it happens, the heat Q transferred to or from a Carnot engine is directly proportional to the absolute temperature T of the reservoir with which it is in contact. That is, for a given Carnot engine,

$$\frac{Q}{T} = \text{constant} \qquad \textit{Carnot engine} \tag{15.6}$$

(The argument that leads to this conclusion is rather long, although it does not involve any physical principles we have not yet met, and so is omitted here.)

According to Eq. (15.6), the ratio Q_2/Q_1 between the amounts of heat ejected and absorbed per cycle by a Carnot engine is equal to the ratio T_2/T_1 between the temperatures of the respective reservoirs. The efficiency of such an engine is therefore

$$\text{Eff} = 1 - \frac{T_2}{T_1} \qquad \textit{Carnot efficiency} \tag{15.7}$$

The smaller the ratio between the absolute temperatures T_2 and T_1, the more efficient the engine (Fig. 15.12). No engine can be 100% efficient because no reservoir can have an absolute temperature of 0 K. (Even if such a reservoir could somehow be created, the exhaust of heat to it by the engine would raise its temperature above 0 K immediately.)

As mentioned earlier, the significance of the Carnot engine is that, being reversible, it has the highest efficiency permitted by the laws of thermodynamics. A real engine is never exactly reversible because of friction and heat losses through the

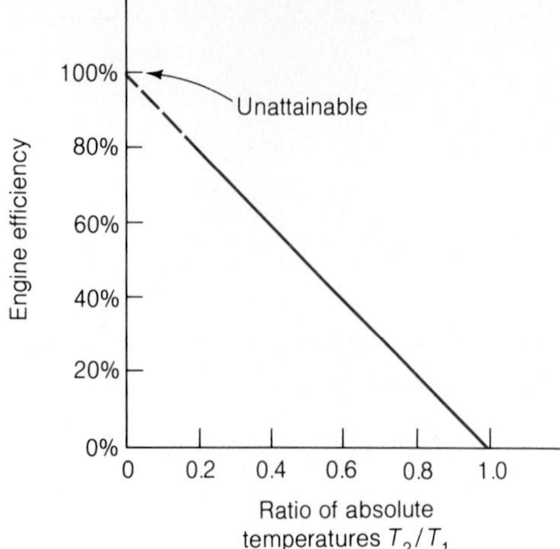

Figure 15.12

The efficiency of a Carnot engine is equal to $1 - T_2/T_1$.

engine walls, so its efficiency is less than that of a Carnot engine. We therefore have a third way to express the second law of thermodynamics:

No engine operating between the absolute temperatures T_1 and T_2 can be more efficient than a Carnot engine operating between the same temperatures, whose efficiency is $1 - T_2/T_1$.

EXAMPLE 15.2

Steam enters a certain steam turbine (Fig. 15.13) at 570°C and emerges into a partial vacuum at 95°C. What is the upper limit to the efficiency of this engine?

SOLUTION The absolute temperatures equivalent to 570°C and 95°C are, respectively, 843 K and 368 K. The efficiency of a Carnot engine operating between these

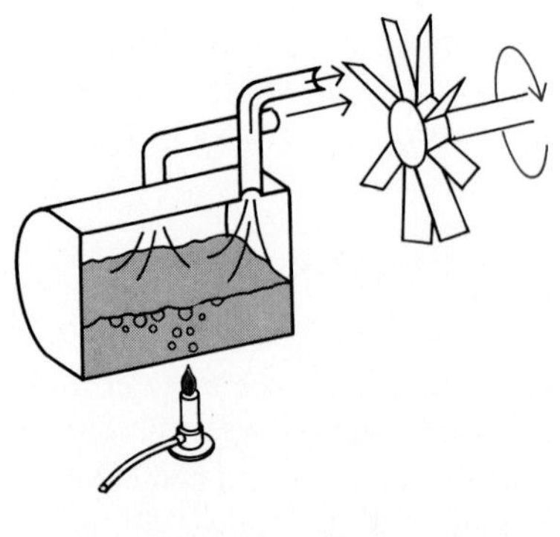

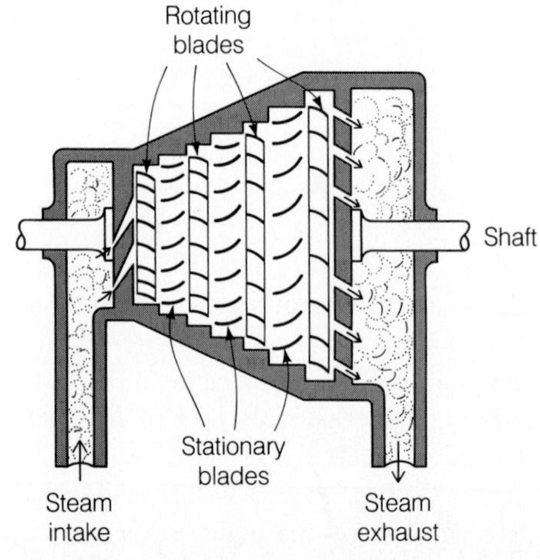

Figure 15.13

(a) A primitive steam turbine. (b) In a modern turbine, steam flows past a dozen or more sets of blades on the same shaft to extract as much power as possible. Stationary blades are interleaved between the moving blades to direct the flow of steam in the most advantageous way.

This steam tubine rotor is 9 m long and turns at 3600 rpm to produce 200 MW of power. Steam enters the turbine at 540° C.

two absolute temperatures is

$$\text{Eff} = 1 - \frac{T_2}{T_1} = 1 - \frac{368\text{ K}}{843\text{ K}} = 0.56 = 56\%$$

The efficiency of a Carnot engine is the highest possible for an engine operating between a given pair of temperatures. An actual steam turbine operating between 570°C and 95°C would have an efficiency of no more than about 40% because of the inevitable presence of friction and heat losses to the atmosphere. Practical problems rule out steam temperatures higher than about 570°C, which would otherwise improve the efficiency. One of these problems is that, at such temperatures, steam can decompose into hydrogen and oxygen. Steel absorbs hydrogen and becomes brittle as a result, which can reduce the lifetime of the steel components of a turbine to an uneconomic extent. ♦

15.6 Internal Combustion Engines

Theory into practice.

An internal combustion engine is relatively efficient because it generates the input heat within the engine itself. In a gasoline engine a mixture of air and gasoline vapor is ignited in each cylinder by a spark plug, and the heat given off is turned into mechanical energy by the pressure of the hot gases on a piston. The greater the ratio between the initial and final volumes of the expanding gases, the greater the engine efficiency. In a gasoline engine this ratio is limited to about 8 to 1, because the gasoline-air mixture in the cylinder will otherwise ignite by itself during its compression before the end of the stroke is reached. The result is a theoretical efficiency of about 55% and an actual efficiency of about 30%.

Figure 15.14 shows the operating cycle of a typical four-stroke gasoline engine. Each cylinder in such an engine has one power stroke for every two shaft revolutions.

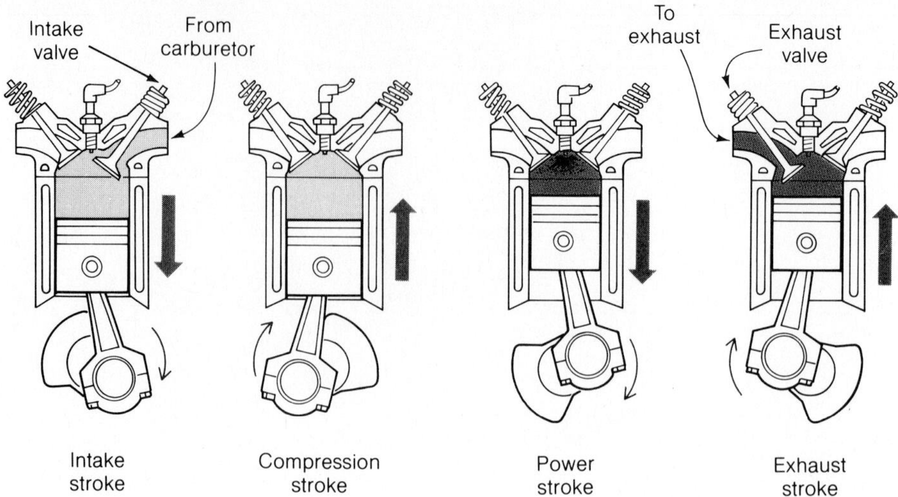

Figure 15.14

The operating cycle of a four-stroke gasoline engine. In the intake stroke, a mixture of gasoline vapor and air from the carburetor is drawn into the cylinder through the intake valve by the suction of the downward-moving piston. In the compression stroke, both valves are closed and the upward-moving piston compresses the fuel-air mixture. At the top of the stroke, the spark plug is fired, which ignites the fuel-air mixture. The burning fuel expands and forces the piston down in the power stroke. At the end of the power stroke, the exhaust valve opens and the upward-moving piston expels the waste gases.

The operating cycle of a gasoline engine (often called the *Otto cycle* after the inventor of the first practical gasoline engine) is shown on a *p-V* diagram in Fig. 15.15. The heat Q_1 enters the engine during the expansion *cd* when the gasoline-air mixture is ignited by the spark plug. The power stroke *de* is an adiabatic expansion, and the heat Q_2 is ejected during *eb* when the exhaust valve is open at the end of the power stroke. (The heat that leaves during the exhaust stroke *ba* is smaller than Q_2 and is ignored in the ideal cycle of the diagram.) The work *W* done in each cycle is equal to the area enclosed by *bcdeb*. The greater the compression ratio, the larger the area and the more work done per cycle.

Diesel engines are more efficient than gasoline engines because of their higher compression ratios. Early ignition is avoided by compressing only air and injecting fuel oil into the hot, compressed air at the instant the piston reaches the top of its travel. No spark plug is required. The compression ratio in a diesel engine is typically 20 to 1, for a theoretical efficiency of as much as 70% and an actual efficiency of about 35%.

The operating cycle of a four-stroke diesel engine is shown on the *p-V* diagram in Fig. 15.16. Air is drawn into the cylinder in the intake stroke *ab* and is compressed adiabatically, which raises its temperature, in the compression stroke *bc*. The heat Q_1 enters the engine in the constant-pressure expansion *cd* during which

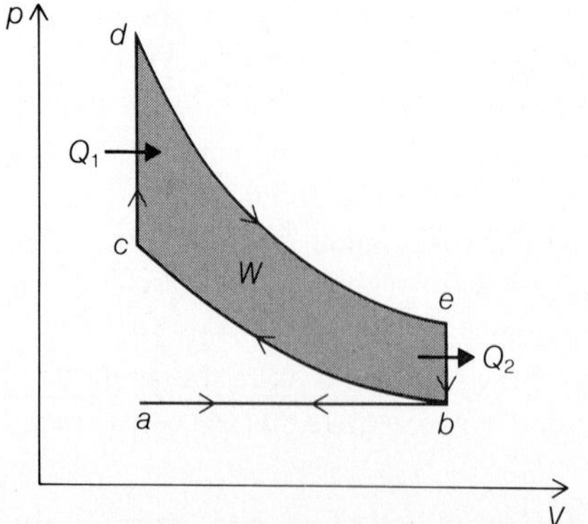

ab: intake of fuel-air mixture
bc: compression stroke
cd: ignition of fuel-air mixture
de: power stroke
e: exhaust valve opens
ba: exhaust stroke

Figure 15.15

Idealized indicator diagram of the Otto cycle of a four-stroke gasoline engine.

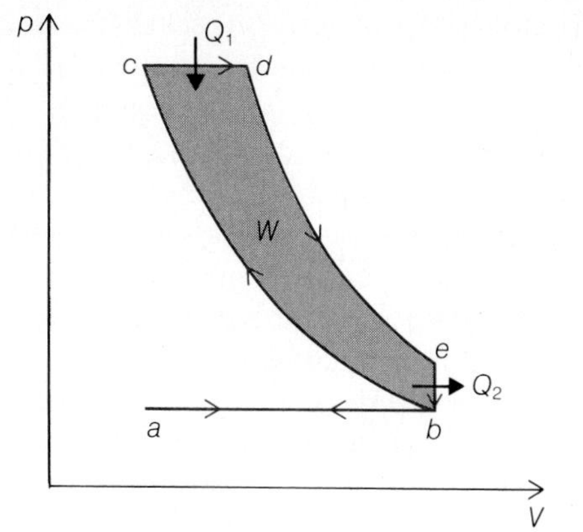

Figure 15.16
Idealized indicator diagram of a four-stroke diesel engine.

fuel is injected into the hot air and burned. Then comes an adiabatic expansion to e when the exhaust valve opens and the heat Q_2 is ejected.

EXAMPLE 15.3

The Mercedes-Benz OM314 is a four-cylinder, four-stroke diesel engine that develops 80 hp at 2600 rpm. The pistons of this engine are 97 mm in diameter, and they travel 128 mm. Find the average pressure on the pistons during each power stroke.

SOLUTION The power output P_c of each cylinder of a gasoline or diesel engine depends on four factors: the average pressure p on the piston during a power stroke; the length L of piston travel; the area A of the piston; and the number N of power strokes per second. Because the force on the piston during a power stroke is $F = pA$, the work done per power stroke is $W = FL = pLA$. Hence

$$\text{Power output per cylinder} = (\text{work done per stroke})\left(\frac{\text{power strokes}}{\text{second}}\right)$$

$$P_c = pLAN$$

Here the area of each piston is

$$A = \frac{\pi d^2}{4} = \frac{\pi(0.097\text{ m})^2}{4} = 0.0074\text{ m}^2$$

In a four-stroke engine a power stroke occurs in each cylinder once every two revolutions, so here there are 1300 power strokes per minute, or 1300/60 = 21.7 per second. Because the engine has four cylinders, its total power output is $P = 4P_c = 4pLAN$ and

$$p = \frac{P}{4LAN} = \frac{(80\text{ hp})(746\text{ W/hp})}{(4)(0.128\text{ m})(0.0074\text{ m}^2)(21.7/\text{s})} = 7.3 \times 10^5\text{ Pa} = 7.3\text{ bars}$$

Of course, the pressure calculated in this way is an average. The actual pressure is greater at the beginning of the stroke just after the fuel is ignited and less at the end of the stroke. ◆

The efficiency of conventional gasoline and diesel engines (nearly as high as that of steam turbines) has to some extent slowed the development of the still more efficient gas turbine. A gas turbine is similar to the steam turbine of Fig. 15.13(b) except that hot gases from the burning fuel pass through its sets of blades instead of steam. A gas turbine is lighter in weight and has fewer moving parts than a piston engine, but the high temperature and high rotational speeds at which it operates present difficulties in manufacture. Gas turbines are nevertheless coming into wider and wider use: "Turboprop" aircraft engines are gas turbines, for instance, and a number of ships are already powered by gas turbines.

The rapidly rotating shaft of a turboprop engine is coupled to a propeller through a reduction gear. In a turbojet engine the propeller is eliminated and the hot gases from the burning fuel are ejected at high speed from the rear of the engine to furnish a reaction force that pushes the aircraft forward (Fig. 15.17). The energy freed by the burning fuel thus goes directly into propulsion with no moving parts except for a turbine that powers the necessary air compressor. Rocket motors are jet engines in which the oxygen or other oxidizing agent needed to burn the fuel comes from an internal reservoir instead of from the atmosphere (see Fig. 5.8). In a solid-fuel rocket, the ultimate in simplicity, both the components required for combustion are combined in a stable mixture whose reaction rate when ignited is relatively slow and steady rather than explosive.

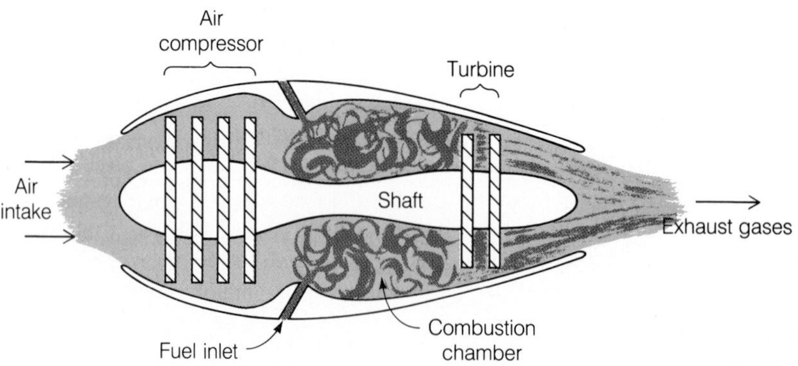

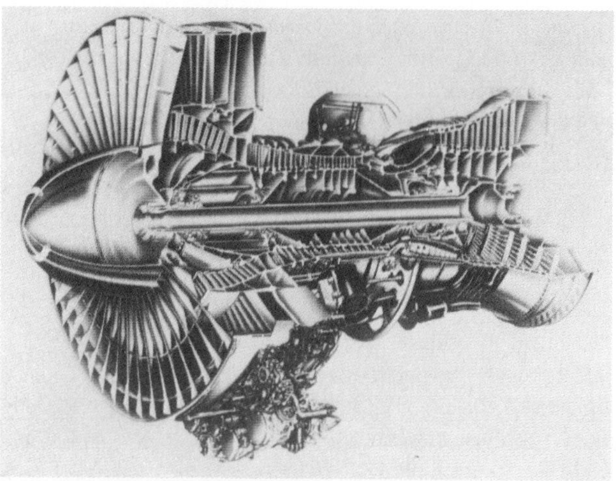

Figure 15.17

Turbojet engine. The reaction force of the exhaust gases pushes the aircraft forward. (Bottom drawing courtesy of United Technologies, Pratt & Whitney.)

Table 15.1 Typical heats of combustion for common fuels. (Gas volumes at 0°C and atmospheric pressure.)

Substance	Heat of Combustion	
Solids	*MJ/kg*	*kcal/kg*
Charcoal	33.9	8,100
Coal	32.6	7,800
Wood	18.8	4,500
Liquids	*MJ/kg*	*kcal/kg*
Diesel oil	44.8	10,700
Domestic fuel oil	45.2	10,800
Ethyl alcohol	32.6	7,800
Gasoline	47.3	11,300
Kerosene	46.0	11,000
Gases	*MJ/m*^{3}	*kcal/m*^{3}
Acetylene	54.0	12,900
Coal gas	18.0	4,300
Hydrogen	10.2	2,445
Natural gas	33–71	8,000–17,000
Propane	86.2	20,600

Heat of Combustion

The **heat of combustion** of a substance is the energy liberated when 1 kg (or 1 m^3) of it is completely burned. Table 15.1 lists heats of combustion for a number of common fuels. If the fuel consumption rate of an engine is known at a certain power output, this table can be used to calculate its efficiency.

EXAMPLE 15.4

The diesel engine of a boat uses 9.3 kg/h of fuel when its power output is 35 kW. Find its efficiency.

SOLUTION The power input to the engine is

$$P_{\text{input}} = \left(9.3\frac{\text{kg}}{\text{h}}\right)\left(44.8 \times 10^6\frac{\text{J}}{\text{kg}}\right)\left(\frac{1}{3600\text{ s/h}}\right) = 1.16 \times 10^5\text{ W} = 116\text{ kW}$$

The engine's efficiency is therefore

$$\text{Eff} = \frac{P_{\text{output}}}{P_{\text{input}}} = \frac{35\text{ kW}}{116\text{ kW}} = 0.30 = 30\%$$

This is less than half its Carnot efficiency. ◆

From Table 15.1 it would seem that hydrogen has a very low heat of combustion, well under those of the other gaseous fuels. However, hydrogen is also much less dense than the other gases, and on a mass basis its heat of combustion of 143 MJ/kg is actually by far the highest of any of the fuels listed. The exceptional heat of combustion of hydrogen is the reason for the use of liquid hydrogen in spacecraft engines, where fuel weight is a critical factor. Liquid hydrogen would be an ideal fuel for many other purposes, except for two factors. First, the safe storage and transport of hydrogen in large quantities is not easy. Second, although the hydrogen in the waters of the earth is almost limitless, a great deal of expensive electric energy is needed to extract it, so the overall efficiency of hydrogen as a fuel is today quite low. Research is under way on methods to use sunlight as the energy source for decomposing water into hydrogen and oxygen. If this research is successful and the safety problems can be overcome, hydrogen may well be widely used in the future.

15.7 Refrigerators

They use outside energy to push heat from a cold reservoir to a hot one.

A refrigerator takes in heat at a low temperature and exhausts it at a higher temperature. According to the second law of thermodynamics, such a process needs an external energy supply to force heat to flow opposite to its normal direction (see Fig. 15.2). In a typical refrigerator two to three times as much heat is taken from the storage chamber as the amount of external energy used.

In nearly all refrigerators the working substance (or **refrigerant**) is a gas that is easily liquefied. Figure 15.18 shows the vaporization curve of Freon 12, one of the most widely used refrigerants. Under conditions of temperature and pressure

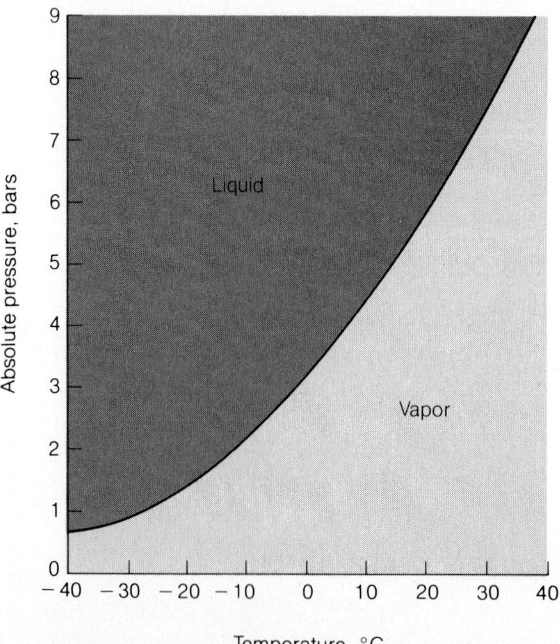

Figure 15.18

Vaporization curve of Freon 12, a common refrigerant. Freon 12 is a trade name for dichlorodifluoromethane, CCl_2F_2.

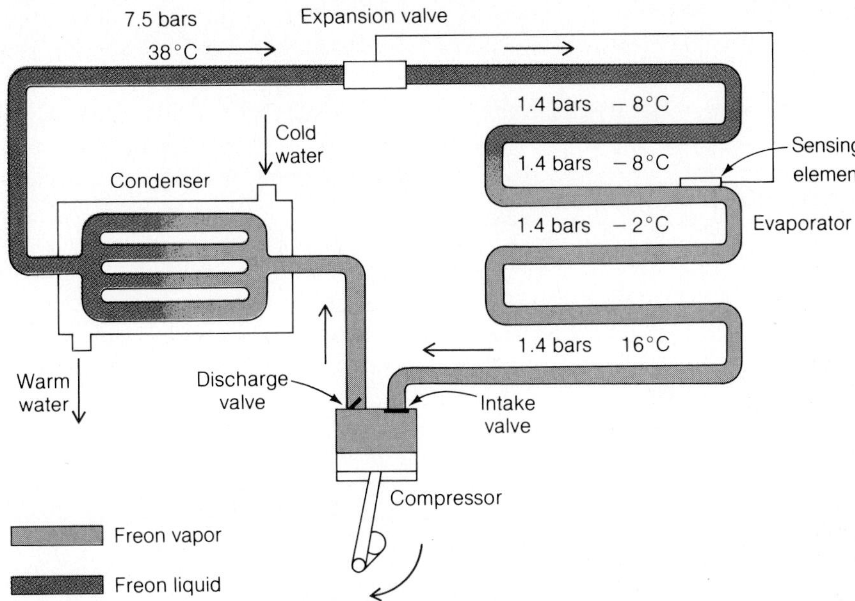

Figure 15.19

A refrigeration system using Freon 12. Gauge pressures are shown. Heat from the refrigerated space enters the system through the evaporator and leaves it through the condenser.

that correspond to points above the curve, only liquid Freon is present, whereas under conditions that correspond to points under the curve, only Freon vapor is present. Other refrigerants have different vaporization curves. The choice of a refrigerant depends on the kind of refrigerator (or air conditioner) involved and the temperatures between which it is to operate.

Let us examine a refrigeration system that uses Freon 12. As shown in Fig. 15.19, this system consists of a **compressor,** a **condenser,** an **expansion valve,** and an **evaporator.** When the piston of the compressor moves downward, Freon vapor at 1.4 bars (gauge pressure) and approximately room temperature is sucked into the cylinder. As the piston reaches the bottom of its stroke and begins to move upward, the intake valve closes and the discharge valve opens. The compressed Freon, which is now at a pressure of 7.5 bars and a high temperature, passes into the condenser in which it is cooled until it liquefies. The condenser may be water cooled, as shown in the sketch, or air cooled. It is in this stage that the heat taken from the refrigerated space is dissipated.

The liquid Freon then goes into the expansion valve from which it emerges at a lower pressure (1.4 bars) and temperature (−8°C). The amount of Freon supplied by the valve is regulated by a sensing element in the refrigerated space; this may be simply a gas-filled bulb that responds to temperature changes by pressure changes that actuate a bellows in the valve. As the cold liquid Freon flows through the evaporator tubes, it absorbs heat from the region being cooled. From Fig. 15.18 we see that, at an absolute pressure of 2.4 bars, the boiling point of Freon 12 is −8°C, so that the heat absorbed by the liquid Freon in the evaporator causes it to vaporize. Farther along in the evaporator, the Freon vapor itself absorbs heat, rising in temperature to perhaps −2°C. Finally the Freon vapor leaves the evaporator and enters the compressor to begin another cycle.

A **heat pump** is a refrigeration system that can take heat from the cold outdoors in winter and deliver it to the interior of a house. In summer the same system can serve as an air conditioner to take heat from the house and exhaust it to the warmer outdoors (Fig. 15.20).

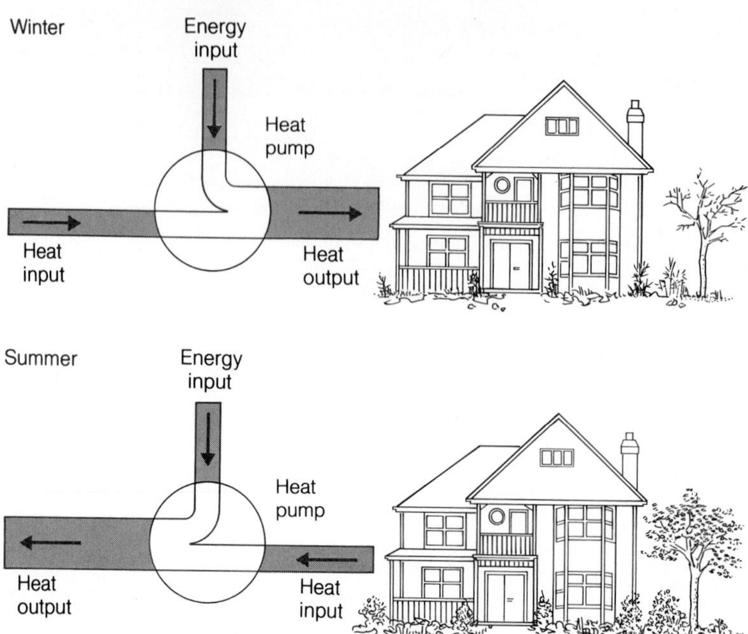

Figure 15.20

A heat pump is a refrigeration system that can be used to heat a house in winter and cool it in summer.

The heat pump is interesting because it transfers more heat than the work done. For example, a heat pump with an energy input of 5 kW might be able to "pump" an additional 15 kW of heat to provide a total of 20 kW. An ordinary furnace would have to burn fuel at a rate of 20 kW and so would require four times as much energy input as the heat pump. However, heat pumps have a cost problem to overcome: Energy in the form of fossil fuel (coal, oil, gas) is still quite a bit cheaper than the electrical energy needed to operate a heat pump, and a heat pump is more expensive than a conventional heating plant. Only when a heat pump is needed for air conditioning as well as for space heating is it a practical proposition.

Freon 12 is one of a family of chlorofluorocarbons (CFCs) widely used in insulating foam and as propellants for aerosol sprays as well as in refrigeration. About a million tons of CFCs are manufactured each year, much of which ends up in the upper atmosphere where its chlorine content acts as a catalyst to break up the ozone (O_3) molecules there. Ozone is an excellent absorber of the short-wavelength ultraviolet radiation that arrives from the sun, radiation that is extremely harmful to living things. The dumping of CFCs into the atmosphere is therefore a serious matter and has already led to a measurable ozone loss. In 1987 an international treaty was signed to stabilize and eventually reduce CFC production, but since CFCs remain in the atmosphere for many decades, we can look forward to a continuing rise in skin cancers and crop damage due to solar ultraviolet radiation.

EXAMPLE 15.5

(a) A Carnot refrigerator is a Carnot engine operating backward. A Carnot refrigerator extracts heat from a storage chamber (reservoir 1) at the absolute temperature T_1 and gives off heat to the outside world (reservoir 2) at the absolute temperature T_2. Find the ratio between the work done on the refrigerator and the heat extracted from the storage chamber. (b) A Carnot refrigerator extracts heat from a freezer at $-5°C$ and exhausts it at 25°C. How much work per joule of heat extracted is needed?

SOLUTION (a) If Q_1 is the heat extracted from the storage chamber and Q_2 is the heat given off to the outside world, the work done is $W = Q_2 - Q_1$. As in a Carnot engine, the heat Q transferred to or from a Carnot refrigerator is proportional to the absolute temperature T at which the transfer takes place, so that

$$\frac{Q_2}{Q_1} = \frac{T_2}{T_1}$$

Hence

$$\frac{W}{Q_1} = \frac{Q_2 - Q_1}{Q_1} = \frac{Q_2}{Q_1} - 1 = \frac{T_2}{T_1} - 1$$

(b) Here $T_1 = 268$ K, $T_2 = 298$ K, and $Q_1 = 1.00$ J. Therefore

$$W = Q_1\left(\frac{T_2}{T_1} - 1\right) = (1.00\text{ J})\left(\frac{298\text{ K}}{268\text{ K}} - 1\right) = (1.00\text{ J})(0.11) = 0.11\text{ J}$$

♦

15.8 Statistical Mechanics

The natural tendency of all systems is toward increasing disorder.

According to the second law of thermodynamics, it is impossible to convert heat completely into any other form of energy. Some of the heat input to an engine *must* be lost. Why? The reason lies in the nature of heat, which is molecular kinetic energy. The temperature of a body is a measure of the average kinetic energy of each of its constituent molecules. Let us see how the notion that matter consists of moving molecules connects with the second law of thermodynamics. Although the discussion will be based on molecules in a gas, the essential ideas hold for matter in any state.

The molecules of a gas are in constant random motion and undergo frequent collisions with one another. Although we cannot hope to follow an individual gas molecule in its wanderings, statistical arguments can tell us what fraction of the time it will have any given amount of kinetic energy. Hence we can find the distribution of molecular energies in a gas sample at a particular temperature. This distribution, which has been confirmed by experiment, has the form shown in Fig. 15.21 and holds for all **equilibrium** conditions in which each molecule has the same average energy over a period of time. A molecule that moves more swiftly than usual at one instant will move less swiftly at a later instant after a number of collisions have taken place.

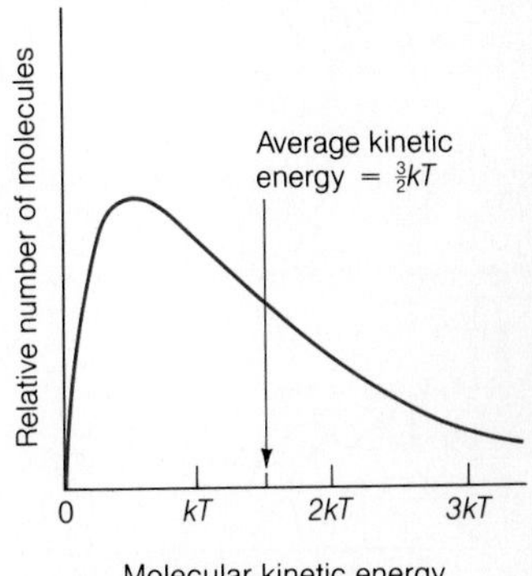

Figure 15.21

The distribution of molecular energies in a gas at a particular temperature.

Equilibrium is the most probable condition according to **statistical mechanics,** a branch of physics that mathematically deduces the behavior of assemblies of

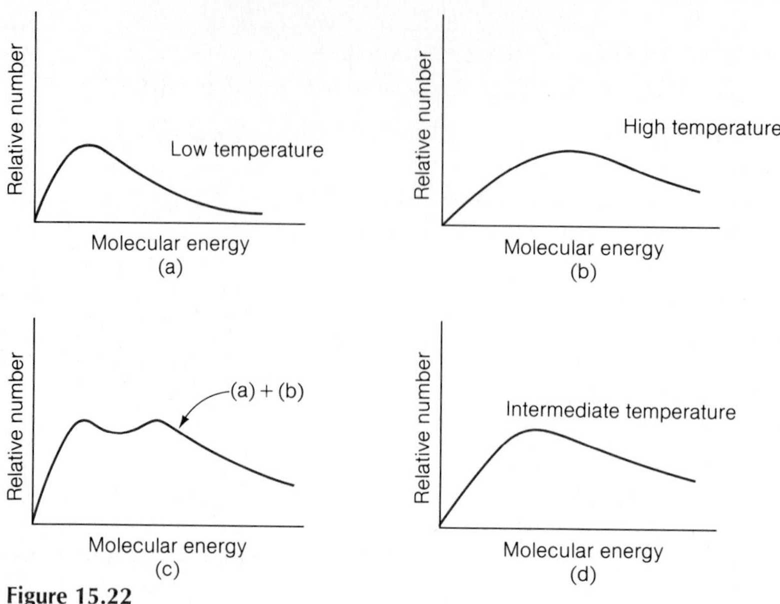

Figure 15.22

The molecular systems (c) and (d) have the same total energy, but the distribution of energies in (d) has the greater probability.

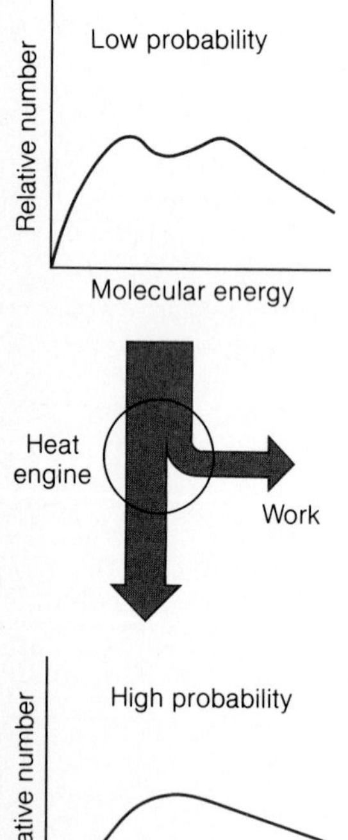

Figure 15.23

Work can be done by a system whose distribution of molecular energies is statistically improbable.

so many particles that deviations from statistically probable behavior are not significant. If we toss two coins, one may well not come up heads and the other tails. However, if we toss a million coins, the percentage deviation from an equal number of heads and tails will be extremely small. We can appreciate why departures from equilibrium are so unlikely if we look at a cubic centimeter—a thimbleful—of air at atmospheric pressure and room temperature. There are 2.7×10^{19} molecules in the cubic centimeter, and each molecule undergoes an average of 4×10^9 collisions per second—almost equivalent to every person in the world colliding with every other person, one at a time, in each second.

Suppose we have a heat reservoir at a high temperature and another at a low temperature. The molecules of each are in equilibrium and have the molecular energy distributions shown in Fig. 15–22(a) and (b). If we consider the two reservoirs as a single system, the molecular energy distribution in the system is like that of (c). This distribution is, in a statistical sense, very improbable. If they were mixed together, the molecules of the two reservoirs would soon blend their energies in collisions to attain the equilibrium distribution of (d), which corresponds to a temperature intermediate between the initial ones of the two reservoirs.

We note the important fact that the total energy contents of the distributions of Fig. 15.22(c) and (d) are identical. The only distinction between them is the way in which the energy is allotted to the molecules on the average. However, it is just this distinction with which the second law of thermodynamics is concerned, because this law states that a system of two heat reservoirs at different temperatures can be made to yield a net work output, whereas a single heat reservoir, no matter how much energy it has, cannot yield any net work. A system of molecules whose energies are distributed in the most probable way is ''dead'' thermodynamically, whereas a system having a different distribution of molecular energies is capable of doing mechanical work as it progresses to an equilibrium state (Fig. 15.23).

BIOGRAPHICAL NOTE

Ludwig Boltzmann

Ludwig Boltzmann (1844–1906) was born in Vienna and attended the university there. He then taught and carried out both experimental and theoretical research at a number of institutions in Austria and Germany, moving from one to another every few years. Boltzmann was interested in poetry, music, and travel as well as in physics; he visited the United States three times, something unusual in those days.

Of Boltzmann's many contributions to physics, the most important were to the kinetic theory of gases and statistical mechanics. The constant k in the formula $KE_{av} = \frac{3}{2}kT$ for the average energy of a gas molecule is named after him in honor of his work on the distribution of molecular energies in a gas. In 1884 Boltzmann derived from thermodynamic considerations the Stefan-Boltzmann law $R = e\sigma T^4$ for the radiation rate of a blackbody. Josef Stefan, who had been one of Boltzmann's teachers, had discovered this law experimentally five years earlier. One of Boltzmann's major achievements was the interpretation of the second law of thermodynamics in terms of order and disorder. A monument to Boltzmann in Vienna is inscribed with his formula $S = k \log W$, which relates the entropy S of a system to its probability W.

Boltzmann was a champion of the atomic theory of matter, still controversial in the late nineteenth century because there was then only indirect evidence of the existence of atoms and molecules. Battles with nonbelieving scientists deeply upset Boltzmann, and in his later years asthma, headaches, and increasingly poor eyesight further depressed his spirits. He committed suicide in 1906, not long after Albert Einstein published a paper on Brownian motion that was to convince the remaining doubters of the atomic theory of its correctness.

Entropy

Although ''disorder'' may not seem a very specific concept, it is possible to define a quantity called **entropy** that is a measure of the disorder of the particles that make up a certain body of matter. For instance, the entropy per kilogram of liquid water is several times greater than that of ice, which reflects the greater disorder of molecules in the liquid state than of those in the solid state.

In terms of entropy the second law of thermodynamics becomes:

The entropy of an isolated system cannot decrease.

If a puddle of water were to rise by itself while turning into ice, the process would conserve energy (and so obey the first law of thermodynamics). However, the process would decrease the entropy of the water, hence it is impossible. The advantage of expressing the second law in terms of entropy is that, because entropy can be determined for a variety of systems, this law can be applied in a much wider context than can the statements of Sections 15.3 and 15.5.

Biological systems might seem to violate the second law. Certainly entropy decreases when a plant turns carbon dioxide and water into leaves and flowers. But this transformation of disorder into order needs the energy of sunlight. If we take into account the increase in the entropy of the sun as it produces the required sunlight, the net result is an increase in the entropy of the universe. There is no way to avoid the second law.

The second law is an unusual physical principle in several respects. It does not apply to individual particles, only to assemblies of many particles. It does not say what can occur, only what cannot occur. And it is unique in that it is closely tied to the direction of time.

Events that involve a few particles are always reversible. Nobody can tell whether a film of billiard balls bouncing around is being run backward or forward. But events that involve many-particle systems are not always reversible. A film of an egg breaking when dropped makes no sense when run backward. The arrow of time always points in the direction of entropy increase. Time never runs backward; a broken egg never reassembles itself; in the universe as a whole, entropy—and the disorder it mirrors—continues to increase.

15.9 The Energy Problem

Unlimited demand, limited supply.

The rise of modern civilization has been paralleled by a steady increase in the world's use of energy. This is no coincidence. All that we do needs energy, and the more energy we have at our command, the better we can meet our desires for food, clothing, shelter, warmth, light, transport, communication, and manufactured goods. In the United States, energy is used at a rate of 11 kW per person.

A single source has provided almost all the energy we use today—the sun. Light and heat arrive directly from the sun; food owes its energy content to photosynthesis; water power exists because solar heat evaporates water that later falls as rain on high ground; and wind power comes from convection in the atmosphere due to unequal heating of the earth's surface by the sun. The fossil fuels coal, oil, and natural gas were formed from plants and animals that lived and stored energy derived from sunlight millions of years ago. Only nuclear energy and heat from sources inside the earth cannot be traced to the sun's rays (Fig. 15.24).

In advanced countries, where the standard of living is already high and populations are stable, energy consumption is unlikely to grow fast. Indeed, it may even decline as energy use becomes more efficient. For more than half the people of the world, however, energy consumption is still low—a rate of less than 1 kW per person. These people seek better lives and their numbers are increasing rapidly, which means much more energy will be needed. Where is it to come from?

The fossil fuels coal, oil, and natural gas, which today furnish by far the greater part of the world's energy, cannot last forever. Oil and natural gas will be the first to be exhausted. At the current rate of consumption, known oil reserves will last only about 35 years more. More oil will certainly be found, and better technology will increase the yield from existing wells, but even so in the next century oil will

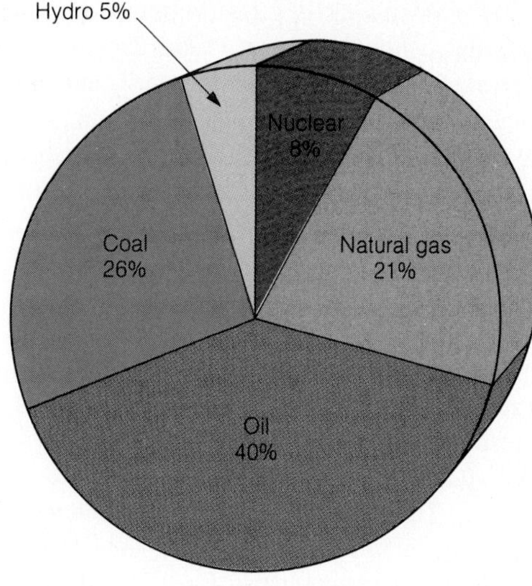

Figure 15.24

A total of nearly 10^{20} J of energy is expected to be produced in the United States in 1995 from the sources shown.

inevitably become scarce. The same is true for natural gas. This is unfortunate because oil and gas burn efficiently and are easy to extract, process, and transport. Half the oil used today goes into fuels that power ships, trains, aircraft, cars, and trucks, and oil and gas are superb feedstocks for synthetic materials of all kinds. Although liquid fuels can be made from coal and coal can serve as the raw material for synthetics, these technologies involve greater expense and greater risk to health and to the environment.

Even though the coal currently consumed each year took about two million years to accumulate, enough remains to last another two or three centuries. But coal is far from being a desirable fuel. Not only does its mining leave large tracts of land unfit for further use, but the air pollution due to coal burning has shortened the lives of millions of people through cancer and respiratory diseases. Coal-burning power plants actually expose their surrounding populations to more radioactivity than do normally operating nuclear plants because of the radon gas they liberate.

Nuclear fuel reserves exceed those of fossil fuels. Properly built and operated plants utilizing this abundant fuel supply are in many respects excellent sources of energy. Nuclear energy provides almost a fifth of the electricity generated in the United States, and in a number of other countries the proportion is even higher; in France it is nearly three-quarters.

To be sure, nuclear energy has serious drawbacks. The potential for large-scale disaster is always present. Two major reactor accidents, at Three Mile Island in Pennsylvania and at Chernobyl in the Soviet Union, have already occurred. Although the overall public-health record of nuclear plants is far better than that of coal-burning ones, above-average cancer and leukemia rates have been reported near badly run nuclear installations here and abroad. Further, a reactor produces many tons of waste each year whose radioactivity will remain high for thousands of years, and the safe disposal of nuclear waste is still unsettled.

The sun and stars obtain their energy from nuclear fusion, and in the long run we may do so as well. As described in Chapter 29, a fusion reactor will get its fuel from the sea, will be safe and nonpolluting, and cannot be adapted for military purposes. But nobody can yet say for sure whether this ultimate energy source will become an everyday reality.

Assuming that fusion power is really on the way, the next question is how to manage until it arrives. Coal can be used more widely, but only at the cost of increased environmental damage and human suffering. And the faster burning of fossil fuels will increase the carbon dioxide in the atmosphere, which, as we saw in Section 14.8, may affect the world's climates. Nuclear energy can help bridge the gap, but memories of Three Mile Island and Chernobyl will have to fade before reactors can be built on a large enough scale.

Renewable Energy Sources

What about the energy of sunlight, of winds and waves, of tides and falling water, of trees and plants? The technology needed to exploit these renewable resources already exists. But even a brief look shows that such energy is unlikely to provide more than a small fraction (though a welcome one) of future needs.

To begin with, a disadvantage of all renewable-energy installations is that they take a lot of space. A city of medium size might use 1000 MW of power. Less than 150 acres is enough for a 1000-MW nuclear plant, whereas 5000 acres might be needed for solar collectors of the same capacity, and over 10,000 acres for windmills. Converting crops into fuel might require 200 square miles of farmland to yield 1000 MW averaged over a year. This is not to say that renewable sources are without value, particularly where local conditions are favorable. Such sources, however, do not seem to be the ultimate answer.

Solar energy Bright sunlight can deliver over 1 kW of power to each square meter it falls on. At this rate a tennis court receives solar energy equivalent to that in a gallon of gasoline every 10 min or so. Although the supply of sunshine varies with location (Fig. 15.25), time of day, season, and weather, plenty is available. One way to use it involves a series of mirrors that track the sun and concentrate its radiation on a boiler whose steam drives a turbine and thence an electric generator. Better in principle are photovoltaic cells that turn light energy directly into electricity. These have no moving parts and avoid thermodynamic limitations because no heat is involved. Most spacecraft rely on solar cells. Unfortunately today's cells are both inefficient and expensive, although better ones are being developed, and a way to store the energy they produce is needed for a continuous supply of electricity.

Wind. A large modern windmill with rotor blades 100 m in diameter might produce 3 MW when the wind speed is 12 m/s (about 27 mi/h). A more typical windmill is

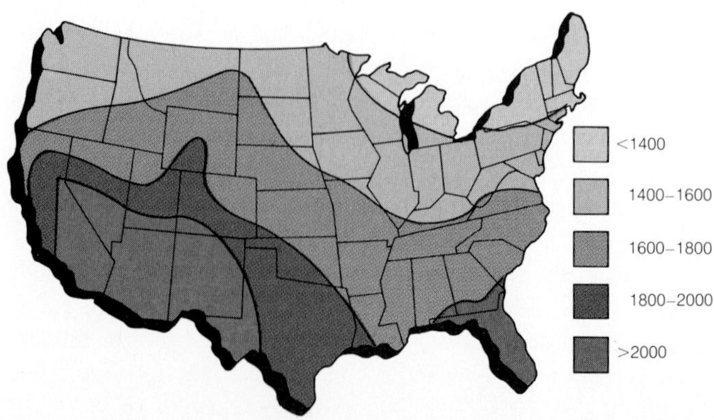

Figure 15.25

Solar energy received per year in the United States expressed in kilowatt-hours per square meter (1 kW·h = 3.6 MJ).

smaller, with an output of perhaps 100 kW. All are noisy. Windmills are cheaper per watt than solar cells, though still more expensive than conventional power stations, and make sense only where winds are steady and strong. Windmills with a total capacity of over 1500 MW are operating in the United States and elsewhere.

Waves. As anybody who has stood in the surf or watched waves dash against a rocky shore knows, waves carry energy in abundance. One method of capturing this energy is to have ocean waves run up a sloping funnellike channel to a reservoir above. Water from the reservoir then powers a turbine generator on its way back to the ocean. Such a system is feasible only where the seabed is so shaped that wave energy is focused on a particular spot on a coast and where the winds that drive the waves are reliable.

Tides. The twice-daily rise and fall of the tides brings corresponding flows of water into and out of bays and river mouths. The energy involved is enormous when the tidal range is great. Tides in the Rance River estuary in northern France have driven 240-MW generators since 1966, and smaller installations are operating in China, Canada, and the Soviet Union. A tidal power project involving a barrier 16 km long has been proposed for the Severn River in England. The project would produce 7200 MW, a fifth of the country's electricity demand. A few other suitable sites on such a scale exist, but the size of the barriers needed is likely to make them impractical.

Hydroelectricity. The power of falling water is responsible for 60,000 MW of electricity in the United States alone. Untapped flows exist here and elsewhere in the world, but it is hard to imagine more than a doubling or tripling of the energy available from this ideal source without running into financial and environmental obstacles.

Biomass. From the time fire came under human control until coal became the dominant fuel a century and a half ago, wood was the chief source of our energy apart from food. Trees grow almost everywhere and ought to be a renewable resource. Over a billion people today in the poorer countries rely on wood for fuel but are burning it up faster than it is being replaced. Wood smoke is even more polluting than coal smoke and is especially high in carcinogens. And wood is too valuable as a material for construction and paper simply to burn it.

In another biomass approach, plant carbohydrates are converted to alcohol for use as vehicle fuel. In Brazil millions of cars run on pure alcohol made from sugar cane, and in the United States a gasoline-alcohol blend is being increasingly used. This technology does not deplete natural resources and causes little pollution. But overall if we include the fuel needed for cultivation and harvesting, it may consume more energy than it provides. In the long run diverting agricultural land from food crops may be unwise: The land needed to run one car on alcohol can provide food for 8 to 16 people.

Geothermal energy. Temperature increases with depth in the earth's crust, but only in a few places is it hot enough near the surface for useful energy to be extracted. One such place is north of San Francisco, where turbines driven by natural steam produce 665 MW of electric power. Similar plants are operating in New Zealand and Italy, but few other regions seem to have much geothermal potential. Although the natural water supply of a geothermal source may eventually run out, the subsurface heat will remain, and water can be pumped down to produce the needed steam.

Ocean thermal energy. In tropical oceans surface water is typically 27°C, whereas at a depth of a kilometer or so it is 5°C or less. As long ago as 1881 warm surface water was proposed as the heat source for vaporizing a suitable fluid that would drive a turbine and then be condensed by cold water drawn from below. Although ''fuel'' in the form of warm surface water is unlimited and free, the thermodynamic efficiency of such an engine would be only a few percent. Hence a very large plant would be needed relative to its capacity. The economics of ocean thermal energy do not seem very promising and a good many engineering problems must be overcome, yet studies continue because the basic idea is so attractive.

Clearly we can expect no magic solution in the near future to the problem of safe, cheap, and abundant energy. The only sensible course is to practice conservation and try to get the best from the various available technologies while pursuing fusion energy as rapidly as possible. As we have seen, each of these technologies has limitations but may be useful in a given situation. If their full potential is realized and population growth slows down, social disaster (starvation, war) and environmental catastrophe (a planet unfit for life) may well be avoided during the wait for fusion.

Even given fusion energy, however, energy consumption cannot rise indefinitely. The second law of thermodynamics is what stands in the way—the inevitable waste heat must go somewhere, and even now its disposal is difficult in many places. In the United States about 10 percent of the flow of all rivers and streams is already being used as cooling water for power stations. If the resultant heating of inland waters increases still further, the biological impact will be severe. Discharging waste heat into the atmosphere through cooling towers on a large scale will alter the weather and climate of the region, not necessarily for the better. The oceans can absorb much waste heat, but locating power plants on their shores poses the problem of carrying their electricity thousands of miles inland. Only a limit to the world's population will prevent a conflict between the desire for more energy and the need to get rid of waste heat.

Important Terms

A **heat engine** is any device that converts heat into mechanical energy or work.

The **first law of thermodynamics** states that the work output of a heat engine is equal to its net heat input.

The **second law of thermodynamics** states that it is impossible to construct an engine, operating in a repeatable cycle, which does nothing other than take energy from a source and perform an equivalent amount of work.

Entropy is a measure of the disorder in a system. The entropy of an isolated system cannot decrease, which is another way to express the second law of thermodynamics.

A **Carnot engine** is an idealized engine that is not subject to such practical difficulties as friction or heat losses by conduction or radiation but that obeys all physical laws. No engine operating between the same two temperatures can be more efficient than a Carnot engine operating between them.

An **isobaric** process takes place at constant pressure, and an **isothermal** process takes place at constant temperature. No heat enters or leaves a system during an **adiabatic** process.

Important Formulas

First law of thermodynamics: $Q = \Delta U + W$

Work done by expanding gas: $\Delta W = p\Delta V$

Engine efficiency: $\text{Eff} = \dfrac{W}{Q_1} = 1 - \dfrac{Q_2}{Q_1}$

Carnot efficiency: $\text{Eff} = 1 - \dfrac{T_2}{T_1}$

MULTIPLE CHOICE

1. A heat engine operates by taking in heat at a particular temperature and
   a. converting it all into work
   b. converting some of it into work and exhausting the rest at a lower temperature
   c. converting some of it into work and exhausting the rest at the same temperature
   d. converting some of it into work and exhausting the rest at a higher temperature

2. The natural direction of heat flow is from a high-temperature reservoir to a low-temperature reservoir, regardless of their respective heat contents. This fact is incorporated in the
   a. first law of thermodynamics
   b. second law of thermodynamics
   c. law of conservation of energy
   d. law of conservation of entropy

3. A type of process that does not need outside energy to reverse is one that occurs at constant
   a. temperature
   b. pressure
   c. volume
   d. speed

4. An adiabatic process in a system is one in which
   a. no heat enters or leaves the system
   b. the system does no work nor is work done on it
   c. the temperature of the system remains constant
   d. the pressure of the system remains constant

5. A process involving a gas that cannot, even in principle, be reversed without work being done is
   a. an isobaric compression
   b. an isothermal compression
   c. an adiabatic compression
   d. an adiabatic expansion

6. In any process the maximum amount of heat that can be converted to mechanical energy
   a. depends only on the intake temperature
   b. depends on the intake and exhaust temperatures
   c. depends on whether kinetic or potential energy is involved
   d. is 100%

7. In any process the maximum amount of mechanical energy that can be converted to heat
   a. depends only on the intake temperature
   b. depends on the intake and exhaust temperatures
   c. depends on whether kinetic or potential energy is involved
   d. is 100%

8. The work output of every heat engine
   a. equals the difference between its heat intake and heat exhaust
   b. equals that of a Carnot engine with the same intake and exhaust temperatures
   c. depends only on its intake temperature
   d. depends only on its exhaust temperature

9. The area enclosed by the p-V graph of a complete heat engine cycle equals
   a. the heat intake per cycle
   b. the heat output per cycle
   c. the work done on the engine per cycle
   d. the work done by the engine per cycle

10. A frictionless heat engine can be 100% efficient only if its exhaust temperature is
   a. equal to its input temperature
   b. less than its input temperature
   c. 0°C
   d. 0 K

11. A process not involved in the operating cycle of a Carnot engine is
   a. an isothermal expansion
   b. an isobaric expansion
   c. an adiabatic compression
   d. an adiabatic expansion

12. A Carnot engine turns heat into work
   a. with 100% efficiency
   b. with 0% efficiency
   c. without itself undergoing a permanent change
   d. with the help of expanding steam

13. A Carnot engine that operates between the absolute temperatures T_1 and T_2
   a. is 100% efficient
   b. has the maximum efficiency possible under these circumstances
   c. has an efficiency of T_2/T_1
   d. has the same efficiency as an actual engine that operates between T_1 and T_2

14. Which of the following engines is normally the least efficient?
   a. gasoline engine
   b. diesel engine
   c. gas turbine
   d. Carnot engine

15. Which of the following engines is the most efficient?
   a. gasoline engine
   b. diesel engine
   c. gas turbine
   d. Carnot engine

16. The fuel in a diesel engine is ignited by
   a. a spark plug
   b. being compressed until it is hot enough
   c. the hot compressed air into which it is injected
   d. exhaust gases remaining from the previous cycle

17. The physics underlying the operation of a refrigerator

most closely resembles the physics underlying

a. a heat engine
b. the melting of ice
c. the freezing of water
d. the evaporation of water

18. A refrigerator

a. produces cold
b. causes heat to vanish
c. removes heat from a region and transports it elsewhere
d. changes heat to cold

19. A refrigerator exhausts

a. less heat than it absorbs from its contents
b. the same amount of heat it absorbs from its contents
c. more heat than it absorbs from its contents
d. any of the above, depending on the circumstances

20. The working substance (or refrigerant) used in most refrigerators is a

a. gas that is easy to liquefy
b. gas that is hard to liquefy
c. liquid that is easy to solidify
d. liquid that is hard to solidify

21. Heat is absorbed by the refrigerant in a refrigerator when it

a. melts
b. vaporizes
c. condenses
d. is compressed

22. A heat pump is basically a

a. pump for hot air
b. pump for hot water
c. heat engine
d. refrigerator

23. When a gas is in equilibrium, its molecules

a. all have the same energy
b. have different energies that remain constant
c. have a certain constant average energy
d. do not collide with one another

24. A system of molecules whose energies are distributed in the most probable way

a. can perform an amount of mechanical work equal to its total energy content
b. can perform an amount of mechanical work that depends on its absolute temperature
c. cannot perform any mechanical work
d. is a Carnot engine

25. The chief source of energy in the world today is

a. coal
b. oil
c. natural gas
d. uranium

26. The source of energy whose reserves are greatest is

a. coal
b. oil
c. natural gas
d. uranium

27. A gas is compressed from 5 L to 3 L under an average pressure of 2 bars. During the compression 0.3 kJ of heat is given off. The internal energy of the gas

a. decreases by 0.1 kJ
b. is unchanged
c. increases by 0.1 kJ
d. increases by 0.4 kJ

28. A Carnot engine absorbs heat at a temperature of 127°C and exhausts heat at a temperature of 77°C. Its efficiency is

a. 13%
b. 39%
c. 61%
d. 88%

29. If a heat engine that exhausts heat at 400 K is to have an efficiency of 33%, it must take in heat at a minimum of

a. 133 K
b. 449 K
c. 532 K
d. 600 K

30. A Carnot engine operates between 800 K and 200 K. If it absorbs 8 kJ of heat in each cycle, the work it does per cycle is

a. 1 kJ
b. 2 kJ
c. 2.7 kJ
d. 6 kJ

QUESTIONS

1. The sun's corona is a very dilute gas at a temperature of about 10^6 K that is believed to extend into interplanetary space at least as far as the earth's orbit. Why can we not use the corona as the high-temperature reservoir of a heat engine in an earth satellite?

2. (a) An attempt is made to cool a kitchen in the summer by switching on an electric fan and closing the kitchen door and windows. What will happen? (b) In another attempt to cool the kitchen, the refrigerator door is left open, again with the kitchen door and windows closed. Now what will happen?

3. A gas sample expands from V_1 to V_2. Does it perform the most work when the expansion takes place at constant pressure, at constant temperature, or adiabatically? In which process does the gas perform the least work? Why?

4. The operation of a steam engine proceeds approximately as follows: (1) Water under pressure is heated to the boiling point; (2) the water turns into steam and expands at constant pressure at its boiling point; (3) the steam enters the cylinder of the engine and expands adiabatically against the piston; (4) the spent steam condenses into water at constant pressure; (5) the water is pumped back into the boiler. Plot the entire cycle on a p-V diagram and indicate in what parts of the cycle heat is absorbed, heat is rejected, and work is done on the outside world.

5. What is the relationship between entropy and the second law of thermodynamics? Between entropy and order? Between entropy and time?

6. What energy sources, if any, cannot be traced to sunlight falling on the earth?

7. Which three fuels provide most of the world's energy today? List their chief advantages and disadvantages.

8. What are some of the disadvantages that have kept renewable energy sources from supplying more than a small fraction of the world's energy?

EXERCISES

15.1 First Law of Thermodynamics

15.2 Work Done by and on a Gas

1. A gas expands by 1.0 L at a constant pressure of 2.0 bars. During the expansion 500 J of heat is added. Find the change in the internal energy of the gas.

2. A gas expands isothermally from 40 L to 60 L while absorbing 5.0 kJ of heat. Find the average pressure on the gas during the expansion.

3. A steam engine built in 1712 had a piston 55 cm in diameter that moved back and forth through a distance of 2.44 m. If the pressure on the piston was a constant 1.0 bar during each expansion, how much work was done?

4. The indicator diagram of a certain engine is shown in Fig. 15.26. (a) Find the work done in each part of the cycle—namely, $a \rightarrow b$, $b \rightarrow c$, $c \rightarrow d$, and $d \rightarrow a$—and identify it as work done by or on the engine. (b) How much net work is done per cycle by the engine?

5. The engine in Exercise 4 is operated in reverse, with each cycle proceeding as *adcba*. Answer the same questions for the engine in this case.

15.4 Carnot Engine

15.5 Engine Efficiency

6. As mentioned in Section 15.9, a typical temperature for surface water in a tropical ocean is 27°C, whereas at a depth of a kilometer or more it is only about 5°C. It has been proposed to operate heat engines using surface water as the hot reservoir and deep water as the cold reservoir. What would the maximum efficiency of such an engine be? Why might such an engine eventually be a practical proposition even with so low an efficiency?

Figure 15.26

Exercise 4.

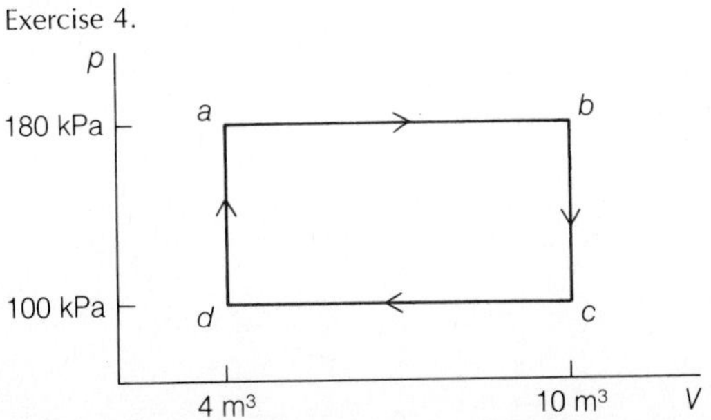

7. An engine operating between 300°C and 50°C is 15% efficient. What would its efficiency be if it were a Carnot engine?

8. An engine is proposed that is to operate between 200°C and 50°C with an efficiency of 35%. Will the engine perform as predicted? If not, what would its maximum efficiency be?

9. One of the most efficient engines ever developed operates between about 2000 K and 700 K. Its actual efficiency is 40%. What percentage of its maximum possible efficiency is this?

10. A steam engine takes in steam at 200°C. What is the maximum temperature at which spent steam can leave the engine if its efficiency is to be 20%?

11. A Carnot engine absorbs 200 kJ of heat at 500K and exhausts 150 kJ. What is the exhaust temperature?

12. Three designs for a heat engine to operate between 450 K and 300 K are proposed. Design *A* is claimed to require a heat input of 2.0 kJ for each kJ of work output, design *B* a heat input of 3.0 kJ, and design *C* a heat input of 4.0 kJ. Which design would you choose and why?

13. A Carnot engine takes in 1.0 MJ of heat from a reservoir at 327°C and exhausts heat to a reservoir at 127°C. How much work does it do?

14. A Carnot engine whose efficiency is 35% takes in heat at 500°C. What must the intake temperature be if the efficiency is to be 50% with the same exhaust temperature?

15.6 Internal Combustion Engines

15. A four-cylinder, four-stroke diesel engine has pistons 108 mm in diameter whose travel is 127 mm. The engine develops 100 kW at 66 rev/s. Find the average pressure on the pistons during the power stroke.

16. The six-cylinder, four-stroke engine of a car has pistons 90 mm in diameter whose travel is 100 mm. The average pressure on the pistons during the power stroke is 5.0 bars. Find the power developed by the engine when it operates at 3000 rpm.

17. A turboprop aircraft engine consumes 250 g/h of kerosene for each horsepower it develops. Find the efficiency of the engine.

18. A 4250-hp aircraft engine is used in a power station in

England to run a 3.0-MW generator to supply electricity during peak-load periods. The engine consumes 370 g of kerosene per kilowatt-hour of energy produced. What is the overall efficiency of the installation?

19. At 800 rpm a diesel engine develops 60 kW and consumes 18 kg of fuel per hour. At 1800 rpm the same engine develops 120 kW and consumes 32 kg of fuel per hour. Find its efficiency at each speed.

20. A certain power station consumes 2000 metric tons (1 metric ton = 10^3 kg) of coal per day. The overall efficiency of the station is 40%. Find the number of megawatts of electricity the station produces.

21. How many cubic meters of propane must be burned to heat 50 L of water from 5°C to 90°C? Assume that 25% of the heat is wasted.

22. A certain jet aircraft uses 10,000 kg of kerosene to travel 1500 km at an average speed of 800 km/h. If the engines of the aircraft develop an average of 11 MW during the flight, what is their efficiency?

15.7 Refrigerators

23. Refrigerators are sometimes rated in **tons,** where 1 ton of refrigeration capacity is that rate of heat removal that can freeze 2000 lb of water at its freezing point to ice per day. (a) Express the refrigeration ton in kW. (b) A $\frac{3}{4}$-ton window air conditioner is found able to maintain a constant temperature of 21°C in an empty room when the outside temperature is 32°C by running half the time. A sitting person liberates about 120 W. How many people can sit reading in the room, each with a 75-W lamp, without exceeding the capacity of the air conditioner?

24. Three designs for a refrigerator to operate between −20°C and 30°C are proposed. Design *A* is claimed to need 100 J per kilojoule of heat extracted, design *B* to need 200 J, and design *C* to need 300 J. Which design would you choose and why?

25. A 500-W refrigerator extracts heat at a rate of 3.6 MJ/h from a storage chamber at −20°C and exhausts heat at 40°C. How much heat per hour would be extracted by a 500-W ideal refrigerator operating between these temperatures?

PROBLEMS

26. A coal-fired electric power plant of 400 MW output operates at an efficiency of 39%, which is about three-quarters of its Carnot efficiency. Water from a river flowing at 50 m^3/s is used to absorb the waste heat. By how much does the river's temperature rise as a result?

27. The total drop of the Wollomombi Falls in Australia is 482 m. What would be the Carnot efficiency of an engine operating between the top and bottom of the falls if the water temperature at the top were 10°C and all the potential energy of the water at the top were converted to heat at the bottom?

28. A Carnot refrigerator is used to make 1.0 kg of ice at −10°C from 1.0 kg of water at 20°C, which is also the temperature of the kitchen. How many joules of work must be done?

29. The coefficient of performance (CP) of a refrigerator is the ratio between the heat it absorbs from a storage chamber and the work required. The higher the CP, the less work is needed to remove a given amount of heat from the storage chamber; typical values of CP range from 2 to 3. (a) Express the CP of a Carnot refrigerator in terms of the absolute temperatures at which it absorbs and rejects heat. (b) Consider a refrigerator whose CP is half that of a Carnot refrigerator and that absorbs heat from a storage chamber at −23°C and rejects heat at 27°C at a rate of 250 kJ/min. Find the rate at which heat is absorbed from the chamber. (c) How much power does the refrigerator need?

30. The resistive forces on a car total 1.2 kN when its speed on a level road is 80 km/h. If the car can go 8.5 km per liter of gasoline at that speed, find the efficiency of its engine and drive train.

Answers to Multiple Choice

1. b	9. d	17. a	25. b
2. b	10. d	18. c	26. d
3. a	11. b	19. c	27. c
4. a	12. c	20. a	28. a
5. a	13. b	21. b	29. d
6. b	14. a	22. d	30. d
7. d	15. d	23. c	
8. a	16. c	24. c	

C H A P T E R

16

ELECTRICITY

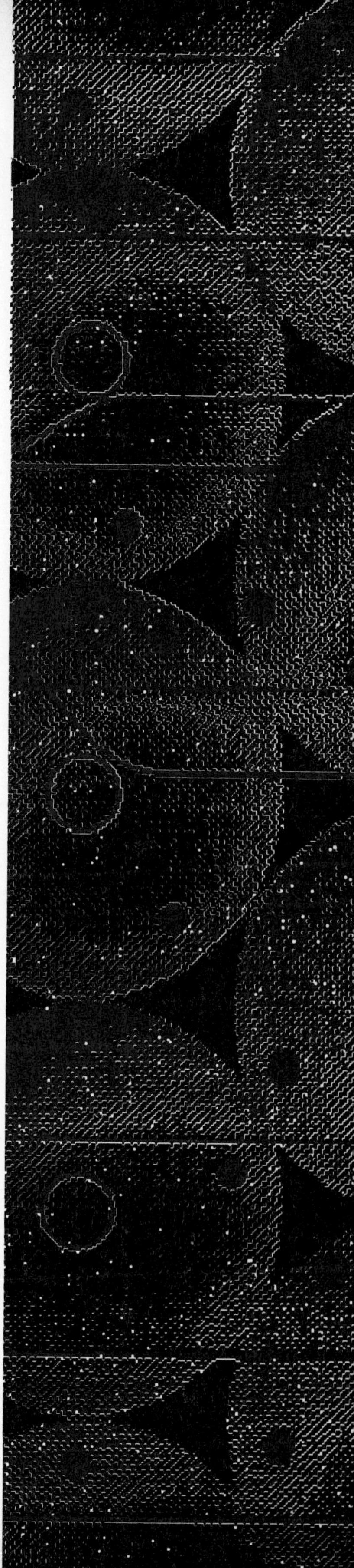

We have now looked into force and motion, mass and energy, temperature and heat. The ideas we have found help make sense of much of the world of our experience, from the paths of the planets across the sky to why heat cannot be entirely turned into work. Are these ideas enough for us to understand how the entire physical universe works?

For an answer all we need do is run a hard rubber comb through our hair on a dry day. Little sparks occur, and afterward the comb can pick up dust and small bits of paper. The attraction is surely not due to gravity, because the gravitational force between comb and paper is far too small. And gravity should not depend on whether the comb is run through our hair or not. What has been revealed by this experiment is an electrical phenomenon, so called after *elektron*, the Greek word for amber, a substance used in the earliest studies of electricity.

16.1 Electric Charge

Like charges repel; unlike charges attract.

Electricity is familiar to all of us as the name for that which causes our light bulbs to glow, many of our motors to turn, our telephones and radios to communicate sounds, our television screens to communicate pictures. But there is more to electricity than its ability to transmit energy and information. Electrical forces hold electrons to nuclei to form atoms, and they hold atoms together to form molecules, solids, and liquids. All of the chief properties of matter in bulk—with the notable exception of mass—can be traced to the electrical nature of its basic particles.

Let us begin our study of electricity by looking at three basic experiments. The first experiment is shown in Fig. 16.1. By convention, we call whatever it is that a rubber rod has by virtue of having been stroked with a piece of fur **negative electric charge.** Part of the negative charge on the rubber rod in Fig. 16.1 passed to the plastic ball when it was touched, and the fact that the ball then flew away from the rod suggests that negative electric charges repel each other.

The next experiment, shown in Fig. 16.2, is very similar. By convention, we call whatever it is that a glass rod has by virtue of having been stroked with a silk cloth **positive electric charge.** Part of the positive charge on the glass rod in

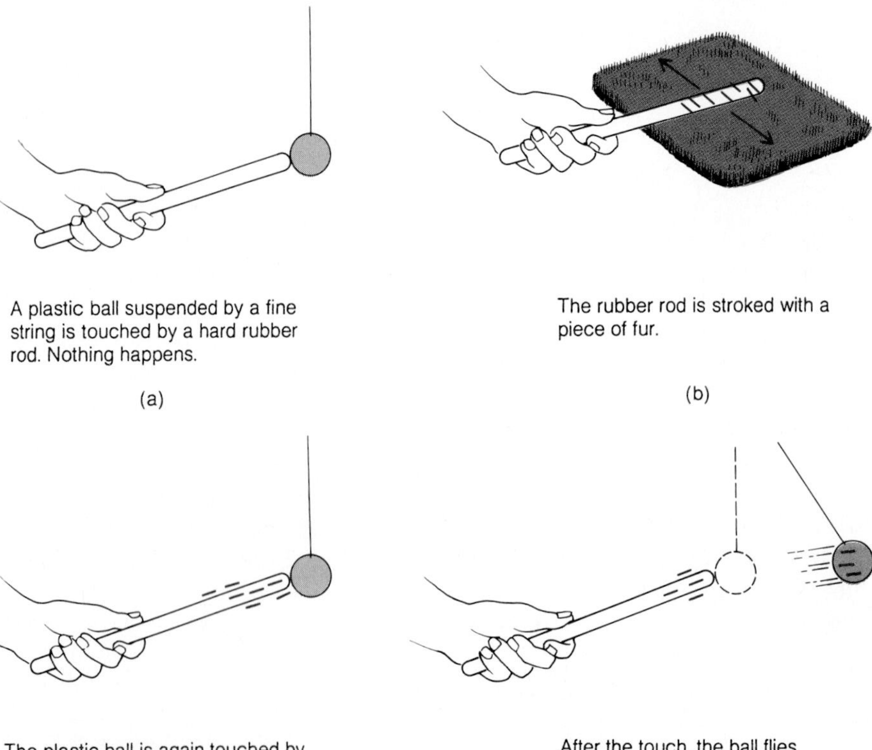

Figure 16.1

A rubber rod stroked with fur becomes negatively charged; two negatively charged objects repel each other.

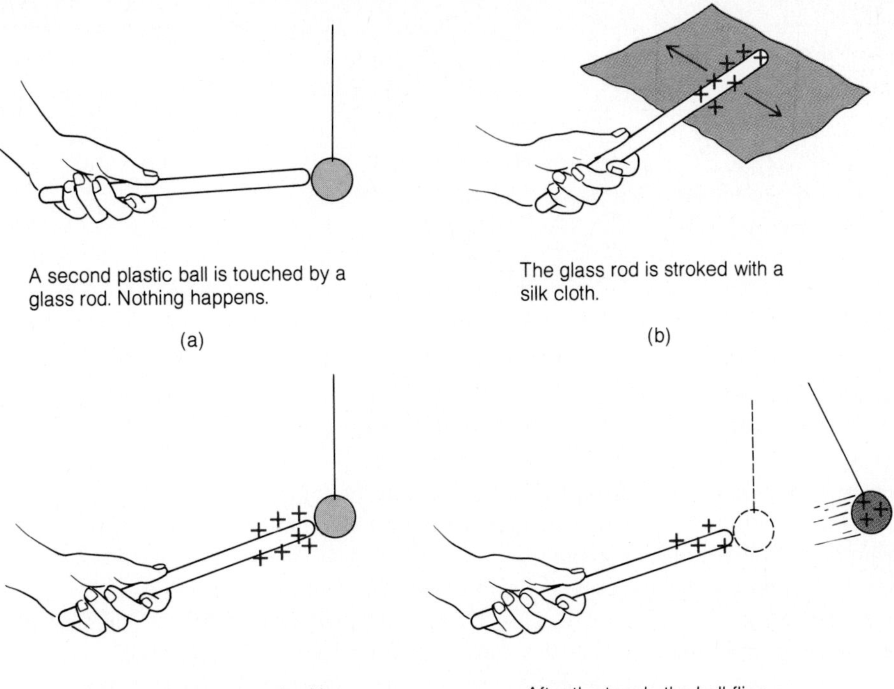

Figure 16.2

A glass rod stroked with silk becomes positively charged; two positively charged objects repel each other.

Fig. 16.2 passed to the plastic ball when it was touched, and the fact that the ball then flew away from the rod suggests that positive electric charges repel each other.

Why is it assumed that the electric charge on the glass rod is different from that on the rubber rod? The reason lies in the result of the third experiment, shown in Fig. 16.3. The attraction of the two plastic balls means (1) that the charges they carry are different, since like charges have already been observed to repel, and (2) that unlike charges attract each other.

The preceding results can be summarized very simply:

Like charges repel; unlike charges attract.

Where do the charges come from when one substance is stroked with another? When we charge one plastic ball with a rubber rod and another with the fur the rod

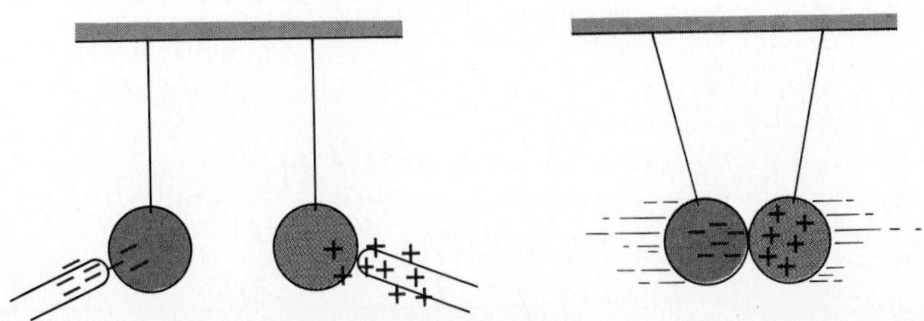

Figure 16.3

Objects with unlike charges attract each other.

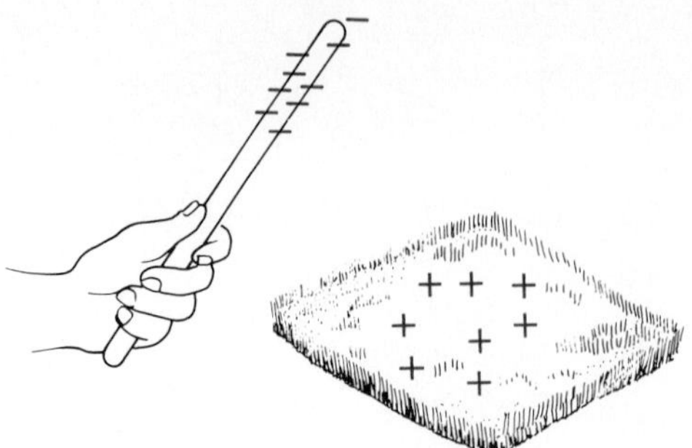

Figure 16.4

The process of stroking a rubber rod with a piece of fur serves to separate charges so that the rod becomes negatively charged and the fur positively charged.

was stroked with, we find that the two balls attract. Since the rubber rod is negatively charged, this experiment shows that the fur is positively charged (Fig. 16.4). A similar experiment with a glass rod and a silk cloth shows that the cloth picks up a negative charge during the stroking. Evidently the process of stroking serves to *separate* charges. We conclude that rubber has an affinity of some kind for negative charges and fur an affinity for positive charges, so that, when rubbed together, each tends to acquire a different kind of charge.

A great many experiments with a variety of substances have shown that there are only these two kinds of electric charge, positive and negative. All electrical phenomena involve either or both kinds of charge. An ''uncharged'' body of matter actually has equal amounts of positive and negative charge, so that appropriate treatment—mere rubbing is sufficient for some substances—can leave an excess of either kind on the body and thereby cause it to show electrical effects.

What is electric charge? All that can be said is that charge, like rest mass, is a fundamental property of certain of the elementary particles of which all matter is composed. Three types of particles are found in atoms, the positively charged **proton,** the negatively charged **electron,** and the neutral (that is, uncharged) **neutron.** The proton and electron have exactly equal amounts of charge, though of opposite sign. Neutrons and protons have nearly the same mass, which is almost 2000 times greater than the electron mass (Table 16.1).

An atom normally contains equal numbers of protons and electrons, so it is electrically neutral unless disrupted in some way. In Figs. 16.1 and 16.2 it is atomic electrons that shift from fur to rubber and from glass to silk to give the observed

Table 16.1 Neutrons, protons, and electrons are the constituents of atoms. An atom contains equal numbers of protons and electrons, so it is electrically neutral.

	Neutron	Proton	Electron
Charge	0	$+e = 1.60 \times 10^{-19}$ C	$-e = -1.60 \times 10^{-19}$ C
	Uncharged	Equal in magnitude, opposite in sign (proton and electron)	
Mass	1.675×10^{-27} kg	1.673×10^{-27} kg	9.11×10^{-31} kg
	Nearly equal (neutron and proton)		Electron mass is 1/1836 of proton mass

charge separations. Figure 16.5 shows a simplified model of a helium atom, whose nucleus consists of two protons and two neutrons. The two electrons that circle the nucleus balance the charges of the protons.

The **principle of conservation of charge** states:

The net electric charge in an isolated system remains constant.

By "net charge" is meant the algebraic sum of the charges present. Net charge can be positive, negative, or zero.

Every known physical process in the universe conserves electric charge. Separating or bringing together charges does not affect their magnitudes, so such rearrangements leave the net charge the same.

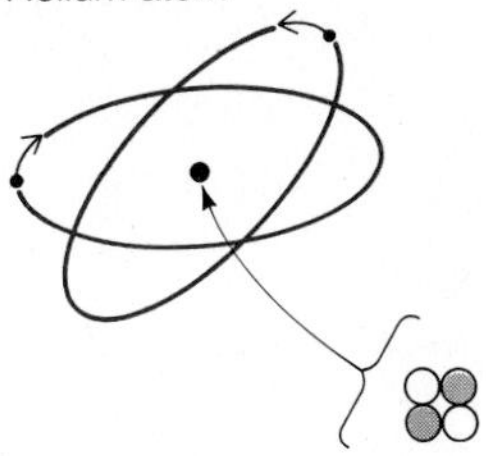

Figure 16.5

A simplified model of the helium atom. The size of the atom is much greater than the size of its nucleus. Nearly all the mass of an atom is concentrated in its nucleus.

16.2 Coulomb's Law

What the force between two electric charges depends on.

In order to arrive at the law of gravitation

$$F = G\frac{m_A m_B}{r^2}$$

Newton had to use astronomical data and an indirect argument, because gravitational forces are appreciable only when at least one of the masses involved is very large. Since electrical forces are much stronger, however, the law they obey can be easily found in the laboratory.

The law of force between charges was first published by the eighteenth-century French scientist Charles Coulomb. If we use the symbol Q for electric charge, **Coulomb's law** for the magnitude F of the force between two charges Q_A and Q_B that are the distance r apart states that

$$F = k\frac{Q_A Q_B}{r^2} \qquad \textit{Coulomb's law} \quad (16.1)$$

The force between two charges is proportional to the product of the charges and is inversely proportional to the square of the distance between them. The quantity k is a constant whose value depends on the units used and on the medium (air, for instance) in which the charges are located. In Eq. (16.1) it is assumed that the charges are either small enough to be considered points or else are spherically symmetric, in which case r is measured from their centers.

Electric force is a vector quantity, and the preceding formula gives only its magnitude. The direction of $\mathbf{F}$ is always along the line joining Q_A and Q_B, and the force is attractive if Q_A and Q_B have opposite signs and repulsive if they have the same signs (Fig. 16.6).

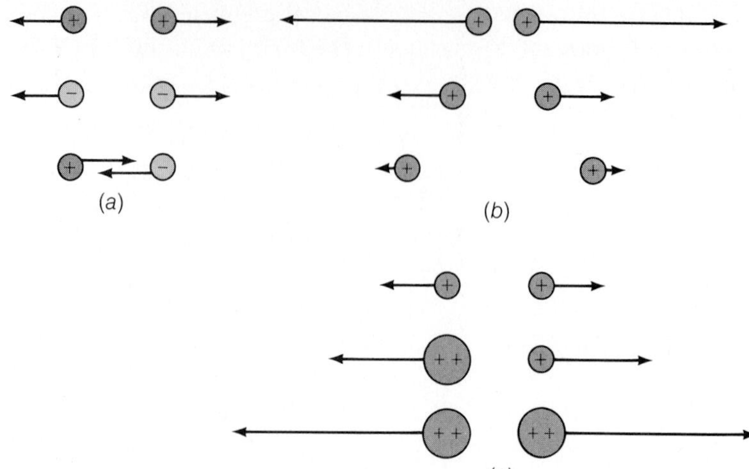

Figure 16.6

(a) Like charges repel, unlike charges attract. (b) The force between two charges varies inversely as the square of their separation. (c) The force is proportional to the product of the charges.

At the beginning of this chapter we noted a familiar observation: A hard rubber comb that has been charged by being passed through someone's hair on a dry day is able to attract small bits of paper. Since the paper bits were originally uncharged, how could the comb exert a force on them?

The explanation depends on Coulomb's law (Fig. 16.7). When the negatively charged comb is brought near the paper, some of the electrons in the paper that are not tightly bound in place move as far away as they can from the comb. Because electrical forces vary inversely with distance, the attraction between the comb and the closer positive charges is greater than the repulsion between the comb and the farther electrons. The paper therefore moves toward the comb. Only a small amount of charge separation actually occurs, and so, with little force available, only very light objects can be attracted in this way.

The Coulomb

The unit of electric charge is the **coulomb** (abbreviated C). The formal definition of the coulomb, given in Chapter 19, is in terms of magnetic forces. A more realistic way to think of the coulomb is in terms of the number of individual

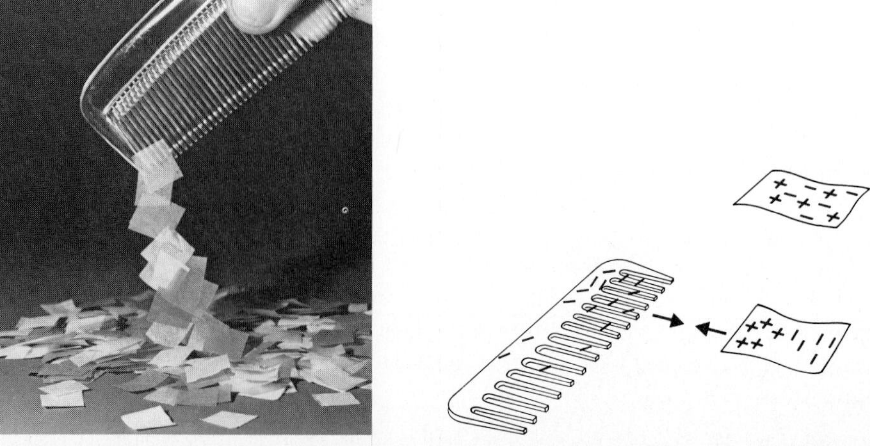

Figure 16.7

How a charged object attracts an uncharged one.

elementary charges that add up to this amount of charge (Fig. 16.8). Because electric charge always occurs in multiples of $\pm$ 1.60 $\times$ 10^{-19} C, this amount of charge has been given a special name, the electron (or electronic) charge, and a special symbol, e:

$$e = 1.60 \times 10^{-19}\ \text{C} \qquad \textit{Electron charge}$$

Thus a charge of $+1.60 \times 10^{-19}$ C is abbreviated $+e$, and one of -3.20×10^{-19} C is abbreviated $-2e$.

In most processes that lead to a net charge on some object, electrons are either added to it or removed from it. Hence we can think of an object whose charge is -1 C as having 6.25×10^{18} electrons more than its normal number, and of an object whose charge is $+1$ C as having 6.25×10^{18} electrons less than its normal number. (By "normal number" is meant a number of electrons equal to the number of protons present, so that the object has no net charge.)

When Q_A and Q_B in Coulomb's law are expressed in coulombs and r in meters, the constant k has the value in vacuum of

$$k = 9.0 \times 10^{9}\ \text{N·m}^2/\text{C}^2$$

The value of k in air is very slightly greater.

The constant k is often written

$$k = \frac{1}{4\pi\epsilon_0}$$

where ϵ_0, called the permittivity of free space, is equal to

$$\epsilon_0 = 8.85 \times 10^{-12}\ \text{C}^2/\text{N·m}^2$$

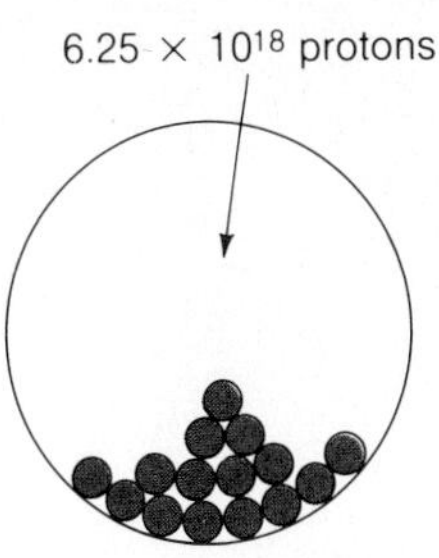

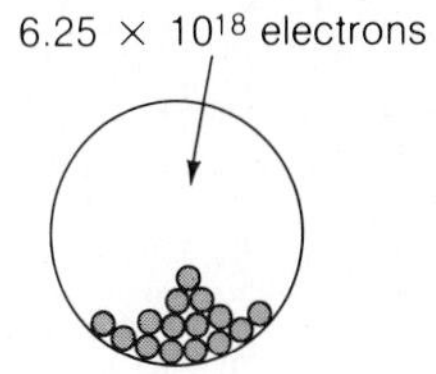

Figure 16.8

Electric charge is not continuous but occurs in multiples of $\pm 1.60 \times 10^{-19}$ C.

EXAMPLE 16.1

Find the magnitude of the force between two charges of 1.0 C each that are 1.0 m apart (Fig. 16.9).

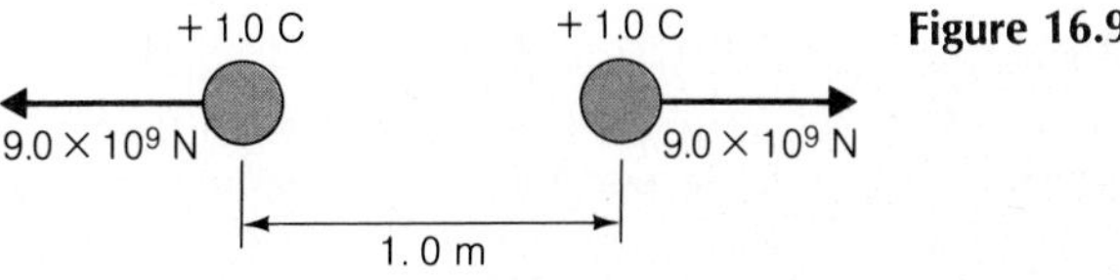

Figure 16.9

SOLUTION From Coulomb's law

$$F = k\frac{Q_AQ_B}{r^2} = \left(9.0 \times 10^9\frac{\text{N·m}^2}{\text{C}^2}\right)\left[\frac{(1.0\ \text{C})(1.0\ \text{C})}{(1.0\ \text{m})^2}\right] = 9.0 \times 10^9\ \text{N}$$

This force is equal to about 2 billion lb. Evidently even the most highly charged objects that can be produced seldom contain more than a tiny fraction of a coulomb of net charge of either sign. ♦

When more than two charges are in the same region, the force on any one of them may be found by adding vectorially the forces exerted on it by each of the others. Usually the component method of vector addition provides the most straightforward means of carrying out the calculation.

EXAMPLE 16.2

Three charges, $Q_1 = +2.0 \times 10^{-9}$ C, $Q_2 = +4.0 \times 10^{-9}$ C, and $Q_3 = -5.0 \times 10^{-9}$ C, are located as shown in Fig. 16.10. Find the magnitude and direction of the net force on Q_1.

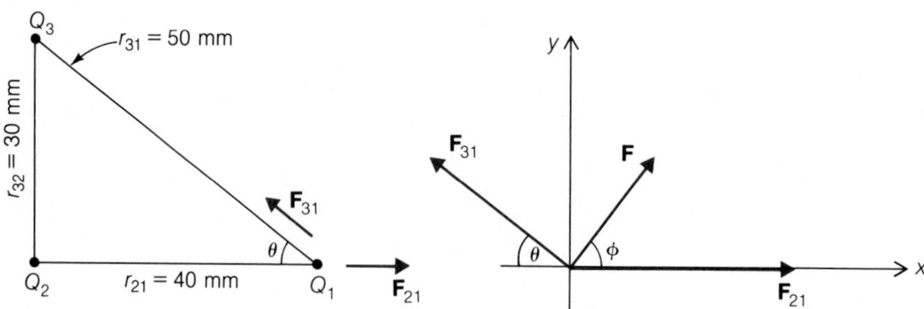

Figure 16.10

SOLUTION Let us call $\mathbf{F}_{21}$ the repulsive force exerted by Q_2 on Q_1 and $\mathbf{F}_{31}$ the attractive force exerted by Q_3 on Q_1. The magnitudes of these forces are

$$F_{21} = \frac{kQ_2Q_1}{r_{21}^2} = \frac{(9.0 \times 10^9\ \text{N}\cdot\text{m}^2/\text{C}^2)(4.0 \times 10^{-9}\ \text{C})(2.0 \times 10^{-9}\ \text{C})}{(4.0 \times 10^{-2}\ \text{m})^2}$$
$$= 4.5 \times 10^{-5}\ \text{N}$$

$$F_{31} = \frac{kQ_3Q_1}{r_{31}^2} = \frac{(9.0 \times 10^9\ \text{N}\cdot\text{m}^2/\text{C}^2)(5.0 \times 10^{-9}\ \text{C})(2.0 \times 10^{-9}\ \text{C})}{(5.0 \times 10^{-2}\ \text{m})^2}$$
$$= 3.6 \times 10^{-5}\ \text{N}$$

The directions of $\mathbf{F}_{21}$ and $\mathbf{F}_{31}$ are parallel to the 40 mm and 50 mm sides of the triangle.

In order to find $\mathbf{F}$, the net force on Q_1, we first resolve $\mathbf{F}_{21}$ and $\mathbf{F}_{31}$ into components. We have, since $\sin\theta = \frac{3}{5}$ and $\cos\theta = \frac{4}{5}$,

$$F_{21x} = F_{21} = 4.5 \times 10^{-5}\ \text{N}$$
$$F_{21y} = 0$$
$$F_{31x} = -F_{31}\cos\theta = -\tfrac{4}{5}F_{31} = -2.9 \times 10^{-5}\ \text{N}$$
$$F_{31y} = F_{31}\sin\theta = \tfrac{3}{5}F_{31} = 2.2 \times 10^{-5}\ \text{N}$$

Hence the components of $\mathbf{F}$ are

$$F_x = F_{21x} + F_{31x} = 1.6 \times 10^{-5}\ \text{N}$$
$$F_y = F_{21y} + F_{31y} = 2.2 \times 10^{-5}\ \text{N}$$

and the magnitude of **F** is accordingly

$$F = \sqrt{F_x^2 + F_y^2} = 2.7 \times 10^{-5}\,\text{N}$$

The direction of **F** can be specified in various ways. If ϕ is the angle between **F** and the $+x$-axis, then

$$\phi = \tan^{-1}\frac{F_y}{F_x} = \tan^{-1} 1.4 = 54° \qquad ◆$$

16.3 Electricity and Matter

Gravitational forces dominate on a large scale, electrical forces on a small scale.

Coulomb's law for the electrical force between charges is very similar in form to Newton's law for the gravitational force between masses. The most striking difference is that electrical forces may be either attractive or repulsive, whereas gravitational forces are always attractive. The latter fact means that matter in the universe tends to come together to form large bodies, such as stars and planets, and these bodies are always found in groups, such as galaxies of stars and families of planets around stars.

There is no comparable tendency for electric charges of a given sign to come together; quite the contrary. Because unlike charges attract strongly, it is hard to separate neutral matter into portions of opposite signs. Furthermore, like charges repel, so it becomes harder and harder to add more charge to an already charged object. Hence the large-scale structure of the universe is largely governed by gravitational forces.

On an atomic scale, though, the relative importance of gravity and electricity is reversed. Elementary particles are so tiny that the gravitational forces between them are insignificant, whereas their electric charges are sufficiently great for electrical forces to govern the structures of atoms, molecules, liquids, and solids.

EXAMPLE 16.3

The hydrogen atom has the simplest structure of all atoms, consisting of a proton and an electron whose average separation is 5.3×10^{-11} m. (For the time being we can think of the electron as circling the proton much as the moon circles the earth, as in Fig. 16.11. A more realistic model of the hydrogen atom—but one that is harder to visualize—will be given later.) Compare the electrical and gravitational forces between the proton and the electron in a hydrogen atom.

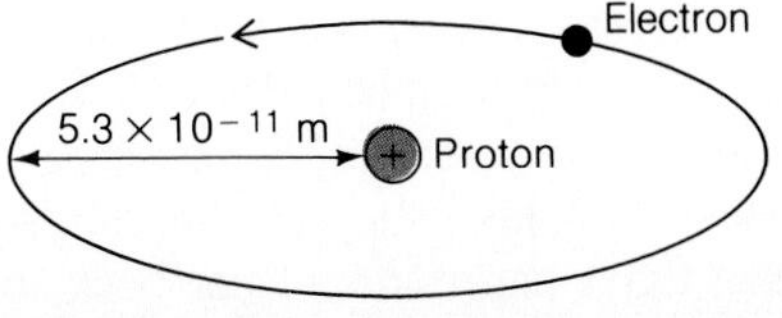

Figure 16.11

A simple model of the hydrogen atom.

SOLUTION The electrical force between the electron and proton is

$$F_e = k\frac{Q_eQ_p}{r^2} = \frac{(9.0 \times 10^9\ \text{N}\cdot\text{m}^2/\text{C}^2)(1.6 \times 10^{-19}\ \text{C})^2}{(5.3 \times 10^{-11}\ \text{m})^2} = 8.2 \times 10^{-8}\ \text{N}$$

while the gravitational force between them is

$$F_g = G\frac{m_em_p}{r^2} = \frac{(6.7 \times 10^{-11}\ \text{N}\cdot\text{m}^2/\text{kg}^2)(9.1 \times 10^{-31}\ \text{kg})(1.7 \times 10^{-27}\ \text{kg})}{(5.3 \times 10^{-11}\ \text{m})^2}$$
$$= 3.7 \times 10^{-47}\ \text{N}$$

The electrical force is over 10^{39} times greater than the gravitational force. Since both forces vary with distance as $1/r^2$, the same ratio holds regardless of how far apart an electron and a proton are. Clearly the electrical forces that subatomic particles exert on one another are so much stronger than their mutual gravitational forces that the latter can be neglected in the atomic world. ◆

16.4 Atomic Structure

An atom is mostly empty space.

Most scientists of the late nineteenth century accepted the idea that the chemical elements consist of atoms, but they knew almost nothing about the atoms themselves. One clue was the discovery that all atoms contain electrons. Since electrons are negatively charged whereas atoms are neutral, atoms must also contain positively charged matter of some kind. But what kind? And arranged in what way?

J. J. Thomson, whose work had led to the identification of the electron, proposed in 1898 that atoms are spheres of positively charged matter that contain embedded electrons, much as a fruitcake is studded with raisins. The reality turned out to be very different.

The most direct way to find out what is inside a fruitcake is simply to plunge a finger into it. In essence this is the classic experiment performed in England in 1911 by Hans Geiger and Ernest Marsden at the suggestion of Ernest Rutherford. The probes they used were fast **alpha particles** emitted by certain radioactive elements. We know today that alpha particles are the nuclei of helium atoms (see Fig. 16.5) and consist of two neutrons and two protons each.

Geiger and Marsden placed a sample of an alpha-emitting substance behind a lead screen with a small hole in it, so that a narrow beam of alpha particles was produced. On the other side of a thin metal foil in the path of the beam, they placed a zinc sulfide screen that gave off a flash of light when struck by an alpha particle, thus showing by how much the alpha particles were scattered from their original direction of motion (Fig. 16.12).

If the charges in an atom were spread more or less evenly throughout its volume, only weak electrical forces would act on alpha particles passing through a thin foil, and they would be deflected by less than 1°. What Geiger and Marsden actually found was that, although most of the alpha particles indeed were not deviated by much, a few were scattered through large angles. Some were even scattered in the backward direction. As Rutherford remarked, "It was almost as incredible as if you fired a 15-inch shell at a piece of tissue paper and it came back and hit you."

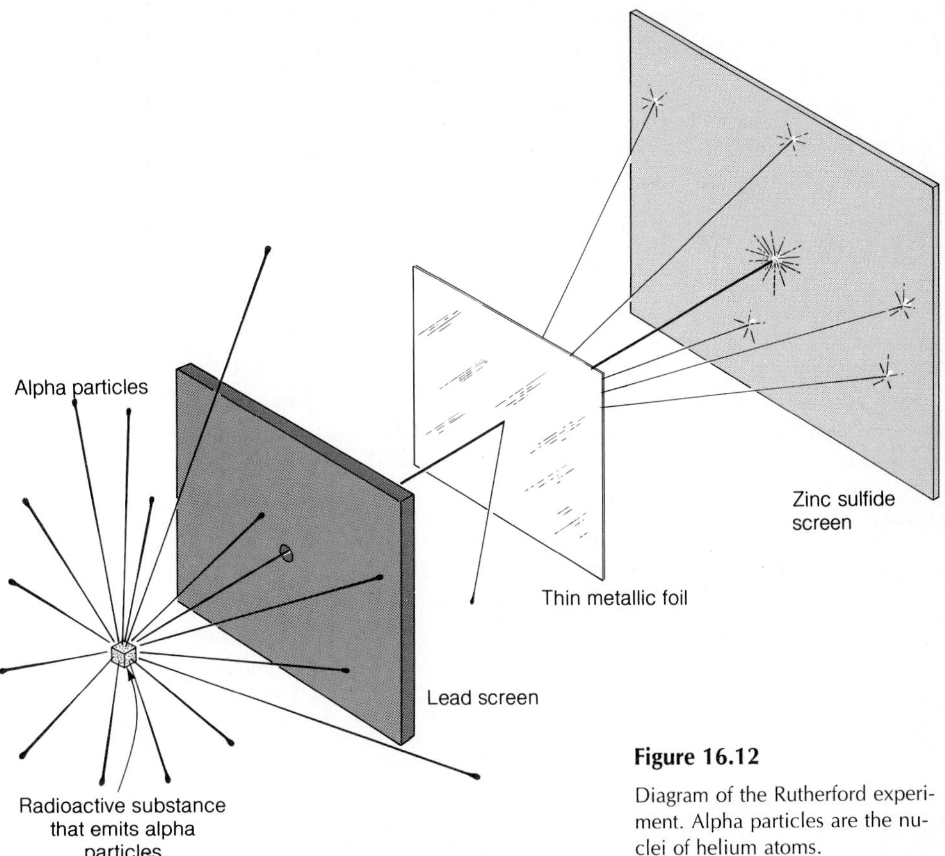

Figure 16.12

Diagram of the Rutherford experiment. Alpha particles are the nuclei of helium atoms.

BIOGRAPHICAL NOTE

Ernest Rutherford

Ernest Rutherford (1871–1937), a native of New Zealand, was on his family's farm digging potatoes when he heard that he had won a scholarship for graduate study at Cambridge University in England. "This is the last potato I will ever dig," he said, throwing down his spade. Thirteen years later he won the Nobel Prize in chemistry.

At Cambridge, Rutherford was a research student under J. J. Thomson, who would soon announce the discovery of the electron. Rutherford's own work was on the newly found phenomenon of radioactivity, and he quickly distinguished between alpha and beta particles, two of the emissions of radioactive materials. In 1898 he went to McGill University in Canada, where he found that alpha particles are the nuclei of helium atoms and that the radioactive decay of an element gives rise to another element. Working with the chemist Frederick Soddy and others, Rutherford traced the successive transformations of radioactive elements, such as uranium and radium, until they end up as stable lead.

In 1907 Rutherford returned to England as professor of physics at Manchester, where in 1911

(continued)

he showed that the nuclear model of the atom was the only one that could explain the observed scattering of alpha particles by thin metal foils. Rutherford's last important discovery, reported in 1919, was the disintegration of nitrogen nuclei into hydrogen and oxygen nuclei when bombarded with alpha particles, the first example of the artificial transmutation of elements into other elements. After other similar experiments, Rutherford suggested that all nuclei contain hydrogen nuclei, which he called protons. He also proposed that a neutral particle was present in nuclei as well.

In 1919 Rutherford became director of the Cavendish Laboratory at Cambridge, where under his stimulus great strides in understanding the nucleus continued to be made. James Chadwick discovered the neutron there in 1932. The Cavendish Laboratory was the site of the first accelerator for producing high-energy particles; with the help of this accelerator, fusion reactions in which light nuclei unite to form heavier ones were observed for the first time.

Horrified by the rise of Nazism in the 1930s, Rutherford helped many Jewish scientists to leave Germany. He died in 1937 of complications of a hernia and was buried near Newton in Westminster Abbey.

Because alpha particles are relatively heavy (almost 8000 times more massive than electrons) and have fairly high initial speeds (typically 2×10^7 m/s), it was clear that strong forces had to be exerted on them to cause such marked deflections. The only atomic model able to account for such forces is one that consists of a tiny **nucleus** in which the atom's positive charge and nearly all of its mass are concentrated, with the electrons some distance away (Fig. 16.13).

With the atom largely empty space, it is easy to see why most alpha particles go right through a thin foil. On the other hand, an alpha particle that comes near a nucleus experiences a strong electrical force and is likely to be scattered through a large angle. (The atomic electrons, being very light, are readily knocked out of the way by alpha particles. The situation is reversed for the nuclei, which are heavier than alpha particles.) Rutherford derived a formula for the scattering of alpha particles by thin foils on the basis of his hypothesis that agreed with the experimental results. He is therefore credited with the discovery of the nucleus.

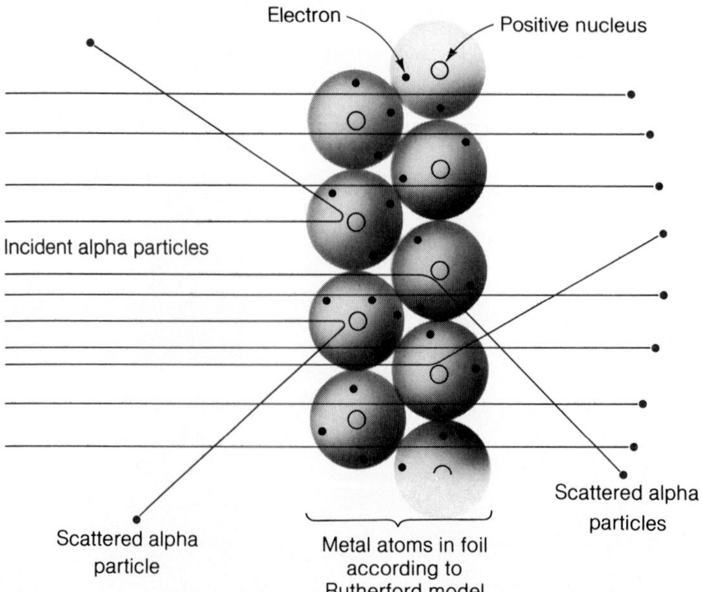

Figure 16.13

According to the Rutherford model of the atom, positive charge is concentrated in a tiny nucleus at its center with electrons some distance away. Strong electric forces can occur within atoms on the basis of this model, and it accordingly predicts considerable deflections of some of the alpha particles striking a thin foil. This prediction agrees with experiment. The spheres shown here represent the regions in which the electrons belonging to the respective atoms circulate; see Chapter 27 for a more complete discussion.

16.5 Electrical Conduction

Most substances conduct electricity either very well or very badly.

An electric current is a flow of charge. Nearly all substances fall into two categories: **conductors,** through which charge can flow easily, and **insulators,** through which charge can flow only with great difficulty. Metals, many liquids, and **plasmas** (gases whose particles are charged) are conductors. Nonmetallic solids, certain liquids, and gases whose molecules are electrically neutral are insulators. Several substances, called **semiconductors,** are intermediate in their ability to conduct charge.

In a solid metal each atom gives up one or more electrons to a common "gas" of freely moving electrons. These electrons can move easily through the crystal structure of the metal, so if one end of a metal wire is given a positive charge and the other end a negative charge, electrons will flow through the wire from the negative to the positive end. This flow, of course, is an electric current. By supplying new electrons to the negative end of the wire and removing electrons from the positive end as they arrive there—which can be done by connecting the wire to a battery or to a generator—a steady current can be maintained in the wire.

In nonmetallic solids, such as salt, glass, rubber, minerals, wood, and plastics, all the atomic electrons are fixed in particular atoms or groups of atoms and cannot move from place to place. Such solids are therefore insulators. Actually, nonmetallic solids do conduct very small amounts of current, but their abilities to do this are much inferior to those of metals. For instance, when identical bars of copper and sulfur are connected to the same battery, about 10^{23} times more current flows in the copper bar.

As mentioned earlier, there are a few substances called semiconductors through which charge flows more readily than through insulators but with more difficulty than through conductors. Thus about 10^{7} times more current flows in a germanium bar connected to a battery than in a sulfur bar of the same size, but this is still about 10^{16} times less current than in a copper bar. The electrical conductivity of solids is discussed in more detail in Chapter 28.

Electrical conduction in liquids and gases is different from that in metals. The current in a metal consists of a flow of electrons past the stationary atoms in its structure. The current in a fluid other than a liquid metal, however, consists of a flow of entire atoms or molecules that are electrically charged. An atom or molecule that carries a net charge is called an **ion**, and both positive and negative ions participate in conduction in liquids and gases.

An atom or molecule becomes a positive ion when it loses one or more of its electrons. If it gains one or more electrons in addition to its usual number, it becomes a negative ion. The fundamental positive charges in matter are protons, which are very tightly bound in the nucleus of every atom. Atomic electrons, however, are held more loosely, and one or two of them can be detached from an atom with relative ease. Thus the oxygen and nitrogen gases in ordinary air become ionized when a spark occurs, in the presence of a flame, and by the passage of X rays or even ultraviolet light. These processes so disturb the air molecules that some electrons are dislodged, leaving behind positive ions. The liberated electrons almost at once become attached to other nearby molecules to create negative ions (Fig. 16.14).

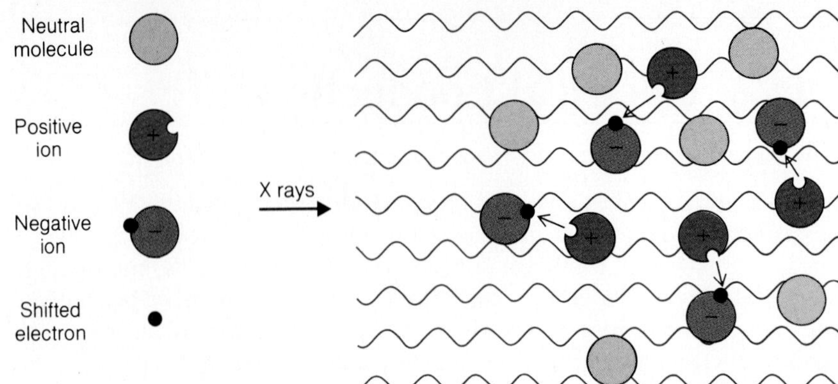

Figure 16.14

Schematic representation of the ionization of air by X rays. A molecule losing an electron becomes a positive ion; a molecule gaining an electron becomes a negative ion.

The electrical attraction between positive and negative charges in time brings the ions together, and the extra electrons on the negative ions return to the positive ions. The gas molecules are then neutral, as they were originally. This **recombination** is rapid at normal atmospheric pressure and temperature.

In the upper atmosphere, where air molecules are so far apart that the recombination of ions is a slow process, the continual bombardment of X rays and ultraviolet light from the sun maintains a certain proportion of ions at all times. The layers of ions in the upper atmosphere constitute the **ionosphere,** and they make possible long-range radio communication by their ability to reflect radio waves (see Fig. 22.9). The ionosphere is an example of a plasma. The behavior of a plasma, unlike that of an ordinary gas, is strongly influenced by electrical and magnetic forces because its particles are charged. Most of the universe is in the plasma state.

The earth as a whole is a fairly good conductor. Hence if a charged object is connected to the earth by a piece of metal, the charge is conducted away from the object to the earth. This convenient way to remove excess charge from an object is called **grounding** the object. As a safety measure, the metal shells of electrical appliances are grounded through special wires that give electric charges in the shells paths to the earth. The round post in the familiar three-prong electric plug is the ground connection.

Superconductivity

Electric conductors, even the best, resist to some extent the flow of charge through them at ordinary temperatures. At very low temperatures, however, certain metals, alloys, and chemical compounds permit currents to pass unimpeded through them. This phenomenon is called **superconductivity.**

Superconductivity was discovered by Kamerlingh Onnes in the Netherlands in 1911. Aluminum becomes superconducting at temperatures below 1.2 K and lead at temperatures below 7.2 K, to give two examples. If a current is set up in a closed wire loop at room temperature, it will die out in less than a second even if the wire is made of a good conductor such as copper or silver. If the wire is made of a superconductor and is kept cold enough, on the other hand, the current will continue indefinitely. Currents have persisted unchanged in superconducting loops for several years.

The effect of electrical resistance is to dissipate as heat some of the energy of an electric current in a conductor. An analogy can be made with the presence of friction in a mechanical system, where kinetic energy is dissipated as heat. About 10% of the electric energy generated in the United States is lost as heat in transmission lines. For this reason superconductivity is of great potential importance for the carrying of electric energy. Because magnetic fields are produced by electric currents (Chapter 19), superconductivity is also important in applications where strong magnetic fields are required. Already laboratory electromagnets with superconducting coils are in use, and experimental electric motors with superconducting windings have been built. There is no fundamental reason why superconducting magnets cannot be used to suspend trains above their tracks and thereby both increase their speeds and reduce their energy needs.

Despite much effort, until 1986 no substance was known whose critical temperature (the temperature below which it is superconducting) was higher than 23 K. In that year Alex Müller and Georg Bednorz, working in Switzerland, studied a class of materials that had never before been suspected of superconducting behavior. They discovered a superconducting ceramic whose critical temperature was 35 K, and soon afterward others extended their approach to produce superconductors with critical temperatures over 100 K. Although still extremely cold by everyday standards (100 K = -173°C), such temperatures are above the 77-K boiling point of liquid nitrogen, which is cheap and readily available. The much greater cost of liquid helium, needed to reach the lower critical temperatures of the period up to 1986, prevented widespread commercial application of superconductivity before then.

The new superconductors are not without problems of their own. In particular, their superconductivity tends to fail when they carry very high currents or in the presence of strong magnetic fields. Also, like other ceramics, the new superconductors are relatively weak mechanically. For these reasons the first large-scale industrial use of superconductivity is likely to be in increasing the operating speeds of computers and in other electronic applications.

A material that is superconducting at room temperature would revolutionize the world's technology. By reducing the waste of electric energy, the rate at which the world's resources are being depleted would also be reduced. Such a material, inconceivable until very recently, is now the subject of an active search.

16.6 What Is a Field?

A region of space with a continuous distribution of a certain property.

Electrical forces, like gravitational ones, act between objects that may be far apart. A **force field** is a model that provides a framework for understanding how forces are transmitted from one object to another across empty space.

A successful model does more than just organize our knowledge of a certain phenomenon. Such a model also enables us to predict hitherto unsuspected effects and relationships, which of course must then be verified by experiment. The creative role of a scientific model is beautifully illustrated by the electromagnetic field model proposed by James Clerk Maxwell over a century ago. With the help of this model, Maxwell predicted in 1864 that electromagnetic waves should exist; he calculated that their speed should be $c = 3 \times 10^8$ m/s; and he suggested that light consists of such waves. In 1887, after Maxwell's death, Heinrich Hertz confirmed Maxwell's hypothesis in the laboratory.

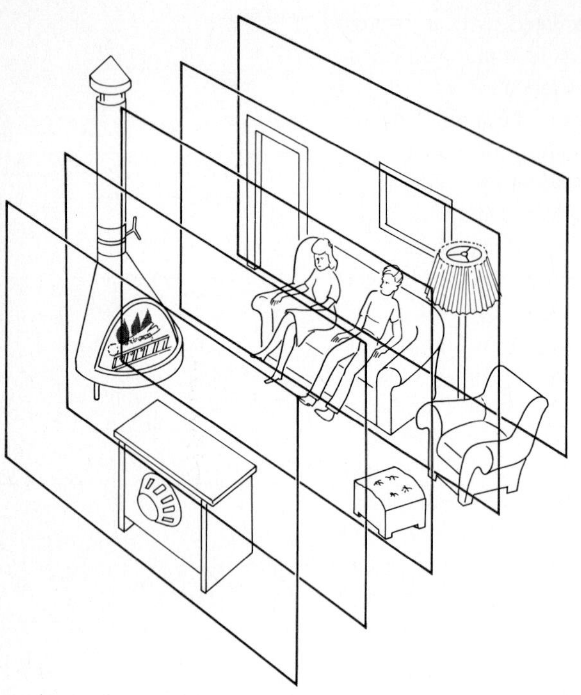

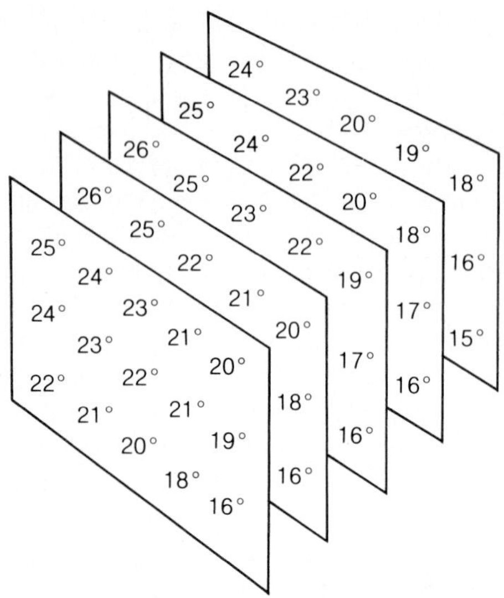

Figure 16.15

The temperature field in a room is a scalar field, because the temperature at a point has no direction associated with it. The temperatures here are expressed in degrees Celsius.

A field—in the sense that physicists use the word—is a region of space in which a certain quantity has a definite value at every point. Thus it is appropriate to speak of the temperature field in a room, of the velocity field of the water in a river, and of the gravitational field around the earth. But it is not appropriate to speak of the ''chair field'' in a room, or of the ''rowboat field'' in a river, or of the ''airplane field'' around the earth, because chairs, rowboats, and airplanes are separate objects. Their presence somewhere is not a property of a region of space that varies continuously throughout that region in the way that temperature, water velocity, and gravitational force vary.

To determine the temperature field in a room, we must measure the temperature at a great many points with a thermometer. The results might be displayed by a series of cross-sectional maps of the room with temperature values written in at each point of measurement, as in Fig. 16.15.

A better way to show the temperature field is to draw a series of lines on each map that connect points having the same temperatures (Fig. 16.16). These lines are called **isotherms** and might be drawn, for example, for temperatures that are 2°C apart. Naturally the actual measurements are not necessarily 16°, 18°, 20°, and so on; we must interpolate between the actual measurements to find the contours of the various isotherms.

Temperature is a scalar quantity because it involves a magnitude only. To say that the temperature somewhere is 20°C describes it completely. The field of a scalar quantity is a **scalar field,** and it can always be pictured by a plot of isolines (lines joining sets of identical values) as in the case of a temperature field. A vector quantity has direction as well as magnitude, so picturing a **vector field** is less easy. In Section 16.8 we will see how to do this for an electric field.

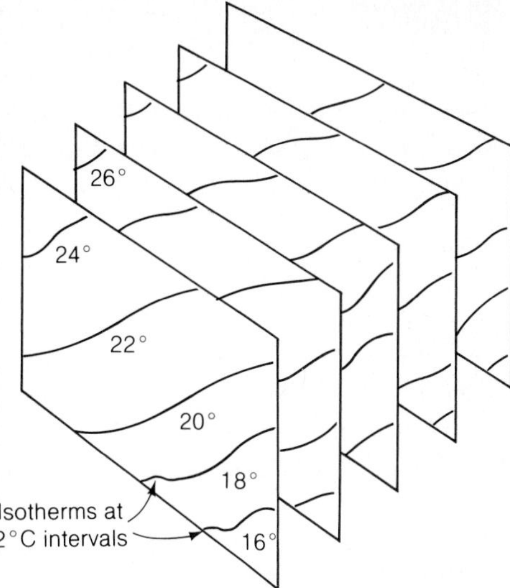

Figure 16.16

An isotherm is a line that joins points whose temperatures are the same. Here the temperature field in the room in Fig. 16.15 is displayed by means of isotherms.

16.7 Electric Field

Its action on a charge leads to a force on the charge.

When an electric charge is present somewhere, the properties of space around it are altered in such a way that another charge brought to this region will experience a force there. The "alteration in space" caused by a charge at rest is called its **electric field,** and any other charge is considered to interact with the field and not directly with the charge that gives rise to it.

All forces, not just electric ones, can be interpreted as arising through the action of a force field of one kind or another. Thus the sun is regarded as having a gravitational field around it, and the forces exerted by this field on the planets are what hold them in their orbits. In the case of the "direct-contact" forces between the solid objects of everyday life, the force fields involved are electric fields.

The electric field **E** at a point in space is defined as the ratio of the force **F** on a charge Q (assumed positive) at that point to the magnitude of Q:

$$\mathbf{E} = \frac{\mathbf{F}}{Q} \qquad \textit{Electric field} \quad (16.2)$$

The units of **E** are newtons per coulomb. Electric field is a vector quantity that possesses both magnitude and direction.

Once we know what the electric field **E** is at some point, from the definition we see that the force that the field exerts on a charge Q at that point is

$$\mathbf{F} = Q\mathbf{E} \qquad \textit{Electric force} \quad (16.3)$$

Force = (charge)(electric field)

EXAMPLE 16.4

The electric field between the electrodes of the gas discharge tube used in a certain neon sign has the magnitude 5.0×10^4 N/C. Find the acceleration of a neon ion of mass 3.3×10^{-26} kg in the tube if it carries a charge of $+e$ (Fig. 16.17).

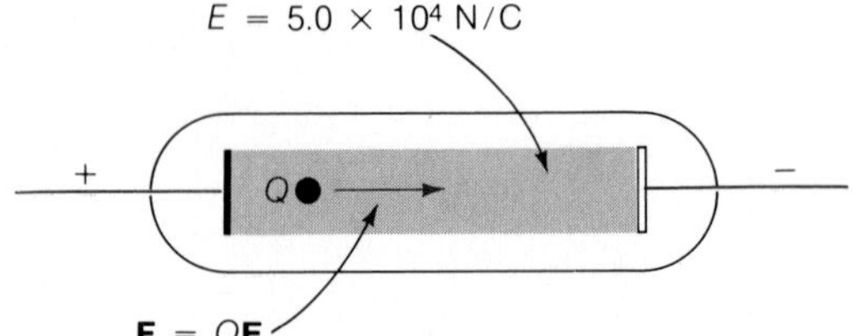

Figure 16.17
A charge Q in an electric field **E** experiences a force equal to Q**E**.

SOLUTION Since $e = 1.6 \times 10^{-19}$ C, the force on the ion has the magnitude

$$F = QE = (1.6 \times 10^{-19}\text{ C})(5.0 \times 10^4\text{ N/C}) = 8.0 \times 10^{-15}\text{ N}$$

Hence the acceleration of the ion is

$$a = \frac{F}{m} = \frac{8.0 \times 10^{-15}\text{ N}}{3.3 \times 10^{-26}\text{ kg}} = 2.4 \times 10^{11}\text{ m/s}^2$$

This is 25 billion times greater than the acceleration of gravity. The speed such an ion actually reaches is limited by frequent collisions with the neon atoms in its path. ◆

We can use Coulomb's law to determine the magnitude of the electric field around a single charge Q. First we find the force F that Q exerts on another charge Q' at the distance r away, which is

$$F = k\frac{QQ'}{r^2}$$

Since the electric field that Q' experiences is $E = F/Q'$ we have

$$E = \frac{F}{Q'} = k\frac{Q}{r^2} \qquad \textit{Electric field of a charge} \quad (16.4)$$

This formula tells us that the electric field magnitude at the distance r from a point charge is directly proportional to the magnitude Q of the charge and is inversely proportional to the square of r.

When more than one charge contributes to the electric field at a point P, the net field **E** is the *vector sum* of the fields of the individual charges. That is,

$$\mathbf{E} = \mathbf{E}_1 + \mathbf{E}_2 + \mathbf{E}_3 + \cdots = \Sigma\mathbf{E}_n \qquad \textit{Total field} \quad (16.5)$$

EXAMPLE 16.5

Two charges, $Q_1 = +2.0 \times 10^{-8}$ C and $Q_2 = +3.0 \times 10^{-8}$ C, are 50 mm apart, as in Fig. 16.18. What is the electric field halfway between them?

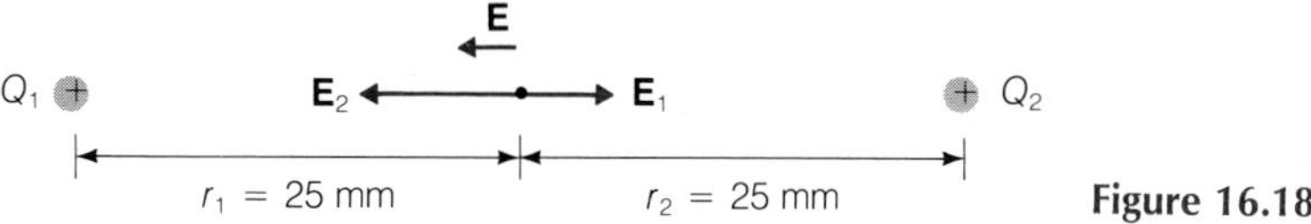

Figure 16.18

SOLUTION At a point halfway between the charges, $r_1 = r_2 = 25$ mm. Here the electric fields of the two charges have the respective magnitudes

$$E_1 = k\frac{Q_1}{r_1^2} = \frac{(9.0 \times 10^9 \text{N}\cdot\text{m}^2/\text{C}^2)(2.0 \times 10^{-8}\text{C})}{(2.5 \times 10^{-2}\text{m})^2} = 2.9 \times 10^5 \text{ N/C}$$

$$E_2 = k\frac{Q_2}{r_2^2} = \frac{(9.0 \times 10^9 \text{N}\cdot\text{m}^2/\text{C}^2)(3.0 \times 10^{-8}\text{C})}{(2.5 \times 10^{-2}\text{m})^2} = 4.3 \times 10^5 \text{N/C}$$

The direction of an electric field at a point is the direction in which a positive charge placed there would move. Hence the direction of $\mathbf{E}_1$ is to the right in Fig. 16.18 and that of $\mathbf{E}_2$ is to the left. Since $\mathbf{E}_1$ and $\mathbf{E}_2$ are in opposite directions, the magnitude of their sum **E** is

$$E = E_2 - E_1 = 1.4 \times 10^5 \text{ N/C}$$

The direction of **E** is toward Q_1.

If the charges had both been negative rather than positive, the field halfway between them would have the same magnitude but would be in the opposite direction. If one charge had been negative and the other positive, $\mathbf{E}_1$ and $\mathbf{E}_2$ would point in the same direction along the line between the charges. The total field **E** there would then have the magnitude $E_1 + E_2$ and its direction would be toward the negative charge. ♦

EXAMPLE 16.6

Where is the electric field equal to zero in the neighborhood of the charges of the previous example?

SOLUTION Since the fields $\mathbf{E}_1$ and $\mathbf{E}_2$ are in opposite directions along a line between Q_1 and Q_2, the point where they cancel each other out must lie on this line. The cancellation occurs where the magnitudes E_1 and E_2 are equal:

$$E_1 = E_2 \qquad k\frac{Q_1}{r_1^2} = k\frac{Q_2}{r_2^2} \qquad \frac{Q_1}{r_1^2} = \frac{Q_2}{r_2^2}$$

This last is an equation with two unknown quantities, r_1 and r_2. We need another equation that relates r_1 and r_2 in order to find their values. Such an equation comes from the fact that the distance between Q_1 and Q_2 is $s = 50$ mm, so that

$$r_1 + r_2 = s \qquad r_2 = (s - r_1)$$

Hence

$$\frac{Q_1}{r_1^2} = \frac{Q_2}{r_2^2} = \frac{Q_2}{(s - r_1)^2}$$

To solve for r_1, we rewrite the equation as

$$\frac{(s - r_1)^2}{r_1^2} = \frac{Q_2}{Q_1}$$

Then we take the square root of both sides:

$$\frac{s - r_1}{r_1} = \sqrt{\frac{Q_2}{Q_1}}$$

Finally we multiply both sides by r_1 and solve for r_1, which gives

$$r_1 = \frac{s}{1 + \sqrt{Q_2/Q_1}}$$

Here $s = 50$ mm and $\sqrt{Q_2/Q_1} = 1.22$, so that

$$r_1 = \frac{s}{1 + \sqrt{Q_2/Q_1}} = \frac{50 \text{ mm}}{1 + 1.22} = \frac{50 \text{ mm}}{2.22} = 23 \text{ mm}$$

The point where $\mathbf{E} = 0$ lies between the charges 23 mm from the charge Q_1. ◆

16.8 Electric Field Lines

Imaginary lines that help us to picture an electric field.

Electric field lines provide a way to give intuitive form to our thinking about an electric field. These lines are constructed as follows: At several points in space we draw arrows in the direction a positive charge there would experience a force (Fig. 16.19a). Then we connect these arrows to form smooth curves, which are the electric field lines (Fig. 16.19b). Two rules should be kept in mind:

1. Electric field lines leave positive charges and enter negative ones.
2. The spacing of field lines is such that they are close together where the field is strong and far apart where the field is weak.

If we place a positive charge in an electric field, it will experience a force in the direction of the field line it is on. The magnitude of the force will be proportional

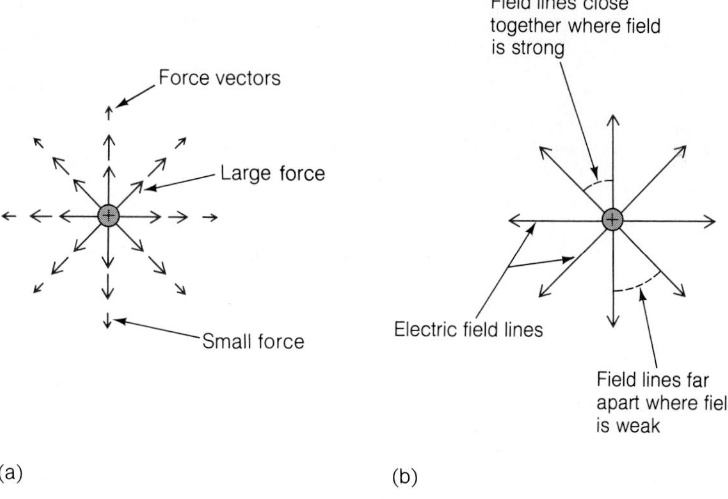

Figure 16.19

How electric field lines around a positive charge are drawn. In (a) the forces whose vectors are shown are those exerted by the charge on a positive test charge.

to the concentration of field lines in its vicinity. Figure 16.20 shows the patterns of field lines around a positive and a negative charge and around pairs of charges.

It is important to keep in mind that field lines do not actually exist as threads in space but are simply a device to help our thinking about force fields. The use of field lines is not limited to electric fields. Gravitational and magnetic fields, for instance, are often described in this way.

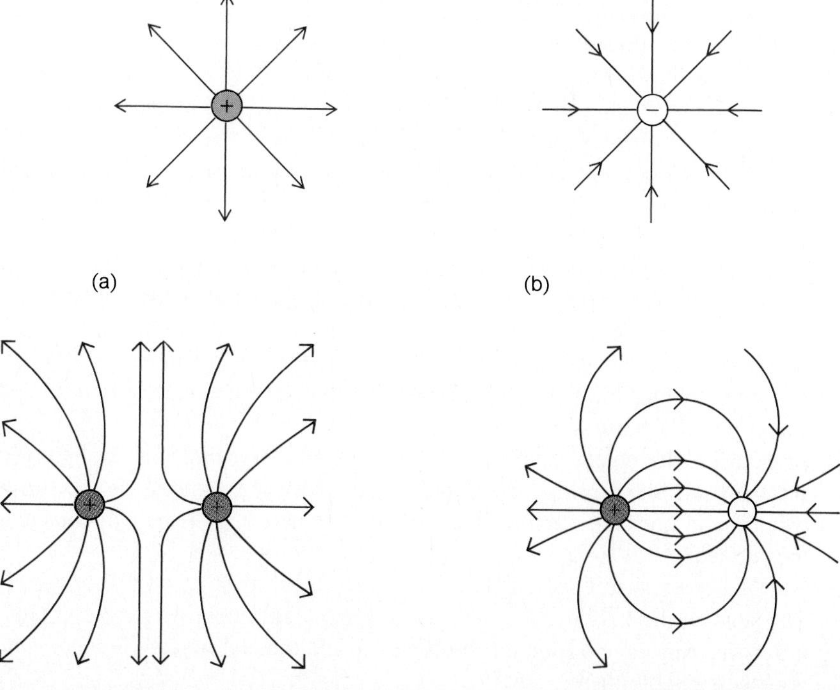

Figure 16.20

Electric field lines near separate charges and near pairs of charges. At large distances relative to the separation of the charges, the field in (c) approaches that of a single charge.

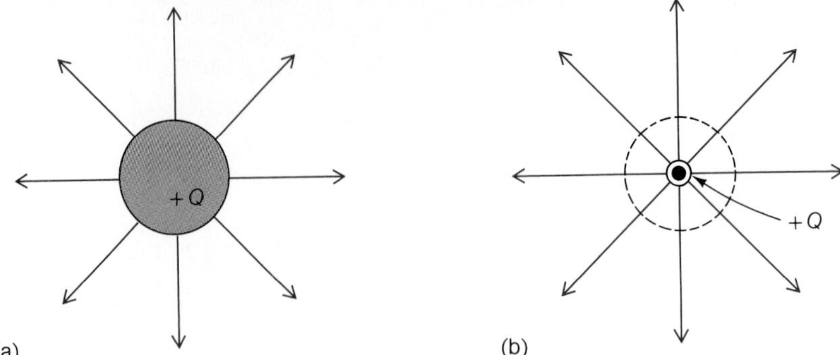

Figure 16.21

(a) Electric field lines around a uniformly charged sphere. (b) Electric field lines around a point charge.

The notion of electric field lines is often helpful as a guide to our thinking when electric fields of charge distributions are concerned. For example, let us inquire into the electric field around a spherical distribution of charge whose total amount is Q. Such a distribution is quite easy to arrange; all we need do is to add the charge Q to a metal sphere. Since metals are good conductors of electricity, the mutual repulsion of the individual charges that make up Q causes them to spread out uniformly over the surface of the sphere so that the charges are as far apart as possible. It does not matter whether the sphere is hollow or solid.

Whatever the number of field lines we choose to represent the electric field around a point charge Q, the same number must be associated with a body that carries the same net charge Q. In the case of a charged sphere, the field lines are symmetrically arranged. Therefore the pattern of field lines outside the sphere is identical with that around a point charge of the same magnitude (Fig. 16.21). The same is true for the electric field **E** itself, of course.

Three important properties of a charged metal object are illustrated in Fig. 16.22:

1. Inside the object the electric field is zero everywhere. The electric fields due to the individual charges on the surface all cancel out in the interior, although outside the object they do not. We can see why this must be true even without a formal calculation. Suppose there *were* an electric field **E** in the interior of the object. Then the charges inside the object that are free to move (electrons in the case of a metal) would move under the influence of the field **E.** But no currents are observed in a charged conducting object except for a moment after the charge is placed on it. Since energy is needed to maintain an electric current, a supply of energy would be needed for currents to continue in such an object. The only conclusion is that the interior of an isolated conducting object is always free of electric field.
2. Just outside the object the electric field **E** must be perpendicular to the object's surface. If **E** had a component parallel to the surface, electrons on the surface would be in constant motion. Because they are not, **E** must be perpendicular to the surface.
3. All the excess charge (of either sign) on the object must be on its surface. If any point in the interior had a net charge, an electric field would be present around it. But there are no electric fields inside a metal object, so there can be no excess charge there.

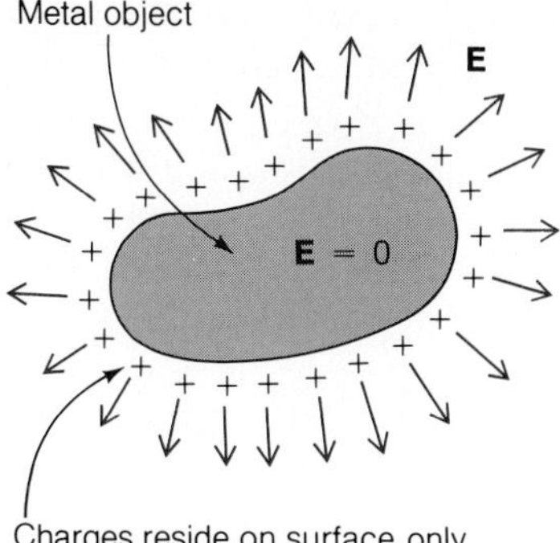

Figure 16.22

There are neither charges nor an electric field inside a charged metal object. Just outside the object the field is perpendicular to its surface.

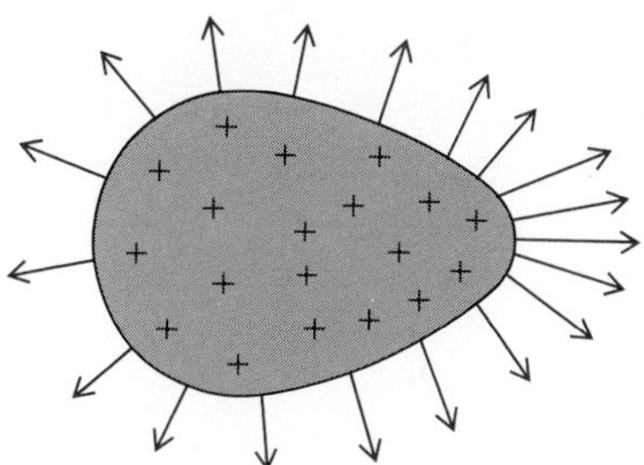

Figure 16.23

The electric field near a charged asymmetric metal object is strongest near the parts of the object that have the smallest radii of curvature.

Electric Field Around an Irregular Metal Body

When a nonspherical metal body is charged, the individual charges do not distribute themselves uniformly on its surface. As a general rule, the more highly curved parts have greater concentrations of charge than the gently curved parts. The electric field is accordingly most intense near the highly curved parts (Fig. 16.23). We can see why this must be true from the preceding discussion of the electric field around a charged metal object. There is no field inside the object, and the field just outside it is perpendicular to its surface. In order to bring this about, the charges must be arranged on the object's surface so that the inward-directed parts of the fields of charges on one side of the object exactly balance the inward-directed parts of the fields on the opposite side. Hence the charges at the small end of the egg-shaped object in Fig. 16.23 must be closer together than those at the large end for their respective fields to cancel out in the interior.

At a metal point the electric field may become so great that it causes a separation of charge in nearby air molecules. The resulting electric "discharge" is visible as a luminous glow. If two oppositely charged pointed rods are brought close together, charge flows between them via disrupted air molecules and a spark occurs. If more gently curved charged bodies, such as large spheres, are similarly brought near each other, their separation must be much smaller before a spark can occur.

Important Terms

Electric charge, like mass, is a fundamental property of certain of the elementary particles of which all matter is composed. There are two kinds of electric charge, **positive charge** and **negative charge;** charges of like sign repel, unlike charges attract. The unit of charge is the **coulomb.** All charges, of either sign, occur in multiples of the fundamental **electron charge** of 1.6×10^{-19} coulomb.

The principle of **conservation of charge** states that the net electric charge in an isolated system remains constant.

Coulomb's law states that the force one charge exerts on another is directly proportional to the magnitudes of the charges and inversely proportional to the square of the distance between them.

An **atom** consists of a tiny, positively charged nucleus surrounded at some distance by electrons. The nucleus consists of protons and neutrons held tightly together by nuclear forces, and the number of electrons equals the number of protons, so the atom as a whole is electrically neutral.

An **ion** is an atom or a group of atoms that carries a net electric charge. An atom or a group of atoms becomes a negative ion when it picks up one or more electrons in addition to its normal number, and becomes a positive ion when it loses one or more of its usual number.

A **force field** is a region of space at every point of which an appropriate test object would experience a force.

An **electric field** exists wherever an electrical force acts on a charged particle. The magnitude of an electric field at a point is equal to the force that would act on a charge of +1 C placed there. The direction of the field is the direction of the force on the charge. The unit of electric field is the newton/coulomb (N/C).

Important Formulas

Coulomb's law: $F = k\dfrac{Q_A Q_B}{r^2}$

Electric field: $\mathbf{E} = \dfrac{\mathbf{F}}{Q}$

Electric field of a charge: $E = k\dfrac{Q}{r^2}$

MULTIPLE CHOICE

1. A rod with an unknown charge attracts a plastic ball. Which one or more of the following might describe them?
- **a.** The rod is positive and the ball is negative.
- **b.** The rod is negative and the ball is positive.
- **c.** The rod is positive and the ball is uncharged.
- **d.** The rod is negative and the ball is uncharged.

2. Electric charge
- **a.** is a continuous quantity that can be subdivided indefinitely
- **b.** is a continuous quantity but it cannot be subdivided into smaller parcels than $\pm 1.6 \times 10^{-19}$ C
- **c.** occurs only in separate parcels, each of $\pm 1.6 \times 10^{-19}$ C
- **d.** occurs only in separate parcels, each of ± 1 C

3. Which one or more of the following statements is not true?
- **a.** The electron and proton have charges of the same magnitude but of opposite sign.
- **b.** The electron and proton have the same mass.
- **c.** The proton and neutron have the same mass.
- **d.** Atomic nuclei contain only protons and neutrons.

4. The electrons in an atom
- **a.** are bound to it permanently
- **b.** are some distance away from the nucleus
- **c.** have more mass than the nucleus
- **d.** may be positively or negatively charged

5. An object has a positive electric charge whenever
- **a.** it has an excess of electrons
- **b.** it has a deficiency of electrons
- **c.** the nuclei of its atoms are positively charged
- **d.** the electrons of its atoms are positively charged

6. The nucleus of an atom
- **a.** contains most of the atom's mass
- **b.** is nearly as large as the atom
- **c.** is electrically neutral
- **d.** always deflects incident alpha particles by less than 1°

7. In the formula $F = kQ_AQ_B/r^2$, the value of the constant k
- **a.** is the same under all circumstances
- **b.** depends on the medium the charges are located in
- **c.** is different for positive and negative charges
- **d.** has the numerical value 1.6×10^{-19}

8. Relative to the electrical force between two protons, the gravitational force between them is
- **a.** weaker
- **b.** equal in magnitude
- **c.** stronger
- **d.** any of the above, depending on how far apart the protons are

9. The electric field at a point in space is equal in magnitude to
- **a.** the number of electric field lines that pass through the point
- **b.** the electric charge there
- **c.** the force a charge of one coulomb would experience there
- **d.** the force an electron would experience there

10. Electric field lines
- **a.** leave negative charges and enter positive ones
- **b.** are close together where E is weak and far apart where E is strong

c. have the same pattern around a charged metal sphere as they do around a point charge
d. physically exist

11. Ten million electrons are placed on a solid copper sphere. The electrons become
a. uniformly distributed on the sphere's surface
b. uniformly distributed in the sphere's interior
c. concentrated at the center of the sphere
d. concentrated at the bottom of the sphere

12. If 10,000 electrons are removed from a neutral plastic ball, its charge is now
a. $+1.6 \times 10^{-15}$ C
b. $+1.6 \times 10^{-23}$ C
c. -1.6×10^{-15} C
d. -1.6×10^{-23} C

13. Two charges of $+Q$ are 1 cm apart. If one of the charges is replaced by a charge of $-Q$, the magnitude of the force between them is
a. zero
b. smaller
c. the same
d. larger

14. A charge of $+q$ is placed 2 cm from a charge of $-Q$. A second charge of $+q$ is then placed next to the first. The force on the charge of $-Q$
a. decreases to half its former magnitude
b. remains the same
c. increases to twice its former magnitude
d. increases to four times its former magnitude

15. Two charges repel each other with a force of 10^{-6} N when they are 10 cm apart. When they are brought closer together until they are 2 cm apart, the force between them becomes
a. 4×10^{-8} N
b. 5×10^{-6} N
c. 8×10^{-6} N
d. 2.5×10^{-5} N

16. Two charges, one positive and the other negative, are initially 2 cm apart and are then pulled away from each other until they are 6 cm apart. The force between them is now smaller by a factor of
a. $\sqrt{3}$
b. 3
c. 9
d. 27

17. The force between two charges of -3.0×10^{-9} C that are 50 mm apart is
a. 1.8×10^{-16} N
b. 3.6×10^{-15} N
c. 1.6×10^{-6} N
d. 3.2×10^{-5} N

18. Two equal charges attract each other with a force of 1.0×10^{-5} N. When they are moved 4 mm farther apart, the force between them becomes 2.5×10^{-6} N. The original separation of the charges was
a. 1 mm
b. 2 mm
c. 4 mm
d. 8 mm

19. The magnitude of the charges in Multiple Choice 18 is
a. 6.7×10^{-11} C
b. 1.3×10^{-10} C
c. 1.1×10^{-9} C
d. 2.1×10^{-9} C

20. The magnitude of the force on an electron in an electric field of 200 N/C is
a. 8.0×10^{-22} N
b. 3.2×10^{-21} N
c. 3.2×10^{-17} N
d. 6.4×10^{-15} N

21. The electric field 20 mm from a certain charge has a magnitude of 1.0×10^5 N/C. The magnitude of **E** 10 mm from the charge is
a. 2.5×10^4 N/C
b. 5.0×10^4 N/C
c. 2.0×10^5 N/C
d. 4.0×10^5 N/C

QUESTIONS

1. (a) When two objects attract each other electrically, must both of them be charged? (b) When two objects repel each other electrically, must both of them be charged?

2. What reasons might there be for the universal belief among scientists that there are only two kinds of electric charge?

3. How can the principle of charge conservation be reconciled with the fact that a rubber rod can be charged by stroking it with a piece of fur?

4. An insulating rod has a charge of $+Q$ at one end and a charge of $-Q$ at the other. How will the rod tend to behave when it is placed near a fixed positive charge that is initially equidistant from the ends of the rod?

5. How will the rod in Question 4 tend to behave when it is placed in a uniform electric field whose direction is (a) parallel to the rod and (b) perpendicular to the rod?

6. How do we know that the inverse square force holding the earth in its orbit around the sun is not an electrical force?

7. Electricity was once regarded as a weightless fluid, an excess of which was "positive" and a deficiency of which was "negative." What phenomena can this hypothesis still explain? What phenomena can it not explain?

8. Nearly all the mass of an atom is concentrated in its nucleus. Where is its charge located?

9. In what ways do the Thomson and Rutherford models of the atom agree? In what ways do they disagree?

10. Most alpha particles pass through gases and thin metal foils with no deflection. To what conclusion regarding atomic structure does this observation lead?

11. What property of the electrons in a metal allows it to conduct electric current readily? What property of the electrons in an insulator prevents it from conducting electric current readily?

12. How does electrical conduction in a metal differ from that in an ionized gas?

13. List several good conductors of electricity and several good insulators. How well do these substances conduct heat? What general relationship between an ability to conduct electricity and an ability to conduct heat can you infer?

14. What aspect of superconductivity has prevented its large-scale application thus far?

15. How could you distinguish between an electric field and a gravitational field?

16. Can electric field lines ever intersect in space? Explain.

17. What can you tell about the force a charged object would experience at a given point in an electric field by looking at a sketch of the field lines?

18. Sketch the pattern of electric field lines in the neighborhood of two charges, $+Q$ and $+2Q$, that are a short distance apart.

19. Sketch the pattern of electric field lines in the neighborhood of two charges, $-Q$ and $+2Q$, that are a short distance apart.

EXERCISES

16.2 Coulomb's Law

1. Find the charge of 1.0 μg of protons.

2. Two charges attract each other with a force of 4.0×10^{-6} N when they are 4.0 mm apart. Find the force between them when their separation is increased to 5.0 mm.

3. Two electric charges originally 80 mm apart are brought closer together until the force between them is greater by a factor of 16. How far apart are they now?

4. Two charges of unknown magnitude and sign are observed to repel one another with a force of 0.10 N when they are 50 mm apart. What will the force be when they are (a) 100 mm apart? (b) 500 mm apart? (c) 10 mm apart?

5. The nucleus of a hydrogen atom is a single proton. Find the force between the two protons in a hydrogen molecule, H_2, that are 7.42×10^{-11} m apart. (The two electrons in the molecule spend more time between the protons than outside them, which leads to attractive forces that balance the repulsion of the protons and permit a stable H_2 molecule; see Chapter 28.)

6. A charge of -5.0×10^{-7} C is 10 cm from a charge of $+6.0 \times 10^{-6}$ C. Find the magnitude and direction of the force on each charge.

7. A charge of $+5.00 \times 10^{-9}$ C is attracted by a charge of -3.00×10^{-7} C with a force of 0.135 N. How far apart are they?

8. Two identical metal spheres, one with a charge of $+2.0 \times 10^{-5}$ C and the other with a charge of -1.0×10^{-5} C, are 10 cm apart. (a) What is the force between them? (b) The two spheres are brought into contact, and then separated again by 10 cm. What is the force between them now?

9. How far apart should two electrons be if the force each exerts on the other is to equal the weight of an electron at sea level?

10. The permittivity ϵ of air is only a trifle greater than ϵ_0, the permittivity of free space, but ϵ is considerably greater for some other materials. For instance, the permittivity of water is $\epsilon = 80\ \epsilon_0$. Find the force, in air and in water, between two charges of $+2.0 \times 10^{-10}$ C that are 1.0 mm apart.

11. A particle carrying a charge of $+6.0 \times 10^{-7}$ C is located halfway between two other charges, one of $+1.0 \times 10^{-6}$ C and the other of -1.0×10^{-6} C, that are 40 cm apart. All three charges lie on the same straight line. What is the magnitude and direction of the force on the $+6.0 \times 10^{-7}$ C charge?

12. A test charge of -5.0×10^{-7} C is placed between two other charges so that it is 50 mm from a charge of -3.0×10^{-7} C and 10 cm from a charge of -6.0×10^{-7} C. The three charges lie along a straight line. What is the magnitude and direction of the force on the test charge?

16.4 Atomic Structure

13. According to a simplified model, the hydrogen atom consists of a proton circled by an electron whose orbit has a radius of 5.3×10^{-11} m. How fast must the electron be moving if the required centripetal force is provided by the electric force exerted by the proton?

14. At what distance apart (if any) are the electrical and gravitational forces between two electrons equal in magnitude? Between two protons? Between an electron and a proton?

16.7 Electric Field

15. Find the electric field 40 cm from a charge of $+7.0 \times 10^{-5}$ C.

16. The average distance between the electron and the proton in a hydrogen atom is 5.3×10^{-11} m. How strong is the electric field the electron experiences?

17. Four charges of $+1.0\ \mu$C are at the corners of a square that measures 1.0 m on each side. Find the electric field at the center of the square.

18. A particle carrying a charge of 1.0×10^{-5} C starts moving from rest in a uniform electric field whose magnitude is 50 N/C. (a) What is the force on the particle? (b) How much kinetic energy will the particle have after it has moved 1.0 m?

19. How strong an electric field is needed to support a proton against gravity at sea level?

20. Two charges of $+4.0\ \mu$C and $+8.0\ \mu$C are 2.0 m apart. What is the electric field halfway between them?

21. Two charges, one of $+1.5\ \mu C$ and the other of $+3.0\ \mu C$, are 20 cm apart. Where is the electric field along the line joining them equal to zero?

22. A charge of $+Q$ is placed on a cubical copper box 20 cm on an edge. What is the magnitude and direction of the electric field at the center of the box?

PROBLEMS

23. Two 3.0-g balloons are suspended from a nail by strings 50 cm long. Each balloon has a charge of $+Q$, and there is an angle of 40° between the strings. Find Q.

24. Suppose the force between the earth and the moon were electrical rather than gravitational, with the earth having a positive charge and the moon having a negative one. If the magnitude of each charge were proportional to the respective body's mass, find the Q/m ratio required for the moon to follow its present orbit of 3.84×10^8-m radius with its period of 27.3 days. The earth's mass is 6.0×10^{24} kg, and the moon's mass is 7.3×10^{22} kg.

25. Two charges, one of -1.0×10^{-6} C and the other of -3.0×10^{-6} C are 40 cm apart. (a) Where should a charge of -1.0×10^{-7} C be placed on the line between the other charges in order that there be no resultant force on it? (b) Where should a charge of $+1.0 \times 10^{-7}$ C be placed in order that there be no resultant force on it?

26. Two charges, one of $+2.0 \times 10^{-8}$ C and the other of $+1.0 \times 10^{-8}$ C are 20 cm apart. (a) Where should a charge of -1.0×10^{-7} C be placed in order that there be no resultant force on it? (b) Where should a charge of $+1.0 \times 10^{-7}$ C be placed in order that there be no resultant force on it?

27. Three charges, $+Q$, $+Q$, and $-Q$, are at the vertexes of an equilateral triangle a long on each side. Find the magnitude and direction of the force on one of the positive charges.

28. Four charges of $+1.0 \times 10^{-8}$ C are at the corners of a square 20 cm on each side. Find the magnitude and direction of the force on one of them.

29. What is the electric field at one vertex of an equilateral triangle whose sides are 1.0 m long if there are charges of $+20\ \mu C$ at the other vertexes?

30. An electric dipole consists of the charges $+Q$ and $-Q$ a distance A apart. Find the electric field produced by a dipole at a distance R from the line joining the charges along the perpendicular bisector of that line, and show that $E = kQA/R^3$ when $R \gg A$.

31. Find the electric field produced by an electric dipole at a distance R from the center of the line joining the two charges along an extension of that line, and show that $E = 2kQA/R^3$ when $R \gg A$. How does the direction of E in this case compare with its direction in Problem 30?

Answers to Multiple Choice

1. a, b, c, d	7. b	13. c	19. b
2. c	8. a	14. c	20. c
3. b, c	9. c	15. d	21. d
4. b	10. c	16. c	
5. b	11. a	17. d	
6. a	12. a	18. c	

CHAPTER

17

ELECTRIC ENERGY

In our study of mechanics we found the related concepts of work and energy to be useful in analyzing a wide variety of situations. These concepts are equally useful in the study of electricity—in particular, electric current. Instead of potential energy itself, a related quantity called potential difference turns out to be especially appropriate in describing electrical phenomena. Later in this chapter we will look into how capacitors store electric energy in their electric fields.

17.1 Electric Potential Energy

A charge located in an electric field has potential energy.

Let us consider the potential energy of a charge in a uniform electric field with the help of a gravitational analogy. At the left in Fig. 17.1(a) is a uniform electric field **E** between two parallel, uniformly charged plates A and B. At the right is a region near the earth's surface in which the gravitational field is also uniform. Now we place a particle of charge Q (assumed positive) in the electric field and a particle of mass m in the gravitational field. As shown in Figure 17.1(b), the charge is acted on by the electrical force

$$\mathbf{F}_{\text{elec}} = Q\mathbf{E}$$

and the mass is acted on by the gravitational force

$$\mathbf{F}_{\text{grav}} = m\mathbf{g}$$

If the charge is on plate A and we want to move it to plate B, we must apply a force of magnitude QE to it because we have to push against a force of this magnitude (and opposite direction) exerted by the electric field. When the charge is at B, we will have done the work

$$\text{Work} = (\text{force})(\text{distance})$$
$$W = QEd$$

on it, where d is the distance the charge has moved (Fig. 17.1c). Similarly, to raise the mass from the ground to a height h, we must apply a force of magnitude mg to it, and the work we do is

$$W = mgh$$

At plate B the charge has the potential energy

$$\text{PE} = QEd \qquad \text{PE } \textit{of charge in uniform electric field} \quad (17.1)$$

with respect to A. If we let it go, the PE will become kinetic energy as the electric field **E** accelerates the charge, and when the charge is back at A, it will have a KE equal to QEd (Fig. 17.1d). In the same way, the mass has PE *with respect to the ground* in its new location. This PE is equal to the work done in raising it through the height h and is

$$\text{PE} = mgh$$

If we let the mass go, it will fall to the ground with a final KE of mgh.

To summarize: The amount of work QEd must be performed to move a charge Q from A to B, the distance d apart, in a uniform electric field **E**. At B the charge accordingly has the potential energy QEd. If the charge is released at B, the

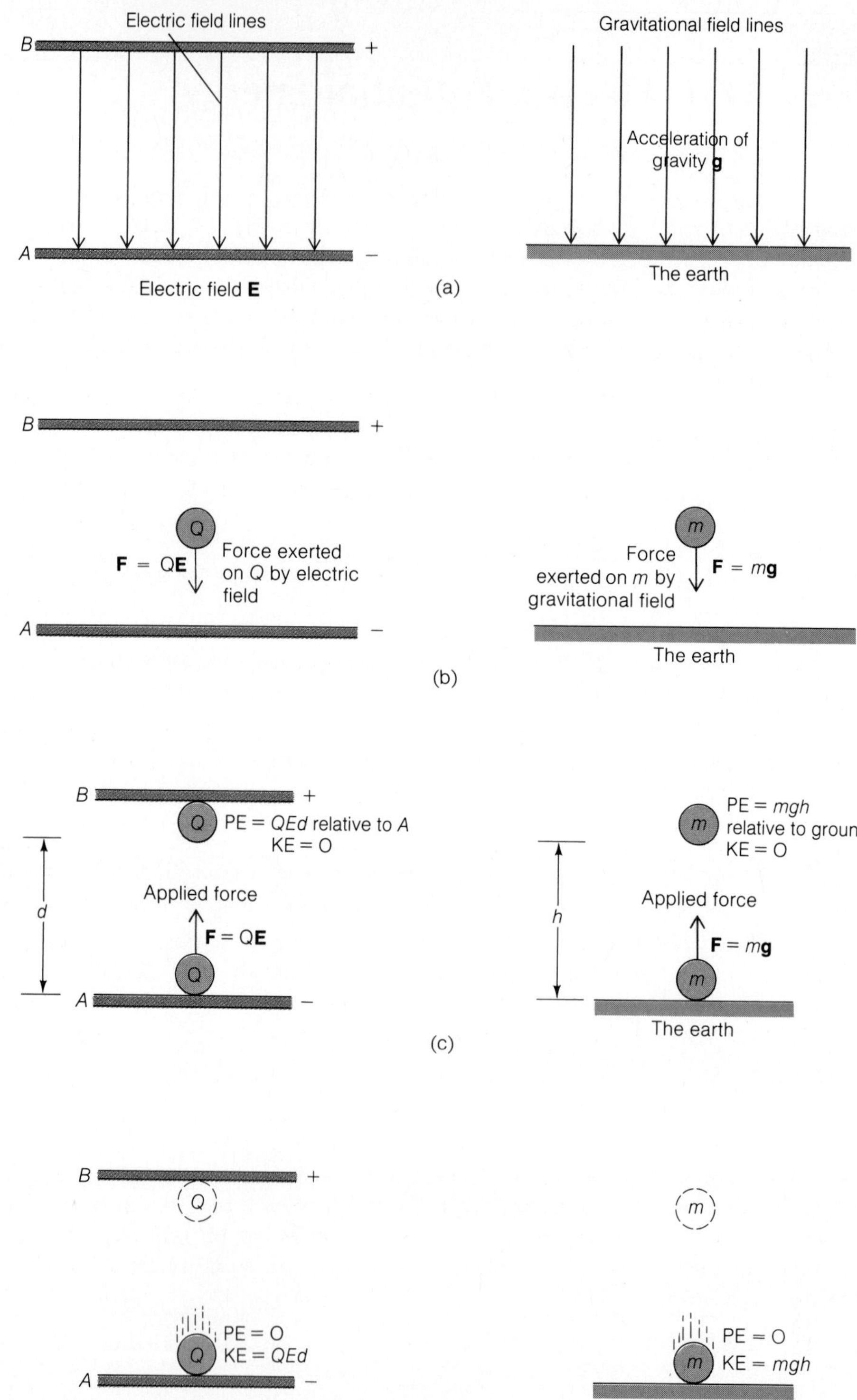

Figure 17.1

Analogy between electric and gravitational potential energy.

force $Q\mathbf{E}$ acting on it produces an acceleration such that the charge has the kinetic energy QEd when it is back at A. Thus the work done in moving the charge in the field becomes PE, which in turn becomes KE when it is released.

There is no change in the energy of a charge moved perpendicular to an electric field, just as there is no change in the energy of a mass moved perpendicular to a gravitational field (for instance, along the earth's surface).

17.2 Potential Energy of Two Charges

It is a property of the system of both charges.

A particle of charge Q_A that is located the distance r from another particle of charge Q_B has electric potential energy there because a force is exerted on it by the electric field of Q_B. When Q_A and Q_B have the same sign, the force is repulsive; when the charges have opposite signs, the force is attractive. In either case, when Q_A is released, it will begin to move and gain KE at the expense of its original PE (Fig. 17.2 shows this effect for two like charges).

We have been assuming that Q_B is fixed in place. If instead Q_A is fixed in place, then we can speak of the PE of Q_B in the electric field of Q_A. This PE is exactly the same as before, since by Newton's third law of motion the force one object exerts on another is equal in magnitude to the force the second object exerts on the first.

In reality, the PE belongs to the *system* of the two particles. If the particles are both released, *both* of them begin to move, and they share the original PE between them. The relative speeds of the two particles will be such as to conserve linear momentum, so the lighter particle will move faster than the heavier one. (The total amount of energy available depends on the charges of the particles, and the division of the energy depends on their masses.)

If one of the particles has a much greater mass than the other has, its speed will be much smaller than that of the other, and we may then legitimately regard it as being fixed in place. For example, the mass of a proton is 1836 times that of an electron, so in a hydrogen atom, which consists of a proton and an electron, we may think of the proton as being stationary and of the electron as having the PE of the system.

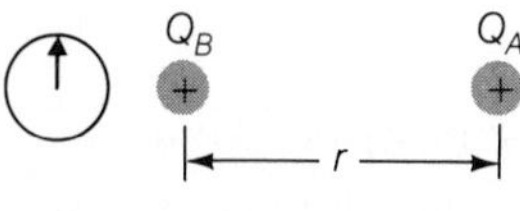

Figure 17.2

A charge has potential energy when it is in the electric field of another charge. Actually, the potential energy belongs to the system of the two charges. Here Q_B is assumed to be fixed in place, so the PE of the system becomes kinetic energy of Q_A when Q_A is released.

A potential energy of any kind must be specified relative to a reference location. Near the earth, for example, the gravitational PE of an object is usually given with the earth's surface as the reference location for PE = 0. In the case of charges interacting with one another, the reference location is chosen to be infinity, since the electric field of a charge falls to zero an infinite distance away.

Because the electric field **E** of the charge Q_B (which we assume to be fixed in place) is not uniform, it is not easy to calculate the PE of Q_A when it is the distance r away. What we must do is start with Q_A at infinity and imagine we move it a short distance toward Q_B. In this short distance we consider **E** as having a constant magnitude, so we can find the work needed for this step from the formula $W = QEd$. Then we move Q_A through another short interval closer to Q_B. Again we consider **E** as having a constant magnitude, though a greater one than before, and we find the work needed for the second step. By continuing this process until Q_A is r away from Q_B and adding up all the amounts of work needed, we can find the PE of Q_A. With the help of calculus this addition yields the result

$$\text{PE} = k\frac{Q_A Q_B}{r} \qquad \textit{Potential energy of system of two charges} \qquad (17.2)$$

If the charges have the same sign, their PE is positive, so a positive PE corresponds to a repulsive force. If the charges have opposite signs, their PE is negative, so a negative PE corresponds to an attractive force. The PE of a charge decreases as it moves away from another charge of the same sign and increases as it moves away from another charge of the opposite sign.

EXAMPLE 17.1

An electron of initial speed 1.0×10^3 m/s is aimed at another electron, whose position is fixed, from a distance of 1.0 mm. How close to the stationary electron will the other one approach before it comes to a stop and reverses its direction?

SOLUTION Figure 17.3 illustrates the situation. Since $Q_A = Q_B = -e$ here, the potential energies of the moving electron at the initial distance r_1 ($= 1.0 \times 10^{-3}$ m) and the final distance r_2 are, respectively,

$$\text{PE}_1 = \frac{ke^2}{r_1} \qquad \text{PE}_2 = \frac{ke^2}{r_2}$$

The potential energies are both positive, corresponding to a repulsive force between the electrons. The difference between the two PE values is equal to the initial kinetic

Figure 17.3

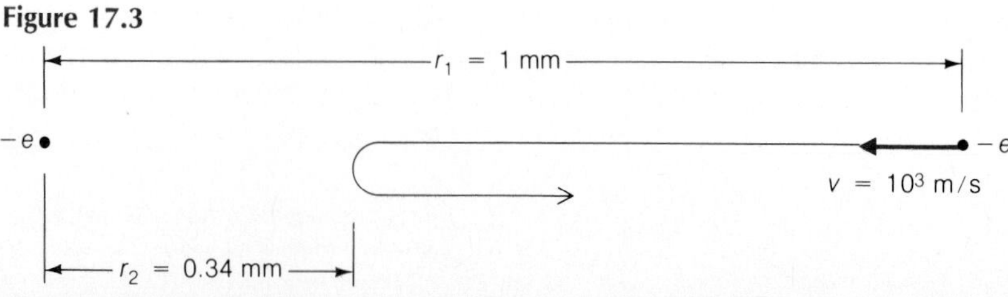

energy $\frac{1}{2}mv^2$ of the moving electron, and therefore

$$\text{KE} = \text{PE}_2 - \text{PE}_1$$

$$\tfrac{1}{2}mv^2 = ke^2\left(\frac{1}{r_2} - \frac{1}{r_1}\right)$$

$$\frac{1}{r_2} = \frac{mv^2}{2ke^2} + \frac{1}{r_1}$$

$$r_2 = \frac{1}{(mv^2/2ke^2) + (1/r_1)} = 3.4 \times 10^{-4}\ \text{m} = 0.34\ \text{mm}$$

◆

17.3 Potential Difference

The volt is the unit of potential difference.

The quantity **potential difference** describes the situation of a charge in an electric field in an especially convenient way. The potential difference V_{AB} between two points A and B is defined as the ratio between the work that must be done to take a charge Q from A to B and the value of Q:

$$V_{AB} = \frac{W_{AB}}{Q} \qquad \textit{Potential difference} \quad (17.3)$$

Potential difference = work per unit charge

The unit of potential difference is the joule per coulomb. Because this quantity is so often used, its unit has been given a name of its own, the **volt** (V). Thus

1 volt = 1 joule/coulomb

The volt is named after the Italian physicist Alessandro Volta (1745–1827), who invented the battery.

In a uniform electric field, $W_{AB} = QEd$, with the result that the potential difference between A and B is QEd/Q or

$$V_{AB} = Ed \qquad \textit{Potential difference in a uniform electric field} \quad (17.4)$$

In a uniform field the potential difference between two points is the product of the field magnitude E and the separation d of the two points in a direction parallel to that of **E** (Fig. 17.4).

From Eq. (17.4) we see that the volt/meter (V/m) is equal to the newton/coulomb (N/C), so that either can be used as the unit of electric field.

A positive potential difference means that the energy of the charge is *greater* at B than at A. A negative potential difference means that its energy is *less* at B than at A. If V_{AB} is positive, then a charge at B tends to return to A. If V_{AB} is negative, the charge tends to move farther away from A.

One advantage of specifying the potential difference between two points in an electric field, rather than the magnitude of the field between them, is that an

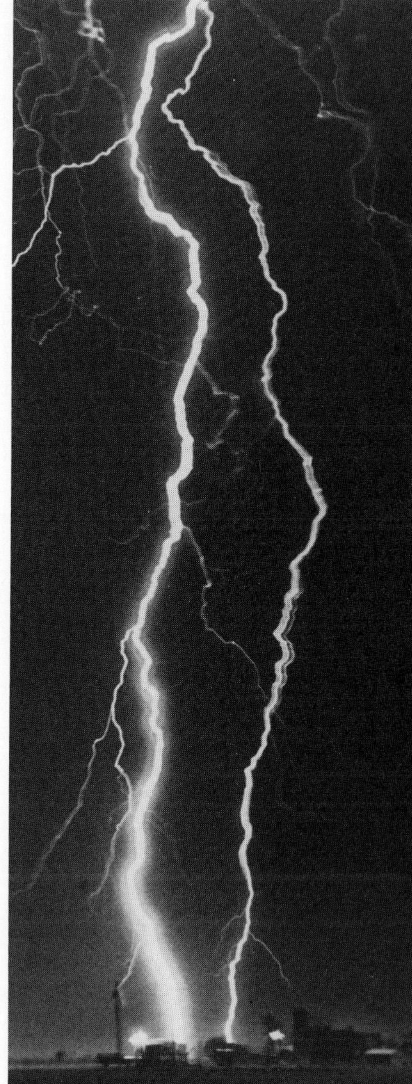

Charge separation begins in a thunderstorm when droplets of water and ice crystals collide, and the separation continues on a larger scale mainly through convection. The result is usually a region of negative charge a few kilometers across in the storm center with a region of positive charge in the clouds above it. The density of separated charge is about a coulomb per cubic kilometer. The charge separation produces a potential difference of up to several hundred million volts, and the electrical discharges that then occur appear as lightning flashes.

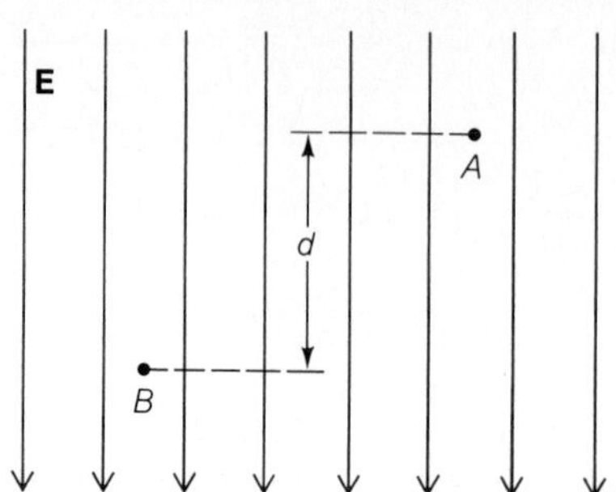

Figure 17.4

The potential difference between two points in a uniform electric field **E** is equal to *Ed*, where *d* is the component parallel to **E** of the distance between the points.

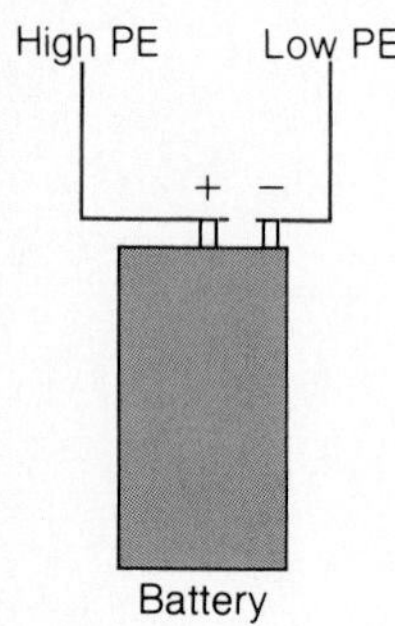

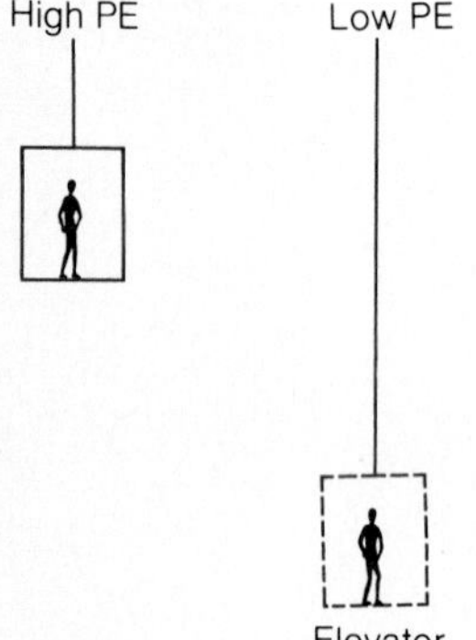

Figure 17.5

A battery is to a positive electric charge as an elevator is to a mass.

electric field is normally created by imposing a difference of potential between two points in space. A battery uses chemical reactions to produce a potential difference between two terminals (Fig. 17.5). A "6-volt" battery is one that has a potential difference of 6 V between its terminals. A generator is another device for producing a potential difference. Batteries and generators are to electric charges as elevators are to masses: All of them increase the potential energy of what they act on.

When a charge Q goes from one terminal of a battery whose potential difference is V with respect to the other, Eq. (17.3) tells us that the work

$$W = QV \qquad \text{Work done on a charge} \qquad (17.5)$$

is done on it regardless of the path taken by the charge and regardless of whether the actual electric field that acted on the charge is strong or weak. Given V, we can find W at once, no matter what the details of the process are. Hence the notion of potential difference simplifies the analysis of electrical phenomena, just as the notion of potential energy simplifies the analysis of mechanical phenomena.

EXAMPLE 17.2

Figure 17.6 shows a tube that has a source of electrons at one end and a metal plate at the other. A 100-V battery is connected between the electron source and the plate, so that there is a potential difference of 100 V between them. The negative terminal of the battery is connected to the electron source. What is the speed of the electrons when they arrive at the metal plate? (The tube is evacuated to prevent collisions between the electrons and air molecules.)

SOLUTION To find the electron's kinetic energy, we note that the work done by the electric field within the tube on an electron is $W = QV = eV$. Since the KE of the electron is equal to the work done on it, its KE after passing through the entire

Figure 17.6

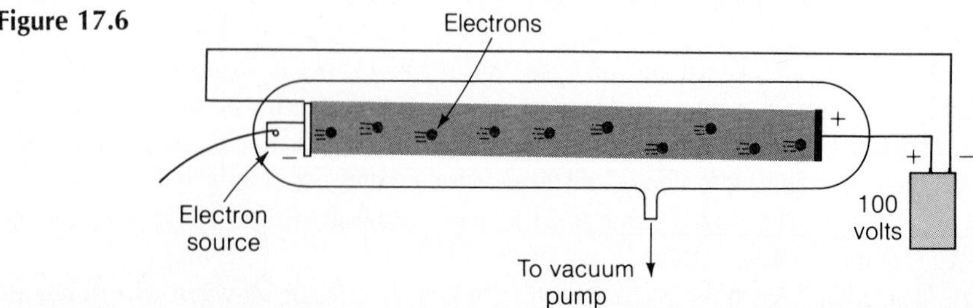

field is eV. Hence

$$\text{KE} = W = eV = \tfrac{1}{2}mv^2$$

$$v = \sqrt{\frac{2eV}{m}} = \sqrt{\frac{2(1.6 \times 10^{-19}\ \text{C})(100\ \text{V})}{9.1 \times 10^{-31}\ \text{kg}}} = 5.9 \times 10^6\ \text{m/s}$$ ♦

EXAMPLE 17.3

The electron source and the positive electrode are 400 mm apart in the tube of the preceding example. (a) What is the magnitude E of the electric field (assumed uniform) between them? (b) If this distance is reduced to 200 mm, what is the new value of E and what effect does this change have on the final speed of the electrons?

SOLUTION (a) The electric field magnitude in a uniform field is given by $E = V/d$ from the definition of potential difference in an electric field. Hence

$$E = \frac{100\ \text{V}}{0.400\ \text{m}} = 250\ \text{V/m}$$

(b) If d is reduced to 200 mm, the field magnitude increases to

$$E = \frac{100\ \text{V}}{0.200\ \text{m}} = 500\ \text{V/m}$$

The energy given to an electron by this field when it travels through the entire 200 mm is the same as the energy given to an electron that travels through the 400-mm length of the previous, weaker field, since in both cases QEd is the same. We reach the same conclusion by noting that KE $= QV$ and the potential difference V in this problem is independent of the spacing of the electrodes. The final electron speed is therefore unchanged. ♦

17.4 The Electron Volt

The energy unit of atomic and nuclear physics.

The **electron volt** (abbreviated eV) is the customary energy unit in atomic and nuclear physics. By definition, 1.00 eV is the energy gained by an electron that has been accelerated through a potential difference of 1.00 volt. Hence

$$W = QV$$

$$1.00\ \text{eV} = (1.60 \times 10^{-19}\ \text{C})(1.00\ \text{V})$$

and so

$$1.00\ \text{eV} = 1.60 \times 10^{-19}\ \text{J} \qquad \textit{Electron volt} \quad (17.6)$$

Typical quantities expressed in electron volts are the ionization energy of an atom (the work needed to remove one of its electrons) and the binding energy of a molecule (the work needed to break it apart into separate atoms). Thus the ionization energy of nitrogen is usually given as 14.5 eV, and the binding energy of the hydrogen molecule, which consists of two hydrogen atoms, is usually given as 4.5 eV.

The eV is too small a unit for nuclear physics, where its multiples the MeV (10^6 eV) and the GeV (10^9 eV) are commonly used. The M and G, respectively, signify *mega* (= 10^6) and *giga* (= 10^9) and are used in connection with other units as well, for instance, the megabuck ($\$10^6$) and the gigawatt ($10^9$ watts). A typical quantity expressed in MeV is the energy liberated when the nucleus of a uranium atom splits into two parts. Such **fission** of a uranium nucleus releases an average of 200 MeV; this is the process that powers nuclear reactors and weapons.

EXAMPLE 17.4

What is the speed of a neutron whose kinetic energy is 50 eV?

SOLUTION The kinetic energy of the neutron is

$$\text{KE} = (50\ \text{eV})(1.6 \times 10^{-19}\ \text{J/eV}) = 8.0 \times 10^{-18}\ \text{J}$$

Since $\text{KE} = \frac{1}{2}mv^2$ and the neutron mass is 1.67×10^{-27} kg, the speed of the neutron is

$$v = \sqrt{\frac{2\ \text{KE}}{m}} = \sqrt{\frac{(2)(8.0 \times 10^{-18}\ \text{J})}{1.67 \times 10^{-27}\ \text{kg}}} = 9.8 \times 10^4\ \text{m/s}$$

◆

EXAMPLE 17.5

A hydrogen atom consists of a proton and an electron an average of 5.30×10^{-11} m apart. Find the potential energy of the electron in eV.

SOLUTION Here the two charges are $-e$ and $+e$, where $e = 1.60 \times 10^{-19}$ C. Hence

$$\text{PE} = k\frac{Q_A Q_B}{r} = k\frac{(-e)(e)}{r} = \left(9.00 \times 10^9 \frac{\text{N·m}^2}{\text{C}^2}\right)\left(\frac{-(1.60 \times 10^{-19}\ \text{C})^2}{5.30 \times 10^{-11}\text{m}}\right)$$
$$= -4.35 \times 10^{-18}\ \text{J}$$

Since $1\ \text{eV} = 1.60 \times 10^{-19}$ J,

$$\text{PE} = \frac{-4.35 \times 10^{-18}\text{J}}{1.60 \times 10^{-19}\ \text{J/eV}} = -27.2\ \text{eV}$$

The minus sign signifies that the force on the electron is directed toward the proton. (The KE of the electron in a hydrogen atom is 13.6 eV, so the total energy of the electron is

$$PE + KE = -27.2 \text{ eV} + 13.6 \text{ eV} = -13.6 \text{ eV}$$

The negative total energy means that the hydrogen atom is a stable system, because work must be done by an outside agency to free the electron from the proton.) ♦

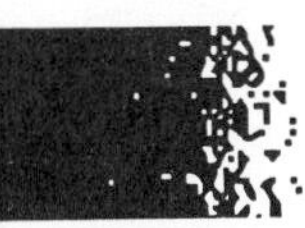

17.5 The Action Potential

Potential differences occur across cell membranes.

Animal cells have an excess of negative charge inside their membranes and an excess of positive charge on the outside. The potential difference across the membrane of a nerve or muscle cell is normally 90 mV (0.090 V). This potential difference is a consequence of the different permeability of the membrane to different ions.

To see how the process works, let us consider two solutions of potassium chloride, KCl, separated by a membrane that permits K^+ ions to pass through it but hinders Cl^- ions from doing so (Fig. 17.7). The solution at the left has the higher concentration of KCl. K^+ ions from both solutions diffuse through the membrane, but because the solution on the left has more of them per unit volume, the number going to the right at first exceeds the number going to the left. As a result the solution on the left soon has a deficiency of K^+ ions relative to its Cl^- ions and hence a net negative charge. At the same time the solution on the right has an excess of K^+ ions and hence a net positive charge. Now an electric field exists across the membrane that favors the diffusion of K^+ ions from right to left. When the influence of the electric field exactly balances the opposite influence of the difference in concentration, as many K^+ ions pass through the membrane in one direction as in the other.

Actual cells do have higher concentrations of K^+ ions inside their membranes compared with the concentration in the fluid around them. The observed difference between these concentrations is consistent with a potential difference of 90 mV. Sodium ions, Na^+, are also present in living tissue, but their concentration is higher

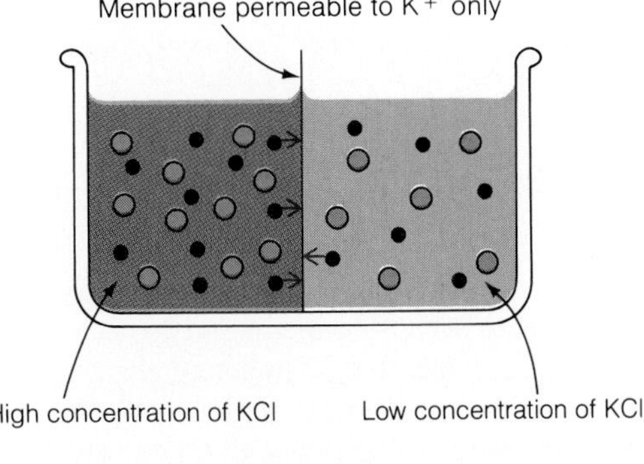

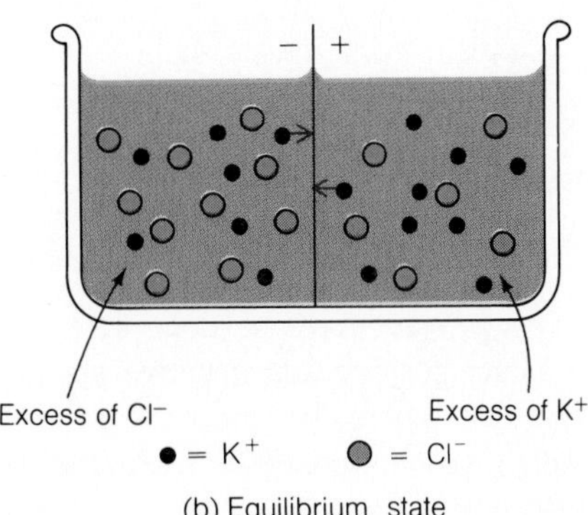

Figure 17.7

(a) Initially the KCl concentration is higher in the solution to the left of the semipermeable membrane, but in each solution the numbers of K^+ and Cl^- ions are equal. Because the concentration of K^+ ions is greater in the solution at left, more K^+ ions diffuse through the membrane to the right than to the left. (b) At equilibrium, the solution at left has an excess of Cl^- ions, and the one at right has an excess of K^+ ions. The rate of diffusion of K^+ ions is now the same in both directions, and there is a potential difference across the membrane.

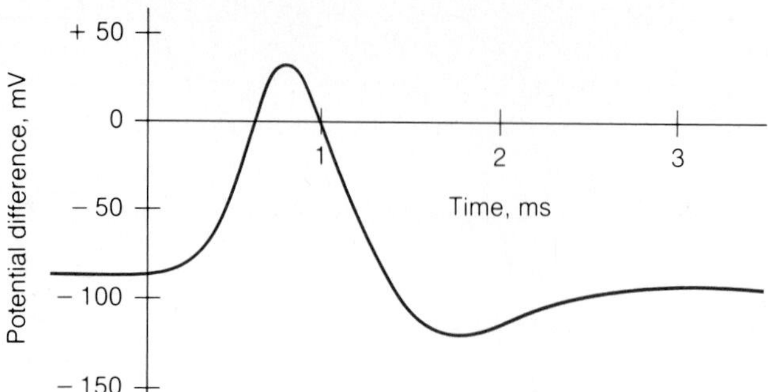

Figure 17.8

Action potential in a nerve cell. The interior of the cell is normally at a potential of −90 mV with respect to the outside.

in the extracellular fluid, and cell membranes are ordinarily impermeable to them. When a nerve or muscle cell is stimulated, its membrane momentarily becomes permeable to Na^+ ions. These ions then flow into the cell to neutralize the excess negative charge there. (Cl^- is not the only negative ion involved.) The membrane, in effect, is short-circuited by the Na^+ ions.

Because the concentration of Na^+ ions is greater outside the cell (the opposite of the case of K^+ ions), in a short time the potential difference across the cell membrane becomes reversed, with the inside at a small positive potential relative to the outside (Fig. 17.8). Then the cell membranes again become impermeable to Na^+ ions, and K^+ diffusion restores the potential difference to its normal value. The entire process involved in producing such an **action potential** takes about a millisecond (0.001 s). A mechanism that is not completely understood then "pumps" the excess Na^+ ions out of the cell into the fluid around it.

Action potentials of the kind just described go from nerve cell to nerve cell at speeds of about 30 m/s. When a nerve impulse arrives at a muscle cell, an action potential triggered in it travels along the muscle fiber and causes it to contract. The action potentials in heart muscle are especially large and are easily detected by electrodes placed on the chest. An **electrocardiogram** is a record of these potentials.

Some fish, notably the giant electric eel, contain specialized cells called **electroplaques** whose chief purpose is to develop action potentials when stimulated. Large numbers of such cells are stacked together. Although each cell is limited to a pulse of not much more than 0.1 V, the total can amount to several hundred volts. Such fish use these potentials to stun their prey. Over two-thirds of the mass of an electric eel consists of 4000 rows of electroplaques, which can produce a 600-V pulse.

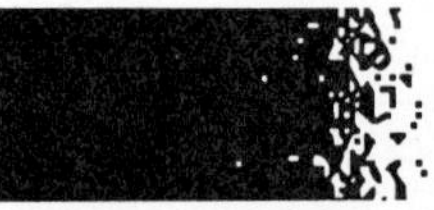

17.6 Electric Field Energy

A charged capacitor stores energy in its electric field.

Work must be done to create an electric field because work is needed to separate positive and negative charges against the Coulomb forces attracting them together. Where does the work go? The answer is that it is stored as potential energy, which we can think of as residing in the electric field between the charges. A system of conductors that stores energy in the form of an electric field is called a **capacitor**.

An example of a capacitor is a pair of parallel metal plates of area A that are a distance d apart, as in Fig. 17.9. Let us imagine that we create an electric field between them by bringing electrons from one plate to the other until a total charge of Q has been transferred. When we are finished, there will be a potential difference of V between the plates.

At the beginning, the potential difference between the plates is small, and little work is needed to move each electron. As the charge builds up, the potential difference becomes greater, and more work is needed per electron. The average potential difference $\overline{V}$ during the charge transfer is

$$\overline{V} = \frac{V_{\text{final}} + V_{\text{initial}}}{2} = \frac{V + 0}{2} = \tfrac{1}{2}V$$

Since the total charge transferred is Q, the work W that was done is the product of Q and the average potential difference V, namely,

$$W = Q\overline{V} = \tfrac{1}{2}QV \qquad \textit{Energy of charged capacitor} \quad (17.7)$$

Because the metal plates are good conductors, the + and − charges on them are spread out evenly, and the field between the plates (except near the edges) is uniform with the magnitude

$$E = 4\pi k\frac{Q}{A} \qquad \textit{Electric field inside capacitor} \quad (17.8)$$

The derivation of this formula, which follows from the considerations discussed in Chapter 16, is not given here because it involves calculus. We can rewrite Eq. (17.8) in the form

$$Q = \frac{AE}{4\pi k} \qquad (17.9)$$

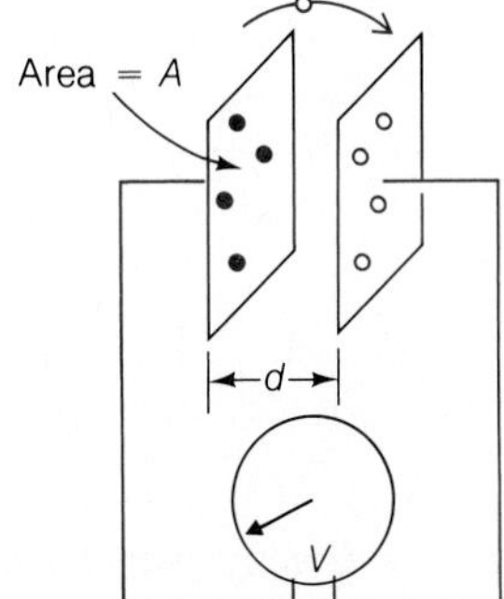

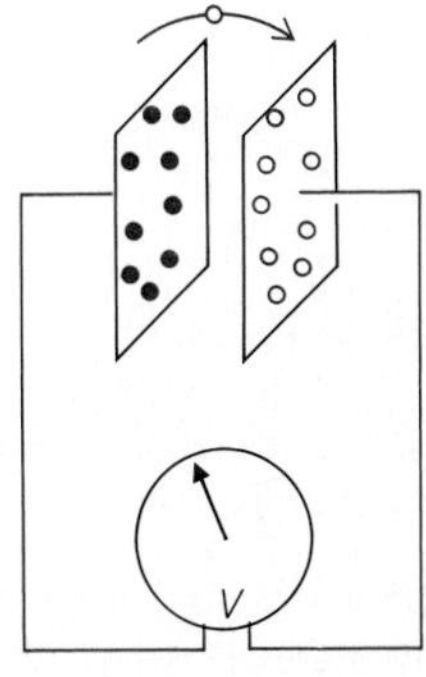

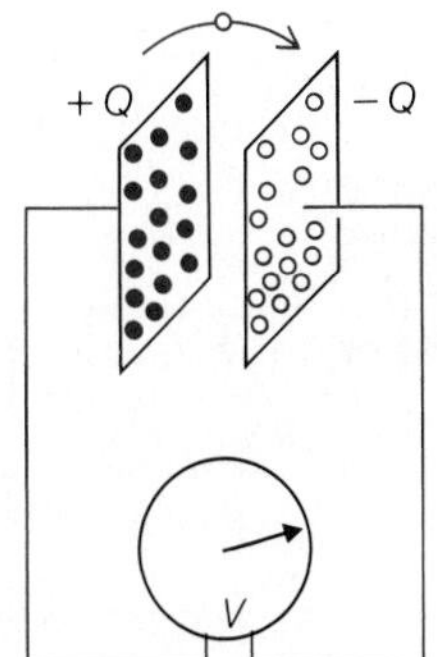

Figure 17.9

As a capacitor is charged, the potential difference between its plates builds up and more work must be done to transfer each successive charge. (In an actual capacitor, the electrons are transferred by means of an electric circuit connecting the plates, not directly.)

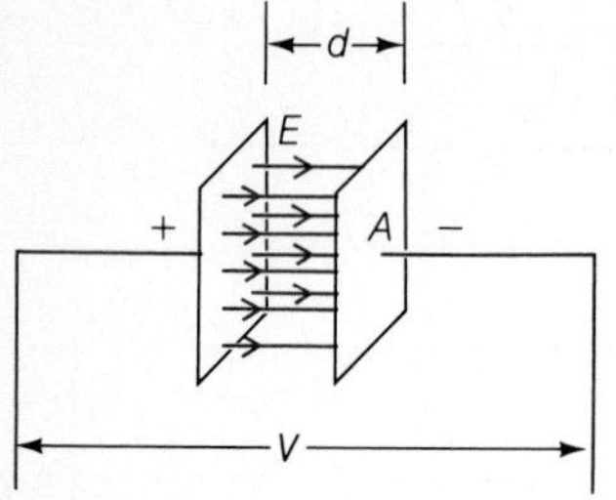

Figure 17.10

The energy density of the electric field in a region is the total energy in the region divided by its volume.

The potential difference between the plates is equal to

$$V = Ed \tag{17.10}$$

Hence the electric potential energy of the system of two charged plates is

$$W = \tfrac{1}{2}QV = \frac{Ad}{8\pi k}E^2 \tag{17.11}$$

We note that Ad, the product of the area of each plate and the distance between them, is the volume occupied by the electric field **E** (Fig. 17.10). If we define the **energy density** w of an electric field as the electric potential energy per unit volume associated with it, $w = W/Ad$, and so

$$w = \frac{E^2}{8\pi k} \qquad \textit{Electric energy density} \tag{17.12}$$

The energy density of an electric field is directly proportional to the square of its magnitude E. Even though this formula was derived for a special situation, it is a completely general result.

EXAMPLE 17.6

Dry air is an insulator provided the electric field in it does not exceed about 3.0×10^6 V/m. What energy density does this correspond to?

SOLUTION Because $E = 3.0 \times 10^6$ V/m,

$$w = \frac{E^2}{8\pi k} = \frac{(3.0 \times 10^6 \text{ V/m})^2}{8\pi(9.0 \times 10^9 \text{ N}\cdot\text{m}^2/\text{C}^2)} = 40 \text{ J/m}^3$$

At sea level 1 m^3 of air has a mass of 1.3 kg, so if this amount of energy were gravitational potential energy, it would correspond to an elevation of about 3 m. ◆

17.7 Capacitance

The ratio of the charge on either conductor of a capacitor to the potential difference between the conductors.

The potential difference V across a capacitor is always directly proportional to the charge Q on either of its plates (Fig. 17.11). The more the charge, the stronger the electric field between the plates, and the greater the potential difference. The ratio between Q and V is therefore a constant for any capacitor and is known as its

capacitance (symbol C):

$$C = \frac{Q}{V} \qquad \textit{Capacitance} \quad (17.13)$$

$$\text{Capacitance} = \frac{\text{charge on either conductor}}{\text{potential difference between conductors}}$$

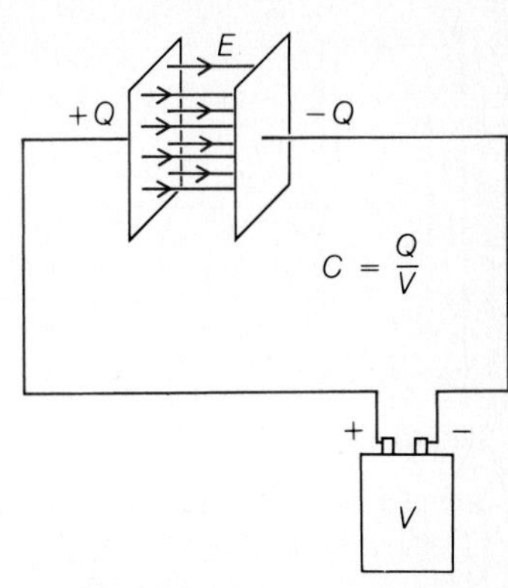

Figure 17.11

The ratio between the charge on a particular capacitor and the potential difference across it is a constant called its *capacitance*.

The unit of capacitance is the **farad**, abbreviated F, where

1 farad = 1 coulomb/volt

The farad is so large a unit that, for practical purposes, it is usually replaced by the **microfarad** (μF) or **picofarad** (pf), whose values are

$$1\,\mu\text{F} = 10^{-6}\,\text{F} \qquad 1\,\text{pF} = 10^{-12}\,\text{F}$$

The capacitance of a pair of separated conductors depends solely on their geometry and on the material between them. In the case of a parallel-plate capacitor in vacuum, from Eqs. (17.9) and (17.10) we have

$$C = \frac{Q}{V} = \frac{AE/4\pi k}{Ed}$$

$$= \frac{1}{4\pi k}\frac{A}{d} \qquad \textit{Parallel-plate capacitor} \quad (17.14)$$

To a good degree of approximation, the same formula can be used for such a capacitor in air as well.

EXAMPLE 17.7

The plates of a parallel-plate capacitor are 10 cm square and 1.0 mm apart. Find its capacitance in air and the charge each plate will have when a potential difference of 100 V is applied.

SOLUTION Here $A = 1.0 \times 10^{-2}\ \text{m}^2$ and $d = 1.0 \times 10^{-3}$ m. In air the capacitance is

$$C = \left[\frac{1}{(4\pi)(9.0 \times 10^9\ \text{N}\cdot\text{m}^2/\text{C}^2)}\right]\left(\frac{1.0 \times 10^{-2}\ \text{m}^2}{1.0 \times 10^{-3}\ \text{m}}\right) = 8.9 \times 10^{-11}\ \text{F}$$

$$= 89\ \text{pF}$$

If a potential difference of 100 V is placed across the plates of this capacitor, each plate will have a charge of

$$Q = CV = (8.9 \times 10^{-11}\ \text{F})(100\ \text{V}) = 8.9 \times 10^{-9}\ \text{C}$$

◆

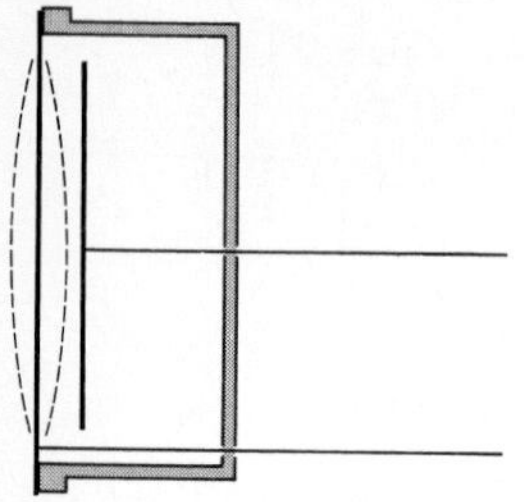

Figure 17.12

Sound waves cause the diaphragm of a capacitor microphone to vibrate, and the resulting changes in its capacitance are transformed into electrical signals.

A parallel-plate capacitor can be used as a microphone if one of its plates is light and flexible enough to respond to the changing air pressure of a sound wave (Fig. 17.12). When the pressure increases, the diaphragm moves closer to the fixed plate, and C increases; when the pressure drops, the diaphragm moves outward, and C decreases. Because the voltage across the capacitor is fixed, the changes in C vary the charge on it, and the result is an electrical output whose variations match those in the incoming sound wave.

The formula $\frac{1}{2}QV$ for the potential energy of a charged capacitor derived in the previous section can be written in three equivalent ways:

$$W = \tfrac{1}{2}QV \qquad \textit{Potential energy of charged capacitor} \qquad (17.15)$$

$$W = \tfrac{1}{2}CV^2 \qquad (17.16)$$

$$W = \frac{1}{2}\frac{Q^2}{C} \qquad (17.17)$$

These equations hold for all capacitors, regardless of their construction.

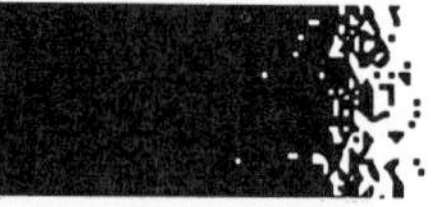

EXAMPLE 17.8

A heart attack often leads to a condition called **fibrillation** in which the heart's actions lose their synchronization and it is unable to pump blood effectively. This condition can often be corrected by an electric shock to the heart that completely stops it for a moment. The heart may then start again spontaneously in its normal rhythm. An appropriate such shock can be provided by the discharge of a 10-μF capacitor that has been charged to a potential difference of 6000 V. (a) How much energy is released in the current pulse? (b) How much charge passes through the patient's body?

SOLUTION

$$\text{(a) } W = \tfrac{1}{2}CV^2 = \tfrac{1}{2}(1.0 \times 10^{-5}\text{ F})(6000\text{ V})^2 = 180\text{ J} = 0.18\text{ kJ}$$

This is the amount of kinetic energy a baseball moving at 50 m/s (113 mi/h) has.

$$\text{(b) } Q = CV = (1.0 \times 10^{-5}\text{ F})(6000\text{ V}) = 0.060\text{ C}$$

◆

17.8 Dielectric Constant

An insulator between its plates reduces the voltage across a charged capacitor.

Let us now see what happens when a slab of an insulating material is placed between the plates of a capacitor.

Although an insulator cannot conduct an appreciable electric current, it can respond to an electric field in another way. The electrons in the molecules of many substances are not uniformly distributed. A molecule of this kind is called a **polar molecule,** and it behaves as though one end were positively charged and the other

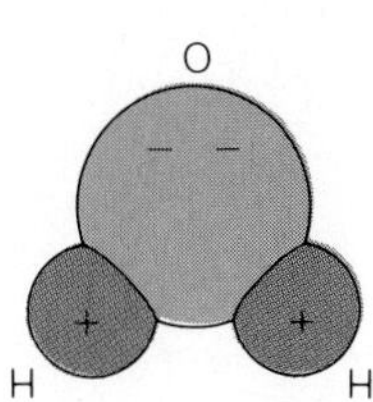

Figure 17.13

The water molecule is polar because the end where the hydrogen atoms are attached behaves as if positively charged and the opposite end behaves as if negatively charged.

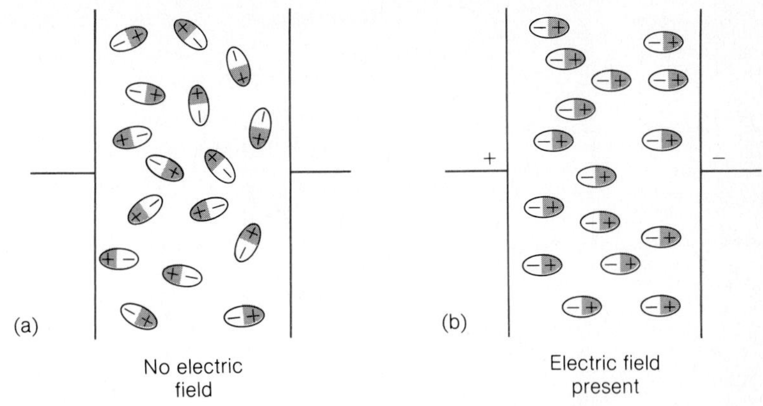

Figure 17.14

An electric field tends to align polar molecules opposite to the field.

negatively charged (Fig. 17.13). In the absence of an external electric field, the molecules in an assembly of polar molecules are oriented randomly, as in Fig. 17.14(a). When an electric field is present, however, the molecules become aligned opposite to the field, as in Fig. 17.14(b).

Although nonpolar molecules ordinarily have symmetric charge distributions, an electric field can distort their arrangements of electrons so that an effective separation of charge takes place (Fig. 17.15). Again the molecules have their charged ends aligned opposite to the external field. In either case, then, the net electric field between the plates of the capacitor is *less* than it would be with nothing between them.

The **dielectric constant,** symbol K, of a substance is a measure of how effective it is in reducing an electric field set up across a sample of it. For a capacitor with a given charge Q, reducing the electric field means reducing V as well. Since $C = Q/V$, this means an increase in capacitance. If the capacitance of a capacitor is C_0 when there is a vacuum between the plates, its capacitance will be

$$C = KC_0 \tag{17.18}$$

when a substance of dielectric constant K is between the plates (Fig. 17.16). Table 17.1 lists dielectric constants for various substances. Water and alcohol molecules are highly polar, and the values of K for water, ice, and ethyl alcohol are accordingly high.

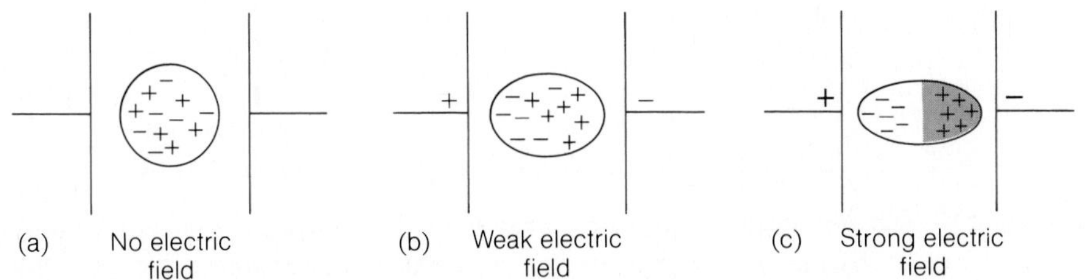

Figure 17.15

An electric field tends to distort the charge distributions in molecules that are ordinarily nonpolar.

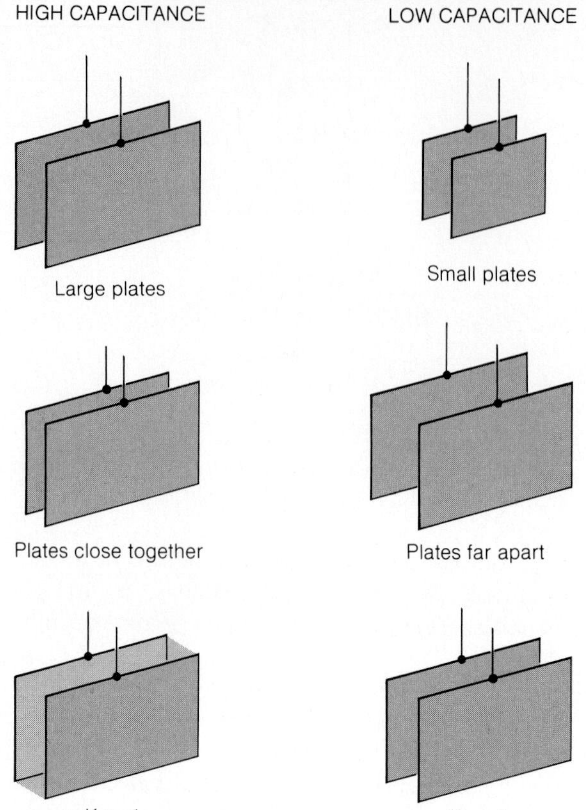

Figure 17.16

The capacitance of a parallel-plate capacitor depends on the area and spacing of its plates and on the dielectric constant K of the medium between them.

Table 17.1 Dielectric constants at 20°C except where noted

Substance	K	Substance	K
Air	1.0006	Mica	2.5–7
Air, liquid (−191°C)	1.4	Neoprene	6.7
Alcohol, ethyl	26	Sulfur	3.9
Benzene	2.3	Teflon	2.1
Glass	5–8	Water	80
Ice (−2°C)	94	Waxed paper	2.2

EXAMPLE 17.9

A capacitor with air between its plates is connected to a 50-V source and then disconnected. The space between the plates of the charged capacitor is filled with Teflon ($K = 2.1$). What is the potential difference across the capacitor now?

SOLUTION The initial charge on the capacitor is $Q = C_1V_1$. When the Teflon dielectric is inserted, the charge remains the same but the capacitance increases to

$C_2 = KC_1$. The new voltage across the capacitor is therefore

$$V_2 = \frac{Q}{C_2} = \frac{C_1 V_1}{KC_1} = \frac{V_1}{K} = \frac{50 \text{ V}}{2.1} = 24 \text{ V}$$

The ratio V_2/V_1 is independent of the original capacitance of the capacitor. ♦

EXAMPLE 17.10

What effect does filling the space between the plates of a capacitor with a material of dielectric constant K have on the energy content of the capacitor? The capacitor is disconnected from the charging circuit before the material is inserted.

SOLUTION A charged capacitor has an energy of $W = Q^2/2C$. The charge Q on each plate does not change when the material is inserted. The capacitance of the capacitor, however, is now $C = KC_0$, where C_0 was its initial value. The new energy W of the capacitor is therefore related to its original energy W_0 by $W = W_0/K$. Because K is greater than 1, energy has been lost, with the missing energy having gone into the alignment and/or distortion of the molecules of the dielectric material. To remove the material would require doing enough work to make up the difference between W and W_0. ♦

Besides its dielectric constant, an important property of a material used between the plates of a capacitor is its **dielectric strength,** which is the maximum electric field that can safely be applied to it before it breaks down and loses its insulating ability. Air has a dielectric strength of about 3×10^6 V/m; thus a voltage of 300 V will produce a spark in an air gap of 0.1 mm. Most materials used as separators in capacitors and to insulate wires have dielectric strengths that exceed 10^7 V/m.

Commercial capacitors consist of many interleaved plates to make possible a high capacitance in a small unit (Fig. 17.17). Solid dielectrics are usually used, both to increase C and to maintain a fixed distance between the plates. With sheets of a solid dielectric such as waxed paper or mica between them, the plates can be of inexpensive metal foil, without the rigidity required to prevent accidental contact if only air separated them.

The capacitance of the variable capacitor in the center can be adjusted by turning its shaft to change the amount of overlap of its metal plates.

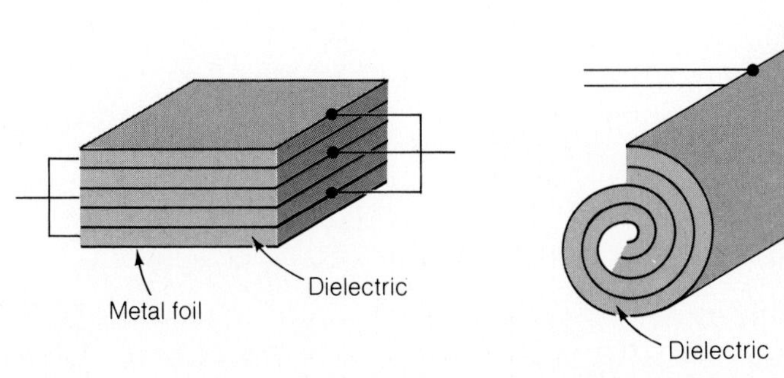

Figure 17.17

Most fixed capacitors are made of interleaved sheets of metal foil separated by layers of dielectric material.

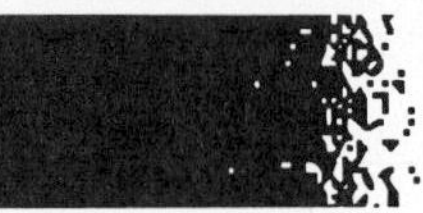

17.9 Capacitors in Combination

Finding the equivalent capacitance of capacitors connected in series and in parallel.

The **equivalent capacitance** of a set of interconnected capacitors is the capacitance of the single capacitor that can replace the set.

Figure 17.18 shows three capacitors in **series,** which means that they are joined end-to-end. Each has charges of the same magnitude Q on its plates, in agreement with the principle of conservation of charge. Hence the potential differences across the capacitors are, respectively,

$$V_1 = \frac{Q}{C_1} \qquad V_2 = \frac{Q}{C_2} \qquad V_3 = \frac{Q}{C_3}$$

If C is the equivalent capacitance of the set and $V = Q/C$ is the potential difference across the set, then

$$V = V_1 + V_2 + V_3$$

$$\frac{Q}{C} = \frac{Q}{C_1} + \frac{Q}{C_2} + \frac{Q}{C_3}$$

$$\frac{1}{C} = \frac{1}{C_1} + \frac{1}{C_2} + \frac{1}{C_3}$$

For any number of capacitors in series,

$$\frac{1}{C} = \frac{1}{C_1} + \frac{1}{C_2} + \frac{1}{C_3} + \cdots \qquad \textit{Capacitors in series} \quad (17.19)$$

The reciprocal of the equivalent capacitance of a series arrangement of capacitors is equal to the sum of the reciprocals of the individual capacitances. Evidently C is smaller than the capacitance of any of the individual capacitors.

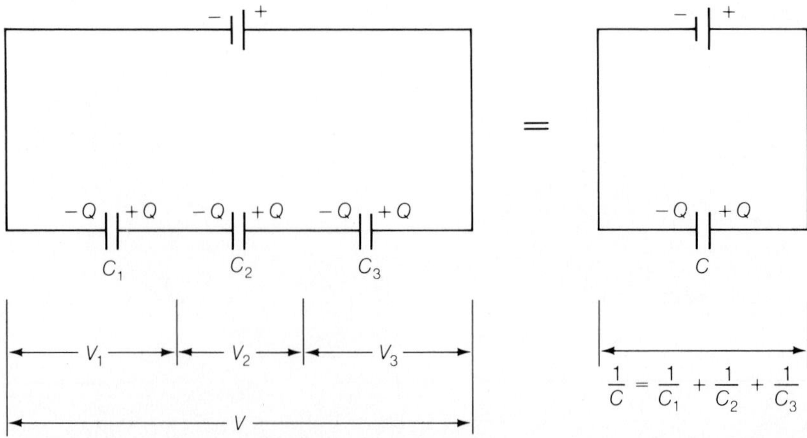

Figure 17.18

Capacitors in series. A charge of the same magnitude Q is present on all the plates of the capacitors. The symbol ⊣⊢ represents a battery and the symbol ⊣⊢ represents a capacitor.

If there are only two capacitors in series,

$$\frac{1}{C} = \frac{1}{C_1} + \frac{1}{C_2} = \frac{C_1 + C_2}{C_1 C_2}$$

and so

$$C = \frac{C_1 C_2}{C_1 + C_2}$$

A calculator with a reciprocal (1/*X*) key makes it easy to solve Eq. (17.19) in an actual problem. If three capacitors are in series, the key sequence would be: [C_1] [1/*X*][+][C_2] [1/*X*][+][C_3][1/*X*][=][1/*X*].

EXAMPLE 17.11

Two capacitors, one of 10 μF and the other of 20 μF, are connected in series across a 12-V battery, as in Fig. 17.19. Find the equivalent capacitance of the combination, the charge on each capacitor, and the potential difference across it.

SOLUTION The equivalent capacitance of the two capacitors is

$$C = \frac{C_1 C_2}{C_1 + C_2} = \frac{(10\ \mu\text{F})(20\ \mu\text{F})}{10\ \mu\text{F} + 20\ \mu\text{F}} = 6.7\ \mu\text{F}$$

and so the charges on them are

$$Q_1 = Q_2 = Q = CV = (6.7 \times 10^{-6}\ \text{F})(12\ \text{V}) = 8.0 \times 10^{-5}\ \text{C}$$

The potential differences across the capacitors are, respectively,

$$V_1 = \frac{Q}{C_1} = \frac{8.0 \times 10^{-5}\ \text{C}}{1.0 \times 10^{-5}\ \text{F}} = 8.0\ \text{V}$$

$$V_2 = \frac{Q}{C_2} = \frac{8.0 \times 10^{-5}\ \text{C}}{2.0 \times 10^{-5}\ \text{F}} = 4.0\ \text{V}$$

The potential difference is greatest across the capacitor of smaller C. The sum of V_1 and V_2 is 12 V, as it should be.

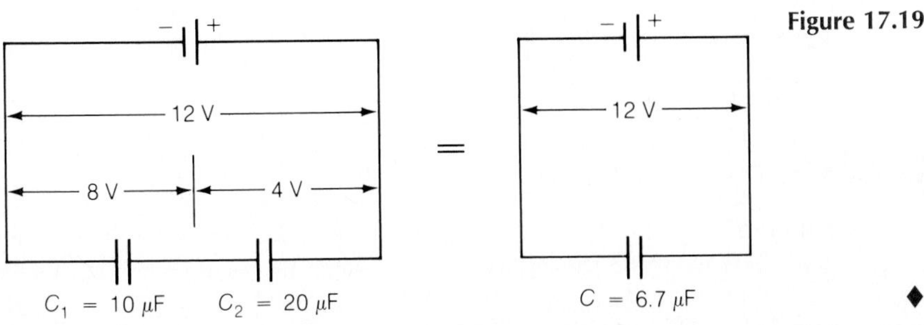

Figure 17.19

◆

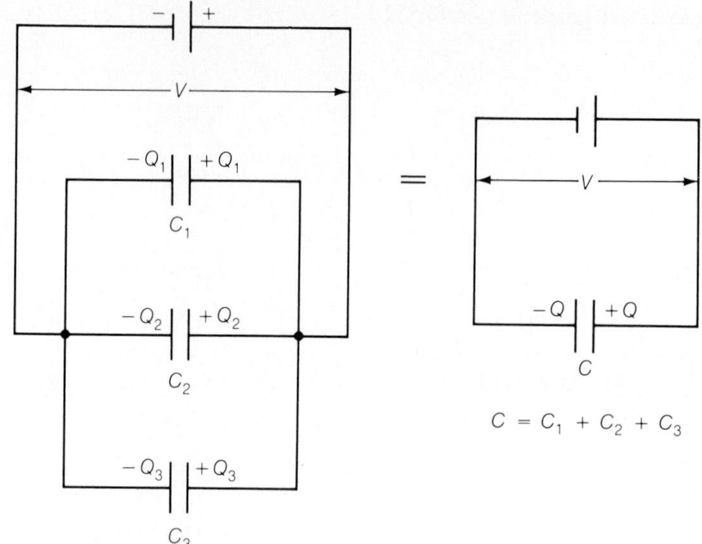

Figure 17.20

Capacitors in parallel. The same potential difference V is across all of them, but each has a charge on its plates whose magnitude is proportional to its capacitance.

Figure 17.20 shows three capacitors connected in **parallel,** which means that each has one end joined to the corresponding end of all the others. The same potential difference is across all of them, so that the charges on their plates have the respective magnitudes

$$Q_1 = C_1V \qquad Q_2 = C_2V \qquad Q_3 = C_3V$$

The total charge $Q_1 + Q_2 + Q_3$ on either the positive or the negative plates of the capacitors is equal to the charge Q on the corresponding plate of the equivalent capacitor. Hence

$$Q = Q_1 + Q_2 + Q_3$$
$$CV = C_1V + C_2V + C_3V$$
$$C = C_1 + C_2 + C_3$$

This result can be generalized to

$$C = C_1 + C_2 + C_3 + \cdots \qquad \textit{Capacitors in parallel} \quad (17.20)$$

More energy is stored in a set of capacitors when they are connected in parallel across a given potential difference than when they are connected in series, because the electric fields in them are stronger in the parallel case.

EXAMPLE 17.12

The capacitors of Fig. 17.19 are reconnected in parallel across the same battery. Compare the energies of the capacitors now with what they were when connected in series.

SOLUTION In the series connection the capacitors had the energies

$$W_1(\text{series}) = \tfrac{1}{2}QV_1 = \tfrac{1}{2}(8.0 \times 10^{-5}\,\text{C})(8.0\,\text{V}) = 3.2 \times 10^{-4}\,\text{J}$$
$$W_2(\text{series}) = \tfrac{1}{2}QV_2 = \tfrac{1}{2}(8.0 \times 10^{-5}\,\text{C})(4.0\,\text{V}) = 1.6 \times 10^{-4}\,\text{J}$$

The capacitor of smaller C had the greater energy. In the parallel connection the capacitors have the same potential difference $V = 12$ V, and their energies are

$$W_1(\text{parallel}) = \tfrac{1}{2}C_1V^2 = \tfrac{1}{2}(1.0 \times 10^{-5}\,\text{F})(12\,\text{V})^2 = 7.2 \times 10^{-4}\,\text{J}$$
$$W_2(\text{parallel}) = \tfrac{1}{2}C_2V^2 = \tfrac{1}{2}(2.0 \times 10^{-5}\,\text{F})(12\,\text{V})^2 = 14 \times 10^{-4}\,\text{J}$$

In this parallel connection the total stored energy is 4.4 times as great as it was in the series connection. Here the capacitor of larger C has the greater energy. ◆

Important Terms

The electric **potential difference** between two points is the work per unit charge that must be done to take a charge from one of the points to the other. The unit of potential difference is the **volt,** which is equal to 1 J/C.

The **electron volt** is the energy acquired by an electron that has been accelerated by a potential difference of 1 V. It is equal to 1.6×10^{-19} J.

The **energy density** of an electric field is the electric potential energy per unit volume associated with it.

A **capacitor** is a device that stores electric energy in the form of an electric field. The ratio between the charge on either plate of a capacitor and the potential difference between the plates is called its **capacitance.** The unit of capacitance is the **farad,** which is equal to 1 coulomb per volt.

The **dielectric constant** K of a particular material is a measure of how effective it is in reducing an electric field set up across a sample of it.

A **polar molecule** is one whose charge distribution is not uniform, so that one end is positive and the other negative even though the molecule as a whole is electrically neutral.

Important Formulas

Potential energy of a charge in uniform field: $\text{PE} = QEd$

Potential energy of two charges: $\text{PE} = k\dfrac{Q_AQ_B}{r}$

Potential difference: $V_{AB} = \dfrac{W_{AB}}{Q} = Ed$ (in uniform electric field)

Energy density of electric field in free space: $w = \dfrac{E^2}{8\pi k}$

Capacitance: $C = \dfrac{Q}{V}$

Parallel-plate capacitor:

$$C = \frac{KA}{4\pi kd} \quad (K = \text{dielectric constant})$$

Potential energy of charged capacitor:

$$W = \tfrac{1}{2}QV = \tfrac{1}{2}CV^2 = \frac{Q^2}{2C}$$

Capacitors in series: $\dfrac{1}{C} = \dfrac{1}{C_1} + \dfrac{1}{C_2} + \dfrac{1}{C_3} + \cdots$

Capacitors in parallel: $C = C_1 + C_2 + C_3 + \cdots$

MULTIPLE CHOICE

1. Of the following quantities, the one that is vector in character is electric

a. charge
b. field
c. energy
d. potential difference

2. The magnitude of the electric field in the region between two parallel oppositely charged metal plates is

a. zero
b. uniform throughout the region
c. greatest near the positive plate
d. greatest near the negative plate

3. From its definition, the unit of electric field E is the N/C. An equivalent unit of E is the

a. V·m b. V·m^2
c. V/m d. V/m^2

4. A system of two charges has a positive potential energy. This signifies that

a. both charges are positive
b. both charges are negative
c. both charges are positive or both are negative
d. one charge is positive and the other is negative

5. An electron and a proton are accelerated through the same potential difference.

a. The electron has the greater KE.
b. The proton has the greater KE.
c. The electron has the greater speed.
d. The proton has the greater speed.

6. The electron volt is a unit of

a. charge
b. potential difference
c. energy
d. momentum

7. The energy content of a charged capacitor resides in its

a. plates b. potential difference
c. charge d. electric field

8. When a slab of insulating material is placed between the plates of a charged capacitor, the electric field there relative to what it was before is

a. less
b. the same
c. more
d. any of the above, depending on the circumstances

9. The farad is not equivalent to which of the following combination of units?

a. C^2/J b. C/V
c. C·V^2 d. J/V^2

10. The plate area of a parallel-plate capacitor of capacitance C is doubled and the distance between the plates is halved, so that the volume of the dielectric remains the same. The new capacitance of the capacitor is

a. $C/4$ b. C
c. $2C$ d. $4C$

11. A parallel-plate capacitor with air between its plates is charged until a potential difference of V appears across it. Another capacitor, having hard rubber (dielectric constant = 3) between its plates but otherwise identical, is also charged to the same potential difference. If the energy of the first capacitor is W, that of the second is

a. $\frac{1}{3}W$ b. W
c. $3W$ d. $9W$

12. An electric field of magnitude 200 V/m can be produced by applying a potential difference of 10 V to a pair of parallel metal plates separated by

a. 20 mm b. 50 mm
c. 20 m d. 2000 m

13. A charge of 10^{-10} C between two parallel metal plates 1 cm apart experiences a force of 10^{-5} N. The potential difference between the plates is

a. 10^{-5} V b. 10 V
c. 10^3 V d. 10^5 V

14. In charging a storage battery, a total of 2.0×10^5 C is transferred from one set of electrodes to another. The potential difference between the electrodes is 12 V. The energy stored in the battery is

a. 1.7×10^4 J b. 2.4×10^6 J
c. 2.4×10^7 J d. 2.9×10^7 J

15. A storage battery is being charged at a rate of 100 W at a potential difference of 13.6 V. The rate at which charge is being transferred between its plates is

a. 0.136 C/s b. 0.54 C/s
c. 7.35 C/s d. 1360 C/s

16. An electron between two metal plates 40 mm apart experiences an acceleration of 1.0×10^{12} m/s. The potential difference between the plates is

a. 0.23 V b. 5.7 V
c. 0.14 kV d. 2.8×10^{12} V

17. An electron whose speed is 1.0×10^7 m/s has a KE of

a. 4.6×10^{-17} eV b. 0.28 keV
c. 0.73 keV d. 5.7 keV

18. An electron whose KE is 150 eV has a speed of

a. 7.3×10^6 m/s b. 5.1×10^7 m/s
c. 2.3×10^8 m/s d. 7.3×10^{13} m/s

19. A capacitor acquires a charge of 0.002 C when connected across a 50-V battery. Its capacitance is

a. 1 μF b. 2 μF
c. 4 μF d. 40 μF

20. A 50-μF capacitor has a potential difference of 8 V across it. Its charge is

a. 4×10^{-3} C b. 4×10^{-4} C
c. 6.25×10^{-5} C d. 6.25×10^{-6} C

21. If a 20-μF capacitor is to have an energy content of 2.5 J, it must be placed across a potential difference of

a. 150 V b. 350 V
c. 500 V d. 0.25 MV

22. A parallel-plate capacitor has a capacitance of 50 pF in air and 110 pF when immersed in turpentine. The dielectric constant of turpentine is

a. 0.45 b. 0.55
c. 1.1 d. 2.2

23. Two 50-μF capacitors are connected in series. The equivalent capacitance of the combination is

a. 25 μF b. 50 μF
c. 100 μF d. 200 μF

24. The capacitor combination in Multiple Choice 23 is connected across a 100-V battery. The potential difference across each capacitor is

a. 25 V b. 50 V
c. 100 V d. 200 V

25. Two 50-μF capacitors are connected in parallel. The equivalent capacitance of the combination is

a. 25 μF b. 50 μF
c. 100 μF d. 200 μF

26. The capacitor combination in Multiple Choice 25 is connected across a 100-V battery. The potential difference across each capacitor is

a. 25 V b. 50 V
c. 100 V d. 200 V

QUESTIONS

1. The potential energy of a certain system of two charges increases as the charges are moved farther apart. What does this tell us about the signs of the charges?

2. The electric potential V at a point can be defined as the potential energy per coulomb of a charge located there relative to its PE at infinity. What is the potential at a distance r from a charge Q?

3. An "electron gun" provides the beam of electrons that traces out the image on the screen of a television picture tube. In an electron gun, electrons are accelerated by a potential difference between a suitable source and an anode (positive electrode) that has a hole through which the beam can pass. What effect does doubling the potential difference have on the speed of the electrons that reach the screen?

4. Why do you think an insulator grows hot when placed in an electric field that reverses its direction rapidly? (This process, called **dielectric heating,** makes it possible to heat a nonmetallic solid to a high temperature evenly with no risk of burning its exterior.)

5. Is there any kind of material that, when inserted between the plates of a capacitor, reduces its capacitance?

6. A sheet of mica whose dielectric constant is 5 is placed between the plates of a charged, isolated parallel-plate capacitor. How is the potential difference across the capacitor affected? How is the charge on the capacitor affected?

EXERCISES

17.2 Potential Energy of Two Charges

1. In the Rutherford experiment described in Section 16.4, alpha particles of charge $+2e$ and kinetic energy 7.7 MeV were directed at a gold foil. The nucleus of a gold atom contains 79 protons, so its charge is $+79e$. Find the closest distance an alpha particle gets to a gold nucleus when it approaches head-on. (The radius of the gold nucleus is about $\frac{1}{6}$ of this distance.)

2. Two electrons are 1.0×10^{-9} m apart when both are released. What is the speed of the electrons when they are 1.0×10^{-8} m apart?

17.3 Potential Difference

3. Twelve joules of work are needed to transfer 2.0 C of charge from one terminal of a storage battery to another. What is the potential difference between the terminals?

4. An electric field stronger than 3.0×10^6 V/m will cause sparks to occur in air. What is the maximum potential difference that can be applied across two metal plates 1.0 mm apart before sparking begins?

5. A potential difference of 50 V is applied across two parallel metal plates, and an electric field of 1.0×10^4 V/m is produced. How far apart are the plates?

6. The potential difference between two parallel metal plates that are 50 mm apart is 1.0×10^4 V. Find the force on an electron located between the plates.

7. A cloud is at a potential of 8.0×10^6 V relative to the ground. A charge of 40 C is transferred in a lightning stroke between the cloud and the ground. Find the energy dissipated.

8. The electrodes in a neon sign are 1.2 m apart, and the potential difference across them is 8000 V. (a) Find the acceleration of a neon ion of mass 3.3×10^{-26} kg and charge $+e$ in the field. (b) If the ion starts at the positive electrode of the sign and moves unimpeded to its negative electrode, how much energy would it gain? (c) Why is it extremely unlikely that the ion would actually acquire this much energy?

9. Two parallel metal plates are 40 mm apart. If the force on an electron between the plates is to be 1.0×10^{-14} N, what should the potential difference between them be?

10. The storage battery of a car is being charged by an alternator at the rate of 10 C/s. The potential difference across the alternator's terminals is 14 V, and the alternator is 90% efficient. Find the power supplied by the car's engine to charge the battery.

11. In charging a 20-kg storage battery, a total of 2.0×10^5 C is transferred from one set of electrodes to another. The potential difference between the electrodes is 14 V. (a) How much energy is stored in the battery? (b) If this energy were used to raise the battery above the ground, how high would it go? (c) If this energy were used to provide the battery with kinetic energy, what would its speed be?

12. A potential difference of 20 kV is used to accelerate electrons in a television picture tube. (The impact of these electrons on the tube's screen produces the flashes of light that make up the picture we see.) How much kinetic energy do the electrons acquire? What is their speed?

17.4 The Electron Volt

13. What is the kinetic energy in electron volts of a potassium atom of mass 6.5×10^{-26} kg when its speed is 1.0×10^6 m/s?

14. What is the kinetic energy in electron volts of an electron whose speed is 1.0×10^6 m/s?

15. What is the speed of an electron whose kinetic energy is 50 eV?

16. What is the speed of a neutron whose kinetic energy is 50 eV?

17. What potential difference is needed between two metal plates 20 mm apart to accelerate an electron to a speed of 5.0×10^6 m/s? What will the KE of the electron be, expressed in electron volts?

17.5 The Action Potential

18. As mentioned in the text, the stimulation of a nerve or muscle cell leaves it with an excess of Na^+ (sodium) ions that must be "pumped" out through the cell wall against a potential difference of 90 mV. If the pumping rate per cell is 1.0×10^9 ions/s, how much power is needed?

17.6 Electric Field Energy

19. A potential difference of 300 V is applied across a pair of parallel metal plates 10 mm apart. What is the energy density of the electric field between the plates?

20. The electric field near the earth's surface is about 100 V/m. How much electric energy is stored in the lowest kilometer of the atmosphere?

17.7 Capacitance

21. A 25-μF capacitor is connected to a source of potential difference of 1000 V. What is the resulting charge on the capacitor? How much energy does it contain?

22. What is the potential difference between the plates of a 20-μF capacitor whose charge is 0.010 C? How much energy does it contain?

23. What potential difference must be applied across a 10-μF capacitor if it is to have an energy content of 1.0 J?

24. A 0.50-μF capacitor has plates 0.50 mm apart in air. What is the maximum charge the capacitor can have before sparks occur between the plates?

25. The plates of a parallel-plate capacitor are 50 cm^2 in area and 1.0 mm apart. (a) What is its capacitance? (b) When the capacitor is connected to a 45-V battery, what is the charge on either plate? (c) What is the energy of the charged capacitor?

26. A parallel-plate capacitor with plates 0.20 mm apart in air has a charge on each plate of 4.0×10^{-8} C when the potential difference is 250 V. Find its capacitance, the area of each plate, and the stored energy.

27. A 200-μF capacitor requires an energy of 50 J to fire a flashlamp. What voltage is needed to charge the capacitor? How much charge passes through the flashlamp?

28. A variable capacitor set at 200 pF is connected to a 100-V battery. The battery is then disconnected, and the capacitor is adjusted to a capacitance of 10 pF. (a) What is the potential difference across the capacitor now? (b) In the absence of friction, how much work had to be done to change the capacitor to the new setting?

17.8 Dielectric Constant

29. The capacitance of a parallel-plate capacitor is increased from 8.0 μF to 50.0 μF when a sheet of glass is inserted between its plates. What is the dielectric constant of the glass?

30. A capacitor with air between its plates is connected to a battery, and each of its plates receives a charge of 1.0×10^{-4} C. While still connected to the battery, the capacitor is immersed in oil, and a further charge of 1.0×10^{-4} C is added to each plate. What is the dielectric constant of the oil?

31. The space between the plates of the capacitor in Exercise

25 is filled with sulfur. Answer the same questions for this case.

17.9 Capacitors in Combination

32. Find the equivalent capacitance of a 20-μF capacitor and a 50-μF capacitor that are connected in series.

33. Find the equivalent capacitance of a 20-μF capacitor and a 50-μF capacitor that are connected in parallel.

34. List the capacitances that can be obtained by combining three 10-μF capacitors in all possible ways.

35. Three capacitors whose capacitances are 5.0, 10.0, and 50.0 μF are connected in series across a 12.0-V battery. Find the charge on each capacitor and the potential difference across it.

36. Three capacitors whose capacitances are 2.0, 4.0, and 5.0 μF are connected in series across a 100-V battery. Find the charge on each capacitor and the potential difference across it.

37. The three capacitors in Exercise 35 are connected in parallel across the same battery. Find the charge on each capacitor and the potential difference across it.

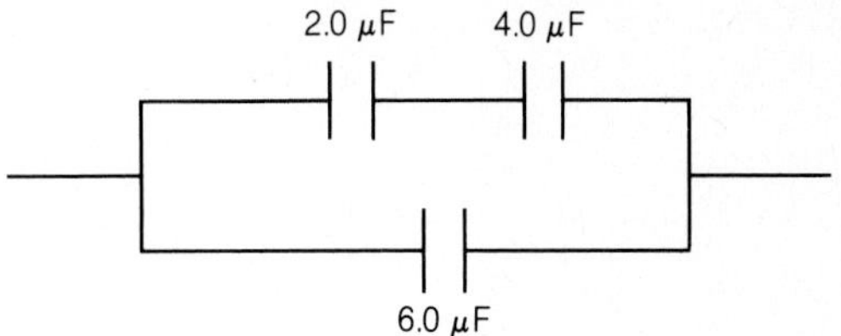

Figure 17.21

Exercise 39.

38. The three capacitors in Exercise 36 are connected in parallel across the same battery. Find the charge on each capacitor and the potential difference across it.

39. (a) Find the equivalent capacitance of the system shown in Fig. 17.21. (b) The dielectric of the 2.0-μF capacitor breaks down and becomes conducting. Find the equivalent capacitance of the system now. (c) If the dielectric of the 6.0-μF capacitor had broken down instead, what would the equivalent capacitance of the system be?

PROBLEMS

40. A potential difference of 1000 V is applied across two parallel metal plates 100 mm apart. An electron leaves the negative plate at the same time as a proton leaves the positive plate. (a) Find the speeds and kinetic energies of the particles when they reach the opposite plates. (b) At what distance from the positive plate do the electron and proton pass each other?

41. A potential difference of 10 V is applied across two parallel metal plates 20 mm apart. An electron is projected at a speed of 1.0×10^7 m/s halfway between the plates and parallel to them. How far will the electron travel before striking the positive plate?

42. The "electron gun" of a television picture tube has an accelerating potential difference of 15 kV and a power rating of 25 W. How many electrons reach the screen per second? At what speed?

43. A parallel-plate capacitor of capacitance C is given the charge Q and then disconnected from the circuit. How much work is required to pull the plates of this capacitor to twice their original separation?

44. A 1.0-μF capacitor and a 2.0-μF capacitor are each charged across a potential difference of 1200 V. The capacitors are disconnected from the source of the potential difference and are then connected with terminals of the same sign together. What is the final charge of each capacitor?

45. The charged capacitors in Problem 44 are instead connected with terminals of opposite sign together. Now what is the final charge of each capacitor?

Answers to Multiple Choice

1. b	**8.** a	**15.** c	**22.** d
2. b	**9.** c	**16.** a	**23.** a
3. c	**10.** d	**17.** b	**24.** b
4. c	**11.** c	**18.** a	**25.** c
5. c	**12.** b	**19.** d	**26.** c
6. c	**13.** c	**20.** b	
7. d	**14.** b	**21.** c	

C H A P T E R 18

ELECTRIC CURRENT

An electric current consists of a flow of charge from one place to another. Currents, not static charges, are involved in nearly all practical applications of electricity. In an electric circuit, current is the means by which energy is transferred from a source such as a battery or a generator to a load. The load may be a lamp, a motor, or any other device that absorbs electric energy and converts it into some other form of energy or into work. In this chapter we consider the chief factors that govern direct currents in simple circuits. In later chapters the magnetic effects of currents and how they are used in technology will be examined. The important topic of alternating current is the subject of Chapter 21.

18.1 Electric Current

A conducting path and a potential difference are both needed for a current to occur.

The magnitude of an electric current, denoted I, is the rate at which charge passes a given point. If the net charge Q goes past in the time interval t, then the average current is

$$I = \frac{Q}{t} \qquad \textit{Electric current} \quad (18.1)$$

$$\text{Current} = \frac{\text{charge}}{\text{time interval}}$$

The unit of electric current is the **ampere** (A), where

$$1 \text{ ampere} = 1 \text{ coulomb/second}$$

The current in a 100-W light bulb is a little less than 1 A.

The direction of a current is, by convention, taken as that in which *positive* charges would have to move in order to produce the same effects as the observed current. Thus a current is always assumed to proceed from the positive terminal of a battery or a generator to its negative terminal in an external circuit (Fig. 18.1).

Despite this convention, actual electric currents in metals consist of flows of electrons, which carry negative charges. However, a current that consists of negative particles moving in one direction is electrically the same as a current that consists of positive particles moving the other way. Since there is no overwhelming reason to prefer one way of designating current to the other, we will follow the usual practice of considering current as a flow of positive electric charge.

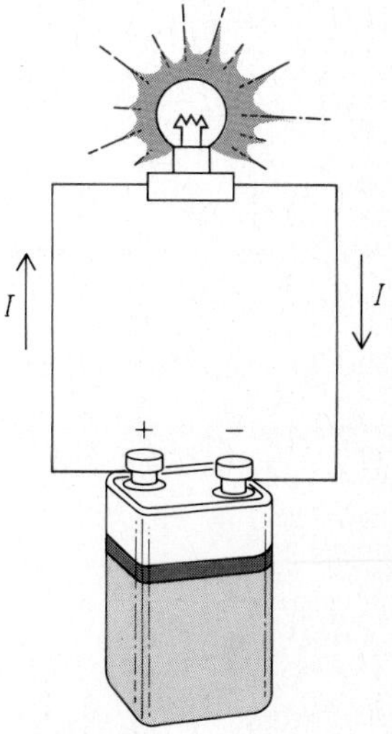

Figure 18.1

By convention, an electric current is assumed to flow from the positive terminal of a battery or generator to its negative terminal in an external circuit. Actual currents in metals consist of electrons that move in the opposite direction. A complete conducting path such as this one is called a *circuit*.

EXAMPLE 18.1

About 10^{20} electrons/cm participate in conducting electric current in a certain wire. That is, about 10^{20} electrons in each centimeter of the wire are moving when the wire is carrying a current. What is the average speed of the electrons when there is a current of 1 A in the wire?

SOLUTION Each electron has a charge of $e = 1.6 \times 10^{-19}$ C. Because there are 10^{20} electrons per centimeter in the wire, the number per meter is 10^{22} and the charge in motion per meter in a length x of the wire is $Q/x = 10^{22}\ e/\text{m} = 1.6 \times 10^{3}$ C/m. The time needed for the electrons in a length x of the wire to pass a given point at the average speed v is $t = x/v$. The current in the wire is therefore

$$I = \frac{Q}{t} = \frac{Qv}{x} \quad \text{and} \quad v = \frac{I}{(Q/x)} = \frac{1\text{ A}}{1.6 \times 10^{3}\text{ C/m}} = 6 \times 10^{-4}\text{ m/s}$$

This is less than 1 mm/s, slower than a tired caterpillar. The average speed is so low because the moving electrons, which are accelerated by the electric field in the wire, collide frequently with atoms in their paths and have to start over again each time. ♦

The following two conditions must be met in order for an electric current to exist between two points:

1. There must be a conducting path between the two points along which charge can flow.
2. There must be a difference of potential between the two points. (A superconductor is an exception to this requirement.) Just as the rate of flow of water between the ends of a pipe depends on the difference of pressure between them, so the rate of flow of charge between two points depends on the difference of potential between them (Fig. 18.2). A large potential difference means a large "push" given to each charge.

A particular conducting path—for instance, a copper wire, a light bulb, an electric heater—is usually called a conductor, even though this is also the name of the class of substances through which current flows readily. The **resistance** of a conductor is the ratio between the potential difference V across it and the resulting current I that flows:

$$R = \frac{V}{I} \qquad \textit{Resistance} \quad (18.2)$$

$$\text{Resistance} = \frac{\text{potential difference}}{\text{current}}$$

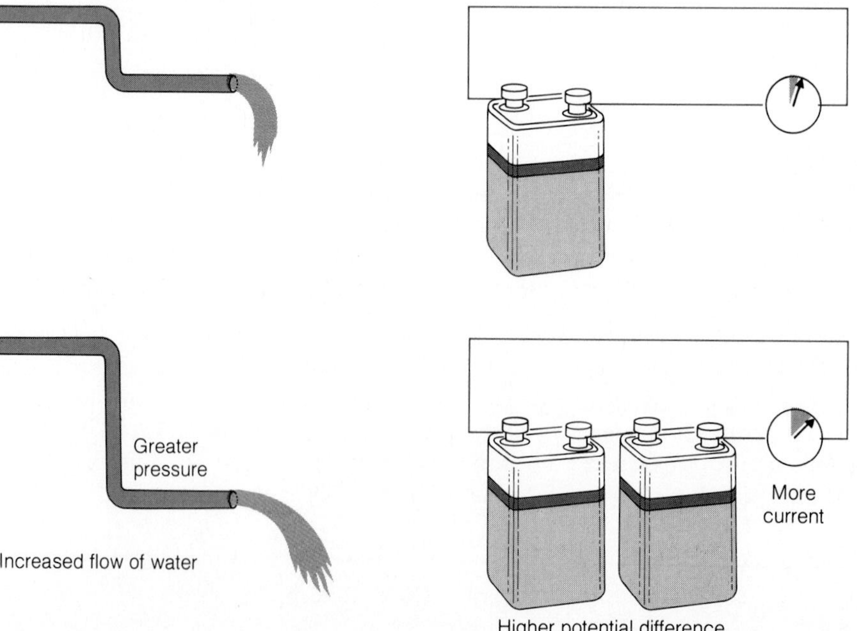

Figure 18.2

The role of potential difference in producing an electric current is analogous to the role of pressure in producing a flow of water.

The unit of resistance is the **ohm** (Ω), where

$$1 \text{ ohm} = 1 \text{ volt/ampere}$$

A conductor in which there is a current of 1 A when a potential difference of 1 V exists across it has a resistance of 1 Ω. (The symbol Ω is the Greek capital letter *omega*.)

EXAMPLE 18.2

A 120-V electric heater draws a current of 15 A. Find its resistance.

SOLUTION From the definition of resistance

$$R = \frac{V}{I} = \frac{120 \text{ V}}{15 \text{ A}} = 8\ \Omega$$

◆

18.2 Ohm's Law

Current is proportional to voltage in many substances.

The resistance of a conductor depends in general both on its properties—its nature and its dimensions—and on the potential difference applied across it. In some conductors R increases when V increases, in others R decreases when V increases, and in still others R depends on the direction of the current (Fig. 18.3).

In a conductor whose resistance is constant, the current is proportional to the voltage:

$$I = \frac{V}{R} \quad (R = \text{constant}) \qquad \textit{Ohm's law} \quad (18.3)$$

This relationship is called **Ohm's law** because it was first verified experimentally by the German physicist Georg Ohm (1787–1854). Despite its name, Ohm's law is not a fundamental physical principle. Rather, it is a relationship that describes the electrical

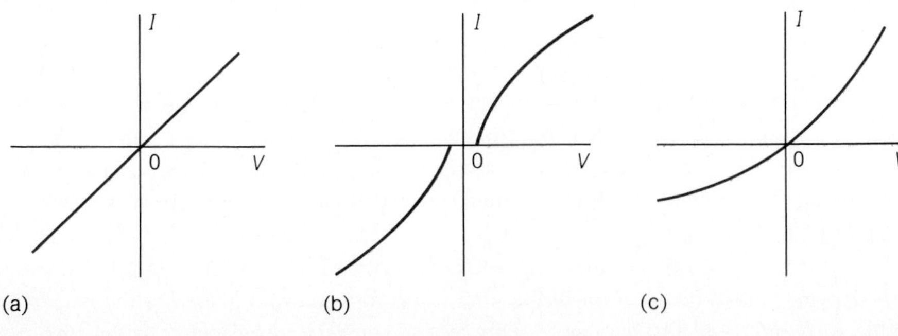

Figure 18.3

The relationship between current and voltage for (a) a metal, (b) a solution of NaCl (table salt), and (c) a semiconductor diode. Only in (a) is I proportional to V, a relationship known as Ohm's law.

behavior of many substances, notably most metals, over a wide range of voltages and currents.

Ohm's law is not the same as the definition of resistance, $R = V/I$. Ohm's law holds only for conductors in which the ratio V/I is constant; such conductors are said to be **ohmic.** On the other hand, the resistance of a conductor is given by $R = V/I$ whether or not it is constant as the voltage is changed.

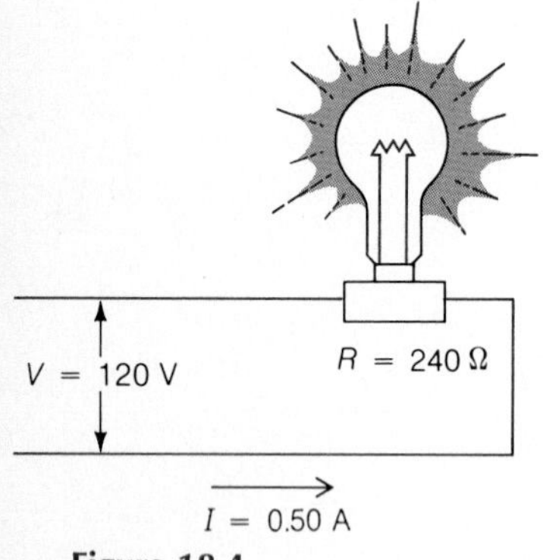

Figure 18.4

EXAMPLE 18.3

A light bulb has a resistance of 240 Ω. Find the current in it when it is placed in a 120-V circuit (Fig. 18.4).

SOLUTION From Ohm's law

$$I = \frac{V}{R} = \frac{120\text{ V}}{240\ \Omega} = 0.5\text{ A}$$

◆

Electrical Hazards

An electric current affects body tissue both by stimulating nerves and muscles and through the heat produced. Tissue is a fairly good conductor because of the ions in solution it contains. Dry skin has more resistance and can protect the rest of the body to some extent in case of accidental exposure to a high voltage, protection that disappears when the skin is wet. A current of as little as 1 mA (0.001 A) can be felt by most people, one of 5 to 10 mA is painful, and one of 15 mA or more causes muscle contractions that may prevent the person involved from letting go of the source of the current. Breathing may then become impossible.

Because a closed conducting path is necessary for a current to occur, touching a single "live" conductor has no effect if the body is isolated. However, if a person at the same time is in contact with a water pipe, is standing on wet soil, or otherwise is grounded, a current will pass through his or her body. The resistance of the human body itself is of the order of magnitude of 1000 Ω, so contact via wet skin with a 120-V line will lead to a current in the neighborhood of $I = 120\text{ V}/1000\ \Omega = 0.12\text{ A} = 120\text{ mA}$. Such a current is extremely dangerous because it is likely to cause the heart muscles to contract rapidly and irregularly and, if allowed to persist, will cause death.

The water in a bathtub is grounded via the tub's drainpipe, so a person in the tub is at risk if he or she touches any electrical device, even a switch; the moisture on a wet finger may be sufficient to provide a conducting path to the interior of the device. Another situation of great potential danger occurs in hospitals where electrical appliances are often attached to patients to monitor various functions or to control them, as in the case of cardiac pacemakers. Even a minor malfunction of such an appliance, or the failure to ground it properly, may have fatal consequences because the conducting path in the body is then short compared with what it is when the contact is made via a finger.

It is sometimes thought that a fuse or a circuit breaker in a circuit eliminates any danger. Such a device "opens" a circuit when the current is greater than a certain value and thus prevents damage to the equipment being used or to the wiring that carries the current. For instance, a 15-A fuse contains a length of special wire that melts when the current is 15 A or more. Since only about 50 mA (1/300 of 15 A) is enough to kill, fuses and circuit breakers do not protect people. A device called a ground-fault circuit interrupter, however, will open a circuit when there is a leakage current to ground of as little as 5 mA, and this does offer useful protection.

18.3 Resistivity

A measure of the ability of a substance to conduct electric current.

The resistance of a conductor that obeys Ohm's law depends on three factors:

1. The material of which it is composed: The ability to carry an electric current varies more than almost any other physical property of a material.
2. Its length L: The longer the conductor, the greater its resistance.
3. Its cross-sectional area A: The thicker the conductor, the less its resistance.

Once again we note the correspondence to water flowing through a pipe. The longer the pipe, the more friction against the pipe wall slows the water, and the wider the pipe, the larger the volume of water that can pass through per second when everything else is the same (Fig. 18.5).

The simple formula

$$R = \rho\frac{L}{A} \qquad \textit{Resistance of ohmic conductor} \quad (18.4)$$

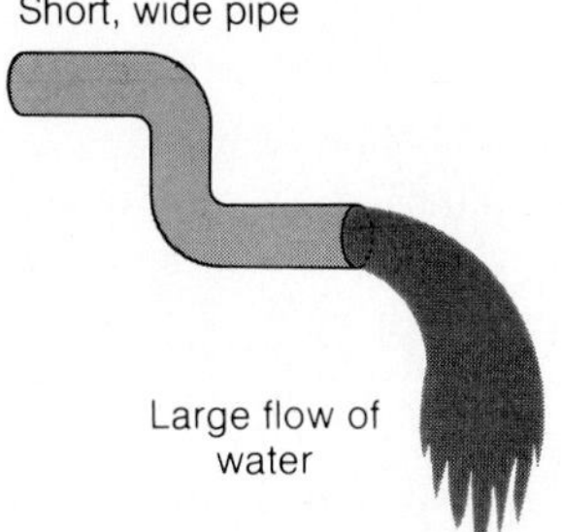

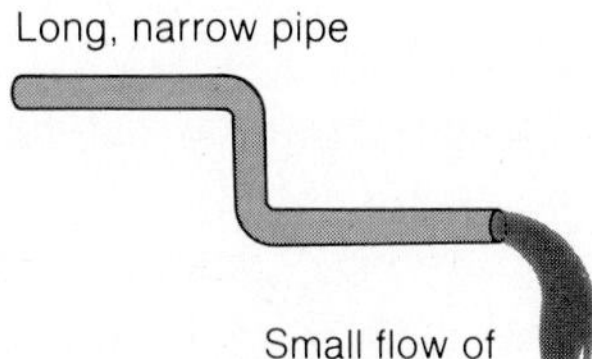

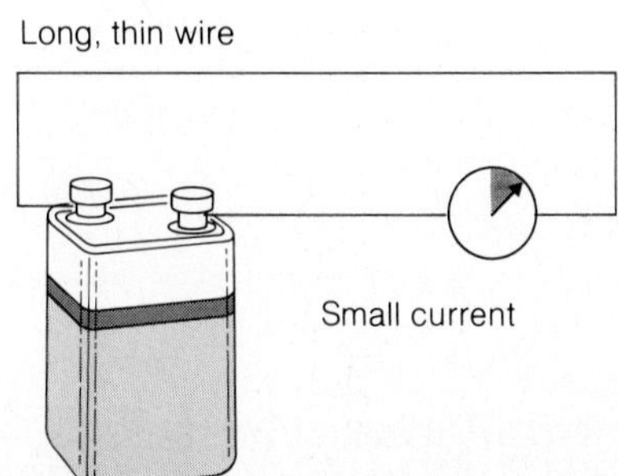

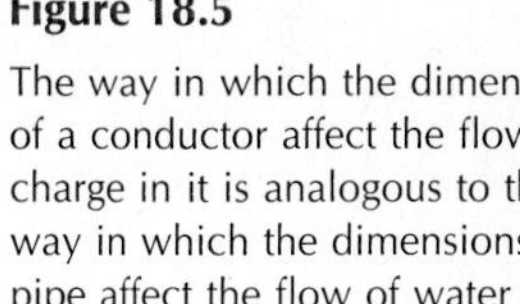

Figure 18.5

The way in which the dimensions of a conductor affect the flow of charge in it is analogous to the way in which the dimensions of a pipe affect the flow of water in it.

Table 18.1 Approximate resistivities (at 20°C) and their temperature coefficients

Substance	ρ ($\Omega\cdot$m)	α (/°C)
Conductors		
Aluminum	2.6×10^{-8}	0.0039
Constantan (60% Cu, 40% Ni)	49×10^{-8}	0.000002
Copper	1.7×10^{-8}	0.0039
Iron	12×10^{-8}	0.0050
Lead	21×10^{-8}	0.0043
Manganin (84% Cu, 12% Mn, 4% Ni)	44×10^{-8}	0.000000
Mercury	98×10^{-8}	0.00088
Platinum	11×10^{-8}	0.0036
Silver	1.6×10^{-8}	0.0038
Semiconductors		
Carbon	3.5×10^{-5}	-0.0005
Germanium	0.5	
Copper oxide (CuO)	1×10^{3}	
Insulators		
Glass	$10^{10} - 10^{14}$	
Quartz	7.5×10^{17}	
Sulfur	10^{15}	

has been found to hold for the resistance of a conductor that obeys Ohm's law. The quantity ρ (the Greek letter *rho*) is called the **resistivity** of the material from which the conductor is made. Table 18.1 lists the resistivities of various substances at room temperature (20°C). Given the nature of a conductor and its dimensions, the value of R can be calculated at once.

In the SI system, lengths are measured in meters and areas in square meters, and the unit of resistivity is accordingly the ohm-meter ($\Omega\cdot$m). Metric wire sizes are usually specified by their cross-sectional areas in square millimeters, so a particular wire might be referred to as 2.5 mm^2, for instance, instead of having its diameter of 1.8 mm quoted.

EXAMPLE 18.4

A water heater draws 30 A from a 120.0-V power source 10 m away. What is the minimum cross section of the wire that can be used if the voltage applied to the heater is not to be lower than 115.0 V?

SOLUTION The permissible voltage drop is 5.0 V, and the resistance that corresponds to this drop when the current is 30 A is

$$R = \frac{V}{I} = \frac{5.0\text{ V}}{30\text{ A}} = 0.167\ \Omega$$

The total length of wire involved is twice the distance between the source and the heater, so $L = (2)(10 \text{ m}) = 20$ m. From Eq. (18.4) we have

$$A = \frac{\rho L}{R} = \frac{(1.7 \times 10^{-8} \Omega\cdot\text{m})(20 \text{ m})}{0.167 \ \Omega} = 2.0 \times 10^{-6} \text{ m}^2 = 2.0 \text{ mm}^2$$ ◆

The resistivities of most substances vary with temperature. In the case of a metal, this is because its atoms vibrate with greater amplitude as the temperature rises. As a result they interfere to a greater extent with the motion of the free electrons that constitute a current in the metal.

If R is the resistance of a conductor at a particular temperature, then the change ΔR in its resistance when the temperature changes by ΔT is approximately proportional to both R and ΔT, and therefore

$$\Delta R = \alpha R \ \Delta T \tag{18.5}$$

The quantity α is the temperature coefficient of resistivity of the material. In Table 18.1 the temperature coefficient of carbon is labeled negative because its resistivity decreases with increasing temperature.

EXAMPLE 18.5

A **resistance thermometer** makes use of the temperature variation of resistivity. When a coil of platinum wire whose resistance at 20°C is 11 Ω is placed in a furnace, its resistance triples. Find the temperature of the furnace, assuming that α remains constant.

SOLUTION Since $R = 11 \ \Omega$ and $\Delta R = 33 \ \Omega - 11 \ \Omega = 22 \ \Omega$,

$$\Delta T = \frac{\Delta R}{\alpha R} = \frac{22 \ \Omega}{(0.0036/°\text{C})(11 \ \Omega)} = 556°\text{C}$$

The temperature of the furnace is $T + \Delta T = 20°\text{ C} + 556°\text{C} = 576°\text{C}$. ◆

18.4 Electric Power

The product of current and voltage.

Electric energy in the form of electric current is converted into heat in an electric stove, into radiant energy in a light bulb, into chemical energy when a storage battery is charged, and into mechanical energy in an electric motor. The widespread use of electric energy is due as much to the ease with which it can be transformed into other kinds of energy as to the ease with which it can be carried through wires.

The work that must be done to take a charge Q through the potential difference V is, by definition,

$$W = QV$$

Since a current I carries the amount of charge $Q = It$ in the time t, the work done is

$$W = IVt \qquad \textit{Electric work} \quad (18.6)$$

The energy input to a device of any kind through which the current I flows when the potential difference V is placed across it is equal to the product of the current, the potential difference, and the time interval.

Power is the rate at which work is being done: $P = W/t$. When the work is done by an electric current, $W = IVt$, so

$$P = IV \qquad \textit{Electric power} \quad (18.7)$$

$$\text{Electric power} = (\text{current})(\text{potential difference})$$

The unit of power is the watt, and when I and V are in amperes and volts, respectively, P will be in watts.

EXAMPLE 18.6

A solar cell 10 cm in diameter produces a current of 2.15 A at 0.45 V in bright sunlight whose intensity is 0.10 W/cm^2. Find the efficiency of the cell.

SOLUTION The area of the cell is $A = \pi d^2/4 = 78.5\ \text{cm}^2$, so it receives solar energy at the rate of $P_{\text{input}} = (0.10\ \text{W/cm}^2)(78.5\ \text{cm}^2) = 7.85\ \text{W}$. The cell's output is

$$P_{\text{output}} = IV = (2.15\ \text{A})(0.45\ \text{V}) = 0.97\ \text{W}$$

The efficiency of the cell is therefore

$$\text{Eff} = \frac{P_{\text{output}}}{P_{\text{input}}} = \frac{0.97\ \text{W}}{7.85\ \text{W}} = 0.12 = 12\%$$

◆

EXAMPLE 18.7

A 1.35-V mercury cell with a capacity of 1.5 A-h is to be used in a cardiac pacemaker. (a) If the power required is 0.10 mW, how long will the cell last? (The capacity of a battery is usually given in terms of the total amount of charge it can circulate, expressed in ampere-hours. Thus a 1.5 A-h battery can supply a current of 1.5 A for 1 h, a current of 0.5 A for 3 h, and so on.) (b) What will the average current be?

SOLUTION (a) The amount of charge that corresponds to 1.5 A-h is

$$Q = (1.5\ \text{A-h})(3600\ \text{s/h}) = 5400\ \text{A-s} = 5400\ \text{C}$$

The total work the cell can perform is therefore

$$W = QV = (5400\ \text{C})(1.35\ \text{V}) = 7290\ \text{J}$$

At a power of $P = 0.10 \text{ mW} = 1.0 \times 10^{-4}$ W, the time involved is

$$t = \frac{W}{P} = 7.29 \times 10^7 \text{ s} = 844 \text{ days} = 2.3 \text{ years}$$

(b) From $P = IV$ the average current delivered by the cell is

$$I = \frac{P}{V} = \frac{1.0 \times 10^{-4} \text{ W}}{1.35 \text{ V}} = 7.4 \times 10^{-5} \text{ A} = 74 \ \mu\text{A}$$ ◆

The power consumed by a resistance that obeys Ohm's law ($I = V/R$) may be expressed in the alternative forms

$$P = IV \qquad \textit{Electric power} \quad (18.7)$$

$$P = I^2R \qquad (18.8)$$

$$P = \frac{V^2}{R} \qquad (18.9)$$

Equation (18.7) holds regardless of the nature of the current-carrying device. Depending on which quantities are known in a specific case, any of these expressions for P may be used.

EXAMPLE 18.8

Find the power consumed by a 240-Ω light bulb when the current through it is 0.50 A.

SOLUTION The formula $P = I^2R$ is easiest to use here. We have

$$P = I^2R = (0.50 \text{ A})^2(240 \ \Omega) = 60 \text{ W}$$ ◆

Owing to the resistance that all conductors offer, electric power is dissipated whenever a current exists, regardless of whether the current also supplies energy that is converted to some other form. Electrical resistance is much like friction: The power consumed in causing a current to flow is dissipated as heat. If too much current flows in a particular wire, it becomes so hot that it may start a fire or even melt. To prevent this from happening, nearly all electric circuits are protected by fuses or circuit breakers, as mentioned earlier, which interrupt the current when I exceeds a safe value. For example, a 15-A fuse in a 120-V power line means that the maximum power that can be carried is

$$P = IV = (15 \text{ A})(120 \text{ V}) = 1800 \text{ W} = 1.8 \text{ kW}$$

The maximum safe current in a wire depends both on how its temperature varies with current and on the nature of its insulation. The larger the diameter of a wire, the more current it can safely carry, partly because its resistance is lower and partly because it has more surface area to dissipate heat. Tables have been compiled that give the allowable currents for copper wires of standard sizes. For example, rubber-covered No. 14 wire (diameter 1.63 mm), the smallest size permitted by the National Electrical Code for use in homes, farms, and industry, is limited to 20 A.

18.5 Resistors in Series

Their equivalent resistance is the sum of the individual resistances.

In analyzing a circuit, it is convenient to group its various components into individual **resistors** that we imagine as joined together by resistanceless wires. The **equivalent resistance** of a set of interconnected resistors is the value of the single resistor that can replace the entire set without affecting the current in the rest of the circuit.

The symbol for a resistor, —⋀⋁⋀⋁—, represents any circuit component that has electrical resistance and obeys Ohm's law. Although there are an unlimited number of ways in which resistors can be put together, many of them are merely combinations of the basic **series** and **parallel** arrangements (Fig. 18.6). Resistors in series are joined end-to-end, so the same current is present in all of them. On the other hand, resistors in parallel have their ends connected together, so the total current is split up among them.

The potential difference V across the ends of a series set of resistors is the sum of the potential differences V_1, V_2, V_3, . . . across each one. This statement follows from the principle of conservation of energy: If V is the work done per coulomb in pushing a charge through the set of resistors, then it must equal the sum $V_1 + V_2 + V_3 + \cdots$ of the work done per coulomb in pushing the charge through each resistor in turn. In the case of the three resistors in series in Fig. 18.7,

$$V = V_1 + V_2 + V_3$$

Because the same current I passes through all the resistors, the individual potential drops are

$$V_1 = IR_1 \qquad V_2 = IR_2 \qquad V_3 = IR_3$$

Figure 18.6

The symbol of a resistor is —⋀⋁⋀⋁—.

Resistors in series

I R_1 I R_2 I R_3 I

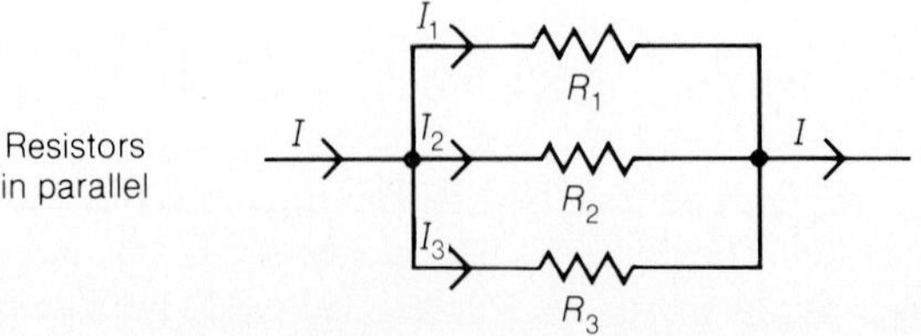

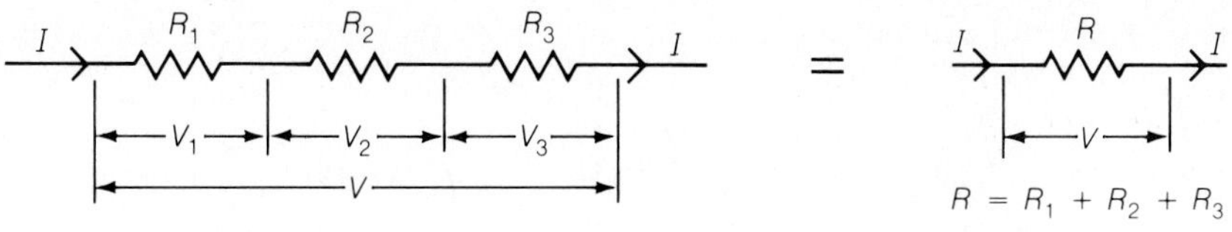

Figure 18.7

Resistors in series. The same current I passes through each resistor.

If we let R be the equivalent resistance of the set, the potential difference across it is

$$V = IR$$

We therefore have

$$V = V_1 + V_2 + V_3$$
$$IR = IR_1 + IR_2 + IR_3$$

Dividing through by the current I gives

$$R = R_1 + R_2 + R_3$$

In general, *the equivalent resistance of a set of resistors connected in series is equal to the sum of the individual resistances*:

$$R = R_1 + R_2 + R_3 + \cdots \qquad \textit{Resistors in series} \quad (18.10)$$

EXAMPLE 18.9

A 5.0-Ω resistor and a 20.0-Ω resistor are connected in series, and a potential difference of 100 volts is applied across them by means of a generator. Find (a) the equivalent resistance of the circuit, (b) the current that flows in it, (c) the potential difference across each resistor, (d) the power dissipated by each resistor, and (e) the power dissipated by the entire circuit.

SOLUTION Successive stages in the solution of this problem are shown in Fig. 18.8.

(a) The equivalent resistance of the two resistors is

$$R = R_1 + R_2 = 5.0\ \Omega + 20.0\ \Omega = 25.0\ \Omega$$

This resistance is more than that of either of the resistors.

(b) The current in the circuit is, from Ohm's law,

$$I = \frac{V}{R} = \frac{100\ \text{V}}{25.0\ \Omega} = 4.0\ \text{A}$$

(c) The potential differences across R_1 and R_2 are

$$V_1 = IR_1 = (4.0\ \text{A})(5.0\ \Omega) = 20\ \text{V}$$
$$V_2 = IR_2 = (4.0\ \text{A})(20\ \Omega) = 80\ \text{V}$$

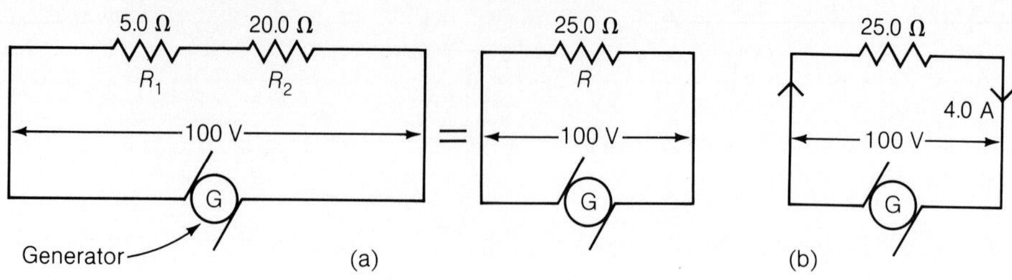

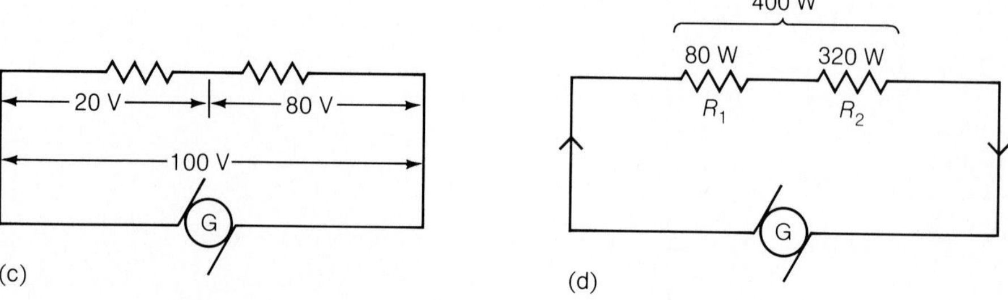

Figure 18.8

We note that the sum of V_1 and V_2 is 100 V, the same as the applied potential difference, as it must be.

(d) The power dissipated by R_1 is

$$P_1 = IV_1 = (4.0\ \text{A})(20\ \text{V}) = 80\ \text{W}$$

We can also find P_1 from the alternative formulas $P = I^2R$ and $P = V^2/R$. In each case we get the same answer:

$$P_1 = I^2R_1 = (4.0\ \text{A})^2(5.0\ \Omega) = 80\ \text{W} \qquad P_1 = \frac{V^2}{R_1} = \frac{(20\ \text{V})^2}{5.0\ \Omega} = 80\ \text{W}$$

The power dissipated by R_2 is

$$P_2 = IV_2 = (4.0\ \text{A})(80\ \text{V}) = 320\ \text{W} = 0.32\ \text{kW}$$

(e) The power dissipated by the entire circuit is

$$P = IV = (4\ \text{A})(100\ \text{V}) = 400\ \text{W} = 0.40\ \text{kW}$$

and is equal to $P_1 + P_2$. The dissipated power appears as heat. ♦

EXAMPLE 18.10

A 20-Ω load whose maximum power rating is 5.0 W is to be connected to a 24-V battery. What is the minimum resistance of the series resistor (Fig. 18.9) that will prevent the power rating from being exceeded?

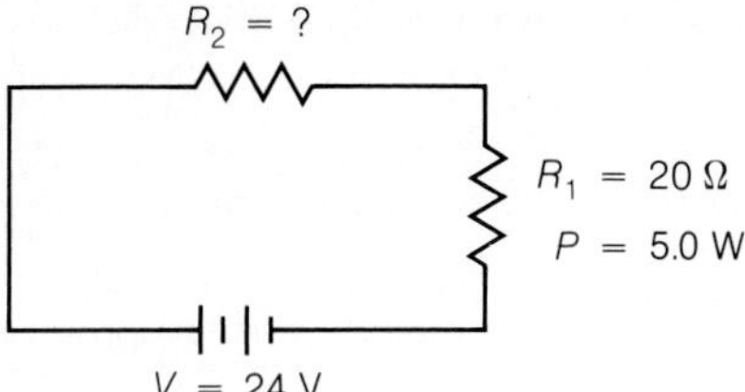

Figure 18.9

SOLUTION The maximum allowable current in the resistor R_1 may be found from the formula $P = I^2R_1$ to be

$$I = \sqrt{\frac{P}{R_1}} = \sqrt{\frac{5.0\ \text{W}}{20\ \Omega}} = 0.50\ \text{A}$$

For a current of 0.50 A, the equivalent resistance R of the circuit should be

$$R = \frac{V}{I} = \frac{24\ \text{V}}{0.50\ \text{A}} = 48\ \Omega$$

Because $R = R_1 + R_2$, the minimum resistance R_2 of the required series resistor is

$$R_2 = R - R_1 = 48\ \Omega - 20\ \Omega = 28\ \Omega$$ ◆

18.6 Resistors in Parallel

Their equivalent resistance is less than any of the individual resistances.

Let us now consider three resistors, R_1, R_2, and R_3, connected in parallel, as in Fig. 18.10. The total current I through the set is the sum of the currents through the separate resistors, so that

$$I = I_1 + I_2 + I_3$$

The potential difference V is the same across all the resistors. By applying Ohm's

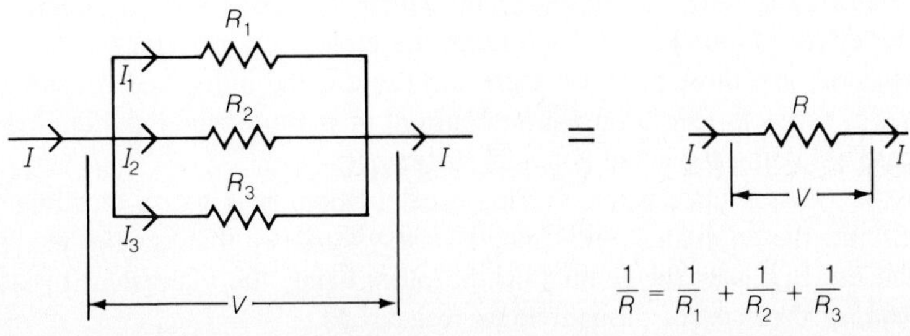

Figure 18.10

Resistors in parallel. The same potential difference V exists across each resistor.

law to each of the resistors, we find that

$$I_1 = \frac{V}{R_1} \qquad I_2 = \frac{V}{R_2} \qquad I_3 = \frac{V}{R_3}$$

The smaller the resistance, the greater the proportion of the total current that flows through it.

The total current flowing through the set of three resistors is given in terms of their equivalent resistance R by

$$I = \frac{V}{R}$$

Hence we have

$$I = I_1 + I_2 + I_3$$

$$\frac{V}{R} = \frac{V}{R_1} + \frac{V}{R_2} + \frac{V}{R_3}$$

The final step is to divide through by V, which gives

$$\frac{1}{R} = \frac{1}{R_1} + \frac{1}{R_2} + \frac{1}{R_3}$$

In general, *the reciprocal of the equivalent resistance of a set of resistors connected in parallel is equal to the sum of the reciprocals of the individual resistances*:

$$\frac{1}{R} = \frac{1}{R_1} + \frac{1}{R_2} + \frac{1}{R_3} + \cdots \qquad \textit{Resistors in parallel} \quad (18.11)$$

The formulas for the equivalent resistances of series and parallel arrangements of resistors are in accord with the basic formula

$$R = \rho\frac{L}{A}$$

for the resistance of a conductor. When several resistors are connected in series, the effect is the same as increasing the length L. When they are connected in parallel, the effect is the same as increasing the cross-sectional area A. Thus a series set of resistors lets through less current than any of the individual resistors, and a parallel set of resistors lets through more current than any of the individual resistors, in each case assuming the same potential difference.

As a check after making a calculation, it is worth recalling that in a series circuit, the equivalent resistance is always *greater* than any of the individual resistances. In a parallel circuit, on the other hand, the equivalent resistance is always *smaller* than any of the individual resistances.

In the case of two resistors in parallel, the equivalent resistance is

$$\frac{1}{R} = \frac{1}{R_1} + \frac{1}{R_2} = \frac{R_1 + R_2}{R_1 R_2}$$

Taking the reciprocal of both sides of this equation gives the convenient result

$$R = \frac{R_1 R_2}{R_1 + R_2} \qquad \textit{Two resistors in parallel} \quad (18.12)$$

When using a calculator with a reciprocal [1/*X*] key, it is easier to use Eq. (18.11) directly, as we saw in Section 17.9 in the case of the similar formula Eq. (17.19).

EXAMPLE 18.11

A 5.0-Ω resistor and a 20.0-Ω resistor are connected in parallel, and a potential difference of 100 volts is applied across them by means of a generator. Find (a) the equivalent resistance of the circuit, (b) the current that flows in each resistor and in the circuit as a whole, (c) the power dissipated by each resistor, and (d) the power dissipated by the entire circuit.

SOLUTION Successive stages in the solution of this problem are shown in Fig. 18.11.

(a) The equivalent resistance of the resistors is

$$R = \frac{R_1 R_2}{R_1 + R_2} = \frac{(5.0\ \Omega)(20.0\ \Omega)}{5.0\ \Omega + 20.0\ \Omega} = 4.0\ \Omega$$

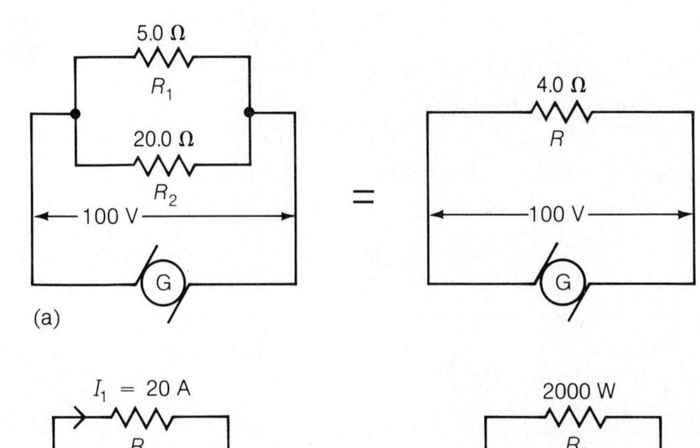

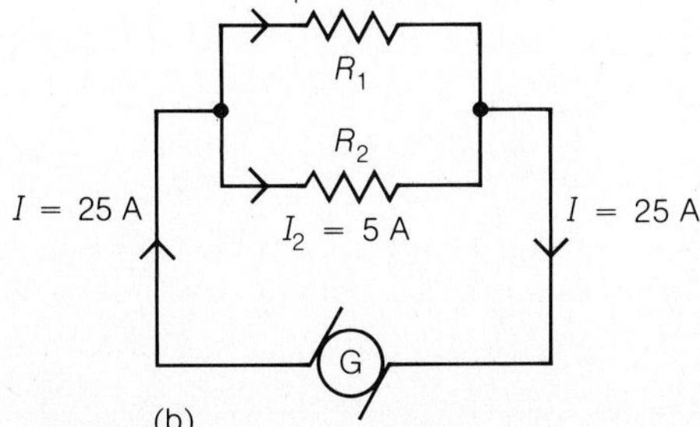

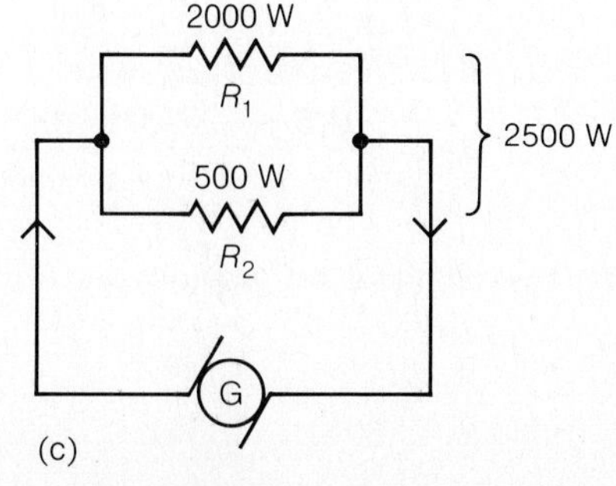

Figure 18.11

(b) The currents that flow in R_1 and R_2 are

$$I_1 = \frac{V}{R_1} = \frac{100\ \text{V}}{5.0\ \Omega} = 20\ \text{A} \qquad I_2 = \frac{V}{R_2} = \frac{100\ \text{V}}{20.0\ \Omega} = 5\ \text{A}$$

The total current in the circuit is

$$I = \frac{V}{R} = \frac{100\ \text{V}}{4.0\ \Omega} = 25\ \text{A}$$

and is equal to $I_1 + I_2$.

(c) The powers dissipated by R_1 and R_2 are

$$P_1 = I_1V = (20\ \text{A})(100\ \text{V}) = 2000\ \text{W} = 2.0\ \text{kW}$$
$$P_2 = I_2V = (5\ \text{A})(100\ \text{V}) = 500\ \text{W} = 0.5\ \text{kW}$$

(d) The power dissipated by the entire circuit is

$$P = IV = (25\ \text{A})(100\ \text{V}) = 2500\ \text{W} = 2.5\ \text{kW}$$

and is equal to $P_1 + P_2$.

We recall from the previous section that, when a 100-V potential difference is applied across the same two resistors connected in series, the power dissipated is only 400 W. The difference is due to the lower equivalent resistance of the parallel combination (4 Ω instead of 25 Ω), which permits more current to flow. Since $P = IV$ and V is the same in both cases, the greater current in the parallel circuit leads to a greater power dissipation. ◆

EXAMPLE 18.12

A circuit has a resistance of 100 Ω. How can it be reduced to 60 Ω?

SOLUTION In order to have an equivalent resistance of $R = 60\ \Omega$, a resistor R_2 must be connected in parallel with the circuit of $R_1 = 100\ \Omega$ (Fig. 18.12). To find the value of R_2 we proceed as follows:

$$\frac{1}{R} = \frac{1}{R_1} + \frac{1}{R_2}$$
$$\frac{1}{R_2} = \frac{1}{R} - \frac{1}{R_1} = \frac{R_1 - R}{R_1R}$$
$$R_2 = \frac{R_1R}{R_1 - R} = \frac{(100\ \Omega)(60\ \Omega)}{100\ \Omega - 60\ \Omega} = 150\ \Omega$$

With a calculator we would work from

$$\frac{1}{R_2} = \frac{1}{60\ \Omega} - \frac{1}{100\ \Omega}$$

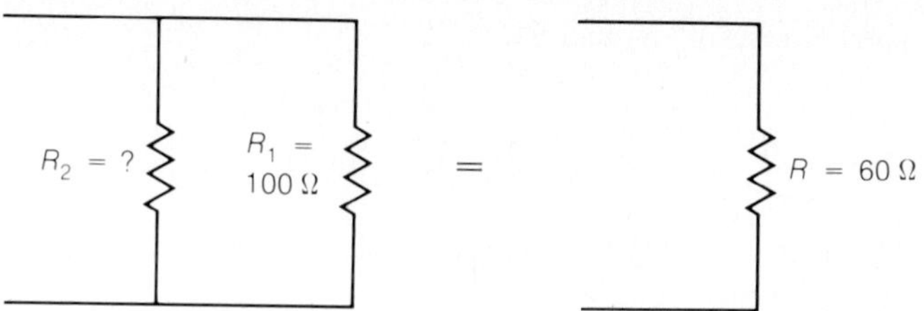

Figure 18.12

and use this key sequence:

[60][1/X][−][100][1/X][=][1/X] ◆

EXAMPLE 18.13

Find the equivalent resistance of the set of resistors shown in Fig. 18.13.

SOLUTION Figure 18.14 shows how the original circuit is broken up into its series and parallel groupings. The first step is to find the equivalent resistance R' of the

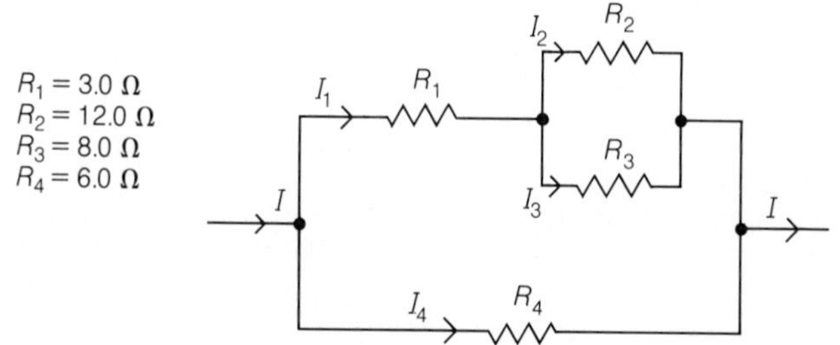

Figure 18.13

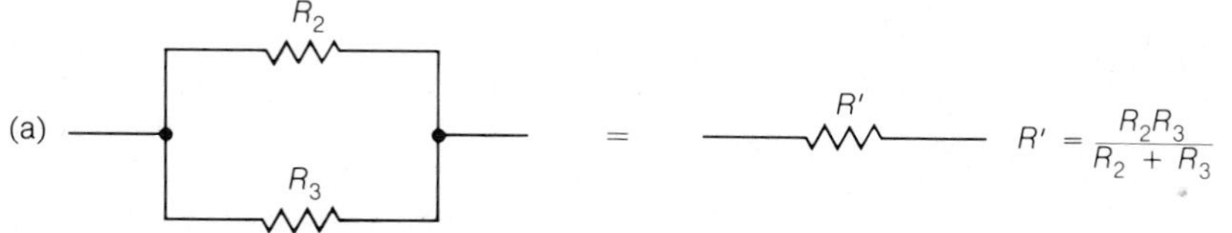

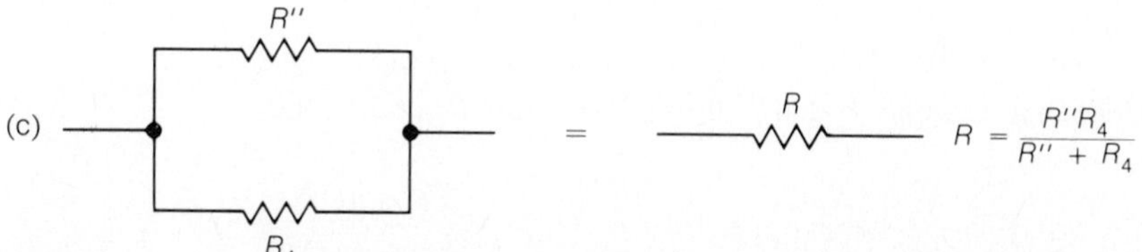

Figure 18.14

Successive steps in determining the equivalent resistance R of the resistor network shown in Fig. 18.13.

parallel resistors R_2 and R_3:

$$R' = \frac{R_2R_3}{R_2 + R_3} = \frac{(12.0\,\Omega)(8.0\,\Omega)}{12.0\,\Omega + 8.0\,\Omega} = 4.8\,\Omega$$

This pair of resistors is in series with R_1. The equivalent resistance R'' of the upper branch of the circuit is therefore

$$R'' = R' + R_1 = 4.8\,\Omega + 3.0\,\Omega = 7.8\,\Omega$$

The upper and lower branches of the circuit are in parallel, which means that the equivalent resistance R of the entire circuit is

$$R = \frac{R''R_4}{R'' + R_4} = \frac{(7.8\,\Omega)(6.0\,\Omega)}{7.8\,\Omega + 6.0\,\Omega} = 3.4\,\Omega$$

◆

EXAMPLE 18.14

A potential difference of 12.0 V is applied across the set of resistors shown in Fig. 18.13. Find the current in each resistor.

SOLUTION The full potential difference is applied across R_4, and so, by Ohm's law, the current I_4 in it is

$$I_4 = \frac{V}{R_4} = \frac{12.0\text{ V}}{6.0\,\Omega} = 2.0\text{ A}$$

The current I_1 in R_1 also flows through the entire upper branch of the circuit. Since the equivalent resistance of this branch is R'',

$$I_1 = \frac{V}{R''} = \frac{12.0\text{ V}}{7.8\,\Omega} = 1.54\text{ A}$$

The potential difference V' across the parallel resistors R_2 and R_3 is equal to the current I_1 through the equivalent resistance R' multiplied by R':

$$V' = I_1R' = (1.54\text{ A})(4.8\,\Omega) = 7.4\text{ V}$$

Another way to find V' is to subtract the potential difference

$$V_1 = I_1R_1 = (1.54\text{ A})(3.0\,\Omega) = 4.6\text{ V}$$

from the total of 12.0 volts to obtain

$$V' = V - V_1 = 12.0\text{ V} - 4.6\text{ V} = 7.4\text{ V}$$

Hence the currents I_2 and I_3 through resistors R_2 and R_3 are

$$I_2 = \frac{V'}{R_2} = \frac{7.4\text{ V}}{12.0\,\Omega} = 0.62\text{ A} \qquad I_3 = \frac{V'}{R_3} = \frac{7.4\text{ V}}{8.0\,\Omega} = 0.92\text{A}$$

◆

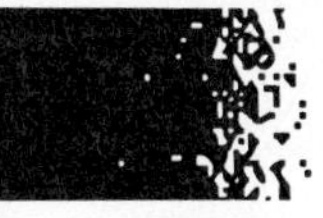

18.7 Electromotive Force

The no-load voltage of a source.

The potential difference across a battery, a generator, or other source of electric energy when it is not connected to any external circuit is called its **electromotive force.** Electromotive force is usually referred to simply as emf, and its symbol is $\mathcal{E}$.

As charges pass through a source of electric energy, work is done on them. The emf of the source is the work done per coulomb on the charges. The emf of a car's storage battery is 12 V, which means that 12 J of work are done on each coulomb of charge that passes through the battery. In the case of a battery, chemical energy is converted into electric energy by means of the work done on the charges in transit through it; in a generator, mechanical energy is converted into electric energy; in a thermocouple, heat energy is converted into electric energy; and so on. The emf of an electric source bears a relationship to its power output similar to that of applied force to mechanical power in a machine, which is the reason for its name.

When a source of electric energy is part of a complete circuit, a current I flows. The potential difference across the terminals of the source is then less than its emf owing to its **internal resistance.** Every electric source has a certain amount of internal resistance r, which means that a potential drop Ir occurs *within* the source. Hence the actual terminal voltage V across a source of emf $\mathcal{E}$ and internal resistance r is

$$V = \mathcal{E} - Ir \qquad \textit{Terminal voltage} \qquad (18.13)$$

Terminal voltage = emf − potential drop within source

If the source is disconnected, no current flows, and $V = \mathcal{E}$. The existence of a current lowers the value of V by an amount proportional to I.

Figure 18.15 shows a battery of emf $\mathcal{E}$ connected to a circuit whose equivalent resistance is R. The total resistance in the entire circuit is R plus the internal resistance r of the battery, so that the current I that flows is

$$I = \frac{\mathcal{E}}{R + r} \qquad (18.14)$$

The actual potential difference across the battery is given by Eq. (18.13).

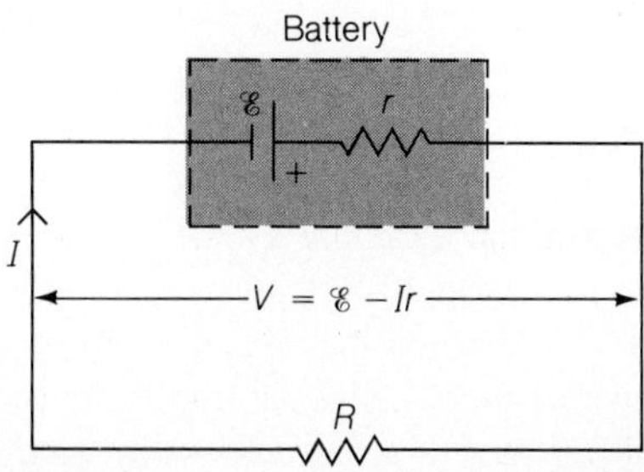

Figure 18.15

The terminal voltage V of a battery in a circuit is always less than its emf $\mathcal{E}$ because of the potential drop Ir within the battery itself.

The internal resistance of a battery governs the maximum current it can supply. One of the major advantages of lead-acid batteries is their low internal resistances, which permit large currents to be drawn for short periods to operate the starting motors of gasoline and diesel engines. The lower the temperature of the electrolyte in a battery, the more slowly its ions move and the higher the internal resistance. As a result the current available for the starting motor in freezing weather may be less than half that available on a warm day.

EXAMPLE 18.15

A "D" cell of emf 1.5 V and internal resistance 0.3 Ω is connected to a flashlight bulb whose resistance is 3.0 Ω. Find the current in the circuit and the terminal voltage of the cell.

SOLUTION The current in the circuit is

$$I = \frac{\mathscr{E}}{R + r} = \frac{1.5\ \text{V}}{3.0\ \Omega + 0.3\ \Omega} = 0.45\ \text{A}$$

The terminal voltage of the cell is

$$V = \mathscr{E} - Ir = 1.5\ \text{V} - (0.45\ \text{A})(0.3\ \Omega) = 1.4\ \text{V}$$ ◆

EXAMPLE 18.16

A storage battery whose emf is 12 V and whose internal resistance is 0.20 Ω is to be charged at a rate of 20 A. What applied voltage is required?

SOLUTION The applied voltage V must exceed the battery's emf $\mathscr{E}$ by the amount Ir to provide the charging current I. Hence

$$V = \mathscr{E} + Ir = 12\ \text{V} + (20\ \text{A})(0.20\ \Omega) = 16\ \text{V}$$ ◆

Figure 18.16

Cells in series and in parallel. In series, terminals of opposite polarity are connected in line. In parallel, terminals of the same polarity are connected together.

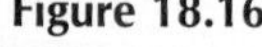

Series

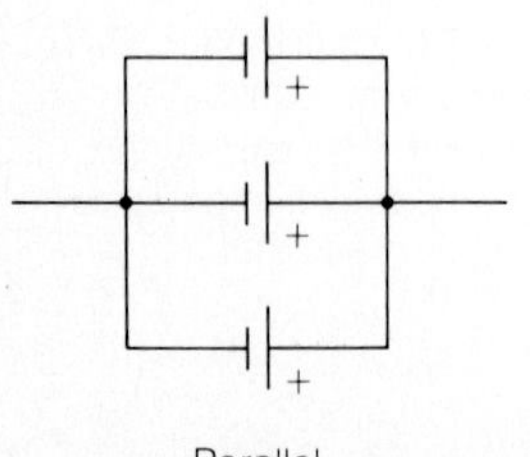

Parallel

Cells in Combination

When the emf of a single cell is too small for an application, two or more cells can be connected in series (Fig. 18.16). The emf of the set is the sum of the emf's of the individual cells, and the internal resistance of the set is the sum of the individual internal resistances:

$$\mathscr{E}_{\text{series}} = \mathscr{E}_1 + \mathscr{E}_2 + \mathscr{E}_3 + \cdots$$
$$r_{\text{series}} = r_1 + r_2 + r_3 + \cdots$$

A familiar example of such an arrangement is the use of six lead-acid cells in series to make the 12-V battery of a car.

When the emf of a cell or a battery is sufficient but its capacity is too small, two or more cells with the same emf can be connected in parallel. (If the cells have different emf's, currents will circulate among them that waste energy.) The emf and internal resistance of the set are now as follows:

$$\mathscr{E}_{\text{parallel}} = \mathscr{E}_1 = \mathscr{E}_2 = \mathscr{E}_3 = \cdots$$
$$\frac{1}{r_{\text{parallel}}} = \frac{1}{r_1} + \frac{1}{r_2} + \frac{1}{r_3} + \cdots$$

If all the cells have the same internal resistance r, r_{parallel} is just r divided by the number of cells. The reduced internal resistance of a parallel set of cells means that more current can be drawn by a given load. When a car's battery is too weak to start the car on a cold day, the remedy is another battery put in parallel with it by using jumper cables, which increases the available current. The ampere-hour capacity of a series set of cells is the same as that of each cell; the capacity of a parallel set is the sum of the capacities of the individual cells.

18.8 Impedance Matching

Energy transfer from one system to another is most efficient when both have the same impedance.

When a source of electric energy is connected to a load, as in Fig. 18.17, the rate at which energy is transferred to the load is a maximum when both source and load have the same resistance. If R is the load resistance and I is the current, then according to Eq. (18.14), $I = \mathscr{E}/(R + r)$, and the power in the load is

$$P = I^2R = \frac{\mathscr{E}^2R}{(R + r)^2} \qquad \textit{Power transfer} \quad (18.15)$$

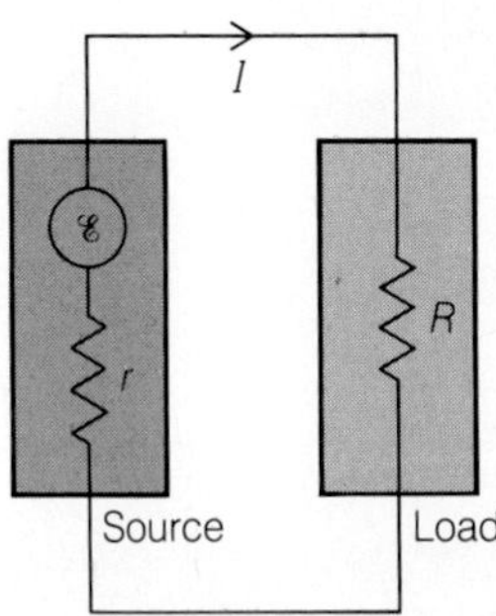

Figure 18.17

What value of the load resistance R will produce the maximum rate of energy transfer in this circuit?

The maximum value of P occurs when $R = r$:

$$P_{\text{max}} = \frac{\mathscr{E}^2R}{(R + R)^2} = \frac{\mathscr{E}^2R}{(2R)^2} = \frac{\mathscr{E}^2}{4R} \qquad (R = r) \quad (18.16)$$

Figure 18.18 shows how P/P_{max} varies with the resistance ratio R/r.

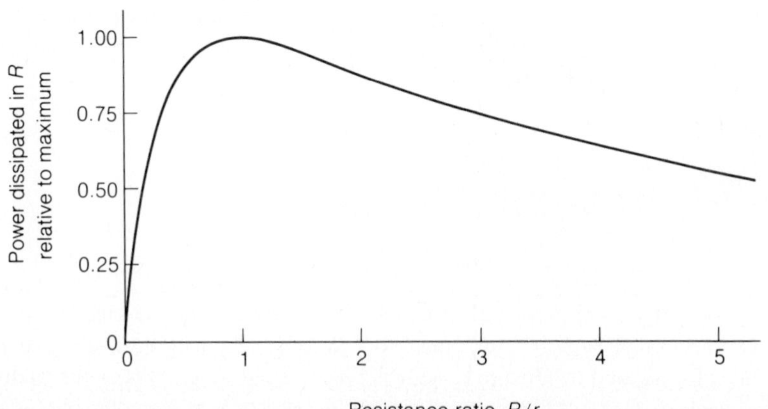

Figure 18.18

The power transferred to an external load of resistance R is a maximum when R equals the internal resistance r of the source.

EXAMPLE 18.17

Verify that the power transfer from a 100-V source of internal resistance $r = 10\ \Omega$ to a load resistance R is a maximum when $R = 10\ \Omega$.

SOLUTION Table 18.2 shows the current I and power P calculated from Eqs. (18.14) and (18.15) for values of R from 2 Ω to 20 Ω. The maximum power of 250 W indeed occurs for $R = r = 10\ \Omega$. Although the current is greater when R is less than r, the smaller R means that P is less than P_{max}. When R is greater than r, the current is then sufficiently low that P is less than P_{max} in this case as well.

Table 18.2 Current I and load power P for various values of R for the circuit of Fig. 18.17 when $\mathcal{E} = 100$ V, $r = 10\ \Omega$

$R(\Omega)$	I(A)	P(W)	$R(\Omega)$	I(A)	P(W)
2	8.3	139	12	4.5	248
4	7.1	204	14	4.2	243
6	6.3	234	16	3.8	237
8	5.6	247	18	3.6	230
10	6.0	250	20	3.3	222

◆

The conclusion reached in the preceding discussion is an example of **impedance matching**. When energy is being transferred from one system to another (here from the battery to the external resistance), the efficiency is greatest when both systems have the same **impedance,** which is a general term for resistance to the flow of energy in whatever form it may take in a particular case.

We met impedance matching earlier, the first time in Section 5.7. There we saw that when a moving object strikes a stationary one, the maximum energy transfer takes place when both have the same mass. In this situation the inertia of each object, as measured by its mass, represents its impedance. The curve of relative energy transfer versus mass ratio in Fig. 5.17 is identical in form to that in Fig. 18.18.

Another example of impedance matching occurs when a pulse or wave moves down a stretched string, as discussed in Section 12.1. When two strings having the same mass per unit length are joined together, a pulse in one passes to the other with no reflection. However, if the second string has either a greater or a smaller mass per unit length, reflection occurs at the junction and not all the energy of the pulse is transmitted (see Figs. 12.5 and 12.6). Though best known for its application to electric circuits, impedance matching is an important concept in many other branches of physics and engineering as well.

18.9 Kirchhoff's Rules

They make it possible to analyze complex circuits.

It is often difficult or impossible to figure out the currents that flow in the various branches of a complex network just by computing equivalent resistances. Two rules formulated by Gustav Kirchhoff (1824–1887) make it possible to find the current in each part of a direct-current circuit, no matter how complicated, if we are given

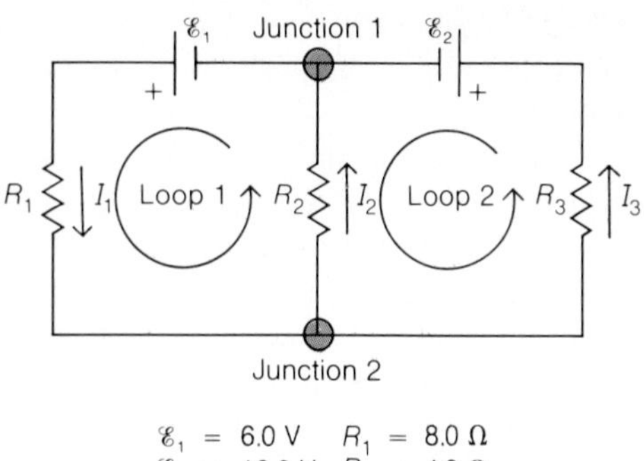

Figure 18.19

A *junction* is a point where three or more wires come together; a *loop* is any closed conducting path in the network. Here the internal resistances of the batteries are included in R_1 and R_3. The directions of the currents I_1, I_2, and I_3 are chosen arbitrarily; the value of I_3 turns out to be negative, meaing that its actual direction is opposite to the one shown.

the emf's of the sources of potential difference and the resistances of the various circuit elements. These rules apply to **junctions,** which are points where three or more wires come together, and to **loops,** which are closed conducting paths that are part of the circuit (Fig. 18.19).

Kirchhoff's first rule follows from the conservation of electric charge. Charge is never found to collect at any point in a circuit, nor can it be created there, and so there cannot be any net current into or out of a junction. Hence the first rule:

---

1. The sum of the currents flowing into a junction is equal to the sum of the currents flowing out of the junction.

The second rule follows from the conservation of energy. The sum of the emf's in a loop equals the amount of work done per coulomb by the sources of emf *on* a charge that moves once around the loop. The work done per coulomb *by* the charge as it moves around the loop equals the sum of the IR potential drops along the way. To conserve energy, the work done on the charge must be the same as the work done by the charge. Thus we have Kirchhoff's second rule:

---

2. The sum of the emf's around a loop is equal to the sum of the *IR* potential drops around the loop.

A definite procedure must be followed when Kirchhoff's rules are applied to a network. Let us call a section of a loop between two junctions a **branch.** The first step is to give the current in each branch a symbol of its own (I_1, I_2, I_3, and so on) and to assume a direction for each current. These symbols and directions should be marked on the circuit diagram, as in Fig. 18.19. The current is the same in all the resistors and sources of emf in a given branch, but of course the currents are different in the different branches.

It does not matter which direction is chosen for each current. If we have guessed correctly about the direction of the current in a branch, the solution of the problem will give a positive value for the current. If we have guessed wrong, a negative value will show that the actual current in the branch is in the opposite direction.

When the currents have been assigned to each branch, we then apply Kirchhoff's first rule to the junctions in the network. This will give us as many equations as the number of junctions. However, we will always find that one of these equations is a combination of the others and so contains no new information. Thus the number of junction equations we can use is always one less than the number of junctions.

The third step is to apply Kirchhoff's second rule to the loops in the network. In going around a loop, we must follow a consistent path, either clockwise or counterclockwise. These paths should be marked on the diagram, as in Fig. 18.19. An emf is reckoned positive if we meet the negative terminal of its source first. If instead we meet the positive terminal first, the emf is reckoned negative. An *IR* drop is considered positive if the assumed current in the resistor is in the same direction as the path we are following. The drop is considered negative if the current is opposite in direction to that of our path. The internal resistance of an emf source is treated as if it were a separate resistor in series with the source.

As in the case of the junction equations, it is important to avoid loop equations that merely duplicate information in the other loop equations. As a guide, the number of useful loop equations will equal the number of separate areas enclosed by branches of the network. (We can think of these areas as adjacent plots of land.)

Thus we have three steps to follow when using Kirchhoff's rules:

1. Assume directions for the branch currents.
2. Apply the first rule to the currents at the junctions.
3. Apply the second rule to the emf's and *IR* drops in the loops.

EXAMPLE 18.18

Find the current in each of the resistors in the network shown in Fig. 18.19.

SOLUTION The assumed directions of the unknown currents I_1, I_2, and I_3 are shown in the figure. Applying Kirchhoff's first rule to junction 1, we obtain

$$I_1 = I_2 + I_3 \qquad \textit{Junction 1}$$

This rule applied to junction 2 yields the same result.

In applying Kirchhoff's second rule to the two loops, we will follow counterclockwise routes. For loop 1,

$$\mathscr{E}_1 = I_1R_1 + I_2R_2 \qquad \textit{Loop 1}$$

and for loop 2,

$$-\mathscr{E}_2 = -I_2R_2 + I_3R_3 \qquad \textit{Loop 2}$$

In loop 2 we consider $\mathscr{E}_2$ as negative because we meet its positive terminal first, and I_2 as negative because its direction is opposite to our counterclockwise path.

There is also a third loop, namely, the outside one in Fig. 18.19, which similarly must obey Kirchoff's second rule. Proceeding counterclockwise yields

$$-\mathscr{E}_2 + \mathscr{E}_1 = I_1R_1 + I_3R_3 \qquad \textit{Outside loop}$$

This last equation is just the sum of the two preceding loop equations. (With two enclosed areas in the network, we expect only two independent loop equations.) Hence we may use the junction equation and any two of the loop equations to solve for the unknown currents. Nothing will be gained by using all three loop equations.

There are three unknown currents, and we have three separate equations relating them. Hence the problem can be solved by routine algebra. One way to do this is to start by substituting $I_2 + I_3$ for I_1 in the first loop equation to obtain

$$\mathscr{E}_1 = I_2R_1 + I_3R_1 + I_2R_2$$

We now solve both this equation and the second loop equation for I_3:

$$I_3 = \frac{\mathscr{E}_1 - I_2R_1 - I_2R_2}{R_1} \qquad I_3 = \frac{-\mathscr{E}_2 + I_2R_2}{R_3}$$

Setting the two expressions for I_3 equal enables us to solve for I_2 as follows:

$$\frac{\mathscr{E}_1 - I_2R_1 - I_2R_2}{R_1} = \frac{-\mathscr{E}_2 + I_2R_2}{R_3}$$

$$\mathscr{E}_1R_3 - I_2R_1R_3 - I_2R_2R_3 = -\mathscr{E}_2R_1 + I_2R_1R_2$$

$$I_2(R_1R_3 + R_2R_3 + R_1R_2) = \mathscr{E}_2R_1 + \mathscr{E}_1R_3$$

$$I_2 = \frac{\mathscr{E}_2R_1 + \mathscr{E}_1R_3}{R_1R_3 + R_2R_3 + R_1R_2}$$

Finally we insert the values of $\mathscr{E}_1$, $\mathscr{E}_2$, R_1, R_2, and R_3 given in Fig. 18.19 to obtain

$$I_2 = 1.05 \text{ A}$$

With the value of I_2 known, we can find the value of I_3 from the second loop equation:

$$I_3 = \frac{-\mathscr{E}_2 + I_2R_2}{R_3} = -0.83 \text{ A}$$

The minus sign means that the direction of I_3 is opposite to that shown in Fig. 18.19. The junction equation finally gives us the value of I_1:

$$I_1 = I_2 + I_3 = 0.22 \text{ A}$$

The same results would have been found had we chosen other directions for the currents or taken other routes around the loops in applying Kirchhoff's second rule. The important thing is not what choice of directions is made but to follow that choice consistently in working out the problem.

The outside loop equation, which was not used in the calculation, gives us a handy way to check the results. We have

$$-\mathscr{E}_2 + \mathscr{E}_1 = I_1R_1 + I_3R_3 \qquad \textit{Outside loop}$$

$$-10 \text{ V} + 6 \text{ V} = (0.22 \text{ A})(8\ \Omega) + (-0.83 \text{ A})(7\ \Omega)$$

$$-4 \text{ V} = -4 \text{ V}$$

In this calculation symbols for the various quantities were used until the very end to make clear the procedure. In working out problems of this kind, however, it is sometimes more convenient to substitute numerical values from the start. ◆

18.10 Capacitive Time Constant

To charge or discharge a capacitor takes time.

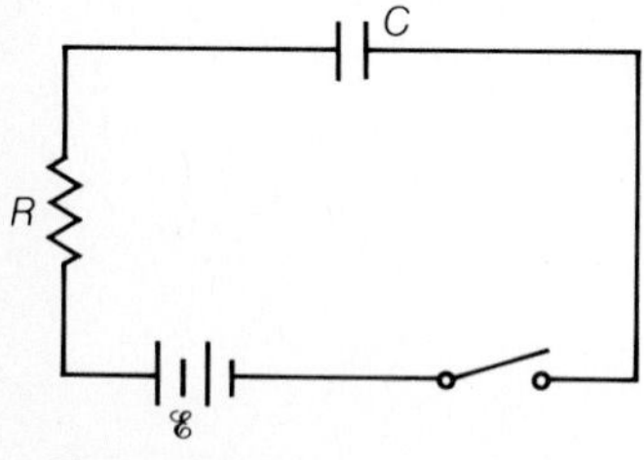

Figure 18.20

A circuit that contains a capacitor, a resistor, and a source of emf in series. When the switch is closed, the charge on the capacitor increases gradually to its ultimate value of $C\mathscr{E}$.

A circuit that contains capacitance as well as resistance does not respond instantly to changes in the applied emf. For instance, when a capacitor is connected to a battery, as in Fig. 18.20, it does not at once become fully charged.

At first the only limit to the current to the capacitor is the resistance R in the circuit. Thus the initial current is $I = \mathscr{E}R$, where $\mathscr{E}$ is the emf of the battery. As the capacitor becomes charged, however, a potential difference appears across it, whose polarity is such as to tend to oppose the further flow of current. When the charge on the capacitor has built up to some value Q, this opposing potential difference is $V = Q/C$. Hence the net potential difference is $\mathscr{E} - Q/C$, and the current is

$$I = \frac{\mathscr{E} - Q/C}{R} \qquad (18.17)$$

$$\text{Current} = \frac{\text{applied emf} - \text{voltage across capacitor}}{\text{resistance}}$$

As Q increases, then, its *rate* of increase drops. This gives a steadily decreasing slope in the curve of Fig. 18.21, which is a graph showing how Q varies with time when a capacitor is being charged; the capacitor is connected to the battery at $t = 0$.

Figure 18.21

The growth of charge in the capacitor of Fig. 18.20 after the switch is closed.

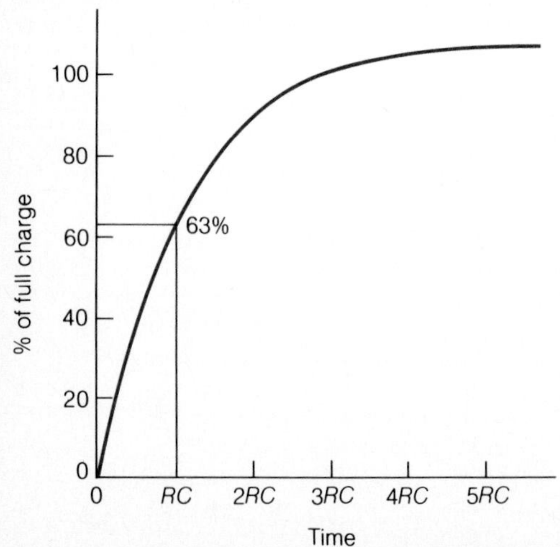

Figure 18.22

The decay of charge in a circuit containing capacitance and resistance when the battery is short-circuited.

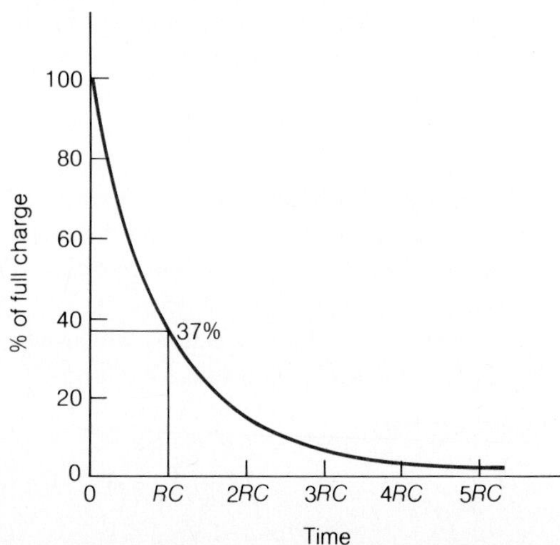

A mathematical analysis of Eq. (18.17) shows that after a time interval of RC (the product of the resistance R in the circuit and the capacitance C of the capacitor), the charge on the capacitor reaches 63% of its ultimate value of $Q_0 = C\mathscr{E}$. The time RC is therefore a convenient measure of how rapidly the capacitor becomes charged and is accordingly called the **time constant** of the circuit:

$$T_C = RC \qquad \textit{Capacitive time constant} \quad (18.18)$$

In principle the capacitor acquires its ultimate charge Q_0 only after an infinite time has elapsed. As we can see from Fig. 18.21, however, this value is very nearly reached after only a few time constants. At $t = 3RC$, the charge is 95% of Q_0, and at $t = 4RC$ it is 98% of Q_0.

If a capacitor with an initial charge is discharged through a resistance, its charge decreases with time as shown in Fig. 18.22. After the time RC the charge on the capacitor is reduced to 37% of its original value, a drop of 63%.

EXAMPLE 18.19

A 5.0-μF capacitor is charged by being connected to a 1.5-V dry cell. The total resistance of the circuit is 2.0 Ω. (a) What is the final charge on the capacitor and how long does it take to reach 63% of this charge? (b) The battery is then disconnected from the capacitor. If the resistance of the dielectric between the capacitor plates is $1.0 \times 10^{10}\ \Omega$, after what period of time will the charge on the capacitor drop to 37% of its initial value?

SOLUTION (a) The ultimate charge on the capacitor is

$$Q_0 = CV = (5.0 \times 10^{-6}\ \text{F})(1.5\ \text{V}) = 7.5 \times 10^{-6}\ \text{C}$$

The time constant of the circuit is

$$T_C = RC = (2.0\ \Omega)(5.0 \times 10^{-6}\ \text{F}) = 1.0 \times 10^{-5}\ \text{s}$$

and so this is the time needed for the capacitor to acquire 63% of its ultimate charge.

(b) When the battery is disconnected from the capacitor, charge gradually leaks through the dielectric of the latter. The time constant for discharge is

$$T_C = RC = (1.0 \times 10^{10}\ \Omega)(5.0 \times 10^{-6}\ \text{F}) = 5.0 \times 10^{4}\ \text{s} = 14\ \text{h}$$

and in this period the charge on the capacitor will drop to 37% of its original value. ◆

Important Terms

A flow of electric charge from one place to another is called an **electric current.** The unit of electric current is the **ampere,** which is equal to a flow of 1 coulomb/second.

Ohm's law states that the current in a metallic conductor is proportional to the potential difference between its ends. The **resistance** of a conductor is the ratio between the potential difference across its ends and the current that flows. The unit of resistance is the **ohm,** which is equal to 1 volt/ampere. The resistance of a conductor is proportional to its length and to the **resistivity** ρ of the material of which it is made, and inversely proportional to its cross-sectional area.

The **equivalent resistance** of a set of interconnected resistors is the value of the single resistor that can be substituted for the entire set without affecting the current that flows in the rest of any circuit of which it is a part. Resistors in **series** are connected consecutively so that the same current flows through all of them, whereas resistors in **parallel** have their terminals connected together so that the total current is split up among them.

The **electromotive force** (emf) of a battery, a generator, or other source of electric energy is the potential difference across its terminals when no current flows. When a current is flowing, the terminal voltage is less than the emf owing to the potential drop in the **internal resistance** of the source.

Kirchhoff's rules for network analysis are: (1) The sum of the currents flowing into a junction of three or more wires is equal to the sum of the currents flowing out of the junction; (2) the sum of the emf's around a closed conducting loop is equal to the sum of the IR potential drops around the loop.

The **capacitive time constant** of a circuit that contains capacitance and resistance is a measure of the time needed for the charge on the capacitor to respond to changes in the applied emf.

Important Formulas

Electric current: $I = \dfrac{Q}{t}$

Resistance: $R = \dfrac{V}{I}$

Ohm's law: $I = \dfrac{V}{R}$ *(holds for most metals)*

Resistance of ohmic conductor: $R = \rho\dfrac{L}{A}$

Power: $P = IV$

$= I^2R = \dfrac{V^2}{R}$ *(ohmic conductor)*

Resistors in series: $R = R_1 + R_2 + R_3 + \cdots$

Resistors in parallel: $\dfrac{1}{R} = \dfrac{1}{R_1} + \dfrac{1}{R_2} + \dfrac{1}{R_3} + \cdots$

$R = \dfrac{R_1R_2}{R_1 + R_2}$ *(two resistors)*

Emf and terminal voltage: $V = \mathscr{E} - IR$

Capacitive time constant: $T_C = RC$

MULTIPLE CHOICE

1. The resistance of a conductor does not depend on its
- **a.** mass
- **b.** length
- **c.** cross-sectional area
- **d.** resistivity

2. A certain wire has a resistance R. The resistance of another wire, identical with the first except for having twice its diameter, is
- **a.** $\frac{1}{4}R$
- **b.** $\frac{1}{2}R$
- **c.** $2R$
- **d.** $4R$

3. Which of the following combinations of length and cross-sectional area will give a certain volume of copper the least resistance?
- **a.** L and A
- **b.** $2L$ and $\frac{1}{2}A$
- **c.** $\frac{1}{2}L$ and $2A$
- **d.** Does not matter, because the volume of copper remains the same

4. The temperature of a copper wire is raised. Its resistance
- **a.** decreases
- **b.** remains the same
- **c.** increases
- **d.** any of the above, depending on the temperatures involved

5. Of the following combinations of units, the one that is not equal to the watt is the
- **a.** J/s
- **b.** AV
- **c.** $A^2\Omega$
- **d.** Ω^2/V

6. A resistor R_1 dissipates the power P when connected to a certain generator. If a resistor R_2 is inserted in series with R_1, the power dissipated by R_1
- **a.** decreases
- **b.** increases
- **c.** remains the same
- **d.** may do any of the above, depending on the values of R_1 and R_2

7. If the resistor R_2 in Multiple Choice 6 is placed in parallel with R_1, the power dissipated by R_1

a. decreases
b. increases
c. remains the same
d. may do any of the above, depending on the values of R_1 and R_2

8. A battery of emf $\mathscr{E}$ and internal resistance r is connected to an external circuit of equivalent resistance R. If $R = r$,

a. the current in the circuit will be a minimum
b. the current in the circuit will be a maximum
c. the power dissipated in the circuit will be a minimum
d. the power dissipated in the circuit will be a maximum

9. A battery is connected to an external circuit. The potential drop within the battery is proportional to

a. the emf of the battery
b. the equivalent resistance of the circuit
c. the current in the circuit
d. the power dissipated in the circuit

10. Which of the following is neither a basic physical law nor derivable from one?

a. Coulomb's law
b. Ohm's law
c. Kirchhoff's first law
d. Kirchhoff's second law

11. If the wrong direction is assumed for a current I in a network being analyzed by Kirchhoff's rules, the value of the current that is obtained will be

a. I b. $-I$
c. incorrect d. 0

12. An electric heater draws a current of 20 A when connected to a 120-V power source. Its resistance is

a. 0.17 Ω b. 6 Ω
c. 8 Ω d. 2400 Ω

13. The 8.0-Ω coil of a loudspeaker carries a current of 0.80 A. The potential difference across its terminals is

a. 0.1 V b. 5.1 V
c. 6.4 V d. 10 V

14. The power rating of an electric motor that draws a current of 3.00 A when operated at 120 V is

a. 40 W b. 0.36 kW
c. 0.54 kW d. 1.08 kW

15. When a 100-W, 240-V light bulb is operated at 200 V, the current that flows in it is

a. 0.35 A b. 0.42 A
c. 0.50 A d. 0.58 A

16. A 115-V, 1-kW electric oven is by mistake connected to a 230-V power line that has a 15-A fuse. The oven will

a. give off less than 1 kW of heat
b. give off 1 kW of heat
c. give off more than 1 kW of heat
d. blow the fuse

17. A 230-V, 1-kW electric oven is by mistake connected to a 115-V power line that also has a 15-A fuse. The oven will

a. give off less than 1 kW of heat
b. give off 1 kW of heat
c. give off more than 1 kW of heat
d. blow the fuse

18. The energy content of a fully charged 12-V storage battery of capacity 50 A-h is

a. 50 J b. 0.60 kJ
c. 36 kJ d. 2.2 MJ

19. A 20-V potential difference is applied across a series combination of a 10-Ω resistor and a 30-Ω resistor. The current in the 10-Ω resistor is

a. 0.5 A b. 0.67 A
c. 1 A d. 2 A

20. The potential difference across the 10-Ω resistor in Multiple Choice 19 is

a. 5 V b. 10 V
c. 15 V d. 20 V

21. The equivalent resistance of a 10-Ω resistor and a 30-Ω resistor connected in parallel is

a. 0.13 Ω b. 7.5 Ω
c. 20 Ω d. 40 Ω

22. A 20-V potential difference is applied across the resistors in Multiple Choice 21. The current in the 10-Ω resistor is

a. 0.5 A b. 1 A
c. 2 A d. 2.67 A

23. The equivalent resistance of a network of three 2-Ω resistors cannot be

a. 0.67 Ω b. 1.5 Ω
c. 3 Ω d. 6 Ω

24. Two identical resistors in parallel have an equivalent resistance of 2 Ω. If the resistors were in series, their equivalent resistance would be

a. 2 Ω b. 4 Ω
c. 8 Ω d. 16 Ω

25. A 12-V potential difference is applied across a series combination of four 6-Ω resistors. The current in each resistor is

a. 0.5 A b. 2 A
c. 8 A d. 18 A

26. A 12-V potential difference is applied across a parallel combination of four 6-Ω resistors. The current in each resistor is

a. 0.5 A b. 2 A
c. 8 A d. 18 A

27. A resistor of unknown resistance is in parallel with a 12-Ω resistor. A battery of emf 24 V and negligible internal resistance is connected across the combination. The battery provides a current of 3 A. The unknown resistance is

a. 8 Ω b. 12 Ω
c. 24 Ω d. 36 Ω

28. A 10-Ω resistor and a 20-Ω resistor are connected in parallel to a battery. If heat is produced in the 10-Ω resistor at the rate P, the rate at which heat is produced in the 20-Ω resistor is

a. $P/4$ b. $P/2$
c. P d. $2P$

29. A 10-Ω resistor and a 20-Ω resistor are connected in series to a battery. If heat is produced in the 10-Ω resistor at the rate P, the rate at which heat is produced in the 20-Ω resistor is

a. $P/4$ b. $P/2$
c. P d. $2P$

30. A 50-V battery is connected across a 10-Ω resistor and a current of 4.5 A flows. The internal resistance of the battery is

a. 0 b. 0.5 Ω
c. 1.1 Ω d. 5 Ω

31. A 4.0-Ω resistor is connected to a battery of emf 20 V and internal resistance 1.0 Ω. The potential difference across the resistor is

a. 4 V b. 16 V
c. 19 V d. 20 V

32. Two batteries of emf 6 V and internal resistance 1 Ω are connected in parallel. The load that will absorb the greatest power from the combination has a resistance of

a. 0.5 Ω b. 1 Ω
c. 2 Ω d. 6 Ω

QUESTIONS

1. Why are two wires used to carry electric current instead of a single one?

2. Which of the I-V graphs shown in Fig. 18.23 correspond to a substance that obeys Ohm's law?

3. When a metal object is heated, both its dimensions and its resistivity increase. Is the increase in resistivity likely to be a consequence of the increase in length?

4. It is sometimes said that an electrical appliance "uses up" electricity. What does such an appliance actually use in its operation?

5. Alice and Fred are discussing whether an electric heater with a large or with a small resistance will yield the greater heat output. Alice favors a small value of R because $P = V_2/R$; Fred favors a large value of R because $P = I^2R$. What is your conclusion?

6. The light bulbs in Fig. 18.24 are identical. Which bulb gives off the most light? The least light?

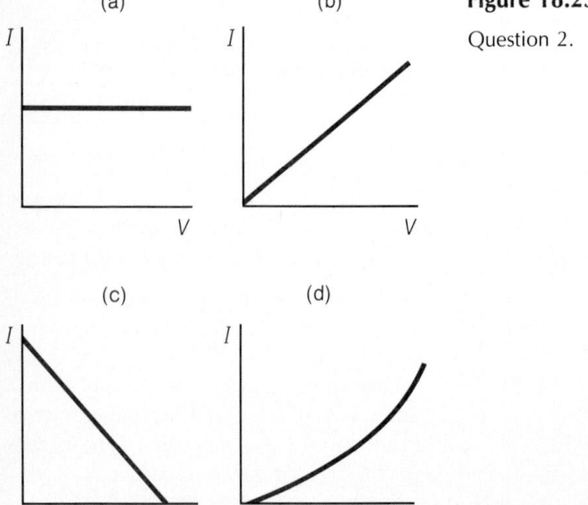

Figure 18.23
Question 2.

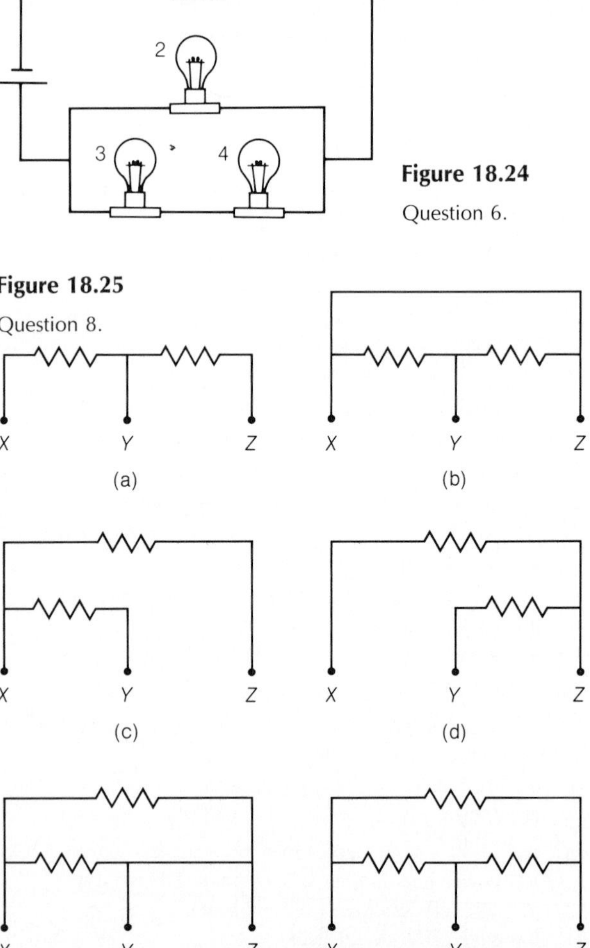

Figure 18.24
Question 6.

Figure 18.25
Question 8.

7. Why is it undesirable to connect cells of different emf in parallel?

8. A black box with three terminals marked X, Y, and Z contains two or three resistors of unknown resistance that may be connected in any of the six ways shown in Fig. 18.25. A battery is connected across the terminals X and Z, and the potential differences V_{xy} and V_{yz} between terminals x and y and between y and z, respectively, are measured. It is found that $V_{xy} = 2$ V and $V_{yz} = 4$ V. (a) Which, if any, of the circuits shown could give this result? (b) If more than one circuit is possible, what further test could tell which of them is correct?

EXERCISES

18.1 Electric Current

1. Sensitive instruments can detect the passage of as few as 60 electrons/s. To what current does this correspond?

2. Cosmic rays are atomic nuclei, mainly protons, that move at high speeds through the Milky Way galaxy, of which the sun is a member. An average of about 1500 of them arrive per square meter per second at the top of the atmosphere; because the earth's magnetic field affects their paths, their intensity varies somewhat around the earth. Assuming that all the cosmic ray particles are protons, find the total cosmic-ray current reaching the earth.

18.2 Ohm's Law

3. Find the resistance of a 120-V electric toaster that draws a current of 8.0 A.

4. A 240-V water heater has a resistance of 24 Ω. What must be the minimum rating of the fuse in the circuit to which the heater is connected?

5. What potential difference must be applied across a 1500-Ω resistor in order that the resulting current be 50 mA?

6. In resistance welding, a pair of electrodes presses two sheets of metal together. Heat from the passage of current melts the metal sheets where they are in contact between the electrodes to form a weld. What potential difference is needed to produce a current of 10 kA when the resistance of the metal between the electrodes is 0.20 mΩ?

18.3 Resistivity

7. How does the resistance per meter of 5.0-mm^2 copper wire compare with that of 2.0-mm^2 copper wire?

8. The resistance of No. 8 copper wire is 0.21 Ω/100 m. What is the resistance per 100 m of No. 8 aluminum wire?

9. American Wire Gage 000 wire has a cross-sectional area of 85 mm^2. Find the resistance of 10 km of copper wire of this size at 20°C.

10. Find the resistance of a carbon cylinder 2.0 mm in diameter and 20 mm long.

11. A silver wire 2.0 m long is to have a resistance of 0.50 Ω. What should its diameter be?

12. A 40-Ω resistor is to be wound from platinum wire 0.10 mm in diameter. How much wire is needed?

13. A metal rod 1.0 m long and 10 mm in diameter is drawn out into a wire 1.0 mm in diameter. (a) What is the length of the wire? (b) Compare the resistance of the rod with the resistance of the wire.

14. A metal bar of length a has a square cross section of length b on each edge. (a) Compare the resistance between its square ends with the resistance between opposite long faces. (b) What is the resistance ratio when a = 20 cm and b = 10 mm?

15. The resistance of a copper wire is 100 Ω at 20°C. Find its resistance at 0°C and at 80°C.

16. Motors, generators, transformers, and other electrical devices should not be operated above certain temperatures, whose values depend on their construction. The temperature in the interior of such a device can be found by measuring the resistance of one of its windings before it is run and after it has been run for some time. If the copper field windings of an electric motor have resistances of 100 Ω at 20°C and of 115 Ω when the motor is operating, what is their temperature under the latter conditions?

17. An iron wire has a resistance of 2.00 Ω at 0°C and a resistance of 2.46 Ω at 45°C. Find the temperature coefficient of resistivity of the wire.

18. A copper wire 1.0 mm in diameter carries a current of 12 A. Find the potential difference between two points in the wire that are 100 m apart.

19. Aluminum wires are sometimes used instead of copper wires to transmit electric power. What is the ratio between the masses of an aluminum and a copper wire of the same length whose cross-sectional areas are such that they have the same resistance?

18.4 Electric Power

20. Currents of 3.0 A flow through two wires, one that has a potential difference of 60 V across its ends and another that has a potential difference of 120 V across its ends. Compare (a) the rates at which charge passes through the wires and (b) the rates at which energy is dissipated in the wires.

21. An electric drill rated at 400 W is connected to a 240-V power line. How much current does it draw?

22. Find the maximum current in a 50-Ω, 10-W resistor if its power rating is not to be exceeded.

23. An electric motor whose power output is 0.50 hp draws 4.0 A at 120 V. What is its efficiency?

24. The alternator on a car's engine is 83% efficient. Find the mechanical power the alternator takes when it is delivering 20 A at 14 V.

25. A 240-V electric motor whose efficiency is 80% draws 10 A when it is used to operate a milling machine. Find the power output of the motor.

26. How much current is drawn by a 0.50-hp electric motor operated from a 120-V source of electricity? Assume that 80% of the electric energy absorbed by the motor is turned into mechanical work.

27. A 240-V clothes dryer draws a current of 15 A. How much energy, in kilowatt-hours and in joules, does it use in 45 min of operation?

28. A fully charged 12-V storage battery has an energy content of 3.5 MJ. What is its capacity in ampere-hours?

29. When a certain 1.5-V battery is used to power a 3.0-W flashlight bulb, it is exhausted after an hour's use. (a) How much charge has passed through the bulb in this period of time? (b) If the battery costs $0.50, what is the cost of a kilowatt-hour of electric energy obtained in this way? How does this compare with the cost of the electric energy supplied to your home?

30. A trolley car of mass 1.0×10^4 kg takes 10 s to reach a speed of 8.0 m/s starting from rest. It operates from a 5.0-kV power line and is 50% efficient. Find the average current it draws during the acceleration.

18.5 Resistors in Series

31. A 200-Ω resistor and a 500-Ω resistor are in series as part of a larger circuit. If the voltage across the 200-Ω resistor is 2.0 V, what is the voltage across the 500-Ω resistor?

32. It is desired to limit the current in an 80-Ω resistor to 0.50 A when it is connected to a 50-V power source. What is the value of the series resistor that is needed?

33. A stingy American moves to Europe, where 240 V is the normal household voltage, and proposes to use his old 120-V light bulbs by connecting them two at a time in series. How good is this idea when both bulbs of each pair have the same power rating? When the bulbs have different ratings—for instance, 10 W and 100 W at 120 V?

34. A set of Christmas tree lights consists of 12 bulbs connected in series to a 120-V power source. Each bulb has a resistance of 5.0 Ω. (a) What is the current in the circuit? (b) How much power is dissipated in the circuit?

35. A 5.0-Ω light bulb and a 10.0-Ω light bulb are connected in series with a 12.0-V battery. (a) What is the current in each bulb? (b) What is the voltage across each bulb? (c) What is the power dissipated by each bulb and the total power dissipated by the circuit?

36. A 20-Ω resistor and a 60-Ω resistor are in series as part of a larger circuit. The two resistors dissipate a total of 20 W. (a) How much power does each resistor dissipate? (b) What is the potential difference across each resistor?

37. A 5.0-Ω resistor and a 10.0-Ω resistor are connected in series. If the power dissipated in the 5.0-Ω resistor is 125 W, what is the potential difference across the combination?

18.6 Resistors in Parallel

38. How can the resistance of a 20.0-Ω circuit be reduced to 5.0 Ω?

39. List the resistances that can be obtained by combining three 100-Ω resistors in all possible ways.

40. Three identical resistors connected in parallel have an equivalent resistance of $1\frac{2}{3}$ Ω. What would their equivalent resistance be if connected in series?

41. A 20-Ω resistor, a 40-Ω resistor, and a 50-Ω resistor are connected in parallel across a 60-V power source. Find the equivalent resistance of the set and the current in each resistor.

42. In a certain bus forty 15-W, 30-V light bulbs are connected in parallel to a 30-V power source. (a) What is the current in each bulb? (b) What is the current provided by the source? (c) How much power is provided by the source?

43. The light bulbs in Exercise 35 are connected in parallel across the same battery. Answer the same questions for this arrangement.

44. What is the potential difference between the points *A* and *B* in the circuit shown in Fig. 18.26?

45. (a) Find the equivalent resistance of the circuit in Fig. 18.27. (b) What is the current in the 8.0-Ω resistor when a potential difference of 12.0 V is applied to the circuit?

46. Each of the resistors in the circuit in Fig. 18.28 can safely dissipate 10 W. What is the maximum power the entire circuit can dissipate?

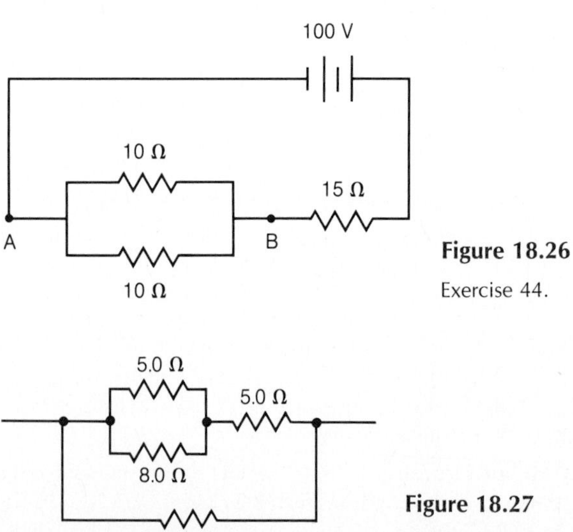

Figure 18.26
Exercise 44.

Figure 18.27
Exercise 45.

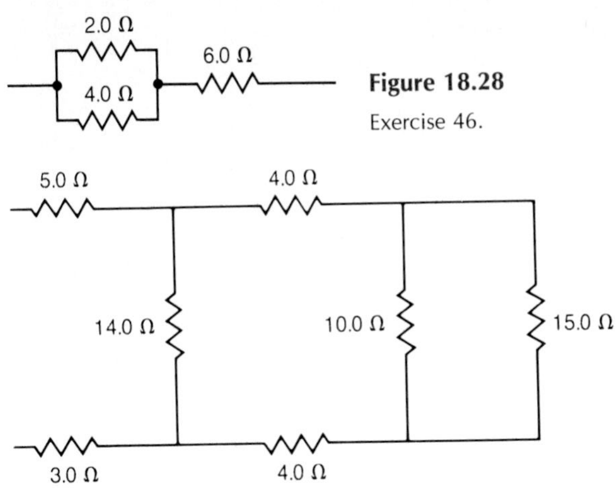

Figure 18.28

Exercise 46.

Figure 18.29

Exercise 47.

47. A 60-V potential difference is applied to the circuit in Fig. 18.29. Find the current in the 10-Ω resistor. (*Hint:* Redraw the circuit to bring out the series and parallel combinations of resistors more clearly.)

48. (a) Find the equivalent resistance of the circuit in Fig. 18.30. (b) What is the current in the 12-Ω resistor when a potential difference of 100 V is applied to the circuit?

18.7 Electromotive Force

49. A 12-V battery of internal resistance 1.5 Ω is connected to an 8.0-Ω resistor. Find the total power produced by the battery and the percentage of this power dissipated as heat within it.

50. A battery of emf 24 V is connected to an 11-Ω resistor. The potential difference across the battery terminals is 22 V. What is the internal resistance of the battery?

51. A battery having an emf of 24 V is connected to a 10-Ω load, and a current of 2.2 A flows. Find the internal resistance of the battery and its terminal voltage.

52. A generator has an emf of 240 V and an internal resistance of 0.30 Ω. When the generator is supplying a current of 20 A, find (a) its terminal voltage, (b) the power supplied to the load, and (c) the power dissipated in the generator itself.

53. The brightness of a light bulb depends on the power dissipated by its filament. As a dry cell ages, its internal resistance increases while its emf remains approximately unchanged at 1.5 V. A fresh No. 6 dry cell might have an internal resistance of 0.05 Ω and an old one an internal resistance of 0.20 Ω. Find the ratio between the powers dissipated in a 0.25-Ω bulb when it is connected to a fresh and to an old dry cell.

54. Four batteries, each of emf 6.0 V and internal resistance 0.30 Ω, are connected in series with a load of 2.0 Ω. Find the current in the load.

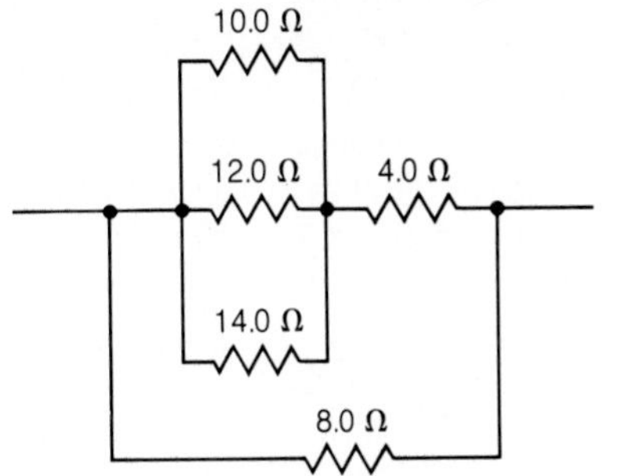

Figure 18.30

Exercise 48.

55. If the batteries in Exercise 54 are connected in parallel with the same load, what is the current in the load?

56. Twelve cells, each of emf 2.1 V and internal resistance 0.20 Ω, are connected as shown in Fig. 18.31. Find the emf and internal resistance of the combination.

57. A 12-V storage battery with an internal resistance of 0.012 Ω delivers 80 A when used to crank a gasoline engine. If the battery mass is 20 kg and it has an average specific heat of 0.84 kJ/kg·°C, what is its rise in temperature during 1 min of cranking the engine?

58. A 5.0-Ω resistor and a 10.0-Ω resistor are connected in parallel. This combination is connected in series with another pair of parallel resistors whose resistances are both 8.0 Ω. (a) What is the equivalent resistance of the network? (b) The network is connected to a 24-V battery whose internal resistance is 1.5 Ω. Find the current in each of the resistors.

18.10 Capacitive Time Constant

59. A reusable flash bulb requires an energy of 100 J for its discharge. A 450-V battery is used to charge a capacitor for this purpose. The resistance of the charging circuit is 15 Ω. (a) What is the required capacitance? (b) What is the time constant of the circuit?

60. A 100-μF electrolytic capacitor has a leakage current of 5 μA when the potential difference across its terminals is 12 V. If the capacitor is connected to a 12-V battery and then removed, how long will it take for the charge to fall to 37% of its original value?

61. A 5-μF capacitor is connected across a 1000-V battery with wires whose resistance is a total of 5000 Ω. (a) What is the time constant of this circuit? (b) What is the initial current that flows when the battery is connected? (c) How long would it take to charge the capacitor if this current remained constant?

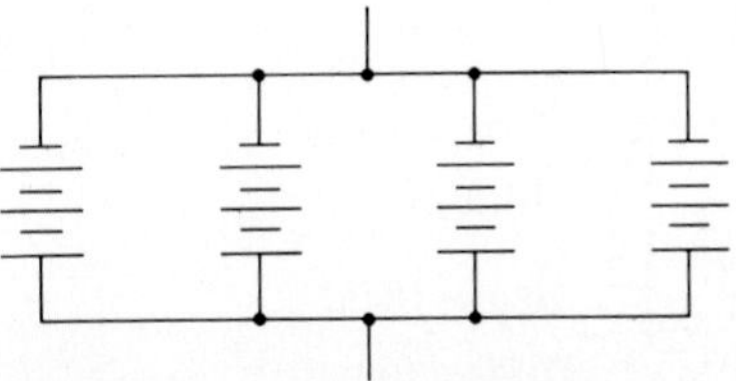

Figure 18.31

Exercise 56.

PROBLEMS

62. A 2-μF capacitor is connected in series with a 6-μF capacitor and a 100-Ω resistor. (a) What is the time constant of the combination? (b) The combination is connected to a 24-V source for a time equal to the time constant and then is disconnected. What is the voltage across each capacitor?

63. A Wheatstone bridge (shown in Fig. 18.32) provides a convenient means for measuring an unknown resistance R in terms of the known resistances A and B and the calibrated variable resistance C. The resistance of C is varied until no current flows through the meter G, in which case the bridge is said to be *balanced*. Show that $R = AC/B$ when the bridge is balanced.

64. Large electromagnets are sometimes wound with copper tubing through which water is circulated to carry away the heat produced. If such an electromagnet has a coil whose resistance is 0.80 Ω and through which 30 L/min of water flows, find the rise in temperature of the water when a current of 300 A is present in the coil. Assume that all the heat produced is absorbed by the water.

65. The insulation in an electric cord is damaged, and a short circuit occurs 2.0 m from a 120-V outlet. The copper wires in the cord are 1.0 mm in diameter. If all the power developed in the wires goes into heating them, how long will it take for the wires to melt? (Neglect the change in resistivity with temperature.)

66. (a) Find the potential difference between points a and b of the circuit in Fig. 18.33. (b) What resistor should be connected to the 16-Ω resistor to make $V_{ab} = 0$? How should the connection be made?

67. Find the values of V_1 and V_2 and the current through the middle resistor of the circuit in Fig. 18.34. The internal resistances of the batteries are included in the values of the resistors.

68. A capacitor is charged by connecting it through a resistance to a battery. Verify that half the work done by the battery is dissipated as heat.

69. A certain 16-cell 32-V storage battery has an emf of 33.5 V when charged to 75% of its 250 A-h capacity. The internal resistance of the battery is 0.1 Ω. It is desired to charge the battery to its full capacity, when its emf will be 34.3 V. (a) What potential difference must be applied to the battery if it is to be charged at the initial rate of 40 A? (b) If this potential difference is held constant, what will be the rate of charge at the end of the process? (c) The emf of the battery arises from the conversion of chemical to electric energy; the higher potential difference is required for charging in order to pass a current through the battery and thereby produce chemical changes that store energy. Find the proportion of the power supplied during the charging process that is stored as chemical energy and the proportion that is dissipated as heat. (Assume an average battery emf during charging of 33.9 V.)

70. Three identical 1.5-V dry cells with internal resistances of 0.15 Ω are connected in parallel with an external 0.5-Ω resistor. How much current flows through the resistor?

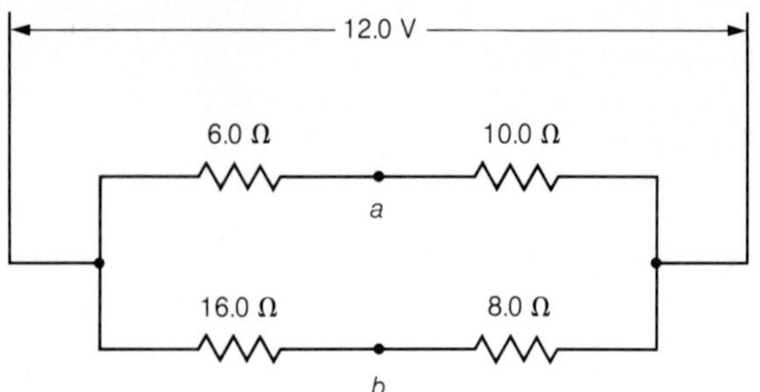

Figure 18.33
Problem 66.

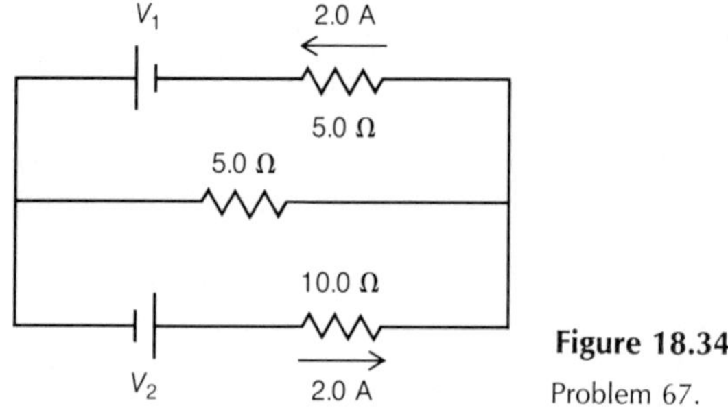

Figure 18.34
Problem 67.

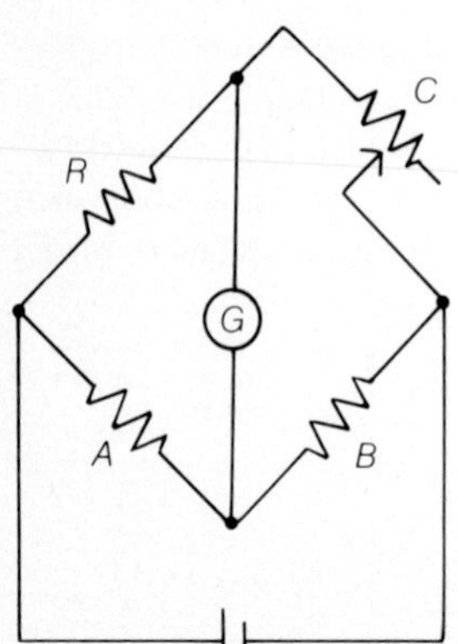

Figure 18.32
Problem 63.

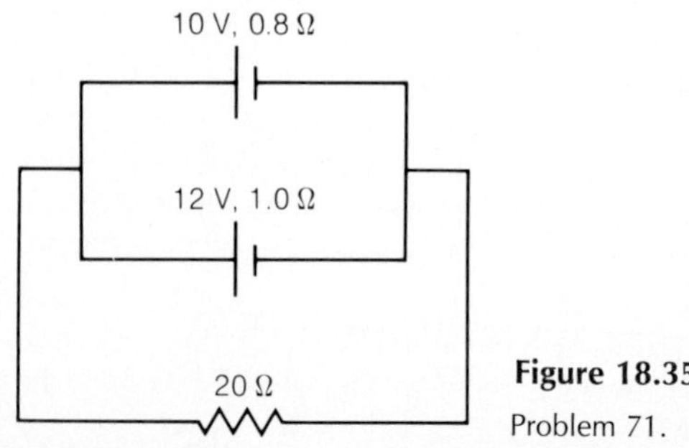

Figure 18.35
Problem 71.

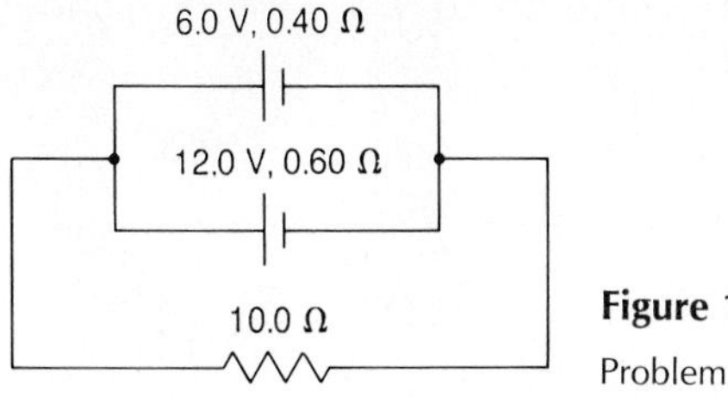

Figure 18.36
Problem 72.

71. Find the current in the 20-Ω resistor of the circuit in Fig. 18.35.

72. Find the current in the 10-Ω resistor of the circuit in Fig. 18.36.

73. Find the values of R_1 and R_2 of the circuit in Fig. 18.37.

74. Find the currents in the three resistors of the circuit in Fig. 18.38. The internal resistances of the batteries are included in the values of the resistors.

75. Find the currents in each of the resistors of the circuit in Fig. 18.39. The internal resistances of the batteries are included in the values of the resistors.

76. (a) Find the current in the 5.0-Ω resistor of the circuit in Fig. 18.40. (b) Find the potential difference between points a and b.

77. Find the potential difference between points a and b and between points a and c of the circuit in Fig. 18.41.

78. If points a and c of the circuit in Fig. 18.41 are connected, find the potential difference between points a and b.

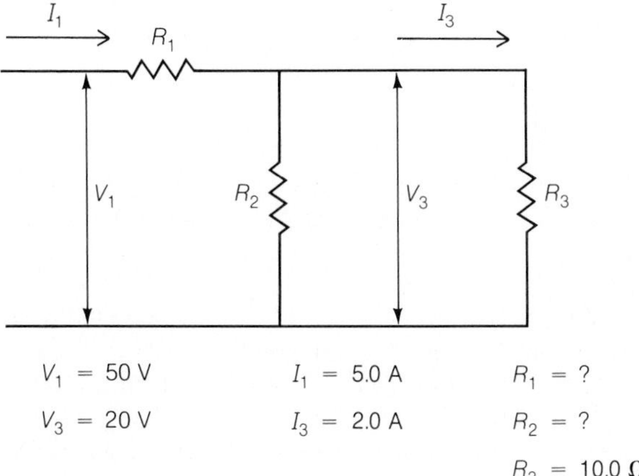

Figure 18.37
Problem 73.

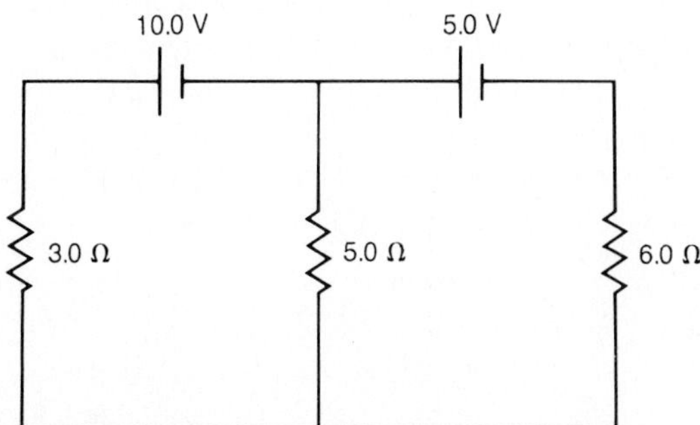

Figure 18.38
Problem 74.

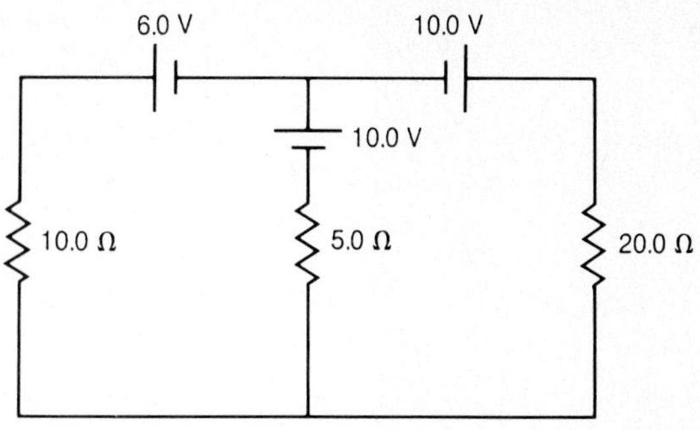

Figure 18.39
Problem 75.

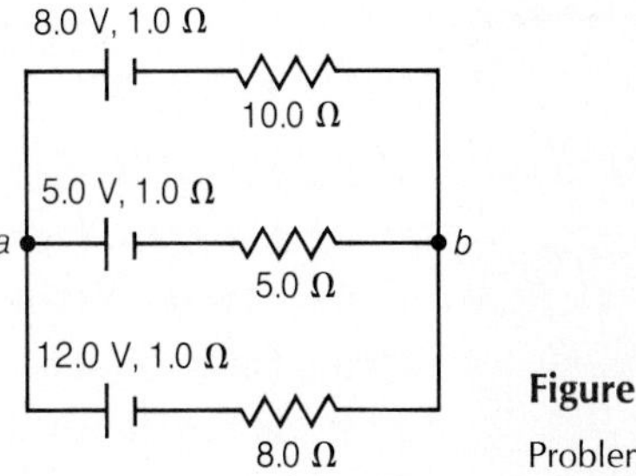

Figure 18.40
Problem 76.

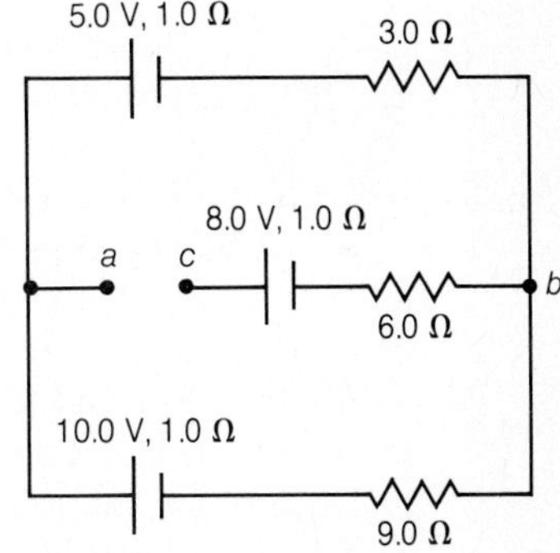

Figure 18.41
Problem 77.

Answers to Multiple Choice

1. a	9. c	17. a	25. a
2. a	10. b	18. d	26. b
3. c	11. b	19. a	27. c
4. c	12. b	20. a	28. b
5. d	13. c	21. b	29. d
6. a	14. b	22. c	30. c
7. c	15. a	23. b	31. b
8. d	16. d	24. c	32. a

CHAPTER 19

MAGNETISM

We are all familiar with magnets and have marveled at their ability to attract nails and other iron or steel objects. Not all of us, however, appreciate how closely related electricity and magnetism are.

Charges at rest relative to an observer appear to exert only electrical forces on one another. When the charges are moving relative to the observer, though, the forces between them are different from before. These differences are by tradition said to arise from "magnetic" forces. Thus magnetic forces represent modifications to electrical forces, modifications that come about because of the motion of charges. Because electrical and magnetic effects are distinct from each other, we can legitimately consider electrical and magnetic forces separately and think in terms of separate electric and magnetic fields.

19.1 The Nature of Magnetism

Magnetic forces arise from the interactions of moving charges.

The forces that moving charges exert on one another are different from those the same charges exert when at rest. For instance, if we place a current-carrying wire parallel to another current-carrying wire, with the currents in the same direction, we find that the wires attract each other (Fig. 19.1). If the currents are in opposite directions, the wires repel each other.

Electrical forces cannot be responsible for these observations, because there is no net charge on a wire when it carries a current. The forces that come into being when electric currents interact are called **magnetic forces.** All magnetic effects can ultimately be traced to currents or, more exactly, to moving charges.

The gravitational force between two masses and the electrical force between two charges at rest are both **fundamental forces** in the sense that they cannot be accounted for in terms of anything else. On the other hand, the force a bat exerts on a ball is not fundamental, because it can be traced to the electrical forces between the atomic electrons of the bat and those of the ball.

What about magnetic forces? It is an important fact that whatever it is in nature that shows itself as an electrical force between stationary charges *must*, according to the theory of relativity (Chapter 25), also show itself as a magnetic force between moving charges. One effect is not possible without the other. Thus there is only a single fundamental interaction between charges, the **electromagnetic interaction,** which has two aspects, electric and magnetic.

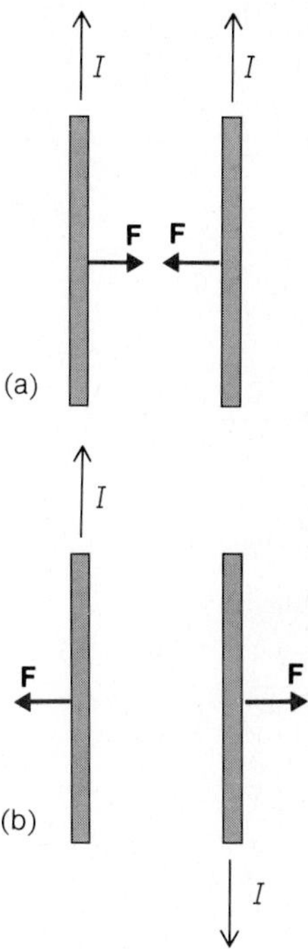

Figure 19.1

(a) Parallel electric currents in the same direction attract each other. (b) When the currents are in opposite directions, they repel each other.

Every electric charge has an electromagnetic field around it. The electric part of this field is always there, but the magnetic part appears only when relative motion is present. In the case of a current-carrying wire, there is only a magnetic field because the wire itself is electrically neutral. The electric fields of the electrons are canceled out by the opposite electric fields of the positive ions in the wire, but the ions are stationary and have no magnetic fields to cancel the magnetic fields of the moving electrons. If we simply move a wire that has no current flowing in it, the electric and magnetic fields of the electrons are canceled out by the electric and magnetic fields of the positive ions.

19.2 Magnetic Field

It is defined in terms of the force it exerts on a moving charge; its unit is the tesla.

We recall from Chapter 16 that the electric field **E** somewhere is defined in terms of the force **F** the field exerts on a stationary positive charge Q located there. Because F is proportional to Q, it made sense to specify the magnitude of **E** by the ratio

$$E = \frac{F}{Q} \qquad \textit{Electric field magnitude}$$

The direction of **E** is that of **F.** Once we know what **E** is, we can easily find the magnitude and direction of the electric force on *any* charge that happens to be there.

The symbol for **magnetic field** is **B.** Because magnetic forces act only on moving charges, **B** is defined in terms of the magnetic force **F** exerted on a positive charge Q whose velocity is **v.** Experiment and theory both show that the magnetic force on a moving charge is proportional to two factors. One is the product Qv. The larger the charge and the faster it moves, the greater the magnetic force.

The other factor concerns direction. When a charge is in a magnetic field, there is always a certain line along which it can move with no magnetic force acting on it. The direction of **B** is taken to lie along this line. When **v** is at the angle θ with respect to **B,** the magnetic force on the charge is found to be proportional to $\sin\theta$. Thus F is a maximum at $\theta = 90°$—the maximum force occurs for motion perpendicular to **B.** Because F depends on $Qv \sin\theta$, the magnitude B of the magnetic field responsible for **F** is defined, by analogy with the definition of electric field $E = F/Q$, as

$$B = \frac{F}{Qv \sin\theta} \qquad \textit{Magnetic field magnitude} \quad (19.1)$$

We note that $v \sin\theta$ is the component of **v** perpendicular to **B.** When **v** is itself perpendicular to **B,** $\sin\theta = \sin 90° = 1$ and

$$B = \frac{F}{Qv} \qquad (\mathbf{v} \perp \mathbf{B}) \qquad (19.2)$$

Although **B** at a point lies along the line through the point where $\mathbf{F} = 0$, two opposite directions are possible along that line. Hence more has to be said about the direction of **B.** Let us consider a charge moving perpendicularly to that line. The direction of **B** is, by convention, given by a right-hand rule (Fig. 19.2):

Open your right hand so that the fingers are together and the thumb sticks out. When your thumb is in the direction of v and your palm faces in the direction of F, your fingers are in the direction of B.

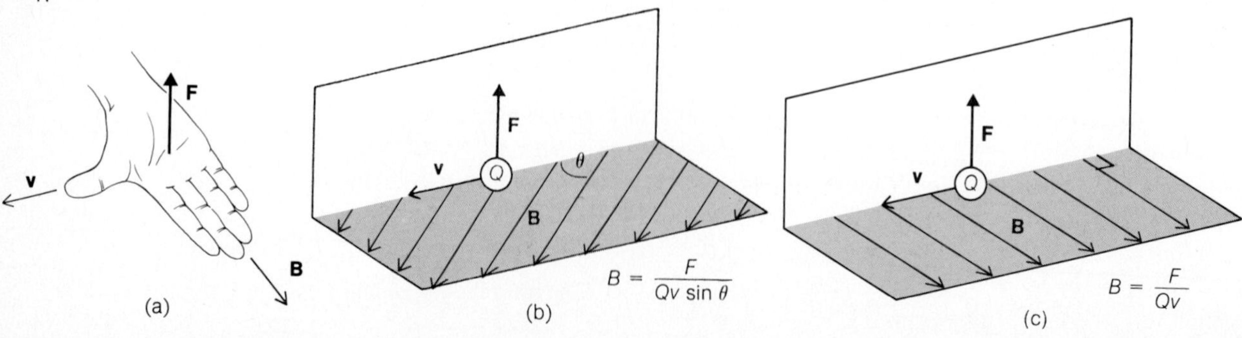

Figure 19.2

(a) Right-hand rule for the direction of the magnetic field **B** that exerts the force **F** on a positively charged particle of velocity **v.** If the particle has a negative charge, the force is in the opposite direction. (b) The magnitude of **B** is defined in terms of F, Q, and v. (c) When $\mathbf{v} \perp \mathbf{B}$, $\sin\theta = \sin 90° = 1$.

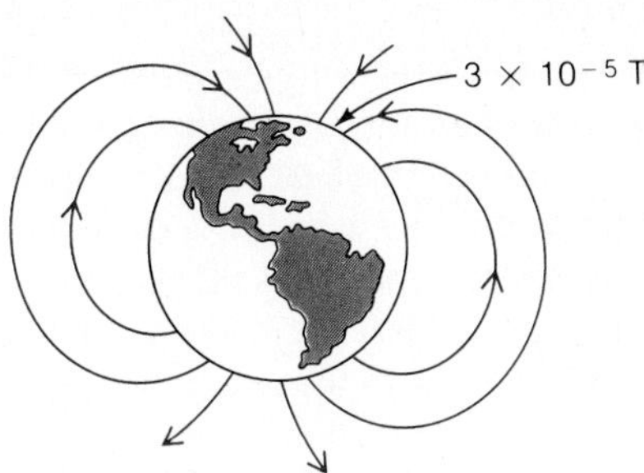

The magnitude of the earth's magnetic field at sea level is about 3×10^{-5} T.

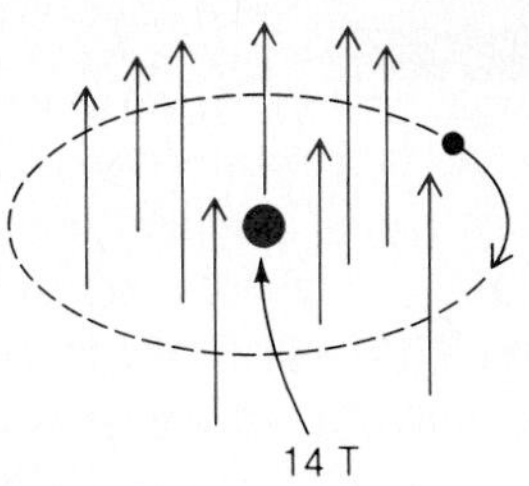

The magnetic field produced at the nucleus of a hydrogen atom by the electron circling around it is about 14 T.

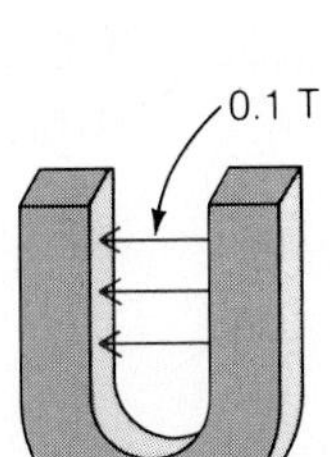

The magnetic field near a strong permanent magnet is about 0.1 T.

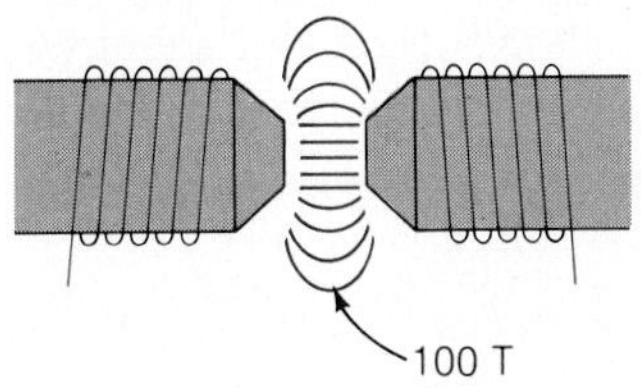

The most powerful magnetic fields achieved in the laboratory have magnitudes in the neighborhood of 100 T.

Figure 19.3

Some representative values of magnetic field. The earth's magnetic field is discussed in Section 19.9.

(An easy way to remember this rule is to associate the outstretched thumb with hitchhiking, and so with velocity; the palm with pushing on something, and so with force; and the parallel fingers with magnetic field lines.)

This completes an operational definition of **B,** since we now can find **B** in any region of space by performing suitable experiments. Such an experiment might use a cathode-ray tube in which the electron beam is deflected by the magnetic field as well as by an electric field, as described in Example 19.1 below.

The unit of magnetic field is, from its definition, the newton/ampere-meter (N/A·m), because the units of Qv are coulomb-meter/second = ampere-meter. The name **tesla,** abbreviated T, has been given to this unit:

$$1 \text{ tesla} = 1 \text{ T} = 1 \text{ N/A·m}$$

Thus a force of 1 N will be exerted on a charge of 1 C when it is moving at 1 m/s perpendicular to a magnetic field whose magnitude is 1 T. Another unit of **B** in common use is the **gauss,** where

$$1 \text{ gauss} = 10^{-4} \text{ T} \quad \text{or} \quad 1 \text{ T} = 10^4 \text{ gauss}$$

Nikola Tesla (1856–1943), a pioneering electrical engineer, was born in what is now Yugoslavia and moved to the United States in 1884; Karl Friedrich Gauss (1777–1855) was a German mathematician and physicist.

Figure 19.3 shows some representative values of magnetic field that may help to give a feeling for the magnitude of the tesla.

EXAMPLE 19.1

Before their nature was understood, electrons were known as "cathode rays." Even today, a tube in which an electron beam can be controlled by electric and/or magnetic fields so as to trace a desired pattern on a fluorescent screen is called a **cathode-ray tube.** The picture tube of a television set is an example. Figure 19.4 shows a cathode-ray tube whose beam consists of electrons that have been accelerated through a potential difference of 1000 V. A magnetic field **B** applied to the region between the deflection plates bends the beam out of the paper, but this effect is canceled out when the electric field E between the plates is 10^4 V/m. Find the direction and magnitude of **B.**

SOLUTION Since electrons have negative charges, the direction of **B** must be opposite to that given by the right-hand rule, so **B** is downward. To find B, we start with the speed v of the electrons. Since the electron kinetic energy $\frac{1}{2}mv^2$ is equal to the energy eV they gain in the electron gun,

$$\frac{1}{2}mv^2 = eV$$

$$v = \sqrt{\frac{2eV}{m}}$$

(We must be careful not to confuse eV, the product of the electron charge e and the potential difference V, with eV, the abbreviation for the electron volt, which is a unit of energy.)

When the electrical force eE on the electron balances the magnetic force evB (since $\theta = 90°$ here),

$$eE = evB$$

$$B = \frac{E}{v}$$

Substituting for v gives

$$B = \frac{E}{v} = E\sqrt{\frac{m}{2eV}} = (10^4 \text{ V/m})\sqrt{\frac{9.1 \times 10^{-31}\text{ kg}}{2(1.6 \times 10^{-19}\text{ C})(10^3\text{ V})}}$$
$$= 5.3 \times 10^{-4}\text{ T}$$

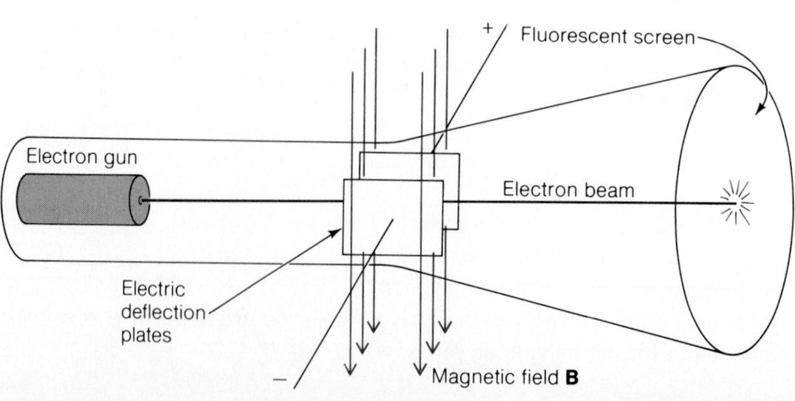

Figure 19.4

Experimental arrangement to determine **B.** The electric field between the plates is adjusted until the electron beam is undeflected. The same procedure can be used with a known magnetic field to determine the charge-to-mass ratio e/m of the electron.

The same experimental arrangement can be used with a known B to find the ratio e/m between the charge and mass of the electron. This was first done in 1897 by the English physicist J. J. Thomson, whose finding that e/m is always the same provided the first definite evidence that "cathode rays" are actually streams of particles. Later work by Robert A. Millikan in the United States showed that electrons all have the same charge $-e = -1.60 \times 10^{-19}$ C, from which the electron mass could be established. ◆

19.3 Magnetic Field of a Current

The stronger the current, the stronger the magnetic field.

Unfortunately it is usually rather difficult to calculate the magnetic field of a given current configuration. The results in the most important cases are given here.

Long, Straight Current

Figure 19.5 shows the form of the magnetic field around a long, straight wire that carries the current I. The magnetic field lines are concentric circles with the current at the center. The sense of the field is given by another right-hand rule:

Grasp the wire with your right hand so that your thumb points in the direction of the current. The curled fingers of that hand point in the direction of the magnetic field.

The magnetic field a distance s from the wire has the magnitude

$$B = \frac{\mu I}{2\pi s} \qquad \textit{Long, straight current} \quad (19.3)$$

The greater the current and the closer one is to it, the stronger the magnetic field.

The constant μ is called the **permeability** of the medium in which the magnetic field exists. In free space

$$\mu_0 = 4\pi \times 10^{-7}\ \text{T·m/A} = 1.257 \times 10^{-6}\ \text{T·m/A}$$

so that

$$\frac{\mu_0}{4\pi} = 1.000 \times 10^{-7}\ \text{T·m/A}$$

The value of μ in air is very close to μ_0; we will assume they are the same here. The unit of permeability is sometimes expressed in other ways—for instance, as N/A^2.

Figure 19.5

The field lines of the magnetic field around a long, straight current consist of concentric circles. The sense of the field is given by the right-hand rule.

EXAMPLE 19.2

Find the magnetic field in air 10 mm from a wire that carries a current of 1.0 A.

SOLUTION Since 10 mm $= 1.0 \times 10^{-2}$ m, we have (Fig. 19.6)

$$B = \frac{\mu_0 I}{2\pi s} = \frac{(4\pi \times 10^{-7}\ \text{T}\cdot\text{m/A})(1\ \text{A})}{(2\pi)(1.0 \times 10^{-2}\ \text{m})} = 2.0 \times 10^{-5}\ \text{T}$$

This is only a little smaller than the magnitude of the earth's magnetic field. For this reason great care is taken aboard ships to keep current-carrying wires away from magnetic compasses.

Figure 19.6

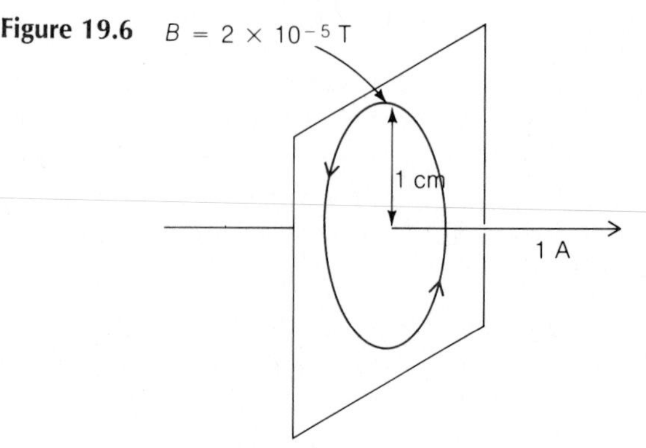

♦

Current Loop

The magnetic field around a circular current loop has the form shown in Fig. 19.7. At the center of the loop **B** is perpendicular to the plane of the loop with the magnitude

$$B = \frac{\mu I}{2r} \qquad \textit{Center of current loop} \quad (19.4)$$

where I is the current in the loop and r is its radius. The direction of **B** is given by

Figure 19.7

(a) The magnetic field around a circular current loop (b) At the center of the loop, **B** is perpendicular to the plane of the loop and its direction is given by the right-hand rule shown. (c) If there are N loops, the magnetic fields of the individual loops add up to give a field N times as strong as each one produces by itself.

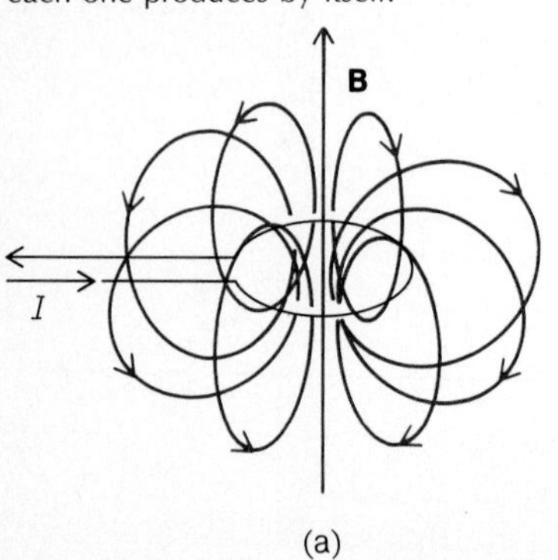

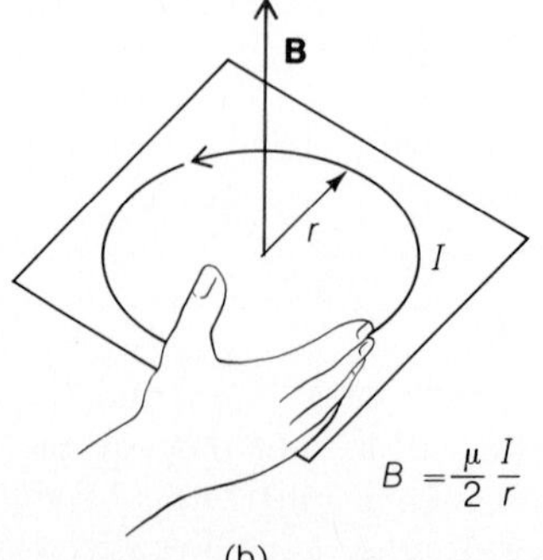

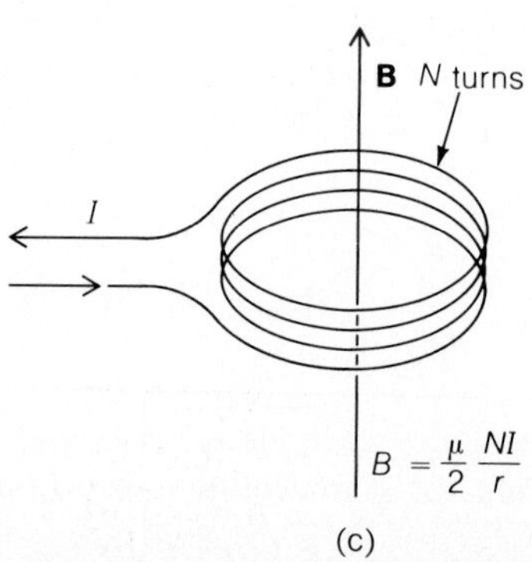

still another right-hand rule:

Grasp the loop so that the curled fingers of your right hand point in the direction of the current; the thumb of that hand then points in the direction of B.

In the case of a flat coil of more than one loop, as in Fig. 19.7(c), the magnetic fields of each individual loop add up to give a proportionately stronger field. If there are N turns, then

$$B = \frac{\mu NI}{2r} \qquad \textit{Center of flat coil} \quad (19.5)$$

Solenoid

A **solenoid** is a coil of wire in the form of a helix (Fig. 19.8). If the turns are close together and the solenoid is long relative to its diameter, then the magnetic field inside it is uniform and parallel to its axis except near the ends. The direction of the field inside a solenoid is given by the same right-hand rule that gives the direction of **B** inside a current loop.

The magnetic field in the interior of a solenoid of the length l that has N turns of wire and carries the current I has the magnitude

$$B = \mu\frac{N}{l}I \qquad \textit{Interior of solenoid} \quad (19.6)$$

The diameter of the solenoid does not matter, provided it is small compared with the length l.

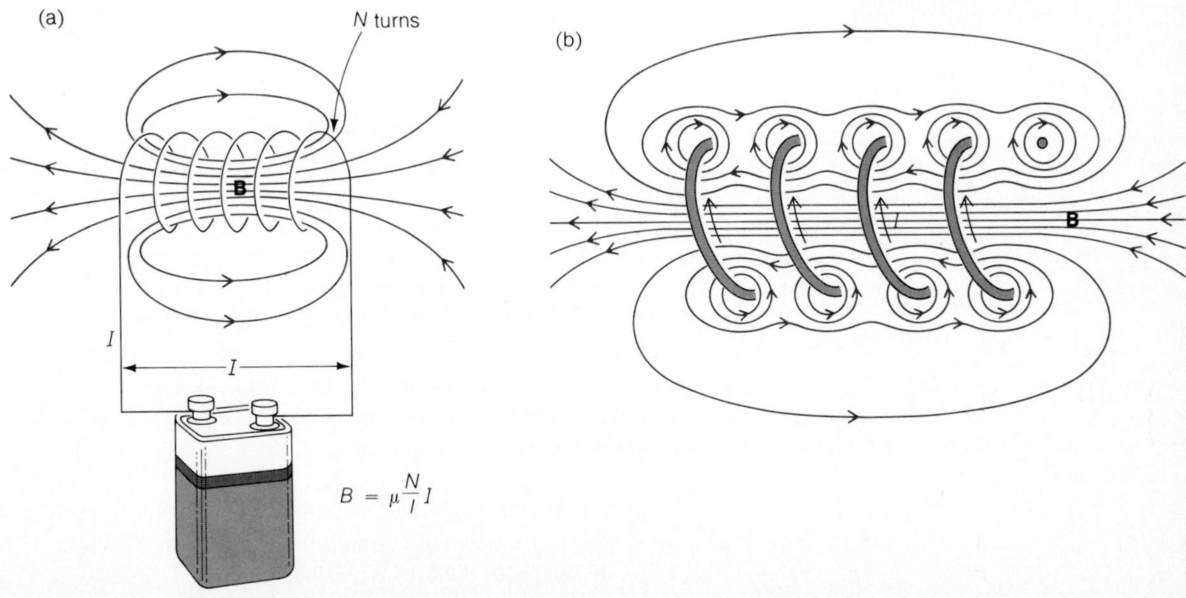

Figure 19.8

(a) The magnetic field inside a solenoid is uniform except near its ends if the solenoid is long relative to its diameter and if its turns are close together. (b) An expanded view of a solenoid showing how the magnetic fields of the individual turns add together to give a uniform field inside it.

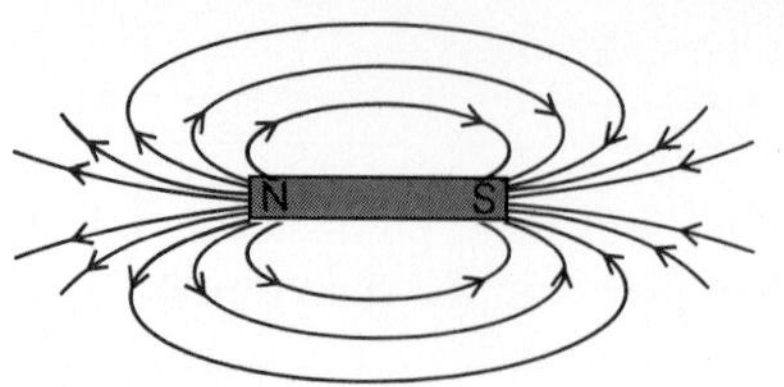

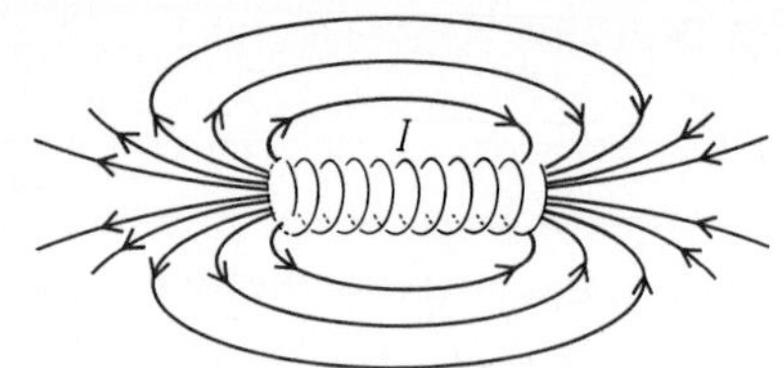

Figure 19.9

The magnetic fields of a bar magnet and of a solenoid are the same.

EXAMPLE 19.3

A solenoid 20 cm long and 40 mm in diameter with an air core is wound with a total of 200 turns of wire. The solenoid's axis is parallel to the earth's magnetic field at a place where the latter is 3.0×10^{-5} T in magnitude. What should the current in the solenoid be for its field to exactly cancel the earth's field inside the solenoid?

SOLUTION In air $\mu = \mu_0$, and so, since $l = 0.20$ m here, the required current is

$$I = \frac{Bl}{\mu_0 N} = \frac{(3.0 \times 10^{-5}\,\text{T})(0.20\,\text{m})}{(4\pi \times 10^{-7}\,\text{T}\cdot\text{m/A})(2.0 \times 10^2)} = 0.024\,\text{A} = 24\,\text{mA}$$

The solenoid diameter here has no significance except as a check that the solenoid is long relative to its diameter. ◆

The magnetic field of a bar magnet is identical with that of a solenoid, which is not surprising since all permanent magnets owe their character to an alignment of atomic current loops no different in principle from the alignment of the current loops in a solenoid (Fig. 19.9). The behavior of permanent magnets is discussed in Section 19.9.

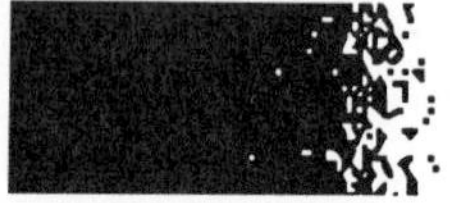

19.4 Magnetic Properties of Matter

Diamagnetism, paramagnetism, and ferromagnetism.

The strength of the magnetic field of a current-carrying solenoid is changed when a rod of almost any material is used as a core. Some materials increase B (for instance, oxygen and aluminum), others decrease B (for instance, mercury and bismuth), but in almost all cases the difference is very small. A few substances, however, give a dramatic increase in B—the new field may be hundreds or thousands of times stronger. Such substances are called **ferromagnetic** (Fig. 19.10). Iron is the most familiar example, but nickel, cobalt, and certain alloys are also ferromagnetic, as are ceramics called **ferrites.**

The magnetic properties of matter can be traced almost entirely to atomic electrons; the contribution of the nucleus is very minor. The magnetic behavior of an atomic electron has two origins:

1. An electron in certain respects resembles a spinning charged sphere, which we may imagine as a series of ultrasmall current loops, as in Fig. 19.11(a). Hence every electron has the magnetic field of a tiny bar magnet.

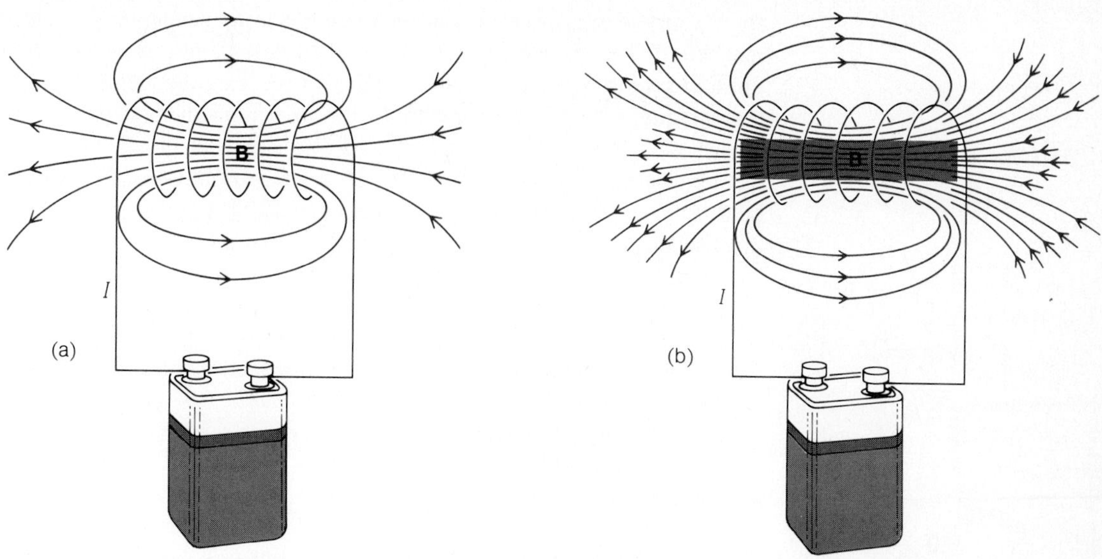

Figure 19.10

(a) Magnetic field of a solenoid with no core. (b) Using a ferromagnetic core increases the magnetic field considerably.

2. When it is part of an atom, an electron may be thought of as circling the nucleus much like a planet circling the sun. As in the case of electron spin, this is a crude model of the actual situation, but it is adequate for many purposes. An orbiting electron is a current loop, and so the result is again the magnetic field of a tiny bar magnet (Fig. 19.11b).

In most substances whose atoms or molecules have an even number of electrons, the various magnetic fields of the electrons cancel each other out in pairs when there is no external magnetic field. (If we shake an even number of small bar magnets together in a box, they will also end up paired off with opposite poles together.) When an external field is present, however, the orbital motions of the electrons are affected. By Lenz's law, described in Section 20.3, the resulting changes in the orbital magnetic fields are such as to oppose the external field, an effect called **diamagnetism.** Diamagnetism thus reduces the strength of a magnetic field.

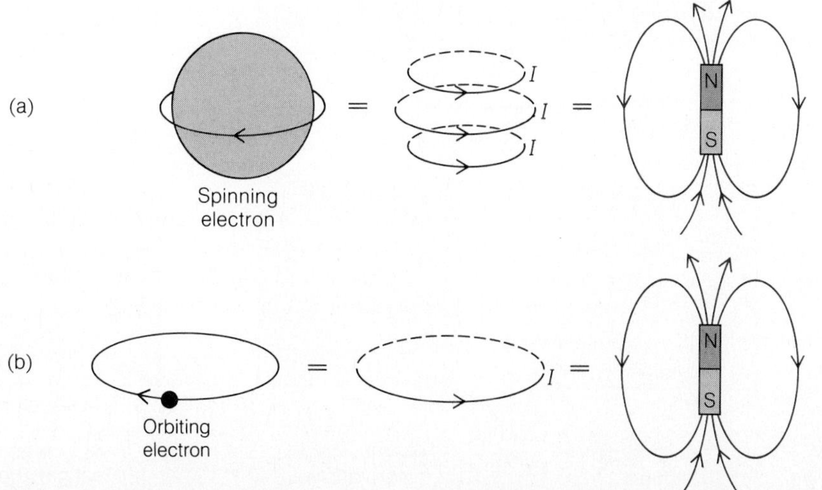

Figure 19.11

Sources of atomic magnetism.

In other substances one or more electrons per atom or molecule have spin magnetic fields that are not canceled out. In an external magnetic field, the fields of these electrons tend to line up to enhance the external field. This effect is called **paramagnetism.** The increase in B due to paramagnetism (when it is present) is always greater than the decrease due to diamagnetism (which is always present), so the net result is an increase in B. The increase is usually small, however, because the constant thermal agitation of the atoms or molecules prevents complete alignment of the spin magnetic fields with the external magnetic field.

Ferromagnetism

In a ferromagnetic material the unpaired electrons in each atom interact strongly with those in adjacent atoms. This causes the unpaired spin magnetic fields in all the atoms to be locked together. Atoms in a ferromagnetic material are accordingly grouped together in assemblies called **domains,** typically about 5×10^{-5} m across and just visible in a microscope. In an unmagnetized sample the directions of magnetization of the domains are random, although in each domain the unpaired electron spins are parallel. When such a sample is placed in an external magnetic field, either the spins within the domains turn to line up with the field or, in pure and homogeneous materials, the domain walls change so that those domains already lined up with the field grow at the expense of the others (Fig. 19.12). When all the unpaired spins in a ferromagnetic sample are lined up, no further increase in B is possible, and the sample is said to be **saturated.**

Above a certain temperature (770°C in the case of iron), the atoms in a domain have enough kinetic energy to overcome the interatomic forces that align their spins. The spins and their magnetic fields then become randomly oriented, and the ferromagnetic material loses its special magnetic properties. Thus heating a ''permanent magnet'' can demagnetize it.

Electromagnet used to load scrap iron and steel.

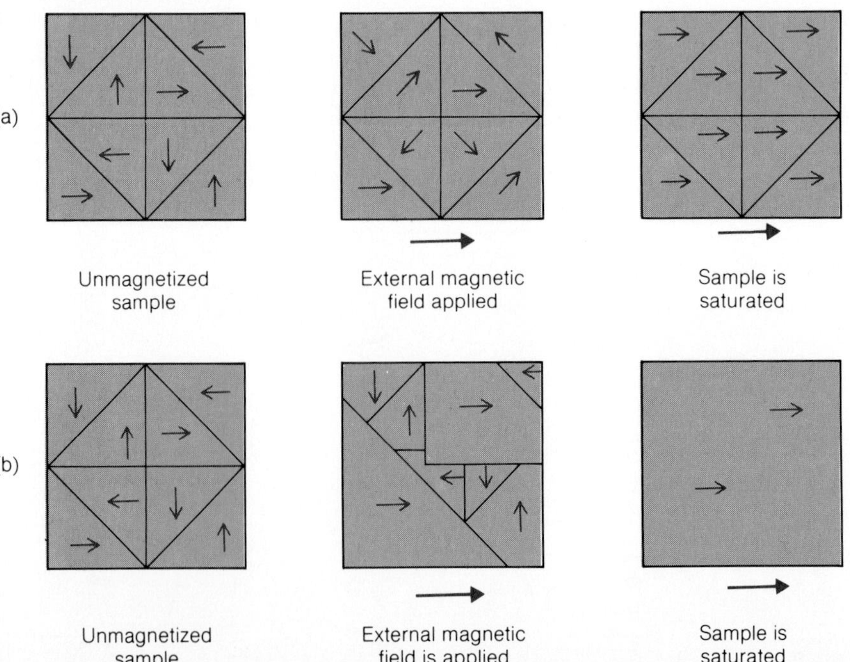

Figure 19.12

(a) Magnetization of a ferromagnetic material by domain alignment. (b) Magnetization by domain growth.

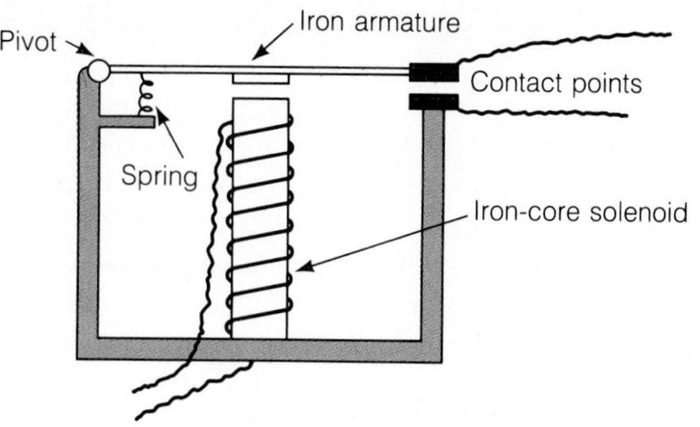

Figure 19.13

A relay is a switch operated by an electromagnet. A small current in the solenoid is enough to pull the armature down against the spring and close the contact points. When the current in the solenoid is cut off, the spring pushes the armature up to break the secondary circuit. The relay of a heating system is actuated by a thermostat whose contacts close when the room temperature drops below a preset value. Because little current is needed to operate the relay, the thermostat can be quite small, and light wiring is sufficient between it and the relay. Large contact points can be provided on the relay to carry the current needed by the oil burner and the blowers or pumps that circulate the hot air or hot water it produces.

Iron, unlike steel (which is an alloy, or mixture, of iron with carbon and other elements), loses nearly all of its magnetization when an external magnetic field is removed. Hence if an iron rod is placed inside a solenoid, we have a very strong magnet that can be turned on and off just by switching the current in the solenoid on and off. Such an **electromagnet** is much stronger than the solenoid itself and can be stronger than a permanent magnet as well. Also, unlike a permanent magnet, its field can be controlled at will by adjusting the current in the solenoid.

Electromagnets are among the most widely used electrical devices. They range in size from the tiny one in a telephone receiver that causes a steel plate to vibrate and thus produce the sounds we hear to the giant electromagnets used to pick up cars in scrap yards. A **relay** is a switch operated by an electromagnet (Fig. 19.13).

Let us see what happens to its magnetic field when the current in an iron-core solenoid is increased.

Figure 19.14 is a plot of B (the total field, including the contribution of the iron core) against B_0 (the field due to the solenoid itself, without the core). Because B is so much greater than B_0, the respective scales on the graph are different. At first, B_0 is too weak to align the domains in the iron to any great extent, so B increases slowly with increasing B_0. When B_0 is made stronger, it is more effective in aligning the domains, and B rises rapidly until it is over 5500 times greater than B_0. Eventually, when the domains are virtually all aligned with B_0, B levels off to a nearly constant saturated value. Clearly μ, the permeability of the iron, is not a fixed quantity but varies with the magnetizing field B_0.

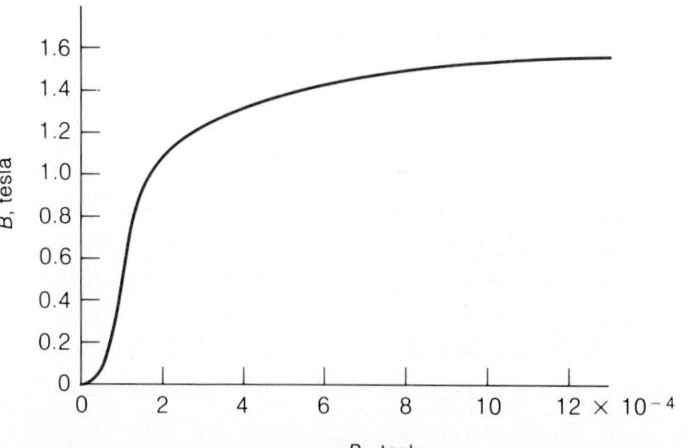

Figure 19.14

The magnetization curve of annealed iron. B_0 is the magnetic field without the iron present, and B is the total field including the effect of the iron.

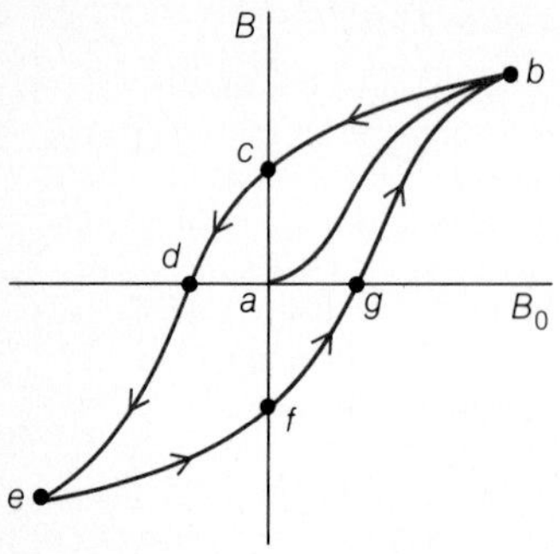

Figure 19.15
Hysteresis loop.

A ferromagnetic material tends to keep some of its magnetization even when the magnetic field that originally aligned its domains is removed. Hence μ does not even have a fixed value at a given B_0 but may take on different values depending on its past history. This phenomenon is called **hysteresis.**

Suppose we vary the current in a ferromagnetic core solenoid from zero to a maximum in one direction, down through zero to a maximum in the other direction, back through zero to the first maximum, and so on. Figure 19.15 is a plot of B versus B_0 for this cycle. When we first turn on the current, the B-B_0 curve is the same as in Fig. 19.14. At the point b we reduce the current, so that B_0 drops, but now B does *not* retrace its original path. From b to c the values of B are higher than they were from a to b at corresponding values of B_0 owing to the magnetic ''memory'' of the core. At c there is no current in the coil and $B_0 = 0$, yet B still has the magnitude B_c. The ferromagnetic sample is now permanently magnetized.

Reversing the direction of B_0 does not at first reverse B but merely reduces it, until finally, at d, B_0 is sufficiently negative to bring B to $B = 0$. A further increase in $-B_0$ takes the B-B_0 curve to e, where B and B_0 are equal in magnitude and opposite in sign to what they were at b. When B_0 is again brought to $B_0 = 0$, B is at f, where $B_f = -B_c$. Increasing B_0 in the positive sense returns B_0 to point b, but along a curve on which B is always less than it was from a to b. Further cycles simply retrace the curve *bcdefgb*.

The hysteresis curve of the iron used in electromagnet cores is narrow, so that ac is short and the field B_c that remains when B_0 is reduced to 0 is very small.

19.5 Force on a Moving Charge

The magnetic force is a sideways push.

A magnetic field can exert a force on an electric current, whether it is a current in a wire, a moving charged particle, or an atomic current such as those in a piece of iron. Whenever we switch on an electric motor, we see this property in action.

According to Eq. (19.1) the force on a particle of charge Q and velocity $\mathbf{v}$ in the magnetic field $\mathbf{B}$ has the magnitude

$$F = QvB \sin \theta \qquad \textit{Force on moving charge} \quad (19.7)$$

where θ is the angle beween $\mathbf{v}$ and $\mathbf{B}$ (Fig. 19.2b). The direction of the force $\mathbf{F}$ is given by the right-hand rule shown in Fig. 19.2(a).

The work done by a force on a body depends on the component of the force in the direction the body moves. Because the force on a charged particle in a magnetic field is perpendicular to its direction of motion, the force does no work on it. Hence the particle keeps the same speed v and energy it had when it entered the field, even though it is deflected. On the other hand, the speed and energy of a charged particle in an *electric* field are always affected by the interaction between the field and the particle, unless $\mathbf{v}$ is perpendicular to $\mathbf{E}$.

A particle of charge Q and velocity $\mathbf{v}$ that is moving in a uniform magnetic field so that $\mathbf{v}$ is perpendicular to $\mathbf{B}$ experiences a force of magnitude

$$F = QvB \qquad (\mathbf{v} \perp \mathbf{B}) \qquad (19.8)$$

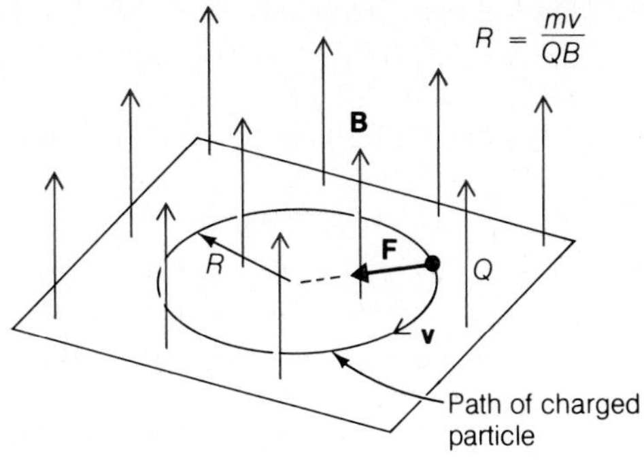

Figure 19.16

The path of a charged particle moving perpendicularly to a uniform magnetic field is a circle.

since $\sin 90° = 1$. This force is perpendicular to both **v** and **B**, so the particle travels in the circular path shown in Fig. 19.16.

To find the radius R of the circular path of the charged particle, we note that the magnetic force QvB provides the particle with the centripetal force mv^2/R that keeps it moving in a circle. Setting equal the magnetic and centripetal forces gives

$$F_{\text{magnetic}} = F_{\text{centripetal}}$$

$$QvB = \frac{mv^2}{R}$$

The radius of the particle's circular path is therefore

$$R = \frac{mv}{QB} \qquad \textit{Orbit radius} \quad (19.9)$$

The greater the momentum mv, the larger the circle, and the stronger the field, the smaller the circle.

A charged particle that moves parallel to a magnetic field experiences no force and is not deflected. The same particle moving perpendicularly to the field follows a circular path. What about the case when the particle moves at any other angle with respect to **B**? If we call $\mathbf{v}_{\|}$ the component of the particle's velocity **v** that is parallel to **B** and $\mathbf{v}_{\perp}$ the component of **v** perpendicular to **B**, then the motion of the particle is the resultant of a forward motion at the velocity $\mathbf{v}_{\|}$ and a circular motion perpendicular to this whose radius is $mv_{\perp}/QB$. As shown in Fig. 19.17, the path is a helix.

Something very interesting happens when a charged particle moving in a magnetic field approaches a region where the field becomes stronger. The magnetic field lines that describe such a field converge, because their spacing is proportional

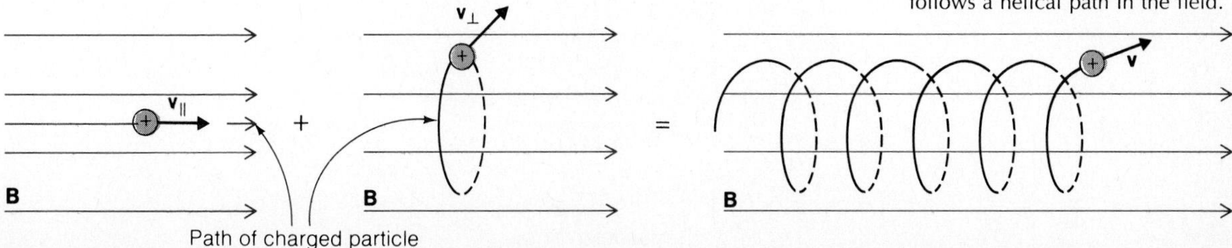

Figure 19.17

A charged particle that has velocity components both parallel and perpendicular to a magnetic field follows a helical path in the field.

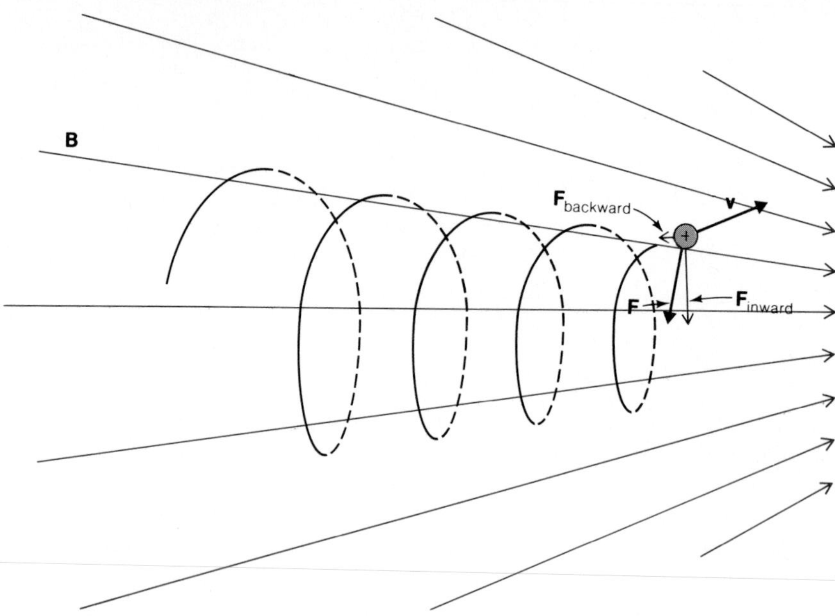

Figure 19.18

The principle of the magnetic mirror.

to the magnitude of the field they describe. The force the particle experiences now has a backward component as well as the inward component that leads to its helical path, as shown in Fig. 19.18. The backward force may be strong enough and extend over a long enough distance to reverse the particle's direction of motion. A converging magnetic field can thus act as a **magnetic mirror.**

Magnetic mirrors are found both in the laboratory and in nature. In the laboratory a pair of them can be used as a "magnetic bottle," as in Fig. 19.19, to contain a hot plasma in research on thermonuclear fusion. If a solid container were used, contact with its walls would contaminate the plasma and also cool it so that the ions would not have enough energy to interact. Magnetic bottles of this kind are somewhat leaky because ions moving along the axis of a magnetic mirror experience no backward force and hence can escape.

The earth's magnetic field traps electrons and protons from space in the **magnetosphere,** a giant doughnut-shaped magnetic bottle that surrounds the earth and extends from about 1000 km above the equator out to perhaps 65,000 km. The magnetosphere contains large numbers of particles with relatively high energies (100 MeV, for instance). Figure 19.20 shows a typical particle path in the magnetosphere.

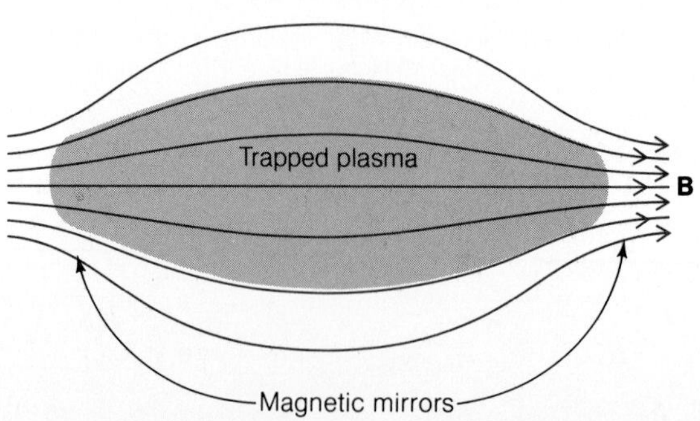

Figure 19.19

A "magnetic bottle."

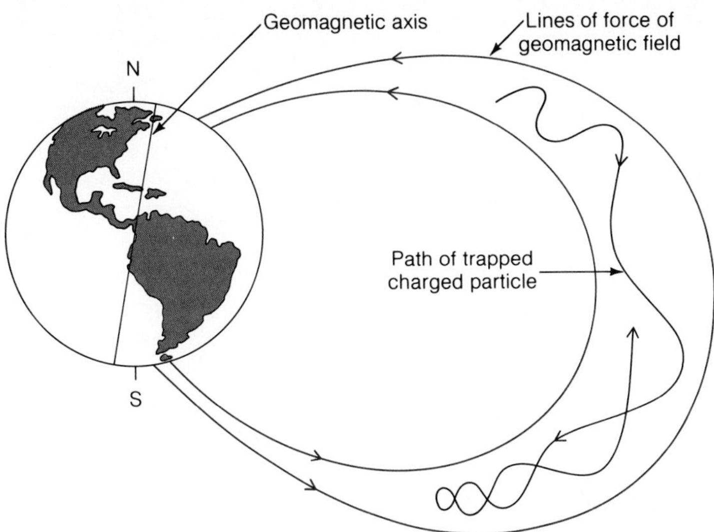

Figure 19.20

Protons and electrons are trapped by the earth's magnetic field.

BIOGRAPHICAL NOTE

André-Marie Ampère

André-Marie Ampère (1775–1836), the son of a merchant who was executed during the French Revolution, was largely self-taught and had mastered advanced mathematics while still in his early teens. Starting as a teacher in local schools near Lyon when he was 21, Ampère rose rapidly to a series of professorships in Paris. In 1808 Napoleon appointed him inspector-general of the French university system, a post he held, in addition to his other jobs, until his death. The loss of his father, the early death of his much-loved first wife, a disastrous second marriage, and financial problems all clouded Ampère's life; the epitaph he chose for his tombstone was *Tandem felix*—"Happy at last."

In contrast to the misfortunes of his personal life were the successes of Ampère's scientific career. His first publications were on mathematics. In one of them he used the theory of probability to prove that a player of a game of chance must eventually lose to another player who starts out with much more money; to quit while you're ahead is more sensible than trying to break the bank. Ampère then went on to chemistry, where he had some excellent ideas. Others, however, always seemed to have arrived at these ideas first and got credit for them, though there was no question that Ampère's work had been done independently.

In 1820 the Danish physicist Hans Christian Oersted (1777–1851) discovered that a magnetic field surrounds every electric current. Upon hearing of this effect, Ampère began a series of experiments of his own, which he analyzed to give "a great theory of these phenomena and of all others known for magnets," as he wrote to his son only two weeks later. Ampère's results included the right-hand rule for magnetic field direction and the law named after him that describes the force between two current-carrying wires of arbitrary relative orientation. This law accounts for the attraction between parallel currents and the repulsion between antiparallel ones.

(continued)

Ampère showed that the magnetic field of a current-carrying solenoid (his word) is the same as that of a bar magnet. He went on to surmise that the magnetism of a material such as iron is the result of permanent loops of electric current in its atoms, a concept very much ahead of his time that was not taken seriously by his contemporaries. The unit of electric current is called the ampere because Ampère was the first to distinguish clearly between current and potential difference in a circuit. The galvanometer, which measures the magnitude of an electric current in terms of the magnetic force the current exerts, is based on a suggestion by Ampère.

Ampère, the "Newton of electricity" to James Clerk Maxwell (who was responsible for the electromagnetic theory of light), summed up his research in electromagnetism in 1827 in *The Mathematical Theory of Electrodynamical Phenomena, Derived Solely from Experiment*. Like Newton's *Principia*, this book was the foundation stone of a whole branch of physics. Ampère did little scientific work afterward, and died of pneumonia at 61.

EXAMPLE 19.4

A long, straight wire carries a current of 100 A. (a) What is the force on an electron moving parallel to the wire, in the opposite direction to the current, at a speed of 1.0 $\times$ 10^7 m/s when it is 10 cm from the wire? (b) What is the force on the electron under the same circumstances when it is moving perpendicularly toward the wire?

SOLUTION (a) The magnetic field 0.10 m from the wire is, from Eq. (19.3),

$$B = \frac{\mu_0 I}{2\pi s} = \frac{(4\pi \times 10^{-7}\,\text{T·m/A})(100\,\text{A})}{(2\pi)(0.10\,\text{m})} = 2.0 \times 10^{-4}\,\text{T}$$

Since $\mathbf{v} \perp \mathbf{B}$ here, the force on the electron has the magnitude

$$\begin{aligned} F = evB &= (1.6 \times 10^{-19}\,\text{C})(1.0 \times 10^7\,\text{m/s})(2.0 \times 10^{-4}\,\text{T}) \\ &= 3.2 \times 10^{-16}\,\text{N} \end{aligned}$$

Because the electron has a negative charge, the direction of the force is opposite to that given by the right-hand rule of Fig. 19.2 and so is toward the wire (Fig. 19.21).

(b) Here $\mathbf{v} \perp \mathbf{B}$ also, so the force has the same magnitude. The direction of the force, however, is now the same as that of the current. ♦

EXAMPLE 19.5

A **mass spectrometer** is an instrument that measures atomic and molecular masses. The first step in the operation of the simple mass spectrometer shown in Fig. 19.22 is to produce positive ions of the substance under study, which is often done by bombardment with electrons. The ions have charges of $+e$ and are then accelerated by an electric field. (Ions with other charges are sometimes present but are easy to take into account.) A pair of slits eliminates ions not moving in the desired direction. Then the beam passes through a **velocity selector** that consists of uniform electric and magnetic fields perpendicular to each other and to the ion beam. Only ions with a specific speed v that depends on the values of E and B are not deflected. Once past the velocity selector, the ions follow curved paths in the magnetic field whose radii depend on their masses. (a) The velocity selector of a mass spectrometer has crossed

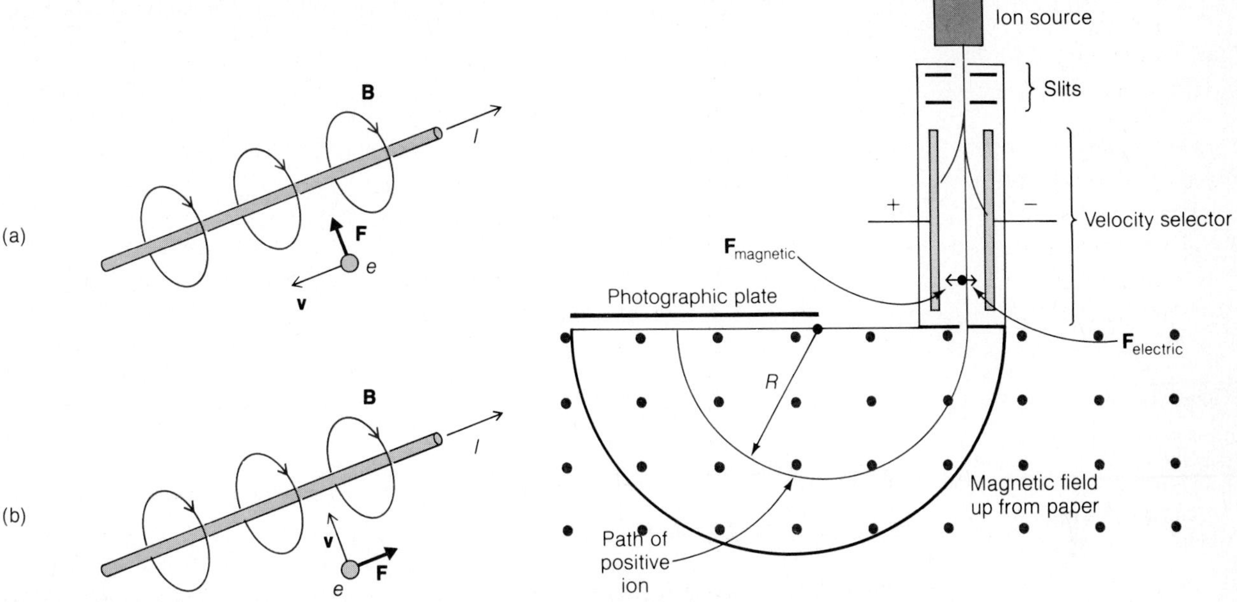

Figure 19.21
Example 19.4.

Figure 19.22
A simple mass spectrometer (Example 19.5). Modern instruments use electrical ion detectors and more complicated fields.

fields of $E = 40.0$ kV/m and $B = 0.0800$ T. What is the speed of the ions that pass through it? (b) Ions of a certain lithium isotope have radii of curvature in the same magnetic field of 390 mm after leaving the velocity selector. What is their mass?

SOLUTION (a) The electric field **E** exerts the force eE on the ions to the right, and the magnetic field **B** exerts the force evB on them to the left. The condition for no deflection is then $F_{\text{electric}} = F_{\text{magnetic}}$, $eE = evB$, and so

$$v = \frac{E}{B} = \frac{4.00 \times 10^4 \text{ V/m}}{8.00 \times 10^{-2} \text{ T}} = 5.00 \times 10^5 \text{ m/s}$$

(b) From Eq. (19.9) the mass of the ions is

$$m = \frac{QBR}{v} = \frac{(1.60 \times 10^{-19} \text{ C})(8.00 \times 10^{-2} \text{ T})(0.390 \text{ m})}{5.00 \times 10^5 \text{ m/s}}$$
$$= 9.98 \times 10^{-27} \text{ kg}$$

◆

19.6 Force on a Current

A magnetic field has the same effect on a current element as on a moving charge.

Since an electric current is a flow of charge, we would expect a current-carrying wire to be affected by a magnetic field in the same way a moving charged particle is. According to Eq. (19.1) the force on a charge Q whose velocity is **v** when

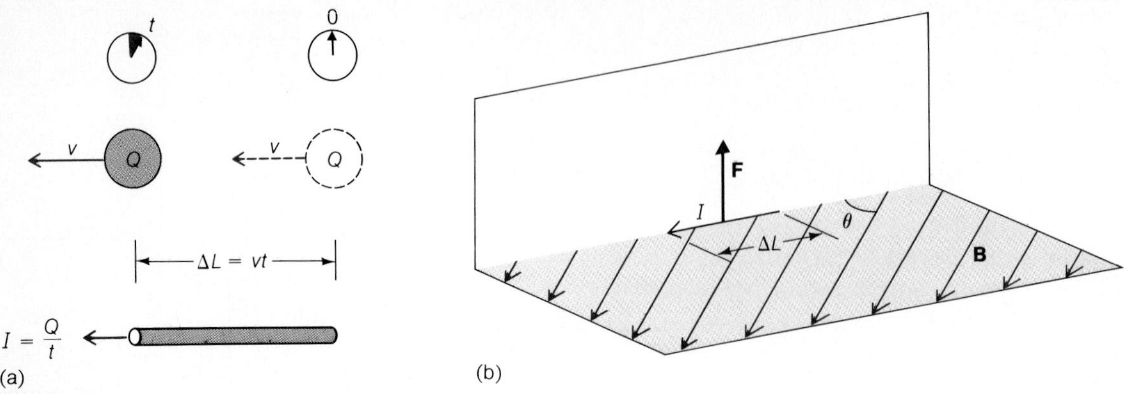

Figure 19.23

The force on a charge Q moving with the speed v in a magnetic field is the same as that on a wire ΔL long carrying the current I, where $I\,\Delta L = Qv$.

it is in the magnetic field **B** has the magnitude

$$F = QvB \sin\theta$$

where θ is the angle between **v** and **B.** What we have to do is replace the Qv of this formula with the quantity appropriate for a current.

Figure 19.23 (a) shows a particle of charge Q and speed v. In the time t the particle travels the distance

$$\Delta L = vt$$

and while it does so, it is equivalent to a current of

$$I = \frac{Q}{t}$$

Hence

$$v = \frac{\Delta L}{t} \quad \text{and} \quad Q = It$$

so that

$$Qv = I\,\Delta L$$

We conclude that the force on an element ΔL long of current I when it is in a magnetic field **B** has the magnitude

$$F = I\,\Delta LB \sin\theta \qquad \textit{Force on current element} \qquad (19.10)$$

where θ is the angle between the direction of I and that of **B** (Fig. 19.23b).

There are two ways to figure out the direction of the force on a current element in a magnetic field. Both give the same result, of course, and deciding which one to use in a particular case is largely a matter of personal preference. The first is

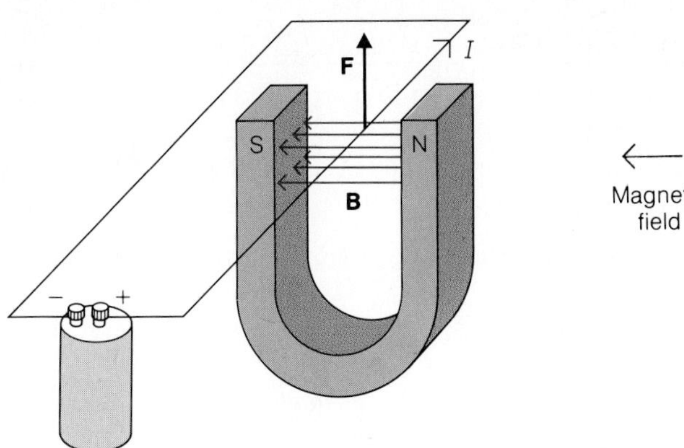

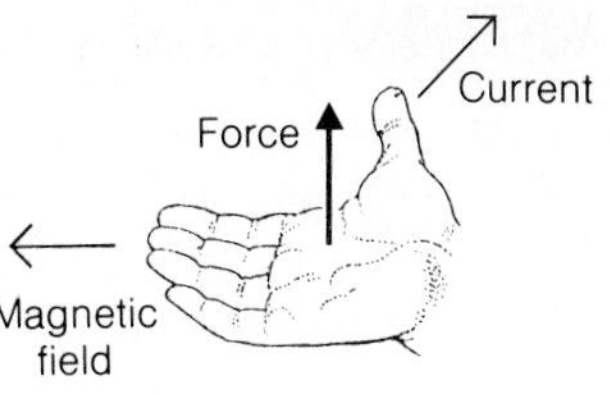

Figure 19.24

The right-hand rule for the direction of the force on a current-carrying wire in a magnetic field.

the same as the right-hand rule used for a moving charge in Section 19.2, except that now the thumb points in the direction of the current. This version of the rule is illustrated in Fig. 19.24.

The other way to find the direction of **F** is based on the pattern of magnetic field lines around a current in a magnetic field. Figure 19.25(a) shows the field lines of a uniform field **B**, and (b) shows the field lines around a wire carrying a current *I*. Since the current is into the paper, the field lines are concentric circles in the clockwise sense. When field (a) is added vectorially to field (b), the resulting pattern of field lines is like that shown in (c): The lines are closer together in the region above the wire where the field of the current is in the same direction as **B** and farther apart under the wire where the field of the current is opposite to **B.** *The direction of the force on the current element is from the region of strong field to the region of weak field,* as though the magnetic field lines were rubber bands that try to straighten out when distorted by the presence of the current.

The pictorial method for establishing the direction of **F** is easy to use and appeals to the intuition. However, we must keep in mind that field lines do not physically exist but are only a way to visualize the magnitude and direction of a force field. It is the field itself that exists in space as a continuous property of the region it occupies, not a series of strings. Despite the artificial nature of field lines, they can still be very helpful in representing various aspects of the interaction between magnetic fields and electric currents, and we can use them for this purpose whenever appropriate.

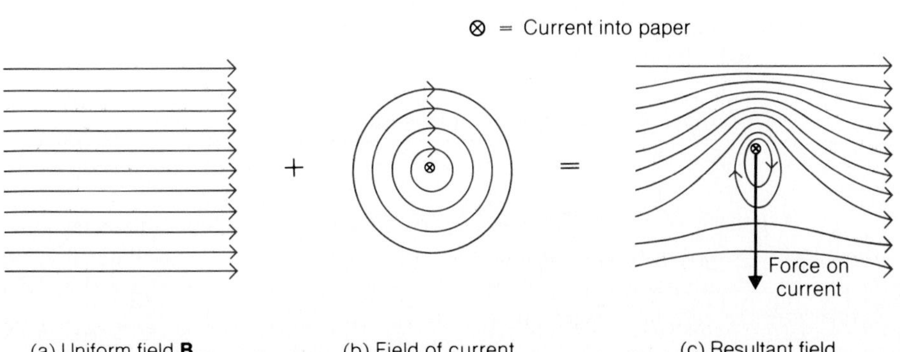

Figure 19.25

The direction of the force on a current element in a magnetic field is from the region of strong field to the region of weak field.

EXAMPLE 19.6

A wire carrying a current of 100 A due west is suspended between two towers 50 m apart. The lines of force of the earth's magnetic field enter the ground there in a northerly direction at a 45° angle; the magnitude of the field at that location is 5.0×10^{-5} T. Find the force on the wire exerted by the earth's field.

SOLUTION The wire is perpendicular to **B**, and so the magnitude of the force is

$$F = I\,\Delta LB \sin\theta = (100\text{ A})(50\text{ m})(5.0 \times 10^{-5}\text{ T})(\sin 90°) = 0.25\text{ N}$$

This force is about an ounce. By either of the methods described above, the force acts downward at a 45° angle with the ground toward the south (Fig. 19.26).

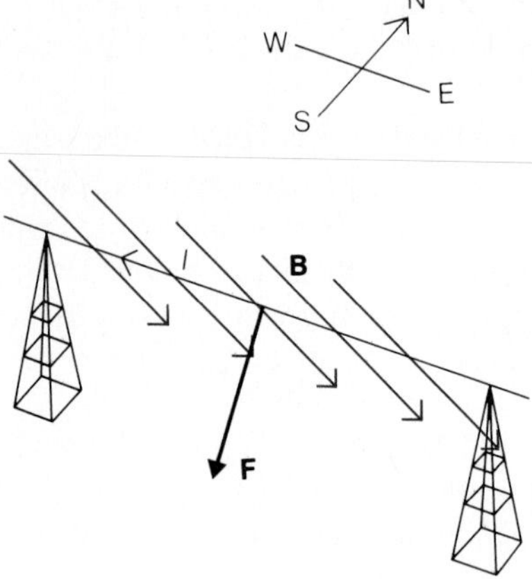

Figure 19.26

◆

EXAMPLE 19.7

A copper wire whose linear density is 10 g/m is stretched horizontally perpendicular to the direction of the horizontal component of the earth's magnetic field at a place where the magnitude of that component is 2.0×10^{-5} T. What must the current in the wire be for its weight to be supported by the magnetic force on it? What do you think would happen to such a wire if this current were passed through it?

SOLUTION Here $\theta = 90°$ and $\sin\theta = 1$, so Eq. (19.10) becomes $F = I\,\Delta L\,B$. The magnetic force per meter on the wire is $F/\Delta L$ and must be equal and opposite to the wire's weight per meter $(m/L)g$. Hence $F/\Delta L = IB = (m/L)g$ and

$$I = \frac{(m/L)g}{B} = \frac{(1.0 \times 10^{-2}\text{ kg/m})(9.8\text{ m/s}^2)}{2.0 \times 10^{-5}\text{ T}} = 4.9 \times 10^3\text{ A}$$

A current of nearly 5000 A is very high for such a small-diameter wire, which leads to the question of whether the wire can carry it without melting. It is left for the reader's amusement to show that the wire would indeed melt a fraction of a second after the current is switched on. ◆

Magnetic forces are being used to support (''levitate'') and propel high-speed experimental trains. By eliminating friction (except for air resistance), a ''maglev'' train has already exceeded 500 km/h, over 300 mi/h. Two systems are currently being developed. In West Germany attractive magnets lift the undercarriage of the train to about a centimeter from a guiding T-shaped track. Very precise control is needed to keep the train floating correctly. In Japan repulsive magnets are used, a more stable method because the closer the train gets to its guiding track, the more the upward push. Stronger magnetic fields are required, however, which means more expensive superconducting electromagnets to produce them. In both systems an alternating current passed through electromagnets under the track creates an attractive force in front of the train's own magnets and a repulsive force behind them. The higher the frequency of the current, the higher the train's speed.

The German *Transrapid* train, using magnetic forces for both support and propulsion, has reached a speed of 413 km/h.

19.7 Force Between Two Currents

Parallel currents attract, antiparallel currents repel.

Every current has a magnetic field around it, and as a result nearby currents exert magnetic forces on one another.

Figure 19.27 shows two parallel wires a distance s apart that carry the currents I_1 and I_2 respectively. The magnetic fields a distance s from each of the wires are, from Eq. (19.3),

$$B_1 = \frac{\mu I_1}{2\pi s} \qquad B_2 = \frac{\mu I_2}{2\pi s}$$

The fields are perpendicular to the wires, which means that $\theta = 90°$ and $\sin\theta = 1$. Therefore the force F_{12} on a length L of current 1 exerted by the magnetic field of current 2 is

$$F_{12} = I_1\, \Delta L B_2 \sin\theta = \frac{\mu I_1 I_2}{2\pi s} L \tag{19.11}$$

The force F_{21} on current 2 exerted by the magnetic field of current 1 has the same

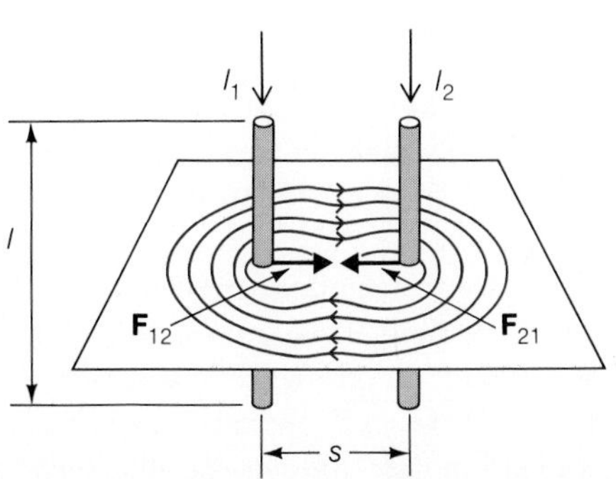

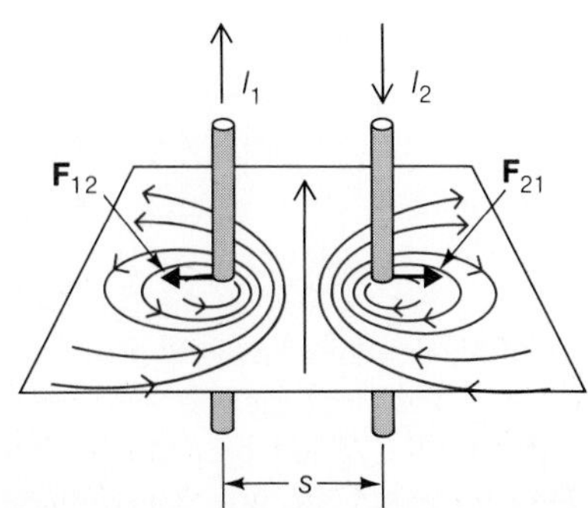

Figure 19.27

Equal and opposite forces are exerted by parallel currents on each other. The forces are attractive when the currents are in the same direction, repulsive when they are in opposite directions.

magnitude. Both forces can be expressed in the form

$$\frac{F}{L} = \frac{\mu I_1 I_2}{2\pi s} \qquad \textit{Force between parallel currents} \qquad (19.12)$$

where F/L is the *force per unit length* each wire exerts on the other by virtue of its magnetic field. From the pattern of lines of force around each wire, it is clear that the forces are always opposite in direction and hence obey Newton's third law of motion, as they must. The forces are attractive when the currents are in the same direction and repulsive when they are in opposite directions.

EXAMPLE 19.8

The cables that connect the starting motor of a car with its battery are 10 mm apart for a distance of 40 cm. Find the forces between the cables when the current in them is 300 A.

SOLUTION The currents in the cables are opposite in direction, and hence the forces are repulsive. Their magnitudes, taking $\mu = \mu_0$, are

$$F = \frac{\mu_0 I^2}{2\pi s} L = \frac{(4\pi \times 10^{-7}\ \text{T·m/A})(300\ \text{A})^2(0.40\ \text{m})}{(2\pi)(1.0 \times 10^{-2}\ \text{m})} = 0.72\ \text{N}$$

This is 0.16 lb. ◆

Equation (19.12) is used to define the ampere: An ampere is that current in each of two parallel wires 1 m apart in free space that produces a force on each wire of exactly 2×10^{-7} N per meter of length. (Thus $\mu_0 = 4\pi \times 10^{-7}$ T·m/A.) In turn, the coulomb is defined in terms of the ampere as that amount of charge transferred per second by a current of 1 A. The ampere is chosen as the primary electrical unit instead of the coulomb because it can be defined in terms of a more direct experiment than would be possible with the coulomb.

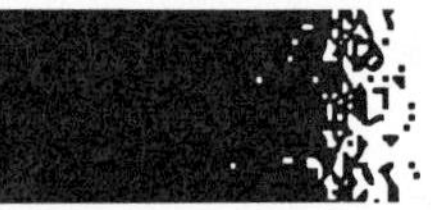

19.8 Torque on a Current Loop

How an electric motor works.

No net force acts on a loop of current in a uniform magnetic field. Instead, a torque occurs that tends to turn the loop to bring its plane perpendicular to **B.** This effect underlies the operation of all electric motors, from the tiniest one in a clock to the many-thousand-horsepower giant in a locomotive.

Let us look at the forces on each side of a rectangular current-carrying wire loop whose plane is parallel to a uniform magnetic field **B,** as in Fig. 19.28(a). The sides A and C of the loop are parallel to **B,** so there is no magnetic force on them. Sides B and D are perpendicular to **B,** however, and each therefore experiences a force. To find the directions of the forces on B and D, we can use the right-hand rule: With the fingers of the right hand in line with **B** and the outstretched thumb in

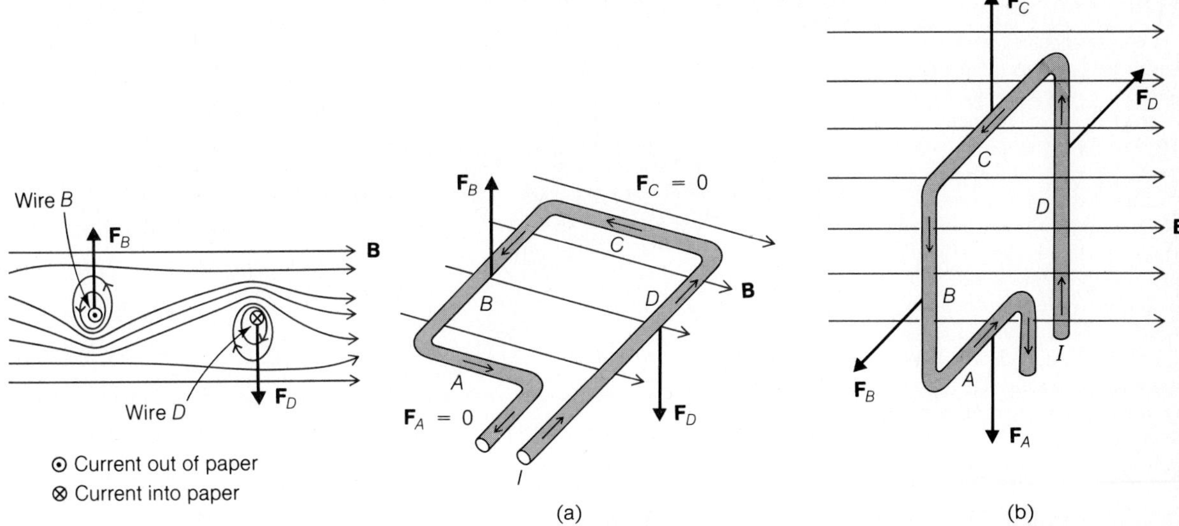

Figure 19.28

(a) A current-carrying wire loop whose plane is parallel to a magnetic field experiences a torque. (b) If the plane of the loop is perpendicular to the magnetic field, there is no torque on the loop. In both cases there is no net force on the loop.

line with **I**, the palm faces the same way as **F**. What we find is that $\mathbf{F}_B$ is opposite in direction to $\mathbf{F}_D$. The same conclusion follows from the pattern of lines of force.

The forces $\mathbf{F}_B$ and $\mathbf{F}_D$ are the same in magnitude, so there is no net force on the current loop. Because $\mathbf{F}_B$ and $\mathbf{F}_D$ do not act along the same line, they exert a torque on the loop that tends to turn it. Although the origin of this torque is easiest to understand in the case of a rectangular loop, the torque occurs for a loop of any shape in a magnetic field when the plane of the loop is not perpendicular to the magnetic field.

If the plane of the loop is perpendicular to the magnetic field instead of parallel to it, there is neither a net force nor a net torque on it. This is easy to verify by applying the right-hand rule in Fig. 19.28(b). Evidently $\mathbf{F}_A$ and $\mathbf{F}_C$ cancel each other out, and $\mathbf{F}_B$ and $\mathbf{F}_D$ also cancel each other out. There is no torque now because $\mathbf{F}_A$ and $\mathbf{F}_C$ have the same line of action, and $\mathbf{F}_B$ and $\mathbf{F}_D$ have the same line of action.

These results can be summarized by saying:

When the plane of a current loop in a magnetic field B is not perpendicular to B, the loop tends to turn so that its plane becomes perpendicular to B.

Galvanometers and Motors

A **galvanometer** is a device based on the effect just described that is used to measure electric current. As in Fig. 19.29, a U-shaped permanent magnet is used to provide a magnetic field, and between its poles is a small coil wound on an iron core to enhance the torque developed when the unknown current is passed through it. The coil assembly is held in place by two bearings, and a pair of hairsprings keeps the pointer at 0 when there is no current in the coil.

When a current flows, there is a torque on the coil because of the interaction between the current and the magnetic field, and the coil rotates as far as it can against the opposing torque of the springs. The more the current, the stronger the torque and

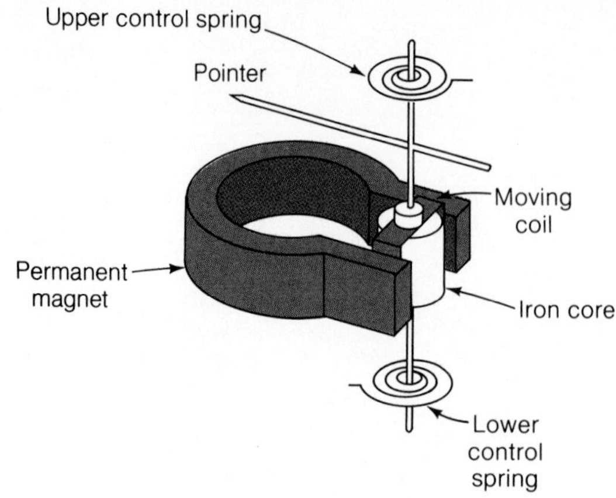

Figure 19.29

The construction of a common type of galvanometer. Such galvanometers can respond to currents as small as 0.1 microampere (10^{-7} A). Even more sensitivity can be attained if the moving coil is suspended by a thin wire to which a small mirror is attached: Bearing friction is avoided in this way, and the mirror deflects a light beam so that the "pointer" may be a meter or more long instead of a few centimeters. Laboratory galvanometers like this can be used to measure currents of 10^{-10} A.

the farther the coil turns. The restoring torque of the hairsprings is proportional to the angle through which they are twisted, and as a result the deflection of the pointer is directly proportional to the current I in the coil.

How does an electric motor achieve continuous motion? The torque a magnetic field exerts on a current loop disappears when the loop turns so that its plane is perpendicular to the field direction. If the loop swings past this position, the torque on it will be in the opposite sense and will return the loop to the perpendicular orientation. In order for a motor to rotate continuously, then, the current in the loop must be reversed each time it turns through 180°. The way this is done is shown in Fig. 19.30. The current is led to the loop by means of graphite rods called **brushes** that press against a split ring called a **commutator.** As the loop rotates, the current is reversed twice per turn as the commutator segments make contact alternately with the brushes. The torque is always in the same direction, except at the moments of switching when it is zero because the loop is perpendicular to the field. However, the angular momentum of the loop carries it past this point so it can go on turning.

While actual direct-current (dc) electric motors, such as the starter motor of a car, are the same in principle as the simple device in Fig. 19.30, they use a number

Stationary windings of a large electric motor.

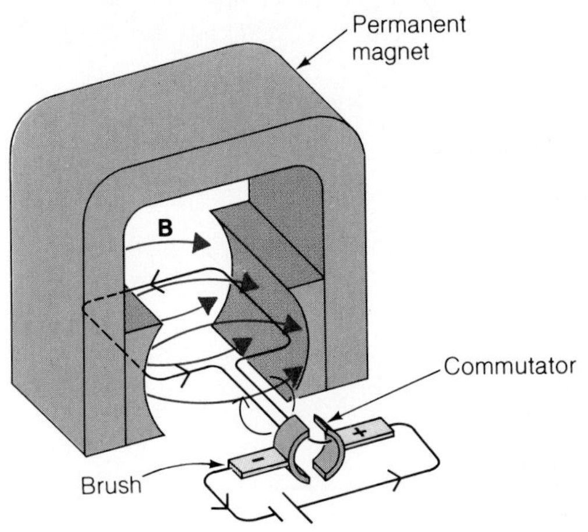

Figure 19.30

A simple dc electric motor. The commutator automatically reverses the current in the rotating loop twice per rotation so that the torque will stay in the same direction.

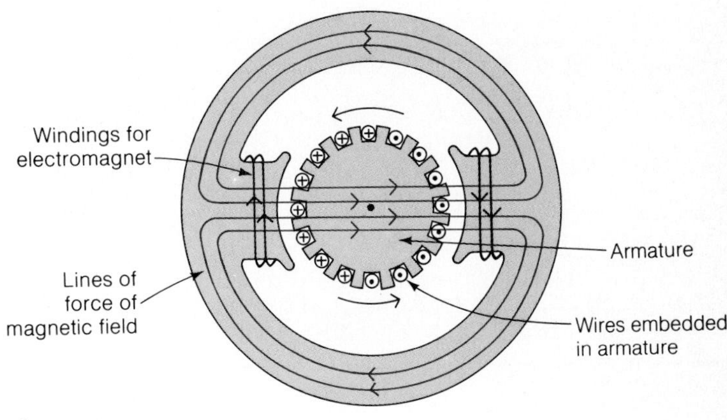

Figure 19.31

Actual dc electric motors employ various means to increase the available torque.

of methods to increase the available torque. Electromagnets rather than permanent magnets provide the field, and there are six or more different coils with many turns each on a slotted iron core called an **armature,** instead of a single loop (Fig. 19.31). A commutator with a pair of segments for each coil is provided so that only those coils approximately parallel to the magnetic field receive current at any time, which means that maximum torque is developed continuously.

Electric energy for industrial and domestic purposes is usually transmitted by **alternating current** (ac), whose direction periodically reverses itself. In the United States the frequency of ordinary ac is 60 Hz, which means that the current changes direction 120 times per second. This frequency is 50 Hz in much of the rest of the world. In ac electric motors, commutators and brushes are not needed because the current itself does the required reversing, which makes such motors easier to build and more reliable than dc motors.

In order to start the armature of an ac electric motor turning, and to ensure that it begins to turn the desired way, an auxiliary stationary winding is used. This winding, together with the operating winding, creates a magnetic field whose direction rotates about the motor's axis. This rotating field pulls the armature around when the motor is switched on. As the armature approaches its normal running speed, the alternating magnetic fields of the operating windings are sufficient to keep it going, and (depending on the motor design) the starting winding may then be cut out of the circuit.

19.9 Magnetic Poles

They come in pairs.

The magnetic field of a bar magnet is the same as that of a solenoid, as we saw in Fig. 19.9. Hence we can use the ideas of this chapter to understand how such magnets behave.

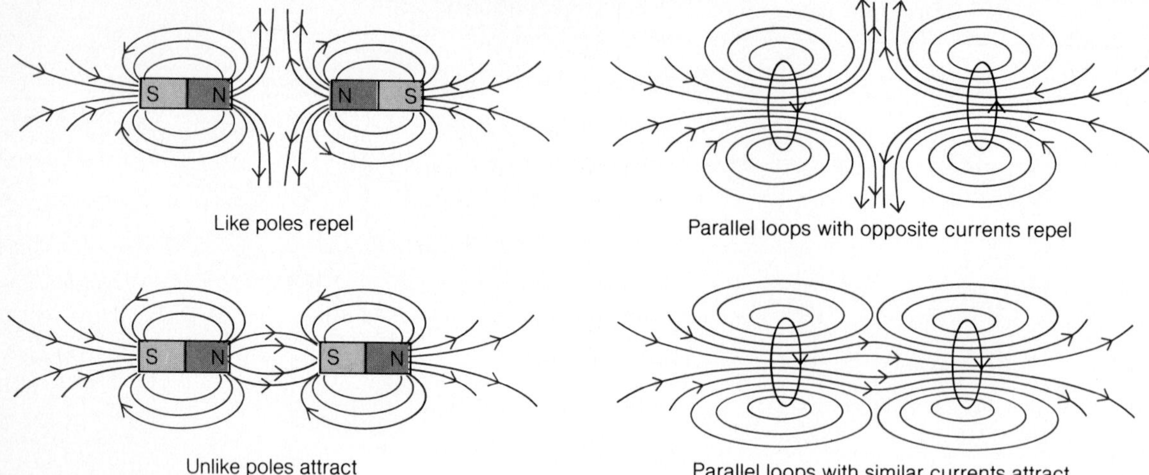

Figure 19.32

Interactions between magnets can be traced to interactions between current loops.

Because the external magnetic field of a bar magnet seems to originate in its ends, these are by custom called its **poles.** At one time it was believed that the poles were ''magnetic charges'' analogous to electric charges, and that the field of the magnet was due to these poles. This belief was reinforced by the repulsion of like poles and the attraction of unlike ones, effects that have their true explanation in the forces between parallel and antiparallel currents (Fig. 19.32).

An important difference between magnetic poles and electric charges is that magnetic poles always occur in pairs of equal strength and opposite polarity. If a magnet is sawed in half, the poles are not separated, but instead two new magnets are created, as in Fig. 19.33. Magnetic poles are therefore not really like electric charges, and all effects they are supposed to cause can be explained in terms of current-carrying solenoids.

Measurements of the earth's magnetic field show that it is very much like the field that would be produced by a powerful current loop whose center is a few hundred kilometers from the earth's center and whose plane is tilted by 11° from the plane of the earth's equator (Fig. 19.34). On the basis of geological evidence, the earth is thought to have a core of molten iron 3470 km (2160 mi) in radius, a little

Figure 19.33

Cutting a magnet in half produces two new magnets.

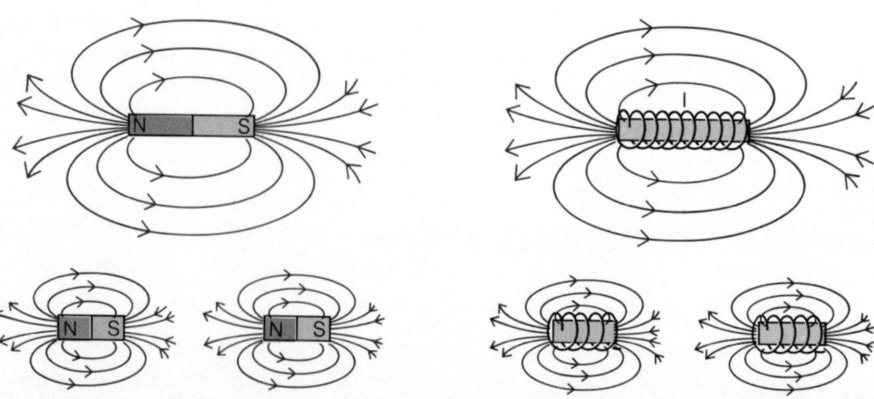

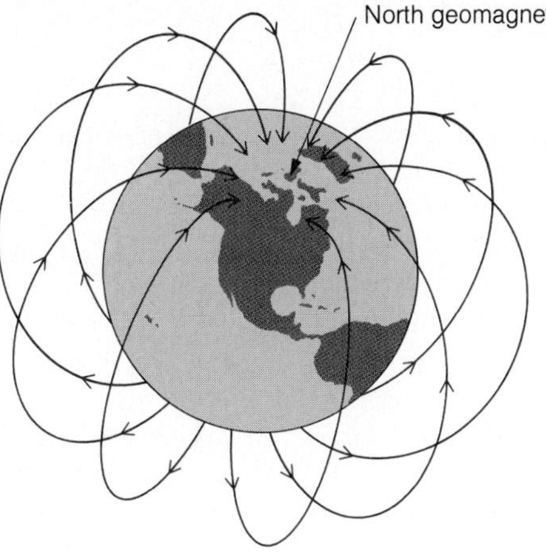

Figure 19.34

The earth's magnetic field originates in currents in its core of molten iron. The magnetic axis is tilted by 11° from the axis of rotation.

over half the earth's radius, and there is no doubt today that electric currents in this core are responsible for the observed geomagnetic field. Exactly how these currents came into being and how they are maintained are still uncertain, however.

As we saw, a current loop tends to turn in a magnetic field until its axis is parallel to the field. A bar magnet does the same. Because of the earth's magnetic field, a magnet suspended by a string therefore turns so as to line up in an approximately north-south direction. A compass consists of a pivoted magnetized iron needle together with a card that permits directions relative to magnetic north to be determined (Fig. 19.35). The end of a freely swinging magnet that points toward the north is called its north-seeking pole, usually shortened to just **north pole.** The other end is its south-seeking pole, or **south pole.** *Magnetic field lines leave the north pole of a magnet and enter its south pole.* (The north geomagnetic pole is thus a south pole, and the south geomagnetic pole is a north pole. This has confused people for several hundred years.)

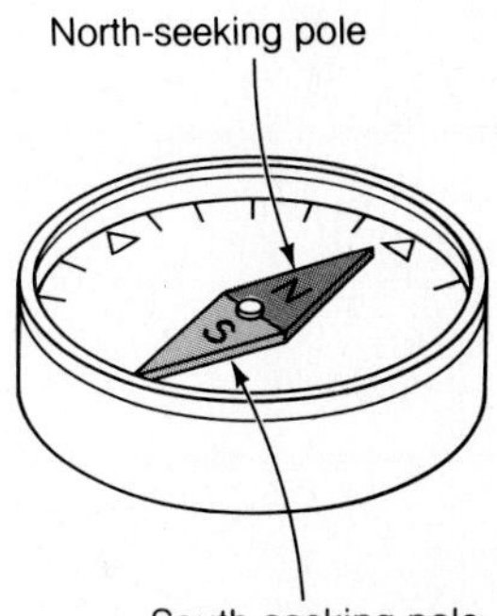

Figure 19.35

A magnetic compass.

The mechanism by which a magnet or a solenoid attracts an iron object (or an object of any other ferromagnetic material such as cobalt or nickel) is very similar to the way in which an electric charge attracts an uncharged object. First the presence of the magnet causes the atomic magnets in the iron object to line up with its field by one of the mechanisms shown in Fig. 19.12, and then the attraction of opposite poles leads to a force on the object that draws it toward the magnet, as in Fig. 19.36.

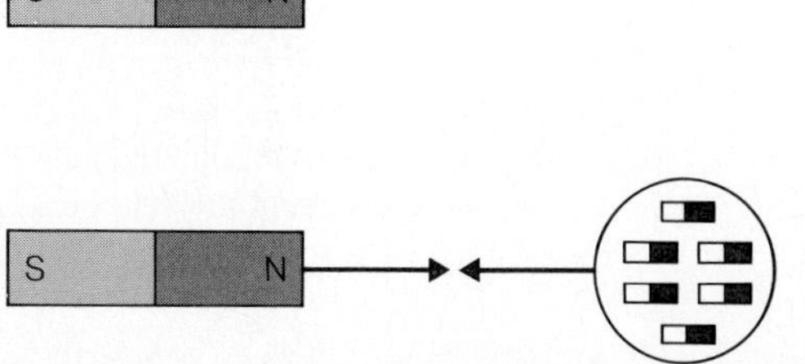

Figure 19.36

How a permanent magnet attracts a ferromagnetic object.

Important Terms

Charged particles in motion relative to an observer exert forces on one another that are different from the electrical forces they exert when at rest. These differences are by custom said to arise from **magnetic forces.** In reality, magnetic forces represent modifications of electrical forces due to the motion of the charges involved, as predicted by the theory of relativity.

A **magnetic field** exists wherever a magnetic force would act on a moving charged particle. The direction of a magnetic field **B** at a point is such that a charged particle would experience no force if it moved in that direction at the point. The magnitude of **B** is numerically equal to the force that would act on a charge of 1 C moving at 1 m/s perpendicularly to **B.** The unit of magnetic field is the **tesla** (T), equal to 1 N/A·m.

When different substances are inserted in a current-carrying wire coil, the magnetic field in its vicinity changes. Those substances that lead to a slight increase in B are called **paramagnetic,** those that lead to a great increase in B are called **ferromagnetic,** and those that lead to a slight decrease in B are called **diamagnetic.** A **permanent magnet** is an object composed of ferromagnetic material whose atomic current loops have been aligned.

The **permeability** of a medium is a measure of its magnetic properties. Paramagnetic and ferromagnetic substances have higher permeabilities than free space, and diamagnetic ones that lower permeabilities. The permeability of a ferromagnetic material in a given magnetizing field B_0 depends on its past history as well as on B_0, a phenomenon called **hysteresis.**

The ends of a permanent magnet are called its **poles.** Magnetic lines of force leave the **north pole** of a magnet and enter its **south pole.**

Important Formulas

Magnetic field: $B = \dfrac{F}{Qv \sin \theta}$

Field around long, straight current: $B = \dfrac{\mu I}{2\pi s}$

Field at center of flat coil: $B = \dfrac{\mu NI}{2r}$

Field inside a solenoid: $B = \mu \dfrac{N}{l} I$

Force on moving charge: $F = QvB \sin \theta$

Force on current element: $F = I \, \Delta L \, B \sin \theta$

Force between parallel currents: $\dfrac{F}{L} = \dfrac{\mu I_1 I_2}{2\pi s}$

MULTIPLE CHOICE

1. All magnetic fields originate in
- **a.** iron atoms
- **b.** permanent magnets
- **c.** magnetic domains
- **d.** moving electric charges

2. An observer moves past a stationary electron. His instruments measure
- **a.** an electric field only
- **b.** a magnetic field only
- **c.** both electric and magnetic fields
- **d.** any of the above, depending on his speed

3. Magnetic fields do not interact with
- **a.** stationary electric charges
- **b.** moving electric charges
- **c.** stationary permanent magnets
- **d.** moving permanent magnets

4. Magnetic field lines provide a convenient way to visualize a magnetic field. Which of the following statements is not true?
- **a.** The path followed by an iron particle moving in a magnetic field corresponds to a field line.
- **b.** The path followed by an electric charge moving in a magnetic field corresponds to a field line.
- **c.** A compass needle in a magnetic field lines up parallel to the field lines around it.
- **d.** Magnetic field lines do not actually exist.

5. A drawing of the field lines of a magnetic field provides information on
- **a.** the direction of the field only
- **b.** the magnitude of the field only
- **c.** both the direction and the magnitude of the field
- **d.** the source of the field

6. In a drawing of magnetic field lines, the stronger the field is,
- **a.** the closer together the field lines are
- **b.** the farther apart the field lines are
- **c.** the more nearly parallel the field lines are
- **d.** the more divergent the field lines are

7. The magnetic field near a strong permanent magnet might be

a. 10^{-9} T **b.** 10^{-5} T
c. 0.1 T **d.** 100 T

8. The earth's magnetic field at sea level typically is

a. 3×10^{-9} T b. 3×10^{-5} T
c. 3×10^{5} T d. 3×10^{9} T

9. A current is flowing east along a power line. If we neglect the earth's field, the direction of the magnetic field below it is

a. north b. east
c. south d. west

10. The magnetic field lines around a long, straight current are in the form of

a. straight lines parallel to the current
b. straight lines that radiate perpendicularly from the current, like the spokes of a wheel
c. concentric circles centered on the current
d. concentric helixes whose axis is the current

11. Inside a solenoid the magnetic field

a. is zero
b. is uniform
c. increases with distance from the axis
d. decreases with distance from the axis

12. The magnitude of the magnetic field inside a solenoid of N turns does not depend on

a. the nature of the medium inside the solenoid
b. the solenoid's length
c. the solenoid's diameter
d. the current in the solenoid

13. When a ferromagnetic substance is inserted in a current-carrying solenoid, the magnetic field is

a. slightly decreased b. greatly decreased
c. slightly increased d. greatly increased

14. Which of the following statements about the magnetic field of a solenoid with an iron core is (are) not always true?

a. Increasing I increases **B**.
b. Decreasing I decreases **B**.
c. $\mathbf{B} = 0$ when $I = 0$.
d. Changing the direction of I changes the direction of **B**.

15. When a magnetized iron bar is strongly heated, its magnetic field

a. becomes weaker
b. becomes stronger
c. reverses its direction
d. is unchanged

16. An electron enters a magnetic field parallel to **B**. The electron's

a. motion is unaffected
b. direction is changed
c. speed is changed
d. energy is changed

17. An electron enters a magnetic field perpendicularly to **B**. The electron's

a. motion is unaffected
b. direction is changed
c. speed is changed
d. energy is changed

18. An ion moves in a circular orbit of radius R in a magnetic field. If the particle's speed is doubled, the orbit radius will become

a. $R/2$ b. R
c. $2R$ d. $4R$

19. An ion that moves through crossed electric and magnetic fields (that is, so that $\mathbf{E} \perp \mathbf{B}$) perpendicularly to both **E** and **B** is not deflected when its speed is equal to

a. EB b. E/B
c. B/E d. B/E^2

20. Two parallel wires carry currents in the same direction.

a. The wires attract each other.
b. The wires repel each other.
c. Torques act on the wires that tend to turn them so that they are perpendicular.
d. Neither forces nor torques act on the wires.

21. A current-carrying wire is in a uniform magnetic field with the direction of the current the same as that of the field.

a. There is a force on the wire that tends to move it parallel to the field.
b. There is a force on the wire that tends to move it perpendicularly to the field.
c. There is a torque on the wire that tends to rotate it until it is perpendicular to the field.
d. There is neither a force nor a torque on the wire.

22. A current-carrying loop in a magnetic field always tends to rotate until the plane on the loop is

a. parallel to the field
b. perpendicular to the field
c. either parallel or perpendicular to the field, depending on the direction of the current
d. at a 45° angle with the field

23. The nature of the force responsible for the operation of an electric motor is

a. electrical
b. magnetic
c. a combination of electrical and magnetic
d. either electrical or magnetic, depending on the design of the motor

24. The magnetic field of a bar magnet most closely resembles the magnetic field of

a. a straight current-carrying wire
b. a stream of electrons moving parallel to one another
c. a current-carrying wire loop
d. a horseshoe magnet

25. A permanent magnet does not exert a force on

a. an unmagnetized iron bar
b. a magnetized iron bar
c. a stationary electric charge
d. a moving electric charge

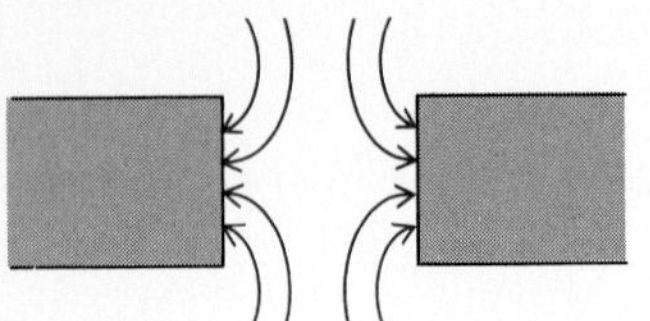

Figure 19.37

Multiple Choice 26.

26. The magnetic field shown in Fig. 19.37 is produced by
  a. a north pole and a south pole
  b. two north poles
  c. two south poles
  d. a south pole and an unmagnetized iron bar

27. At different places on the earth's surface, the earth's magnetic field
  a. is the same in direction and magnitude
  b. may be different in direction but not in magnitude
  c. may be different in magnitude but not in direction
  d. may be different in both magnitude and direction

28. The magnetic field 2 cm from a long, straight wire is 10^{-6} T. The current in the wire is
  a. 0.01 A   b. 0.1 A
  c. 1 A   d. 10 A

29. The magnetic field inside a 100-turn solenoid 20 mm long that carries a 10-A current is
  a. 0.00063 T   b. 0.0013 T
  c. 0.063 T   d. 0.13 T

30. In a certain electric motor, wires that carry a current of 6 A are perpendicular to a magnetic field of 0.5 T. The force per centimeter on these wires is
  a. 0.03 N   b. 0.2 N
  c. 3 N   d. 0.3 kN

31. Two parallel wires in free space are 10 cm apart and carry currents of 10 A each in the same direction. The force each wire exerts on the other per meter of length is
  a. 2×10^{-7} N, attractive
  b. 2×10^{-7} N, repulsive
  c. 2×10^{-4} N, attractive
  d. 2×10^{-4} N, repulsive

QUESTIONS

1. An experimenter is able to measure electric, magnetic, and gravitational fields. Which will be detected when (a) a proton moves past her, and (b) when she moves past a proton?

2. When you face the screen of a TV tube, in what direction is the magnetic field of the electron beam?

3. Figure 19.38 shows a current-carrying wire and a compass. In what direction does the compass needle point?

4. A solenoid has two windings, one on top of the other, each with the same length, number of turns, and resistance. The windings are such that their magnetic fields are in the same direction. Compare the magnetic field inside the solenoid when the windings are connected (a) in series and (b) in parallel to the same battery of negligible internal resistance.

5. An alternating current whose variation with time follows a sine curve is sent through an iron-core solenoid. Does the resulting magnetic field also vary sinusoidally with time? Explain.

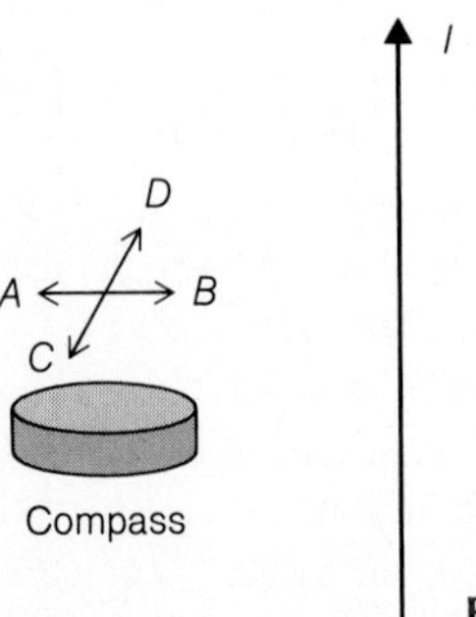

Figure 19.38

Question 3.

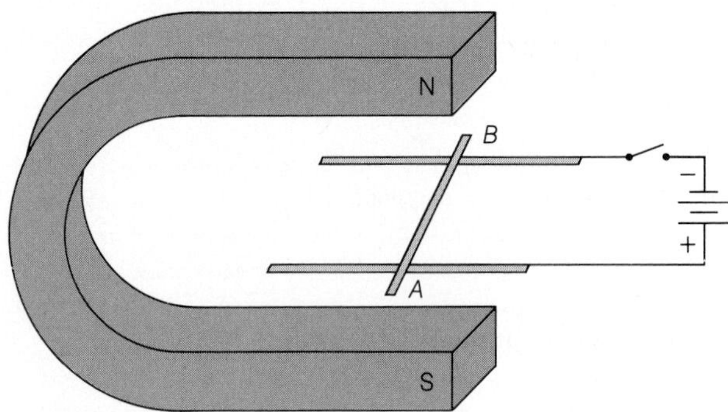

Figure 19.39

Question 10.

6. What aspect of the magnetic field of a "magnetic mirror" enables it to reflect approaching charged particles?

7. An electron is moving vertically upward when it enters a magnetic field directed to the east. In what direction is the force on the electron?

8. What path is followed by an electron sent into a current-carrying solenoid along its axis? What is its path when the electron enters the solenoid at an angle relative to its axis?

9. A current-carrying wire is in a magnetic field. (a) What angle should the wire make with **B** for the force on it to be zero? (b) What should the angle be for the force to be a maximum?

10. A length of copper wire *AB* rests across a pair of parallel copper wires that are connected to a battery through a switch, as in Fig. 19.39. The arrangement is placed between the poles

of a magnet, and the switch is closed. Does the wire move to the left or to the right, or rise upward, or press downward?

11. A stream of protons is moving parallel to a stream of electrons. Is the force between the two streams necessarily attractive? Explain.

12. A beam of protons, initially moving slowly, is accelerated to higher and higher speeds. What happens to the diameter of the beam during this process?

13. The electron beam in a cathode-ray tube is aimed parallel to a nearby wire carrying a current in the same direction. Is the electron beam deflected? If so, in what direction?

14. What, if anything, happens to the length of a helical spring when a current is passed through it?

15. A current is passed through a loop of highly flexible wire. What shape does the loop assume? Why?

16. What is the direction of the magnetic field that causes the wire loop in Fig. 19.40 to experience a clockwise torque about the z-axis?

17. A current-carrying wire loop is in a uniform magnetic field. Under what circumstances, if any, will there be no torque on the loop? No net force? Neither torque nor net force?

18. A bar magnet is placed in a uniform magnetic field so that it is parallel to the field with its north pole in the field direction. Sketch the field lines of the resulting magnetic field **B** around the magnet. Indicate where **B** = 0.

19. The bar magnet in Question 18 is turned so that its north pole is now opposite to the direction of the uniform field. Sketch the field lines of the resulting magnetic field **B** around the magnet for this situation. Indicate where **B** = 0.

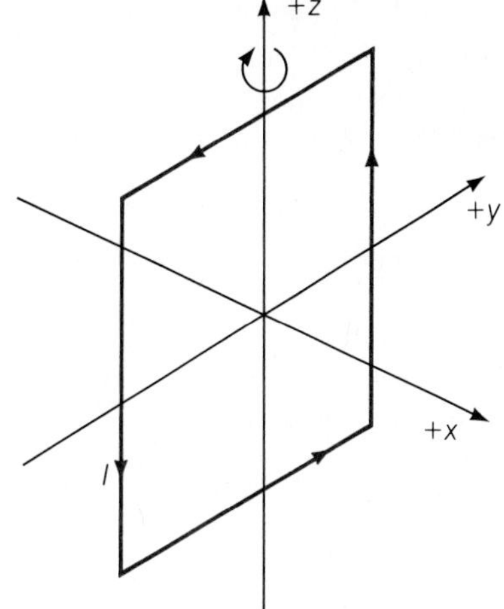

Figure 19.40
Question 16.

20. Why is a piece of iron attracted by *either* pole of a magnet?

21. Would you expect a compass to be more accurate near the equator or in the polar regions? Why?

22. What is believed to be the origin of the earth's magnetic field?

EXERCISES

19.3 Magnetic Field of a Current

1. At what distance from a long, straight wire carrying a current of 12 A does the magnetic field equal that of the earth, approximately 3.0×10^{-5} T?

2. A power line 10 m above the ground carries a current of 5000 A. Find the magnetic field magnitude of the current on the ground directly under the cable.

3. Two parallel wires 100 mm apart carry currents in the same direction of 8.0 A. Find the magnetic field magnitude midway between the wires.

4. Two parallel wires 100 mm apart carry currents in opposite directions of 8.0 A. Find the magnetic field magnitude midway between the wires.

5. Two parallel wires 200 mm apart carry currents in the same direction of 5.0 A. Find the magnetic field magnitude between the wires 50 mm from one of them and 150 mm from the other.

6. Two parallel wires 200 mm apart carry currents in opposite directions of 5.0 A. Find the magnetic field magnitude between the wires 50 mm from one of them and 150 mm from the other.

7. What should the current be in a wire loop 10 mm in diameter if the magnetic field magnitude at the center of the loop is to be 0.0010 T?

8. A 10-turn circular coil of radius 20 mm carries a current of 0.50 A. Find the magnetic field magnitude at its center.

9. A long solenoid is wound with 30 turns/cm. What is the magnetic field magnitude inside the solenoid when the current is 3.8 A?

10. A solenoid 10 cm long is meant to have a magnetic field magnitude of 0.0020 T inside it when the current is 3.0 A. How many turns are needed?

11. The magnetic field inside a long current-carrying spring has the magnitude B. What is the field there when the spring is pulled out to twice its normal length?

12. A toroidal coil is a solenoid bent around into a doughnut shape with its axis in a circle of radius R. Find the magnetic field magnitude inside a toroidal coil of R = 100 mm that has 800 turns when the current is 0.50 A.

13. A tube 200 mm long is wound with 400 turns of wire in one layer and with 200 turns in a second layer in the opposite sense to the first. A current of 0.10 A is passed through the coils. What is the magnetic field magnitude in the interior of the tube?

14. A solenoid is wound with 18 turns/cm and carries a current of 1.2 A. Another layer of turns is wound over the solenoid with 10 turns/cm, and a current of 5.0 A is passed through the new coil, opposite to the direction of the current in the solenoid. What is the magnetic field magnitude in the interior of the solenoid?

15. A current of 0.50 A is passed through a solenoid wound with 20 turns/cm. A single loop of wire 10 mm in radius is bent around the middle of the solenoid. What should the current in this loop be in order for the magnetic field at its center to cancel out the field of the solenoid there?

19.5 Force on a Moving Charge

16. An electron in a television picture tube travels at 3×10^7 m/s and is acted on both by gravity and by the earth's magnetic field. Which exerts the greater force on the electron?

17. What is the radius of the path of an electron whose speed is 1.0×10^7 m/s in a magnetic field of magnitude 0.020 T when the electron's path is perpendicular to the field?

18. Find the minimum radius of curvature in the earth's magnetic field at sea level, assuming $B = 3.0 \times 10^{-5}$ T, of (a) a proton whose speed is 2.0×10^7 m/s and (b) an electron of the same speed.

19. A charge of $+1.0 \times 10^{-6}$ C is moving at 500 m/s along a path parallel to a long, straight wire and 100 mm from it. The wire carries a current of 2.0 A in the same direction as that of the charge. What is the magnitude and direction of the force on the charge?

20. A charge of $+2.0 \times 10^{-6}$ C is moving at 1.0×10^3 m/s at a distance of 120 mm away from a straight wire carrying a current of 4.0 A. Find the magnitude and direction of the force on the charge when it is moving parallel to the wire (a) in the same direction as the current and (b) in the opposite direction to the current.

21. An electron is moving at 6.0×10^7 m/s at a distance of 50 mm from a long, straight wire carrying a current of 40 A. Find the magnitude and direction of the force on the electron when it is moving parallel to the wire (a) in the same direction as the current and (b) in the opposite direction to the current.

22. The charge in Exercise 20 is moving in a direction perpendicular to the same wire. Find the magnitude and direction of the force on the charge when it is moving (a) toward the wire and (b) away from the wire.

23. The electron in Exercise 21 is moving in a direction perpendicular to the same wire. Find the magnitude and direction of the force on the electron when it is moving (a) toward the wire and (b) away from the wire.

24. An electron moving through an electric field of 500 V/m and a magnetic field of 0.10 T experiences no force. The two fields and the electron's direction of motion are all mutually perpendicular. What is the speed of the electron?

25. A velocity selector uses a magnet to produce a 0.050-T magnetic field and a pair of parallel metal plates 10 mm apart to produce a perpendicular electric field. What potential difference should be applied to the plates to permit singly charged ions of speed 5.0×10^6 m/s to pass through the selector?

26. A mass spectrometer has a velocity selector that consists of a magnetic field of 0.0400 T perpendicular to an electric field of 5.00×10^4 V/m. The same magnetic field is then used to deflect the ions. Find the radius of curvature of singly charged lithium ions of mass 1.16×10^{-26} kg in this spectrometer.

19.6 Force on a Current

27. A wire 1.0 m long is perpendicular to a magnetic field of 0.10 T. (a) What is the force on the wire when it carries a current of 10 A? (b) What is the force if the wire is parallel to the magnetic field?

28. A horizontal wire 600 mm long whose mass is 4.0 g is to be supported magnetically against the force of gravity. The current in the wire is 12.0 A and goes from south to north. Find the magnitude and direction of the magnetic field of least magnitude that will support the wire.

29. A horizontal north-south wire 5.0 m long is in a 0.020-T magnetic field whose direction is northeast. (a) What is the magnitude and direction of the force on the wire when a 4.0-A current flows north in it? (b) When the same current flows south in it?

30. A vertical wire 2.0 m long is in a 1.0×10^{-2} T magnetic field whose direction is northeast. (a) What is the magnitude and direction of the force on the wire when a 5.0-A current flows upward in it? (b) When the same current flows downward in it?

19.7 Force Between Two Currents

31. The parallel wires in a lamp cord are 2.0 mm apart. What is the force per meter between them when the cord is used to supply power to a 120-V, 200-W light bulb?

32. A certain electric transmission line consists of two wires 4.0 m apart that carry currents of 1.0×10^4 A. If the towers supporting the wires are 200 m apart, how much force does each current exert on the other between the towers?

PROBLEMS

33. Two long, parallel wires a distance s apart each carry the current I in opposite directions. Verify that the magnetic field at a point equidistant from the wires and x away from the plane in which they lie is given by $2\mu_0 Is/\pi(4x^2 + s^2)$.

34. The table below lists corresponding values of B_0 and B for a type of carbon steel, where B_0 is the magnetic field of a current-carrying coil with no core and B is the magnetic field of the same coil with a carbon steel core. (a) Plot a graph of B versus B_0 for this material. (b) Plot a graph of the relative permeability μ/μ_0 of this material versus B_0. Approximately what is the maximum value of μ/μ_0, and at approximately what value of B_0 is this maximum reached?

B_0	B
0.0×10^{-5} T	0.0 T
4.2	0.2
6.3	0.4
7.7	0.6
9.0	0.8
12	1.0
20	1.2
36	1.4
75	1.6

35. According to the Bohr model, a hydrogen atom consists of an electron that circles a proton at a speed of 2.2×10^6 m/s in an orbit of radius 5.3×10^{-11} m. (a) Find the magnetic field the proton experiences as a result of the motion of the electron around it. (b) Find the magnetic field that would be needed by an electron with this speed to move in an orbit of this radius.

36. Prove that the time required for a charged particle in a magnetic field to make a complete revolution is independent of its speed and the radius of its orbit.

37. In a typical loudspeaker a permanent magnet creates a radial magnetic field in which a wire coil attached to the apex of a paper cone can move perpendicularly to the field (Fig. 19.41). An alternating current in the coil causes the cone to oscillate and thereby produce sound waves. If the 40-turn coil of a certain loudspeaker is 8.0 mm in radius and is located in a 0.40-T magnetic field, what is the force on the coil when the current in it is 0.050 A?

Figure 19.41

A loudspeaker (Problem 37).

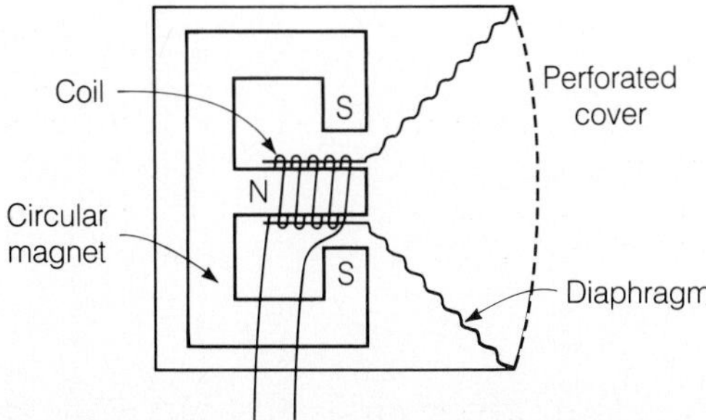

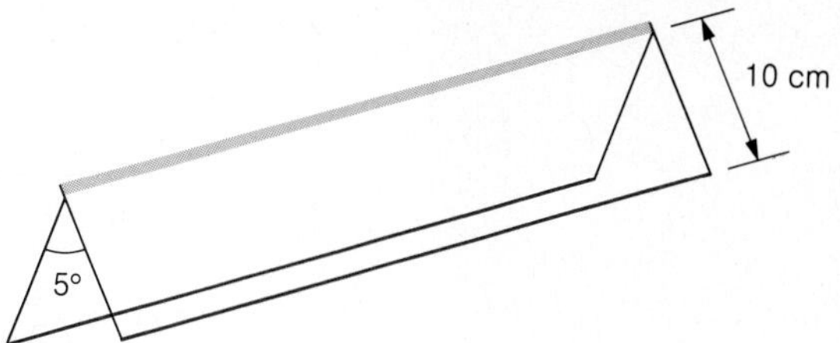

Figure 19.42

Problem 38.

38. Two parallel wires 100 cm long each of mass 20 g are suspended by strings 10 cm long from an overhead rod (Fig. 19.42). When a current I is passed through the wires in opposite directions, the wires swing out so that the strings are 5° apart. Find I.

39. (a) The current loop in Fig. 19.28(a) has the width x and the length y, as in Fig. 19.43. Verify that the magnetic torque on the loop about the axis shown in $\tau = IAB$, where $A = xy$ is the area enclosed by the loop. (In fact, this is a general formula that holds for a loop of any shape, provided it lies in a plane parallel to **B**.) (b) If θ is the angle between the plane of the loop and **B**, verify that $\tau = IAB \cos \theta$.

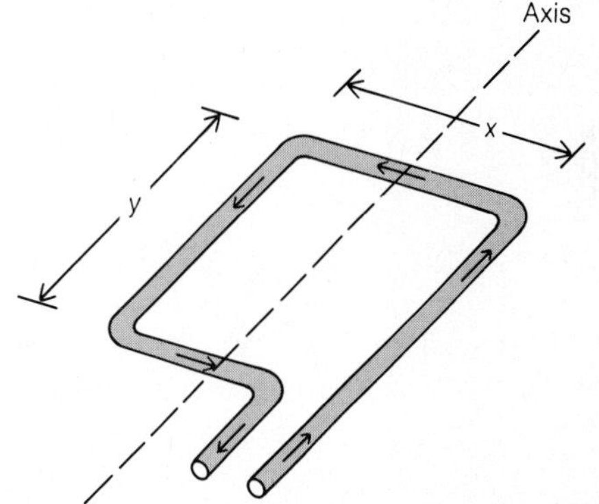

Figure 19.43

Problem 39.

Answers to Multiple Choice

1. d	**9.** a	**17.** b	**25.** c
2. c	**10.** c	**18.** c	**26.** c
3. a	**11.** b	**19.** b	**27.** d
4. b	**12.** c	**20.** a	**28.** b
5. c	**13.** d	**21.** d	**29.** c
6. a	**14.** a, b, c, d	**22.** b	**30.** a
7. c	**15.** a	**23.** b	**31.** c
8. b	**16.** a	**24.** c	

C H A P T E R 20

ELECTROMAGNETIC INDUCTION

We have seen that one way in which electricity and magnetism are related is the presence of a magnetic field around every electric current. Are there any other ways? In particular, since a current produces a magnetic field, can a magnetic field in its turn somehow produce a current? Many of the early workers in electricity and magnetism tackled this problem without success until, in 1831, Michael Faraday in England and Joseph Henry in the United States independently discovered electromagnetic induction.

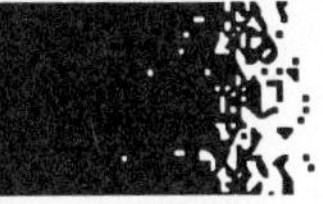

20.1 Electromagnetic Induction

An emf is produced when there is relative motion between a conductor and a magnetic field.

A galvanometer connected to a stationary wire in an unchanging magnetic field shows no sign of a current. What Faraday and Henry found is that *moving* the wire produces a current (Fig. 20.1). It does not matter whether the wire or the source of the magnetic field is being moved, provided that a component of the motion is perpendicular to the field. The origin of the current lies in the *relative motion* between a conductor and a magnetic field. This effect is known as **electromagnetic induction.**

Electromagnetic induction occurs for the following reason:

An electromotive force (emf) is produced in a conductor whenever there is relative motion between the conductor and a magnetic field, provided the motion has a component across the field direction.

It is this emf that leads to the current that flows whenever there is relative motion across the field lines between a conductor and a magnetic field. In fact, no actual motion of either a wire or a source of magnetic field is needed, because a magnetic field that changes in strength has moving field lines associated with it.

Moving a wire *parallel* to a magnetic field does not give rise to a current. Electromagnetic induction occurs only when there is a component of the wire's velocity perpendicular to the field.

Electromagnetic induction should not come as a surprise to us because it is in accord with the force a magnetic field exerts on a moving charge. As we saw in the previous chapter, a wire carrying a current is pushed sideways in a magnetic field because of the forces on the moving electrons. In Faraday's and Henry's experiments, electrons are also moved through a magnetic field, but now by shifting the entire wire. As before, the electrons are pushed to the side, which here moves them along the wire. The motion of these electrons along the wire is the electric current we find.

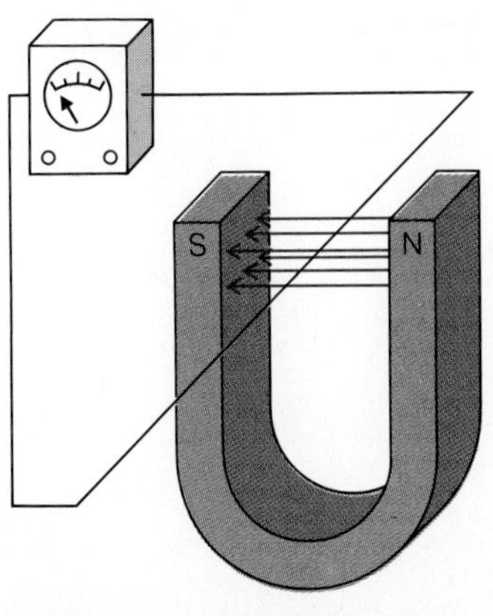

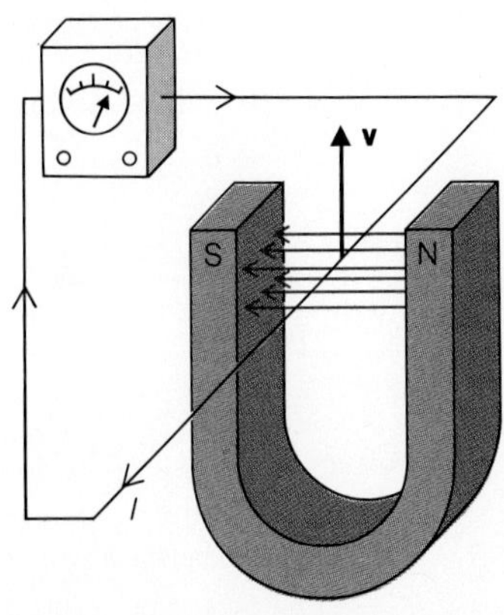

Figure 20.1

(a) There is no current in a stationary wire in a magnetic field. (b) When the wire is moved across the field, a current is produced. Reversing the direction of motion also reverses the direction of the current.

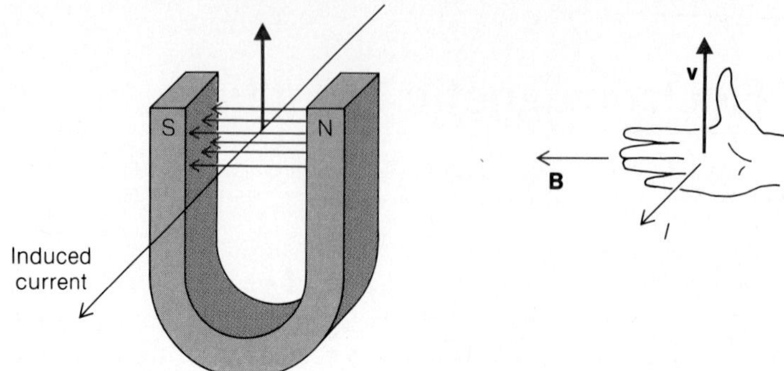

Figure 20.2

Right-hand rule for the direction of an induced current.

The right-hand rule for the force on a moving positive charge in a magnetic field can also be used to give the direction of the induced current in a wire moving across a magnetic field. Hold your right hand so that the fingers point in the direction of **B** and the outstretched thumb is in the direction of **v**. The palm then faces in the same direction as the force on the positive charges in the wire, and hence in the direction of the conventional current (Fig. 20.2).

In the arrangement of Fig. 20.1, a current will also flow if the magnet is moved vertically past the wire, instead of the wire being moved past the magnet. This effect is the basis of the playing head of a tape recorder (Fig. 20.3). The tape

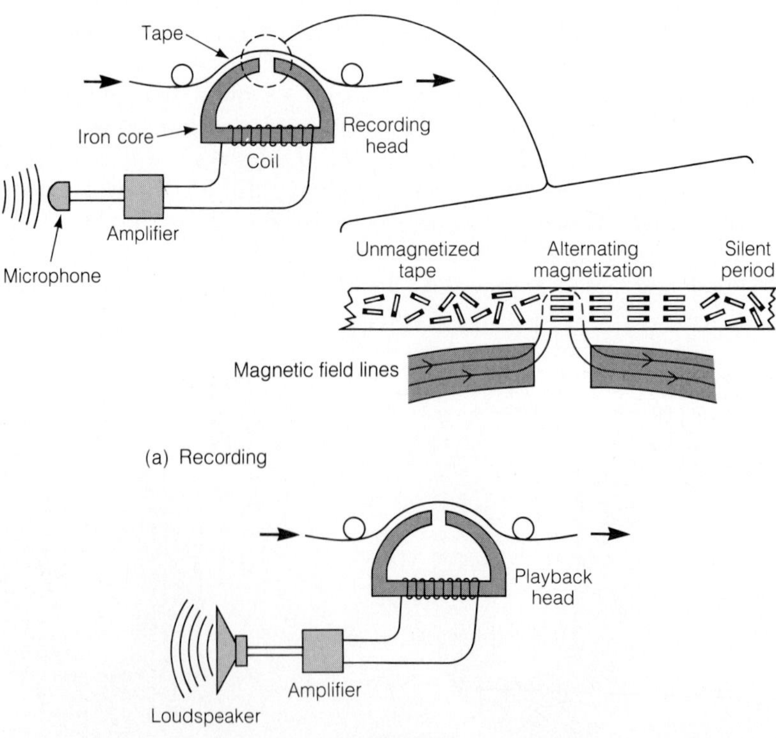

Figure 20.3

A tape recorder. (a) Recording. (b) Playback. The polarity and degree of magnetization of the ferromagnetic coating on the tape correspond to the pressure fluctuations of the original sound wave.

used is a plastic film with a ferromagnetic coating, and it is magnetized by being passed close to an electromagnet whose current varies with the signal to be recorded. For instance, the recording head would be connected via an amplifier to a microphone if music or other sound were to be registered on the tape. The playing head contains a small coil, and the current induced in it by the changing magnetic fields of the moving tape is amplified and fed into a loudspeaker to reconstruct the original sound.

20.2 Moving Wire in a Magnetic Field

The emf depends on the wire's length and velocity and on the field strength.

When a wire moves through a magnetic field **B** with a velocity **v** that is not parallel to **B**, an induced emf $\mathscr{E}_i$ comes into being between the ends of the wire. It is not hard to figure out what $\mathscr{E}_i$ is in terms of **B**, **v**, and the length L of the wire.

Let us assume that the wire is perpendicular to **B** and that **v** has the component $\mathbf{v}_\perp$ perpendicular to **B**, as in Fig. 20.4. From Eq. (19.8) the force on a positive charge Q in the moving wire has the magnitude

$$F = Qv_\perp B$$

The direction of this force is along the wire, as shown.

Now we consider a charge Q that moves from one end of the wire to the other under the influence of the magnetic force $Qv_\perp B$. The work done on the charge by whatever moves the wire is

$$\text{Work} = (\text{force})(\text{distance})$$
$$W = FL = Qv_\perp BL$$

since the wire is L long. By definition the potential difference between two points is the work done in moving a unit charge between these points. The potential difference between the ends of the wire is therefore W/Q or

$$\mathscr{E}_i = BLv_\perp \qquad \textit{Motional emf} \quad (20.1)$$

This potential difference is the emf induced in the wire by its motion through the magnetic field (Fig. 20.5).

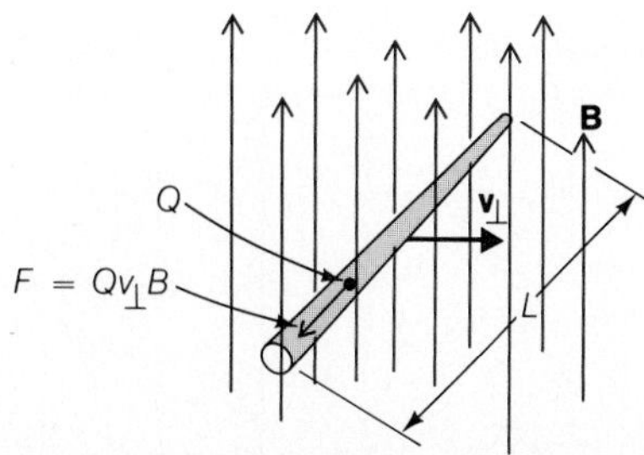

Figure 20.4

The force on a charge Q within a conductor when the conductor is moved through a magnetic field is $Qv_\perp B$, where $v_\perp$ is the component of the conductor's velocity perpendicular to **B**.

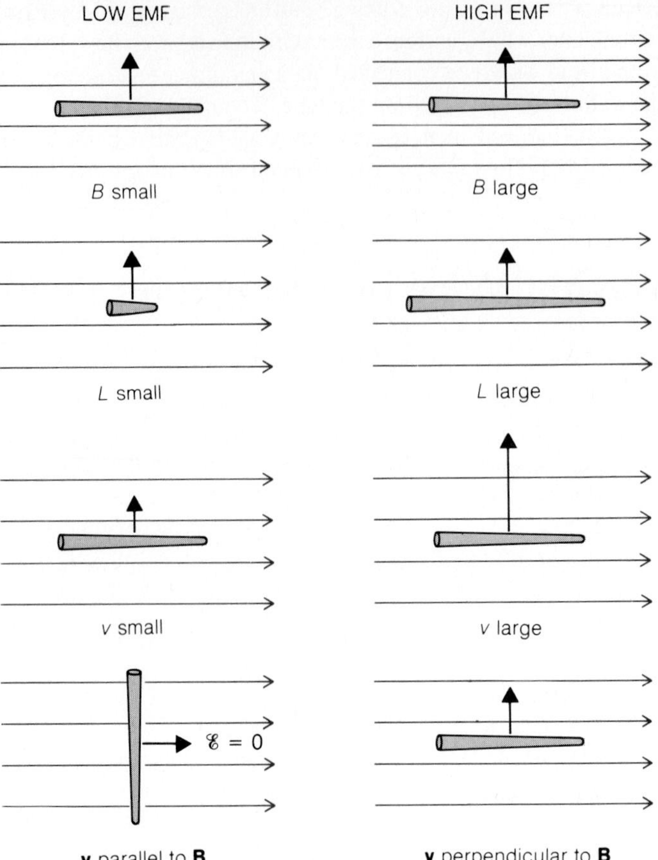

Figure 20.5

The emf induced in a conductor moving through a magnetic field depends on the magnetic field strength, on the length and speed of the conductor, and on the direction of **v** relative to **B.** The conductors shown here are oriented perpendicular to **B;** that is, perpendicular to the page.

If the moving wire is part of a complete circuit, how much current will flow? The answer is given by Ohm's law, since an induced emf is like any other emf in its ability to cause a current. If the total resistance in the circuit is R, the resulting current will be $I = \mathscr{E}/R$ as long as the emf exists.

EXAMPLE 20.1

Find the potential difference between the wing tips of a jet airplane induced by its motion through the earth's magnetic field. The total wing span of the airplane is 40 m, and its speed is 300 m/s in a region where the vertical component of the earth's field has the magnitude 3.0×10^{-5} T (Fig. 20.6).

Figure 20.6

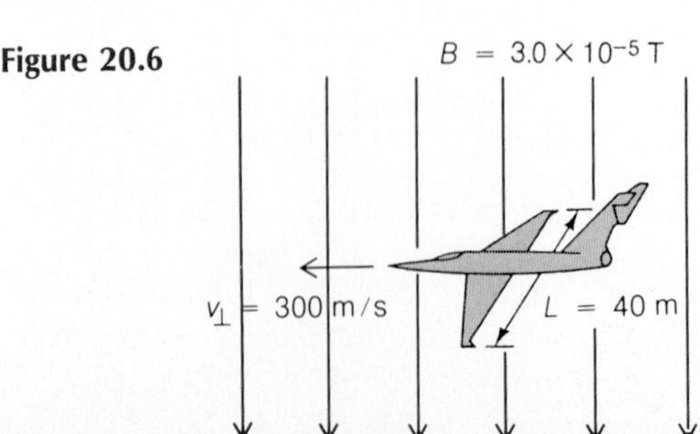

SOLUTION Substituting the given values of $v_\perp$, L, and B in Eq. (20.1) for the induced emf yields

$$\mathscr{E}_i = BLv_\perp = (3.0 \times 10^{-5}\ \text{T})(40\ \text{m})\left(300\ \frac{\text{m}}{\text{s}}\right) = 0.36\ \text{V}$$

♦

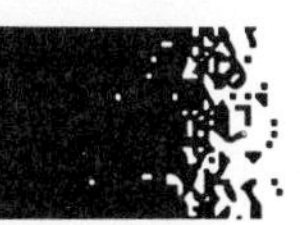

20.3 Faraday's Law

The induced emf in a loop equals the rate of change of the magnetic flux through it.

A convenient way to look at electromagnetic induction makes use of the idea of **magnetic flux.** As in Fig. 20.7, we consider a flat wire loop of area A in a magnetic field **B**, where θ is the angle between a perpendicular to the plane of the loop and the direction of **B.** The loop can have any shape. The total magnetic flux Φ (Greek capital letter *phi*) through the loop is defined as

$$\Phi = B_\perp A = BA\cos\theta \qquad \textit{Magnetic flux} \quad (20.2)$$

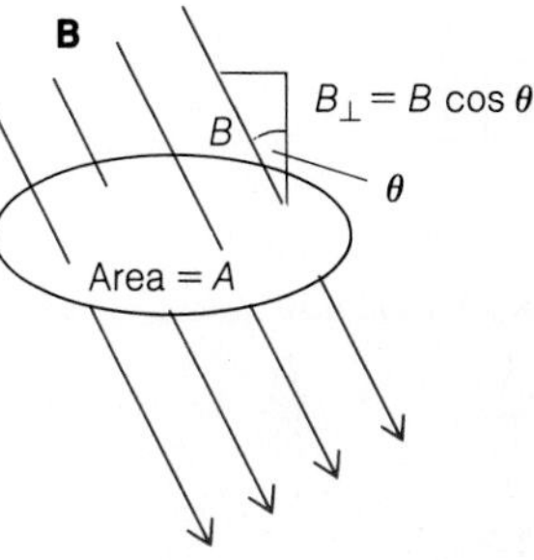

Figure 20.7

The magnetic flux Φ through a flat wire loop of area A is proportional to A and to the component $B_\perp$ of the magnetic field perpendicular to the plane of the loop. The emf induced in the loop is equal to the rate at which the flux Φ it encloses is changing.

where $B_\perp = B\cos\theta$ is the component of **B** perpendicular to the plane of the loop. The unit of flux is the **weber** (Wb), where 1 Wb = 1 T·m^2.

Faraday found that the electromotive force $\mathscr{E}_i$ in such a wire loop is *equal to the rate of change of the flux through it*. That is,

$$\mathscr{E}_i = -\frac{\Delta\Phi}{\Delta t} \qquad \textit{Induced emf} \quad (20.3)$$

Here $\Delta\Phi$ is the change in the flux Φ that takes place during a period of time Δt. Equation (20.3) is **Faraday's law of electromagnetic induction.** The reason for the minus sign is Lenz's law, which will be described shortly.

The induced emf in a wire loop is equal to the potential difference that would be found between the ends of an identical *open* wire loop. The notion of emf is useful here because it is meaningless to speak of the potential difference around a closed circuit. The emf in a closed circuit is the potential difference that would be found between the ends of the circuit if it were cut anywhere.

If a coil of N turns replaces the single loop of Fig. 20.7, the induced emf's in the turns are added together, and the total emf in the entire coil is

$$\mathscr{E}_i = -N\frac{\Delta\Phi}{\Delta t} \qquad (20.4)$$

Lenz's Law

The minus sign in Eqs. (20.3) and (20.4) follows from the law of conservation of energy. If the sign of $\mathscr{E}$ were the same as that of $\Delta\Phi/\Delta t$, the induced electric current would be in such a direction that its own magnetic field would *add* to that of the external field **B.** This additional changing field would then increase the rate of

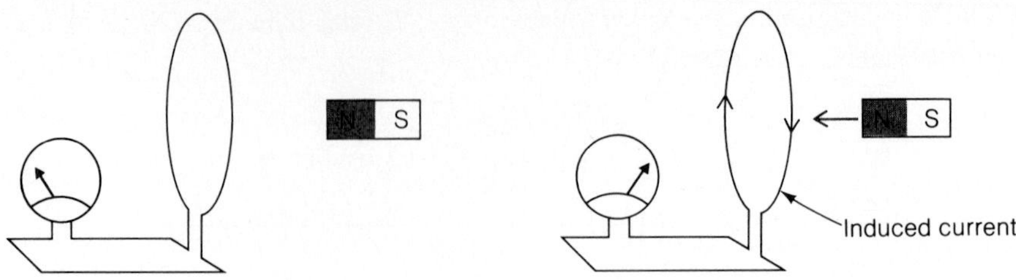

Figure 20.8

The current induced in the wire loop is such that its own magnetic field opposes the field change due to the motion of the permanent magnet, in accordance with Lenz's law.

change of the flux Φ, and more and more current would flow even if the original $\Delta\Phi/\Delta t$ were to stop. But energy is associated with every current, and no current can increase without energy being supplied. The only possibility, then, is that $\mathscr{E}$ be opposite in sign to $\Delta\Phi/\Delta t$, which means:

The direction of an induced current is always such that its own magnetic field opposes the change in flux responsible for producing it.

This observation is known as **Lenz's law.** An example of Lenz's law is illustrated in Fig. 20.8.

EXAMPLE 20.2

A 12-turn coil 100 mm in diameter has its axis parallel to a magnetic field of 0.50 T that is produced by a nearby electromagnet. The current in the electromagnet is cut off, and as the field collapses, an average emf of 8.0 V is induced in the coil. What is the length of time required for the field to disappear?

SOLUTION The change in the flux through the coil is

$$\Delta\Phi = BA = B\pi r^2 = (-0.50\ \text{T})(\pi)(0.050\ \text{m})^2 = -0.0039\ \text{Wb}$$

Hence the flux drops to zero in the time

$$\Delta t = -\frac{N\Delta\Phi}{\mathscr{E}_i} = -\frac{(12)(-0.0039\ \text{Wb})}{8.0\ \text{V}} = 0.0059\ \text{s}$$

♦

EXAMPLE 20.3

A metal disk that rotates through a magnetic field, as shown in Fig. 20.9, constitutes a **homopolar generator.** Its operation can be understood if we imagine the disk replaced by a bicycle wheel, with an emf being induced in each spoke as it moves across the field. In the case of the disk, what moves across the field is the line on the disk between the electrical contact and the shaft at any instant. (a) Derive a formula for the emf of such a generator in which the disk has the radius r and makes f turns per second through a uniform magnetic field **B.** (b) If r = 15 cm and B = 1.0 T, what must f be in order to produce an emf of 1.0 V?

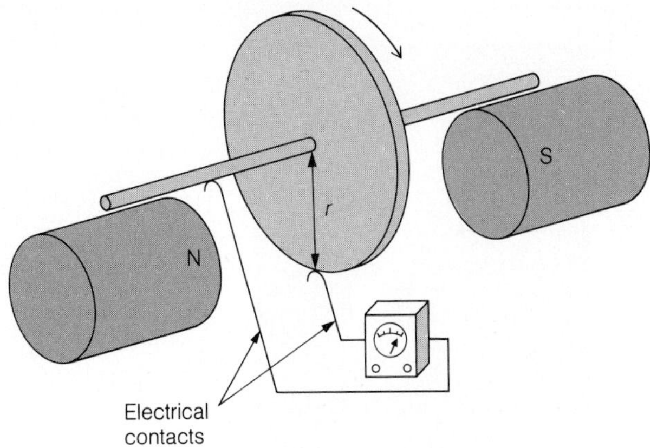

Figure 20.9

A homopolar generator.

SOLUTION (a) In each revolution all the radii of the disk move across the magnetic field. The situation is the same as though one radius had moved through a magnetic field that passed through the entire disk. The area of the disk is πr^2, hence the magnetic flux through which each radius moves in each revolution is $\Delta\Phi = BA = \pi r^2 B$, and the emf is $\mathscr{E}_i = \Delta\Phi/\Delta t = \pi r^2 f B$.

$$\text{(b) } f = \frac{\mathscr{E}_i}{\pi r^2 B} = \frac{1.0\text{ V}}{\pi(0.15\text{ m})^2(1.0\text{ T})} = 14\text{ rev/s}$$

◆

A Derivation of Faraday's Law

It is not hard to derive Faraday's law of electromagnetic induction from the formula for the emf induced in a wire moved across a magnetic field. Figure 20.10 shows a moving wire that slides across the legs of a U-shaped metal frame, so that the frame completes the loop. At the moment (say $t = 0$) when the moving wire is

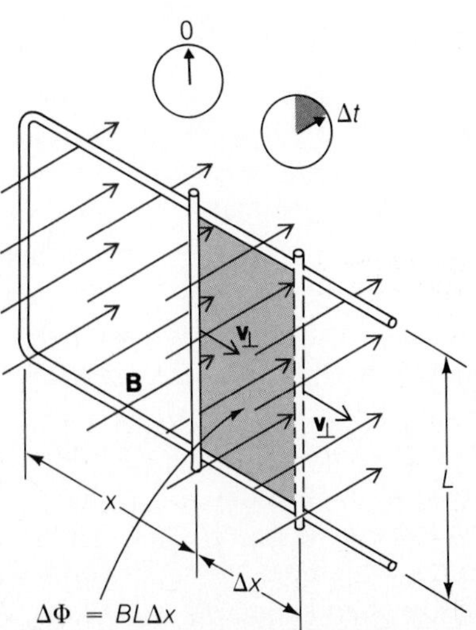

Figure 20.10

The flux enclosed by the wire frame and movable wire increases as the wire moves to the right. An emf is induced in the wire by its motion through the magnetic field, which is equal to the rate of change of the enclosed flux, in agreement with Faraday's law of electromagnetic induction.

BIOGRAPHICAL NOTE

Michael Faraday

Michael Faraday (1791–1867), the son of a blacksmith who lived near London, had little education and at 13 was apprenticed to a bookbinder. Encouraged by his employer, Faraday read as well as bound books, especially books on science. In 1812 a customer gave Faraday tickets to public lectures by the noted chemist Humphrey Davy. Faraday took careful notes, bound them nicely, and presented them to Davy with a request for a job. Faraday later wrote that he "wanted to escape from trade which I thought vicious and selfish and enter the service of science, which I imagined made its pursuers amiable and liberal. . . . [Davy] smiled at my notion of the superior moral feelings of philosophic men, and said he would leave to me the experience of a few years to set me right on that matter."

Davy hired Faraday as his assistant and took him along on a tour of Europe the next year, during which Faraday met the foremost scientists of the time. Faraday learned a great deal on this tour, and on his return began an extraordinarily fruitful scientific career. He started in chemistry and soon discovered benzene, liquefied a number of gases for the first time, developed superior steels, and prepared a new type of glass that led to improvements in optical instruments. The pioneering work of Davy on electrolysis was extended by Faraday, whose careful studies led to what are today called Faraday's laws of electrolysis.

Although Faraday was involved mainly in chemistry until about 1830, he had been intrigued in 1821 by Oersted's discovery that electric currents can lead to magnetic forces, and used this concept to build the first electric motor. He felt there must be a reciprocal process whereby a magnet could produce an electrical effect. He tried various approaches, but all failed until in 1831 he found that the key to electromagnetic induction was relative motion between a wire and "lines of magnetic force which would be detected by iron filings." The modern concept of a force field developed from Faraday's lines of force, which he invented to enable him to describe phenomena he could not capture in equations because of his ignorance of mathematics. (The American physicist Joseph Henry had actually demonstrated electromagnetic induction earlier, but he published his results later than Faraday.)

Faraday realized the implications of electromagnetic induction at once and soon had generators and transformers working in his laboratory as well as electric motors. Asked by a politician what good these devices were, Faraday replied, "At present I do not know, but one day you will be able to tax them." With his full attention now on electricity and magnetism, Faraday made a number of other important contributions in the next few years. Among them were the use of field lines to help interpret electrical effects and the idea of capacitance (its unit, the farad, is named after him).

During much of his life Faraday suffered from spells of headaches, dizziness, and memory loss. Rest led to temporary improvement, but in 1839 the failure of his memory was so severe that he left his laboratory for a long period to recover. At that time other scientists had similar symptoms, which may have been the result of mercury poisoning. In 1845 Faraday returned to work and established that the plane of polarization of polarized light is rotated when the light passes through a piece of special glass located in a strong magnetic field. This led Faraday to speculate that light has some sort of electromagnetic nature, a notion that Maxwell was to turn into a detailed theory 18 years later. Another important discovery that Faraday made was diamagnetism, the property certain substances have of being repelled rather than attracted by a magnet.

In the 1850s Faraday's activity again slackened because of lapses of memory. He died in 1867 at 76, leaving behind laboratory notebooks with over 16,000 entries that testify to his remarkable originality, intuition, skill, and diligence.

the distance x from the closed end of the frame, the flux enclosed by the loop is

$$\Phi = BA = BLx$$

At the later time $t = \Delta t$, the moving wire is $x + \Delta x$ from the closed end of the frame, so the flux now enclosed by the loop is

$$\Phi + \Delta\Phi = BLx + BL\,\Delta x$$

Hence the increase in the enclosed flux during the time Δt is

$$\Delta\Phi = BL\,\Delta x$$

The speed of the moving wire is $v_\perp$, and so

$$\Delta x = v_\perp\,\Delta t$$

and

$$\Delta\Phi = BLv_\perp\,\Delta t$$

$$\frac{\Delta\Phi}{\Delta t} = BLv_\perp$$

But $BLv_\perp$ is the emf induced in the moving wire. Hence we have, inserting a minus sign to remind ourselves of Lenz's law,

$$\mathscr{E}_i = -\frac{\Delta\Phi}{\Delta t}$$

which is Faraday's law.

20.4 The Generator

Electromagnetic induction underlies its operation.

Almost all the world's electric energy is produced by electromagnetic induction, something to keep in mind whenever we switch on a lamp. In a **generator** a coil of wire is turned in a magnetic field so that the flux through the coil changes constantly. The resulting emf in the coil causes a current to flow in an external circuit, and cables can carry this current for long distances from its origin.

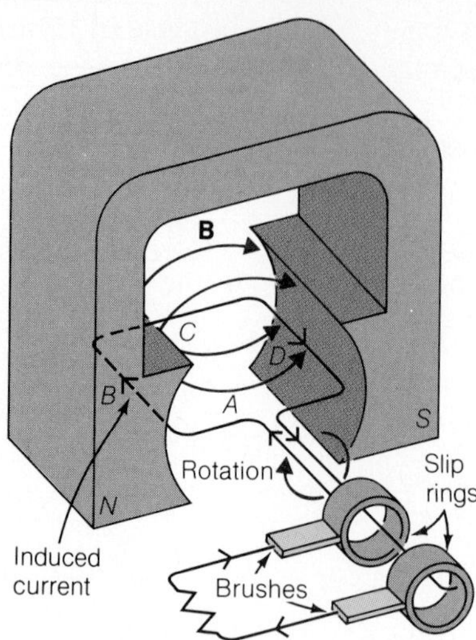

Figure 20.11

A simple ac generator.

Despite its name, a generator does not *create* electric energy. What it does is *convert* mechanical energy into electric energy, just as a battery converts chemical energy into electric energy.

The construction of an idealized simple generator is shown in Fig. 20.11. A wire loop is rotated in a magnetic field, and the ends of the loop are connected to two **slip rings** on the shaft. Brushes pressing against the slip rings join the loop to an external circuit. Only sides *B* and *D* of the loop contribute to the induced emf.

Because the sides of the loop reverse their direction of motion through the magnetic field twice per rotation, the induced current is also reversed twice per rotation and for this reason is called an **alternating current.** The emf of this generator varies with time as in Fig. 20.12. The shape of the curve is that of a cosine (or sine) curve. In a complete cycle, which corresponds to one turn of the loop, the emf increases to a maximum, falls to zero, continues to a negative maximum, and then returns to zero. If the loop rotates *f* times per second, the generator output will go through *f* cycles per second. An ac generator is often called an **alternator.**

Figure 20.12

The variation with time of the emf of a simple ac generator.

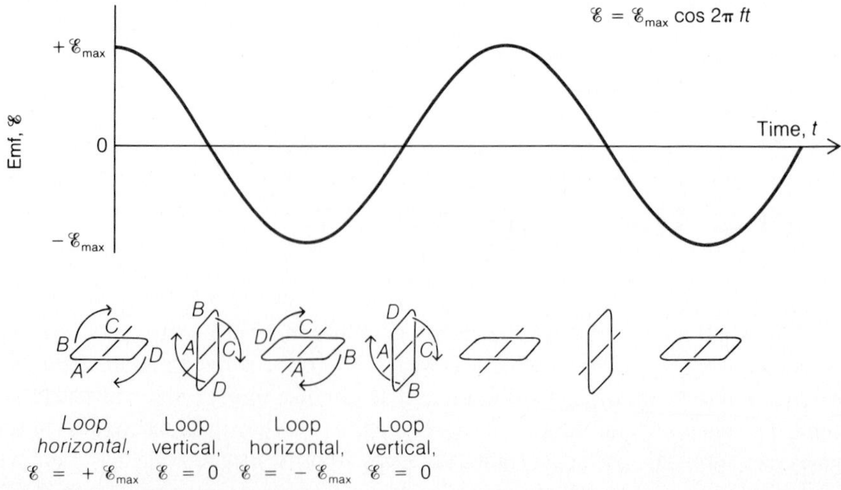

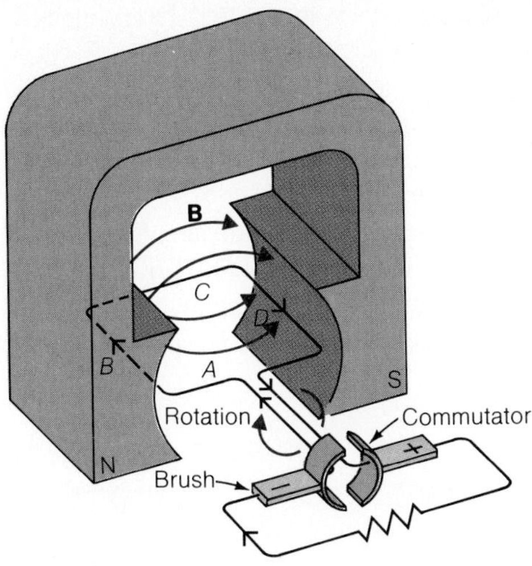

Figure 20.13

A simple dc generator. The current in the brushes is always in the same direction.

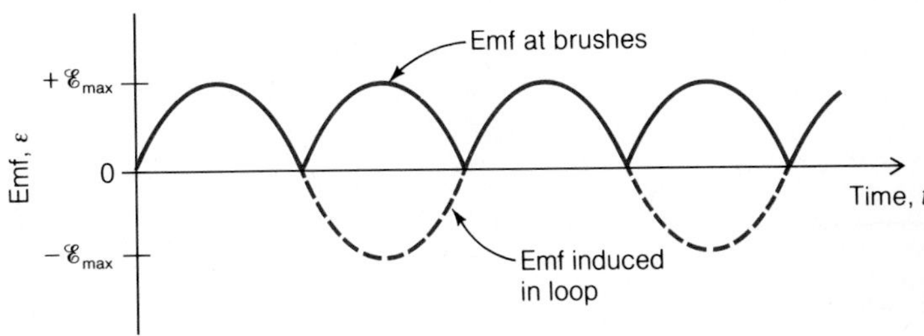

Figure 20.14

The variation with time of the emf of a simple dc generator.

The ac output of a simple generator can be changed to dc by substituting a commutator for the slip rings that connect the rotating coil with the external circuit (Fig. 20.13). The emf of such a dc generator varies with time as in Fig. 20.14. The emf is always in the same direction, but it rises to a maximum and drops to zero twice per complete rotation. The resemblance between the dc generator in Fig. 20.13 and the dc motor in Fig. 19.30 is not an accident: One is the inverse of the other. If the output terminals of a dc generator are connected to a battery, the generator will run as a motor.

Another approach is to use a **rectifier** in the output circuit of an alternator. A rectifier is a device that permits current to pass through it in only one direction; the operation of semiconductor rectifiers is described in Chapter 28. Because alternators are simpler to construct than dc generators and are more reliable, they are often used with semiconductor rectifiers to produce the direct current needed in cars and other vehicles.

In nearly all actual generators, electromagnets produce the magnetic field. (If a permanent magnet is used, the generator is called a **magneto**.) In an alternator the armature and field windings are often exchanged so that, in effect, the field rotates inside a stationary armature. This makes no difference to the physics involved because there is still relative movement of a conductor and a magnetic field. The advantage of a stationary armature is that a high-voltage output can be taken directly from it without any moving electrical contacts while the field current is supplied at a modest voltage through such contacts.

Back Emf

As we just saw, a coil rotated in a magnetic field is a source of emf, which implies that an emf is generated in the rotating armature of an electric motor. This induced emf occurs because wires are being moved through a magnetic field. The presence of a current of external origin in the armature at the same time, which is what causes the armature to turn in the first place, does not matter. Every electric motor is also a generator.

By Lenz's law all induced emf's are such as to oppose the changes that bring them into being. This means that the induced emf in a motor armature is opposite in direction to the external voltage applied to the armature. The effect of this **back emf** (or **counter emf**) is to reduce the current in the armature.

Back emf is equal to $-\Delta\Phi/\Delta t$ and so is proportional to the angular speed of the armature. It is therefore zero when the motor is turned on and increases as the speed increases. As a result the current is a maximum when the motor starts, and it drops afterward. The existence of back emf in a motor helps to keep its speed constant. At the normal speed,

$$V = \mathscr{E}_b + IR \tag{20.5}$$

Applied voltage = back emf + potential drop in armature

If the speed falls, $\mathscr{E}_b$ decreases and I goes up, and the greater current means more torque on the armature to restore the speed to its original value. If the speed rises above normal, $\mathscr{E}_b$ increases and I goes down, with the result that the torque is reduced and the motor loses speed until it assumes its normal value.

EXAMPLE 20.4

A 120-V dc electric motor has its armature and field windings connected in parallel. (Such a motor is said to be ''shunt wound.'' When armature and field are in series, the motor is said to be ''series wound.'') The armature resistance is 2.0 Ω and the field resistance is 200 Ω. When the motor is at its operating speed, the total current is 3.0 A. Find (a) the back emf of the motor at its operating speed, (b) the efficiency of the motor, and (c) the current in the motor at the moment its switch is turned on.

SOLUTION (a) The current in the field winding is

$$I_f = \frac{V}{R_f} = \frac{120\ \text{V}}{200\ \Omega} = 0.6\ \text{A}$$

Since the total current is 3.0 A, the armature current is

$$I_a = I - I_f = 3.0\ \text{A} - 0.6\ \text{A} = 2.4\ \text{A}$$

The magnitude of the back emf is therefore

$$\mathscr{E}_b = V - I_aR_a = 120\ \text{V} - (2.4\ \text{A})(2.0\ \Omega) = 115\ \text{V}$$

The direction of the back emf is, of course, opposite to that of the impressed potential difference V.

(b) The power lost as heat in the field and armature windings is

$$P_{\text{loss}} = I_f^2R_f + I_a^2R_a = (0.6\ \text{A})^2(200\ \Omega) + (2.4\ \text{A})^2(2.0\ \Omega) = 83.5\ \text{W}$$

The electrical power input of the motor is equal to IV, the total current multiplied by the applied potential difference:

$$P_{\text{input}} = IV = (3.0\text{ A})(120\text{ V}) = 360\text{ W}$$

Hence the motor's efficiency is

$$\text{Eff} = \frac{P_{\text{input}} - P_{\text{loss}}}{P_{\text{input}}} = \frac{360\text{ W} - 83.5\text{ W}}{360\text{ W}} = 0.77 = 77\%$$

(c) At the moment the motor is started, there is no back emf, because the armature is still stationary. The total resistance of the armature and field windings is

$$R = \frac{R_a R_f}{R_a + R_f} = \frac{(2.0\ \Omega)(200\ \Omega)}{2.0\ \Omega + 200\ \Omega} = 1.98\ \Omega$$

The current is therefore

$$I = \frac{V}{R} = \frac{120\text{ V}}{1.98\ \Omega} = 61\text{ A}$$

The starting current is more than 20 times greater than the operating current! ◆

20.5 The Transformer

Stepping voltage up and down.

Earlier we noted that a current can be induced in a wire or other conductor by a changing magnetic field as well as by relative motion between a wire and a constant magnetic field. The essential condition for an induced emf is that magnetic field lines cut across the wire (or vice versa). It does not matter exactly how this comes about.

A change in the current in a wire loop is accompanied by a change in its magnetic field. If another wire loop is nearby, an emf will therefore be induced in it. If an alternating current flows in the first loop, the magnetic field around it will vary periodically and an alternating emf will be induced in the second loop. A **transformer** is a device based on this effect that is used to produce an alternating emf in a secondary circuit larger or smaller than the alternating emf in the primary circuit. A transformer thus permits electric energy to be transferred from one ac circuit to another without a direct connection between them.

With the help of Fig. 20.15, we can see how a changing current in a wire loop induces a changing current in another wire loop not connected to it.

(a) The switch connecting loop 1 with the battery is open, and no current is in either loop.

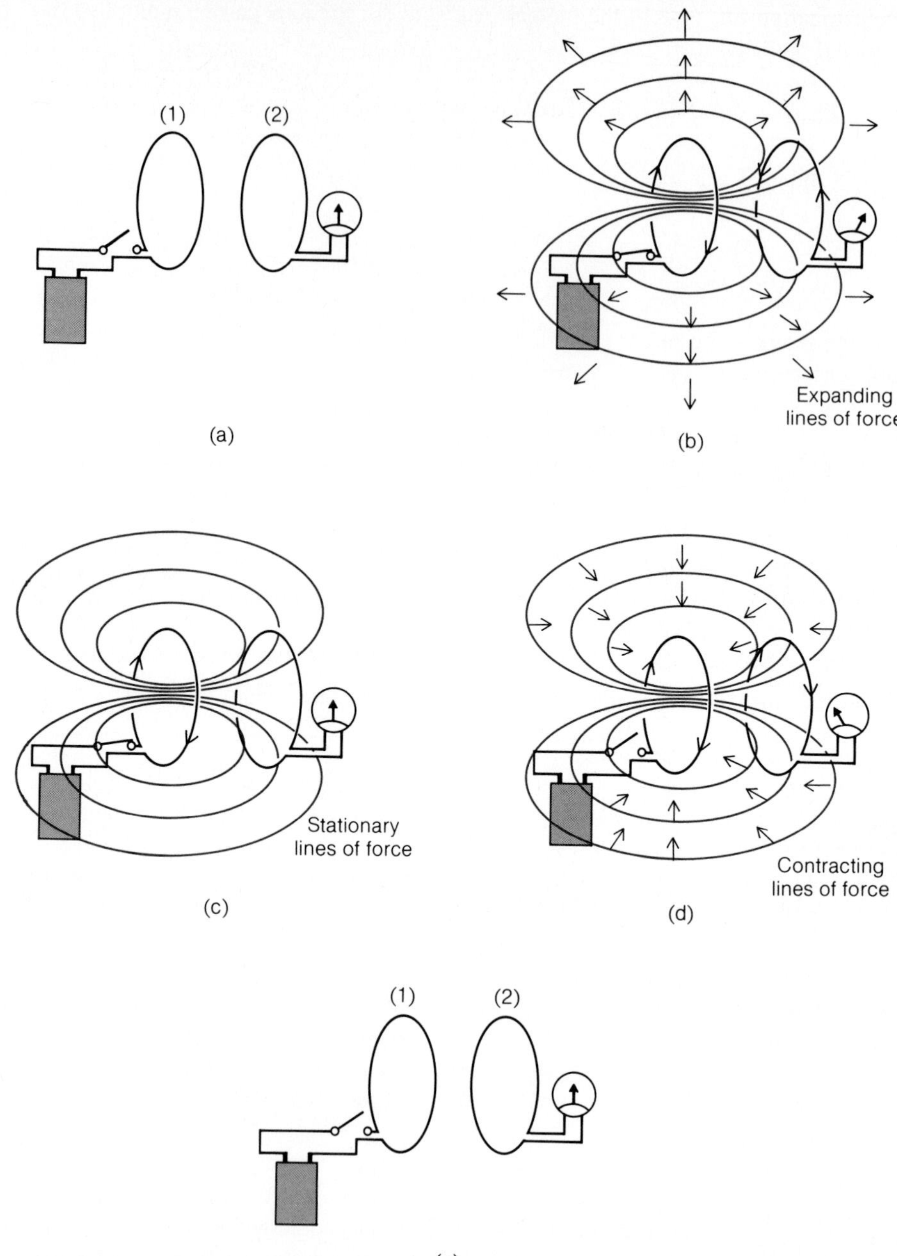

Figure 20.15

When the current in a wire loop is turned on or off, the magnetic field around it changes. The changing field can induce a current in another nearby loop, even though both loops are stationary.

(b) The switch has just been closed, and the expanding field lines resulting from the increasing current in loop 1 cut across loop 2, inducing a current in it. The current in loop 2 is in the *opposite* direction to that in loop 1 because of Lenz's law: The direction of an induced current is always such that its own magnetic field opposes the change in flux that is inducing it.

(c) Now there is a constant current in loop 1, and since the flux through it does not change, no current is induced in loop 2.

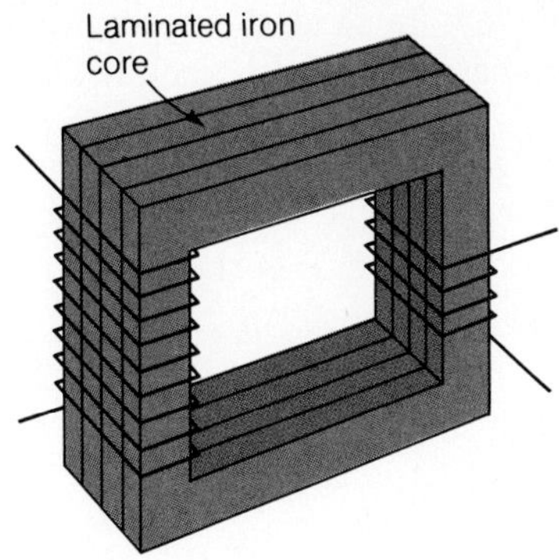

Figure 20.16
A simple transformer.

(d) The switch has just been opened, and the contracting field lines resulting from the decreasing current in loop 1 cut across loop 2, inducing a current in it. The current in loop 2 is now in the *same* direction as that in loop 1. This is because it is the decreasing current in loop 1 that leads to the current in loop 2, and Lenz's law requires that an induced current oppose the change in flux that brings it about.
(e) Finally the current in loop 1 disappears, and again no current is in either loop.

Instead of being switched on and off, the current input to a transformer is alternating current that reverses its direction regularly. The construction of a simple transformer is shown in Fig. 20.16. Coils are used instead of single loops, and they are wound together on an iron core so that the alternating magnetic flux set up by an alternating current in one coil induces an alternating emf in the other coil. The **primary winding** of a transformer is the coil that is fed with an alternating current, and the **secondary winding** is the coil to which power is transferred via the changing magnetic flux. Either winding may be the primary, and the other is then the secondary. A transformer may have more than one secondary winding so that it can provide several different secondary emf's.

Step-up and Step-down Transformers

An ideal transformer has no "leakage" of flux outside the iron core and no losses within it, so that $\Delta\Phi/\Delta t$ is the same for each of the turns of both windings. The emf *per turn* is therefore the same in both windings, and the total emf in each winding is proportional to the number of turns it contains. Hence

$$\frac{\mathscr{E}_1}{\mathscr{E}_2} = \frac{N_1}{N_2} \tag{20.6}$$

$$\frac{\text{Primary emf}}{\text{Secondary emf}} = \frac{\text{total primary turns}}{\text{total secondary turns}}$$

When the secondary winding has more turns than the primary winding, $\mathscr{E}_2$ is greater than $\mathscr{E}_1$. The result is a **step-up transformer.** When the reverse is the case and $\mathscr{E}_2$ is less than $\mathscr{E}_1$, the result is a **step-down transformer.**

In an ideal transformer the power output is equal to the power input. If the primary current is I_1 and the second current is I_2, then

$$\mathscr{E}_1 I_1 = \mathscr{E}_2 I_2$$
$$\text{Power input} = \text{power output}$$

and

$$\frac{I_1}{I_2} = \frac{\mathscr{E}_2}{\mathscr{E}_1} = \frac{N_2}{N_1} \qquad \textit{Transformer} \quad (20.7)$$

Increasing the voltage with a step-up transformer means a decrease in the current, whereas dropping the voltage leads to a greater current.

EXAMPLE 20.5

A transformer connected to a 120-V ac power line has 200 turns in its primary winding and 50 turns in its secondary winding. The secondary is connected to a 100-Ω light bulb. How much current is drawn from the 120-V power line?

SOLUTION The voltage across the secondary of the transformer is

$$V_2 = \frac{N_2}{N_1} V_1 = \left(\frac{50 \text{ turns}}{200 \text{ turns}}\right)(120 \text{ V}) = 30 \text{ V}$$

and so the current in the secondary circuit is

$$I_2 = \frac{V_2}{R} = \frac{30 \text{ V}}{100 \ \Omega} = 0.30 \text{ A}$$

The current in the primary circuit is

$$I_1 = \frac{N_2}{N_1} I_2 = \left(\frac{50 \text{ turns}}{200 \text{ turns}}\right)(0.30 \text{ A}) = 0.075 \text{ A}$$

◆

Transformers at Morro Bay, California, step up electric power generated at 18,000 V to 230,000 V for efficient transmission.

Alternating current owes its wide use largely to the ability of transformers to change the voltage at which it is transmitted. Electricity generated at perhaps 11,000 V is stepped up by transformers at the power station to as much as 750,000 V (or even more) for transmission, and then is stepped down by transformers at substations near the point of use to less than a thousand volts. Local transformers then step this down further to 240 or 120 V. High voltages are desirable because, since $P = IV$, the higher the voltage, the smaller the current for a given amount of power. Power is dissipated as heat at the rate I^2R, so the smaller the current, the less the power lost due to resistance in the transmission line.

EXAMPLE 20.6

Find the power lost as heat when a 10.0-Ω cable is used to transmit 1000 W of electricity at 240 V and at 240,000 V.

SOLUTION The current in the cable in each case is

$$I_a = \frac{P}{V_a} = \frac{1000\ \text{W}}{240\ \text{V}} = 4.17\ \text{A}$$

$$I_b = \frac{P}{V_b} = \frac{1000\ \text{W}}{240{,}000\ \text{V}} = 4.17 \times 10^{-3}\ \text{A}$$

The respective rates of heat production per kilowatt are therefore

$$I_a^2R = (4.17\ \text{A})^2(10.0\ \Omega) = 174\ \text{W}$$

$$I_b^2R = (4.17 \times 10^{-3}\ \text{A})^2(10.0\ \Omega) = 1.74 \times 10^{-4}\ \text{W}$$

a difference of a factor of 10^6. Transmission at 240 V therefore means a million times more power lost as heat than does transmission at 240,000 V. ♦

Eddy Currents

When a piece of metal is placed in a coil that carries an alternating current, the changing magnetic flux causes **eddy currents** to flow in the metal. The energy of these currents is dissipated as heat, and the metal becomes hot as a result. This effect is called **induction heating.** Induction furnaces heat metals rapidly and are often used in foundries to melt casting metal before it is poured into molds.

Eddy currents in a transformer core are wasteful partly because of the power lost as heat and partly because the flux they establish interferes with the proper operation of the transformer. To minimize eddy currents, actual transformer cores are usually laminated, with thin sheets arranged parallel to the magnetic field, as in Fig. 20.16. The laminations are insulated from each other by natural oxide layers on their surfaces or by varnish coatings. Only very weak eddy currents are induced in a core of this kind. In small transformers an alternative technique is to use a core of compressed powdered iron, so that each grain of iron is fairly well insulated from the others.

20.6 Inductance

A self-induced emf occurs whenever the current in a wire loop changes.

Let us look at what happens when a wire loop is connected to a battery. As the current starts to flow, a magnetic field begins to build up. This changing magnetic field induces an emf in the *same* wire loop whose current led to the field in the first place. By Lenz's law the **self-induced emf** is such as to oppose the change in flux responsible for it. Thus an increasing current produces a self-induced emf opposite to the current, which slows down its rise. If instead we switch off an existing current in the loop, the self-induced emf will be in the same direction as the current, which slows down its fall.

The magnitude of the self-induced emf $\mathscr{E}_i$ depends on the rate of change $\Delta\Phi/\Delta t$ of the flux through the loop. For a loop whose geometry is fixed, Φ is proportional to the current I in the loop, and $\Delta\Phi/\Delta t$ is proportional to $\Delta I/\Delta t$, the rate at which the current is changing. Thus we can write

$$\mathscr{E}_i = -L\frac{\Delta I}{\Delta t} \qquad \textit{Self-induced emf} \quad (20.8)$$

where L, the constant of proportionality, is called the **inductance** of the loop. The minus sign, as in Eq. (20.3), follows from Lenz's law. A high value of L means a large $\mathscr{E}_i$ for a given rate of change $\Delta I/\Delta t$, and a low value of L means a small $\mathscr{E}_i$.

The unit of inductance is the **henry** (H), where

$$1\ \text{H} = 1\frac{\text{J}}{\text{A}^2} = 1\frac{\text{V}\cdot\text{s}}{\text{A}}$$

The henry, like the farad, is usually too large a unit for convenience. Inductances are accordingly often expressed in *millihenries* (mH) or *microhenries* (μH), where

$$1\ \text{mH} = 10^{-3}\ \text{H} \qquad 1\ \mu\text{H} = 10^{-6}\ \text{H}$$

EXAMPLE 20.7

An average self-induced emf of -0.75 V is produced in a 25-mH coil when the current in it falls to 0 in 0.010 s. What was the original current in the coil?

SOLUTION From Eq. (20.8)

$$\Delta I = \frac{-\mathscr{E}_i\,\Delta t}{L} = \frac{-(-0.75\ \text{V})(0.010\ \text{s})}{2.5\times 10^{-2}\ \text{H}} = 0.30\ \text{A}$$

Since $\Delta I = I_0 - 0$, the original current was $I_0 = 0.30$ A. ◆

A circuit element in which a self-induced emf accompanies a changing current is called an **inductor.** A wire loop is an example of an inductor. The inductance of

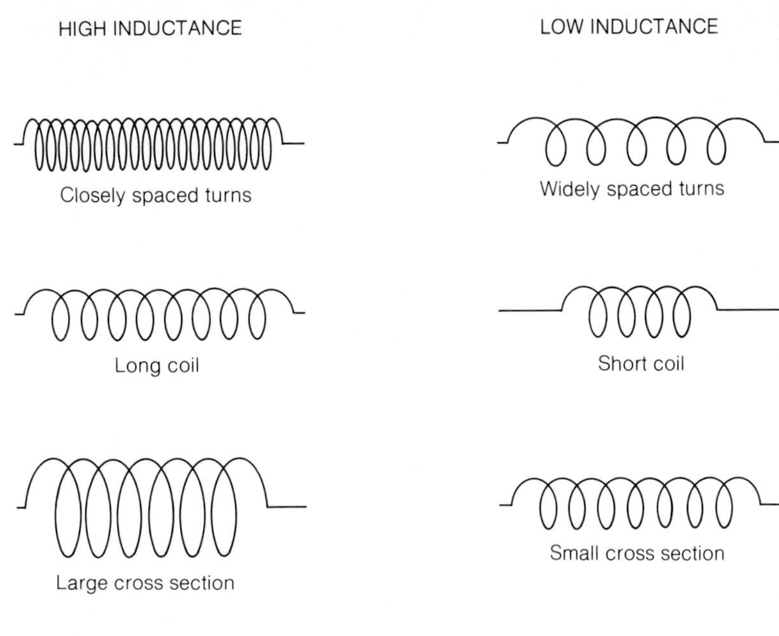

Figure 20.17

The inductance of a coil depends on how closely wound it is, on its length, its cross-sectional area, and on the nature of its core.

an inductor depends on its geometry and on the presence of a core with magnetic properties. A coil with many turns has more inductance than one with few turns because its magnetic field is stronger for a given current (Fig. 20.17). The longer the coil and the larger its cross-sectional area, the more its inductance, because it has a larger volume of magnetic field.

A core in a coil changes its inductance by changing the flux through the coil (see Section 19.4). Thus a diamagnetic core decreases the inductance slightly from its value in free space or air, a paramagnetic core increases it slightly, and a ferromagnetic core increases it considerably—by a factor of as much as 10^4 in some cases—but by an amount that is not constant.

EXAMPLE 20.8

Find the inductance of an N-turn solenoid of length l and cross-sectional area A.

SOLUTION From Eq. (20.4) the emf induced in a coil of N turns through which the magnetic flux changes at the rate $\Delta\Phi/\Delta t$ is

$$\mathscr{E}_i = -N\frac{\Delta\Phi}{\Delta t}$$

Since the magnetic field inside a solenoid is perpendicular to its turns, $\Phi = BA$ and

$$\mathscr{E}_i = -NA\frac{\Delta B}{\Delta t}$$

In the case of a solenoid, according to Eq. (19.6), $B = \mu(N/l)I$, so that

$$\frac{\Delta B}{\Delta t} = \mu \frac{N}{l}\frac{\Delta I}{\Delta t}$$

and

$$\mathscr{E}_i = -\mu N^2 \frac{A}{l}\frac{\Delta I}{\Delta t}$$

Comparing this result with Eq. (20.8), which defines L, gives

$$L = \mu N^2 \frac{A}{l} \qquad \textit{Inductance of a solenoid} \quad (20.9)$$

for the inductance of a solenoid. ◆

EXAMPLE 20.9

Find the inductance in air of a 1000-turn solenoid 10 cm long that has a cross-sectional area of 20 cm^2.

SOLUTION Since 20 cm^2 = 2.0×10^{-3} m^2 and $\mu = \mu_0$ in air (very nearly),

$$L = \mu_0 N^2 \frac{A}{l} = \left(1.26 \times 10^{-6} \frac{\text{T·m}}{\text{A}}\right)(10^3)^2\left(\frac{2.0 \times 10^{-3}\ \text{m}^2}{1.0 \times 10^{-1}\ \text{m}}\right)$$
$$= 2.5 \times 10^{-2}\ \text{H} = 25\ \text{mH}$$ ◆

20.7 Inductive Time Constant

The current in an inductor needs time to build up or to disappear.

When a circuit containing inductance is connected to a battery, the current in the circuit does not rise at once to its final value $I = \mathscr{E}/R$, where $\mathscr{E}$ is the emf of the battery and R is the total resistance in the circuit. As the switch in Fig. 20.18 is closed, the current I starts to grow. The induced emf $-L(\Delta I/\Delta t)$ then comes into being in the opposite direction to the battery emf $\mathscr{E}$. The net emf acting to establish current in the circuit is therefore $\mathscr{E} - L(\Delta I/\Delta t)$, and the current reaches its final value of I_0 in a gradual manner.

When the current in the circuit of Fig. 20.18 is I and is changing at the rate $\Delta I/\Delta t$,

$$\mathscr{E} - L\frac{\Delta I}{\Delta t} = IR \qquad (20.10)$$

Applied emf − induced emf = net voltage

Hence the rate at which the current is increasing is

$$\frac{\Delta I}{\Delta t} = \frac{\mathscr{E} - IR}{L}$$

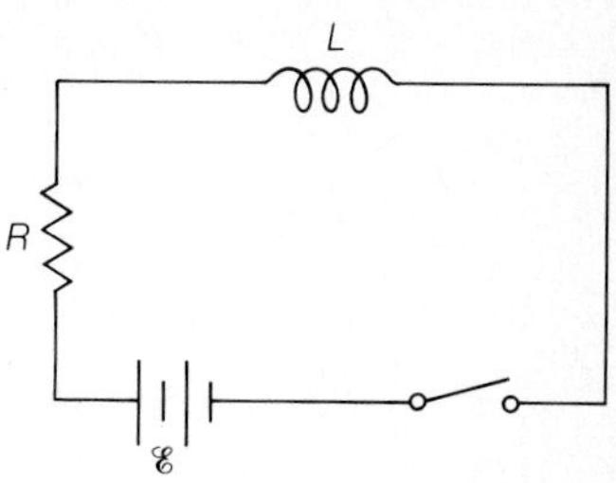

Figure 20.18

A circuit with an inductor, a resistor, and a source of emf in series. When the switch is closed, the current increases gradually to its ultimate value of $\mathscr{E}/R$ because of the opposing self-induced emf in the inductor.

The larger the inductance L, the more slowly the current increases. At the moment the switch is closed, $I = 0$ and $\Delta I/\Delta t$ has its maximum value of

$$\left(\frac{\Delta I}{\Delta t}\right)_{\text{max}} = \frac{\mathscr{E}}{L}$$

Eventually the current reaches its final value of I_0 and $\Delta I/\Delta t = 0$. From then on

$$\frac{\Delta I}{\Delta t} = 0 = \frac{\mathscr{E} - I_0 R}{L}$$

$$I_0 = \frac{\mathscr{E}}{R}$$

Thus the effect of having inductance in the circuit is to delay the establishment of the final current.

The graph in Fig. 20.19 shows how I varies with time when a current is being established in a circuit containing inductance. A mathematical analysis shows that, after a time interval of L/R, the current reaches 63% of I_0. The time $T_L = L/R$ is therefore a convenient measure of how rapidly a current rises in a circuit containing inductance. T_L is called the **inductive time constant** of the circuit:

$$T_L = \frac{L}{R} \qquad \textit{Inductive time constant} \quad (20.11)$$

When the battery in Fig. 20.18 is replaced by a wire, the current I drops slowly in the manner shown in Fig. 20.20, since the induced emf now tends to maintain the existing current. The current falls to 37% of its original value in the time L/R after the battery is removed.

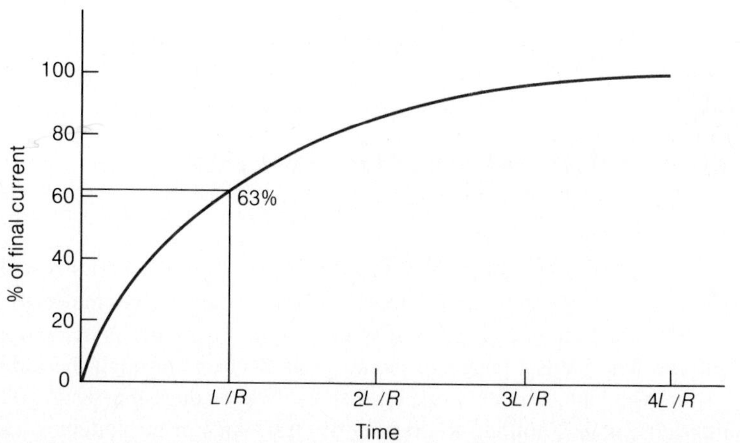

Figure 20.19

The growth of current in a circuit containing inductance and resistance.

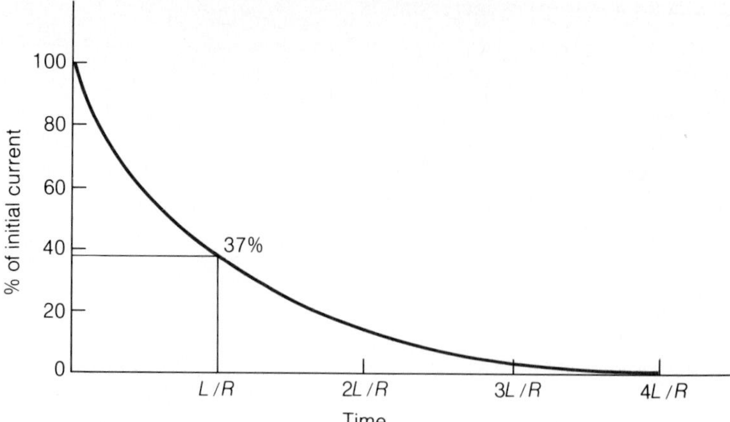

Figure 20.20

The decay of current in a circuit containing inductance and resistance when the battery is replaced by a wire.

EXAMPLE 20.10

A 2.0-H inductor whose resistance is 100 Ω is connected to a 24-V battery of negligible internal resistance. Find the initial current in the circuit and the initial rate at which the current is increasing, the time required for the current to reach 63% of its ultimate value, and the magnitude of the final current.

SOLUTION The initial current is 0, and the initial rate of current increase is

$$\frac{\Delta I}{\Delta t} = \frac{\mathscr{E}}{L} = \frac{24\ \text{V}}{2.0\ \text{H}} = 12\ \text{A/s}$$

The time needed for the current to rise to 63% of the final value is

$$T_L = \frac{L}{R} = \frac{2.0\ \text{H}}{100\ \Omega} = 0.02\ \text{s}$$

and the final current will be

$$I_0 = \frac{\mathscr{E}}{R} = \frac{24\ \text{V}}{100\ \Omega} = 0.24\ \text{A}$$

◆

20.8 Magnetic Energy

Energy is stored in the magnetic field of an inductor.

Energy is associated with magnetic as well as with electric fields. To appreciate this, we return to what happens when an inductor is connected to a battery.

By Lenz's law the self-induced emf $\mathscr{E}_i$ in the inductor opposes the rising current. The battery therefore has to push electrons through the inductor against $\mathscr{E}_i$, so that the battery does work against $\mathscr{E}_i$. Where does the work go? Since the self-induced emf has nothing to do with the resistance of the inductor, the work must go into the magnetic field of the current that is starting to flow. When the current in the inductor reaches its final value, there is no more self-induced emf. The work done to establish the inductor's magnetic field has become magnetic energy.

Let us see how much magnetic energy is contained in an inductor when a steady current I flows through it. This energy is equal to the work W done against $\mathscr{E}_i$ to establish the current starting from $I = 0$.

If we suppose the current rises at a uniform rate from 0 to I in a time interval Δt, then the average current during Δt is $\bar{I} = \frac{1}{2}I$. In this time the total charge Q that passes through the inductor is

$$Q = \bar{I}\Delta t = \tfrac{1}{2}I\,\Delta t$$

The work done to build up the current (and thus to create the final magnetic field) is

$$W = Q\mathscr{E} = \tfrac{1}{2}I\mathscr{E}\,\Delta t$$

where $\mathscr{E}$ is the applied emf. Since $\mathscr{E} = -\mathscr{E}_i$, from Eq. (20.8) we have

$$\mathscr{E} = -\mathscr{E}_i = L\frac{\Delta I}{\Delta t}$$

Here the change in current ΔI in the time Δt is the final current I, so $\Delta I = I$. The work done is therefore

$$W = \tfrac{1}{2}I\left(L\frac{I}{\Delta t}\right)\Delta t$$

and so

$$W = \tfrac{1}{2}LI^2 \qquad \textit{Energy of inductor} \quad (20.12)$$

A circuit of inductance L has the magnetic energy of $\frac{1}{2}LI^2$ stored in it when a current I is present in it. When the potential difference responsible for the current is removed, the energy $\frac{1}{2}LI^2$ is what powers the self-induced emf that slows down the drop in current. The gradual rise of current in a circuit containing inductance may be thought of as the result of absorbing $\frac{1}{2}LI^2$ of magnetic energy, and its gradual drop may be thought of as the result of the transfer of the inductor's energy to the circuit.

EXAMPLE 20.11

How much current is needed in a 50-mH coil for it to have 1.0 J of magnetic energy?

SOLUTION From Eq. (20.12) we have

$$I = \sqrt{\frac{2W}{L}} = \sqrt{\frac{(2)(1.0\text{ J})}{50 \times 10^{-3}\text{ H}}} = 6.3\text{ A}$$

◆

Magnetic Energy Density

In Section 17.6 the energy density w of an electric field of magnitude E was shown to be

$$w = \frac{E^2}{8\pi k} \qquad \textit{Electric energy density}$$

Let us now find the corresponding formula for the energy density of a magnetic field.

Equation (20.9) gives the inductance of a solenoid as $L = \mu N^2 A/l$. Therefore the energy content of the solenoid when the current I flows in it is

$$W = \tfrac{1}{2}LI^2 = \frac{\mu}{2}N^2\frac{A}{l}I^2$$

The magnetic field inside the solenoid is $B = \mu(N/l)I$, according to Eq. (19.6). Hence we can express W in terms of B instead of in terms of I by substituting $I = Bl/\mu N$. We find that

$$W = \frac{\mu}{2}N^2\frac{A}{l}\left(\frac{B^2l^2}{\mu^2N^2}\right) = \frac{lA}{2\mu}B^2$$

The quantity lA is the volume occupied by the magnetic field **B,** because it is the product of the length of the solenoid and its cross-sectional area. (We are considering an ideal solenoid whose length is large relative to its diameter so we can ignore the magnetic field escaping from the ends.) The **magnetic energy density** of the field is therefore

$$w = \frac{W}{lA} = \frac{B^2}{2\mu} \qquad \textit{Magnetic energy density} \quad (20.13)$$

The energy density of a magnetic field is directly proportional to the square of its magnitude B. This formula holds for all magnetic fields, not just for those inside solenoids.

EXAMPLE 20.12

The earth's magnetic field in a certain region has the magnitude 6.0×10^{-5} T. Find the magnetic energy per cubic kilometer in this region.

SOLUTION The magnetic energy density corresponding to $B = 6.0 \times 10^{-5}$ T is, since here $\mu = \mu_0$,

$$w = \frac{B^2}{2\mu_0} = \frac{(6.0 \times 10^{-5}\,\text{T})^2}{2(1.26 \times 10^{-6}\,\text{T·m/A})} = 1.4 \times 10^{-3}\,\text{J/m}^3$$

A cubic kilometer is equal to $(10^3\ \text{m})^3 = 10^9\ \text{m}^3$, so the total magnetic energy in a cubic kilometer is

$$W = wV = (1.4 \times 10^{-3}\ \text{J/m}^3)(10^9\ \text{m}^3) = 1.4 \times 10^6\ \text{J}$$

This amount of energy is enough to raise the 365,000-ton Empire State Building by 0.42 mm. ♦

20.9 Electrical Oscillations

An LC circuit is the electrical analog of a harmonic oscillator.

Let us connect a charged capacitor to an inductor, as in Fig. 20.21. The following sequence of events occurs:

(a) Each of the capacitor's plates starts with the charge Q and its electric energy is $\frac{1}{2}Q^2/C$, whereas the inductor has no energy since $I = 0$.

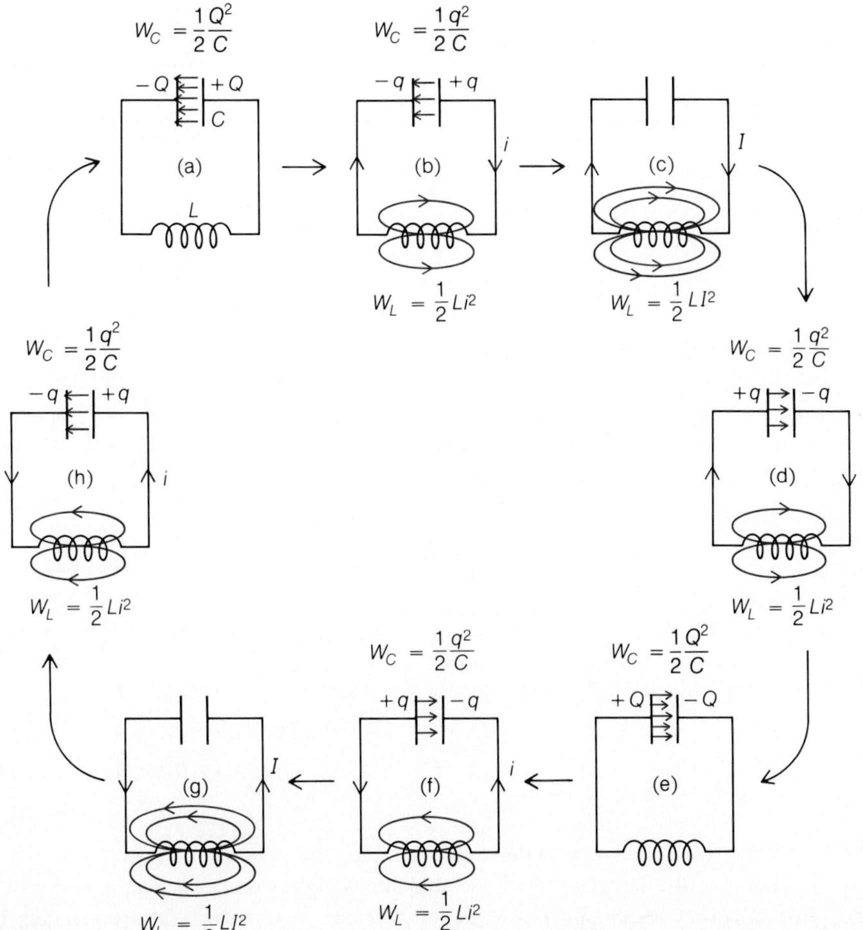

Figure 20.21

The various stages in the oscillation of a circuit containing the inductance L and the capacitance C. The maximum charge on the capacitor is Q, and the maximum current in the circuit is I.

(b) The capacitor is partly discharged, and the current in the inductor leads to a magnetic field whose energy is $\frac{1}{2}Li^2$.
(c) No charge is left on the capacitor's plates, and all the energy of the circuit is magnetic energy $\frac{1}{2}LI^2$.
(d) Now charge begins to build up on the capacitor's plates with polarities the reverse of the original ones, and the electric energy of the capacitor grows at the expense of the magnetic energy of the inductor as the current drops.
(e) At the instant the capacitor is fully charged again, the current is zero and the energy of the circuit is once more entirely electric.
(f) Next the capacitor begins to discharge, and eventually the entire circuit returns to its original state at (a). The cycle will continue to repeat itself indefinitely if no resistance is present.

This sequence is wholly analogous to the behavior of a harmonic oscillator. As we learned in Chapter 11, in a harmonic oscillator energy is constantly interchanged between KE and PE. Here the same sort of interchange occurs, with the electric energy of a charged capacitor corresponding to PE and the magnetic energy of an inductor through which a current flows corresponding to KE.

A harmonic oscillator whose mass is m and whose spring constant is k has the potential and kinetic energies

$$\text{PE} = \tfrac{1}{2}kx^2 \qquad \text{KE} = \tfrac{1}{2}mv^2 \qquad \textit{Harmonic oscillator}$$

where x is the displacement of the object from its equilibrium position and the speed v is the rate of change of x. The electric and magnetic energies in an electrical oscillator are

$$W_e = \tfrac{1}{2}\frac{Q^2}{C} \qquad W_m = \tfrac{1}{2}LI^2 \qquad \textit{Electrical oscillator}$$

where the current I is the rate of change of the charge Q.

In the above formulas we note that $1/C$ corresponds to k, since both determine the amount of energy present at the moment when nothing is moving in the respective oscillators. Similarly L corresponds to m. Both are measures of inertia—a large inductance tends to slow down changes in current, and a large m tends to slow down changes in speed. The frequency of a harmonic oscillator is

$$f = \frac{1}{2\pi}\sqrt{\frac{k}{m}}$$

and so it is not surprising that the frequency of an electrical oscillator is

$$f = \frac{1}{2\pi\sqrt{LC}} \qquad \textit{Oscillator frequency} \qquad (20.14)$$

The larger the inductance L and the capacitance C, the lower the frequency.

The parallel between electrical and mechanical oscillations is a very close one. Just as every mechanical system has certain specific natural frequencies of vibration that depend on its properties, so every electric circuit has a natural frequency of oscillation that depends on L and C. Just as the energy of every actual mechanical vibration is eventually dissipated as heat owing to the inevitable presence of friction, so the energy of every electrical oscillation is eventually dissipated as heat owing to the inevitable presence of resistance in the wires of the circuit.

Electrical resonance is much like mechanical resonance. If a source of emf that alternates at the natural frequency of a circuit is connected to the circuit, energy is fed into the oscillations that can maintain or increase their amplitude. At any other frequency the energy absorbed is small.

Until recently all radio communication was based on the ability of an LC circuit to oscillate only at the frequency given by Eq. (20.14). Such a circuit was used to "tune" a transmitter so that only a single frequency of radio waves was produced, and another similar circuit "tuned" a receiver to the same frequency. Today a different method is often used that is based on the multiplication of a standard frequency by special circuits.

EXAMPLE 20.13

A 5.0-μF capacitor with a charge of 1.0×10^{-4} C is connected to a 20-mH coil. (a) What is the frequency of the oscillations that result? (b) What is the maximum current in the circuit?

SOLUTION (a) From Eq. (20.14) the charge in the circuit will oscillate back and forth through the inductor at the frequency

$$f = \frac{1}{2\pi\sqrt{LC}} = \frac{1}{2\pi\sqrt{(20 \times 10^{-3}\,\text{H})(5.0 \times 10^{-6}\,\text{F})}} = 503\ \text{Hz}$$

(b) The maximum current occurs when all the charge has left the plates of the capacitor, which corresponds to (c) or (g) in Fig. 20.21. At such times the magnetic energy $\frac{1}{2}LI^2$ of the inductor equals the electric energy $\frac{1}{2}Q^2/C$ of the fully charged capacitor, so that

$$\tfrac{1}{2}LI^2 = \tfrac{1}{2}\frac{Q^2}{C}$$

$$I = \frac{Q}{\sqrt{LC}} = \frac{1.0 \times 10^{-4}\,\text{C}}{\sqrt{(20 \times 10^{-3}\,\text{H})(5.0 \times 10^{-6}\,\text{F})}} = 0.32\ \text{A}$$

♦

Important Terms

Electromagnetic induction refers to the production of an electric field in a conductor whenever magnetic field lines move across it.

Lenz's law states that the direction of an induced current must be such that its own magnetic field opposes the changes in flux that are inducing it.

A **generator** is a device that converts mechanical energy into electric energy.

The direction of an **alternating current** reverses itself periodically.

A **back** (or **counter**) **emf** is induced in the rotating coils of an electric motor and is opposite in direction to the external voltage applied to them.

An alternating current flowing in the primary coil of a **transformer** induces another alternating current in the secondary coil. The ratio of the emf's is proportional to the ratio of turns in the coils.

The **inductance** L of a circuit is the ratio between the magnitude of the self-induced emf $\mathscr{E}$ due to a changing current in the circuit and the rate of change $\Delta I/\Delta t$ of the current. The unit of inductance is the **henry.**

The inductive time constant of a circuit that contains inductance and resistance is a measure of the time needed for a current to be established or to disappear in the circuit.

The **energy density** of a magnetic field is the magnetic energy per unit volume associated with it.

Important Formulas

Motional emg: $\mathscr{E}_i = BLv_\perp$

Magnetic flux: $\Phi = B_\perp A = BA \cos\theta$

Induced emf: $\mathscr{E}_i = -N\dfrac{\Delta\Phi}{\Delta t}$

Transformer: $\dfrac{I_1}{I_2} = \dfrac{\mathscr{E}_2}{\mathscr{E}_1} = \dfrac{N_2}{N_1}$

Self-induced emf: $\mathscr{E}_i = -L\dfrac{\Delta I}{\Delta t}$

Inductive time constant: $T_L = \dfrac{L}{R}$

Magnetic energy of inductor: $W = \frac{1}{2}LI^2$

Magnetic energy density: $w = \dfrac{B^2}{2\mu}$

Oscillator frequency: $f = \dfrac{1}{2\pi\sqrt{LC}}$

MULTIPLE CHOICE

1. The emf produced in a wire by its motion across a magnetic field does not depend on
- **a.** the length of the wire
- **b.** the diameter of the wire
- **c.** the orientation of the wire
- **d.** the flux density of the field

2. A bar magnet is passed through a coil of wire. The induced current is greatest when
- **a.** the magnet moves slowly, so that it is inside the coil for a long time
- **b.** the magnet moves fast, so that it is inside the coil for a short time
- **c.** the north pole of the magnet enters the coil first
- **d.** the south pole of the magnet enters the coil first

3. The magnetic flux through a wire loop in a magnetic field **B** does not depend on
- **a.** the area of the loop
- **b.** the shape of the loop
- **c.** the angle between the plane of the loop and the direction of **B**
- **d.** the magnitude B of the field

4. The unit of magnetic flux is the weber, where 1 Wb =
- **a.** 1 T·m^2
- **b.** 1 T/m^2
- **c.** 1 A·m^2
- **d.** 1 A/m^2

5. A wire loop is moved parallel to a uniform magnetic field. The induced emf in the loop
- **a.** depends on the area of the loop
- **b.** depends on the shape of the loop
- **c.** depends on the magnitude of the field
- **d.** is 0

6. According to Lenz's law an induced current
- **a.** occurs whenever a moving conductor has a component of its velocity parallel to a magnetic field
- **b.** occurs whenever a moving conductor has a component of its velocity perpendicular to a magnetic field
- **c.** gives rise to a magnetic field of its own that reinforces the flux change that induces it
- **d.** gives rise to a magnetic field of its own that opposes the flux change that induces it

7. Lenz's law is a consequence of the law of conservation of
- **a.** charge
- **b.** momentum
- **c.** field lines
- **d.** energy

8. When a wire loop is rotated in a magnetic field, the direction of the induced emf changes once in each
- **a.** $\frac{1}{4}$ revolution
- **b.** $\frac{1}{2}$ revolution
- **c.** 1 revolution
- **d.** 2 revolutions

9. The back emf in an electric motor is a maximum when the motor
- **a.** is switched on
- **b.** is increasing in speed
- **c.** has reached its maximum speed
- **d.** is decreasing in speed

10. The alternating current in the secondary coil of a trans-

former is induced by
   a. a varying electric field
   b. a varying magnetic field
   c. the iron core of the transformer
   d. motion of the primary coil

11. The ratio between primary and secondary currents in a transformer does not depend on the
   a. ratio of turns in the two windings
   b. resistance of the windings
   c. nature of the core
   d. primary voltage

12. Eddy currents occur only in
   a. ferromagnetic materials
   b. insulators
   c. conductors
   d. coils

13. The unit of inductance is the henry, where 1 H =
   a. 1 J·A^2 b. 1 J/A^2
   c. 1 V·A   d. 1 V/A

14. A large increase in the inductance of a coil can be achieved by using a core that is
   a. diamagnetic   b. paramagnetic
   c. ferromagnetic   d. polar

15. The time constant of an *RL* circuit is the time needed for the current to reach which percentage of its final value?
   a. 50%   b. 63%
   c. 90%   d. 100%

16. Magnetic fields invariably contain
   a. a ferromagnetic material
   b. inductance
   c. electric current
   d. energy

17. A wire coil carries the current *I*. The magnetic energy of the coil does not depend on
   a. the value of *I*
   b. the number of turns in the coil
   c. whether the coil has an iron core or not
   d. the resistance of the coil

18. The horizontal aluminum main boom of a sailboat sailing downwind at 5.0 m/s is 7.0 m long. The boom is at an angle of 75° relative to the boat's direction of motion. If the vertical component of the earth's magnetic field there is 4.0×10^{-5} T, the potential difference between the ends of the boom is
   a. 0.03 mV   b. 0.36 mV
   c. 1.35 mV   d. 1.40 mV

19. A square wire loop 5 cm on a side is perpendicular to a magnetic field of 0.08 T. If the field drops to 0 in 0.2 s, the average emf induced in the loop during that time is
   a. 0.04 mV   b. 0.5 mV
   c. 1 mV   d. 8 V

20. A 400-turn coil of resistance 6.0 Ω encloses an area of 30 cm^2. How rapidly should a magnetic field parallel to the coil axis change in order to induce a current of 0.30 A in the coil?
   a. 0.0045 T/s   b. 0.25 T/s
   c. 0.67 T/s   d. 1.5 T/s

21. The primary winding of a transformer has 200 turns and its secondary winding has 50 turns. If the current in the secondary winding is 40 A, the current in the primary is
   a. 10 A   b. 80 A
   c. 160 A   d. 8.0 kA

22. A transformer has 100 turns in its primary winding and 300 turns in its secondary. If the power input to the transformer is 60 W, the power output is
   a. 20 W   b. 60 W
   c. 0.18 kW   d. 0.54 kW

23. A 60-W, 24-V light bulb is to be powered from a 120-V ac line. Which of the following combinations of primary and secondary transformer windings is needed?
   a. 20 turns primary, 100 turns secondary
   b. 50 turns primary, 100 turns secondary
   c. 100 turns primary, 20 turns secondary
   d. 100 turns primary, 50 turns secondary

24. The self-induced emf in a 0.10-H coil when the current in it is changing at the rate of 200 A/s is
   a. 10 V   b. 20 V
   c. 0.10 kV   d. 2.0 kV

25. The current in a circuit falls to 0 from 16 A in 0.010 s. The average emf induced in the circuit during the drop is 64 V. The inductance of the circuit is
   a. 0.032 H   b. 0.040 H
   c. 0.25 H   d. 4.0 H

26. A 0.20-H inductor of resistance 8.0 Ω is connected to a 12-V battery of 1.0 Ω internal resistance. The initial rate at which the current in the inductance increases is
   a. 2.4 A/s   b. 7.7 A/s
   c. 30 A/s   d. 60 A/s

27. The time needed for the current in the inductor of Multiple Choice 26 to reach 63% of its final value is
   a. 0.022 s   b. 0.025 s
   c. 1.6 s   d. 1.8 s

28. A 2.0-mH coil carries a current of 10 A. The energy stored in its magnetic field is
   a. 0.050 J   b. 0.10 J
   c. 1.0 J   d. 0.10 kJ

29. The energy contained in a cubic meter of space in which the magnetic field is 1.0 T is
   a. 6.3×10^{-7} J   b. 4.0×10^{-5} J
   c. 4.0×10^{5} J   d. 7.9×10^{5} J

30. The inductance needed with a capacitor of 200 pF in order to have an *LC* circuit that oscillates at 8.0 kHz is
   a. 2.0 mH   b. 4.0 mH
   c. 12 mH   d. 2.0 H

QUESTIONS

1. A car is traveling from New York to Florida. Which of its wheels are positively charged and which are negatively charged?

2. One end of a bar magnet is thrust into a coil, and the induced current is clockwise as seen from the side of the coil into which it was thrust. (a) Is the end of the magnet its north or south pole? (b) What will be the direction of the induced current when the magnet is withdrawn?

3. A bar magnet held vertically with its north pole downward is dropped through a wire loop whose plane is horizontal. Is the current induced in the loop just before the magnet enters it clockwise or counterclockwise as seen by an observer above? What is the direction of the current when the magnet passes through the center of the loop? Just after it leaves the loop?

4. A loop of copper wire is rotated in a magnetic field about an axis along a diameter. (a) Why does the loop resist this rotation? (b) If the loop were made of aluminum wire, would the resistance to rotation be different?

5. Explain how the loudspeaker in Fig. 19.41 can also be used as a microphone to convert sound waves into alternating current.

6. The following are basic electromagnetic phenomena: (a) An electric current produces a magnetic field; (b) a current-carrying wire experiences a force in a magnetic field (unless aligned with the field); and (c) an emf is induced in a wire when there is relative motion between it and a magnetic field. Which of these phenomena is involved in the operation of the following devices: an electric motor, a generator, a transformer, a relay, a tape-recording head, a tape-playback head?

7. Why is it easy to turn the shaft of a generator when it is not connected to an outside circuit, but much harder when such a connection is made?

8. Why does an electric motor require more current when it is turned on than when it is running continuously?

9. A coil is rotated 100 times per second in a magnetic field. What time interval separates successive instants when the induced emf is zero?

10. For which of the following can alternating current be used without first being rectified into direct current: an electromagnet, a light bulb, an electric heater, the charging of a storage battery?

11. How is the back emf in the armature of a dc motor at a given speed of rotation related to the emf developed in the same armature when the motor is used as a generator at the same speed of rotation and with the same current in its field coils?

12. In a series-wound generator the magnetic field is produced by an electromagnet whose windings are connected in series with the armature windings. What happens to the emf in such a generator when the current drawn by the external circuit increases? When the external circuit current decreases? Why?

13. In a shunt-wound generator the electromagnet windings are connected in parallel with the armature windings. What happens to the emf in such a generator when the current drawn by the external circuit increases? When the current decreases? Why?

14. What would happen if the primary winding of a transformer were connected to a battery?

15. What acts on the secondary winding of a transformer to cause an alternating potential difference to occur across its ends even though the primary and secondary windings are not connected?

16. Why are transformer cores made from thin sheets of steel? Would there be any advantage in making permanent magnets in this way?

17. What is the purpose of the air gaps in the recording and playback heads of a tape recorder?

18. The greater its capacitance, the less energy is stored in a capacitor when it is given a certain charge. On the other hand, the greater its inductance, the more energy is stored in an inductor when a certain current is present in it. Explain the difference.

19. What is the direction of the self-induced emf in a coil when the current in it increases? When the current decreases? What is the reason in each case?

20. What becomes of the work done against the back emf in an inductive circuit when a current is being established in it?

EXERCISES

20.2 Moving Wire in a Magnetic Field

1. A car is traveling at 90 km/h in a region where the horizontal component of the earth's magnetic field has the magnitude 2.0×10^{-5} T. Find the emf induced in the car's vertical radio antenna, which is 1.2 m long.

2. A train is traveling at 130 km/h in a region where the vertical component of the earth's magnetic field is 3.0×10^{-5} T. The railway is standard gauge with its rails 1.435 m apart. Find the potential difference between the wheels on each axle of the train's cars.

3. A car is traveling at 30 m/s on a road where the vertical component of the earth's magnetic field is 3.0×10^{-5} T. What is the potential difference between the ends of its axles, which are 2.0 m long?

4. A potential difference of 1.8 V is found between the ends of a 2.0-m wire moving in a direction perpendicular to a

magnetic field at a speed of 12 m/s. What is the magnitude of the field?

20.3 Faraday's Law

5. A rectangular wire loop 50 mm $\times$ 100 mm in size is perpendicular to a magnetic field of 1.0×10^{-3} T. (a) What is the flux through the loop? (b) If the magnetic field drops to zero in 3.0 s, what is the potential difference induced between the ends of the loop during that period?

6. The wire loop in Exercise 5 is turned so that a perpendicular to its plane makes an angle of 40° with **B**. Answer the same questions for this situation.

7. A wire loop is 50 mm in diameter and is oriented with its plane perpendicular to a magnetic field. How rapidly should the field change if a potential difference of 1.0 V is to appear across the ends of the loop?

8. A flat 100-turn coil that encloses an area of 50 cm^2 is turned so that its plane goes from being parallel to a magnetic field of 1.0×10^{-3} T to being perpendicular to the field in 0.040 s. Find the average emf induced in the coil.

9. The magnetic field inside a solenoid 15 mm in radius is increasing at a rate of 0.060 T/s. Find the emf induced in a 400-turn coil wound around the solenoid.

10. A circular wire loop 10 cm in radius with a resistance of 0.40 Ω is in a magnetic field of 0.60 T whose direction is perpendicular to the plane of the loop. The loop is grasped at opposite ends of a diameter and pulled out into a rectangle 30 cm long and 1.4 cm wide in a time of 0.20 s. Find the average current in the wire while it is being deformed.

11. A square wire loop 10 cm on a side is oriented with its plane perpendicular to a magnetic field. The resistance of the loop is 5.0 Ω. How rapidly should the magnetic field change if a current of 2.0 A is to flow in the loop?

20.4 The Generator

12. A 1.50-kW dc generator has a full-load potential difference of 35.0 V and a no-load potential difference of 38.0 V. How much power is dissipated as heat in the armature at full load?

13. The armature of a dc generator has a resistance of 0.20 Ω. The terminal potential difference of the generator is 120 V with no load and 115 V at full load. How much power is delivered at full load?

14. A shunt-wound 120-V, 3.0-kW generator has an armature resistance of 0.10 Ω and a field resistance of 120 Ω. Find its efficiency.

20.5 The Transformer

15. A transformer has 100 turns in its primary winding and 500 turns in its secondary winding. If the primary voltage and current are, respectively, 120 V and 3.0 A, what are the secondary voltage and current? (Assume 100% efficiency in Exercises 15–18.)

16. An electric welding machine employs a current of 400 A. The device uses a transformer whose primary winding has 400 turns and that draws 4.0 A from a 200-V power line. (a) How many turns are there in the secondary winding of the transformer? (b) What is the potential difference across the secondary?

17. A transformer rated at a maximum power of 10 kW is used to couple a 5000-V transmission line to a 240-V circuit. (a) What is the ratio of turns in the windings of the transformer? (b) What is the maximum current in the 240-V circuit?

18. A transformer connected to a 120-V power line has 100 turns in its primary winding and 40 turns in its secondary winding. The primary current is 0.24 A when a light bulb is connected to the secondary winding. What is the resistance of the bulb?

20.6 Inductance

19. The current in a circuit drops from 5.0 A to 1.0 A in 0.10 s. If an average emf of 2.0 V is induced in the circuit while this is happening, what is the inductance of the circuit?

20. What is the self-induced emf in a 0.40-H coil when the current in it is changing at a rate of 500 A/s?

21. Find the inductance of an air-core coil 40 cm long and 40 mm in diameter that has 1000 turns of wire.

22. An air-core solenoid 20 cm long and 20 mm in diameter has an inductance of 0.178 mH. How many turns of wire does it contain?

23. An inductor consists of an iron ring 50 mm in diameter and 1.0 cm^2 in cross-sectional area that is wound with 1000 turns of wire. (Such an inductor is essentially a solenoid bent into a circle.) The permeability of the iron is constant at 400 times that of free space at the magnetic fields at which the inductor will be used. What is the inductance of the inductor?

20.7 Inductive Time Constant

24. Show that the unit of L/R is the second.

25. What is the inductance of a coil whose resistance is 14 Ω and whose time constant is 0.10 s?

26. A 50-mH coil with a resistance of 20 Ω is connected to a 90-V battery of negligible internal resistance. What is the time constant of the circuit?

20.8 Magnetic Energy

27. A magnetic field of 1.0×10^2 T is very strong. (a) How much energy is contained in 1.0 liter of such a field? (b) What electric field would have the same energy density?

28. A 20-mH coil carries a current of 0.20 A. (a) How much energy is stored in its magnetic field? (b) What should the current be in order that it contain 1.0 J of energy?

29. A solenoid 200 mm long and 24.0 mm in diameter is wound with 1200 turns of wire whose total resistance is 40.0 Ω. The solenoid is connected to a 12.0-V battery whose internal resistance is 2.00 Ω. Find the inductance of the solenoid, the final current in it, and its energy content when the final current flows.

30. The solenoid in Exercise 29 is immersed in liquid oxygen whose permeability is 1.0049 times greater than that of free space. Find the inductance of the solenoid, the final current in it, and its energy content when the final current flows.

31. A 2.0-H coil carries a current of 0.50 A. (a) How much energy is stored in it? (b) In how much time should the current drop to 0 if an emf of 100 V is to be induced in it?

20.9 Electrical Oscillations

32. Show that the unit of $1/2\pi\sqrt{LC}$ is the hertz.

33. In the antenna circuit of a certain radio receiver, the inductance is fixed at 4.0 mH but the capacitance can be varied. When the capacitance is 10 pF, what is the frequency of the radio waves the receiver responds to?

34. What inductance is needed in a circuit in which $C = 60$ μF if its natural frequency is to be 30 Hz?

35. What capacitance is needed in a circuit in which $L =$ 2 H if its natural frequency is to be 200 Hz?

36. The frequencies used in the commerical radio broadcasting range from 550 to 1600 kHz. What range of capacitance should a variable capacitor have if it is connected to a coil of inductance 1.0 mH in a circuit designed to respond to frequencies in this band?

37. A 1.0-μF capacitor is charged by being connected to a 10-V battery. The battery is then removed and the capacitor connected to a 10-mH coil. (a) Find the frequency of the resulting oscillations. (b) Find the maximum value of the charge on the capacitor. (c) Find the maximum current that flows through the inductor.

38. A capacitor with the charge Q is connected to an inductor, and electrical oscillations with the frequency f occur as a result. How is the maximum current I that flows in the circuit related to Q and f?

PROBLEMS

39. A metal rod of mass m slides down a pair of parallel metal rails L apart that are at the angle θ with the horizontal, as in Fig. 20.22. The rails are joined by a resistor of resistance R. A magnetic field **B** directed vertically downward is present. As the rod slides down, a current is induced in it that interacts with the magnetic field to produce a force with a component that opposes the rod's downward motion. As a result an equilibrium occurs in which the rod moves downward at a constant speed v. Derive a formula for v.

40. An air-core solenoid 20 cm long and 20 mm in radius is wound with 1000 turns and has a secondary coil of 300 turns wound over it. (a) If the current in the solenoid is changing at a rate of 0.50 A/s, what is the emf in the secondary coil? (b) An iron core with a constant permeability of $\mu = 400\mu_0$ is placed in the solenoid. What is the induced emf now?

41. In a **betatron** (Fig. 20.23) electrons are accelerated to high speeds through the action of an increasing magnetic field. As **B** increases, the energy of an electron moving in a circular path increases by an amount per revolution equal to its increase in energy when moving through an emf given by Eq. (20.4). Thus a circular path of this kind can be regarded as though it were a loop of wire as far as electromagnetic induction is concerned. As the electron goes faster and faster, it needs a stronger magnetic field to stay in the same orbit, which the

Figure 20.22
Problem 39.

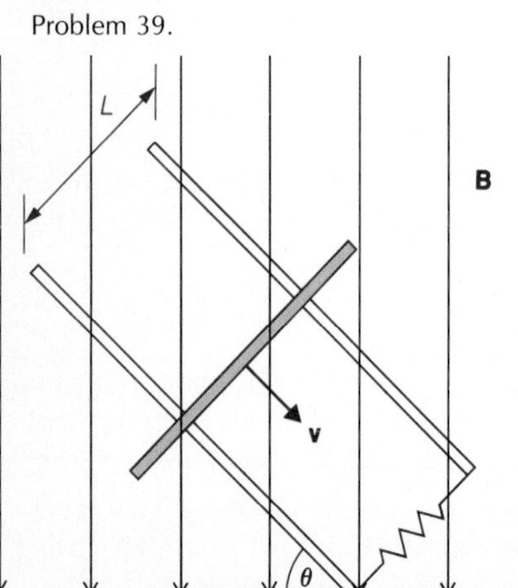

Figure 20.23
A betatron (Problem 41).

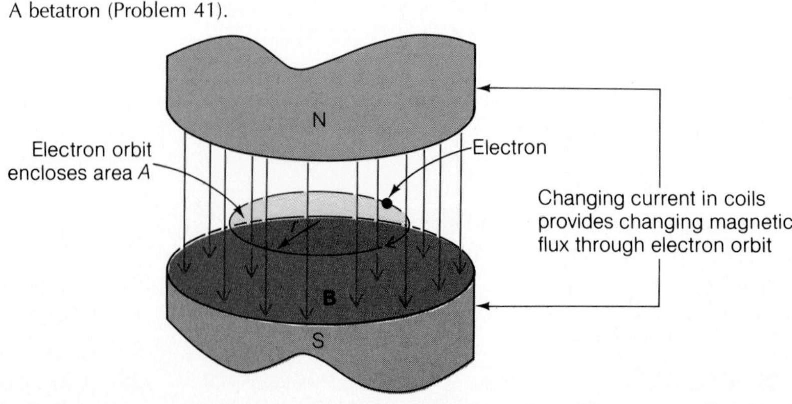

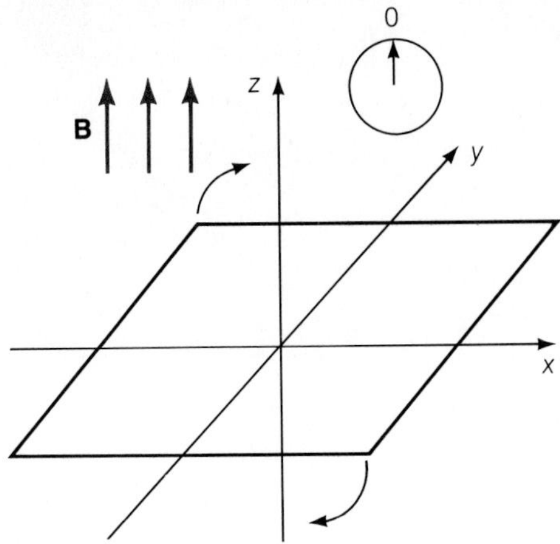

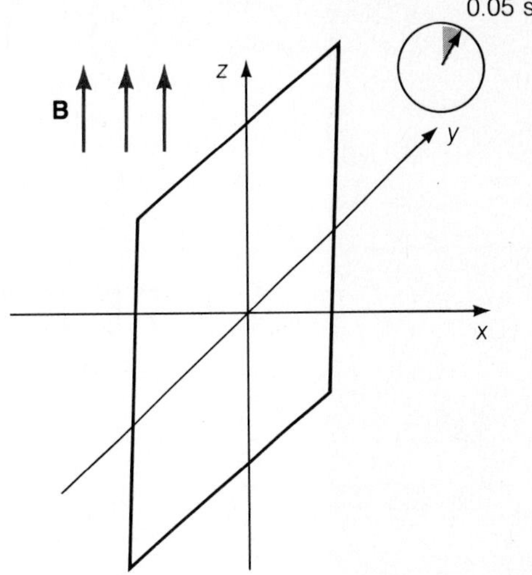

Figure 20.24
Problem 42.

increase in **B** that accelerates the electron can accomplish. Find the number of revolutions an electron must make in a betatron to reach an energy of 100 MeV when its orbit radius is 1.00 m and B is changing at 100 T/s.

42. A square wire loop 100 mm on a side rotates at a constant angular velocity in a uniform magnetic field **B** so that it turns through 90° in 0.050 s, as shown in Fig. 20.24. Initially the loop lies in the xy-plane and afterward in the yz-plane. The field **B** has a magnitude of 0.40 T and points in the $+z$-direction. (a) Find the average emf $\mathscr{E}_{av}$ induced in the loop during the turn. (b) In what part of the turn is $\mathscr{E}$ a minimum? A maximum? What are the values of $\mathscr{E}_{min}$ and $\mathscr{E}_{max}$?

43. A shunt-wound 120-V dc electric motor has an armature resistance of 1.0 Ω and a field resistance of 200 Ω. The motor has a power input of 420 W at 2000 rpm. (a) Find the power output and efficiency of the motor. (b) What is the starting current? (c) What series resistance is required if the starting current is to be limited to 25 A?

44. Find the power output, the power input, and the efficiency of the motor in Problem 43 when it is operating at 1800 rev/min.

45. A series-wound 24-V dc electric motor has an armature resistance of 0.10 Ω and a field resistance of 0.40 Ω. The motor draws 20 A at its normal operating speed. Find the power output, the power input, and the efficiency of the motor.

46. Show that the total inductance L of three inductors in parallel is given by $1/L = 1/L_1 + 1/L_2 + 1/L_3$. (*Hint:* The self-induced emf is the same for all the inductors, because they are connected in parallel.)

47. Show that the total inductance L of three inductors in series is given by $L = L_1 + L_2 + L_3$. (*Hint:* The total self-induced emf of the inductors is equal to the sum of the self-induced emf's of the individual inductors.)

48. A coil 20 cm long and 30 mm in diameter is tightly wound with one layer of copper wire 1.0 mm in diameter. Find its time constant.

49. A potential difference of 50 V is suddenly applied across a 12.0-mH, 8.0-Ω coil. Find (a) the initial current and the initial rate of change of current, (b) the current when the rate of change of current is 2000 A/s, and (c) the final current and final rate of change of current.

50. A potential difference of 100 V is suddenly applied across a 0.50-H, 20.0-Ω inductor. Find (a) the initial rate of increase of current, (b) the rate of increase of current when the current is 3.0 A, (c) the final current, and (d) the energy content of the inductor when the final current flows in it.

51. A 2.0-μF capacitor is charged by being connected to a 24-V battery. The battery is then removed and the capacitor connected to a 50-mH coil. After $\frac{1}{8}$ cycle the initial energy will be equally divided between the electric field in the capacitor and the magnetic field in the coil, as in Fig. 20.21(b). (a) Find the time interval required for this to occur. (b) Find the charge on the capacitor and the current in the inductor at this moment.

Answers to Multiple Choice

1. b	**9.** c	**17.** d	**25.** b
2. b	**10.** b	**18.** c	**26.** d
3. b	**11.** d	**19.** c	**27.** a
4. a	**12.** c	**20.** d	**28.** b
5. d	**13.** b	**21.** a	**29.** c
6. d	**14.** c	**22.** b	**30.** d
7. d	**15.** b	**23.** c	
8. b	**16.** d	**24.** b	

CHAPTER

21

ALTERNATING CURRENT

Nearly all the world's electric energy is carried by alternating current. Mainly this is due to the economy of high-voltage transmission, which minimizes I^2R heat losses. With step-up transformers at the production end and step-down transformers at the consumption end, the transmission voltage is limited only by insulation and atmospheric discharge problems. Alternating current is also preferred in industry because ac electric motors are, as a class, cheaper, more durable, and less in need of maintenance than dc motors. (Dc motors, however, have certain characteristics, such as better speed regulation, that make them more suitable for specialized duties.) Alternating currents are also involved in all aspects of modern communication: The electrical equivalent of a sound wave of a certain frequency is an alternating current of that frequency, and radio waves are produced by antennas fed with high-frequency ac. Alternating current behaves in a circuit in a very different way from direct current, and some insight into ac circuit behavior is needed to understand much of modern technology.

21.1 Effective Current and Voltage

Each is less than the corresponding maximum value.

A direct current is described in terms of its direction and magnitude. In a simple circuit there might be a current of 6 A from the positive terminal of a battery through a resistance network to its negative terminal. An alternating current has neither a constant direction nor a constant magnitude. How do we describe it?

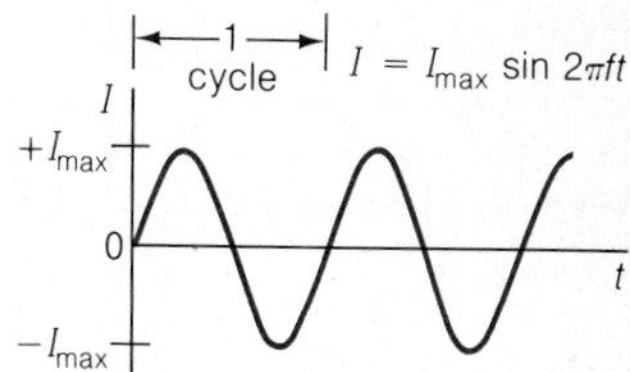

Figure 21.1

The variation of an alternating current with time. The frequency of the current is the number of cycles that occur per second.

Since alternating current flows back and forth in a circuit, it has no "direction" in the same sense as a direct current has. However, the oscillations have a certain frequency in each case, and the value of this frequency—that is, how many times per second the current goes through a complete cycle—is part of the description of the current.

An alternating current varies with time as shown in Fig. 21.1, and it would seem natural to specify its maximum value, I_{max}, as well as its frequency. The trouble with doing this is that I_{max} is not a measure of the ability of a current to do work or produce heat. A 6-A direct current is not equivalent to an alternating current in which $I_{max} = 6$ A. A better procedure is to define an **effective current** I_{eff} such that a direct current of this magnitude produces heat in a resistor at the same rate as the alternating current.

An alternating current varies with time according to the formula

$$I = I_{max} \sin 2\pi ft \tag{21.1}$$

where f is the frequency of the current. Here the current is assumed to be $I = 0$ and is increasing when $t = 0$. Figure 21.1 is a graph of this formula. The rate at which heat is dissipated in a resistance R by an alternating current is, at any time t, given by

$$I^2R = I^2_{max}R \sin^2 2\pi ft$$

The average value of I^2R over a complete cycle is

$$[I^2R]_{av} = I^2_{eff}R = I^2_{max}R[\sin^2 2\pi ft]_{av} \tag{21.2}$$

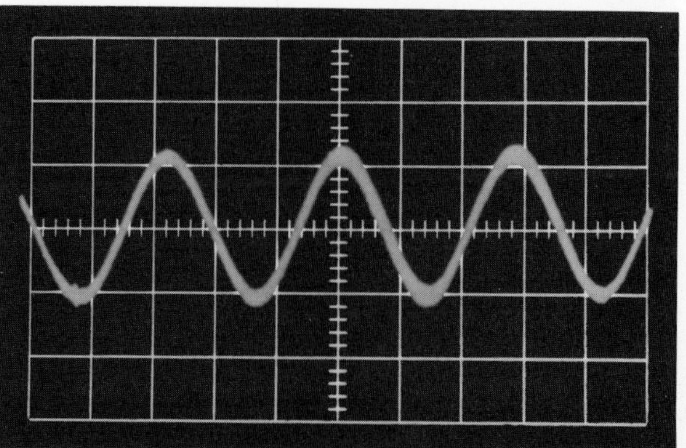

Oscilloscope trace of the signal produced by a tuning fork vibrating at 440 Hz.

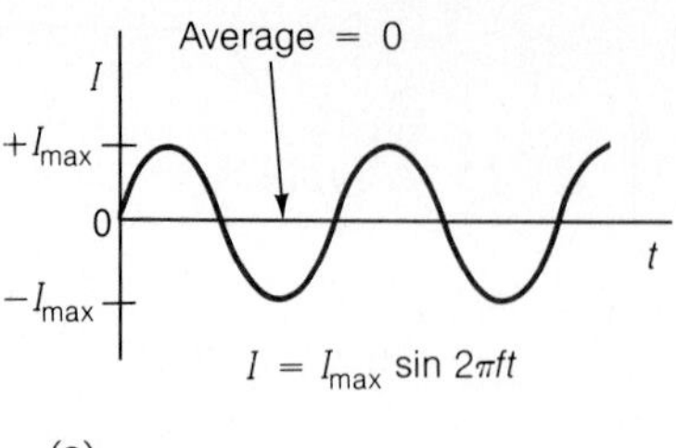

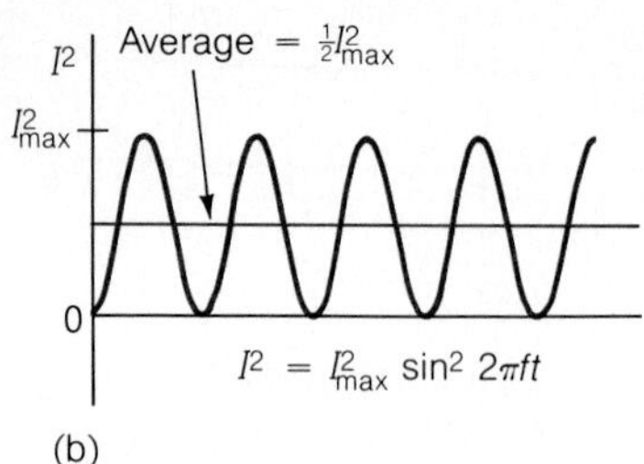

Figure 21.2

(a) The average value of I in an ac circuit is zero. (b) The average value of I^2 in an ac circuit is $\frac{1}{2}I^2_{max}$.

What we must find is the average value of $\sin^2 2\pi ft$ over a complete cycle. (The average value of I over a cycle is 0, since I is positive for half the cycle and negative for the other half. However, I^2 is always positive, and its average is a positive quantity. See Fig. 21.2.)

We begin with the trigonometric identity

$$\sin^2\theta = \tfrac{1}{2}(1 - \cos 2\theta)$$

The average value of $\sin^2 \theta$ over a complete cycle is therefore

$$[\sin^2 \theta]_{av} = \tfrac{1}{2}[1 - \cos 2\theta]_{av} = \tfrac{1}{2} - \tfrac{1}{2}[\cos 2\theta]_{av} = \tfrac{1}{2}$$

because, over a complete cycle, the average value of $\cos 2\theta$ is 0 by the same reasoning as in the case of $\sin \theta$. As a result

$$I^2_{eff}R = I^2_{max}R[\sin^2 2\pi ft]_{av} = \tfrac{1}{2}I^2_{max}R$$

and

$$I_{eff} = \frac{I_{max}}{\sqrt{2}} = 0.707 I_{max} \qquad \textit{Effective current} \quad (21.3)$$

The effective magnitude of an alternating current is 70.7% of its maximum value.

In a similar way the effective voltage in an ac circuit turns out to be

$$V_{eff} = \frac{V_{max}}{\sqrt{2}} = 0.707\ V_{max} \qquad \textit{Effective voltage} \quad (21.4)$$

It is customary to express currents and voltages in ac circuits in terms of their effective values. Thus the potential difference across a ''120-volt, 60-Hz'' power line actually varies from

$$+\,V_{max} = +\frac{V_{eff}}{0.707} = +\frac{120 \text{ volts}}{0.707} = +\,170 \text{ volts}$$

through 0 to -170 volts and back to $+170$ volts a total of 60 times per second (Fig. 21.3).

In what follows, when current, potential difference, and emf values are given for an ac circuit without other qualification, they will refer to the effective magnitudes of these quantities.

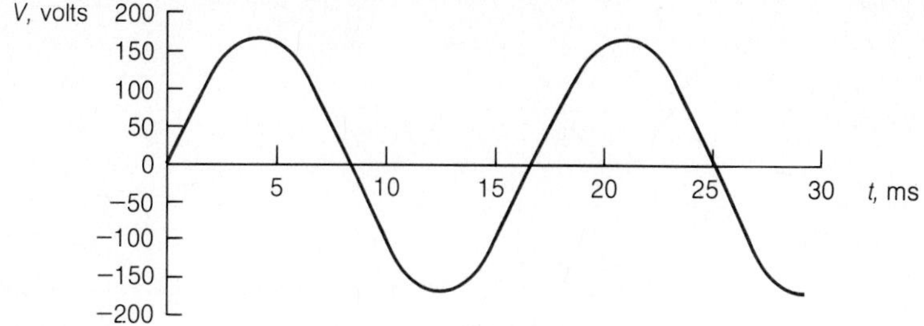

Figure 21.3

The voltage in a "120-V, 60-Hz" power line actually varies from −170 V to +170 V. Each complete cycle takes 16.7 ms.

By analogy with circular motion, angular frequency ω (in radians per second) is often used instead of frequency f (in hertz) in discussing alternating currents, where

$$\omega = 2\pi f$$

In this notation, which will not be used here, the instantaneous current in an ac circuit is written

$$I = I_{max} \sin \omega t$$

What are called here "effective" values of current and voltage are elsewhere sometimes called "root-mean-square" or "rms" values from their definitions as the square roots of the average values of I^2 and V^2.

21.2 Phasors

Rotating vectors used to represent alternating currents or voltages.

A convenient way to represent ac currents and voltages is in terms of **phasors.** This scheme is based on the fact that the component in any direction of a uniformly rotating vector varies sinusoidally—that is, as $\sin \theta$ varies with θ—with time. In the case of an alternating current, the length of the phasor $\mathbf{I}_{max}$ corresponds to I_{max}, and we imagine it to rotate f times per second in a counterclockwise sense (Fig. 21.4).

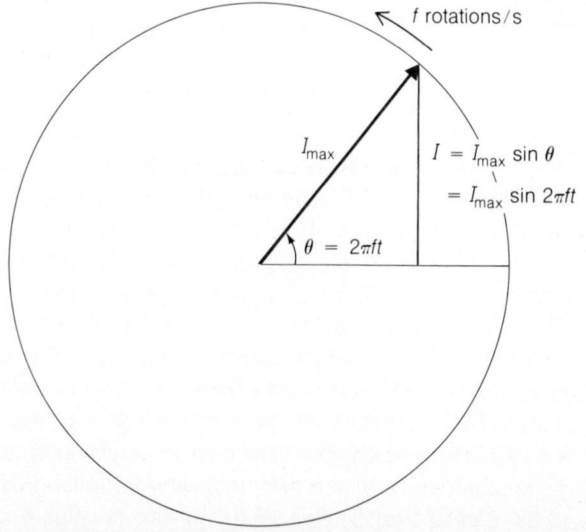

Figure 21.4

Phasor representation of I_{max} and I in an alternating current of frequency f. The vertical component of the phasor I_{max} at any time t is equal to the instantaneous value I of the current at that time.

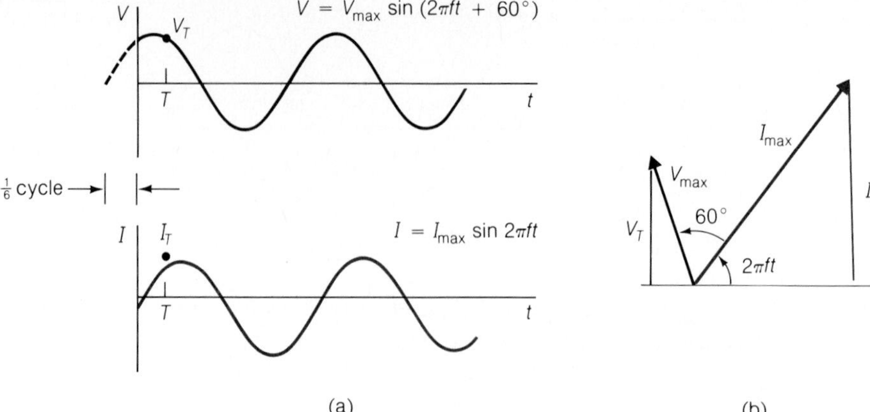

Figure 21.5

(a) In a certain ac circuit, the oscillations in V occur earlier than those in I by $\frac{1}{6}$ cycle. (b) In a phasor diagram, the phasor V_{max} leads the phasor I_{max} by 360°/6 = 60°. The orientations of the phasors correspond to the time T in the graphs of part (a).

Since the angle θ is equal to $2\pi ft$, the vertical component of the phasor at any time t corresponds to the instantaneous current I of Eq. (21.1):

$$I = I_{max} \sin \theta = I_{max} \sin 2\pi ft$$

In a similar way a phasor $\mathbf{V}_{max}$ can be used to represent an ac voltage V.

Phasors are helpful because the current and voltage in an ac circuit or circuit element always have the same frequency but may differ in phase, so that the peaks in each quantity do not occur at the same times. Suppose, for instance, that the voltage in a certain circuit leads the current by $\frac{1}{6}$ cycle, as in Fig. 21.5(a). (That is, the voltage peaks occur $\frac{1}{6}$ cycle earlier than the current peaks do.) In a phasor diagram, such as Fig. 21.5(b), this situation is shown by having the phasor $\mathbf{V}_{max}$ at an angle of 60° counterclockwise from the phasor $\mathbf{I}_{max}$, since 360°/6 = 60°. We must imagine the two phasors rotating together, always with the same angle of 60° between them, and generating the changing values of V and I shown in Fig. 21.5(a) as times goes on.

21.3 Phase Relationships

In a resistor, V is in phase with I; in an inductor, V leads I; in a capacitor, V lags behind I.

All actual electric circuits exhibit resistance, capacitance, and inductance to some degree. When direct current flows in a circuit, only its resistance is significant, but all three properties of the circuit affect the flow of alternating current.

A pure-resistance ac circuit is an idealized circuit whose capacitance and inductance are negligible. The instantaneous values of the voltage and current are in the same phase at all times in such a circuit. Both V and I are zero at the same time, both V and I pass through their maximum values at the same time, and so on (Fig. 21.6). The phasors for V and I therefore remain together.

A pure-inductance ac circuit is an idealized circuit whose resistance and capacitance are negligible. There is no IR potential drop across the inductance, and the potential difference across it is therefore proportional to $\Delta I/\Delta t$, the rate of change of the current (see Section 20.6). In this situation V and I cannot be in phase with

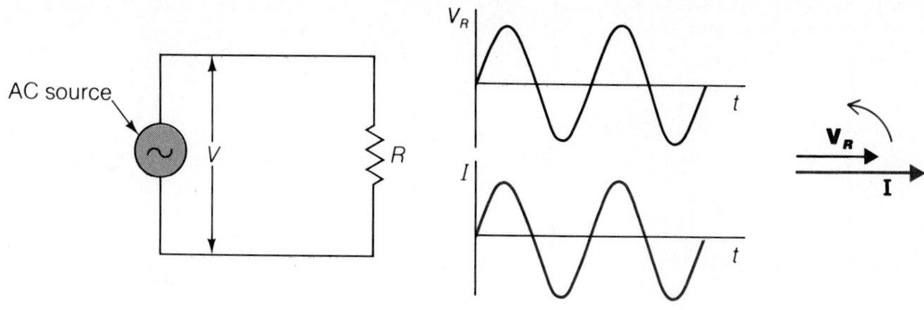

Figure 21.6
Current and voltage are in phase in a pure resistance.

each other. The current I changes most rapidly when $I = 0$, so $V = \pm V_{max}$ when $I = 0$. On the other hand, $\Delta I/\Delta t = 0$ at I_{max}, so $V = 0$ when $I = \pm I_{max}$. As shown in Fig. 21.7,

---

The voltage across a pure inductor leads the current in the inductor by $\frac{1}{4}$ cycle.

That is, the variations in voltage occur $\frac{1}{4}$ cycle *earlier than* the corresponding variations in current.

In a phasor diagram a difference in phase of $\frac{1}{4}$ cycle means that the angle between the phasors **V** and **I** is 360°/4 = 90°. Since **V** leads **I** in a pure inductor, **V** is 90° counterclockwise from **I**, as in Fig. 21.7.

The potential difference across a capacitor depends on the amount of charge stored on its plates. The charge is a maximum when $I = 0$, which is when the current is about to reverse direction and carry away the stored charge. The potential difference across a capacitor is

$$V = \frac{Q}{C}$$

and so $V = \pm V_{max}$ when $I = 0$. The stored charge is 0 when $I = \pm I_{max}$, because at these times the former stored charge is all gone and charge of the opposite sign is about to build up. Hence $V = 0$ when $I = \pm I_{max}$. As shown in Fig. 21.8,

---

The voltage across a pure capacitor lags behind the current into and out of the capacitor by $\frac{1}{4}$ cycle.

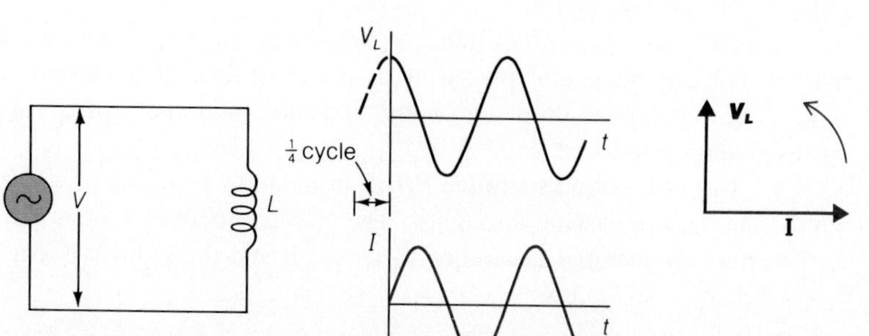

Figure 21.7
The voltage across a pure inductor leads the current in the inductor by $\frac{1}{4}$ cycle.

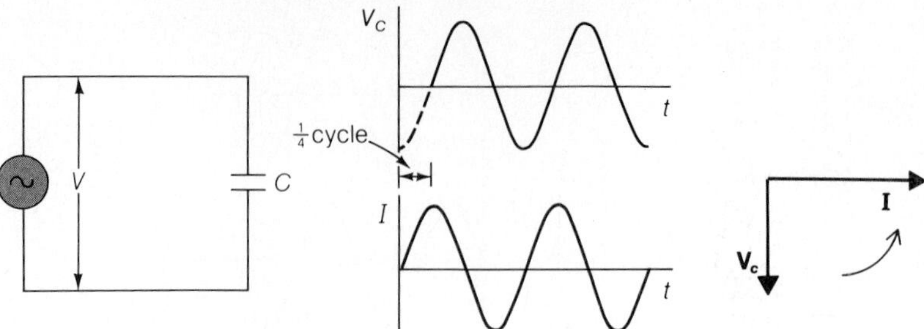

Figure 21.8

The voltage across a pure capacitor lags behind the current into and out of the capacitor by $\frac{1}{4}$ cycle.

That is, the variations in the voltage occur $\frac{1}{4}$ cycle *later than* the corresponding variation in the current. In a phasor diagram **V** is 90° clockwise from **I,** as in Fig. 21.8.

In a pure-capacitance circuit the phase relationship between current and voltage is different from what it is in pure-resistance and pure-inductance circuits. Although alternating current does not flow *through* a capacitor, it does flow *into one plate and out of the other* because changes in the amount of charge stored on one of the plates are mirrored by changes in the charge stored on the other plate. A flow of $+Q$ into one plate means that $+Q$ flows out of the other plate to leave the latter with a net charge of $-Q$. If direct current were involved, eventually enough charge would accumulate to prevent the arrival of any more, and the current would stop. In the case of an alternating current, however, the current always stops and reverses itself periodically anyway, so the presence of a series capacitor does not prevent ac from flowing in the circuit.

21.4 Inductive Reactance

A measure of the opposition of an inductor to the flow of alternating current.

Resistors, inductors, and capacitors all impede the flow of alternating current in a circuit. The effect of resistance is to dissipate part of the electric energy that passes through it into heat. In those conductors in which Ohm's law holds for direct current, it holds for alternating current as well, and

$$I = \frac{V_R}{R} \tag{21.5}$$

Here V_R is the effective potential difference across the resistance R, and I is the effective current through it.

The opposition of an inductor to the flow of alternating current comes from the self-induced back emf produced in it by the changing current. The back emf represents a potential drop across the inductor, and the current in the circuit is correspondingly reduced.

The **inductive reactance** X_L of an inductor is a measure of its effect on an alternating current passing through it. The effective current I in an inductor is related to the effective potential difference V_L across it and the inductive reactance X_L by

$$I = \frac{V_L}{X_L} \tag{21.6}$$

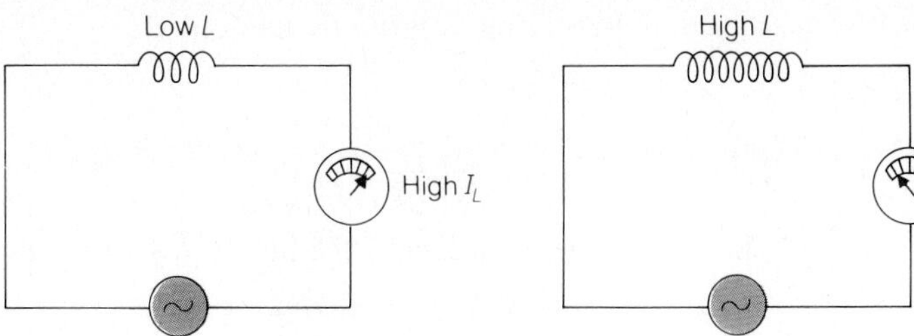

Figure 21.9

At a given frequency, the lower the inductance L, the lower the inductive reactance X_L and the higher the current I_L.

The unit of inductive reactance is the ohm. Equation (21.6) is analogous to Ohm's law, but there is a basic distinction between reactance and resistance in that there is no power loss in an inductor, whereas power is dissipated as heat in a resistor.

The inductive reactance of an inductor is given by the formula

$$X_L = 2\pi fL \qquad \textit{Inductive reactance} \quad (21.7)$$

where f is the frequency of the current in hertz (cycles/second) and L is the inductance in henries. The direct dependence of X_L on f and L is reasonable. The self-induced back emf, which is what opposes the current, is proportional to both $\Delta I/\Delta t$ and L, and so the more rapidly the current changes and the larger the value of L, the greater the back emf (Fig. 21.9).

EXAMPLE 21.1

What is the current in a coil of negligible resistance and inductance 0.40 H when it is connected to a 120-V, 60-Hz power line?

SOLUTION The reactance of the coil is

$$X_L = 2\pi fL = (2\pi)(60 \text{ Hz})(0.4 \text{ H}) = 151 \ \Omega$$

and the current in it is

$$I = \frac{V_L}{X_L} = \frac{120 \text{ V}}{151 \ \Omega} = 0.80 \text{ A}$$

Both the 120-V and 0.80-A figures represent effective values. ♦

21.5 Capacitive Reactance

A measure of the opposition of a capacitor to the flow of alternating current.

The extent to which a capacitor opposes the flow of alternating current depends on its **capacitive reactance** X_C. If V_C is the effective potential difference across a capacitor whose reactance is X_C, the effective current into and out of the capacitor is

$$I = \frac{V_c}{X_c} \qquad (21.8)$$

The capacitive reactance of a capacitor is given by the formula

$$X_C = \frac{1}{2\pi fC} \qquad \textit{Capacitive reactance} \quad (21.9)$$

If f is in hertz and the capacitance C in farads, the unit of X_C is the ohm.

EXAMPLE 21.2

A capacitor whose reactance is 80 Ω at 50 Hz is used in a 60-Hz circuit. What is its reactance in the latter circuit?

SOLUTION The capacitance of the capacitor, from Eq. (21.9), is

$$C = \frac{1}{2\pi f X_C} = \frac{1}{(2\pi)(50\text{ Hz})(80\ \Omega)} = 4 \times 10^{-5}\text{ F}$$

which is 40 μF. Its reactance at 60 Hz is

$$X_C = \frac{1}{2\pi fC} = \frac{1}{(2\pi)(60\text{ Hz})(4 \times 10^{-5}\text{ F})} = 66\ \Omega$$

◆

A capacitor impedes the flow of alternating current through the reverse potential difference that appears across it as charge builds up on its plates. Thus there is a potential drop across a capacitor in an ac circuit that affects the current just as the potential drop in a resistor does.

The inverse dependence of X_C on f and C can be understood from the following argument. If the charge on the plates of a capacitor changes by ΔQ in the time Δt, then the instantaneous current is $I_{\text{inst}} = \Delta Q/\Delta t$. If V_C is the change in the potential difference across the capacitor that leads to ΔQ, then $\Delta Q = C(\Delta V_C)$, and $I_{\text{inst}} = C(\Delta V_C/\Delta t)$. The higher the frequency, the greater the rate of change $\Delta V_C/\Delta t$ of the potential difference and the greater the current. The larger C is, also, the greater the current (Fig. 21.10). Since a high current means a low reactance, increasing f and C decreases X_C.

Capacitive and inductive reactances vary differently with frequency, as Fig. 21.11 shows: X_C decreases with increasing f, whereas X_L increases with increasing f.

In the limit of $f = 0$, which means direct current, $X_L = 0$ and $X_C = \infty$. When the current is steady, there is no self-induced back emf in an inductor, and no

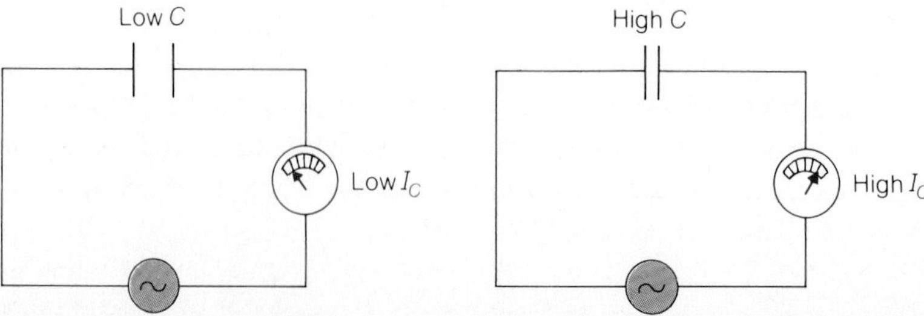

Figure 21.10

At a given frequency, the higher the capacitance C, the lower the capacitive reactance X_C and the higher the current I_C.

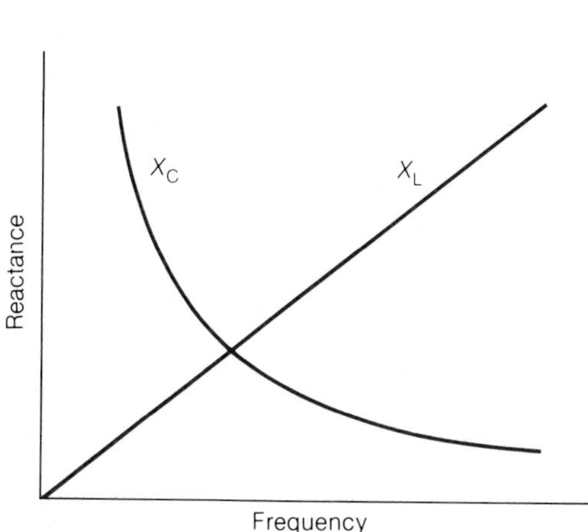

Figure 21.11

The reactance X_L of an inductor increases with frequency, whereas the reactance X_C of a capacitor decreases with frequency. In the dc limit of $f = 0$, $X_L = 0$ and $X_C = \infty$.

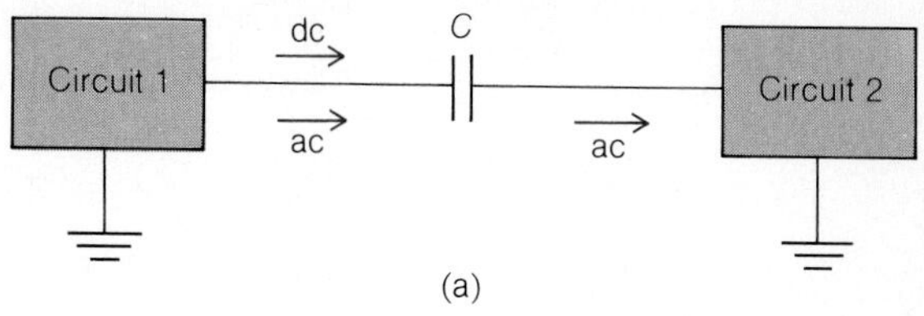

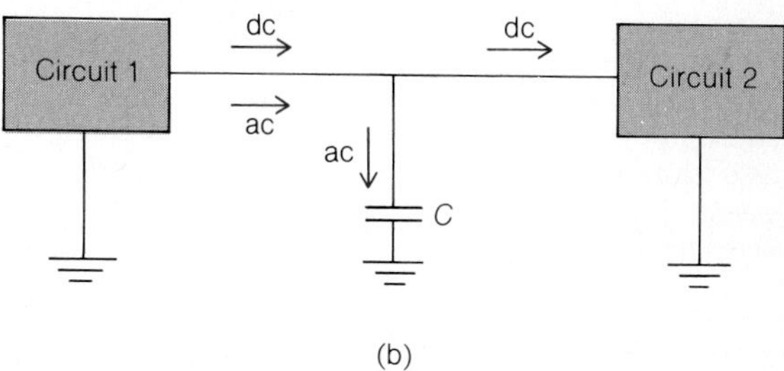

Figure 21.12

A capacitor can be used to favor the passage of either ac or dc between two circuits, depending on how it is connected.

inductive reactance to impede current. On the other hand, a capacitor completely obstructs direct current because the charge that builds up on its plates remains there instead of surging back and forth as it does when an ac potential is applied.

Because capacitors pass ac but not dc, they can be used to favor the transmission of signals of one kind or the other between two circuits. In Fig. 21.12(a) the capacitor permits an ac signal to go from circuit 1 to circuit 2 while stopping any direct current. In Fig. 21.12(b) a direct current can pass between the circuits, but the capacitor provides a path to ground for alternating current, and most of the ac will take this path if the reactance is low.

21.6 Impedance

The ac equivalent of resistance in a dc circuit.

A series circuit that contains resistance, inductance, and capacitance can be represented as in Fig. 21.13, where each of these circuit properties is considered as lumped into a single resistor, inductor, and capacitor. If an ac source of emf is connected to the circuit, at any instant the applied voltage V is equal to the sum of the voltage drops across the various circuit elements:

$$V = V_R + V_L + V_C \qquad \textit{Instantaneous voltage} \quad (21.10)$$

However, V_R, V_L, and V_C are *out of phase with one another*. The voltage across the resistor, V_R, is always in phase with the current I, but V_L is $\frac{1}{4}$ cycle ahead of I, and V_C is $\frac{1}{4}$ cycle behind I. This situation is shown in Fig. 21.13.

While Eq. (21.10) is always correct when V, V_R, V_L, and V_C refer to the instantaneous values of the various voltages, it is *not* correct when effective (or

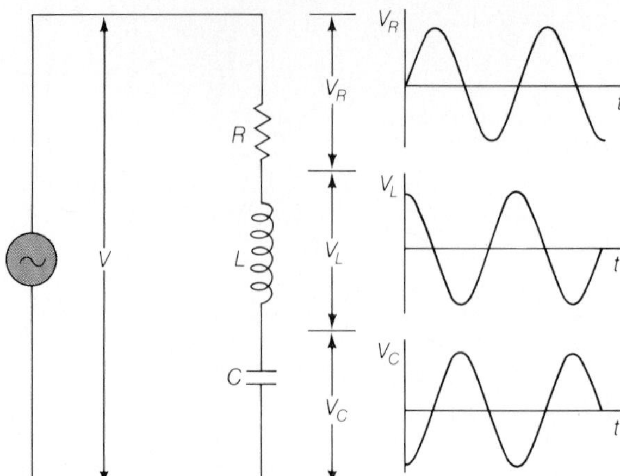

Figure 21.13

The instantaneous value of the voltage V applied to a series RLC circuit is equal to the sum of the instantaneous values of V_R, V_L, and V_C. This relationship does not hold for the effective values of the various voltages because of the phase differences among them.

maximum) values are involved. We must use a vectorial approach to take the phase differences into account.

Figure 21.14 is a phasor diagram that shows the phase differences between $\mathbf{V}_R$, $\mathbf{V}_L$, and $\mathbf{V}_C$. We see that $\mathbf{V}_L$ is 90° ahead of $\mathbf{V}_R$ (hence 90° counterclockwise from $\mathbf{V}_R$, since phasors rotate counterclockwise) and $\mathbf{V}_C$ is 90° behind $\mathbf{V}_R$. The vector sum $\mathbf{V}$ of $\mathbf{V}_R$, $\mathbf{V}_L$, and $\mathbf{V}_C$ represents the effective voltage across the terminals of the circuit. The magnitude of $\mathbf{V}$ is

$$V = \sqrt{V_R^2 + (V_L - V_C)^2} \qquad \textit{Effective voltage} \quad (21.11)$$

The angle ϕ between $\mathbf{V}$ and $\mathbf{V}_R$ is called the **phase angle** because it is a measure of how much the voltage in the circuit leads or lags behind the current. The direction of the effective current phasor $\mathbf{I}$ is always the same as the direction of $\mathbf{V}_R$.

In Fig. 21.14 V_L is greater than V_C, and $(V_L - V_C)$ is a positive quantity. If instead V_L is the smaller quantity, $(V_L - V_C)$ is negative and the vector $\mathbf{V}$ is below the x-axis (Fig. 21.15). However, Eq. (21.11) still applies, since $(V_L - V_C)^2 = (V_C - V_L)^2$. The phase angle in either case is specified by

$$\tan \phi = \frac{V_L - V_C}{V_R} \qquad \textit{Phase angle} \quad (21.12)$$

If $V_C > V_L$, the result will be a negative value for ϕ, which signifies that $\mathbf{V}$ lags behind $\mathbf{V}_R$ (and behind the current $\mathbf{I}$).

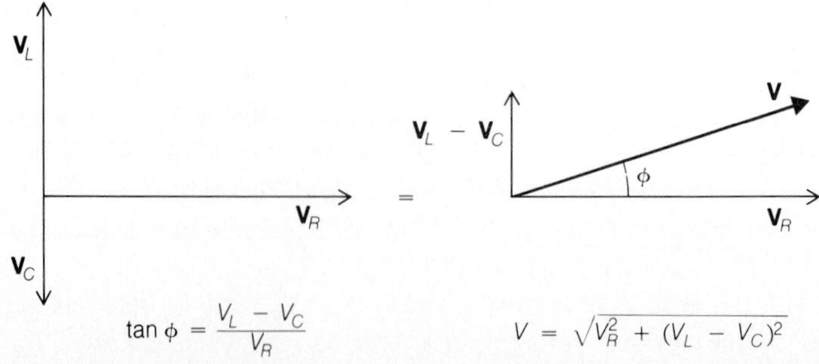

Figure 21.14

Phasor diagram of the effective voltages across the resistor, inductor, and capacitor of Fig. 21.13. The magnitude of the vector sum of the voltage phasors equals the effective voltage across the entire circuit. The angle ϕ is the phase angle. The direction of the effective current phasor $\mathbf{I}$ is always the same as that of $\mathbf{V}_R$.

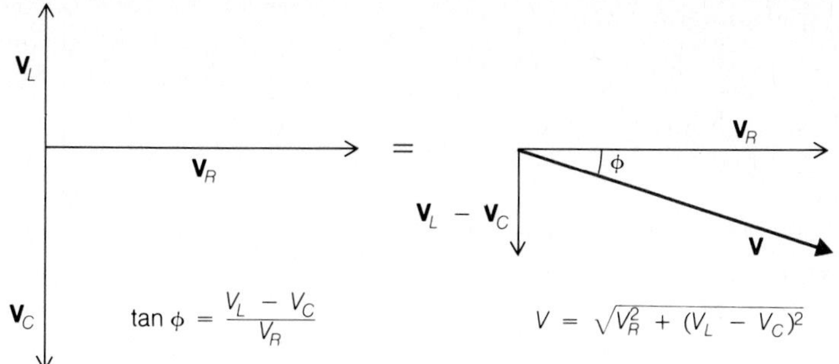

Figure 21.15

The procedures for finding **V** and ϕ are the same whether $\mathbf{V}_L$ is greater than $\mathbf{V}_C$ or the reverse.

EXAMPLE 21.3

A resistor, a capacitor, and an inductor are connected in series across an ac power source. The effective voltages across the circuit components are $V_R = 5$ V, $V_C = 10$ V, and $V_L = 12$ V. Find the effective voltage of the source and the phase angle in the circuit.

SOLUTION (a) From Eq. (21.11)

$$V = \sqrt{V_R^2 + (V_L - V_C)^2} = \sqrt{(5\text{ V})^2 + (12\text{ V} - 10\text{ V})^2} = 5.4\text{ V}$$

We note that V_C and V_L are both greater in magnitude than the source voltage in this case.

(b) Since

$$\tan \phi = \frac{V_L - V_C}{V_R} = \frac{12\text{ V} - 10\text{ V}}{5\text{ V}} = 0.4$$

we have for the phase angle $\phi = 22°$. ♦

Because

$$V_R = IR \qquad V_L = IX_L \qquad \text{and} \qquad V_C = IX_C$$

we can rewrite Eq. (21.11) in the form

$$V = I\sqrt{R^2 + (X_L - X_C)^2}$$

The quantity

$$Z = \sqrt{R^2 + (X_L - X_C)^2} \qquad \textit{Impedance} \quad (21.13)$$

is known as the **impedance** of a series circuit containing resistance, inductance, and capacitance. The unit of impedance is evidently the ohm. Impedance in an ac circuit plays the same role that resistance does in a dc circuit. In an ac circuit

$$I = \frac{V}{Z} \qquad \textit{Current in ac circuit} \quad (21.14)$$

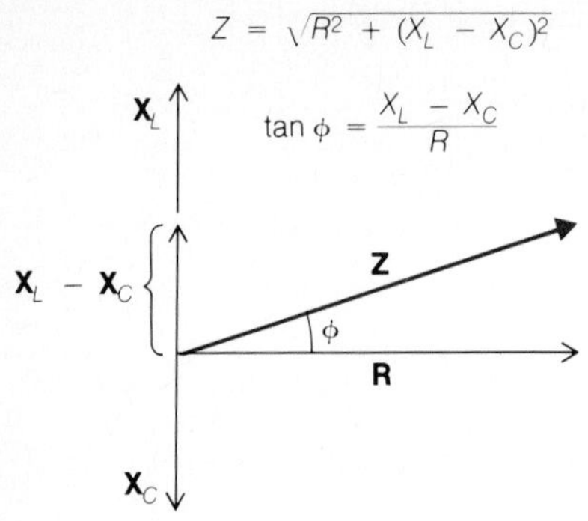

Figure 21.16

The phasor impedance diagram that corresponds to the phasor voltage diagram shown in Fig. 21.14.

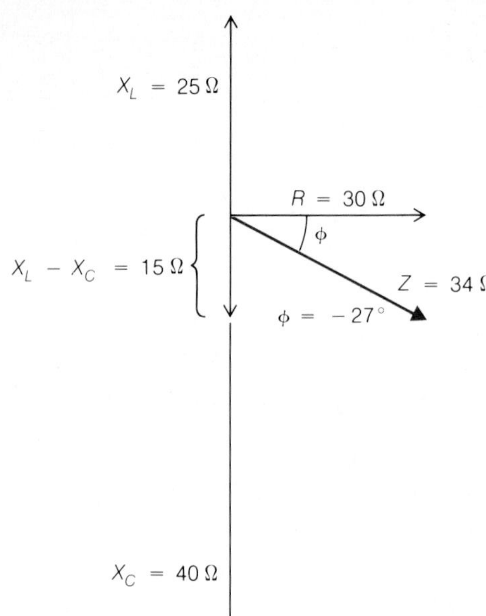

Figure 21.17

A negative phase angle occurs when X_C is greater than X_L. This signifies that the voltage in the circuit lags behind the current (Example 21.4).

is the effective current that flows when the effective voltage V is applied. It is important to keep in mind that Z not only depends on the circuit parameters R, L, and C but also varies with the frequency f.

The current I is the same in all parts of the circuit at all times, and

$$Z = \frac{V}{I} \qquad R = \frac{V_R}{I} \qquad X_L = \frac{V_L}{I} \qquad X_C = \frac{V_C}{I}$$

Hence the phasor voltage diagram of Fig. 21.14 can be replaced by the phasor impedance diagram of Fig. 21.16. The phase angle ϕ is the same in both cases, of course, and can be calculated from the relation

$$\tan \phi = \frac{X_L - X_C}{R} \qquad \textit{Phase angle} \quad (21.15)$$

EXAMPLE 21.4

Analyze a series circuit that consists of a 10-mH inductor, a 10-μF capacitor, and a 30-Ω resistor connected to a source of 100-V, 400-Hz alternating current.

SOLUTION The reactances of the inductor and capacitor at 400 Hz are

$$X_L = 2\pi fL = (2\pi)(400\ \text{Hz})(0.010\ \text{H}) = 25\ \Omega$$

$$X_C = \frac{1}{2\pi fC} = \frac{1}{(2\pi)(400\ \text{Hz})(1.0 \times 10^{-5}\ \text{F})} = 40\ \Omega$$

The phasor impedance diagram for this circuit is shown in Fig. 21.17. The impedance

Z is

$$Z = \sqrt{R^2 + (X_L - X_C)^2} = \sqrt{(30\ \Omega)^2 + (25\ \Omega - 40\ \Omega)^2} = 34\ \Omega$$

The phase angle ϕ is found as follows:

$$\tan\phi = \frac{X_L - X_C}{R} = -\frac{15\ \Omega}{30\ \Omega} = -0.50 \qquad \phi = -27°$$

A negative phase angle signifies that the voltage lags behind the current. Here the lag is 27°, which is $\frac{27}{360}$ or 0.075 of a complete cycle.

The current in the circuit is

$$I = \frac{V}{Z} = \frac{100\ \text{V}}{34\ \Omega} = 2.94\ \text{A}$$

The effective potential difference across each of the circuit elements is

$$V_L = IX_L = (2.94\ \text{A})(25\ \Omega) = 74\ \text{V}$$
$$V_C = IX_C = (2.94\ \text{A})(40\ \Omega) = 118\ \text{V}$$
$$V_R = IR = (2.94\ \text{A})(30\ \Omega) = 88\ \text{V}$$

We see again that the effective voltage across an inductor or a capacitor in an ac circuit can exceed the effective voltage applied to the entire circuit.

The arithmetic sum of the above voltages is 280 V, but this sum means nothing because the voltages are not in phase with one another. The vector sum of the voltages, which takes into account the phase differences among them, is

$$V = \sqrt{V_R^2 + (V_L - V_C)^2} = \sqrt{(88\ \text{V})^2 + (74\ \text{V} - 118\ \text{V})^2} = 98\ \text{V}$$

The difference between this figure and the applied voltage of 100 V is due to the rounding of V_L, V_C, and V_R when they were calculated. ♦

21.7 Resonance

Impedance is a minimum when $X_L = X_C$.

When an ac voltage is applied to a series circuit, the current depends on the frequency. The greatest current flows when the impedance Z is a minimum. Since

$$Z = \sqrt{R^2 + (X_L - X_C)^2}$$

the condition for minimum impedance in a given circuit is that the frequency be such that the inductive and capacitive reactances are equal. When this is true,

$$X_L = X_C \quad \text{and} \quad 2\pi f_0 L = \frac{1}{2\pi f_0 C}$$

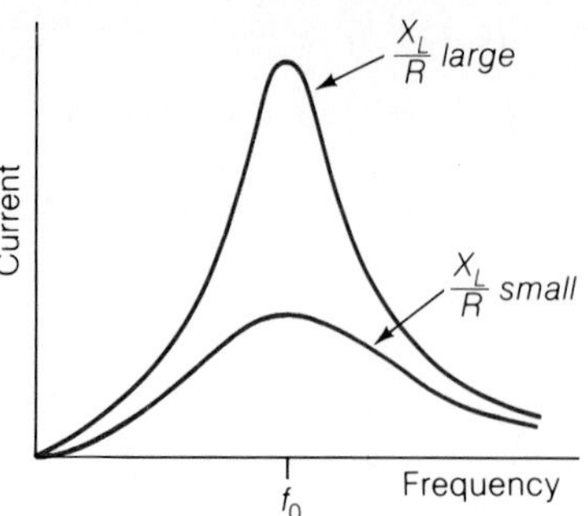

Figure 21.18

The variation with frequency of the current in a series ac circuit. The maximum current occurs at the resonance frequency f_0.

Thus

$$f_0 = \frac{1}{2\pi\sqrt{LC}} \qquad \textit{Resonance frequency} \quad (21.16)$$

When the applied voltage has this frequency, the current is a maximum and is limited only by the resistance R. The frequency f_0 is called the **resonance frequency** of the circuit. **Resonance** occurs in a series circuit when the impressed voltage oscillates with the resonance frequency.

We note that Eq. (21.16) is the same as Eq. (20.14) for the ''natural'' frequency at which an LC circuit will oscillate if the capacitor is initially charged. When a circuit is in resonance with an applied voltage, the energy alternately stored in and discharged from the capacitor is the same as the energy stored in and discharged from the inductor. At other frequencies the energy contents of the capacitor and the inductor are different. The difference interferes with the back-and-forth flow of power and gives rise to an impedance that exceeds the resistance R. The current that flows in an RLC circuit at resonance is

$$I = \frac{V}{R}$$

just as in a dc circuit.

Figure 21.18 shows how the current in a series circuit varies with frequency. The less the resistance R relative to the inductive reactance X_L, the sharper the peak in the curve. The antenna circuit of one kind of radio receiver is tuned to respond to a particular frequency of radio waves by adjusting a variable capacitor or inductor until the resonance frequency of the circuit is equal to the signal frequency. A circuit in which X_L/R is large has a narrow response curve and can separate signals from two stations very close together in frequency.

At resonance, $X_L = X_C$, and there is no phase difference between current and voltage. Another way to describe resonance, then, is to say that current and voltage are in phase at all times, just as in a pure-resistance circuit.

EXAMPLE 21.5

The antenna circuit of a certain radio receiver consists of a 10-mH coil and a variable capacitor; the resistance in the circuit is 50 Ω. An 880-kHz (kilohertz) radio wave produces a potential difference of 1.0×10^{-4} V across the circuit. Find the capacitance required for resonance and the current at resonance.

SOLUTION The capacitance required for resonance at 880 kHz is

$$C = \frac{1}{(2\pi f)^2 L} = \frac{1}{(2\pi \times 8.8 \times 10^5\ \text{Hz})^2(1.0 \times 10^{-2}\ \text{H})}$$
$$= 3.3 \times 10^{-12}\ \text{F} = 3.3\ \text{pF}$$

At resonance, inductive and capacitive reactances cancel each other out, and the current that flows is

$$I = \frac{V}{R} = \frac{1.0 \times 10^{-4}\ \text{V}}{50\ \Omega} = 2.0 \times 10^{-6}\ \text{A} = 2.0\ \mu\text{A}$$

◆

21.8 Power in AC Circuits

Because inductors and capacitors do not absorb power, the apparent power IV may exceed the actual power.

No power is consumed in a pure inductor or capacitor in an ac circuit. These elements act merely as temporary reservoirs of energy and return whatever energy they absorb in one quarter of a cycle to the circuit in the next quarter of that cycle. No power is therefore needed to maintain an ac current in the inductive and capacitive parts of a circuit. The resistance in the circuit, however, dissipates power as heat at the rate $P = IV_R$, where, as usual, I and V_R are effective values.

From Fig. 21.14 we see that

$$V_R = V \cos \phi \tag{21.17}$$

since $V \cos \phi$ is the component of **V** that is in phase with the current phasor **I.** Hence the effective power delivered to an ac circuit is

$$P = IV \cos \phi \qquad \textit{Power in ac circuit} \tag{21.18}$$

The quantity $\cos \phi$ is called the **power factor** of the circuit. The power factor is equal to 1 only at resonance, when current and voltage are in phase; then $\phi = 0$ and $\cos \phi = 1$. Under other circumstances the voltage is not in phase with the current, and the actual power is less than IV.

There is an interesting analogy between Eq. (21.18) and the mechanical power $P = Fv \cos \theta$ developed by a force **F** that acts on an object moving at the velocity **v** when the angle between **F** and **v** is θ. When **F** is parallel to **v,** $\cos \theta = \cos 0 = 1$ and the power is a maximum. When **F** is perpendicular to **v,** $\cos \theta = \cos 90° = 0$ and $P = 0$. In the electrical case, we can think of V as corresponding to F and of I as corresponding to v.

From Fig. 21.16 we see that

$$\cos \phi = \frac{R}{Z} = \frac{R}{\sqrt{R^2 + (X_L - X_C)^2}} \qquad \textit{Power factor} \tag{21.19}$$

The power factor of a circuit is the ratio between its resistance and its impedance. Often power factors are expressed as percentages rather than as decimals or fractions. Thus a phase angle of 60° means a power factor of

$$\cos 60° = 0.50 = 50\%$$

Apparent Power

Instruments called **wattmeters** respond directly to the effective product of V and I. An ac wattmeter connected in a circuit gives a lower value for the power than the product of the effective values of V and I obtained from a separate voltmeter and ammeter in the same circuit (except at resonance). This is because the separate meters are not affected by the phase difference between current and voltage. Alternating-current generators, transformers, and power lines are usually rated in **volt-amperes,** the product of effective voltage and current without regard to actual power. This is because higher values of V and/or I must be supplied to a circuit whose power factor is less than 1 than is reflected in its power consumption. We can think of the power factor as having the unit of watts per volt-ampere.

Alternating-current devices of various kinds—for example, ac electric motors and fluorescent lamps—may have net inductive or capacitive reactances and so have power factors of less than 100%. A power factor of 70%, for instance, means that 1 kVA (kilovolt-ampere) of apparent power must be supplied for every 700 W of power actually consumed. This is an uneconomical situation, partly because of the additional generator capacity required, and partly because of the additional heat losses in the transmission lines as the unused power circulates between the generator and the device. The remedy is to introduce capacitors or inductors into the power line to increase the power factor to an acceptable figure.

EXAMPLE 21.6

An inductive load connected to a 24-V, 400-Hz power source draws a current of 2.0 A and dissipates 40 W. (a) Find the power factor of the load and the phase angle ϕ. (b) What series capacitance is needed to make the power factor 100%? (c) What would the current be? (d) How much power would the load then dissipate?

SOLUTION (a) Since $P = IV \cos \phi$, the power factor is

$$\cos \phi = \frac{P}{IV} = \frac{40 \text{ W}}{(2.0 \text{ A})(24 \text{ V})} = 0.83 = 83\%$$

The phase angle is therefore $\phi = 34°$.

(b) The power factor will be 100% when $X_C = X_L$. In the original circuit $X_C = 0$, and so, from Eq. (21.15), $X_L = R \tan \phi$. The resistance in the circuit is

$$R = \frac{P}{I^2} = \frac{40 \text{ W}}{(2.0 \text{ A})^2} = 10 \ \Omega$$

Hence the inductive reactance is

$$X_L = R \tan \phi = (10 \ \Omega)(\tan 34°) = 6.75 \ \Omega$$

Since the required capacitive reactance X_C must equal X_L when $f = 400$ Hz,

$$C = \frac{1}{2\pi f X_C} = \frac{1}{(2\pi)(400\ \text{Hz})(6.75\ \Omega)} = 5.9 \times 10^{-5}\ \text{F} = 59\ \mu\text{F}$$

(c) When $X_C = X_L$, $Z = R$ and

$$I = \frac{V}{Z} = \frac{V}{R} = \frac{24\ \text{V}}{10\ \Omega} = 2.4\ \text{A}$$

(d) $P = I^2R = (2.4\ \text{A})^2(10\ \Omega) = 57.6$ W. ♦

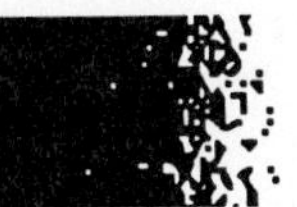

21.9 Impedance Matching

Power transfer is a maximum between circuits with the same impedance.

In Section 18.8 we saw that the maximum power transfer between two dc circuits occurs when their resistances are equal. The same applies to ac circuits with the impedance of the circuits rather than their resistances as the quantities to be matched. An additional consideration with ac circuits is that an impedance mismatch may result in a distorted signal by altering the properties of the circuits—for instance, by changing their resonance frequencies.

Ac circuits have the advantage that a transformer can be used to correct impedance mismatches. A transformer is often used for this purpose in an audio system where the amplifier circuit might have an impedance of several thousand ohms whereas the voice coil of the loudspeaker has an impedance of only a few ohms. Connecting the voice coil directly to the amplifier would be grossly inefficient.

Let us calculate the ratio of turns N_1/N_2 between the primary and secondary windings of a transformer needed to couple a circuit of impedance Z_1 to a circuit of impedance Z_2. If the voltages and currents in the circuits are, respectively, V_1, I_1 and V_2, I_2, then from Eq. (20.7)

$$\frac{V_1}{V_2} = \frac{N_1}{N_2} \quad \text{and} \quad \frac{I_2}{I_1} = \frac{N_1}{N_2}$$

Since $Z_1 = V_1/I_1$ and $Z_2 = V_2/I_2$,

$$\frac{Z_1}{Z_2} = \frac{V_1 I_2}{V_2 I_1} = \left(\frac{N_1}{N_2}\right)^2$$

and the ratio of the turns is

$$\frac{N_1}{N_2} = \sqrt{\frac{Z_1}{Z_2}} \qquad \textit{Impedance matching} \quad (21.20)$$

Thus if a loudspeaker whose voice coil has an impedance of 10 Ω is to be used with an amplifier whose load impedance is 9000 Ω, a transformer should be used whose ratio of turns is

$$\frac{N_1}{N_2} = \sqrt{\frac{Z_1}{Z_2}} = \sqrt{\frac{9000\ \Omega}{10\ \Omega}} = 30$$

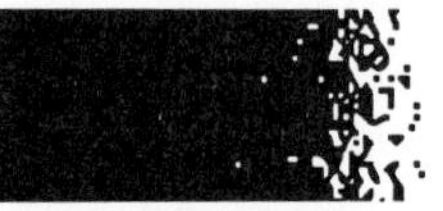

21.10 Parallel AC Circuits and Filters

The branch currents in a parallel circuit are not in phase.

When a resistor, an inductor, and a capacitor are connected in parallel, as in Fig. 21.19, the potential difference is the same across each circuit element:

$$V = V_R = V_L = V_C$$

The total instantaneous current is the sum of the instantaneous currents in each branch, as in the case of a dc parallel circuit. However, this is not true of the total effective current, because the branch currents are not in phase. The current I_R is always in phase with the voltage V, but I_C leads V by 90° and I_L lags behind V by 90°.

A phasor diagram of the situation is shown in Fig. 21.20. The magnitudes of the currents in the branches are

$$I_R = \frac{V}{R} \qquad I_C = \frac{V}{X_C} \qquad I_L = \frac{V}{X_L}$$

and their vector sum is

$$I = \sqrt{I_R^2 + (I_C - I_L)^2} \qquad \textit{Effective current} \quad (21.21)$$

The phase angle ϕ specified by

$$\tan \phi = \frac{I_C - I_L}{I_R} \qquad \textit{Phase angle} \quad (21.22)$$

Figure 21.19

A parallel *RLC* circuit. The potential differences across the circuit elements are the same.

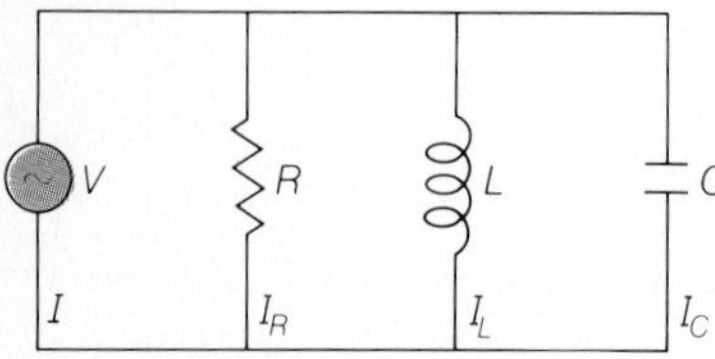

Figure 21.20

A phasor diagram of the effective currents in the resistor, inductor, and capacitor of Fig. 21.19. The magnitude *I* of the vector sum of the current vectors is equal to the effective current through the entire circuit.

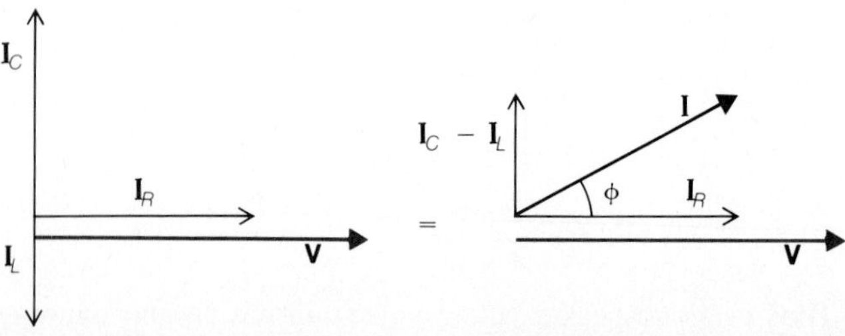

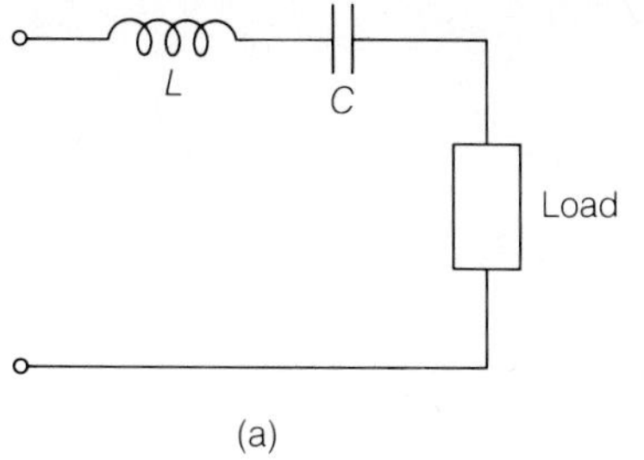

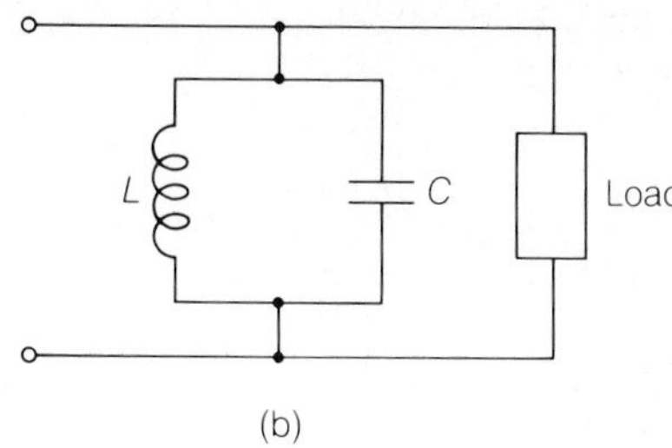

Figure 21.21

Band-pass filters. The impedance Z of a series LC circuit is a minimum near f_0, whereas Z is a maximum near f_0 for a parallel LC circuit.

is the angle between the current and the voltage. A positive phase angle means that the current leads the voltage, a negative phase angle means that the current lags behind the voltage. The power dissipated in a parallel alternating-current circuit is given by the same formula as in the case of a series circuit, namely, $P = IV \cos \phi$.

In a series RLC circuit, the impedance is a minimum at resonance and increases at higher and lower frequencies than f_0. At resonance, $X_L = X_C$, $Z = R$, and $I = V/R$. In a parallel RLC circuit, resonance again corresponds to $X_L = X_C$, $Z = R$, and $I = V/R$, but now the impedance is a *maximum* at f_0 since at higher and lower frequencies some current can pass through the inductor and capacitor as well as through the resistor. Thus a series circuit can be used as a selector to favor a particular frequency, and a parallel circuit with the same L and C can be used as a selector to discriminate against the same frequency. Such circuits form the basis of **band-pass** and **band-reject** filters.

Figure 21.21 shows two kinds of band-pass filters. At frequencies near f_0 the impedance of the series LC circuit of Fig. 21.21(a) is low, so the load current is high. (From Fig. 21.18 we can see why a band of frequencies rather than a single one is favored.) At frequencies much above or below f_0 the impedance is high, so the load current is low. In Fig. 21.21(b) the impedance of the parallel LC circuit is high for frequencies near f_0, and most of the current flows through the load. At other frequencies current is diverted through the LC circuit and thus reduces the load current. The band-reject filters of Fig. 21.22 behave in just the opposite ways to minimize the load current for frequencies near f_0.

Band-pass filters make it possible to send a number of different signals at the same time through a single pair of wires. What is done is to have each signal modulate a "carrier" alternating current of constant frequency, so that the resulting variations in the amplitude of the carrier correspond to the signal. This procedure is illustrated in Fig. 22.6 for radio transmission. Modulated carriers of different frequencies can be transmitted at the same time from one place to another along the

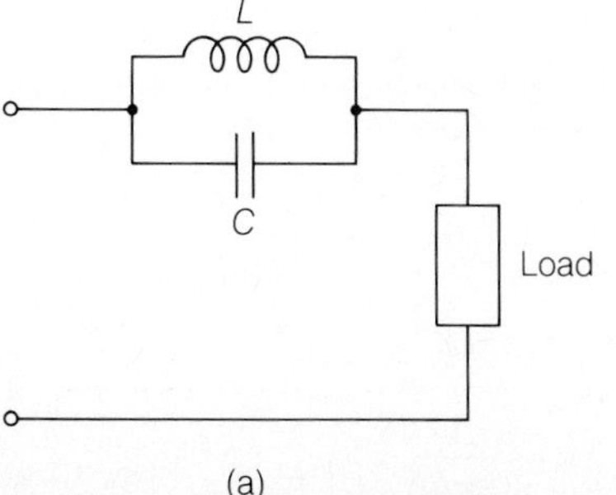

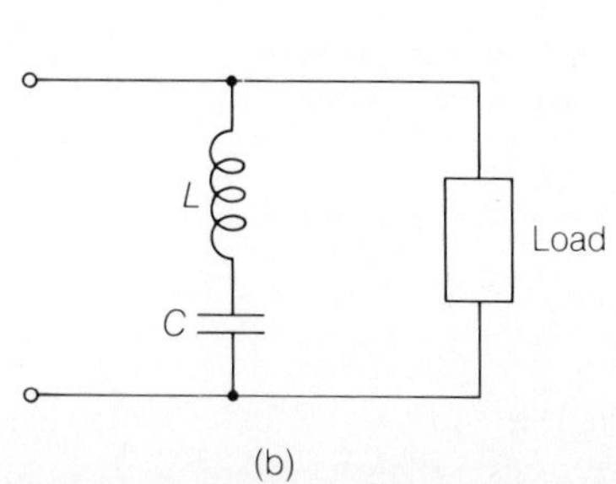

Figure 21.22

Band-reject filters.

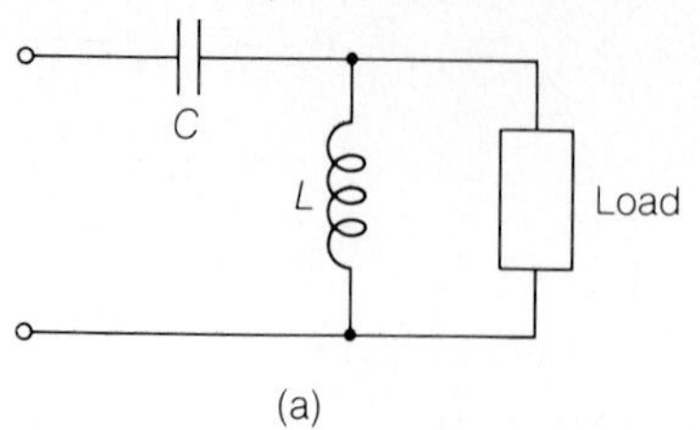

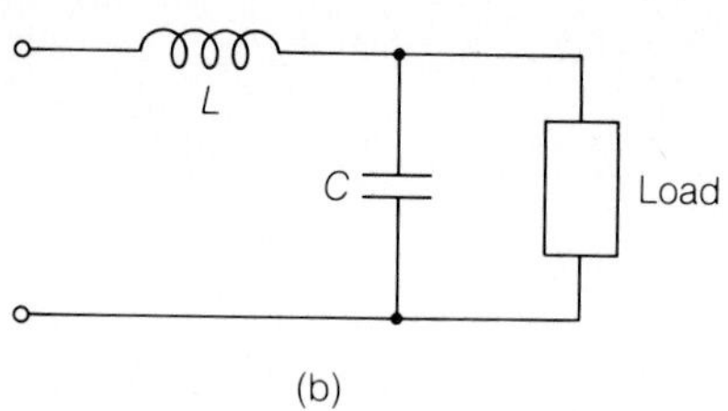

Figure 21.23

(a) High-pass filter. At low frequencies, X_C is large and X_L is small. (b) Low-pass filter. At high frequencies, X_C is small and X_L is large.

same pair of wires. At the receiving end band-pass filters separate the carriers and special circuits then demodulate them to reconstruct the original signals. Telephone systems have used this technique for over 60 years. Up to 32 simultaneous telephone conversations can be carried on an ordinary pair of wires, and a coaxial cable, in which one of the conductors is in the form of a tube insulated from a central wire, can carry thousands of them.

Other arrangements of inductors and capacitors can act as **high-pass** filters to discriminate against low-frequency currents, and as **low-pass** filters to favor them. In the high-pass filter in Fig. 21.23(a), since $X_C = 1/(2\pi fC)$, the lower the frequency, the greater the reactance of the capacitor. In addition, since $X_L = 2\pi fL$, the lower the frequency, the less the reactance of the inductor and the more the current diverted through it instead of passing through the load. These effects are used in the opposite ways in the low-pass filter in Fig. 21.23(b), which is widely used to smooth out ripples in the output of alternating-current to direct-current rectifiers (see Section 28.9). More elaborate circuits than those of Figs. 21.21 to 21.23 can be designed that provide sharper cutoffs between the passed and rejected frequency bands.

Important Terms

The **effective value** of an alternating current is such that a direct current of this magnitude produces heat in a resistor at the same rate as the alternating current.

A **phasor** is a rotating vector whose projection can represent either current or voltage in an ac circuit.

The **phase relationships** between the instantaneous voltage and instantaneous current in ac circuit components are as follows: The voltage across a pure resistor is in phase with the current; the voltage across a pure inductor leads the current by $\frac{1}{4}$ cycle; the voltage across a pure capacitor lags behind the current by $\frac{1}{4}$ cycle.

The **inductive reactance** X_L of an inductor is a measure of its effect on an alternating current. The **capacitive reactance** X_C of a capacitor is a measure of its effect on an alternating current. Both X_L and X_C vary with the frequency of the current.

The **impedance** Z of an ac circuit is analogous to the resistance of a dc circuit. The **resonance frequency** of an ac circuit is that frequency for which the impedance is a minimum.

The **power factor** of an ac circuit is the ratio between the power consumed in the circuit and the product of the effective current and voltage there; this ratio is equal to that between the resistance and the impedance of the circuit, and is less than 1, except at resonance. The unit of the apparent power $V_{eff}I_{eff}$ is the **volt-ampere,** as distinct from the watt, which is the unit of consumed power.

A **filter** is a circuit that discriminates against the passage of currents in a certain frequency range.

Important Formulas

Effective current: $I_{eff} = \dfrac{I_{max}}{\sqrt{2}} = 0.707\, I_{max}$

Effective voltage: $V_{eff} = \dfrac{V_{max}}{\sqrt{2}} = 0.707\, V_{max}$

Inductive reactance: $X_L = 2\pi f L$

Capacitive reactance: $X_C = \dfrac{1}{2\pi fC}$

Impedance: $Z = \sqrt{R^2 + (X_L - X_C)^2}$

Current in ac circuit: $I = \dfrac{V}{Z}$

Phase angle: $\tan \phi = \dfrac{X_L - X_C}{R}$ $\quad \cos \phi = \dfrac{R}{Z}$

Resonance frequency: $f_0 = \dfrac{1}{2\pi\sqrt{LC}}$

Power in ac circuit: $P = IV \cos \phi$

Impedance-matching transformer: $\dfrac{N_1}{N_2} = \sqrt{\dfrac{Z_1}{Z_2}}$

MULTIPLE CHOICE

1. In an ac circuit the voltage
- **a.** leads the current
- **b.** lags behind the current
- **c.** is in phase with the current
- **d.** is any of the above, depending on the circumstances

2. The voltage cannot be exactly in phase with the current in a circuit that contains
- **a.** only resistance
- **b.** only inductance
- **c.** inductance and capacitance
- **d.** inductance, capacitance, and resistance

3. The voltage lags behind the current by $\frac{1}{4}$ cycle in
- **a.** a pure capacitor
- **b.** a pure inductor
- **c.** a pure resistor
- **d.** a circuit with capacitance and inductance

4. The unit of inductive reactance is the
- **a.** henry **b.** tesla
- **c.** weber **d.** ohm

5. When voltage and current are in phase in an ac circuit, the
- **a.** impedance is 0 **b.** reactance is 0
- **c.** resistance is 0 **d.** phase angle is 90°

6. The impedance of a circuit does not depend on
- **a.** I **b.** f
- **c.** R **d.** C

7. A resistor, a capacitor, and an inductor are connected in series to a source of ac power. If the inductance is decreased, the impedance of the circuit
- **a.** decreases
- **b.** increases
- **c.** decreases or increases
- **d.** decreases, increases, or remains the same

8. The power dissipated as heat in an ac circuit depends on its
- **a.** resistance
- **b.** inductive reactance
- **c.** capacitive reactance
- **d.** impedance

9. A coil of inductance L has an inductive reactance of X_L in an ac circuit in which the effective current is I. The coil is made from a superconducting material and has no resistance. The rate at which power is dissipated in the coil is
- **a.** 0 **b.** IX_L
- **c.** I^2X_L **d.** IX_L^2

10. The power factor of a circuit is equal to
- **a.** RZ **b.** R/Z
- **c.** X_L/Z **d.** X_C/Z

11. At resonance, it is not true that
- **a.** $R = Z$ **b.** $X_L = 1/X_C$
- **c.** $P = IV$ **d.** $I = V/R$

12. The impedance of a parallel *RLC* circuit at resonance is
- **a.** less than R
- **b.** equal to R
- **c.** more than R
- **d.** any of the above, depending on the circumstances

13. Impedance is a maximum at resonance in
- **a.** a series *RLC* circuit
- **b.** a parallel *RLC* circuit
- **c.** all *RLC* circuits
- **d.** no *RLC* circuits

14. The power factor of a circuit in which $X_L = X_C$
- **a.** is 0
- **b.** is 1
- **c.** depends on the ratio X_L/X_C
- **d.** depends on the value of R

15. A voltmeter across an ac circuit reads 50 V, and an ammeter in series with the circuit reads 5 A. The power consumption of the circuit
- **a.** is less than or equal to 250 W
- **b.** is exactly equal to 250 W
- **c.** is equal to or more than 250 W
- **d.** may be less than, equal to, or more than 250 W

16. The power factor of a certain circuit in which the voltage lags behind the current is 80%. To increase the power factor to 100%, it is necessary to add to the circuit additional
- **a.** resistance **b.** capacitance
- **c.** inductance **d.** impedance

17. A simple low-pass filter can be made by using
- **a.** an inductor in series with the load
- **b.** a capacitor in series with the load
- **c.** a resistor in series with the load
- **d.** an inductor in parallel with the load

18. The inductive reactance of a 1.0-mH coil in a 5.0-kHz circuit is

a. 3.1 Ω b. 6.3 Ω
c. 10 Ω d. 31 Ω

19. A current of 0.50 A flows in an electromagnet of negligible resistance when it is connected to a source of 120-V, 60-Hz alternating current. The inductance of the electromagnet is

a. 0.011 mH b. 0.16 H
c. 0.64 H d. 0.24 kH

20. The capacitive reactance of a 5.0-μF capacitor in a 20-kHz circuit is

a. 0.63 Ω b. 1.6 Ω
c. 5.0 Ω d. 16 Ω

21. A 2.0-μF capacitor is connected to a 50-V, 400-Hz power source. The current that flows is

a. 0.20 mA b. 0.25 A
c. 2.5 A d. 3.5 A

22. A resistor, a capacitor, and an inductor are connected in series to an ac source of frequency f. The effective voltages across the circuit components are V_R = 10 V, V_C = 20 V, and V_L = 14 V. The effective voltage of the source is

a. 8 V b. 12 V
c. 16 V d. 35 V

23. The phase angle in the circuit in Multiple Choice 22 is

a. −31° b. +31°
c. −53° d. +53°

24. The resonance frequency of the circuit in Multiple Choice 22 is

a. less than f
b. equal to f
c. greater than f
d. any of the above, depending on the values of the components

25. In a series ac circuit R = 10.0 Ω, X_L = 8.0 Ω, and X_C = 6.0 Ω when the frequency is f. The impedance at this frequency is

a. 10.2 Ω b. 12 Ω
c. 24 Ω d. 104 Ω

26. The phase angle in the circuit in Multiple Choice 25 is

a. 0.2° b. 2°
c. 11° d. 45°

27. The resonance frequency of the circuit in Multiple Choice 25 is

a. less than f
b. equal to f
c. greater than f
d. any of the above, depending on the applied voltage

28. The resonance frequency of a circuit that contains a 50-mH inductor and a 0.20 μF capacitor is

a. 16 kHz b. 10 kHz
c. 63 kHz d. 16 MHz

29. A coil connected to a 120-V ac source draws a current of 0.50A and dissipates 50 W. If a capacitor were connected in series to bring the power factor to 1, the circuit would dissipate

a. 50 W b. 60 W
c. 72 W d. 100 W

30. A 2.00-μF capacitor, a 5.00-mH inductor, and a 30.0-Ω resistor are connected in series across a 100-V, 1.00-kHz power source. The current in the circuit is

a. 1.76A b. 2.65A
c. 3.33A d. 5.50A

31. The power dissipated in the circuit in Multiple Choice 30 is

a. 93 W b. 176 W
c. 202 W d. 211 W

32. The minimum rating in volt-amperes needed by the power source of the circuit in Multiple Choice 30 is

a. 93 V·A b. 176 V·A
c. 265 V·A d. 333 V·A

33. What should the capacitance of the circuit in Multiple Choice 30 be for it to have a resonance frequency of 1.00 kHz?

a. 5.07 μF b. 31.8 μF
c. 50.7 μF d. 71 μF

34. With that capacitance, the power dissipated in the circuit in Multiple Choice 30 would be

a. 144 W b. 176 W
c. 230 W d. 333 W

QUESTIONS

1. What happens to the reactance of a coil when an iron core is inserted in it?

2. What happens to the reactance of a capacitor with air between its plates when the space between its plates is filled with a substance of dielectric constant $K > 1$?

3. A coil has the inductive reactance A when an ac voltage of frequency F is applied to it. What is the reactance when the frequency is changed to $F/2$? to $2F$?

4. A capacitor has the capacitive reactance B when an ac voltage of frequency F is applied to it. What is the reactance when the frequency is changed to $F/2$? To $2F$?

5. What properties of an ac circuit are described by its resistance, capacitive reactance, inductive reactance, and impedance? What are the similarities and differences among these quantities?

6. You have dc and ac sources of the same emf and dc and ac ammeters. How would you use them to decide whether a circuit element in a black box with two terminals is a resistor, a capacitor, an inductor, or a rectifier?

7. The frequency of the alternating potential difference applied to a series *RLC* circuit is halved. What happens to the resistance, the inductive reactance, and the capacitive reactance of the circuit? What further information is needed to establish what happens to the impedance of the circuit?

8. An alternating potential difference whose frequency is higher than the resonance frequency of a series *RLC* circuit is applied to it. Does the voltage in the circuit lead or lag behind the current?

9. What is the significance of the power factor of a circuit? Is it independent of frequency? Under what circumstances (if any) can it be zero? Under what circumstances (if any) can it be 100%?

10. What kind of filter is shown in Fig. 21.24?

11. What kind of filter is shown in Fig. 21.25?

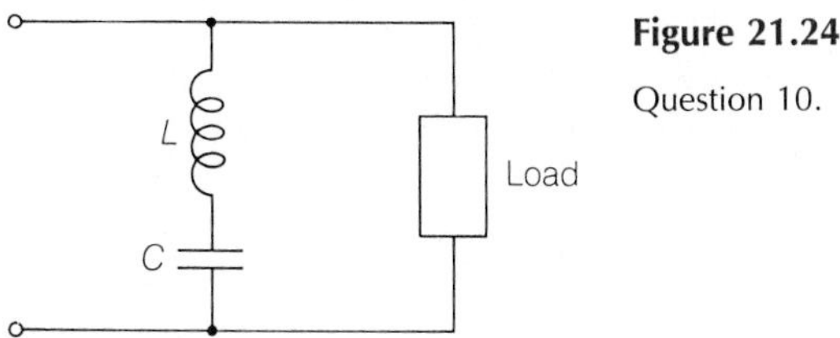

Figure 21.24

Question 10.

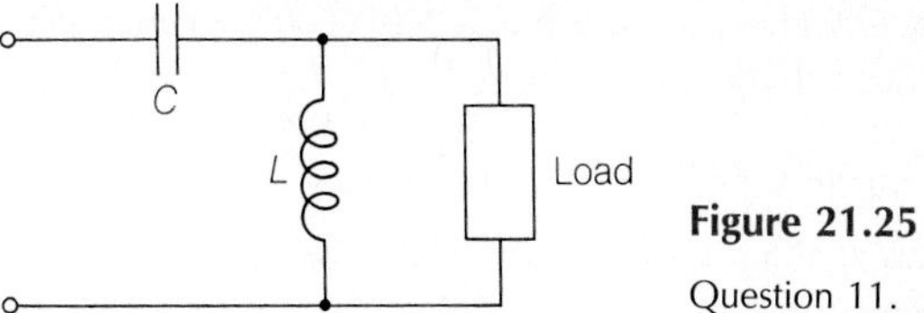

Figure 21.25

Question 11.

12. Loudspeakers are limited in their frequency responses, with large speakers ("woofers") providing more accurate sound reproduction at low frequencies and small ones ("tweeters") at high frequencies. A crossover circuit that divides the output of an audio amplifier into a low-frequency and a high-frequency component is needed when a pair of such speakers is used. Which output of the circuit shown in Fig. 21.26 should be connected to the woofer and which to the tweeter?

Figure 21.26

Question 12.

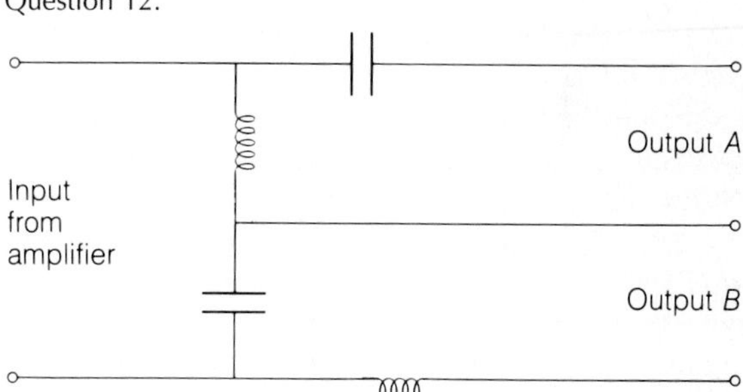

EXERCISES

21.1 Effective Current and Voltage

1. The graph in Fig. 21.27 shows how the current in a certain ac circuit varies with time. What is the frequency of the current?

2. What is the effective current in the circuit shown in Fig. 21.27?

3. An ammeter in series with an ac circuit reads 10 A and a voltmeter across the circuit reads 60 V. (a) What is the maximum current in the circuit? (b) What is the maximum potential difference across the circuit? (c) Does the maximum current necessarily occur at the same moments as the maximum voltage?

Figure 21.27

Exercise 1.

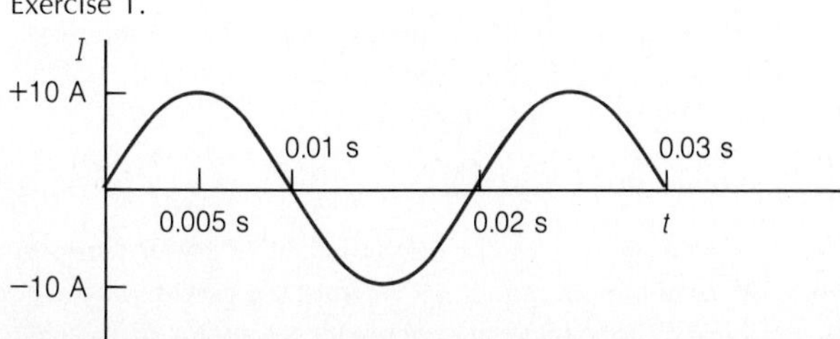

4. The dielectric used in a certain capacitor breaks down at a voltage of 300 V. Find the highest effective sinusoidal ac voltage that can be applied to it.

5. The potential difference across a source of alternating current varies sinusoidally with time with $V_{max} = 100$ V. Find the value of the instantaneous potential difference (a) $\frac{1}{8}$ cycle, (b) $\frac{1}{4}$ cycle, (c) $\frac{3}{8}$ cycle, and (d) $\frac{1}{2}$ cycle after $V = 0$.

6. A certain 250-Hz ac power source has a maximum potential difference of 24 V. Find the value of the instantaneous potential difference (a) 0.0005 s, (b) 0.001 s, (c) 0.002 s, (d) 0.0025 s, (e) 0.004 s, and (f) 0.005 s after $V = 0$.

21.4 Inductive Reactance

7. What is the reactance of a 5.0-mH inductor at 10 Hz? At 10 kHz?

8. The reactance of an inductor is 80 Ω at 500 Hz. Find its inductance.

9. Find the current that flows when a 3.0-mH inductor is connected to a 15-V, 5.0-kHz power source.

10. A current of 0.80 A flows through a 50-mH inductor of negligible resistance that is connected to a 120-V power source. What is the frequency of the source?

11. An inductor of negligible resistance whose reactance is

120 Ω at 200 Hz is connected to a 240-V, 60-Hz power line. What is the current in the inductor?

21.5 Capacitive Reactance

12. What is the reactance of an 80-pF capacitor at 10 kHz? At 10 MHz?

13. Find the current that flows when a 10-μF capacitor is connected to a 15-V, 5.0-kHz power source.

14. The reactance of a capacitor is 50 Ω at 200 Hz. What is its capacitance?

15. A capacitor of unknown capacitance is found to have a reactance of 120 Ω at 200 Hz. The capacitor is connected to a 240-V, 60-Hz power line. (a) What is the current in the circuit? (b) Why is it not the same as that in Exercise 11?

21.6 Impedance

16. A pure capacitor, a pure inductor, and a pure resistor are connected in series across an ac power source. A voltmeter placed in turn across each circuit element reads 15 V, 20 V, and 20 V, respectively. What is the potential difference of the source?

17. A 5.0-μF capacitor is connected in series with a 300-Ω resistor, and a 120-V, 50-Hz potential difference is applied. Find the current in the circuit and the power dissipated.

18. A 30-mH, 60-Ω inductor is connected to a 20-V, 400-Hz power source. Find the current in the inductor and the power dissipated in it.

19. The current in a resistor is 2.0 A when it is connected across a 240-V, 50-Hz line. How much inductance should be connected in series with the resistor to reduce the current to 1.0 A?

20. The current in a resistor is 2.0 A when it is connected across a 240-V, 50-Hz line. How much capacitance should be connected in series with the resistor to reduce the current to 1.0 A?

21. A capacitor of unknown capacitance is connected in series with an 80-Ω resistor. The combination is found to draw 0.50 A when connected to a 120-V, 60-Hz power source. Find (a) the capacitance of the capacitor, (b) the power dissipated in the capacitor, and (c) the power dissipated in the resistor.

22. A coil of unknown inductance and resistance is observed to draw 50 mA when a dc potential difference of 5.0 V is applied. When the coil is connected to a source of 400-Hz ac, however, a potential difference of 8.0 V is required to yield the same current. (a) Find the inductance and resistance of the coil. (b) Find the power dissipated in the coil in each situation.

23. A coil of unknown inductance and resistance is observed to draw 8.0 A when a 40-V dc potential difference is applied, and 5.0 A when a 40-V, 60-Hz ac potential difference is applied. (a) Find the inductance and resistance of the coil. (b) Find the power dissipated in the coil in each situation.

24. A series circuit has a resistance of 40 Ω, an inductive reactance of 30 Ω, and a capacitive reactance of 50 Ω when connected to a certain ac source. Find (a) the impedance of the circuit, (b) the phase angle, and (c) the potential difference required for a current of 1.5 A to flow.

25. A pure capacitor, a pure inductor, and a pure resistor are connected in series across a 60-V ac power source. The potential difference across the capacitor is 60 V, and that across the inductor is also 60 V. What is the potential difference across the resistor?

26. The voltage leads the current in a certain ac circuit by 30°. The effective current in the circuit is 5.0 A. (a) Is the capacitive reactance greater or less than the inductive reactance? (b) What is the instantaneous value of I when the instantaneous value of V is 0?

27. A series circuit has a resistance of 20 Ω, an inductive reactance of 20 Ω, and a capacitive reactance of 20 Ω when connected to an ac source. Find (a) the impedance of the circuit, (b) the phase angle, and (c) the potential difference required for a current of 20 A to flow.

28. A 10-μF capacitor, a 0.10-H inductor, and a 60-Ω resistor are connected in series across a 120-V, 60-Hz power line. Find (a) the current in the circuit, (b) the power dissipated in it, and (c) the potential difference across each of the circuit elements.

29. The circuit in Exercise 28 is connected across a 120-V, 30-Hz power line. Answer the same questions for this case.

21.7 Resonance

30. In the antenna circuit of a radio receiver that is tuned to a certain station, $R = 5.0\ \Omega$, $L = 5.0$ mH, and $C = 5.0$ pF. (a) Find the frequency of the station. (b) If the voltage applied to the circuit is 0.50 mV at this frequency, what is the resulting current? (c) What should the capacitance be in order to receive an 800-kHz radio signal?

31. In a series ac circuit, $R = 20\ \Omega$, $X_L = 10\ \Omega$, and $X_C = 25\ \Omega$ when the frequency is 400 Hz. Find the resonance frequency of the circuit.

32. In a series ac circuit $R = 10\ \Omega$, $X_L = 20\ \Omega$, and $X_C = 30\ \Omega$ when the applied frequency is 1.0 kHz. Find the resonance frequency of the circuit.

33. A resistor, a capacitor, and an inductor are in series with an ac power source of frequency f. The effective voltages across the components are $V_R = 96$ V, $V_C = 64$ V, and $V_L = 136$ V. Find the effective voltage of the source and the phase angle in the circuit. Is the resonance frequency of the circuit less than f, equal to f, or greater than f?

21.8 Power in AC Circuits

34. A 10-kW electric motor has an inductive power factor of 70%. What minimum rating in kVA must the power line have? A capacitor is connected in series with the motor to increase the power factor to 100%. How does this affect the required rating of the power line?

35. An inductive load connected to a 120-V, 60-Hz power source draws a current of 5.0 A and dissipates 450 W. (a) Find the capacitance needed in series with the load to increase the power factor to 100%. (b) Find the power dissipated when this capacitor is connected.

36. A 10-μF capacitor, a 30-mH inductor, and a 15-Ω resistor are connected in series with a 10-V, 250-Hz power source. Find (a) the impedance of the circuit, (b) the current in it, (c) the power factor, (d) the power dissipated, and (e) the minimum apparent-power rating of the source. (f) What current will flow if the circuit is connected to a 10-V power line whose frequency is equal to its resonance frequency?

37. A 60-μF capacitor, a 0.30-H inductor, and a 50-Ω resistor are connected in series with a 120-V, 60-Hz power source. Find (a) the impedance of the circuit, (b) the current in it, (c) the power factor, (d) the power dissipated, and (e) the minimum apparent-power rating of the source. (f) What current will flow if the circuit is connected to a 120-V power line whose frequency is equal to its resonance frequency?

21.9 Impedance Matching

38. A microphone whose impedance is 20 Ω is to be used with an amplifier whose input impedance is 50,000 Ω. What should the ratio of turns be in the required transformer?

21.10 Parallel AC Circuits and Filters

39. In a parallel ac circuit R = 40 Ω, X_L = 20 Ω, and X_C = 30 Ω when the circuit is connected to a 60-V, 1.0-kHz power source. What is the total current in the circuit?

40. A 10-Ω resistor and an 8.0-μF capacitor are connected in parallel across a 10-V, 1.0-kHz power source. Find the current in each component, the total current, the impedance of the circuit, the phase angle, and the power dissipated by the circuit.

41. A 10-Ω resistor and a 2.0-mH inductor are connected in parallel across a 10-V, 1.0-kHz power source. Answer the same questions for this circuit.

42. A 10-Ω resistor, an 8.0-μF capacitor, and a 2.0-mH inductor are connected in parallel across a 10-V, 1.0-kHz power source. Answer the same questions for this circuit.

PROBLEMS

43. A circuit that contains inductance and resistance has an impedance of 50 Ω at 100 Hz and an impedance of 100 Ω at 500 Hz. What are the values of the inductance and the resistance of the circuit?

44. A circuit that contains capacitance and resistance has an impedance of 30 Ω at 80 Hz and an impedance of 14 Ω at 240 Hz. What are the values of the capacitance and resistance of the circuit?

45. An inductor dissipates 75 W of power when it draws 1.0 A from a 120-V, 60-Hz power line. (a) What is its power factor? (b) What capacitance should be connected in series with it to increase its power factor to 100%? (c) How much current would the circuit then draw? (d) How much power would it then dissipate? (e) What should the minimum volt-ampere rating of the power line be then?

46. A circuit consisting of a capacitor in series with a resistor draws 3.6 A from a 50-V, 100-Hz power line. The circuit dissipates 120 W. (a) What is the precise phase relationship between voltage and current in the circuit? (b) What series inductance should be inserted in the circuit if the current and voltage are to be in phase? (c) How much current would the circuit then draw? (d) How much power would it dissipate?

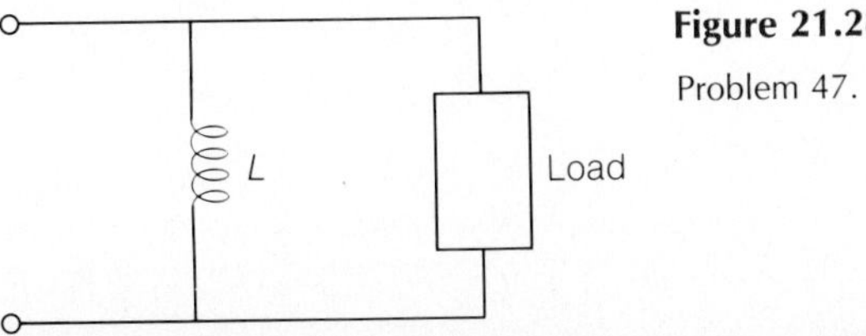

Figure 21.28
Problem 47.

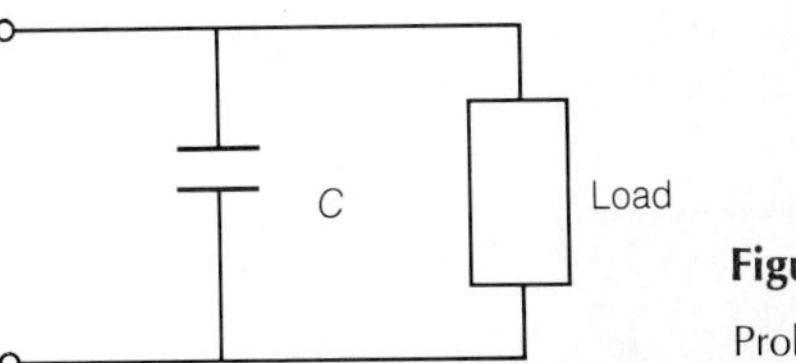

Figure 21.29
Problem 48.

47. Figure 21.28 shows a simple high-pass filter. (a) What inductance is needed to discriminate against frequencies under 20 kHz when the load is 50 Ω? (b) What proportion of the total current passes through the load when the frequency of the applied signal is 2.0 kHz? (c) When it is 200 kHz?

48. Figure 21.29 shows a simple low-pass filter. (a) What capacitance is needed to discriminate against frequencies above 20 kHz when the load is 50 Ω? (b) What proportion of the total current passes through the load when the frequency of the applied signal is 2.0 kHz? (c) When it is 200 kHz?

Answers to Multiple Choice

1. d	**10.** b	**19.** c	**28.** a
2. b	**11.** b	**20.** b	**29.** c
3. a	**12.** b	**21.** b	**30.** a
4. d	**13.** b	**22.** b	**31.** a
5. b	**14.** b	**23.** a	**32.** b
6. a	**15.** a	**24.** c	**33.** a
7. c	**16.** c	**25.** a	**34.** d
8. a	**17.** a	**26.** c	
9. a	**18.** d	**27.** a	

CHAPTER

22

LIGHT

The discovery that light consists of electromagnetic waves is among the most noteworthy achievements of science. The existence of electromagnetic waves was proposed in 1864 by the British physicist James Clerk Maxwell, who went on to suggest that light waves were of this kind. Experimental confirmation came some years later. Although the wave behavior of light can be analyzed directly from Maxwell's theory, such calculations require advanced mathematics. A simpler, though less comprehensive, approach to optics had been devised in 1678 by the Dutch physicist Christian Huygens. In this chapter the reflection and refraction of light at plane surfaces are considered with the help of Huygens' method.

22.1 Electromagnetic Waves

Coupled electric and magnetic fields that move with the speed of light.

From the earlier work of Faraday, Maxwell knew that a changing magnetic field causes charges to move in a wire loop. An electric field can also set charges in motion, which means that a changing magnetic field is equivalent in its effects to an electric field (Fig. 22.1a). Maxwell proposed the converse (Fig. 22.1b):

A changing electric field is equivalent in its effects to a magnetic field.

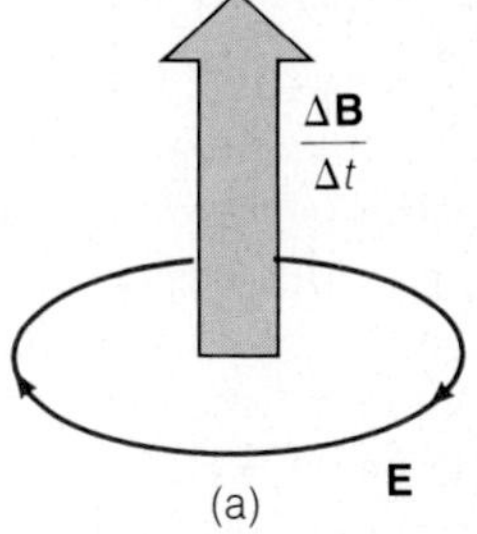

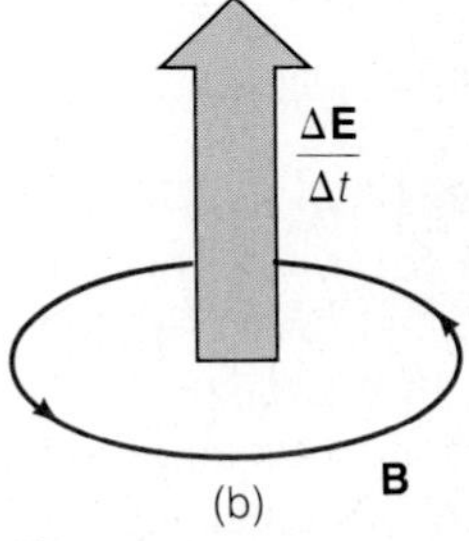

Figure 22.1

(a) According to Faraday's law, a changing magnetic field gives rise to an electric field. (b) According to Maxwell's hypothesis, a changing electric field gives rise to a magnetic field.

The electric fields produced by electromagnetic induction are easy to observe because metals offer little resistance to the flow of electrons. Even a weak electric field can lead to a measurable current in a metal. Weak magnetic fields are much harder to detect, however. Maxwell's hypothesis was based on an indirect argument rather than on experimental findings, unlike Faraday's hypothesis of electromagnetic induction.

To appreciate Maxwell's train of thought, imagine a capacitor being charged by a battery. Current flows in the connecting wires, and electric charges of opposite sign build up on the capacitor plates. As the charges grow, the electric field E increases with them. The greater the current, the faster the field changes. The strength I of the current is thus mirrored by the rate of change $\Delta E/\Delta t$ of the electric field. What Maxwell did was to extend this correspondence to include magnetic effects. Just as a current I has a magnetic field associated with it, said Maxwell, so does a changing electric field $\Delta E/\Delta t$. The result was a particularly symmetric set of four equations describing electricity and magnetism that remain valid to this day.

If Maxwell was right, electromagnetic (or simply em) waves should occur in which fluctuating electric and magnetic fields are coupled together both by electromagnetic induction and by the converse mechanism he had proposed. The energy of the waves would be shared by their electric and magnetic fields. Maxwell was able to show that the speed c in free space of em waves is given by

$$c = \frac{1}{\sqrt{\epsilon_0 \mu_0}} \qquad \textit{Speed of light} \quad (22.1)$$

$$= \frac{1}{\sqrt{(8.85 \times 10^{-12}\ \mathrm{C^2/N{\cdot}m^2})(1.26 \times 10^{-6}\ \mathrm{T{\cdot}m/A})}}$$

$$= 3.00 \times 10^8\ \mathrm{m/s}$$

where ϵ_0 is the permittivity of free space and μ_0 is its permeability. This is the same speed that had been measured for light waves! The agreement was too great to be accidental, and Maxwell concluded that light consists of em waves. To five significant figures the value of c is 2.9979×10^8 m/s.

During Maxwell's lifetime the idea of em waves remained without direct experimental support. Finally, in 1888 the German physicist Heinrich Hertz showed that em waves indeed do exist and have just the properties Maxwell had predicted.

BIOGRAPHICAL NOTE

James Clerk Maxwell

James Clerk Maxwell (1831–1879) was born in Scotland at about the time Faraday discovered electromagnetic induction. At 19 he entered Cambridge University to study physics and mathematics. While still a student, he investigated the physics of color vision and later used his ideas to make the first color photograph. Maxwell became known to the scientific world at 24 when he showed theoretically that the rings of Saturn could not be solid or liquid but must consist of separate small bodies. At about this time Maxwell became interested in electricity and magnetism and grew convinced that the wealth of phenomena Faraday and others had discovered were not isolated effects but had an underlying unity of some kind. Maxwell's initial step in establishing that unity came in 1856 with the paper "On Faraday's Lines of Force," in which he developed a mathematical description of electric and magnetic fields.

Maxwell left Cambridge in 1856 to teach at a college in Scotland and later at King's College in London. In this period he expanded his ideas on electricity and magnetism to create a single comprehensive theory of electromagnetism. The fundamental equations he arrived at remain the foundations of the subject today. From these equations Maxwell predicted that electromagnetic waves should exist that travel with the speed of light, described the properties the waves should have, and surmised that light consisted of electromagnetic waves. Sadly, he did not live to see his work confirmed in the experiments of the German physicist Heinrich Hertz (1857–1884).

Maxwell's contributions to kinetic theory and statistical mechanics were on the same profound level as his contributions to electromagnetism. His calculations showed that the viscosity of a gas ought to be independent of its pressure, a surprising result that Maxwell, with the help of his wife, confirmed experimentally. They also found that the viscosity was proportional to the absolute temperature of the gas. Maxwell's explanation for this proportionality gave him a way to estimate the size and mass of molecules, which until then could only be guessed at. Maxwell shares with Boltzmann credit for the equation that gives the distribution of molecular energies in a gas.

In 1865 Maxwell returned to his family's home in Scotland. There he continued his research and also composed a treatise on electromagnetism that was to be the standard text on the subject for many decades. It was still in print a century later. In 1871 Maxwell went back to Cambridge to establish and direct the Cavendish Laboratory, named in honor of the pioneering physicist Henry Cavendish and financed by the Duke of Devonshire, a member of the Cavendish family. Maxwell died of abdominal cancer at the age of 48 in 1879, the year in which Albert Einstein was born. Maxwell had been the greatest theoretical physicist of the nineteenth century; Einstein was to be the greatest theoretical physicist of the twentieth century.

Hertz generated the waves by applying an alternating emf to an air gap between two metal balls. The width of the gap was small enough for a spark to occur each time the emf reached a peak. A wire loop with a small gap was the detector; em waves set up oscillations in the loop that gave rise to sparks in the gap. Hertz determined the wavelength and speed of the waves he generated, showed that they had both electric and magnetic components, and found that they exhibited the same optical behavior as light waves.

To see how an em wave can come into being, imagine connecting a pair of metal rods to an oscillator that produces an alternating emf, as in Fig. 22.2. Suppose that at first the upper rod has a single positive charge that can move, and that the lower rod has a single negative charge that can move. (Only electrons move in actual metals, to be sure, and a great many of them do so in even a small metal rod. The model used here leads to the same result that a more realistic model would yield, but is easier to visualize.)

(a) When the oscillator is switched on, the positive charge in the upper rod begins to move upward and the negative charge in the lower rod begins to move downward. The electric field lines around the charges are indicated by the colored lines, and the magnetic field lines due to the motion of the charges (which are coaxial circles perpendicular to the paper) are indicated by crosses when their direction is into the paper and by dots when their direction is out of the paper. (The dots represent arrowheads and the crosses represent the tail feathers of arrows.)

(b) The charges have reached the limit of their motion and have stopped, so that they no longer produce a magnetic field. The outer magnetic field lines do not disappear but continue to travel outward.

(c) The emf of the oscillator now begins to decrease, and the charges move toward each other. The result is a magnetic field in the opposite direction to the earlier field. The electric field is in the same direction as before.

(d) The emf has passed through 0 and begun to increase in the opposite sense. As a result there is now a negative charge in the upper rod and a positive one in the lower rod that begin to move apart. The electric field is therefore opposite in direction to the earlier field, but the magnetic field is in the same direction because magnetically a positive charge moving downward is equivalent to a negative charge moving upward.

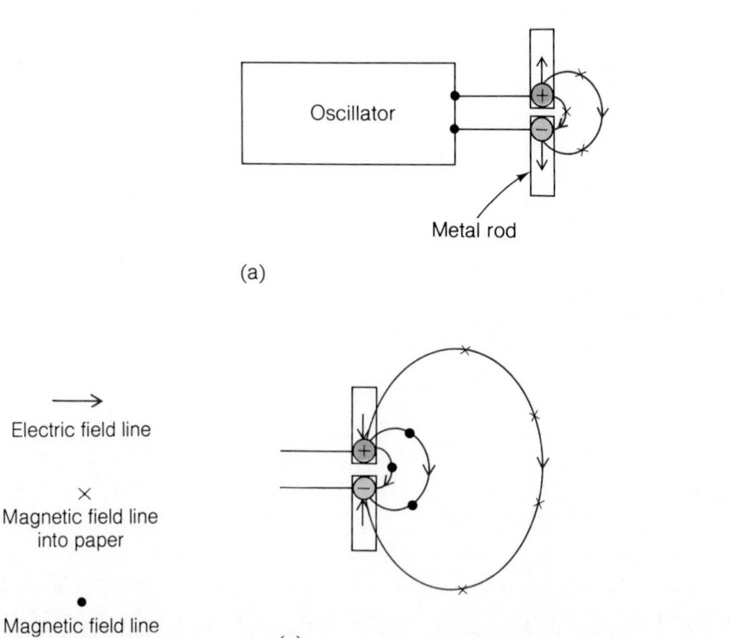

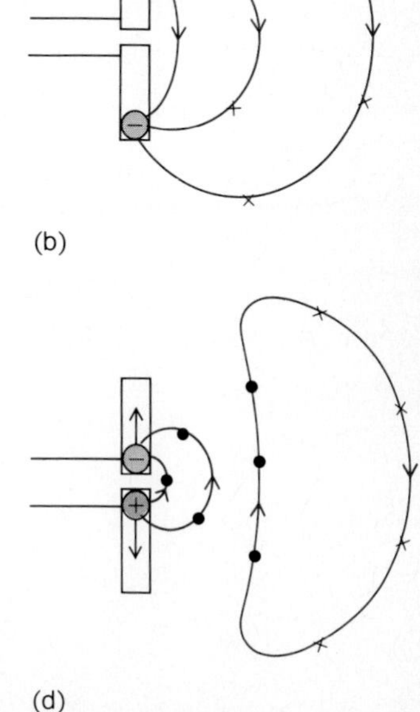

Figure 22.2

A pair of metal rods connected to an electrical oscillator emit electromagnetic waves.

Owing to this sequence of changes in the fields, the outermost electric and magnetic field lines, respectively, form into closed loops. These loops, which lie in perpendicular planes, are divorced from the oscillating charges that gave rise to them and continue moving outward, making up an em wave. As the charges continue oscillating back and forth, further associated loops of electric and magnetic field lines are emitted that form an expanding pattern of loops.

Figure 22.3(a) shows the form of the electric and magnetic fields that spread outward from a pair of oscillating charges. The actual fields are three-dimensional, so that the magnetic field lines form loops in planes perpendicular to the line joining the charges.

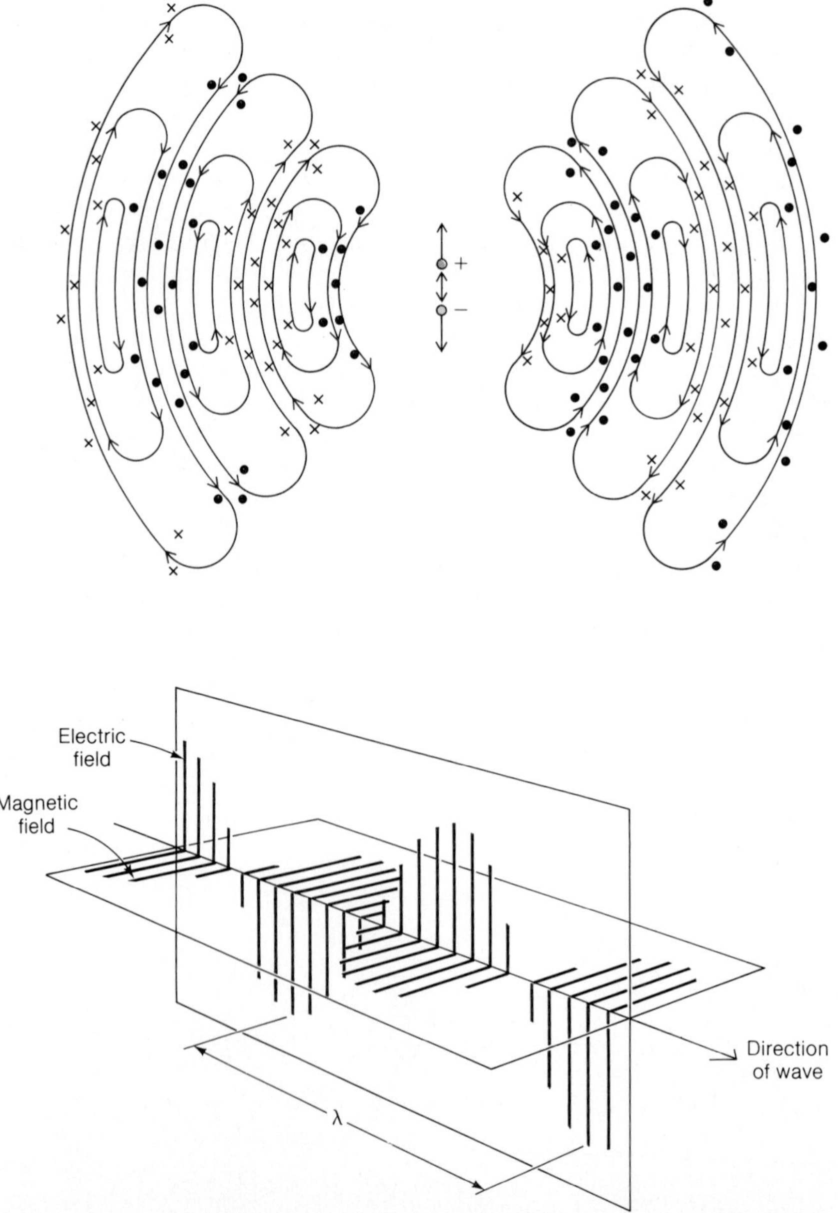

Figure 22.3

(a) The configuration of electric and magnetic fields that spread outward from a pair of oscillating charges. (b) The electric and magnetic fields of an electromagnetic wave far from its source vary as shown here. The field directions are perpendicular to each other and to the direction of propagation.

Three properties of em waves are worth noting:

1. The variations occur simultaneously in both fields (except close to the oscillating charges), so that the electric and magnetic fields have maxima and minima at the same times and in the same places.
2. The directions of the electric and magnetic fields are perpendicular to each other and to the direction in which the waves are moving. Light waves are therefore transverse.
3. The speed of the waves depends only on the electric and magnetic properties of the medium they travel in, not on the amplitudes of the field variations.

Figure 22.3(b) is an attempt to illustrate properties 1 and 2 in terms of field lines of **E** and **B** a long distance from a source of electromagnetic waves. Closer to the source the field lines are curved, as in Fig. 22.3(a). It is worth keeping in mind that, unlike the other types of waves considered in Chapter 12—waves in a stretched string, water waves, sound waves—nothing material moves in the path of an em wave. The only changes are in electric and magnetic fields.

We have been looking at a special kind of em wave source. Actually, *all accelerated charges radiate em waves,* regardless of how the acceleration occurs. For instance, a charge that moves in a curved path is accelerated and so gives off em waves, losing energy as it does so.

22.2 Types of Em Waves

X rays and radio waves as well as light consist of em waves.

Light is not the only example of an em wave. Although all em waves share certain basic properties, other features of their behavior depend on their frequencies. Light waves themselves span a short frequency interval, from about 4.3×10^{14} Hz for red light to about 7.5×10^{14} Hz for violet light. Em waves with frequencies between these limits are the only ones that the eye responds to, and specialized instruments of various kinds are required to detect waves with higher and lower frequencies. Figure 22.4 shows the em wave **spectrum** from the low frequencies used in radio communication to the high frequencies found in X rays and gamma rays (which are considered in later chapters). The wavelengths corresponding to the various frequencies are also shown.

As we recall, the product of the frequency f of a wave and its wavelength λ is just the wave speed, here c. Given the wavelength or frequency of a particular electromagnetic wave, we can immediately find the other quantity.

EXAMPLE 22.1

Find the wavelengths of yellow light whose frequency is 5.00×10^{14} Hz and of radio waves whose frequency is 1.00 MHz. (1 MHz = 1 megahertz = 10^6 Hz.)

SOLUTION The wavelength of the yellow light is

$$\lambda = \frac{c}{f} = \frac{3.00 \times 10^8 \text{ m/s}}{5.00 \times 10^{14} \text{ Hz}} = 6.00 \times 10^{-7} \text{ m} = 600 \text{ nm}$$

which is less than 1/1000 of a millimeter. (1 nm = 1 nanometer = 10^{-9} m. The nanometer is often used for expressing the wavelengths found in light.) The width of a human hair is about 70 of these wavelengths. By the same procedure we find that a 1.00-MHz radio wave has a wavelength of 300 m. ♦

Shortly after Hertz's experiments with em waves, Guglielmo Marconi conceived the idea of using the waves for communication. In Hertz's work only a few meters separated transmitter and receiver. Marconi was able to extend the range to many kilometers, and in 1899 sent a radio message from France to England over the English Channel. Two years later he sent signals from England across the Atlantic Ocean to Newfoundland with the help of kites to raise his antennas. Morse code was used initially; radio communication using speech was first achieved in 1906.

The two chief ways to incorporate information in an em wave are **amplitude modulation** and **frequency modulation.** (The term *modulation* means alteration or

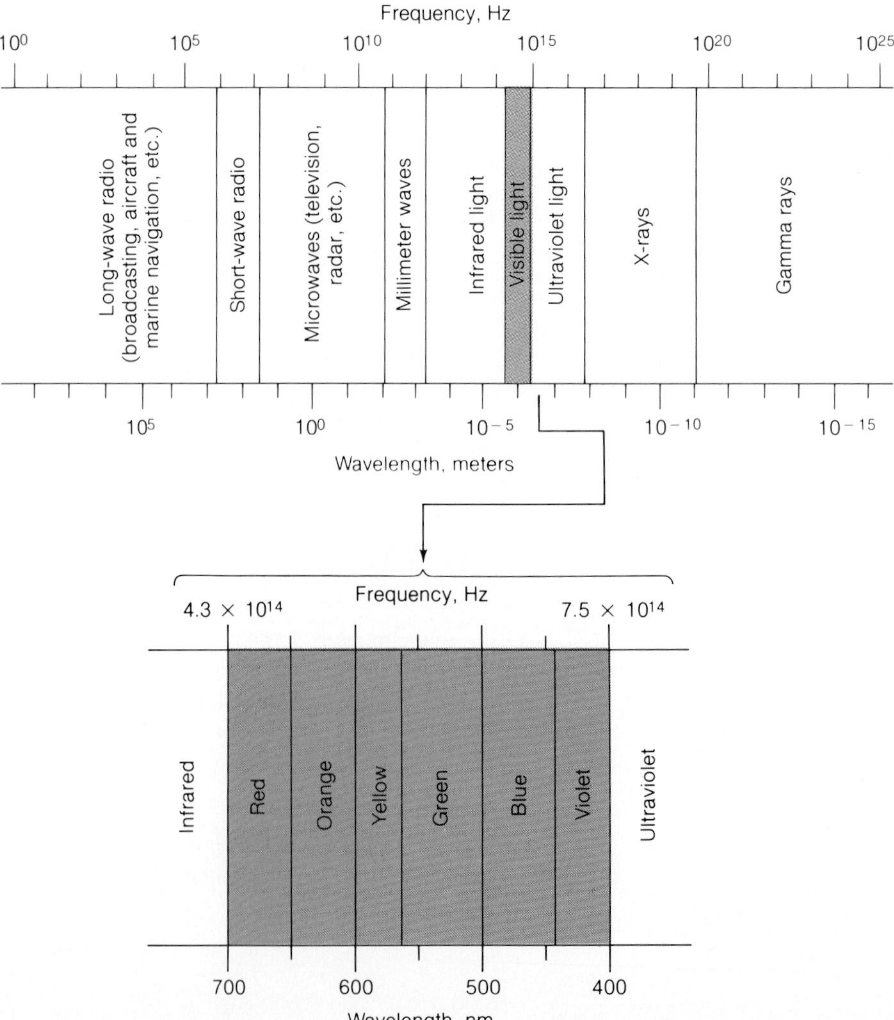

Figure 22.4

The electromagnetic wave spectrum. The boundaries of the various categories are not sharp. (1 nm = 1 nanometer = 10^{-9} m.)

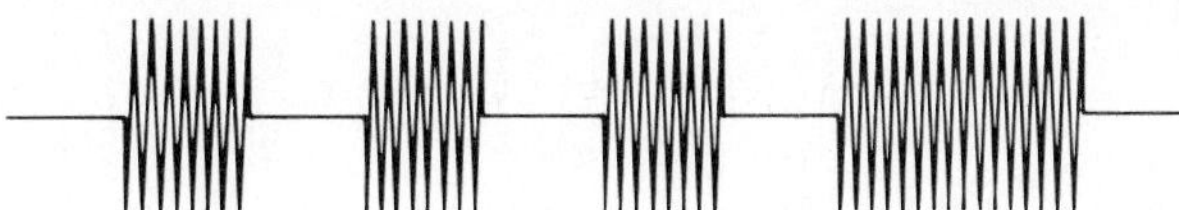

Figure 22.5

In radiotelegraphy, a constant-frequency electromagnetic wave is switched on and off in a coded sequence to transmit information. The letter *v*, which is · · · — in the Morse code, is shown.

change.) In amplitude modulaton (AM) a "carrier wave" of constant frequency is varied in amplitude, with the variations constituting the signal. The simplest example is a flashlight switched on and off to give a series of dots and dashes. Another example is the radiotelegraph, in which coded sequences of dots and dashes are used to represent letters of the alphabet, numbers, or other specific information (Fig. 22.5).

A carrier wave can also be modulated so that the amplitude variations correspond to sound waves, as in Fig. 22.6. A microphone converts sound waves to an equivalent electric signal, which is then amplified and combined with the carrier. The final wave is broadcast from an antenna, and at the receiving station the carrier

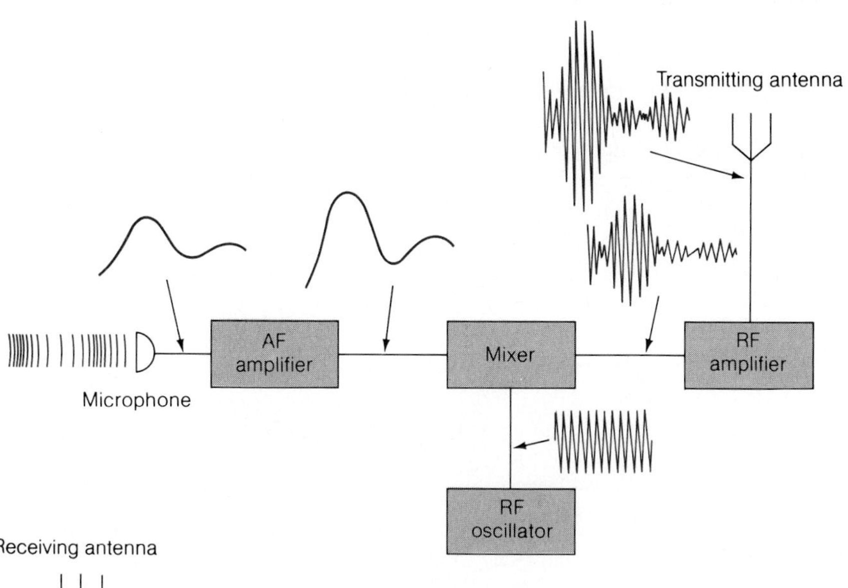

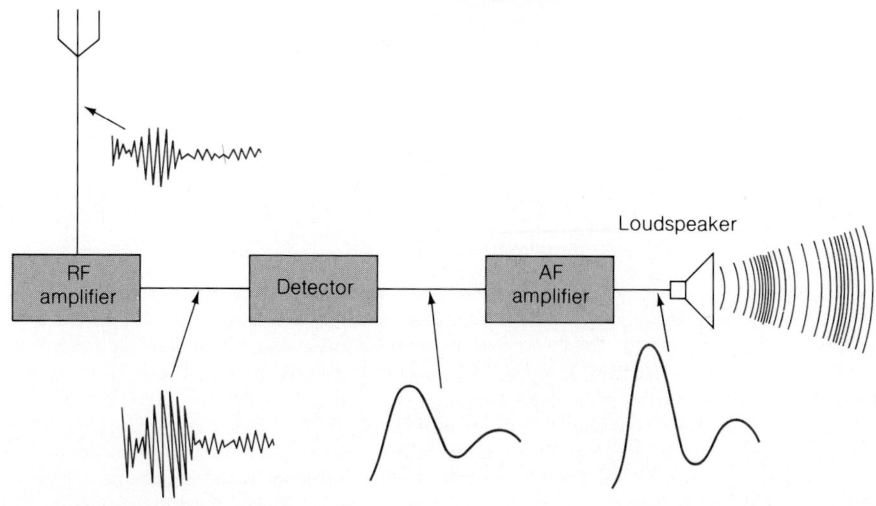

Figure 22.6

In amplitude modulation, the amplitude of a constant-frequency carrier wave is varied in accordance with an audio signal. The transmitting and receiving systems shown are highly simplified.

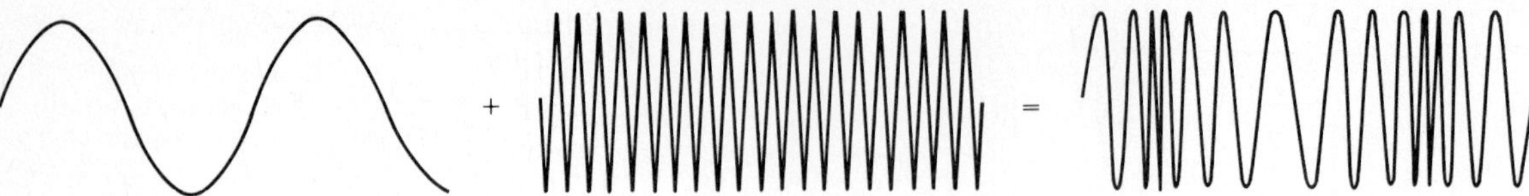

Figure 22.7

The information content of a frequency-modulated wave resides in its frequency variations rather than in its amplitude variations.

is removed to leave an audio signal. The latter is then amplified and used to generate sound waves in a loudspeaker.

One problem with amplitude modulation is that such sources of random em waves as electric storms and electric machinery can interfere with the broadcast waves to cause "static." In frequency modulation (FM) the frequency, not the amplitude, of the carrier is varied in accordance with the audio signal (Fig. 22.7). The information content of a frequency-modulated em wave is virtually immune to disturbance.

In television the desired scene is focused on the light-sensitive screen of a special tube, which is then scanned in a zig-zag fashion by an electron beam. The signal extracted from the screen varies with the image brightness at each successive point that is scanned. This signal is then used to modulate a carrier wave for broadcasting. At the receiver, suitable circuits demodulate the wave. In the picture tube the image is reproduced with the help of an electron beam that moves in step with the electron beam in the camera tube (Fig. 22.8).

Radio waves in the short-wave band of Fig. 22.4 have the useful property of being reflected from the earth's ionosphere (Section 16.5), a region high in the

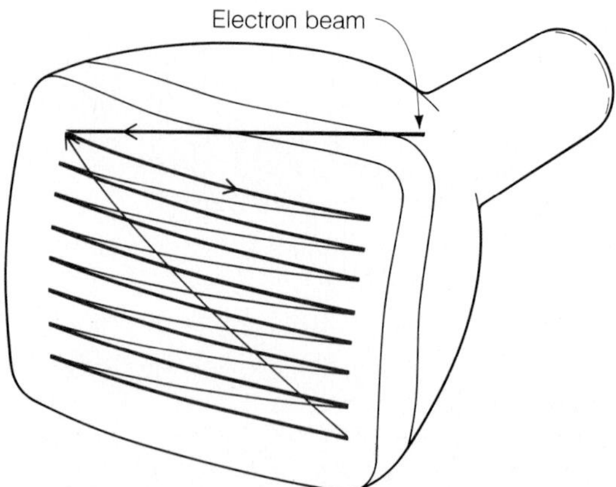

Figure 22.8

The screen of a television picture tube is coated with a phosphor, a substance that emits light when struck by electrons and then continues to glow for a short time afterward. This permits a picture to be built up that lasts long enough to be perceived as a whole. In the tube a beam of electrons is accelerated through 5 kV to 50 kV and then is deflected by magnetic fields to cover the screen in a pattern of horizontal lines starting at the upper left. The returns from right to left take about 1/10 as long as the sweeps from left to right that produce the image. In the United States the image is divided into 525 lines, but elsewhere more lines are used to give better picture quality. A complete image is built up by two scans of the screen, each covering alternate lines, a process that takes place 30 times per second. During the scans the beam intensity is varied to produce the variations in brightness that constitute the visible image.

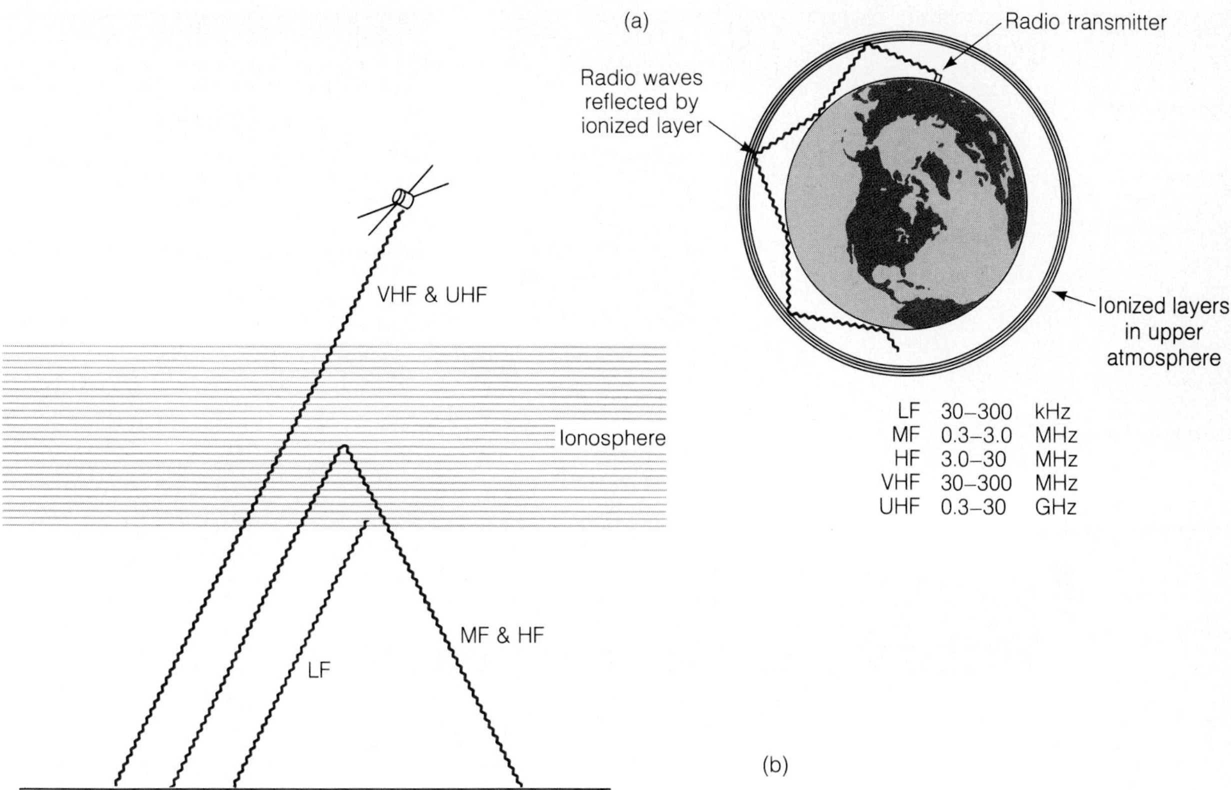

Figure 22.9

(a) The ionosphere is a region in the upper atmosphere whose ionized layers make possible long-range radio communication by their ability to reflect radio waves. (b) Only radio waves in the medium- and high-frequency (MF and HF) bands are reflected by the ionosphere. Low-frequency (LF) waves are absorbed there. Very-high- and ultrahigh-frequency (VHF and UHF) waves pass through the ionosphere and so are used to communicate with satellites and spacecraft.

atmosphere where ions are relatively abundant. Without the ionosphere, radio communication would be limited to short distances because em waves travel in straight lines and would be shielded from remote receivers by the curvature of the earth. However, radio waves of appropriate frequencies can bounce one or more times between the ionosphere and the earth's surface. Long-range transmission is therefore possible, even to the opposite side of the earth (Fig. 22.9).

The ion content of the ionosphere and the heights of the individual ion layers vary with time of day and are influenced by streams of fast protons and electrons emitted by the sun during periods of sunspot activity. Hence no absolute rules can be given for which frequencies are best for which distances. Generally speaking, though, the higher the frequency, the greater the range. Thus a frequency of 10 MHz would ordinarily give good results between 300 and 2400 km, whereas 20 MHz would be better for 2400 to 11,000 km.

The higher-frequency waves used for FM and television broadcasting are not reflected by the ionosphere, so their reception is limited to the line of sight between transmitting and receiving antennas. When the frequencies are up around 10^{10} Hz, corresponding to wavelengths in the centimeter range, the waves can readily be focused into narrow beams. Such beams are reflected by objects such as ships and airplanes, which is the basis of **radar** (from *ra*dio *d*etection *a*nd *r*anging). A rotating antenna is used to send out a pulsed beam, and the distance of a particular target is established by the time needed for the echo to return to the antenna. A picture tube is used to display the echoes, with the center of the image representing the position of the antenna.

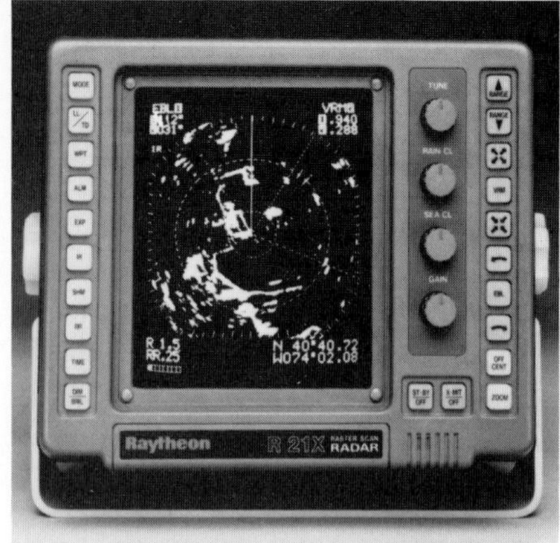

The rotating scanner of a radar set sends out pulses of microwaves in a narrow beam and then detects their reflections from objects in their paths. These reflections are then displayed on a screen. The center of the display corresponds to the position of the scanner.

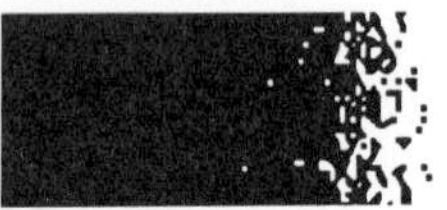

22.3 Poynting Vector

Em waves carry energy and momentum.

The **Poynting vector S** describes the transport of energy by an electromagnetic wave. The direction of **S** is the direction of the wave, and its magnitude S is equal to the rate at which energy is being carried by the wave per square meter of area perpendicular to the wave direction.

We recall from Eq. (17.12) that the energy density of an electric field E is

$$w_e = \frac{E^2}{8\pi k}$$

Since in free space $k = 1/4\pi\epsilon_0$, we can also write this as

$$w_e = \frac{\epsilon_0 E^2}{2}$$ *Electric energy density*

The energy density of a magnetic field B in free space is, from Eq. (20.13),

$$w_m = \frac{B^2}{2\mu_0}$$ *Magnetic energy density*

Figure 22.10 shows an em wave passing through an imaginary box of area A and thickness Δx. If E and B are the magnitudes of the electric and magnetic fields in the box at a certain time, the total energy density in the box is

$$w = w_e + w_m = \frac{\epsilon_0 E^2}{2} + \frac{B^2}{2\mu_0}$$

One of Maxwell's findings was that the electric and magnetic energy densities in an em wave are equal, so that we can write

$$w = 2w_e = \epsilon_0 E^2$$

The total energy ΔW in the box is $\Delta W = wV$, where $V = A\Delta x$ is the volume of the box. Hence

$$\Delta W = wV = \epsilon_0 E^2 A\Delta x$$

The wave carries this energy at the speed c, and in the time $\Delta t = \Delta x/c$ the energy will have passed through the box. Thus the rate P at which energy flows through the box is

$$P = \frac{\Delta W}{\Delta t} = \frac{\epsilon_0 E^2 A\Delta x}{\Delta x/c} = c\epsilon_0 E^2 A$$

Since the Poynting vector S is the rate of energy flow per unit area,

$$S = \frac{P}{A} = c\epsilon_0 E^2$$

The unit of S is the watt/m^2.

If E_{max} is the maximum value of the electric field of the wave, $S_{max} = c\epsilon_0 E_{max}^2$. Because E in the wave varies sinusoidally with time, the same argument used in Section 21.1 to find the average current in an ac circuit gives for the average value $\overline{S}$ of the Poynting vector

$$\overline{S} = \frac{c\epsilon_0 E_{max}^2}{2} \qquad \textit{Average Poynting vector} \quad (22.2)$$

We can also express $\overline{S}$ in terms of B_{max}, the maximum value of the magnetic field of the wave. Since the energy densities w_e and w_m are equal,

$$\frac{\epsilon_0 E^2}{2} = \frac{B^2}{2\mu_0} \qquad E = \frac{B}{\sqrt{\epsilon_0 \mu_0}}$$

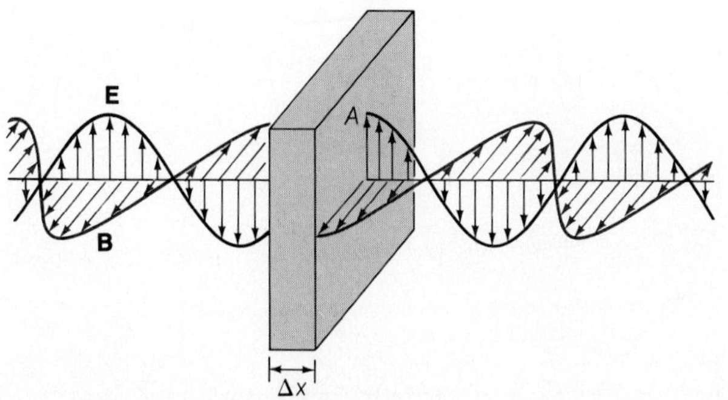

Figure 22.10

Em wave passing through an imaginary box of area A and thickness Δx.

But $1/\sqrt{\epsilon_0\mu_0} = c$, the speed of light! Thus

$$E = cB \qquad \textit{Em wave} \quad (22.3)$$

The electric and magnetic fields in an em wave are proportional to each other, with c as the constant of proportionality. We therefore have

$$\overline{S} = \frac{cB_{max}^2}{2\mu_0} \qquad \textit{Average Poynting vector} \quad (22.4)$$

EXAMPLE 22.2

Em radiation from the sun arrives at the earth at the rate of about 1.4 kW per m^2 of area perpendicular to the sun's rays (Fig. 22.11). Find the maximum electric and magnetic field magnitudes of a single em wave (as distinct from the mixture of different em waves that comes from the sun) whose Poynting vector is 1.4 kW/m^2.

SOLUTION The maximum electric field magnitude is

$$\begin{aligned} E_{max} &= \sqrt{\frac{2S}{c\epsilon_0}} \\ &= \sqrt{\frac{(2)(1.4 \times 10^3\ \text{W/m}^2)}{(3.0 \times 10^8\ \text{m/s})(8.85 \times 10^{-12}\ \text{C}^2/\text{N}\cdot\text{m}^2)}} \\ &= 1.03 \times 10^3\ \text{V/m} = 1.0\ \text{kV/m} \end{aligned}$$

which is appreciable. The corresponding value of B_{max} is

$$B_{max} = \frac{E_{max}}{c} = \frac{1.03 \times 10^3\ \text{V/m}}{3.0 \times 10^8\ \text{m/s}} = 3.4 \times 10^{-6}\ \text{T}$$

which is quite small. The electric field of an em wave is responsible for most of the effects the wave produces when it interacts with matter, even though the electric and magnetic energy densities are the same.

Figure 22.11

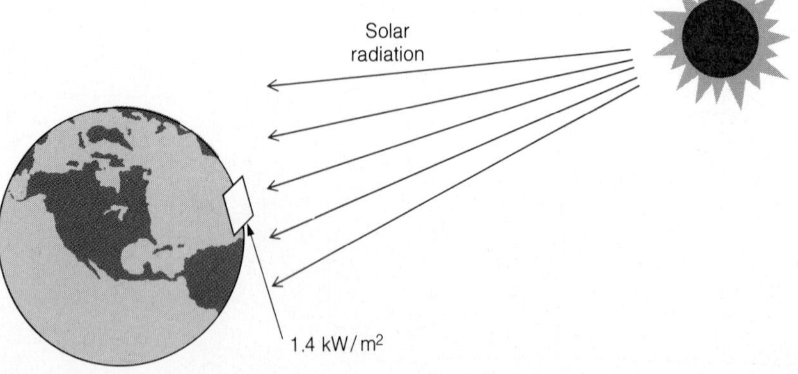

♦

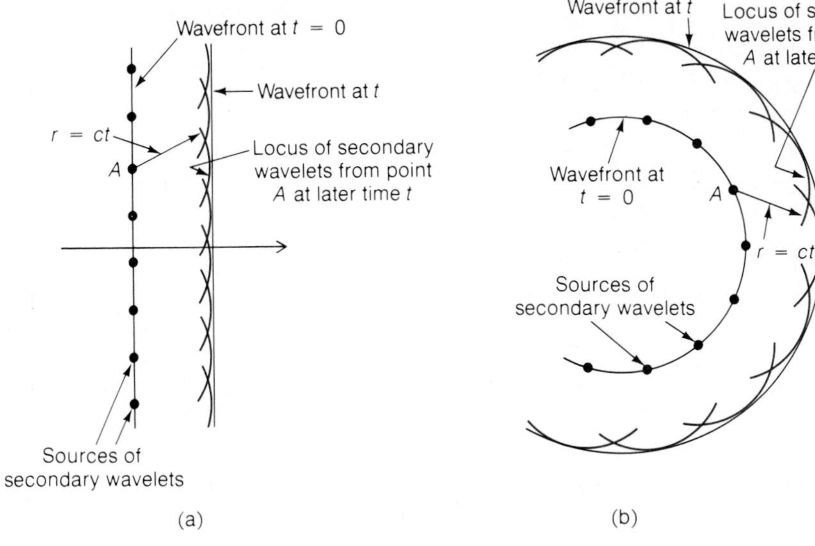

Figure 22.15

Huygens' principle applied to the propagation of (a) plane and (b) spherical wavefronts.

into the "shadow" region). The motion of wavefronts is what counts, although we can both properly and conveniently use rays to summarize our conclusions.

22.5 Reflection

Mirror, mirror on the wall.

Every object reflects some of the light that falls on it, and this reflected light enables us to see it. In most cases the surface of the object has irregularities that spread out an initially parallel beam of light in all directions to produce **diffuse reflection** (Fig. 22.16a). A surface so smooth that any irregularities in it are small relative to the wavelength of the light falling on it behaves differently. When a parallel beam of light is directed at such a surface, it is **specularly reflected** in only one direction (Fig. 22.16b). Reflection from a brick wall is diffuse, whereas reflection from a mirror is specular. Often a mixture of both kinds of reflection occurs, as in the case of a surface coated with varnish or glossy enamel. Our concern here is with specular reflection only.

When we see an object, what enters our eyes are light waves reflected from its surface. What we perceive as the object's color therefore depends on two things: the kind of light falling on it, and the nature of its surface. If white light is used to

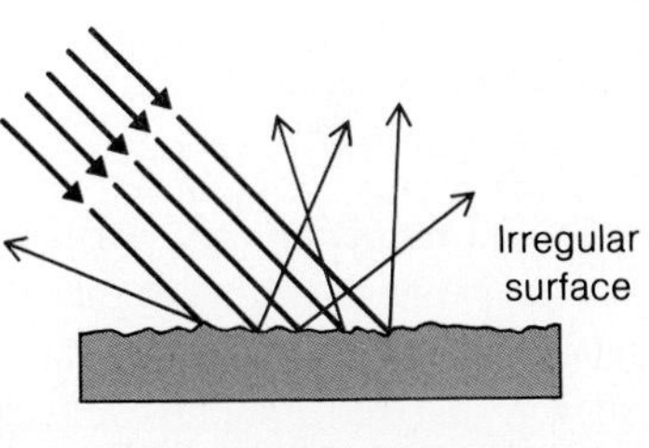

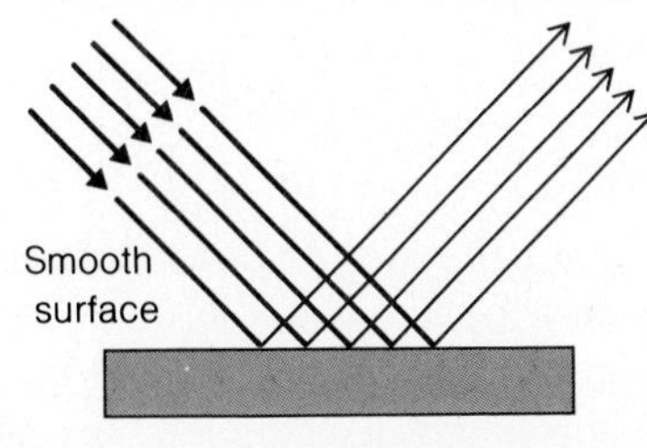

Figure 22.16

Light reflecting from irregular and smooth surfaces.

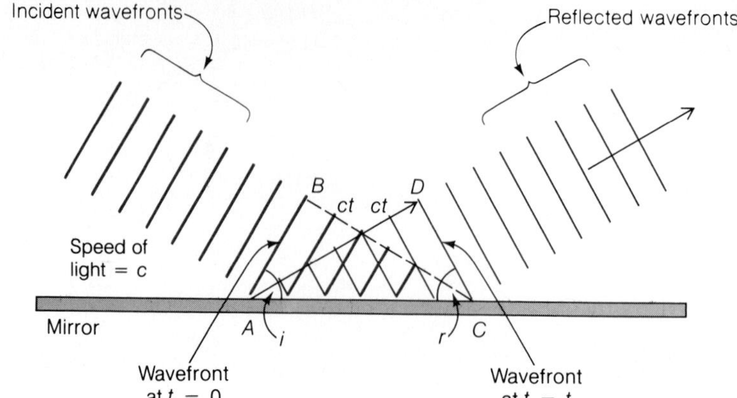

Figure 22.17

The behavior of successive wavefronts during reflection from a plane mirror. The angle of incidence *i* and the angle of reflection *r* are equal.

illuminate an object that absorbs all colors other than red, the object will appear red. If green light is used instead, the object will look black because it absorbs green light. A white object reflects light of all wavelengths equally well, and its apparent color is the same as the color of the light reaching it. A black object, on the other hand, absorbs light of all wavelengths, and it looks black no matter what color light reaches it.

Figure 22.17 shows the specular reflection of a series of plane wavefronts, which corresponds to a parallel beam of light. At $t = 0$ the wavefront AB just touches the mirror, and secondary wavelets start to spread out from A. After a time t the end B of the wavefront reaches the mirror at C. The secondary wavelets from A are now at the point D. The wavefront that was AB at $t = 0$ is therefore CD at the later time t.

The angle i between an approaching wavefront and the reflecting surface is called the **angle of incidence,** and angle r between a receding wavefront and the reflecting surface is called the **angle of reflection.** Here

$$\sin i = \frac{BC}{AC} \qquad \text{and} \qquad \sin r = \frac{AD}{AC}$$

The wavelet that comes from A takes the time t to reach D. Hence $AD = ct$, where c is the velocity of light. The point B of the wavefront AB also takes the time t to reach C, and so $BC = ct$ as well. Therefore $AD = BC$ and $\sin i = \sin r$, from which we conclude that the angles i and r are equal. Thus we have the basic law of reflection:

The angle of reflection of a plane wavefront with a plane mirror is equal to the angle of incidence.

Figure 22.18

Ray picture of reflection. The angles of incidence and reflection are measured with respect to the normal to the reflecting surface, which is a line perpendicular to it. The incident ray, the normal, and the reflected ray all lie in the same plane.

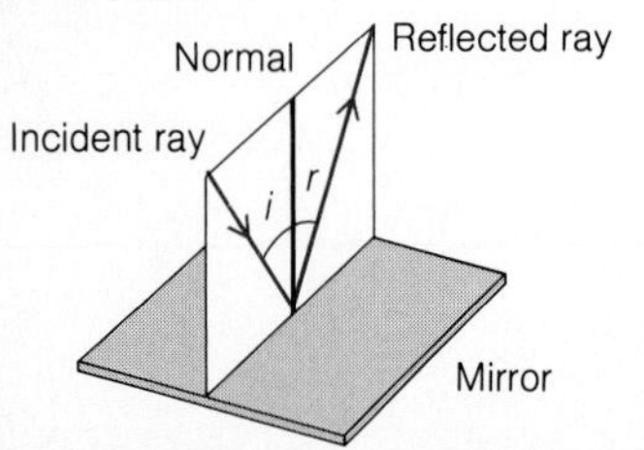

In terms of rays, the angles of incidence and reflection are measured with respect to the **normal** to the reflecting surface at the point where the light strikes it (Fig. 22.18). The normal is a line drawn perpendicular to the surface at that point. In this representation, too, the angles of incidence and reflection are equal. The incident ray, the reflected ray, and the normal all lie in the same plane.

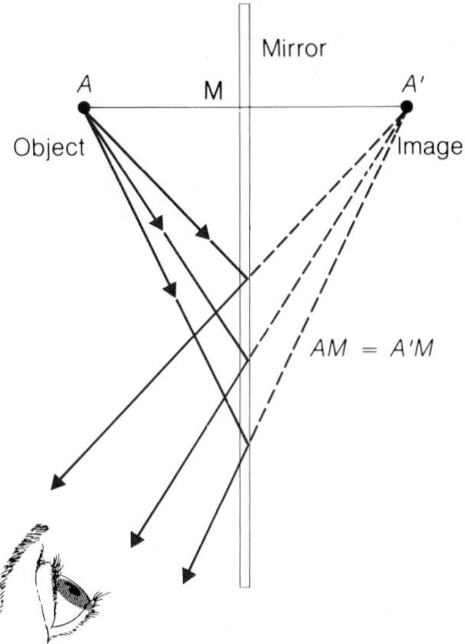

Figure 22.19

Rays from a point object at A appear to come from A' after reflection from a plane mirror. The image is as far behind the mirror as the object is in front of it.

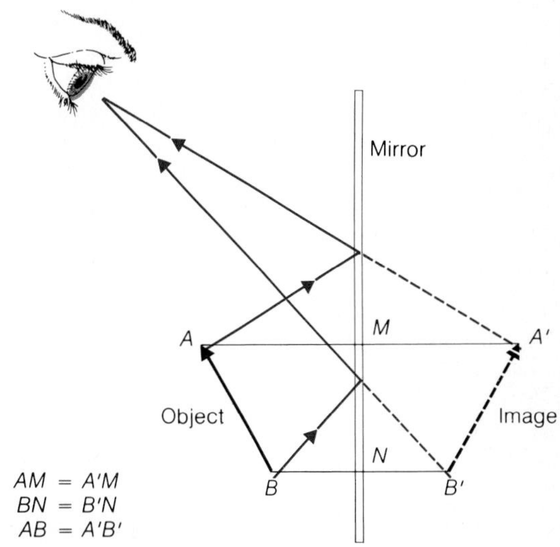

Figure 22.20

Image formation by a plane mirror. The image is erect and is the same size and shape as the object.

Mirror Image

When we look at the image of an object in a plane mirror, the image seems to be the same size as the object and as far behind the mirror as the object is in front of it. Why?

In Fig. 22.19 three light rays from a point object at A are reflected by a mirror. To the eye these rays apparently come from the point A' behind the mirror. The rays that seem to come from the image do not actually pass through it, and for this reason the image is said to be **virtual.** (A **real image** is formed by light rays that pass through it.) From the figure it is clear that $AM = A'M$, so that the object and the image are the same distance on either side of the mirror.

In Fig. 22.20 an actual object is reflected by a plane mirror. Again simple geometry shows that the object and its virtual image are the same size. Every point in the image is directly behind the corresponding point on the object, and so the orientation of the image is the same as that of the object. The image is therefore **erect.** However, left and right are interchanged in a mirror image, as in Fig. 22.21, because front and back have been reversed by the reflection. Thus a printed page appears backward in a mirror, and what seems to be one's left hand in a mirror is actually one's right hand.

Figure 22.21

Left and right are interchanged in reflection.

22.6 Refraction

The bending of light when it goes from one medium to another.

Experience tells us that a light ray that passes at a slanting angle from one medium to another, say from air to water, is deflected at the surface between the two

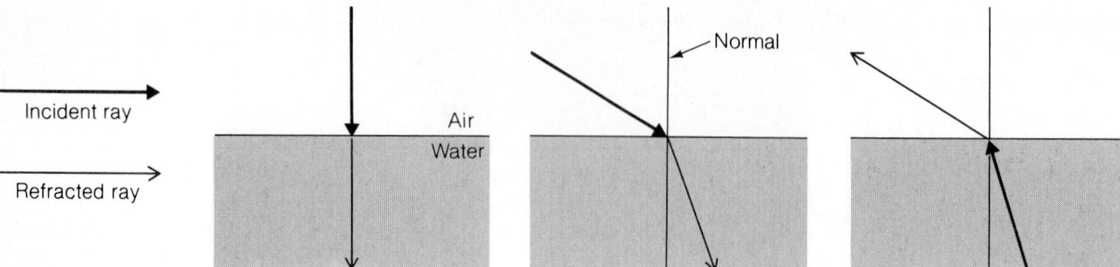

Figure 22.22

The bending of a light beam when it passes obliquely from one medium to another is called *refraction*.

media. This phenomenon is called **refraction.** Refraction gives rise to such familiar effects as the apparent distortion of objects partly submerged in water.

Several important aspects of refraction are illustrated in Fig. 22.22. When a light ray goes from air into water along the normal to the surface between them, it simply continues along the same path. However, when it enters the water at any other angle, it is bent toward the normal. The paths are reversible; so a light ray emerging from the water is bent *away* from the normal as it enters the air.

Refraction occurs because light travels at different speeds in the two media. Let us see what happens to the plane wavefront *AB* in Fig. 22.23 when it passes obliquely from a medium in which its speed is v_1 to a medium in which its speed is v_2, where v_2 is less than v_1. At $t = 0$ the wavefront *AB* just touches the interface between the two media. After a time t, the end *B* of the wavefront reaches the interface at *C*, and the secondary wavelets from *A* are now at *D*. Since v_2 is less than v_1, the secondary wavelets generated at the interface go a shorter distance in the same time interval than do wavelets in the first medium. The distance *AD* is therefore shorter than *BC*, and the refracted wavefronts move in a different direction from that of the incident wavefronts.

Something similar to refraction occurs when a tracked vehicle such as a bulldozer makes a turn. To turn a bulldozer to the right, which corresponds to the situation in Fig. 22.23, the right-hand track is slowed down, and the greater speed of the left-hand track then swings the bulldozer to the right.

The angle i between an approaching wavefront and the interface between two media is called the *angle of incidence* of the wavefront. The angle r between a receding wavefront and the interface it has passed through is called the **angle of**

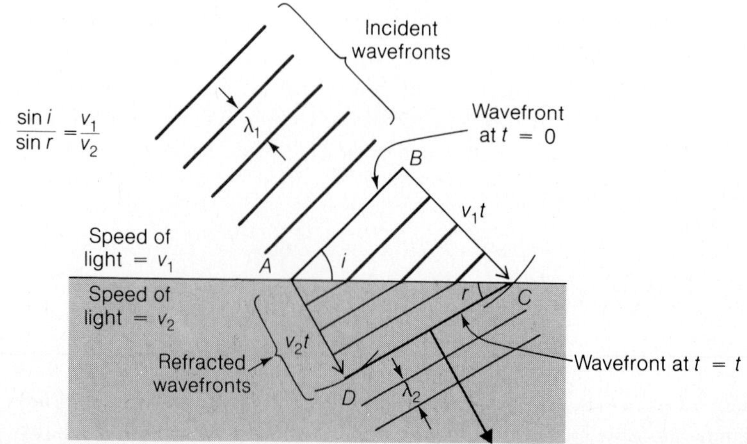

Figure 22.23

The behavior of successive wavefronts during refraction. The speed of light v_1 in the first medium is greater than that in the second medium.

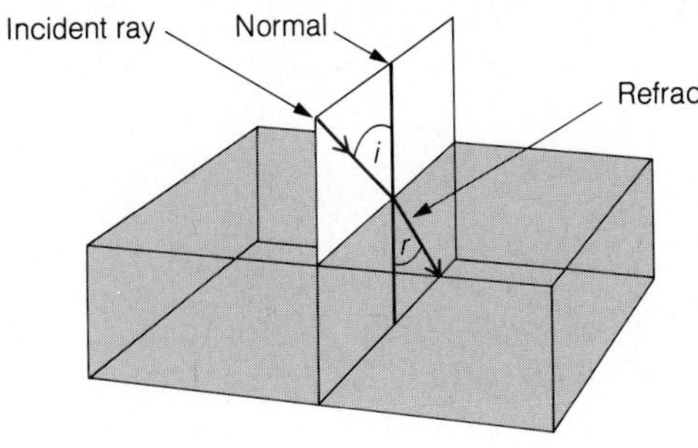

Figure 22.24

Ray picture of refraction. The incident ray, the normal, and the refracted ray all lie in the same plane.

refraction. From Fig. 22.23 we see that

$$\sin i = \frac{BC}{AC} \qquad \text{and} \qquad \sin r = \frac{AD}{AC}$$

Because $BC = v_1 t$ and $AD = v_2 t$,

$$\frac{\sin i}{\sin r} = \frac{BC}{AD} = \frac{v_1 t}{v_2 t}$$

and so we have

$$\frac{\sin i}{\sin r} = \frac{v_1}{v_2} \qquad \textit{Snell's law} \quad (22.7)$$

This useful result is known as *Snell's law* after its discoverer, the seventeenth-century Dutch astronomer Willebrord Snell. It states:

The ratio of the sines of the angles of incidence and refraction is equal to the ratio of the speeds of light in the two media.

Refraction, like reflection, can be described in terms of the ray model of light. As in Fig. 22.24, the angles i and r are taken with respect to the normal to the interface at the point where the rays meet it. The incident ray, the normal, and the refracted ray all lie in the same plane.

Why does light travel more slowly in a material medium than in a vacuum? The medium consists of atoms that contain electrons. When a light wave arrives at an atom, its electrons absorb energy from the wave as they begin to oscillate at the same frequency. These oscillations then reradiate light waves with this frequency. Because of the inertia of the electrons and because they are bound to atoms, their oscillations lag slightly behind those of the incoming wave. The result is a wave speed in the medium that is less than its speed in free space.

Refraction is not confined to light waves. For example, water waves that approach a sloping beach are refracted. Regardless of the direction of the waves in open water, their direction becomes more and more perpendicular to the shoreline as they come nearer (Fig. 22.25). The reason is that the speed of a water wave decreases in shallow water because of friction with the bottom, so a wavefront moving at an

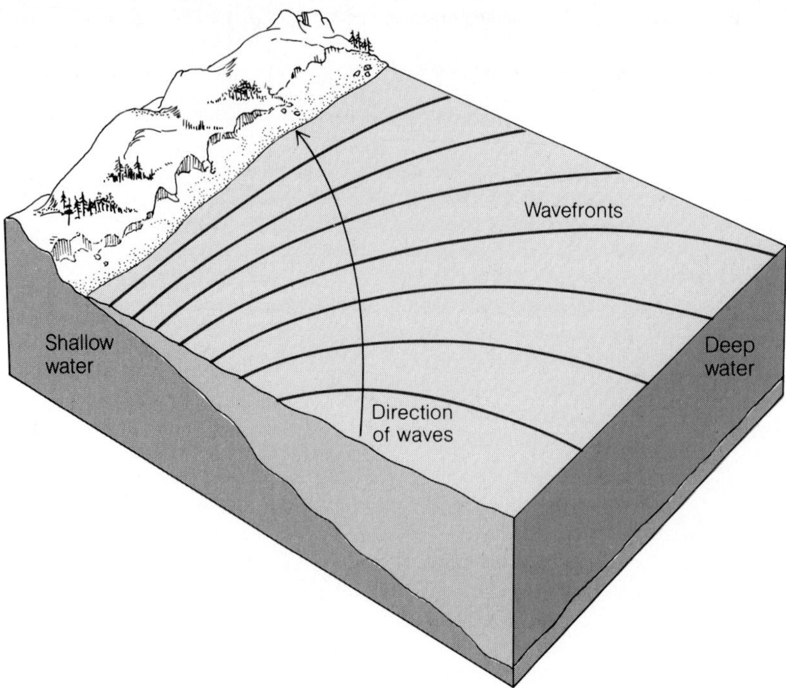

Figure 22.25

Water waves that approach the shore at an angle are refracted because they move more slowly in shallow water.

angle shoreward is progressively affected by the change in speed and swings around until it is parallel to the shore.

22.7 Index of Refraction

The greater its value in a medium, the more slowly light travels there.

The ratio between the speed of light c in free space and its speed v in a particular medium is called the **index of refraction** of the medium. The greater the index of refraction, the greater the extent to which a light beam is deflected on entering or leaving the medium (Fig. 22.26). The symbol for index of refraction is n, so that

$$n = \frac{c}{v} \qquad \textit{Index of refraction} \quad (22.8)$$

Table 22.1 is a list of the values of n for a number of substances.

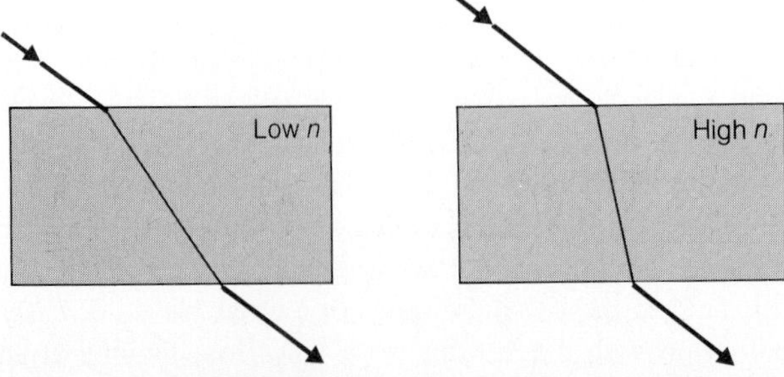

Figure 22.26

The greater the index of refraction n of a medium, the greater the deflection of a light beam on entering or leaving it.

Table 22.1 Indexes of refraction. The values of n vary slightly with the frequency of the light.

Substance	n	Substance	n
Air	1.0003	Glass, flint	1.63
Benzene	1.50	Ice	1.31
Carbon disulfide	1.63	Lucite™ and Plexiglas™	1.51
Diamond	2.42	Quartz	1.46
Ethyl alcohol	1.36	Water	1.33
Glass, crown	1.52	Zircon	1.92

It is easy to rewrite Snell's law in terms of the indexes of refraction n_1 and n_2 of two successive media. In these media light has the respective speeds

$$v_1 = \frac{c}{n_1} \quad \text{and} \quad v_2 = \frac{c}{n_2}$$

and so Snell's law in Eq. (22.7) becomes

$$\frac{\sin i}{\sin r} = \frac{v_1}{v_2} = \frac{c/n_1}{c/n_2} = \frac{n_2}{n_1}$$

This is usually written in the form

$$n_1 \sin i = n_2 \sin r \qquad \textit{Snell's law} \qquad (22.9)$$

EXAMPLE 22.3

A beam of parallel light enters a block of ice at an angle of incidence of 30° (Fig. 22.27). What is the angle of refraction in the ice?

SOLUTION The indexes of refraction of air and ice are, respectively, 1.00 and 1.31. From Eq. (22.9)

$$\sin r = \frac{n_{\text{air}}}{n_{\text{ice}}} \sin i = \frac{1.00}{1.31} \sin 30^\circ = 0.382$$

$$r = 22^\circ$$

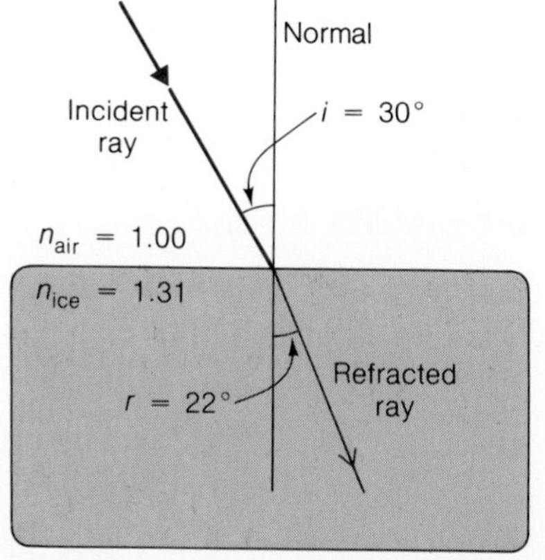

Figure 22.27

◆

EXAMPLE 22.4

A light ray is incident at an angle of 45° on one side of a glass plate of index of refraction 1.6. Find the angle at which the ray emerges from the other side of the plate.

SOLUTION The geometry of this problem is shown in Fig. 22.28. The angle of refraction r_1 at the first side of the plate is found as follows:

$$\sin r_1 = \frac{n_{\text{air}}}{n_{\text{glass}}} \sin i_1 = \frac{1.0}{1.6} \sin 45° = 0.442$$

$$r_1 = 26°$$

Because the sides of the plate are parallel, the angle of incidence i_2 at the second side is equal to r_1. If we let r_2 be the angle of refraction at which the ray emerges from the plate,

$$\sin r_2 = \frac{n_{\text{glass}}}{n_{\text{air}}} \sin i_2 = \frac{1.6}{1.0} \sin 26° = 0.701$$

$$r_2 = 45°$$

The ray leaves the glass plate parallel to its original direction but displaced to one side. This is a general result for parallel-sided plates that holds for all angles of incidence and all indexes of refraction; of course, when $i_1 = 0$, the ray is not displaced.

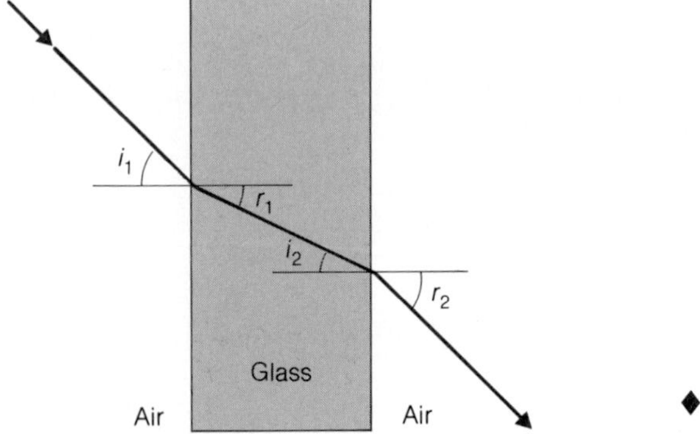

Figure 22.28

Light entering a glass plate with parallel sides emerges parallel to its original direction but is displaced to one side.

◆

EXAMPLE 22.5

A light ray enters the glass prism shown in Fig. 22.29 parallel to its base. At what angle of refraction does the ray leave the prism? The index of refraction of the glass is 1.5.

SOLUTION Because the sum of the interior angles of a triangle is 180°, each of the base angles θ of the prism is 65°. From the figure $i_1 + \theta = 90°$, so the first angle of incidence is $i_1 = 90° - 65° = 25°$. The first angle of refraction is found

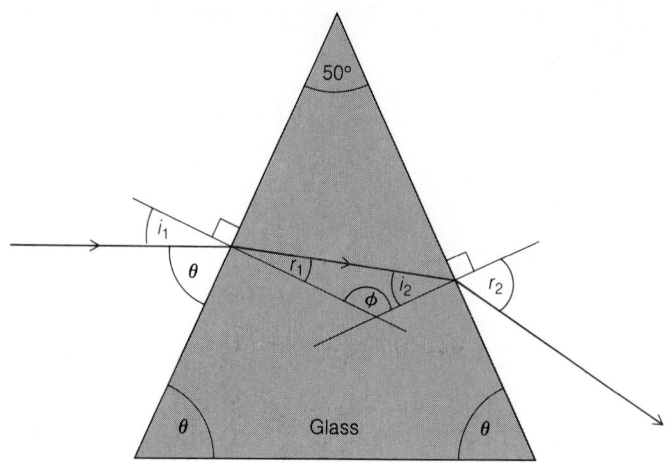

Figure 22.29

in the usual way:

$$\sin r_1 = \frac{n_{\text{air}}}{n_{\text{glass}}} \sin i_1 = \frac{1.0}{1.5} \sin 25° = 0.282$$

$$r_1 = 16°$$

From the geometry of Fig. 22.29 the angle ϕ between the two normals is $\phi = 180° - 50° = 130°$. Hence

$$r_1 + i_2 + \phi = 180° \qquad i_2 = 180° - 16° - 130° = 34°$$

and

$$\sin r_2 = \frac{n_{\text{glass}}}{n_{\text{air}}} \sin i_2 = \frac{1.5}{1.0} \sin 34° = 0.839$$

$$r_2 = 57°$$

◆

Dispersion

The index of refraction of a medium depends to some extent on the frequency of the light involved, with the highest frequencies having the highest values of n. In ordinary glass the index of refraction for violet light is about 1% greater than that for red light, for example.

Because a different index of refraction means a different deflection when a light beam enters or leaves a medium, a beam containing more than one frequency is split into a corresponding number of different beams when it is refracted. This effect, called **dispersion,** is illustrated in Fig. 22.30, which shows the result of directing a narrow pencil of white light at one face of a glass prism. The initial beam separates into beams of various colors, from which we conclude that white light is actually a mixture of light of these different colors. The band of colors that emerges from the prism is known as a **spectrum.**

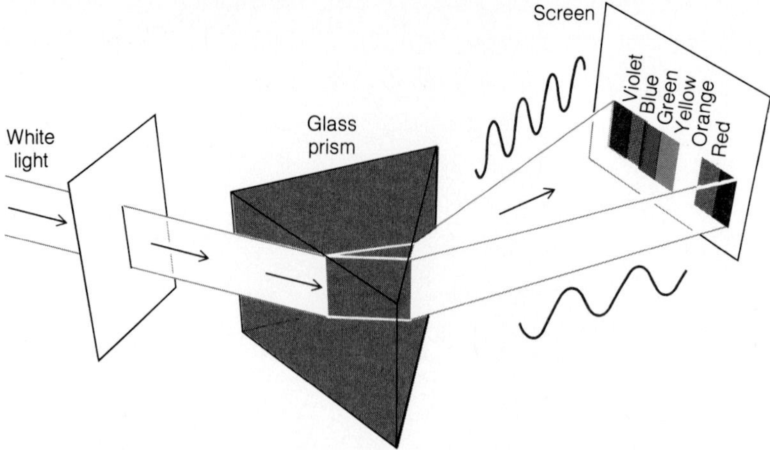

Figure 22.30

Dispersion of white light by a prism. The various colors blend into one another smoothly.

Dispersion is especially conspicuous in diamond, which is the reason for the vivid play of color when white light shines on a cut diamond. The sparkle of a cut diamond is due partly to its high refractive index and partly to the way it is cut (Fig. 22.31).

Dispersion in water droplets is responsible for rainbows, which are seen when the sun is behind an observer who is facing falling rain. Figure 22.32 shows what happens when a ray of sunlight enters a raindrop. The ray is first refracted, then reflected at the far surface, and finally refracted again when it emerges. Dispersion occurs at each refraction, with the result that the angles between the incoming sunlight and the violet and red ends of the spectrum are, respectively, 40° and 42°. Thus the light that reaches the observer comes from a ring in the sky between 40° and 42° from the direction of the sunlight. In this ring red light is on the outside because, being deviated most, it originates in droplets farther away. Between the red outside of the rainbow and the violet inside appear all the other colors of the spectrum. A person in an airplane can see the entire ring, but from the ground only the upper part appears to give the familiar arc-shaped spectrum.

Light arriving at raindrops at a larger angle than that shown in Fig. 22.32 can undergo two reflections inside each drop before emerging, which leads to an

Figure 22.31

(a) A diamond cut in the "brilliant" style has 33 facets in its upper part and 25 in its lower part. The proportions of the facets are critical in giving the maximum of sparkle. (b) Dispersion gives a cut diamond its fire.

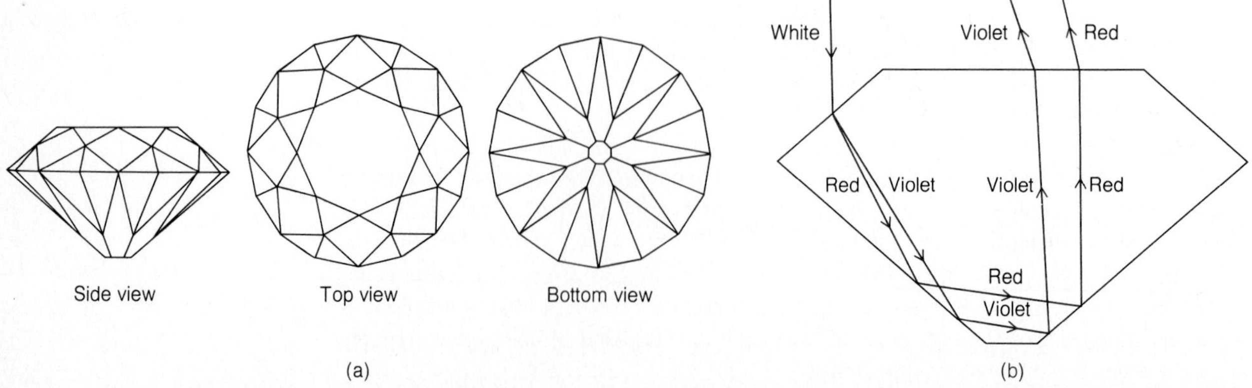

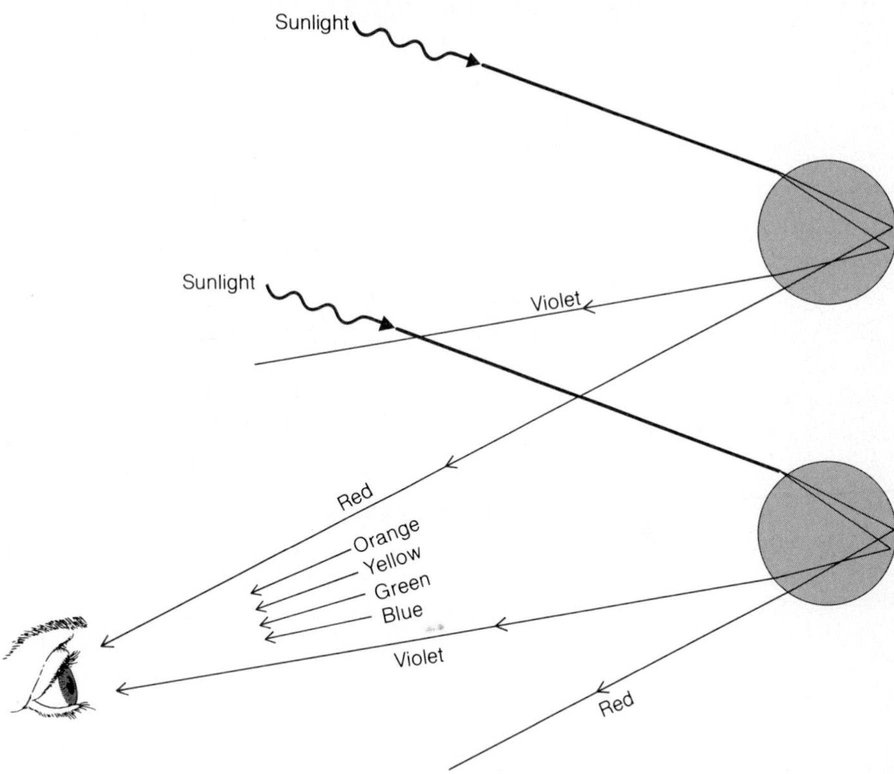

Figure 22.32

Rainbows are created by the dispersion of sunlight by raindrops. Red light arrives at the eye of the observer from the upper drop shown here; violet light arrives from the lower drop. Other raindrops yield the other colors and produce a continuous arc in the sky.

outer secondary rainbow whose colors are in the reverse order. Because some light is refracted out of a water drop at each reflection, the double reflection makes the secondary rainbow much fainter than the primary one.

22.8 Apparent Depth

Submerged objects seem closer to the surface than they actually are.

Dip an oar in a lake, and it will seem bent upward where it enters the water because its submerged portion appears to be closer to the surface than it actually is. Figure 22.33 shows how this effect comes about. Light leaving the tip of the oar at B is bent away from the normal upon entering the air. To an observer above, who instinctively interprets what he or she sees in terms of the straightline propagation of light, the tip of the oar is a C, and the submerged part of the oar seems to be AC and not AB.

It is not hard to relate the apparent and actual depths of a submerged object. Figure 22.34 shows a fish at the point F, an actual depth h below the surface of a body of water. To an observer in the air the fish is at F', only h' below the surface. If we restrict ourselves to rays that are nearly vertical,

$$\frac{\sin i}{\sin r} \approx \frac{\tan i}{\tan r} \qquad \textit{Small-angle approximation} \quad (22.10)$$

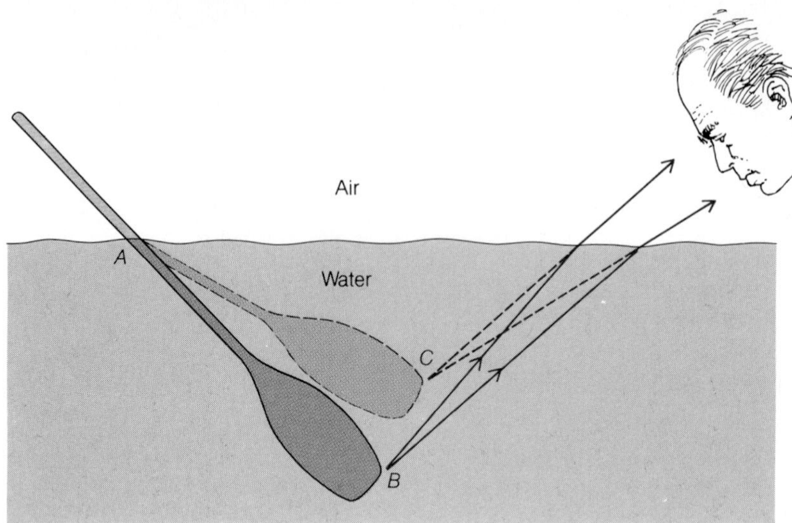

Figure 22.33

An oar appears bent when partly immersed in water because of refraction.

because $\sin \theta \approx \tan \theta$ when θ is small. From Fig. 22.34 we see that

$$\tan i = \frac{x}{h} \qquad \text{and} \qquad \tan r = \frac{x}{h'}$$

where x is the horizontal distance between the position of the fish and the point Q

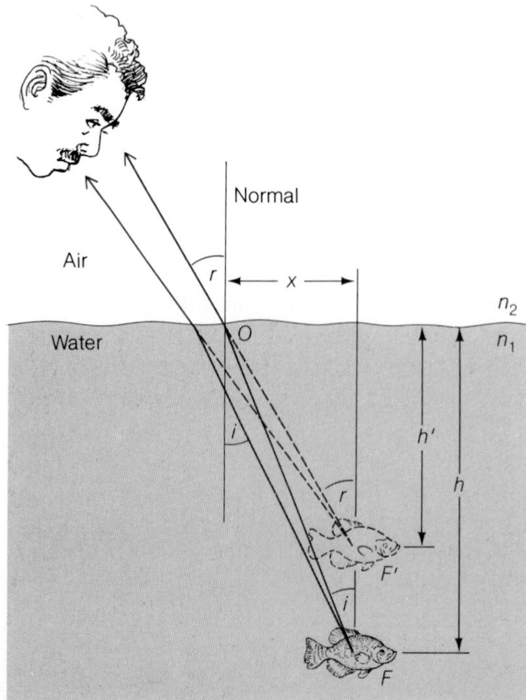

Figure 22.34

Submerged bodies seem closer to the surface than they actually are.

where the ray under consideration leaves the water. Hence

$$\frac{\tan i}{\tan r} = \frac{x/h}{x/h'} = \frac{h'}{h}$$

and so, from Eq. (22.10) and Snell's law,

$$\frac{h'}{h} = \frac{n_2}{n_1} \qquad \textit{Apparent depth} \quad (22.11)$$

EXAMPLE 22.6

To a sailor standing on the deck of a yacht, the water depth appears to be 3.0 m. If this estimate is correct in terms of what she sees, what is the true depth?

SOLUTION Here, with $n_1 = 1.33$ and $n_2 = 1.00$ (the indexes of refraction of water and air, respectively),

$$\frac{h'}{h} = \frac{1.00}{1.33} = 0.752$$

The apparent depth h' is only about three-quarters of the true depth h. Hence

$$h = \frac{h'}{0.752} = \frac{3.0\text{ m}}{0.752} = 4.0\text{ m}$$

♦

22.9 Total Internal Reflection

Past the critical angle, light is reflected rather than refracted.

An interesting effect called **total internal reflection** can occur when light goes from one medium to another whose index of refraction is lower—for instance, from glass or water to air.

In a situation of this kind, the angle of refraction is greater than the angle of incidence, and a light ray is bent *away* from the normal (ray 2 in Fig. 22.35). As the angle of incidence is increased, a certain **critical angle** i_c is reached for which the angle of refraction is 90°. The "refracted" ray now travels along the interface between the two media and cannot escape (ray 3). A ray approaching the boundary at an angle of incidence past the critical angle is reflected back into the medium it comes from, with the angle of reflection being equal to the angle of incidence as in any other instance of reflection (ray 4).

To find the value of the critical angle, we set $i = i_c$ and $r = 90°$ in Snell's law. Because $\sin 90° = 1$, we obtain

$$n_1 \sin i_c = n_2 \sin 90°$$

$$\sin i_c = \frac{n_2}{n_1} \qquad \textit{Critical angle} \quad (22.12)$$

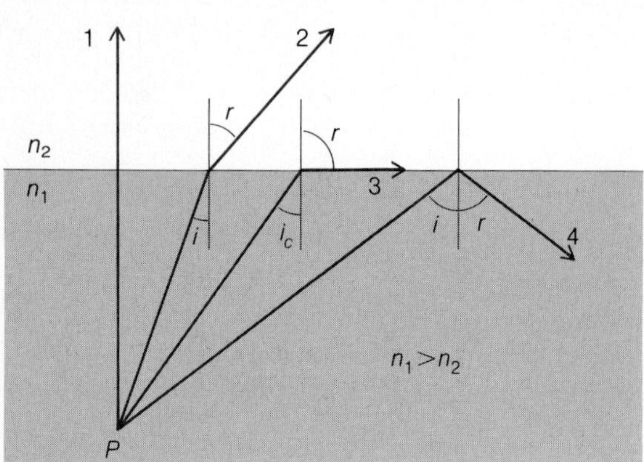

Figure 22.35

Total internal reflection occurs when the angle of refraction of a light ray going from one medium to another of lower index of refraction equals or exceeds 90°.

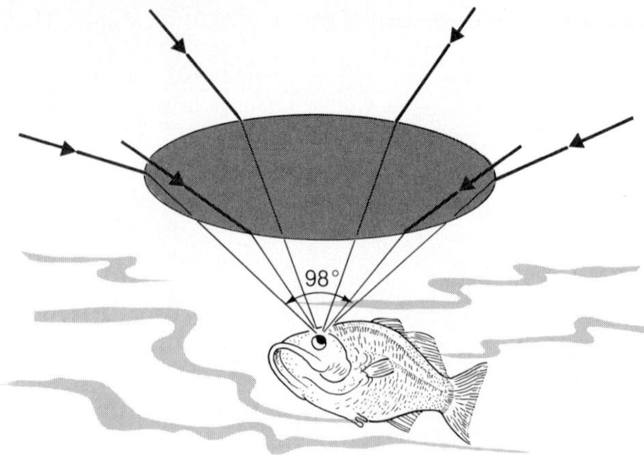

Figure 22.36

An underwater observer sees a circle of light at the surface. All the light reaching him or her is concentrated in a cone 98° wide.

For a ray going from water to air,

$$\sin i_c = \frac{n_{\text{air}}}{n_{\text{water}}} = \frac{1.00}{1.33} = 0.752$$
$$i_c = 49°$$

What does a person (or a fish) who is underwater see when looking upward? By reversing all the rays of light in Fig. 22.35 (the paths taken by light rays are always reversible), it is clear that light from everywhere above the water's surface reaches the eyes through a circle on the surface. The rays are all concentrated in a cone whose angular width is twice i_c, which is 98°. Outside this cone is darkness (Fig. 22.36).

The sharpness and brightness of a light beam are better preserved by total internal reflection than by reflection from an ordinary mirror. Optical instruments accordingly use prisms instead of mirrors whenever light is to be changed in direction. Figure 22.37 shows three types of totally reflecting prisms in common use.

Figure 22.37

Three applications of total internal reflection.

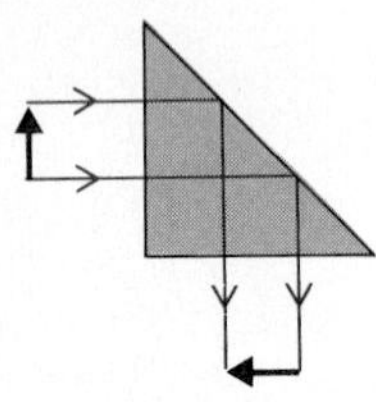

90° deflection

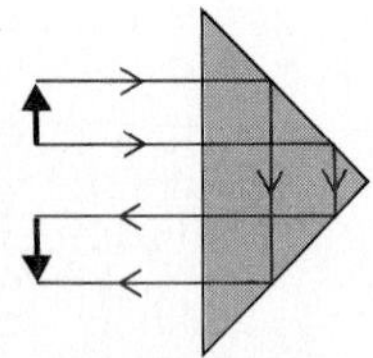

180° deflection

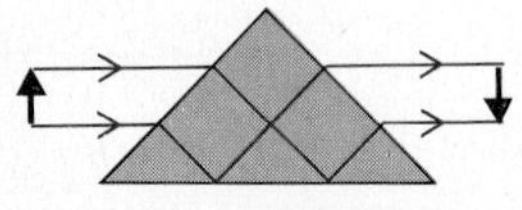

Image inversion

Fiber Optics

Total internal reflection makes it possible to ''pipe'' light from one place to another with a rod of glass or transparent plastic. As in Fig. 22.38, successive internal reflections occur at the surface of the rod, and nearly all the light entering at one end emerges at the other. If a cluster of narrow glass fibers is used instead of a single thick rod, an image can be transferred from one end to the other with each fiber carrying a part of the image. Because a tube of glass fibers is flexible, it can be used for such purposes as examining a person's stomach by passing the tube in through the mouth. Some of the fibers are used to provide light for illumination, and the others carry the reflected light back outside for viewing.

Glass fibers are coming into increasing use in telephone systems, where they have a number of advantages over electric wires for carrying information. In a fiber-optic system, the electric signals that would normally be sent directly through copper

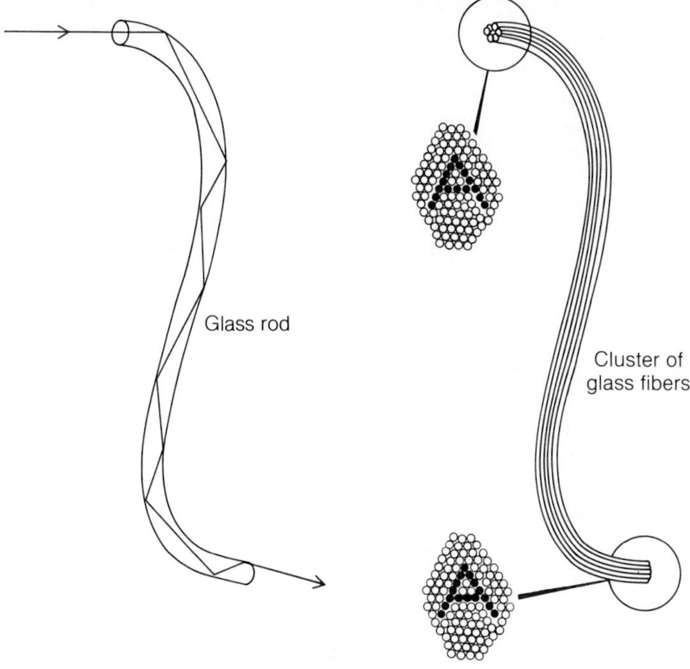

Figure 22.38

Light can be "piped" from one place to another by means of successive internal reflections in a glass rod.

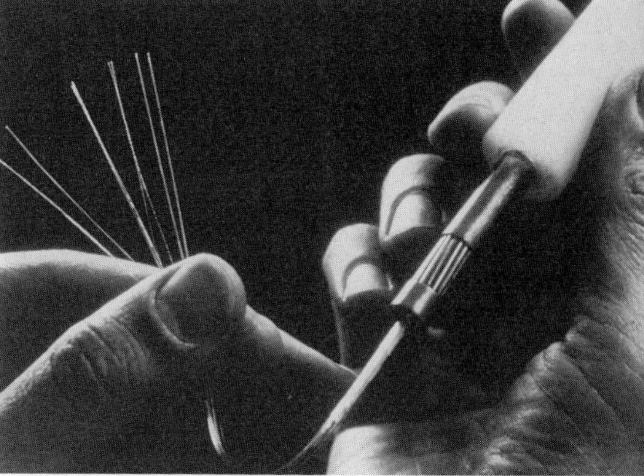

Each of the thin glass fibers in this cable can carry thousands of telephone conversations in the form of coded flashes of light.

wires are first converted into a series of pulses according to a suitable code. The pulses are then transmitted as flashes of light down a thin glass fiber (5–50 μm in diameter), and at the other end the flashes are reconverted into normal electric signals.

Modern electronic methods permit a maximum of 32 telephone conversations to be carried at the same time by a pair of wires, but the theoretical maximum for a glass fiber is in the millions. As a practical matter, several thousand conversations can today be carried on the same fiber. For instance, a six-fiber transatlantic cable over 5500 km long that links the United States and Europe can carry 40,000 conversations at the same time. Optical fibers have the further advantage that signals in them fade less rapidly than electric ones do in copper wires, and so fewer amplifiers are needed over long distances. Also, electrical interference from nearby circuits is impossible.

Important Terms

Maxwell's hypothesis is that a changing electric field produces a magnetic field.

Electromagnetic waves consist of coupled electric and magnetic field oscillations. Radio waves, microwaves, light waves, X rays, and gamma rays are all electromagnetic waves differing only in their frequency.

The **Poynting vector S** describes the flow of energy in an em wave. Its direction is that of the wave, and its magnitude is equal to the rate at which energy is being carried by the wave per unit cross-sectional area. When em waves strike a surface, they exert **radiation pressure** on it because of their momenta.

In **amplitude modulation,** information is contained in variations in the amplitude of a constant-frequency carrier wave. In **frequency modulation,** information is contained in variations in the frequency of the carrier wave.

A **wavefront** is an imaginary surface that joins points where all the waves from a source are in the same phase of oscillation. According to **Huygens' principle,** every point on a wavefront can be considered as a point source of secondary wavelets that spread out in all directions with the wave speed of the medium. The wavefront at any time is the envelope of these wavelets.

In **diffuse reflection** an incident beam of parallel light is spread out in many directions, whereas in **specular reflection** the angle of reflection is equal to the angle of incidence. A **mirror** is a specular reflecting surface of regular form that can produce an image of an object placed before it.

A **real image** of an object is formed by light rays that pass through the image; the image would therefore appear on a properly placed screen. A **virtual image** can only be seen by the eye because the light rays that seem to come from the image actually do not pass through it.

The bending of a light beam when passing from one medium to another is called **refraction.** The quantity that governs the degree to which a light beam will be deflected in entering a medium is its **index of refraction,** defined as the ratio between the speed of light in free space and its speed in the medium. **Dispersion** refers to the splitting up of a beam of light containing different frequencies by passage through a substance whose index of refraction varies with frequency.

Snell's law states that the ratio between the sine of the angle of incidence of a light ray on an interface between two media and the sine of the angle of refraction is equal to the ratio of the speeds of light in the two media.

In **total internal reflection,** light arriving at a medium of lower index of refraction at an angle greater than the **critical angle** of incidence is reflected back into the medium it came from.

Important Formulas

Poynting vector: $\bar{S} = \dfrac{E_{max}B_{max}}{2\mu_0} = \dfrac{cB_{max}^2}{2\mu_0} = \dfrac{E_{max}^2}{2c\mu_0}$

Radiation pressure: $p = \dfrac{\bar{S}}{c}$ (absorbed radiation)

$p = \dfrac{2\bar{S}}{c}$ (reflected radiation)

Index of refraction: $n = \dfrac{c}{v}$

Snell's law: $n_1 \sin i = n_2 \sin r$

Apparent depth: $\dfrac{h'}{h} = \dfrac{n_2}{n_1}$

Critical angle: $\sin i_c = \dfrac{n_2}{n_1}$

MULTIPLE CHOICE

1. A changing electric field gives rise to
a. a magnetic field **b.** electromagnetic waves
c. sound waves **d.** nothing in particular

2. In an electromagnetic wave the electric field is
a. parallel to both the magnetic field and the wave direction
b. perpendicular to both the magnetic field and the wave direction
c. parallel to the magnetic field and perpendicular to the wave direction
d. perpendicular to the magnetic field and parallel to the wave direction

3. In a vacuum the speed of an electromagnetic wave
a. depends on its frequency
b. depends on its wavelength
c. depends on its electric and magnetic fields
d. is a universal constant

4. The magnetic field of an em wave at a certain place and time has the magnitude B. The corresponding magnitude of the electric field of the wave is
a. proportional to B
b. proportional to B^2
c. proportional to $1/B$
d. unrelated to B

5. The highest frequencies are found in
a. X rays **b.** ultraviolet light
c. radio waves **d.** radar waves

6. The quality in sound that corresponds to color in light is
a. amplitude **b.** resonance
c. waveform **d.** pitch

7. Most of the effects an em wave produces when it interacts with matter are due to its
a. electric field
b. magnetic field
c. speed
d. spectrum

8. All real images
a. are erect
b. are inverted
c. can appear on a screen
d. cannot appear on a screen

9. When you look at yourself in a plane mirror, what you see is a

a. real image behind the mirror
b. real image in front of the mirror
c. virtual image behind the mirror
d. virtual image in front of the mirror

10. When a beam of light enters one medium from another, a quantity that never changes is its

a. direction b. speed
c. frequency d. wavelength

11. The index of refraction of a material medium

a. is less than 1
b. is equal to 1
c. is greater than 1
d. may be any of the above

12. Relative to the angle of incidence, the angle of refraction

a. is smaller
b. is the same
c. is larger
d. may be any of the above

13. A light ray enters one medium from another along the normal. The angle of refraction

a. is 0
b. is 90°
c. equals the critical angle
d. depends on the indexes of refraction of the two media

14. Light goes from medium *A* to medium *B* at an angle of incidence of 40°. The angle of refraction is 30°. The speed of light in *B*

a. is less than that in *A*
b. is the same as that in *A*
c. is greater than that in *A*
d. may be any of the above, depending on the specific medium

15. Figure 22.39 shows a ray of light going from water to air. The refracted ray is

a. *A* b. *B*
c. *C* d. *D*

16. Which of the diagrams in Fig. 22.40 could represent the path of a light ray through a glass block in air?

a. *A* b. *B*
c. *C* d. *D*

Figure 22.39

Multiple Choice 15.

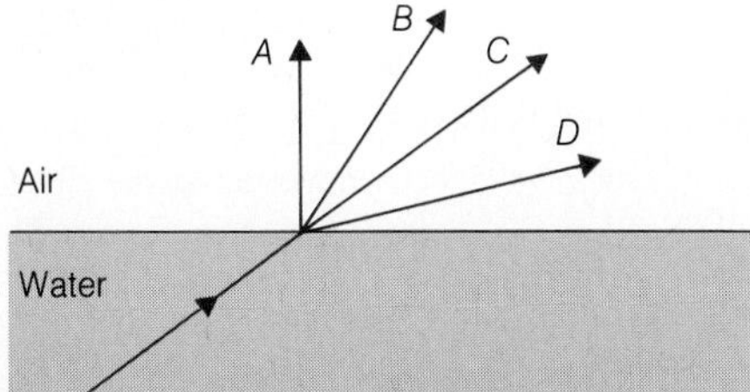

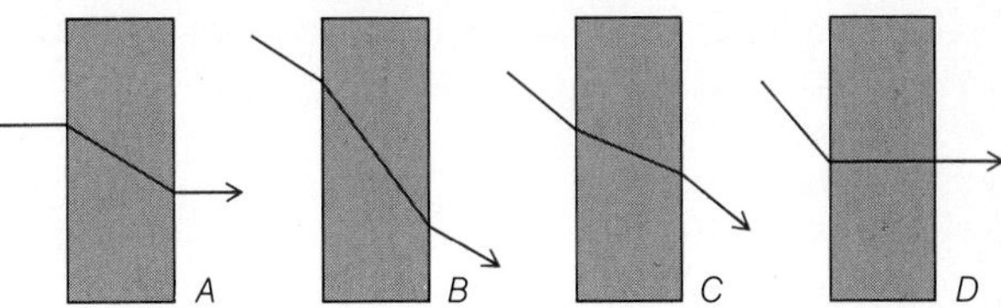

Figure 22.40

Multiple Choice 16.

17. Dispersion is the term used to describe

a. the splitting of white light into its component colors in refraction
b. the propagation of light in straight lines
c. the bending of a beam of light when it goes from one medium to another
d. the bending of a beam of light when it strikes a mirror

18. The depth of an object submerged in a transparent liquid

a. always seems less than its actual depth
b. always seems more than its actual depth
c. may seem less or more than its actual depth, depending on the index of refraction of the liquid
d. may seem less or more than its actual depth, depending on the angle of view

19. Total internal reflection can occur when light passes from one medium to another

a. that has a lower index of refraction
b. that has a higher index of refraction
c. that has the same index of refraction
d. at less than the critical angle

20. When a light ray approaches a glass-air interface from the glass side at the critical angle, the angle of refraction is

a. 0
b. 45°
c. 90°
d. equal to the angle of incidence

21. Which of the diagrams in Fig. 22.41 could represent the path of a light ray through a glass prism in air? The critical angle in glass is about 42°.

a. *A* b. *B*
c. *C* d. *D*

22. All the radiant energy from a 1.0-kW light bulb (10% efficient) is concentrated in the beam of a movie projector. If

Figure 22.41

Multiple Choice 21.

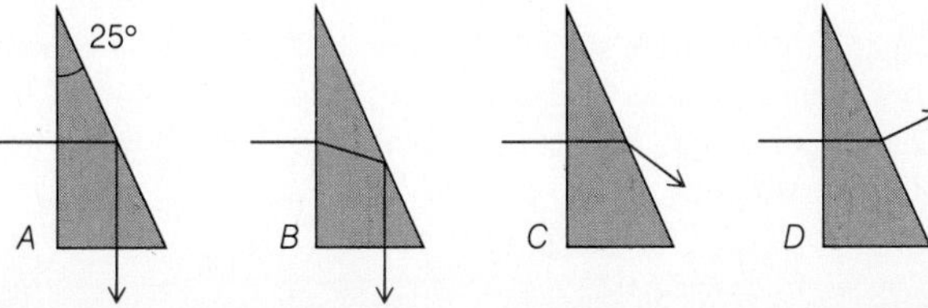

the beam is entirely reflected by the screen, the force on the screen is

a. 6.7×10^{-11} N
b. 6.7×10^{-9} N
c. 6.7×10^{-7} N
d. 6.7×10^{-5} N

23. Light enters a glass plate whose index of refraction is 1.6 at an angle of incidence of 30°. The angle of refraction is

a. 18° b. 19°
c. 48° d. 53°

24. Light leaves a slab of transparent material whose index of refraction is 2.0 at an angle of refraction of 0°. The angle of incidence is

a. 0° b. 30°
c. 45° d. 90°

25. Light enters a glass plate at an angle of incidence of 40° and is refracted at an angle of refraction of 25°. The index of refraction of the glass is

a. 0.625 b. 0.66
c. 1.52 d. 1.6

26. An underwater swimmer shines a flashlight beam upward at an angle of incidence of 40°. The angle of refraction is 60°. The index of refraction of water is

a. 0.67 b. 0.74
c. 1.3 d. 1.5

27. A fish frozen in the clear ice (n = 1.3) of a frozen lake appears to be 40 cm below the surface. Its actual depth is

a. 31 cm b. 46 cm
c. 52 cm d. 80 cm

28. The critical angle of incidence for light going from crown glass (n = 1.5) to ice (n = 1.3) is

a. 12° b. 42°
c. 50° d. 60°

QUESTIONS

1. Why was electromagnetic induction discovered much earlier than its converse, the production of a magnetic field by a varying electric field?

2. Under what circumstances does a charge radiate electromagnetic waves?

3. Why are light waves able to travel through a vacuum, whereas sound waves cannot?

4. Light is said to be a transverse wave phenomenon. What is it that varies at right angles to the direction in which a light wave travels?

5. What is the relationship between the directions of the electric and magnetic fields of an em wave? What is the relationship between the field magnitudes?

6. What is the direction of the Poynting vector of an em wave headed vertically upward with its electric field in the north-south direction? What is the direction of the magnetic field of the wave?

7. A radio transmitter has a vertical antenna. Does it matter whether the receiving antenna is vertical or horizontal? Explain.

8. What is a wavefront? What is the relationship between a light ray and the wavefronts whose motion it is used to describe?

9. Huygens' principle is useful for studying the behavior of light only in certain situations, whereas the em theory of light has a much broader scope. Huygens' principle, however, has one major advantage over the em theory of light. What do you think it is?

10. What is the significance of the ionosphere for radio communication?

11. Of the following colors, which corresponds to the lowest frequency? To the shortest wavelength? Red, yellow, blue, green.

12. The Swedish flag has a yellow cross on a blue background. How does it appear in red light? In blue light?

13. The Danish flag has a red cross on a white background. How does it appear in red light? In blue light?

14. Under what circumstances do the incident ray, the reflected ray, and the normal all lie in the same plane?

15. What is the difference between a real image and a virtual image?

16. Do electromagnetic waves have the same speed in all transparent media? If not, how do their speeds in a transparent medium compare with their speed in free space?

17. Explain why a cut diamond held in white light shows flashes of color. What would happen if it were held in red light?

18. Why is a beam of white light not dispersed into its component colors when it passes perpendicularly through a pane of glass?

19. A glass of water (n = 1.33) and a glass of ethyl alcohol (n = 1.36) seem to be filled to the same depth when viewed from above. Is the liquid level actually the same in both glasses? If not, which liquid is deeper?

20. In Fig. 22.30 the incident light is refracted at the right-hand face of the prism. Why is it not internally reflected, as in the cases shown in Fig. 22.37?

EXERCISES

22.2 Types of Em Waves

1. The marine radiotelephone station on Tahiti transmits at the frequency 4390.2 kHz. What wavelength does this correspond to?

2. A radar sends out 0.050-μs pulses of microwaves whose wavelength is 2.5 cm. What is the frequency of these microwaves? How many waves does each pulse contain?

3. A nanosecond is 10^{-9} s. (a) What is the frequency of an electromagnetic wave whose period is 1.0 ns? (b) What is its wavelength? (c) To what class of electromagnetic waves does it belong?

22.3 Poynting Vector

4. A marine radar operates at a wavelength of 3.2 cm and a power of 7.0 kW per pulse. (a) What is the frequency of the radar waves? (b) How much reaction force is exerted on the antenna when each pulse is emitted?

5. The 50-W light bulb of a slide projector is 8% efficient. If all of its luminous output is concentrated on a slide 50 mm square and the slide absorbs half the light, what is the force on the slide?

6. A radio wave exerts a pressure of 1.0×10^{-8} Pa on a reflecting surface. What are the maximum magnitudes of its electric and magnetic fields?

7. For good reception a radio wave should have a maximum electric field of at least 1.0×10^{-4} V/m when it arrives at a receiving antenna. (a) What is the maximum magnetic field of such a wave? (b) What is the magnitude of its Poynting vector? (c) What pressure does it exert when it is absorbed?

8. An aircraft is being designed to be supported against gravity by radiation pressure from a searchlight on the ground. The lower surface of the aircraft is a perfect reflector, and the entire luminous output of the searchlight is focused on it. If the aircraft's mass is 1000 kg, what is the required luminous power output of the searchlight? Does the horizontal area of the aircraft matter? Neglect practical considerations such as aiming the beam.

22.5 Reflection

9. A plane mirror is mounted on the back of a truck that is traveling at 30 km/h. How fast does the image of a man standing in the road behind the truck seem to be moving away from him?

10. What is the height of the smallest mirror in which a woman 160 cm tall can see herself at full length? Does it matter how far from the mirror she stands?

22.6 Refraction

22.7 Index of Refraction

11. Using the indexes of refraction given in Table 22.1 and taking the speed of light in free space as 3.0×10^8 m/s, find the speed of light in (a) air, (b) diamond, (c) crown glass, and (d) water.

12. The frequency of the light in a particular beam depends solely on its source, whereas the wavelength depends on the speed of light in the medium it travels through. Find the ratio between the wavelengths of a light beam that passes from one medium to another in terms of their indexes of refraction.

13. A flashlight is frozen into a block of ice. If its beam strikes the surface of the ice at an angle of incidence of 37°, what is the angle of refraction?

14. What is the angle of refraction of a beam of light that enters the surface of a lake at an angle of incidence of 50°?

15. A beam of light strikes a pane of glass at an angle of incidence of 60°. If the angle of refraction is 35°, what is the index of refraction of the glass?

16. A beam of light enters a liquid of unknown composition at an angle of incidence of 30° and is deflected by 5° from its original path. Find the index of refraction of the liquid.

17. The index of refraction of ordinary crown glass for red light is 1.51 and for violet light, 1.53. A beam of white light falls on a cube of such glass at an incident angle of 40°. What is the difference between the angles of refraction of the red and the violet light?

18. A ray of light passes through a plane boundary separating two media whose indexes of refraction are $n_1 = 1.5$ and $n_2 = 1.3$. (a) If the ray goes from medium 1 to medium 2 at an angle of incidence of 45°, what is the angle of refraction? (b) If the ray goes from medium 2 to medium 1 at the same angle of incidence, what is the angle of refraction?

19. A double-glazed window consists of two panes of glass separated by an air space for better insulation. (a) Assuming that the glass has an index of refraction of 1.5, find the angle at which a ray of sunlight enters a room if it arrives at the window at an angle of incidence of 70°. (b) The window leaks, and the space between the panes becomes filled with water. What is the angle at which the ray enters the room now?

20. A layer of benzene is floating on water. Find the angle of refraction in the water of a light ray whose angle of incidence from air to the benzene is 45°.

21. A glass prism has an apex angle of 60° and an index of refraction of 1.45. If a light ray enters the prism parallel to its base, at what angle of refraction does it leave the prism?

22. The prism in Exercise 21 is placed in water. Find the angle of refraction in this case.

22.8 Apparent Depth

23. The olive in a martini cocktail ($n = 1.35$) appears to be 40 mm below the surface. What is the actual depth of the olive?

24. A glass paperweight in the form of a 60-mm cube is placed on a letter. How far underneath the top of the cube does the letter appear if the index of refraction of the glass is 1.5?

25. A barrel of ethyl alcohol is 80 cm high. How high does it appear to somebody looking into it from above?

26. A glass block has the dimensions 10 cm × 10 cm × 16 cm. When it rests on one of its 10-cm × 10-cm faces and is examined from above, the block appears to be a 10-cm cube. What is the index of refraction of the glass?

22.9 Total Internal Reflection

27. Find the critical angle for total internal reflection in a diamond when the diamond is (a) in air, and (b) immersed in water.

28. The critical angle for total internal reflection in Lucite is 41°. Find its index of refraction.

29. Snell's law also applies to the refraction of sound waves. (a) Can sound waves be totally reflected when they reach an air-water interface from the air side? From the water side? (b) Find the critical angle for the total reflection of sound waves that arrive at an air-seawater interface. The speed of sound is 343 m/s in air and 1531 m/s in seawater.

PROBLEMS

30. The intensity of solar radiation varies inversely with the square of the distance from the sun. The earth is an average of 1.5×10^8 km from the sun. (a) How close to the sun should the earth be for the solar radiation pressure to be 1.0 Pa? Assume the radiation arrives as a single wave and is totally absorbed. (b) What would the corresponding maximum electric and magnetic energy densities be for the wave at that distance?

31. A ray of light strikes a glass plate at an angle of incidence of 55°. Some of the light is reflected and the rest is refracted. If the reflected and refracted rays are perpendicular to each other, what is the index of refraction of the glass?

32. A light ray enters one end of a glass fiber at an angle of incidence of θ, as in Fig. 22.42. (a) If the index of refraction of the glass is n, what is the maximum value of θ that will permit the ray to be totally reflected from the wall of the fiber? (b) What is the value of θ_{max} for $n = 1.35$?

33. A light ray enters a prism of apex angle 65° and index of refraction 1.5 at an angle of incidence of 40°, as in Fig. 22.43. What is the total deviation θ_D of the ray?

34. A light bulb is on the bottom of a swimming pool 1.8 m below the water surface. A person on a diving board above the pool sees a circle of light. What is its diameter?

Figure 22.42

Problem 32.

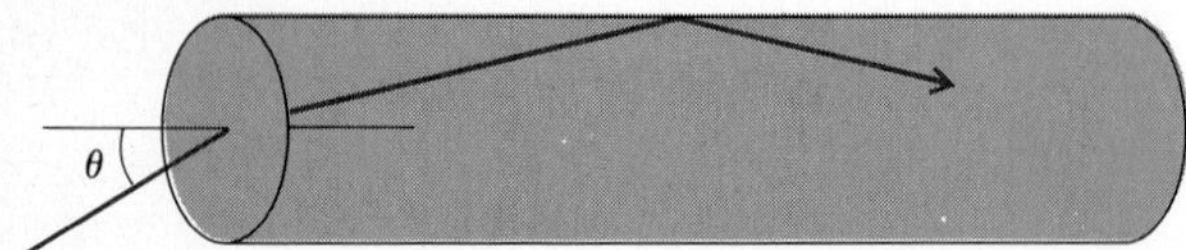

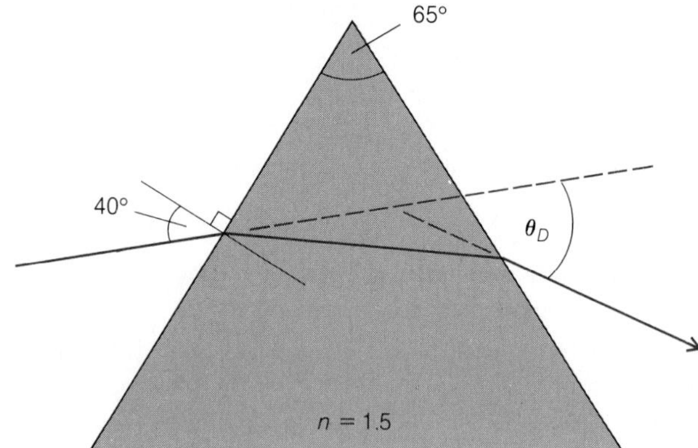

Figure 22.43

Problem 33.

35. Prisms are used in optical instruments such as binoculars instead of mirrors because total internal reflection better preserves the sharpness and brightness of a light beam. Find the minimum index of refraction of the glass to be used in a prism that is meant to change the direction of a light beam by 90°.

Answers to Multiple Choice

1. a	8. c	15. d	22. c
2. b	9. c	16. c	23. a
3. d	10. c	17. a	24. a
4. a	11. c	18. a	25. c
5. a	12. d	19. a	26. c
6. d	13. a	20. c	27. c
7. a	14. a	21. c	28. d

CHAPTER

23

LENSES AND MIRRORS

A lens is a piece of glass or other transparent material so shaped that it can produce an image by refracting light that comes from an object. The image may be real or virtual; erect or inverted; and larger, smaller, or the same size as the object. Lenses have many uses: in eyeglasses to improve vision, in cameras to record scenes, in projectors to show images on a screen, in microscopes to enable us to see small things, in telescopes to enable us to see distant things, and so forth. Mirrors that have properly curved reflecting surfaces also can form images of objects. As we will find, the optics of lenses and those of mirrors have many similarities.

23.1 Lenses

A converging lens has a positive focal length, a diverging lens has a negative focal length.

A simple way to appreciate how a lens produces its effects is to begin with a pair of prisms, as in Fig. 23.1(a). When the prisms have their bases together, parallel light that approaches is deviated so that the various rays intersect. They do not intersect at a single focal point, however, because each prism merely changes the direction of the rays without affecting the shape of the wavefronts. If properly curved rather than flat surfaces are employed, as in Fig. 23.1(b), the result is called a **converging lens,** and it acts to bring an incoming parallel beam of light to a single focal point F. Because the rays pass through F, the focal point is real.

A **diverging lens** may similarly be thought of as a development of a pair of prisms with their apexes together (Fig. 23.2). The prisms spread incoming parallel light outward, but the diverging rays cannot be projected back to a single point. The corresponding lens has curved surfaces so shaped that the diverging rays appear to originate from a single point, the virtual focal point F.

Figure 23.3 shows how lenses affect wavefronts. The speed of light in glass (or other transparent material) is less than in air, and light takes more time to pass through a certain thickness of glass than through the same thickness of air. Light that enters the center of a converging lens is thus retarded more than light that enters toward the edge. The result is that the wavefronts converge after passing through the lens. Exactly the opposite effect occurs with a diverging lens.

A lens with spherical surfaces, or one plane and one spherical surface, gives an undistorted image of something in front of it, provided the lens is relatively thin.

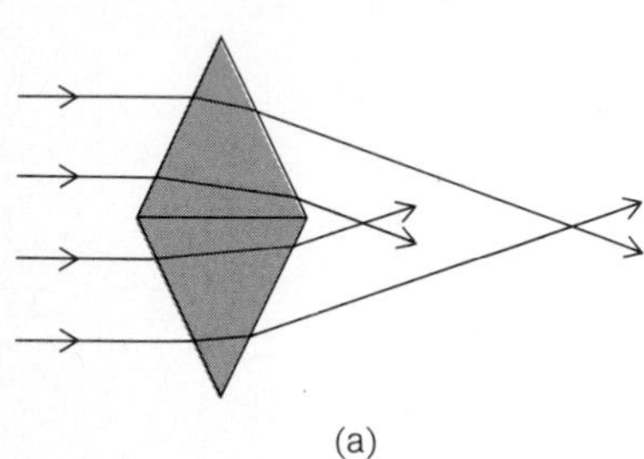

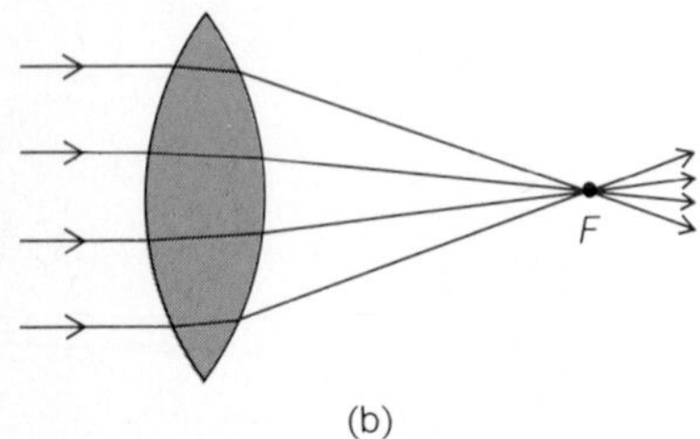

Figure 23.1

A converging lens may be thought of as a development of two prisms placed base to base. Such a lens brings parallel rays to the real focal point F, which is real because the refracted rays pass through it.

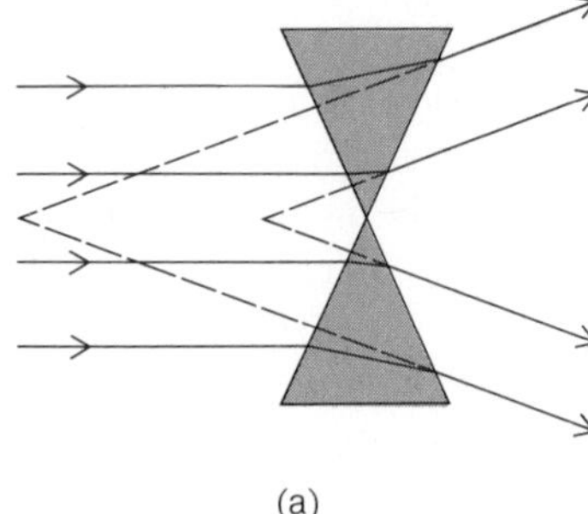

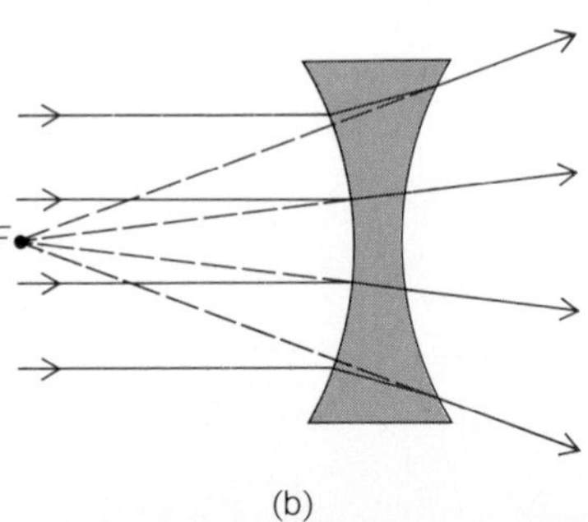

Figure 23.2

A diverging lens may be thought of as a development of two prisms placed apex to apex. Such a lens spreads out parallel rays as though they came from the virtual focal point F, which is virtual because the refracted rays do not pass through it.

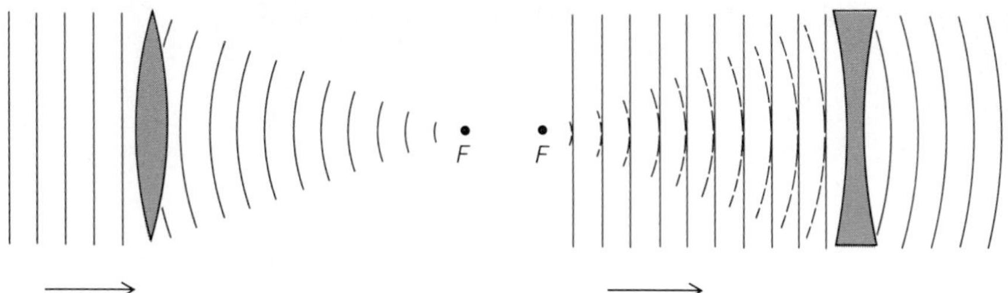

Figure 23.3
How converging and diverging lenses affect plane wavefronts.

Such lenses are also the easiest to manufacture. **Aspherical** (nonspherical) lens surfaces are used in special applications to avoid certain kinds of distortion that occur with spherical lenses. The various forms of simple lenses are shown in Fig. 23.4. Converging lenses are always thickest in the center, whereas diverging lenses are always thinnest in the center. A meniscus lens has one concave and one convex surface.

Focal Length

The distance from a lens to its focal point is called its **focal length** f. The focal length of a particular lens depends both on the index of refraction n of its material relative to that of the medium it is in and on the radii of curvature R_1 and R_2 of its surfaces.

A **thin lens** is one whose thickness is small compared with R_1 and R_2. The **lensmaker's equation** holds for such a lens:

$$\frac{1}{f} = (n - 1)\left(\frac{1}{R_1} + \frac{1}{R_2}\right) \qquad \textit{Lensmaker's equation} \quad (23.1)$$

In using this equation, it does not matter which surface of the lens is considered as 1 and which as 2. However, the sign given to each radius of curvature *is* important (Table 23.1). We will consider a radius as + if the surface is convex (curved outward), and as − if the surface is concave (curved inward). A plane surface has, in effect, an infinite radius of curvature, and $1/R$ for a plane surface is therefore 0. In Fig. 23.4(a) both radii are +; in (b) the first radius is infinite and the second is +; in (c) the first radius is − and the second is +; in (d) both radii are −; and so on.

Depending on its shape, a lens may have a positive or a negative focal length. A positive focal length signifies a converging lens, and a negative one signifies a diverging lens.

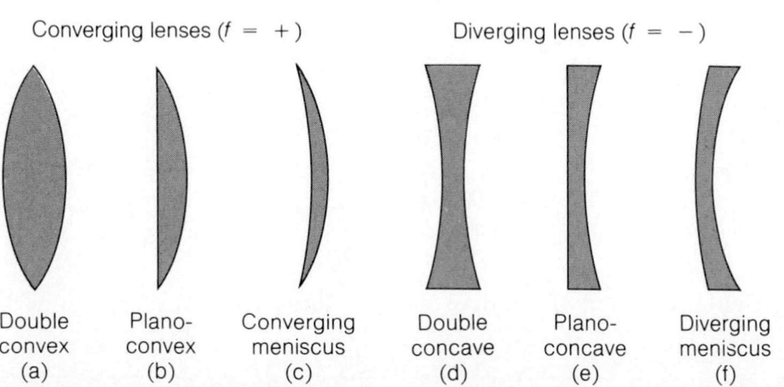

Figure 23.4
Some simple lenses. The focal length of a converging lens is reckoned positive, that of a diverging lens is reckoned negative.

Table 23.1 Sign conventions for the lensmaker's equation. A plane surface is considered to have an infinite radius, so for it $1/R = 0$.

Quantity	Positive	Negative
Radius	Convex	Concave
Focal length	Converging lens	Diverging lens

EXAMPLE 23.1

A meniscus lens has a convex surface whose radius of curvature is 25 cm and a concave surface whose radius of curvature is 15 cm. The index of refraction is 1.52. Find the focal length of the lens and whether it is converging or diverging.

SOLUTION Here $R_1 = +25$ cm and $R_2 = -15$ cm, in accord with the sign convention. From the lensmaker's equation, Eq. (23.1),

$$\frac{1}{f} = (n-1)\left(\frac{1}{R_1} + \frac{1}{R_2}\right) = (1.52 - 1.00)\left(\frac{1}{25\text{ cm}} - \frac{1}{15\text{ cm}}\right)$$
$$= (0.52)(0.040 - 0.067)\text{ cm}^{-1} = -0.014\text{ cm}^{-1}$$

and so

$$f = -71\text{ cm}$$

The negative focal length indicates a diverging lens. ♦

EXAMPLE 23.2

The lens in Example 23.1 is placed in water, whose index of refraction is 1.33. Find its focal length there.

SOLUTION The index of refraction of the glass relative to water is

$$n' = \frac{\text{index of refraction of glass}}{\text{index of refraction of water}} = \frac{1.52}{1.33} = 1.14$$

From the lensmaker's equation, since R_1 and R_2 are the same in both air and water, the ratio between the focal length f' of the lens in water and its focal length f in air is

$$\frac{f'}{f} = \frac{n-1}{n'-1} = \frac{1.52 - 1}{1.14 - 1} = 3.7$$

Hence

$$f' = 3.7f = (3.7)(-71\text{ cm}) = -263\text{ cm}$$

The focal length of *any* lens made of this glass is 3.7 times longer in water than in air. ♦

23.2 Image Formation

How to find an image by tracing rays through a lens.

A scale drawing gives us a simple way to find the position, size, and nature of an image formed by a lens. What we do is consider two different light rays that come from a certain point on an object and then trace their paths until they (or their extensions) come together again after being refracted by the lens.

It is worth noting that a lens has *two* focal points, one on each side the distance f from its center. The focal point on the side of the lens from which the light comes is the **near focal point,** and the one on the other side of the lens is the **far focal point.**

The three easiest rays to trace, as shown in Fig. 23.5, are:

1. A ray that leaves the object parallel to the lens axis. When this ray is refracted by the lens, it passes through the far focal point of a converging lens or seems to come from the near focal point of a diverging lens.
2. A ray that leaves the object and passes through the near focal point of a converging lens or is aimed at the far focal point of a diverging lens. When this ray is refracted, it continues parallel to the axis.
3. A ray that leaves the object and goes through the center of the lens. If the lens is thin relative to the radii of curvature of its surfaces, this ray is not deviated.

Ordinarily only two of these rays are enough to establish the image of an object. Figure 23.6 shows how the properties of the image produced by a converging lens depend on the position of the object. As usual, solid lines represent the actual paths taken by light rays, and dashed lines represent virtual paths.

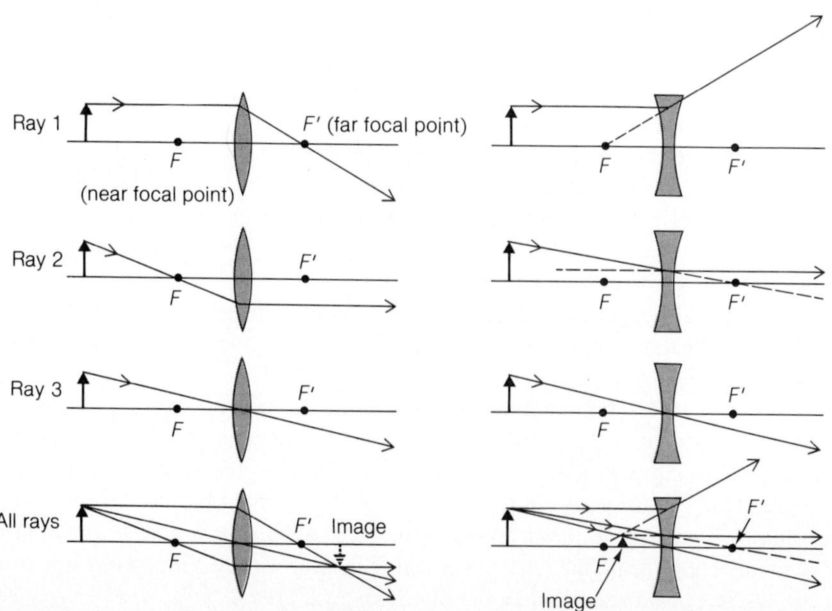

Figure 23.5

The position and size of an image produced by a thin lens can be determined by tracing any two of the rays shown. In tracing the rays, any deviations produced by the lens are assumed to occur at its central plane.

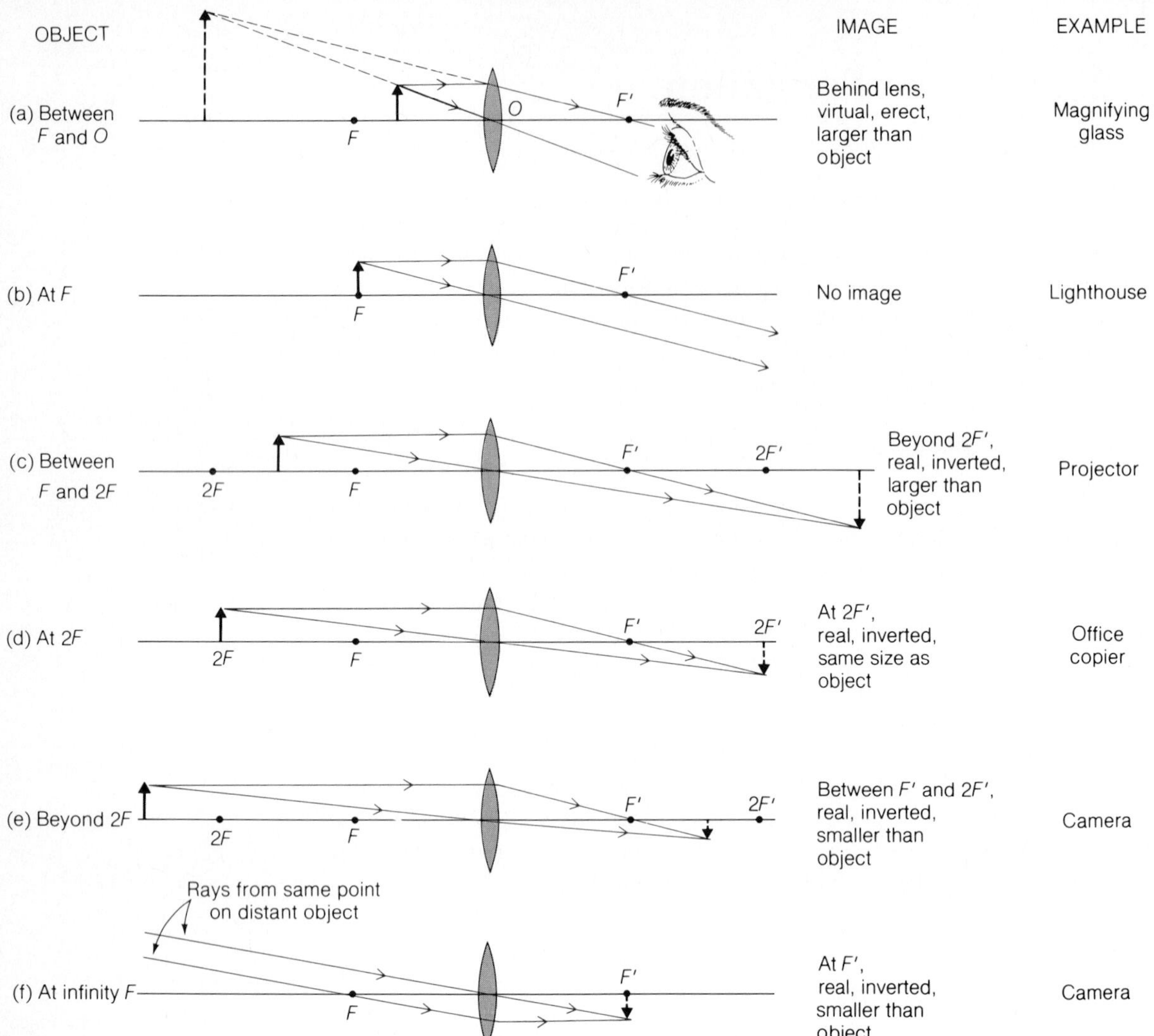

Figure 23.6

Image formation by a converging lens.

When the object is closer to the lens than F, the near focal point, as in (a), the image is erect, enlarged, and virtual. The image seems to be behind the lens because the refracted rays diverge as though coming from a point behind it. (A virtual image can be seen by the eye, but it cannot appear on a screen because no light rays actually pass through such an image.)

No image is formed of an object at the focal point, because the refracted rays are all parallel and hence never meet, as in (b). We can think of the image in this case as being at infinity.

An object farther from the lens than F always has an inverted real image that may be smaller than, the same size as, or larger than the object, depending on whether the object distance is between f and $2f$, as in (c); equal to $2f$, as in (d); or greater than $2f$, as in (e) and (f).

In contrast to the diversity of image sizes, natures, and locations produced by a converging lens, the image of a real object formed by a diverging lens is always virtual, erect, and smaller than the object (Fig. 23.7).

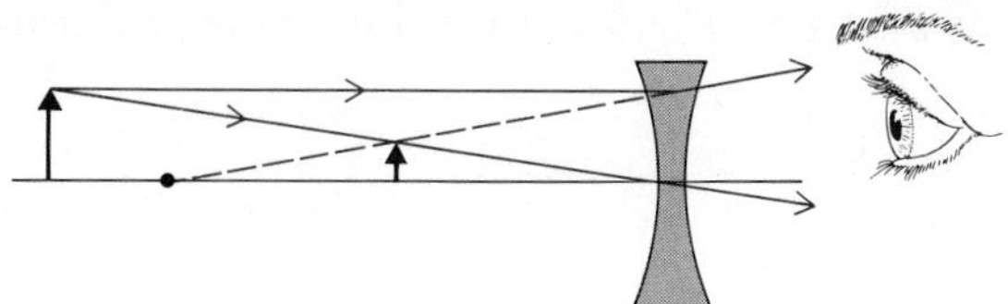

Figure 23.7

The image of a real object formed by a diverging lens is always virtual, erect, and smaller than the object.

23.3 The Lens Equation

How object and image distances are related to the focal length of a lens.

A simple formula relates the positions of the image and the object of a thin lens to the lens's focal length f. In terms of the symbols

p = distance of object from lens

q = distance of image from lens

f = focal length of lens

this formula is

$$\frac{1}{p} + \frac{1}{q} = \frac{1}{f} \qquad \textit{Lens equation} \quad (23.2)$$

When any two of the three quantities are known, the third can be found. The sign conventions to be observed when using the lens equation are as follows:

1. The focal length f is considered + for a converging lens, − for a diverging lens.
2. Object distance p and image distance q are considered + for real objects and images, − for virtual objects and images. (When the virtual image produced by a lens is used as the object of another lens, the image is said to be the virtual object of the second lens.)

The lens equation can be derived geometrically with the help of Fig. 23.8. We observe that the triangles ABO and $A'B'O$ are similar, which means that corresponding sides of these triangles are proportional. Hence

$$\frac{A'B'}{AB} = \frac{A'O}{AO} = \frac{q}{p} \qquad (23.3)$$

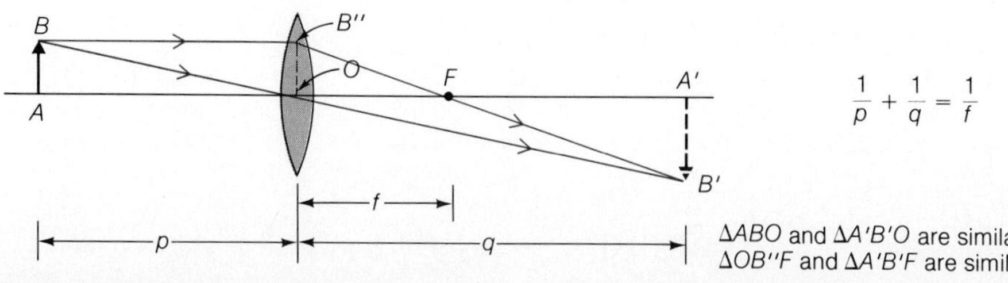

Figure 23.8

A ray diagram for deriving the lens equation.

The triangles $OB''F$ and $A'B'F$ are also similar, and

$$\frac{A'B'}{OB''} = \frac{A'F}{OF} = \frac{q - f}{f} = \frac{q}{f} - 1 \tag{23.4}$$

Since the light ray BB'' is parallel to the axis,

$$OB'' = AB$$

Therefore

$$\frac{A'B'}{OB''} = \frac{A'B'}{AB}$$

which means the right-hand sides of Eqs. (23.3) and (23.4) are equal:

$$\frac{q}{f} - 1 = \frac{q}{p}$$

Dividing each term of this equation by q gives

$$\frac{1}{f} - \frac{1}{q} = \frac{1}{p}$$

and, rearranging terms, we obtain the lens equation, Eq. (23.2):

$$\frac{1}{p} + \frac{1}{q} = \frac{1}{f}$$

The preceding derivation used a converging lens with an object distance greater than its focal length. Equation (23.2) is perfectly general, however, and holds for any object distance and for diverging lenses as well.

It is often convenient to solve the lens equation for p or q before using it in a calculation. We find that

$$\frac{1}{p} = \frac{1}{f} - \frac{1}{q} = \frac{q - f}{qf}$$

$$p = \frac{qf}{q - f} \tag{23.5}$$

and, in a similar way, that

$$q = \frac{pf}{p - f} \tag{23.6}$$

$$f = \frac{pq}{p + q} \tag{23.7}$$

If a calculator with a reciprocal [1/*X*] key is employed, Eq. (23.2) can be used directly to solve a numerical problem. The sequence of key strokes to find f, for instance, would be $(p)[1/X][+](q)[1/X][=][1/X]$.

EXAMPLE 23.3

A planoconvex lens of focal length 5.0 cm is used in a reading lamp to focus light from a bulb on a book (Fig. 23.9). If the lens is 60.0 cm from the book, how far should it be from the bulb's filament?

SOLUTION Here the focal length of the lens is +5.0 cm and the image distance is +60.0 cm. From the thin-lens equation the object distance is

$$p = \frac{qf}{q - f} = \frac{(60.0\text{ cm})(5.0\text{ cm})}{60.0\text{ cm} - 5.0\text{ cm}} = 5.5\text{ cm}$$

Figure 23.9

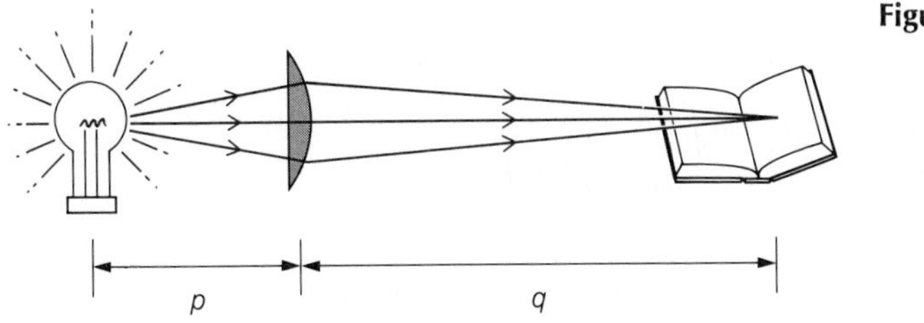

◆

EXAMPLE 23.4

Movie directors often use diverging lenses to see how scenes would appear on the screen without having to move the camera into place first. If a double-concave lens of focal length −50 cm is used for this purpose, where does the image of a scene 10 m away seem to be located?

SOLUTION Since $f = -0.50$ m and $p = 10$ m,

$$q = \frac{pf}{p - f} = \frac{(10\text{ m})(-0.50\text{ m})}{10\text{ m} - (-0.50\text{ m})} = -0.48\text{ m}$$

The image is located 48 cm behind the lens. The negative image distance signifies a virtual image, which is always on the same side of a lens as its object. This situation was pictured in Fig. 23.7. ◆

23.4 Magnification

The greater the ratio of object and image distances, the greater the magnification.

The **linear magnification** m of an optical system of any kind is the ratio between the sizes of the image and of the object. A magnification of exactly 1 means that the image and the object are the same size; a magnification of more than 1 means

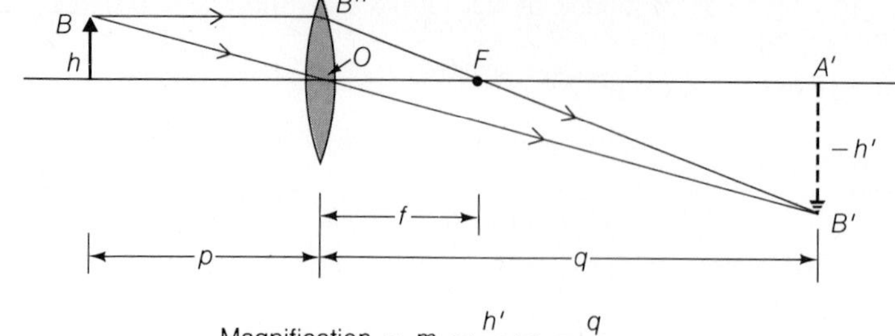

Figure 23.10

The magnification of a lens is equal to minus the ratio of image and object distances. A positive magnification means an erect image; a negative magnification means an inverted image.

that the image is larger than the object; and a magnification of less than 1 means that the image is smaller than the object. By ''size'' is meant any transverse linear dimension—for instance, height or width. Thus we can write

$$m = \frac{h'}{h}$$

$$\text{Linear magnification} = \frac{\text{image height}}{\text{object height}}$$

where h = height of object and h' = height of image. A positive m corresponds to an erect image, a negative m to an inverted image.

Figure 23.10 shows a converging lens that forms the image $A'B'$ of the object AB. The object height is $AB = h$, and that of the image is $A'B' = -h'$ because the image is inverted. The triangles ABO and $A'B'O$ are similar, and their corresponding sides are proportional. Hence

$$\frac{A'B'}{AB} = \frac{A'O}{AO}$$

Because $A'B' = -h'$, $AB = h$, $A'O = q$, and $AO = p$, we find that

$$\frac{h'}{h} = -\frac{q}{p}$$

and the magnification produced by the lens is

$$m = \frac{h'}{h} = -\frac{q}{p} \qquad \textit{Linear magnification of lens} \qquad (23.8)$$

$$\text{Linear magnification} = \frac{\text{image height}}{\text{object height}} = -\frac{\text{object distance}}{\text{image distance}}$$

This formula is a general one that holds for diverging as well as converging lenses and for any object distance.

A useful feature of Eq. (23.8) is that it automatically tells us whether an image is erect (m positive) or inverted (m negative). Table 23.2 summarizes the various sign conventions we have been using.

Table 23.2 Sign conventions for lenses

Quantity	Positive	Negative
Focal length f	Converging lens	Diverging lens
Object distance p	Real object	Virtual object
Image distance q	Real image	Virtual image
Magnification m	Erect image	Inverted image

EXAMPLE 23.5

A "magnifying glass" is a converging lens held less than its focal length from an object being examined (see Fig. 23.6a). How far should a double-convex lens whose focal length is 15 cm be held from an object to produce an erect image three times larger?

SOLUTION The required magnification is +3 since an erect image is required. From Eq. (23.8)

$$m = -\frac{q}{p} \quad \text{or} \quad q = -mp = -3p$$

Using this value of q in the thin-lens formula gives for $f = +15$ cm

$$\frac{1}{f} = \frac{1}{q} + \frac{1}{p} = -\frac{1}{3p} + \frac{1}{p} = \frac{2}{3p}$$

$$p = \frac{2}{3}f = \frac{2}{3}(15\text{ cm}) = 10\text{ cm}$$

When the lens is held 10 cm from the object, the image will be virtual, erect, and enlarged three times. ♦

EXAMPLE 23.6

A slide projector is to be used with its lens 6.0 m from a screen. If a projected image 1.5 m square of a slide 50 mm square is desired, what should the focal length of the lens be?

SOLUTION Projectors use converging lenses to produce enlarged, inverted, real images of transparencies that are illuminated from behind (Fig. 23.11). Here the

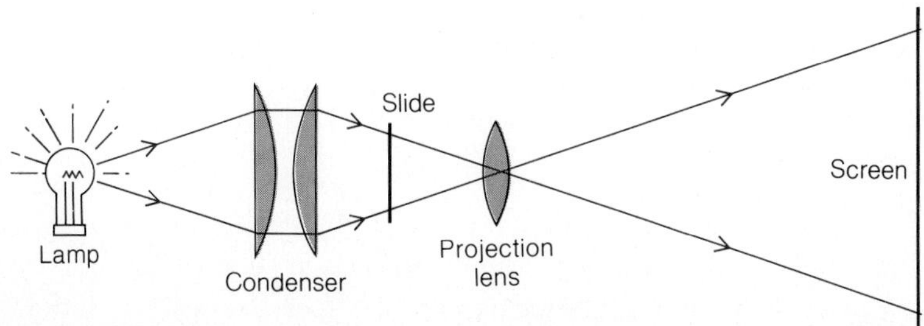

Figure 23.11

An optical projection system. The condenser causes the slide to be evenly illuminated, thereby making possible an image of uniform brightness on the screen.

image distance is given as $q = 6.0$ m, and the required magnification is

$$m = \frac{h'}{h} = \frac{-1.5\text{ m}}{0.050\text{ m}} = -30$$

The image height is considered negative as it is inverted. From Eq. (23.8)

$$p = -\frac{q}{m} = -\frac{6.0\text{ m}}{-30} = 0.20\text{ m}$$

From Eq. (23.7) we find that the lens should have a focal length of

$$f = \frac{pq}{p+q} = \frac{(0.20\text{ m})(6.0\text{ m})}{0.20\text{ m} + 6.0\text{ m}} = 0.19\text{ m} = 19\text{ cm}$$ ◆

23.5 The Camera

The larger the f-number, the smaller the aperture and the less the light reaching the film.

A camera uses a converging lens to form an image on a light-sensitive photographic film (Fig. 23.12). The distance between the lens and the film is adjustable to permit bringing an object at any distance to a sharp focus on the film. The film is exposed by opening the shutter for a fraction of a second. The shorter the exposure time, the better we can "freeze" a moving object. The longer the exposure time, the more light reaches the film and hence the better we can capture a dimly lit scene.

For a given shutter speed, the amount of light reaching the film from a scene depends on the area of the lens aperture, which is regulated by an iris diaphragm (Fig. 23.13). Apertures are specified in terms of the focal length f of the lens. Thus $f/8$ means that the diameter d of the opening is $\frac{1}{8}$ of f. The "8" in this case is referred to as the "f-number" of the aperture, so that

$$f\text{-number} = \frac{\text{focal length}}{\text{aperture diameter}} = \frac{f}{d}$$

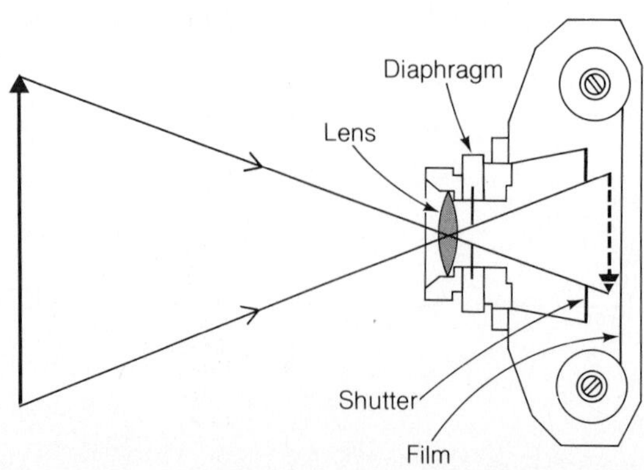

Figure 23.12

A camera. The light-sensitive film is exposed by opening the shutter for a fraction of a second. The adjustable diaphragm permits the intensity of the light entering the camera to be varied to suit the shutter speed and film used.

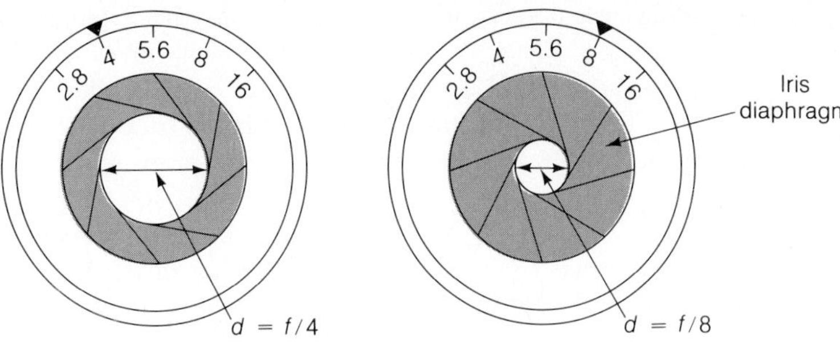

Figure 23.13

The larger the *f*-number of a camera diaphragm, the smaller the aperture. The sequence of values is such that changing the aperture by one *f*-number changes its area and the amount of light reaching the film by a factor of 2. An aperture of *f*/4 admits four times as much light as one of *f*/8.

The larger the *f*-number, the smaller the aperture and the less the light that reaches the film.

The amount of light per unit area arriving at the film depends on the ratio between the area $\pi d^2/4$ of the aperture and q^2, the square of the image distance q. (If we double q, the light that enters the camera is spread over four times the original area.) Because q is usually quite close to f in a camera, an aperture of a given *f*-number gives practically the same light per unit area on the film for a given scene whatever the focal length of the lens. This is the reason for the *f*-number scheme. The diaphragm of a camera lens is calibrated using a sequence of *f*-number values (for instance, 2.8, 4, 5.6, 8, 11, 16) such that their squares are approximately in the ratios 1:2:4:8 and so on. This means that changing the setting by one *f*-number changes the illumination of the film by the factor 2. The same amount of light reaches the film of a camera during exposures of $\frac{1}{50}$ s at *f*/8, $\frac{1}{100}$ s at *f*/5.6, and $\frac{1}{200}$ s at *f*/4.

Besides helping to govern how much light reaches the film, the size of a camera's lens aperture affects the **depth of field,** which is the range of object distances at which reasonably sharp images are produced. As Fig. 23.14 shows, reducing the aperture (increasing the *f*-number) leads to an increased depth of field. Thus more of a scene will appear sharp if a slow shutter speed and small aperture are used (provided motion in the scene is not a problem, of course).

Figure 23.14

Reducing the aperture of a lens increases the depth of field. Here arrowhead *A* is in focus on the film, whereas arrowhead *B* is in focus behind the film because it is closer to the lens. (a) With a large aperture, the image of *B* on the film is very indistinct. (b) With a small aperture, the image of *B* on the film is sharper. When a 50-mm lens is set to a distance of 5 m, objects from 4.5 to 5.6 m are in reasonably sharp focus at *f*/1.8, but at *f*/16 the range is 2.6 to 138 m.

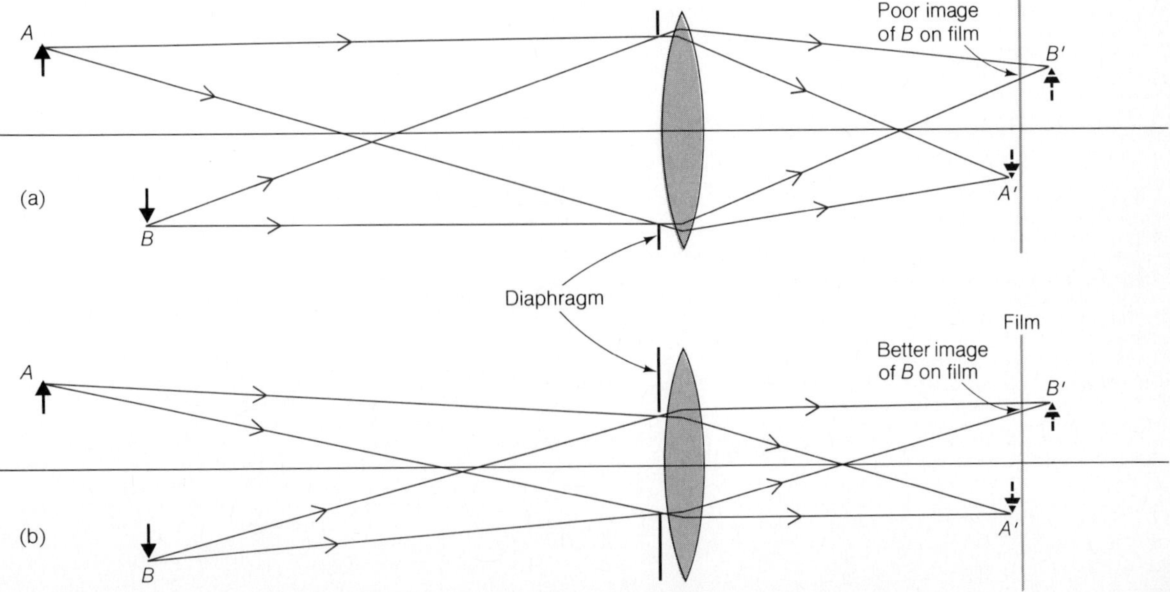

The film size and the desired angle of view are what determine the appropriate focal length of a camera lens. A "normal" lens usually gives an angle of view of about 45°. This means a focal length of 50 mm for a camera using 35-mm film on which the image size is 24 × 36 mm. A lens of shorter focal length is a **wide-angle lens** because it captures more of a given scene, though at the expense of reducing the sizes of details in the scene. Typical wide-angle lenses for a 35-mm camera have focal lengths of 35 and 28 mm, which give angles of view, respectively, of 63° and 75°. A **telephoto lens** has a long focal length to give larger images of distant objects, but this means a reduced angle of view. A 135-mm telephoto lens for a 35-mm camera has an angle of view of 18°, and it is only 8° for a 300-mm telephoto lens.

EXAMPLE 23.7

(a) What range of motion should the 50-mm lens of a 35-mm camera have if the camera is to be able to photograph objects as close as 50 cm from the lens? (b) How wide would the image on the film be of a flower 60 mm across when photographed from a distance of 50 cm? (c) What is the aperture diameter when the diaphragm is set at *f*/2.8? At *f*/16?

SOLUTION (a) Here the image distance q for an object distance of $p = 50$ cm $=$ 500 mm is

$$q = \frac{pf}{p-f} = \frac{(500 \text{ mm})(50 \text{ mm})}{500 \text{ mm} - 50 \text{ mm}} = 56 \text{ mm}$$

An object at infinity is brought to a sharp focus on the film when the image distance is equal to the focal length (see Fig. 23.6f), which in this case is 50 mm. Hence a range of adjustment of 6 mm will permit the camera to photograph objects from 50 cm away to infinity.

(b) From Eq. (23.8) the linear magnification here is

$$m = -\frac{q}{p} = -\frac{56 \text{ mm}}{500 \text{ mm}} = -0.11$$

Hence the width of the flower on the film would be

$$h' = mh = (-0.11)(60 \text{ mm}) = -6.6 \text{ mm}$$

The minus sign means that the image is inverted.

(c) At *f*/2.8 the aperture diameter of the lens is

$$d = \frac{f}{2.8} = \frac{50 \text{ mm}}{2.8} = 18 \text{ mm}$$

and at *f*/16 it is

$$d = \frac{50 \text{ mm}}{16} = 3 \text{ mm}$$

◆

23.6 The Eye

A remarkable optical instrument.

The structure of the human eye is shown in Fig. 23.15. Its diameter is typically a little less than 3 cm, and its more or less spherical shape is maintained by internal pressure. Incoming light first passes through the **cornea,** the transparent outer membrane, and then enters a liquid called the **aqueous humor.** The **lens** of the eye is a jellylike assembly of tiny transparent fibers that slide over one another when the shape of the lens is altered by the **ciliary muscle** to which it is attached by ligaments. In front of the lens is the colored **iris** whose aperture is the **pupil.** Behind the lens is a cavity filled with another liquid, the **vitreous humor,** and the lining of this cavity is the **retina.**

The retina has millions of tiny structures called **rods** and **cones** that are sensitive to light. The cones are responsible for vision in bright conditions, the rods for vision in dim conditions. Because the sensation of color is produced by the cones, colors are difficult to distinguish in a faint light. The rods and cones respond to the image formed by the lens on the retina and transmit the information to the brain through the **optic nerve.**

No photoreceptors are near the optic nerve, so that region is known as the "blind spot." The blind spot covers a field of view about 8° high and 6° wide. We are not normally aware of the blind spot for two reasons: The blind spots of the two eyes obscure different fields of vision, and the eyes are never completely at rest but are in constant scanning motion. Vision is most acute for the image focused on the **fovea,** a small region near the center of the retina. The rest of the field of view appears less distinct. Since the fovea contains only cones, it is easier to see something in a dim light by looking a bit to one side instead of directly at it.

In bright light the pupil contracts to reduce the amount of light reaching the retina, and it widens in faint light. A fully opened pupil lets in about 16 times as much light as a fully contracted one, since the respective pupil diameters are about 8 and 2 mm for a ratio of 4:1. The retina itself can also cope with a considerable range of brightnesses.

The sensitivity of the eye varies with wavelength, so light of the same intensity but of different wavelengths will give different impressions of brightness. Maximum sensitivity in bright light occurs in the middle of the visible spectrum at about $\lambda = 555$ nm, which corresponds to green (Fig. 23.16). Maximum sensitivity in dim light, when the rods rather than the cones are chiefly involved, occurs at about

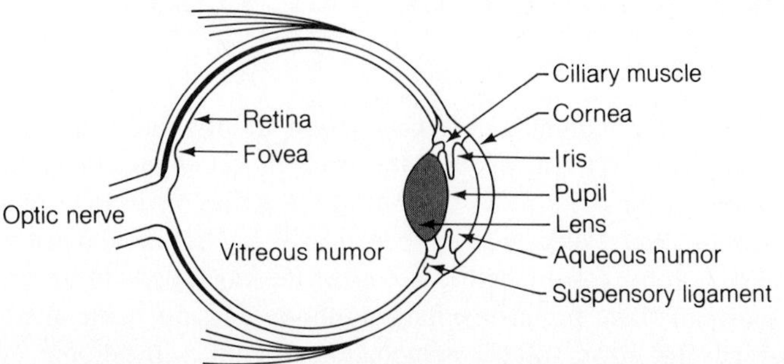

Figure 23.15

The human eye slightly larger than life-size.

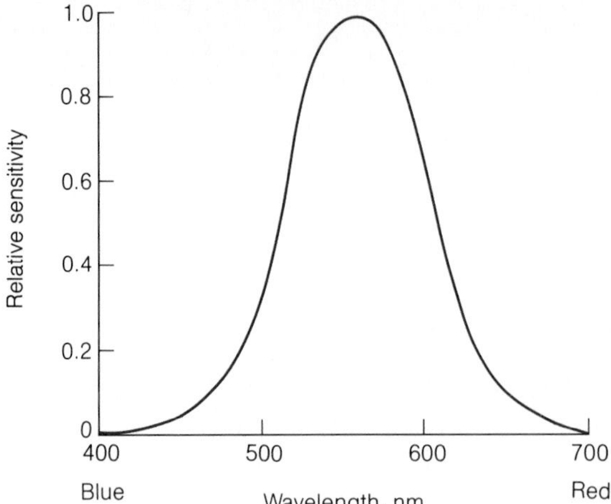

Figure 23.16

How the sensitivity of a normal human eye to bright light varies with wavelength. In dim light the wavelength of maximum sensitivity shifts from about 555 nm (green light) to about 510 nm.

510 nm. This wavelength is closer to the blue end of the spectrum and accounts for the reduced ability of the eye to respond to a weak red light.

Refraction at the cornea is responsible for most of the focusing power of the eye. The lens is less effective because its index of refraction is not very different from those of the aqueous and vitreous humors on both sides of it. The shape of the lens is controlled by the ciliary muscle. When this relaxes, the lens brings objects at infinity to a sharp focus on the retina. To permit a closer object to be viewed, the ciliary muscle forces the lens into a more convex shape whose focal length is appropriate for the object distance involved. The image distance in this situation, which is the lens-retina distance, is fixed, and focusing is thus accomplished by changing the focal length of the lens. This process is called **accommodation.** In the case of a camera, on the other hand, the focal length of the lens is fixed, and focusing is done by changing the image distance.

The limited range of accommodation of the lens is not enough to make up for the loss of refractive power that occurs when the eye is immersed in water. This loss is considerable because the index of refraction of the cornea (1.38) is close to that of water (1.33). Hence underwater objects cannot be brought to a sharp focus unless a face mask or goggles are used to keep water away from the cornea. Fish are able to see clearly when submerged because the lenses of their eyes are spherical and have high indexes of refraction. Focusing in most fish eyes is carried out by shifting the lens closer to or farther from the retina, in the same way a camera lens is focused.

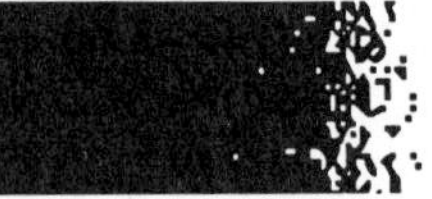

23.7 Defects of Vision

How corrective lenses work.

Two common defects of vision are **myopia** (nearsightedness) and **hyperopia** (farsightedness). In myopia, the eyeball is too long, and light from something at infinity comes to a focus in front of the retina, as in Fig. 23.17(b). Accommodation permits nearby objects to be seen clearly, but not more distant ones. A diverging lens of the proper focal length can correct this condition. In hyperopia, the eyeball is too short, and light from something at infinity does not come to a focus within the eyeball at all. Its power of accommodation permits a hyperopic eye to focus on distant

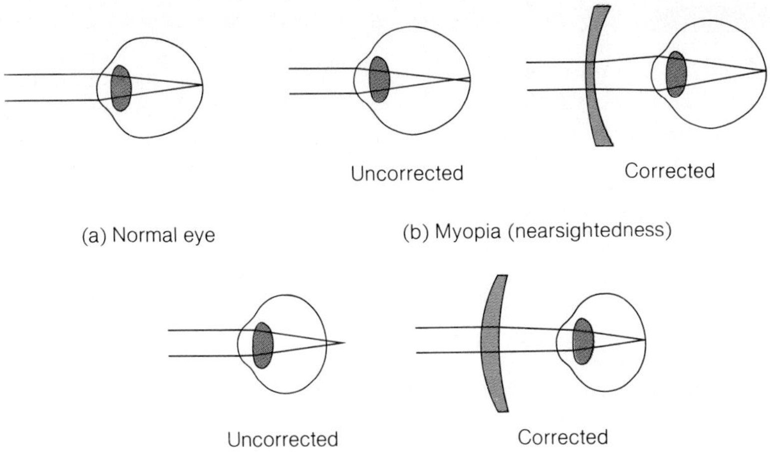

Figure 23.17

Myopia and hyperopia are common defects of vision that can be remedied with corrective lenses.

objects, but the range of accommodation is not enough for nearby objects to be seen clearly. The correction for hyperopia is a converging lens, as in Fig. 23.17(c).

Opticians often use the *power D* of a lens, expressed in **diopters,** in place of its focal length. If the focal length f is given in meters, then

$$D\,(\text{diopters}) = \frac{1}{f} \qquad \textit{Dioptric power} \quad (23.9)$$

Thus a converging lens of $f = +50$ cm $= +0.50$ m could also be described as having a power of $+2$ diopters.

EXAMPLE 23.8

A certain nearsighted eye cannot see objects distinctly when they are more than 25 cm away. Find the power in diopters of a correcting lens that will enable this eye to see distant objects clearly.

SOLUTION The purpose of the lens is to form an image 25 cm in front of the eye of an object that is infinitely far away. Because the image is to be on the same side of the lens as the object, the image must be virtual, so $q = -25$ cm. With an object distance of $p = \infty$,

$$\frac{1}{f} = \frac{1}{p} + \frac{1}{q} = \frac{1}{\infty} = \frac{1}{25\text{ cm}} = 0 - \frac{1}{25\text{ cm}}$$

$$f = -25\text{ cm} = -0.25\text{ m}$$

The minus sign means that the lens is diverging. Its power in diopters is

$$D = \frac{1}{f} = \frac{-1}{0.25\text{ m}} = -4\text{ diopters}$$

◆

EXAMPLE 23.9

A certain farsighted eye cannot see objects distinctly when they are closer than 1.00 m away. Find the power in diopters of a correcting lens that will enable this eye to read a letter 25 cm away.

SOLUTION The purpose of the lens is to form an image 1.00 m in front of the eye of an object that is 25 cm away. Since the image is to be on the same side of the lens as the object, the image must be virtual, and so $q = -1$ m. With an object distance of $p = 25$ cm $= 0.25$ m,

$$f = \frac{pq}{p+q} = \frac{(0.25\text{ m})(-1.00\text{ m})}{0.25\text{ m} + (-1.00\text{ m})} = +0.33\text{ m}$$

The plus sign means that the lens is converging. Its power in diopters is

$$D = \frac{1}{f} = \frac{1}{0.33\text{ m}} = +3\text{ diopters}$$

◆

The range of accommodation decreases with age as the lens hardens, a condition known as **presbyopia.** Distance vision is not affected, but the "near point" of the eye (the closest position at which objects can be seen distinctly) gets increasingly far away. From perhaps 7 cm at age 10, the near point recedes until it is often 2 m or more at age 60. The correction for presbyopia is a converging lens. If the range of accommodation is severely limited, more than one set of corrective lenses may be required. These are often combined in "bifocal" and "trifocal" lenses.

The closer an object is to the eye, the larger the object seems and more of its detail can be made out. Since the eye cannot focus on objects closer than the near point, this sets a limit to the magnification of which the eye is capable. The distance of most distinct vision is usually taken as 25 cm (about 10 in.), which is a comfortable object distance for most people. For purposes of calculation, an optical instrument such as a microscope or a telescope is therefore assumed to form a virtual image 25 cm behind the lens (or lens system) nearest the eye.

Astigmatism is a defect of vision caused by the cornea (or sometimes the lens) having different curvatures in different planes. When light rays that lie in one plane are in focus on the retina of an astigmatic eye, those in other planes will be in focus either in front or in back of the retina. As a result only one of the bars of a cross can be in focus at the same time (Fig. 23.18). Astigmatism is a source of eyestrain because the mechanism of accommodation continually varies the focus of the lens in an effort to produce a completely sharp image. The remedy is a corrective lens that has a cylindrical curvature, as in Fig. 23.19.

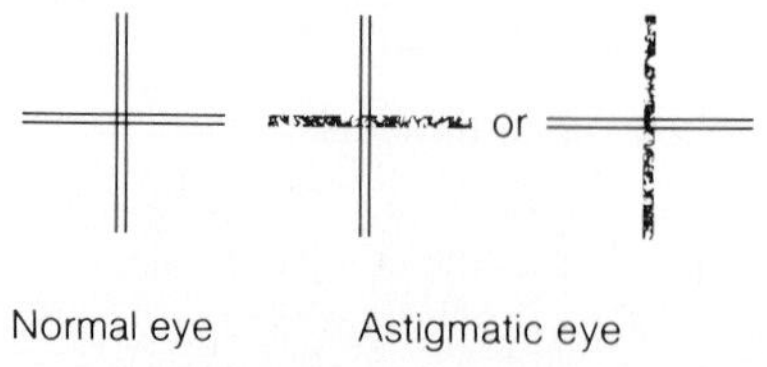

Figure 23.18

How a cross is seen by a normal eye and by an astigmatic eye.

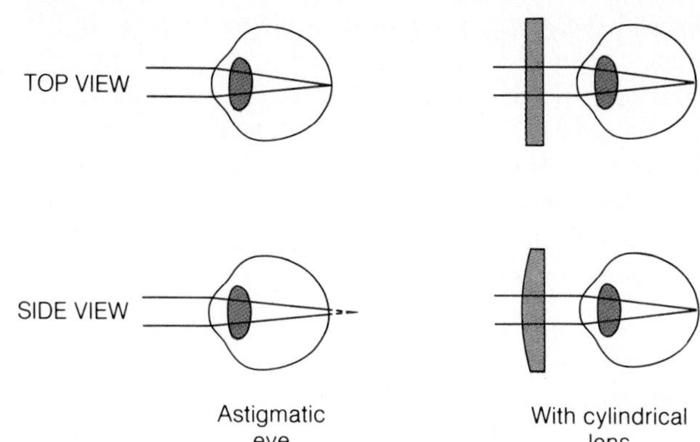

Figure 23.19

How a cylindrical lens can improve the image formed by an astigmatic eye.

23.8 The Microscope

It produces enlarged images of nearby objects.

In many applications a system of two or more lenses is better than a single lens. An ordinary magnifying glass, for instance, cannot produce images enlarged beyond 3 × or so without severe distortion. However, two converging lenses arranged as a **microscope** can give good images enlarged hundreds of times.

The optical system of a microscope is shown in Fig. 23.20. The **objective** is a lens of short focal length that forms an enlarged real image of the object. This image is further enlarged by the **eyepiece,** which acts as a simple magnifier to form a virtual final image. The image produced by the objective is thus the object of the eyepiece, and the final image has been magnified twice.

The total magnification produced by a two-lens system such as a microscope is the product of the magnification m_1 of the first lens (the objective) and the magnification m_2 of the second lens (the eyepiece):

$$m = m_1 m_2 \qquad \textit{Total magnification} \quad (23.10)$$

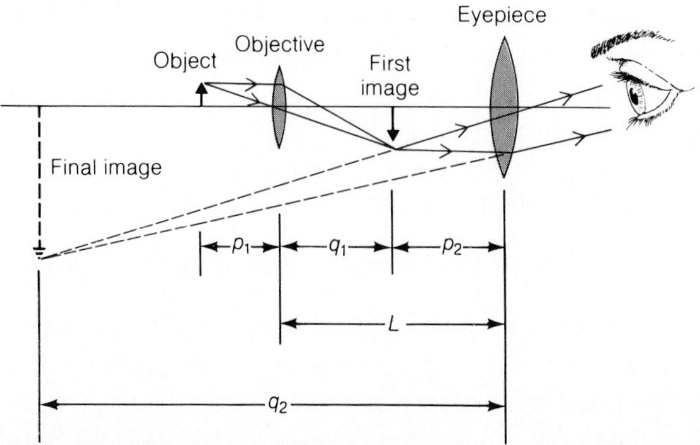

Figure 23.20

In a microscope the image formed by the objective is further magnified by the eyepiece. In this diagram the rays used to locate the final image are not the same as those used to locate the first image.

Typical objectives yield magnifications of 10× to 100×, and standard eyepiece magnifications are 5×, 10×, and 15×. Total magnifications of up to 1500 × are therefore possible with a good laboratory microscope. As discussed in the next chapter, however, the wave nature of light limits the useful magnification of a microscope to a maximum of perhaps 500×. Higher magnifications give larger images but do not show finer details. Actual microscopes use compound lenses that consist of two to six elements in place of the single lenses shown in Fig. 23.20, but their optical behavior is the same.

EXAMPLE 23.10

A microscope has an objective of focal length 4.00 mm and an eyepiece of focal length 20.0 mm. The image distance of the objective is 160 mm and that of the eyepiece is 250 mm (which are the usual figures for these quantities). (a) Find the magnification produced by each lens and by the entire microscope. (b) What is the distance between the objective and the eyepiece?

SOLUTION (a) The object distance of the objective is

$$p_1 = \frac{q_1 f_1}{q_1 - f_1} = \frac{(160\text{ mm})(4.00\text{ mm})}{160\text{ mm} - 4.00\text{ mm}} = 4.10\text{ mm}$$

and so its magnification is

$$m_1 = -\frac{q_1}{p_1} = -\frac{160\text{ mm}}{4.10\text{ mm}} = -39.0$$

The minus sign means the image is inverted.

(b) With $q_2 = -250$ mm and $f_2 = 20.0$ mm, the same procedure yields for the eyepiece

$$p_2 = \frac{q_2 f_2}{q_2 - f_2} = \frac{(-250\text{ mm})(20.0\text{ mm})}{-250\text{ mm} - 20.0\text{ mm}} = 18.5\text{ mm}$$

$$m_2 = -\frac{q_2}{p_2} = -\frac{-250\text{ mm}}{18.5\text{ mm}} = 13.5$$

(c) The magnification of the microscope is

$$m = m_1 m_2 = (-39.0)(13.5) = -527$$

The distance between the objective and the eyepiece (see Fig. 23.20) is

$$L = q_1 + p_2 = 160\text{ mm} + 18.5\text{ mm} = 178.5\text{ mm}$$

which is 179 mm to three significant figures. ♦

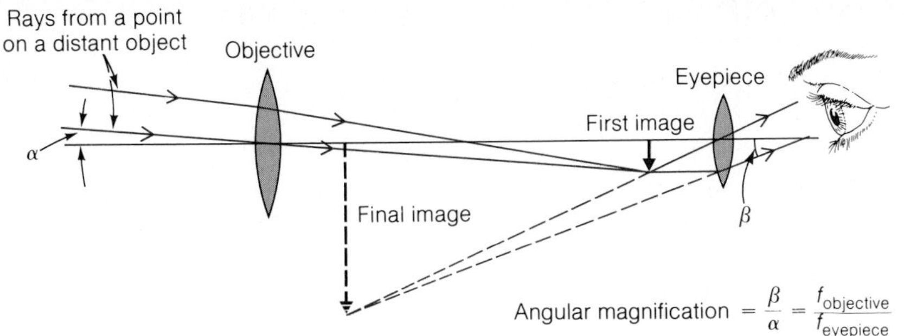

Figure 23.21

Ray diagram of a simple telescope. The rays used to locate the final image are not the same as those used to locate the first image.

23.9 The Telescope

Its angular magnification equals the ratio of focal lengths of objective and eyepiece.

A **telescope** is a lens system used to examine distant objects. As in a microscope, two lenses are involved, with an eyepiece to enlarge the image produced by the objective. A telescope objective, however, has a long focal length, whereas that of a microscope is very short.

Figure 23.21 shows a simple telescope. The image produced by the objective is real, inverted, and smaller than the object. The eyepiece then acts as a simple magnifier to form an enlarged virtual image whose object is the initial image. In practice the object distance is usually very long relative to the focal length of the objective, and the final image is smaller than the object itself. But the image seen by the eye is larger than it would be without the telescope, so the effect is the same as if the object were closer to the eye than it actually is.

Telescopes are described in terms of the **angular magnification** they produce. This quantity is the ratio between the angle β subtended at the eye by the image and the angle α subtended at the eye by the object seen directly. For distant objects ($p \gg f$) the angular magnification of a telescope is simply the ratio between the focal lengths of its objective and eyepiece:

$$m_{\text{ang}} = \frac{f_{\text{objective}}}{f_{\text{eyepiece}}} = \frac{\beta}{\alpha} \qquad \textit{Angular magnification} \quad (23.11)$$

EXAMPLE 23.11

A birdwatcher uses a telescope with an objective of focal length 60 cm and an eyepiece of focal length 1.5 cm to examine a hummingbird 5.0 cm long that is 20 m away. If the image distance of the eyepiece is 25 cm (the distance of most distinct vision), what is the apparent length of the hummingbird?

SOLUTION The angular magnification of the telescope is

$$m_{\text{ang}} = \frac{f_{\text{objective}}}{f_{\text{eyepiece}}} = \frac{60\text{ cm}}{1.5\text{ cm}} = 40$$

The angle α subtended by the bird from the location of the telescope is

$$\alpha = \frac{\text{object length}}{\text{object distance}} = \frac{0.05 \text{ m}}{20 \text{ m}} = 0.0025 \text{ radian}$$

If L is the length of the bird's image as seen through the telescope, the angle β this image subtends is

$$\beta = \frac{\text{image length}}{\text{image distance}} = \frac{L}{25 \text{ cm}}$$

Since the angular magnification of the telescope is 40, from Eq. (23.11) we have

$$\beta = m_{\text{ang}}\alpha$$

$$\frac{L}{25 \text{ cm}} = (40)(0.0025 \text{ radian})$$

$$L = 2.5 \text{ cm}$$

The hummingbird seems to be 2.5 cm long, half its actual length, and to be located 25 cm from the viewer's eye. ◆

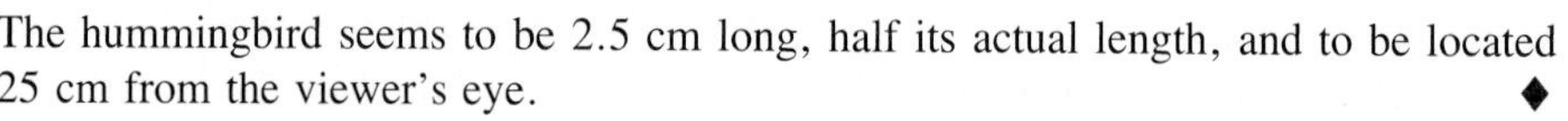

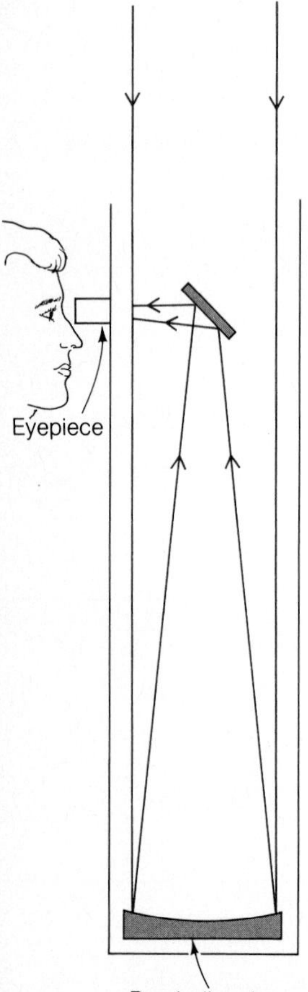

Figure 23.22
One type of reflecting telescope.

The higher the magnification of a telescope, the greater in diameter the objective must be in order to gather in enough light for the image to be visible. This sets a limit to the useful magnification of a refracting telescope, since a large glass lens tends to distort under its own weight. The largest refracting telescope in the world is the 1.02-m-diameter instrument at Yerkes Observatory in Wisconsin, whose objective has a focal length of nearly 20 m.

Modern astronomical telescopes use concave parabolic mirrors as their objectives. Such a mirror produces a real image of a distant object (Section 23.11) and can be supported from behind. Further advantages are that a parabolic mirror is not subject to the aberrations mentioned in the next section and that it needs only one surface to be ground to a precise shape. A small secondary mirror reflects the image outside the telescope tube for viewing or, more often, for photographing (Fig. 23.22). When used as a camera, a telescope needs no eyepiece—its objective lens or mirror simply acts as a giant telephoto lens.

The largest single-mirror telescope, located in the Soviet Union, has a mirror 6 m in diameter, but technical problems have prevented it from reaching its full potential. Next in size is a 5-m reflector at Mount Palomar in California that has been enlarging the horizons of astronomy since 1948. The latest generation of astronomical telescopes does not rely on single large mirrors, which practical difficulties limit to at most 8 m. Instead, a number of individual mirrors are linked to produce a single image. In one approach the Keck telescope in Hawaii uses hexagonal segments to give a collecting surface 10 m across. In another approach separate circular mirrors are used. A proposed new American telescope will have four 7.5-m mirrors mounted in one structure for the equivalent of a 15-m mirror. A third scheme, which is being developed in Europe, uses four individual 8-m telescopes whose images are added together electronically.

EXAMPLE 23.12

The diameter of the planet Mars is 6.8×10^6 m. What focal length must a telescope objective have in order to produce a photographic image of Mars 1.0 mm in diameter at a time when Mars is 8.0×10^{10} m from the earth?

SOLUTION Here $h = 6.8 \times 10^6$ m and $h' = 1.0 \text{ mm} = 1.0 \times 10^{-3}$ m, so the required magnification is

$$m = \frac{h'}{h} = -\frac{1.0 \times 10^{-3}\text{ m}}{6.8 \times 10^6\text{ m}} = -1.5 \times 10^{-10}$$

A minus sign is used because the image will be inverted. The object distance p is the Mars-earth distance of 8×10^{10} m. Since the focal length f is going to be much smaller than p, $p - f \approx p$ and the image distance q is

$$q = \frac{pf}{p - f} \approx f$$

The formula $m = -q/p$ therefore becomes $m = -f/p$ here, and

$$f = -mp = -(-1.5 \times 10^{-10})(8.0 \times 10^{10}\text{ m}) = 12\text{ m}$$

The telescope objective should have a focal length of 12 m. Of course, further enlargement of the image can be made from the negative. ◆

In stellar astronomy the purpose of a big telescope is not magnification, for the stars are too far away to ever appear as more than points of light. The virtue here of a large telescope lies in its light-gathering ability, which can reveal faint objects that would otherwise be invisible.

A telescope gives an erect image if a third lens is used between the objective and the eyepiece (Fig. 23.23). The disadvantage is that the telescope tube is longer. A better scheme uses a pair of prisms that both invert the image so that it is erect and shorten the instrument length (Fig. 23.24). This is the arrangement used in binoculars.

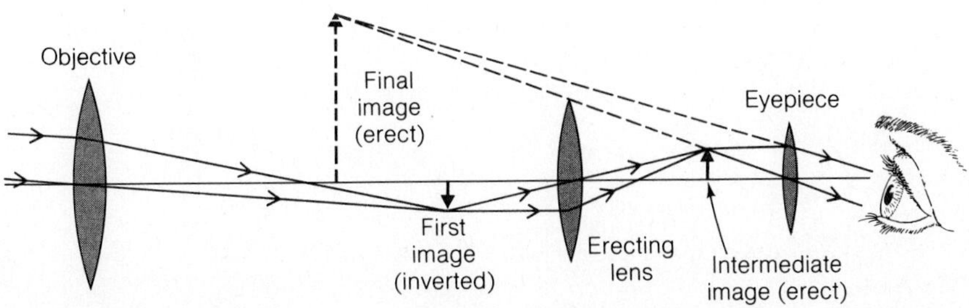

Figure 23.23
In a terrestrial telescope an intermediate lens is used to produce an erect image.

Figure 23.24

A pair of prisms is used to erect the image in each half of a prism binocular.

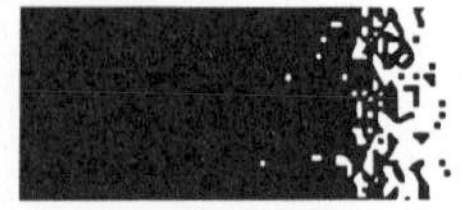

23.10 Lens Aberrations

Simple lenses do not give perfect images.

The image formed by a simple lens is never an exact replica of its object. Of the variety of imperfections to which such an image is subject, the most familiar is the presence of fringes of color around whatever is being looked at. This **chromatic aberration** occurs because the index of refraction of glass varies with wavelength. As a result the focal length of a lens is slightly different from light of different colors (Fig. 23.25a). The remedy is to combine a converging and a diverging lens made of different glass so that the dispersion produced by one is canceled by the other while leaving a net converging or diverging power. Such an **achromatic lens** is illustrated in Fig. 23.25(b).

The lens equation was derived on the basis of light rays that made only small angles with the axis. When a simple lens is used to form images of objects some distance from the axis, a variety of aberrations arise even if the light used is monochromatic. One of them is **spherical aberration,** in which rays passing near the lens rim come to a focus closer to the lens than rays near the axis (Fig. 23.26). Another is distortion of the image because the magnification of a simple lens varies with the distance of an object from the axis. By using several lenses of different types of glass and different curvatures to replace a single lens, the various aberrations can be minimized. Compound lenses that consist of two or more elements are always used in high-performance optical systems such as those in microscopes, prism binoculars, and quality cameras.

Figure 23.25

A compound lens made of different types of glass can correct for chromatic aberration.

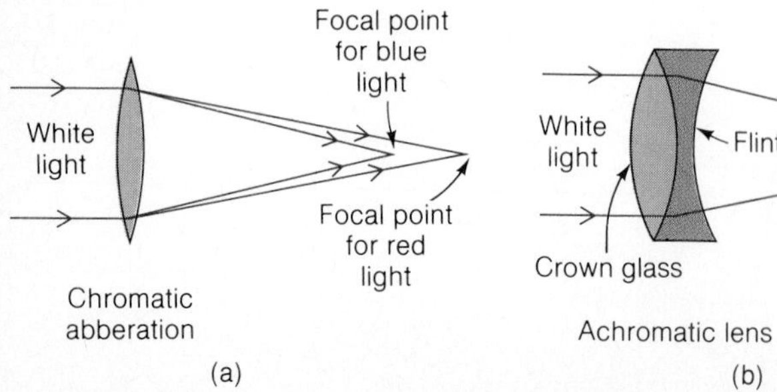

Figure 23.26

Spherical aberration.

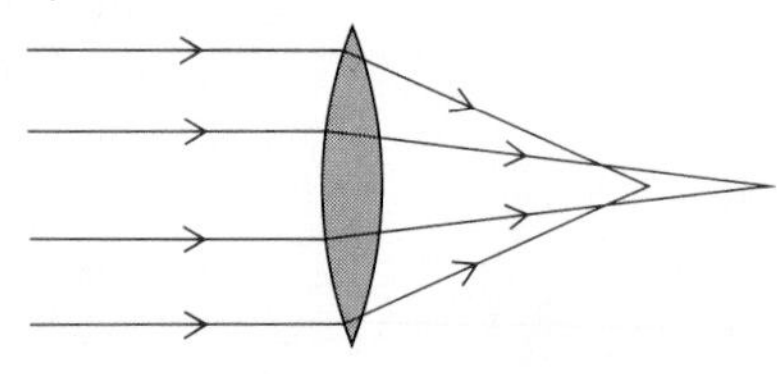

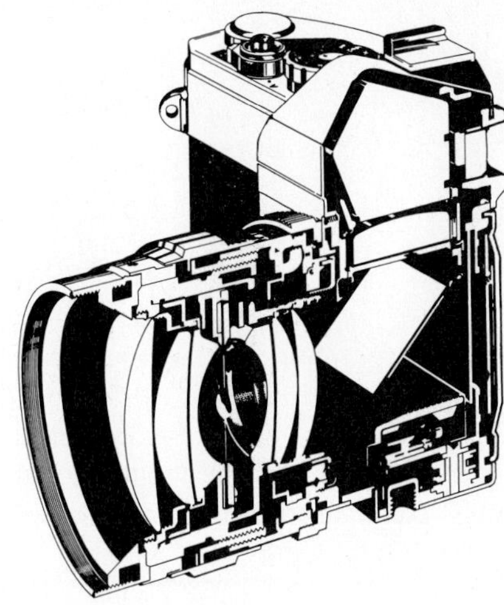

A compound lens that consists of several individual lenses, as in this camera, minimizes aberrations to produce a better image than a single lens can.

23.11 Spherical Mirrors

Concave mirrors are converging, convex mirrors are diverging.

Mirrors with spherical reflecting surfaces form images in much the same way that lenses do. Figure 23.27(a) shows how a concave mirror converges a parallel beam of light to a real focal point, and Fig. 23.27(b) shows how a convex mirror diverges such a beam so that the reflected rays seem to originate in a virtual focal point behind the mirror. If R is the radius of the reflecting surface in each case, the focal lengths of these mirrors are

$$f = \frac{R}{2} \qquad \textit{Concave mirror} \quad (23.12)$$

$$f = -\frac{R}{2} \qquad \textit{Convex mirror} \quad (23.13)$$

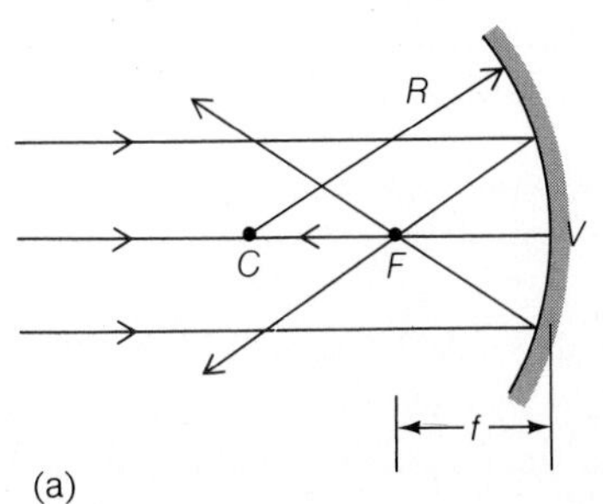

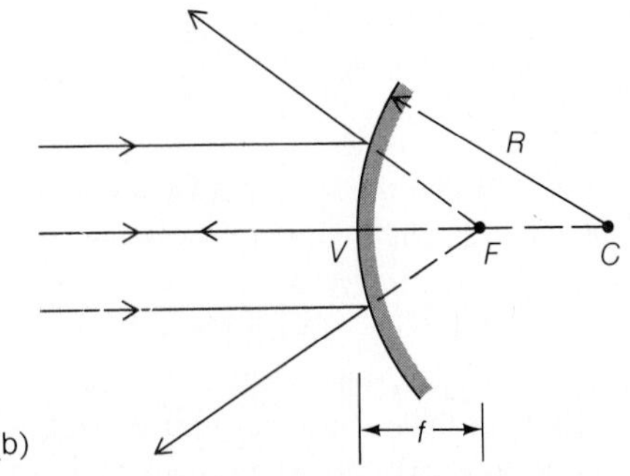

R = radius of curvature
V = vertex
C = center of curvature
F = focal point
f = focal length

Figure 23.27

(a) A concave spherical mirror converges a parallel beam of incident light. (b) A convex spherical mirror diverges a parallel beam of incident light.

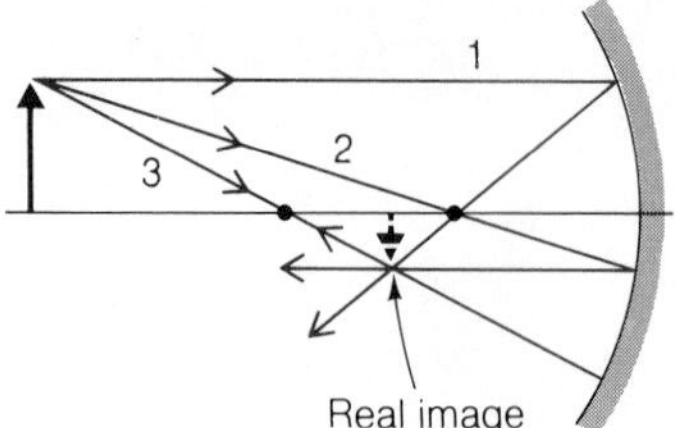

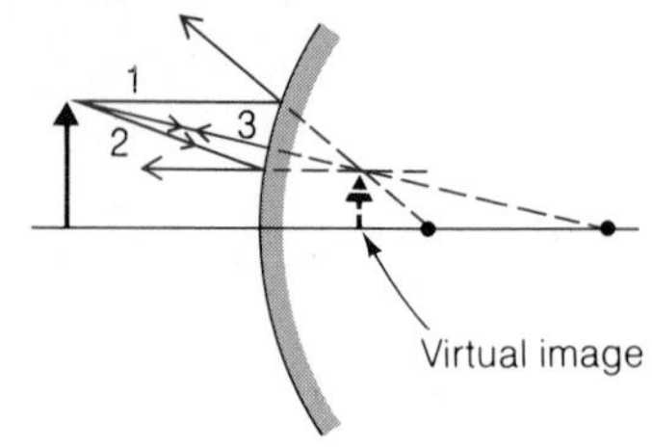

Figure 23.28

The position and size of an image produced by a spherical mirror can be determined by tracing any two of the rays shown.

The position, size, and nature of the image produced by a spherical mirror of an object in front of it can be found with the help of a scale drawing. As in the case of a lens, what we do is consider two different light rays that come from a certain point on the object and trace their paths until they (or their backward extensions) come together again after reflection. The three easiest rays to trace, as shown in Fig. 23.28, are:

1. A ray that leaves the object parallel to the mirror axis. When this ray is reflected, it passes through the focal point of a concave mirror or seems to come from the focal point of a convex mirror.
2. A ray that leaves the object and passes through the focal point of a concave mirror, or is aimed at the focal point of a convex mirror. When this ray is reflected, it proceeds parallel to the axis.
3. A ray that leaves the object along a radius of the mirror. When this ray is reflected, it returns along its original path.

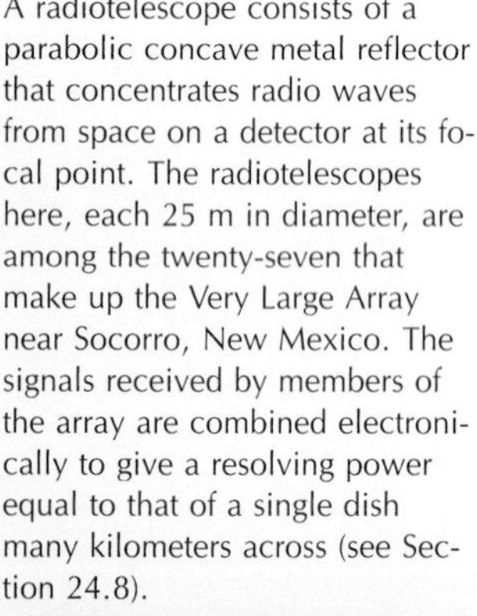
A radiotelescope consists of a parabolic concave metal reflector that concentrates radio waves from space on a detector at its focal point. The radiotelescopes here, each 25 m in diameter, are among the twenty-seven that make up the Very Large Array near Socorro, New Mexico. The signals received by members of the array are combined electronically to give a resolving power equal to that of a single dish many kilometers across (see Section 24.8).

Only two of these rays are needed to locate the image of a reflected object. Figure 23.29 shows how the properties of the image produced by a concave mirror vary with the position of the object. Solid lines represent the actual paths taken by light rays, and dashed lines represent virtual paths.

When the object is closer to the mirror than the focal point F, as in (a), the image is erect, enlarged, and virtual. The image seems to be behind the mirror because the reflected light rays from the mirror diverge as though coming from a point behind the mirror.

When the object is at the focal point, as in (b), no image is formed because the reflected rays are all parallel and so do not meet. We can say that the image in this case is at infinity.

An object between the focal point F and the center of curvature C, as in (c), has an inverted, enlarged image that is real. The image would appear on a screen placed at its position.

When the object is at C, as in (d), its image is at the same place and is the same size but is inverted. An object past C, as in (e) and (f), has a real image that is smaller and is inverted.

The image formed by a convex mirror of a real object is always erect, smaller than the object, and virtual (Fig. 23.30). The field of view of a convex mirror is wider than that of a plane mirror, which accounts for a number of its applications, such as at blind corners in roads and in rear-view mirrors in cars.

The object and image distances p and q of a spherical mirror are related to its focal length by the same formula that holds for a thin lens:

$$\frac{1}{p} + \frac{1}{q} = \frac{1}{f} \qquad \textit{Mirror equation} \quad (23.14)$$

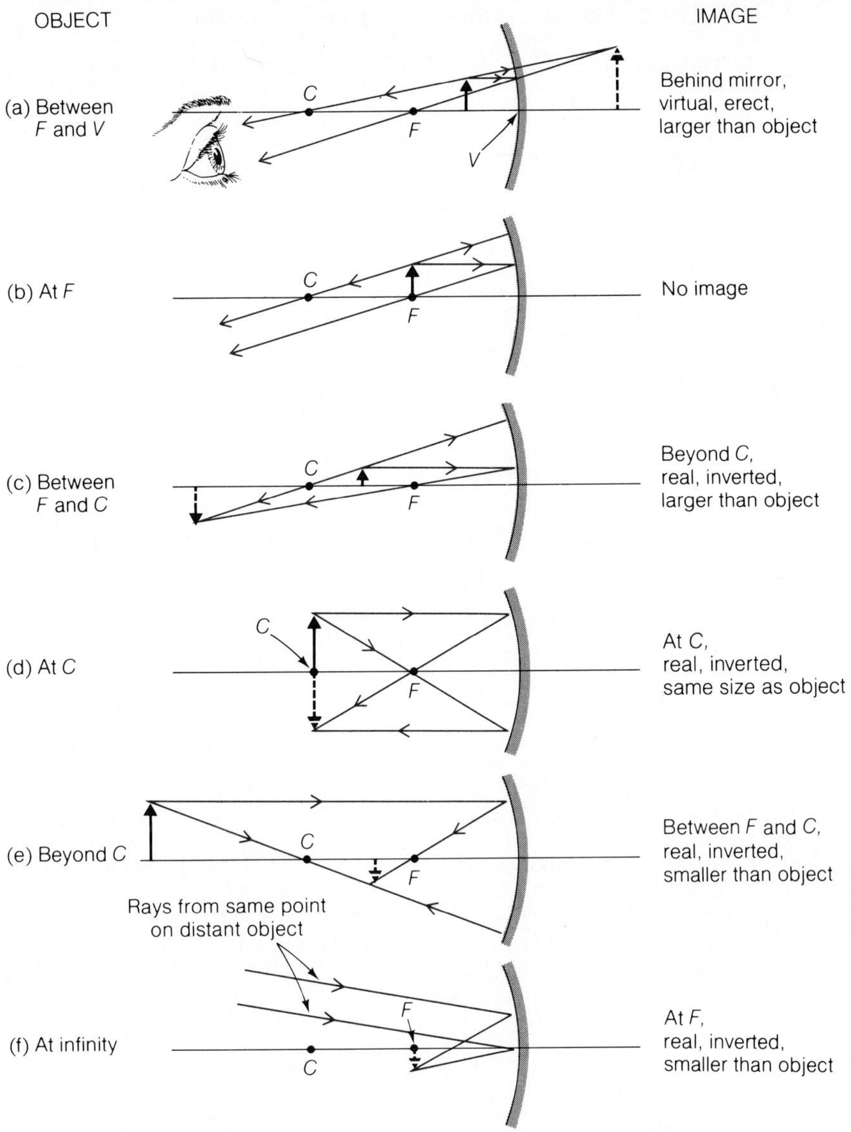

OBJECT

IMAGE

(a) Between *F* and *V*

C F V

Behind mirror, virtual, erect, larger than object

(b) At *F*

C F

No image

(c) Between *F* and *C*

C F

Beyond *C*, real, inverted, larger than object

(d) At *C*

C F

At *C*, real, inverted, same size as object

(e) Beyond *C*

C F

Between *F* and *C*, real, inverted, smaller than object

Rays from same point on distant object

(f) At infinity

F C

At *F*, real, inverted, smaller than object

Figure 23.29

Image formation by a concave mirror.

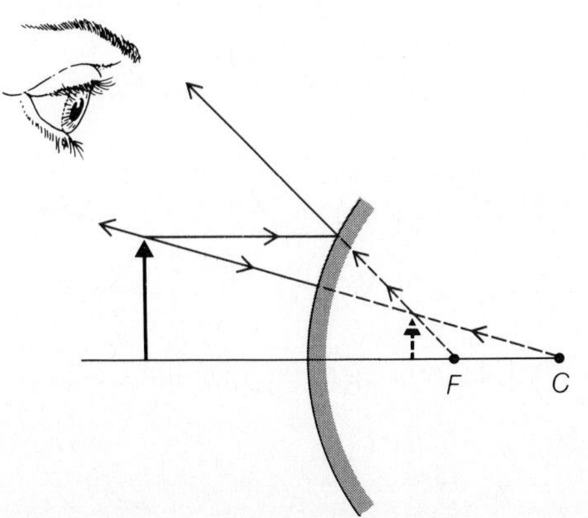

Figure 23.30

The image of a real object formed by a convex mirror is always virtual, erect, and smaller than the object.

Table 23.3 Sign conventions for spherical mirrors

Quantity	Positive	Negative
Focal length f	Concave mirror	Convex mirror
Object distance p	Real object	Virtual object
Image distance q	Real image	Virtual image
Magnification m	Erect image	Inverted image

Equations (23.5), (23.6), and (23.7) hold here as well as for lenses. The linear magnification of a mirror also follows the same formula as for a lens:

$$m = \frac{h'}{h} = -\frac{q}{p} \qquad \textit{Magnification of a mirror} \quad (23.15)$$

The sign conventions for a mirror are given in Table 23.3.

EXAMPLE 23.13

A candle 5.0 cm high is placed 40 cm from a concave mirror whose radius of curvature is 60 cm. Find the position, size, and nature of the image.

SOLUTION The focal length of the mirror is

$$f = \frac{R}{2} = \frac{60\text{ cm}}{2} = 30\text{ cm}$$

The situation therefore corresponds to that shown in Fig. 23.29(c). The image distance q is

$$q = \frac{pf}{p-f} = \frac{(40\text{ cm})(30\text{ cm})}{40\text{ cm} - 30\text{ cm}} = 120\text{ cm}$$

The image distance is positive, so the image is a real one. From Eq. (23.15) the size of the image is

$$h' = -h\frac{q}{p} = (-5.0\text{ cm})\left(\frac{120\text{ cm}}{40\text{ cm}}\right) = -15\text{ cm}$$

The image of the candle is three times as large as the candle itself and is inverted. ◆

EXAMPLE 23.14

A concave shaving mirror has a focal length of 40 cm. How far away from it should one's face be for the reflected image to be erect and twice its actual size?

SOLUTION Here the magnification is +2 since the image is erect. Hence

$$m = -\frac{q}{p} = 2 \qquad q = -2p$$

The negative image distance signifies that the image is virtual; see Fig. 23.29(a). Now we substitute $f = 40$ cm and $q = -2p$ into the mirror equation and solve for the object distance p:

$$\frac{1}{p} + \frac{1}{q} = \frac{1}{f}$$

$$\frac{1}{p} - \frac{1}{2p} = \frac{1}{40 \text{ cm}}$$

$$\frac{1}{2p} = \frac{1}{40 \text{ cm}}$$

$$p = 20 \text{ cm}$$

When one's face is 20 cm from the mirror, the virtual image one sees is magnified twice and is erect. ♦

Mirrors do not suffer from chromatic aberration but do exhibit spherical aberration. In the case of a concave mirror, rays reflected from the outer part of the mirror cross the axis closer to the vertex than rays reflected from the central part (Fig. 23.31). When a high-quality image is required, as in the case of an astronomical telescope (see Fig. 23.22), a concave mirror whose cross-sectional shape is a parabola is the answer. Despite spherical aberration, however, spherical mirrors are common because they are the easiest to manufacture.

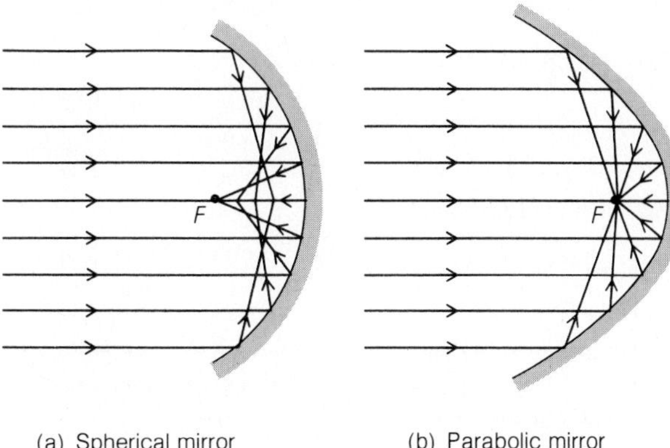

Figure 23.31

(a) Spherical aberration. The more highly curved the mirror, the less sharp the image. (b) A mirror with a parabolic cross section avoids spherical aberration.

Important Terms

A **lens** is a transparent object that can produce an image of an object placed before it. A **converging lens** brings parallel light to a single **real focal point,** whereas a **diverging lens** deviates parallel light outward as though it originated at a single **virtual focal point.** The distance from a lens to its focal point is its **focal length.**

The **magnification** of an optical system is the ratio between the size of the image and that of the object.

A **microscope** is a lens system used to produce enlarged images of nearby objects. A **telescope** is a lens system used to produce larger images of distant objects than would be seen by the unaided eye, although the image itself may be smaller than the actual object.

A **concave mirror** curves inward toward its center and converges parallel light to a single real focal point.

A **convex mirror** curves outward toward its center and diverges parallel light as though the reflected light came from a single virtual focal point behind the mirror. The distance from a mirror to its focal point is the focal length of the mirror.

Important Formulas

Lensmaker's equation:

$$\frac{1}{f} = (n - 1)\left(\frac{1}{R_1} + \frac{1}{R_2}\right)$$ (R is + for convex, − for concave surface)

Focal length of concave mirror: $f = \frac{R}{2}$

Focal length of convex mirror: $f = -\frac{R}{2}$

Lens and mirror equation: $\frac{1}{p} + \frac{1}{q} = \frac{1}{f}$

Alternative forms of lens and mirror equation:

$$p = \frac{qf}{q - f} \qquad q = \frac{pf}{p - f} \qquad f = \frac{pq}{p + q}$$

Linear magnification: $m = \frac{h}{h'} = -\frac{q}{p}$

$$m = m_1 m_2$$

Angular magnification of telescope: $m_{ang} = \frac{f_{objective}}{f_{eyepiece}}$

MULTIPLE CHOICE

1. A negative focal length corresponds to which one or more of the following?
- a. Double-convex lens
- b. Planoconcave lens
- c. Convex mirror
- d. Concave mirror

2. An object farther from a converging lens than its focal point always has an image that is
- a. inverted
- b. virtual
- c. the same in size
- d. smaller in size

3. An object closer to a converging lens than its focal point always has an image that is
- a. inverted
- b. virtual
- c. the same in size
- d. smaller in size

4. When the distance of a real object from a converging lens equals the focal length of the lens,
- a. the image is virtual, erect, and larger than the object
- b. the image is real, inverted, and larger than the object
- c. the image is real, inverted, and the same size as the object
- d. no image is formed

5. Relative to its object, a real image formed by a lens is always
- a. erect
- b. inverted
- c. smaller
- d. larger

6. Relative to its object, a real image formed by a spherical mirror is always
- a. erect
- b. inverted
- c. smaller
- d. larger

7. The image of a real object produced by a diverging lens is never
- a. real
- b. virtual
- c. erect
- d. smaller than the object

8. A negative magnification corresponds to an image that is
- a. erect
- b. inverted
- c. smaller than the object
- d. larger than the object

9. A negative image distance signifies an image that is
- a. real
- b. virtual
- c. erect
- d. inverted

10. The image a camera forms on the film is
- a. always real
- b. always virtual
- c. sometimes real and sometimes virtual
- d. neither real nor virtual

11. Enlarging the lens aperture of a camera
- a. increases the *f*-number
- b. increases the depth of field
- c. enlarges the image
- d. permits a faster shutter speed

12. Which of the following combinations of shutter speed and lens opening will admit the most light to the film of a camera?
- a. 1/125 s at *f*/8
- b. 1/125 s at *f*/16
- c. 1/250 s at *f*/4
- d. 1/250 s at *f*/5.6

13. The lens of the eye forms an image on the retina that is
- a. real and erect
- b. real and inverted
- c. virtual and erect
- d. virtual and inverted

14. When a converging lens of focal length F is used as a magnifying glass, the object distance must be

a. less than F b. equal to F
c. between F and $2F$ d. more than $2F$

15. Four lenses with the listed focal lengths are being considered for use as a microscope objective. The one that will produce the greatest magnification with a given eyepiece has the focal length

a. −5 mm b. +5 mm
c. −5 cm d. +5 cm

16. Four lenses with the listed focal lengths are being considered for use as a telescope objective. The one that will produce the greatest magnification with a given eyepiece has the focal length

a. −1 m b. +1 m
c. −2 m d. +2 m

17. A concave mirror produces an erect image when the object distance is

a. less than f b. equal to f
c. between f and $2f$ d. greater than $2f$

18. The image formed by a concave mirror is larger than the object

a. when p is less than $2f$
b. when p is more than $2f$
c. for no values of p
d. for all values of p

19. The image formed by a convex mirror is larger than the object

a. when p is less than $2f$
b. when p is more than $2f$
c. for no values of p
d. for all values of p

20. An object is located 10 cm from a converging lens of focal length 12 cm. The image distance is

a. +5.5 cm b. −5.5 cm
c. +60 cm d. −60 cm

21. An object is located 12 cm from a converging lens of focal length 10 cm. The image distance is

a. +5.5 cm b. −5.5 cm
c. +60 cm d. −60 cm

22. The image of an object 10 cm from a lens is located 10 cm behind the object. The focal length of the lens is

a. +6.7 cm b. −6.7 cm
c. +20 cm d. −20 cm

23. A pencil 10 cm long is placed 70 cm in front of a lens of focal length +50 cm. The image is

a. 4 cm long and erect
b. 4 cm long and inverted
c. 25 cm long and erect
d. 25 cm long and inverted

24. A pencil 10 cm long is placed 100 cm in front of a lens of focal length +50 cm. The image is

a. 5 cm long and erect
b. 5 cm long and inverted
c. 10 cm long and erect
d. 10 cm long and inverted

25. A pencil 10 cm long is placed 175 cm in front of a lens of focal length +50 cm. The image is

a. 4 cm long and erect
b. 4 cm long and inverted
c. 25 cm long and erect
d. 25 cm long and inverted

26. A magnifying glass is to be used at the fixed object distance of 10.0 mm. If it is to produce an erect image magnified 5.00 times, its focal length should be

a. +2.0 mm b. +8.0 mm
c. +12.5 mm d. +50.0 mm

27. A lens placed 9.00 cm from a postage stamp produces a virtual image of it 3.00 cm from the lens. The lens has a focal length of

a. −2.25 cm b. +2.25 cm
c. −4.50 cm d. +4.50 cm

28. An object 40 cm from a converging lens has an image 40 cm away from the lens on the other side. The focal length of the lens is

a. 20 cm b. 40 cm
c. 60 cm d. 80 cm

29. A lens held 20.0 cm from an object forms a virtual image of it 10.0 cm from the lens. The focal length of the lens is

a. −6.7 cm b. +6.7 cm
c. −20 cm d. +20 cm

30. A projector is intended to produce an image 140 cm wide of a slide 35.0 mm wide. If the projector is to be located 5.00 m from the screen, the focal length of the projector's lens should be

a. +122 mm b. +125 mm
c. −128 mm d. +5.13 m

31. A convex mirror is ground with a radius of curvature of 12 cm. Its focal length is

a. 6 cm b. 24 cm
c. −6 cm d. −24 cm

32. A pencil 10 cm long is placed 30 cm in front of a mirror of focal length +50 cm. The image is

a. 2.5 cm long and erect
b. 25 cm long and erect
c. 0.25 m long and erect
d. 25 cm long and inverted

33. A pencil 10 cm long is placed 100 cm in front of a mirror of focal length +50 cm. The image is

a. 3 cm long and erect
b. 10 cm long and erect
c. 3 cm long and inverted
d. 10 cm long and inverted

34. A pencil 10 cm long is placed 30 cm in front of a mirror of focal length −50 cm. The image is
   a. 25 cm long and erect
   b. 6.3 cm long and erect
   c. 25 cm long and inverted
   d. 6.3 cm long and inverted

QUESTIONS

1. A fortune-teller's crystal ball is 15 cm in diameter. (a) Would you expect the lensmaker's equation to hold for the focal length of this ball? (b) Would you expect the ball to form undistorted images?

2. Does a spherical bubble of air in a volume of water act to converge or to diverge light passing through it?

3. A double-convex lens made of crown glass is placed in a tank of benzene. Will it act as a converging or as a diverging lens there?

4. Under what circumstances, if any, is a light ray that passes through a converging lens not deflected? Under what circumstances, if any, is a light ray that passes through a diverging lens not deflected?

5. Under what circumstances, if any, will a diverging lens form an inverted image of a real object? Under what circumstances, if any, will a converging lens form an erect image of a real object?

6. Is the mercury column in a thermometer wider or narrower than it appears?

7. Is there any way in which a diverging lens, used by itself, can form a real image of a real object? Is there any way in which a converging lens, used by itself, can form a virtual image of a real object?

8. When it emerges from the lens, which path will the incident ray shown in Fig. 23.32 follow?

9. If the screen is moved farther away from a projector, how should the projector's lens be moved to keep the image in focus?

10. A gnu moves toward a camera. Should the camera lens be moved toward or away from the film to keep the gnu in focus?

11. What can you say about the properties of an image when the magnification is (a) less than −1? (b) between −1 and 0? (c) between 0 and +1? (d) greater than +1?

12. What are the characteristics of the image formed on the retina by the lens of the eye?

13. Is there any way in which a convex mirror, used by itself, can form a real image of a real object? Is there any way in which a concave mirror, used by itself, can form a virtual image of a real object?

Figure 23.32

Question 8.

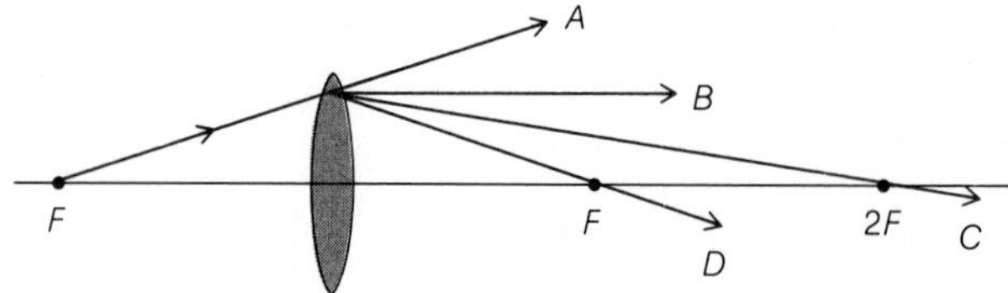

EXERCISES

23.1 Lenses

1. A double-concave lens has surfaces whose radii of curvature are both 400 mm. The lens is made from flint glass whose index of refraction is 1.55. Find its focal length.

2. A meniscus lens has a concave surface of radius 30 cm and a convex surface of radius 25 cm. The index of refraction of the glass is 1.50. Find the focal length of the lens.

3. A converging meniscus lens has surfaces whose radii of curvature are 20 cm and 30 cm. The lens is made from crown glass whose index of refraction is 1.50. Find its focal length.

4. A diverging meniscus lens has surfaces whose radii of curvature are 20 cm and 30 cm. The lens is made from crown glass whose index of refraction is 1.50. Find its focal length.

5. The index of refraction of crown glass is 1.523 for blue light and 1.517 for red light. How far apart are the focal points for blue and red light of a planoconvex lens of crown glass whose radius of curvature is 200 mm?

6. A double-convex lens whose focal length is 35 cm has surfaces whose radii of curvature are 25 cm and 50 cm, respectively. Find the index of refraction of the glass.

7. Both surfaces of a double-concave lens whose focal length is −18 cm have radii of 20 cm. Find the index of refraction of the glass.

8. A planoconcave lens of focal length 90 mm is to be ground from quartz of index of refraction 1.55. Find the required radius of curvature.

9. A converging lens made from glass of $n = 1.60$ has a focal length in air of 150 mm. (a) Is it converging or diverging when immersed in water? (b) What is its focal length in water?

10. A diverging lens made from glass of $n = 1.55$ has a focal length of −80 mm in air. (a) Is it converging or diverging when immersed in carbon disulfide ($n = 1.63$)? (b) What is its focal length in carbon disulfide?

23.3 The Lens Equation

11. A lens is to be used to focus sunlight on a piece of paper to ignite it. What kind of lens should be used? If the focal length of the lens is 80 mm, how far should it be held from the paper?

12. In order to photograph herself for a passport picture, a woman stands beside her camera 60 cm from a mirror. If the focal length of the camera lens is 40 mm, how far from the film should it be?

13. A lens whose focal length is -60 cm is used to look at a duck 4.00 m away. Where is the image located? Is it real or virtual?

14. A lens held 20 cm from a sardine produces a virtual, erect image of it that appears to originate 10 cm in front of the sardine. What is the focal length of the lens? Is it converging or diverging?

15. A lens held 20 cm from a sardine produces a real, inverted image of it 30 cm on the other side. What is the focal length of the lens? Is it converging or diverging?

23.4 Magnification

16. A candle is placed with its flame 10 cm from a lens whose focal length is $+15$ cm. Find the location of the image of the flame. Is the image larger or smaller than the actual flame? What is the character of the image—that is, is it erect or inverted, real or virtual?

17. The flame of the candle in Exercise 16 is 15 cm from the lens. Answer the same questions for this case.

18. The flame of the candle in Exercise 16 is 25 cm from the lens. Answer the same questions for this case.

19. The flame of the candle in Exercise 16 is 30 cm from the lens. Answer the same questions for this case.

20. The flame of the candle in Exercise 16 is 50 cm from the lens. Answer the same questions for this case.

21. A movie director holds a diverging lens 5.00 m from an actress and sees her one-tenth of her normal size. What is the focal length of the lens?

22. A camera whose lens has a focal length of 90 mm is used to photograph a wall 4.00 m away. (a) How far in front of the film should the lens be placed? (b) If the image on the film covers a square 40 mm on a side, how much of the wall does it include?

23. An aerial camera whose lens has a focal length of 1.0 m is used to photograph a military base from an altitude of 7.0 km. How long is the image on the film of a tank 9.0 m long?

24. The moon's diameter is 3476 km and its average distance from the earth is 3.8×10^5 km. (a) What is the diameter of its image when a 35-mm camera with a 300-mm telephoto lens is used to photograph it? (b) What would the image diameter be if a camera using 9×12-cm film were used with the same lens?

25. A magnifying glass of 10-cm focal length is held 80 mm from a stamp. What is the actual length of a feature of the stamp that appears to be 10 mm long?

26. A magnifying glass of 50-mm focal length is held 35 mm from a spider egg 0.50 mm long. What is the apparent size of the egg?

23.5 The Camera

27. What is the diameter of an *f*/8 telephoto lens whose focal length is 300 mm?

28. The lens of a 35-mm camera is marked *f*/1.8 and is 15 mm in diameter. What is its focal length? Is it a wide-angle, normal, or telephoto lens?

29. An exposure meter indicates that a camera should be set for a shutter speed of $\frac{1}{125}$ s and an aperture of *f*/8 to photograph a certain scene. If the shutter speed is changed to $\frac{1}{500}$ s, what should the aperture be changed to?

30. The settings of a camera are changed from $\frac{1}{125}$ s at *f*/16 to $\frac{1}{250}$ s at *f*/5.6. What is the difference in the amount of light reaching the film?

23.6 The Eye

23.7 Defects of Vision

31. The image distance in a certain normal eye is 20 mm. What range of focal lengths does the eye have if its near point is 20 cm?

32. A myopic eye cannot bring to a focus objects farther than 15 cm away. What type of lens is needed to permit this eye to see clearly objects at infinity, and what focal length should such a lens have?

33. A presbyopic eye has a near point 100 cm away. What type of lens is needed to permit the eye to see clearly objects 25 cm away, and what focal length should such a lens have?

34. A hyperopic eye has a near point of 60 cm. What is its near point when a correcting lens of $+3.33$ diopters is used?

35. A myopic person whose eyes have far points of 60 cm is lent a pair of glasses whose power is -1.5 diopters. How far can she see clearly with these glasses?

23.8 The Microscope

36. In a microscope an objective of 10.0-mm focal length is used with a 5× eyepiece. (a) What is the magnification of the microscope? (b) How far should the objective be from the specimen being examined? Assume that the image distances of the objective and the eyepiece are, respectively, 160 mm and 250 mm here and in Exercise 37.

37. In a microscope an objective of 4.0-mm focal length is used with a 10× eyepiece. (a) What is the magnification of the microscope? (b) How far should the objective be from the specimen being examined?

23.9 The Telescope

38. A telescope with an objective of focal length 90 cm and an eyepiece of focal length 30 mm is used to look at the moon, whose angular diameter as seen from the earth is about 0.5°. If the moon appears to be 25 cm in front of the eyepiece, what is its apparent diameter?

39. A telescope has an objective whose focal length is 60 cm. (a) Find the eyepiece focal length needed for an angular magnification of 20. (b) The telescope is used to watch a young zebra 500 m away. If the zebra's image is 20 mm long and is located 25 cm in front of the eyepiece, find the zebra's actual length.

23.11 Spherical Mirrors

40. A convex mirror has a radius of curvature of 40 cm. What is its focal length? Where is its focal point?

41. A butterfly is 20 cm in front of a concave mirror whose focal length is 40 cm. Find the location of the image. Is the image larger or smaller than the butterfly? What is the character of the image—that is, is it erect or inverted, real or virtual?

42. A dime is 40 cm in front of the mirror in Exercise 41. Answer the same questions for the image of the dime.

43. A peanut is 50 cm in front of the mirror in Exercise 41. Answer the same questions for the image of the peanut.

44. A caterpillar is 80 cm in front of the mirror in Exercise 41. Answer the same questions for the image of the caterpillar.

45. A button is 100 cm in front of the mirror in Exercise 41. Answer the same questions for the image of the button.

46. An object 6 cm high is 30 cm in front of a convex mirror whose focal length is 50 cm. What are the height and the character of the image?

47. A dentist's concave mirror has a diameter of 10 mm and a focal length of 25 mm. What magnification does it produce when held 18 mm from a tooth?

48. A coin 15.0 mm in diameter is placed 15.0 cm from a spherical mirror. The coin's image is 5.0 mm in diameter and is erect. Is the mirror concave or convex? What is its radius of curvature?

49. The coin in Exercise 48 is placed 15.0 cm from another spherical mirror. The coin's image is now 30.0 mm in diameter and is again erect. Is the new mirror concave or convex? What is its radius of curvature?

50. A man stands 6.0 m from a concave mirror, and an inverted image of himself of the same height is formed on a screen beside him. What is the radius of curvature of the mirror?

51. A worm crawls toward a polished metal ball 60 cm in diameter lying on a lawn. How far from the surface of the ball is the worm when its image appears to be 10 cm behind the surface?

52. A mirror in an amusement park produces an erect image four times enlarged of anyone standing 3.0 m away. (a) Is the mirror concave or convex? (b) What is its radius of curvature?

53. What should the radius of curvature of a convex mirror be if it is to produce an image one-fifth the size of an object 150 cm away?

54. The moon is 3476 km in diameter. What radius of curvature should a concave mirror have if it is to produce a lunar image 10 mm in diameter when the moon is 3.84×10^5 km away?

PROBLEMS

55. A 150-mm camera lens, originally focused on a scene 10 m away, is moved 5 mm farther from the film. Scenes at what distance are now in focus?

56. Verify that the effective focal length f of two thin lenses of focal lengths f_1 and f_2 that are in contact is given by

$$\frac{1}{f} = \frac{1}{f_1} + \frac{1}{f_2}$$

(To do this, let the image produced by the first lens be the object of the second.)

57. A converging lens 30 cm from a candle produces an inverted image of the candle on a screen 30 cm from the other side of the lens. A diverging lens is then placed midway between the candle and the other lens. When the candle is moved back 10 cm from its original position, its image reappears on the screen. Find the focal lengths of the two lenses.

58. A frog is 80 cm in front of a lens whose focal length is +50 cm. Another lens, of focal length +20 cm, is 100 cm behind the first lens. Where is the image of the frog located? Is it real or virtual?

59. A photographic enlarger is being designed to produce prints 40 cm × 50 cm from negatives 10 cm × 12.5 cm. (a) If the maximum distance from negative to print paper is to be 60 cm, what should the focal length of the lens be? (b) How far would such a negative be from the print paper using the same lens if the enlargement size were 20 cm × 25 cm?

60. The actual size of each frame of 16-mm motion picture film is 7.5 mm × 10.5 mm. A projector whose lens has a focal length of 25 mm is to be used with a screen 1.5 m wide. How far from the projector should the screen be located?

61. An object should be about 25 cm from a normal eye for maximum distinctness of vision. (a) Find a formula for the magnification of a converging lens in terms of its focal length when the lens is used as a magnifying glass with an image distance of -25 cm; see Fig. 23.6(a). (b) Find the magnification of a lens whose focal length is 50 mm.

62. A myopic eye with near and far points of 12 and 20 cm, respectively, is given a corrective lens that permits distant vision. What is the new near point?

63. A hyperopic person whose eyeglass lenses are converging with a power of $+2.5$ diopters can see objects distinctly as close as 25 cm away. What is the distance of the near point without the eyeglasses?

64. A nearsighted person whose eyeglass lenses are diverging with a power of -2.5 diopters can see distant objects distinctly. What is the distance of the far point without the eyeglasses?

65. A telescope has an objective of 750-mm focal length and an eyepiece of 25-mm focal length. The telescope is focused on a distant albatross. (a) What is the angular magnification? (b) How far apart are the lenses? Assume that the final image is 25 cm in front of the eyepiece.

66. A telescope with an objective of focal length 1.0 m and an eyepiece of focal length 5.0 cm is used to examine a penguin 40 cm high. Find the apparent height of the penguin when it is (a) 50 m, and (b) 5.0 m away from the telescope. Assume that the image is 25 cm in front of the eyepiece.

67. A Galilean telescope has a converging lens as its objective and a diverging lens as its eyepiece. Its advantages are that it produces an erect image and is short in length; its main disadvantage is a narrow field of view that limits it to low magnifications. Inexpensive opera glasses consist of a pair of Galilean telescopes. Show that the distance between the objective and the eyepiece in a Galilean telescope when it is focused on a distant object with the final image at infinity is the difference between the absolute values $|f_1|$ and $|f_2|$ of their focal lengths (absolute value is the value without regard to sign).

68. A virtual image 60 mm long is formed of a paper clip 20 mm long placed 100 mm in front of a concave mirror of unknown curvature. Where else can the paper clip be placed for an image 60 mm long to be formed? What is the character of the image in the latter case?

69. A pawn 50 mm high is placed in front of a concave mirror whose radius of curvature is 100 cm. What are the two object distances that will lead to images 200 mm high? What is the character of the image in each case?

Answers to Multiple Choice

1. b, c	10. a	19. c	28. a
2. a	11. d	20. d	29. c
3. b	12. c	21. c	30. a
4. d	13. b	22. c	31. c
5. b	14. a	23. d	32. b
6. b	15. b	24. d	33. d
7. a	16. d	25. b	34. b
8. b	17. a	26. c	
9. b	18. a	27. c	

C H A P T E R 24

PHYSICAL OPTICS

In the previous chapter the wave nature of light played only a small role, because lenses and mirrors are easier to analyze with the help of rays. But other aspects of optics directly involve the wave nature of light. The study of such phenomena is called **physical optics,** whereas the study of effects that can be treated by using rays is called **geometrical optics.** Since the laws of reflection and refraction follow from Huygens' principle, geometrical optics is really an approximation of physical optics. In this chapter we will find that a purely wave approach is necessary to account for the interference, diffraction, polarization, and scattering of light.

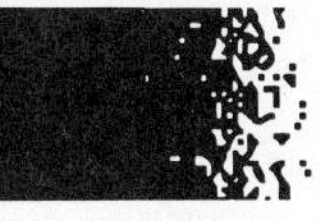

24.1 Interference of Light

What happens when light waves from different sources meet.

When light waves from one source are mixed with those from another source, the two groups of waves are said to **interfere.** We recall from Chapter 12 the principle of superposition, which governs interference: When two or more waves of the same nature go past a point at the same time, the instantaneous amplitude at that point is the sum of the instantaneous amplitudes of the individual waves. Constructive interference refers to the reinforcement of waves in phase (in step) with one another, and destructive interference refers to the partial or complete cancellation of waves out of phase with one another (Fig. 24.1).

Anyone with a pan of water can see how interference between water waves can lead to a water surface disturbed in a variety of characteristic patterns. Two people who hum fairly pure tones slightly different in frequency will hear beats as the result of interference in the sound waves. But if we shine light from two flashlights at the same place on a screen, there is no evidence of interference: The region of overlap is merely uniformly brighter.

It is hard to observe interference in light for two reasons. First, the wavelengths in light are very short, about 1% of the width of a human hair. Second, every natural source of light emits light waves only as short groups of random phase, so that any interference that occurs is usually averaged out during even the briefest period of observation by the eye or photographic film unless special procedures are used. Interference in light is nevertheless just as real as interference in water or sound waves. One example of it is familiar to everybody—the bright colors of a thin film of oil spread out on a water surface.

Coherence

Two sources of waves are said to be **coherent** if there is a fixed phase relationship between the waves they emit during the time the waves are being observed. It does not matter whether the waves are exactly in step when they leave the sources, or exactly out of step, or anything in between. The important thing is that the phase relationship stays the same. If the sources shift back and forth in relative phase while the observation is made, the phase differences average out, and there will be no interference pattern. Such sources are **incoherent.**

The question of coherence is especially significant for light waves because an excited atom radiates for 10^{-8} s or less, depending on its environment. Therefore

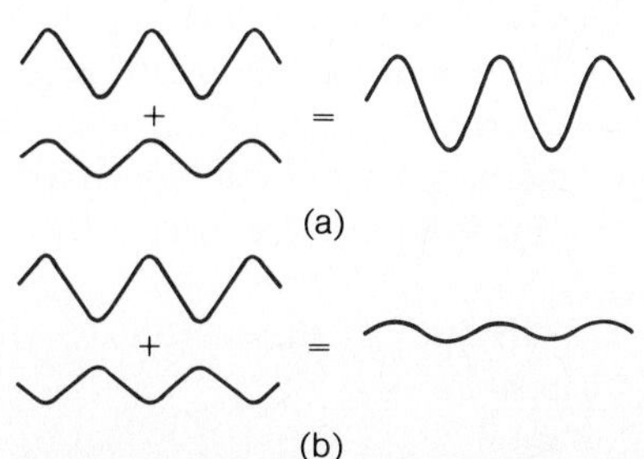

Figure 24.1

(a) Constructive interference. (b) Destructive interference.

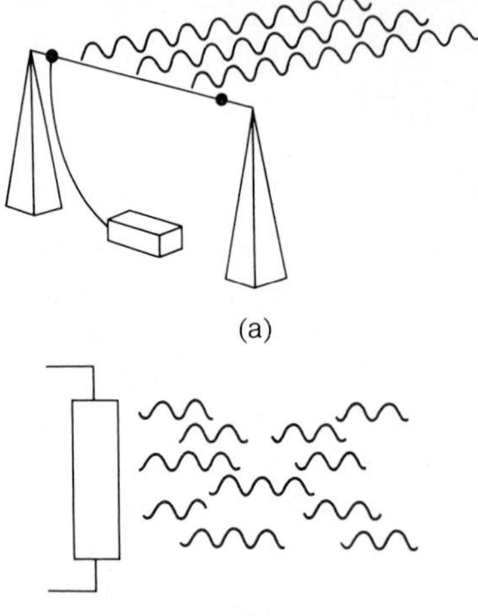

Figure 24.2

(a) Radio waves from an antenna are coherent. (b) Light waves from a gas discharge tube are incoherent.

a monochromatic light source such as a gas discharge tube (a neon sign is an example) does not emit a continuous series (''train'') of waves as a radio antenna does but instead emits separate short wave groups whose phases are random. The light from such a tube actually comes from a great many individual, uncoordinated sources, namely the gas atoms, which are in effect being switched on and off rapidly and irregularly (Fig. 24.2).

Suppose we have two point sources of monochromatic light—for instance, a discharge tube with a cover that has two pinholes close together. Different atoms are behind each pinhole, so they are independent sources. Therefore we have at most 10^{-8} s to observe the interference of waves from the two sources. If our detecting instruments are fast enough, as some modern electronic devices are, interference can be demonstrated and the sources can be considered coherent. If we are limited to the eye and to photographic film, which average light signals over times far greater than 10^{-8} s, no interference can be observed in the light from the two sources. The sources must then be considered incoherent. Like beauty, coherence lies in the eye of the beholder.

Does the brief lifetime of an excited atom mean that interference patterns can never be actually seen but can only be recorded by instruments? Not at all. There are three ways to construct separate sources of light coherent for long enough periods of time to produce visible interference patterns:

1. Illuminate two (or more) slits with light from one slit behind them. Then the light waves from the secondary slits are automatically coordinated.
2. Obtain coherent virtual sources from a single source by reflection or refraction. This is how interference is produced by thin films of oil.
3. Coordinate the radiating atoms in each separate source so that they always have the same phase even though different atoms are radiating at successive instants. This is done in the **laser.**

The first two of these methods will be discussed in this chapter and the third in Chapter 27.

24.2 Double Slit

It produces an interference pattern of light and dark lines.

The interference of light waves was demonstrated in 1801 by Thomas Young, who used an arrangement like that shown in Fig. 24.3. A source of monochromatic light (that is, light consisting of only a single wavelength) is placed behind a narrow slit S in a barrier screen. Another barrier with two similar slits A and B is placed on the other side. Light from S passes through both A and B and then to the viewing screen.

If light were not a wave phenomenon, we would expect to find the viewing screen completely dark, since no light ray can reach it from the source along a straight path. What actually happens is that each slit acts as a source of secondary wavelets—we recall Huygens' principle from Section 22.4—so that the entire screen is illuminated. Even the point S', separated from S by the barrier between the slits A and B, turns out to be bright rather than dark.

Owing to interference the screen is not evenly lit but shows a pattern of alternate bright and dark lines. Light waves from slits A and B are exactly in phase,

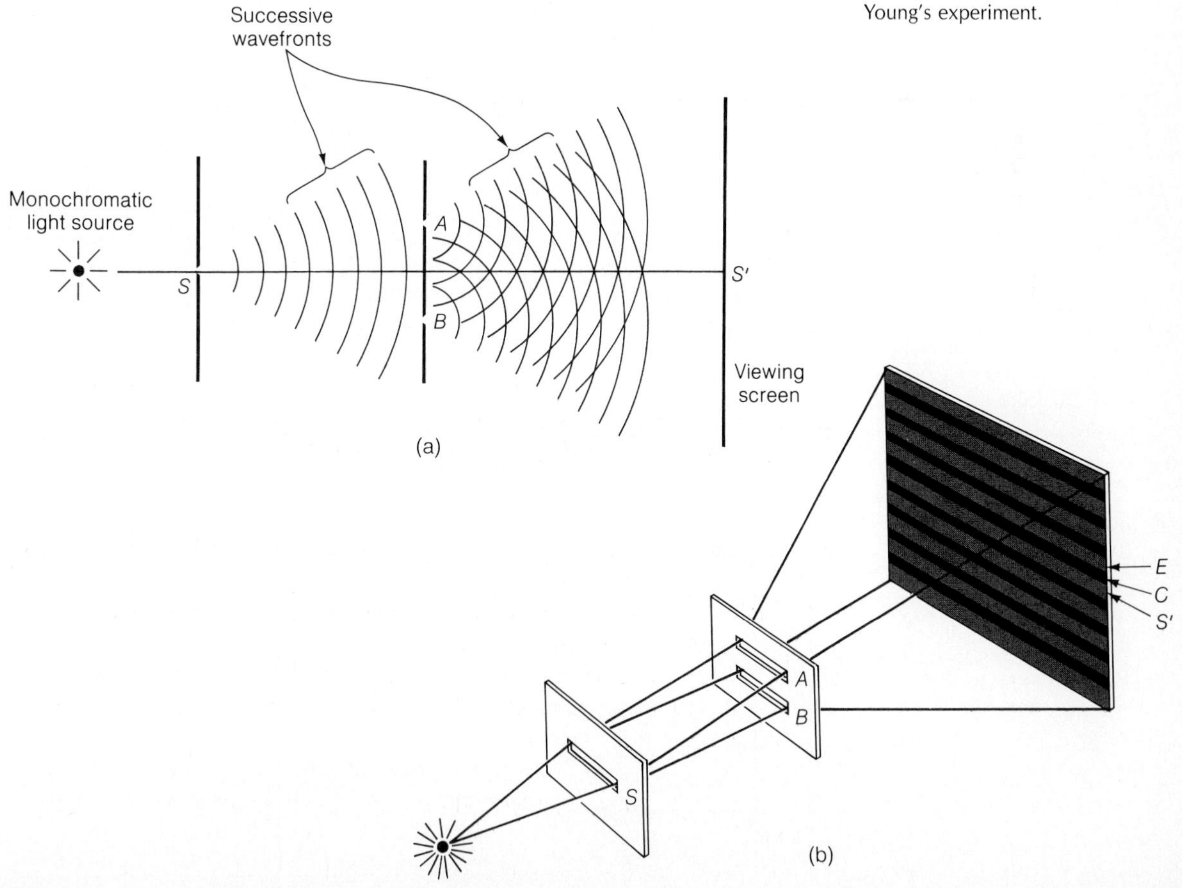

Figure 24.3

(a) Young's double-slit experiment. In accord with Huygens' principle, each slit acts as a source of secondary wavelets. (b) The appearance of the screen in Young's experiment.

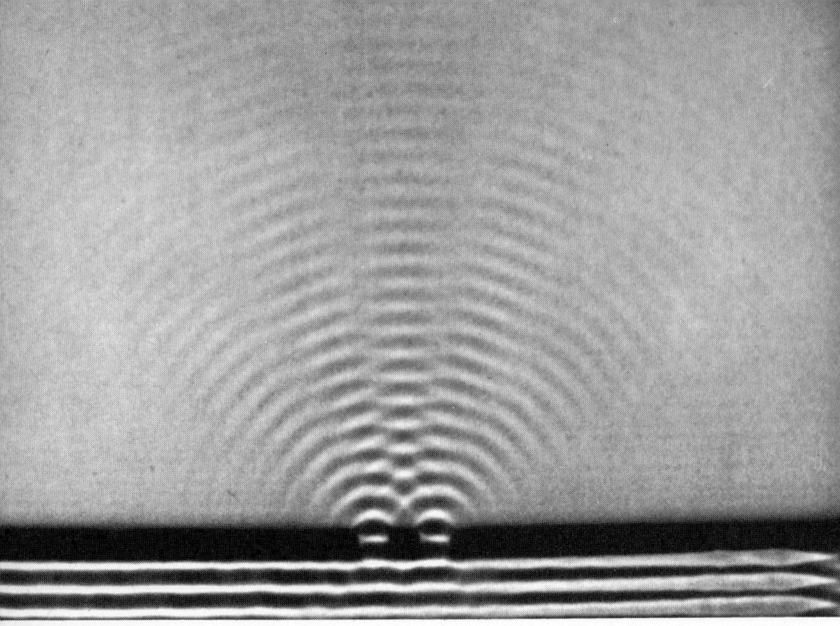

Double-slit interference pattern in water waves.

since A and B are the same distance from S (Fig. 24.4). The centerline S' of the screen is equally distant from A and B, so light waves from these slits interfere constructively there to produce a bright line.

What happens at the position C on the screen located to one side of S'? The distance BC is longer than the distance AC by the amount BD, which is equal to exactly half a wavelength of the light being used. That is,

$$BD = \tfrac{1}{2}\lambda \qquad \textit{Dark line}$$

When a crest from A reaches C, this difference in path length means that a trough from B arrives there at the same time, since $\frac{1}{2}\lambda$ separates a crest and a trough in the same wave. The two cancel each other out, the light intensity at C is zero, and the result is a dark line.

If we go past C on the screen, we will come to a point E such that the distance BE is greater by exactly one wavelength than the distance AE. That is, the difference BF between BE and AE is

$$BF = \lambda \qquad \textit{Bright line}$$

When a crest from A reaches E, a crest from B also arrives there. Waves arriving at E from both slits are always in the same part of their cycles, and they constructively interfere to produce a bright line at E.

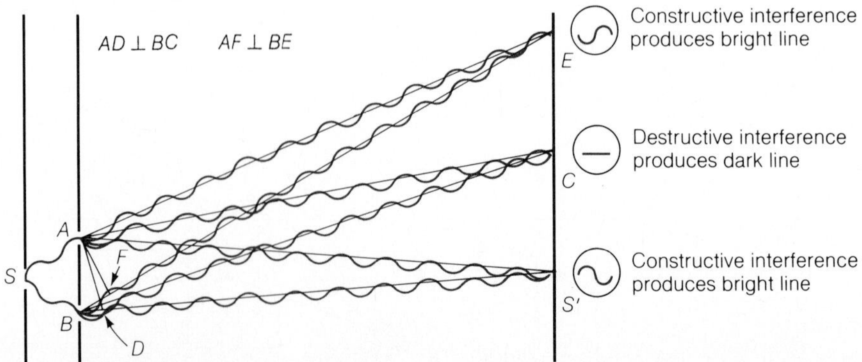

Figure 24.4

Origin of the double-slit interference pattern.

In this way we find that the alternate bright and dark lines on the screen correspond, respectively, to locations where constructive and destructive interference occurs. Waves reaching the screen from A and B along paths that are equal or that differ by a whole number of wavelengths (λ, 2λ, 3λ, and so on) reinforce. Waves whose paths differ by an odd number of half wavelengths ($\frac{1}{2}\lambda$, $\frac{3}{2}\lambda$, $\frac{5}{2}\lambda$, and so on) cancel. At intermediate locations on the screen the interference is only partial, so that the light intensity on the screen varies gradually between the bright and dark lines.

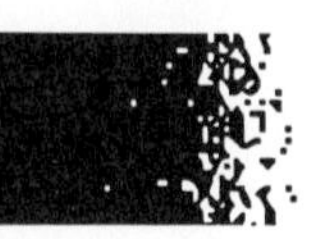

24.3 Wavelength of Light

How to measure it with a double slit.

The interference pattern produced by a double slit gives us a way to find the wavelength of the light being used.

In Fig. 24.5 d is the separation of the slits, L is the distance from the slits to the screen, and y is the distance from the central point S' on the screen to the point Q whose illumination we are observing. The waves that travel from slit B to Q must travel s farther than those from slit A. As we saw in Section 24.2, when the path difference s is

$$s = 0, \lambda, 2\lambda, 3\lambda, \ldots \qquad \textit{Constructive interference} \quad (24.1)$$

where λ is the wavelength of the light from the source, waves from A and B arrive at Q in the same stage of their cycles and reinforce to produce a bright line. On the other hand, when the path difference s is

$$s = \tfrac{1}{2}\lambda, \tfrac{3}{2}\lambda, \tfrac{5}{2}\lambda, \ldots \qquad \textit{Destructive interference} \quad (24.2)$$

waves from A and B arrive at Q in the opposite stages of their cycles and cancel to produce a dark line.

From Fig. 24.5 we see that the triangles ABD and $E'QS'$ are similar, since each is a right triangle with two sides perpendicular to two sides of the other. Corresponding sides of similar triangles are proportional, hence

$$\frac{s}{y} = \frac{d}{E'Q} \qquad s = \frac{dy}{E'Q}$$

In an actual experiment d and y are much smaller than L, so that $E'Q$ is very nearly

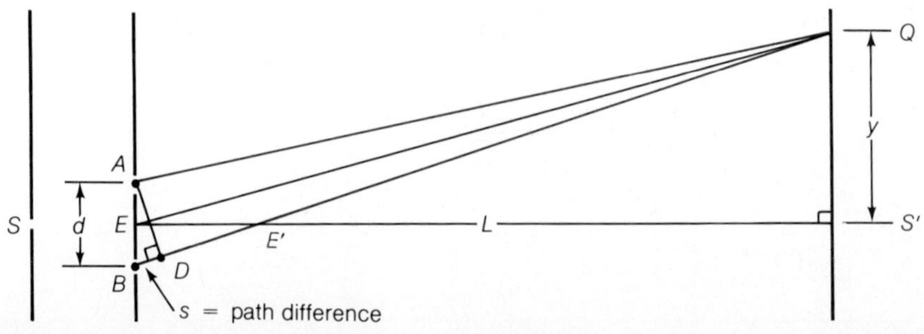

Figure 24.5

A diagram of the double-slit experiment.

equal to L. Taking them as equal gives

$$s = \frac{dy}{L} \tag{24.3}$$

Combining Eqs. (24.1) and (24.2) with Eq. (24.3) leads to the conditions for bright and dark lines to occur at Q:

$$y = 0, \frac{L\lambda}{d}, \frac{2L\lambda}{d}, \frac{3L\lambda}{d}, \ldots \qquad \textit{Bright lines} \tag{24.4}$$

$$y = \frac{L\lambda}{2d}, \frac{3L\lambda}{2d}, \frac{5L\lambda}{2d}, \ldots \qquad \textit{Dark lines} \tag{24.5}$$

According to Eq. (24.4) there is a bright line when $y = 0$, corresponding to the center of the screen S', a bright line on either side of S' a distance $L\lambda/d$ from it, another bright line on either side a distance $2L\lambda/d$ from S', and so on (see Fig. 24.3). Similarly there is a dark line on either side of S' a distance $L\lambda/2d$ from it, another dark line on either side a distance $3L\lambda/2d$ from it, and so on.

EXAMPLE 24.1

Monochromatic yellow light illuminates two narrow slits 1.00 mm apart. The viewing screen is 1.00 m from the slits, and the distance from the central bright line to the bright line nearest it is found to be 0.589 mm (Fig. 24.6). Find the wavelength of the light.

SOLUTION For the second bright line Eq. (24.4) gives $y = L\lambda/d$. Hence

$$\lambda = \frac{yd}{L} = \frac{(5.89 \times 10^{-4}\,\text{m})(1.00 \times 10^{-3}\,\text{m})}{1.00\,\text{m}} = 5.89 \times 10^{-7}\,\text{m} = 589\,\text{nm}$$

Figure 24.6

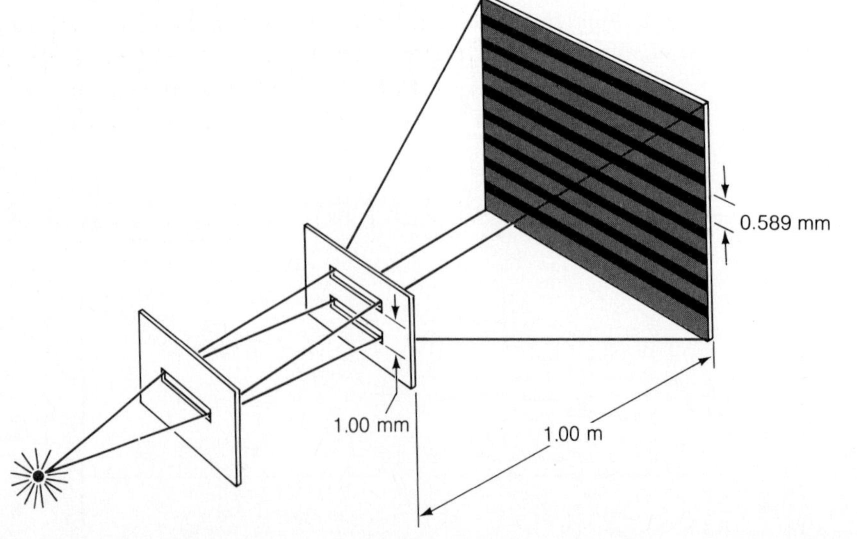

◆

24.4 Diffraction Grating

It produces sharper and brighter interference patterns than a double slit.

There are two difficulties in using a double slit for measuring wavelengths. First, the "bright" lines on the screen are actually quite faint, and an intense light source is therefore needed. Second, the lines are relatively broad, and it is hard to locate their centers accurately. A **diffraction grating** that consists, in essence, of a large number of parallel slits overcomes both of these difficulties.

Gratings are made by ruling grooves on a glass or metal plate with a diamond; the clear bands between the grooves are the "slits" (Fig. 24.7). Replica gratings, made by allowing a transparent liquid plastic to harden in contact with an original grating, are ordinarily used in practice. Replica gratings are often given a thin coating of silver or aluminum and produce their characteristic diffraction patterns by the interference of reflected rather than transmitted light. A phonograph record or compact disk held at a glancing angle acts as a reflection grating by virtue of its closely spaced grooves. Opals are natural diffraction gratings whose constituent particles (tiny spheres of silicon dioxide) are spaced about 250 nm apart in regular arrays.

Gratings are ruled with from 200 to 1000 lines/mm, and a lens is used to focus the light from the slits between them on a screen. The effect of phase differences among the rays from the various slits is enhanced by their great number, and the intensity of light on the screen falls rapidly on either side of the center of each bright line. The bright lines are therefore sharp, and, because there are so many slits, they are also bright in a literal sense. Very accurate wavelength determinations can be made with the help of a grating, and wavelengths that are close together can be resolved. The analysis of a light beam in terms of the particular wavelengths it contains is today almost invariably carried out by using a grating.

Figure 24.8 shows a diffraction grating that forms a bright line on a screen at a deviation angle of θ from the original beam of light. The condition for a bright line is that $s = n\lambda$, where $n = 1, 2, 3$, and so on. Since $s = d \sin \theta$, where d is the spacing of the slits, bright lines occur at those angles for which

$$\sin \theta = n\frac{\lambda}{d} \qquad n = 0, 1, 2, 3, \ldots$$

Bright lines (24.6)

When polychromatic light is incident on a grating, a series of spectra is formed on each side of the original beam corresponding to $n = 1$, $n = 2$, and so on. The **first-order spectrum** contains bright lines for which $n = 1$, the **second-**

(a)

(b)

Figure 24.7

(a) A diffraction grating. Each space between the ruled grooves acts as a slit. (b) The greater the number of slits, the sharper the line a single wavelength produces.

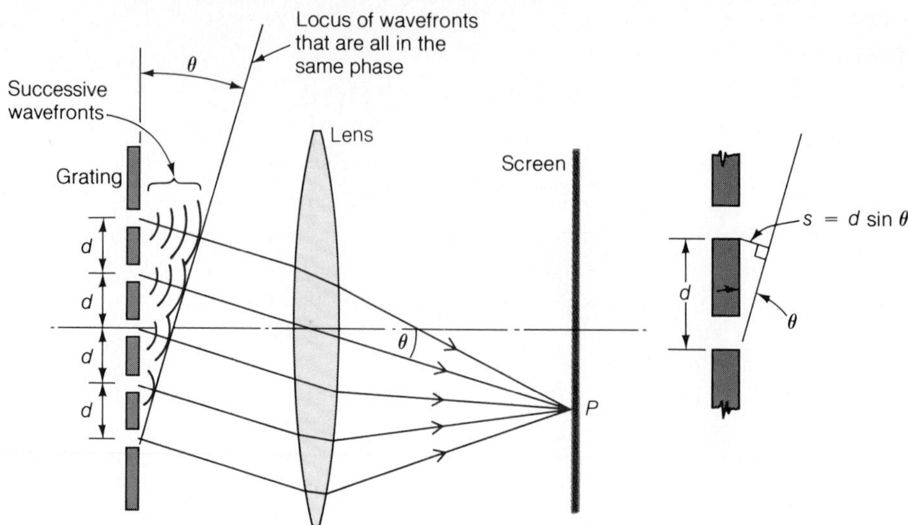

Figure 24.8
The plane diffraction grating.

order spectrum contains bright lines for which $n = 2$, and so on (Fig. 24.9). In some gratings the higher-order spectra overlap, so that, for example, the blue end of the third-order spectrum may be deviated by less than the red end of the second-order spectrum. According to Eq. (24.6) the angle θ increases with wavelength. Hence blue

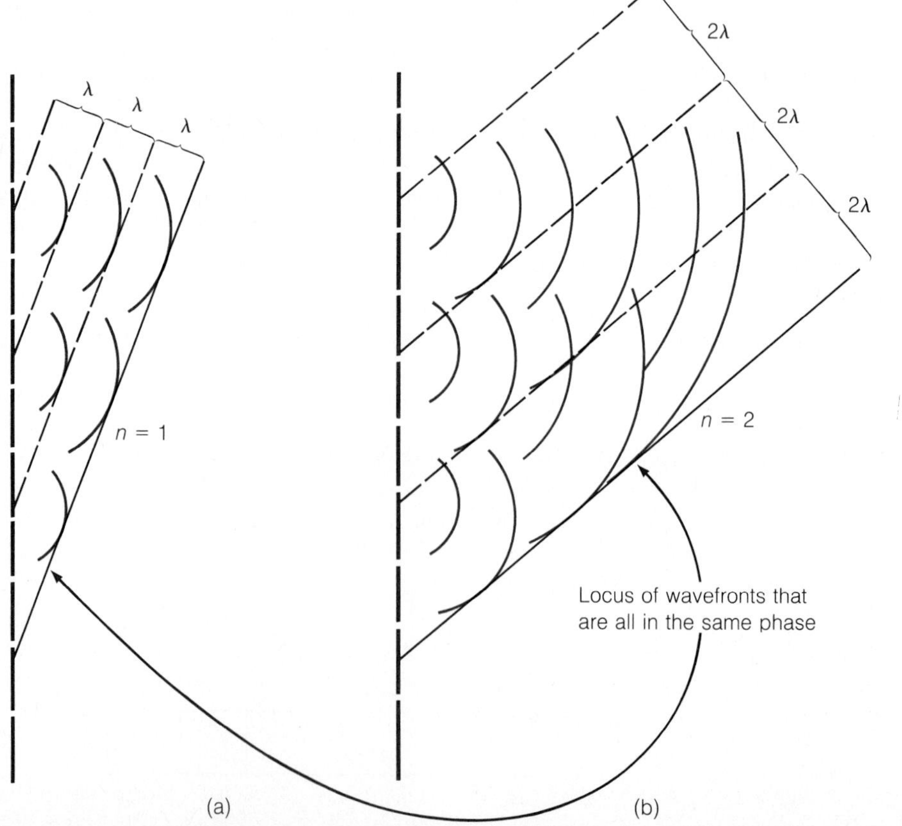

Figure 24.9
(a) In a first-order spectrum, the paths of diffracted rays from successive slits differ by λ in length. (b) In a second-order spectrum, the path differences are 2λ.

light is deviated least and red light most, which is the reverse of what occurs when a prism is used to form a spectrum.

EXAMPLE 24.2

Visible light includes wavelengths from approximately 400 nm (violet light) to 700 nm (red light). Find the angular width of the first-order spectrum produced by a grating ruled with 8000 lines/cm.

SOLUTION The slit spacing d that corresponds to 8000 lines/cm is

$$d = \frac{10^{-2}\ \text{m/cm}}{8.00 \times 10^{3}\ \text{lines/cm}} = 1.25 \times 10^{-6}\ \text{m}$$

Since $n = 1$ for a first-order spectrum, the angular deviations of violet and red light, respectively, are given by Eq. (24.6) as

$$\sin\theta_v = \frac{\lambda_v}{d} = \frac{4.00 \times 10^{-7}\ \text{m}}{1.25 \times 10^{-6}\ \text{m}} = 0.320 \qquad \theta_v = 19°$$

$$\sin\theta_r = \frac{\lambda_r}{d} = \frac{7.00 \times 10^{-7}\ \text{m}}{1.25 \times 10^{-6}\ \text{m}} = 0.560 \qquad \theta_r = 34°$$

The total width of the spectrum is therefore $34° - 19° = 15°$ (Fig. 24.10).

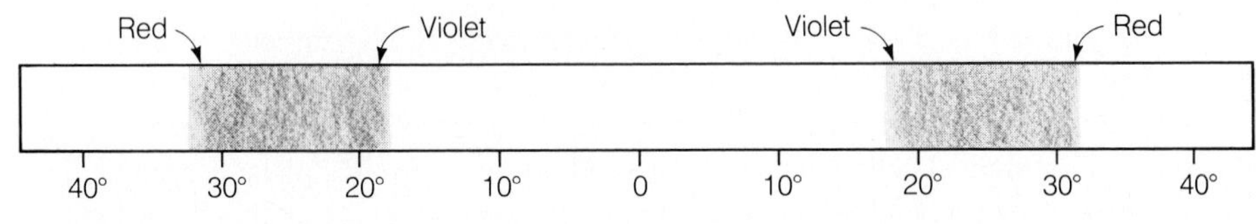

Figure 24.10
First-order spectra of white light produced by a grating ruled with 800 lines/mm.

◆

EXAMPLE 24.3

A radio transmitter operating at 15 MHz has a number of vertical antennas 50 m apart on an east-west line. How many intensity maxima in the horizontal plane are there? In what directions?

SOLUTION The line of antennas corresponds to a diffraction grating whose spacing is $d = 50$ m. The wavelength of the radio waves is

$$\lambda = \frac{c}{f} = \frac{3.0 \times 10^{8}\ \text{m/s}}{15 \times 10^{6}\ \text{Hz}} = 20\ \text{m}$$

Hence $\lambda/d = 0.40$. There is an intensity maximum at θ_0, which means in the north and south directions. Maxima also occur at angles such that $\sin\theta_n = n\lambda/d \leq 1$, since the highest value $\sin\theta$ can have is 1. When $n = 1$, Eq. (24.6) gives

$$\sin\theta_1 = \frac{\lambda}{d} = 0.40 \qquad \theta_1 = 24°$$

and when $n = 2$,

$$\sin\theta_2 = 2\frac{\lambda}{d} = 0.80 \qquad \theta_2 = 53^\circ$$

Because $3\lambda/d = 1.2$, there are only these 2 maxima on either side of θ_0. With 5 maxima to the north and 5 to the south, the total number of intensity maxima is 10. ♦

24.5 X-Ray Diffraction

How to determine crystal structures.

The atoms of a solid are arranged in regular patterns, and so its surface ought to act as a reflection grating. Then why does every mirror not produce a spectrum?

Light waves are reflected when atomic electrons at the surface of a reflecting material are caused to oscillate by the incoming em waves. The reradiated waves have their central maximum ($n = 0$) at an angle of reflection equal to the angle of incidence. Because atoms in a solid are only about 0.1 nm apart, however, λ/d equals several thousand for visible light. Thus spectra do not occur, because $\sin\theta$ cannot be more than 1. But wavelengths comparable to atomic spacings in solids are found in X rays, and indeed X-ray diffraction is the means whereby the structures of many solids have been determined. The double-helix form of the DNA molecule was discovered with the help of X-ray interference patterns.

Figure 24.11 shows a beam of monochromatic X rays directed at a crystal. Most of them go right through without interacting, but some are scattered by atoms in their paths. The conditions for a scattered ray to emerge are the same as those for the reflection of light: The incident and scattered rays must lie in a plane perpendicular to the layers of atoms, and the angle of incidence must equal the angle of scattering. In addition, unless all of the X rays scattered in a given direction are in phase, destructive interference will result, and they will cancel one another out.

We can derive the condition for constructive interference by considering the scattering of a beam of X rays from two successive layers of atoms a distance a apart, as in Fig. 24.12. Ray II must travel the distance $a \sin\theta$ farther than ray I to reach its scattering atom, and then a further $a \sin\theta$ to leave the crystal. Thus ray II goes

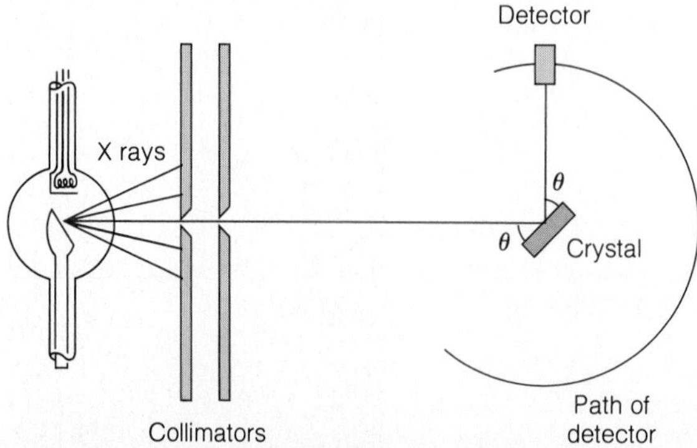

Figure 24.11

A simple X-ray diffraction apparatus.

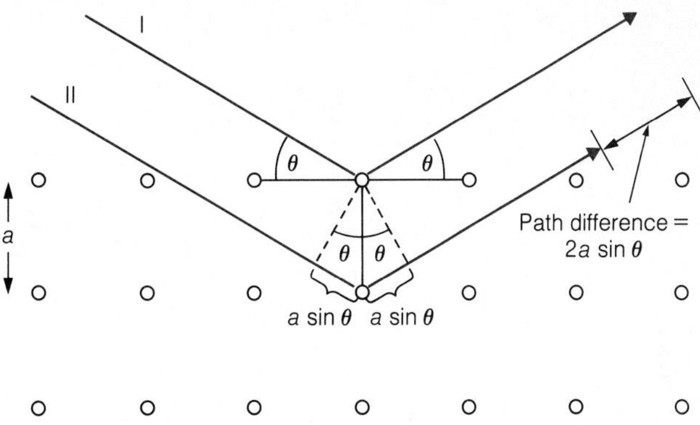

Figure 24.12

X-ray scattering from a crystal with evenly spaced atoms.

$2a$ sin θ farther than ray I. If this path difference is a whole number of X-ray wavelengths, the two rays will reinforce each other. The condition for constructive interference is therefore

$$2a \sin \theta = n\lambda \qquad n = 1, 2, 3, \ldots \tag{24.7}$$

Path difference = whole number of wavelengths

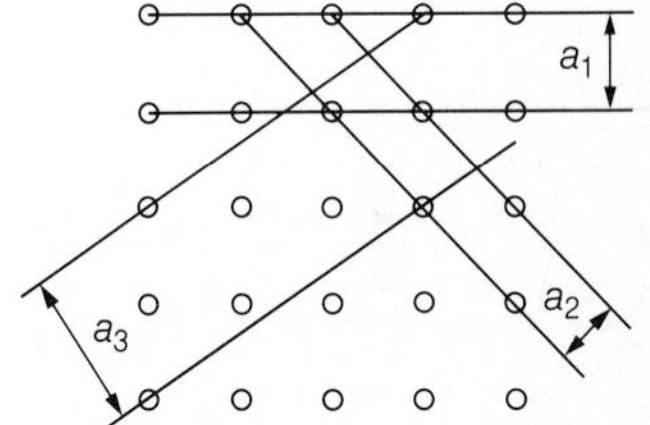

Figure 24.13

The atoms in a crystal fall into several series of parallel layers.

By directing a beam of X rays of known wavelength at a crystal and varying the angle of incidence to find the angles at which scattering occurs, the spacing a between adjacent layers of atoms can be established. Of course, the atoms fall into more than one series of parallel layers, as in Fig. 24.13, and a different spacing will be found for each of them. Most crystals have more complicated three-dimensional structures than the simple one shown in Fig. 24.12, and developments of the procedure just described have been able to establish what the structures are.

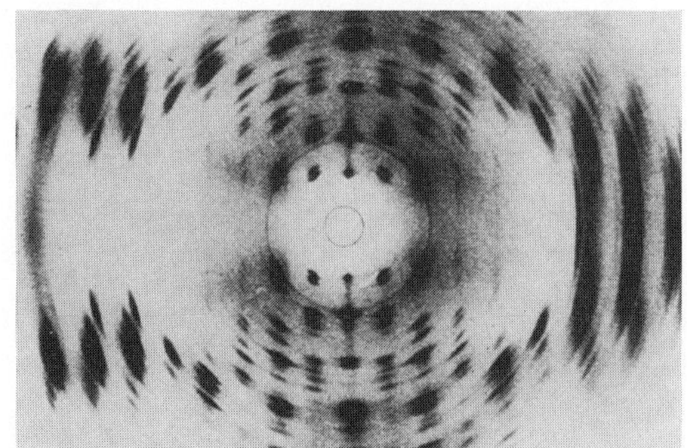

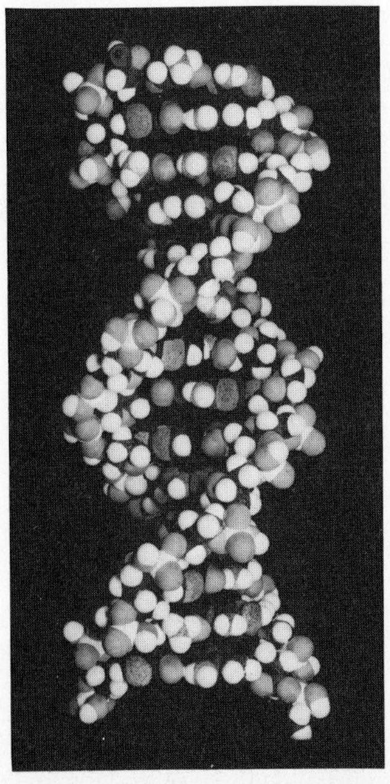

X-ray diffraction photographs such as this one led to the discovery of the double-helix form of the DNA molecule. A model of a small part of a DNA molecule is shown. A human DNA molecule, which is normally coiled and folded into a microscopic package called a chromosome, would be a meter or so long if stretched out. The development and functioning of every living organism is controlled by the DNA in its cells. When the organism reproduces, copies of its DNA are passed on to the new generation.

24.6 Thin Films

Where the colors come from.

We have all seen the marvelous rainbow colors that appear in soap bubbles and thin oil films. Some of us may also have observed the patterns of light and dark bands that occur when two glass plates are almost (but not quite) in perfect contact. Both effects owe their origins to a combination of reflection and interference.

Let us consider a beam of monochromatic light that strikes a thin film of soapy water. Figure 24.14 shows a ray picture of what happens. We notice that some reflection takes place at both the air-soap and soap-air interfaces. This is a general result: Waves are always partially reflected when they go from one medium to another in which their speed is different. (See the discussion in Section 12.1 on the reflection of pulses in a string under similar circumstances.)

A light ray consists of a succession of wavefronts. Figure 24.15 is the same diagram with the wavefronts drawn in. In (a) the two reflected wave trains are out of phase, and they interfere destructively to cancel each other partially or completely. Most or all of the light reaching this part of the soap bubble therefore passes right through.

Another part of the soap film may have a different thickness. When the film is a little thinner than in (a), the waves in the two reflected trains are exactly in phase, and they interfere constructively to reinforce one another, as in Fig. 24.15(b). Light reaching this part of the soap bubble is strongly reflected. Shining monochromatic light on a soap bubble therefore gives a pattern of light and dark that results from the varying thickness of the bubble.

When the white light shines on a soap bubble, light waves of each wavelength present pass through the soap film without reflection at those places where the film

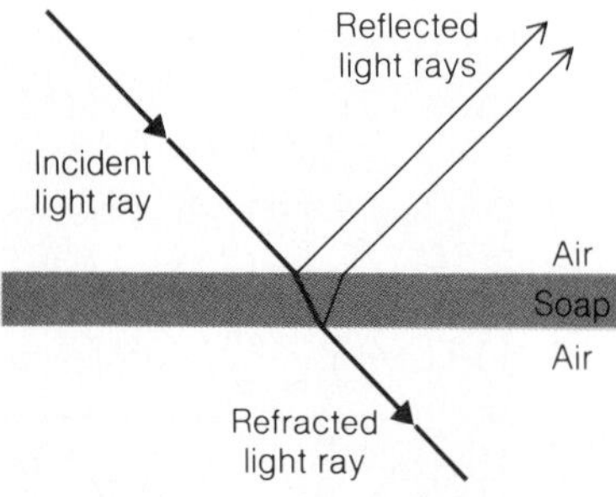

Figure 24.14

Reflection occurs at both surfaces of a soap film.

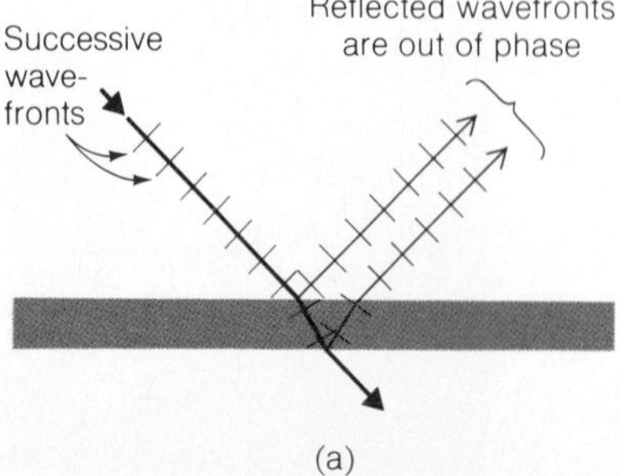

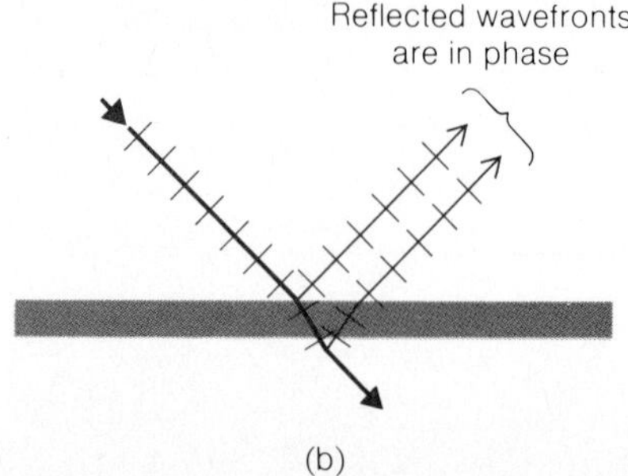

Figure 24.15

(a) Destructive and (b) constructive interference in a thin film.

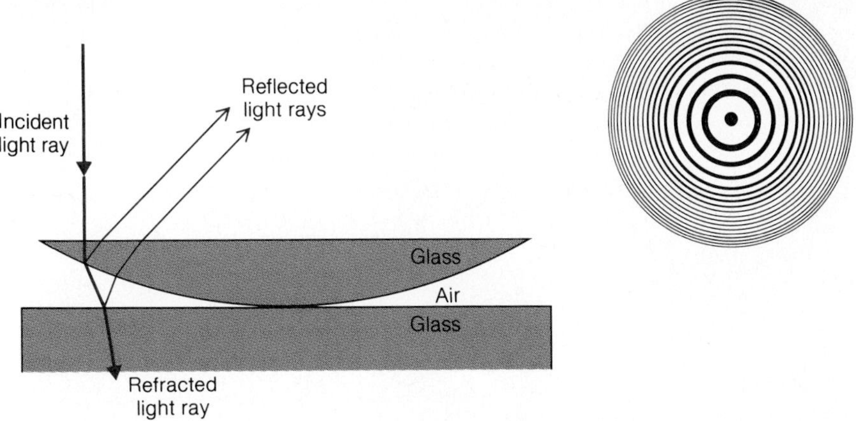

Figure 24.16
Newton's rings.

is exactly the right thickness for the two reflected rays to interfere destructively. Light waves of the other wavelengths are reflected to at least some extent and give rise to the vivid colors seen. The varying thickness of the bubble means that the color of the light reflected from the bubble changes from place to place. Exactly the same effect is responsible for the coloration of thin oil films. Generally speaking, soap or oil films whose thickness is comparable to the wavelengths in visible light give rise to the most striking color effects.

A thin film of air between two sheets of glass or transparent plastic also gives a pattern of colored bands when illuminated with white light. A notable example is **Newton's rings,** which occur when a slightly curved lens is placed on a flat glass plate (the curvature is exaggerated in Fig. 24.16). Again reflection takes place at both the top and bottom of the film, and again the result is constructive or destructive interference, depending on the film thickness. Because the thickness of the air film increases with distance from the central point of contact, the pattern of light and dark bands consists of concentric circles.

The dark spot at the center of a set of Newton's rings is not what we might expect to find (Fig. 24.17). At the center, where the air film between the pieces of

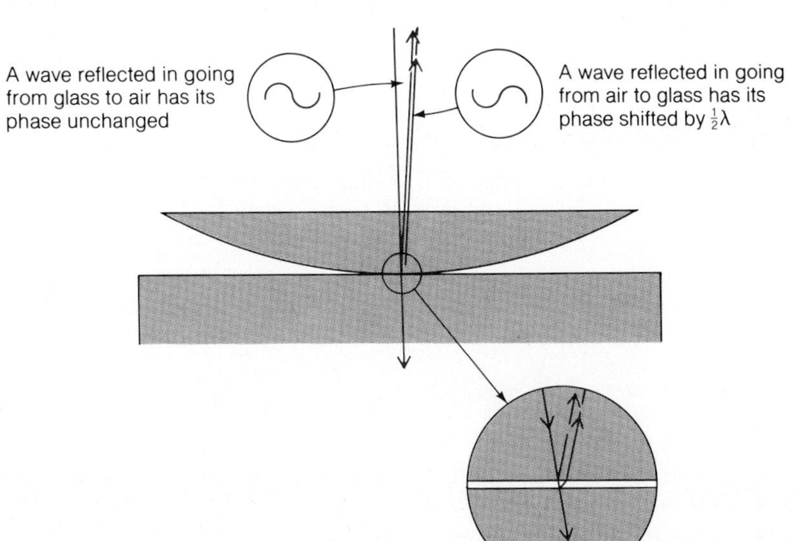

Figure 24.17
Origin of dark spot at center of Newton's rings.

Interference of light in soap bubbles gives rise to patterns of light and dark.

glass is minute, the path difference between the waves reflected from the upper and lower surfaces of the film is negligible. Hence there ought to be constructive interference and reinforcement of the light to yield a bright spot. What this analysis overlooks is the fact that a wave reflected at the surface of a new medium in which its speed is less (in optical terms, a medium of higher index of refraction) is shifted by half a wavelength. That is, a positive displacement of the wave variable is reflected as a negative one, and vice versa. The same effect was noted in the discussion of pulses in a stretched string and is shown in Fig. 12.5. Thus a $\frac{1}{2}\lambda$-shift occurs when light waves are reflected in going from air to glass, but not in going from glass to air. In consequence the two wave trains reflected at the center of a Newton's ring pattern exactly cancel each other out to give the dark spot actually observed.

EXAMPLE 24.4

Two flat glass plates 12 cm long are separated at one edge by a piece of foil 0.020 mm thick, as shown in exaggerated fashion in Fig. 24.18. (a) How far apart are the interference bands when the arrangement is perpendicularly illuminated by red light of $\lambda = 680$ nm? (b) How far apart are the bands if the space between the plates is filled with water?

SOLUTION (a) Light reflected from the upper surface of the lower plate is shifted in phase by $\frac{1}{2}\lambda$, whereas light reflected by the lower surface of the upper plate is not shifted in phase. As a result a dark band will occur when the path difference between light rays reflected by the two surfaces is 0, λ, 2λ, 3λ, and so on. At a point where the plates are d apart, the path difference is $2d$, so dark bands occur when $2d = m\lambda$,

$$d = m\frac{\lambda}{2} \qquad m = 0, 1, 2, 3, \ldots$$

Since corresponding sides of similar triangles are proportional to each other,

$$\frac{d}{D} = \frac{l}{L} \qquad l = \frac{dL}{D}$$

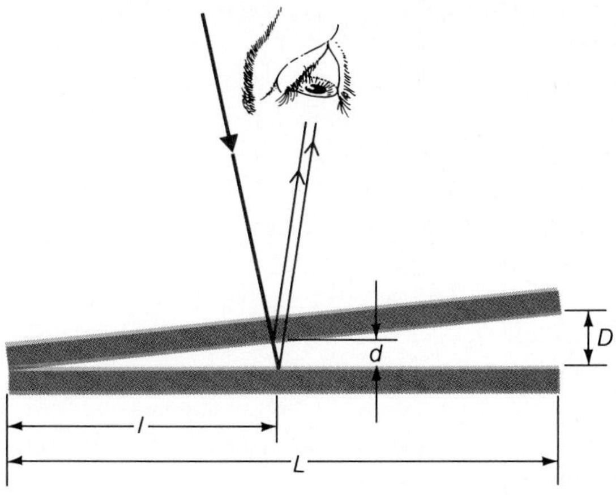

Figure 24.18

Substituting $d = m\lambda/2$, $\lambda = 680$ nm, $L = 12$ cm, and $D = 0.020$ mm yields

$$l = m\frac{\lambda L}{2D} = m\frac{(680 \times 10^{-9}\text{ m})(0.12\text{ m})}{2(2.0 \times 10^{-5}\text{ m})} = m(0.0020\text{ m})$$

$$= m(2.0\text{ mm}) \qquad m = 0, 1, 2, 3 \ldots$$

This result means that the first dark line is at $l = 0$, the second at $l = 2.0$ mm, the third at $l = 4.0$ mm, and so on, so that the bands are 2.0 mm apart. The bright bands are similarly spaced, of course.

(b) From Table 22.1 we see that the index of refraction of water is 1.33, which is less than that of glass. Hence the phase shift of $\frac{1}{2}\lambda$ at the upper surface of the lower glass plate still takes place, and the only difference is that the wavelength in the water film is reduced to λ/n. Substituting λ/n for λ in the formula for l in part (a) gives

$$l = m\left(\frac{2.0\text{ mm}}{1.33}\right) = m(1.5\text{ mm})$$

The dark bands are now 1.5 mm apart. ♦

Coated Lenses

About 4% of the light striking a glass-air interface is reflected. This is not a lot, but there may be many glass-air interfaces in an optical instrument, and the total amount of light lost through reflection may be considerable. There are ten such interfaces in each of the optical systems in a pair of binoculars, for instance, so only about two-thirds of the incoming light actually gets through to the observer's eyes. Even in a camera, where there are fewer glass-air interfaces, reflections are a nuisance because they may lead to secondary images that blur the picture.

To reduce reflections at a glass-air interface, the glass can be coated with a very thin layer of a transparent substance (usually magnesium fluoride) whose index of refraction is intermediate between those of glass and of air. If the layer is exactly $\frac{1}{4}\lambda$ thick, light reflected at its bottom will have traveled $\frac{1}{2}\lambda$ farther when it rejoins light reflected at the top of the layer, and the two will cancel out exactly, as in Fig. 24.15(a). (Light waves reflected at both the air-fluoride and fluoride-glass surfaces are shifted in phase by $\frac{1}{2}\lambda$, so these shifts have no effect on the cancellation.)

But the cancellation just described is exact only for a particular wavelength λ, whereas white light contains a range of wavelengths. What is therefore done is to choose a wavelength in the middle of the visible spectrum, which corresponds to green light, so that at least partial cancellation occurs over a wide range. The red and violet ends of the spectrum are accordingly least affected, and the light reflected from a coated lens is a mixture of these colors, a purplish hue. The average reflectivity of a glass surface coated in this way is only about 1%. Multiple coatings are sometimes used to reduce reflection even further; a triple coating brings the average reflectivity below 0.5%. Despite their lack of perfection at suppressing reflections, coated lenses transmit appreciably more light than uncoated ones, and they are universally used in fine optical instruments.

24.7 Diffraction

Why shadows are never completely dark.

Waves bend around the edge of an obstacle in their path, a behavior called **diffraction.** We all have heard sound that originated around the corner of a building from where we were standing, for example. These sound waves cannot have traveled in a straight line to our ears. Figure 24.19 shows water waves being diffracted at a gap in a barrier. The waves on the far side of the gap spread out into the geometrical "shadow" of the gap's edges. The diffracted waves spread out as though they originated at the gap, in accord with Huygen's principle.

Diffraction in light is harder to observe because the wavelengths in visible light are so short, less than 10^{-6} m, and the extent of the bending into the shadow zone is correspondingly small. (In contrast, a typical audible sound wave might have a wavelength of 1 m, and a typical wave in a pan of water might have a wavelength of 10 cm; it is easy to notice diffraction effects with such waves.) In fact, because he could not discover any diffraction with his relatively crude apparatus, Newton felt sure that light could not consist of waves.

Figure 24.20 shows what the shadow of a razor blade looks like with the use of monochromatic light from a pinhole. Such patterns of light and dark fringes are the result of interference between secondary wavelets from different parts of the same wavefront, not from different sources as in Young's double-slit experiment. The

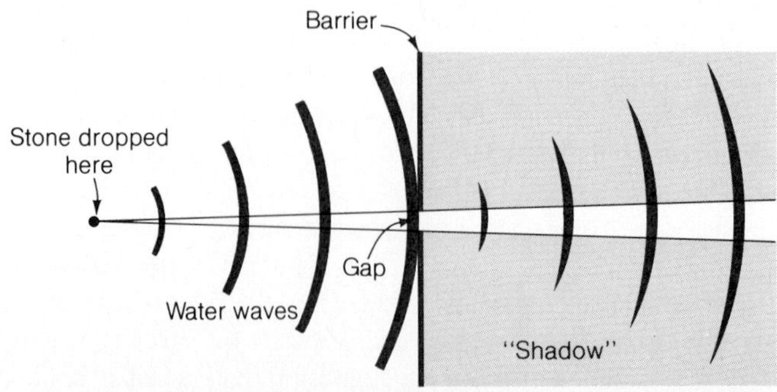

Figure 24.19

Diffraction in water waves. The waves on the far side of the gap spread out as though they had originated at the gap.

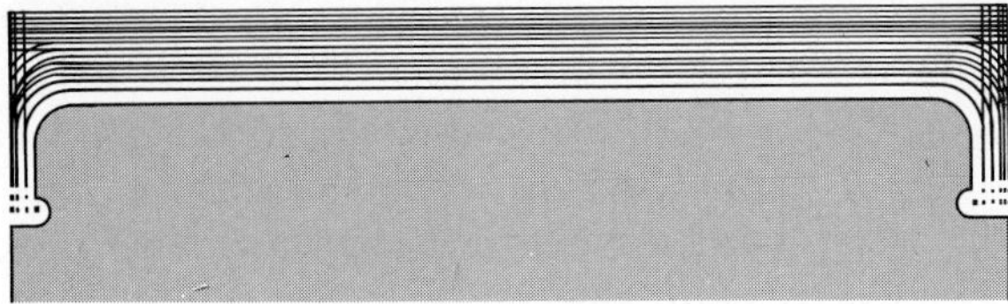

Figure 24.20

The shadow of a razor blade.

wavefronts in a beam of unobstructed light produce secondary wavelets that interfere in such a way as to produce new wavefronts exactly like the old ones (see Fig. 22.15). By obstructing part of the wavefronts, points in the shadow region are not reached by secondary wavelets from the entire initial wavefronts but only from part of them, and the result is an interference pattern.

Single Slit

Because wavelets from different parts of the same wavefront interfere with one another, light passing through a single slit can give rise to an interference pattern. Figure 24.21 shows what happens when light of wavelength λ falls on a slit a wide. The intensity of the light that reaches the screen varies as indicated to give the appearance of a series of bright and dark lines. Let us see how this pattern arises.

When $\theta = 0$, wavelets from the slit all travel the same distance to the screen, where they are in phase and produce a bright central band. As θ increases, however, path differences between wavelets grow until destructive interference is complete to give the first dark lines at the angle θ_1 on either side of the central bright band. This occurs when the wavelet from A, at the top of the slit, travels λ farther than the wavelet from E, at the bottom of the slit, as in Fig. 24.22(a).

To see why, we consider the wavelets that leave A and C, halfway down. These wavelets are $\lambda/2$ different in path length and so cancel out. Next we consider wavelets that leave the slit just below the ones from A and C. These too are exactly out of phase and also cancel out. Going down the slit in this way to wavelets from

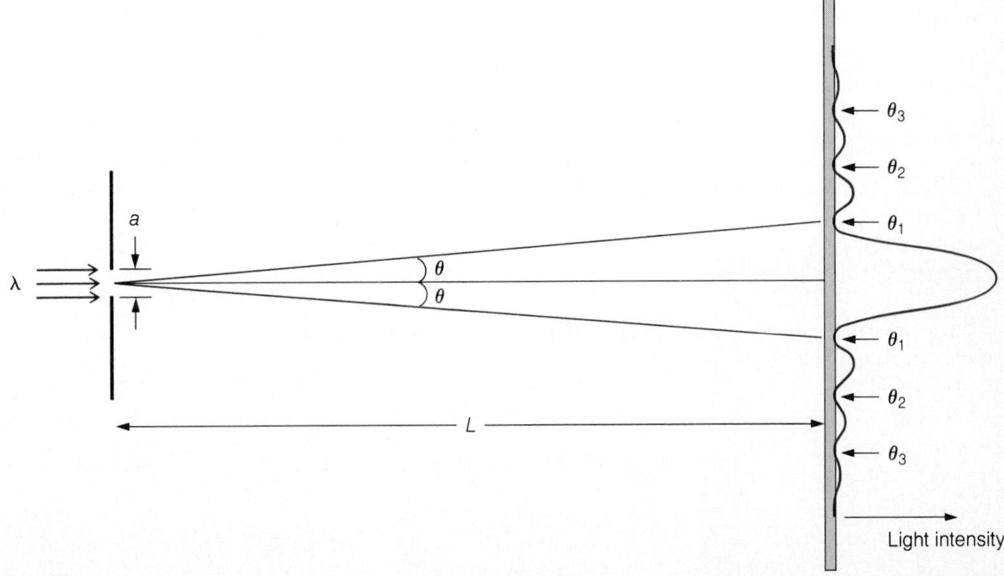

Figure 24.21

Single-slit diffraction. Dark lines appear on the screen at the angles given by the formula $\sin \theta_m = m\,\lambda/a$, where a is the width of the slit and $m = 1, 2, 3, \ldots$

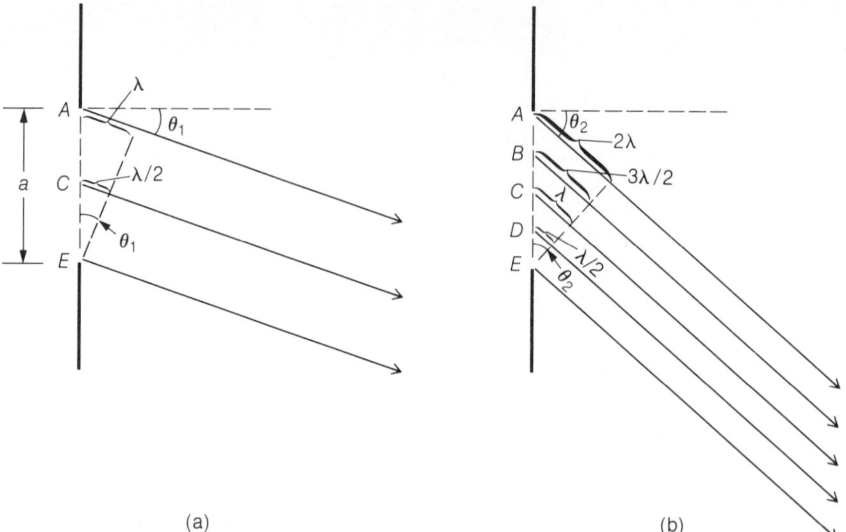

Figure 24.22

Origin of the single-slit diffraction pattern. (a) For the first dark line at θ_1, the path difference between wavelets from the top and bottom of the slit is λ. (b) For the second dark line at θ_2, the path difference is 2λ.

C and E, we find that all the wavelets from the slit in this direction cancel out in pairs. From Fig. 24.22(a) it is clear that

$$\sin\theta_1 = \frac{\lambda}{a} \tag{24.8}$$

is the condition for the first diffraction minimum, which appears as a dark line on the screen.

Cancellation is less and less complete as θ increases past θ_1. After a maximum is reached, cancellation again begins to occur and eventually gives a second minimum at θ_2. As in Fig. 24.22(b), this minimum corresponds to a path difference between wavelets from A and E of 2λ. To verify this, we now imagine the slit divided into four parts. Wavelets from A and B cancel each other, as do wavelets from C and D, because the members of each pair are $\lambda/2$ different in path length. As before, we consider wavelets from below A and B and from below C and D, which cancel out in pairs to give a dark line on the screen. Here $\sin\theta_2 = 2\lambda/a$ is the condition for the diffraction minimum. Continuing this process gives the general formula

$$\sin\theta_m = \frac{m\lambda}{a} \qquad m = 1, 2, 3, \ldots \qquad \textit{Single-slit diffraction minima} \tag{24.9}$$

Single-slit diffraction pattern in water waves. The motion is from left to right.

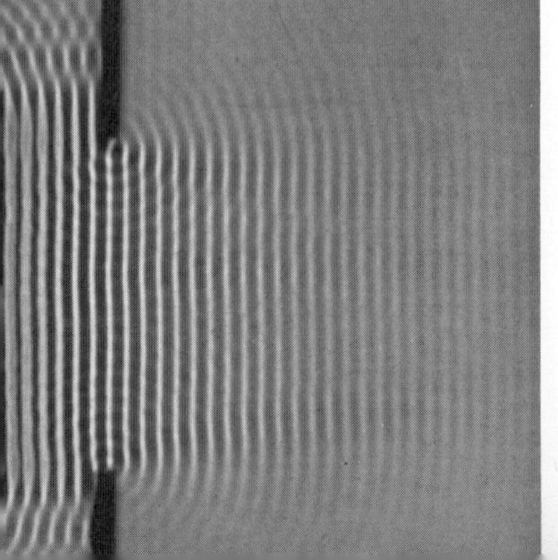

The smaller the slit width a relative to the wavelength λ, the wider the bright bands.

EXAMPLE 24.5

Monochromatic light of wavelength λ illuminates a slit 60 cm from a screen. Find the width of the central maximum for slit widths of 2λ and 10λ.

SOLUTION (a) Here $\lambda/a = 0.5$, so that

$$\sin\theta_1 = \frac{\lambda}{a} = 0.5 \qquad \theta_1 = 30°$$

The distance x_1 from the center of the diffraction pattern to the first minimum on either side is, with the help of Fig. 24.21,

$$x_1 = L \tan \theta_1$$

Hence the total width of the central maximum is

$$2x_1 = 2L \tan \theta_1 = (2)(60 \text{ cm})(\tan 30°) = 69 \text{ cm}$$

(b) Here $\lambda/a = 0.1$ and

$$\sin \theta_1 = \frac{\lambda}{a} = 0.1 \qquad \theta_1 = 6°$$

Hence

$$2x_1 = (2)(60 \text{ cm})(\tan 6°) = 12 \text{ cm}$$

◆

24.8 Resolving Power

Diffraction limits the useful magnification of an optical system.

No matter how perfect a lens, a mirror, or an optical system, the image of a point source of light it produces is always a tiny disk of light with bright and dark interference fringes around it. The diffraction pattern of light of wavelength λ from a point source that passes through a circular aperture of diameter D has its first minimum when

$$\sin \theta = 1.22 \frac{\lambda}{D} \qquad \textit{Circular aperture} \quad (24.10)$$

This is a slightly larger angle than that given by Eq. (24.8) for the first minimum in a single-slit diffraction pattern. We can use Eq. (24.10) to estimate the **resolving power** of an optical system, which refers to its ability to keep separate the images of two objects that are close together.

Let us consider the diffraction patterns on a screen produced by light from two sources near each other that has passed through a circular aperture. If the patterns overlap sufficiently, it will be impossible to distinguish them—we will see a single fringed blob of light instead of two. The criterion for just being able to resolve the sources was chosen by the English physicist Lord Rayleigh (1842–1919) as the angular separation θ_0 between them such that the first minimum of one pattern on the image falls on the central maximum of the other pattern (Fig. 24.23). This is the angular separation of Eq. (24.10). Since θ_0 is always very small, we can replace $\sin \theta_0$ by θ_0 (in radians) to give

$$\theta_0 = 1.22 \frac{\lambda}{D} \qquad \textit{Angular resolving power} \quad (24.11)$$

Figure 24.23

According to Rayleigh's criterion, the angular limit of resolution θ_0 of two light sources corresponds to the central maximum of the diffraction pattern of one source overlapping the first minimum of the pattern of the other source.

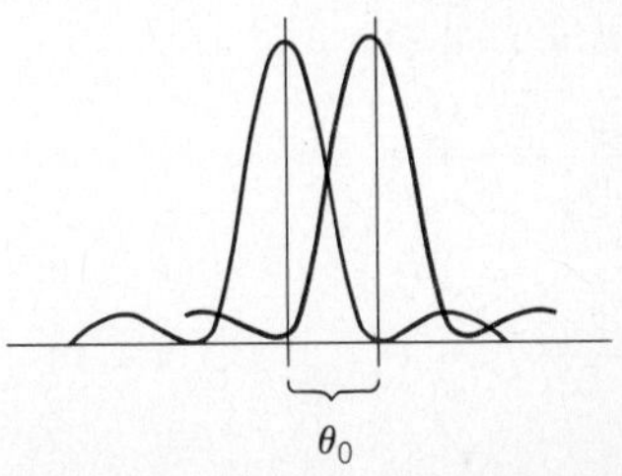

Large lens

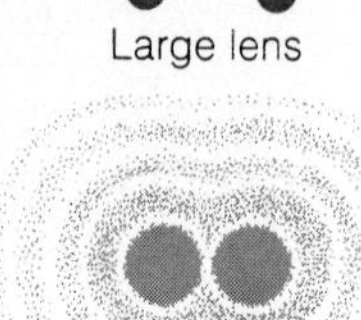

Small lens

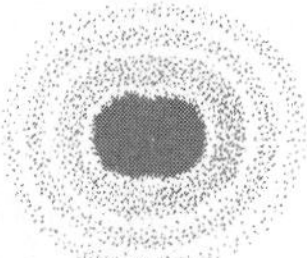

Very small lens

Figure 24.24

A large lens or mirror is better able to resolve nearby objects than a small one.

Two sources less than θ_0 apart cannot be resolved, no matter how high the magnification (Fig. 24.24).

Although Eq. (24.11) was obtained for the images of point sources of light, it is a reasonable approximation of the resolving power of a lens or mirror of diameter D used to examine actual objects. In the case of a microscope or telescope, D is taken as the diameter of the objective lens. There is no advantage to using a higher magnification than will just reveal features that subtend the angle θ_0 at the lens.

If two objects d apart are the distance L from an observer, the angle between them, in radians, is

$$\theta = \frac{d}{L}$$

Hence Eq. (24.11) can be rewritten

$$d_0 = 1.22\frac{\lambda L}{D} \qquad \textit{Linear resolving power} \quad (24.12)$$

In this formula

d_0 = minimum separation of objects that can be resolved

λ = wavelength of the light used

L = object distance

D = diameter of object lens or mirror

EXAMPLE 24.6

The pupils of a person's eyes under ordinary conditions of illumination are about 3 mm in diameter, and the distance of most distinct vision is 25 cm for most people. What is the resolving power of the eye at this distance under the assumption that it is limited only by diffraction?

SOLUTION Using $\lambda = 550$ nm, which is in the middle of the visible spectrum, Eq. (24.12) gives

$$d_0 = 1.22\frac{\lambda L}{D} = \frac{(1.22)(5.5 \times 10^{-7}\text{ m})(0.25\text{ m})}{0.003\text{ m}} = 6 \times 10^{-5}\text{ m} = 0.06\text{ mm}$$

The photoreceptors in the retina are not quite close enough together to permit this degree of resolution, and 0.1 mm is a more realistic figure under good conditions. In terms of angular resolving power, $\theta_0 \approx 5 \times 10^{-4}$ radians for the human eye. ♦

The dependence of resolving power on λ/D poses a severe problem for radiotelescopes, which are antennas designed to receive radio waves from astronomical objects such as stars, galaxies, and gas clouds. Because radio waves from space might have wavelengths as much as a million times greater than those in visible light, an antenna would have to be tens or hundreds of kilometers across to be able to separate sources as close together in the sky as optical telescopes can. In fact, modern electronics permits such resolution to be achieved by combining the signals received by a series of widely spaced antennas. Radio astronomers are currently setting up such an antenna array extending from the Virgin Islands across the Continental United States to Hawaii, a span of about 8000 km—nearly the diameter of the earth. This array should provide better resolution than is possible with even the largest optical telescope.

Numerical Aperture

The **numerical aperture** NA of a microscope objective is a measure of its resolving power. If α is the half-angle of the cone of light rays that reach an objective from an object, as in Fig. 24.25, and n is the index of refraction of the medium between the objective and the object, then by definition

$$\text{NA} = n \sin \alpha \qquad \textit{Numerical aperture} \quad (24.13)$$

The resolving power of the objective is

$$d_0 = 1.22 \frac{\lambda}{2\text{NA}} \qquad \textit{Microscope objective} \quad (24.14)$$

The greater the diameter of an objective relative to its focal length and the higher the index of refraction n, the greater the NA and the finer the detail that can be resolved.

For an objective used in air, the angle α has a maximum possible value of 90°, which would mean NA = 1 and $d_0 = 0.61\ \lambda$. In practice, α seldom exceeds 75°, which corresponds to an NA of 0.97. However, if a liquid of high index of refraction is used between the objective and the object, the NA can be increased. Thus an immersion oil of $n = 1.55$, together with a lens for which $\alpha = 75°$, gives an NA of 1.5. The higher the magnification of an objective, the higher the NA it requires in order that the enlarged details be sharp enough to be distinguished. An NA of 0.25 is sufficient for a 10× objective, for instance, whereas an NA of at least 1.25 is usual for a 100× objective.

The eyepiece of a microscope provides the user with an image 25 cm in front of the eye, and so the limit to the detail one can see clearly is about 0.1 mm, as mentioned in Example 24.6. The finest detail the microscope itself can resolve is

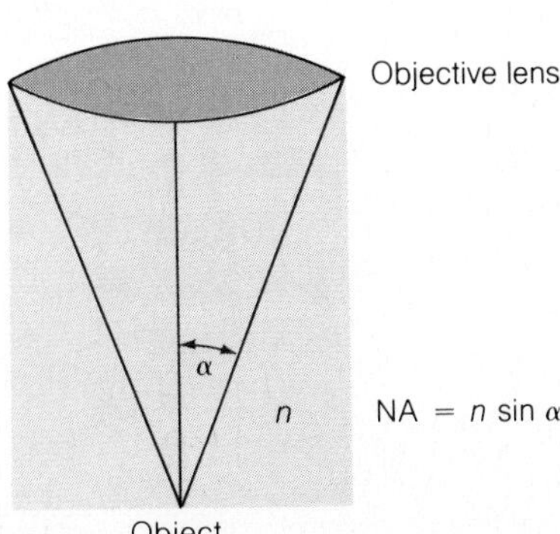

Figure 24.25

The numerical aperture of a microscope objective depends on the angular width of the cone of light that reaches it from the object and also on the index of refraction of the medium between the objective and the object.

d_0, which for $\lambda = 550$ nm and NA $= 1.5$ is

$$d_0 = 1.22\frac{\lambda}{2\text{NA}} = \frac{(1.22)(5.5 \times 10^{-7}\text{ m})}{(2)(1.5)} = 2.2 \times 10^{-7}\text{ m}$$

This is about half a wavelength of the light being used!

The ratio between 0.1 mm and d_0 is

$$M = \frac{10^{-4}\text{ m}}{2.2 \times 10^{-7}\text{ m}} = 455$$

This represents the maximum useful magnification of the microscope in the sense that higher magnifications will give larger images but will not reveal any more detail. A larger image is nevertheless useful when examining small objects in order to reduce eyestrain, but magnifications much beyond 1000× serve no purpose. For really high magnifications an electron microscope must be used. The electrons in such a device behave in certain respects like waves with very short wavelengths (see Section 26.5), and so greater resolving powers are possible.

24.9 Polarization

Polarized transverse waves lie in a single plane of polarization.

A **polarized** beam of transverse waves is one whose vibrations occur in only a single direction perpendicular to the direction in which the beam travels. The entire wave motion is confined to a plane called the **plane of polarization** (Fig. 24.26).

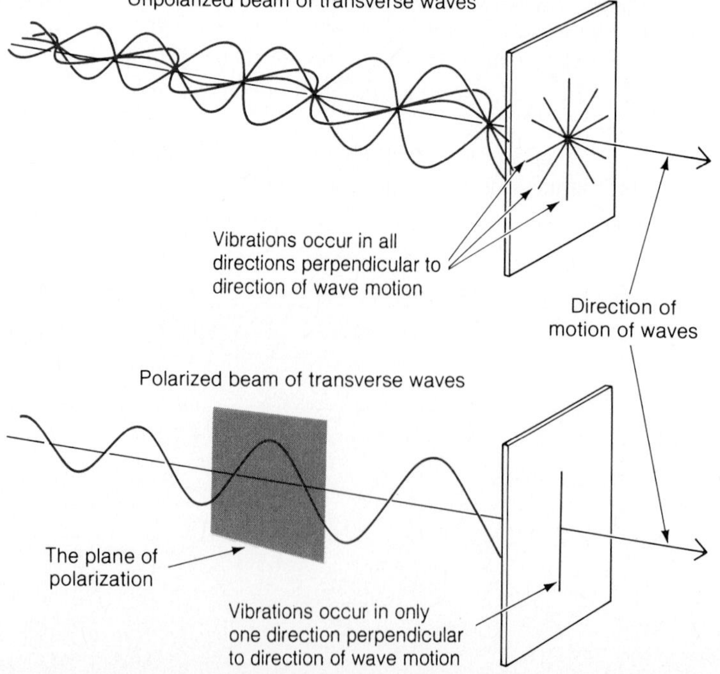

Figure 24.26

An unpolarized and a polarized beam of transverse waves.

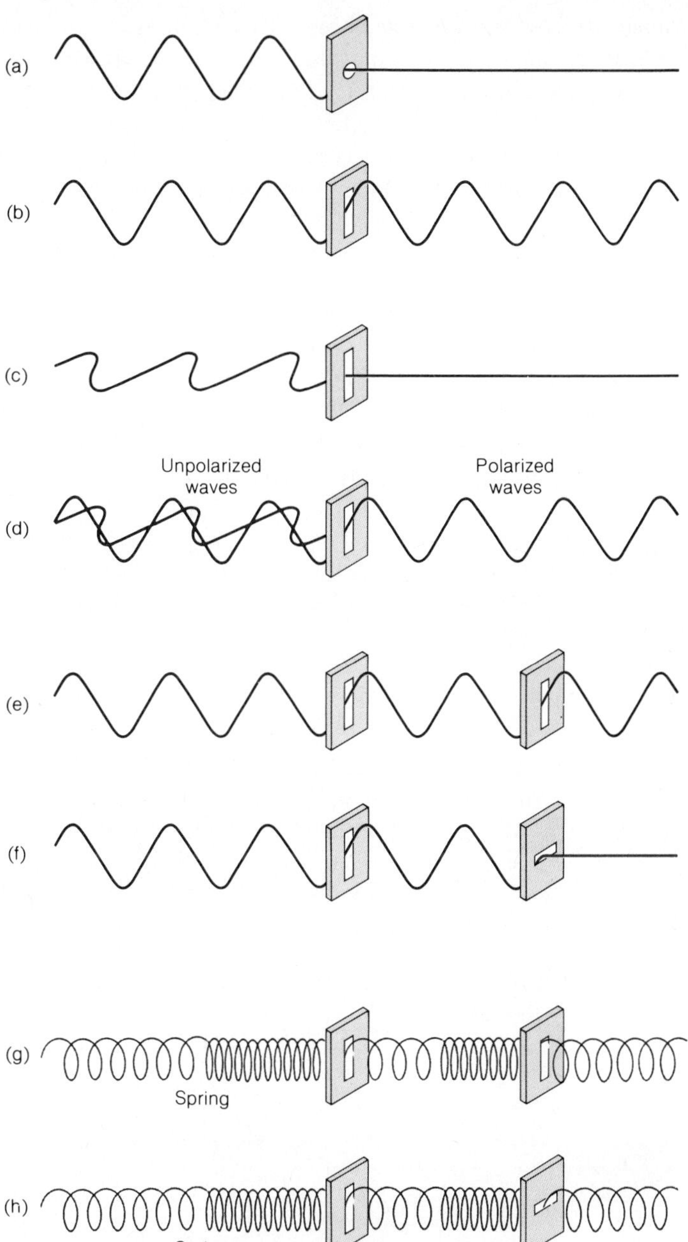

Figure 24.27

Mechanical analogies of fundamental polarization phenomena.

When many different directions of polarization are present in a beam of transverse waves, vibrations occur equally often in all directions perpendicular to the direction of motion, and the beam is then said to be **unpolarized.** Since the vibrations that make up longitudinal waves can take place in only one direction—namely, that in which the waves travel—longitudinal waves cannot be polarized.

Light waves are transverse, and it is possible to produce and detect polarized light. To clarify the ideas involved, let us first consider the behavior of transverse waves in a stretched string. If the string passes through a hole in a barrier, as in Fig. 24.27(a), waves traveling down the string are stopped because the string cannot vibrate there. When the hole is replaced by a vertical slot, waves whose vibrations are vertical can get through the barrier, but waves with vibrations in other directions cannot; see Fig. 24.27(b) and (c). In a situation in which several waves vibrating in different directions move down the string, the slot stops all but vertical vibrations, as in (d). An initially unpolarized series of waves has become polarized.

Such an approach can be used to determine whether a particular wave can be polarized or not. In the case of a stretched string, what we do is put another barrier a short distance from the first, as in Fig. 24.27(e). If the slot in the new barrier is also vertical, those waves that can get through the first one can also get through the second. If the slot in the new barrier is horizontal, however, it will stop all waves that reach it from the first barrier, as in (f).

Should longitudinal waves (say in a spring) go through the barrier, it is possible that their amplitudes might decrease in passing through the slots, but the relative alignments of the slots would not matter; see Fig. 24.27(g) and (h). On the other hand, the alignment of the slots is the critical factor in the case of transverse waves.

This chain of reasoning made it possible for the polarization of light waves to be demonstrated in the last century. A number of substances—for instance, quartz, calcite, and tourmaline—have different indexes of refraction for light with different planes of polarization relative to their crystal structures. Prisms can be made from these substances that transmit light in only a single plane of polarization. When a beam of unpolarized light is incident upon such a prism, only those of its waves whose planes of polarization are parallel to a particular plane in the prism emerge from the other side. The remainder of the waves are absorbed or deflected.

Polaroid is an artificially made polarizing material in wide use that transmits light with only a single plane of polarization. To exhibit the transverse nature of light waves, we first place two Polaroid disks in line so that their axes of polarization are parallel (Fig. 24.28a) and note that all light passing through one disk also passes through the other. Then we rotate the second disk. As we do so, less and less light gets through. Finally, when the axis of polarization of the second disk is perpendicular to that of the first, as in Fig. 24.28(b), all the light that passes through the first disk is stopped by the second.

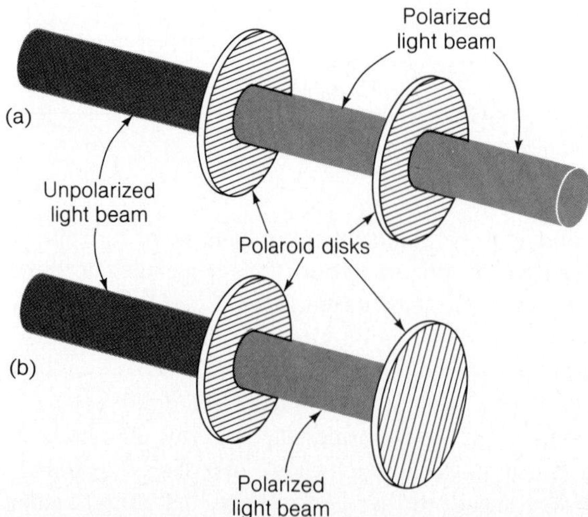

Figure 24.28

Experiment showing the transverse nature of light waves.

Overlapping Polaroid sheets whose directions of polarization are perpendicular.

Just what is it whose vibrations are aligned in a beam of polarized light? As we know, light waves consist of oscillating electric and magnetic fields perpendicular to each other. Because it is the electric fields of light waves whose interactions with matter produce nearly all common optical effects, the plane of polarization of a light wave is considered to be the plane in which lie both the direction of its electric field and the direction of the wave [the vertical plane in Fig. 22.3(b)]. Even though nothing material moves during the passage of a light wave, it is possible to establish its transverse nature and identify its plane of polarization.

Polarization by Reflection

Light can be polarized when it is reflected from the interface between two media whose indexes of refraction are different. Figure 24.29 shows a beam of unpolarized light arriving at such an interface at the angle of incidence p. Two planes

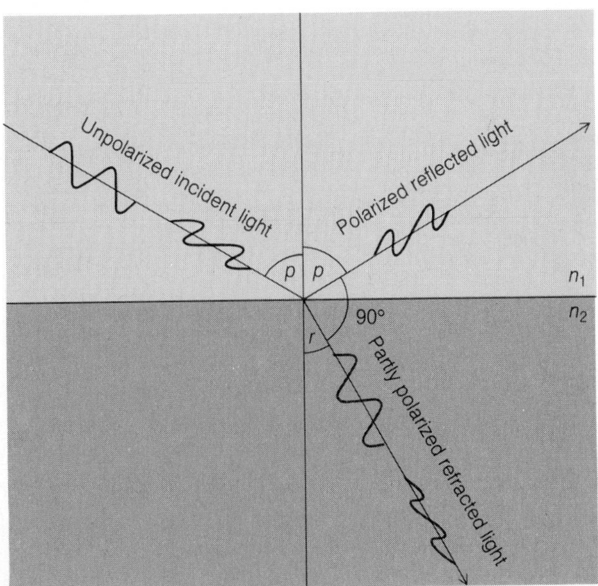

Figure 24.29

Polarization by reflection.

of polarization in the incoming beam are indicated, one in the plane of the paper and one perpendicular to it. (We can think of the beam as having its various polarizations resolved into these components.) Part of the beam is reflected at the angle p, and the rest is refracted at the angle r. If the angle between the reflected and refracted beams is a right angle, none of the waves whose planes of polarization are parallel to the paper is reflected. Thus the reflected wave is completely polarized and the refracted wave is partly polarized.

To find the angle of incidence for which the reflected beam is polarized, we use the fact that $p + r = 90°$ together with Snell's law (Eq. 22.9):

$$n_1 \sin p = n_2 \sin r = n_2 \sin (90° - p) = n_2 \cos p$$

Hence

$$\frac{\sin p}{\cos p} = \frac{n_2}{n_1}$$

$$\tan p = \frac{n_2}{n_1} \qquad \textit{Polarizing angle} \qquad (24.15)$$

The angle p is called **Brewster's angle.**

EXAMPLE 24.7

At what angle of incidence is sunlight reflected from the surface of a lake fully polarized?

SOLUTION Here $n_1 = n_{air} = 1.00$ and $n_2 = n_{water} = 1.33$, so

$$\tan p = \frac{n_2}{n_1} = 1.33 \qquad p = 33°$$

The polarization of sunlight reflected by a lake can be verified by looking at the image of the sun through Polaroid sunglasses while rotating them. ◆

24.10 Scattering

Why the sky is blue.

As mentioned earlier, the oscillating electric field in a light wave causes electrons in the atoms and molecules it encounters to oscillate at the same frequency. The electrons then reradiate light waves with this frequency. As a result, some of the incoming light is **scattered** to the side of its original path.

In general, the intensity of light of wavelength λ scattered by something small compared with λ is proportional to λ^{-4}. The shorter the wavelength, the greater the proportion of the incoming light that is scattered. This is the reason the sky is blue. When we look at the sky, what we see is light from the sun that has been scattered by molecules in the upper atmosphere. Blue light, which consists of the

BIOGRAPHICAL NOTE

Lord Rayleigh

Lord Rayleigh (1842–1919) was born John William Strutt to a wealthy English family and inherited his title on the death of his father. After being educated at home, he went on to be an outstanding student at Cambridge University and then spent some time in the United States. On his return Rayleigh set up a laboratory in his home. There he carried out both experimental and theoretical research except for a five-year period when he directed the Cavendish Laboratory at Cambridge following Maxwell's death in 1879.

For much of his life Rayleigh's work concerned the behavior of waves of all kinds, and he made many contributions to acoustics and optics. One of the types of wave an earthquake produces is named after him. In 1871 Rayleigh explained the blue color of the sky in terms of the preferential scattering of short-wavelength sunlight in the atmosphere. The formula for the resolving power of an optical instrument is another of his achievements.

At the Cavendish Laboratory, Rayleigh completed the standardization of the volt, the ampere, and the ohm, a task Maxwell had begun. Back at home, he found that nitrogen prepared from air is very slightly denser than nitrogen prepared from nitrogen-containing chemical compounds. Together with the chemist William Ramsay (1852–1916), Rayleigh showed that the reason for the discrepancy was a hitherto unknown gas that made up about 1% of the atmosphere. They called the gas argon, from the Greek word for "inert," because argon did not react with other substances. Ramsay went on to discover the other inert gases neon ("new"), krypton ("hidden"), and xenon ("stranger"). He was also able to isolate the lightest inert gas, helium, which had 30 years earlier been identified in the sun by its spectral lines (see Chapter 27); *helios* means "sun" in Greek. Rayleigh and Ramsay won Nobel prizes in 1904 for their work on argon.

What was possibly Rayleigh's greatest contribution to science came after the discovery of argon and took the form of an equation that did not agree with experiment. The problem was accounting for the spectrum of blackbody radiation, that is, the relative intensities of the different wavelengths present in such radiation. Rayleigh calculated the shape of this spectrum; because the astronomer James Jeans pointed out a small error Rayleigh had made, the result is called the Rayleigh-Jeans formula. The formula follows directly from the laws of physics known at the end of the nineteenth century—and it is hopelessly incorrect, as Rayleigh and Jeans were aware. (For instance, the formula predicts that a blackbody should radiate energy at an infinite rate.) The search for a correct blackbody formula led to the founding of the quantum theory of radiation by Max Planck and Albert Einstein, a theory that was to completely revolutionize physics.

Despite the successes of quantum theory and of Einstein's theory of relativity that followed soon afterward, Rayleigh, after a lifetime devoted to classical physics, never really accepted them. He died in 1919.

---

shortest wavelengths, is scattered about ten times more readily than red light, so the scattered light is chiefly blue in color (Fig. 24.30a). At sunrise or sunset, when sunlight must make a very long passage through the atmosphere to reach an observer, much of its short-wavelength content is scattered out along the way, and the sun

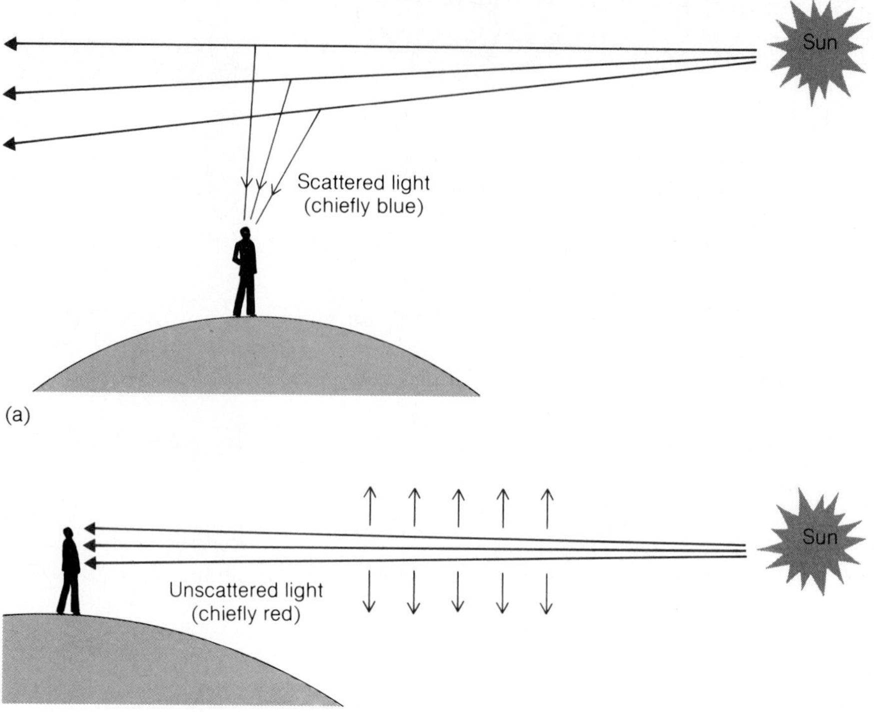

Figure 24.30

(a) Blue light is scattered the most by the earth's atmosphere, red light the least. Hence the sky appears blue. (b) Why the sun appears red at sunrise and sunset.

accordingly appears red in color (Fig. 24.30b). The water droplets and ice crystals in clouds are larger than λ, and the scattering they produce is independent of λ. Hence clouds do not appear colored. Above the atmosphere the sky appears black, and astronauts can see the moon, stars, and planets in the daytime.

Skylight not only is blue but is partly polarized as well. To check this, all one has to do is hold up a piece of Polaroid against the sky and rotate it. Sunglasses are often made with Polaroid lenses so oriented as to discriminate against the polarized skylight, which reduces glare while affecting other light to a lesser extent. The human eye itself responds equally to all states of polarization. The eyes of bees, however, can detect the polarization of scattered sunlight from a patch of blue sky. This enables bees to find the direction of the sun for navigation when the sun itself is obscured by clouds.

Figure 24.31 shows why scattered light is polarized. A beam of unpolarized light heading in the $+z$-direction strikes some air molecules and is scattered. In this process electrons in the molecules are set in vibration by the electric fields of the light waves, and the vibrating electrons then reradiate. Because the electric field of an electromagnetic wave is perpendicular to its direction of motion, the electric fields in the initial beam lie in the xy-plane only. A scattered wave that proceeds downward, which is the $-y$-direction in Fig. 24.31, can have its electric field in the x-direction only, so it is polarized. Only two planes of polarization for the incident light are shown for clarity. In the case of intermediate planes of polarization, the components of **E** in the x-direction can also lead to scattered waves traveling downward that are similarly polarized. Because skylight arrives at our eyes from a variety of directions, the polarization is not complete, but enough occurs to be easily demonstrated.

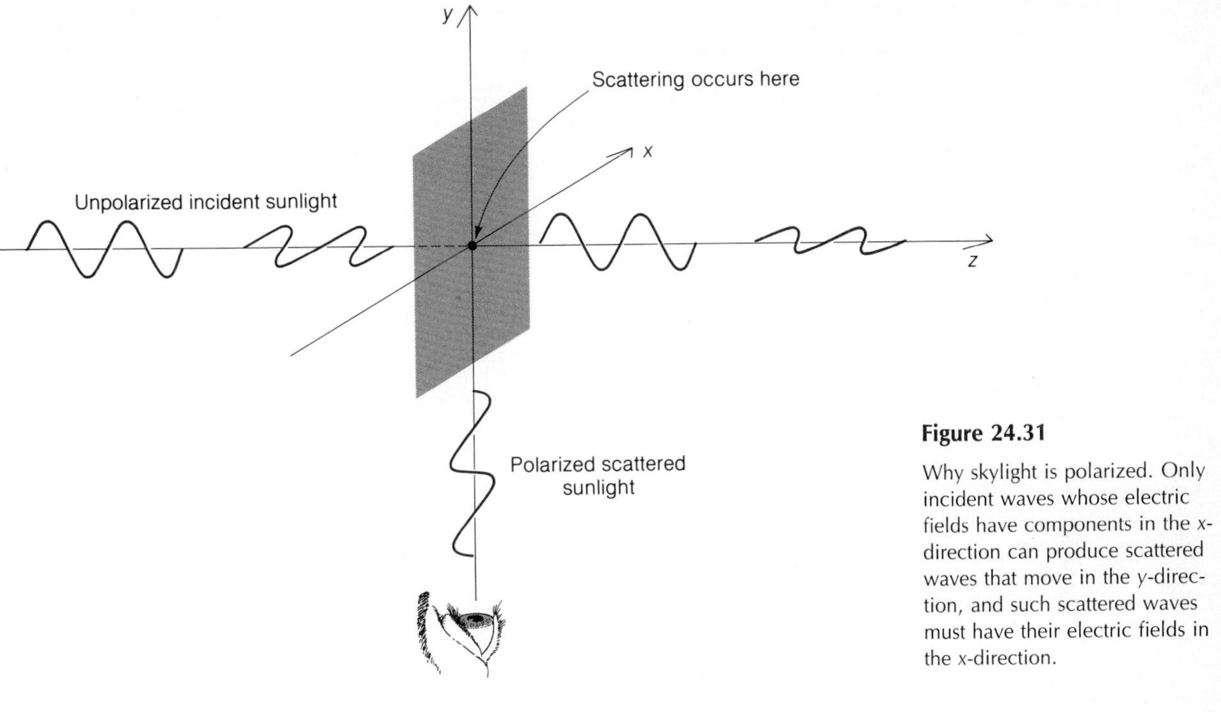

Figure 24.31

Why skylight is polarized. Only incident waves whose electric fields have components in the x-direction can produce scattered waves that move in the y-direction, and such scattered waves must have their electric fields in the x-direction.

Important Terms

Two sources of waves are **coherent** if there is a fixed phase relationship between the waves they emit during the time the waves are being observed. Interference can be observed only in waves from coherent sources.

The ability of waves to bend around the edges of obstacles in their paths is called **diffraction.** A **diffraction grating** is a series of parallel slits that produces a spectrum through the interference of light that is diffracted by them.

The **resolving power** of an optical system refers to its ability to produce separate images of nearby objects; resolving power is limited by diffraction, and the larger the objective lens of an optical system, the greater its resolving power.

A **polarized** beam of transverse waves is one whose vibrations occur in only a single direction perpendicular to the direction in which the beam travels, so that the entire wave motion is confined to a plane called the **plane of polarization.** An **unpolarized** beam of transverse waves is one whose vibrations occur equally often in all directions perpendicular to the direction of motion.

Important Formulas

Double slit:

$$y = 0, \frac{L\lambda}{d}, \frac{2L\lambda}{d}, \frac{3L\lambda}{d}, \cdots \quad \text{(bright lines)}$$

Diffraction grating: $\sin\theta = n\frac{\lambda}{d}$ (bright lines)

X-ray diffraction: $2a \sin\theta = n\lambda$

Single slit: $\sin\theta_m = \frac{m\lambda}{a}$ (dark lines)

Resolving power: $\theta_0 = 1.22\frac{\lambda}{D}$

$$d_0 = 1.22\frac{\lambda L}{D}$$

Polarizing angle: $\tan p = \frac{n_2}{n_1}$

MULTIPLE CHOICE

1. In order to produce an interference pattern, the waves used must
   a. all have the same wavelength
   b. be in a narrow beam
   c. be coherent
   d. be electromagnetic

2. Coherent electromagnetic waves are not emitted by
   a. two antennas connected to the same radio transmitter
   b. two pinholes in an opaque shield over a sodium-vapor lamp
   c. a pinhole in an opaque shield over a sodium-vapor lamp and its reflection in a mirror
   d. two lasers

3. In a double-slit experiment the maximum intensity of the first bright line on either side of the central one occurs on the screen at locations where the arriving waves differ in path length by
   a. $\lambda/4$ b. $\lambda/2$
   c. λ d. 2λ

4. When a spectrum is formed by a double slit or a grating, the light deviated most
   a. is red
   b. is green
   c. is blue
   d. depends on the slit spacing

5. A diffraction grating is superior to a double slit for wavelength measurements for which one or more of the following reasons?
   a. The line of a single wavelength is sharper.
   b. The line of a single wavelength is brighter.
   c. Light of a larger range of wavelengths can be analyzed.
   d. The angular width of the spectrum is much wider.

6. The greater the number of lines that are ruled on a grating of given width,
   a. the shorter the wavelengths that can be diffracted
   b. the longer the wavelengths that can be diffracted
   c. the narrower the spectrum that is produced
   d. the broader the spectrum that is produced

7. X rays are used to determine crystal structures because
   a. their wavelengths are much shorter than atomic spacings in crystals
   b. their wavelengths are comparable to atomic spacings in crystals
   c. their wavelengths are much longer than atomic spacings in crystals
   d. they are easily polarized

8. Thin films of oil and soapy water owe their brilliant colors to a combination of reflection and
   a. refraction b. interference
   c. diffraction d. polarization

9. A factor not involved in the production of the colored patterns exhibited when white light is incident on a thin oil film is
   a. the presence of different wavelengths in white light
   b. the varying thickness of the oil film
   c. interference between incident and reflected light
   d. reflection at both the upper and lower surfaces of the oil film

10. Coating a glass surface with a thin layer of a substance whose index of refraction is between those of air and glass completely prevents the reflection of light
   a. of one wavelength only
   b. of two wavelengths only
   c. of purple color only
   d. of all wavelengths

11. In a single-slit diffraction pattern
   a. all the bright bands have the same width
   b. the central bright band is the widest
   c. the central bright band is the narrowest
   d. any of the above, depending on the ratio between the wavelength of the light and the slit width

12. One way to improve the resolving power of a lens is to increase the
   a. wavelength of the light used
   b. brightness of the light used
   c. diameter of the lens
   d. object distance

13. The numerical aperture of a microscope objective lens is a measure of its
   a. diameter b. focal length
   c. magnification d. resolving power

14. An unpolarized beam of transverse waves is one whose vibrations
   a. are confined to a single plane
   b. occur in all directions
   c. occur in all directions perpendicular to their direction of motion
   d. have not passed through a Polaroid disk

15. It is impossible to polarize
   a. white light b. radio waves
   c. X rays d. sound waves

16. The sky is blue because
   a. air molecules are blue
   b. the lens of the eye is blue
   c. the scattering of light is more efficient the shorter its wavelength
   d. the scattering of light is more efficient the longer its wavelength

17. Light scattered perpendicularly to an incident beam is

partly

a. diffracted b. polarized
c. monochromatic d. coherent

18. Seen from above the atmosphere, the sky is

a. blue b. red
c. white d. black

19. Polarized light cannot be produced by

a. reflection
b. scattering
c. diffraction
d. selective absorption

20. Monochromatic green light of wavelength 500 nm illuminates a pair of narrow slits 1.0 mm apart. The separation of bright lines on the interference pattern formed on a screen 2.0 m away is

a. 0.10 mm b. 0.25 mm
c. 0.40 mm d. 1.0 mm

21. The wavelength of light that is deviated in the first order by 15° by a 5000-line/cm grating is

a. 52 nm b. 259 nm
c. 518 nm d. 773 nm

22. A pair of binoculars designated "7 × 50" has a magnification of 7 and an objective lens diameter of 50 mm. The smallest detail that in principle can be perceived with such an instrument when viewing an object 1 km away in light of wavelength 500 nm has a linear dimension of approximately

a. 1 mm b. 1 cm
c. 10 cm d. 1 m

QUESTIONS

1. Can light from incoherent sources interfere? If so, then why is a distinction made between coherent and incoherent sources?

2. What becomes of the energy of the light waves whose destructive interference leads to dark lines in an interference pattern?

3. What advantages has a diffraction grating over a double slit for determining wavelengths of light?

4. What governs the angular width of the first-order spectrum of white light produced by a grating?

5. What is the difference between the first-order and the second-order spectra produced by a grating? Which is wider? Does a prism produce spectra of different orders?

6. Why is visible light not useful for studying the structure of a crystal, whereas X rays are?

7. As a soap bubble is blown up, its wall becomes thinner and thinner. Just before the bubble breaks, the thinnest part of its wall turns black. Why?

8. What do diffraction and interference have in common? How do they differ? In Young's double-slit experiment, which effects are due to diffraction and which to interference?

9. Radio waves diffract pronouncedly around buildings, whereas light waves, which are also electromagnetic waves, do not. Why?

10. Explain the peculiar appearance of a distant light source when seen through a piece of finely woven cloth.

11. Give two advantages a large-diameter telescope objective has over a small-diameter one.

12. A camera can be made by using a pinhole instead of a lens. What happens to the sharpness of the picture if the hole is too large? If it is too small?

13. Since light consists of transverse waves, why is not every light beam polarized?

14. What is the relationship between the plane of polarization of a transverse wave and its direction of propagation?

15. Which of the following can occur in (a) transverse waves and (b) longitudinal waves: refraction, dispersion, interference, diffraction, polarization?

EXERCISES

24.2 Double Slit

24.3 Wavelength of Light

1. Two parallel slits 0.10 mm apart are illuminated by light of wavelength 546 nm. A viewing screen is 0.80 m from the slits. (a) How far from the central bright line is the next bright line? (b) How far is the bright line after that? (c) How far is the third dark line?

2. Two parallel slits 0.12 mm apart are illuminated by light of wavelength 500 nm. A viewing screen is 1.5 m from the slits. (a) How far from the central bright line is the next bright line? (b) How far is the first dark line? (c) How far is the fifth dark line?

3. Two parallel slits 0.25 mm apart are illuminated by light of two wavelengths, 500 and 600 nm. A viewing screen is 2.0 m away from the slits. How far are the first-order bright lines of one wavelength from those of the other?

4. Monochromatic light is used to illuminate a pair of narrow slits 0.30 mm apart, and the interference pattern is observed on a screen 90 cm away. The second dark band appears 3.0 mm from the center of the pattern. What is the wavelength of the light?

5. Light of unknown wavelength is used to illuminate two parallel slits 1.0 mm apart. Adjacent bright lines on the interference pattern that results on a screen 1.5 m away are 0.65 mm apart. What is the wavelength of the light?

6. Two parallel slits are illuminated by light of two wavelengths, one of which is 580 nm. On a viewing screen an unknown distance from the slits, the fourth dark line of the light of the known wavelength coincides with the fifth bright line of the light of the unknown wavelength. Find the unknown wavelength.

24.4 Diffraction Grating

7. Light of wavelength 750 nm is directed on a grating ruled with 4000 lines/cm. What is the angular deviation of this light in (a) the first order and (b) the third order?

8. A 5500-line/cm diffraction grating produces an image deviated by 27° in the second order. Find the wavelength of the light.

9. The antennas of Example 24.3 are moved so that they are 15 m apart on the same east-west line. How many intensity maxima in the horizontal plane are there now? In what directions?

10. Light containing wavelengths of 500 and 550 nm is directed at a 2000-line/cm grating. How far apart are the lines formed by these wavelengths on a screen 4.0 m away in the second order?

11. White light that contains wavelengths from 400 to 700 nm is directed at a 3000-line/cm grating. How wide is the first-order spectrum on a screen 2.0 m away?

24.5 X-Ray Diffraction

12. A beam of X rays with wavelengths down to 20 pm is directed at a crystal whose atomic planes are 0.20 nm apart at an angle of 7°. X rays of what wavelengths will be diffracted?

13. The distance between atomic planes in calcite is 0.30 nm. What is the smallest angle between these planes and an incident beam of 30-pm X rays at which diffracted X rays can be detected?

24.7 Diffraction

14. Light of wavelength λ illuminates a single slit. Under what, if any, circumstances will no diffraction minima occur?

15. Light of wavelength 600 nm illuminates a slit and produces a diffraction pattern on a screen 50 cm away whose central maximum has a total width of 60 cm. How wide is the slit?

16. Light of wavelength 500 nm illuminates a slit 0.0020 mm wide. A screen is 100 cm away. Find the distances from the center of the screen of the first, second, and third dark lines.

17. Monochromatic light of unknown wavelength illuminates a slit 0.0030 mm wide and produces a central maximum 20° wide. Find the wavelength of the light.

24.8 Resolving Power

18. A radar has a resolving power of 30 m at a range of 1.0 km. What is the minimum width of its antenna if its operating frequency is 9500 MHz?

19. The Hale telescope at Mount Palomar in California has a concave mirror 5.0 m in diameter. How many meters apart must two features of the moon's surface be in order to be resolved by this telescope? Take the distance from the earth to the moon as 3.8×10^5 km and use 500 nm for the wavelength of the light.

20. An astronaut circles the earth in a satellite at an altitude of 150 km. If the diameter of his pupils is 2.0 mm and the average wavelength of the light reaching his eyes is 550 nm, is it conceivable that he can distinguish sports stadiums on the earth's surface? Private houses? Cars?

21. The Jodrell Bank radiotelescope has a parabolic reflecting "dish" 76 m in diameter. (a) What is the angular diameter in degrees of a point source of radio waves of wavelength 21 cm as seen by this telescope? (b) How does this figure compare with the angular width of the moon, whose diameter is 3476 km and whose average distance from the earth is 3.8×10^5 km?

22. According to a famous battle command, "Don't fire until you see the whites of their eyes." (a) If the diameter of the white of an eye is 20 mm, the diameter of the pupil in bright sunlight is 2.0 mm, and the light has a wavelength of 500 nm, what is this distance on the basis of the theoretical resolution formula? (b) The actual resolving power of the eye, which is limited by the structure of the retina, is about 5.0×10^{-4} rad. Find the maximum distance at which the white of an eye can be distinguished on the basis of this figure.

23. At night the pupils of a person's eyes are 8.0 mm in diameter. (a) How many km away from a car facing a woman will she be able to distinguish its headlights from each other? (b) If her pupils were 4.0 mm in diameter (say at twilight), how far away from the car could she distinguish its headlights from each other? Assume the headlights are 1.5 m apart, that the average wavelength of their light is 600 nm, and that her eyes are capable of attaining their theoretical resolving power.

PROBLEMS

24. Interference can be demonstrated by using a single light source together with a mirror, as in Fig. 24.32. Light from the source S illuminates the screen directly, and the combination of this light with light reflected from the mirror produces an interference pattern. Find the condition for interference maxima on the screen for light of wavelength λ in terms of the distances shown in the figure. Note that the reflection introduces a phase shift of $\lambda/2$.

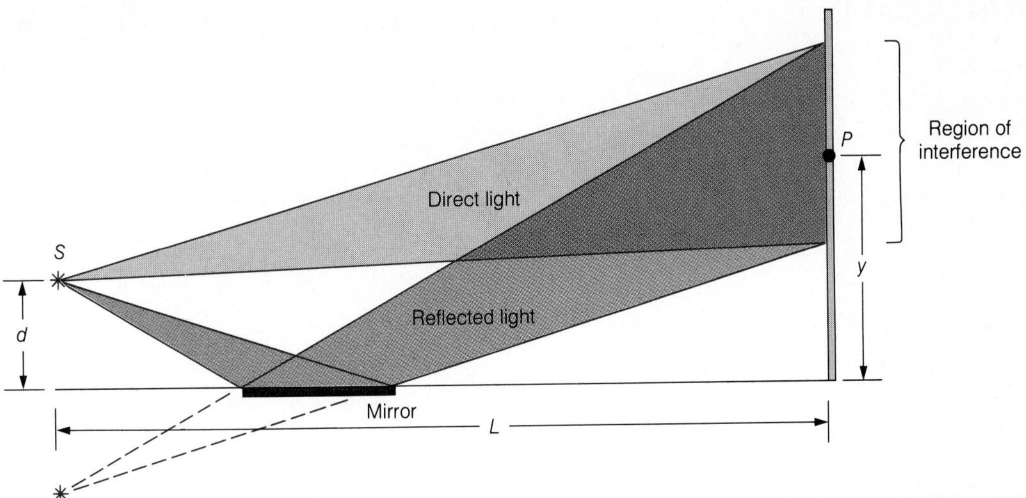

Figure 24.32

Problem 24. Both d and y are actually small compared with L.

25. How many diffracted images are formed on either side of the central image when radiation of wavelength 600 nm falls on a 4000-line/cm grating?

26. White light that contains wavelengths from 400 to 700 nm is directed at a 5000-line/cm grating. Do the first- and second-order spectra overlap? The second- and third-order spectra? If there is an overlap, will the use of a grating with a different number of lines/cm change the situation?

27. The index of refraction of magnesium fluoride is 1.38. How thick should an antireflection coating of this material be for maximum cancellation at 550 nm, which is the middle of the visible spectrum?

28. The index of refraction of a certain soap bubble illuminated by white light is $n = 1.35$. If the bubble appears orange ($\lambda = 630$ nm) at a point nearest the viewer, what is the minimum thickness of the bubble at that point? Be sure to take into account any phase shifts on reflection.

29. A 50-mm^3 drop of oil spreads out uniformly over an area of 3000 cm^2 on the surface of a body of water. When light of wavelength 500 nm is directed perpendicularly on the oil film, the film appears black. The index of refraction n of the oil is greater than that of water. What is the minimum possible value of n?

30. A certain laser produces a beam of monochromatic light whose wavelength is 550 nm and whose initial diameter is 1.0 mm. (a) In what distance will diffraction have caused the beam to double its diameter? (b) If the initial diameter of the beam were 10 mm, would the doubling distance be different? If so, what would it be?

31. The smaller the aperture of a camera lens, the greater the depth of field. However, a small aperture means reduced resolution. The criterion for an enlarged print to show sharp detail from a small negative is that the image of a point object on the negative be not more than 0.010 mm across. Find the maximum f-number of a camera lens for this criterion to be met for a distant object in 550-nm light.

Answers to Multiple Choice

1. a	**7.** b	**13.** d	**19.** c
2. b	**8.** b	**14.** c	**20.** d
3. c	**9.** c	**15.** d	**21.** c
4. a	**10.** a	**16.** c	**22.** b
5. a, b	**11.** b	**17.** b	
6. d	**12.** c	**18.** d	

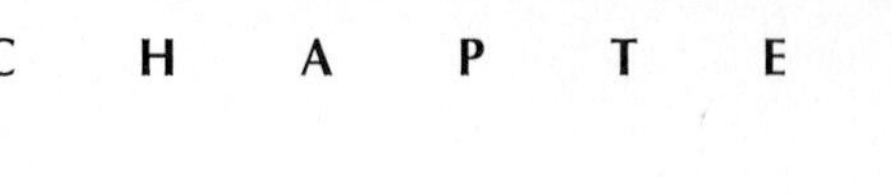

CHAPTER 25

RELATIVITY

Few physical theories represent so drastic an assault on traditional habits of thought as the theory of relativity. Relativity links time and space, matter and energy, electricity and magnetism—and for all the seeming magic of its conclusions, most of them can be reached with the simplest of mathematics. The theory of relativity was proposed in 1905 by Albert Einstein, and little of physical science since then has remained unaffected by his ideas.

25.1 Special Relativitiy

All motion is relative.

Until now nothing much has been said about how such quantities as length, time, and mass are measured. It was simply assumed that these quantities could be determined in some way. Because standard units have been established for each quantity, it would not seem to matter who does the job—everybody ought to get the same figure. There is certainly no question of principle associated with, say, finding the length of an airplane on the ground: All we need do is place one end of a tape measure at the airplane's nose and note the number on the tape at the airplane's tail.

But what if the airplane is in flight and we are on the ground? It is not hard to find the length of a distant object with the help of a surveyor's transit to measure angles, a tape measure to establish a baseline, and a knowledge of trigonometry. Because the airplane is moving, however, things become more complicated. Now we must take into account the fact that light does not travel instantaneously from one place to another but does so at a definite, fixed speed—and light is the means by which information is carried from a distant object to our instruments. When a careful analysis is made of the problem of measuring physical quantities when there is relative motion between the instruments and whatever is being observed, many surprising results emerge.

In Chapter 1 the notion of **frame of reference** was introduced. When we see something moving, what we actually detect is that its position relative to something else is changing (Fig. 25.1). A passenger moves relative to a train; the train moves relative to the earth; the earth moves relative to the sun; the sun moves relative to the galaxy of stars (the "Milky Way") of which it is a member; and so on. In each case a frame of reference is part of the description of the motion. It means nothing to say that something is moving unless we know the frame of reference with respect to which the motion occurs.

There is no universal frame of reference that can be used everywhere. If we see something changing its position with respect to us at constant velocity, we have no way of knowing whether *it* is moving or *we* are moving. If we were isolated from the rest of the universe, we could not find out if we were moving at constant velocity or not—indeed, the question would make no sense. All motion is relative to the observer, and there is no such thing as "absolute motion."

The theory of relativity deals with the consequences of the lack of a universal frame of reference. **Special relativity,** proposed in 1905 by Albert Einstein, treats problems that involve frames of reference moving at constant velocity with respect to one another. **General relativity,** put forward ten years later by Einstein, treats problems that involve frames of reference accelerated with respect to one another. Special relativity has had an enormous impact on almost all of physics, and its chief results will be examined here.

Two principles underlie special relativity. The principle of relativity states:

The laws of physics are the same in all frames of reference moving at constant velocity with respect to one another.

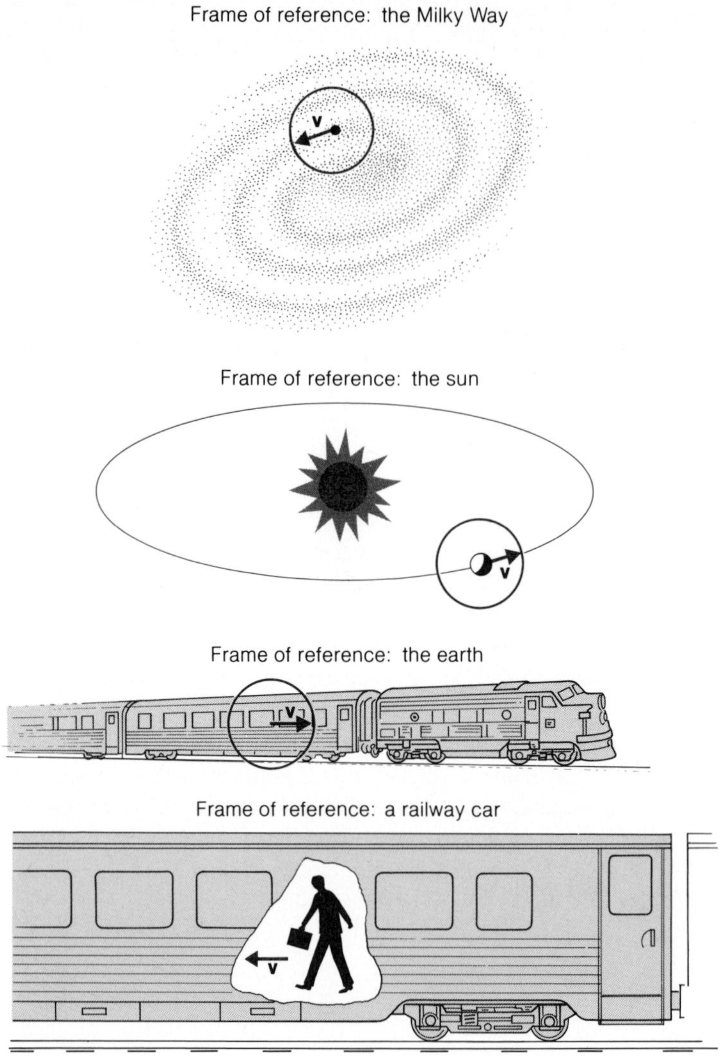

Figure 25.1

Some frames of reference.

This principle follows from the absence of a universal frame of reference. If the laws of physics were different for different observers in relative motion, the observers could tell from these differences which of them were "stationary" in space and which were "moving." But such a distinction does not exist in nature, and the principle of relativity expresses this fact.

Thus experiments of any kind performed, for instance, in an elevator climbing at a constant velocity give exactly the same results as they would when the elevator is at rest or is falling at a constant velocity. On the other hand, an isolated observer *can* detect accelerations, as any elevator passenger can verify.

The second principle is based on the results of a great many experiments:

The speed of light in free space has the same value for all observers, regardless of their state of motion or the state of motion of the source.

To appreciate how remarkable this principle is, let us look at a hypothetical experiment basically no different from actual ones that have been carried out in a number of ways. Suppose I turn on a searchlight just as you take off in a spacecraft at a speed of 250,000 km/s (Fig. 25.2). We both measure the speed of the light waves from the searchlight by using identical instruments. From the ground I find their speed to be 300,000 km/s, as usual. "Common sense" tells me you ought to find a speed of (300,000 − 250,000) km/s, or only 50,000 km/s, for the same light waves. But you also find their speed to be 300,000 km/s, even though to me you seem to be moving parallel to the waves at 250,000 km/s. As so often, common sense is wrong.

There is only one way to account for the apparent discrepancy between these results without violating the principle of relativity. It must be true that measurements of space and time are not absolute but depend on the relative motion of the observer and that which is observed. If I were to measure from the ground the rate at which your clock ticks and the length of your meter stick, I would find that the clock ticks more slowly than it did on the ground and that the meter stick is shorter in the direction of motion of the spacecraft. To you, your clock and meter stick are the same as they were on the ground before you took off. To me they are different because of the relative motion, but in such a way that the speed of light you measure

Figure 25.2

The speed of light is the same to all observers

v = 250,000 km/s

c = 300,000 km/s

(a)

(b)

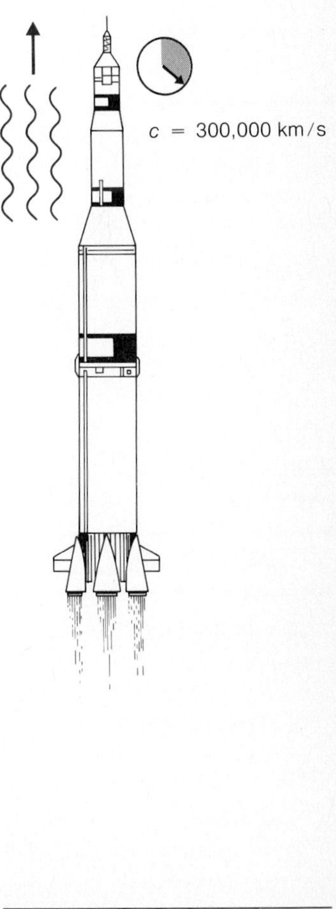

is the same 300,000 km/s that I measure. Time intervals and lengths are relative quantities, but the speed of light in free space is the same to all observers.

Before Einstein's work, a conflict had existed between the principles of mechanics, which were then based on Newton's laws of motion, and those of electricity and magnetism, which had been developed into a unified theory by Maxwell. Newtonian mechanics had worked well for over two centuries. Maxwell's theory not only covered all that was known at the time about electric and magnetic phenomena but had also predicted the existence of electromagnetic waves and identified light as an example of them. But the equations of Newtonian mechanics and those of electromagnetism differ in the way they relate measurements made in one frame of reference with those made in another frame in relative motion. Einstein showed that Maxwell's theory is consistent with special relativity whereas Newtonian mechanics is not, and his modification of mechanics brought these branches of physics into accord. As we will find, relativistic and Newtonian mechanics agree for relative speeds much lower than the speed of light, which is why the latter theory seemed correct for so long. However, at higher speeds Newtonian mechanics fails and must be replaced by a relativistic formulation.

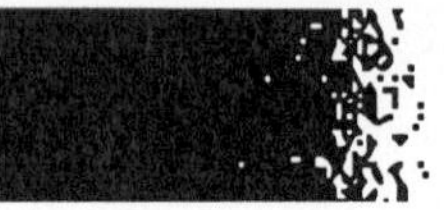

25.2 Relativity of Time

A moving clock ticks more slowly than a clock at rest.

Measurements of time intervals are affected by relative motion between an observer and what is observed. As a result, a clock moving with respect to an observer ticks more slowly than it does without such motion, and all processes (including those of life) occur more slowly to an observer when they take place in a frame of reference in relative motion.

We begin with the particularly simple clock shown in Fig. 25.3. In this clock a pulse of light is reflected back and forth between two mirrors. Whenever the light strikes the lower mirror, an electric signal is produced that marks the recording tape. Each mark corresponds to the tick of an ordinary clock.

Let us consider two of these clocks, one of them at rest in a laboratory and another in a spaceship moving at the speed v relative to the laboratory. An observer in the laboratory watches both clocks: Does she find that they tick at the same rate?

Figure 25.4 shows the laboratory clock in action. The mirrors are L apart, and the time interval between ticks is t_0. Hence the time needed for the light pulse to travel the distance L between the mirrors at the speed c is $t_0/2$, and so

$$L = c\left(\frac{t_0}{2}\right) \qquad t_0 = \frac{2L}{c}$$

Figure 25.5 shows the moving clock with its mirrors parallel to the direction of the relative velocity as seen from the laboratory. The time interval between ticks is t. Because the clock is moving, the pulse of light follows a zigzag path in which it travels the distance $ct/2$ in going from one mirror to the other in the time $t/2$. From the Pythagorean theorem,

$$\left(\frac{ct}{2}\right)^2 = L^2 + \left(\frac{vt}{2}\right)^2$$

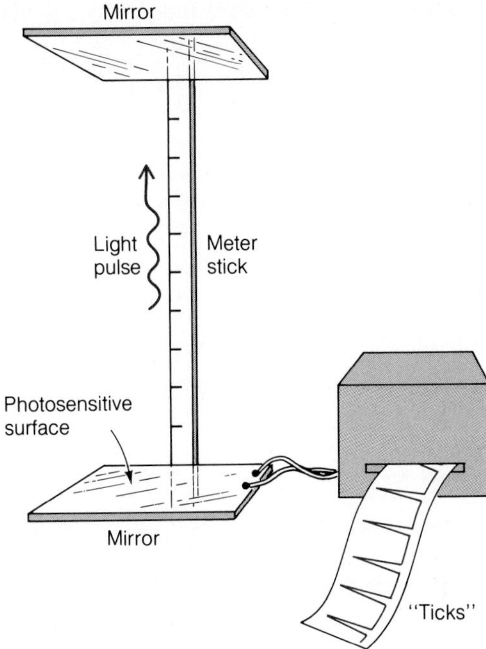

Figure 25.3

A light-pulse clock.

How is t related to t_0? To find out, we first solve the preceding equation for t:

$$\frac{t^2}{4}(c^2 - v^2) = L^2$$

$$t^2 = \frac{4L^2}{(c^2 - v^2)} = \frac{(2L)^2}{c^2(1 - v^2/c^2)}$$

$$t = \frac{2L/c}{\sqrt{1 - v^2/c^2}}$$

Figure 25.4

Light-pulse clock in the laboratory as seen by an observer in the laboratory. The dial represents a conventional clock in the laboratory.

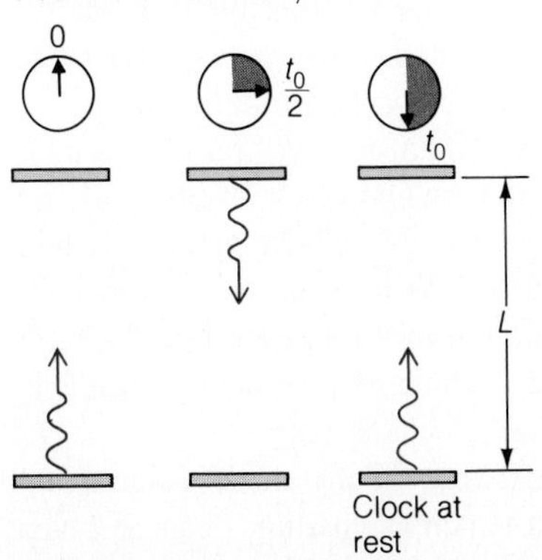

Figure 25.5

Light-pulse clock in the spaceship as seen by an observer in the laboratory. The dial represents a conventional clock in the laboratory.

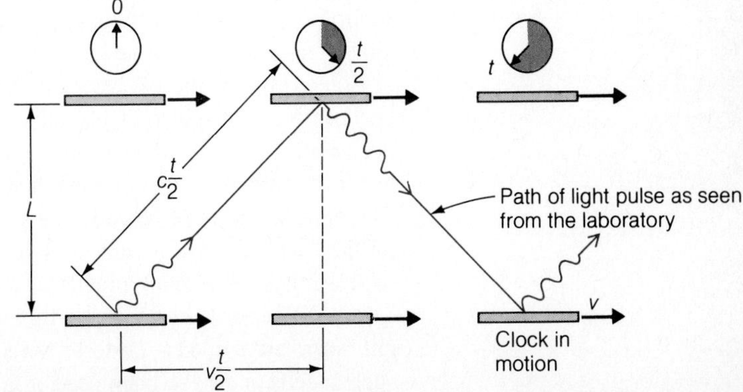

But the quantity $2L/c$ is the time interval t_0 between ticks in the laboratory clock, as we saw. Hence

$$t = \frac{t_0}{\sqrt{1 - v^2/c^2}} \qquad \textit{Time dilation} \quad (25.1)$$

Here is a reminder of what the symbols in this formula represent:

t_0 = time interval on clock at rest relative to observer

t = time interval on clock in relative motion to observer as determined by observer

v = speed of relative motion

c = speed of light

Because the quantity $\sqrt{1 - v^2/c^2}$ is always smaller than 1 for a moving object, t is always greater than t_0. *A clock moving with respect to an observer ticks more slowly than a clock that is stationary with respect to the same observer.* This effect is referred to as **time dilation** (to dilate is to become larger).

Now let us turn the situation around. What does an observer in a spacecraft find when he compares his clock with one on the ground? The only change needed in the preceding argument is the direction of motion: If the person on the ground sees the spacecraft heading east, the person in the spacecraft sees the laboratory on the ground heading west. To the person in the spacecraft, the light pulse of the ground clock follows a zigzag path that requires a total time per round trip of

$$t = \frac{t_0}{\sqrt{1 - v^2/c^2}}$$

whereas the light pulse in his own clock takes t_0 for the round trip. Thus to the person in the spacecraft, the clock on the ground ticks at a slower rate than his own clock does. A clock moving relative to an observer *always* is slower than a clock at rest relative to him, regardless of where the observer is located.

A light clock is a more exotic timepiece than most of us are familiar with. What if a stationary cuckoo clock and a moving one are compared: Do we again find that the moving clock runs more slowly?

The principle of relativity makes it easy for us to predict the result. Suppose cuckoo clocks tick at exactly the same rate to all observers, whether there is relative motion or not. We put a cuckoo clock and a light-pulse clock (which *does* tick more slowly in motion) on a spacecraft. On the ground they show the same time.

In flight the two clocks show different times to an observer on the ground, because the light-pulse clock ticks slower whereas the cuckoo clock (by hypothesis) does not. To an observer in the spacecraft, however, the two clocks agree, because to him the clocks are stationary and it is the ground that is moving away from him. Therefore the laws of physics that govern the operation of the clocks must be different on the spacecraft from what they are on the ground—which contradicts the principle of relativity. *All* moving clocks tick more slowly than clocks at rest.

We must keep in mind that the slowing down of a moving clock is significant only at relative speeds close to the speed of light. Elementary particles can be given such speeds and have been used in most of the experiments that have confirmed time dilation. Today's spacecraft are far too slow for this. For instance, the highest speed relative to the earth reached by the *Apollo II* spacecraft on its way to the moon was only about 10.8 km/s, or 0.0036% of the speed of light. At this speed clocks on the spacecraft differ from those on the earth by less than 1 part in 10^9.

EXAMPLE 25.1

Find the speed relative to the earth of a spacecraft whose clock runs exactly 1 s slow per day compared with a terrestrial clock.

SOLUTION Here $t_0 = (24\text{ h})(60\text{ min/h})(60\text{ s/min}) = 86{,}400\text{ s}$ is the time interval on the earth, and $t = 86{,}401\text{ s}$ is the time interval on the spacecraft. We begin by solving Eq. (25.1) for v and then substitute the values of t_0, t, and c:

$$t = \frac{t_0}{\sqrt{1 - v^2/c^2}}$$

$$1 - v^2/c^2 = (t_0/t)^2$$

$$v = c\sqrt{1 - (t_0/t)^2} = (3.00 \times 10^8\text{ m/s})\sqrt{1 - \left(\frac{86{,}400\text{ s}}{86{,}401\text{ s}}\right)^2}$$

$$= 1.44 \times 10^6\text{ m/s}$$

This is more than a thousand times faster than existing spacecraft. ◆

25.3 Muon Decay

Time dilation permits muons to reach the earth before decaying.

A good illustration of time dilation occurs in the decay of unstable elementary particles called **muons,** which are further described in Chapter 30.

Muons have masses 207 times that of the electron and may have positive or negative electric charges. A muon decays into a positron (positively charged electron) or an electron soon after it comes into being. Muons are created at high altitudes as an ultimate result of collisions between the nuclei of atoms in the earth's atmosphere and fast cosmic-ray particles, which are largely protons, that reach the earth from space. A muon passes through each square centimeter of the earth's surface a little more often than once a minute.

Muon speeds are about 2.994×10^8 m/s, or $0.998c$. But in $t_0 = 2.2 \times 10^{-6}$ s (2.2 μs), the average muon lifetime, they can travel a distance of only

$$vt_0 = (2.994 \times 10^8\text{ m/s})(2.2 \times 10^{-6}\text{ s}) = 6.6 \times 10^2\text{ m} = 0.66\text{ km}$$

whereas they actually come into being at elevations ten or more times greater than this.

The key to resolving this paradox is to note that the average muon lifetime of 2.2 μs is what an observer at rest with respect to a muon would find. If we could collect some muons at the instant of their creation and time their decays when they are at rest, we would find an average of $t_0 = 2.2\ \mu s$.

However, when we are on the ground and the muons are hurtling toward us at $0.998c$, we find instead that their lifetimes have been extended by time dilation to

$$t = \frac{t_0}{\sqrt{1 - v^2/c^2}} = \frac{2.2 \times 10^{-6}\ \text{s}}{\sqrt{1 - \dfrac{(0.998c)^2}{c^2}}} = 34.8 \times 10^{-6}\ \text{s}$$

The fast muons have lifetimes almost 16 times longer than those at rest. In a time interval of $t = 34.8 \times 10^{-6}$ s, a muon whose speed is $0.998c$ can cover the distance

$$vt = (2.994 \times 10^8\ \text{m/s})(34.8 \times 10^{-6}\ \text{s}) = 1.04 \times 10^4\ \text{m} = 10.4\ \text{km}$$

Despite its brief life span of $t_0 = 2.2\ \mu s$ in its own frame of reference, a muon is able to reach the ground from high altitudes because in the frame of reference in which these altitudes are measured, the muon lifetime is $t = 34.8\ \mu s$.

25.4 Length Contraction

Faster means shorter.

As we have seen, the fact that the lives of cosmic-ray muons are short (in their frame of reference) does not stop them from reaching the ground, because these lives are increased 16-fold (in our frame of reference) by the relative motion. But what if somebody could go with the muons at $0.998c$, so that to her the muons are at rest? Both the muons and the observer are now in the same frame of reference, and the muon lifetime is only $2.2\ \mu s$ in this frame. The question is, does the moving observer find that the muons reach the ground, or does the observer find that they decay beforehand?

The principle of relativity states that the laws of physics are the same in all frames of reference moving at constant velocity with respect to one another. An observer on the ground and an observer moving with the muons are in relative motion at a constant velocity. If we on the ground find that the muons reach our apparatus before they decay, then the moving observer must also find the same thing. Though the appearance of an event may be different to different observers, the fact of the event's occurrence is the same for all of them.

The only way somebody in a muon's frame of reference can reconcile its arrival at sea level with the lifetime of $2.2\ \mu s$ she finds is if the distance the muon travels is shortened by virtue of its motion (Fig. 25.6). The principle of relativity tells us by how much—it must be by the same factor of $\sqrt{1 - v^2/c^2}$ that the muon lifetime is extended from the point of view of a stationary observer. Thus a distance we on the ground measure to be L_0 will appear to the muon as the shorter distance

$$L = L_0 \sqrt{1 - v^2/c^2}$$

In our frame of reference the average distance a muon can go at the speed $v = 0.998c$ before it decays is $L_0 = 10.4$ km. The corresponding distance in the

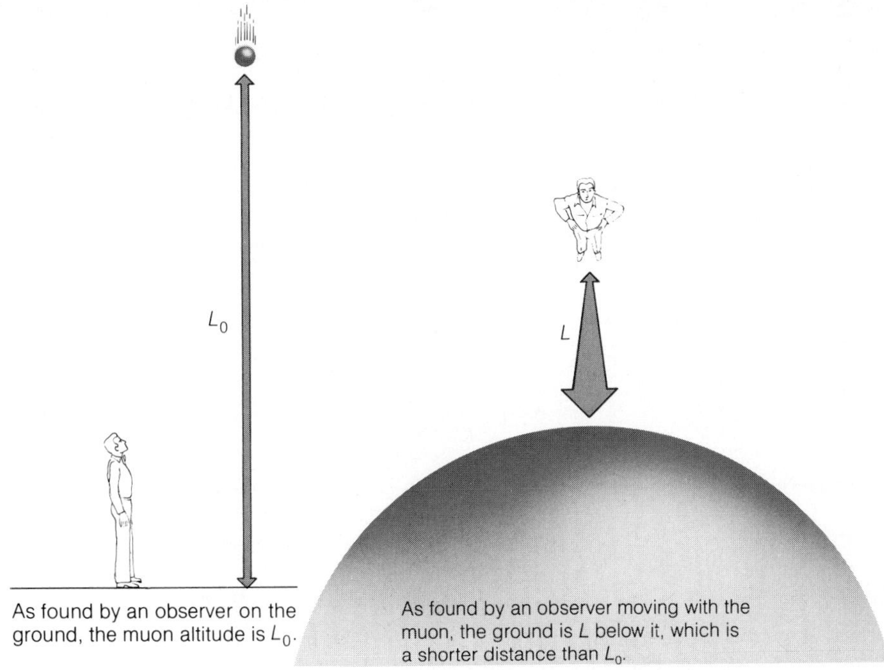

Figure 25.6
The muon size is greatly exaggerated here; in fact, the muon may actually be a point particle with no extension in space.

muon's frame of reference is

$$L = L_0\sqrt{1 - v^2/c^2} = (1.04 \times 10^4\ \text{m})\sqrt{1 - \frac{(0.998c)^2}{c^2}}$$
$$= 6.6 \times 10^2\ \text{m} = 0.66\ \text{km}$$

This is just how far a muon traveling at $0.998c$ can go in 2.2 μs. Both points of view—from the frame of reference of someone on the ground, to whom the muon lifetime is increased, and from the frame of reference of the muon itself, to which the distance to the ground is reduced—give the same result.

The relativistic shortening of lengths is described by the formula

$$L = L_0\sqrt{1 - v^2/c^2} \qquad \textit{Length contraction} \qquad (25.2)$$

The symbols in this formula have these meanings:

L_0 = length measured when the object is at rest
L = length measured when the object is in relative motion
v = speed of relative motion
c = speed of light

The length of an object moving with respect to an observer is shorter to the observer than when it is at rest with respect to him. This shortening works both ways. To a person in a spacecraft, objects on the earth are shorter than they were when he was on the ground, and someone on the ground finds that the spacecraft is shorter in flight

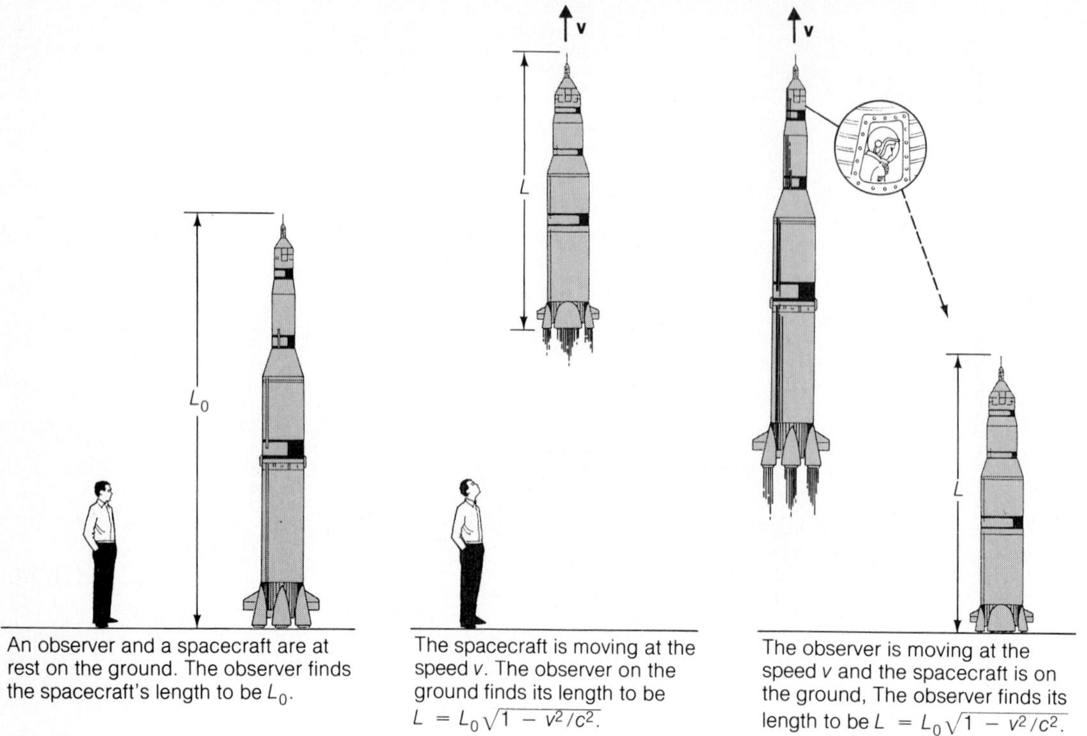

An observer and a spacecraft are at rest on the ground. The observer finds the spacecraft's length to be L_0.

The spacecraft is moving at the speed v. The observer on the ground finds its length to be $L = L_0\sqrt{1 - v^2/c^2}$.

The observer is moving at the speed v and the spacecraft is on the ground, The observer finds its length to be $L = L_0\sqrt{1 - v^2/c^2}$.

Figure 25.7

Relativistic length contraction.

than when it was at rest (Fig. 25.7). (To the person in the spacecraft, its length is the same whether on the ground or in flight, since it is always at rest with respect to him.) The length of an object is a maximum in a reference frame in which it is stationary, and its length is less in a reference frame with respect to which it is moving.

Only lengths in the direction of motion undergo contractions. Thus to the outside observer a spacecraft is shorter in flight than on the ground, but it is not narrower.

The relativistic length contraction is negligible for ordinary speeds but is an important effect at speeds close to the speed of light. A speed of 1000 km/s seems fast to us, yet it results in a shortening in the direction of motion to only

$$\frac{L}{L_0} = \sqrt{1 - (v^2/c^2)} = \sqrt{1 - \frac{(1000 \text{ km/s})^2}{(300{,}000 \text{ km/s})}} = 0.999994 = 99.9994\%$$

of the length at rest. On the other hand, something traveling at nine-tenths the speed of light is shortened to

$$\frac{L}{L_0} = \sqrt{1 - \frac{(0.9c)^2}{c^2}} = 0.44 = 44\%$$

of its length at rest, a major change.

25.5 Velocity Addition

Two and two may not equal four.

One of the principles of special relativity holds that all observers, regardless of their relative motion, will find the same value for c, the speed of light in free space. But if we throw a ball forward at 10 m/s from a car moving at 20 m/s relative to a road, everyday experience tells us the ball's speed relative to the road will be 30 m/s. Hence we would expect a pulse of light emitted by a spacecraft moving at a relative speed v toward the earth to have a speed of $c + v$ relative to the earth, which is greater than c. However, everyday experience may not be an adequate guide to events remote from that experience.

Because measurements of length and time are different in frames of reference in relative motion, it follows that measurements of speeds, too, will be different in such frames. The correct formula for velocity addition when the velocities are along the same straight line turns out to be

$$V = \frac{V' + v}{1 + vV'/c^2} \qquad \textit{Velocity addition} \quad (25.3)$$

In this formula, V' is the speed of something with respect to a frame of reference that itself is moving at the speed v relative to an observer, and V is the speed the observer measures. (If $\mathbf{V}'$ is opposite to $\mathbf{v}$, then $-V'$ is used in place of V'.)

When V' and v are small compared with the speed of light c, $V \approx V' + v$, which is the case for a ball thrown from a moving car. For a pulse of light emitted from a spacecraft, $V' = c$, and an observer on the earth would find the pulse to have the speed

$$V = \frac{V' + v}{1 + vV'/c^2} = \frac{c + v}{1 + vc/c^2} = \frac{c + v}{1 + v/c} = \frac{c(c + v)}{c + v} = c$$

All observers, regardless of their relative motion, find c for the speed of light in free space.

EXAMPLE 25.2

Spacecraft *Alfa* has a speed of $0.900c$ with respect to the earth. If spacecraft *Bravo* is to pass *Alfa* in the same direction at a relative speed of $0.500c$, what speed must *Bravo* have with respect to the earth?

SOLUTION According to classical mechanics, *Bravo* would have to travel at $0.900c + 0.500c = 1.400c$ with respect to the earth. By using Eq. (25.3) with $V' = 0.500c$ and $v = 0.900c$ we find instead that the required speed is only

$$V = \frac{V' + v}{1 + vV'/c^2} = \frac{0.500c + 0.900c}{1 + (0.900c)(0.500c)/c^2} = 0.966c$$

which is less than c. Spacecraft *Bravo* must go only 6.6% faster than a spacecraft traveling at $0.900c$ (both speeds measured from the earth) in order to pass it at a relative speed of $0.500c$ (as measured from either spacecraft). ◆

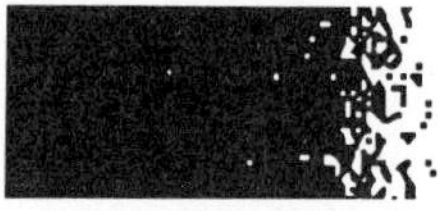

25.6 Twin Paradox

A longer life, but it will not seem longer.

Since life processes have regular rhythms, a person is a biological clock and must behave like any other clock when in motion relative to an observer. There is no difference in principle between heartbeats and the ticks of a clock. The slowing down of a moving clock means that the life processes of a person in a moving spacecraft appear to a ground observer to occur less rapidly than they do on the ground, so he or she ages more slowly than somebody on the ground does.

The celebrated case of the twins Dick and Jane illustrates time dilation in space travel. Dick is 20 years old when he embarks on a space voyage at a speed of 297,000 km/s, which is 99% of the speed of light. To Jane, who has stayed behind, the pace of Dick's life is slower than her own by a factor of

$$\sqrt{1 - v^2/c^2} = \sqrt{1 - (0.99)c^2/c^2} = 0.141 \approx \tfrac{1}{7}$$

To Jane, Dick's heart beats only once for every seven beats of Jane's heart; Dick takes only one breath for every seven of Jane's; Dick thinks only one thought for every seven of Jane's. Eventually Dick returns after 70 years have elapsed by Jane's calendar—but Dick is only 30 years old, whereas Jane, the stay-at-home twin, is 90 years old (Fig. 25.8).

Jane is baffled by the youth of her astronaut brother. ''After all,'' she argues, ''according to the principle of relativity, *my* life processes should have appeared seven times slower to Dick, so by the same reasoning I ought to be 30 and Dick ought to be 90.''

But advanced age has dulled Jane's powers of reasoning. The two situations are not interchangeable. Dick was accelerated when he started out, when he reversed

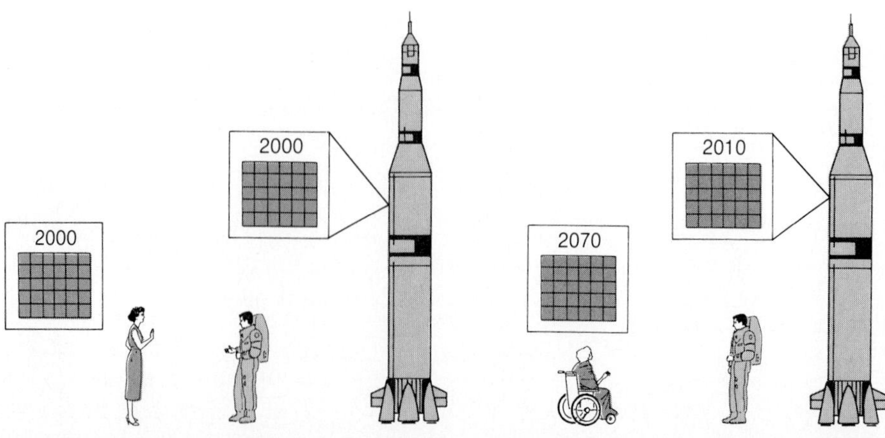

Figure 25.8

According to special relativity, an astronaut who returns from a space voyage will be younger than his twin who remains on the earth. Speeds close to the speed of light are needed for this effect to be conspicuous.

his direction to head for home, and when he landed on the earth. In each acceleration he changed from one constant-velocity frame of reference to another. Jane, however, stayed in the same constant-velocity frame of reference at all times (the various accelerations of the earth are negligible here). Jane is therefore entitled to apply the time-dilation formula to Dick's entire voyage, but Dick is not.

If we want to look at Dick's journey from his own point of view, we must take into account that the distance L he covers is shortened by the fraction

$$\frac{L}{L_0} = \sqrt{1 - v^2/c^2} = 0.141 \approx \tfrac{1}{7}$$

relative to the distance L_0 measured by Jane from the earth. To Dick, time passes at the usual rate, and his voyage has taken 10 years because he has traveled the shorter distance L. Thus Dick's life span has not been extended *to him*, since regardless of his sister's 70-year wait, he has spent only 10 years on the trip by his own reckoning.

The nonsymmetric aging of the twins has been verified by experiments in which accurate clocks were taken on airplane trips around the world and then compared with identical clocks that had been left behind.

25.7 Electricity and Magnetism

Relativity is the bridge.

One of the puzzles that set Einstein on the trail of special relativity was the connection between electricity and magnetism. The ability of his theory to clarify the nature of this connection is one of its triumphs.

It is not obvious that, when we pick up a nail with a magnet, we are witnessing a consequence of relative motion. Most relativistic effects do not appear in everyday life because the speeds of things around us are so small compared with the speed of light. Even though experiments show that there are moving electrons in the atoms of the nail and the magnet, their speeds are nowhere near that of light. The puzzle is underscored when we consider that the effective speeds of the electrons that carry a current in a wire are less than 1 mm/s—less than the speed of a caterpillar. Yet current-carrying wires do give rise to appreciable magnetic effects, as anyone who has seen an electric motor knows.

If we think about the matter for a moment, though, the idea that electricity and magnetism are connected via relativity becomes less unlikely. To begin with, electrical forces are very strong, so even a small change in their character due to relative motion (which is what magnetic forces represent) may have large consequences. As we saw in Chapter 16, the electrical force between the electron and proton in a hydrogen atom is more than 10^{39} times greater than the gravitational force between them. Second, although the individual charges involved in magnetic forces usually do move slowly, there may be so many of them that the total effect is not negligible. For example, even a modest current in a wire involves the motion of 10^{20} electrons in each centimeter of the wire.

To see how relativity accounts for the magnetic forces between moving charges, let us look into how the forces between two parallel currents come into being. In doing so, we must keep in mind that, like the speed of light, **electric charge**

is relativistically invariant. A charge whose magnitude is found to be Q in one frame of reference will be found to be Q in all other frames of reference regardless of their relative velocities.

Figure 25.9(a) shows two parallel conductors with no current in them. We imagine them to contain equally spaced positive and negative charges that are at rest. The conductors are electrically neutral.

In Fig. 25.9(b) we see the same conductors when they carry currents in the same direction. The positive charges move to the right at the speed u and the negative

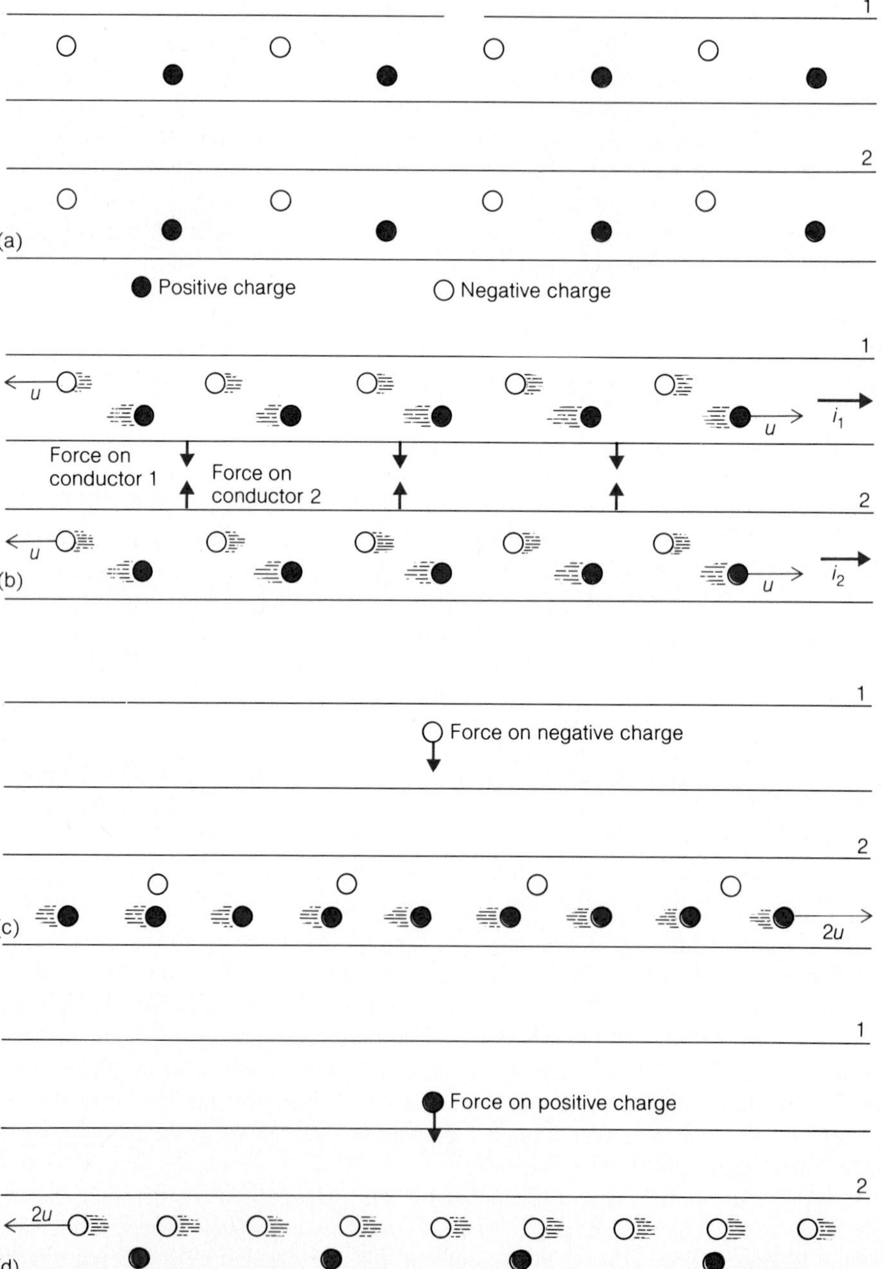

Figure 25.9

(a) Two idealized conductors that contain equal numbers of positive and negative charges. (b) The conductors attract each other when they carry currents in the same direction. (c) Conductor 2 as seen by a negative charge in conductor 1 has a net positive charge. (d) Conductor 2 as seen by a positive charge in conductor 1 has a net negative charge. The various length contractions are greatly exaggerated here.

BIOGRAPHICAL NOTE

Albert Einstein

Albert Einstein (1879–1955), bitterly unhappy with the rigid discipline of the schools of his native Germany, completed his education in Switzerland and got a job examining patent applications at the Swiss Patent Office. Then, in 1905, he published three short papers that were to change decisively the course not only of physics but of modern civilization as well.

The first paper, on the photoelectric effect, proposed that light has a dual character with particle as well as wave properties. The subject of the second paper was Brownian motion, the irregular zigzag motion of particles of suspended matter, such as pollen grains in water. Einstein showed that Brownian motion results from the bombardment of the particles by randomly moving molecules in the fluid in which they are suspended. This provided the long-awaited definite link with experiment that convinced the remaining doubters of the molecular theory of matter. The third paper introduced the special theory of relativity. Even the most unexpected of Einstein's conclusions were soon confirmed, and the development of what is now called modern physics began in earnest.

Einstein's general theory of relativity, published in 1915, related gravity to the structure of space and time. In this theory the force of gravity can be thought of as arising from a warping of space-time around a body of matter so that a nearby mass tends to move toward it, much as a marble rolls toward the bottom of a saucer-shaped hole. From general relativity came a number of remarkable predictions, such as that light should be subject to gravity, all of which were verified experimentally. The later discovery that the universe is expanding fit neatly into the theory. In 1917 Einstein introduced the idea of stimulated emission of radiation, an idea that bore fruit 40 years later in the invention of the laser.

The development of quantum mechanics in the 1920s disturbed Einstein, who never accepted its probabilistic rather than deterministic view of events on an atomic scale. "God does not play dice with the world," he said, but for once his physical intuition seemed to be leading him in the wrong direction.

Einstein left Germany in 1933 after Hitler came to power and spent the rest of his life at the Institute for Advanced Study in Princeton, New Jersey, thereby escaping the fate of millions of other European Jews at the hands of the Germans. His last years were spent in an unsuccessful search for a theory that would bring gravitation and electromagnetism together into a single picture, a problem worthy of his gifts but one that remains unsolved to this day.

---

charges move to the left at the same speed u, as seen from the laboratory frame of reference. The spacing of the charges is smaller than before by the factor $\sqrt{1 - u^2/c^2}$ because of the relativistic length contraction. The charges of both signs have the same speed in the idealized situation we are considering, hence the contractions in their spacings are the same. Thus the conductors are still neutral to an observer in the laboratory frame of reference. There is an attractive force between the conductors: How does it arise?

We begin by looking at conductor 2 from the frame of reference of one of the negative charges in conductor 1, as in Fig. 25.9(c). To this charge, the negative charges in conductor 2 are at rest, since they are (as we see the situation from the outside) all moving at the same speed u as it is. The spacing of the negative charges is not contracted, as it is to an observer in the laboratory, so they are the same distance apart, as in Fig. 25.9(a). However, in this frame the positive charges in conductor 2 are moving at the speed $2u$, and their spacing accordingly exhibits a greater contraction. Conductor 2 therefore appears positively charged to the negative charge in conductor 1, and there is an attractive electrical force on this charge in its own frame of reference.

From the frame of reference of one of the positive charges in conductor 1, as in Fig. 25.9(d), the positive charges in conductor 2 are at rest. Their spacing, in the absence of any relativistic length contraction, is the same as in Fig. 25.9(a). The negative charges in conductor 2 have the speed $2u$ relative to the positive charge in conductor 1, and they accordingly appear closer together than they do in the laboratory frame of reference. Thus there is a net negative charge on conductor 2 as seen by a positive charge in conductor 1, and this charge is attracted electrically to conductor 2.

An identical argument shows that both the negative and the positive charges in conductor 2 are attracted to conductor 1. To any of the charges in either conductor, the force on it is an "ordinary" electrical force that occurs because the charges of opposite sign in the other conductor are closer together than the charges of the same sign. The result is a net attractive force. To an observer in the laboratory both conductors are electrically neutral. He or she therefore finds it natural to ascribe the force to a special "magnetic" interaction between the currents in the conductors.

As in everything else where there is relative motion, the frame of reference from which a phenomenon is viewed is a basic part of the description of the phenomenon. Although for many purposes it is convenient to think of magnetic forces as something different from electrical ones, we must keep in mind that both are manifestations of a single electromagnetic interaction between charges.

A similar approach accounts for the repulsive force between parallel currents in opposite directions. Again, the "magnetic force" turns out to be a consequence of Coulomb's law, charge invariance, and the principles of special relativity.

Currents in metal wires actually consist of flows of electrons only, with the positive ions remaining in place. The advantage of considering the idealized currents described above, which are electrically equivalent to actual currents, is that they are easier to analyze. The results are exactly the same in all cases.

25.8 Relativity of Mass

Rest mass is least.

Another important finding of the theory of relativity is that the mass of an object is not the same to all observers but depends on its speed with respect to each observer. The variation of mass with speed is given by the formula

$$m = \frac{m_0}{\sqrt{1 - v^2/c^2}} \qquad \textit{Mass increase} \quad (25.4)$$

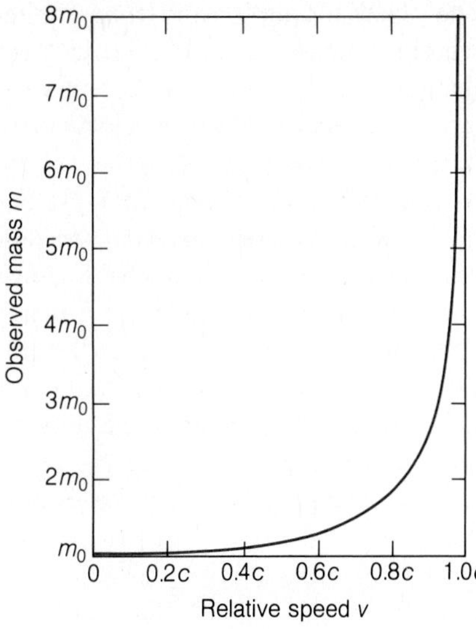

Figure 25.10

The relativity of mass. The closer the relative speed v is to the speed of light c, the greater the observed mass m is compared with the rest mass m_0.

whose symbols have these meanings:

m_0 = mass measured when object is at rest (''rest mass'')
m = mass measured when object is in relative motion
v = speed of relative motion
c = speed of light

Since the denominator of Eq. (25.4) is always less than 1, an object will always appear more massive when in relative motion than when at rest. This mass increase acts both ways. To a measuring device on the rocket ship in flight, its twin ship on the ground also appears to have a mass m greater than its own mass m_0.

Relativistic mass increases are significant only at speeds approaching that of light (Fig. 25.10). The rest mass of the *Apollo II* spacecraft was about 63,070 kg. On its way to the moon, the spacecraft's speed was about 10,840 m/s relative to the earth. Its mass in flight, as measured by an observer on the earth therefore increased to

$$m = \frac{m_0}{\sqrt{1 - \frac{v^2}{c^2}}} = \frac{63{,}070 \text{ kg}}{\sqrt{1 - \frac{(1.084 \times 10^4 \text{ m/s})^2}{(2.998 \times 10^8 \text{ m/s})^2}}} = 63{,}070.000041 \text{ kg}$$

This is not much of a change.

Smaller objects can be given much higher speeds. It is not very hard to accelerate electrons (rest mass 9.109×10^{-31} kg) to speeds of, say, $0.9999c$. The mass of such an electron is

$$m = \frac{m_0}{\sqrt{1 - \frac{v^2}{c^2}}} = \frac{9.109 \times 10^{-31} \text{ kg}}{\sqrt{1 - \frac{(0.9999c)^2}{c^2}}} = 644 \times 10^{-31} \text{ kg}$$

The electron's mass is 71 times its rest mass! The mass increases predicted by the relativistic mass formula, even such remarkable ones as this, agree with experimental findings.

Equation (25.4) has something to say about the greatest speed an object can have. The closer v approaches c, the closer v^2/c^2 approaches 1, and the closer $\sqrt{1 - (v^2/c^2)}$ approaches zero. As the denominator of Eq. (25.4) becomes smaller, the mass m becomes larger. If the relative speed v actually were equal to the speed of light, the object's mass would be infinite. The idea of an infinite mass is, of course, nonsense in many ways. Such a mass would have needed an infinite force to reach c, for instance. In addition, its length in the direction of motion would be zero by Eq. (25.2) so that its volume would be zero, and it would exert an infinite gravitational force on all other bodies in the universe. Hence we interpret Eq. (25.4) to mean that no material body can ever reach the speed of light.

Provided that linear momentum is defined as

$$m\mathbf{v} = \frac{m_0\mathbf{v}}{\sqrt{1 - v^2/c^2}} \qquad \textit{Relativistic momentum} \qquad (25.5)$$

conservation of momentum is just as valid in special relativity as it is in classical physics. However, Newton's second law of motion is correct only in the form

$$\text{Force} = \text{rate of change of momentum} \qquad \textit{Relativistic second law} \qquad (25.6)$$

$$\mathbf{F} = \frac{\Delta(m\mathbf{v})}{\Delta t}$$

where $m\mathbf{v}$ is given by Eq. (25.5). This is *not* the same as

$$\text{Force} = \text{(mass)(rate of change of velocity)}$$

$$\mathbf{F} = m\frac{\Delta\mathbf{v}}{\Delta t} = m\mathbf{a}$$

because m as well as $\mathbf{v}$ changes when an object is accelerated. Relativistic mechanics is therefore more complicated than classical mechanics.

25.9 Mass and Energy

$E = mc^2$.

The most far-reaching conclusion of special relativity is that mass and energy are closely related—so closely that matter can be converted into energy and energy into matter. The **rest energy** of a body is the energy equivalent of its mass. If a body has the mass m_0 when it is at rest, its rest energy is

$$E_0 = m_0c^2 \qquad \textit{Rest energy} \qquad (25.7)$$

where c is the speed of light.

Why are we not aware of rest energy as we are aware of kinetic and potential energies? After all, a 1-kg object—such as this book—contains the rest energy

$$m_0c^2 = (1\ \text{kg})(3 \times 10^8\ \text{m/s})^2 = 9 \times 10^{16}\ \text{J}$$

which is enough energy to send a payload of a million tons to the moon. How can so much energy be bottled up without revealing itself in some manner?

In fact, all of us *are* familiar with processes in which rest energy is liberated, only we do not usually think of them in these terms. In every chemical reaction in which energy is given off—for instance, a fire—matter is converted into energy in the form of heat, which is molecular kinetic energy. But the amount of matter that vanishes in such reactions is so small that it escapes our notice. When 1.0 kg of dynamite explodes, 6.0×10^{-11} kg of matter is transformed into energy. The lost mass is so tiny a fraction of the total mass that it is impossible to detect directly (hence the ''law'' of conservation of mass in chemistry). However, the result is the evolution of

$$m_0c^2 = (6.0 \times 10^{-11}\ \text{kg})(3.0 \times 10^8\ \text{m/s})^2 = 5.4 \times 10^6\ \text{J}$$

of energy, which is very hard not to detect.

EXAMPLE 25.3

Solar energy reaches the earth at the rate of about 1.4 kW per square meter of surface perpendicular to the direction of the sun. By how much does the mass of the sun decrease per second owing to this energy loss? The mean radius of the earth's orbit is 1.5×10^{11} m.

SOLUTION The surface area of a sphere of radius r is

$$A = 4\pi r^2$$

so the total power radiated by the sun, which is equal to the power received by a sphere whose radius is that of the earth's orbit, is given by

$$P = \left(\frac{P}{A}\right)(A) = \left(\frac{P}{A}\right)(4\pi r^2)$$
$$= (1.4 \times 10^3\ \text{W/m}^2)(4\pi)(1.5 \times 10^{11}\ \text{m})^2 = 4.0 \times 10^{26}\ \text{W}$$

Thus the sun loses $E_0 = 4.0 \times 10^{26}$ J of rest energy per second, which means that its rest mass decreases by

$$m_0 = \frac{E_0}{c^2} = \frac{4.0 \times 10^{26}\ \text{J}}{(3 \times 10^8\ \text{m/s})^2} = 4.4 \times 10^9\ \text{kg}$$

per second. Since the sun's mass is 2.0×10^{30} kg, it is in no immediate danger of running out of matter. The mechanism by which rest mass is converted into energy in the sun and other stars is discussed in Chapter 29. ♦

Energy of a Moving Object

An object of rest mass m_0 moving relative to an observer at the speed v has the total energy

$$E = mc^2 = \frac{m_0c^2}{\sqrt{1 - v^2/c^2}} \qquad \textit{Total energy} \quad (25.8)$$

where m is the mass the observer measures. This total energy is the sum of the object's rest energy E_0 and kinetic energy KE:

$$E = E_0 + \text{KE} = m_0c^2 + \text{KE} \qquad \textit{Total energy} \quad (25.9)$$

Since the zero level of potential energy is arbitrary, PE is not included here. If the PE of the object changes, its total energy will change by the same amount.

From Eqs. (25.8) and (25.9) we see that the kinetic energy of a moving object is given by

$$\text{KE} = mc^2 - m_0c^2 = \frac{m_0c^2}{\sqrt{1 - v^2/c^2}} - m_0c^2 \qquad \textit{Kinetic energy} \quad (25.10)$$

This formula is rather different from the kinetic energy formula $\frac{1}{2}m_0v^2$ we have been using thus far. However, it is not hard to verify that the relativistic formula for KE reduces to $\frac{1}{2}m_0v^2$ when v is small compared with c.

We start by noting that when x is small, the binomial expansion of algebra shows that $1/\sqrt{1 - x}$ can be approximated by

$$\frac{1}{\sqrt{1 - x}} \approx 1 + \frac{x}{2} \qquad x << 1$$

The symbol $<<$ means "much smaller than." Here we let $x = v^2/c^2$, so that

$$\frac{1}{\sqrt{1 - v^2/c^2}} \approx 1 + \frac{1}{2}\frac{v^2}{c^2} \qquad v << c$$

and

$$\text{KE} = \frac{m_0c^2}{\sqrt{1 - v^2/c^2}} - m_0c^2 \approx \left(1 + \frac{1}{2}\frac{v^2}{c^2}\right)m_0c^2 - m_0c^2 \approx \frac{1}{2}m_0v^2 \qquad v << c$$

The relativistic formula is correct for all speeds, and the formula $\text{KE} = \frac{1}{2}m_0v^2$ is the low-speed approximation to it. The greater the speed v, the more the formula $\frac{1}{2}m_0v^2$ understates the true KE of an object.

The degree of accuracy required is what determines whether the approximation $\frac{1}{2}m_0v^2$ is appropriate. For instance, when $v = 0.033c$ (1×10^7 m/s), the approximation understates the true KE by only 0.08%; when $v = 0.1c$ (3×10^7 m/s), it understates the true KE by 0.8%; but when $v = 0.5c$, the understatement is a significant 19%, and when $v = 0.999c$, the approximation is too small by 4300%.

According to KE $= \frac{1}{2}m_0v^2$, an object would need a kinetic energy of $\frac{1}{2}m_0c^2$, half its rest energy, to move at the speed of light. According to Eq. (25.10) it would need an infinite kinetic energy, and so such a speed is unattainable.

EXAMPLE 25.4

How many times greater than its rest mass is the mass of an electron whose kinetic energy is exactly 1 GeV? The rest energy of the electron is 0.511 MeV.

SOLUTION Since 1 GeV $= 10^9$ eV $= 10^3$ MeV,

$$\frac{m}{m_0} = \frac{mc^2}{m_0c^2} = \frac{m_0c^2 + \text{KE}}{m_0c^2} = \frac{0.511\text{ MeV} + 1000.000\text{ MeV}}{0.511\text{ MeV}} = 1958$$

Electrons with energies well in excess of 1 GeV occur naturally in the cosmic radiation in space and have been produced in the laboratory as well. ♦

EXAMPLE 25.5

An object at rest explodes into two fragments whose rest masses are both 1.0 g. The fragments move apart at speeds of $0.60c$ relative to the original object. What was the rest mass of the original object?

SOLUTION Let us call the rest mass of the original object m_0 and those of the fragments m_{01} and m_{02}, with the speeds of the latter being respectively v_1 and v_2. The total energy m_0c^2 of the original object must equal the sum of the total energies m_1c^2 and m_2c^2 of the fragments, so

$$m_0c^2 = m_1c^2 + m_2c^2 = \frac{m_{01}c^2}{\sqrt{1 - v_1^2/c^2}} + \frac{m_{02}c^2}{\sqrt{1 - v_2^2/c^2}}$$

Since $m_{01} = m_{02} = 1.0$ g and $v_1 = v_2 = 0.60c$, dividing through by c^2 and substituting these values gives

$$m_0 = \frac{2m_{01}}{\sqrt{1 - v_1^2/c^2}} = \frac{2(1.0\text{ g})}{\sqrt{1 - (0.60)^2}} = 2.5\text{ g}$$

Thus 20% of the original object's mass became kinetic energy of the fragments in the explosion. ♦

25.10 General Relativity

Gravity is due to a warping of space-time.

Special relativity shows us how to interpret what we observe in frames of reference that move at constant velocity with respect to us. The laws of physics are valid in all such frames of reference, but measurements of some quantities—notably

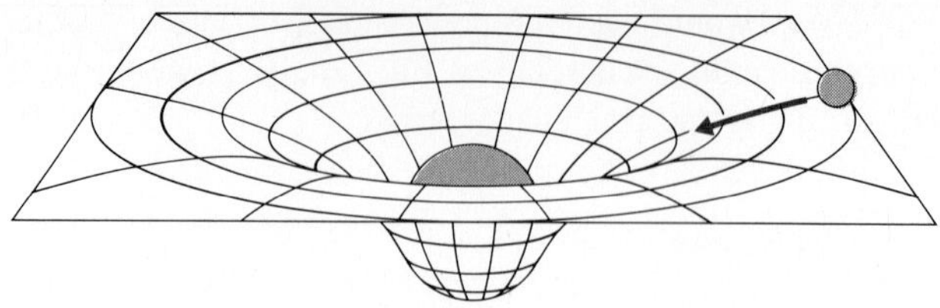

Figure 25.11

General relativity pictures gravity as a warping of the structure of space and time due to the presence of a body of matter. An object nearby experiences an attractive force as a result of this distortion in space-time, much as a marble rolls toward the bottom of a saucer-shaped hole in the ground.

time intervals, lengths, and masses—are affected by relative motion. Measurements of other quantities—notably the speed of light and electric charge—are not affected.

Einstein's **general theory of relativity** goes further by including the effects of accelerated motion on what we observe. In this theory gravitation turns out to be related to the structure of space and time. What is meant by "the structure of space and time" can be given a precise meaning mathematically, but unfortunately not on the level we have been using here. The essential conclusion is that the force of gravity arises from a warping of space-time around a body of matter so that other bodies tend to move toward it, much as a marble rolls toward the bottom of a saucer-shaped hole in the ground (Fig. 25.11). It may seem merely as though one abstract concept is replacing another, but in fact the new point of view has led to a variety of remarkable discoveries.

One of the basic ideas of general relativity is the **principle of equivalence:** An observer in a closed laboratory cannot ditinguish between the effects produced by a gravitational field and those produced by an acceleration of the laboratory. This principle is another way to express the experimental finding that the inertial mass of an object, which governs its acceleration when a force acts on it, is always equal to its gravitational mass, which governs the gravitational pull another object exerts on it. (The two masses are actually proportional; the constant of proportionality is set equal to 1 by an appropriate choice of the gravitational constant G.)

It follows from the principle of equivalence that light should be subject to gravity. If a light beam is directed across an accelerated laboratory, as in Fig. 25.12, its path relative to the laboratory will be curved. This means that if the light beam were subject to the gravitational field the acceleration is equivalent to, it would follow the same curved path.

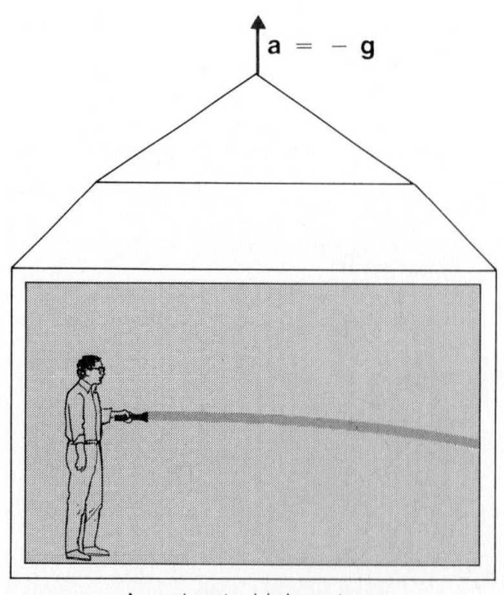

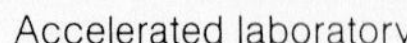

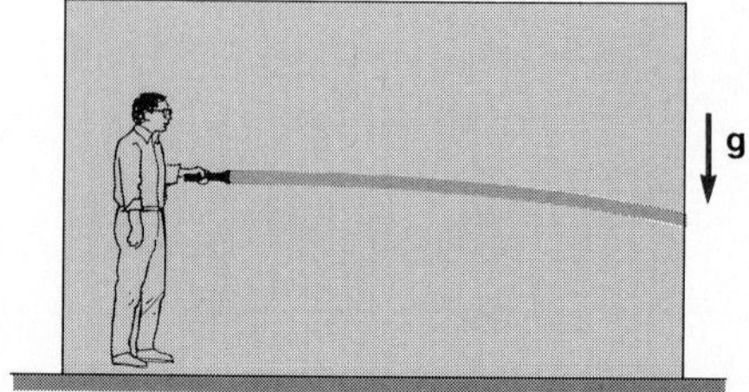

Figure 25.12

According to the principle of equivalence, events that take place in an accelerated laboratory cannot be distinguished from those that take place in a gravitational field. Hence the deflection of a light beam relative to an observer in an accelerated laboratory means that light must be similarly deflected in a gravitational field.

According to general relativity, light rays that pass near the sun should have their paths bent toward it by 0.0005°—the diameter of a dime seen from a mile away. This prediction was first confirmed by photographs of stars that appeared in the sky near the sun during an eclipse in 1919, when they could be seen because the sun's disk was covered by the moon. These photographs were then compared with photographs of the same region of the sky taken when the sun was far away (Fig. 25.13). More recently laboratory experiments based on the quantum theory of light (Chapter 26) have independently verified that light is indeed subject to gravity.

The more massive a star and the smaller it is, the stronger the pull of gravity at its surface and the higher the escape speed needed for something to leave the star. In the case of the sun, it is 617 km/s (as we saw in Chapter 6, the escape speed from the earth is 11.2 km/s.) If the ratio M/R between the mass and radius of a star is large enough, the escape speed is more than the speed of light, and nothing, not even light, can ever get out. The star cannot radiate and so is invisible, a **black hole** in space. For a star with the sun's mass, the critical radius is 3 km, a quarter of a million times smaller than the sun's present radius. Anything passing near a black hole will be sucked into it, never to return to the rest of the universe.

Since it is invisible, how can a black hole be detected? A black hole that is a member of a double-star system (double stars are quite common) will reveal its presence by its gravitational effect on the other member, which will rotate about the center of mass of the system. In addition, the black hole will attract matter from the other star. This matter will be compressed and heated to such high temperatures that X rays will be emitted in profusion. One of a number of invisible X-ray-emitting objects that astronomers believe to be a black hole is known as Cygnus X–1, whose mass is about eight times that of the sun but whose radius is thought to be only about 10 km. More speculative, but with much support, is the idea that many, if not all, galaxies of stars have massive black holes at their centers. Such a black hole might contain as much matter as a hundred million suns.

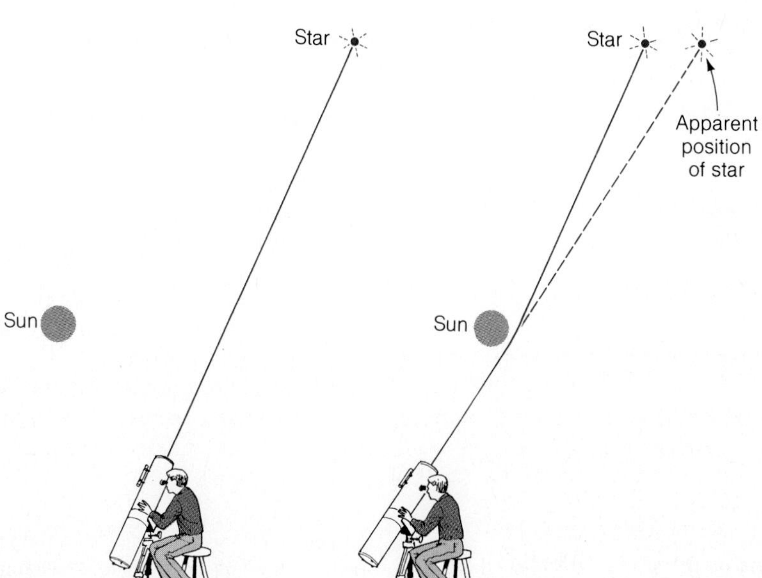

Figure 25.13

Starlight passing near the sun is deflected by its strong gravitational field. The deflection can be measured during a solar eclipse when the sun's disk is obscured by the moon.

Important Terms

The **special theory of relativity** relates measurements made on an object or phenomenon from frames of reference moving at constant velocity with respect to one another.

The relativistic **time dilation** refers to the fact that a clock moving with respect to an observer appears to tick less rapidly than it does to an observer traveling with the clock.

The relativistic **length contraction** refers to the decrease in the measured length of an object when it is moving relative to an observer.

The **relativity of mass** refers to the increase in the measured mass of an object when it is moving relative to an observer. The object's **rest mass** is its mass measured when it is at rest relative to the observer.

Rest energy is the energy an object has by virtue of its mass alone.

The **general theory of relativity** concerns frames of reference accelerated with respect to one another. One of the conclusions of general relativity is that light is affected by gravitational fields.

Important Formulas

Time dilation: $t = \dfrac{t_0}{\sqrt{1 - v^2/c^2}}$

Length contraction: $L = L_0\sqrt{1 - v^2/c^2}$

Mass increase: $m = \dfrac{m_0}{\sqrt{1 - v^2/c^2}}$

Rest energy: $E_0 = m_0c^2$

Kinetic energy: $\mathrm{KE} = mc^2 - m_0c^2$
$\approx \frac{1}{2} m_0v^2 \qquad v << c$

MULTIPLE CHOICE

1. The theory of relativity is in conflict with
- **a.** experiment
- **b.** Newtonian mechanics
- **c.** electromagnetic theory
- **d.** ordinary mathematics

2. Which, if any, of the following quantities have the same value to all observers?
- **a.** The speed of light
- **b.** The charge of the muon
- **c.** The mass of the muon
- **d.** The average lifetime of the muon

3. According to the principle of relativity, the laws of physics are the same in all frames of reference
- **a.** at rest with respect to one another
- **b.** moving toward or away from one another at constant velocity
- **c.** moving parallel to one another at constant velocity
- **d.** all of the above

4. The lifetime of a muon in motion relative to an observer appears to him to be
- **a.** shorter than its lifetime at rest
- **b.** the same as its lifetime at rest
- **c.** longer than its lifetime at rest
- **d.** any of the above, depending on the relative velocity

5. A spacecraft has left the earth and is moving toward Mars. An observer on the earth finds that, relative to measurements made when it was at rest, the spacecraft's
- **a.** length is greater
- **b.** mass is smaller
- **c.** clocks tick faster
- **d.** momentum is greater

6. A spacecraft headed toward the earth at the speed v sends out a pulse of light. The pulse travels toward the earth at a relative speed of
- **a.** c
- **b.** $c + v$
- **c.** $c/\sqrt{1 - v^2/c^2}$
- **d.** $c\sqrt{1 - v^2/c^2}$

7. Alphonse makes a round trip to Pluto while his twin brother Gaston stays at home. When Alphonse returns,
- **a.** Alphonse is younger
- **b.** Gaston is younger
- **c.** they are the same age
- **d.** any of the above, depending on the spacecraft's speed

8. The formula $\mathrm{KE} = \frac{1}{2}m_0v^2$ for kinetic energy
- **a.** is the correct formula if v is properly interpreted
- **b.** always gives too high a value
- **c.** is the low-speed approximation to the correct formula
- **d.** is the high-speed approximation to the correct formula

9. A gravitational field
- **a.** always attracts light
- **b.** attracts light only if the field is strong enough
- **c.** may attract or repel light, depending on circumstances
- **d.** neither attracts nor repels light

10. No light can emerge from a black hole because
 a. its electric field is too strong
 b. its magnetic field is too strong
 c. its gravitational field is too strong
 d. it is surrounded by thick clouds

11. When an object whose length is 1 m when at rest relative to an observer approaches the speed of light relative to the observer, its length approaches
 a. 0 b. 0.5 m
 c. 2 m d. ∞

12. When an object whose rest mass is 1 kg relative to an observer approaches the speed of light relative to the observer, its mass approaches
 a. 0 b. 0.5 kg
 c. 2 kg d. ∞

13. A man 6.0 ft tall lies along the axis of a space vehicle traveling at $0.90c$. His height as measured by a stationary observer is
 a. 1.9 ft b. 2.6 ft
 c. 6.0 ft d. 14 ft

14. Which of the following speeds must an object have if its mass is to be double its rest mass?
 a. $c/2$ b. $\sqrt{3}c/2$
 c. c d. $2c$

15. A particle of rest mass 1.00 g is moving relative to an observer at 0.600 c. The observer measures its mass to be
 a. 0.80 g b. 1.25 g
 c. 1.56 g d. 1.58 g

16. The mass equivalent of 6.0 MJ is
 a. 6.7×10^{-11} kg b. 5.4×10^{-9} kg
 c. 6.7×10^{-3} kg d. 2.0×10^{-2} kg

QUESTIONS

1. Which of the following questions can an observer in a closed laboratory answer? (a) Is the laboratory at rest? (b) Is it in motion at constant velocity? (c) Is it accelerated in a particular direction? (d) Is it rotating?

2. The electron beam in a television picture tube can move across the screen faster than the speed of light. Why does this not violate the special theory of relativity?

3. If the speed of light were smaller than it is, would relativistic phenomena be more or less conspicuous than they are now?

4. Does a laboratory at rest on the earth's surface constitute a nonaccelerated frame of reference? If not, where on the surface is the acceleration greatest? Where is it least? Is it zero anywhere in the earth?

5. The length of a rod is measured by several observers, one of whom is stationary with respect to the rod. What must be true of the figure obtained by the latter observer?

6. Is the density of a moving object less than, the same as, or more than it appears to an observer when the object is at rest?

7. An object moving in a circle has an average velocity of zero. Would you expect it to have a relativistic mass increase?

8. The hydrogen molecule contains two protons and two electrons. The electrons are moving considerably faster than the protons. The most accurate measurements to date indicate that the hydrogen molecule is electrically neutral to at least 1 part in 10^{20}. What significance has this result with respect to the question of whether electric charge is an invariant quantity, like the speed of light in free space, or depends on relative motion, like mass, length, and the duration of a time interval?

9. An object of length L_0, mass m_0, and charge Q_0 that contains a clock whose ticks are t_0 apart, all measured at rest, is now moving away from an observer. Which of these properties does the observer find smaller than before? Larger?

10. Is the mass of an object the same whether it is moving toward an observer or away from him at the same speed? Is it the same whether it moves toward the observer or the observer moves toward it at the same speed? What is the connection between the answers to these questions and the principle of relativity?

11. The potential energy of a golf ball in a hole is negative relative to the ground. Under what circumstances (if any) is its kinetic energy negative? Its rest energy?

12. Light cannot escape from black holes, yet they have been identified with the help of the X rays emitted as matter is sucked into them. How do you think X rays can escape whereas light cannot?

EXERCISES

25.2 Relativity of Time

1. How fast would a spacecraft have to go for each year on the spacecraft to correspond to two years on the earth?

2. A woman leaves the earth in a spacecraft that makes a round trip to the nearest star, 4.00 light-years distant, at a speed of $0.900c$ relative to the earth. How much younger is she on her return than her twin sister who remained behind? (A light-year is the distance light travels in free space in a year. It is equal to 9.46×10^{15} m.)

3. A certain process requires 1.00×10^{-6} s to occur in an atom at rest in the laboratory. To an observer in the laboratory, how much time will this process require when the atom is moving at a speed of 5.00×10^{7} m/s?

4. Find the speed of a spacecraft whose clock runs 1.0 s slow per hour relative to a clock on the earth.

25.4 Length Contraction

5. An atomic nucleus 5.00×10^{-15} m in diameter is moving at a speed of 1.00×10^8 m/s. What is its thickness in its direction of motion as measured by a laboratory observer?

6. An astronaut is standing in a spacecraft parallel to its direction of motion. An observer on the earth finds that the spacecraft speed is $0.60c$ and the astronaut is 1.3 m tall. What is the astronaut's height as measured when she is at rest?

7. A meter stick appears only 300 mm long to an observer. (a) What is its relative speed? (b) An observer times the passage of the meter stick past a point in her own frame of reference. What does she find?

25.8 Relativity of Mass

8. An electron is moving at 2.98×10^8 m/s. What is its mass?

9. Find the mass of an object whose rest mass is 1000 g when it is traveling at 10%, 90%, and 99% of the speed of light.

10. What speed must a particle have if its mass is to be triple its rest mass?

11. A man has a mass of 100 kg on the ground. When he is in a spacecraft in flight, an observer on the earth measures his mass to be 101 kg. How fast is the spacecraft moving?

25.9 Mass and Energy

12. The combustion of 1.00 kg of gasoline liberates 47.3 MJ of energy. How much mass is lost when 1.00 kg of gasoline is burned in the engine of a car? Would you expect to be able to measure the change in mass?

13. The source of the sun's energy (and therefore, directly or indirectly, of nearly all energy available on earth) is the conversion of hydrogen to helium. As described in Chapter 29, the nuclei of four hydrogen atoms, each of mass 1.673×10^{-27} kg, join together in a series of separate reactions to yield a helium nucleus of mass 6.646×10^{-27} kg. How much energy is liberated each time a helium nucleus is formed? How many helium nuclei are formed to produce the 1.00×10^7 J a moderately active person requires per day?

14. Typical chemical reactions absorb or release energy at the rate of several eV per molecular change. What change in mass is associated with the absorption or release of 1.00 eV?

15. The rest mass of a particle called the neutral pion is 2.40×10^{-28} kg. Find the particle's rest energy in MeV.

16. How many joules of energy per kilogram of rest mass are required to bring a spacecraft from rest to a speed of $0.900c$?

17. What is the kinetic energy in MeV of a neutron whose mass is double its rest mass? The rest energy of the neutron is 940 MeV.

PROBLEMS

18. Two observers, *A* on earth and *B* in a rocket ship whose speed is 2.00×10^8 m/s, both set their watches to the same time when the ship is abreast of the earth. (a) How much time must elapse by *A*'s reckoning before the watches differ by 1.00 s? (b) To *A*, *B*'s watch seems to run slow. To *B*, does *A*'s watch seem to run fast, run slow, or keep the same time as his own watch?

19. An airplane is flying at 300 m/s (672 mi/h). How much time must elapse before a clock in the airplane and one on the ground differ by 1.0 s? (*Hint*: See the approximation used in Section 25.9.)

20. A woman on the moon sees two spacecraft, *A* and *B*, coming toward toward her from opposite directions at the respective speeds of $0.800c$ and $0.900c$. (a) What does a man on *A* measure for the speed with which he is approaching the moon? For the speed with which he is approaching *B*? (b) What does a man on *B* measure for the speed with which he is approaching the moon? For the speed with which he is approaching *A*?

21. An object moving at $0.50c$ with respect to an observer disintegrates into two fragments that move in opposite directions relative to their center of mass along the same line of motion as the original object. One fragment has a speed of $0.60c$ in the backward direction relative to the center of mass, and the other has a speed of $0.50c$ in the forward direction relative to the center of mass. What speeds will the observer find?

22. An electron has a kinetic energy of 100 keV. Find its speed according to classical and relativistic mechanics. The rest energy of the electron is 511 keV.

23. How much mass does a proton gain when it is accelerated to a kinetic energy of 500 MeV? The rest mass of the proton is 1.67×10^{-27} kg, and its rest energy is 938 MeV.

24. An electron whose speed relative to an observer in a laboratory is $0.80c$ is also being studied by an observer moving in the same direction as the electron at a speed of $0.50c$ relative to the laboratory. What is the kinetic energy, in MeV, of the electron to each observer?

Answers to Multiple Choice

1. b	5. d	9. a	13. b
2. a, b	6. a	10. c	14. b
3. d	7. a	11. a	15. b
4. d	8. c	12. d	16. a

CHAPTER 26

PARTICLES AND WAVES

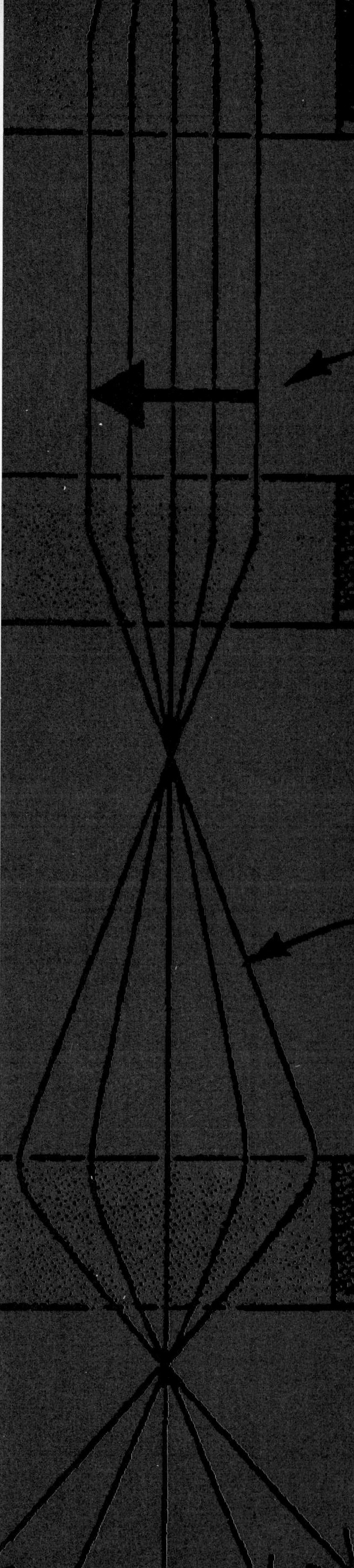

Particles and waves are separate concepts in everyday life. A stone thrown into a lake and the ripples that spread out from where it lands resemble each other only in their ability to carry energy from one place to another. But in the world of the atom the situation is different. Electromagnetic waves—which, as we know, show such typical wave behavior as diffraction and interference—turn out to have important properties in common with particles. And electrons—whose particle nature is evident whenever we look at a television picture tube—turn out to have important properties in common with waves. On a very small scale of size, a wave-particle duality replaces the distinction between waves and particles so obvious on a large scale. This duality, as we will see, is the key to understanding the world of the atom.

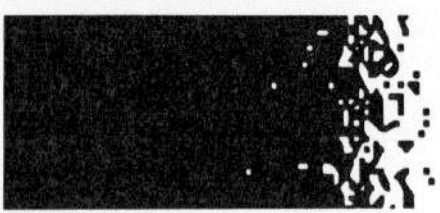

26.1 Photoelectric Effect

The energies of electrons liberated by light depend on the frequency of the light.

When light (particularly ultraviolet light) falls on a metal surface, electrons are found to be given off (Fig. 26.1). This phenomenon is known as the **photoelectric effect.** The photoelectric cell that measures light intensity in a camera, the solar cell that produces electric current when sunlight falls on it, and the television camera tube that converts the image of a scene into an electric signal are all based on the photoelectric effect.

Since light is electromagnetic in nature and carries energy, there seems nothing unlikely about the photoelectric effect; the energy absorbed by the metal may somehow concentrate on individual electrons and reappear as kinetic energy. The situation should be like water waves dislodging pebbles from a beach. But three experimental findings show that no such simple explanation is possible:

1. The electrons (called **photoelectrons**) are emitted at once, even when the light is faint. However, because the energy in an em wave is supposed to be spread across the wavefronts, a period of time should elapse before an individual electron accumulates enough energy to leave the metal. Several months might be needed for a really weak light beam.
2. A bright light yields more photoelectrons than a dim light, but their average KE remains the same. The em theory of light, on the contrary, predicts that the more intense the light, the greater the KE of the electrons.
3. The higher the frequency of the light, the more KE the electrons have. Blue light yields faster electrons than red light (Fig. 26.2). At frequencies below a critical one f_0 (which is characteristic of the metal used), no electrons are given off. Above f_0 the photoelectrons range in energy from 0 to a maximum value that increases with increasing frequency (Fig. 26.3).

Until the discovery of the photoelectric effect a century ago, the em theory of light had been completely successful in explaining the behavior of light. But no

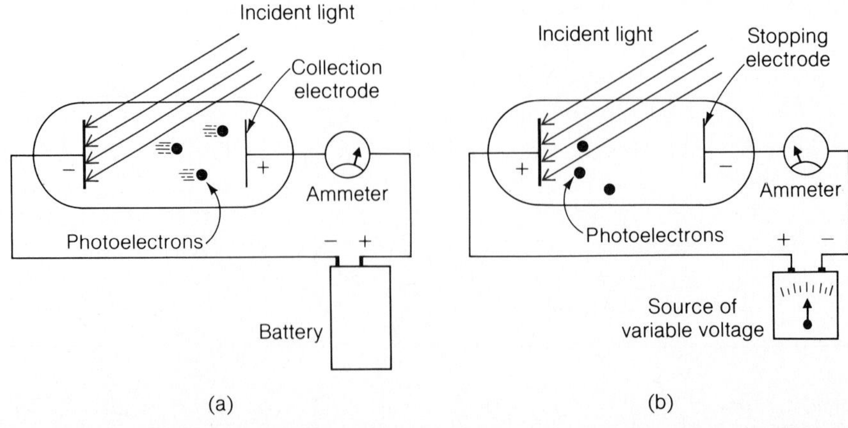

Figure 26.1

(a) A method of detecting the photoelectric effect. The photoelectrons ejected from the irradiated metal plate are attracted to the positive collection electrode at the other end of the tube, and the current that results is measured with an ammeter. (b) A method of detecting the maximum energy of the photoelectrons. Note polarity opposite to that in (a). As the stopping electrode is made more negative, the slower photoelectrons are repelled before they can reach it. Finally a voltage will be reached at which no photoelectrons whatever are received at the stopping electrode, as indicated by the current dropping to zero, and this voltage corresponds to the maximum photoelectron energy.

Each photon has the energy hf. Hence

$$\text{Number of photons} = \frac{\text{detectable energy}}{\text{energy/photon}} = \frac{1 \times 10^{-18}\ \text{J}}{(6.63 \times 10^{-34}\ \text{J}\cdot\text{s})(5 \times 10^{14}\ \text{Hz})} = 3\ \text{photons}$$ ◆

EXAMPLE 26.2

The photoelectric work function for copper is 4.5 eV. Find the maximum energy of the photoelectrons when ultraviolet light of 1.5×10^{15} Hz falls on a copper surface.

SOLUTION The quantum energy of the incident photons in eV is

$$hf = \frac{(6.63 \times 10^{-34}\ \text{J}\cdot\text{s})(1.5 \times 10^{15}\ \text{Hz})}{1.6 \times 10^{-19}\ \text{J/eV}} = 6.2\ \text{eV}$$

Hence the maximum photoelectron energy is

$$\text{KE}_{\text{max}} = hf - hf_0 = (6.2 - 4.5)\text{eV} = 1.7\ \text{eV}$$ ◆

What Is Light?

The notion that light travels as a series of little packets of energy is directly opposed to the wave theory of light (Fig. 26.5). Both views have strong experimental support, as we have seen. According to the wave theory, light waves move away from a source with the energy they carry spread out continuously through the wave pattern. According to the quantum theory, light is a series of individual photons, each small enough to be absorbed by a single electron. Yet, despite the particle picture of light it presents, the quantum theory needs the frequency f of the light to describe the photon energy.

Which theory are we to believe? A great many scientific ideas have had to be revised or discarded when they were found to disagree with new data. Here, for the first time, two different theories are both needed to explain a single phenomenon. This situation is not the same as it is, say, in the case of relativistic versus Newtonian mechanics, where one turns out to be an approximation of the other. The connection between the wave theory of light and the quantum theory of light is rather different.

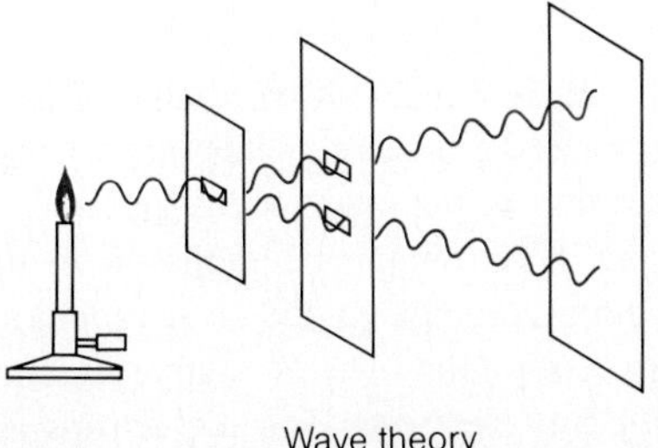

Wave theory of light

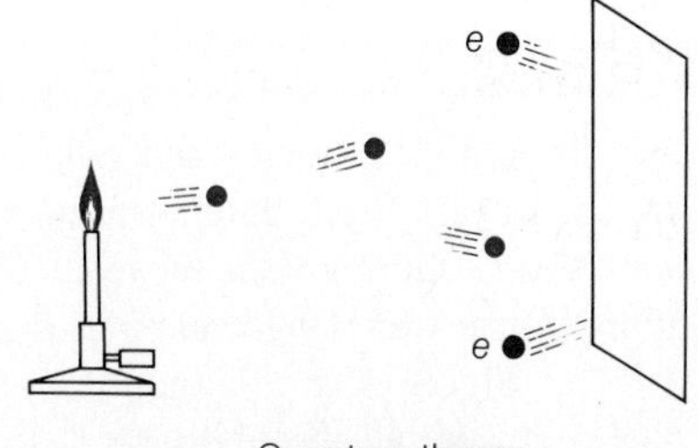

Quantum theory of light

Figure 26.5

The wave theory of light is needed to explain diffraction and interference, which the quantum theory cannot explain. The quantum theory of light is needed to explain the photoelectric effect, which the wave theory cannot explain.

To appreciate this connection, let us consider the formation of a double-slit interference pattern on a screen, as in Fig. 26.5. In the wave model, the light intensity at a place on the screen depends on the value of the Poynting vector S there, as discussed in Section 22.3. In the particle model, this intensity depends instead on Nhf, where N is the number of photons per second per unit area that reach the same place on the screen. Both descriptions must give the same value for the intensity, so N is proportional to S. If N is large enough, somebody looking at the screen would see the usual double-slit interference pattern and would have no reason to doubt the wave model. If N is small—perhaps so small that only one photon at a time reaches the screen—the observer would find a series of apparently random flashes and would assume that he or she is seeing a quantum phenomenon.

If the observer keeps track of the flashes for long enough, though, the pattern they form will be the same as when N is large. Thus the observer is entitled to conclude that the *probability* of finding a photon at a certain place and time depends on the value of S there. If we regard each photon as somehow having a wave associated with it, the intensity of this wave at a given place on the screen determines the likelihood that a photon will arrive there. When it passes through the slits, light is behaving as a wave does. When it strikes the screen, light is behaving as a particle does.

Evidently light has a dual character. The wave theory and the quantum theory complement each other. Either theory by itself is only part of the story and can explain only certain effects. The "true nature" of light includes both wave and particle characters, even though there is nothing in everyday life like that to help us visualize it.

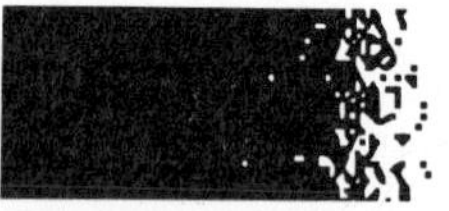

26.3 X Rays

They consist of high-energy photons.

If photons of light can give up their energy to electrons, can the kinetic energy of moving electrons be converted into photons? The answer is that such a transformation not only is possible but had in fact been discovered (though not understood) prior to the work of Planck and Einstein.

In 1895 the German physicist Wilhelm Roentgen found that a mysterious, highly penetrating radiation is emitted when high-speed electrons impinge on matter. The X rays (so called because their nature was then unknown) caused phosphorescent substances to glow, exposed photographic plates, traveled in straight lines, and were not affected by electric or magnetic fields. The more energetic the electrons, the more penetrating the X rays. The greater the number of electrons, the greater the intensity of the resulting X-ray beam.

After more than ten years of study it was finally established that X rays exhibit both interference and polarization effects, leading to the conclusion that they are em waves. From the interference experiments their wavelengths were found to be very short, shorter than those in ultraviolet light. Electromagnetic radiation in the approximate wavelength interval from 0.01 to 10 nm is today classed as X-radiation.

Figure 26.6 is a diagram of an idealized X-ray tube. The dc source A sends a current through the filament, heating it until it emits electrons. These electrons are then accelerated toward a metal target by the potential difference V provided by the

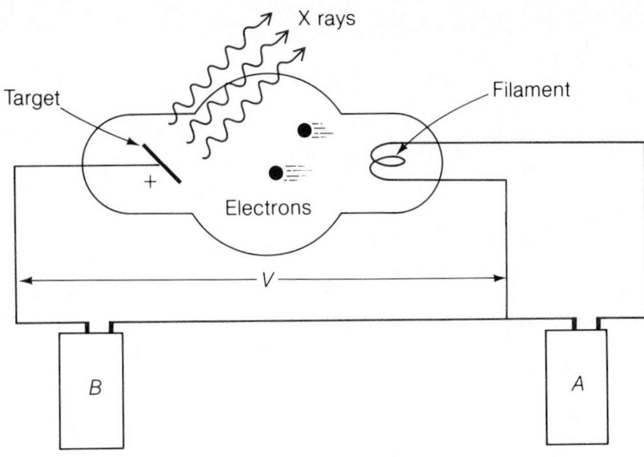

Figure 26.6

An idealized X-ray tube. *A* and *B* are sources of direct current.

dc source *B*. The tube is evacuated to permit the electrons to reach the target freely. The impact of the electrons causes the evolution of X rays from the target.

What is the physical process involved in the production of X rays? It is known that charged particles emit em waves when accelerated, and so we may reasonably identify X rays as the radiation that accompanies the slowing down of fast electrons when they strike matter. The great majority of the electrons, to be sure, lose their kinetic energy too gradually for X rays to be given off, and merely heat the target. (Hence the targets in X-ray tubes are made of metals with high melting points, and a means for cooling the target is often provided.) A few electrons, however, lose much or all of their energy in single collisions with target atoms, and this is the energy that appears as X rays. In other words, we may regard X-ray production as an inverse photoelectric effect.

Since the threshold energy hf_0 needed to remove an electron from a metal is only a few electron volts, whereas the accelerating potential V in an X-ray tube usually exceeds 10,000 volts, we can neglect the work function hf_0 here. The highest frequency f_{max} found in an X-ray beam should therefore correspond to a quantum energy of hf_{max}, where hf_{max} equals the kinetic energy KE $= Ve$ of an electron that has been

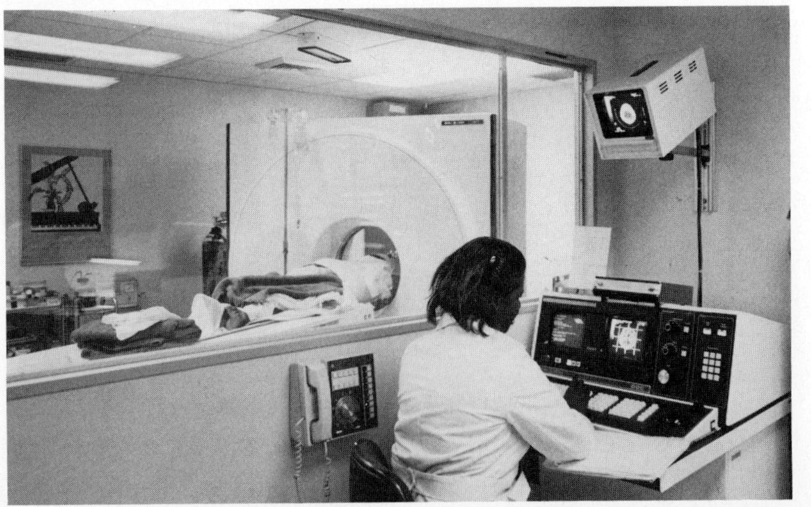

In a CAT (computerized axial tomography) scanner, a series of X-ray exposures of a patient taken from different directions are combined by a computer to give cross-sectional images of the part of the body being examined. The computer, in effect, slices up the tissues on the basis of the X-ray exposures, and any desired slice can be displayed.

accelerated through a potential difference of V. We conclude that

$$hf_{\max} = Ve \qquad \textit{X-ray energy} \quad (26.3)$$

in the operation of an X-ray tube.

EXAMPLE 26.3

Find the highest frequency present in the radiation from an X-ray machine whose operating potential is 50,000 volts.

SOLUTION From Eq. (26.3) we find that

$$f_{\max} = \frac{Ve}{h} = \frac{(5.0 \times 10^4\ \text{V})(1.6 \times 10^{-19}\ \text{C})}{6.6 \times 10^{-34}\ \text{J}\cdot\text{s}} = 1.2 \times 10^{19}\ \text{Hz}$$

The corresponding wavelength is $\lambda = c/f = 2.5 \times 10^{-11}\ \text{m} = 0.025\ \text{nm}$. ♦

26.4 Photons as Particles

Further confirmation of the photon model.

According to the quantum theory of light, photons behave like particles except for their lack of rest mass. How far can this analogy be carried? For instance, can we consider a collision between a photon and an electron as if both were billiard balls?

Figure 26.7 shows a collision in which a photon of frequency f strikes an electron at rest and is scattered away from its original path, while the electron receives an impulse and moves off. In the impact the photon loses an amount of energy equal to the kinetic energy KE gained by the electron, so the scattered photon has a lower frequency f', where

$$\text{Change in photon energy} = \text{electron kinetic energy}$$
$$hf - hf' = \text{KE}$$

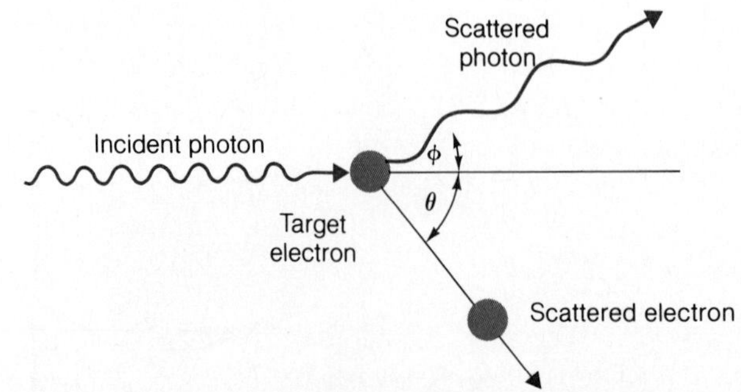

Figure 26.7

The scattering of a photon by an electron is called the Compton effect. Energy and momentum are conserved in such an event, with the result that the scattered photon has a lower frequency (longer wavelength) than the incident photon.

As we saw in Section 22.3, em waves carry momentum as well as energy, hence photons must do so also. To find the momentum of a photon of frequency f, we can take its energy hf as equivalent to the total energy mc^2 of a particle of mass m, even though the photon has no rest mass. On this basis $hf = mc^2$, and the photon "mass" is

$$m_p = \frac{hf}{c^2} \qquad \textit{Photon "mass"} \quad (26.4)$$

Because the photon travels at the speed of light, its momentum is $p = m_p c$, so that

$$p = \frac{hf}{c} \qquad \textit{Photon momentum} \quad (26.5)$$

A collision between a photon and an electron can be analyzed with the help of special relativity, starting from conservation of energy and conservation of linear momentum. The result is a formula that relates the change in photon frequency to the angle ϕ between the original photon direction and the scattered photon direction:

$$\frac{1}{f'} - \frac{1}{f} = \frac{h}{m_0 c^2}(1 - \cos \phi)$$

This formula is easy to check by experiment: A beam of X rays of frequency f is directed at a target, and the frequencies of the scattered X rays are measured at various angles ϕ. The greater the scattering angle, the greater should be the change in frequency. Theory and experiment agree, which signifies that photons do indeed possess the momentum $p = hf/c$ and do indeed behave like particles in collisions. Photon scattering is called the **Compton effect** in honor of its discoverer, the American physicist Arthur H. Compton (1892–1962). The Compton effect is the chief means by which X rays lose energy as they pass through matter.

Photons and Gravity

As we have seen, a photon behaves in collisions as if it had the inertial mass hf/c^2 even though it lacks rest mass. According to the principle of equivalence discussed in Section 25.10, the gravitational and inertial masses of a material object are the same. If this principle applies to photons—which is reasonable since light is subject to gravity—then a photon ought to behave as if it had a gravitational mass of hf/c^2.

When we drop a stone of mass m from a height H, the gravitational pull of the earth accelerates it as it falls and the stone gains the energy mgH. The stone's final kinetic energy $\frac{1}{2}mv^2$ is equal to mgH, so its final speed is $\sqrt{2gH}$ (Fig. 26.8). But a photon has the speed of light and cannot go any faster. However, a photon's initial quantum energy hf will increase by $m_p gH$ if it is affected by gravity. This means that the photon's frequency ought to increase after it has "fallen" in a gravitational field.

In a laboratory experiment the frequency change is so small that we can neglect the corresponding change in the photon's "mass" m_p. If f' is the new

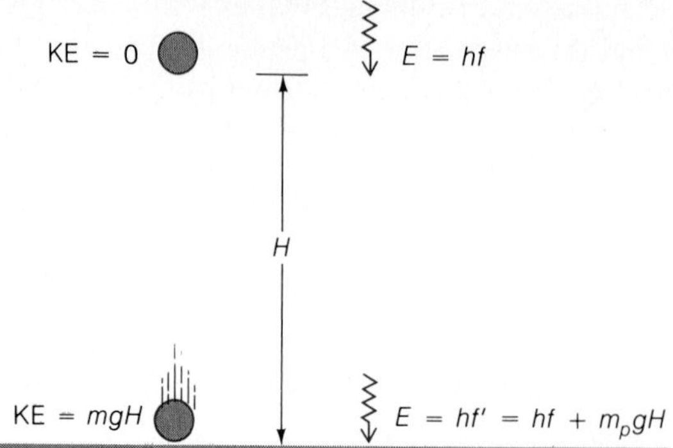

Figure 26.8

A photon falling in a gravitational field gains energy, just as a stone does. This gain of energy appears as an increase in frequency.

frequency, then,

$$\text{Final photon energy} = \text{initial photon energy} + \text{increase in energy}$$
$$hf' = hf + m_p gH$$

Because $m_p = hf/c^2$,

$$hf' = hf + \frac{hfgH}{c^2} = hf\left(1 + \frac{gH}{c^2}\right)$$

For $H = 20$ m, a reasonable height for a laboratory experiment, the relative change in frequency is

$$\frac{\Delta f}{f} = \frac{f' - f}{f} = \frac{gH}{c^2} = \frac{(9.8 \text{ m/s}^2)(20 \text{ m})}{(3.0 \times 10^8 \text{ m/s})^2} = 2.2 \times 10^{-15}$$

Such a frequency change can be detected, and in experiments using gamma rays, which are em waves of higher frequency than X rays, the effect has been confirmed. In the case of visible light, the frequency change is a few hertz for $H = 20$ m.

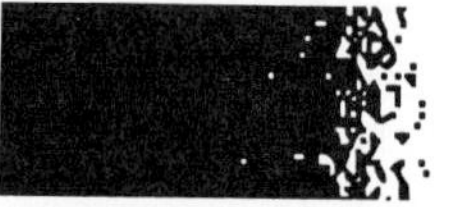

26.5 Matter Waves

A moving body behaves in certain ways as though it had a wave nature.

As we have seen, em waves under certain circumstances behave like particles. Is it possible that what we normally think of as particles have wave properties, too? This speculation was first made by Louis de Broglie in 1924. Soon afterward de Broglie's idea was taken up and developed by a number of other physicists (notably Heisenberg. Schrödinger, Born, Pauli, and Dirac) into the elaborate, mathematically difficult—but very beautiful—theory called **quantum mechanics.**

The advent of quantum mechanics did more than provide an accurate and complete description of atomic phenomena. It also altered the way in which physicists approach nature, so that they now think in terms of probabilities instead of in terms of certainties. The atomic world is closer in many respects to a roulette wheel than to a clock—but it is a roulette wheel that obeys certain rules, otherwise known as laws of physics.

De Broglie started with Eq. (26.5) for the linear momentum of a photon. Since $\lambda f = c$, this momentum can be expressed in terms of wavelength as $p = h/\lambda$. Hence for a photon

$$\lambda = \frac{h}{p} \qquad \textit{Photon wavelength} \quad (26.6)$$

De Broglie suggested that this equation for wavelength is a perfectly general one, applying to material objects as well as to photons. In the case of an object $p = mv$, and so its **de Broglie wavelength** is

$$\lambda = \frac{h}{mv} \qquad \textit{de Broglie wavelength} \quad (26.7)$$

The more momentum an object has, the shorter its wavelength. The relativistic formula $m = m_0/\sqrt{1 - v^2/c^2}$ must be used for m to find the de Broglie wavelength of a fast particle.

How can de Broglie's hypothesis be tested? Only waves can be diffracted and can reinforce and cancel each other by interference. A few years after de Broglie's work, Clinton Davisson and Lester Germer, in the United States, and G. P. Thomson, in England, independently demonstrated that streams of electrons are diffracted when they are scattered from crystals. The diffraction patterns they observed were in complete accord with the electron wavelengths predicted by de Broglie's formula.

In certain aspects of its behavior, a moving object resembles a wave, and in other aspects it resembles a particle. Which type of behavior is most conspicuous depends on how the object's de Broglie wavelength compares with its dimensions and with the dimensions of whatever it interacts with. Two examples will help us appreciate this statement.

EXAMPLE 26.4

In one of their experiments Davisson and Germer aimed a beam of 54-eV electrons at a nickel crystal (Fig. 26.9). Compare the de Broglie wavelength of these electrons with the spacing of the atomic planes in the crystal, which is 0.91×10^{-10} m.

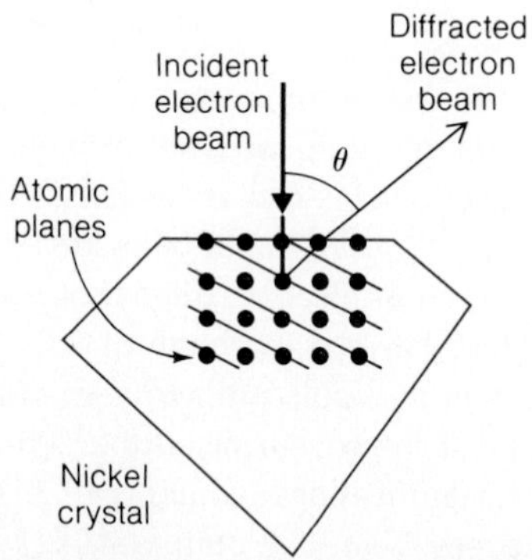

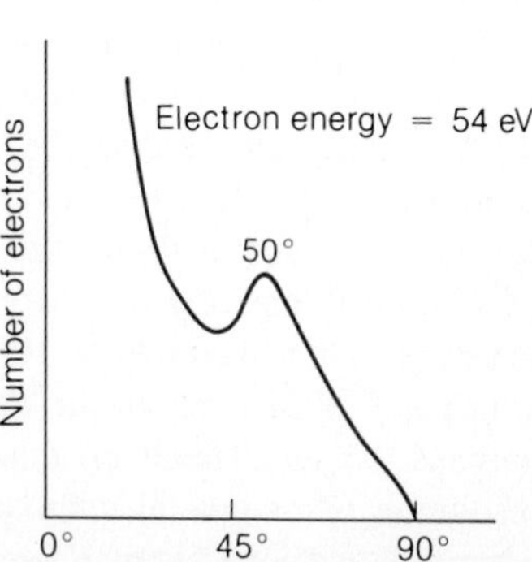

Figure 26.9

The Davisson-Germer experiment. The peak in the number of diffracted 54-eV electrons at θ = 50° is in agreement with de Broglie's formula for the wavelength of a moving particle.

SOLUTION The kinetic energy of a 54-eV electron is

$$\text{KE} = (54\ \text{eV})(1.6 \times 10^{-19}\ \text{J/eV}) = 8.6 \times 10^{-18}\ \text{J}$$

The momentum of such an electron can be calculated nonrelativistically. Here KE = $\frac{1}{2}mv^2$, $mv = \sqrt{2m\text{KE}}$, and so, from Eq. (26.7),

$$\lambda = \frac{h}{mv} = \frac{h}{\sqrt{2m\text{KE}}} = \frac{6.6 \times 10^{-34}\ \text{J}\cdot\text{s}}{\sqrt{2(9.1 \times 10^{-31}\ \text{kg})(8.6 \times 10^{-18}\ \text{J})}} = 1.7 \times 10^{-10}\ \text{m}$$

This is the same order of magnitude as the spacing of the atoms in the nickel crystal. We recall that diffraction is prominent only when the wavelength of the waves involved is comparable to the spacing of the scattering centers. Davisson and Germer found that the scattered 54-eV electrons were concentrated in just the direction ($\theta = 50°$) predicted by the theory of diffraction for waves of $\lambda = 1.7 \times 10^{-10}$ m, instead of the more even distribution expected for purely billiard-ball scattering. They interpreted the result as support for de Broglie's hypothesis. ♦

EXAMPLE 26.5

Find the de Broglie wavelength of a 1500-kg car whose speed is 30 m/s.

SOLUTION The car's wavelength is

$$\lambda = \frac{h}{mv} = \frac{6.63 \times 10^{-34}\ \text{J}\cdot\text{s}}{(1.5 \times 10^3\ \text{kg})(30\ \text{m/s})} = 1.5 \times 10^{-38}\ \text{m}$$

The wavelength is so small compared with the car's dimensions that no wave behavior is to be expected. ♦

As we know, the resolving power of any optical system is proportional to the wavelength of the light used. This limits ordinary microscopes to useful magnifications of less than 500×. Moving electrons, however, may have wavelengths much shorter than those of light waves. X rays also have short wavelengths, but it is difficult to focus them. On the other hand, electrons, by virtue of their charge, are easily controlled by electric and magnetic fields.

In an **electron microscope** a beam of electrons passes through a very thin specimen in a vacuum chamber, to prevent scattering of the beam and hence blurring of the image. The electron beam is then brought to a focus on a fluorescent screen or a photographic plate by a system of magnetic fields that act as lenses (Fig. 26.10). Because the numerical aperture of a magnetic "lens" cannot be made as high as that of an optical lens, the full theoretical resolution of electron waves cannot be realized in practice. For example, a 100-keV electron has a wavelength of 3.7×10^{-12} m, which is 0.0037 nm, but the actual resolution available ordinarily would not exceed perhaps 0.1 nm. However, this is still a great improvement on the 200-nm or more resolution of an optical microscope, and magnifications of 100,000 × or more are commonly achieved with electron microscopes.

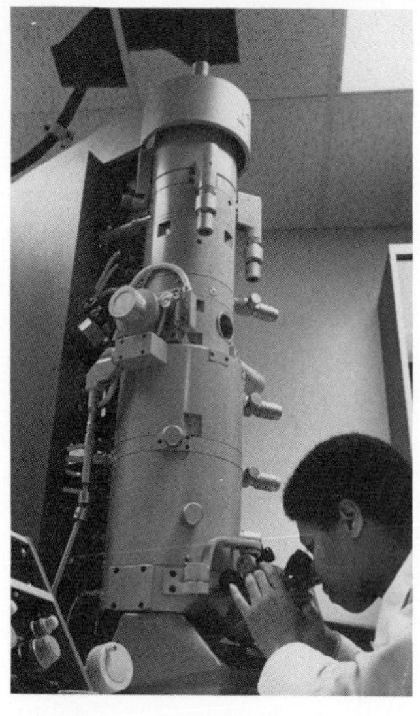

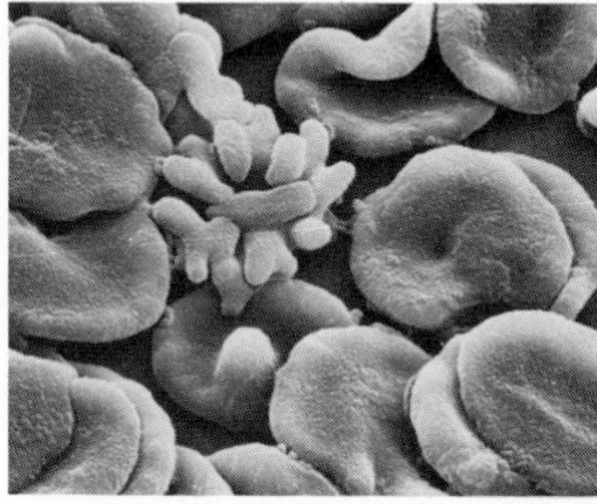

An electron microscope and an electron micrograph of red blood cells.

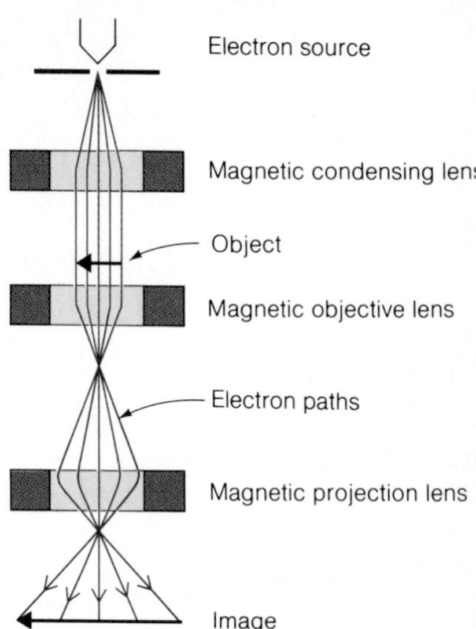

Figure 26.10

An electron microscope uses electrons instead of light waves to produce an enlarged image. Because electron de Broglie wavelengths are shorter than optical wavelengths, an electron microscope has a higher resolving power and hence can be used to produce higher magnifications. The magnetic "lenses" are current-carrying coils whose magnetic fields focus an electron beam, just as the glass lenses in an ordinary microscope focus a light beam.

26.6 Waves of What?

Waves of probability.

In water waves, the quantity that varies periodically is the height of the water surface. In sound waves, it is pressure in the medium the waves travel through. In light waves, electric and magnetic fields vary. What is it that varies in the case of matter waves?

The quantity whose variations constitute the matter waves of a moving object is known as its **wave function.** The symbol for wave function is ψ, the Greek letter *psi*.

The value of ψ^2 for a particular object at a certain place and time is proportional to the probability of finding the object at that place at that time.

For this reason the quantity ψ^2 is called the **probability density** of the object. A large value of ψ^2 means that it is likely to be found at the specified place and time; a small value of ψ^2 means that it is unlikely to be found at that place and time. Thus matter waves may be regarded as waves of probability.

Why does the probability of finding an object depend on ψ^2 and not on the wave function ψ itself? The answer is subtle. The amplitude of every wave varies from $-A$ to $+A$ to $-A$ to $+A$, and so on, where A is the maximum absolute value of whatever the wave variable is. But a negative probability is meaningless: The probability that an object be found must lie between 0 (the object is definitely not there) and 1 (the object is definitely there). An intermediate probability, say 0.4, means that there is a 40% chance of finding the object. A probability of -0.4,

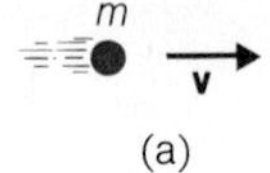

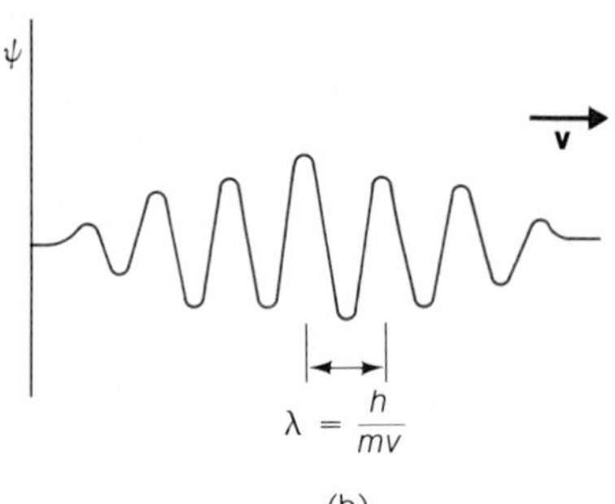

Figure 26.11

(a) Particle description of moving object. (b) Wave description of moving object.

however, makes no sense at all. Methods are known that allow us to find the wave function ψ for moving objects in a great many situations, and each value of ψ must then be squared to get a positive quantity that can be compared with experiment. (Actually, ψ sometimes turns out to have an imaginary component that contains the factor $\sqrt{-1}$, and in such cases ψ^2 is obtained in another way.)

A group or packet of matter waves is associated with every moving object. The packet travels with the same speed as the object does. The waves in the packet have the average wavelength $\lambda = h/mv$ given by de Broglie's formula (Fig. 26.11). Even though we cannot visualize what is meant by ψ and so cannot form a mental image of matter waves, the agreement between theory and experiment signifies that the notion of matter waves is a meaningful way to describe moving objects.

26.7 Particle in a Box

Why a trapped particle can have only certain energies.

When an object is confined to a certain region instead of being able to move freely, its wave properties lead to some remarkable consequences. Let us see what these consequences are in the simplest case—that of a particle trapped in an imaginary box of width L whose walls are such that the particle does not lose energy as it bounces back and forth along a straight line (Fig. 26.12). We will assume that the particle's speed v is small, so relativistic considerations can be ignored.

The wave equivalent of a particle in a box is a standing de Broglie wave. The reason for this is the same as in the case of standing waves in a stretched string (Section 12.5): The wave variable—transverse displacement in the case of a string, wave function ψ in the case of a moving particle—must be 0 at the walls of the box in order that the waves be reflected there. Hence the possible de Broglie wavelengths depend on the width L of the box (Fig. 26.13). The longest possible wavelength is $\lambda = 2L$, next is $\lambda = L$, then $\lambda = 2L/3$, and so on; in general, the permitted wavelengths are given by

Figure 26.12

A moving particle in a box of width L.

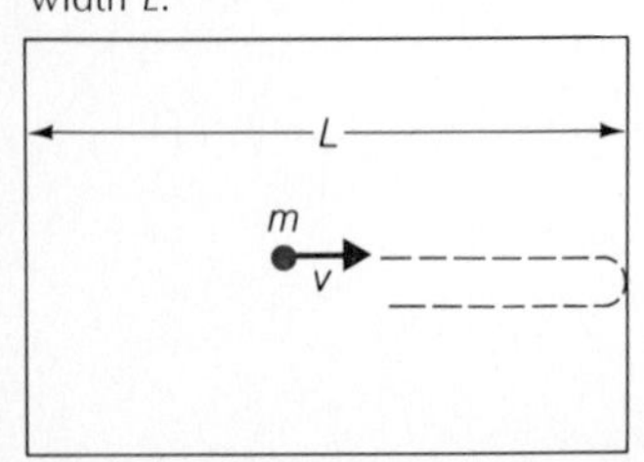

$$\lambda_n = \frac{2L}{n} \qquad n = 1, 2, 3, \ldots \qquad \textit{Permitted wavelengths} \quad (26.8)$$

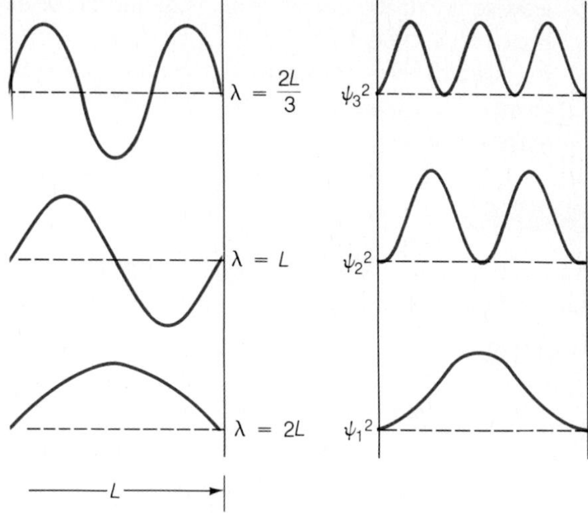

Figure 26.13

Wave functions ψ and probability densities ψ^2 of a particle in a box.

The wavelength of a particle of mass m and speed v is

$$\lambda = \frac{h}{mv}$$

so the restrictions on λ imposed by the size of the box amount to restrictions on the speed v the particle can have. These in turn lead to restrictions on its energy. The kinetic energy of a moving particle of wavelength λ is

$$\text{KE} = \tfrac{1}{2}mv^2 = \frac{1}{2m}(mv)^2 = \frac{h^2}{2m\lambda^2}$$

Our trapped particle has only kinetic energy, and because the only wavelengths it can have are $\lambda = 2L/n$, the only energies it can have are

$$E_n = \frac{n^2h^2}{8mL^2} \qquad n = 1, 2, 3, \ldots \qquad \textit{Energy levels} \quad (26.9)$$

Each permitted energy is called an **energy level,** and the integer n that corresponds to a given energy level is called its **quantum number.**

We can draw three important general conclusions from Eq. (26.9). These conclusions apply to *any* particle confined to a certain region of space, such as an atomic electron held captive by the attraction of the positively charged nucleus, and not just to the artificial situation of a particle in a box.

1. A trapped particle can have only certain specific energies and no others. The energy of such a particle is said to be **quantized,** and the magnitudes of the energy levels depend on the way in which the particle's motion is restricted.
2. A trapped particle cannot have zero energy; it must have a certain minimum amount E_1. Since the de Broglie wavelength of a particle is $\lambda = h/mv$, a speed of $v = 0$ means an infinite wavelength. But there is no way to reconcile an infinite wavelength with a trapped particle, so such a particle must possess at

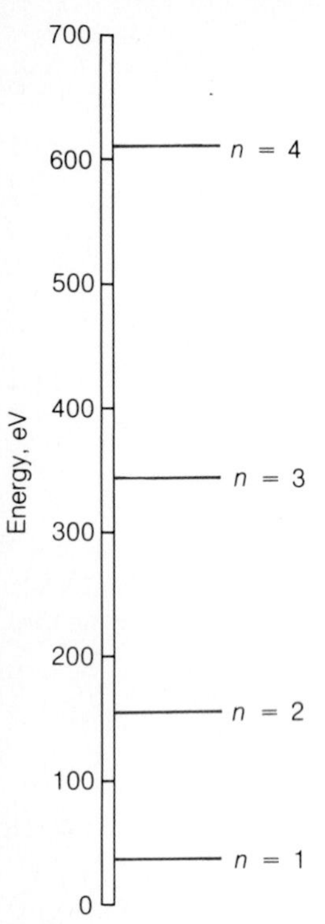

Figure 26.14

Energy levels of an electron confined to a box 10^{-10} m wide. A trapped particle cannot have zero energy.

least some kinetic energy. This is the origin of the "zero-point" energy mentioned in Section 13.6.

3. Because Planck's constant is so small—only 6.63×10^{-34} J·s—quantization of energy is conspicuous only when m and L are also small. Two examples will make this clear.

EXAMPLE 26.6

An electron is in a box 1.0×10^{-10} m across, which is the order of magnitude of atomic dimensions. Find its permitted energies.

SOLUTION From Eq. (26.9)

$$E_n = \frac{n^2h^2}{8mL^2} = \frac{n^2(6.6 \times 10^{-34}\ \text{J}\cdot\text{s})^2}{8(9.1 \times 10^{-31}\ \text{kg})(1.0 \times 10^{-10}\ \text{m})^2}$$
$$= 6.0 \times 10^{-18}\, n^2\ \text{J} = 38n^2\text{eV} \qquad n = 1, 2, 3, \ldots$$

The energies the electron can have are therefore 38 eV, 152 eV, 342 eV, 608 eV, and so on (Fig. 26.14). If such a box existed, the quantization of a trapped electron's energy would be a prominent feature of the system. (And indeed energy quantization *is* prominent in the case of an atomic electron, as we will find in Chapter 27.) ◆

EXAMPLE 26.7

A 10-g marble is in a box 10 cm across. Find its permitted energies.

SOLUTION Here

$$E_n = \frac{n^2h^2}{8mL^2} = \frac{(6.6 \times 10^{-34}\ \text{J}\cdot\text{s})^2 n^2}{8(1.0 \times 10^{-2}\ \text{kg})(0.10\ \text{m})^2} = 5.4 \times 10^{-64}\, n^2\ \text{J}$$
$$n = 1, 2, 3, \ldots$$

Here energy quantization cannot be detected. When $n = 1$, the marble's speed is 3.3×10^{-31} m/s, a speed no experiment could tell from zero, and the spacing between the higher energy levels is too small for measurement. ◆

On a macroscopic scale of size, then, quantum effects are unobservable, which is why classical physics is adequate on this scale. But on the scale of the atom, quantum effects are dominant, and the concepts and principles of classical physics must be replaced by others of a less familiar character. In fact, classical physics turns out to be just an approximation of quantum physics, which is perfectly general in its range of application.

26.8 Uncertainty Principle

We cannot know the future because we cannot know the present.

To regard a moving object as a wave packet suggests that there are fundamental limits to the accuracy with which we can measure such "particle" properties as its position and momentum.

The object whose wave packet is shown in Fig. 26.11 may be located anywhere within the packet at a given time. Of course, the probability density ψ^2 is a maximum in the middle of the packet, so it is most likely to be found there. But we may still find the object anywhere that ψ^2 is not actually zero.

The narrower the wave packet, the more precisely an object's position can be specified (Fig. 26.15a). However, the wavelength of the waves in a narrow packet is not well defined. There are just not enough waves to measure λ accurately. Because $\lambda = h/mv$, the particle's momentum mv is therefore not a precise quantity. If we make a series of momentum measurements, we will find a broad range of values.

On the other hand, a wide wave packet, such as that in Fig. 26.15(b), has a clearly defined wavelength. The momentum that corresponds to this wavelength is therefore a precise quantity, and a series of measurements will give a narrow range of values. But where is the particle located? The width of the packet is now too great for us to be able to say exactly where it is at a given time.

Thus we have the **uncertainty principle:**

It is impossible to know both the exact position and the exact momentum of an object at the same time.

This principle, which was discovered by Werner Heisenberg in 1927, is one of the most significant of physical laws.

The uncertainty principle can be put on a quantitative basis by regarding a wave packet as the result of the superposition of many trains of sinusoidal waves, as discussed in Chapter 12, with each train having a different wavelength (Fig. 26.16). The narrower the wave packet, the greater the spread of wavelengths involved and hence the greater the uncertainty Δmv. A smaller range of wavelengths is needed for a wide wave packet, but now Δx is large. The advantage of this approach is that it permits a precise analysis of the uncertainty principle, with the result that

$$\Delta x \Delta mv \geq \frac{h}{2\pi} \qquad \textit{Uncertainty principle} \quad (26.10)$$

In words, the uncertainty principle states that

$$\begin{pmatrix}\text{Uncertainty} \\ \text{in position}\end{pmatrix}\begin{pmatrix}\text{uncertainty} \\ \text{in momentum}\end{pmatrix} \quad \textit{is equal to or greater than} \quad \frac{\text{Planck's constant}}{2\pi}$$

These uncertainties have nothing to do with the instruments we use to make measurements. Of course, poor instruments will give poor results, but even perfect ones could not give exact results for both position and momentum at the same time. Since we cannot know where an object is right now and what its momentum is, we cannot say anything for sure about where it will be in the future or how fast it will be moving then. We cannot know the future because we cannot know the present. But our ignorance is not total. We can still say that the object is more likely to be in one place than in another and that its momentum is more likely to have a certain value than to have another. Because h is so small, the uncertainty principle is significant only on an atomic scale—for objects larger than molecules the probabilities are so great as to be practically certainties.

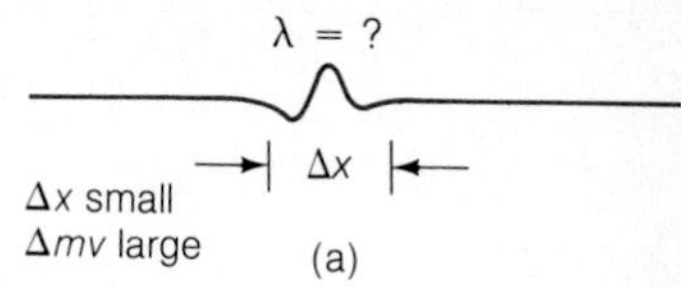

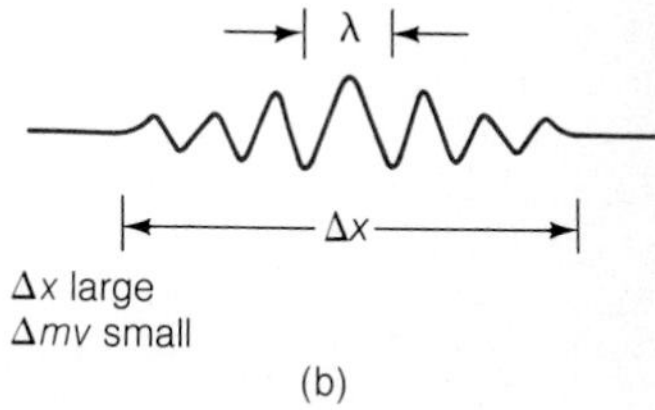

Figure 26.15

(a) A narrow wave packet means a small uncertainty Δx in the position of a moving object but a large uncertainty in its wavelength and hence momentum. (b) A wide wave packet means a large uncertainty in the position of a moving object but a small uncertainty in its wavelength and hence momentum. It is impossible to have a narrow wave packet with a well-defined wavelength.

Figure 26.16

(a) If a moving object can be represented by a wave with a single wavelength, its position cannot be established at all. (b) A wave packet is the result of superposing waves of different wavelengths. The greater the range of wavelengths, the narrower the packet.

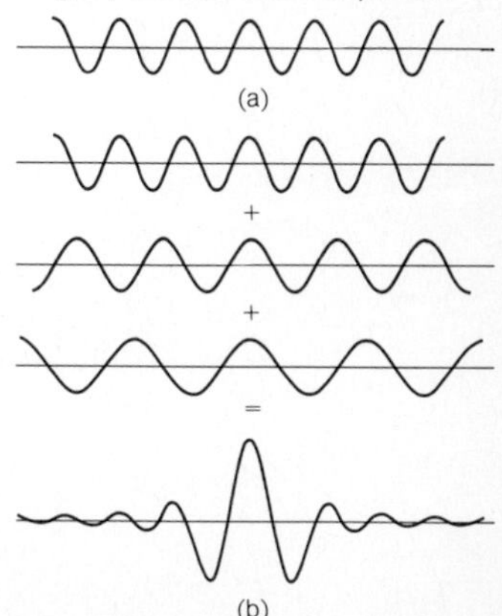

EXAMPLE 26.8

A measurement locates a proton with an accuracy of $\pm 1.0 \times 10^{-11}$ m. Find the uncertainty in the proton's position 1.0 s later. Assume $v << c$.

SOLUTION Let us call the uncertainty in the proton's position Δx_0 at the time $t = 0$. The uncertainty in its momentum at this time is

$$\Delta mv \geq \frac{h}{2\pi \Delta x_0}$$

Since $v << c$, the uncertainty in the proton's speed is

$$\Delta v = \frac{\Delta mv}{m} \geq \frac{h}{2\pi m \Delta x_0}$$

The distance x the proton covers in the time t cannot be predicted more accurately than

$$\Delta x = t\Delta v \geq \frac{ht}{2\pi m \Delta x_0}$$

Hence Δx is inversely proportional to Δx_0. The *more* we know about the proton's position at $t = 0$, the *less* we know about its later position at $t = t$. The value of Δx at $t = 1.0$ s is

$$\Delta x \geq \frac{(6.63 \times 10^{-34}\ \text{J·s})(1.0\ \text{s})}{(2\pi)(1.67 \times 10^{-27}\ \text{kg})(1.0 \times 10^{-11}\ \text{m})} \geq 6300\ \text{m}$$

This is 6.3 km—nearly 4 mi! ◆

The uncertainty principle is not merely a negative statement. Here is an illustration of how it can give us valuable information in a simple way.

EXAMPLE 26.9

The radius of the hydrogen atom is 5.3×10^{-11} m. Assuming that the uncertainty Δx in the location of its single electron has the same value, estimate the minimum energy of the electron.

SOLUTION From the uncertainty principle

$$\Delta mv \geq \frac{h}{2\pi \Delta x} \geq \frac{6.63 \times 10^{-34}\ \text{J·s}}{(2\pi)(5.3 \times 10^{-11}\ \text{m})} \geq 2.0 \times 10^{-24}\ \text{kg·m/s}$$

This momentum uncertainty corresponds to a kinetic energy uncertainty of

$$\Delta \text{KE} = \frac{(\Delta mv)^2}{2m} = \frac{(2.0 \times 10^{-24}\ \text{kg·m/s})^2}{2(9.1 \times 10^{-31}\ \text{kg})} = 2.2 \times 10^{-18}\ \text{J}$$

which is

$$\Delta KE = \frac{2.2 \times 10^{-18}\ \mathrm{J}}{1.6 \times 10^{-19}\ \mathrm{J/eV}} = 14\ \mathrm{eV}$$

This is the lower limit to the kinetic energy of the electron, and in fact it is equal to the kinetic energy of an electron in the lowest energy level of a hydrogen atom.

When a similar calculation is made of the minimum energy of an electron in an atomic nucleus, whose radius is typically 5×10^{-15} m, the result turns out to be over 20 MeV. (Such a calculation must be made relativistically.) However, the electrons associated with nuclei, even unstable nuclei, never have more than a small fraction of this energy, and we conclude that electrons cannot be present in atomic nuclei. ♦

26.9 Uncertainty Principle: Alternative Approach

Considering an object as a particle gives the same result.

The uncertainty principle makes sense from the point of view of the particle properties of waves as well as from the point of view of the wave properties of particles. Suppose we wish to measure the position and momentum of an object at a certain moment. To do so, we must touch it with something that will carry the required information back to us. That is, we must poke it with a stick, shine light on it, or perform some similar act. The measurement process itself thus requires that the body be interfered with in some way. If we consider this interference in detail, we are led to the same uncertainty principle as before even without taking into account the wave nature of moving objects.

Suppose we look at an electron with the help of light whose wavelength is λ, as in Fig. 26.17. Each photon of this light has the momentum h/λ. When one of these photons bounces off the electron (which must occur if we are to "see" it), the electron's original momentum will be changed. The exact amount of the change Δmv

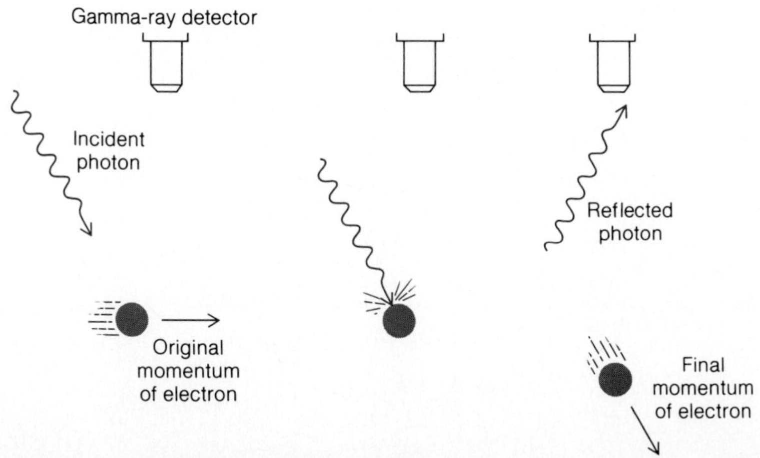

Figure 26.17
An electron cannot be observed without affecting its momentum. The more accurately the electron's position is determined, the greater the change in momentum.

cannot be predicted, but it will be of the same order of magnitude as the photon momentum h/λ. Hence

$$\Delta mv \approx \frac{h}{\lambda}$$

The *larger* the wavelength of the observing photon, the smaller the uncertainty in momentum.

Because light is a wave phenomenon as well as a particle phenomenon, we cannot expect to determine the electron's location with perfect accuracy even with the best of instruments. An estimate of the minimum uncertainty Δx in the measurement might be one photon wavelength. That is,

$$\Delta x \geq \lambda$$

The *smaller* the wavelength, the smaller the uncertainty in location. Hence if we use light of short wavelength to increase the accuracy of our position measurement, there will be a corresponding decrease in the accuracy of our momentum measurement because the higher photon energy will disturb the electron's motion to a greater extent. Light of long wavelength will yield an accurate momentum but an inaccurate position.

Combining the two preceding formulas gives

$$\Delta x \, \Delta mv \geq h$$

A detailed calculation shows that somewhat better accuracy is possible, so that the limit of the product $\Delta x \, \Delta mv$ is $h/2\pi$ as before, instead of just h.

Energy and Time

A form of the uncertainty principle that involves energy and time is easy to obtain. Suppose we are measuring the energy E emitted in some process during the period of time Δt. If the energy is in the form of electromagnetic waves, the time Δt limits the accuracy with which we can measure the wave frequency f. The minimum uncertainty in the number of waves we count in a given wave train is of the order of one wave. Because

$$\text{Frequency} = \frac{\text{number of waves}}{\text{time interval}}$$

the uncertainty in frequency Δf of our determination is

$$\Delta f \geq \frac{1}{\Delta t}$$

The corresponding uncertainty in energy ΔE is $\Delta E = h \, \Delta f$ and so

$$\Delta E \geq \frac{h}{\Delta t} \quad \text{or} \quad \Delta E \, \Delta t \geq h$$

A more detailed calculation changes this result to

$$\Delta E\,\Delta t \geq \frac{h}{2\pi} \qquad \textit{Uncertainties in energy and time} \qquad (26.11)$$

Equation (26.11) states that the product of the uncertainty in an energy measurement and the time available for the measurement is at best equal to $h/2\pi$. This form of the uncertainty principle can be derived in other ways as well and is a general conclusion that is not limited to electromagnetic waves. Equation (26.11) will be used in Chapter 30 to estimate the mass of a certain type of elementary particle.

Important Terms

The **photoelectric effect** is the emission of electrons from a metal surface when light shines on it.

The **quantum theory of light** states that light travels in tiny bursts of energy called **quanta** or **photons.** The quantum theory of light is required to account for the photoelectric effect.

X rays are high-frequency electromagnetic waves emitted when fast electrons impinge on matter.

A moving body behaves as though it had a wave character. The waves representing such a body are **matter waves,** also called **de Broglie waves.** The wave variable in a matter wave is its **wave function,** whose square is the **probability density** of the body. The value of the probability density of a particular body at a certain place and time is proportional to the probability of finding the body at that place and time. Matter waves may thus be regarded as waves of probability.

Because of its wave nature, a particle restricted to a definite region of space can have only certain specific energies, each of which is called an **energy level** and corresponds to a **quantum number.**

The **uncertainty principle** is an expression of the limit set by the wave nature of matter on finding both the position and the state of motion of a moving body.

Important Formulas

Quantum energy: $E = hf$

Photoelectric effect: $hf = \text{KE}_{\text{max}} + hf_0$

X-ray energy: $hf_{\text{max}} = Ve$

de Broglie wavelength: $\lambda = \dfrac{h}{mv}$

Uncertainty principle: $\Delta x\,\Delta mv \geq \dfrac{h}{2\pi}$

$$\Delta E\,\Delta t \geq \frac{h}{2\pi}$$

MULTIPLE CHOICE

1. A surface emits photoelectrons only when the light shone on it exceeds a certain

a. frequency b. wavelength
c. intensity d. speed

2. When light is directed on a metal surface, the kinetic energies of the photoelectrons

a. vary with the intensity of the light
b. vary with the frequency of the light
c. vary with the speed of the light.
d. are random

3. An increase in the brightness of the light directed at a metal surface causes an increase in the photoelectrons'

a. wavelength b. speed
c. energy d. number

4. The photoelectric effect can be understood on the basis of

a. the electromagnetic theory of light
b. the special theory of relativity
c. the principle of superposition
d. none of the above

5. In a vacuum all photons have the same
a. frequency
b. wavelength
c. energy
d. speed

6. The rest mass of a photon
a. is zero
b. is the same as that of an electron
c. depends on its frequency
d. depends on its energy

7. An increase in the voltage applied to an X-ray tube causes an increase in the X rays'
a. wavelength
b. speed
c. energy
d. number

8. The Compton effect refers to the scattering of
a. photons by electrons
b. photons by crystals
c. de Broglie waves by electrons
d. de Broglie waves by crystals

9. A photon that has moved downward in the earth's gravitational field has, relative to the original value,
a. a higher speed
b. a higher frequency
c. a longer wavelength
d. less energy

10. The description of a moving body in terms of matter waves is legitimate because
a. it is based on common sense
b. matter waves have actually been seen
c. the analogy with electromagnetic waves is plausible
d. theory and experiment agree

11. The wave packet that corresponds to a moving particle
a. has the same size as the particle
b. has the same speed as the particle
c. has the speed of light
d. consists of X rays

12. A moving body is described by the wave function ψ at a certain time and place; ψ^2 is proportional to the body's
a. electric field
b. speed
c. energy
d. probability of being found

13. The dimensions of the region in which a particle is trapped govern
a. its minimum mass
b. its minimum wavelength
c. its minimum energy
d. the number of possible energies it can have

14. A particle trapped in a certain region of space can have
a. only certain energies above a certain minimum
b. any energy above a certain minimum
c. only certain energies below a certain maximum
d. any energy below a certain maximum

15. The maximum de Broglie wavelength a particle trapped in a box L wide can have is
a. $L/4$
b. $L/2$
c. L
d. $2L$

16. The wider the wave packet of a particle is,
a. the larger it is
b. the longer its wavelength is
c. the more precisely its position can be determined
d. the more precisely its momentum can be determined

17. If Planck's constant were larger than it is,
a. moving bodies would have shorter wavelengths
b. moving bodies would have higher energies
c. moving bodies would have higher momenta
d. the uncertainty principle would be significant on a larger scale of size

18. Light of wavelength 500 nm consists of photons whose energy is
a. 1.1×10^{-48} J
b. 1.3×10^{-27} J
c. 4.0×10^{-19} J
d. 1.7×10^{-15} J

19. An X-ray photon has an energy of 6.6×10^{-15} J. The frequency that corresponds to this energy is
a. 4.4×10^{-48} Hz
b. 1.0×10^{-19} Hz
c. 1.0×10^{15} Hz
d. 1.0×10^{19} Hz

20. A lamp emits light of frequency 5.0×10^{15} Hz at a power of 25 W. The number of photons given off per second is
a. 1.3×10^{-19}
b. 8.3×10^{-17}
c. 7.6×10^{18}
d. 1.9×10^{50}

21. The de Broglie wavelength of an electron whose speed is 1.0×10^8 m/s is
a. 5.9×10^{-56} m
b. 1.5×10^{-19} m
c. 7.3×10^{-12} m
d. 1.4×10^{11} m

22. The speed of an electron whose de Broglie wavelength is 1.0×10^{-10} m is
a. 6.6×10^{-24} m/s
b. 3.8×10^3 m/s
c. 7.3×10^6 m/s
d. 1.0×10^{10} m/s

23. A particle trapped in a certain box has an energy of 8 eV in the $n = 2$ state. The lowest energy the particle can have in the box is
a. 0.5 eV
b. 1 eV
c. 2 eV
d. 4 eV

QUESTIONS

1. Why do you think the wave aspect of light was discovered earlier than its particle aspect?

2. If Planck's constant were smaller than it is, would quantum phenomena be more or less conspicuous than they are now in everyday life?

3. What happens to the frequency and wavelength of the X rays emitted from a metal surface when the speed of the electrons that strike it is increased? What happens to the speed, energy, and number per second of the X-ray photons?

4. How does the speed of the wave packet that corresponds to a moving particle compare with (a) the particle's speed and (b) the speed of light?

5. Must a particle have an electric charge in order for matter waves to be associated with its motion?

6. A photon and a proton have the same wavelength. What can be said about how their linear momenta compare? About how the photon's energy compares with the particle's kinetic energy?

7. What is the simplest experimental procedure that can distinguish between a gamma ray whose wavelength is 10^{-11} m and an electron whose de Broglie wavelength is also 10^{-11} m?

8. Can the rest mass of a moving particle be determined by measuring its de Broglie wavelength?

9. A proton and an electron have the same de Broglie wavelength. How do their speeds compare?

10. The uncertainty principle applies to all objects, yet its consequences are significant only for such tiny particles as electrons, protons, and neutrons. Why?

EXERCISES

26.2 Quantum Theory of Light

1. Find the energy of a 700-nm photon.

2. Find the frequency and wavelength that correspond to a 5.0-MeV photon.

3. How many photons per second are emitted by a 150-W amateur radio transmitter operating on a wavelength of 20 m?

4. A detached retina is being ''welded'' back by using 20-ms pulses from an 0.50-W laser operating at a wavelength of 643 nm. How many photons are in each pulse?

5. What is the maximum wavelength of light that will lead to photoelectric emission from platinum, whose work function is 5.6 eV? In what part of the spectrum is such light?

6. The threshold frequency for photoelectric emission in calcium is 7.7×10^{14} Hz. Find the maximum energy in electron volts of the photoelectrons when light of frequency 1.20×10^{15} Hz is directed on a calcium surface.

7. A silver ball is suspended by a string in a vacuum chamber, and ultraviolet light of wavelength 200 nm is directed at it. What electric potential will the ball acquire as a result? The work function of silver is 4.7 eV.

26.3 X Rays

8. The operating voltage of an X-ray tube is 50 kV. Find the highest frequency found in the X rays it emits.

9. Electrons are accelerated in television tubes through potential differences of about 10 kV. Find the highest frequency of the em waves that are emitted when these electrons strike the screen of the tube. What type of waves are these?

10. What voltage must be applied to an X-ray tube for it to emit X rays with a minimum wavelength of 3.0×10^{-11} m?

26.5 Matter Waves

11. What is the de Broglie wavelength of a 1.0-mg grain of sand blown by the wind at a speed of 20 m/s?

12. Find the de Broglie wavelength of a 1.0-MeV proton. Since the rest energy of the proton is 938 MeV, the calculation may be made nonrelativistically.

13. Calculate the de Broglie wavelength of (a) an electron whose speed is 1.0×10^8 m/s and (b) an electron whose speed is 2.0×10^8 m/s. Use relativistic formulas.

14. Show that the de Broglie wavelength of an oxygen molecule in thermal equilibrium in the atmosphere at 20°C is smaller than its diameter of about 4×10^{-10} m.

15. Green light has a wavelength of about 550 nm. Through what potential difference must an electron be accelerated to have this wavelength?

26.7 Particle in a Box

16. The lowest energy possible for a certain particle in a certain box is 1 eV. What are the next two higher energies the particle can have?

17. Derive a formula for the energy levels (in MeV) of a neutron confined to a one-dimensional box 1.0×10^{-14} m wide. What is its minimum energy? (The diameter of an atomic nucleus is of this order of magnitude.)

26.8 Uncertainty Principle

26.9 Uncertainty Principle: Alternative Approach

18. (a) An electron is confined in a box 1.0×10^{-9} m in width. What is the uncertainty in its speed? (b) A proton is confined in the same box. What is the uncertainty in its speed?

19. An atom in an energy level above its lowest one gives up its excess energy by emitting one or more photons, each with a characteristic frequency, as described in Chapter 27. Usually about 1.0×10^{-8} s elapses between the excitation of an atom and the time it emits a photon. Find the uncertainty in the photon frequency in hertz.

20. From the formula for the possible energies of a particle trapped in a box, find the minimum momentum the particle can have and compare it with the prediction of the uncertainty principle.

21. The position and momentum of a 1.0-keV electron are determined at the same time. If the position is found with an uncertainty of 1.0×10^{-10} m, what is the percentage of uncertainty in the momentum?

22. The position of a proton, initially at rest, is to be determined without giving it more than 1.0 keV of KE. Find the maximum accuracy with which the proton's position can be determined.

PROBLEMS

23. Light from the sun reaches the earth at the rate of about 1.4×10^3 W/m^2 of area perpendicular to the direction of the light. Assume sunlight is monochromatic with a frequency of 5.0×10^{14} Hz. (a) How many photons fall per second on each square meter of the earth's surface directly facing the sun? (b) How many photons are present in each cubic meter near the earth on the sunlit side?

24. Light of wavelength 420 nm falls on the cesium surface of a photoelectric cell at the rate of 5.0 mW. Given that one photoelectron is emitted for every 10^4 incident photons, find the current produced by the cell. The work function of cesium is 1.9 eV.

25. A metal surface illuminated by 8.5×10^{14} Hz light emits electrons whose maximum energy is 0.52 eV. The same surface illuminated by 12.0×10^{14} Hz light emits electrons whose maximum energy is 1.97 eV. From these data find Planck's constant and the work function of the surface.

26. (a) How much time is needed to measure the KE of an electron whose speed is 10 m/s with an uncertainty of no more than 0.1%? How far will the electron have traveled in this time interval? (b) Make the same calculations for a 1.0-g insect with the same speed. What do these sets of figures indicate?

27. Verify that the uncertainty principle can be expressed in the form $\Delta L\ \Delta\theta \geq h/2\pi$, where ΔL is the uncertainty in the angular momentum of a body and $\Delta\theta$ is the uncertainty in radians in its angular position. To do this, consider a particle of mass m moving in a circle.

Answers to Multiple Choice

1. a	7. c	13. c	19. d
2. b	8. a	14. a	20. c
3. d	9. b	15. d	21. c
4. d	10. d	16. d	22. c
5. d	11. b	17. d	23. c
6. a	12. d	18. c	

CHAPTER 27

THE ATOM

When Ernest Rutherford discovered in 1911 that an atom consists of a tiny, positively charged nucleus surrounded by electrons, he solved one problem but brought into focus a whole set of new ones. To begin with, what keeps the electrons out there? How are the electrons arranged in the atoms of different elements? What happens when two or more atoms join to form a molecule? Why are some combinations of atoms stable (the water molecule H_2O, for instance), whereas others (HO_2, for instance) never occur? How do atoms and molecules interact to form liquids and solids? As we will see in this chapter and the next, quantum concepts provide the basis for understanding atomic structure.

27.1 The Hydrogen Atom

Standing waves in the atom.

Our starting point will be the hydrogen atom, the simplest of all, which has a single electron outside a nucleus that consists of only a proton. The electron cannot be at rest, because the electric attraction of the nucleus would pull it in at once. If the electron is moving, however, a stable orbit like that of a planet around the sun would seem to be possible.

Experiments indicate that 13.6 eV of work is needed to break apart a hydrogen atom into a proton and an electron that go their separate ways (Fig. 27.1). With the help of calculations based on what we already know about mechanics and electricity (see Section 27.4), this figure leads to an orbital radius of $r = 5.3 \times 10^{-11}$ m and an orbital speed of $v = 2.2 \times 10^6$ m/s for the electron in the hydrogen atom. The proton mass is 1836 times the electron mass, so we can consider the proton as stationary with the electron revolving around it, the required centripetal force being provided by the electrical force exerted by the proton (Fig. 27.2).

What such an analysis overlooks is that, according to electromagnetic theory, all accelerated electric charges radiate electromagnetic waves—and an electron moving in a circular path is certainly accelerated. Thus an atomic electron circling its nucleus should radiate away energy, spiraling inward until it is swallowed up by the nucleus (Fig. 27.3). Clearly the ordinary laws of physics cannot explain the stability of the hydrogen atom, whose electron must be whirling around the nucleus to keep from being pulled into it yet must be losing energy all the time.

The first successful theory of the hydrogen atom was put forward in 1913 by Niels Bohr, a Dane. Bohr applied then-new quantum ideas to atomic structure to come up with a model that, even though later replaced by a theory of greater accuracy and scope, still provides a useful mental picture of the atom for many purposes. The concept of matter waves leads in a natural way to the Bohr model, and this is the

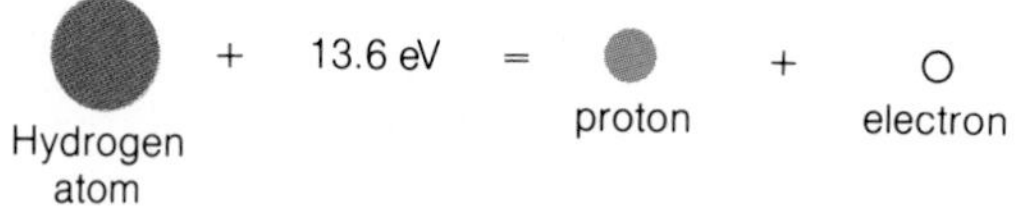

Figure 27.1

To break a hydrogen atom apart requires 13.6 eV.

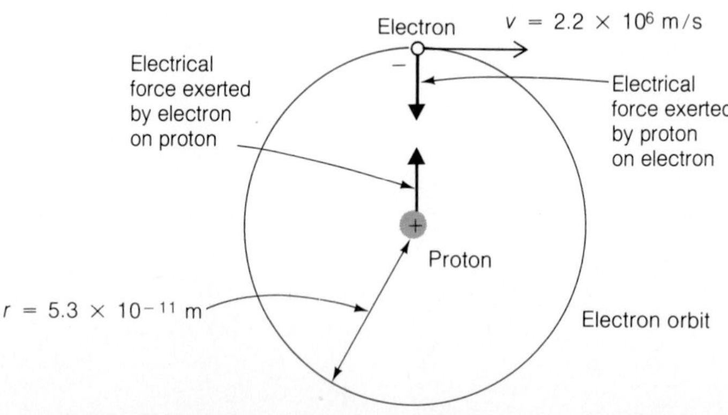

Figure 27.2

The hydrogen atom consists of an electron circling a proton. The electrical force exerted by the proton on the electron provides the centripetal force required to hold it in a circular path. The proton is 1836 times as heavy as the electron, so its motion under the influence of the electrical force that the electron exerts is small.

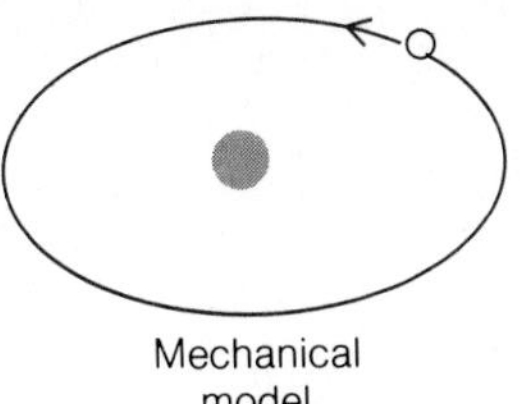

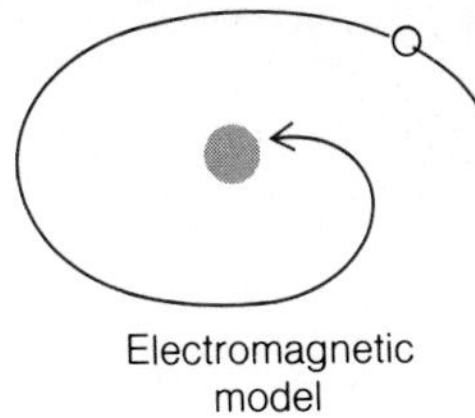

Figure 27.3

The mechanical and electromagnetic models of the hydrogen atom are in conflict. Only quantum theory can account for the stability of atoms.

route we will follow here. Bohr himself used a different approach because matter waves were unknown at the time, which makes his achievement all the more remarkable. The results are exactly the same in both cases.

Let us begin by looking into the wave properties of the electron in the hydrogen atom. The de Broglie wavelength λ of an object of mass m and speed v is $\lambda = h/mv$ where h is Planck's constant. The speed of the electron in a hydrogen atom is $v = 2.2 \times 10^6$ m/s, and so the wavelength of its matter waves is

$$\lambda = \frac{h}{mv} = \frac{6.63 \times 10^{-34}\ \text{J}\cdot\text{s}}{(9.1 \times 10^{-31}\ \text{kg})(2.2 \times 10^6\ \text{m/s})} = 3.3 \times 10^{-10}\ \text{m}$$

This is an exciting result, because the electron's orbit has a circumference of exactly

$$2\pi r = (2\pi)(5.3 \times 10^{-11}\ \text{m}) = 3.3 \times 10^{-10}\ \text{m}$$

We therefore conclude that *the orbit of the electron in a hydrogen atom corresponds to one complete electron wave joined on itself* (Fig. 27.4).

Figure 27.4

(a) The electron orbit in a ground-state hydrogen atom is exactly one de Broglie wavelength in circumference. (b) The orbit corresponds to a complete electron wave joined on itself.

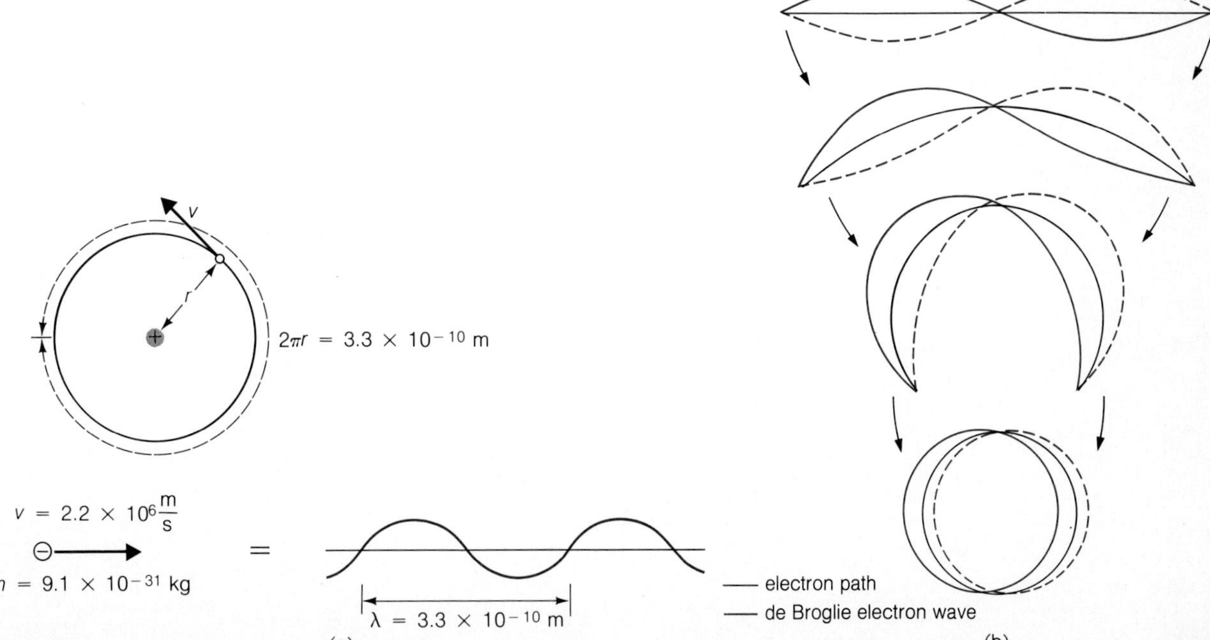

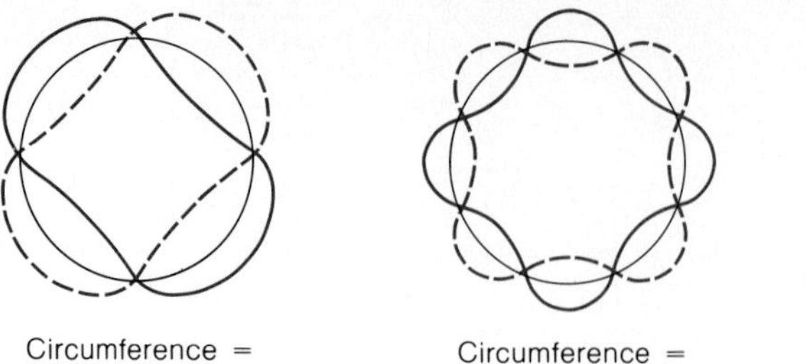

Figure 27.5

Three possible modes of vibration of a wire loop.

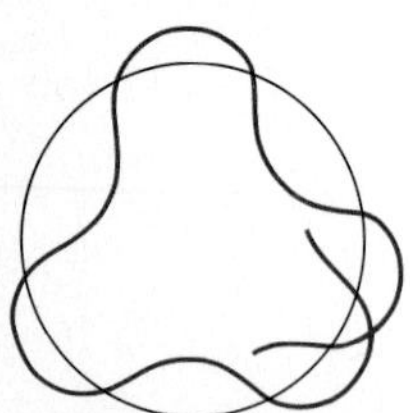

Figure 27.6

Unless a whole number of wavelengths fits into the wire loop, destructive interference causes the vibrations to die out rapidly.

The fact that the electron orbit in a hydrogen atom is one electron wavelength in circumference is just the clue we need to construct a theory of the atom. If we look at the vibrations of a wire loop, as in Fig. 27.5, we see that their wavelengths always fit a whole number of times into the loop's circumference, each wave joining smoothly with the next. With no dissipative effects, such vibrations would go on forever.

Why are these the only vibrations possible in a wire loop? A fractional number of wavelengths cannot be fitted into the loop and still allow each wave to join smoothly with the next (Fig. 27.6). The result would be destructive interference as the waves travel around the loop, and the vibrations would die out rapidly.

By considering the behavior of electron waves in the hydrogen atom as analogous to standing waves in a wire loop, then, we may postulate the following:

An electron can circle an atomic nucleus only if its orbit is a whole number of electron wavelengths in circumference.

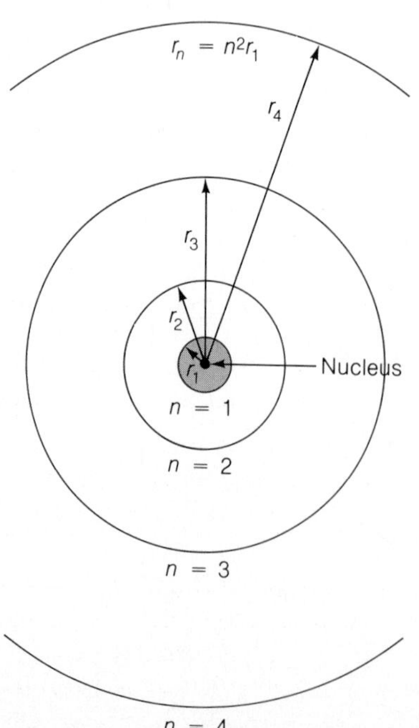

Figure 27.7

Electron orbits in the hydrogen atom according to the Bohr model. The orbit radii are proportional to n^2.

This postulate is the decisive one in our understanding of the atom. We note that it combines both the particle and the wave characters of the electron into a single statement. Although we can never observe these antithetical characters at the same time in an experiment, they are inseparable in nature.

It is easy to express in a formula the condition that a whole number of electron wavelengths fit into the electron's "orbit." The circumference of a circular orbit of radius r is $2\pi r$, and so the condition for orbit stability is

$$n\lambda = 2\pi r_n \qquad n = 1, 2, 3, \ldots \qquad \textit{Condition for orbit stability} \quad (27.1)$$

where r_n designates the radius of the orbit that contains n wavelengths. The quantity n is called the **quantum number** of the orbit.

A straightforward calculation (given in Section 27.4) shows that the stable electron orbits are those whose radii are given by the formula

$$r_n = n^2 r_1 \qquad n = 1, 2, 3, \ldots \qquad \textit{Orbital radii in Bohr atom} \quad (27.2)$$

where $r_1 = 5.3 \times 10^{-11}$ m is the radius of the innermost orbit (Fig. 27.7).

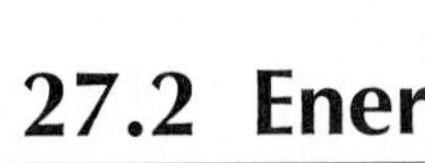

27.2 Energy Levels

Each orbit has a characteristic energy.

The total energy of a hydrogen atom is not the same in the various permitted orbits. The energy E_n of a hydrogen atom whose electron is in the nth orbit is given by

$$E_n = \frac{E_1}{n^2} \qquad n = 1, 2, 3, \ldots \qquad \textit{Energy levels of hydrogen atom} \quad (27.3)$$

where $E_1 = -13.6$ eV $= -2.18 \times 10^{-18}$ J is the energy corresponding to the innermost orbit. The energies specified by this formula are called the **energy levels** of the hydrogen atom and are the only ones possible. The situation is rather like that of a person on a ladder, who can stand only on its steps, not in between them.

The energy levels of the hydrogen atom are all less than zero, which means that the electron does not have enough energy to escape from the atom. The lowest energy level E_1, corresponding to the quantum number $n = 1$, is called the **ground state** of the atom. The higher levels E_2, E_3, E_4, and so on are called **excited states** (Fig. 27.8).

As the quantum number n increases, the energy E_n approaches zero. In the limit of $n = \infty$, $E_\infty = 0$ and the electron is no longer bound to the proton to form an atom. An energy greater than zero corresponds to an unbound electron. Because such an electron has no closed orbit that must satisfy quantum conditions, it may have any positive energy whatever.

The work needed to remove an electron from an atom in its ground state is called its **ionization energy.** The ionization energy is therefore equal to $-E_1$, the energy that must be provided to raise an electron from its ground state to an energy of $E = 0$, when it is free. In the case of hydrogen, the ionization energy is 13.6 eV, because the ground-state energy of the hydrogen atom is -13.6 eV.

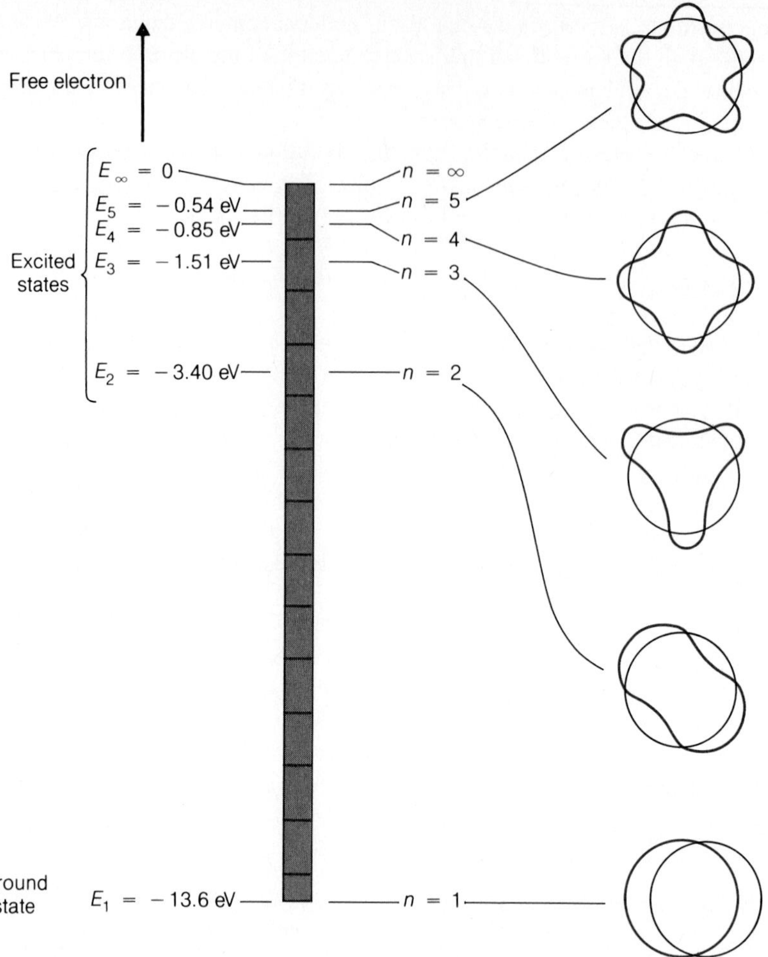

Figure 27.8

Energy levels of the hydrogen atom.

EXAMPLE 27.1

An electron collides with a hydrogen atom in its ground state and excites it to a state of $n = 3$. How much energy was given to the hydrogen atom in this inelastic (KE not conserved) collision?

SOLUTION From Eq. (27.3) the energy change of a hydrogen atom that goes from an initial state of quantum number n_i to a final state of quantum number n_i is

$$\Delta E = E_f - E_i = \frac{E_1}{n_f^2} - \frac{E_1}{n_i^2} = E_1\left(\frac{1}{n_f^2} - \frac{1}{n_i^2}\right)$$

Here $n_i = 1$ and $n_f = 3$, and $E_1 = -13.6$ eV, so

$$\Delta E = E_1\left(\frac{1}{3^2} - \frac{1}{1^2}\right) = -13.6\left(\frac{1}{9} - 1\right)\text{eV} = 12.1 \text{ eV}$$

◆

Quantization in the Atomic World

The presence of energy levels in an atom—which is true for all atoms, not just the hydrogen atom—is another example of the fundamental graininess of physical quantities on a microscopic scale. In the everyday world, matter, electric charge, energy, and so on seem continuous and capable of being cut up, so to speak, into parcels of any size we like. In the world of the atom, however, matter consists of elementary particles of fixed rest masses that join together to form atoms of fixed rest masses; electric charge aways comes in multiples of $+e$ and $-e$; energy in the form of electromagnetic waves of frequency f always comes in separate photons of energy hf; and stable systems of particles, such as atoms, can have only certain energies and no others.

Other quantities in nature are also grainy, or **quantized,** and it has turned out that this graininess is the key to understanding how the properties of matter we are familiar with in everyday life originate in the interactions of elementary particles. In the case of the atom, the quantization of energy follows from the wave nature of moving bodies: In an atom the electron wave functions can occur only in the form of standing waves, much as a violin string can vibrate only at those frequencies that give rise to standing waves.

27.3 Atomic Spectra

Each element has a characteristic line spectrum.

When an electric current passes through a glass tube filled with neon gas, the gas gives off a bright orange light. We have all seen neon signs based on this effect, perhaps without being aware of how closely related the color of the light is to the way in which the electrons in each neon atom are arranged.

Every element has a characteristic spectrum that contains certain wavelengths only. Figure 27.9 shows how a spectrometer spreads out the light from an "excited" gas or vapor into a series of bright lines, each of a single wavelength. These lines make up the **emission spectrum** of the element (Fig. 27.10). Because some of the lines are more intense than the rest, the light usually gives the impression of being a specific color (orange in the case of neon) even though other colors are present as

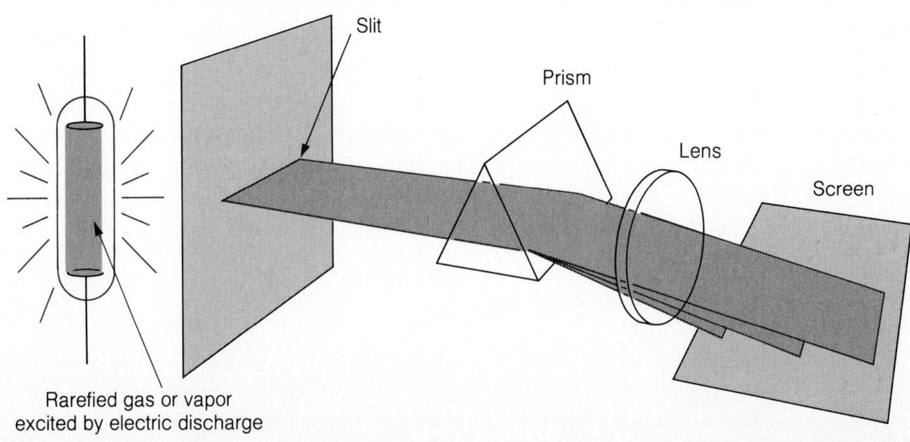

Figure 27.9

The spectrum of a substance may be obtained by "exciting" a sample of its vapor in a tube by means of an electric discharge and then directing the light it gives off through a prism or grating.

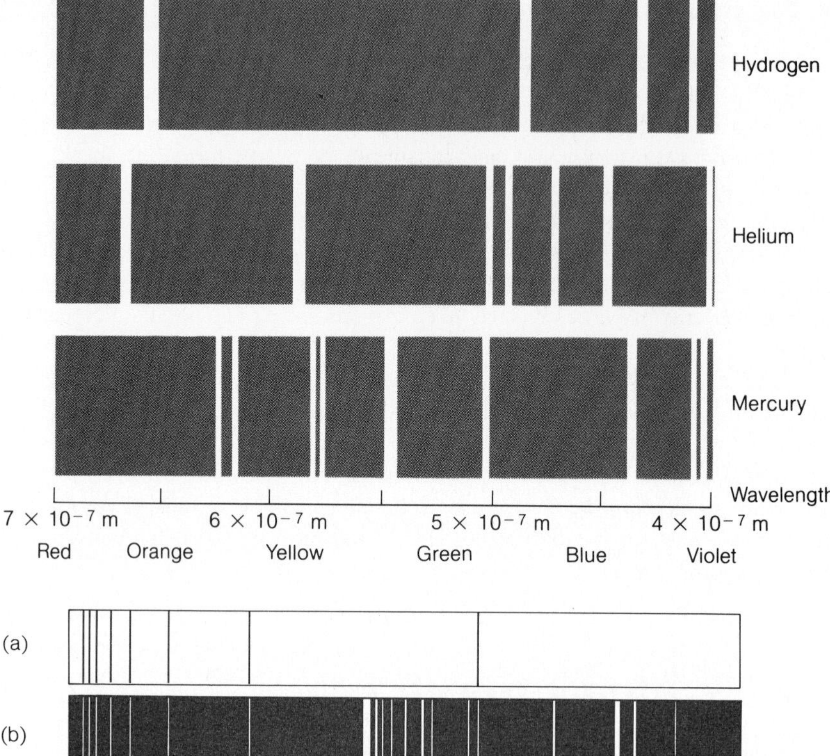

Figure 27.10

The most prominent lines in the emission spectra of hydrogen, helium, and mercury in the visible region.

Figure 27.11

(a) Absorption spectrum of sodium vapor. (b) Emission spectrum of sodium vapor. Each dark line in the absorption spectrum corresponds to a bright line in the emission spectrum.

Streams of protons and electrons from the sun excite atoms in the upper atmosphere to produce the eerie glow of an aurora. The green colors originate in oxygen, the reds in both oxygen and nitrogen. The earth's magnetic field deflects the paths of the solar ions so that most auroras occur in belts around each pole.

well. Spectral lines are found in the infrared and ultraviolet regions as well as in the visible region. An emission spectrum is different from a **continuous spectrum**—that is, the rainbow band produced when light from a hot object passes through a spectrometer—because a continuous spectrum contains all wavelengths and not just a few.

Spectra of another kind, **absorption spectra,** occur when light from a hot source passes through a cool gas before entering the spectrometer. The light source alone would give a continuous spectrum, but the gas absorbs certain wavelengths out of the light that goes through it. Hence the continuous spectrum appears to be crossed by dark lines, each line corresponding to one of the wavelengths absorbed by the gas.

If the bright-line spectrum of an element is compared with its absorption spectrum, the dark lines in the latter have the same wavelengths as a number of the bright lines in the emission spectrum (Fig. 27.11). Thus a cool gas absorbs some of the wavelengths of the light that it emits when excited. The spectrum of sunlight contains dark lines because the luminous part of the sun, which radiates much like a blackbody heated to 5800 K, has around it an envelope of cooler gas. Because every element has a unique spectrum, the spectrometer is a valuable tool in chemistry and astronomy. Helium was discovered in the solar atmosphere through its spectrum before being identified on the earth (*helios* is Greek for ''sun'').

Origin of Spectral Lines

A century ago the wavelengths in the spectrum of an element were found to fall into sets called **spectral series.** The spectral series of hydrogen are shown in Fig. 27.12, and a simple formula turns out to relate the wavelengths in each series.

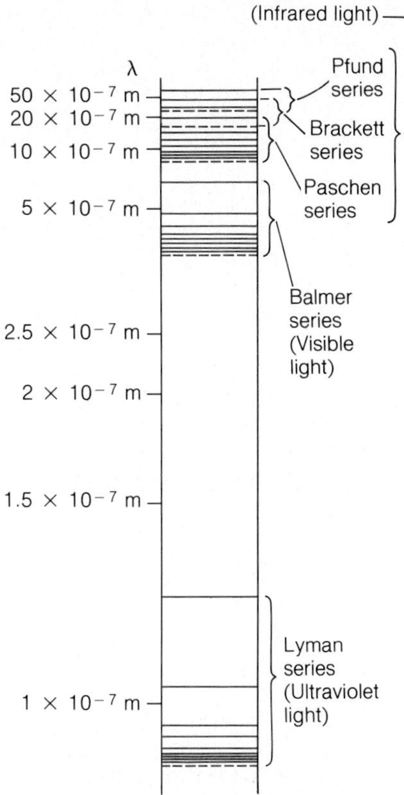

Figure 27.12

The line spectrum of hydrogen with the various series of spectral lines indicated. The wavelength scale is not linear in order to cover the entire spectrum.

The presence of separate energy levels in the hydrogen atom suggests a connection with line spectra. Let us suppose that when an electron in an excited state drops to a lower state, the difference in energy between the states is emitted as a single photon of light. Because electrons cannot exist in an atom except in certain specific energy levels, a jump from one level to the other, with the energy difference being given off all at once in a photon rather than gradually, fits in well with this model.

If the quantum number of the initial (higher energy) state is n_i and the quantum number of the final (lower energy) state is n_f, what we assert is that

$$E_i - E_f = hf \qquad \textit{Origin of spectral lines} \qquad (27.4)$$

Initial energy − final energy = quantum energy

where f is the frequency of the emitted photon and h is Planck's constant. This formula agrees with the experimental data on the spectrum of hydrogen.

Figure 27.13 is an energy-level diagram of the hydrogen atom showing the possible transitions from initial quantum states to final ones. Each transition—or "jump"—involves a characteristic amount of energy and hence a photon of a certain characteristic frequency. The larger the energy difference between initial and final energy levels, as indicated by the lengths of the arrows, the higher the frequency of the emitted photon. The origins of the various series of spectral lines are marked on the diagram.

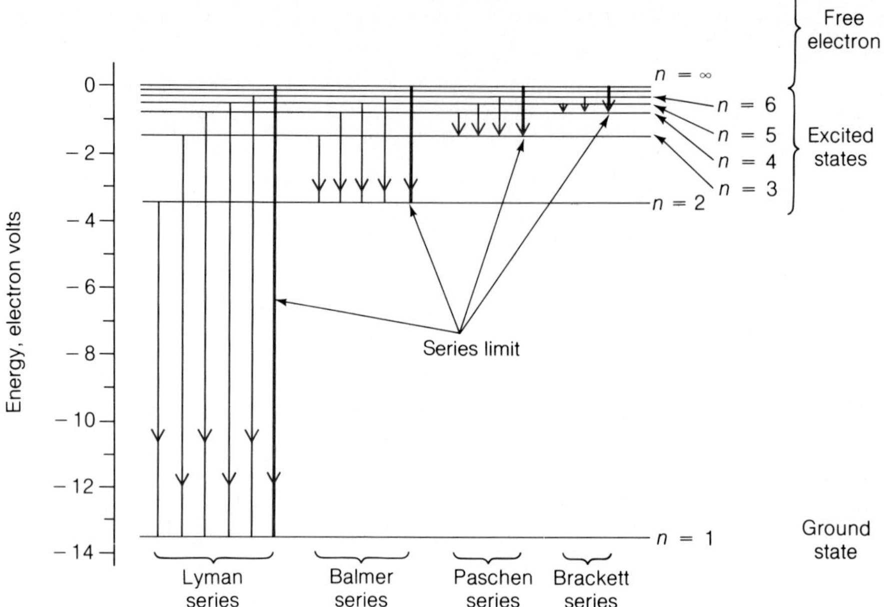

Figure 27.13

Energy-level diagram of the hydrogen atom with some of the transitions that give rise to spectral lines indicated.

EXAMPLE 27.2

Derive a formula for the wavelength of the photon emitted when a hydrogen atom in an excited state of quantum number n_i drops to a state of lower quantum number n_f.

SOLUTION Since $hf = E_i - E_f$ and $E_n = E_1/n^2$,

$$hf = E_1\left(\frac{1}{n_i^2} - \frac{1}{n_f^2}\right)$$

$$f = \frac{E_1}{h}\left(\frac{1}{n_i^2} - \frac{1}{n_f^2}\right) = -\frac{E_1}{h}\left(\frac{1}{n_f^2} - \frac{1}{n_i^2}\right)$$

Because $\lambda = c/f$, $1/\lambda = f/c$ and

$$\frac{1}{\lambda} = -\frac{E_1}{ch}\left(\frac{1}{n_f^2} - \frac{1}{n_i^2}\right) = R\left(\frac{1}{n_f^2} - \frac{1}{n_i^2}\right)$$

where $R = -E_1/ch = 1.097 \times 10^7\ \text{m}^{-1}$. We recall that E_1 is a negative quantity (-13.6 eV, in fact), so R is a positive quantity. The formula for $1/\lambda$ was originally obtained from studies of the spectra themselves, and its derivation was a triumph for Bohr's theory of the hydrogen atom. The various spectral series correspond to different values of n_f, as is clear from Fig. 27.13. The Lyman series is specified by $n_f = 1$, the Balmer series by $n_f = 2$, and so on. ◆

EXAMPLE 27.3

Find the longest wavelength present in the Balmer series of hydrogen.

SOLUTION In the Balmer series the quantum number of the final state is $n_f = 2$. The longest wavelength in this series corresponds to the smallest energy difference between energy levels. Hence the initial state must be $n_i = 3$:

$$\frac{1}{\lambda} = R\left(\frac{1}{n_f^2} - \frac{1}{n_i^2}\right) = R\left(\frac{1}{2^2} - \frac{1}{3^2}\right) = 0.139R$$

$$\lambda = \frac{1}{0.139R} = \frac{1}{0.139(1.097 \times 10^7\ \text{m}^{-1})} = 6.56 \times 10^{-7}\ \text{m}$$

$$= 656\ \text{nm}$$

This wavelength is near the red end of the visible spectrum. ♦

27.4 The Bohr Model

A mixture of classical and quantum concepts.

Let us see how the various results of the Bohr theory of the hydrogen atom can be obtained. We start with the classical picture of the hydrogen atom in which the electron is assumed to circle the proton in a stable orbit. If the orbit radius is r, the centripetal force $F_c = mv^2/r$ on the electron is provided by the electrical force $F_e = ke^2/r^2$ between the electron and the proton, so that

$$\frac{mv^2}{r} = k\frac{e^2}{r^2}$$

The electron speed is therefore related to its orbit radius r by the formula

$$v = e\sqrt{\frac{k}{mr}} \tag{27.5}$$

The total energy E of a hydrogen atom is the sum of the electron's kinetic energy KE $= \frac{1}{2}mv^2$ and the electric potential energy PE $= -ke^2/r$ of the system of electron and proton. The latter formula was discussed in Section 17.2. Hence

$$E = \text{KE} + \text{PE} = \frac{mv^2}{2} - k\frac{e^2}{r}$$

In view of Eq. (27.5),

$$E = k\left(\frac{e^2}{2r} - \frac{e^2}{r}\right) = -\frac{ke^2}{2r} \tag{27.6}$$

The total energy of the hydrogen atom is negative, as it must be for the electron to be bound to the proton. If E were greater than zero, the electron would not follow a closed orbit about the proton.

Experiments show that 13.6 eV of work must be done to break a hydrogen atom apart into a proton and an electron. Thus the total energy of the hydrogen atom in its ground state is $E = -13.6$ eV and is made up of 13.6 eV of KE and -27.2 eV of PE. We can find the orbital radius of the electron from Eq. (27.6) for E. Because 13.6 eV $= 2.18 \times 10^{-18}$ J, we have

$$r = -\frac{k e^2}{2E} = -\frac{(9.0 \times 10^9\ \text{N}\cdot\text{m}^2/\text{C}^2)(1.6 \times 10^{-19}\ \text{C})^2}{2(-2.18 \times 10^{-18}\ \text{J})}$$
$$= 5.3 \times 10^{-11}\ \text{m}$$

The electron's speed is, from Eq. (27.5),

$$v = e\sqrt{\frac{k}{mr}} = 1.6 \times 10^{-19}\ \text{C}\sqrt{\frac{9.0 \times 10^9\ \text{N}\cdot\text{m}^2/\text{C}^2}{(9.1 \times 10^{-31}\ \text{kg})(5.3 \times 10^{-11}\ \text{m})}}$$
$$= 2.2 \times 10^6\ \text{m/s}$$

This speed is well below that of light ($c = 3 \times 10^8$ m/s), and so our nonrelativistic calculation is justified.

To derive a formula for the radii r_n of the possible electron orbits in the hydrogen atom, we start from the quantum condition of Eq. (27.1),

$$n\lambda = 2\pi r_n \qquad n = 1, 2, 3, \ldots$$

BIOGRAPHICAL NOTE

Niels Bohr

Niels Bohr (1885–1962) was born and spent most of his life in Copenhagen, Denmark. His family was a distinguished one: Bohr's father was professor of physiology at the University of Copenhagen, his younger brother Harald became a noted mathematician, and his son Aage, like Bohr himself, would win a Nobel Prize in physics. In 1911 Bohr received his doctorate in physics and went to England to broaden his scientific horizons. At Rutherford's laboratory in Manchester, Bohr was introduced to the just-discovered nuclear model of the atom, which was in conflict with existing principles of mechanics and electromagnetism. Bohr felt that the quantum theory of light must somehow be the key to understanding atomic structure.

Back in Copenhagen in 1913, a friend suggested to Bohr that Balmer's formula for some of the spectral lines of hydrogen might be relevant to his quest. "As soon as I saw Balmer's formula the whole thing was immediately clear to me," Bohr said later. To construct his theory, Bohr began with two revolutionary ideas. The first was that an atomic electron can circle its nucleus only in certain orbits, and the other was that an atom emits or absorbs a photon of light when an electron jumps from one permitted orbit to another.

What is the condition for a permitted orbit? To find out, Bohr used as a guide what became known as the correspondence principle: When quantum numbers are very large, quantum effects should not be conspicuous, and the quantum theory must then give the same results as would classical physics. Applying this principle showed that the electron in a permitted orbit must have an angular momentum that is a multiple of $h/2\pi$. A decade later Louis de Broglie would explain this quantization of angular momentum in terms of the wave nature of a moving electron.

Bohr's daring mixture of classical and quantum concepts accounted for all the spectral series of hydrogen, not just the Balmer series, but the publication of the theory aroused great controversy. Some well-known physicists, among them Albert Einstein, were enthusiastic supporters of Bohr. Others felt differently: Otto Stern and Max von Laue were so upset that they said they would quit physics if Bohr were right. (They later changed their minds.) Bohr and others tried to extend his model to many-electron atoms with occasional success—for instance, the correct prediction of the properties of the then-unknown element hafnium—but real progress had to wait for Wolfgang Pauli's exclusion principle of 1925.

In 1916 Bohr returned to Rutherford's laboratory, where he stayed until 1919. Then an Institute of Theoretical Physics was created for him in Copenhagen, and he directed it until his death. The institute was a magnet for quantum theoreticians from all over the world, who were stimulated by the exchange of ideas at regular meetings there. Bohr's last important work came in 1939, when he used an analogy between a large nucleus and a liquid drop to explain why nuclear fission, which had just been discovered, occurs in certain nuclei but not in others. During World War II Bohr worked on the development of the atomic bomb at Los Alamos, New Mexico. After the war, Bohr returned to Copenhagen, where he died in 1962.

The formula for the de Broglie wavelength of the electron is $\lambda = h/mv$. Using Eq. (27.5) for v with r_n in place of r gives

$$\lambda = \frac{h}{me}\sqrt{\frac{mr_n}{k}} = \frac{h}{e}\sqrt{\frac{r_n}{mk}}$$

and so the quantum condition becomes

$$n\lambda = \frac{nh}{e}\sqrt{\frac{r_n}{mk}} = 2\pi r_n$$

Finally we square both sides and solve for r_n to obtain

$$\frac{n^2h^2}{e^2}\frac{r_n}{mk} = 4\pi^2 r_n^{\,2}$$

$$r_n = \frac{h^2n^2}{4\pi^2kme^2} \qquad n = 1, 2, 3, \ldots \tag{27.7}$$

We can write this as

$$r_n = r_1 n^2 \qquad n = 1, 2, 3, \ldots \tag{27.8}$$

where r_1 is the radius of the first (innermost) orbit. This result was given as Eq. (27.2).

The total energy E_n of a hydrogen atom whose electron is in the nth orbit depends only on the radius r_n of that orbit because

$$E = -\frac{k}{2}\frac{e^2}{r}$$

for all orbits. Hence the possible energies of a hydrogen atom are limited to

$$E_n = -\frac{ke^2}{2h^2n^2/4\pi^2kme^2} = -\frac{2\pi^2k^2me^4}{h^2n^2} \qquad n = 1, 2, 3, \ldots \tag{27.9}$$

We can write this as

$$E_n = \frac{E_1}{n^2} \qquad n = 1, 2, 3, \ldots \tag{27.10}$$

where $E_1 = -13.6$ eV is the energy of the innermost orbit. This result was given as Eq (27.3).

It is worth noting that the formula for E_n contains only the directly measurable quantities k, m, e, and h, all of which are constants of nature. The formula is in complete agreement with the experimentally determined values of E_n.

27.5 Quantum Theory of the Atom

Four quantum numbers for each electron.

The Bohr theory agrees nicely with experiment, but it has some serious limitations. For example, it cannot be applied successfully to the atoms of elements other than hydrogen, that is, to atoms with more than one electron. These limitations are absent from the more general **quantum theory of the atom** that was developed in the 1920s. Not only has this theory provided a framework for understanding the structures of all atoms, however complex, but it has also furnished key insights that explain how and why atoms join together to form molecules, liquids, and solids.

In the Bohr model an atomic electron is supposed to move in a circular orbit, and all that changes as it revolves around its nucleus is its position on this circle. A single quantum number is all that is needed to describe the physical state of such an electron.

In the quantum theory of the atom, an electron is not limited to a fixed orbit but moves about in three dimensions. We can think of the electron as circulating in a probability cloud specified by a wave function ψ. Where the cloud is most dense (ψ^2 has the highest value), the electron is most likely to be found. Where the cloud is least dense (ψ^2 has the lowest value), the electron is least likely to be found. Figure 27.14 shows a cross section of the probability cloud for the ground state of the hydrogen atom, which is the state in which the electron has the lowest energy and is the one usually occupied.

Three quantum numbers, not the single one of the Bohr theory, turn out to be needed to specify the size and shape of the probability cloud of an atomic electron. One of them, the **principal quantum number** n, can have the values 1, 2, 3, . . .

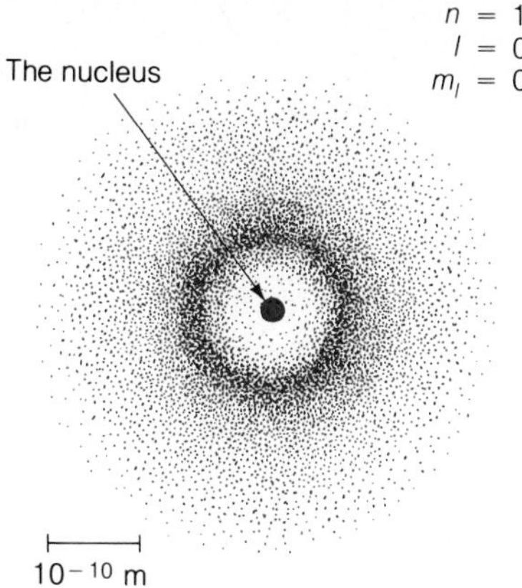

Figure 27.14

Cross section of the probability cloud for the ground state of the hydrogen atom.

This quantum number is the chief factor that governs the electron's energy and its most probable distance from the nucleus. The larger n is, the greater the energy (that is, the less negative E_n is) and the farther the electron tends to be from the nucleus.

The other two quantum numbers, l and m_l, together govern the electron's angular momentum **L** and the form of its probability cloud. According to quantum theory, angular momentum as well as energy is quantized in an atom. The possible values of the magnitude L of the angular momentum are determined by l, the **orbital quantum number.** An electron whose principal quantum number is n can have an orbital quantum number of 0 or any whole number up to $n - 1$. For instance, if $n = 3$, the values l can have are 0, 1, or 2.

Angular momentum is a vector quantity, and to describe it completely requires that its direction as well as its magnitude L be specified. This is the role of the **magnetic quantum number** m_l. An electron whose orbital quantum number is l can have a magnetic quantum number that is 0 or any whole number between $-l$ and $+l$. Thus if $l = 2$, the values m_l can have are -2, -1, 0, $+1$, and $+2$.

An electron with the angular momentum **L** due to its motion around a nucleus is equivalent to an electric current loop and so can interact with an external magnetic field **B.** The magnetic quantum number m_l determines the direction of **L** relative to **B** and so controls the potential energy of the atom in the magnetic field. As a result, in a magnetic field the energy levels of an atom are split into sublevels, each of different m_l. The spectral lines of that element when it is in a magnetic field are similarly split into components whose spacing varies with B (Fig. 27.15). This phenomenon is called the **Zeeman effect.** It is especially valuable in astronomy, where it was responsible for, among other findings, the discoveries of magnetic fields in the sun and other stars and of the magnetic nature of sunspots.

Figure 27.15

An example of the Zeeman effect. The stronger the magnetic field, the greater the separation of the components into which each spectral line is split.

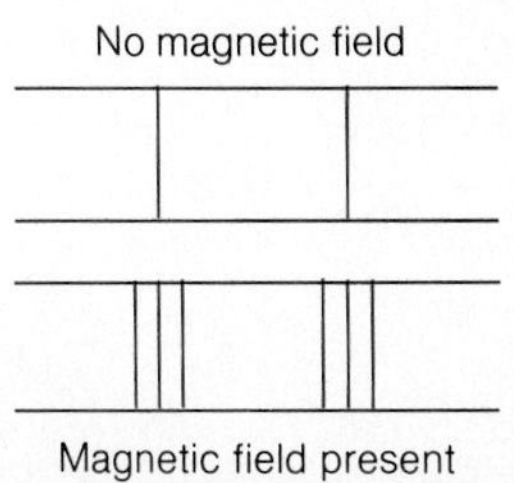

Still another quantum number is needed to describe completely an atomic electron, the **spin magnetic quantum number** m_s. An electron behaves as though it were a spinning charged sphere, which means it is, in effect, a tiny bar magnet. The angular momentum of the spin is the same for all electrons; what m_s determines is the direction of the spin. An electron with $m_s = +\frac{1}{2}$ ("spin up") aligns itself with an external magnetic field **B,** and an electron with $m_s = -\frac{1}{2}$ ("spin down") aligns itself opposite to **B** (Fig. 27.16).

Electron spin is supported both by experiment and by theory. The British physicist Paul A. M. Dirac was able to show, on the basis of a relativistic version of quantum mechanics, that the electron *must* have spin. On the other hand, the notion

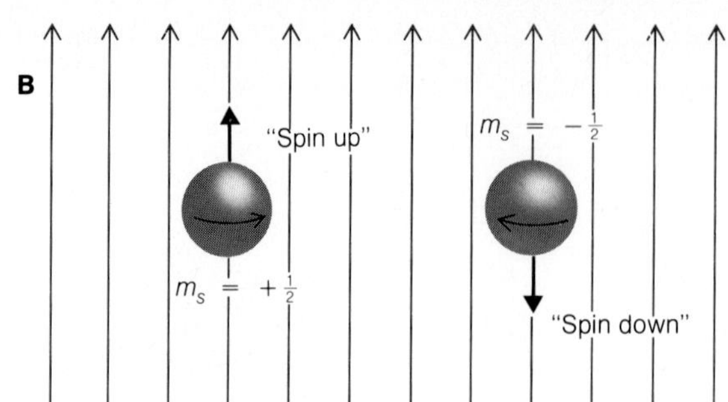

Figure 27.16

The spin magnetic quantum number m_s of an atomic electron has two possible values, $+\frac{1}{2}$ and $-\frac{1}{2}$. Electrons behave in certain respects like spinning charged spheres, but experiments suggest they are more likely to be point particles with no extension in space.

of an electron as a spinning charged sphere cannot be literally correct. For one thing, observations of the scattering of electrons by other electrons at high energies indicate that the electron must be less than 10^{-18} m in diameter. With such a small diameter, the electron's equator would have to move faster than the speed of light to give it the angular momentum associated with spin. But even if we cannot picture an electron, we can still say for sure that it has a certain rest mass, charge, angular momentum, and magnetic behavior.

27.6 Exclusion Principle

A different set of quantum numbers for each electron in an atom.

The four quantum numbers needed to describe each state an atomic electron can have are listed in Table 27.1. There are $2n^2$ possible quantum states for each value of the principal quantum number n. For example, when $n = 3$, there are 18 states, as shown in Table 27.2.

In a hydrogen atom the electron is normally in its ground state of $n = 1$. What about more complex atoms? Are all 92 electrons of a uranium atom in the same $n = 1$ state, jammed into a single probability cloud? Many lines of evidence make this idea unlikely.

Table 27.1 Quantum numbers of an atomic electron

Name	Symbol	Possible values	Quantity determined
Principal	n	$1, 2, 3, \ldots$	Electron energy
Orbital	l	$0, 1, 2, \ldots, n-1$	Magnitude of angular momentum
Magnetic	m_l	$-l, \ldots, 0, \ldots, +l$	Direction of angular momentum
Spin magnetic	m_s	$-\frac{1}{2}, +\frac{1}{2}$	Direction of electron spin

Table 27.2 Quantum states of an electron of principal quantum number $n = 3$

	$m_l = 0$	$m_l = -1$	$m_l = +1$	$m_l = -2$	$m_l = +2$	
$l = 0$:	⇃↾					$\uparrow m_s = +\frac{1}{2}$
$l = 1$:	⇃↾	⇃↾	⇃↾			$\downarrow m_s = -\frac{1}{2}$
$l = 2$:	⇃↾	⇃↾	⇃↾	⇃↾	⇃↾	

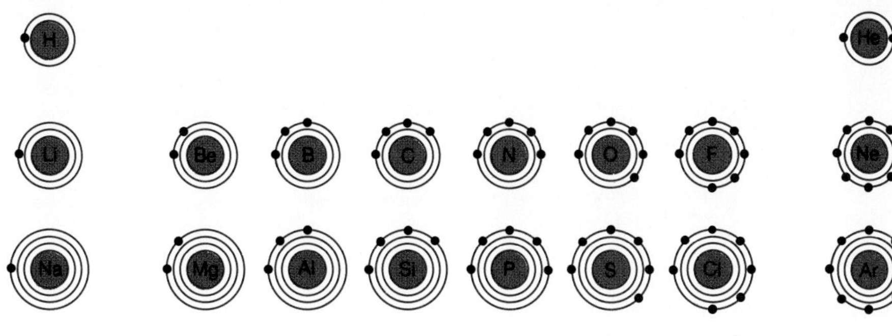

Figure 27.17

Schematic representation of the electron structures of the 18 lightest elements. The electrons in the filled inner shells of the second and third rows of atoms are not shown. Metals are at the left, nonmetals at the right; hydrogen is in a category by itself.

An example of such evidence is the great difference in chemical behavior shown by certain elements whose atomic structures differ by just one electron. Thus the elements that have the atomic numbers 9, 10, and 11 are, respectively, the corrosive gas fluorine, the inert gas neon, and the metal sodium. Because the electron structure of an atom controls how it interacts with other atoms, it makes no sense that the chemical properties of the elements should change so sharply with a small change in atomic number if all the electrons in an atom were in the same probability cloud.

In 1925 Wolfgang Pauli solved the problem of the electron arrangement in an atom with more than one electron. His **exclusion principle** states:

No two electrons in an atom can exist in the same quantum state.

That is, each electron in a complex atom must have a different set of the quantum numbers n, l, m_l, and m_s. The exclusion principle can be generalized to refer to the electrons in *any* small region of space, regardless of whether or not they constitute an atom, as we will find in Chapter 28.

The electrons in an atom that share the same quantum number n are said to occupy the same **shell.** These electrons average about the same distance from the nucleus and usually have comparable, though not identical, energies. Since the number of electrons in a shell of given n is limited to $2n^2$, there can be at most 2 electrons in the $n = 1$ shell, 8 in the $n = 2$ shell, 18 in the $n = 3$ shell, and so on. To minimize the atom's energy, the electrons in an atom in its ground state occupy the innermost shells to the maximum extent possible. For example, the chlorine atom, which contains a total of 17 electrons, has 2 electrons in its $n = 1$ shell, 8 in its $n = 2$ shell, and 7 in its $n = 3$ shell. Figure 27.17 is a highly schematic representation of the electron structures of the 18 lightest elements.

27.7 Atomic Excitation

The origins of emission and absorption spectra.

There are two principal ways in which an atom may be excited to an energy level above that of its ground state and thereby become able to radiate. One of these ways is by a collision with another atom during which part of their kinetic energy is

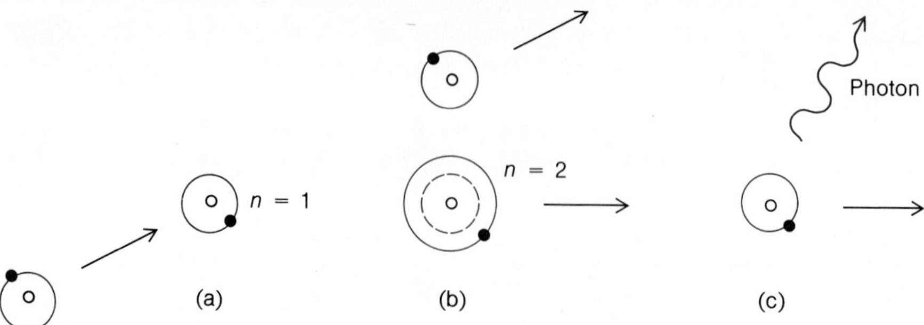

Figure 27.18

In (a) both atoms are in their ground states. During the collision some kinetic energy is transformed into excitation energy, and in (b), the target atom is in an excited state. In (c) the target atom has returned to its ground state by emitting a photon.

transformed into electron energy within either or both of the participating atoms. An atom excited in this way will then lose its excitation energy by emitting one or more photons in the course of returning to its ground state (Fig. 27.18). In an electric discharge in a gas, an electric field accelerates electrons and ions until their kinetic energies are great enough to excite atoms with which they happen to collide. A neon sign is a familiar example of this effect.

Another excitation mechanism is the absorption by an atom of a photon of light whose energy is just the right amount to raise it to a higher energy level. A photon of wavelength 121.7 nm is emitted when a hydrogen atom in the $n = 2$ state drops to the $n = 1$ state. The absorption of a photon of wavelength 121.7 nm by a hydrogen atom initially in the $n = 1$ state will therefore bring it up the $n = 2$ state. This process explains the origin of absorption spectra (Fig. 27.19).

When white light (which contains all wavelengths) is passed through hydrogen gas, photons of those wavelengths that correspond to transitions between hydrogen energy levels are absorbed. The resulting excited hydrogen atoms reradiate their excitation energy almost at once, but these photons come off in random directions, not all in the same direction as in the original beam of white light (Fig. 27.20). The dark lines in an absorption spectrum are therefore never totally dark, but only appear so by contrast with the bright background of transmitted light. We would expect the lines in the absorption spectrum of a substance to be the same as lines in its emission spectrum, which is what is found. As Fig. 27.11 shows, only a few of the emission lines appear in absorption. The reason is that reradiation is so rapid that essentially

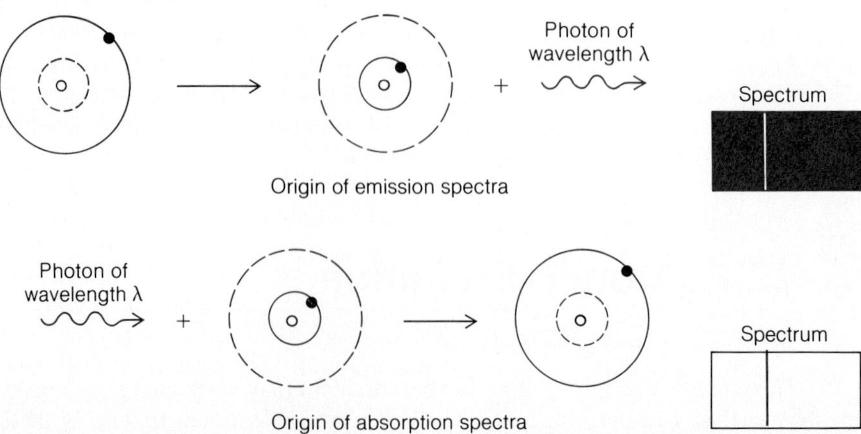

Figure 27.19

The origins of emission and absorption spectra.

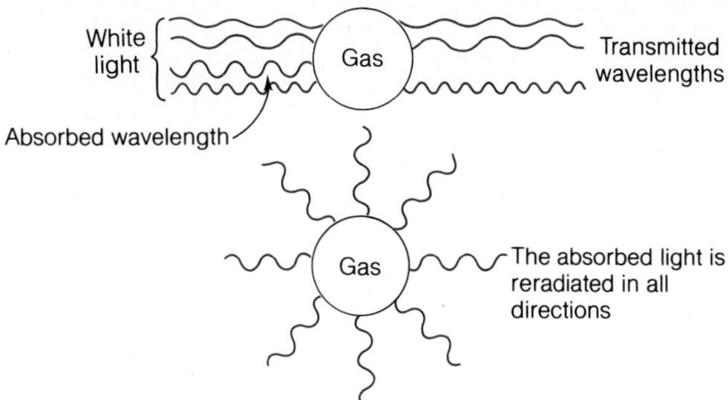

Figure 27.20

The dark lines in an absorption spectrum are never totally dark.

all the atoms in an absorbing gas or vapor are initially in their lowest energy states and therefore take up and reemit only photons that represent transitions to and from that state.

27.8 The Laser

How to produce light waves all in step.

The **laser** is a device that produces a light beam with some remarkable properties:

1. The beam is extremely intense, more intense by far than the light from any other source. To achieve an energy density equal to that in some laser beams, a hot object would have to be at a temperature of 10^{30} K.
2. There is very little divergence, so that a laser beam from the earth that was reflected by a mirror left on the moon during the *Apollo 11* expedition remained narrow enough to be detected upon its return to the earth—a total journey of more than 750,000 km.
3. The light is essentially monochromatic.
4. The light is coherent, with the waves all exactly in phase with one another (Fig. 27.21). It is possible to obtain interference patterns not only by placing two slits in a laser beam but also by using beams from two separate lasers.

The term *laser* stands for *l*ight *a*mplification by *s*timulated *e*mission of *r*adiation. Let us look into how a laser works.

Left to itself, an atom always stays in its ground state, the quantum state in which it has the least energy. When an atom is raised to an excited state, it usually falls to its ground state almost at once, sometimes first dropping to an intermediate excited state. A photon is normally emitted in each transition. The lifetime of most excited states is only about 10^{-8} s, but certain states are **metastable** (temporarily stable). An atom may remain in a metastable state for 10^{-3} s or more before radiating (Fig. 27.22), provided it does not undergo a collision in the meantime in which its excitation energy would be lost. The operation of lasers depends on the existence of metastable states in atoms.

Figure 27.21

A laser produces a beam of light whose waves all have the same frequency and are in phase with one another. Such a beam is said to be monochromatic and coherent. The beam is also narrow and spreads out very little even over a long distance.

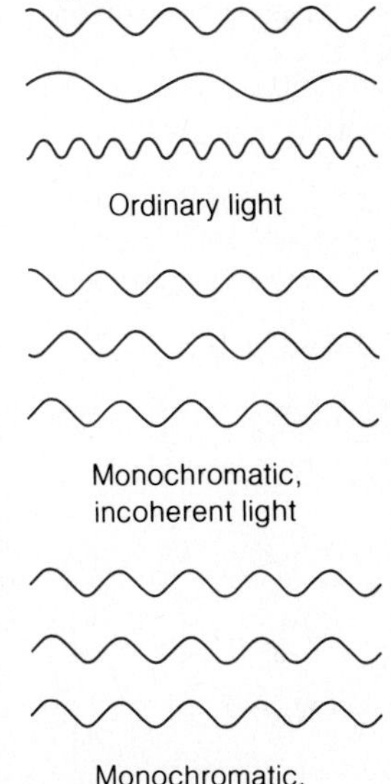

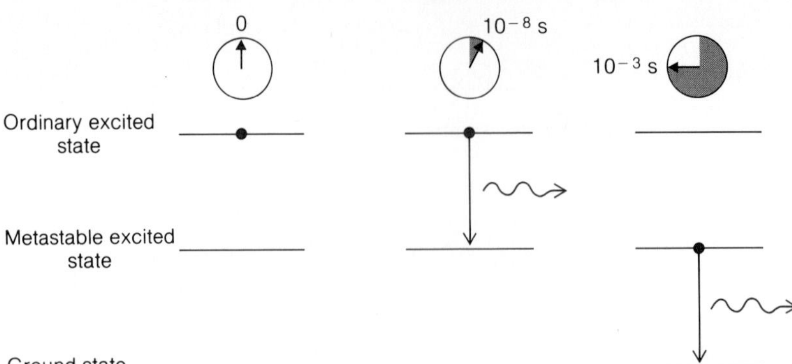

Figure 27.22

An atom can exist in a metastable energy level for a longer time before radiating than it can in an ordinary energy level.

Three kinds of transition involving electromagnetic radiation are possible between two energy levels in an atom whose energy difference is $E_2 - E_1 = hf$ (Fig. 27.23):

1. **Induced absorption** occurs when the atom is in the lower level and is raised to the upper level by absorbing a photon of energy hf.
2. **Spontaneous emission** occurs when the atom is in the upper level and falls by itself to the lower level by emitting a photon of energy hf.
3. **Induced emission** occurs when the atom is in the upper level and radiation of frequency f causes a transition from the upper level to the lower level. In induced emission the emitted light waves are exactly in phase with the incoming ones, so that the result is an enhanced beam of coherent light.

Now suppose we have a group of atoms of some kind that have metastable states of excitation energy hf. If we can raise most of the atoms to the metastable level and then shine light of frequency f on the group, more induced emission than

A laser produces an intense beam of monochromatic, coherent light from the cooperative radiation of excited atoms or molecules. Shown is part of a laser system that can deliver thousands of joules of energy to a target the size of a grain of sand in a pulse lasting less than 10^{-9} s for experiments in controlled fusion energy.

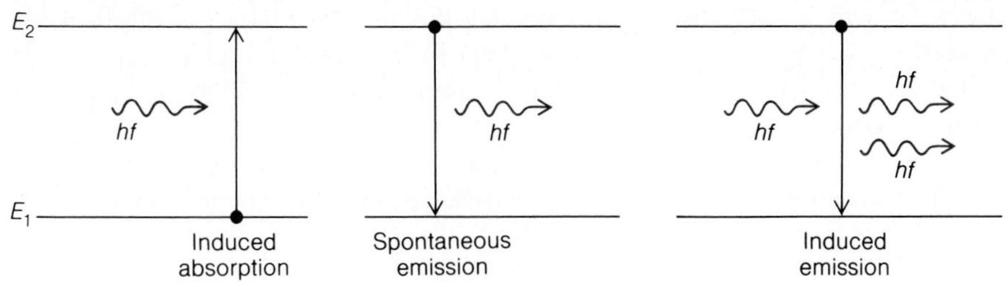

Figure 27.23

The three types of transition between energy levels in an atom.

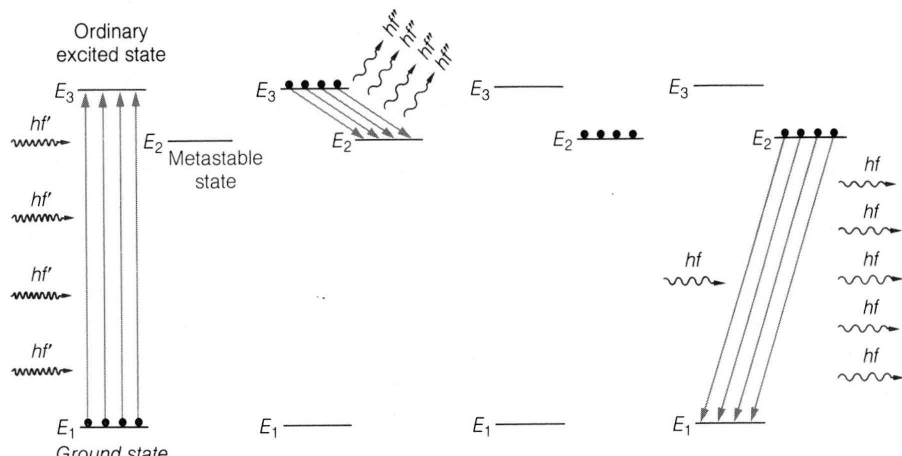

Figure 27.24

How a laser operates.

absorption will occur and the original light will be amplified (Fig. 27.24). This is the concept that underlies the operation of the laser. The term **population inversion** is given to an assembly of atoms most of which are in excited states, because normally the ground states are the more highly populated.

Population inversions can be produced in a number of ways. One of them, called optical pumping, uses an external light source some of whose photons have the right frequency to raise the atoms to excited states that decay into the desired metastable ones. This is the method used in the ruby laser, in which the chromium ions Cr^{3+} in a ruby crystal are excited by light from a xenon-filled flash lamp (Fig. 27.25). (A ruby is a crystal of Al_2O_3 a small number of whose Al^{3+} ions are replaced by Cr^{3+} ions, which are responsible for the reddish color. Such ions are Cr atoms that have lost three electrons each.)

The Cr^{3+} ions are raised in the pumping process to a level E_3 from which they decay to the metastable level E_2 by losing energy to other atoms in the crystal. This metastable level has a lifetime of about 0.003 s. Because there are comparatively

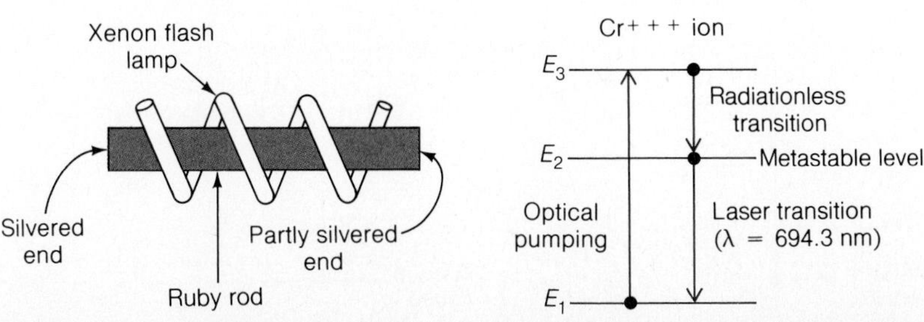

Figure 27.25

The ruby laser produces pulses of light.

few Cr^{3+} ions in the ruby rod and a xenon flash lamp emits a great deal of light, most of the Cr^{3+} ions can be pumped to the E_2 level in this way. A few Cr^{3+} ions in the E_2 level then spontaneously fall to the E_1 level, and some of the resulting photons bounce back and forth between the reflecting ends of the ruby rod. The presence of light of the right frequency now stimulates the other Cr^{3+} ions in the E_2 level to radiate, and the result is an avalanche that produces a large pulse of red light that emerges from the partly silvered end of the rod. The length of the rod is made to be exactly a whole number of half-wavelengths long, so the radiation trapped in it forms an optical standing wave. Since the induced emissions are stimulated by the standing wave, their waves are all in step with it.

The helium-neon gas laser achieves a population inversion in a different way. A mixture of helium and neon is placed in a glass tube that has parallel mirrors at both ends. An electric discharge is then produced in the gas by electrodes connected to a source of high-frequency alternating current. Collisions with electrons from the discharge then excite He and Ne atoms to metastable states respectively 20.61 and 20.66 eV above their ground states (Fig. 27.26). Some of the excited He atoms give their energy to ground-state Ne atoms in collisions, with the 0.05 eV of additional energy coming from the kinetic energy of the atoms. The purpose of the He atoms is thus to help produce a population inversion in the Ne atoms.

An excited Ne atom emits a photon of wavelength $\lambda = 632.8$ nm in the transition to another excited state, which is the laser transition. Then another photon is spontaneously emitted, which produces only incoherent radiation because it is not induced. The rest of the excitation energy is lost in collisions with the tube walls. Because the electron impacts that produce the population inversion occur all the time, an He-Ne laser operates continuously, unlike a ruby laser.

Many other types of laser have been constructed, including solid-state and chemical lasers. Solid-state lasers are small and have limited power outputs, but they are ideal for use in fiber-optic transmission lines in which signals are carried in the form of modulated light beams instead of as modulated electric currents. Such transmission lines are widely used between telephone exchanges. In a chemical laser, molecules in metastable excited states are produced by chemical reactions, and the overall efficiency of the system is thereby improved.

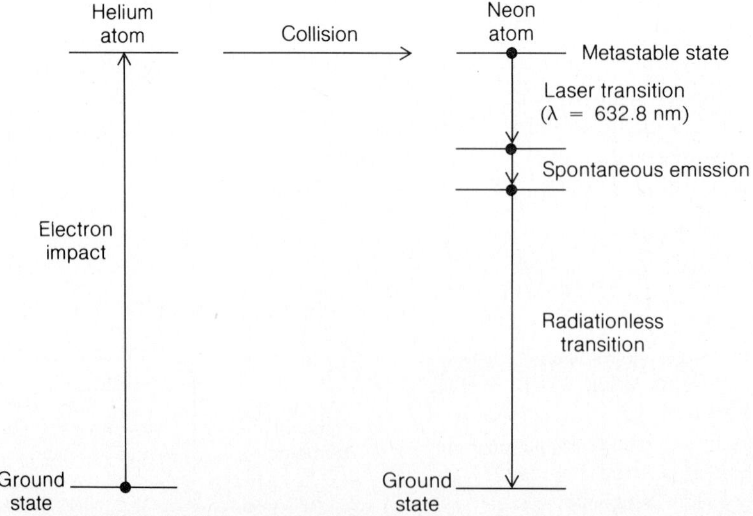

Figure 27.26

Energy levels involved in the helium-neon laser, which operates continuously.

Important Terms

An **emission spectrum** consists of the various wavelengths of light emitted by an excited substance. An **absorption spectrum** consists of the various wavelengths of light absorbed by a substance when white light is passed through it.

According to the **Bohr theory of the atom,** an electron can circle an atomic nucleus indefinitely without radiating energy if the circumference of its orbit is a whole number of electron wavelengths. The number of wavelengths that fit into a particular permitted orbit is called the **quantum number** of that orbit. The electron energies corresponding to the various quantum numbers constitute the **energy levels** of the atom, of which the lowest is the **ground state** and the rest are **excited states.**

When an electron in an excited state drops to a lower state, the difference in energy between the states is emitted as a single photon of light; when a photon of the same wavelength is absorbed, the electron goes from the lower to the higher state. These two processes account for the properties of **emission** and **absorption spectra,** respectively.

According to the **quantum theory of the atom,** each electron in an atom is described by its **principal quantum number** n, which governs its energy, its **orbital quantum number** l, which governs the magnitude of its angular momentum, and its **magnetic quantum number** m_l, which governs the orientation of its angular momentum. To each set of quantum numbers there is a corresponding **probability cloud** that governs the likelihood the electron thus described will be found in any particular location in an atom.

Every electron has a certain intrinsic amount of angular momentum called its **spin.** The **spin magnetic quantum number** m_s of an atomic electron has two possible values, $+\frac{1}{2}$ and $-\frac{1}{2}$. Owing to its spin, every electron acts like a tiny bar magnet.

According to the Pauli **exclusion principle,** no two electrons in an atom can exist in the same quantum state.

The electrons in an atom that have the same total quantum number n are said to occupy the same **shell.**

A **laser** is a device for producing a narrow, monochromatic, coherent beam of light. The term stands for *l*ight *a*mplification by *s*timulated *e*mission of *r*adiation. Laser operation depends on the existence of **metastable states,** which are excited atomic states that can persist for unusually long periods of time.

Important Formulas

Condition for orbit stability: $n\lambda = 2\pi r_n$

Energy levels of hydrogen atom:

$$E_n = \frac{E_1}{n^2} \qquad n = 1, 2, 3, \ldots$$

Origin of spectral lines: $E_i - E_f = hf$

MULTIPLE CHOICE

1. Classical physics is unable to account for
- **a.** the presence of electrons in atoms
- **b.** why atomic electrons are in motion
- **c.** why atoms can radiate electromagnetic waves
- **d.** why atoms can be stable

2. In the Bohr model of the hydrogen atom, the electron revolves around the nucleus in order to
- **a.** emit spectral lines
- **b.** produce X rays
- **c.** form energy levels that depend on its speed
- **d.** keep from falling into the nucleus

3. In the Bohr model, an electron can revolve around the nucleus of a hydrogen atom indefinitely if its orbit is
- **a.** a perfect circle
- **b.** sufficiently far from the nucleus to avoid capture
- **c.** less than one de Broglie wavelength in circumference
- **d.** exactly one de Broglie wavelength in circumference

4. The electron of a ground-state hydrogen atom
- **a.** has left the atom
- **b.** is at rest
- **c.** is in its orbit of lowest energy
- **d.** is in its orbit of highest energy

5. The energy difference between adjacent energy levels
- **a.** is the same for all quantum numbers
- **b.** is smaller for small quantum numbers
- **c.** is larger for small quantum numbers
- **d.** shows no regularity

6. An atom emits a photon when one of its electrons
- **a.** collides with another of its electrons
- **b.** is removed from the atom
- **c.** shifts to a quantum state of lower energy
- **d.** shifts to a quantum state of higher energy

7. The bright-line spectrum produced by the excited atoms

of an element contains wavelengths that
  a. are the same for all elements
  b. are characteristic of the particular element
  c. are evenly distributed throughout the entire visible spectrum
  d. are different from the wavelengths in its dark-line spectrum

8. A neon sign does not produce
  a. a line spectrum
  b. an emission spectrum
  c. an absorption spectrum
  d. photons

9. In an emission spectral series,
  a. the lines are evenly spaced
  b. the excited electrons all started from the same energy level
  c. the excited electrons all return to the same energy level
  d. the excited electrons leave the atoms

10. The sun's spectrum consists of a light background crossed by dark lines. This suggests that the sun
  a. is a hot object surrounded by a hot atmosphere
  b. is a hot object surrounded by a cool atmosphere
  c. is a cool object surrounded by a hot atmosphere
  d. is a cool object surrounded by a cool atmosphere

11. The quantum number n of the lowest energy state of a hydrogen atom
  a. is 0
  b. is 1
  c. depends on the orbit size
  d. depends on the electron speed

12. Most excited states of an atom have lifetimes of about
  a. 10^{-8} s b. 10^{-3} s
  c. 1 s d. 10 s

13. Which of the following transitions in a hydrogen atom emits the photon of highest frequency?
  a. $n = 1$ to $n = 2$ b. $n = 2$ to $n = 1$
  c. $n = 2$ to $n = 6$ d. $n = 6$ to $n = 2$

14. Which of the following transitions in a hydrogen atom absorbs the photon of highest frequency?
  a. $n = 1$ to $n = 2$ b. $n = 2$ to $n = 1$
  c. $n = 2$ to $n = 6$ d. $n = 6$ to $n = 2$

15. Which of the following transitions in a hydrogen atom emits the photon of lowest frequency?
  a. $n = 1$ to $n = 2$ b. $n = 2$ to $n = 1$
  c. $n = 2$ to $n = 6$ d. $n = 6$ to $n = 2$

16. The energy of a quantum state of an atom depends chiefly on its
  a. principal quantum number n
  b. orbital quantum number l
  c. magnetic quantum number m_l
  d. spin magnetic quantum number m_s

17. The angular momentum of an atomic electron is
  a. not quantized
  b. quantized in magnitude only
  c. quantized in direction only
  d. quantized in both magnitude and direction

18. A large value of the probability density ψ^2 of an atomic electron at a certain place and time signifies that the electron
  a. is likely to be found there
  b. is certain to be found there
  c. has a great deal of energy there
  d. has a great deal of charge there

19. The number of quantum numbers needed to determine the size and shape of the probability cloud of an atomic electron is
  a. 1 b. 2
  c. 3 d. 4

20. Electrons
  a. have no magnetic properties
  b. have the same magnetic behavior as particles of iron
  c. behave like tiny bar magnets of different strengths
  d. behave like tiny bar magnets of the same strength

21. According to the exclusion principle, no two electrons in an atom can have the same
  a. spin b. speed
  c. orbit d. set of quantum numbers

22. Each probability cloud in an atom can be occupied by
  a. one electron
  b. two electrons with spins in the same direction
  c. two electrons with spins in opposite directions
  d. any number of electrons

23. An atom can be raised to an excited state by which one or more of the following processes?
  a. A collision with another atom or other particle
  b. The absorption of a photon
  c. The spontaneous emission of a photon
  d. The induced emission of a photon

24. The operation of the laser is based on
  a. the uncertainty principle
  b. the interference of de Broglie waves
  c. induced emission of radiation
  d. spontaneous emission of radiation

25. Which one or more of the following properties are characteristic of the light waves from a laser?
  a. The waves all have the same frequency.
  b. The waves are all in step with one another.
  c. The waves form a narrow beam.
  d. The waves have higher photon energies than light waves of the same frequency from an ordinary source.

QUESTIONS

1. How are the Bohr and Rutherford models of the hydrogen atom related? In the Bohr model, why is the electron pictured as revolving around the nucleus?

2. In the Bohr model the electron is in constant motion. How can such an electron have a negative amount of energy?

3. What kind of spectrum is observed in (a) light from the hot filament of a light bulb; (b) light from a sodium-vapor highway lamp; (c) light from an electric light bulb surrounded by cool sodium vapor?

4. When radiation with a continuous spectrum is passed through a volume of hydrogen gas whose atoms are all in the ground state, which spectral series will be present in the resulting absorption spectrum?

5. Explain why the spectrum of hydrogen has many lines, although a hydrogen atom contains only one electron.

6. Would you expect the fact that the atoms of an excited gas are in rapid random motion to have any effect on the sharpness of the spectral lines they produce?

7. What quantity is governed by each of the quantum numbers n, l, m_l, and m_s?

8. Which quantum number is not involved in describing the probability cloud of an atomic electron?

9. Under what circumstances do electrons exhibit spin? Do electrons in inner shells have more spin angular momentum than those in outer shells?

10. What significance does electron spin have in atomic structure?

11. Under what circumstances can two electrons share the same probability cloud in an atom?

12. According to the Bohr model, the radius of an atomic electron's orbit is proportional to n^2. Is there any aspect of the probability-cloud model that might correspond to this relationship?

13. Why is the length of the optical cavity of a laser so important?

EXERCISES

27.1 The Hydrogen Atom

27.2 Energy Levels

1. How much energy is needed to ionize a hydrogen atom when it is in the $n = 4$ state?

2. Find the average kinetic energy per molecule in a gas at room temperature (20°C) and show that this is much less than the energy required to raise a hydrogen atom from its ground state ($n = 1$) to its first excited state ($n = 2$).

3. To what temperature must a hydrogen gas be heated if the average molecular kinetic energy is to equal the binding energy of the hydrogen atom?

4. Find the quantum number of the Bohr orbit in a hydrogen atom that is 1 mm in radius. What is the energy (in electron volts) of an atom is this state? Why is such an orbit unlikely to be occupied?

5. (a) Find the de Broglie wavelength of the earth. (b) What is the quantum number that characterizes the earth's orbit about the sun? (The earth's mass is 6.0×10^{24} kg, its orbital radius is 1.5×10^{11} m, and its orbital speed is 3.0×10^4 m/s.)

27.3 Atomic Spectra

6. Find the wavelength of the spectral line that corresponds to a transition in hydrogen from the $n = 6$ state to the $n = 3$ state. In what part of the spectrum is this?

7. Find the wavelength of the spectral line that corresponds to a transition in hydrogen from the $n = 10$ state to the ground state. In what part of the spectrum is this?

8. A proton and an electron, far apart and at rest initially, combine to form a hydrogen atom in the ground state. A single photon is emitted in this process. What is its wavelength?

9. What is the shortest wavelength present in the Brackett series of spectral lines?

10. What is the shortest wavelength present in the Paschen series of spectral lines?

11. A beam of electrons is used to bombard gaseous hydrogen. What is the minimum energy in electron volts the electrons must have if the second line of the Paschen series, corresponding to a transition from the $n = 5$ state to the $n = 3$ state, is to be emitted?

12. An excited hydrogen atom emits a photon of wavelength λ in returning to the ground state. (a) Derive a formula that gives the quantum number of the initial excited state in terms of λ and R. (b) Use this formula to find n_i for a 102.55-nm photon.

13. (a) An object of mass 0.10 g undergoes simple harmonic motion with a frequency of 20 Hz and an amplitude of 2.0 mm. If this oscillation were to emit a quantum of radiation at the same frequency, by what percentage would its energy be reduced? (b) An object of mass 1.0×10^{-26} kg undergoes simple harmonic motion with a frequency of 2.0×10^{15} Hz and an amplitude of 2.0×10^{-12} m. If this oscillation were to emit a quantum of radiation at the same frequency, by what percentage would its energy be reduced? (c) What conclusion can we draw from these results?

27.4 The Bohr Model

14. A negative muon ($m = 207\ m_e$, $Q = -e$) can be captured by a proton to form a **muonic atom.** What is the radius of the first Bohr orbit of such an atom? What is the ionization energy of such an atom? Assume that the proton remains stationary as the muon revolves around it (in reality, both revolve around a common center of mass).

27.5 Quantum Theory of the Atom

27.6 Exclusion Principle

15. What are the possible values of the magnetic quantum number of an atomic electron with the orbital quantum number $l = 4$?

16. What are the possible values of the orbital and magnetic quantum numbers of an atomic electron with the principal quantum number $n = 3$?

17. How many elements would there be if atoms with occupied electron shells up through $n = 6$ could exist?

PROBLEMS

18. (a) Derive a formula for the frequency of revolution of an electron in the nth orbit of the Bohr atom. (b) Find the frequencies of revolution of the electron when it is in the $n = 1$ and $n = 2$ orbits. (c) An electron spends about 10^{-8} s in an excited state before it drops to a lower state by giving up energy in the form of a photon. How many revolutions does an electron in the $n = 2$ state of the Bohr atom make before dropping to the $n = 1$ state? How does this compare with the number of revolutions the earth has made around the sun in the 4.5×10^9 years of its existence?

19. Show that the angular momentum of an electron in the nth orbit of a Bohr atom is equal to $nh/2\pi$. (In fact, this was part of Bohr's original formulation of his theory of the hydrogen atom, which he carried out before de Broglie waves had been discovered.)

20. A beam of electrons whose energy is 13.0 eV is used to bombard gaseous hydrogen. What series of wavelengths will be emitted?

21. Repeat the derivation of the Bohr theory for a one-electron ion whose nuclear charge is Ze and show that the energy of the electron is proportional to Z^2.

22. Calculate the radius and speed of an electron in the ground state of doubly ionized lithium and compare them with the radius and speed of the electron in the ground state of the hydrogen atom. (Li^{2+} has a nuclear charge of $3e$.)

23. Steam at 100°C can be thought of as an excited state of water at 100°C. Suppose that a laser could be built based on the transition from steam to water, with the energy lost per molecule of steam appearing as a photon. What would the frequency of such a photon be? To what region of the spectrum does this correspond? The heat of vaporization of water is 2260 kJ/kg, and its molar mass is 18.02 g/mole. Ignore the work done when the water expands as it becomes steam.

Answers to Multiple Choice

1. d	**8.** c	**15.** d	**22.** c
2. d	**9.** c	**16.** a	**23.** a, b
3. d	**10.** b	**17.** d	**24.** c
4. c	**11.** b	**18.** a	**25.** a, b, c
5. c	**12.** a	**19.** c	
6. c	**13.** b	**20.** d	
7. b	**14.** a	**21.** d	

C H A P T E R 28

ATOMS IN COMBINATION

Individual atoms are rare on the earth and in its lower atmosphere; only the inert gas atoms (such as those of helium and argon) occur by themselves. All other atoms are found linked in small groups called molecules or in larger ones as liquids or solids. Some of these groups consist of atoms of the same element; others consist of atoms of different elements. In every case the arrangement is favored over separate atoms because of interactions between the atoms that reduce the total energy of the system. The quantum theory of the atom accounts for these interactions in a natural way, with no special assumptions, which is further testimony to the power of this approach.

28.1 The Hydrogen Molecule

Electrical forces hold atoms together to form molecules.

A molecule is a group of atoms that stick together strongly enough to act as a single particle. A molecule always has a certain definite structure. Hydrogen molecules always have two hydrogen atoms each, for instance, and water molecules always consist of one oxygen atom and two hydrogen atoms each. A piece of iron is not a molecule, because even though its atoms stay together, any number of them do so to form an object of any size or shape.

A molecule of a given kind is complete in itself with little tendency to gain or lose atoms. If one of its atoms is somehow removed or another atom is somehow attached, the result is a molecule of a different kind with different properties. A liquid or a solid, on the other hand, can gain or lose additional atoms of the kinds already there without changing its character.

Atoms are bound together to form molecules by sharing electrons. An example is the hydrogen molecule, H_2, which consists of two hydrogen atoms. In this molecule the two protons are 7.42×10^{-11} m apart, and the two electrons, one from each atom, belong to the entire molecule rather than to their parent protons. Because the electrons spend more time on the average between the protons than they do on the outside, there is effectively a net negative charge between the protons (Fig. 28.1). The attractive force of this charge on the protons is more than enough to counterbalance the repulsion between them. If the protons are too close together, however, their repulsion becomes dominant.

The balance between attractive and repulsive forces occurs at a separation of 7.42×10^{-11} m, where the total energy of the H_2 molecule is -4.5 eV. Hence 4.5 eV of work must be done to break an H_2 molecule into two H atoms:

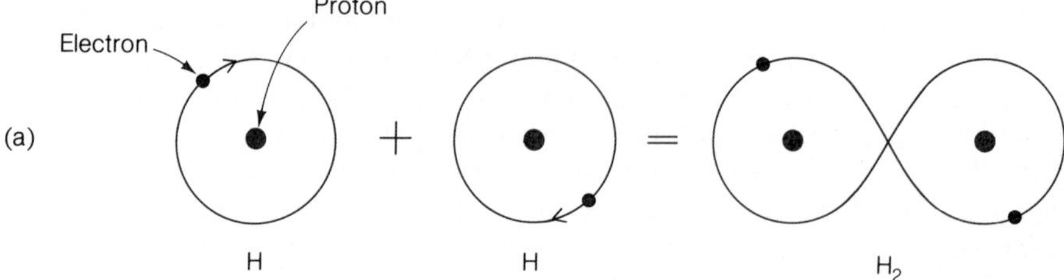

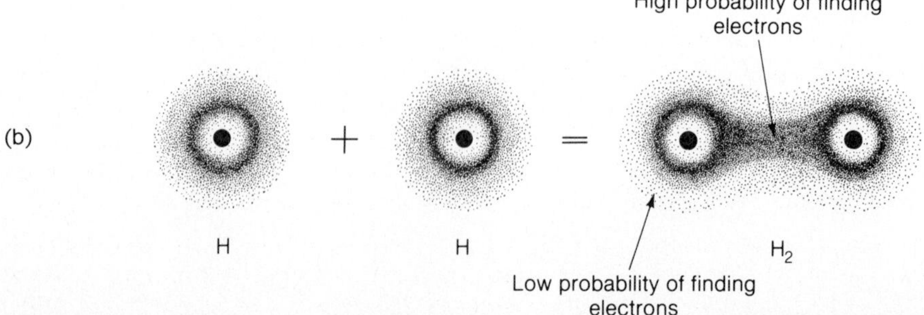

Figure 28.1

(a) Orbit model of the hydrogen molecule. (b) Probability-cloud model of the hydrogen molecule. In both models the shared electrons spend more time on the average between the protons, which leads to an attractive force.

$$H_2 + 4.5 \text{ eV} \rightarrow H + H$$

By comparison, the binding energy of the hydrogen atom is 13.6 eV:

$$H + 13.6 \text{ eV} \rightarrow p^+ + e^-$$

This is an example of the general rule that it is easier to break up a molecule than to break up an atom.

An argument based on the uncertainty principle lets us see, even without a calculation, why a hydrogen molecule should have less energy (and hence more stability) than two separate hydrogen atoms. According to the uncertainty principle, the larger the uncertainty in a particle's position, the less the uncertainty in its momentum. Therefore the larger the region of space to which an electron is confined, the less its momentum and hence energy need be. An electron shared by two protons has more room in which to move about than an electron bound to a single proton to form a hydrogen atom, so its energy is less.

Molecules and the Exclusion Principle

Why do only two hydrogen atoms join together to form a molecule? Why not three, or four, or a hundred? The basic reason for the limited size of molecules is the exclusion principle, which does not allow more than one electron in an atomic system to have the same set of quantum numbers. If a third H atom were to be brought up to an H_2 molecule, its electron would have to leave the $n = 1$ shell and go to the $n = 2$ shell in order for an H_3 molecule to be formed, since only two electrons can occupy the $n = 1$ shell (see Fig. 27.17). But an $n = 2$ electron in the hydrogen atom has over 10 eV more energy than an $n = 1$ electron, and the binding energy in H_2 is only 4.5 eV. Hence an H_3 molecule, if it could somehow be put together, would immediately break apart into $H_2 + H$ with the release of energy. Similar considerations hold for other molecules.

The exclusion principle is also responsible for the fact that the inert gases do not occur as molecules. The electron shells of the inert gas atoms are all filled to capacity, so in order for two of them to share an electron pair, one of the electrons would have to go into an empty shell of higher energy. The increase in energy involved would be more than the energy decrease produced by sharing the electrons, and therefore no such molecules as He_2 or Ar_2 occur.

28.2 Covalent Bond

An electron pair shared by two atoms.

The mechanism by which electron sharing holds atoms together to form molecules is known as **covalent bonding.** We can think of the atoms as being held together by **covalent bonds,** with each shared pair of electrons making up a bond.

More complex atoms than hydrogen also join together to form molecules by sharing electrons. Depending on the electron structures of the atoms involved, there may be one, two, or three covalent bonds—shared electron pairs—between the atoms. For example, in the oxygen molecule O_2 there are two bonds between the O atoms, and in the nitrogen molecule N_2 there are three bonds between the N atoms. Covalent bonds are often represented by a dash for each shared pair of electrons. Thus the H_2,

O_2, and N_2 molecules can be represented as follows:

$$\mathrm{H-H} \qquad \mathrm{O=O} \qquad \mathrm{N\equiv N}$$

Covalent bonds are not limited to atoms of the same element or to only two atoms per molecule. Here are two examples of the bonding in more complicated molecules:

```
                                        H
             H                          |
Water, H2O   |       Ammonia, NH3       N—H
             O—H                        |
                                        H
```

We see that oxygen participates in two bonds in H_2O and nitrogen in three bonds in HN_3, whereas H participates in only one bond in both cases. This behavior agrees with the fact that the hydrogen, oxygen, and nitrogen molecules have one, two, and three bonds, respectively, between their atoms.

Carbon atoms tend to form four covalent bonds at the same time, because they have four electrons in their outer shells and these shells lack four electrons to be complete. The structures of the common covalent molecules methane, carbon dioxide, and acetylene illustrate the different bonds in which carbon atoms can participate to form molecules:

```
   H
   |
H—C—H        O=C=O         H—C≡C—H
   |
   H

Methane     Carbon dioxide     Acetylene
```

Carbon atoms are so versatile in forming covalent bonds with each other as well as with other atoms that millions of carbon compounds are known. The molecules in some carbon compounds contain tens of thousands of atoms. Such compounds were once thought to originate only in living things, and their study is accordingly known even today as organic chemistry.

In a molecule composed of different atoms, the shared electrons favor one or another of the kinds of atoms. The resulting molecule may have a nonuniform distribution of electric charge, with some parts of the molecule being positively charged and other parts negatively charged. A molecule of this kind is called a **polar molecule,** whereas a molecule whose charge distribution is symmetric is called a **nonpolar molecule.** Thus the water molecule is highly polar because the two O—H bonds are 104.5° apart and the two shared electron pairs spend more time near the O atom than near the H atoms.

28.3 Solids and Liquids

Long- and short-range order.

The covalent bonds that can tie a fixed number of atoms together into a molecule can also tie an unlimited number of them together into a solid or a liquid. Other bonding mechanisms are found in solids and liquids as well, one of which is responsible for the ability of metals to conduct electric current. Although only a tiny part of the universe is in the solid state and even less is in the liquid state, matter in these forms makes up much of the physical world of our experience. Modern technology is to a large extent based on the unique characteristics of various solid materials.

Most solids are crystalline in nature, with their atoms or molecules arranged in regular, repeated patterns. A **crystal** is thus characterized by the presence of **long-range order** in its structure. Every crystal of a given kind, large or small, has the same geometrical form. The word *crystal* suggests salt and sugar grains, mineral samples, and sparkling gem stones. But metals and snowflakes are crystalline, as are graphite (the soft ''lead'' in pencils), the fibers of asbestos, and the clear, flat plates of mica. Clay is composed of tiny crystals that can trap water between them to give an easily shaped material.

Other solids lack the definite arrangements of atoms and molecules so conspicuous in crystals. They can be thought of as liquids whose stiffness is due to an exaggerated viscosity. Examples of such **amorphous** (''without form'') solids are pitch, glass, and many plastics. The structures of amorphous solids exhibit **short-range order** only.

Some substances—for instance, B_2O_3—can exist in either crystalline or amorphous forms. In both cases each boron atom is surrounded by three larger oxygen atoms, which represents a short-range order. In a B_2O_3 crystal a long-range order is also present, as shown in a two-dimensional representation in Fig. 28.2. Amorphous B_2O_3, a glassy material, lacks this additional regularity.

The lack of long-range order in amorphous solids means that the bonds vary in strength. When an amorphous solid is heated, the weakest bonds break at lower temperatures than the others, and the solid softens gradually. In a crystalline solid the bonds break simultaneously, and melting has a sudden onset.

Liquids have more in common with solids than with gases, even though liquids share with gases the ability to flow from place to place. Because the density of a given liquid is never far from that of the same substance in solid form, the bonding mechanism must be similar in both cases. When a solid is heated to its melting point, its atoms or molecules pick up enough energy to shift the bonds holding them together so that they form into separate clusters. Not until the liquid is heated to its vaporization point are the atoms or molecules able to break loose completely and form a gas. This interpretation of the liquid state is confirmed by X-ray studies that reveal clusters of atoms or molecules in a liquid (that is, short-range order like that in amorphous solids). Unlike the permanent arrangements in a solid, however, the clusters in a liquid are constantly shifting their arrangements.

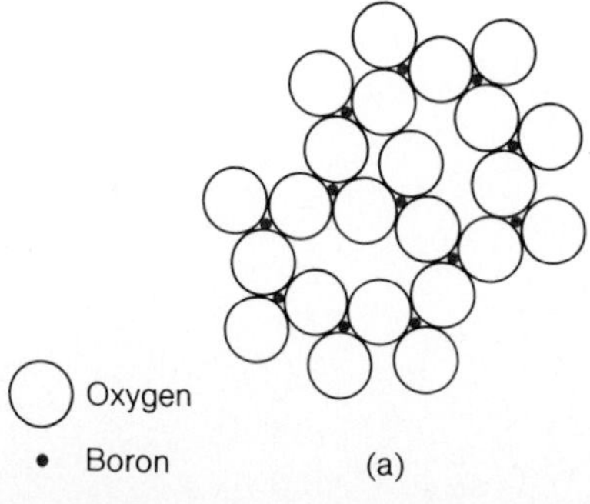

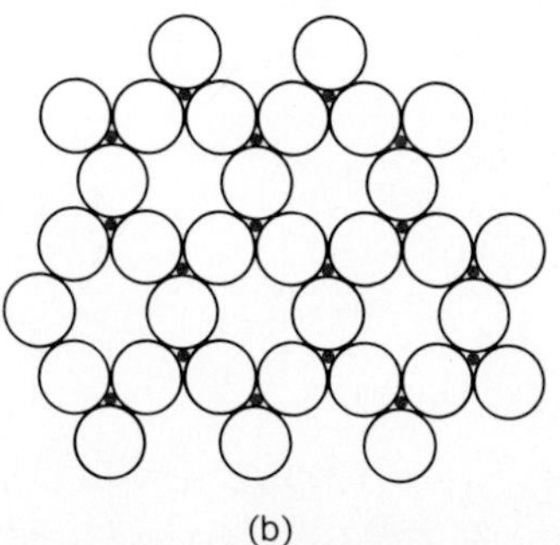

Figure 28.2

Structure of B_2O_3. (a) Amorphous B_2O_3, a glass, has short-range order only in its structure. (b) Crystalline B_2O_3 has long-range order as well.

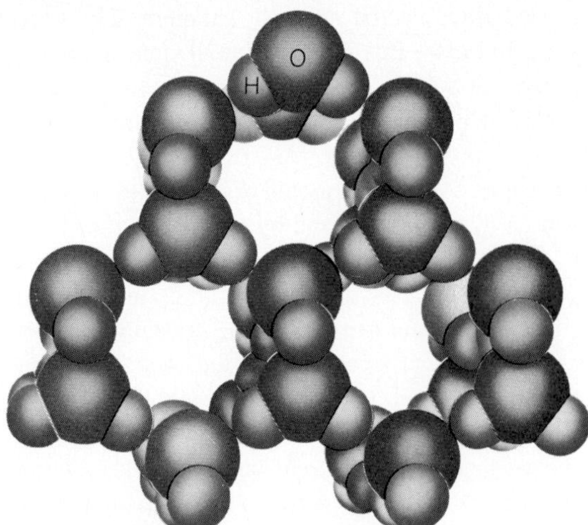

Figure 28.3

The structure of an ice crystal, showing the open hexagonal arrangement of the H_2O molecules. There is less order in liquid water, which allows the molecules to be closer together on the average than they are in ice. Thus density of ice is less than that of water, and ice floats.

The unusual behavior of water near the freezing point that was mentioned in Section 13.2 can be traced to the effect just described. Ice crystals have very open structures because each H_2O molecule can participate in only four bonds with other H_2O molecules (Fig. 28.3). In other solids each atom or molecule may have as many as 12 nearest neighbors. This allows the assemblies to be more compact. Because clusters of molecules are smaller and less stable in the liquid state, water molecules are on the average packed more closely together than are ice molecules, and water has the higher density. Hence ice floats. The density of water increases from 0°C to a maximum at 4°C as large clusters of H_2O molecules break up into smaller ones that occupy less space. Only above 4°C does the normal thermal expansion of a liquid show up as a decreasing density with increasing temperature (see Fig. 13.6).

28.4 Covalent and Ionic Solids

Electron sharing and electron transfer in crystals.

Molecular bonds are all essentially covalent, but in solids four bonding mechanisms are found: covalent, ionic, van der Waals, and metallic.

Diamond is an example of a covalent crystal. Figure 28.4 shows the array of carbon atoms in diamond, with each carbon atom sharing electron pairs with the

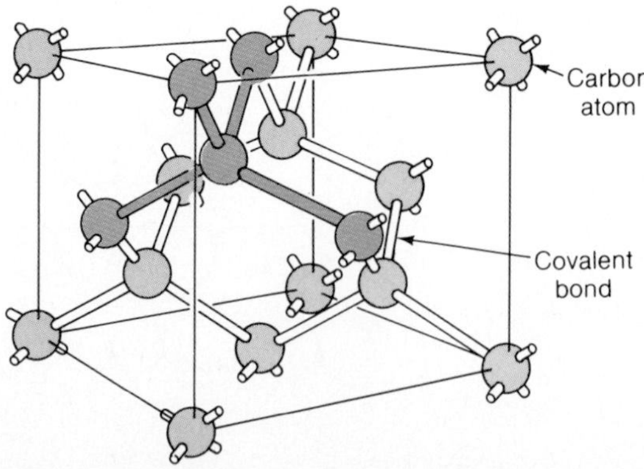

Figure 28.4

Covalent structure of diamond. Each carbon atom shares electron pairs with four other carbon atoms.

The mineral calcite is the chief constituent of limestone and marble. Calcite crystals consist of calcium (Ca^{2+}) and carbonate (CO_3^{2-}) ions held together by ionic bonds.

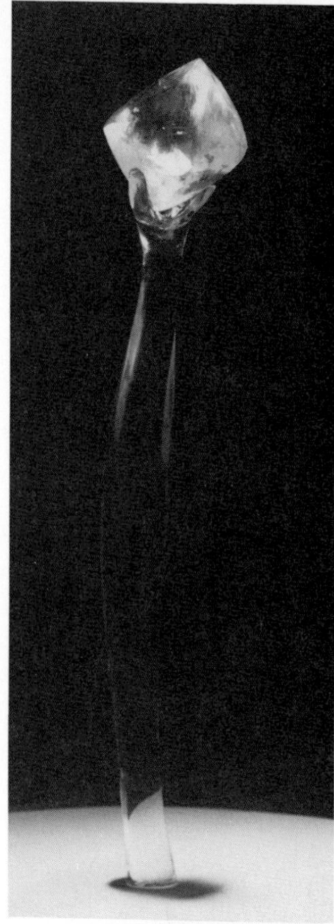

The covalent bonds between its carbon atoms give diamond its characteristic hardness and high melting point. The mass of a diamond gemstone is customarily given in carats, where 1 carat is about 0.2 g. This stone has a mass of 253.7 carats.

four carbon atoms nearest it. All the electrons in the outer shells of the carbon atoms participate in the bonding, and as a result diamond is extremely hard and must be heated to over 3500°C before its crystal structure is disrupted and it melts. Purely covalent crystals are relatively few in number. In addition to diamond, some examples are silicon, germanium, and silicon carbide (''Carborundum''). Like diamond, all are hard and have high melting points.

In covalent bonding, two atoms share one or more pairs of electrons. In **ionic bonding,** one or more electrons from one atom transfer to another atom, producing a positive ion and a negative ion that attract each other. Let us see how ionic bonding functions in the case of NaCl, whose crystals constitute ordinary table salt. We will consider a hypothetical NaCl ''molecule'' for convenience. Salt crystals are actually aggregates of Na and Cl atoms that, although they do not pair off into individual molecules, do interact through ionic bonds whose nature can be most easily examined in terms of a single Na-Cl unit.

Figure 28.5 shows schematically an electron shifted from a sodium atom, which becomes an Na^+ ion, to a chlorine atom, which becomes a Cl^- ion. The ions then attract each other electrically. The combination is stable because more energy is needed to pull the Na^+ and Cl^- ions apart than will be supplied by the return of the

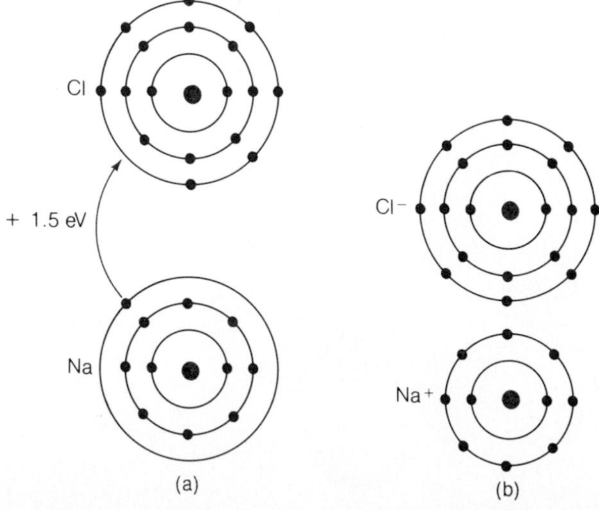

Figure 28.5

(a) Electron transfer in NaCl. (b) The resulting ions attract each other electrically.

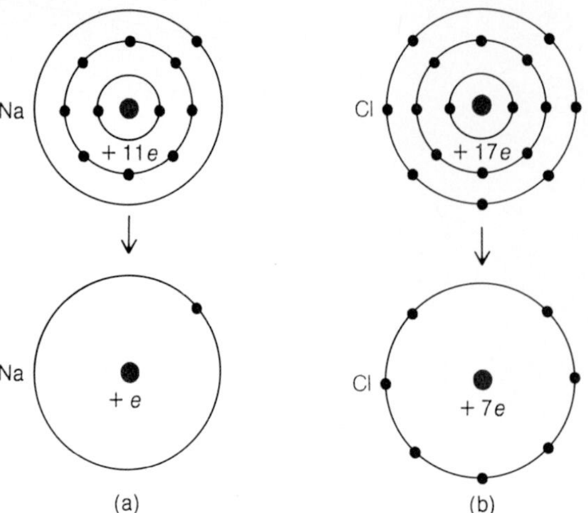

Figure 28.6

(a) A sodium atom. The presence of ten electrons in the $n = 1$ and $n = 2$ shells effectively shields the outer $n = 3$ electron from all but $+e$ of the nuclear charge. In consequence an Na atom tends to lose its outer electron to become an Na^+ atom. (b) A chlorine atom. The inner electrons leave unshielded $+7e$ of the nuclear charge. In consequence a Cl atom tends to pick up another electron to become a Cl^- ion.

shifted electron from the Cl^- ion to the Na^+ ion. Not all pairs of atoms can interact by electron transfer to form stable combinations. Let us see why Na and Cl can do so.

The single outer electron of the Na atom is relatively easy to detach because the ten inner electrons shield it from all but $+e$ of the actual nuclear charge of $+11e$ (Fig. 28.6a). In the Cl atom, on the other hand, the ten inner electrons leave $+7e$ of the nuclear charge unshielded, so the attractive force on the outer electrons is much greater (Fig. 28.6b). Accordingly, a Cl atom tends to pick up an additional electron to become a Cl^- ion. Thus Na and Cl are an ideal match: One readily loses an electron; the other readily gains an electron. The same is true for the partners in other ionic solids, which owe these tendencies to their electron structures.

The structure of a NaCl crystal is shown in Fig. 28.7. Each ion behaves much like a point charge and thus tends to attract to itself as many ions of opposite sign as can fit around it. In a NaCl crystal each Na^+ ion is surrounded by six Cl^- ions, and vice versa. In crystals having different structures, the number of "nearest neighbors" around each ion may be 3, 4, 6, 8, or 12. Ionic bonds are usually fairly

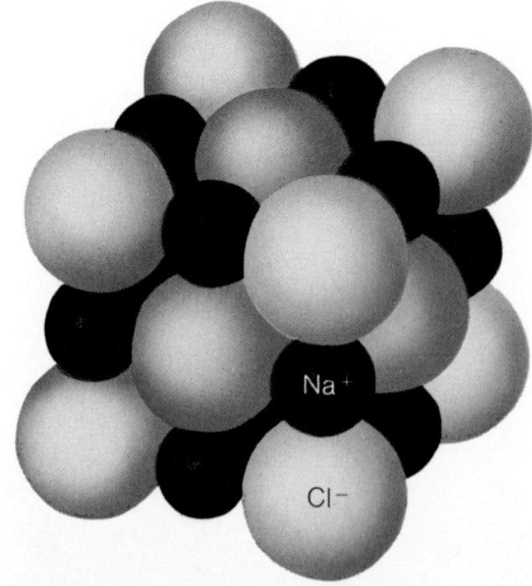

Figure 28.7

A NaCl crystal is composed of Na^+ and Cl^- ions in an array such that each ion is surrounded by six ions of the other kind. Na^+ ions are small because they have lost their outer electrons.

strong, and consequently ionic crystals are strong and hard and have high melting points.

Many crystalline bonds are partly ionic and partly covalent in origin. An example of such mixed bonding is quartz, SiO_2.

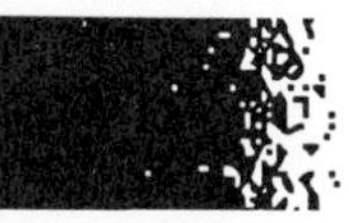

28.5 Van der Waals Bonds

Weak but everywhere.

A number of molecules and nonmetallic atoms have electron structures that do not lend themselves to either covalent or ionic bonding. In this category are the inert gas atoms, which have filled outer shells, and molecules such as methane (Fig. 28.8), whose outer electrons are fully involved in covalent bonds. Even these virtually noninteracting substances, however, condense into liquids and solids at low enough temperatures through the action of **van der Waals forces,** which were proposed a century ago by the Dutch physicist Johannes van der Waals. Such familiar effects as friction, surface tension, adhesion, and cohesion also arise from these forces.

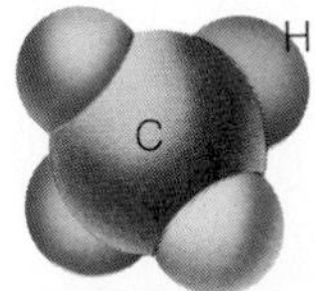

Figure 28.8

Model of the methane molecule, CH_4.

A type of van der Waals force arises between polar molecules, whose charge distributions are asymmetric. When one such molecule comes near another, the ends of opposite polarity attract each other to hold the molecules together (Fig. 28.9).

A somewhat similar effect occurs when a polar molecule is near a nonpolar molecule. The electric field of the polar molecule distorts the initially symmetric charge distribution of the nonpolar molecule, as in Fig. 28.10, and the two then attract each other in the same way as any other pair of polar molecules. This phenomenon is like that involved in the attraction of bits of paper by a charged rubber comb shown in Fig. 16.7.

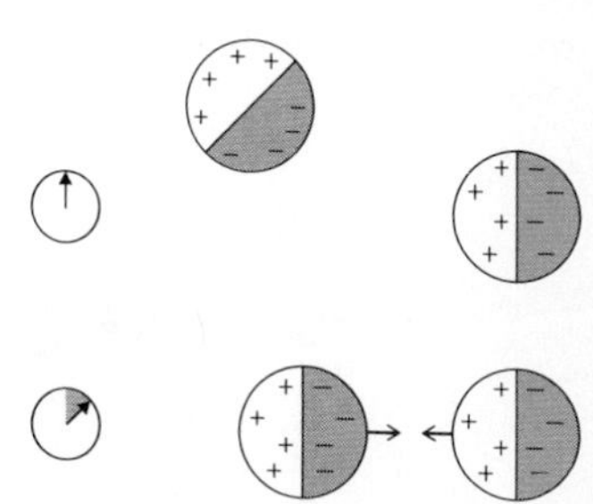

Figure 28.9

Polar molecules attract each other electrically.

Nonpolar molecules attract one another in much the same way that polar molecules attract nonpolar molecules. In a nonpolar molecule the electrons are distributed symmetrically *on the average*. However, *at any moment* one part of the molecule contains more electrons than usual and the rest of the molecule contains fewer. Thus every molecule (and atom) behaves as though it were polar, though the direction and magnitude of the polarization vary constantly. The fluctuations in the charge distributions of nearby nonpolar molecules keep in step through the action of electrical forces, and these forces also hold the molecules together (Fig. 28.11).

Figure 28.10

A polar molecule attracts a nonpolar one by first distorting the latter's originally symmetric charge distribution.

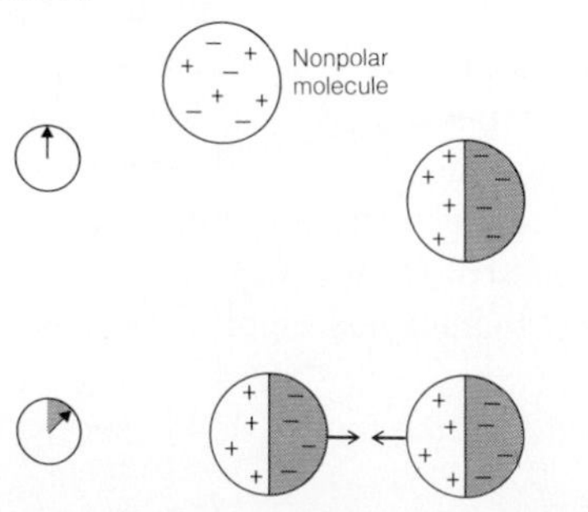

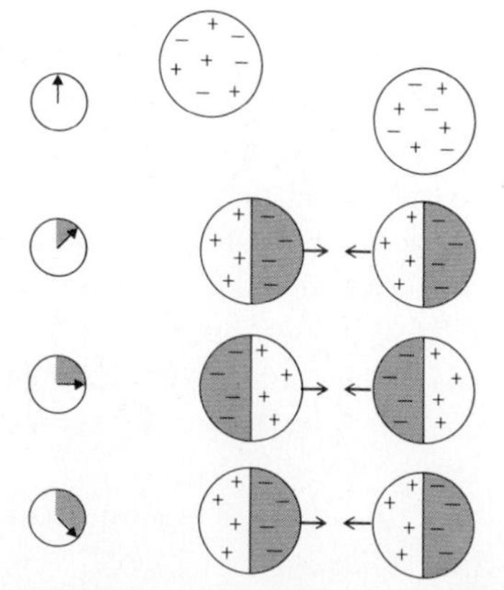

Figure 28.11

Nonpolar molecules have, on the average, symmetric charge distributions, but at any instant the distributions are not necessarily symmetric. The fluctuations in the charge distributions of adjacent nonpolar molecules keep in step, which leads to an attractive force between them.

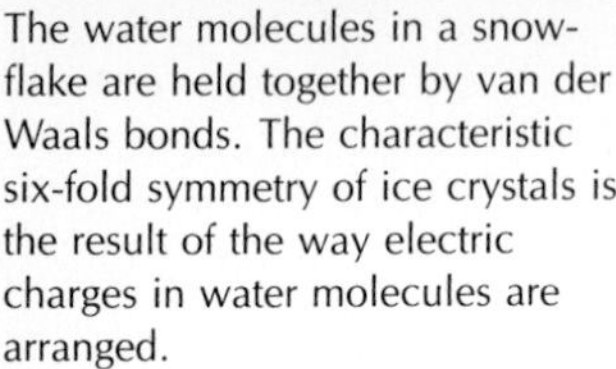

The water molecules in a snowflake are held together by van der Waals bonds. The characteristic six-fold symmetry of ice crystals is the result of the way electric charges in water molecules are arranged.

In general, van der Waals bonds are considerably weaker than ionic, covalent, and metallic bonds. Usually less than 1% as much energy is needed to remove an atom or a molecule from a van der Waals solid as is required in the case of ionic or covalent crystals. As a result the inert gases and compounds with symmetric molecules liquefy and vaporize at rather low temperatures. For example, the boiling point of argon is $-186°C$, and the boiling point of methane is $-161°C$. Molecular solids such as ordinary ice and dry ice (solid CO_2), which consist of whole molecules rather than of atoms or ions, generally have little mechanical strength.

28.6 Metallic Bond

The electron "gas" that bonds metals also makes them good conductors.

Metal atoms have only a few electrons in their outer shells (see Fig. 27.17), and these electrons are not very securely attached, as we saw in the discussion of ionic bonding. When metal atoms come together to form a solid, their outer electrons are given up to a common "gas" of electrons that move relatively freely through the resulting assembly of metal ions. The negatively charged electron gas acts to hold together the positively charged metal ions.

The electron-gas picture of the metallic bond accounts nicely for the properties of metals. A metal conducts heat and electricity well because of the ease with which the free electrons can move about. (In other kinds of solids, all the electrons are bound to particular atoms or pairs of atoms.) Because the free electrons respond readily to electromagnetic waves, metals are opaque to light and have reflective surfaces. Also, because the atoms in a metal are not linked by specific bonds, the properties of alloys—mixtures of different metals—do not depend critically on the relative proportions of each kind of atom, provided their sizes are similar. The copper and tin in bronze, for instance, need not be present in an exact ratio. In sodium chloride, on the other hand, the ratio between sodium and chlorine is always the same.

Table 28.1 summarizes the characteristics of the four basic kinds of crystalline solids.

Table 28.1 Types of crystalline solids. The cohesive energy E_c is the work needed to remove an atom (or molecule) from the crystal and so indicates the strength of the bonds holding it in place.

Type	Covalent	Ionic	Molecular	Metallic
Lattice	Shared electrons	Negative ion; Positive ion	Instantaneous charge separation in molecule	Metal ion; Electron gas
Bond	Shared electrons	Electric attraction	Van der Waals forces	Electron gas
Properties	Very hard; high melting point; soluble in very few liquids	Hard; high melting point; may be soluble in polar liquid such as water	Soft; low melting and boiling points; soluble in covalent liquids	Ductile; metallic luster; ability to conduct heat and electric current readily
Example	Diamond, C $E_c = 7.4$ eV/atom	Sodium chloride, NaCl $E_c = 3.3$ eV/atom	Methane, CH_4 $E_c = 0.1$ eV/molecule	Sodium, Na $E_c = 1.1$ eV/atom

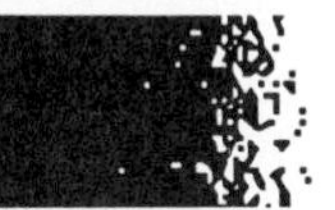

28.7 Energy Bands

Its energy band structure determines whether a solid will be a conductor, an insulator, or a semiconductor.

The notion of energy bands provides a useful framework for understanding the electrical behavior of solids. The nature and properties of semiconductors in particular are clarified with the help of an energy-band analysis.

When atoms are brought as close together as those in a crystal, they interact with one another to such an extent that their outer electron shells constitute a single system of electrons common to the entire array of atoms. The exclusion principle forbids more than two electrons (one with each spin) in any energy level of a system. This principle is obeyed in a crystal because the energy levels of the outer electron shells of the various atoms are all slightly altered by their mutual interaction. (The inner shells do not interact and therefore do not change.) As a result of the shifts in the energy levels, an **energy band** exists in a crystal in place of each sharply defined energy level of its component atoms (Fig. 28.12). Although these bands are actually a multitude of individual energy levels, as many as there are atoms in the crystal, the levels are so near one another as to form a nearly continuous distribution.

The energy bands in a crystal correspond to energy levels in an atom, and an electron in a crystal can only have an energy that falls within one of these bands. The various energy bands in a crystal may or may not overlap, depending on the composition of the crystal (Fig. 28.13). If they do not overlap, the gaps between them represent energies that electrons in the crystal cannot have. The gaps are accordingly known as **forbidden bands.**

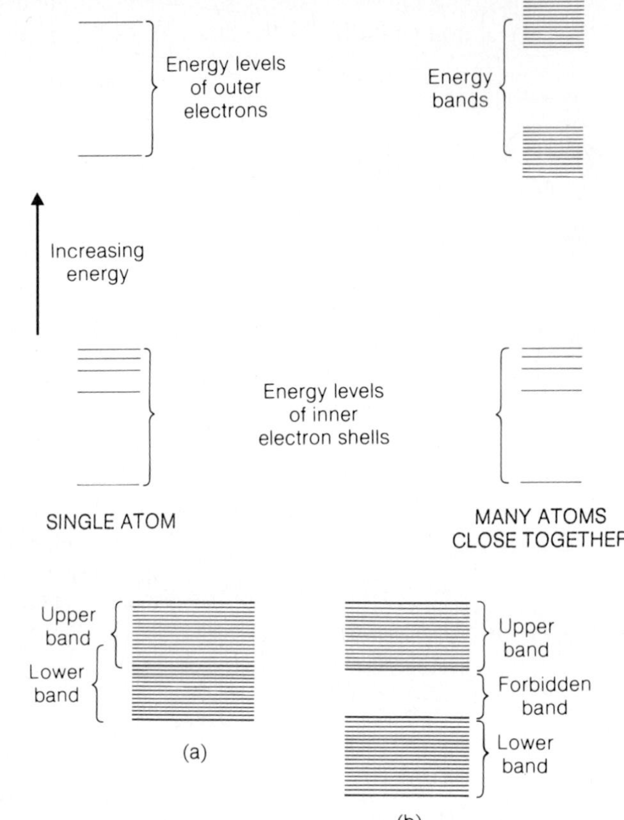

Figure 28.12

Energy bands replace energy levels of outer electrons in an assembly of atoms that are close together.

Figure 28.13

(a) Overlapping energy bands. (b) The gap between energy bands that do not overlap is called a forbidden band.

The energy bands contain all the energies that an outer electron in a crystalline solid can have. The electrical properties of such a solid depend both on its energy-band structure and on the way in which the bands are normally occupied by electrons.

Figure 28.14

Energy levels and bands in solid sodium, which is a metal. (Not to scale.)

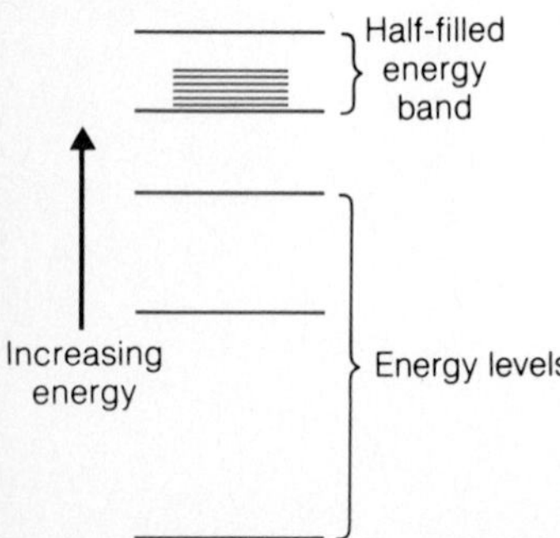

Figure 28.14 shows the energy bands of solid sodium. A sodium atom has only one electron in its outer shell. This means that the upper energy band in a sodium crystal is only half filled with electrons, because each level within the band, like each level in the atom, is capable of containing *two* electrons. When an electric field is set up in a sodium crystal, electrons readily pick up the small additional energy they need to move up in their energy band. The additional energy is in the form of kinetic energy, and the moving electrons constitute an electric current. Sodium is therefore a good conductor, as are other crystalline solids with energy bands that are only partly filled. Such solids are metals.

Figure 28.15 shows the energy-band structure of diamond. The two lower energy bands are completely filled, and there is a gap of 6 eV between the top of the higher of these bands and the empty band above it. Hence a minimum of 6 eV of additional energy must be given to an electron in a diamond crystal if it is to be capable of free motion, because it cannot have an energy lying in the forbidden band. Such an energy boost cannot be readily given to an electron in a crystal by an electric field. Diamond, like other solids with similar energy-band structures, is therefore an electrical insulator.

Silicon has a crystal structure similar to that of diamond, and, as in diamond, a gap separates the top of a filled energy band from an empty higher band. However,

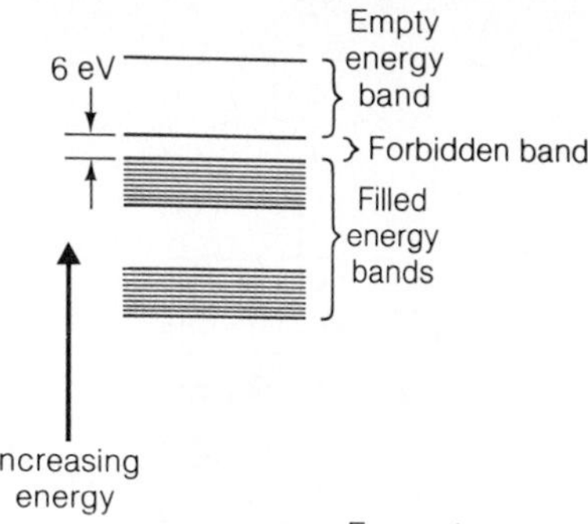

Figure 28.15

Energy levels and bands in diamond, which is an electrical insulator. (Not to scale.)

whereas the gap is 6 eV wide in diamond, it is only 1.1 eV wide in silicon. At very low temperatures silicon is hardly better than diamond as a conductor, but at room temperature a few of its electrons have enough kinetic energy of thermal origin to exist in the higher band. These few electrons are enough to permit a small amount of current to flow when an electric field is applied. Thus silicon has a resistivity intermediate between those of conductors (such as sodium) and those of insulators (such as diamond), and is classified as a **semiconductor.**

The optical properties of solids and their energy-level structures are closely related. Photons of visible light have energies from about 1 to 3 eV. A free electron in a metal can readily absorb such a photon because its allowed energy band is only partly filled. Metals are accordingly opaque, as mentioned before. The characteristic luster of a metal is due to the reradiation of light absorbed by its free electrons. If the metal surface is smooth, the reradiated light appears as a reflection of the original incident light.

For an electron in an insulator to absorb a photon, on the other hand, the photon energy must be more than 3 eV if the electron is to jump across the forbidden band to the next allowed band. Insulators therefore cannot absorb photons of visible light and are transparent. To be sure, most samples of insulating materials do not appear transparent, but this is usually due to the scattering of light by irregularities in their structures. Insulators are opaque to ultraviolet light, whose higher frequencies mean high enough photon energies to allow electrons to cross the forbidden band.

Because the forbidden bands in semiconductors are comparable in width to the photon energies of visible light, they are usually opaque to visible light. However, they are transparent to infrared light whose lower frequencies mean photon energies too low to be absorbed. Thus infrared lenses can be made from the semiconductor germanium, whose appearance is that of a solid metal.

28.8 Impurity Semiconductors

Electrons carry current in n-*type semiconductors; holes carry current in* p-*type semiconductors.*

Small amounts of impurity can have a marked effect on the conductivity of a semiconductor. Suppose we mix a little arsenic with silicon. Silicon atoms have four electrons in their outer shells; arsenic atoms have five. When an arsenic atom replaces a silicon atom in a silicon crystal, four of its outer electrons join in covalent bonds with its nearest neighbors. The extra electron needs very little energy to be

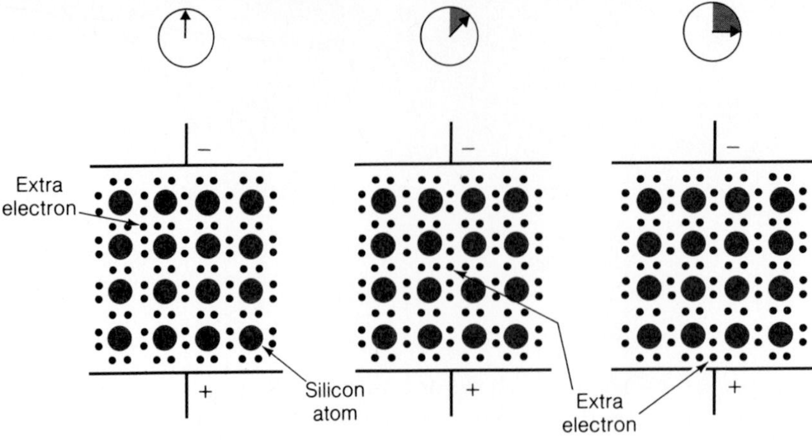

Figure 28.16

Current in an *n*-type semiconductor is carried by surplus electrons that do not fit into the electron-bond structure of the crystal.

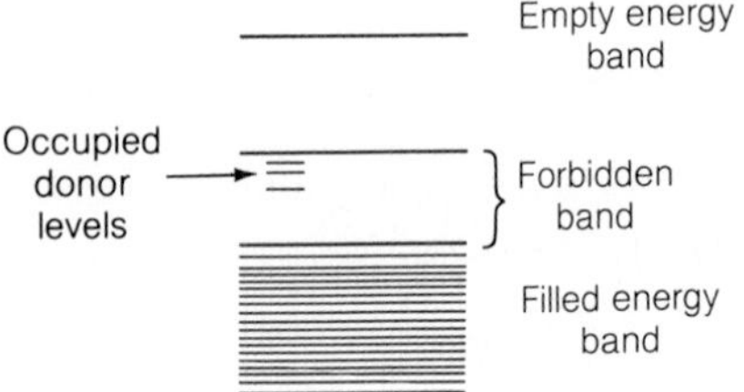

Figure 28.17

Donor levels due to arsenic atoms in a silicon crystal.

detached and move about freely in the crystal. Such a solid is called an ***n*-type** semiconductor because electric current in it is carried by the motion of negative charges (Fig. 28.16). In an energy-band diagram, as in Fig. 28.17, the effect of arsenic as an impurity is to supply occupied energy levels just below an empty energy band. These levels are called **donor levels.**

If we add gallium atoms to a silicon crystal, a different effect occurs. Gallium atoms have only three electrons in their outer shells, and their presence leaves vacancies called **holes** in the electron structure of the crystal. An electron requires little energy to move into a hole, but as it does so, it leaves a new hole behind. When an electric field is applied to a silicon crystal that has gallium as an impurity, electrons move toward the positive electrode by successively filling holes. The flow of current here is best described in terms of the motion of the holes, which behave as though they were positive charges since they move toward the negative electrode (Fig. 28.18). A material of this kind is therefore called a ***p*-type** semiconductor. In the energy-band diagram of Fig. 28.19 we see that the effect of gallium as an impurity is to provide energy levels, called **acceptor levels,** just above the highest filled band. Electrons that enter these levels leave behind unoccupied levels in the formerly filled band that allow the conduction of current.

Adding an impurity to a semiconductor is called **doping.** Phosphorus, antimony, and bismuth as well as arsenic have atoms with five outer electrons and so can be used as donor impurities in doping silicon and germanium to yield an *n*-type semiconductor. Similarly indium and tellurium as well as gallium have atoms with three outer electrons and so can be used as acceptor impurities. A tiny amount of impurity can lead to a dramatic change in the conductivity of a semiconductor. For instance, 1 part of a donor impurity per 10^8 parts of germanium increases its conduc-

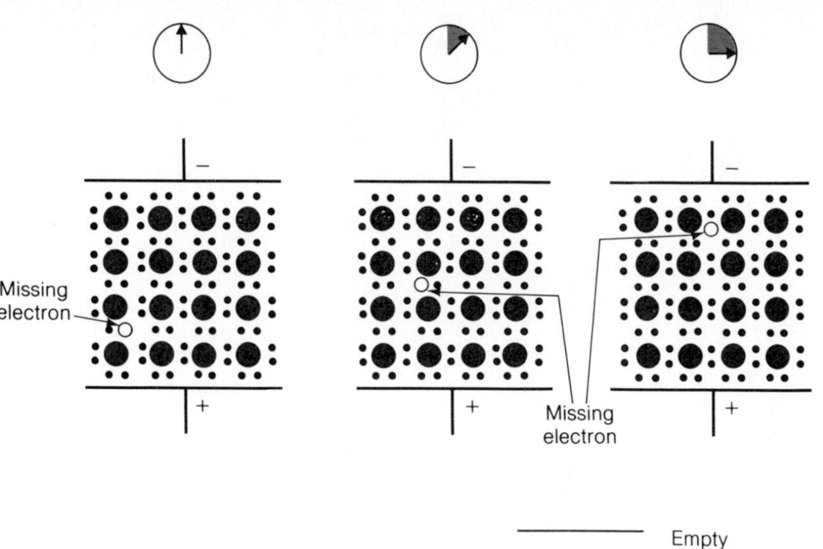

Figure 28.18

Current in a p-type semiconductor is carried by the motion of "holes," which are sites of missing electrons. Holes move toward the negative electrode as a succession of electrons move into them.

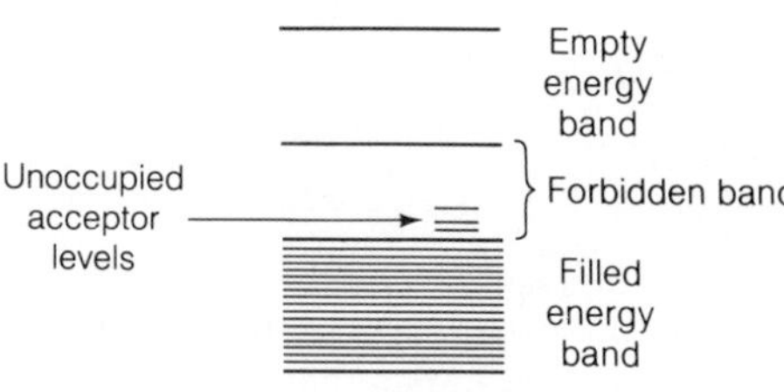

Figure 28.19

Acceptor levels due to gallium atoms in a silicon crystal.

tivity by a factor of 12. Silicon and germanium are not the only semiconducting materials with practical applications. Another important class of semiconductors consists of such compounds as GaAs, GaP, InSb, InAs, and InP.

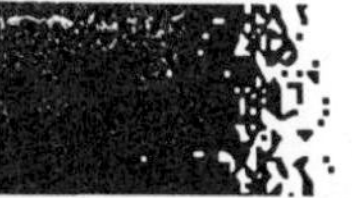

28.9 Semiconductor Devices

The properties of the p-n *junction are responsible for the microelectronics industry.*

The operation of most of the semiconductor devices that have revolutionized modern electronics is based on the nature of junctions between p- and n-type materials. Such junctions can be made in several ways. A method widely used to manufacture integrated circuits involves diffusing impurities in vapor form into a semiconductor wafer in regions defined by masks. A series of diffusion steps using donor and acceptor impurities is part of the procedure that results in circuits with as many as 2 million resistors, capacitors, diodes, and transistors on a chip 5 mm square.

Diodes

A characteristic property of a p-n semiconductor junction is that electric current can pass through it much more readily in one direction than in the other. In the crystal shown in Fig. 28.20, the left-hand end is a p-type region (current carried by holes) and the right-hand end is an n-type region (current carried by electrons). When a voltage is applied across the crystal so that the p-end is negative and the

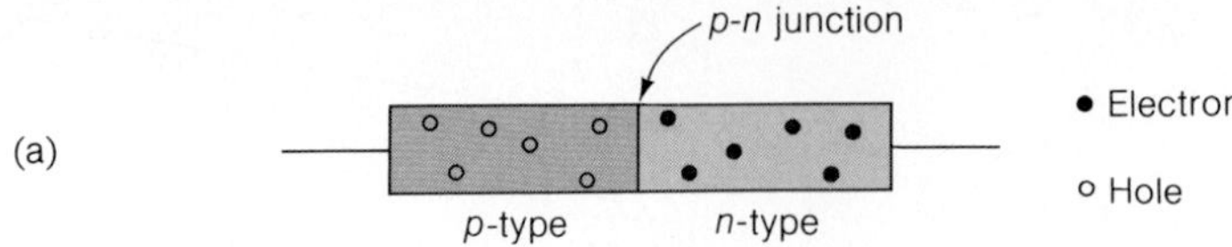

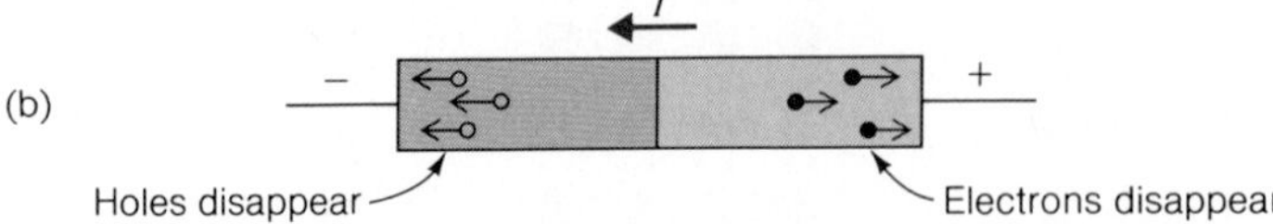

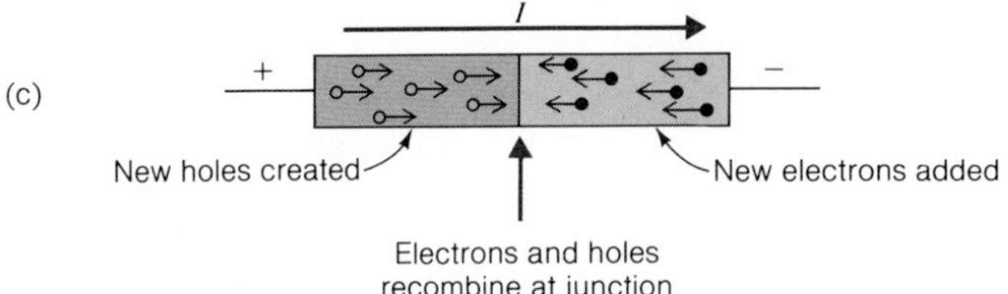

Figure 28.20

(a) Current is carried in a *p*-type semiconductor by the motion of holes and in an *n*-type semiconductor by the motion of electrons. (b) Reverse bias, little current flows. (c) Forward bias, much current flows.

n-end positive, the holes in the *p*-region migrate to the left and the electrons in the *n*-region migrate to the right. Only a limited number of holes and electrons are in the respective regions, and new ones appear spontaneously at only a very slow rate. The current through the entire crystal is therefore negligible. This case is called **reverse bias.**

Figure 28.20(c) shows the same crystal with the connections changed so that the *p*-end is positive and the *n*-end negative, which is **forward bias.** Now new holes are created all the time by the removal of electrons at the *p*-end while new electrons are fed into the *n*-end of the crystal. The holes migrate to the right and the electrons to the left to produce a net positive current flowing from + to −. The electrons and holes meet at the junction between the *p*- and *n*-regions and recombine there: A hole is a missing electron, and when electrons and holes come together, they disappear into the regular structure of the crystal and can no longer act as current carriers. Thus current can flow readily through a semiconductor junction from the *p*- to the *n*-region, but hardly at all in the opposite direction. Figure 28.21 shows how I varies with V at a *p-n* junction. Devices using such junctions, called **diodes,** are widely employed.

The simplest way to use a diode to rectify alternating current is shown in Fig. 28.22(a). Because half the ac input is wasted in such a **half-wave** rectifier circuit, a **full-wave** circuit is more common. A full-wave circuit usually employs four rectifiers in the "bridge" arrangement of Fig. 28.22(b). Current flows through one pair of rectifiers when it is in one direction and through the other pair when it is in the opposite direction to give the pulsating output shown. Ripple-free direct current can be obtained by using a combination of capacitors and inductors (see Section 21.10) to absorb energy at the top of each half cycle and then release it at the bottom.

Energy is needed to create an electron-hole pair, and this energy is released when an electron and a hole recombine. In silicon and germanium the recombination energy is absorbed by the crystal as heat, but in certain other semiconductors, notably gallium arsenide, a photon is emitted when recombination occurs. This is the basis of the **light-emitting diode.** Solid-state lasers have also been made that utilize this phenomenon.

Figure 28.21

The relationship between current and voltage in a semiconductor diode.

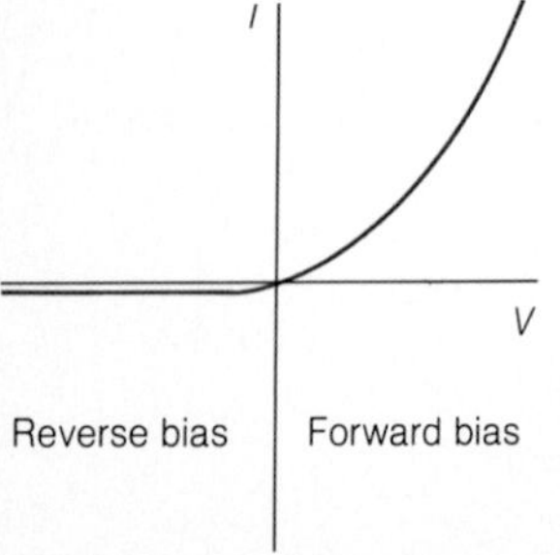

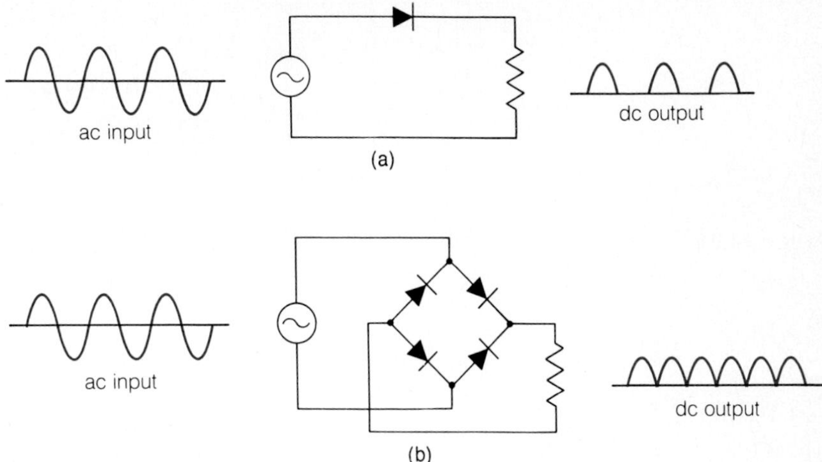

Figure 28.22

(a) Half-wave rectifier circuit. The symbol of a rectifier is

(b) Full-wave rectifier circuit.

Transistors

Transistors are semiconductor devices that can amplify weak signals into strong ones. Figure 28.23 shows an *n-p-n* junction transistor, which consists of a *p*-type material sandwiched between *n*-type materials. The *p*-type region is called the **base,** and the two *n*-type regions are the **emitter** and the **collector.**

In the figure the transistor is connected as a simple amplifier. There is a forward bias across the emitter-base junction in this circuit, so electrons pass readily from the emitter to the base. Depending on the rate at which holes are produced in the base, which in turn depends on its potential relative to the emitter and hence on the signal input, a certain proportion of the electrons from the emitter will recombine there. The rest of the electrons migrate across the base to the collector to complete the emitter-collector circuit. If the base is at a high positive potential relative to the emitter, many holes are produced in the base to recombine with electrons from the emitter. Little current can then flow through the transistor. If the relative positive potential of the base is low, the electrons arriving from the emitter will outnumber the holes being formed in the base, and the surplus electrons will continue across the base to the collector. Changes in the input-circuit current are thus mirrored by changes in the output-circuit current, which is only a few percent smaller.

The ability of the transistor in Fig. 28.23 to produce amplification comes about because the reverse bias across the base-collector junction permits a much higher voltage in the output circuit than that in the input circuit. Because electric

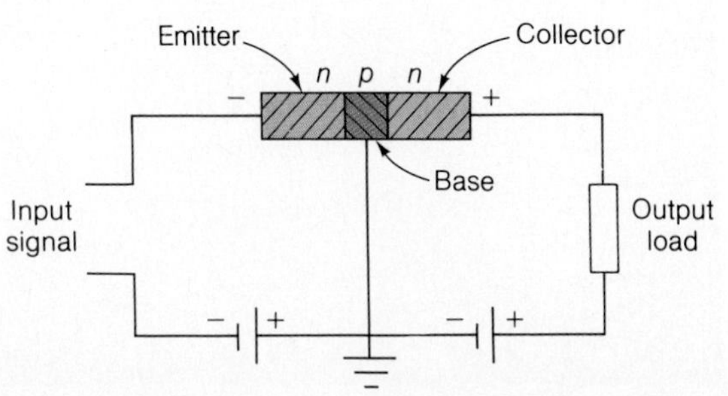

Figure 28.23

A simple transistor amplifier.

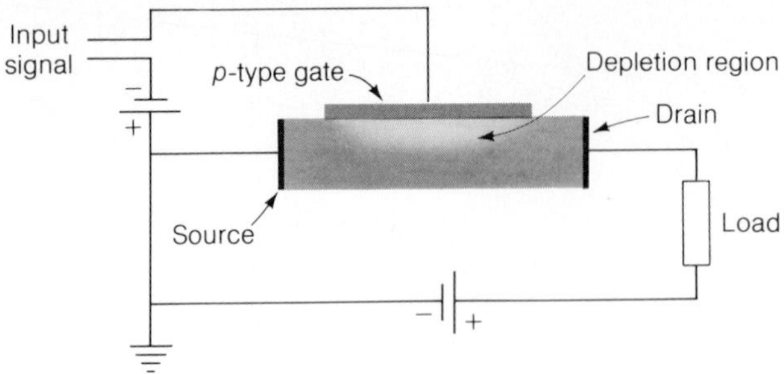

Figure 28.24
A field-effect transistor.

power = (current)(voltage), the power of the output signal can greatly exceed the power of the input signal.

One problem with the junction transistor is that it is difficult to incorporate large numbers of them in an integrated circuit. The **field-effect transistor** (FET) lacks this disadvantage and is widely used today. As in Fig. 28.24, an *n*-channel field-effect transistor consists of a strip of *n*-type material with contacts at each end together with a strip of *p*-type material, called the **gate,** on one side. When connected as shown, electrons move from the **source** terminal to the **drain** terminal through the *n*-type channel. The *p*-*n* junction is given a reverse bias, and as a result both the *n* and *p* materials near the junction are depleted of charge carriers (see Fig. 28.20b). The higher the reverse potential on the gate, the larger the depleted region in the channel, and the fewer the electrons available to carry the current. Thus the gate voltage controls the channel current.

In a metal oxide semiconductor field-effect transistor (MOSFET), the semiconductor gate is replaced by a metal film separated from the channel by an insulating layer of silicon dioxide. The metal film is thus capacitively coupled to the channel, and its potential controls the drain current through the number of induced charges in the channel. A MOSFET is easier to make than a FET and occupies only a few percent of the area needed for a junction transistor.

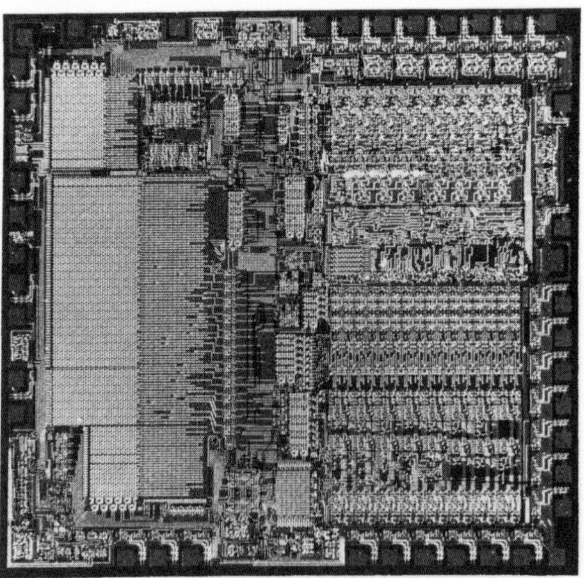

This silicon chip, about 6 mm square, is a microprocessor designed for telephone applications that contains tens of thousands of resistors, capacitors, diodes, and transistors.

Important Terms

A **molecule** is a group of atoms that stick together strongly enough to act as a single particle. A molecule of a given compound always has a certain definite structure and is complete in itself, with little tendency to gain or lose atoms.

In a **covalent bond** between atoms in a molecule or a solid, the atoms share one or more electron pairs.

Solids whose constituent atoms or molecules are arranged in regular, repeated patterns are called **crystalline.** When only short-range order is present, the solid is **amorphous.**

In an **ionic bond,** electrons are transferred from one atom to another, and the two then attract each other electrically.

Van der Waals forces arise from the electric attraction between asymmetric charge distributions in atoms and molecules.

The **metallic bond** that holds metal atoms together in the solid state arises from a ''gas'' of freely moving electrons that pervades the entire metal.

Because the atoms in a crystal are so close together, the energy levels of their outer electron shells are altered slightly to produce **energy bands** characteristic of the entire crystal, in place of the individual sharply defined energy levels of the separate atoms. Gaps between energy bands represent energies that electrons in the crystal cannot have and are called **forbidden bands.**

Semiconductors are intermediate between conductors and insulators in their ability to carry electric current. An ***n*-type semiconductor** is one in which electric current is carried by the motion of electrons. A ***p*-type semiconductor** is one in which electric current is carried by the motion of **holes,** which are vacancies in the electron structure that behave like positive charges.

A semiconductor **diode** permits current to flow through it in only one direction. Signals can be amplified by means of a **transistor,** another semiconductor device.

MULTIPLE CHOICE

1. Relative to the energy needed to separate the electrons of an atom from its nucleus, the energy needed to separate the atoms of a molecule is

a. smaller b. about the same
c. larger d. much larger

2. The hydrogen atoms in a hydrogen molecule stick together because

a. their electrons spend all their time between the nuclei
b. their electrons spend more time between the nuclei than outside them
c. their electrons spend more time outside the nuclei than between them
d. their nuclei attract each other

3. A hydrogen molecule contains only two hydrogen atoms because

a. of the exclusion principle
b. of the uncertainty principle
c. only two such atoms can fit together
d. with more atoms, the electrons would collide with one another

4. In a molecule other than H_2 that consists of two atoms,

a. only a single pair of electrons can be shared
b. more than one pair of electrons can be shared
c. one of the atoms must be a hydrogen atom
d. one of the atoms must be a carbon atom

5. In a covalent bond,

a. electrons are shifted from one atom to another
b. only atoms of the same element are present
c. there must be at least one carbon atom
d. adjacent atoms share electron pairs

6. Covalent bonds never involve the sharing of

a. 2 electrons b. 3 electrons
c. 4 electrons d. 6 electrons

7. The number of covalent bonds a hydrogen atom forms when it combines chemically is

a. 1 b. 2
c. 3 d. 4

8. The number of covalent bonds a carbon atom usually forms when it combines chemically is

a. 1 b. 2
c. 3 d. 4

9. When two or more atoms join to form a stable molecule,

a. energy is absorbed
b. energy is given off
c. there is no energy change
d. any of the above, depending on the circumstances

10. Short-range order is never found in

a. crystalline solids
b. amorphous solids
c. liquids
d. gases

11. A crystalline solid always

a. is transparent
b. is held together by covalent bonds
c. is held together by ionic bonds
d. has long-range order in its structure

12. An amorphous solid is closest in structure to
- a. a covalent crystal
- b. an ionic crystal
- c. a semiconductor
- d. a liquid

13. An amorphous solid
- a. has its particles arranged in a regular pattern
- b. is held together by ionic bonds
- c. does not melt at a definite temperature but softens gradually
- d. consists of nonpolar molecules

14. The lowest melting points are usually found in solids held together by
- a. covalent bonds
- b. ionic bonds
- c. metallic bonds
- d. van der Waals bonds

15. Van der Waals forces arise from
- a. electron transfer
- b. electron sharing
- c. symmetric charge distributions
- d. asymmetric charge distributions

16. A polar molecule can attract
- a. only ions
- b. only other polar molecules
- c. only nonpolar molecules
- d. all of the above

17. The particles that make up the lattice of the van der Waals crystal of a compound are
- a. electrons b. atoms
- c. ions d. molecules

18. The particles that make up the lattice of an ionic crystal are
- a. electrons b. atoms
- c. ions d. molecules

19. The particles that make up the lattice of a covalent crystal are
- a. electrons b. atoms
- c. ions d. molecules

20. A property of metals that is not due to the electron "gas" that pervades them is their unusual ability to
- a. conduct electricity b. conduct heat
- c. reflect light d. form oxides

21. Ice is an example of
- a. a covalent solid b. an ionic solid
- c. a metallic solid d. a molecular solid

22. Diamond is an example of
- a. a covalent solid b. an ionic solid
- c. a metallic solid d. a molecular solid

23. A crystal whose upper energy band is partly occupied by electrons is
- a. a conductor
- b. an insulator
- c. an n-type semiconductor
- d. a p-type semiconductor

24. A crystal with a wide forbidden band between a filled lower band and an empty upper band is
- a. a conductor
- b. an insulator
- c. an n-type semiconductor
- d. a p-type semiconductor

25. A solid with occupied donor levels below an empty energy band is
- a. an n-type semiconductor
- b. a p-type semiconductor
- c. a transistor
- d. a diode

26. Diamond is transparent because
- a. its atoms are far apart
- b. it is held together by covalent bonds
- c. the forbidden band above its highest filled energy band is narrow relative to the photon energies in visible light
- d. the forbidden band above its highest filled energy band is wide relative to the photon energies in visible light

27. A hole in a p-type semiconductor is
- a. a surplus electron
- b. a missing electron
- c. a missing atom
- d. a donor level

28. Current in a p-type semiconductor is carried by
- a. electrons b. holes
- c. positive ions d. negative ions

29. Current in an n-type semiconductor is carried by
- a. electrons b. holes
- c. positive ions d. negative ions

30. A junction between n- and p-type semiconductors
- a. acts as an insulator for currents in both directions
- b. conducts current in both directions equally well
- c. conducts current readily when the p-end is positive and the n-end is negative
- d. conducts current readily when the p-end is negative and the n-end is positive

31. A light-emitting diode gives off a photon of light when
- a. two holes collide
- b. two electrons collide
- c. a hole is created by the freeing of an electron
- d. a hole is filled by recombination with an electron

QUESTIONS

1. What is wrong with the model of a hydrogen molecule in which the two electrons are supposed to follow figure-eight orbits that encircle the two protons?

2. What must be true of the spins of the two electrons in the H_2 molecule?

3. The energy needed to detach the electron from a hydrogen atom is 13.6 eV, but the energy needed to detach an electron from a hydrogen molecule is 15.7 eV. Why do you think the latter energy is greater?

4. Why do the inert gas atoms almost never participate in covalent bonds?

5. What property of carbon atoms allows them to form so many varied and complex molecules?

6. The atoms in a molecule are said to share electrons, yet some molecules are polar. Explain.

7. Why are electrons much more readily liberated from lithium when it is irradiated with ultraviolet light than from fluorine?

8. Why are Cl atoms more active chemically than Cl^- ions?

9. Why are Na atoms more active chemically than Na^+ ions?

10. The separation between Na^+ and Cl^- ions in an NaCl crystal is 2.8×10^{-10} m, whereas it is 2.4×10^{-10} m in an NaCl molecule such as might exist in the gaseous state. Why is it reasonable for these separations to be different?

11. Van der Waals forces can hold inert gas atoms together to form solids, but they cannot hold such atoms together to form molecules in the gaseous state. Why not?

12. The temperature of a gas falls when it passes slowly from a full container to an empty one through a porous plug. Since the expansion is into a rigid container, no mechanical work is done. What is the origin of the fall in temperature?

13. Lithium atoms, like hydrogen atoms, have only a single electron in their outer shells, yet lithium atoms do not join together to form Li_2 molecules the way hydrogen atoms form H_2 molecules. Instead, lithium is a metal with each atom part of a crystal structure. Why?

14. What is the connection between the ability of a metal to conduct electricity and its ability to conduct heat?

15. Does the "gas" of freely moving electrons in a metal include all the electrons present? If not, which electrons are members of the "gas"?

16. What is the basic reason that energy bands rather than specific energy levels exist in a solid?

17. What kind of solid is one whose upper energy band is partly filled with electrons?

18. Why are some solids transparent to visible light and others opaque?

19. Semiconductors are usually opaque to visible light but transparent to infrared light. Why is this so?

20. The energy gap in silicon is 1.1 eV, and in diamond it is 6 eV. How transparent would you expect these substances to be to visible light?

21. How does the energy-band structure of a solid determine whether it is a conductor, a semiconductor, or an insulator of electricity?

22. The forbidden energy band in germanium that lies between the highest filled band and the empty band above it has a width of 0.7 eV. Compare the conductivity of germanium with that of silicon, in which the gap is 1.1 eV, at (a) very low temperatures and (b) room temperature.

23. What is the distinction between *n*- and *p*-type semiconductors?

24. Does the addition of a small amount of indium to germanium result in the formation of an *n*- or a *p*-type semiconductor? An indium atom has three electrons in its outermost shell; a germanium atom has four.

25. At a *p-n* junction which direction of current represents forward bias and which reverse bias? In which direction does current flow more readily?

26. Is energy absorbed or given off when an electron and a hole recombine? If energy is absorbed, where does it come from? If energy is given off, where does it go?

Answers to Multiple Choice

1. a	**9.** b	**17.** d	**25.** a
2. b	**10.** d	**18.** c	**26.** d
3. a	**11.** d	**19.** b	**27.** b
4. b	**12.** d	**20.** d	**28.** b
5. d	**13.** c	**21.** d	**29.** a
6. b	**14.** d	**22.** a	**30.** c
7. a	**15.** d	**23.** a	**31.** d
8. d	**16.** d	**24.** b	

C H A P T E R 29

THE NUCLEUS

The behavior of atomic electrons is responsible for the chief properties (except mass) of atoms, molecules, solids, and liquids. For this reason, until now we have not had to know more about the nucleus of an atom other than that it is a tiny, positively charged bit of matter that provides the atom with most of its mass and holds its electrons in place. The nucleus nevertheless turns out to be of supreme importance. To begin with, the various elements exist only because nuclei can contain more than one proton despite their mutual repulsion. And the continuing evolution of the universe is powered in part by energy that comes from nuclear reactions and transformations in the sun and other stars.

29.1 Nuclear Composition

Atomic nuclei of an element have the same numbers of protons but may have different numbers of neutrons.

The electron structure of the atom was understood before even the composition of its nucleus was known. The reason is that the nucleus is held together by forces vastly stronger than the electrical forces that hold the electrons to the nucleus, and it is correspondingly harder to break apart a nucleus to find out what is inside. Changes in the electron structure of an atom, such as those that occur in the emission of photons or in the formation of chemical bonds, involve energies of only several eV. Changes in nuclear structure, however, involve energies of several MeV, a million times more.

As we know, the simplest nucleus, that of the hydrogen atom, usually consists of a single proton, whose charge is $+e$ and whose mass is

$$m_{\text{proton}} = 1.673 \times 10^{-27} \text{ kg}$$

The proton mass is 1836 times that of the electron, so nearly all of the hydrogen atom's mass resides in its nucleus. This is true of all other atoms as well.

Elements more complex than hydrogen have nuclei that contain **neutrons** as well as protons. The neutron, as its name suggests, is uncharged. The neutron mass is slightly more than that of the proton:

$$m_{\text{neutron}} = 1.675 \times 10^{-27} \text{ kg}$$

Neutrons and protons are jointly called **nucleons.**

Every neutral atom contains the same number of protons and electrons. This number is the **atomic number** of the element. Thus the atomic number of hydrogen is 1, of helium 2, of lithium 3, and of uranium 92. Atomic numbers of the elements are listed in Appendix C. Except in the case of ordinary hydrogen atoms, the number of neutrons in a nucleus equals or, more often, exceeds the number of protons. The compositions of atoms of the four lightest elements are illustrated in Fig. 29.1.

Although all atoms of a given element have nuclei with the same numbers of protons, the numbers of neutrons may not be the same. For instance, even though

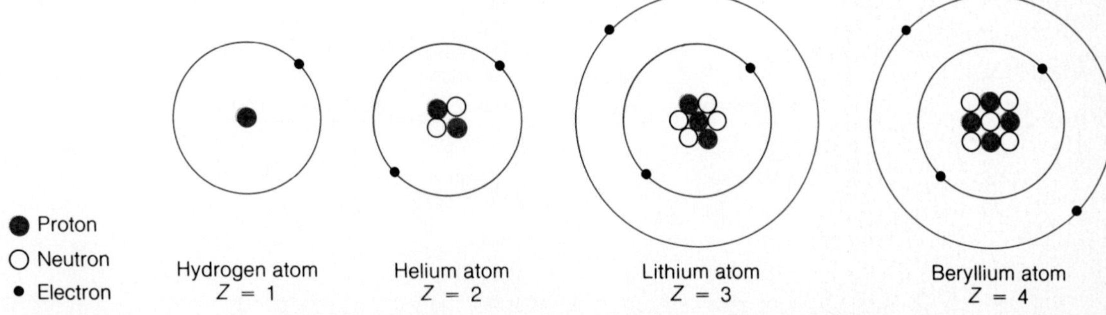

Figure 29.1
The electronic and nuclear compositions of hydrogen, helium, lithium, and beryllium atoms.

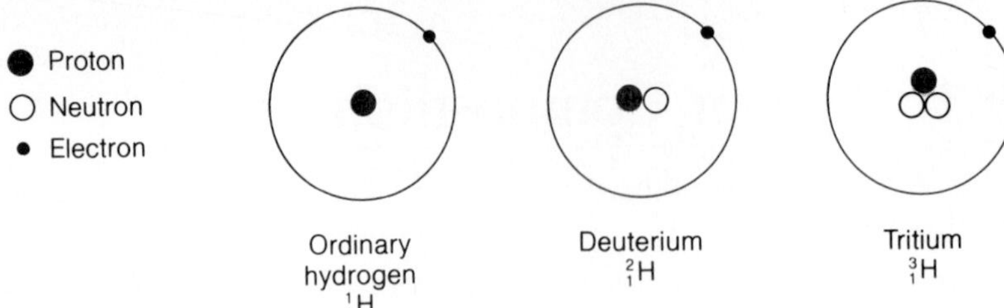

Figure 29.2

The three isotopes of hydrogen.

over 99.9% of hydrogen nuclei are just single protons, a few also contain a neutron as well, and a very few two neutrons (Fig. 29.2). The varieties of an element that differ in the numbers of neutrons their nuclei contain are called its **isotopes.** Table 29.1 describes the isotopes of hydrogen and chlorine. The **atomic mass unit** u used in the table has the value

$$1 \text{ u} = 1.660 \times 10^{-27} \text{ kg}$$

The hydrogen isotope deuterium is stable, but tritium is radioactive (Section 29.5), and a sample of it gradually changes to an isotope of helium. The flux of cosmic rays from space continually replenishes the earth's tritium by nuclear reactions in the atmosphere. Only about 2 kg of tritium of natural origin is present on the earth's surface, nearly all of it in the oceans.

All elements have isotopes. The atomic number of an element determines how many electrons its atoms have and how they are arranged, which in turn governs the chemical behavior of the element. The isotopes of the element have the same atomic number, so it is not surprising that the two isotopes of chlorine, for instance, have the same yellow color, the same suffocating odor, and the same efficiency as poisons and bleaching agents. Because boiling and freezing points depend somewhat on atomic mass, they are slightly different for the two isotopes, as is their density. Other physical properties of isotopes may vary more drastically with their compositions. An example is the radioactivity of tritium, in contrast to the stability of ordinary hydrogen and deuterium.

Table 29.1 The isotopes of hydrogen and chlorine found in nature

	Properties of element		Properties of isotope			
Element	**Atomic number**	**Average atomic mass, u**	**Protons in nucleus**	**Neutrons in nucleus**	**Atomic mass, u**	**Relative abundance**
Hydrogen	1	1.008	1	0	1.008	99.985%
			1	1	2.014	0.015%
			1	2	3.016	Very small
Chlorine	17	35.46	17	18	34.97	75.53%
			17	20	36.97	24.47%

The conventional symbols for nuclear species, or **nuclides,** all follow the pattern ${}^{A}_{Z}X$, where

X = chemical symbol of the element

Z = atomic number of the element
= number of protons in the nucleus

A = mass number of the nuclide
= number of protons and neutrons in the nucleus

Hence ordinary hydrogen is designated ${}^{1}_{1}H$, since its atomic number and mass number are both 1, while tritium is designated ${}^{3}_{1}H$. The two isotopes of chlorine in Table 29.1 are designated ${}^{35}_{17}Cl$ and ${}^{37}_{17}Cl$, respectively.

29.2 Size and Stability

Nuclei are very small, and their ratios of neutrons to protons fall in a narrow range.

The Rutherford scattering experiment (Section 16.4) gives information on nuclear dimensions as well as on atomic structure. When their energies are not too great, as was the case in Rutherford's experiment, alpha particles do not get close enough to the target nuclei for the scattering to differ from that due to nuclei of infinitely small size. At higher energies, though, discrepancies appear between Rutherford's theory and the data then can be used to estimate nuclear dimensions.

EXAMPLE 29.1

The scattering of 7.7-MeV alpha particles (which are ${}^{4}_{2}He$ nuclei) by a gold foil follows the predictions of Coulomb's law. Use this observation to set an upper limit to the size of the gold nucleus.

SOLUTION An alpha particle comes closest to a nucleus when it is headed directly toward it. At the distance of closest approach r_0, the initial KE of the particle is entirely converted to electric potential energy PE, and thus

$$\text{KE} = \text{PE} = \frac{kQ_AQ_B}{r_0}$$

$$r_0 = \frac{kQ_AQ_B}{\text{KE}}$$

Here $Q_A = 2e$ is the charge of an alpha particle, $Q_B = 79e$ is the charge of a gold nucleus, and

$$\text{KE} = (7.7 \times 10^6 \text{ eV})(1.6 \times 10^{-19} \text{ J/eV}) = 1.2 \times 10^{-12} \text{ J}$$

is the alpha-particle energy. Since $k = 9.0 \times 10^9 \text{ N}\cdot\text{m}^2/\text{C}^2$, we have for the upper limit to the radius of the gold nucleus

$$r_0 = \frac{(9.0 \times 10^9\ \text{N}\cdot\text{m}^2/\text{C}^2)(2)(79)(1.6 \times 10^{-19}\ \text{C})^2}{1.2 \times 10^{-12}\ \text{J}} = 3.0 \times 10^{-14}\ \text{m}$$

This is less than 1/10,000 of the radius of the gold atom. The actual radius of the gold nucleus is still smaller, about 7.0×10^{-15} m. ♦

More recent experiments with high-energy electrons, protons, and neutrons give more precise figures for nuclear radii, which range from about 1.1×10^{-15} m for the proton to about 7.4×10^{-15} m for the $^{235}_{92}U$ nucleus. Like the atom, the nucleus has a fuzzy boundary, not a sharp one.

The density of nuclear matter is about 2.4×10^{17} kg/m^3, which is equivalent to 4 billion tons per cubic inch. Certain stars, called **neutron stars,** are made up of atoms that have been so compressed that most of their protons and electrons have fused into neutrons, which are the most stable form of matter under enormous pressures. The densities of neutron stars are comparable to those of nuclei: A neutron star packs the mass of one or two suns into a sphere only about 10 km in radius. If the earth were this dense, it would fit into a large apartment house. Neutron stars are the remnants of supernova explosions (Section 29.12). **Pulsars** are believed to be neutron stars that rotate rapidly and have magnetic fields whose axes are not aligned with the axes of rotation. These fields trap tails of ionized gas that radiate light, radio waves, and X rays, which arrive at the earth in bursts typically a fraction of a second apart as the pulsars spin. Thus a pulsar is like a lighthouse whose flashes are due to a rotating beam of light.

Neutron/Proton Ratio

Only certain combinations of protons and neutrons form stable nuclei. In the lightest nuclei there are about as many neutrons as protons, whereas in heavier ones there are more neutrons than protons. Figure 29.3 is a plot of neutron number N versus Z for stable nuclei. Evidently the number of neutrons in the nuclei of a given element is a fairly critical quantity. Let us see why.

There are two opposing tendencies in a nucleus. The first is a tendency for N to equal Z. Just like an electron in an atom, a nucleon in a nucleus can have only certain specific energies, and, also like an electron, it has spin. Because neutrons and protons obey the exclusion principle, at most two of each kind of nucleon (one whose spin is "up" and one whose spin is "down") can occupy each quantum state. As with atomic energy levels, nuclear energy levels are filled in sequence to give nuclei of minimum energy and hence maximum stability. Thus the boron isotope $^{12}_{5}B$ has more energy than the carbon isotope $^{12}_{6}C$ because one of its neutrons is in a higher energy level, and $^{12}_{5}B$ is accordingly unstable (Fig. 29.4). If created in a nuclear reaction, a $^{12}_{5}B$ nucleus changes by beta decay into a $^{12}_{6}C$ nucleus in a fraction of a second (see Section 29.5).

The other tendency in a nucleus is for the number of neutrons to exceed the number of protons. This is because of the electric repulsion exerted by the protons

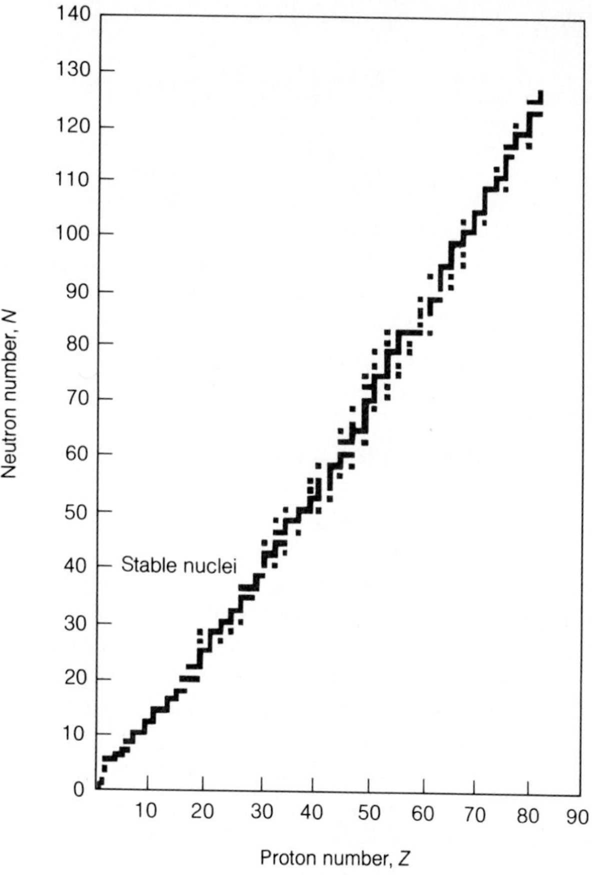

Figure 29.3

The number of neutrons versus the number of protons in stable nuclei. The larger the nucleus, the greater the proportion of neutrons.

on one another, which must be balanced by the attractive nuclear forces that act between nucleons. The repulsive electrical forces increase more rapidly with Z than the attractive nuclear forces increase with A, the total number of nucleons. Hence a greater proportion of neutrons, which produce only attractive forces, is needed for stability in large nuclei.

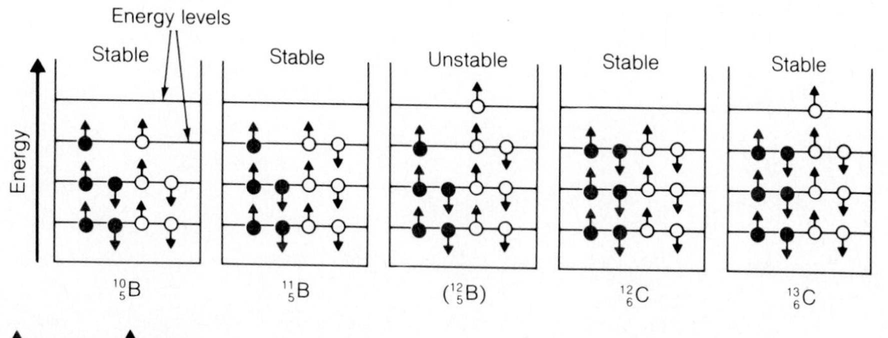

Figure 29.4

Each nuclear energy level can contain two protons of opposite spins and two neutrons of opposite spins. In the light nuclei, energy levels are filled in sequence so the resulting configuration is one of minimum energy.

29.3 Binding Energy

The energy needed to break up a nucleus into its neutrons and protons.

The nucleus of a deuterium atom consists of a proton and a neutron. Thus we would expect that the mass of a deuterium atom, ${}^2_1H$, should be equal to the mass of an ordinary hydrogen atom, ${}^1_1H$, plus the mass of a neutron. However, it turns out that the mass of ${}^2_1H$ is 0.0024 u *less* than the combined masses of a ${}^1_1H$ atom and a neutron (Fig. 29.5).

The case of deuterium is an example of a general observation:

Atomic nuclei always have less mass than the combined masses of their constituent particles.

The energy equivalent of the "missing" mass of a nucleus is called its **binding energy.** In order to break up the nucleus, this amount of energy must be supplied either by colliding with another particle or by absorbing a photon. Binding energies arise from the action of the nuclear forces that hold nuclei together, just as ionization energies of atoms, which must be supplied to remove electrons from them, are due to the action of the electrical forces that hold them together.

The usual energy unit in nuclear physics is the MeV, where

$$1 \text{ MeV} = 10^6 \text{ eV} = 1.60 \times 10^{-13} \text{ J}$$

In terms of MeV, the energy equivalent of 1 u of mass is

$$E_u = 931.5 \text{ MeV}$$ *Energy per atomic mass unit*

Table 29.2 gives the rest masses and their energy equivalents for several particles.

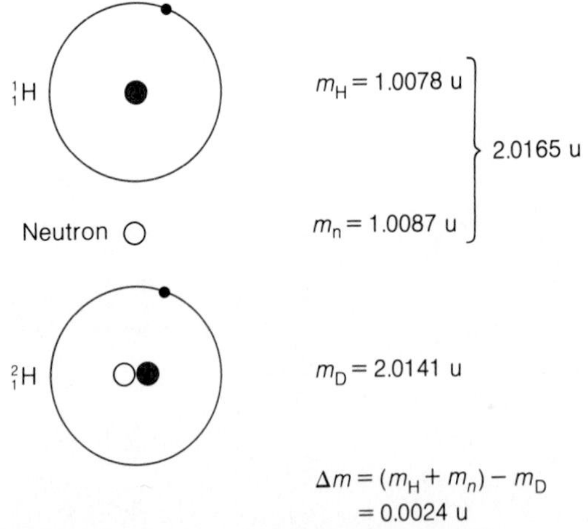

Figure 29.5

The mass of every atom is less than the total of the masses of its constituent neutrons, protons, and electrons. This phenomenon is illustrated here for the deuterium atom.

Table 29.2 Rest masses and energy equivalents for selected particles

Particle	Rest mass, u	Equivalent energy, MeV
Electron	0.000549	0.511
Proton	1.007277	938.3
Neutron	1.008665	939.6
Hydrogen atom	1.007825	938.8
Deuterium atom	2.014102	1876.1
Helium ($^4_2$He) atom	4.002603	3728.4

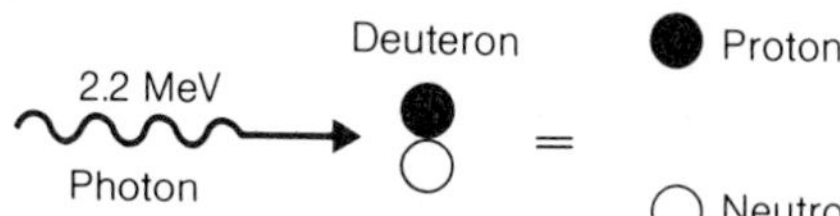

Figure 29.6

The binding energy of the deuteron is 2.2 MeV, which is the energy difference between the deuteron's mass and the combined mass of its constituent neutron and proton. Absorbing 2.2 MeV—for instance, by being struck by a 2.2-MeV gamma-ray photon—gives a deuteron enough additional mass to split into a neutron and a proton.

When calculating binding energies, the masses of neutral atoms and of electrons, protons, and neutrons are used. The ionization energies of the electrons, which are very small compared with nuclear binding energies, usually can be neglected.

Since the "missing" mass in a deuteron (as the deuterium nucleus is called) is 0.0024 u, its binding energy is

$$(0.0024 \text{ u})(931.5 \text{ MeV/u}) = 2.2 \text{ MeV}$$

This figure is confirmed by experiments that show that the minimum energy a photon must have in order to disrupt a deuteron is 2.2 MeV (Fig. 29.6).

Nuclear binding energies are strikingly high. The range for stable nuclei is from 2.2 MeV for $^2_1$H (deuterium) to 1640 MeV for $^{209}_{83}$Bi (an isotope of the metal bismuth). Larger nuclei are all unstable and decay radioactively, as described in Sections 29.5 and 29.6. To appreciate how high binding energies are, we can compare them with more familiar energies in terms of kilojoules of energy per kilogram of mass. In these units a typical binding energy is 8×10^{11} kJ/kg—800 billion kJ/kg. By contrast, to boil water involves a heat of vaporization of a mere 2260 kJ/kg, and even the heat given off by burning gasoline is only 4.7×10^4 kJ/kg, 17 million times smaller.

Binding Energy per Nucleon

The **binding energy per nucleon** of a nucleus is its binding energy divided by the number A of neutrons and protons it contains. Figure 29.7 is a plot of binding energy per nucleon versus mass number A. Except for the high peak for $^4_2$He, the curve is a quite regular one. Nuclei of intermediate size have the highest binding energies per nucleon, which means that their nucleons are held together more securely than the nucleons in both heavier and lighter nuclei. The maximum in the curve is 8.8 MeV/nucleon at $A = 56$, which corresponds to the iron nucleus $^{56}_{26}$Fe.

A remarkable feature of nuclear structure is illustrated by this curve. Suppose that we split the nucleus $^{235}_{92}$U, whose binding energy is 7.6 MeV/nucleon, into two parts. Each fragment will be the nucleus of a much ligher element and therefore will

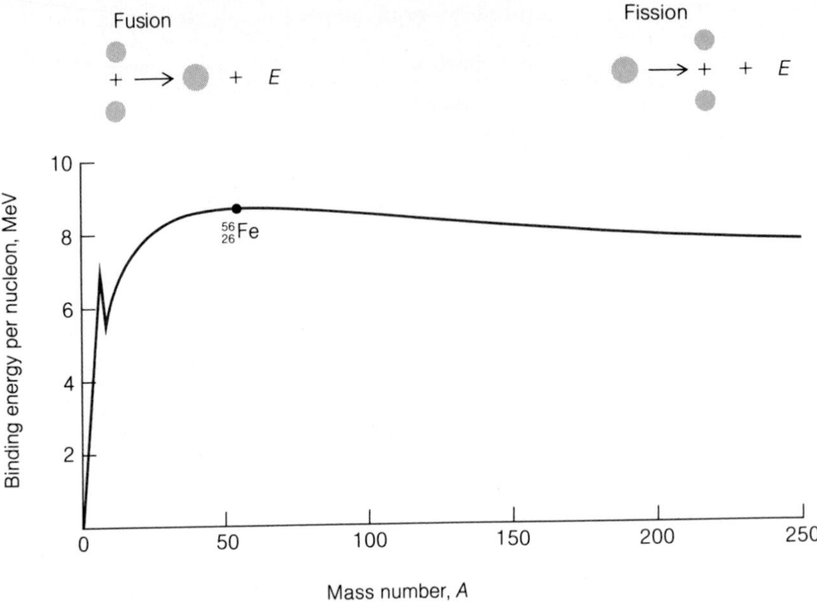

Figure 29.7

Binding energy per nucleon versus mass number. The higher the binding energy per nucleon, the more stable the nucleus. When a heavy nucleus is split into two lighter ones, a process called *fission*, the greater binding energy of the product nuclei causes the liberation of energy. When two very light nuclei join to form a heavier one, a process called *fusion*, the greater binding energy of the product nucleus again causes the liberation of energy.

have a higher binding energy per nucleon than the uranium nucleus. The difference is about 0.8 MeV/nucleon, and so, if such **nuclear fission** were to take place, an energy of

$$\left(0.8\,\frac{\text{MeV}}{\text{nucleon}}\right)(235\text{ nucleons}) = 188\text{ MeV}$$

would be given off per splitting. This is a truly immense amount of energy to be produced in a single atomic event. As a comparison, chemical processes involve energies of the order of magnitude of 1 electron volt per reacting atom, 10^{-8} the energy involved in fission.

Figure 29.7 also shows that if two light nuclei join to form a heavier one, the higher binding energy of the resulting nucleus will also result in the evolution of energy. For instance, if two deuterons combined to make a $^4_2$He nucleus, more than 23 MeV would be released. This process is known as **fusion,** and, together with fission, it promises to be the source of more and more of the world's energy as reserves of fossil fuels are depleted. Nuclear fusion is the means by which the sun and stars obtain their energy.

The graph of Fig. 29.7 has a good claim to being the most significant in all of science. The fact that binding energy exists at all means that nuclei more complex than the single proton of hydrogen can be stable. Such stability in turn accounts for the existence of the various elements and so for the existence of the many and diverse forms of matter around us. Because the curve peaks in the middle, we have the explanation for the energy that powers, directly or indirectly, the evolution of much of the universe—it comes from the fusion of protons and neutrons to form complex nuclei. And the harnessing of nuclear fission in reactors and weapons has irreversibly changed modern civilization.

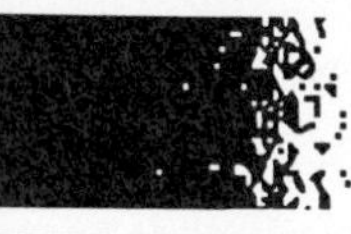

29.4 Strong Nuclear Interaction

The fundamental force that holds nuclei together.

The interaction between nucleons that leads to complex nuclei cannot be electric, because neutrons are uncharged and the positive charges of protons lead to repulsive forces only. It cannot be gravitational, because gravitational forces are far too weak to be able to overcome the repulsive forces between protons. Thus we have a third fundamental interaction, the **strong nuclear interaction,** which is responsible for the existence of atomic nuclei more complex than the single proton of ^1_1H.

Two properties of the strong interaction stand out. First, it is by far the strongest of all the fundamental interactions, as we can tell from nuclear binding energies. To pull apart the neutron and proton in a deuterium nucleus takes 2.22 MeV, but to pull the electron in a deuterium atom away from the nucleus takes only 13.6 eV, more than 100,000 times less work (Fig. 29.8).

The second noteworthy aspect of the strong interaction is its short range. Electrical and gravitational forces fall off with distance as $1/r^2$ and are effective at considerable distances. For example, although the planet Pluto averages 6×10^{12} m from the sun, it is kept in orbit by the gravitational pull of the sun. But nuclear forces operate only over a few nucleon diameters. Up to a separation of about 3×10^{-15} m, the nuclear attraction between two protons is about 100 times stronger than the electric repulsion between them, but beyond this distance the nuclear force dies out rapidly. The nuclear interactions between protons and protons, between protons and neutrons, and between neutrons and neutrons appear to be the same.

Now we can see why the number of stable elements is limited. The larger a nucleus, the stronger the electric repulsion on each of its protons. However, the attractive nuclear forces on each nucleon cannot increase indefinitely, because only a certain number of other nucleons are close enough to interact with it. The largest stable nucleus is the bismuth isotope $^{209}_{83}Bi$, and nuclei larger than the uranium isotope $^{238}_{92}U$ are too unstable to have survived on earth since their formation before the solar system came into being.

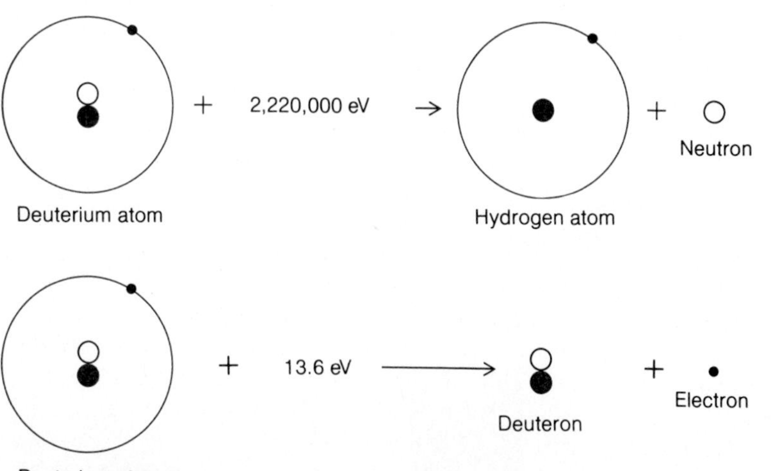

Figure 29.8

Nuclear binding energies are much greater than atomic binding energies.

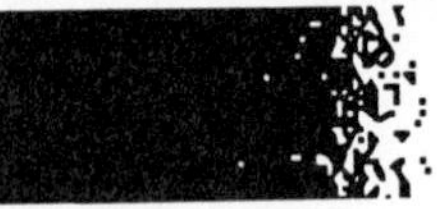

29.5 Radioactive Decay

How unstable nuclei change into stable ones.

Not all atomic nuclei are stable. At the beginning of the twentieth century research by Henri Becquerel, Marie and Pierre Curie, and others, showed that some nuclei spontaneously transform themselves into other nuclear species with the emission of radiation. Such nuclei are said to be **radioactive.**

Unstable nuclei emit three kinds of radiation:

1. **Alpha particles,** which are the nuclei of ${}^{4}_{2}\text{He}$ atoms;
2. **Beta particles,** which are electrons or positrons (positively charged electrons);
3. **Gamma rays,** which are photons of high energy.

The early experimenters identified these radiations with the help of a magnetic field, as in Fig. 29.9. Gamma rays carry no charge and are not affected by the magnetic field. Figure 29.10 indicates the penetrating powers of the different radiations.

To understand why nuclei decay, let us return to Fig. 29.3, which shows the number of neutrons versus the number of protons in stable nuclei. Clearly an element of a given proton number Z has only a very narrow range of possible neutron numbers if it is to be stable.

Suppose now that a nucleus has too many neutrons for stability relative to the number of protons present. If one of the excess neutrons changes into a proton, this will both reduce the number of neutrons and increase the number of protons (Fig. 29.11). To conserve electric charge, such a change means that a negative electron must be emitted. Therefore

$$n^0 \rightarrow p^+ + e^- \qquad \textit{Electron emission} \quad (29.1)$$

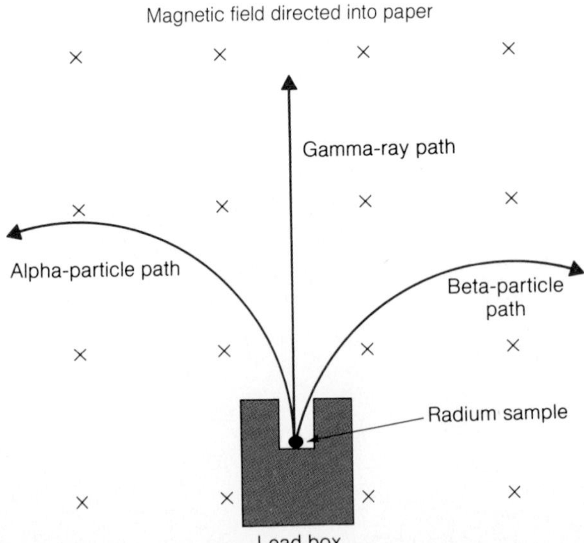

Figure 29.9

The radiations from a radium sample may be analyzed with the help of a magnetic field. In the figure the direction of the field is into the paper; hence the positively charged alpha particles (which are helium nuclei) are deflected to the left and the negatively charged beta particles (which are electrons), to the right. Gamma rays (which are energetic photons) carry no charge and are not affected by the magnetic field.

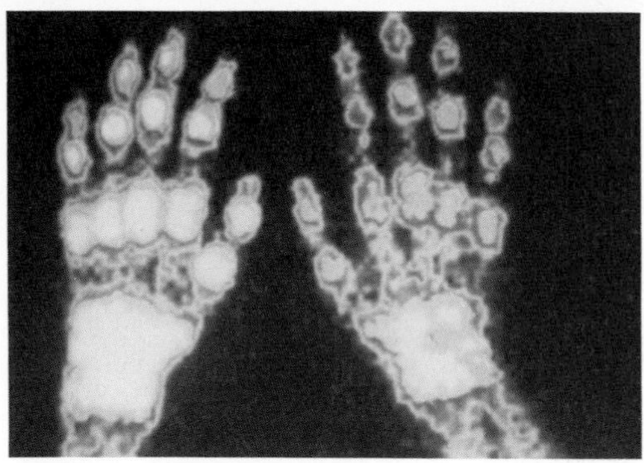

Radionuclides can be traced in living tissue by the radiation they emit. In this image of the hands of a person suffering from rheumatoid arthritis different levels of gamma-ray intensity are shown by different shades of gray. The higher the intensity, the lighter the shade. The radionuclides injected into the bloodstream of the person were absorbed most readily by the inflamed tissue around the arthritic joints. The left hand and wrist are the most severely affected.

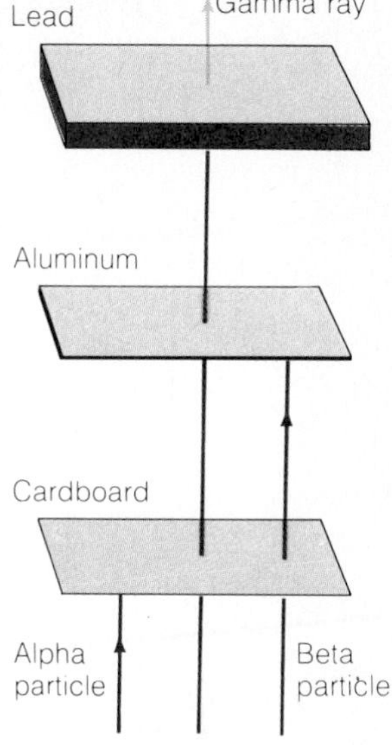

Figure 29.10

Alpha particles are stopped by a piece of cardboard. Beta particles penetrate the cardboard but are stopped by a sheet of aluminum. Even a thick slab of lead may not stop all the gamma rays.

The electron leaves the nucleus and is detectable as a "beta particle." The final nucleus may be left with extra energy due to its shifted binding energy, and this energy is given off in the form of gamma rays. Sometimes more than one such **beta decay** is needed for a particular unstable nucleus to become stable. (In Section 30.2 we will see that this is not the whole story of beta decay.)

Should the nucleus have too few neutrons, the inverse process

$$p^+ \rightarrow n^0 + e^+ \qquad \textit{Positron emission} \quad (29.2)$$

may take place. Here a proton becomes a neutron with the emission of a **positron** (positive electron). This is also called beta decay, because it resembles the emission of negative electrons.

A process that competes with positron emission is the capture by a nucleus with too small a neutron/proton ratio of one of the electrons in its innermost atomic shell. The electron is absorbed by a nuclear proton, which then becomes a neutron.

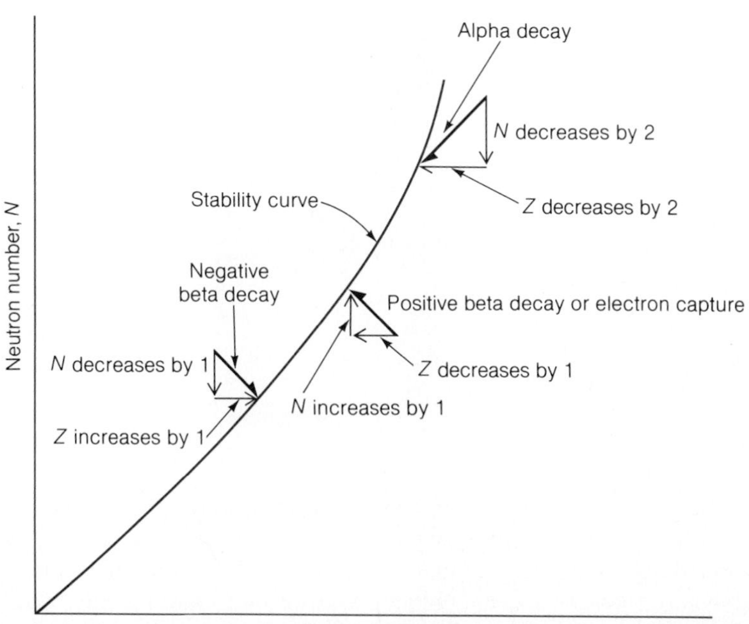

Figure 29.11

How alpha and beta decays tend to bring an unstable nucleus to a stable configuration.

Table 29.3 Radioactive decay. The asterisk (*) denotes an excited nuclear state and γ denotes a gamma-ray photon.

Decay	Nuclear transformation	Example
Alpha	${}^{A}_{Z}X \rightarrow {}^{A-4}_{Z-2}Y + {}^{4}_{2}\text{He}$	${}^{238}_{92}\text{U} \rightarrow {}^{234}_{90}\text{Th} + {}^{4}_{2}\text{He}$
Electron emission	${}^{A}_{Z}X \rightarrow {}_{Z+1}{}^{A}Y + e^-$	${}^{14}_{6}\text{C} \rightarrow {}^{14}_{7}\text{N} + e^-$
Positron emission	${}^{A}_{Z}X \rightarrow {}_{Z-1}{}^{A}Y + e^+$	${}^{64}_{29}\text{Cu} \rightarrow {}^{64}_{28}\text{Ni} + e^+$
Electron capture	${}^{A}_{Z}X + e^- \rightarrow {}_{Z-1}{}^{A}Y$	${}^{64}_{29}\text{Cu} + e^- \rightarrow {}^{64}_{28}\text{Ni}$
Gamma	${}^{A}_{Z}X^* \rightarrow {}^{A}_{Z}X + \gamma$	${}^{87}_{38}\text{Sr}^* \rightarrow {}^{87}_{38}\text{Sr} + \gamma$

This process can be expressed as

$$p^+ + e^- \rightarrow n^0 \qquad \textit{Electron capture} \quad (29.3)$$

Electron capture does not lead to the emission of a particle. However, it can be detected by the X-ray photon produced when one of the atom's outer electrons falls into the vacancy left by the absorbed electron.

Another way certain nuclei can achieve stability is by **alpha decay,** in which an alpha particle consisting of two neutrons and two protons is emitted. Thus negative beta decay increases the number of protons by one and decreases the number of neutrons by one; positive beta decay and electron capture decrease the number of protons by one and increase the number of neutrons by one; and alpha decay decreases both the number of protons and the number of neutrons by two. These processes are summarized in Table 29.3 and shown schematically in Fig. 29.12.

BIOGRAPHICAL NOTE

Marie Sklodowska Curie

Marie Sklodowska Curie (1867–1934) was born in Poland, at that time under the oppressive domination of Russia. Following high school, she worked as a governess until she was 24 so that she could study science in Paris, where she had barely enough money to survive. In 1894 Marie married Pierre Curie, who was eight years older than she and already a noted physicist. In 1897, just after the birth of her first daughter Irene (who was to win a Nobel Prize in physics herself in 1935), Marie began to investigate the newly discovered phenomenon of radioactivity—her word—for her doctoral thesis.

The year before, Antoine Becquerel (1852–1908) had found that uranium emitted a mysterious radiation. Marie, after a search of all the known elements, learned that thorium did so as well. She then examined various minerals for radioactivity. Her studies showed that the uranium ore pitchblende was far more radioactive than its uranium content would suggest. Marie and Pierre together went on to identify first polonium, named for her native Poland, and then radium as the sources of the additional activity. With the primitive facilities that were all they could afford (they had to use their own money), they had succeeded by 1902 in purifying a tenth of a gram of radium from several tons of ore, a task that involved immense physical as well as intellectual labor.

Together with Becquerel, the Curies shared the 1903 Nobel Prize in physics. Pierre ended his acceptance lecture with these words: "One may also imagine that in criminal hands radium might become very dangerous, and here one may ask if humanity has anything to gain by learning the secrets of nature, if it is ready to profit from them, or if this knowledge is not harmful. . . . I am among those who think . . . that humanity will obtain more good than evil from the new discoveries."

In 1906 Pierre was struck and killed by a horse-drawn carriage in a Paris street. Marie continued work on radioactivity, still in an inadequate laboratory, and won the Nobel Prize in chemistry in 1911. Not until her scientific career was near an end did the French government provide her with proper research facilities. Even before Pierre's death, both Curies had suffered from ill health because of their exposure to radiation, and much of Marie's later life was marred by radiation-induced ailments, from which she died. The destructive effects of radiation had a good side, however, and not long after its isolation, radium became widely used to control certain cancers.

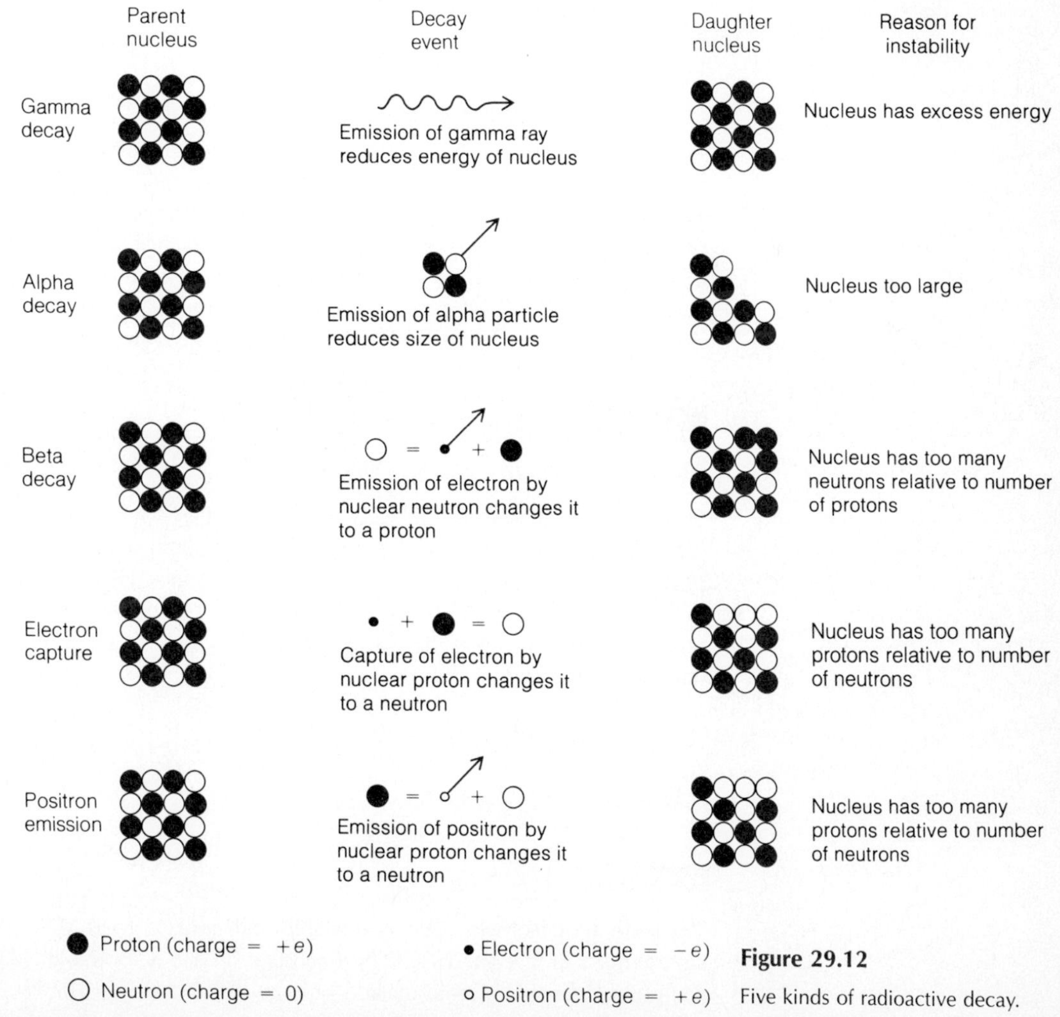

Figure 29.12

Five kinds of radioactive decay.

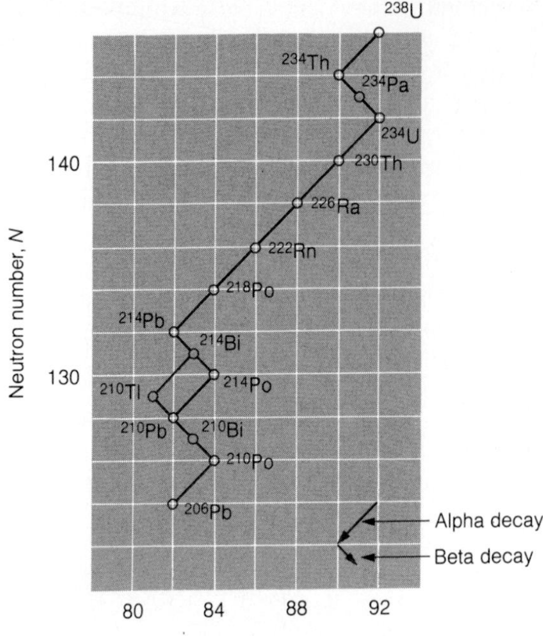

Figure 29.13

The uranium isotope $^{238}_{92}$U undergoes 14 alpha and beta decays to reach the stable lead isotope $^{206}_{82}$Pb. The decay of $^{214}_{83}$Bi can proceed either by alpha emission and then beta emission or in the reverse order.

Often a succession of alpha and beta decays, plus gamma decays to carry off excess energy, is needed before a nucleus reaches stability. An example of such a **decay series** is shown in Fig. 29.13. Here the unstable uranium isotope $^{238}_{92}$U becomes the stable lead isotope $^{206}_{82}$Pb in 14 steps. The series branches at $^{214}_{83}$Bi, which can decay either by alpha emission followed by beta emission or in the reverse order.

EXAMPLE 29.2

The nucleus $^{6}_{2}$He is unstable. What kind of decay would you expect it to undergo?

SOLUTION The most stable helium nucleus is $^{4}_{2}$He, as shown by its position on the binding-energy curve of Fig. 29.7 where it is responsible for the peak at the left of the curve. Because $^{6}_{2}$He has four neutrons whereas $^{4}_{2}$He has only two, the instability of $^{6}_{2}$He must arise from an excess of neutrons. Thus it would seem likely that $^{6}_{2}$He undergoes negative beta decay to become $^{6}_{3}$Li, whose neutron/proton ratio is more consistent with stability:

$$^{6}_{2}\text{He} \rightarrow {}^{6}_{3}\text{Li} + e^{-}$$

This is in fact the manner in which $^{6}_{2}$He decays. ♦

EXAMPLE 29.3

The polonium isotope $^{210}_{84}$Po is unstable and emits a 5.30-MeV alpha particle. The atomic mass of $^{210}_{84}$Po is 209.9829 u and that of $^{4}_{2}$He is 4.0026 u. Identify the daughter nuclide and find its atomic mass.

SOLUTION (a) The daughter nuclide will have an atomic number of 84 − 2 = 82 and a mass number of 210 − 4 = 206. From Table C.1 in the Appendix we see that $Z = 82$ corresponds to lead, so the symbol of the daughter nuclide is ${}^{206}_{82}\text{Pb}$.

(b) The mass equivalent of 5.30 MeV is

$$m_E = \frac{5.30 \text{ MeV}}{931.5 \text{ MeV/u}} = 0.0057 \text{ u}$$

The mass lost by ${}^{210}_{84}\text{Po}$ in its decay equals the atomic mass of the alpha particle ${}^{4}_{2}\text{He}$ plus m_E, which is (4.0026 + 0.0057) u = 4.0083 u. Hence

$$\text{Mass of } {}^{210}_{84}\text{Po} - (\text{mass of } {}^{4}_{2}\text{He} + m_E) = \text{mass of } {}^{206}_{82}\text{Pb}$$

$$209.9829 \text{ u} - 4.0083 \text{ u} = 205.9746 \text{ u}$$

◆

Weak Interaction

The strong interaction that holds nucleons together to form nuclei cannot account for beta decay, positron emission, and electron capture. Another fundamental interaction turns out to be responsible: the **weak interaction.** Its range is so short ($\sim 10^{-17}$ m) that it operates only *within* certain elementary particles and leads to their transformation into other particles. The four fundamental interactions—gravitational, electromagnetic, strong, and weak—are discussed further in the next chapter. The name "weak interaction" arose because the other short-range force that affects nucleons is stronger. The gravitational interaction is actually weaker than the weak interaction.

29.6 Alpha Decay

Tunneling out of a nucleus.

Nuclei that contain more than 209 nucleons are so large that the short-range forces holding them together are barely able to overcome the long-range electrical repulsive forces of their protons. Such a nucleus can reduce its bulk and become more stable by emitting an alpha particle, which decreases its mass number A by 4.

Why do heavy nuclei give off alpha particles instead of, for example, individual protons or ${}^{3}_{2}\text{He}$ nuclei? The reason is the high binding energy of the alpha particle. As a result, an alpha particle has significantly less mass than four separate nucleons. Because of this small mass, an alpha particle can be ejected by a heavy nucleus with energy to spare. Thus the alpha particle released in the decay of ${}^{232}_{92}\text{U}$ has a kinetic energy of 5.4 MeV, whereas 6.1 MeV would have to be supplied from the outside for this nucleus to release a proton and 9.6 MeV to release a ${}^{3}_{2}\text{He}$ nucleus.

Even though alpha decay may be energetically possible in a particular nucleus, it is not obvious just how the alpha particle is able to break away from the nuclear forces that bind it to the rest of the nucleus. Typically an alpha particle has available about 5 MeV of energy with which to escape. However, an alpha particle near the nucleus but outside the range of its nuclear forces has an electric PE of perhaps 25 MeV. That is, if released from here it will have a KE of 25 MeV when

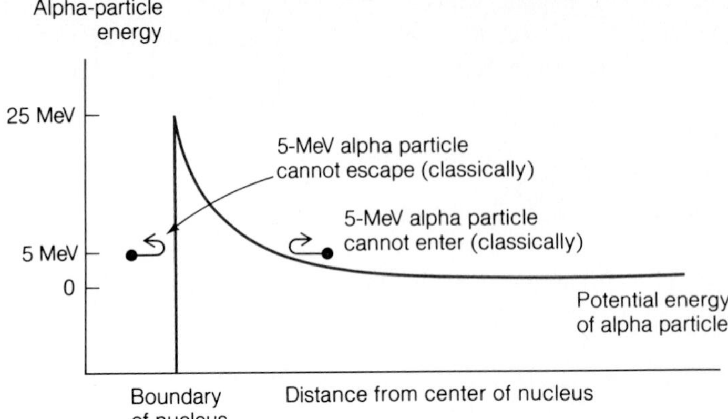

Figure 29.14

The variation in alpha-particle potential energy near a typical heavy nucleus. The alpha particle in this nucleus has a kinetic energy of 5 MeV.

it is infinitely far away as a result of electric repulsion (Fig. 29.14). An alpha particle inside the nucleus therefore should need at least 25 MeV, five times more than is available, in order to break loose.

The alpha particle, then, is in a box whose walls are of such a height that an energy of 25 MeV is needed to climb over them, while the particle itself has only 5 MeV for the purpose.

Quantum mechanics provides the answer to the paradox of alpha decay. Two assumptions are needed: (1) An alpha particle can exist as such inside a nucleus, and (2) it is in constant motion there.

According to quantum theory, a moving particle has a wave character, so the proper classical analog of an alpha particle in a nucleus is a light wave trapped between mirrors and not a particle bouncing back and forth between solid walls. Now in order for a light wave to be reflected from a mirror, it must penetrate the reflecting surface for a short distance (Fig. 29.15). If the mirror is thick, all of the incident light is reflected. If the mirror is very thin, however, some of the incident light can pass right through the mirror, as shown. The theory of this partial transmission, with only minor changes, accounts nicely for alpha decay. The very existence of alpha decay, in fact, further confirms quantum ideas, because the principles of physics that follow from Newton's laws of motion forbid such decay.

Of course, a 25-MeV energy barrier is not very "transparent" to a 5-MeV alpha particle. A large nucleus might be 1.5×10^{-14} m across, and an alpha particle inside it might go back and forth at 2×10^{7} m/s. Hence the alpha particle knocks at the nuclear wall nearly 10^{21} times per second but may have to wait as much as 10^{10} years to escape from certain nuclei.

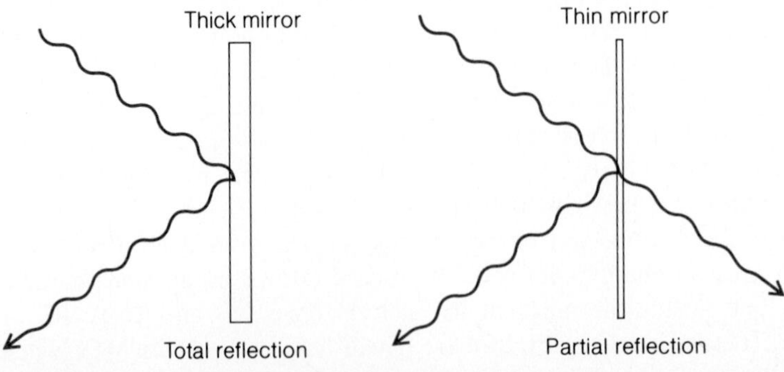

Figure 29.15

In reflection, a wave penetrates the reflecting surface for a short distance and may pass through it if the mirror is sufficiently thin.

The quantum-mechanical penetration of a barrier is known as the **tunnel effect** because the particle escapes *through* the barrier and not over it.

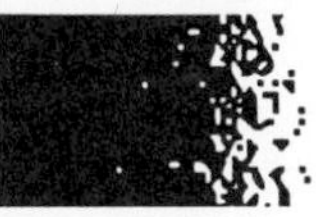

29.7 Half-Life

The shorter the half-life, the more likely a given nucleus will decay in a given time.

The rate R at which a sample of a radioactive nuclide decays is called its **activity.** If ΔN nuclei decay in the time Δt, then

$$R = \frac{\Delta N}{\Delta t} \qquad \textit{Activity} \quad (29.4)$$

The SI unit of activity is named after the French physicist Henri Becquerel, who discovered radioactivity in 1896:

$$1 \text{ becquerel} = 1 \text{ Bq} = 1 \text{ event/s}$$

The traditional unit of activity, still in common use, is the **curie.** Originally the curie was defined as the activity of 1 g of radium ($^{226}_{88}$Ra), and its precise value accordingly changed as measuring techniques improved. For this reason the curie is now defined arbitrarily as

$$1 \text{ curie} = 3.70 \times 10^{10} \text{ events/s} = 37 \text{ GBq}$$

The activity of 1 g of radium is a few percent smaller. A luminous watch dial contains several microcuries of $^{226}_{88}$Ra; ordinary potassium has an activity of about 1 millicurie/kg owing to the presence of the radioactive isotope $^{40}_{19}$K; "cobalt-60" sources of 1 or more curies are widely used in medicine for radiation therapy and industrially for the inspection of metal castings and welded joints.

One of the characteristics of all types of radioactivity is that the rate at which the nuclei in a given sample decay always follows a curve whose shape is like that shown in Fig. 29.16. If we start with a sample whose activity is, say, 100 Bq, it will not continue to decay at that rate but instead fewer and fewer disintegrations will occur in each successive second.

Some radioactive nuclides decay faster than others, but in each case a certain definite time is required for half of an original sample to decay. This time is called the **half-life** of the nuclide. For instance, the radon isotope $^{222}_{86}$Rn undergoes alpha decay to the polonium isotope $^{218}_{84}$Po with a half-life of 3.8 days. Should we start with 1 mg of radon in a closed container (since it is a gas), $\frac{1}{2}$ mg will remain undecayed after 3.8 days; $\frac{1}{4}$ mg will remain undecayed after 7.6 days; $\frac{1}{8}$ mg will remain undecayed after 11.4 days; and so on (Fig. 29.17).

As we would expect, the smaller the half-life $t_{1/2}$ of a nuclide and the greater the number N_0 of its nuclei in a sample, the higher the activity of the sample. The exact relationship turns out to be

$$R = \frac{\Delta N}{\Delta t} = \frac{0.693}{t_{1/2}} N_0 \qquad \textit{Activity} \quad (29.5)$$

Figure 29.16

The rate at which a sample of a radioactive substance decays is not constant but varies with time in the manner shown in the curve.

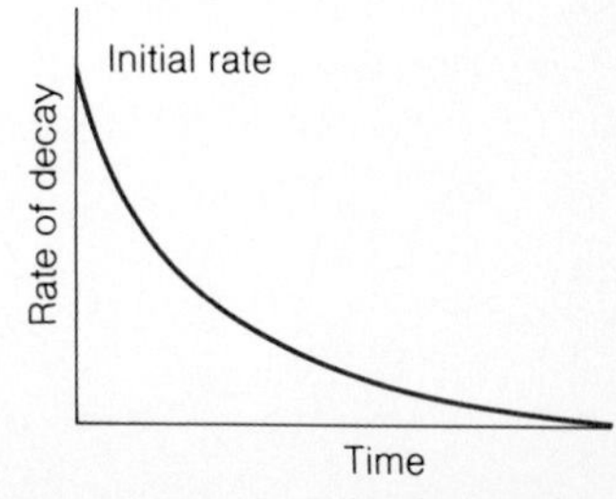

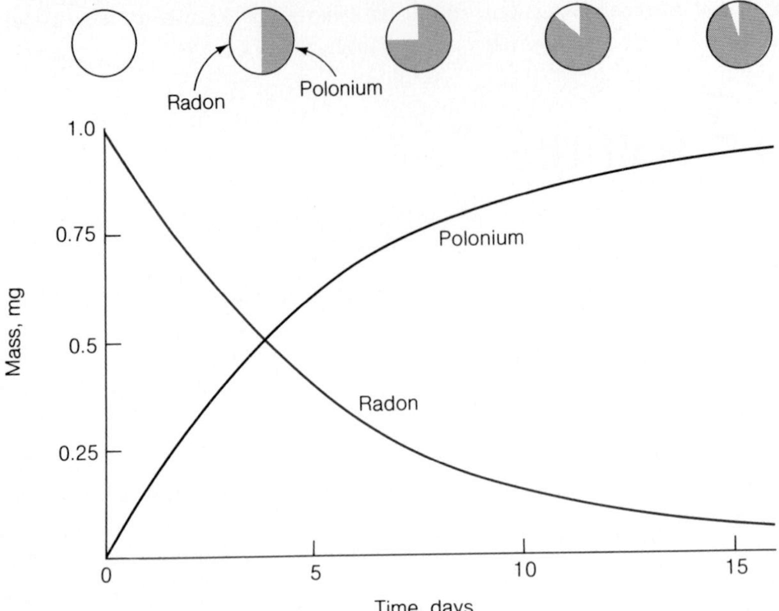

Figure 29.17

The alpha decay of radon ($^{222}_{86}Rn$) to polonium ($^{218}_{84}Po$) has a half-life of 3.8 days. The sample of radon whose decay is graphed here had an initial mass of 1.0 mg.

At a later time t the number of undecayed nuclei will have decreased to N, where

$$N = \frac{N_0}{2^{t/t_{1/2}}} \qquad \textit{Radioactive decay} \quad (29.6)$$

Thus after $t = 1$ half-life, $t/t_{1/2} = 1$ and $N = N_0/2$; after 2 half-lives, $t/t_{1/2} = 2$ and $N = N_0/2^2 = N_0/4$; after 3 half-lives, $t/t_{1/2} = 3$ and $N = N_0/2^3 = N_0/8$; and so on. This is just what was given above for the decay of radon. As N decreases, so does R.

For times that are not multiples of $t_{1/2}$, the calculation of $2^{t/t_{1/2}}$ can be made by using logarithms (see Appendix B.5). Since $\log a^b = b \log a$ in general,

$$\log 2^{t/t_{1/2}} = \frac{t}{t_{1/2}} \log 2 = 0.301 \frac{t}{t_{1/2}}$$

and we have

$$2^{t/t_{1/2}} = \log^{-1}\left(0.301 \frac{t}{t_{1/2}}\right)$$

where $\log^{-1}$ signifies the antilogarithm. To find the antilogarithm of a number x on an electronic calculator, simply enter the number and then press the [10^x] or [INV LOG] key.

EXAMPLE 29.4

The hydrogen isotope tritium, ^3_1H, is radioactive and emits an electron with a half-life of 12.5 years. (a) What does ^3_1H become after beta decay? (b) What percentage of an original sample of tritium will remain undecayed 5 years after its preparation?

SOLUTION (a) When a nucleus emits an electron, its atomic number increases by 1 (corresponding to an increase in nuclear charge of $+e$) and its mass number is unchanged. Helium has the atomic number 2, and so

$$^{3}_{1}\text{H} \rightarrow {}^{3}_{2}\text{He} + e^{-}$$

(b) Here $t/t_{1/2} = (5 \text{ years}/12.5 \text{ years}) = 0.4$ and

$$2^{t/t_{1/2}} = 2^{0.4} = \log^{-1}(0.301 \times 0.4) = \log^{-1} 0.120 = 1.32$$

Hence from Eq. (29.6)

$$\frac{N}{N_0} = \frac{1}{2^{t/t_{1/2}}} = \frac{1}{1.32} = 0.758 = 75.8\%$$

of the original amount of tritium is left after 5 years. The activity of the sample will also have fallen to 75.8% of its original value after 5 years. ◆

EXAMPLE 29.5

Find the activity of 1.0 mg of radon, $^{222}_{86}\text{Rn}$, whose half-life is 3.8 days.

SOLUTION The mass of an atom is very close to its mass number expressed in atomic mass units. Hence the number N_0 of atoms in 1.0 mg of radon is

$$N_0 = \frac{1.0 \times 10^{-6}\,\text{kg}}{(222\,\text{u/atom})(1.66 \times 10^{-27}\,\text{kg/u})} = 2.7 \times 10^{18}\,\text{atoms}$$

The number of nuclei is the same, of course. The half-life of radon in seconds, which is what we need here, is

$$t_{1/2} = (3.8\,\text{days})(86{,}400\,\text{s/day}) = 3.3 \times 10^{5}\,\text{s}$$

From Eq. (29.5) the activity of the sample is

$$R = \frac{0.693}{t_{1/2}} N_0 = \frac{(0.693)(2.7 \times 10^{18})}{3.3 \times 10^{5}\,\text{s}} = 5.7 \times 10^{12}\,\text{Bq} = 5.7\,\text{TBq}$$

This is the number of alpha decays per second that occur in 1 mg of radon. ◆

Half-lives range from billionths of a second to billions of years. Samples of radioactive isotopes decay in the manner shown in Fig. 29.16 because a great many nuclei are involved, each with a certain probability of decaying. The fact that radon has a half-life of 3.8 days signifies that every radon nucleus has a 50% chance of decaying in any 3.8-day period. Because a nucleus does not have a memory, this does *not* mean that a radon nucleus has a 100% chance of decaying in 7.6 days: The likelihood of decay of a given nucleus stays the same until it actually does decay. Thus a half-life of 3.8 days means a 75% probability of decay in 7.6 days, an 87.5% probability of decay in 11.4 days, and so on, because in each interval of 3.8 days the probability is 50%.

This discussion suggests that radioactive decay involves events that take place within individual nuclei, rather than collective processes that involve more than one nucleus in interaction. This idea is confirmed by experiments that show that the half-life of a particular isotope is invariant under changes of pressure, temperature, electric and magnetic fields, and so on, which might, if strong enough, affect internuclear phenomena.

29.8 Radiometric Dating

A clock based on radioactivity.

Radioactivity makes it possible to find the ages of many geological and biological specimens. The ratio between the amounts of a radionuclide and of its stable daughter in a specimen depends on the half-life of the nuclide and the age of the specimen. Let us see how **radiocarbon,** the beta-active carbon isotope $^{14}_{6}C$, can be used to date objects of biological origin.

Cosmic rays are high-energy atomic nuclei, mainly protons, that circulate through the Milky Way galaxy of which the sun is a member. About 10^{18} of them arrive at the earth's atmosphere each second, where they disrupt the nuclei of atoms they encounter to produce showers of secondary particles. Among these secondaries are neutrons that can react with nitrogen nuclei in the atmosphere to form radiocarbon with the emission of a proton:

$$^{14}_{7}N + ^{1}_{0}n \rightarrow ^{14}_{6}C + ^{1}_{1}H$$

The proton picks up an electron and becomes a hydrogen atom. The $^{14}_{6}C$ isotope has too many neutrons for stability and beta-decays to become $^{14}_{7}N$ with a half-life of 5760 years:

$$^{14}_{6}C \rightarrow ^{14}_{7}N + e^{-}$$

About 90 tons of radiocarbon is distributed around the world, an amount kept constant by the cosmic-ray bombardment.

Shortly after their formation, radiocarbon atoms combine with oxygen molecules to form carbon dioxide (CO_2) molecules. Green plants convert water and carbon dioxide into carbohydrates by photosynthesis, which means that every plant contains some radiocarbon. Animals eat plants, and so they too contain radiocarbon. Thus every living thing on earth has a very small proportion of radiocarbon in its tissues. Because the mixing of radiocarbon is relatively efficient, living plants and animals all have the same ratio of radiocarbon to ordinary carbon, which is $^{12}_{6}C$.

When a plant or animal dies, it no longer takes in radiocarbon, and the radiocarbon already present continues to decay. After 5760 years a dead organism has left only half as much radiocarbon as it had while alive, after 11,520 years only a quarter as much, and so on. The ratio of radiocarbon to ordinary carbon therefore provides the ages of ancient objects and remains of organic origin (Fig. 29.18). This technique permits the dating of mummies, wooden implements, cloth, leather, charcoal from campfires, and similar artifacts from ancient civilizations as much as 70,000 years old, about 12 half-lives of $^{14}_{6}C$.

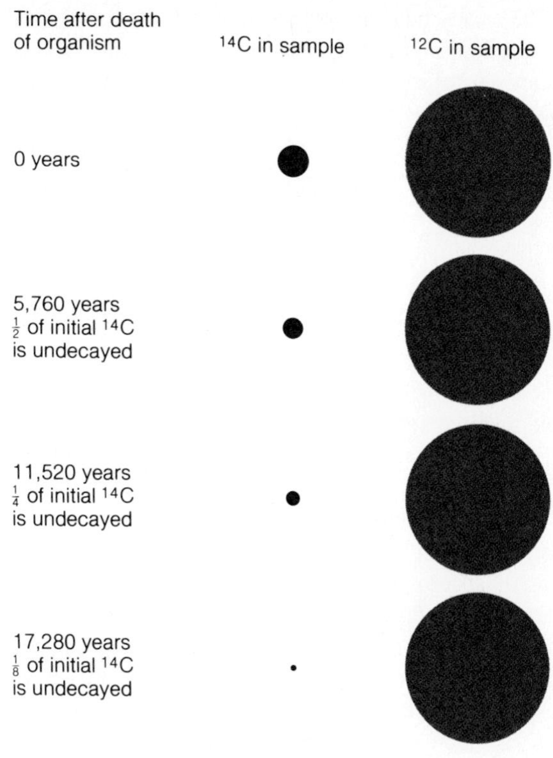

Figure 29.18

The radioactive $^{14}_{6}C$ content of a specimen of plant or animal tissue decreases steadily, whereas its $^{12}_{6}C$ content does not change. Hence the $^{14}_{6}C/^{12}_{6}C$ ratio indicates the time that has passed since the organism died. The half-life of $^{14}_{6}C$ is 5760 years.

Because the earth's history goes back 4.5 or so billion years, geologists use radioactive nuclides of much longer half-lives than that of radiocarbon to date rocks (Table 29.4). The oldest rocks whose ages have been found in this way are in Greenland and are believed to have solidified 3.8 billion years ago. Meteorites and lunar rocks as well as terrestrial rocks have been dated by the methods of Table 29.4.

Most of the energy responsible for the geological history of the earth can be traced to the decay of the radionuclides it contains. The earth is thought to have come into being about 4.5 billion years ago as a cold aggregate of smaller bodies of metallic iron and silicate minerals that had been orbiting the sun. Heat of radioactive origin accumulated inside the infant earth and in time led to partial melting. Gravity then caused the iron to migrate inward to form a molten core; the geomagnetic field comes from electric currents in this core. The lighter minerals rose to form the mantle and crust (see Fig. 6.21). Most of the earth's radioactivity is now concentrated in its upper mantle and crust, where its heat escapes and cannot collect to remelt the earth.

Table 29.4 Geological dating methods

Method	Parent nuclide	Daughter nuclide	Half-life (billion years)
Potassium-argon	$^{40}_{19}K$	$^{40}_{18}Ar$	1.3
Rubidium-strontium	$^{87}_{37}Rb$	$^{87}_{38}Sr$	47
Thorium-lead	$^{232}_{90}Th$	$^{208}_{82}Pb$	14
Uranium-lead	$^{235}_{92}U$	$^{207}_{82}Pb$	0.7
Uranium-lead	$^{238}_{92}U$	$^{206}_{82}Pb$	4.5

The steady outward stream of heat is what powers the geological processes that cause volcanoes and earthquakes, raise mountains, and shift the giant plates into which the earth's surface is divided to produce the drift of the continents.

29.9 Radiation Hazards

Invisible but dangerous.

Like X rays, the various radiations from radioactive nuclei ionize matter through which they pass. All ionizing radiation is harmful to living tissue, although if the damage is slight, the tissue can often repair itself with no permanent effect. It is easy to underestimate radiation hazards because there is usually a delay, sometimes of many years, between an exposure and some of its possible consequences, which include cancer, leukemia, and genetic changes that may lead to children handicapped in various ways.

Radiation dosage is measured in *sieverts* (Sv), where 1 Sv is the amount of any radiation that has the same biological effects as that produced when 1 kg of body tissue absorbs 1 joule of X or gamma rays. (An older unit, the *rem,* is equal to 0.01 Sv.) Although radiobiologists disagree about the exact relationship between radiation exposure and the likelihood of developing cancer, there is no question that such a link exists. Natural sources of radiation lead to a dosage rate per person of about 3 mSv/y averaged over the U.S. population (1 mSv = 0.001 Sv). Other sources of radiation add 0.6 mSv/y, with medical X rays prominent. The total is about 3.6 mSv/y, about the dose received from 25 chest X rays.

Figure 29.19 shows the relative contributions to the radiation dosage received by an average person in the United States. The most important single source is the radioactive gas radon, a decay product of radium whose own origin traces back to the decay of uranium. Uranium is found in many common rocks, notably granite. Hence radon, colorless and odorless, is present nearly everywhere, though usually in

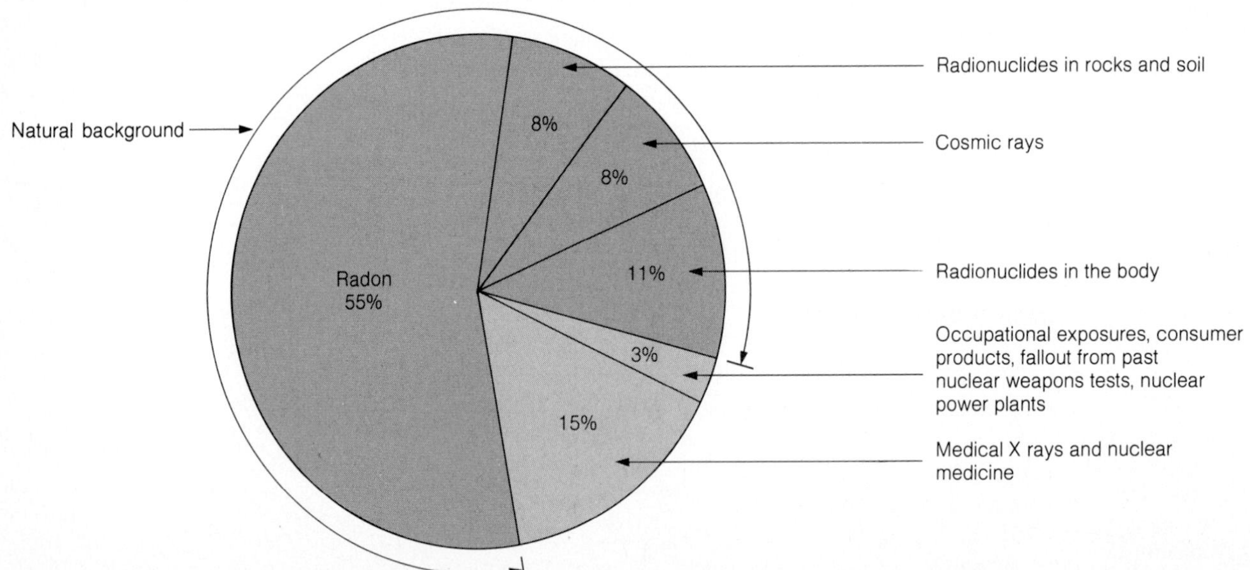

Figure 29.19

Sources of radiation dosage for an average person in the United States. The total is about equivalent to 25 chest X rays per year. Actual dosages vary widely. For instance, radon concentrations are not the same everywhere; some people receive more medical X rays than others; cosmic rays are more intense at high altitudes; and so on. Nuclear power stations are responsible for only 0.08% of the total, although accidents can raise the amount in affected areas to dangerous levels.

amounts too small to endanger health. Problems arise when houses are built in uranium-rich regions, because it is impossible to prevent radon from entering such houses from the ground under them. Surveys show that millions of American homes have radon concentrations high enough to pose a small but definite cancer risk. As a cause of lung cancer, radon is second only to cigarette smoking. Of the various approaches to reducing radon levels in an existing house in a hazardous region, the most effective seems to be to extract air from underneath the ground floor and disperse it into the atmosphere before it can enter the house.

Other natural sources of radiation dosage include cosmic rays and radionuclides in rocks and soil. The human body itself contains small amounts of radioisotopes of such elements as potassium and carbon.

Many useful processes involve ionizing radiation. Some employ such radiation directly, as in the X rays and gamma rays used in medicine and industry. Smoke detectors and false teeth are among the consumer products that contain traces of radionuclides. Sometimes the radiation is unwanted but inescapable, notably in the operation of nuclear reactors and in the disposal of their wastes. It is not always easy to find an appropriate balance between risk and benefit where radiation is concerned. This seems particularly true for medical X rays, which constitute the second largest source of radiation dosage for an average person in the United States.

It seems to be an unfortunate fact that some X-ray exposures are made for no strong reason and do more harm than good. In this category are "routine" chest X rays upon hospital admission, "routine" X rays as part of regular physical examinations, and "routine" dental X rays. The once-routine mass screening by X ray of symptomless young women for breast cancer is now generally thought to have increased, not decreased, the overall death rate due to cancer. Especially dangerous is the X-raying of pregnant women, until recently another "routine" procedure, which dramatically increases the chances of cancer in their children.

Since the carcinogenic properties of X rays have been known since 1902, all this is hard to excuse. X rays unquestionably are valuable in medicine. The point is that every exposure should have a definite justification that outweighs the risk.

29.10 Nuclear Fission

Divide and conquer.

As we saw earlier, a great deal of binding energy will be released if we can break a large nucleus into smaller ones. But nuclei are usually not at all easy to split apart. What we need is a way to disrupt a heavy nucleus without using more energy than we get back from the process.

The solution came in 1938 with the discovery that absorbing a neutron causes a nucleus of the uranium isotope ${}^{235}_{92}U$ to undergo fission. It is not the impact of the neutron that has this effect. What happens is that the ${}^{235}_{92}U$ nucleus becomes ${}^{236}_{92}U$, which is so unstable that it explodes almost at once into two **fission fragments** (Fig. 29.20). Later several other heavy nuclides were found to be fissionable in similar processes.

Because stable light nuclei have proportionately fewer neutrons than do heavy nuclei, the fragments are unbalanced when they are formed and release one or two neutrons each. Usually the fragments are still unstable and undergo radioactive decay to achieve appropriate neutron-proton ratios. The products of fission, such as the

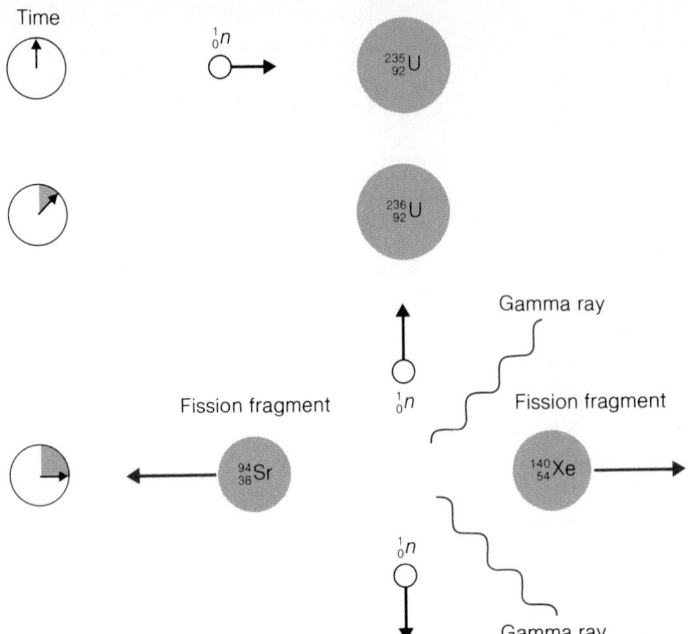

Figure 29.20

In nuclear fission an absorbed neutron causes a heavy nucleus to split into two parts, with the emission of several neutrons and gamma rays.

wastes from a nuclear reactor and the fallout from a nuclear bomb burst, are accordingly highly radioactive. Since some of these products have long half-lives, their radioactivity will persist for many generations and their disposal remains a problem without a really satisfactory solution as yet.

A variety of nuclides may appear as fission fragments. A typical fission reaction is that shown in Fig. 29.20:

$$^{235}_{92}\text{U} + {}^1_0n \rightarrow {}^{236}_{92}\text{U} \rightarrow {}^{140}_{54}\text{Xe} + {}^{94}_{38}\text{Sr} + {}^1_0n + {}^1_0n + 200\text{ MeV} \qquad (29.7)$$

About 84% of the total energy liberated during fission appears as kinetic energy of the fission fragments, about 2.5% as kinetic energy of the neutrons, and about 2.5% in the form of instantaneous emitted gamma rays, with the remaining 11% being given off in the decay of the fission fragments.

Because each fission event liberates neutrons, a self-sustaining series of fissions is possible (Fig. 29.21). For such a **chain reaction** to occur, at least one neutron produced in each fission must, on the average, cause another fission. If too few neutrons initiate fissions, the reaction will slow down and stop. If exactly one neutron per fission causes another fission, energy will be given off at a constant rate, which is the case in a **nuclear reactor.** If the rate of fissioning increases, the energy release will be so rapid that an explosion will occur, which is the case in an **atomic bomb.** If two neutrons from each fission induce further fissions and take 10^{-8} s to do so, a chain reaction starting with a single fission will give off 2×10^{13} J of energy in less than 10^{-6} s.

The destructiveness of nuclear weapons does not stop with their detonation but continues long afterward through the radioactive debris that is produced and dispersed. Tens of thousands of such weapons now exist in a number of countries. The explosion of even a fraction of them might, by causing dust clouds that block

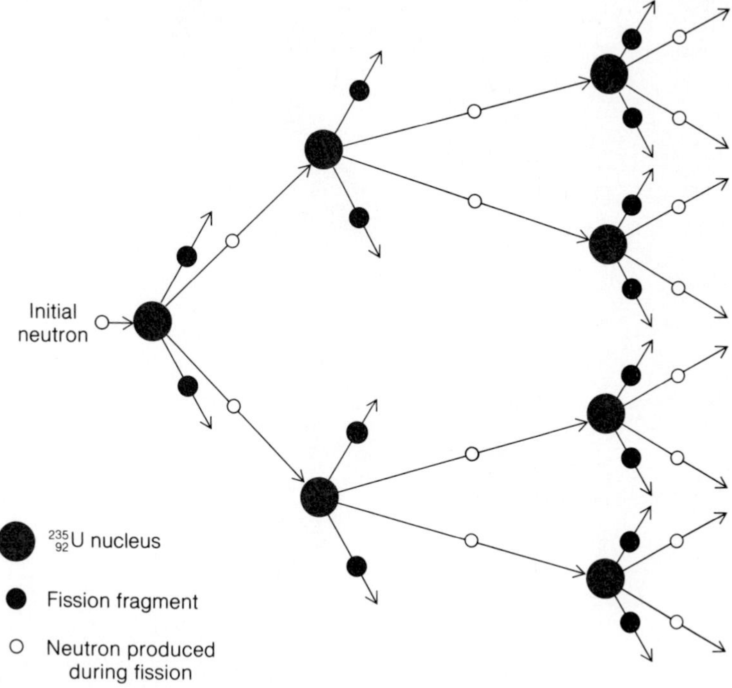

Figure 29.21

Sketch of a chain reaction. If at least one neutron from each fission on the average induces another fission, the reaction is self-sustaining. If more than one neutron per fission on the average induces another fission, the reaction is explosive.

sunlight, lead to months of darkness and bitter cold that could doom many of those living things that survived the blasts themselves and the radioactive fallout from them.

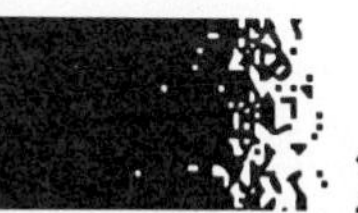

29.11 Nuclear Reactors

From uranium to electricity.

A nuclear reactor is a very efficient source of energy. An output of about 1 MW (1000 kW) is produced by the fission of 1 g of $^{235}_{92}U$ per day, as compared with the burning of 2.6 tons of coal per day for this output in a conventional power plant.

The energy liberated in a nuclear reactor appears as heat, and it is extracted by circulating a coolant liquid or gas. The hot coolant can then be used as the heat source of a conventional steam turbine, which in turn may power an electric generator, a ship, or a submarine. In 1990 over 500 reactors in 26 countries were generating about 200,000 MW of electric power—equivalent to nearly 10 million barrels of oil per day. A number of countries, such as France, obtained over half their electricity from reactors in that year. In the United States in 1990, nuclear energy was responsible for about 20% of the electricity generated, not far from the world average, in second place behind coal (Fig. 29.22).

Each fission in $^{235}_{92}U$ frees an average of 2.5 neutrons, so no more than 1.5 neutrons per fission can be lost in a reactor. Natural uranium, however, contains only 0.7% of the fissionable isotope $^{235}_{92}U$. The rest is $^{238}_{92}U$, which readily captures the fast neutrons emitted during fission but does not undergo fission afterward. The neutrons absorbed by $^{238}_{92}U$ are therefore wasted, and a chain reaction cannot take place in a solid lump of natural uranium.

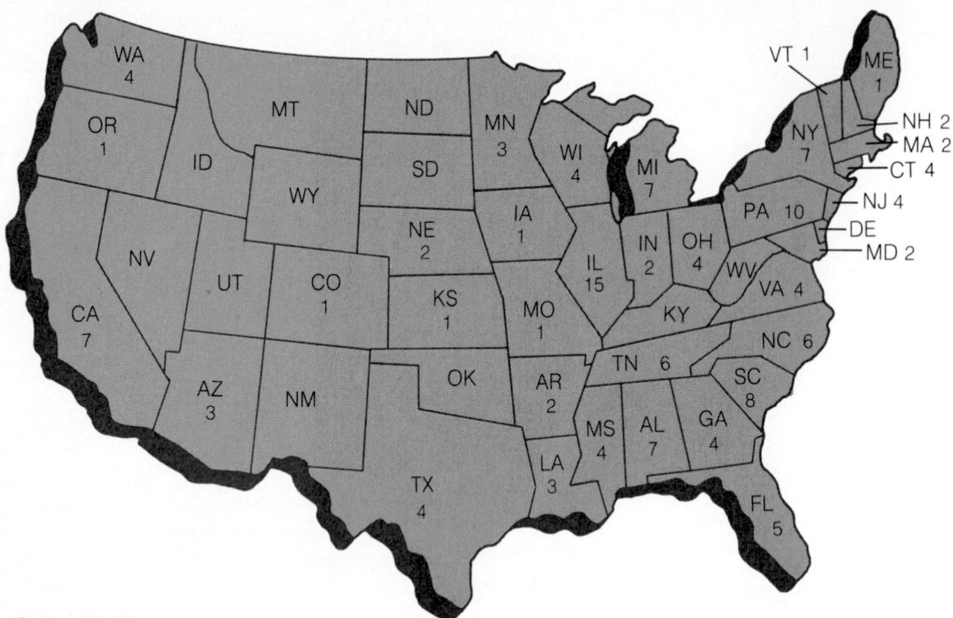

Figure 29.22

Nuclear power plants in the United States. The total number is 142. No new plants have been ordered since 1979, when failures in its cooling system disabled one of the reactors at Three Mile Island in Pennsylvania, and a certain amount of radioactive material escaped. Public unease with the safety record of the nuclear industry, together with a leveling off of demand for electricity, led to a halt in plans to expand nuclear energy further in the United States. A much worse accident destroyed a reactor at Chernobyl in the Soviet Union in 1986, producing many fatalities and widespread contamination in many parts of Europe.

There is a way around this problem. As it happens, $^{238}_{92}U$ has little ability to capture *slow* neutrons, whereas slow neutrons are more likely to induce fission in $^{235}_{92}U$ than fast ones. Slowing down the fast neutrons from fission will thus prevent them from being lost to $^{238}_{92}U$ and at the same time promote further fissions in $^{235}_{92}U$. To slow down fission neutrons, the uranium fuel in a reactor is mixed with a **moderator,** a substance whose nuclei absorb energy from fast neutrons in collisions. From Section 5.7 we know that the more nearly equal in mass colliding particles are, the more energy is transferred. Since hydrogen nuclei are protons with nearly the same mass as neutrons, hydrogen is widely used as a moderator in the form of water, H_2O, whose molecules each contain two hydrogen atoms. The water also serves as the coolant in such reactors.

Unfortunately a neutron that strikes a proton has a certain chance of sticking to it to form a deuteron, $^{2}_{1}H$. Neutron losses in a water-moderated reactor make it necessary to use **enriched uranium** whose $^{235}_{92}U$ content has been increased to about 3%. Most enriched uranium is produced by the gaseous diffusion process in which uranium hexafluoride (UF_6) gas is exposed to a succession of semipermeable barriers.

Loading fuel rods in the nuclear reactor of a power plant in Connecticut. The rods are metal tubes filled with pellets of uranium oxide.

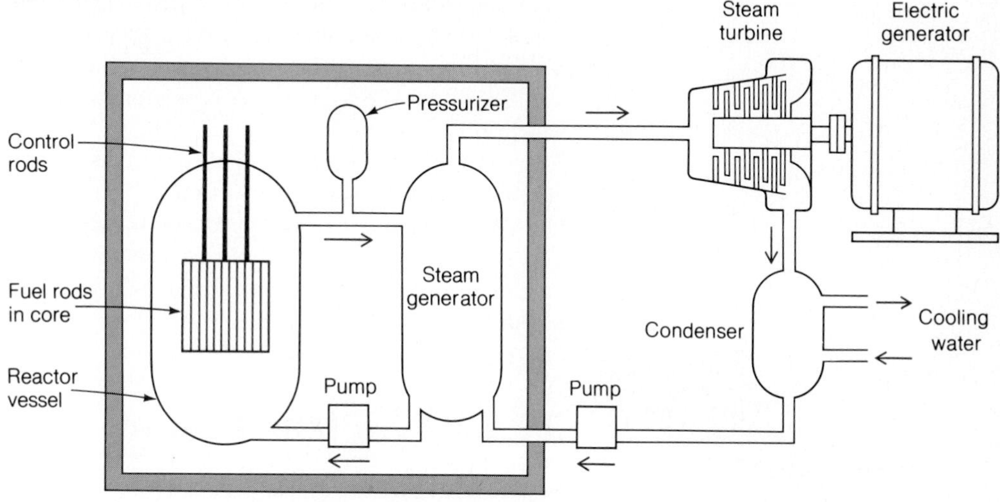

Figure 29.23

The basic design of a pressurized-water nuclear reactor. Water serves both as moderator and as coolant for the core.

Because of their smaller mass, molecules of ${}^{235}_{92}UF_6$ are slightly more likely to diffuse through each barrier than molecules of ${}^{238}_{92}UF_6$, and any desired degree of enrichment can be achieved in this way. Another method is to use gas centrifuges for the separation. Either procedure needs a large plant and much energy, so it is expensive.

The fuel for a water-cooled reactor consists of uranium oxide (UO_2) pellets sealed in long, thin zirconium alloy tubes assembled into a core that is enclosed in a steel pressure vessel. A typical pressure vessel is 10 m high, has an inside diameter of 3 m, and has walls 20 cm thick. In a **pressurized-water reactor** (PWR), the most common type, the water that circulates through the core is kept at a sufficiently high pressure, about 155 atm, to prevent boiling. The water enters the pressure vessel at perhaps 280°C and leaves at 320°C, passing through a heat exchanger to produce steam that drives a turbine (Fig. 29.23). Such a PWR used to produce electricity might contain 70 tons of UO_2 and operate at 2700 MW to yield perhaps 1000 MW of electric power. To control the rate of the chain reaction, the reactor has movable rods of cadmium or boron, which are good absorbers of slow neutrons. As these rods are inserted farther and farther into the reactor, the reaction rate is progressively reduced.

Another type of reactor uses ''heavy'' water, whose molecules contain deuterium (${}^{2}_{1}H$) atoms, instead of ''light'' water because deuterons are less likely than protons to capture neutrons. Natural uranium is the fuel in a reactor of this kind. When graphite, a form of pure carbon, is the moderator, the reactor can be cooled by a gas such as helium or carbon dioxide and operate at much higher temperatures for greater thermodynamic efficiency.

Breeder Reactors

Some nonfissionable nuclides can become fissionable ones by absorbing neutrons. A notable example is ${}^{238}_{92}U$, which becomes ${}^{239}_{92}U$ when it captures a fast neutron. This uranium isotope then beta-decays into ${}^{239}_{93}Np$, an isotope of the element **neptunium,** which is also beta-active. The decay of ${}^{239}_{93}Np$ yields ${}^{239}_{94}Pu$, a fissionable isotope of **plutonium** that can support a chain reaction. The entire sequence is shown in Fig. 29.24.

Figure 29.24

The nonfissionable uranium isotope ${}^{238}_{92}U$ becomes the fissionable plutonium isotope ${}^{239}_{94}Pu$ after absorbing a neutron and undergoing two beta decays. Transformations like this are the basis of the breeder reactor, which produces more fissionable material than it uses up.

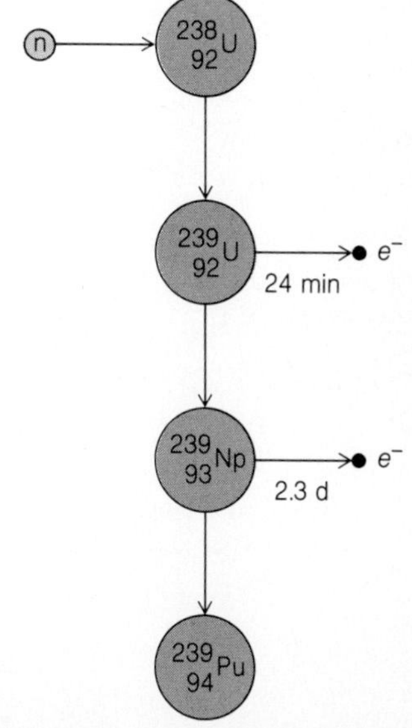

Both neptunium and plutonium are **transuranic elements,** none of which are found on the earth because their half-lives are too short for them to have survived even if they had been present when the earth came into being 4.5 billion years ago. Transuranic elements up to $Z = 105$ have been produced by the neutron bombardment of lighter nuclides.

A **breeder reactor** is one designed to produce as much or more fissionable material than the ${}^{235}_{92}U$ it consumes, either from ${}^{238}_{92}U$ or from some similarly "fertile" nuclide. On the one hand, the widespread use of breeder reactors would mean that known reserves of nuclear fuel would last for centuries to come. On the other hand, since plutonium (unlike the slightly enriched uranium that fuels ordinary reactors) can also be used in nuclear weapons, such reactors would also complicate arms control.

29.12 Nuclear Fusion

How the sun and stars get their energy.

The basic energy-producing process in stars—and hence the source, direct or indirect, of nearly all the energy in the universe—is the fusion of hydrogen nuclei into helium nuclei. This can take place under stellar conditions in two different series of nuclear reactions. In stars like the sun, the **proton-proton** cycle predominates:

$$\begin{aligned} &{}^{1}_{1}H + {}^{1}_{1}H \rightarrow {}^{2}_{1}H + e^{+} + 0.4\,\text{MeV} \\ &{}^{1}_{1}H + {}^{2}_{1}H \rightarrow {}^{3}_{2}He + 5.5\,\text{MeV} \\ &{}^{3}_{2}He + {}^{3}_{2}He \rightarrow {}^{4}_{2}He + 2\,{}^{1}_{1}H + 12.9\,\text{MeV} \end{aligned}$$

Proton-proton cycle

The first two of these reactions must each occur twice for every synthesis of ${}^{4}_{2}He$, so the total energy liberated is 24.7 MeV.

The other reaction sequence, the **carbon cycle,** predominates in stars hotter than the sun. In this cycle, a ${}^{12}_{6}C$ nucleus absorbs three protons, with two positive beta decays along the way, until it becomes the nitrogen nucleus ${}^{15}_{7}N$. When a ${}^{15}_{7}N$ nucleus reacts with a fourth proton, the result is the formation of a ${}^{4}_{2}He$ nucleus and the reappearance of a ${}^{12}_{6}C$ nucleus. Thus the original ${}^{12}_{6}C$ nucleus acts as a kind of catalyst for the process, since it returns at the end ready to start another cycle.

The energy liberated by nuclear fusion is often called **thermonuclear energy.** High temperatures and densities are needed for fusion reactions to occur often enough for a substantial amount of thermonuclear energy to be produced. The high temperature means that the initial light nuclei have enough thermal energy to overcome their mutual electric repulsion and come close enough together to react. The high density means that such collisions are frequent. A further condition for these cycles is a large reacting mass, such as that of a star, because several steps are involved in each cycle and much time may go by between the initial fusion of one proton with another and the ultimate production of an alpha particle.

Fusion reactions that produce helium are not the only ones that occur in stars. In fact, nuclei up to ${}^{56}_{26}Fe$ are produced, this being the nucleus with the highest binding energy per nucleon (see Fig. 29.7). The formation of nuclei heavier than ${}^{56}_{26}Fe$ by fusion requires a net energy input, whereas the formation of lighter nuclei gives off energy and so is favored under ordinary stellar conditions.

Then where do the heavier elements in the universe come from? In a star the inward pull of gravity is balanced by the pressure due to heat given off in fusion reactions. As a star ages, its fuel supply becomes used up, and it contracts. A star whose mass is comparable to that of the sun eventually shrinks down to about the size of the earth to become a **white dwarf.** A star much larger than the sun has a more violent fate, with a sudden collapse followed by a huge explosion. The exploding star is called a **supernova.** When one occurs, it outshines the other stars, galaxies, and planets that we can see for some days or weeks. Nuclei heavier than ${}^{56}_{26}\mathrm{Fe}$ are believed to have been created in supernova explosions, where sufficient energy is available to synthesize them from lighter nuclei. The heavy nuclei become dispersed in interstellar matter and are incorporated in new stars and planets that come into being from this matter.

Fusion Energy on the Earth

Here on the earth, where any reacting mass must be very limited in size, an efficient fusion process cannot involve more than a single step. The fusion reaction that is the basis of current research is the combination of a deuterium nucleus and a tritium nucleus to form a helium nucleus and a neutron:

$${}^{2}_{1}\mathrm{H} + {}^{3}_{1}\mathrm{H} \rightarrow {}^{4}_{2}\mathrm{He} + {}^{1}_{0}n + 17.6\,\mathrm{MeV}$$

Most of the energy given off is carried away by the neutron. To recover this energy, the reaction chamber could be surrounded with liquid lithium, which would absorb energy from the neutrons and carry it to a steam generator. The steam would then power a turbine connected to an electric generator, as in fossil-fuel and fission power plants.

About 1 part in 5000 of the hydrogen in the waters of the world is deuterium, which adds up to over 10^{15} tons. (A gallon of seawater has the potential for fusion energy equal to the chemical energy in 300 gallons of gasoline.) Seawater contains too little tritium for economic recovery, but neutrons react with both isotopes of natural lithium to give tritium and helium:

$${}^{6}_{3}\mathrm{Li} + {}^{1}_{0}n \rightarrow {}^{3}_{1}\mathrm{H} + {}^{4}_{2}\mathrm{He}$$
$${}^{7}_{3}\mathrm{Li} + {}^{1}_{0}n \rightarrow {}^{3}_{1}\mathrm{H} + {}^{4}_{2}\mathrm{He} + {}^{1}_{0}n$$

Once a fusion reactor is given an initial load of tritium, then, it can make enough for its further operation.

The big problem in exploiting fusion energy is to achieve the required combination of temperature, density, and containment time for a deuterium-tritium mixture to react sufficiently to produce a net energy yield. Such a combination occurs in the explosion of a fission ("atomic") bomb, and including the ingredients for fusion in such a bomb leads to a much more destructive weapon, the "hydrogen" bomb.

Two approaches to the controlled release of fusion energy are being explored. In one, strong magnetic fields are used to contain the reacting nuclei, which are in the form of a hot, dense plasma (fully ionized gas). Four decades of research have led to larger and larger experimental magnetic fusion reactors that have brought to light no reasons why eventual success should not be possible. But the engineering problems to be overcome are substantial.

The other approach involves using energetic beams to both heat and compress tiny deuterium-tritium pellets to produce what are, in effect, miniature hydrogen-

The Tokamak reactor at Princeton University uses strong magnetic fields to confine a very hot ionized gas of deuterium and tritium while nuclear fusion reactions take place. A larger-scale reactor of this kind may produce more energy than its operation consumes.

bomb explosions. A number of beams would strike each pellet from all sides to keep it in place as it is squeezed together. If ten pellets the size of a grain of sand are ignited per second, the energy output could provide the electric power needed by a city of 175,000 people. Laser beams are being tried for this purpose as well as beams of particles such as electrons and protons.

Although practical fusion reactors are still in the future (how far, nobody knows), the advantages of thermonuclear energy—cheap, essentially limitless fuel; much less hazard than for fission energy; relatively minor production of radioactive wastes; no possibility of military use—are such that eventually it will almost certainly supply a large part of the world's energy needs.

Important Terms

The **atomic number** of an element is the number of electrons in each of its atoms or, equivalently, the number of protons in each of its atomic nuclei.

The **neutron** is an electrically neutral particle, slightly heavier than the proton, that is present in nuclei together with protons. Neutrons and protons are jointly called **nucleons.** The **mass number** of a nucleus is the number of nucleons it contains.

Isotopes of an element have the same atomic number but different mass numbers. Symbols of **nuclides** (nuclear species) follow the pattern

$$^{A}_{Z}X$$

where X is the chemical symbol of the element, Z its atomic number, and A the mass number of the particular nuclide.

The **binding energy** of a nucleus is the energy equivalent of the difference between its mass and the sum of the masses of its individual constituent nucleons. This amount of energy must be supplied to the nucleus if it is to be completely disintegrated.

Radioactive nuclei spontaneously transform themselves into other nuclear species with the emission of radiation. The radiation may consist of **alpha particles,** which are the nuclei of helium atoms, or **beta particles,** which are positive or negative electrons. Positive electrons are known as **positrons.** Electron capture by a nucleus is an alternative to positron emission. The emission of **gamma rays,** which are energetic photons, permits an excited nucleus to lose its excess energy.

The time required for half of a given sample of a radioactive nuclide to decay is called its **half-life.**

In **nuclear fission** the absorption of neutrons by certain heavy nuclei causes them to split into smaller **fission fragments.** Because each fission also liberates several neutrons, a rapidly multiplying sequence of fissions called a **chain reaction** can occur if a sufficient amount of the proper material is assembled. A **nuclear reactor** is a device in which a chain reaction can be initiated and controlled.

In **nuclear fusion** two light nuclei combine to form a heavier one with the emission of energy. The energy liberated in the process when it takes place on a large scale is called **thermonuclear energy.**

MULTIPLE CHOICE

1. The chemical behavior of an atom is determined by its
- a. atomic number
- b. mass number
- c. binding energy
- d. number of isotopes

2. All isotopes of a given element have the same
- a. number of protons
- b. number of neutrons
- c. number of nucleons
- d. binding energy

3. The number of neutrons in the nucleus of the aluminum isotope $^{27}_{13}$Al is
- a. 13
- b. 14
- c. 27
- d. 40

4. Which of the following is not an isotope of hydrogen?
- a. $^{1}_{0}$H
- b. $^{1}_{1}$H
- c. $^{2}_{1}$H
- d. $^{3}_{1}$H

5. Relative to the sum of the masses of its constituent nucleons, the mass of a nucleus is
- a. greater
- b. the same
- c. smaller
- d. sometimes greater and sometimes smaller

6. The ionization energy of an atom relative to the binding energy of its nucleus is
- a. greater
- b. the same
- c. smaller
- d. sometimes greater and sometimes smaller

7. The binding energy per nucleon is
- a. the same for all nuclei
- b. greatest for very small nuclei
- c. greatest for nuclei of intermediate size
- d. greatest for very large nuclei

8. The element whose nuclei contain the most tightly bound nucleons is
- a. helium
- b. carbon
- c. iron
- d. uranium

9. A consequence of the limited range of the strong nuclear interaction is
- a. the existence of isotopes
- b. the magnitude of nuclear binding energies
- c. the instability of large nuclei
- d. the ratio of atomic size to nuclear size

10. In a stable nucleus the number of neutrons is always
- a. less than the number of protons
- b. less than or equal to the number of protons
- c. more than the number of protons
- d. more than or equal to the number of protons

11. The half-life of a radioactive substance is
- a. half the time needed for a sample to decay entirely
- b. half the time a sample can be kept before it begins to decay
- c. the time needed for half a sample to decay
- d. the time needed for the remainder of a sample to decay after half of it has already decayed

12. As a sample of a radioactive nuclide decays, its half-life
- a. decreases
- b. remains the same
- c. increases
- d. any of the above, depending on the nuclide

13. When a nucleus undergoes radioactive decay, its new mass number is
- a. always less than its original mass number
- b. always more than its original mass number
- c. never less than its original mass number
- d. never more than its original mass number

14. A nucleus with an excess of neutrons may decay radioactively with the emission of
- a. a neutron
- b. a proton
- c. an electron
- d. a positron

15. Electron capture has the same effect on a nucleus as
- a. alpha decay
- b. negative beta decay
- c. positive beta decay
- d. gamma decay

16. Which of these types of radiation has the greatest ability to penetrate matter?
- a. alpha particles
- b. beta particles
- c. gamma rays
- d. X rays

17. Which of these types of radiation has the least ability to penetrate matter?
- a. alpha particles
- b. beta particles
- c. gamma rays
- d. X rays

18. A suitable object for radiocarbon dating would be one whose age is
- a. 20,000 years
- b. 200,000 years
- c. 2 million years
- d. 20 million years

19. The largest amount of radiation received by an average person in the United States comes from
- a. medical X rays
- b. nuclear reactors
- c. fallout from past weapons tests
- d. natural sources

20. By "chain reaction" is meant
- a. the joining together of protons and neutrons to form atomic nuclei
- b. the joining together of light nuclei to form heavy ones

c. the successive fissions of heavy nuclei induced by neutrons emitted in the fissions of other heavy nuclei
d. the burning of uranium in a special type of furnace called a nuclear reactor

21. In a nuclear power plant the nuclear reactor itself is used to supply
a. neutrons b. electricity
c. steam d. heat

22. Most of the energy liberated during fission appears
a. as KE of the neutrons that are emitted
b. as KE of the fission fragments
c. in the decay of the fission fragments
d. as gamma rays

23. Enriched uranium is a better fuel for nuclear reactors than natural uranium because it has a greater proportion of
a. slow neutrons b. deuterium
c. $^{235}_{92}U$ d. $^{238}_{92}U$

24. The purpose of the moderator in a nuclear reactor is to
a. slow down the neutrons emitted during fission
b. absorb the neutrons emitted during fission
c. absorb the gamma rays emitted during fission
d. control the rate at which energy is produced

25. The sun's energy comes from
a. nuclear fission
b. radioactivity
c. the conversion of hydrogen to helium
d. the conversion of helium to hydrogen

26. Fusion reactions on the earth are likely to use as fuel
a. ordinary hydrogen b. deuterium
c. plutonium d. uranium

27. The mass of a $^{7}_{3}Li$ nucleus is 0.042 u less than the sum of the masses of 3 protons and 4 neutrons. The binding energy per nucleon in $^{7}_{3}Li$ is
a. 5.6 MeV b. 10 MeV
c. 13 MeV d. 39 MeV

28. The product of the alpha decay of the bismuth isotope $^{214}_{83}Bi$ is
a. $^{210}_{79}Au$ b. $^{210}_{81}Tl$
c. $^{210}_{83}Bi$ d. $^{218}_{85}At$

29. The helium isotope $^{6}_{2}He$ undergoes beta decay with the emission of an electron. The product of the decay is
a. $^{6}_{1}H$ b. $^{5}_{2}He$
c. $^{7}_{2}He$ d. $^{6}_{3}Li$

30. The product of the positive beta decay of the nitrogen isotope $^{13}_{7}N$ is
a. $^{12}_{6}C$ b. $^{13}_{6}C$
c. $^{14}_{7}N$ d. $^{13}_{8}O$

31. The product of the gamma decay of the aluminum isotope $^{27}_{13}Al$ is
a. $^{27}_{12}Mg$ b. $^{26}_{13}Al$
c. $^{27}_{13}Al$ d. $^{27}_{14}Si$

32. When the bismuth isotope $^{213}_{83}Bi$ decays into the polonium isotope $^{213}_{84}Po$, it emits
a. an alpha particle b. an electron
c. a positron d. a gamma ray

33. The half-life of a certain radioactive isotope is 6 h. If we start out with 10.0 g of the isotope, after 1 day there will be
a. none left b. 0.6 g left
c. 1.6 g left d. 2.5 g left

34. After 10 years 75 g of an original sample of 100 g of a certain radioactive isotope has decayed. The half-life of the isotope is
a. 5 years b. 7.5 years
c. 20 years d. 40 years

QUESTIONS

1. In what ways are the isotopes of an element similar to one another? In what ways are they different?

2. Why don't stable nuclei have more protons than neutrons?

3. What limits the size of a stable nucleus?

4. If the early universe contained protons, neutrons, and electrons but the strong interaction did not exist, what kinds of matter would eventually fill the universe?

5. Radium undergoes spontaneous decay into helium and radon. Why is radium regarded as an element rather than as a chemical compound of helium and radon?

6. What happens to the atomic number and mass number of a nucleus that emits a gamma ray? What happens to the actual mass of the nucleus?

7. List as many aspects of radioactivity as you can that cannot be explained on the basis of classical (prerelativity and prequantum) physics.

8. (a) Under what circumstances does a nucleus emit an electron? A positron? (b) The nuclei $^{14}_{8}O$ and $^{19}_{8}O$ both undergo beta decay in order to become stable nuclei. Which would you expect to emit a positron and which an electron?

9. Radiocarbon dates are never completely reliable. Can you think of any reasons (other than measurement error) why this should be so?

10. (a) What is the basis of the radiocarbon dating procedure? (b) What steps would you take to find the age of an ancient piece of wood by radiocarbon dating?

11. What are the two basic conditions that must be met by a radionuclide in order that it be useful in dating a particular kind of rock?

12. What is the limitation on the fuel that can be used in a reactor whose moderator is ordinary water? Why is the situation different if the moderator is heavy water?

13. Why might a gas-cooled reactor with a graphite moderator be safer than a reactor that uses water as coolant and moderator?

14. What are the differences and similarities between nuclear fission and nuclear fusion? Where does the evolved energy come from in each case?

15. Why are the conditions in the interior of a star favorable for nuclear fusion reactions?

EXERCISES

29.1 Nuclear Composition

1. State the number of neutrons and protons in each of the following nuclei: ${}^{10}_{5}Be$; ${}^{22}_{10}Ne$; ${}^{36}_{16}S$; ${}^{88}_{38}Sr$; ${}^{180}_{72}Hf$.

2. State the number of neutrons and protons in each of the following nuclei: ${}^{6}_{3}Li$; ${}^{13}_{6}C$; ${}^{31}_{15}P$; ${}^{94}_{40}Zr$; ${}^{137}_{56}Ba$.

3. Complete the following nuclear reactions:

$$
\begin{aligned}
&{}^{6}_{3}Li + ? \rightarrow {}^{7}_{4}Be + {}^{1}_{0}n \\
&{}^{10}_{5}B + ? \rightarrow {}^{7}_{3}Li + {}^{4}_{2}He \\
&{}^{35}_{17}Cl + ? \rightarrow {}^{32}_{16}S + {}^{4}_{2}He
\end{aligned}
$$

4. A nucleus of ${}^{15}_{7}N$ is struck by a proton. A nuclear reaction can take place with the emission of either a neutron or an alpha particle. Give the atomic number, mass number, and chemical name of the remaining nucleus in each of these cases.

5. A reaction often used to detect neutrons occurs when a neutron strikes a ${}^{10}_{5}B$ nucleus, with the subsequent emission of an alpha particle. What are the atomic number, mass number, and chemical name of the remaining nucleus?

6. Ordinary boron is a mixture of the ${}^{10}_{5}B$ and ${}^{11}_{5}B$ isotopes and has a composite atomic mass of 10.82 u. Find the percentage of each isotope present in ordinary boron. The atomic masses of ${}^{10}_{5}B$ and ${}^{11}_{5}B$ are, respectively, 10.01 u and 11.01 u.

7. Using the figures in Table 29.1, verify that the average atomic mass of chlorine is 35.46 u.

29.3 Binding Energy

8. The binding energy of ${}^{20}_{10}Ne$ is 160.64 MeV. Find its atomic mass.

9. The atomic mass of ${}^{56}_{26}Fe$ is 55.934939 u. Find its binding energy and binding energy per nucleon.

10. The neutron decays in free space into a proton and an electron. What must be the minimum binding energy contributed by a neutron to a nucleus in order that the neutron not decay inside the nucleus? How does this figure compare with the observed binding energies per nucleon in stable nuclei?

11. The nuclear reaction

$$ {}^{6}_{3}Li + {}^{2}_{1}H \rightarrow 2\,{}^{4}_{2}He $$

evolves 22.4 MeV. Calculate the atomic mass of ${}^{6}_{3}Li$ in u.

29.5 Radioactive Decay

12. What is the daughter nuclide of the alpha decay of the polonium isotope ${}^{210}_{84}Po$?

13. The nucleus ${}^{233}_{90}Th$ undergoes two negative beta decays in becoming an isotope of uranium. What is the symbol of the isotope?

14. The copper isotope ${}^{64}_{29}Cu$ can decay by either electron capture or positron emission. What does it become in each case?

15. A ${}^{80}_{35}Br$ nucleus can decay by electron emission, positron emission, or electron capture. What is the daughter nucleus in each case?

16. The lead isotope ${}^{214}_{82}Pb$ decays into the bismuth isotope ${}^{214}_{83}Bi$. What kind of decay is involved?

17. The thorium isotope ${}^{232}_{90}Th$ decays through the successive emission of 6 alpha particles and 4 electrons. What stable nuclide does it finally become?

18. The nuclide ${}^{232}_{92}U$ (mass 232.0372 u) alpha decays into ${}^{228}_{90}Th$ (mass 228.0287 u). (a) Find the amount of energy released in the decay. (b) Is it possible for ${}^{232}_{92}U$ to decay into ${}^{231}_{92}U$ (mass 231.0364 u) by emitting a neutron? Why? (c) Is it possible for ${}^{232}_{92}U$ to decay into ${}^{231}_{91}Pa$ (mass 231.0359 u) by emitting a proton? Why? (The above masses refer to neutral atoms.)

19. The atomic mass of ${}^{226}_{88}Ra$ is 226.0254 u, and the energy liberated in its alpha decay is 4.87 MeV. (a) Identify the daughter nucleus and find its atomic mass. (b) The alpha particle emitted in the decay is observed to have a KE of 4.78 MeV. Where do you think the other 0.09 MeV goes?

29.7 Half-Life

20. How many disintegrations per second occur in a 25-millicurie sample of thorium?

21. The half-life of radium is 1600 years. How long will it take for $\frac{15}{16}$ of a given sample of radium to decay?

22. The activity of a sample of ${}^{227}_{90}Th$ is observed to be 12.5% of its original amount 54 days after the sample was prepared. What is the half-life of ${}^{227}_{90}Th$?

23. How long does it take for 60% of a sample of radon to decay? Radon has a half-life of 3.8 days.

24. The activity of 1.0 g of $^{226}_{88}\text{Ra}$ is about 1.0 curie. Find its half-life.

25. The potassium isotope $^{40}_{19}\text{K}$ undergoes beta decay with a half-life of 1.83 billion years. Find the activity of 1.0 g of this nuclide.

29.10 Nuclear Fission

26. (a) How much mass is lost per day by a nuclear reactor operated at a 1.0-GW (10^9-W) power level? (b) If each fission in $^{235}_{92}\text{U}$ releases 200 MeV, how many fissions must occur per second to yield a power level of 1.0 GW?

27. $^{235}_{92}\text{U}$ loses about 0.1% of its mass when it undergoes fission. (a) How much energy is released when 1 kg of $^{235}_{92}\text{U}$ undergoes fission? (b) One ton of TNT releases about 4×10^9 J when it is detonated. How many tons of TNT are equivalent in destructive power to a bomb that contains 1 kg of $^{235}_{92}\text{U}$?

29.12 Nuclear Fusion

28. Fusion reactions involving deuterons can take place when the deuterons have energies of about 10 keV or more. Find the temperature of a deuterium plasma in which the average KE of the deuterons is 10 keV.

29. Old stars obtain part of their energy by the fusion of three alpha particles to form a $^{12}_{6}\text{C}$ nucleus, whose atomic mass is 12.0000 u. How much energy is evolved in each such reaction?

PROBLEMS

30. The atomic masses of $^{15}_{7}\text{N}$, $^{15}_{8}\text{O}$, and $^{16}_{8}\text{O}$ are, respectively, 15.0001, 15.0030, and 15.9949 u. (a) Find the average binding energy per nucleon in $^{16}_{8}\text{O}$. (b) How much energy is needed to remove one proton from $^{16}_{8}\text{O}$? (c) How much energy is needed to remove one neutron from $^{16}_{8}\text{O}$? (d) Why are these figures different from one another?

31. The distance between the two protons in a $^{3}_{2}\text{He}$ nucleus is roughly 1.7×10^{-15} m. (a) Calculate the electric potential energy ke^2/r of these protons. (b) Show that this energy is of the right order of magnitude to account for the difference in binding energy between $^{3}_{1}\text{H}$ and $^{3}_{2}\text{He}$. What conclusion can we draw from this result about the dependence of nuclear forces on electric charge? The atomic mass of $^{3}_{1}\text{H}$ is 3.016050 u and that of $^{3}_{2}\text{He}$ is 3.016030 u.

32. Compare the electric potential energy of two protons 5.0×10^{-15} m apart with the binding energy per nucleon of nuclei with $A \approx 60$. Such nuclei are about 5×10^{-15} m in radius.

33. In free space the neutron is unstable and beta-decays into a proton and an electron with a half-life of 10.8 min. What proportion of the original number of neutrons in a beam will remain undecayed after 1 min? After 1 h?

34. The energy of the alpha particles emitted by the polonium isotope $^{210}_{84}\text{Po}$ is 5.30 MeV. The half-life of $^{210}_{84}\text{Po}$ is 138 days. (a) What mass of this nuclide is needed to power a thermoelectric cell of 1.0-W output if the efficiency of energy conversion is 8.0%? (b) What would the power output be after one year?

35. The sun's mass is 2.0×10^{30} kg and its power output is 4.0×10^{26} W. If we assume that the sun today contains only protons and electrons and that only the proton-proton cycle occurs, for roughly how many years can it continue to radiate at this rate?

Answers to Multiple Choice

1. a	10. d	19. d	28. b
2. a	11. c	20. c	29. d
3. b	12. b	21. d	30. b
4. a	13. d	22. b	31. c
5. c	14. c	23. c	32. b
6. c	15. c	24. a	33. b
7. c	16. c	25. c	34. a
8. c	17. a	26. b	
9. c	18. a	27. a	

C H A P T E R 30

ELEMENTARY PARTICLES

The building blocks of ordinary matter are atoms, of which over a hundred different kinds are known. Are atoms elementary particles in the sense that they have no internal structure? The answer, of course, is no, since every atom consists of a central nucleus with one or more electrons around it. Electrons indeed seem to be elementary particles, but nuclei are not. As we have seen, they are made up of nucleons—protons and neutrons. What about nucleons? Because they cannot be broken down into anything else, for a long time they were considered to be true elementary particles. Today, however, it appears that nucleons are actually composed of unusual particles called quarks. A great many other subatomic particles have been discovered in recent years, all of which are unstable and decay into more stable particles soon after being formed in high-energy collisions. Most of these particles, too, seem to be made up of various combinations of quarks. Not only are these findings interesting in themselves, but the theories that describe them also allow us to picture the evolution of the universe starting from 10^{-43} s of the moment when it came into being.

30.1 Particle Detection

The ions they produce give them away.

Photons and subatomic particles can be detected in a variety of ways. Nearly all of them are based on the ionization that results, directly or indirectly, from the passage of a photon or charged particle through matter.

A heavy charged particle, such as a proton or an alpha particle, loses energy chiefly by electric interactions with atomic electrons. The electrons either are raised to excited states or, more often, are pulled away entirely from their parent atoms. The ejected electrons often ionize other atoms themselves. The incoming particle slows down gradually until it comes to a stop or reacts with a nucleus in its path.

Electrons, too, lose energy by ionization, but for them another mechanism may also be important. As we know, electromagnetic radiation is given off whenever an electric charge is accelerated. Energy loss by radiation is more significant for electrons than for heavier particles because they are more violently accelerated when passing near nuclei in their paths. The faster the electron and the greater the nuclear charge of the atoms it encounters, the more it radiates.

Because neutrons are uncharged, they do not interact with atomic electrons in their paths but only with nuclei. Most such interactions are "billiard ball" collisions in which part of the neutron's kinetic energy is given to the target nucleus. As discussed in Section 5.7 the closer the mass of the target nucleus is to the neutron mass, the more the energy that will be transferred. Thus a 2-MeV neutron in a nuclear reactor on the average needs 18 collisions with hydrogen nuclei to reach equilibrium at room temperature (when its most probable energy is 0.025 eV), 25 collisions with deuterium nuclei, 114 with carbon nuclei, and 2150 with uranium nuclei.

The three principal ways in which X and gamma rays lose energy when they pass through matter, as shown in Fig. 30.1, are:

1. **The photoelectric effect.** A photon gives up all its energy to an atomic electron in the absorbing material.
2. **Compton scattering.** A photon loses part of its energy to an atomic electron; the new photon has a lower frequency.
3. **Pair production.** A photon whose energy is at least 1.02 MeV (since m_0c^2 is 0.51 MeV for the electron) can materialize into an electron-positron pair when it passes near a nucleus, as described in Section 30.3.

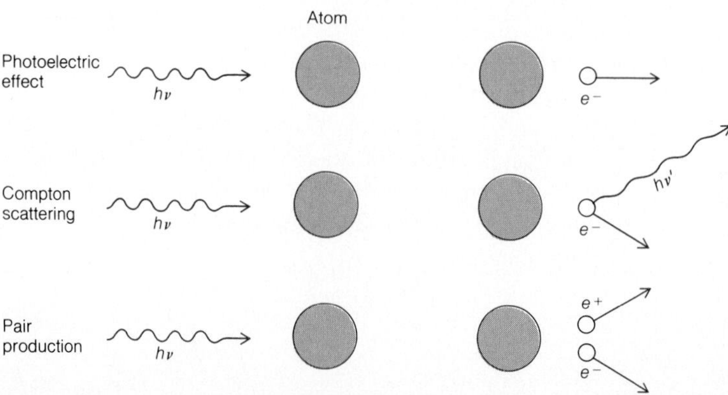

Figure 30.1

X and gamma rays interact with matter chiefly through the photoelectric effect, Compton scattering, and pair production. Pair production requires a photon energy of at least 1.02 MeV.

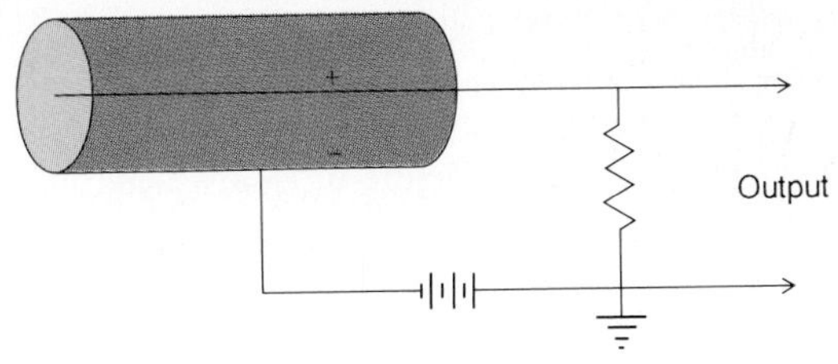

Figure 30.2

A gas-filled detector of this kind can be used either as an ionization chamber or as a Geiger counter. When it is to be an ionization chamber, the applied voltage is low, and the output pulse that occurs when a photon or charged particle passes through the tube is a measure of the amount of ionization produced in the gas. When it is to be a Geiger counter, the applied voltage is high enough to give the initial ions enough energy of their own to cause further ionization, and the result is a large output pulse that is the same for all incident photons or particles.

In each case photon energy is given to electrons, which in turn lose energy mainly by exciting or ionizing atoms in the absorbing material. At low photon energies the photoelectric effect dominates, to be replaced by Compton scattering at higher energies. As the photon energy increases beyond the minimum energy of 1.02 MeV required, pair production becomes more and more important. Gamma rays from radioactive decay and X rays interact with matter largely through Compton scattering.

Ionization Detectors

A number of devices can be used to detect the ionization produced by an energetic photon or charged particle. Some, such as the **ionization chamber** and the **Geiger counter,** are gas-filled tubes with electrodes that attract the electrons and ions to produce voltage pulses in external circuits (Fig. 30.2). Another device, the **scintillation counter,** is based on the flash of light emitted by certain substances (such as those used to make the screen of a television picture tube) when struck by ionizing radiation. The light is then picked up by a sensitive photomultiplier tube (Fig. 30.3). **Semiconductor detectors** make use of the fact that a thin layer on both sides of a *p-n* junction is normally depleted in charge carriers when the crystal has an external voltage across it. Electrons and holes produced at the junction by an ionizing particle lead to a measurable pulse.

Neutrons do not themselves produce ionizaton and so cannot be detected directly. However, a counter whose sensitive volume is rich in hydrogen will respond to fast neutrons through the ionization produced by protons recoiling after collisions with the neutrons. In the case of slow neutrons, one method is to employ the nuclear reaction

$$ {}_{0}^{1}n + {}_{5}^{10}\mathrm{B} \rightarrow {}_{2}^{4}\mathrm{He} + {}_{3}^{7}\mathrm{Li} $$

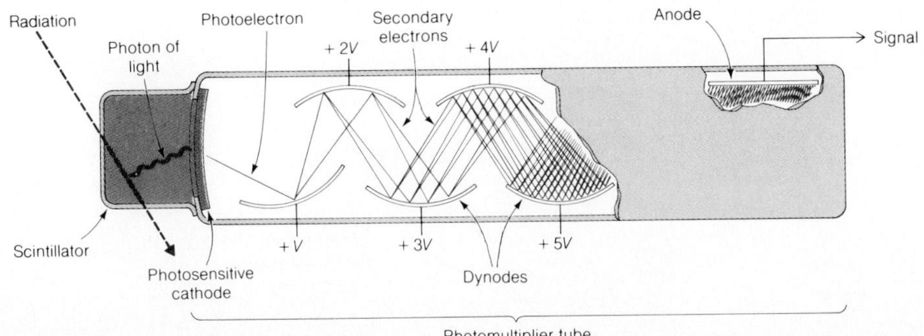

Figure 30.3

Essential features of a scintillation counter. The scintillator may be a crystalline or plastic solid or a liquid. The number of dynodes in the photomultiplier tube may be ten or more, with each dynode several hundred volts higher in potential than the one before it. Each dynode produces 2 to 5 secondary electrons per incident electron for a total amplification of as much as 10^6 or 10^7. A scintillation counter is faster in its operation than an ionization chamber or a Geiger counter.

in which a neutron reacts with a nucleus of the boron isotope ${}^{10}_{5}B$ to produce an alpha particle and a nucleus of the lithium isotope ${}^{7}_{3}Li$. If ${}^{10}_{5}B$ is in the sensitive volume of a suitable counter (the gas BF_3 is often used), ionization that corresponds to an alpha particle signifies the capture of a slow neutron.

Track Detectors

Instruments of the kinds just described are useful when photons or particles of a certain kind that arrive at a certain place are to be counted or have their energies measured. If the interactions and decays of elementary particles are being studied, however, a way to display their paths is needed. The **bubble chamber** is widely used for this purpose. In a bubble chamber a suitable liquid is heated to a temperature above its normal boiling point while under enough pressure to keep it from boiling. The pressure is then reduced suddenly, which leaves the liquid in a superheated condition. A superheated liquid is unstable, and bubbles of vapor form around any ions present. A charged particle moving through the liquid at just this time will leave a track of bubbles that can be photographed. A magnetic field is usually applied to a bubble chamber to permit finding the charges and momenta of the various particles involved in an event that takes place inside it (Fig. 30.4). Bubble chambers for elementary particle research are usually filled with hydrogen, whose nuclei are protons, at a temperature of $-246°C$ and a pressure of 5 bars.

In some experiments a **spark chamber** is preferred to a bubble chamber. A spark chamber consists of a series of parallel metal plates in a gas-filled container. A high voltage is applied between each pair of plates. A charged particle passing through the chamber leaves behind a trail of ions that increases the conductivity of the gas, and sparks occur along this trail. The string of sparks can be photographed, and, as with a bubble chamber, a magnetic field allows the charge and momentum of the particle to be found from the curvature of the track.

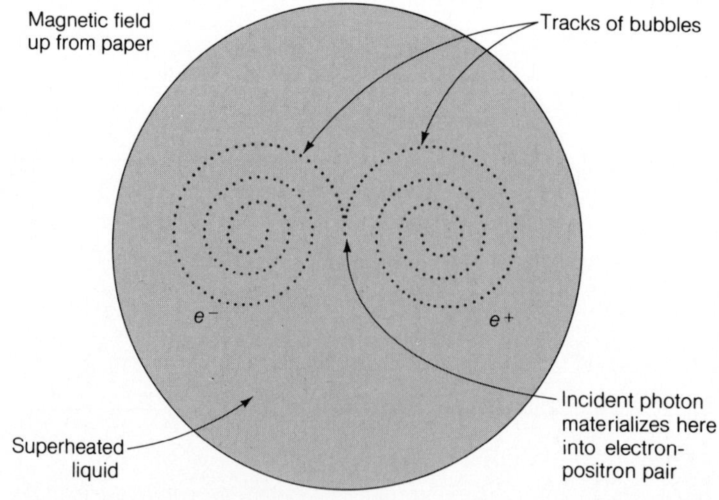

Figure 30.4

How pair production ideally appears in a bubble chamber placed in a magnetic field. Bubbles form along the ion trails of the electron and positron in the superheated liquid. As the particles slow down, the radii of their paths decrease in accord with Eq. (19.9), and the result is a pair of spiral tracks.

30.2 The Neutrino

A particle with neither mass nor charge.

All elementary particles, stable and unstable, fall into two broad categories that depend on their response to the strong nuclear interaction. **Leptons** are not affected by this interaction and show no hints of internal structure. They are as close to being point particles as present measurements can establish. The electron is a lepton. **Hadrons** are subject to the strong interaction and have definite sizes (they are about 10^{-15} m across) and internal structures. The proton and neutron are hadrons.

Another lepton besides the electron is the **neutrino.** Its story began with the discovery that, in beta decay, energy, linear momentum, and angular momentum are all apparently not conserved.

In the case of alpha and gamma decays, the released energy corresponds exactly to the mass difference between the original nucleus and the products of the decay. Instead of all having the same energy, however, the emitted electrons from a particular beta-active nuclide have a variety of energies, as Fig. 30.5 shows. What becomes of the missing energy? Linear momentum, too, is not conserved in beta decays. The emitted electron and the daughter nucleus do not in general move apart in opposite directions, so their momenta cannot cancel out to equal the initial momentum of zero.

As for angular momentum, the neutron, proton, electron, and positron all have spin quantum numbers of magnitude $\frac{1}{2}$. When a neutron turns into a proton and an electron, or a proton into a neutron and a positron, inside a nucleus, the total angular momenta before and after apparently cannot match.

In order to account for these discrepancies without giving up three of the most fundamental and otherwise well-established physical principles, Wolfgang Pauli in 1931 proposed the existence of the neutrino, symbol ν_e (Greek letter *nu*). The

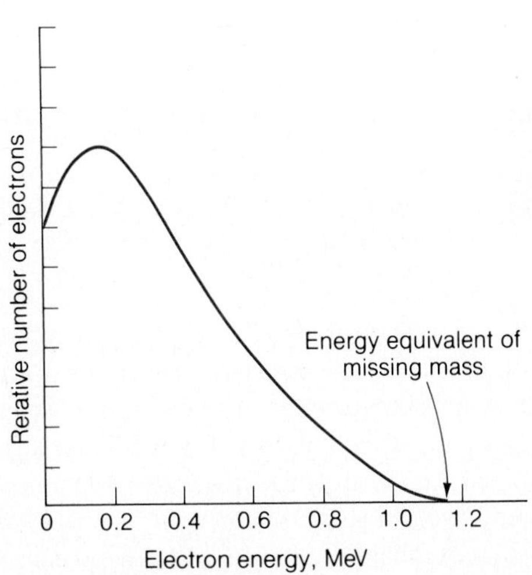

Figure 30.5
The spread of electron energies found in the beta decay of $^{210}_{83}\text{Bi}$. The maximum electron energy is equal to the energy equivalent of the mass lost by the decaying nucleus minus the electron mass.

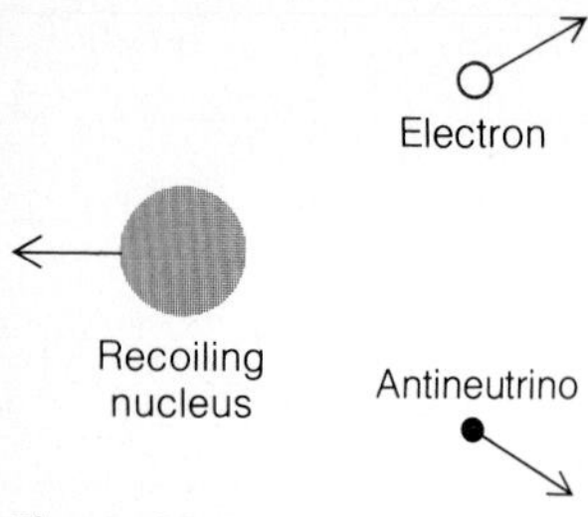

Figure 30.6

An electron and an antineutrino are simultaneously emitted in the beta decay of a nucleus. This makes possible the conservation of energy, momentum, and angular momentum in the process.

neutrino was supposed to have no charge and no mass but could possess both energy and momentum and have an intrinsic spin. According to this idea, an electron and a neutrino are simultaneously emitted in beta decay (Fig. 30.6), which permits energy and both kinds of momentum to be conserved. (In reality, an antineutrino rather than a neutrino is emitted in beta decay; see Section 30.3.)

Lacking charge and mass, and not electromagnetic in nature as is the photon (another massless particle), the neutrino remained undetected for a quarter of a century. So strong was the indirect evidence for it, however, that hardly any physicists doubted its existence. Finally, in 1956 a reaction that, in theory, could be caused only by a neutrino was found to take place. A neutrino that strikes a proton has a small probability of being absorbed to give a neutron and a positron:

$$\nu_e + p \rightarrow n + e^+$$

By placing a sensitive detecting chamber containing hydrogen near a nuclear reactor, in which a great many beta decays take place, the simultaneous appearance of a neutron and a positron would be registered each time the above reaction occurred. When the number of reactions per second that were found agreed with the number predicted, there could be no doubt that neutrinos indeed exist. Some recent theories suggest that the neutrino should have a very small mass rather than none at all, but if so, experiments show that it must be less than about 10^{-5} of the electron mass.

Neutrinos are able to travel unimpeded through vast amounts of matter because they are limited to the weak interaction. On the average, a neutrino must pass through 100 light-years of solid iron before being absorbed—and a light-year, the distance light travels in a year, is 9.5×10^{15} m.

In the sequence of nuclear reactions by which hydrogen is converted to helium in the sun and other stars, two beta decays occur for each helium nucleus formed. A vast number of neutrinos is therefore produced in the sun at all times. Because neutrinos travel freely through matter, almost all of these neutrinos escape into space and take with them 6% to 8% of the total energy generated by the sun. The flood of neutrinos from the sun is such that every cubic centimeter on the earth contains several neutrinos at any moment. The considerable energy carried by the neutrinos created in the sun and the other stars is apparently lost forever from the universe in the sense that it is no longer available for conversion into other forms of energy, such as matter.

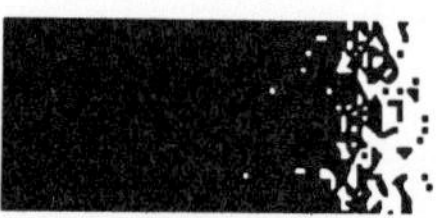

30.3 Antiparticles

The same but different.

Another particle which had to wait a long time after its prediction to be discovered is the negative proton, or **antiproton,** whose symbol is $\overline{p}$. This particle has the same properties as the proton except that its electric charge is negative. The existence of antiprotons was predicted largely on the basis of symmetry arguments. Since the electron has a positive counterpart in the positron, why should the proton not have a negative counterpart as well? Actually, as sophisticated theories show, this is an excellent argument, and few physicists were surprised when the antiproton was actually found.

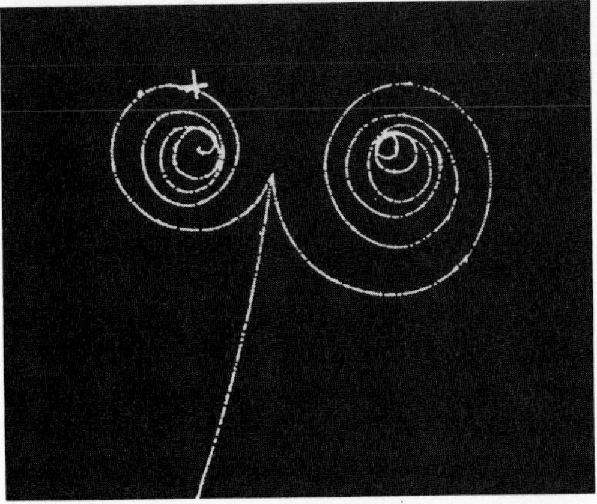

Bubble-chamber photograph of electron-positron pair formation in liquid hydrogen. A high-energy gamma-ray photon created the pair and also gave energy to an atomic electron. A magnetic field in the chamber caused the paths of the electrons to curve in opposite directions.

The reason positrons and antiprotons are so hard to find is that they are readily **annihilated** on contact with ordinary matter (Fig. 30.7). When a positron is near an electron, they attract each other electrically, come together, and then both vanish simultaneously. The missing mass appears in the form of two gamma-ray photons:

$$e^{+} + e^{-} \rightarrow \gamma + \gamma$$

The total mass of the two particles is equivalent to 1.02 MeV, so each photon has an energy of 0.51 MeV. (Their energies must be equal and they must be emitted in opposite directions in order that momentum be conserved.)

While the similar reaction

$$p + \bar{p} \rightarrow \gamma + \gamma$$

can occur when a proton and an antiproton undergo annihilation, it is more usual for the vanished mass to reappear in the form of several mesons, particles that we will consider in the next section.

Electron
Positron
0.51-MeV photon
0.51-MeV photon

Figure 30.7

The mutual annihilation of an electron and a positron.

The reverse of annihilation can also take place, with a photon materializing into a positron and an electron or, if it is energetic enough, into a proton and an antiproton (Fig. 30.8). This phenomenon, known as **pair production,** requires the presence of a nucleus in order that momentum as well as energy be conserved. Any photon energy besides that needed for the mass of the created particles (1.02 MeV for a positron-electron pair, 1872 MeV for a proton-antiproton pair) appears as kinetic energy.

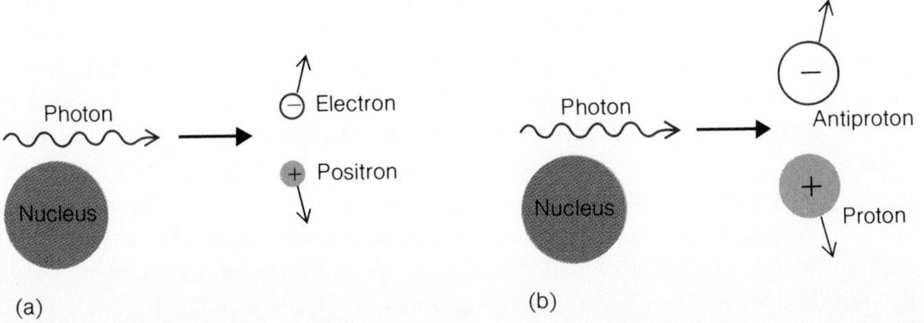

Figure 30.8

The production of (a) an electron-positron pair and (b) a proton-antiproton pair by the materialization of sufficiently energetic photons. Pair production can occur only in the presence of a nucleus.

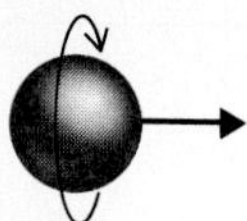

Neutrino

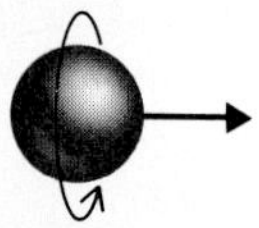

Antineutrino

Figure 30.9

The spins of the neutrino and antineutrino are in opposite senses with respect to their directions of motion.

Antineutrons (symbol $\overline{n}$) and antineutrinos (symbol $\overline{\nu}_e$) have also been identified. The antineutrino differs from the neutrino in that, while the spin axes of both are parallel to their directions of motion, the spin of $\overline{\nu}_e$ is clockwise and that of $\overline{\nu}_e$ is counterclockwise when viewed from behind. A moving neutrino may be thought of as resembling a left-handed screw, and a moving antineutrino as resembling a right-handed screw (Fig. 30.9). An antineutrino is released during a beta decay in which an electron is emitted, and a neutrino is released during a beta decay in which a positron is emitted or an electron is captured. Thus the correct equations of beta decay are

$$n \rightarrow p + e^- + \overline{\nu}_e \qquad \textit{Electron emission} \quad (30.1)$$

$$p \rightarrow n + e^+ + \nu_e \qquad \textit{Positron emission} \quad (30.2)$$

$$p + e^- \rightarrow n + \nu_e \qquad \textit{Electron capture} \quad (30.3)$$

Ordinary atoms are composed of neutrons, protons, and electrons. There is apparently no reason why atoms composed of antineutrons, antiprotons, and positrons should not be stable and behave in every way like ordinary atoms. It would seem to be an attractive notion that equal amounts of matter and **antimatter** came into being at the origin of the universe and became segregated into separate galaxies. The spectra of the light emitted by the members of antimatter galaxies would be exactly the same as the spectra of the light emitted by the members of galaxies of ordinary matter. Thus we have no way to distinguish between the two—except when antimatter comes into contact with ordinary matter. Mutual annihilation would then occur with the release of an immense amount of energy. (A postage stamp of antimatter annihilating a postage stamp of matter would give enough energy to send the space shuttle into orbit.) But the gamma rays of characteristic energies that such an event would create have never been observed, so it seems the universe consists entirely of ordinary matter.

30.4 Particles that Carry Force

Particle exchange can lead to either attraction or repulsion.

In 1935 the Japanese physicist Hideki Yukawa suggested that the strong nuclear interaction could arise from the exchange of particles between nucleons. Today these particles are called **pions.** Pions may be charged or neutral. Those with charges of $+e$ or $-e$ have rest masses of 273 times the electron mass, while neutral pions have rest masses of 264 times the electron mass. The neutral pion, like the photon, is its own antiparticle. Pions are members of a class of elementary particles called **mesons**—the word *pion* is a contraction of the original name π-meson.

The rough analogy illustrated in Fig. 30.10 may help in understanding how particle exchange can lead to both attractive and repulsive forces. Each child in the figure has a pillow. When the children exchange pillows by snatching them from each other's grasp, the effect is like that of a mutually attractive force. On the other hand, the children may also exchange pillows by throwing them at each other. Here conservation of momentum requires that the children move apart, just as if a repulsive force acted between them.

According to Yukawa's theory, nearby nucleons constantly exchange mesons without themselves being altered. Of course, the emission of a meson by a nucleon at rest that does not lose a corresponding amount of mass violates the law of conser-

Figure 30.10

Particle exchange can lead to attractive or repulsive forces.

vation of energy. However, the law of conservation of energy, like all physical laws, deals only with measurable quantities. Because the uncertainty principle restricts the accuracy with which we can perform certain combinations of measurements, it limits the range of application of physical laws such as that of energy conservation.

In Section 26.9 we saw that the uncertainty principle could be expressed as

$$\Delta E\,\Delta t \geqslant \frac{h}{2\pi} \qquad (30.4)$$

This suggests that a process can take place in which energy is temporarily not conserved by an amount ΔE *provided* that the time interval Δt in which the process occurs is not more than $h/2\pi\Delta E$. Thus the creation, transfer, and disappearance of a meson do not conflict with the conservation of energy if the sequence takes place fast enough. Equation (30.4) gives us a way to estimate the mass of the pion.

Let us assume that the temporary energy discrepancy ΔE is of the same magnitude as the rest energy mc^2 of the pion, and that the pion travels at nearly the speed of light c as it goes from one nucleon to another. (These assumptions are crude because the KE of the pion is ignored, but all we are after is an approximate figure for m.) The time Δt the pion spends between its creation in one nucleon and its absorption in another cannot be greater than R/c, where R is the maximum distance that can separate interacting nucleons (Fig. 30.11).

We therefore have, using the symbol $\approx$ to indicate that the result is only an approximation,

$$\Delta E\,\Delta t \approx \frac{h}{2\pi}$$

$$(mc^2)\left(\frac{R}{c}\right) \approx \frac{h}{2\pi}$$

$$m \approx \frac{h}{2\pi Rc} \qquad \textit{Pion mass estimate} \qquad (30.5)$$

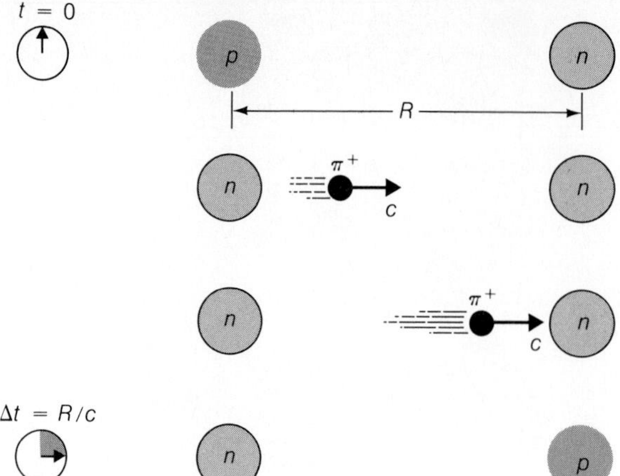

Figure 30.11

A meson can come into being, travel to another nucleon, and disappear if the entire process occurs rapidly enough.

The strong interaction responsible for the attractive forces between nucleons has a range of about 1.7×10^{-15} m. This value for R in the above formula gives $m \approx 2.1 \times 10^{-28}$ kg for the pion mass, which is about 230 electron masses. Of course, such a calculation is hardly rigorous, but if Yukawa's theory has any validity, the pions he postulated should have masses somewhere in this vicinity—as they do. Heavier mesons than the pion have also been discovered, some over a thousand times more massive than the electron. The contribution of these mesons to nuclear forces is, from Eq. (30.5), limited to shorter distances than those characteristic of pions.

How can the meson hypothesis be verified? A sufficiently energetic nuclear collision should be able to liberate mesons without violating energy conservation. Such collisions occur in nature only when fast cosmic-ray particles from space strike oxygen and nitrogen nuclei in the atmosphere, and pions were first identified in cosmic-ray experiments in the 1940s. Since then, high-energy accelerators have permitted pions and other mesons to be studied in the laboratory.

Other Force Carriers

Even before Yukawa's work, particle exchange had been proposed as the mechanism responsible for electromagnetic forces. In this case the particles are photons that, being massless, are not limited in range by Eq. (30.5). However, the farther apart two charges are, the smaller must be the energies of the photons that pass between them (and hence the less the momenta of the photons and the weaker the resulting force) in order that the uncertainty principle not be violated. For this reason electrical forces decrease with distance.

Because the photons exchanged in the interactions of electric charges cannot be detected, they are called **virtual photons.** As in the case of pions, they can become actual photons if enough energy is somehow supplied to free them from the energy conservation constraint. The idea of photons as carriers of electromagnetic forces is attractive on many counts. An obvious one is that it explains why such forces are transmitted with the speed of light and not, say, instantaneously. The full theory is called **quantum electrodynamics.** Its conclusions have turned out to be in extraordinarily precise agreement with the data on such phenomena as the photoelectric effect, pair production and annihilation, and the emission of photons by excited atoms and by accelerated charges.

What about the weak and gravitational interactions? Since the weak interaction has a very short range, the **intermediate bosons** that carry it should have correspondingly large masses—over 30 times the proton mass, in fact. One kind, called W, has a charge of $+e$ or $-e$ and is involved in ordinary beta decays. Another, called Z, is electrically neutral and is heavier than the W; its effects are confined to certain high-energy events. Both the W and the Z particles have been observed experimentally. (A boson is a particle not subject to the exclusion principle.)

Because the range of a gravitational field is unlimited, its carrier, the **graviton,** should have no mass and, like the photon, should travel at the speed of light. It should also be hard to detect since the gravitational interaction is so feeble. There is as yet no definite evidence either for or against the existence of the graviton.

30.5 Leptons

Three pairs plus their antiparticles.

The electron and the neutrino ν_e emitted during beta decay, together with their antiparticles, are leptons, which are particles not affected by the strong interaction. Are there any others?

Outside of nuclei the pion is not stable but decays into a lighter particle, the **muon,** which is a lepton. The decays of the three kinds of pion are shown in Fig. 30.12. Muons are among the decay products of other unstable elementary particles as well. Muons themselves are also unstable and decay into electrons and neutrinos, as in Fig. 30.13.

Since muons decay relatively slowly and are not subject to the strong interaction, they penetrate considerable amounts of matter. As noted in Chapter 25, the great majority of cosmic-ray secondary particles at sea level are muons, and they have been found deep underground.

The neutrinos involved in pion decay have been denoted ν_μ, whereas those involved in beta decay have been denoted ν_e. Until 1962 only a single kind of neutrino was known to exist. In that year an experiment was performed in which pions were produced by bombarding a metal target with high-energy protons from an accelerator. The pion decays liberated neutrinos, and the interactions of these neutrinos with

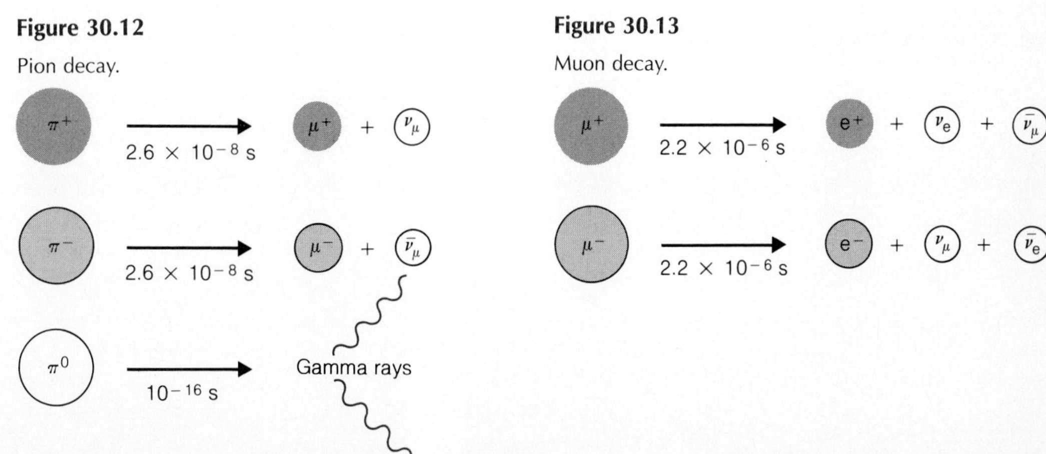

Figure 30.12
Pion decay.

Figure 30.13
Muon decay.

matter was studied. The only inverse reactions found led to the production of muons; no electrons whatever were created. Hence the neutrinos set free in pion decay are different from those set free in beta decay.

A third pair of leptons are the recently discovered **tau** particle and the neutrino associated with it. The mass of the tau is nearly twice that of the proton; like the muon, it is unstable. There is strong evidence for the belief that these three pairs of particles—the electron, the muon, and the tau, together with their neutrinos—are the only leptons.

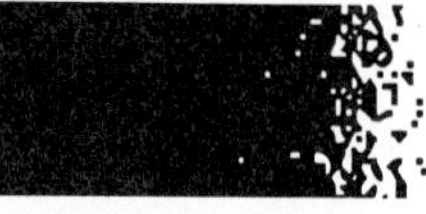

30.6 Hadrons

They are composed of quarks, which have fractional charges and a property called "color."

Several hundred elementary particles have been discovered thus far that are subject to the strong interaction and so are hadrons. Some hadrons, like the pion, are mesons that mediate the strong interaction. They have no spin. Most hadrons are **baryons,** nucleons and heavier particles that have spins of 1/2 and 3/2. Baryons other than nucleons seem to play no role in the behavior of ordinary matter. They decay in less than a billionth of a second, often in a series of steps, into stable particles: protons, electrons, neutrinos, and photons. Many hadrons have several different ways to decay. Here is one sequence that the Ω^- baryon can follow in its decay:

$$\begin{aligned}\Omega^- \rightarrow \Xi^0 &+ \pi^- \\ \hookrightarrow \Lambda^0 &+ \pi^0 \\ \hookrightarrow p^+ &+ \pi^-\end{aligned}$$

The Ω^- (*omega*) baryon's mass is about 1.8 m_{proton}, the Ξ^0 (*xi*) baryon's mass is about 1.4 m_{proton}, and the Λ^0 (*lambda*) baryon's mass is about 1.2 m_{proton}. The π^- and π^0 pions decay as in Fig. 30.12, and the muons from the decay of the π^- pion go on to decay as in Fig. 30.13. Thus the final result of the decay of the Ω^- is a proton, two electrons, four neutrinos, and two photons.

The Tevatron at the Fermi Laboratory in Illinois is 2 km in diameter and has accelerated protons to energies of 1 TeV, over a thousand times their rest energy.

Circulating protons in the Tevatron lower ring are confined to narrow beams by 4.4-tesla superconducting magnets. Collisions of these protons create showers of particles, mostly short-lived, that provide information on the ultimate nature of matter.

Although stable inside a nucleus, the neutron beta decays in free space with an average lifetime of about 15 min:

$$n^0 \rightarrow p^+ + e^- + \bar{\nu}_e$$

Thus the proton is the only stable hadron, since all mesons and all other baryons are unstable.

Quarks

What is known about the various hadrons suggests that they are composite objects whose constituent particles are known as **quarks.** Quarks are thought to be elementary in the same sense as leptons, but unlike leptons they have never been isolated experimentally. The present status of quarks is much like that of neutrinos for 25 years after they were proposed: A wealth of indirect evidence supports their existence, but something stands in the way of detecting them. The parallel may not be accurate, however, because the elusiveness of the neutrino is merely due to its feeble interaction with matter. On the other hand, there may well be a basic reason why quarks cannot occur independently outside hadrons.

Quarks have a number of remarkable properties. One is their possession of fractional electric charges, either $\pm\frac{1}{3}e$ or $\pm\frac{2}{3}e$, something unknown elsewhere in nature. Only six quarks and their antiquarks account for all known hadrons. These quarks seem to fall into three families, or "generations," as shown in Table 30.1, which also lists the leptons associated with each generation.

The quark model of hadrons was proposed in 1963 by Murray Gell-Mann and, independently, by George Zweig. In this model every baryon consists of three quarks and every meson of a quark and an antiquark. Quarks have spins of $\frac{1}{2}$, which accounts for the observed spins of baryons and mesons. Figure 30.14 shows the quark

Table 30.1 The three generations of quarks and leptons. Each particle has an antiparticle of opposite sign. Symbols for antiparticles have a bar over them, for instance, $\bar{u}$.

Generation	Quark	Symbol	Charge, e
1	Up	u	$+\frac{2}{3}$
	Down	d	$-\frac{1}{3}$
2	Charm	c	$+\frac{2}{3}$
	Strange	s	$-\frac{1}{3}$
3	Top	t	$+\frac{2}{3}$
	Bottom	b	$-\frac{1}{3}$
Generation	**Lepton**	**Symbol**	**Charge, e**
1	Electron	e^-	-1
	e-Neutrino	ν_e	0
2	Muon	μ^-	-1
	μ-Neutrino	ν_u	0
3	Tau	τ^-	-1
	τ-Neutrino	ν_r	0

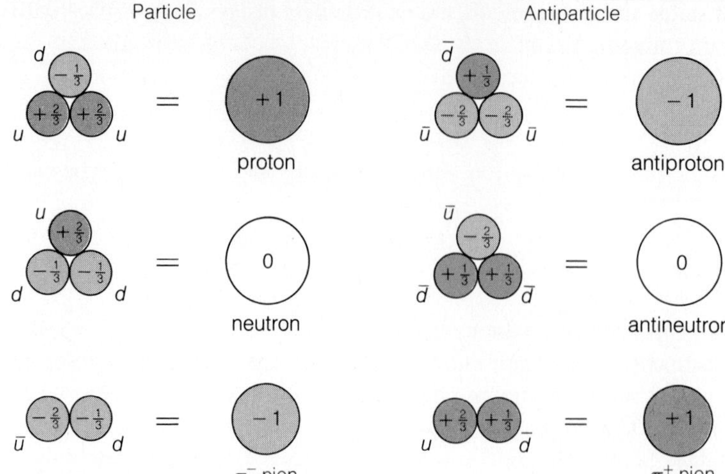

Figure 30.14

Compositions of the proton, neutron, and negative pion, together with their antiparticles, according to the quark model of hadrons. Electric charges are in units of e.

compositions of the proton, neutron, and π^- pion, together with those of their antiparticles. The proton consists of two u and one d quarks, which gives a total charge of $(+\frac{2}{3} + \frac{2}{3} - \frac{1}{3})e = +e$, and the neutron of one u and two d quarks, which gives a total charge of $(+\frac{2}{3} - \frac{1}{3} - \frac{1}{3})e = 0$. The π^+ pion, a meson, consists of a u quark and a $\bar{d}$ antiquark, which gives a total charge of $(+\frac{2}{3} + \frac{1}{3})e = +e$.

Of the three generations of quarks and leptons in Table 30.1, only the first have any connection with ordinary matter. Thus only two leptons and two quarks are

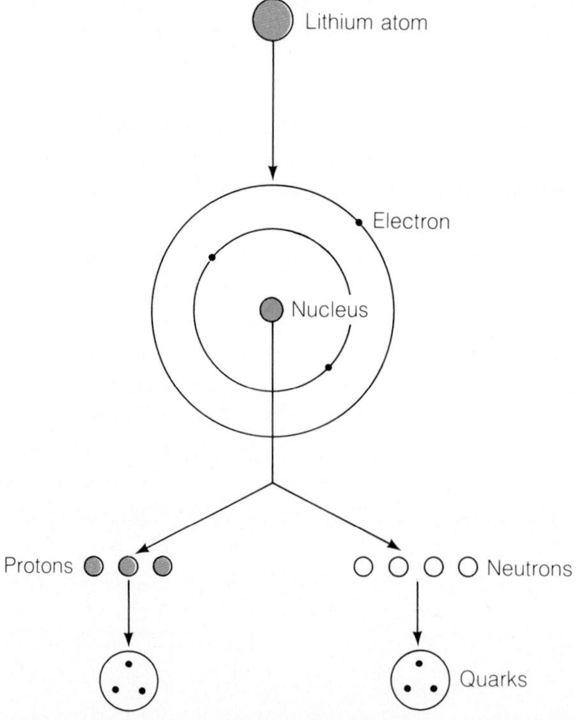

Figure 30.15

The search for truly elementary particles has led to the discovery of particles within particles. Today all ordinary matter seems to be made up of electrons and quarks. Shown are the various levels of organization of a lithium atom.

enough—apparently—to account for all the properties of ordinary matter (Fig. 30.15). The other generations of quarks and leptons are involved only in the unstable particles created in high-energy collisions and in their decays.

Quark Color

A serious objection to the idea that hadrons are composed of quarks was that the presence of more than one quark of the same kind in a single particle (for instance, two *u* quarks in a proton) violates the exclusion principle, to which quarks ought to be subject. To take care of this objection, it was suggested that quarks and antiquarks have an additional property of some kind that can be manifested in different ways. An analogy is electric charge, a property that can be manifested in the two ways that have come to be called positive and negative. In the case of quarks, the new property was whimsically called **color,** and its modes of expression were termed red, green, and blue. The antiquark colors are antired, antigreen, and antiblue.

According to this scheme, all three quarks in a baryon have different colors, which satisfies the exclusion principle. Such a combination can be thought of as white by analogy with the way red, green, and blue light combine to make white light. (Of course, there is no connection between quark colors and actual visual colors.) A meson is supposed to consist of a quark of one color and an antiquark of the corresponding anticolor, thereby canceling out the color. As a result both baryons and mesons are colorless: Quark color is a property that is significant within hadrons but cannot be observed in the outside world.

The notion of quark color has turned out to be more than just a way to get around the exclusion principle. In particular, it is possible to consider the strong interaction as being based on quark color, in the same sense as the electromagnetic interaction is based on electric charge. The resulting theory, called **quantum chromodynamics,** has had a number of successes in explaining elementary-particle phenomena.

30.7 The Fundamental Interactions

Their number grows smaller.

Only four fundamental interactions between elementary particles apparently govern the structure and behavior of the entire physical universe on all scales of size, from atomic nuclei to galaxies of stars. These interactions were described earlier; their basic properties are summarized in Table 30.2. The relative strengths of the interactions vary enormously with distances. Thus the strong force between nearby nucleons completely dominates the gravitational force between them, but when the nucleons are even a millimeter apart, the reverse is true. The existence of nuclei is a consequence of the strong interaction; the existence of atoms is a consequence of the electromagnetic interaction. Because matter in bulk is electrically neutral and the strong and weak interactions are severely limited in range, the gravitational interaction, which is completely negligible on a small scale, becomes the dominant one on a large scale. The function of the weak interaction in the structure of matter seems to be limited to allowing nuclei with inappropriate neutron/proton ratios to undergo corrective beta decays.

If hadrons are indeed composed of quarks, as appears likely, then the strong interaction between hadrons ought to have its origin in an interaction between quarks. The particles that quarks exchange to produce their interaction are called **gluons.** Gluons are supposed to be massless and to move at the speed of light, and each one carries a color and an anticolor. When a quark emits or absorbs a gluon, its color changes accordingly, just as the emission of a charged meson by a nucleon changes its electric charge. For instance, a blue quark that emits a blue-antired gluon becomes a red quark, and a red quark that absorbs this gluon becomes a blue quark. Quantum chromodynamics, the theory of how quarks interact with each other, not only is able to explain the behavior of quarks within hadrons but also has predicted certain effects that were afterward observed in high-energy particle experiments. However, this theory has not yet been developed to the point where it can account in more than a general way for the longer-range interaction between quarks that is observed as the strong interaction between hadrons.

Studies independently carried out in the 1960s by Steven Weinberg and Abdus Salam showed that the weak and electromagnetic interactions are in reality different manifestations of the same essential phenomenon. Supported by experiment, this conclusion is now widely accepted. More recently it has become apparent that the strong interaction can be linked to the unified weak and electromagnetic ones as well. Leptons and quarks find natural places in such a grand unified theory, and it is possible in this way to explain, among other things, why the electron (a lepton) and the proton (a composite of quarks) have electric charges of exactly the same magnitude.

Part of the new theory is an interaction between leptons and quarks that allows a member of one class to be transformed into a member of the other. This implies the instability of protons, hitherto regarded as completely stable, which ought eventually to decay into less massive leptons. The lepton-quark interaction is extremely feeble, to be sure, so the estimated mean life of the proton is around 10^{30} years; by comparison, the universe is less than 10^{11} years old. Nevertheless, proton decay with such a lifetime is within reach of experiment, but the results thus far have all been negative.

Table 30.2 The four fundamental interactions.

Interaction	Particles affected	Range	Relative strength	Particles exchanged	Role in universe
Strong	Quarks	$\sim 10^{-15}$ m	1	Gluons	Holds quarks together to form nucleons
	Hadrons			Mesons	Holds protons and neutrons together to form atomic nuclei
Electromagnetic	Charged particles	∞	$\sim 10^{-2}$	Photons	Determines structures of atoms, molecules, solids, and liquids; is important factor in astronomical universe
Weak	Hadrons and leptons	$\sim 10^{-17}$ m	$\sim 10^{-13}$	Intermediate bosons	Helps determine compositions of atomic nuclei; mediates transformations of quarks and leptons
Gravitational	All	∞	$\sim 10^{-40}$	Gravitons	Assembles matter into planets, stars, and galaxies

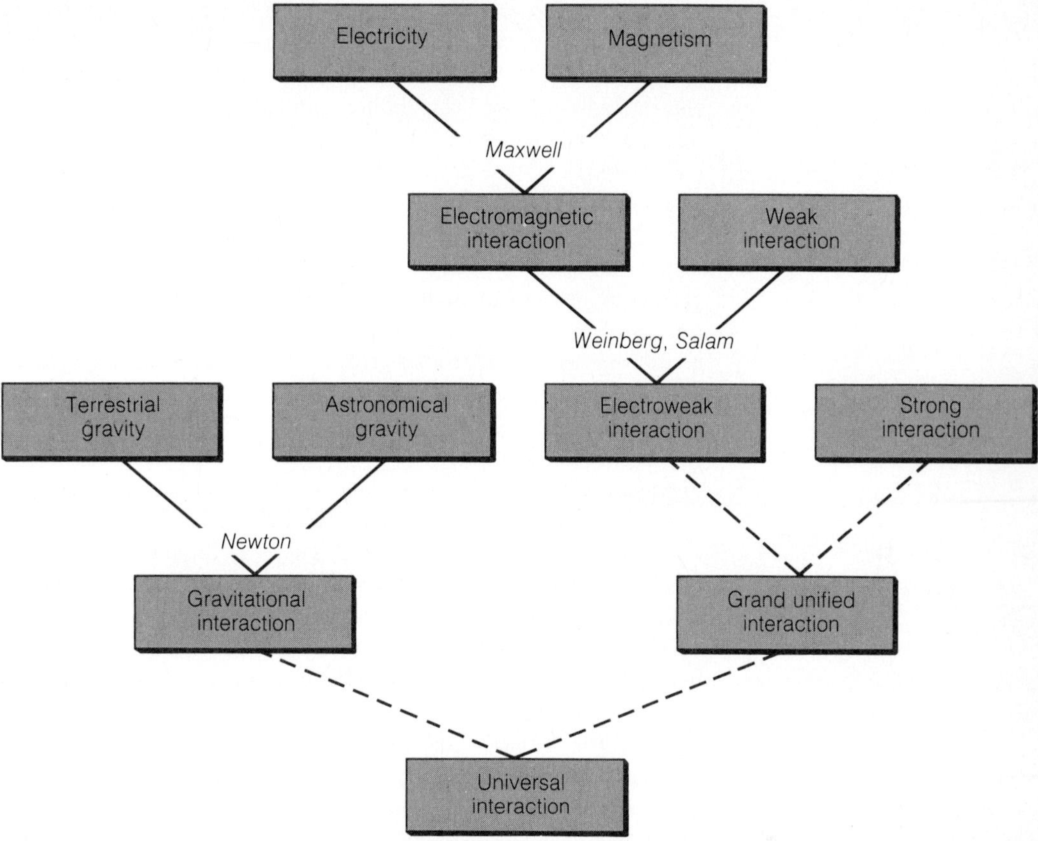

Figure 30.16

One of the goals of physics is a single theoretical picture that unites all the ways in which particles of matter interact with each other. Much progress has been made, but the task is not finished.

Although much remains to be done, and no doubt further surprises are in store, at least the outline of a unified picture of the strong, weak, and electromagnetic interactions is in existence—a remarkable achievement. Still more remarkable would be a way to weave the gravitational interaction into a single skein with the others, and there are hints that this goal is not beyond reach (Fig. 30.16).

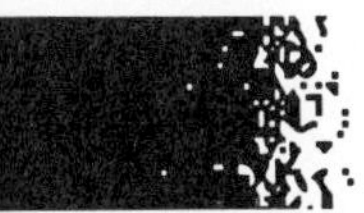

30.8 History of the Universe

It began with a bang.

The observed uniform expansion of the universe points to an origin in a Big Bang about 15 billion years ago, as mentioned in Section 12.8. In the absence of a quantum theory of gravity, nothing can be said about what happened immediately after the Big Bang. From 10^{-43} s on, however, the grand unified theory that ties together the strong, electromagnetic, and weak interactions, even though incomplete, allows a general picture to be sketched of what may well have happened.

Shortly after the Big Bang, the universe was a tiny, immensely hot fireball of matter and energy. As the fireball expanded and cooled, it underwent a series of transitions at specific temperatures. An analogy can be made with the cooling of steam, which becomes water and then ice as its temperature falls. The first transition seems to have occurred at $t = 10^{-35}$ s, when quarks and leptons assumed their present identities. At this point particles outnumbered antiparticles, perhaps a billion and one particles for every billion antiparticles of the same kind. As time went on, matter and antimatter annihilated each other to leave a universe that seems to contain only matter.

Somewhere around $t = 10^{-6}$ s, the quarks joined together to form hadrons. At about 1 s neutrino energies became too small for them to interact with the hadron-lepton soup. The existing neutrinos and antineutrinos remained in the universe but ceased to participate in its evolution. From then on protons could no longer be transformed into neutrons by the inverse beta-decay events, but the free neutrons could beta-decay into protons. However, nuclear reactions were then starting to occur that incorporated many of the neutrons into helium nuclei before they decayed. These reactions stopped at $t = 3$ min when, according to theory, the ratio of protons to ${}^4_2He$ nuclei should have been about 3:1, which in fact is the ratio found in most of the universe today.

From $t = 3$ min to $t \approx 100{,}000$ y, the universe consisted of a plasma of electrons and hydrogen and helium nuclei in thermal equilibrium with radiation. At the latter time the temperature had fallen below 10,000 K; this permitted hydrogen atoms to form without being disrupted immediately by collisions (see Exercise 3 of Chapter 27). Since photons interact much more readily with charged particles than with neutral atoms, matter and radiation were now largely decoupled and the universe became transparent.

The radiation that remained should then have continued to spread out with the rest of the universe, undergoing Doppler shifts to longer and longer wavelengths. We would expect this remnant radiation to come equally strongly from all directions today, and calculations predict for it a spectrum like that of a blackbody at 2.7 K—and such radiation has actually been found. The universe is bathed in a sea of radio waves whose ultimate origin was the primeval fireball that followed the Big Bang. Like the 3:1 hydrogen:helium ratio, this radiation supports our reconstruction of the early history of the universe.

Radio waves that originated early in the history of the universe were first detected with a sensitive receiver attached to this 15-m-long antenna at Holmdel, New Jersey.

Once matter and radiation were decoupled, gravity became a major factor in the evolution of the universe. One or two billion years after the big bang, the hydrogen and helium of the universe began to accumulate in separate clouds under the influence of gravity. As the young galaxies contracted, local concentrations of gas occurred that became the first stars. Some of the early stars went through their life cycles rapidly to end as supernovas (Section 29.12). The explosions of these supernovas dumped elements heavier than helium into the remaining galactic gas. As a result the matter from which later stars condensed was a mixture of all the elements, not just hydrogen and helium. By the time our sun came into being, the loose material of our Milky Way galaxy contained between 1% and 2% of the heavier elements, often in the form of small, solid dust grains. The planets apparently formed at the same time as the sun, and living things developed on at least one of them, which brings us to the present day.

And the future? One possibility is that the universe is ''open'' and will continue to expand forever. It is also possible—the data are insufficient to decide at this time—that enough matter is present in the universe for it to be ''closed.'' If this is the case, gravity will sooner or later stop the expansion, and the universe will begin to contract. The progression of events will then be the reverse of those that took place after the Big Bang, with an eventual Big Crunch. And after that another Big Bang? Then the universe would be cyclic, with no real beginning or end. Enough is not yet known to enable us to say which fate is the more likely.

Important Terms

The **neutrino** is an unchanged particle of small or zero mass that is emitted in the decay of certain elementary particles. It can possess energy and both linear and angular momentum.

Nearly every elementary particle has an **antiparticle** counterpart whose electric charge and certain other properties have the opposite sign. Thus the antiparticle of the electron is the positron (charge $+e$) and that of the proton is the antiproton (charge $-e$). When a particle and an antiparticle of the same kind come together, they **annihilate** each other, with the vanished mass reappearing as photons or mesons.

The four categories of elementary particles are the photons, the leptons, the mesons, and the baryons. **Leptons** are point particles with no internal structures; examples are the electron and the neutrino. **Mesons** are particles that can be regarded as ''carriers'' of the strong interaction; an example is the pion. **Baryons** include nucleons and heavier unstable particles. Mesons and baryons, jointly known as **hadrons,** are subject to the strong interaction and are thought to be composed of **quarks,** which are particles with fractional electric charges that have not as yet been experimentally isolated.

The four **fundamental interactions,** in order of strength, are the strong, the electromagnetic, the weak, and the gravitational. The electromagnetic and weak interactions, and probably the strong as well, are closely related.

MULTIPLE CHOICE

1. Gamma rays do not lose energy in passing through matter by

- **a.** materializing in electron-positron pairs
- **b.** giving up all their energy to atomic electrons in the photoelectric effect
- **c.** giving up part of their energy to atomic electrons in the Compton effect
- **d.** giving up part of their energy to neutrinos

2. The neutrino does not possess

- **a.** charge
- **b.** linear momentum
- **c.** angular momentum
- **d.** energy

3. The mass of the neutrino is

- **a.** zero or nearly so
- **b.** equal to that of the electron

c. between that of the electron and that of the muon
d. greater than that of the muon

4. Neutrinos were shown to actually exist through
a. bubble-chamber pictures of their paths
b. the ionization they produce in a scintillation counter
c. Compton-type scattering by electrons
d. nuclear reactions they cause

5. A particle that is its own antiparticle is the
a. neutron b. neutrino
c. photon d. quark

6. When an electron and a positron annihilate each other, two gamma-ray photons rather than just one are created in order to conserve
a. charge b. energy
c. mass d. linear momentum

7. The process of pair production cannot produce
a. a proton and an antiproton
b. a neutron and an antineutron
c. a neutron and a neutrino
d. an electron and a positron

8. Experimental evidence suggests that the universe contains
a. little or no antimatter
b. some antimatter but less than the amount of matter
c. equal amounts of matter and antimatter
d. more antimatter than matter

9. The strong interaction between nucleons is believed to arise through the exchange of
a. neutrinos b. mesons
c. leptons d. baryons

10. Cosmic-ray particles from space, which are largely energetic protons, interact with atoms in the upper atmosphere to give secondary particles. Most of the secondaries that reach sea level are
a. gamma rays b. electrons and positrons
c. muons d. neutrinos

11. Of the following particles, the one that most closely resembles the muon is the
a. antiproton b. electron
c. pion d. photon

12. Particles unrelated to the strong interaction are
a. leptons b. hadrons
c. quarks d. mesons

13. The particles with the greatest masses are the
a. photons b. leptons
c. mesons d. baryons

14. The only one of the following particles that is a true elementary particle is the
a. neutron b. pion
c. muon d. alpha particle

15. The only one of the following particles that does not decay in free space is the
a. neutrino b. neutron
c. muon d. pion

16. All leptons are
a. stable
b. composed of quarks
c. subject to the weak interaction
d. subject to the strong interaction

17. Quarks and antiquarks have charges of
a. $0, \pm e$ b. $\pm\frac{1}{3}e$
c. $\pm\frac{1}{3}e, \pm\frac{2}{3}e$ d. $0, \pm\frac{1}{3}e, \pm\frac{2}{3}e$

18. Quarks are not present in
a. electrons b. protons
c. neutrons d. hadrons

19. A particle composed of a quark and an antiquark is the
a. muon b. pion
c. proton d. neutrino

20. A particle composed of three quarks is the
a. muon b. pion
c. proton d. neutrino

21. Which of the following would be violated if the quarks in a hadron had the same color?
a. Uncertainty principle
b. Exclusion principle
c. Conservation of energy
d. Conservation of charge

22. The matter in the early universe that eventually condensed into galaxies and then into stars consisted of
a. only hydrogen
b. hydrogen and helium in an approximately 3:1 ratio by mass
c. hydrogen and helium in equal amounts
d. hydrogen and helium in an approximately 1:3 ratio by mass

23. The elements heavier than hydrogen and helium of which the earth and the other planets are composed probably were formed in
a. the sun
b. supernova explosions
c. the Big Bang
d. the Big Crunch

QUESTIONS

1. What is the chief way in which heavy charged particles, such as protons and alpha particles, lose energy in passing through matter?

2. In what ways do electrons lose energy in passing through matter?

3. In what ways do gamma rays lose energy in passing through matter?

4. Discuss the similarities and differences between the photon and the neutrino.

5. What enables neutrinos to travel immense distances through matter without interacting?

6. What must be true of the directions of the two gamma-ray photons into which a neutral pion at rest decays? Why?

7. No particle of fractional charge has yet been observed. If none is found in the future either, does this necessarily mean that the quark hypothesis is wrong?

8. Distinguish between quarks and gluons.

9. A member of the Σ (*sigma*) group of particles consists of two u quarks and an s quark. What is its charge?

10. The Λ (*lambda*) particle consists of a u quark, a d quark, and an s quark. What is its charge?

11. The gravitational interaction is the weakest of all, yet it alone governs the motions of the planets around the sun. Why?

Answers to Multiple Choice

1. d	7. c	13. d	19. b
2. a	8. a	14. c	20. c
3. a	9. b	15. a	21. b
4. d	10. c	16. c	22. b
5. c	11. b	17. c	23. b
6. d	12. a	18. a	

APPENDIX A USEFUL MATHEMATICS

A–1 ALGEBRA

Algebra is a generalized arithmetic in which symbols are used in place of numbers. Algebra thus provides a language in which general relationships can be expressed among quantities whose numerical values need not be known in advance.

The arithmetical operations of addition, subtraction, multiplication, and division have the same meanings in algebra. The symbols of algebra are normally letters of the alphabet. If we have two quantities a and b and add them to give the sum c, we would write

$$a + b = c$$

If we subtract b from a to give the difference d, we would write

$$a - b = d$$

Multiplying a and b together to give e may be written in any of these ways:

$$a \times b = e \qquad ab = e \qquad (a)(b) = e$$

Whenever two algebraic quantities are written together with nothing between them, it is understood that they are to be multiplied.

Dividing a by b to give the quotient f is usually written

$$\frac{a}{b} = f$$

but it may sometimes be more convenient to write

$$a/b = f$$

which has the same meaning.

Parentheses and brackets are used to show the order in which various operations are to be performed. Thus

$$\frac{(a + b)c}{d} - e = f$$

means that, in order to find f, we are first to add a and b together, then multiply their sum by c and divide by d, and finally subtract e.

A–2 EQUATIONS

An *equation* is simply a statement that a certain quantity is equal to another one. Thus

$$7 + 2 = 9$$

which contains only numbers, is an arithmetical equation, and

$$3x + 12 = 27$$

which contains a symbol as well, is an algebraic equation. The symbols in an algebraic equation usually cannot have any arbitrary values if the equality is to hold. Finding the possible values of these symbols is called *solving* the equation. The *solution* of the latter equation above is

$$x = 5$$

since only when x is 5 is it true that $3x + 12 = 27$.

In order to solve an equation, a basic principle must be kept in mind.

Any operation performed on one side of an equation must be performed on the other.

An equation therfore remains valid when the same quantity, numerical or otherwise, is added to or subtracted from both sides, or when the same quantity is used to multiply or divide both sides. Other operations, for instance squaring or taking the square root, also do not alter the equality if the same thing is done to both sides. As a simple example, to solve $3x + 12 = 27$, we first subtract 12 from both sides:

$$3x + 12 - 12 = 27 - 12$$
$$3x = 15$$

To complete the solution we divide both sides by 3:

$$\frac{3x}{3} = \frac{15}{3}$$
$$x = 5$$

To check a solution, we substitute it back in the original equation and see whether the equality is still true. Thus we can check that $x = 5$ by reducing the original algebraic equation to an arithmetical one:

$$3x + 12 = 27$$
$$(3)(5) + 12 = 27$$
$$15 + 12 = 27$$
$$27 = 27$$

Two helpful rules follow directly from the principle stated above. The first is,

Any term on one side of an equation may be transposed to the other side by changing its sign.

To verify this rule, we subtract b from each side of the equation

$$a + b = c$$

to obtain

$$a + b - b = c - b$$
$$a = c - b$$

We see that b has disappeared from the left-hand side and $-b$ is now on the right-hand side.

The second rule is,

A quantity which multiplies one side of an equation may be transposed in order to divide the other side, and vice versa.

To verify this rule, we divide both sides of the equation

$$ab = c$$

by b. The result is

$$\frac{ab}{b} = \frac{c}{b}$$
$$a = \frac{c}{b}$$

We see that b, a multiplier on the left-hand side, is now a divisor on the right-hand side.

EXAMPLE Solve the following equation for x:

$$4\,(x - 3) = 7$$

SOLUTION The above rules are easy to apply here:

$$4\,(x - 3) = 7$$
$$x - 3 = \frac{7}{4}$$
$$x = \frac{7}{4} + 3 = 1.75 + 3 = 4.75$$

When each side of an equation consists of a fraction, all we need to do to remove the fractions is to *cross multiply*:

$$\frac{a}{b} = \frac{c}{d}$$
$$ad = bc$$

What was originally the denominator (lower part) of each fraction now multiplies the numerator (upper part) of the other side of the equation.

EXAMPLE Solve the following equation for y:

$$\frac{5}{y + 2} = \frac{3}{y - 2}$$

SOLUTION First we cross multiply to get rid of the fractions, and then solve in the usual way:

$$5(y - 2) = 3(y + 2)$$
$$5y - 10 = 3y + 6$$
$$5y - 3y = 6 + 10$$
$$2y = 16$$
$$y = 8$$

EXERCISES

1. Solve each of the following equations for x. The answers are given at the end of Appendix A.

(a) $\dfrac{x + 7}{6} = x + 2$

(b) $\dfrac{3x - 42}{9} = 2(7 - x)$

(c) $\dfrac{1}{x + 1} = \dfrac{1}{2x - 1}$

(d) $\dfrac{8}{x} = \dfrac{1}{4 - x}$

(e) $\dfrac{x}{2x - 1} = \dfrac{5}{7}$

A–3 SIGNED NUMBERS

The rules for multiplying and dividing positive and negative quantities are straightforward. First, perform the indicated operation on the absolute value of the quantity (the absolute value of -7 is 7, for instance). If the quantities are both positive or both negative, the result is positive:

$$(+a)(+b) = (-a)(-b) = +ab$$
$$\frac{+a}{+b} = \frac{-a}{-b} = +\frac{a}{b}$$

If one quantity is positive and the other negative the result is negative:

$$(-a)(+b) = (+a)(-b) = -ab$$
$$\frac{-a}{+b} = \frac{+a}{-b} = -\frac{a}{b}$$

A few examples might be helpful:

$$(-6)(-3) = 18 \qquad \frac{-20}{-5} = 4$$
$$(4)(-5) = -20 \qquad \frac{6}{-2} = -3$$
$$(-10)(7) = -70 \qquad \frac{-30}{15} = -2$$

EXAMPLE Find the value of

$$z = \frac{xy}{x - y}$$

when $x = -12$ and $y = 4$.

SOLUTION We begin by evaluating xy and $x - y$, which are

$$xy = (-12) \times 4 = -48$$
$$x - y = (-12) - 4 = -16$$

Hence

$$z = \frac{xy}{x - y} = \frac{-48}{-16} = 3$$

EXERCISES

2. Evaluate the following:

(a) $\dfrac{3(x + y)}{2}$ when $x = 5, y = -2$

(b) $\dfrac{1}{x - y} - \dfrac{1}{x + y}$ when $x = 3, y = -5$

(c) $\dfrac{3(x + 7)}{y + 2}$ when $x = 3, y = -6$

(d) $\dfrac{x + y}{2z} + \dfrac{z}{x - y}$ when $x = -2, y = 2, z = 4$

(e) $\dfrac{x + z}{y} - \dfrac{xy}{2}$ when $x = 2, y = -8, z = 10$

A–4 EXPONENTS

It is often necessary to multiply a quantity by itself a number of times. This process is indicated by a superscript number called the exponent, according to the following scheme:

$$A = A^1$$
$$A \times A = A^2$$
$$A \times A \times A = A^3$$
$$A \times A \times A \times A = A^4$$
$$A \times A \times A \times A \times A = A^5$$

We read A^2 as "A squared" because it is the area of a square of length A on a side; similarly A^3 is called "A cubed" because it is the volume of a cube each of whose sides is A long. More generally we speak of A^n as "A to the nth power." Thus A^5 is read as "A to the fifth power."

When we multiply a quantity raised to some particular power (say A^n) by the same quantity raised to another power (say A^m), the result is that quantity raised to a power equal to the sum of the original exponents. That is,

$$A^n A^m = A^{(n+m)}$$

For example,

$$A^2 A^5 = A^7$$

which we can verify directly by writing out the terms:

$$(A \times A)(A \times A \times A \times A \times A)$$
$$= A \times A \times A \times A \times A \times A \times A = A^7$$

From the above result we see that when a quantity raised to a particular power (say A^n) is to be multiplied by itself a total of m times, we have

$$(A^n)^m = A^{nm}$$

For example,

$$(A^2)^3 = A^6$$

since

$$(A^2)^3 = A^2 \times A^2 \times A^2 = A^{(2+2+2)} = A^6$$

Reciprocal quantities are expressed in a similar way with the addition of a minus sign in the exponent, as follows:

$$\frac{1}{A} = A^{-1} \quad \frac{1}{A^2} = A^{-2} \quad \frac{1}{A^3} = A^{-3} \quad \frac{1}{A^4} = A^{-4}$$

Exactly the same rules as before are used in combining quantities raised to negative powers with one another and with some quantity raised to a positive power. Thus

$$A^5A^{-2} = A^{(5-2)} = A^3$$
$$(A^{-1})^{-2} = A^{-1(-2)} = A^2$$
$$(A^3)^{-4} = A^{-4\times 3} = A^{-12}$$
$$AA^{-7} = A^{(1-7)} = A^{-6}$$

It is important to remember that any quantity raised to the zeroth power, say A^0, is equal to 1. Hence

$$A^2A^{-2} = A^{(2-2)} = A^0 = 1$$

This is more easily seen if we write A^{-2} as $1/A^2$:

$$A^2A^{-2} = A^2 \times \frac{1}{A^2} = \frac{A^2}{A^2} = 1$$

A–5 ROOTS

The *square root* of A, $\sqrt{A}$, is that quantity which, when multiplied by itself, is equal to A:

$$\sqrt{A} \times \sqrt{A} = A$$

The square root has that name because the length of each side of a square of area A is given by $\sqrt{A}$. Some examples of square roots are as follows:

$\sqrt{1} = 1$ because $1 \times 1 = 1$

$\sqrt{16} = 4$ because $4 \times 4 = 16$

$\sqrt{100} = 10$ because $10 \times 10 = 100$

$\sqrt{42.25} = 6.5$ because $6.5 \times 6.5 = 42.25$

$\sqrt{9A^2} = 3A$ because $3A \times 3A = 9A^2$

The square root of a number less than 1 is larger than the number itself:

$\sqrt{0.01} = 0.1$ because $0.1 \times 0.1 = 0.01$

$\sqrt{0.49} = 0.7$ because $0.7 \times 0.7 = 0.49$

The square root of a quantity may be either positive or negative because $(+A)(+A) = (-A)(-A) = A^2$. Whether the positive or negative root (or even both) is correct in a given problem depends upon the details of the problem and is usually obvious. Using exponents we see that, because

$$(A^{1/2})^2 = A^{2\times(1/2)} = A^1 = A$$

we can express square roots by the exponent $\frac{1}{2}$:

$$\sqrt{A} = A^{1/2}$$

Other roots may be expressed similarly. The *cube root* of a quantity A, written $\sqrt[3]{A}$, when multiplied by itself twice equals A. That is,

$$\sqrt[3]{A} \times \sqrt[3]{A} \times \sqrt[3]{A} = A$$

which may be more conveniently written

$$(A^{1/3})^3 = A$$

where $\sqrt[3]{A} = A^{1/3}$.

In general the nth root of a quantity, $\sqrt[n]{A}$, may be written $A^{1/n}$, which is a more convenient form for most purposes. Some examples may be helpful:

$$\sqrt{A^4} = (A^4)^{1/2} = A^{1/2\times 4} = A^2$$
$$(\sqrt{A})^4 = (A^{1/2})^4 = A^{4\times 1/2} = A^2$$
$$\sqrt{\sqrt{A}} = (A^{1/2})^{1/2} = A^{1/2\times 1/2} = A^{1/4}$$
$$(A^{1/4})^{-7} = A^{-7\times(1/4)} = A^{-7/4}$$
$$(A^3)^{-1/3} = A^{-(1/3)\times 3} = A^{-1}$$
$$(A^{1/4})^{1/4} = A^{(1/4)\times(1/4)} = A^{1/16}$$

Although procedures exist for finding roots arithmetically with pencil and paper, in practice it is much easier to use an electronic calculator. Calculators for science and engineering usually have a square root [$\sqrt{}$] key. For other roots, such as $A^{1/3}$, and for nonintegral exponents in general, such as $A^{3.27}$, logarithms must be used, as described in Appendix B-5. A calculator with [LOG] and [10^x] or [INV LOG] keys makes such quantities easy to evaluate.

A–6 EQUATIONS WITH POWERS AND ROOTS

An equation that involves powers or roots or both is subject to the same basic principle that governs the manipulation of simpler equations: Whatever is done to one side must be done to the other. Hence the following rules:

An equation remains valid when both sides are raised to the same power, that is, when each side is multiplied by itself the same number of times as the other side.

An equation remains valid when the same root is taken of both sides.

EXAMPLE Newton's law of gravitation states that the force F between two bodies the distance r apart whose masses are m_A and m_B is given by the formula

$$F = G\frac{m_A m_B}{r^2}$$

where G is a universal constant. Solve this formula for r.

SOLUTION We begin by transposing the r^2 to give

$$Fr^2 = Gm_A m_B$$

Next we transpose the F to give

$$r^2 = \frac{Gm_A m_B}{F}$$

and finally take the square root of both sides:

$$r = \sqrt{r^2} = \sqrt{\frac{Gm_A m_B}{F}}$$

The positive value of the root is correct here since a negative separation r between the bodies is meaningless.

Quadratic equations, which have the general form

$$ax^2 + bx + c = 0$$

are often encountered in algebraic calculations. In such an equation, a, b, and c are constants. A quadratic equation is satisfied by the values of x given by the formulas

$$x_+ = \frac{-b + \sqrt{b^2 - 4ac}}{2a}$$

$$x_- = \frac{-b - \sqrt{b^2 - 4ac}}{2a}$$

By looking at these formulas, we can see that the nature of the solution depends upon the value of the quantity $b^2 - 4ac$. When $b^2 = 4ac$, $\sqrt{b^2 - 4ac} = 0$ and the two solutions are equal to just $x = -b/2a$. When $b^2 > 4ac$, $\sqrt{b^2 - 4ac}$ is a real number and the solutions x_+ and x_- are different. When $b^2 < 4ac$, $\sqrt{b^2 - 4ac}$ is the square root of a negative number, and the solutions are different. The square root of a negative number is called an *imaginary number* because squaring a real number, whether positive or negative, always gives a positive number: $(+2)^2 = (-2)^2 = +4$, hence $\sqrt{-4}$ cannot be either $+2$ or -2. Imaginary numbers do not occur in the sort of physical problems treated in this book.

EXAMPLE Solve the quadratic equation

$$2x^2 - 3x - 9 = 0$$

SOLUTION Here $a = 2$, $b = -3$, and $c = -9$, so that

$$b^2 - 4ac = (-3)^2 - 4(2)(-9) = 9 + 72 = 81$$

The two solutions are

$$x_+ = \frac{-b + \sqrt{b^2 - 4ac}}{2a}$$

$$= \frac{-(-3) + \sqrt{81}}{2(2)} = \frac{3 + 9}{4} = \frac{12}{4} = 3$$

$$x_- = \frac{-b - \sqrt{b^2 - 4ac}}{2a}$$

$$= \frac{-(-3) - \sqrt{81}}{2(2)} = \frac{3 - 9}{4} = \frac{-6}{4} = -\frac{3}{2}$$

EXAMPLE Solve the quadratic equation

$$x^2 - 4x + 4 = 0$$

SOLUTION Here $a = 1$, $b = -4$, and $c = 4$, so that

$$b^2 - 4ac = (-4)^2 - 4(1)(4) = 16 - 16 = 0$$

Hence the only solution is

$$x = \frac{-b}{2a} = \frac{-(-4)}{2} = 2$$

EXERCISES

3. Solve each of the following equations for x.

(a) $9x^2 + 12x + 4 = 0$

(b) $x^2 + x - 2 = 0$

(c) $2x^2 - 5x + 2 = 0$

(d) $x^2 + x - 20 = 0$

(e) $4x^2 - 4x - 11 = 0$

A–7 TRIGONOMETRIC FUNCTIONS

A *right triangle* is a triangle two sides of which are perpendicular. Such triangles are frequently encountered in physics, and it is necessary to know how their sides and angles

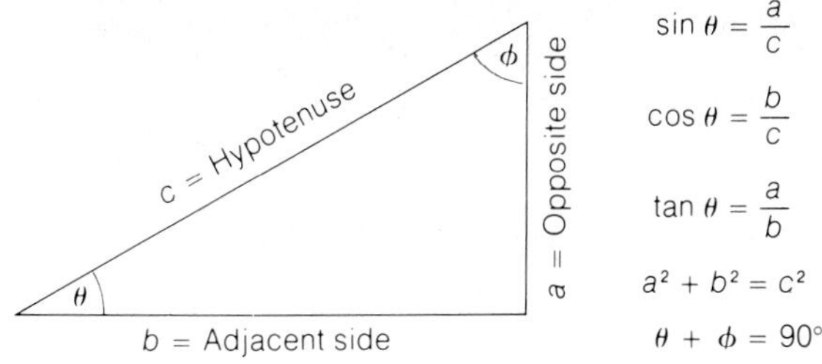

FIG. A–1 A right triangle

are related. The *hypotenuse* of a right triangle is the side opposite the right angle, as in Fig. A–1; it is always the longest side. The three basic trigonometric functions, the *sine, cosine,* and *tangent* of an angle, are defined as follows:

$$\sin\theta = \frac{a}{c} = \frac{\text{opposite side}}{\text{hypotenuse}}$$

$$\cos\theta = \frac{b}{c} = \frac{\text{adjacent side}}{\text{hypotenuse}}$$

$$\tan\theta = \frac{a}{b} = \frac{\text{opposite side}}{\text{adjacent side}}$$

From these definitions we see that

$$\frac{\sin\theta}{\cos\theta} = \frac{a/c}{b/c} = \frac{a}{b} = \tan\theta$$

Numerical values of sin θ, cos θ, and tan θ for angles form 0° to 90° are given in Appendix C. These figures may be used for angles from 90° to 360° with the help of this table:

	θ	90° + θ	180° + θ	270° + θ
sin	sin θ	cos θ	− sin θ	− cos θ
cos	cos θ	− sin θ	− cos θ	sin θ
tan	tan θ	$-\frac{1}{\tan\theta}$	tan θ	$-\frac{1}{\tan\theta}$

For example, if we require the value of sin 120°, we first note that 120° = 90° + 30°. Then, since

$$\sin(90° + \theta) = \cos\theta$$

we have

$$\sin 120° = \sin(90° + 30°) = \cos 30° = 0.866$$

Similarly, to find tan 342°, we begin by noting that 342° = 270° + 72°. Then, since

$$\tan(270° + \theta) = -\frac{1}{\tan\theta}$$

we have

$$\tan 342° = \tan(270° + 72°) = -\frac{1}{\tan 72°} = -\frac{1}{3.078}$$

$$= -0.325$$

The *inverse* of a trigonometric function is the angle whose function is given. For instance, the inverse of sin θ is the angle θ. If sin θ = *x*, then the angle θ may be designated either as θ = arcsin *x* or as θ = $\sin^{-1} x$. It is important to keep in mind that an expression such as $\sin^{-1} x$ does *not* mean 1/sin *x*. The inverse trigonometric functions are as follows:

$$\sin\theta = x$$
$$\theta = \arcsin x = \sin^{-1} x = \text{angle whose sine is } x$$
$$\cos\theta = y$$
$$\theta = \arccos y = \cos^{-1} y = \text{angle whose cosine is } y$$
$$\tan\theta = z$$
$$\theta = \arctan z = \tan^{-1} z = \text{angle whose tangent is } z$$

To find an inverse function, say the angle whose cosine is 0.907, the procedure with some calculators is to enter 0.907 and then press in succession the [ARC] and [COS] keys. The result is θ = 25° to the nearest degree. With other calculators, a shift key must first be pressed that converts the [SIN], [COS], and [TAN] keys, respectively, to $\sin^{-1}$, $\cos^{-1}$, and $\tan^{-1}$, and then the [COS] key is pressed. With a table of cosines, we look for the value nearest to 0.907 in the body of the table, which is 0.906, and then read across to find the corresponding angle.

EXERCISES

4. Express the following functions in terms of angles between 0 and 90°.

(a) sin 100°

(b) sin 300°

(c) cos 150°

(d) cos 350°

(e) tan 180°

(f) tan 220°

A–8 SOLVING A RIGHT TRIANGLE

To solve a given triangle means to find the values of any unknown sides or angles in terms of the values of the known sides and angles. A triangle has three sides and three angles, and we must know the values of at least three of these six quantities, including one of the sides, to solve the triangle for the others. In a right triangle, one of the angles is always 90°, and so all we need here are the lengths of any two of its sides or the length of one side and the value of one of the other angles to find the remaining sides and angles.

Suppose we know the length of the side b and the angle θ in the right triangle of Fig. A–1. From the definitions of sine and tangent we see that

$$\tan\theta = \frac{a}{b}$$

$$a = b\tan\theta$$

$$\sin\theta = \frac{a}{c}$$

$$c = \frac{a}{\sin\theta}$$

This gives us the two unknown sides a and c. To find the unknown angle ϕ, we can use any of these formulas:

$$\phi = \sin^{-1}\frac{b}{c} \qquad \phi = \cos^{-1}\frac{a}{c} \qquad \phi = \tan^{-1}\frac{b}{a}$$

Alternatively, we can use the fact that the sum of the angles in any triangle is 180°. Because one of the angles in a right triangle is 90°, the sum of the other two must be 90°:

$$\theta + \phi = 90°$$

Hence $\phi = 90° - \theta$ here.

Another useful relationship in a right triangle is the Pythagorean theorem, which states that the sum of the squares of the sides of such a triangle adjacent to the right angle is equal to the square of its hypotenuse. For the triangle of Fig. A–1,

$$a^2 + b^2 = c^2$$

Thus we can always express the length of any of the sides of a right triangle in terms of the other sides:

$$a = \sqrt{c^2 - b^2}$$

$$b = \sqrt{c^2 - a^2}$$

$$c = \sqrt{a^2 + b^2}$$

EXAMPLE In the triangle of Fig. A–1, a = 7 cm and b = 10 cm. Find c, θ, and ϕ.

SOLUTION From the Pythagorean theorem,

$$c = \sqrt{a^2 + b^2} = \sqrt{7^2 + 10^2}\,\text{cm}$$

$$= \sqrt{149}\,\text{cm} = 12.2\,\text{cm}$$

Since $\tan\theta = a/b$,

$$\theta = \tan^{-1}\frac{a}{b} = \tan^{-1}\frac{7\,\text{cm}}{10\,\text{cm}}$$

$$= \tan^{-1}0.7 = 35°$$

The value of the other angle ϕ is given by

$$\phi = 90° - \theta = 90° - 35° = 55°$$

EXERCISES

5. Find the values of the unknown sides and angles in the right triangles for which the following data are known:

(a) $\theta = 45°, a = 10$

(b) $\theta = 15°, b = 4$

(c) $\theta = 25°, c = 5$

(d) $a = 3, b = 4$

(e) $a = 5, c = 13$

ANSWERS

1. (a) -1
(b) 8
(c) 2
(d) 32/9
(e) 5/3

2. (a) 4.5
(b) 0.625
(c) -3
(d) -1
(e) 6.5

3. (a) $-2/3$
(b) $1, -2$
(c) 2, 0.5
(d) $4, -5$
(e) $0.5 \pm \sqrt{3} = 2.232, -1.232$

4. (a) cos 10°
(b) $-\cos 30°$
(c) $-\sin 60°$
(d) sin 80°
(e) $-1/\tan 90°$ or tan 0. Since $\tan 90° = \infty$ and $\tan 0 = 0$, both expressions give the same value.
(f) tan 40°

5. (a) $b = 10, c = 14.1, \phi = 45°$
(b) $a = 1.07, c = 4.14, \phi = 75°$
(c) $a = 2.11, b = 4.53, \phi = 65°$
(d) $c = 5, \theta = 37°, \phi = 53°$
(e) $b = 12, \theta = 23°, \phi = 67°$

APPENDIX B POWERS OF TEN AND LOGARITHMS

B–1 POWERS OF TEN

Very small and very large numbers are common in physics. For example, the mass of an electron is 0.000, 000,000,000,000,000,000,000,000,000,910,9 kilogram, and the mass of the earth is 5,983,000,000,000, 000,000,000,000 kilograms. Such numbers in ordinary decimal form are clumsy to write and to make calculations with, and it is hard to appreciate their precise magnitudes because of the sea of zeros.

A better method for expressing numbers makes use of powers-of-ten notation. This method is based on the fact that all numbers may be represented by a number between 1 and 10 multiplied by a power of 10. In powers-of-ten notation the mass of an electron is written simply as 9.109×10^{-31} kilogram and the mass of the earth is written as 5.983×10^{24} kilogram.

Table B-1 contains powers of 10 from 10^{-10} to 10^{10}. Evidently positive powers of 10 (which cover numbers greater than 1) follow this pattern:

TABLE B-1

10^{-10} = 0.000,000,000,1	10^{0} = 1
10^{-9} = 0.000,000,001	10^{1} = 10
10^{-8} = 0.000,000,01	10^{2} = 100
10^{-7} = 0.000,000,1	10^{3} = 1000
10^{-6} = 0.000,001	10^{4} = 10,000
10^{-5} = 0.000,01	10^{5} = 100,000
10^{-4} = 0.000,1	10^{6} = 1,000,000
10^{-3} = 0.001	10^{7} = 10,000,000
10^{-2} = 0.01	10^{8} = 100,000,000
10^{-1} = 0.1	10^{9} = 1,000,000,000
10^{0} = 1	10^{10} = 10,000,000,000

$$10^0 = 1. \quad = 1 \quad = 1 \text{ with decimal point moved 0 places,}$$

$$10^1 = 1.0 \quad = 10 \quad = 1 \text{ with decimal point moved 1 place to the right,}$$

$$10^2 = 1.00 \quad = 100 \quad = 1 \text{ with decimal point moved 2 places to the right,}$$

$$10^3 = 1.000 \quad = 1{,}000 \quad = 1 \text{ with decimal point moved 3 places to the right,}$$

$$10^4 = 1.0000 \quad = 10{,}000 \quad = 1 \text{ with decimal point moved 4 places to the right,}$$

$$10^5 = 1.00000 \quad = 100{,}000 \quad = 1 \text{ with decimal point moved 5 places to the right,}$$

$$10^6 = 1.000000 = 1{,}000{,}000 = 1 \text{ with decimal point moved 6 places to the right, and so on.}$$

The exponent of the 10 indicates how many places the decimal point is moved *to the right* from 1.000 · · ·

A similar pattern is followed by negative powers of 10, whose values always lie between 0 and 1:

$$10^0 = 1. \quad = 1 \quad = 1 \text{ with decimal point moved 0 places,}$$
$$10^{-1} = 01. \quad = 0.1 \quad = 1 \text{ with decimal point moved 1 place to the left,}$$
$$10^{-2} = 001. \quad = 0.01 \quad = 1 \text{ with decimal point moved 2 places to the left,}$$
$$10^{-3} = 0001. \quad = 0.001 \quad = 1 \text{ with decimal point moved 3 places to the left,}$$
$$10^{-4} = 00001. \quad = 0.000,1 \quad = 1 \text{ with decimal point moved 4 places to the left,}$$
$$10^{-5} = 000001. \quad = 0.000,01 \quad = 1 \text{ with decimal point moved 5 places to the left,}$$
$$10^{-6} = 0000001. \quad = 0.000,001 \quad = 1 \text{ with decimal point moved 6 places to the left, and so on.}$$

The exponent of the 10 now indicates how many places the decimal point is moved *to the left* from 1.

Here are a few examples of powers-of-ten notation:

$$600 = 6 \times 100 = 6 \times 10^2$$
$$7940 = 7.94 \times 1000 = 7.94 \times 10^3$$
$$93{,}000{,}000 = 9.3 \times 10{,}000{,}000 = 9.3 \times 10^7$$
$$0.023 = 2.3 \times 0.01 = 2.3 \times 10^{-2}$$
$$0.000{,}035 = 3.5 \times 0.000{,}01 = 3.5 \times 10^{-5}$$

EXERCISES

1. Express the following numbers in decimal notation.

(a)	2×10^5	(e)	1.003×10^{-6}
(b)	8×10^{-2}	(f)	10^{-10}
(c)	7.819×10^2	(g)	9.56×10^{-5}
(d)	4.51×10^8		

2. Express the following numbers in powers-of-ten notation.

(a)	70	(e)	1,000,000
(b)	0.14	(f)	0.007,890
(c)	3.81	(g)	351,600
(d)	8400		

B–2 USING POWERS OF TEN

Let us see how to make calculations using numbers written in powers-of-ten notation. To add or subtract numbers written in powers-of-ten notation, they must be expressed in terms of the *same* power of ten.

$$7 \times 10^4 + 2 \times 10^5 = 0.7 \times 10^5 + 2 \times 10^5 = 2.7 \times 10^5$$
$$5 \times 10^{-2} + 3 \times 10^{-4} = 5 \times 10^{-2} + 0.03 \times 10^{-2} = 5.03 \times 10^{-2}$$
$$8 \times 10^{-3} - 7 \times 10^{-4} = 8 \times 10^{-3} - 0.7 \times 10^{-3} = 7.3 \times 10^{-3}$$
$$4 \times 10^5 - 1 \times 10^6 = 4 \times 10^5 - 10 \times 10^5 = -6 \times 10^5$$

To multiply powers of ten together, add their exponents:

$$(10^n)(10^m) = 10^{n+m}$$

Be sure to take the sign of each exponent into account.

$$(10^2)(10^3) = 10^{2+3} = 10^5$$
$$(10^7)(10^{-3}) = 10^{7-3} = 10^4$$
$$(10^{-2})(10^{-4}) = 10^{-2-4} = 10^{-6}$$

To multiply numbers written in powers-of-ten notation, multiply the decimal parts of the numbers together and add the exponents to find the power of ten of the product:

$$(A \times 10^n)(B \times 10^m) = AB \times 10^{n+m}$$

If necessary, rewrite the result so the decimal part is a number between 1 and 10.

$$(3 \times 10^2)(2 \times 10^5) = (3 \times 2) \times 10^{2+5} = 6 \times 10^7$$
$$(8 \times 10^{-5})(3 \times 10^7) = (8 \times 3) \times 10^{-5+7} = 24 \times 10^2 = 2.4 \times 10^3$$
$$(1.3 \times 10^{-3})(4 \times 10^{-5}) = (1.3 \times 4) \times 10^{-3-5} = 5.2 \times 10^{-8}$$
$$(-9 \times 10^{17})(6 \times 10^{-18}) = (-9 \times 6) \times 10^{17-18} = -54 \times 10^{-1} = -5.4$$

To divide one power of ten by another, subtract the exponent of the denominator from the exponent of the numerator:

$$\frac{10^n}{10^m} = 10^{n-m}$$

Be sure to take the sign of each exponent into account.

$$\frac{10^5}{10^3} = 10^{5-3} = 10^2$$

$$\frac{10^{-2}}{10^4} = 10^{-2-4} = 10^{-6}$$

$$\frac{10^{-3}}{10^{-7}} = 10^{-3-(-7)} = 10^{-3+7} = 10^4$$

To divide a number written in powers-of-ten notation by another number written that way, divide the decimal parts of the numbers in the usual way and use the above rule to find the exponent of the power of ten of the quotient:

$$\frac{A \times 10^n}{B \times 10^m} = \frac{A}{B} \times 10^{n-m}$$

If necessary, rewrite the result so the decimal part is a number between 1 and 10:

$$\frac{6 \times 10^5}{3 \times 10^2} = \frac{6}{3} \times 10^{5-2} = 2 \times 10^3$$

$$\frac{2 \times 10^{-7}}{8 \times 10^4} = \frac{2}{8} \times 10^{-7-4} = \frac{1}{4} \times 10^{-11}$$

$$= 0.25 \times 10^{-11} = 2.5 \times 10^{-12}$$

$$\frac{-7 \times 10^5}{10^{-2}} = -7 \times 10^{5-(-2)} = -7 \times 10^{5+2}$$

$$= -7 \times 10^7$$

$$\frac{5 \times 10^{-2}}{-2 \times 10^{-9}} = -\frac{5}{2} \times 10^{-2-(-9)} = -2.5 \times 10^{-2+9}$$

$$= -2.5 \times 10^7$$

To find the reciprocal of a power of ten, change the sign of the exponent:

$$\frac{1}{10^n} = 10^{-n}$$

$$\frac{1}{10^{-m}} = 10^m$$

$$\frac{1}{10^5} = 10^{-5}$$

$$\frac{1}{10^{-3}} = 10^3$$

Hence the prescription for finding the reciprocal of a number written in powers-of-ten notation is

$$\frac{1}{A \times 10^n} = \frac{1}{A} \times 10^{-n}$$

For example,

$$\frac{1}{2 \times 10^3} = \frac{1}{2} \times 10^{-3} = 0.5 \times 10^{-3} = 5 \times 10^{-4}$$

$$\frac{1}{4 \times 10^{-8}} = \frac{1}{4} \times 10^8 = 0.25 \times 10^8 = 2.5 \times 10^7$$

The powers-of-ten method of writing large and small numbers makes arithmetic involving such numbers relatively easy to carry out. Here is a calculation that would be very tedious if each number were kept in decimal form.

$$\frac{(3800)(0.0054)(0.000{,}001)}{(430{,}000{,}000)(73)}$$

$$= \frac{(3.8 \times 10^3)(5.4 \times 10^{-3})(10^{-6})}{(4.3 \times 10^8)(7.3 \times 10^1)}$$

$$= \frac{(3.8)(5.4)}{(4.3)(7.3)} \times \frac{(10^3)(10^{-3})(10^{-6})}{(10^8)(10^1)}$$

$$= 0.65 \times 10^{(3-3-6-8-1)} = 0.65 \times 10^{-15}$$

$$= 6.5 \times 10^{-16}$$

EXERCISES

3. Perform the following additions and subtractions.

(a) $3 \times 10^2 + 4 \times 10^3$
(b) $7 \times 10^{-2} + 2 \times 10^{-3}$
(c) $4 \times 10^{-5} + 5 \times 10^{-3}$
(d) $6.32 \times 10^2 + 5$
(e) $4 \times 10^3 - 3 \times 10^2$
(f) $3.2 \times 10^{-4} - 5 \times 10^{-5}$
(g) $7 \times 10^4 - 2 \times 10^{-5}$
(h) $4.76 \times 10^{-3} - 4.81 \times 10^{-3}$

4. Evaluate the following reciprocals.

(a) $\frac{1}{10^2}$ (c) $\frac{1}{10^{-2}}$

(b) $\frac{1}{2 \times 10^2}$ (d) $\frac{1}{4 \times 10^{-4}}$

5. Perform the following calculations.

(a) $\frac{(500{,}000)(18{,}000)}{9{,}000{,}000}$

(b) $\frac{(30)(80{,}000{,}000{,}000)}{0.0004}$

(c) $\dfrac{(30{,}000)(0.000{,}000{,}6)}{(1000)(0.02)}$

(d) $\dfrac{(0.002)(0.000{,}000{,}05)}{0.000{,}004}$

(e) $\dfrac{(0.06)(0.0001)}{(0.000{,}03)(40{,}000)}$

(f) $\dfrac{(3 \times 10^4)(5 \times 10^{-12})}{10^3}$

(g) $\dfrac{9 \times 10^{12}}{9 \times 10^{-12}}$

(h) $\dfrac{(8 \times 10^{10})(3)}{6 \times 10^{-4}}$

(i) $\dfrac{10^{-3}}{(5 \times 10^4)(2 \times 10^2)}$

(j) $\dfrac{(5 \times 10^5)(2 \times 10^{-18})}{4 \times 10^4}$

B–3 POWERS AND ROOTS

To square a power of ten, multiply the exponent by 2; to cube a power of ten, multiply the exponent by 3:

$$(10^n)^2 = 10^{2n}$$
$$(10^n)^3 = 10^{3n}$$

In general, to raise a power of ten to the mth power, multiply the exponent by m:

$$(10^n)^m = 10^{m \times n}$$

Be sure to take the sign of each exponent into account, as in these examples:

$$(10^3)^2 = 10^{2 \times 3} = 10^6$$
$$(10^{-2})^5 = 10^{5 \times -2} = 10^{-10}$$
$$(10^{-4})^{-2} = 10^{-2 \times -4} = 10^8$$

To raise a number written in powers-of-ten notation to the mth power, multiply the decimal part of the number by itself m times and multiply the exponent of the power of ten by m:

$$(A \times 10^n)^m = A^m \times 10^{m \times n}$$

If necessary, rewrite the result so the decimal part is a number between 1 and 10:

$$(2 \times 10^5)^2 = 2^2 \times 10^{2 \times 5} = 4 \times 10^{10}$$

$$(3 \times 10^{-3})^3 = 3^3 \times 10^{3 \times -3} = 27 \times 10^{-9}$$
$$= 2.7 \times 10^{-8}$$

$$(5 \times 10^{-2})^{-4} = 5^{-4} \times 10^{-4 \times -2} = \frac{1}{5^4} \times 10^8$$
$$= \frac{1}{625} \times 10^8 = 0.0016 \times 10^8$$
$$= 1.6 \times 10^5$$

To take the mth root of a power of ten, divide the exponent by m:

$$\sqrt[m]{10^n} = (10^n)^{1/m} = 10^{n/m}$$

Thus

$$\sqrt{10^4} = (10^4)^{1/2} = 10^{4/2} = 10^2$$
$$\sqrt[3]{10^9} = (10^9)^{1/3} = 10^{9/3} = 10^3$$
$$\sqrt[3]{10^{-9}} = (10^{-9})^{1/3} = 10^{-9/3} = 10^{-3}$$

In powers-of-ten notation, the exponent of the 10 must be an integer. Hence in taking the mth root of a power of ten, the exponent should be an integral multiple of m. Instead of, for example,

$$\sqrt{10^5} = (10^5)^{1/2} = 10^{2.5}$$

which, while correct, is hardly useful, we would write

$$\sqrt{10^5} = \sqrt{10^1 \times 10^4}$$
$$= \sqrt{10} \times \sqrt{10^4}$$
$$= 3.16 \times 10^2$$

Here are two other examples:

$$\sqrt{10^{-3}} = \sqrt{10^1 \times 10^{-4}} = \sqrt{10} \times \sqrt{10^{-4}}$$
$$= 3.16 \times 10^{-2}$$
$$\sqrt[3]{10^8} = \sqrt[3]{10^2 \times 10^6} = \sqrt[3]{100} \times \sqrt[3]{10^6}$$
$$= 4.64 \times 10^2$$

To take the mth root of a number expressed in powers-of-ten notation, first write the number so the exponent of the 10 is an integral multiple of m. Then take the mth root of the decimal part of the number and divide the exponent by m to find the power of ten of the result:

$$(A \times 10^n)^{1/m} = \sqrt[m]{A} \times 10^{n/m}$$

Thus

$$\sqrt{9 \times 10^4} = \sqrt{9} \times 10^{4/2} = 3 \times 10^2$$
$$\sqrt{9 \times 10^{-6}} = \sqrt{9} \times 10^{-6/2} = 3 \times 10^{-3}$$

$$\sqrt{4 \times 10^7} = \sqrt{40 \times 10^6} = \sqrt{40} \times 10^{6/2}$$
$$= 6.32 \times 10^3$$
$$\sqrt[3]{3 \times 10^{-5}} = \sqrt[3]{30 \times 10^{-6}} \quad \sqrt[3]{30} \times 10^{-6/3}$$
$$= 3.11 \times 10^{-2}$$

EXERCISES

6. Evaluate the following powers.

(a) $(2 \times 10^7)^2$ (c) $(3 \times 10^{-8})^2$

(b) $(2 \times 10^7)^{-2}$ (d) $(5 \times 10^{-4})^{-3}$

7. Evaluate the following roots. Assume $\sqrt{4} = 2$, $\sqrt{40} = 6.3$, $\sqrt[3]{4} = 1.6$, $\sqrt[3]{40} = 3.4$, and $\sqrt[3]{400} = 7.4$.

(a) $(4 \times 10^6)^{1/2}$ (f) $(4 \times 10^{13})^{1/3}$

(b) $(4 \times 10^7)^{1/2}$ (g) $(4 \times 10^{14})^{1/3}$

(c) $(4 \times 10^{-4})^{1/2}$ (h) $(4 \times 10^{-6})^{1/3}$

(d) $(4 \times 10^{-5})^{1/2}$ (i) $(4 \times 10^{-7})^{1/3}$

(e) $(4 \times 10^{-12})^{1/3}$ (j) $(4 \times 10^{-8})^{1/3}$

B–4 SIGNIFICANT FIGURES

An advantage of powers-of-ten notation is that it gives no false impression of the degree of accuracy with which a number is stated. For instance, the equatorial radius of the earth is 6378 km, but it is often taken as 6400 km for convenience in making rough calculations. To indicate the approximate character of the latter figure, all we have to do is write

$$r = 6.4 \times 10^3 \text{ km}$$

whose meaning is

$$r = (6.4 \pm 0.05) \times 10^3 \text{ km}$$

With this method, how large the number is and how accurate it is are both clear. The accurately known digits, plus one uncertain digit, are called *significant figures;* in the above case, r has two significant figures, 6 and 4. If we require greater accuracy, we would write

$$r = 6.38 \times 10^3 \text{ km}$$

which contains three significant figures, or

$$r = 6.378 \times 10^3 \text{ km}$$

which contains four significant figures.

Sometimes one or more zeros in a number are significant figures, and it is proper to retain them when expressing the number in powers-of-ten notation. There is quite a difference between 3×10^5 and 3.00×10^5:

$$3 \times 10^5 = (3 \pm 0.5) \times 10^5$$
$$3.00 \times 10^5 = (3 \pm 0.005) \times 10^5$$

When quantities are combined arithmetically, the result is no more accurate than the quantity with the largest uncertainty. Suppose a 75-kg person picks up a 0.23-kg apple. The total mass of person plus apple is still 75 kg because all we know of the person's mass is that it is somewhere between 74.5 and 75.5 kg, which means an uncertainty greater than the apple's mass. If the person's mass is instead quoted as 75.0 kg, the mass of person plus apple is 75.2 kg; if it is quoted as 75.00 kg, the mass of person plus apple is 75.23 kg. Thus

$$75 \text{ kg} + 0.23 \text{ kg} = 75 \text{ kg}$$
$$75.0 \text{ kg} + 0.23 \text{ kg} = 75.2 \text{ kg}$$
$$75.00 \text{ kg} + 0.23 \text{ kg} = 75.23 \text{ kg}$$

Significant figures must be taken into account in multiplication and division also. For example, if we divide 1.4×10^5 by 6.70×10^3, we are not justified in writing

$$\frac{1.4 \times 10^5}{6.70 \times 10^3} = 20.89552 \cdots$$

We may properly retain only two significant figures, corresponding to the two significant figures in the numerator, and so the correct answer is just 21.

In a calculation with several steps, however, it is a good idea to keep an extra digit in the intermediate steps and to round off the result only at the end. As an example,

$$\frac{5.7 \times 10^4}{3.3 \times 10^{-2}} + \sqrt{1.8 \times 10^{12}} = 1.73 \times 10^6 + 1.34 \times 10^6$$
$$= 3.07 \times 10^6 = 3.1 \times 10^6$$

If the intermediate results had been rounded off to two digits, however, the result would have been the incorrect

$$1.7 \times 10^6 + 1.3 \times 10^6 = 3.0 \times 10^6$$

B–5 LOGARITHMS

The logarithm of a number N is the exponent n to which a given base number a must be raised in order that $a^n = N$. That is, if

$$N = a^n, \qquad \text{then} \qquad n = \log_a N$$

Here are some examples using the base $a = 10$:

$$1000 = 10^3, \quad \text{therefore } \log_{10} 1000 = 3;$$

$$5 = 10^{0.699}, \quad \text{therefore } \log_{10} 5 = 0.699;$$

$$0.001 = 10^{-3}, \text{therefore } \log_{10} 0.001 = -3.$$

Since the decimal system of numbers has the base 10, this is a convenient number to use as the base of a system of logarithms. Logarithms to the base 10 are called *common logarithms* and are denoted simply as "log N." Another widely used system of logarithms uses $e = 2.718 \cdots$ as the base. Such logarithms are called *natural logarithms* (because they arise in a natural way in calculus) and are denoted "ln N." To go from one system to the other these formulas are needed:

$$\log N = 0.43429 \ln N$$

$$\ln N = 2.3026 \log N$$

Logarithms are defined only for positive numbers because the quantity a^n is positive whether n is positive, negative, or zero. Since n is the logarithm of a^n, n can describe only a positive number.

Let us consider a number N that is the product of two numbers x and y, so that $N = xy$. If $x = 10^n$ and $y = 10^m$, then

$$N = xy = 10^n \times 10^m = 10^{n+m}$$

$$\log N = n + m$$

Since $n = \log x$ and $m = \log y$, $n + m = \log x + \log y$, and so

$$\log N = \log x + \log y$$

Thus we have the general rule for the logarithm of a product:

$$\log xy = \log x + \log y$$

Similar reasoning gives the additional rules

$$\log \frac{x}{y} = \log x - \log y$$

$$\log x^n = n \log x$$

Before the days of electronic calculators, logarithms were widely used to simplify arithmetical work because they permit replacing multiplication and division by addition and subtraction, which are easier to do and less prone to error. Today, of course, calculators are used for such routine arithmetic. However, logarithms are still needed for finding powers and roots, with calculators replacing tables of logarithms. For example, to find $(2.13)^4$ we proceed as follows:

$$\log (2.13)^4 = 4 \log 2.13 = (4)(0.3284) = 1.314$$

$$(2.13)^4 = \log^{-1} 1.314 = 20.6$$

To obtain log 2.13 with a calculator, enter 2.13 and press the [LOG] key. To obtain $\log^{-1}$ 1.314 (the *antilogarithm* of 1.314, which is the number whose logarithm is 1.314), the 1.314 is entered and the [10^x] key pressed, since 10^x is the number whose logarithm is x. This key is alternatively designated [INV LOG].

The same procedure is followed for negative exponents, for instance $1/\sqrt[4]{7}$:

$$\frac{1}{\sqrt[4]{7}} = \frac{1}{7^{1/4}} = 7^{-1/4}$$

$$\log 7^{-1/4} = -\tfrac{1}{4} \log 7 = -\frac{1}{4} \times 0.8451 = -0.2113$$

$$7^{-1/4} = \log^{-1}(-0.2113) = 0.615$$

The logarithms of numbers written in powers-of-ten notation can be found with a calculator without first converting them to decimal notation. Let us consider 6.04×10^9. Since

$$\log xy = \log x + \log y$$

we have

$$\log (6.04 \times 10^9) = \log 6.04 + \log 10^9$$

But from the definition of logarithms to the base 10, $\log 10^9 = 9$. We therefore have

$$\log (6.04 \times 10^9) = 0.7810 + 9 = 9.7810$$

The same procedure holds for a number smaller than 1, for instance 2.4×10^{-5}:

$$\log (2.4 \times 10^{-5}) = \log 2.4 + \log 10^{-5}$$
$$= 0.3802 - 5 = -4.6198$$

EXERCISES

7. Evaluate the following with the help of logarithms.

(a) $0.0181^{1.5}$ (d) $\sqrt[4]{156}$

(b) $62.2^{7.13}$ (e) $(6.24 \times 10^{-4})^{1/3}$

(c) $(8.15 \times 10^{14})^6$ (f) $(2.71 \times 10^5)^{1/8}$

ANSWERS

1. (a) 200,000 (e) 0.000,001,003

(b) 0.08 (f) 0.000,000,000,1

(c) 781.9 (g) 0.000,0956

(d) 451,000,000

2. (a) 7×10^1
(b) 1.4×10^{-1}
(c) 3.81
(d) 8.4×10^3
(e) 1×10^6
(f) 7.890×10^{-3}
(g) 3.516×10^5

3. (a) 4.3×10^3
(b) 7.2×10^{-2}
(c) 5.04×10^{-3}
(d) 6.37×10^2
(e) 3.7×10^3
(f) 2.7×10^{-4}
(g) -1.3×10^5
(h) -5×10^{-5}

4. (a) $10^{-2} = 0.01$
(b) $5 \times 10^{-3} = 0.005$
(c) $10^2 = 100$
(d) $2.5 \times 10^3 = 2500$

5. (a) 1×10^3
(b) 6×10^{15}
(c) 9×10^{-4}
(d) 2.5×10^{-5}
(e) 5×10^{-6}
(f) 1.5×10^{-10}
(g) 1×10^{24}
(h) 4×10^{14}
(i) 1×10^{-10}
(j) 2.5×10^{-17}

6. (a) 2×10^3
(b) 6.3×10^3
(c) 2×10^{-2}
(d) 6.3×10^{-3}
(e) 1.6×10^{-4}
(f) 3.4×10^4
(g) 7.4×10^4
(h) 1.6×10^{-2}
(i) 7.4×10^{-3}
(j) 3.4×10^{-3}

7. (a) 2.44×10^{-3}
(b) 6.16×10^{12}
(c) 2.93×10^{89}
(d) 3.53
(e) 8.55×10^{-2}
(f) 4.78

APPENDIX C
TABLES

TABLE C–1
The Elements

Atomic number	*Element*	*Symbol*	*Atomic mass**	*Atomic number*	*Element*	*Symbol*	*Atomic mass**
1	Hydrogen	H	1.008	36	Krypton	Kr	83.80
2	Helium	He	4.003	37	Rubidium	Rb	85.47
3	Lithium	Li	6.941	38	Strontium	Sr	87.62
4	Beryllium	Be	9.012	39	Yttrium	Y	88.91
5	Boron	B	10.81	40	Zirconium	Zr	91.22
6	Carbon	C	12.01	41	Niobium	Nb	92.91
7	Nitrogen	N	14.01	42	Molybdenum	Mo	95.94
8	Oxygen	O	16.00	43	Technetium	Tc	(97)
9	Fluorine	F	19.00	44	Ruthenium	Ru	101.1
10	Neon	Ne	20.18	45	Rhodium	Rh	102.9
11	Sodium	Na	22.99	46	Palladium	Pd	106.4
12	Magnesium	Mg	24.31	47	Silver	Ag	107.9
13	Aluminum	Al	26.98	48	Cadmium	Cd	112.4
14	Silicon	Si	28.09	49	Indium	In	114.8
15	Phosphorus	P	30.97	50	Tin	Sn	118.7
16	Sulfur	S	32.06	51	Antimony	Sb	121.8
17	Chlorine	Cl	35.46	52	Tellurium	Te	127.6
18	Argon	Ar	39.95	53	Iodine	I	126.9
19	Potassium	K	39.10	54	Xenon	Xe	131.3
20	Calcium	Ca	40.08	55	Cesium	Cs	132.9
21	Scandium	Sc	44.96	56	Barium	Ba	137.3
22	Titanium	Ti	47.90	57	Lanthanum	La	138.9
23	Vanadium	V	50.94	58	Cerium	Ce	140.1
24	Chromium	Cr	52.00	59	Praseodymium	Pr	140.9
25	Manganese	Mn	54.94	60	Neodymium	Nd	144.2
26	Iron	Fe	55.85	61	Promethium	Pm	(145)
27	Cobalt	Co	58.93	62	Samarium	Sm	150.4
28	Nickel	Ni	58.70	63	Europium	Eu	152.0
29	Copper	Cu	63.55	64	Gadolinium	Gd	157.3
30	Zinc	Zn	65.38	65	Terbium	Tb	158.9
31	Gallium	Ga	69.72	66	Dysprosium	Dy	162.5
32	Germanium	Ge	72.59	67	Holmium	Ho	164.9
33	Arsenic	As	74.92	68	Erbium	Er	167.3
34	Selenium	Se	78.96	69	Thulium	Tm	168.9
35	Bromine	Br	79.90	70	Ytterbium	Yb	173.0

*The unit of mass is the u. Elements whose atomic masses are given in parentheses have not been found in nature but have been produced by nuclear reactions in the laboratory. The atomic mass in such a case is the mass number of the longest-lived radioactive isotope of the element.

TABLE C–1
The Elements (continued)

Atomic number	Element	Symbol	Atomic mass*	Atomic number	Element	Symbol	Atomic mass*
71	Lutetium	Lu	175.0	89	Actinium	Ac	(227)
72	Hafnium	Hf	178.5	90	Thorium	Th	232.0
73	Tantalum	Ta	180.9	91	Protactinium	Pa	231.0
74	Tungsten	W	183.9	92	Uranium	U	238.0
75	Rhenium	Re	186.2	93	Neptunium	Np	(237)
76	Osmium	Os	190.2	94	Plutonium	Pu	(244)
77	Iridium	Ir	192.2	95	Americium	Am	(243)
78	Platinum	Pt	195.1	96	Curium	Cm	(247)
79	Gold	Au	197.0	97	Berkelium	Bk	(247)
80	Mercury	Hg	200.6	98	Californium	Cf	(251)
81	Thallium	Tl	204.4	99	Einsteinium	Es	(254)
82	Lead	Pb	207.2	100	Fermium	Fm	(257)
83	Bismuth	Bi	209.0	101	Mendelevium	Md	(258)
84	Polonium	Po	(209)	102	Nobelium	No	(255)
85	Astatine	At	(210)	103	Lawrencium	Lr	(260)
86	Radon	Rn	222	104	Rutherfordium	Rf	(257)
87	Francium	Fr	(223)	105	Hahnium	Ha	(260)
88	Radium	Ra	226.0				

TABLE C–2
Natural Trigonometric Functions

Angle					Angle				
Degree	Radian	Sine	Cosine	Tangent	Degree	Radian	Sine	Cosine	Tangent
0°	.000	0.000	1.000	0.000					
1°	.017	.018	1.000	.018	16°	.279	.276	.961	.287
2°	.035	.035	0.999	.035	17°	.297	.292	.956	.306
3°	.052	.052	.999	.052	18°	.314	.309	.951	.325
4°	.070	.070	.998	.070	19°	.332	.326	.946	.344
5°	.087	.087	.996	.088	20°	.349	.342	.940	.364
6°	.105	.105	.995	.105	21°	.367	.358	.934	.384
7°	.122	.122	.993	.123	22°	.384	.375	.927	.404
8°	.140	.139	.990	.141	23°	.401	.391	.921	.425
9°	.157	.156	.988	.158	24°	.419	.407	.914	.445
10°	.175	.174	.985	.176	25°	.436	.423	.906	.466
11°	.192	.191	.982	.194	26°	.454	.438	.899	.488
12°	.209	.208	.978	.213	27°	.471	.454	.891	.510
13°	.227	.225	.974	.231	28°	.489	.470	.883	.532
14°	.244	.242	.970	.249	29°	.506	.485	.875	.554
15°	.262	.259	.966	.268	30°	.524	.500	.866	.577

TABLE C–2
Natural Trigonometric Functions (continued)

Angle					Angle				
Degree	*Radian*	*Sine*	*Cosine*	*Tangent*	*Degree*	*Radian*	*Sine*	*Cosine*	*Tangent*
31°	.541	.515	.857	.601	61°	1.065	.875	.485	1.804
32°	.559	.530	.848	.625	62°	1.082	.883	.470	1.881
33°	.576	.545	.839	.649	63°	1.100	.891	.454	1.963
34°	.593	.559	.829	.675	64°	1.117	.899	.438	2.050
35°	.611	.574	.819	.700	65°	1.134	.906	.423	2.145
36°	.628	.588	.809	.727	66°	1.152	.914	.407	2.246
37°	.646	.602	.799	.754	67°	1.169	.921	.391	2.356
38°	.663	.616	.788	.781	68°	1.187	.927	.375	2.475
39°	.681	.629	.777	.810	69°	1.204	.934	.358	2.605
40°	.698	.643	.766	.839	70°	1.222	.940	.342	2.747
41°	.716	.658	.755	.869	71°	1.239	.946	.326	2.904
42°	.733	.669	.743	.900	72°	1.257	.951	.309	3.078
43°	.751	.682	.731	.933	73°	1.274	.956	.292	3.271
44°	.768	.695	.719	.966	74°	1.292	.961	.276	3.487
45°	.785	.707	.707	1.000	75°	1.309	.966	.259	3.732
46°	.803	.719	.695	1.036	76°	1.326	.970	.242	4.011
47°	.820	.731	.682	1.072	77°	1.344	.974	.225	4.331
48°	.838	.743	.669	1.111	78°	1.361	.978	.208	4.705
49°	.855	.755	.656	1.150	79°	1.379	.982	.191	5.145
50°	.873	.766	.643	1.192	80°	1.396	.985	.174	5.671
51°	.890	.777	.629	1.235	81°	1.414	.988	.156	6.314
52°	.908	.788	.616	1.280	82°	1.431	.990	.139	7.115
53°	.925	.799	.602	1.327	83°	1.449	.993	.122	8.144
54°	.942	.809	.588	1.376	84°	1.466	.995	.105	9.514
55°	.960	.819	.574	1.428	85°	1.484	.996	.087	11.43
56°	.977	.829	.559	1.483	86°	1.501	.998	.070	14.30
57°	.995	.839	.545	1.540	87°	1.518	.999	.052	19.08
58°	1.012	.848	.530	1.600	88°	1.536	.999	.035	28.64
59°	1.030	.857	.515	1.664	89°	1.553	1.000	.018	57.29
60°	1.047	.866	.500	1.732	90°	1.571	1.000	.000	∞

TABLE C–3
The Greek Alphabet

Α	α	Alpha
Β	β	Beta
Γ	γ	Gamma
Δ	δ	Delta
Ε	ϵ	Epsilon
Ζ	ζ	Zeta
Η	η	Eta
Θ	θ	Theta
Ι	ι	Iota
Κ	κ	Kappa
Λ	λ	Lambda
Μ	μ	Mu
Ν	ν	Nu
Ξ	ξ	Xi
Ο	ο	Omicron
Π	π	Pi
Ρ	ρ	Rho
Σ	σ	Sigma
Τ	τ	Tau
Υ	υ	Upsilon
Φ	ϕ	Phi
Χ	χ	Chi
Ψ	ψ	Psi
Ω	ω	Omega

ANSWERS TO ODD-NUMBERED QUESTIONS AND EXERCISES

CHAPTER 1

Questions

1. Centimeter; yard; kilometer. **3.** Yes; an object dropped from a great height compared with one dropped from a smaller height. **5.** C; D. **7.** No.

Exercises

1. 0.112 km. **3.** 31 mi/h. **5.** 4047 m^2; 247 acre/km^2. **7.** 144 in^3; $(1/12)ft^3 = 0.0833\ ft^3$; 2360 cm^3. **9.** 0.45 s. **11.** 0.029 s. **13.** 3×10^{-15} m. **15.** 338 km/h. **17.** 0.50 km. **19.** 200 km. **21.** 1.2 m/s^2; $-2.0\ m/s^2$. **23.** 5.0 s. **25.** 0.60 m/s^2; $-0.80\ m/s^2$. **27.** The airplane's speed is 500 km/h from 4:00 P.M. to 5:41 P.M., 620 km/h from 5:41 P.M. to 7:54 P.M., and 520 km/h from 7:54 P.M. to 10:00 P.M. **29.** 90.4 m. **31.** (a) 8.8 s; 88 m. (b) 8.8 s; 0.61 km. (c) 1.4 km. **33.** $32g$. **35.** 1.0×10^6 m. **37.** Yes; 47 m/s. **39.** Yes. **41.** 27 m/s. **43.** 78 m. **45.** 15 m/s (up); 15 m/s (down). **47.** -19.8 m/s (down); -29.6 m/s (down). **49.** 13 m. **51.** 14 m/s. **53.** 11 m/s; 5.7 s. **55.** 1.2 s; 12 m/s. **57.** 19 s. **59.** 20 m/s. **61.** 46 m; 78 m. **63.** 1.3 m/s. **65.** 44 mi/h. **67.** The car had an initial speed of 6 m/s and accelerated at 0.1 m/s^2 for 3 min to 24 m/s. It remained at 24 m/s for 4 min and then slowed to a stop at 0.2 m/s^2 in 2 min. **69.** 12.8 m. **71.** 50 s. **73.** 3.3 s; 22 m/s. **75.** 12 s. **77.** See Fig. A.1.

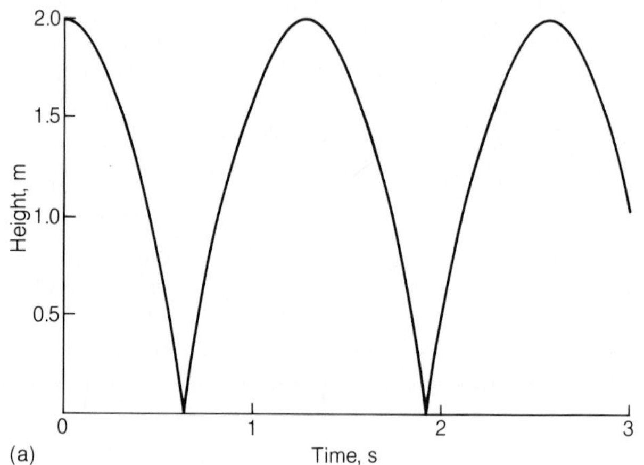

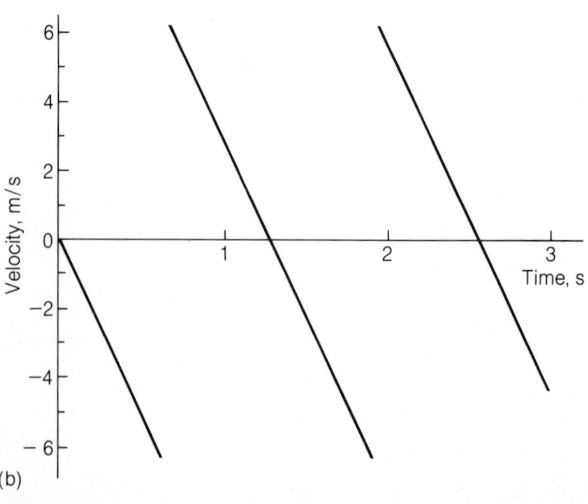

Figure A.1
Exercise 77, Chapter 1.

CHAPTER 2

Questions

1. No. The distinction between vector and scalar quantities is simply that vector quantities have directions associated with them. Both kinds of quantity are found in the physical world. **3.** Yes. **5.** 30 N; 0. **7.** (a) No. *B* and *C* reach the ground together; *A* reaches it later. (b) No. *B* has the lowest speed; *A* and *C* have the same higher speed.

9. The squirrel should stay where it is. If it lets go, it will fall with the same acceleration as the bullet and so will be struck.

Exercises

1. 0.11 kN. **3.** 81 m, 30° above horizontal. **5.** 43 km; 172 km. **7.** 7.6 kn; 68° west of north. **9.** 1.7 m/s; 0.76 m/s to the east. **11.** (a) 8.0 min. (b) 19° upstream of a line directly across the river. (c) 8.5 min. **13.** 94 km/h; 94 km/h. **15.** 68 N. **17.** 453 N. **19.** 8.4 m/s. **21.** 85 km/h; 181 km/h; 128 km/h. **23.** 7°. **25.** 182 km; northwest. **27.** 111 km/h, 26° south of east; 111 km/h, 26° north of west. **29.** 10 cm. **31.** 179 m/s. **33.** 12 m/s. **35.** 0.71 km. **37.** 36 m/s; 33° below the horizontal. **39.** 10 m/s. **41.** 20 m/s. **43.** 200 m; 44 m/s in both cases. **45.** 5.0 m/s; 1.3 m/s. **47.** 15 N; 37° south of east. **49.** 19.5 N; 37° clockwise from $-x$ direction. **51.** (a) 14 cm; 8° counterclockwise from the $+x$ direction. (b) 14 cm; 8° counterclockwise from the $+y$ direction. (c) 14 cm; 8° counterclockwise from the $-y$ direction. **53.** 133°. **55.** 23 m/s; 33°; 70°. **57.** 32 m/s; 31 m; 21 m/s; 29 m. **59.** (a) $H = (v_0^2/2g)\sin\theta$. (b) $4/(\tan\theta)$; 4. **61.** (a) 8.0 km; 31 s; 1.1 km. (b) 60°; 53 s; 3.4 km.

CHAPTER 3

Questions

1. No. Other forces may come into being that cancel out the applied force. **3.** The deceleration of a person falling on loose earth is more gradual than if he falls on concrete; hence the force acting on him is less. **5.** $a = g$. **7.** Yes, by applying a downward force in addition to the downward force of gravity. **9.** No. Action and reaction forces act on different objects, so a single force can certainly act on an object. **11.** A propeller works by pushing backward on the air, whose reaction force in turn pushes the propeller itself forward. No air means no reaction force, so the idea is no good. **13.** (a) For it to fly, the bird's wings must exert a downward force on the air in the box equal to the bird's weight and this force is transmitted to the box. Hence the scale reading is the same whether the bird stands or flies. (b) Now the force exerted on the air by the bird's wings is not completely transmitted to the cage, and the scale reading will be smaller when the bird flies. Because some force is still transmitted to the bottom of the cage, the reading when the bird flies will be somewhat greater than that corresponding to the empty cage. **15.** Yes; no.

Exercises

1. 1.5×10^{-5} N. **3.** 0.67 m/s². **5.** 29 m/s. **7.** $1.9g$; 2.2×10^5 N. **9.** 9.6 kN. **11.** 0.50 kN. **13.** 10 kN; the goat is likely to be killed. **15.** Upper string, 147 N; lower string, 98 N. **17.** 0.82 kg; 25 m/s². **19.** 2.0 m/s² each, the first upward and the second downward. **21.** 8.8 kN; 10.8 kN. **23.** 4.0 kN = 4.1 mg. **25.** Down; no; yes, 1.4 m/s². **27.** 19 m/s². **29.** 0.3 kN. **31.** 2.0 N. **33.** 0.041. **35.** 36 m. **37.** 0.4 μg; 0.6 μg; μg. **39.** 0.23; 4.5 N. **41.** 6.9 m/s²; 56 m. **43.** 0.18. **45.** 0.12 kN; up the ramp. **47.** 3.3 kg. **49.** Upper string, 93 N; lower string, 53 N. **51.** $v_t = \sqrt{mg/k}$. **53.** 4.9 m/s²; 3.9 m/s². **55.** 21 N; 26 N. **57.** No; 9.5×10^5 N. **59.** 2.1 s. **61.** 24 m/s; 0.28 km. **63.** (c) 76°.

CHAPTER 4

Questions

1. No work is done by a net force acting on a moving object when the force is perpendicular to the direction of the object's motion. **3.** Yes; no. **5.** The work done is the same for both twins. **7.** The center of mass of the stored water drops to half its original height; hence half the original PE is lost.

Exercises

1. None. **3.** 5.0 kJ; 4.0 kJ; 1.0 kJ. **5.** 78 N; 1.2 kJ. **7.** 8.8 kJ. **9.** 1.0 kJ; 1.0 kJ. **11.** 24 J. **13.** 4.0 kW. **15.** 3.9 kW. **17.** 23 kN. **19.** 8.0 m/s. **21.** 78%. **23.** 65 W. **25.** 0.37 km/h. **27.** 3.2 m/s. **29.** 2.8 kJ. **31.** 7.1 m/s. **33.** 8.8 s. **35.** 36 kW. **37.** 40 m. **39.** 1.3×10^5 N. **41.** 90 cm. **43.** 29.4 MW; 29,400. **45.** 80%. **47.** 5.0 m. **49.** 24 J; 9.3 J. **51.** 1.7×10^5 J. **53.** 0.14 kJ. **55.** 40 J. **57.** 0.62 kJ. **59.** 0.46 m/s. **61.** 2.3 m/s. **63.** (a) 5.5 kJ; 0.39 kW. (b) 5.5 kJ; 0.71 kW. (c) 5.5 kJ; 78 W. **65.** 2.7 kN; 48 kW. **67.** F_{roll} = 2.5 N; k = 0.20 N/(m/s)².

CHAPTER 5

Questions

1. Yes; yes. **3.** $p_2 = p_1/\sqrt{2}$. **5.** The definition of mass and the first law of motion are both included in the principle of conservation of linear momentum. This principle goes further because it can be applied to systems consisting of any number of objects that interact with one another in any manner at all, and it holds in three dimensions, not just in one. **7.** (a) Yes. (b) In the opposite direction to that in which the man walks. (c) The car also comes to a stop. **9.** In such a collision the total momentum of the objects is initially zero, and the objects stick together afterward. The total momentum remains zero.

Exercises

1. 5.0 kg·m/s. **3.** $KE_A = 1.4 \times 10^5$ J; $KE_B = 6.9 \times 10^4$ J; $p_A = p_B = 1.7 \times 10^4$ kg·m/s. **5.** 3.9×10^7 kg·m/s; 114 s. **7.** 7.8×10^5 N. **9.** 8.1 kg; 8.0 kg. **11.** 8. **13.** 1.0 m/s; 0.9 m/s. **15.** 13 s. **17.** 24 m/s in the original direction. **19.** No; yes; yes. **21.** 4.0 km. **23.** 720 kg/s. **25.** Initial, 5.2 m/s²; final, 40 m/s². **27.** 3.3×10^4 km/s. **29.** 1.4 m/s. **31.** 1.3 m/s; 2.0 J. **33.** 56 km/s; 37 kJ. **35.** 0.99 m/s; 27° north of west. **37.** 28 m/s. **39.** 1.6 m/s; 3.6 m/s; 33%. **41.** $KE_s/KE_d = 1.2$; $p_s/p_d = 2.1$. The KEs are comparable, but the momenta are very different. This result suggests that the work done, rather than the impulse given, is similar for the athletes. **43.** Second law of motion; work-energy; impulse-momentum; 6.8×10^5 N. **45.** 17 m/s in the same plane as the other fragments, bisecting the angle between the backward extensions of their paths. **47.** (a) $v = (1 + M/m)\sqrt{2gh}$. (b) 561 m/s. **51.** 3.8 m/s; 3.2 m/s.

CHAPTER 6

Questions

1. Under no circumstances. **3.** At the bottom of the circle, where the string must support all the ball's weight as well as provide centripetal force. **5.** The mass of a sprinter and the force his or her legs can exert are both unchanged on the moon. Hence the acceleration is unchanged and the time needed would not be different. **7.** In both cases the centripetal force is provided by the object's weight mg. **9.** Shorter. **11.** Such a vehicle does not have to depend only on its initial KE for the work needed to leave the earth; hence it need not reach the escape speed.

Exercises

1. 52 cm/s; 1.8 m/s². **3.** 2.3 km. **5.** 0.38 rev/s. **7.** 0.38 kN. **9.** 8.9 kN. **11.** 20 m/s. **13.** 33.3, no; 78, yes. **15.** 2.0 km. **17.** 94 cm. **19.** 2.8 m/s; same speed. **21.** 11 m/s. **23.** 2.0 s. **25.** (a) 6.7×10^{10} N. (b) 6.7×10^{10} N. (c) 3.3×10^{-10} m/s²; 1.3×10^{-10} m/s². **27.** 2.0×10^{30} kg. **29.** 1.4 mm. The moon's "fall" is a deflection from a straight line into an orbit. **31.** $g/2$. **33.** $\sqrt{10}\, r_e$. **35.** 6.3 m/s². **37.** 7.7 km/s; 5.6×10^3 s (about 93 min). **39.** 11 m/s²; 25 km/s. **41.** $\sqrt{3.0}\, g$. **43.** 225 days. **45.** 7.8 N. **47.** $(2\pi/T)^2 r_e \cos\theta$. **49.** 0.94 kN. **51.** 3.8×10^7 m from the moon. **53.** $M_{sun}/M_{earth} = 3.4 \times 10^5$. **55.** Sun, 5.1×10^{-7} N; moon, 11.4×10^{-7} N. **57.** (a) Yes, to the given precision. (b) 5.67×10^{26} kg.

CHAPTER 7

Questions

1. Such a mass distribution maximizes I and so increases the flywheel's KE at a given angular speed. **3.** The rectangle. **5.** The solid cylinder reaches the bottom first, because its moment of inertia is smaller and hence less of its initial PE becomes KE of rotation. **7.** They will reach the bottom at the same time. **9.** The location of the pivot point. **11.** Reducing I means that a given torque leads to a faster turn, making it possible to complete more somersaults before reaching the water. **13.** The length of the day will increase. The earth's moment of inertia will be greater when the water from the ice caps becomes uniformly distributed and hence, from conservation of angular momentum, the angular speed must decrease.

Exercises

1. 0.698 rad. **3.** 2.9×10^{-4} rad; 0.073 mm. **5.** 34 cm. **7.** Hour: 1.45×10^{-4} rad/s; minute: 1.75×10^{-3} rad/s; second: 0.105 rad/s. **9.** 15.3 rad/s. **11.** 94.5 m/s; 3.0×10^3 rpm. **13.** 226 km/h. **15.** 11 rad/s. **17.** 4 m/s; 2 m/s; 0. **19.** 0.80 J. **21.** $(2/3)MR^2$; $\sqrt{2/5}R$. **23.** (a) 2.6×10^{29} J. (b) Because the earth's actual moment of inertia is less than that found in (a), its actual rotational energy is less than that in (a). **25.** One-half. **27.** $v(\text{solid})/v(\text{hollow}) = 1.09$. **29.** 12 rad/s; 5.8 m/s; 50 J; 101 J. **31.** 0.50 rad/s². **33.** 60 rev. **35.** 35 rad/s². **37.** 0.29 kN. **39.** 100 kg·m². **41.** 4.5 s; 450 J. **43.** 8.0 N·m. **45.** 235 kW; 6.7 rad/s. **47.** 0.12 kN. **49.** 2.1 kW. **51.** mvs; no. **53.** $(v/R)(1 + M/2m)$. **55.** 7.2×10^6 m/s. **57.** $\sqrt{3g/L}$. **59.** $v = \sqrt{2gh/(1 + M/2m)}$. **63.** 2.5 m/s². **65.** 537 kg. **67.** 3.5 rev/s; 49 J, 34 J; the decrease in KE is due to the work done in stretching the string. **69.** (a) 50 rad/s. (b) 2.5 kJ of KE were lost, mainly to heat. (c) 50 rad/s each. (d) No KE was lost in separation.

CHAPTER 8

Questions

1. (a) The forces that act on the object. (b) The forces the object exerts on anything else. (c) Yes. (d) Yes. **3.** (a) Perpendicular to the wall. (b) The force on the ground is always greater in magnitude than the weight of the ladder, because the force has a horizontal component equal to the reaction force of the wall on the ladder as well as a vertical component equal to the ladder's weight. **5.** Negative; positive. **7.** About whatever point makes the calculations easiest; the results will be the same regardless of where the point is. **9.** Greater than 1; less than 1. **11.** 4.

Exercises

1. 0.16 kN. **3.** 0.28 kN. **5.** 8.7 kg. **7.** 94 N; 31° above the horizontal. **9.** 0.39 kN; 0.27 kN. **11.** 197 kg. **13.** 0.92 kg. **15.** 69 kN; 59 kN. **17.** 1.0 m from the center. **19.** μh. **21.** 0.25 kN. **23.** 36 cm. **25.** 1.6 m. **27.** 0.31 kN; 1.3 m. **29.** 1.00 m. **31.** 9.0 kN. **33.** Forepaws, 0.41 kN; hind paws, 0.33 kN. **35.** 0.62 kN. **37.** 88 cm. **39.** 20 cm to the left of its original position. **41.** 67 mm to the right of the cross member. **43.** 6.1 kN; 400 m. **45.** 4.5 kN. **47.** 0.18 kN. **49.** $1/(2 \tan \theta)$. **51.** 0.53 kN each, directed along the leg. **53.** 12 kN. **55.** $\mu L/\sqrt{1 + \mu^2}$. **57.** 17 kN. **59.** 0.24 kN; 0.42 kN. **61.** 150 N down; 55 N away from the wall. **63.** Top: 203 N; 9.5° from vertical. Bottom: 33 N; horizontal.

CHAPTER 9

Questions

1. The second procedure is more likely to break the string because here the tension in the string is twice as great with the reaction force exerted by the tree being equal and opposite to the pull of the two boys. **3.** All are under the same stress. The strain varies as $1/Y$; hence the aluminum leg experiences the greatest strain and the steel leg the least strain. **5.** B stretches twice as much as A. **7.** Inside the building, on the lower part of the floor; on the balcony, on the upper part of the floor.

Exercises

1. 78 kg. **3.** 55 L; 140 m^3. **5.** (a) 5.52×10^3 kg/m^3. (b) The interior must consist of denser materials than those at the surface; no. **7.** 0.77 g/cm^3. **9.** 1.8×10^3 L/min. **11.** 78 mm. **13.** 9 kN; 49 kN. **15.** The compressive stress is nearly 12 times less than the ultimate strength. **17.** They are about the same. **19.** 1.2×10^{11} Pa. **21.** 0.29 mm. **23.** 0.02 mm. **25.** 9.2×10^{10} Pa. **27.** 0.14 mm. **29.** 6.1×10^{-7} m. **31.** 3.3 kPa. **33.** 7.1 mm. **35.** 40 mm^3. **37.** 1.4×10^{-4}%. **39.** 9.3 m/s. **41.** 5.2 m/s^2. **43.** (a) Steel, 65%; copper, 35%. (b) d(copper)/d(steel) = 1.35. **45.** 2.4×10^{12} Pa.

CHAPTER 10

Questions

1. As the steam inside the can condenses, the internal pressure falls below atmospheric pressure. **3.** Atmospheric pressure. **5.** Airliners fly at high altitudes where there is a large pressure difference between their pressurized interiors and the outside atmosphere. Light aircraft do not have pressurized interiors, so the pressures on both sides of their windows are the same; thus the windows have no net force to withstand in normal flight, even if they are large. **7.** The pressures are the same. **9.** There is no buoyant force because there is no water under the block to provide an upward force. **11.** The load on the bridge does not change because the boat displaces a volume of water whose weight equals its own weight. **13.** The water level falls. The floating canoe displaced a volume of water V whose weight equalled its own weight. The sunken canoe, however, displaces a volume of water equal to the volume of its aluminum shell which is smaller than V, because aluminum is denser than water. **15.** A ship is a hollow shell, so its average density is less than that of water. If the ship fills with water, its average density increases until it is greater than that of water and then the ship sinks. **17.** Ice on the wings interferes with the intended flow of air past them and so decreases the lift they produce. Ice on the fuselage increases the weight of the airplane and increases air resistance, but these effects are usually less significant than the reduction in lift. **19.** No. For example, water at 20°C has a greater density than ethyl alcohol but a smaller viscosity.

Exercises

1. 0.44 kPa; 0.87 kPa; 1.74 kPa. **3.** 1.4 N. **5.** 3.2 kPa. **7.** 63 atm. **9.** 0.16 N. **11.** 12.5 N; 8.0 cm. **13.** The forces are the same. **15.** 1.10×10^8 Pa. **17.** 1.03 m. **19.** 55 torr. **21.** 1.9×10^5 N. **23.** 700 kg/m^3. **25.** 0.89 N. **27.** 10 mm. **29.** 880 kg/m^3. **31.** 0.89 kN. **33.** 2.4×10^5 kg. **35.** 90 g; 20 cm^3. **37.** 4.5 mm. **39.** 7.5 W. **41.** 1.9 kg/s. **43.** 1.5 m. **45.** 1.8 m. **47.** 0.52 kPa/m. **49.** 0.67 mm/s; about 10^{10}. **51.** 2.8 m/s. **53.** 0.74 mm. **57.** 1053 kg/m^3. **59.** 22 g. **61.** 36 kN·m. **63.** 200 N. **65.** 9.8 kPa. **67.** (a) 1.3 m/s; 2.3 bars. (b) 1.3 m/s; 1.9 bars. **69.** 0.88 m/s; 3.5 m/s; 55 kPa. **71.** (a) $L^3T^{-1} = M^{a+c}L^{b-a-2c}T^{-a-2c}$. (b) $a + c = 0$; $b - a - 2c = 3$; $a + 2c = 1$. (c) $a = -1$; $b = 4$; $c = 1$. **73.** 15 mm.

CHAPTER 11

Questions

1. At either extreme of its motion the energy is entirely PE; at the midpoint the energy is entirely KE. **3.** When the object has vertical sides, so the restoring force is proportional to the displacement of the object above or below its equilibrium level. **5.** PE_{max} and KE_{max} for a harmonic oscillator are always equal. **7.** When the string is vertical. **9.** 0; $1/T$; $1/T$. **11.** In the ax's head so that there will be no reaction force on the hands of the user.

Exercises

1. 1.0 kN/m. **3.** 1.4 m/s. **5.** 4.2 J. **7.** 10 mm; 3.6 Hz. **9.** 9.9 kN/m. **11.** 2.3 s. **13.** 0.57 s. **15.** 3.1 m/s. **17.** 11 Hz. **19.** 4.1 m/s; 1.7×10^4 m/s^2 = 1.8×10^3 g. **21.** At half the amplitude. **23.** 39 N/m; 0.31 m/s; 2.0 m/s^2. **25.** 4.9 s. **27.** 94 mm. **29.** $\sqrt{2gL(1 - \cos\theta}$ **33.** 1.5 s; 53 cm. **35.** 1.5 J. **37.** (a) 8 cm; 12 cm. (b) 0.25 s. **39.** The force would be directed inward and be proportional to the distance from the equilibrium position (the center of the earth). The period would be 5.1×10^3 s, or about 1 h 25 min. **41.** 159 cm; 9.55 m/s^2. **43.** (a) On the far side of the wheel's axis from the piston (the 3 o'clock position). (b) 20 cm; 25 rad/s. (c) 5.0 m/s; 1.3×10^2 m/s^2.

CHAPTER 12

Questions

1. The energy dissipates as heat, and the string becomes warmer as a result. **3.** It increases fourfold because the rate of flow of energy is proportional to A^2. **5.** In solids, because their constituent particles are more tightly coupled together than those of gases or liquids. **9.** Frequency; amplitude; harmonic structure. **11.** The other fork could also have a frequency of 246 Hz to give 10 beats per second. **13.** (a) The short-wavelength sound will be 9 times more intense than the other one because the intensity of a mechanical wave is proportional to f^2. (b) The short-wavelength sound will seem louder, but less than 9 times louder, because the response of the ear is roughly logarithmic.

Exercises

3. 3.0 kHz. **5.** 0.86 Hz; 1.2 s. **7.** 8.8 m/s; 1.3 m/s. **9.** 255 m/s. **11.** (a) 600 m/s. (b) 600 Hz; 900 Hz; 1200 Hz. **13.** 0.60 g/cm. **15.** 805 Hz; 1611 Hz; 2416 Hz. **17.** 1.36 s. **19.** 29 cm, which is not far from the width of a typical head. **21.** 1.6 km. **23.** 440 Hz, 3.4 m; 440 Hz, 78 cm. **25.** 1.15. **27.** 26 cm. **29.** 3.4 m. **31.** $\Delta f = f_1/2$, so f_1 and f_2 are consonant. **33.** 9.8 beats/s. **35.** 408 Hz. **37.** 518 Hz; 483 Hz. **39.** 1.3 m/s. **41.** 5.78×10^{-7} m. **43.** 10; 100; 10,000. **45.** 67 dB. **47.** 1.0 W/m^2. **49.** 253 Hz. **51.** $2fu/(v - u)$. **53.** (a) $f_r = f_s(v - v_b)/(v + v_b)$. (b) $v_b = v(f_s - f_r)/(f_s + f_r)$. (c) 31 mm/s. **55.** 57 dB. **57.** 3.2. **59.** 85 dB.

CHAPTER 13

Questions

1. The glass bulb heats up and expands before the mercury inside does. **3.** Glass expands less than copper with a rise in temperature, so the bulb would crack as it heats up. Lead-in wires must have the same coefficient of thermal expansion as glass. **5.** Because of the volume occupied by its molecules. **7.** No. The only temperature scale on which such comparisons might make any sense is the absolute scale. **9.** Gas molecules undergo frequent collisions with one another, which considerably increases the time needed for a particular molecule to get from one place to another. **11.** A relative humidity of 60% at 22°C means that the air contains 60% as much water vapor as the maximum possible at that temperature. Increasing T will lower the relative humidity and decreasing T will raise it. **13.** The moisture content of the room is the same after it has been heated; therefore the relative humidity has decreased.

Exercises

1. 626°F; 2138°F. **3.** −40°C = −40°F. **5.** −80°C. **7.** 9×10^{-5} m. **9.** 8.1×10^{-6}/°C. **11.** 39.985 m. **13.** 1.25 cm^3. **15.** 10.8 g/cm^3. **17.** $0.12. **21.** 11 mm; 1.9×10^5 N. **23.** 2.5 m^3. **25.** 9.5 kg. **27.** 20 strokes. **29.** p/T = constant; 87°C. **31.** 2.2 bars. **33.** 4.0×10^5 Pa; 2.7 m^3; 2.7 m^3. **35.** 533°C. **37.** 2.99×10^{-25} kg. **39.** 2.92×10^{24} atoms. **41.** 627°C. **43.** 0.40 km/s. **45.** H_2, 3.5 km/s; He, 2.5 km/s; N_2, 0.94 km/s; O_2, 0.88 km/s. These are 32%, 22%, 8.4%, and 7.9%, respectively, of the escape speed, so an appreciable fraction of the H_2 molecules and He atoms originally in the atmosphere would have escaped. **47.** 4×10^9 collision/s. **49.** 110 moles. **51.** 44.6 moles; 2.69×10^{25} molecules. **53.** 64.1 g/mole = 64.1 u/molecule. **55.** 5.1 L. **57.** 26 g. **59.** 230°C. **61.** 33 mm. **63.** The water level rises; 23°C. **65.** 10 m. **67.** 13 mm. **69.** 6.6 kg/m^3. **71.** 0.36 kg. **73.** 6.2×10^{23} molecules/mole.

CHAPTER 14

Questions

1. The specific heat capacity of alcohol is less than that of water, hence its ability to carry heat from the engine to the radiator is less. **3.** The ice is more effective because of its heat of fusion. **5.** See Fig. A.2. **7.** A liquid. **9.** The wood particles in sawdust are mixed with air and so are not in as close contact with one another as they are in solid wood. **11.** In this location convection is facilitated, which increases the rate of heat transfer to the water. **13.** (a) Conduction, convection, and radiation. (b) The same mechanisms, but radiation is less significant in this case. **15.** The rate of heat loss will decrease, because polished copper has a lower emissivity. **17.** The vacuum prevents heat transfer by conductor or convection, and the silvered surfaces are poor radiators.

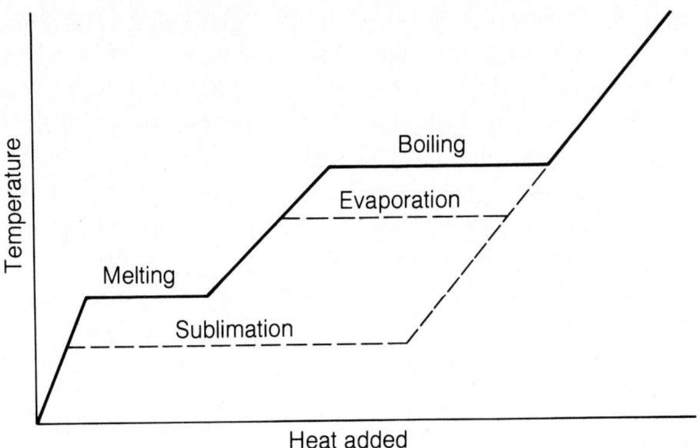

Figure A.2

Question 5, Chapter 14.

Exercises

1. 71 kJ. **3.** 1.7 kJ/kg·°C. **5.** 5.1 kJ; 0.34%. **7.** 14 kg/h. **9.** 8.9×10^3 kg; 45 drums. **11.** 0.0023*h*°C/m. **13.** 2.5 min. **15.** 34°C. **17.** 21°C. **19.** 0.18 kg/h. **21.** 0.35 kg. **23.** 2.7 MJ; 3.1 MW. **25.** 0.82 km/s. **27.** 0.33 km/s. **29.** 2.0 m; 1.5 m; 33 cm. **31.** 5.2×10^{11} J. **33.** 1.9 kW. **35.** 6.8×10^5 W; the rate will decrease. **37.** 303°C. **39.** 0.14 kg/h. **41.** 9.4 min. **43.** 646°C. **45.** The radiation rate of the cooler surface is 55% that of the warmer surface. **47.** 51 W. **49.** 22°C. **51.** 0.85. **53.** 26 g. **55.** 0.74 kg. **57.** A mixture of steam and water at 100°C; 94 g of steam has condensed. **59.** 30 kJ/kg; 3.0 kJ/kg·°C. **61.** 61 mm. **63.** 0.74 kg/h.

CHAPTER 15

Questions

1. There is no suitable low-temperature reservoir in space. **3.** The work done at constant pressure is $p(V_2 - V_1)$ and is a maximum; the final temperature would exceed the initial temperature. At constant temperature, since pV/T = constant, the pressure would drop during the expansion and less work would be done. If the expansion were adiabatic, the temperature would drop, and accordingly the final pressure would be lower than in either of the other cases. Hence the least work is done during an adiabatic expansion. **5.** (a) The statement that the entropy of an isolated system cannot decrease is equivalent to the second law. (b) The higher the entropy of a system, the more disordered are the particles that compose it. (c) Entropy increases with time; thus both have the same direction. **7.** Oil, coal, and natural gas. Advantages: All are easy to use and relatively cheap; in the case of coal, considerable resources. Disadvantages: Mining and burning coal damages the environment; burning all three adds CO_2 to the atmosphere and enhances the greenhouse effect; oil and gas resources are limited.

Exercises

1. It increases by 300 J. **3.** 58 kJ. **5.** (a) 0; 0.60 MJ by the engine; 0; 1.08 MJ on the engine. (b) -0.48 MJ (net work is done *on* the engine). **7.** 44%. **9.** 62%. **11.** 375 K. **13.** 0.33 MJ. **15.** 6.5 bars. **17.** 23%. **19.** 27%; 30%. **21.** 0.28 m³. **23.** 3.5 kW; about 7 persons. **25.** 7.6 MJ/h. **27.** 0.4%. **29.** $T_1/(T_2 - T_1)$; 179 kJ/min; 1.2 kW.

CHAPTER 16

Questions

1. No; yes. **3.** An equal amount of charge of the opposite sign appears on the fur. **5.** It will not move; it will rotate until aligned with the field. **7.** It can explain phenomena that involve charges whose magnitudes are large compared with the electron charge. It cannot explain the behavior of ions or elementary particles or the structure of matter on an atomic scale. **9.** They agree that the positive matter in the atom has much greater mass than does the negative matter. In the Thomson model, however, the positive matter is assumed to occupy the entire atomic volume, whereas in the Rutherford model it is assumed to occupy only a tiny region at the atom's center. **11.** (a) The ability of certain of the electrons to move freely through the metal, rather like molecules in a gas. (b) None of the electrons is free to move through the insulator. **13.** All metals are good electrical conductors, and most nonmetallic solids are good insulators. Liquids may be either. Gases are good insulators if not ionized. Among solids, good electrical conductors are good heat conductors and vice-versa. **15.** Compare the forces that each field exerts on a charged and on an uncharged object. **17.** The direction of the force is given by the direction of the field lines at that point. The magnitude of the force is proportional to the density of field lines there.

Exercises

1. 96 C. **3.** 20 mm. **5.** 4.2×10^{-8} N. **7.** 10 mm. **9.** 5.08 m. **11.** 0.27 N, toward the negative charge. **13.** 2.2×10^6 m/s. **15.** 3.9×10^6 N/C. **17.** $E = 0$. **19.** 1.02×10^{-7} N/C. **21.** 83 mm from the $+1.5$ μC charge. **23.** 0.37 μC. **25.** 146 mm from the smaller charge; in the same place. **27.** kQ^2/a^2; parallel to the line joining the other positive charge and the negative charge; toward the negative charge. **29.** 3.1×10^5 N/C. **31.** $(2kQAR)/[R^2 - (A/2)^2]^2 \to 2kQA/R^3$ for $R \gg A$. The field direction is from the negative charge to the positive charge, which is opposite to the direction in Problem 30.

CHAPTER 17

Questions

1. The charges have different signs. **3.** The electron speed increases by a factor of $\sqrt{2}$. **5.** No.

Exercises

1. 3.0×10^{-14} m. **3.** 6.0 V. **5.** 5.0 mm. **7.** 3.2×10^{8} J. **9.** 2.5 kV. **11.** 2.8 MJ; 14 km; 0.53 km/s. **13.** 2.0×10^{5} eV. **15.** 4.2×10^{6} m/s. **17.** 71 V; 71 eV. **19.** 4.0×10^{-3} J/m³. **21.** 25 mC; 12.5 J. **23.** 0.45 kV. **25.** 44 pF; 2.0×10^{-9} C; 4.5×10^{-8} J. **27.** 707 V; 0.14 C. **29.** 6.25. **31.** 1.7×10^{-10} F; 7.8×10^{-9} C; 1.7×10^{-7} J. **33.** 70 μF. **35.** 37.5 μC on each; 7.50 V; 3.75 V; 0.75 V. **37.** 60 μC; 120 μC; 600 μC; 12 V across each. **39.** 7.3 μF; 10 μF; infinite. **41.** 15 cm. **43.** $Q^2/2C$. **45.** 4×10^{-4} C; 8×10^{-4} C.

CHAPTER 18

Questions

1. With a single wire, charge would be permanently transferred from one end to the other, and soon so large a charge separation would occur that the electric field that produced the current would be canceled out. With two wires, charge can be sent on a round trip, so to speak, and a net flow of energy from one place to another can occur without a net transfer of charge. **3.** The increase in resistivity cannot be due to the expansion of the metal, because its cross-sectional area expands faster than its length and so the resistivity should decrease on this basis. Also, the coefficient of linear expansion is too much smaller than the temperature coefficient of resistivity for the expansion of the metal to have any major effect on resistivity. **5.** If the heater is to be connected to a power source of fixed voltage, which is normally the case, then $P = V^2/R$ is the appropriate formula and a small R will give the greater rate of heat output. **7.** Some of the power produced by the cell of high emf will be dissipated in the cell of low emf.

Exercises

1. 9.6×10^{-18} A. **3.** 15 Ω. **5.** 75 V. **7.** 40% as great. **9.** 2.0 Ω. **11.** 0.29 mm. **13.** 100 m; the resistance of the wire is 10^4 that of the rod. **15.** 92 Ω; 123 Ω. **17.** 5.1×10^{-3}/°C. **19.** 0.46. **21.** 1.67 A. **23.** 78%. **25.** 1.9 kW. **27.** 2.7 kWh = 9.7 MJ. **29.** 7.2×10^{3} C; $167, which is quite a bit more expensive. **31.** 5.0 V. **33.** (a) Idea is all right. (b) The small bulb will be brighter than at 120 V but will soon burn out, because it must dissipate 3.3 times as much power as it was designed to do; the large bulb will be much dimmer than before. This idea is not so good. **35.** (a) 0.80 A in each bulb. (b) 4.0 V; 8.0 V. (c) 3.2 W; 6.4 W; 9.6 W. **37.** 75 V. **39.** 33 Ω; 67 Ω; 150 Ω; 300 Ω. **41.** 10.5 Ω; 3.0 A; 1.5 A; 1.2 A. **43.** (a) 2.4 A; 1.2 A. (b) 12.0 V; 12.0 V. (c) 28.8 W; 14.4 W; 43.2 W. **45.** 2.2 Ω; 0.57 A. **47.** 1.2 A. **49.** 15 W; 16%. **51.** 0.9 Ω; 22 V. **53.** 2.25. **55.** 2.9 A. **57.** 0.27°C. **59.** 988 μF; 15 ms. **61.** 25 ms; 0.2 A; 25 ms. **65.** 9.2×10^{-2} s. **67.** $V_1 = 10$ V; $V_2 = 20$ V; $I = 0$. **69.** 37.5 V; 32 A; 89%, 11%. **71.** 0.53 A. **73.** $R_1 = 6.0$ Ω; $R_2 = 6.7$ Ω. **75.** From left to right: 0.86 A down; 1.49 A up; 0.63 A down. **77.** 6.4 V; 1.6 V. **79.** 1.5×10^{-4} s; 3.8 V, 11.3 V.

CHAPTER 19

Questions

1. All three fields are detected in both cases. **3.** *C*. **5.** No. Because of hysteresis in the iron core, B is not proportional to I, and so the variations of B with time do not exactly follow the variations of I with time. **7.** South. **9.** 0°; 90°. **11.** The electric force between the streams is attractive and the magnetic force is repulsive. Which force predominates depends on the speeds of the particles because the magnetic force increases with speed whereas the electric force does not change. **13.** The electron beam is deflected away from the wire. **15.** A circle, because the forces that act are repulsive and a circle is the configuration in which each part of the loop is as far as possible from the other parts. **17.** There is never a net force on the loop in such a field. There is no torque on the loop when its plane is perpendicular to the field. **19.** See Fig. A.3. **21.** A magnetic compass is more accurate near the equator, where the horizontal component of the earth's magnetic field is strongest. In the polar regions the field has only a very small horizontal component and so can exert relatively little torque on a compass needle.

Exercises

1. 80 mm. **3.** $\mathbf{B} = 0$. **5.** 1.3×10^{-5} T. **7.** 8.0 A. **9.** 0.014 T. **11.** $B/2$. **13.** 1.3×10^{-4} T. **15.** 20 A. **17.** 2.8 mm. **19.** 2.0×10^{-9} N, toward the wire. **21.** (a) 1.5×10^{-15} N, away from the wire. (b) same magnitude, toward the wire. **23.** (a) 1.5×10^{-15} N, in the direction of the current. (b) same magnitude, opposite to the direction of the current. **25.** 2.5 kV. **27.** 1.0 N; 0. **29.** 0.28 N, down; 0.28 N, up.

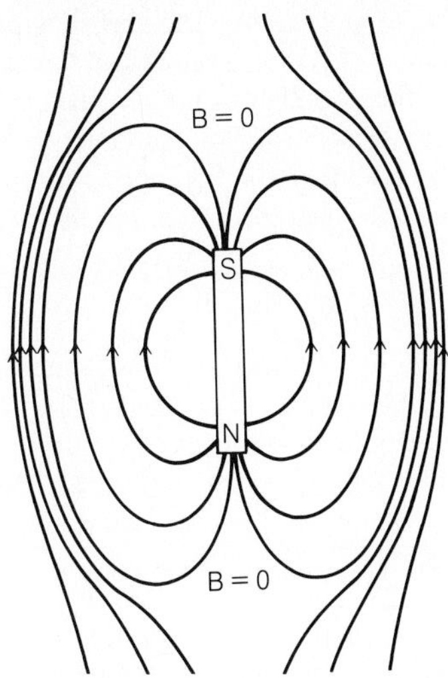

Figure A.3

Question 19, Chapter 19.

31. 2.8×10^{-4} N/m. **35.** 13 T; 2.4×10^5 T. **37.** 0.040 N.

CHAPTER 20

Questions

1. The left-hand wheels have positive charges, and the right-hand wheels have negative charges. **3.** Counterclockwise; no current; clockwise. **5.** As the diaphragm vibrates in response to sound waves, the coil moves back and forth in the field of the magnet. This induces a current in the coil whose frequency and amplitude correspond to those of the sound wave. **7.** When the generator is connected to an outside circuit, work must be done to turn its shaft as the current produced delivers energy to the circuit. **9.** The induced emf is zero when the plane of the coil is perpendicular to the magnetic field, which happens twice per rotation. Hence $T = 0.005$ s. **11.** The emf's are the same. **13.** (a) The emf decreases because the greater I is, the greater the IR drop in the armature and the less the potential difference across the generator are. Hence the current in the magnet windings drops and, as the emf is proportional to B, it too decreases. (b) A similar argument explains the increase in the emf when the current drops. **15.** The changing magnetic field produced by an alternating current in the primary winding. **17.** The air gap in the recording head permits the magnetic field of its coil to reach the tape. The air gap in the playback head permits the magnetic field of the tape to enter the iron core of its coil. **19.** (a) From Lenz's law, the self-induced emf will be opposite to the direction of the current. (b) For the same reason, the emf here will be in the same direction as that of the current.

Exercises

1. 0.60 mV. **3.** 1.8 mV. **5.** 5.0×10^{-6} Wb; 1.7 μV. **7.** 5.1×10^2 T/s. **9.** 17 mV. **11.** 1.0×10^3 T/s. **13.** 2.9 kW. **15.** 600 V; 0.60 A. **17.** 21; 42 A. **19.** 50 mH. **21.** 3.9 mH. **23.** 0.32 H. **25.** 1.4 H. **27.** 4.0 MJ; 3×10^{10} V/m. **29.** 4.09 mH; 0.286 A; 1.67×10^{-4} J. **31.** 0.25 J; 0.01 s. **33.** 0.80 MHz. **35.** 0.32/μF. **37.** 1.6 kHz; 1.0×10^{-5} C; 0.10 A. **39.** $v = (mgR \sin\theta)/(BL \cos\theta)^2$. **41.** 3.18×10^5 revolutions. **43.** 340 W, 81%; 120 A; 3.8 Ω. **45.** 280 W; 480 W; 58%. **49.** 4.17×10^3 A/s; 3.25 A; 6.25 A. **51.** 2.5×10^{-4} s; 34 μC, 0.11 A.

CHAPTER 21

Questions

1. It increases. **3.** $A/2$; $2A$. **5.** The resistance of an ac circuit determines how much of the electrical energy that passes through it is dissipated as heat. The capacitive reactance and inductive reactance, respectively, determine the extent to which the capacitors and inductors in the circuit oppose the flow of current without dissipating energy. X_L and X_C vary with frequency, whereas R does not, and their combined effects depend on $X_L - X_C$, so their analogy with resistance is very limited. The impedance determines the current in the circuit when a potential difference V of a certain frequency is applied according to $I = V/Z$; thus Z has the same role as R does in a dc circuit with respect to I, although the rate of energy dissipation is equal to I^2R in both cases. **7.** $R = R_0$; $X_L = X_{L0}/2$; $X_C = 2X_C$. In order to determine what happens to Z, the values of X_{L0} and X_{C0} must be known. **9.** The power factor of a circuit is the ratio between its resistance and impedance and governs the rate at which power is absorbed by the circuit for a given I and V; it varies with frequency. The power factor can be 0 only when $R = 0$; it is 100% at resonance, when $Z = R$. **11.** High-pass filter.

Exercises

1. 50 Hz. **3.** 14 A; 85 V; no. **5.** 71 V; 100 V; 71 V; 0. **7.** 0.31 Ω: 0.31 kΩ. **9.** 0.16 A. **11.** 6.7 A. **13.** 4.7 A. **15.** 0.60 A; X_L and X_C vary differently with frequency. **17.** 0.17 A; 8.7 W. **19.** 0.66 H. **21.** 12 μF; 0; 20 W. **23.** 17 mH, 5.0 Ω; 320 W, 125 W. **25.** 60 V. **27.** 20 Ω; 0; 400 V. **29.** 0.23 A; 3.3 W; 14 V (resistor), 4.4 V (inductor), 124 V (capacitor). **31.** 0.63 kHz. **33.** 120 V; +37°; less than f. **35.** 1.7×10^{-4} F; 800 W. **37.** 85 Ω; 1.4 A; 59%; 99 W; 0.17 kVA; 2.4 A. **39.** 1.8 A. **41.** 1.0 A (resistor), 0.80 A (inductor), 1.3 A (total); 7.8 Ω; −39°; 10 W.

43. 28 mH; 47 Ω. **45.** 63%; 28 μF; 1.6 A; 0.19 kW; 0.19 kV·A. **47.** 0.40 mH; 10.0%; 99.5%.

CHAPTER 22

Questions

1. The electric fields produced by electromagnetic induction are readily detected through the currents they produce in metal wires, but no such simple means is available for detecting the weak magnetic fields. **3.** Light waves consist of coupled electric and magnetic field fluctuations and need no material medium in which to occur. Sound waves consist of pressure fluctuations in a material medium and hence cannot occur in a vacuum. **5.** Perpendicular to each other; proportional to each other. **7.** A vertical antenna will give better reception because the direction of the electric field of the radiation is vertical; therefore the induced emf in a vertical antenna will be a maximum. **9.** No elaborate mathematics is needed to use Huygens's principle. **11.** Red; blue. **13.** All red; as a black cross on a blue background. **15.** A real image is formed by light rays that pass through it; a screen placed at the location of the image would show the image. A virtual image is formed by the backward extension of light rays that were diverted by refraction or reflection. The light rays that seem to come from it do not actually pass through a virtual image, and so it cannot appear on a screen although it can be seen by the eye. **17.** The flashes of color are the result of dispersion. In red light the flashes would be red only. **19.** The water is deeper.

Exercises

1. 68.3 m. **3.** 1.0×10^9 Hz; 30 cm; microwaves. **5.** 7×10^{-9} N. **7.** 3.3×10^{-13} T; 1.3×10^{-11} W/m^2; 4.4×10^{-20} Pa. **9.** 60 km/h. **11.** 3.00×10^8 m/s; 1.24×10^8 m/s; 1.97×10^8 m/s; 2.26×10^8 m/s. **13.** 52°. **15.** 1.51. **17.** 0.35°. **19.** 70°; 70°. **21.** 68°. **23.** 54 mm. **25.** 59 cm. **27.** 24°; 33°. **29.** Yes; no; 13°. **31.** 1.43. **33.** 48°. **35.** 1.41.

CHAPTER 23

Questions

1. No; no. **3.** Crown glass has a smaller index of refraction than benzene. The immersed lens will act as a diverging lens because wavefronts passing through it are retarded less going through its center where the path length in the medium of higher light speed is greater. Thus the wavefronts are affected as in the right-hand diagram of Fig. 23.3, even though the lens is convex. **5.** Under no circumstances; when the object is between the lens and the focal point. **7.** No; yes, if the object is placed between the lens and the focal point. **9.** The lens should be moved toward the slide. **11.** (a) Inverted and larger than the object. (b) Inverted and smaller than the object. (c) Erect and smaller than the object. (d) Erect and larger than the object. **13.** (a) No. (b) A concave mirror forms a virtual image of any object that is between the mirror and its focal point.

Exercises

1. −364 mm. **3.** 120 cm. **5.** 4.4 mm. **7.** 1.56. **9.** Converging; 443 mm. **11.** Converging; 80 mm. **13.** The image is virtual and 52 cm from the lens toward the duck. **15.** 12 cm; converging. **17.** No image is formed. **19.** 30 cm on the other side of the lens; same size; inverted; real. **21.** −556 mm. **23.** 1.3 mm. **25.** 2.0 mm. **27.** 37.5 mm. **29.** f/4. **31.** 20 mm to 18 mm. **33.** Converging; +33 cm. **35.** 6.0 m. **37.** 390; 4.1 mm. **39.** 30 mm; 2.0 m. **41.** 40 cm behind the mirror; erect; virtual; twice as large. **43.** 200 cm in front of the mirror; inverted; real; four times as large. **45.** 67 cm in front of the mirror; inverted; real; 2/3 as large. **47.** 3.6. **49.** Concave; 60.0 cm. **51.** 30 cm. **53.** 75 cm. **55.** 3.2 m. **57.** +15.0 cm; −37.5 cm. **59.** +96 mm; 432 mm. **61.** (a) $m = (25 \text{ cm}/f) + 1$. (b) 6.0. **63.** 67 cm. **65.** 30; 773 mm. **69.** 625 mm, real, inverted; 375 mm, virtual, erect.

CHAPTER 24

Questions

1. Light from the incoherent sources can interfere, but the resulting interference pattern shifts continually because there is no definite phase relationship between the beams from the sources. When the sources are coherent, there is such a relationship, and the pattern is stable and therefore discernible. **3.** The bright lines formed by a diffraction grating are narrower and brighter than those formed by a double slit. **5.** The first-order spectrum contains bright lines for which $\sin\theta = \lambda/d$, where d is the spacing of the slits, and the second-order spectrum contains bright lines for which $\sin\theta = 2\lambda/d$. The second-order spectrum is deviated by more than the first-order spectrum, and is wider as well (twice as wide, in fact). A prism produces only a single spectrum. **7.** The light waves reflected by the outer surface of the soap film are shifted in phase by $\lambda/2$, whereas those reflected by the inner surface of the film are not shifted in phase; hence destructive interference occurs in the reflected light where the film is extremely thin. **9.** The wavelengths in visible light are very small relative to the dimensions of a building, whereas those in radio waves are more nearly comparable. **11.** Higher resolving power; greater light-gathering ability and hence ability to form images of faint objects.

13. The individual waves in an ordinary light beam are polarized, but their planes of polarization are random so the beam itself is unpolarized. **15.** All; all except polarization.

Exercises

1. 4.4 mm; 8.7 mm; 11 mm. **3.** 0.80 mm. **5.** 4.3 $\times 10^{-7}$ m. **7.** 17°; 64°. **9.** Two; one to the north and one to the south. **11.** 19 cm. **13.** 2.9°. **15.** 1.2 $\times 10^{-6}$ m. **17.** 5.2 $\times 10^{-7}$ m. **19.** 46 m. **21.** 0.39°; $\theta_{moon} = 0.52°$. **23.** 16 km; 8 km. **25.** 4. **27.** 99.6 nm. **29.** 1.5. **31.** *f*/15.

CHAPTER 25

Questions

1. c, d. **3.** More conspicuous. **5.** The rod appears longest to the stationary observer. **7.** Yes, because its speed is not zero. **9.** The length is shorter, the mass and time interval are larger, the charge is unchanged. **11.** Kinetic and rest energies are always positive quantities.

Exercises

1. 2.6 $\times 10^8$ ms. **3.** 1.01 μs. **5.** 4.71 $\times 10^{-15}$ m. **7.** 2.86 $\times 10^8$ m/s; 1.05 $\times 10^{-9}$ s. **9.** 1005 g; 2294 g; 7089 g. **11.** 4.2 $\times 10^7$ m/s. **13.** 4.1 $\times 10^{-12}$ J; 2.42 $\times 10^{18}$. **15.** 135 MeV. **17.** 940 MeV. **19.** 2.00 $\times 10^{12}$ s (more than 60,000 years). **21.** $-0.14c$; $0.80c$. **23.** 8.90 $\times 10^{-28}$ kg.

CHAPTER 26

Questions

1. Such wave phenomena as diffraction and interference are easier to demonstrate than are quantum phenomena. **3.** The frequency increases and the wavelength decreases; the speed is unchanged, the energy increases, and the number per second is unchanged. **5.** No. **7.** Apply an electric or magnetic field and look for a deflection. **9.** The electron has the higher speed.

Exercises

1. 1.77 eV. **3.** 1.5 $\times 10^{28}$ photons/s. **5.** 220 nm; ultraviolet. **7.** 1.5 V. **9.** 2.4 $\times 10^{18}$ Hz; X rays. **11.** 3.3 $\times 10^{-29}$ m. **13.** 6.9 $\times 10^{-12}$ m; 2.7 $\times 10^{-12}$ m. **15.** 5.0 μV. **17.** $E_n = n^2E_1$; $E_1 = 2.1$ MeV. **19.** 1.6 $\times 10^7$ Hz. **21.** 6.2%. **23.** 4.2 $\times 10^{21}$ photons/m^2·s; 1.4 $\times 10^{13}$ photons/m^3. **25.** $h = 6.6 \times 10^{-34}$ J·s; 3.0 eV.

CHAPTER 27

Questions

1. (a) The Rutherford model indicates the division of the hydrogen atom into a nucleus and a relatively distant electron, and the Bohr model goes on from there to specify the motion and energy of the electron. (b) In order not to be pulled into the nucleus by the electric force exerted by the nucleus. **3.** (a) Continuous emission spectrum. (b) Emission line spectrum. (c) Absorption line spectrum. **5.** A hydrogen sample contains a great many atoms, each of which can undergo a variety of possible transitions. **7.** The quantum number n governs the total energy of an atomic electron; l governs the magnitude of its orbital angular momentum; m_l governs the direction of its orbital angular momentum; and m_s governs the direction of its spin angular momentum. **9.** Spin is an intrinsic property of electrons, which they always exhibit. All electrons have the same spin. **11.** When their spins are opposite. **13.** The length must be an integral number of half-wavelengths so that a standing wave is set up in the cavity that causes further emissions to be exactly in phase with the previous ones, thus producing a coherent beam.

Exercises

1. 0.85 eV. **3.** 1.05 $\times 10^5$ K. **5.** 3.7 $\times 10^{-63}$ m; 2.6 $\times 10^{74}$. **7.** 92 nm; ultraviolet. **9.** 1.46 μm. **11.** 13.06 eV. **13.** 4.2 $\times 10^{-25}$ %; 42%: Quantum phenomena are conspicuous only on a very small scale of size. **15.** $-4, -3, -2, -1, 0, 1, 2, 3, 4$. **17.** 182. **23.** 1.02 $\times 10^{14}$ Hz; infrared.

CHAPTER 28

Questions

1. The existence of such definite orbits is not consistent with the uncertainty principle. **3.** The additional attractive force of the two protons exceeds the mutual repulsion of the electrons to increase the binding energy per electron. **5.** Their ability to bond with each other as well as with other atoms. **7.** Li atoms have a single electron outside a closed shell, and this electron is held relatively weakly since the inner electrons shield most of the full nuclear charge. In fluorine, the outer shells lack an electron of completion, and the electrons in these shells are held very tightly to the nucleus since little of the nuclear charge is shielded from them. **9.** Na^+ ions have closed shells, whereas each Na atom has a single outer electron that can be detached relatively easily. **11.** Van der Waals forces are too weak to hold inert gas atoms together against the forces exerted during collisions in the gaseous state. **13.** A maximum of two electrons can exist in the first shell of an atom, and so only two H atoms can join at a time. A maximum of eight electrons can exist in the second shell of an atom, so there is no limit to the number of Li atoms that can join to form an array of atoms. As a result lithium is a metallic solid under ordinary conditions, whereas hydrogen is a diatomic gas. **15.** Only the outermost electrons in the atoms of a metal are members of its ''gas'' of free electrons. **17.** A metal. **19.** The forbidden bands in semiconductors are comparable in width to the photon energies in visible light; hence semiconductors absorb visible light and are opaque to it. The lower frequencies in infrared light mean that photon energies are too low to be absorbed; hence semiconductors are transparent to this light. **21.** A solid with a partly filled upper energy band is a conductor because electrons in this band can readily absorb small amounts of energy from an electric field and move through the solid. A solid with a filled upper band and a wide forbidden band above it is an insulator because electrons in the filled band require very large amounts of energy in order to jump the forbidden band to the empty band above it. A solid with a filled upper band and a narrow forbidden band is a semiconductor because a few electrons have enough thermal energy to enter the empty band above the forbidden band, and these electrons can move through the solid under the influence of an electric field. **23.** In an *n*-semiconductor, current is carried by electrons. In a *p*-type semiconductor, current is carried by holes, which are sites of missing electrons. **25.** (a) *p-n*, forward bias; *n-p*, reverse bias. (b) Forward bias.

CHAPTER 29

Questions

1. The isotopes have the same atomic number and hence the same electron structures; therefore they have the same chemical behavior. They have different numbers of neutrons, hence different atomic masses. **3.** The limited range of the strong interaction. **5.** Helium and radon cannot be combined chemically to form radium, nor can radium be broken down into helium and radon by chemical means. **7.** (a) Alpha and beta decays lead to the transmutation of an element into another element. (b) Radioactivity involves the conversion of matter into energy. (c) Radioactivity is a statistical process, so the decay of an individual nucleus cannot be predicted. There is no cause–effect relationship as in classical physics. (d) An alpha particle escapes through a PE barrier, not over it. **9.** The basic reason is that the $^{14}C/^{12}C$ ratio in the atmosphere has not been constant in the past. In the past century and a half, for instance, much ^{12}C has been added to the atmosphere by the burning of the fossil fuels coal and oil whose original ^{14}C content has long ago decayed. More recently, tests of nuclear weapons have increased the ^{14}C concentration. In addition, there is no reason to suppose that cosmic rays have always reached the earth at their present intensity. **11.** The radionuclide must be found in the rock, and the radionuclide's half-life must be within a few orders of magnitude of the age of the rock. **13.** If there is a leak in a water-cooled reactor, the coolant escapes and the fuel rods may then melt with potentially diastrous results. In a gas-cooled reactor, the heat capacity of the graphite moderator will reduce the temperature rise if the coolant gas stops circulating. **15.** The density in the interior of a star is extremely high, hence there are many collisions between nuclei there; the temperature is also extremely high, hence many of the colliding nuclei have high enough energies to interact.

Exercises

1. 5*n*, 5*p*; 12*n*, 10*p*; 20*n*, 16*p*; 50*n*, 38*p*; 108*n*, 72*p*. **3.** ^2_1H; ^1_0n; ^1_1H. **5.** $Z = 3$, $A = 7$; lithium. **9.** 492.3 MeV; 8.79 MeV/nucleon. **11.** 6.015 u. **13.** $^{233}_{92}U$. **15.** $^{80}_{36}Kr$; $^{80}_{34}Se$; $^{80}_{34}Se$. **17.** $^{208}_{82}Pb$. **19.** $^{222}_{86}Rn$; 222.0176 u; KE of the recoiling nucleus. **21.** 6400 yr. **23.** 5.0 days. **25.** 1.8×10^5 Bq. **27.** 9×10^{13} J; 2×10^4 tons. **29.** 7.274 MeV. **31.** 0.85 MeV; 0.76 MeV; therefore nuclear forces seem to be very nearly independent of electric charge. **33.** 94%; 2.1%. **35.** 3.0×10^{18} s (about 100 billion years).

CHAPTER 30

Questions

1. Such particles lose energy mainly by electric interactions with the electrons of atoms in their paths. Usually the electrons are pulled away from their parent atoms to produce ion-electron pairs. **3.** Photoelectric effect, Compton scattering, and pair production. **5.** Neutrinos are subject only to the weak interaction, which is both feeble and acts only over extremely short distances. **7.** No; it is possible there is a reason why quarks cannot exist except in combination with other quarks as hadrons. **9.** $+e$. **11.** The strong and weak interactions have very short ranges. The range of the electromagnetic interaction is unlimited, but the repulsive forces between charges of the same sign and the attractive forces between charges of opposite sign are both so strong that matter in bulk is always electrically neutral.

PHOTO CREDITS

Chapter 1

p. 19, AP/Wide World Photos; p. 20, AP/Wide World Photos; p. 21, AIP/Niels Bohr Library, Physics Today Collection

Chapter 2

p. 38, © David Weintraub, Photo Researchers, Inc.; p. 51, AP/Wide World Photos

Chapter 3

p. 61, AP/Wide World Photos; p. 66, AP/Wide World Photos; p. 68, AIP/Niels Bohr Library

Chapter 4

p. 92, AP/Wide World Photos; p. 103, AP/Wide World Photos; p. 108, AIP/Niels Bohr Library, E. Scott Barr Collection

Chapter 5

p. 118, Tim Davis, Photo Researchers, Inc.; p. 125, NASA

Chapter 6

p. 145, Abram G. Schoenfeld, Photo Researchers, Inc.; p. 149, Six Flags Over Georgia; p. 152, AIP/Niels Bohr Library; p. 154, NASA; p. 160, NASA; p. 167, NOAO

Chapter 7

p. 171, Black & Decker U.S. Inc., Hunt Valley, MD 21030; p. 186, AP/Wide World Photos; p. 188, Fundamental Photographs, New York

Chapter 8

p.199, Fundamental Photographs, New York; p. 214, The Bettmann Archive; p. 216, Pamela Clark Photography

Chapter 9

p. 231, Courtesy of the Geological Survey of Canada (GSC-120457); p. 243, Manchester (NH) Historical Associations

Chapter 10

p. 254, John Deere 590D hydraulic excavator. Photo courtesy Deere & Company; p. 256, © Ron Church, Photo Researchers, Inc.; p. 262, Sears/Zemansky/Young, *College Physics,* 7th Ed., © 1991, Addison-Wesley Publishing Co., Inc., Reading, MA. Fig 13-27(c). Reprinted with permission of publisher; p. 265, Takeshi Takahara, Photo Researchers, Inc.; p. 270, The Bettmann Archive

Chapter 11

p. 301, Rijksmuseum voorde Geschiedenis der Wetchschappen/AIP Niels Bohr Library

Chapter 12

p. 322 (top), U.S. Coast Guard Photograph; p. 322 (bottom), AP/Wide World Photos; p. 327 (left), AP/Wide World Photos; p. 327 (right), AP/Wide World Photos; p. 331, 1970 © Estate of Harold E. Edgerton. Courtesy of Palm Press; p. 333, © Sherry Suris, Photo Researchers, Inc.; p. 339, Palomar Observatory Photograph

Chapter 13

p. 353, Courtesy of the Triborough Bridge and Tunnel Authority; p. 358, AIP/Niels Bohr Library; p. 360, © Bruce Roberts, Photo Researchers, Inc.

Chapter 14

p. 405, © Dr. R. Clark & M. Goff, Photo Researchers, Inc.

Chapter 15

p. 427, Courtesy of Electric Power Research Institute; p. 437, University of Vienna/AIP Niels Bohr Library

Chapter 16

p. 452, Fundamental Photographs, New York; p. 457, UK Atomic Energy Authority. Courtesy AIP Niels Bohr Library

Chapter 17

p. 479, AP/Wide World Photos; p. 491, Fundamental Photographs, New York

Chapter 19

p. 546, Bill Bachman, Photo Researchers, Inc.; p. 551, The Bettmann Archive; p. 557, Transrapid International/ © 1988 Discover Publications

Chapter 20

p. 578, Pacific Gas & Electric; p. 586, AIP/Niels Bohr Library

Chapter 21

p. 605, Fundamental Photographs, New York

Chapter 22

p. 632, AIP/Niels Bohr Library; p. 640 (left), Raytheon Company; p. 640 (right), Raytheon Company; p. 643, AP/Wide World Photos; p. 659, Courtesy of AT&T Archives

Chapter 23

p. 689, Eugene Hecht, © 1987, Addison-Wesley Publishing Co., Inc., Reading, Massachusetts. Fig. 5.110. Reprinted with permission of publisher; p. 690, Ken Briggs, Photo Researchers, Inc.

Chapter 24

p. 704, by permission of DC Heath & Co.; p. 711 (left), © Science Source, Photo Researchers, Inc.; p. 711 (right), © Omikron, Photo Researchers, Inc.; p. 714, Fundamental Photographs, New York; p. 718, by permission of DC Heath & Co.; p. 725, Fundamental Photographs, New York; p. 727, AIP/Niels Bohr Library, Physics Today Collection

Chapter 25

p. 749, AIP/Niels Bohr Library

Chapter 26

p. 767, © Guy Gillette, Photo Researchers, Inc.; p. 773 (left), Rapho, Photo Researchers, Inc.; p. 773 (right), Biophoto Associates, Photo Researchers, Inc.

Chapter 27

p. 792, Jack Finch, Photo Researchers, Inc.; p. 796, Niels Bohr Institute, Courtesy of AIP/Niels Bohr Library; p. 804, Courtesy of the Lawrence Livermore National Laboratory

Chapter 28

p. 817 (left), Jerome Wyckoff; p. 817 (right), Smithsonian Institution Photo No. 312-A; p. 820 (left), © R.B. Hoit, Photo Researchers, Inc.; p. 820 (right), © Carl Zeiss, Photo Researchers, Inc.; p. 828, Courtesy of AT&T Archives

Chapter 29

p. 843, © CNRI/Science Photo Library, Photo Researchers, Inc.; p. 858, © Nancy Pierce, Photo Researchers, Inc.; p. 862, Princeton University; p. 844, AIP Niels Bohr Library, W.F. Meggers Collection

Chapter 30

p. 873, Lawrence Berkeley Laboratory, University of California; p. 878 (left), Fermilab Visual Media Services; p. 878 (right), Fermilab Visual Media Services; p. 884, Courtesy of AT&T Archives

INDEX

CONVERSION FACTORS

TIME

1 hour = 60 min = 3600 s
1 day = 1440 min = 8.64×10^4 s
1 year = 365.2 days = 3.156×10^7 s

ANGLE

1 radian (rad) = 57.30° = 57°18′
1° = 0.01745 rad (180° = π rad)
1 rev/min (rpm) = 0.1047 rad/s
1 rad/s = 9.549 rpm

LENGTH

1 meter (m) = 100 cm = 39.37 in. = 3.281 ft
1 centimeter (cm) = 10 millimeters (mm)
= 0.3937 in.
1 kilometer (km) = 1000 m = 0.6214 mi
1 foot (ft) = 12 in. = 0.3048 m = 30.48 cm
1 inch (in.) = 2.540 cm
1 mile (mi) = 5280 ft = 1.609 km
1 nautical mile (nmi) = 6076 ft. = 1.152 mi
= 1.852 km

AREA

1 m^2 = 10^4 cm^2 = 10.76 ft^2
1 cm^2 = 10^{-4} m^2 = 0.1550 in.^{2}
1 ft^2 = 144 in.^{2} = 9.290×10^{-2} m^2 = 929.0 cm^2
1 in.^{2} = 6.452 cm^2
1 hectare (ha) = 10^4 m^2 = 2.471 acres
1 acre = 43,560 ft^2 = 0.4049 ha

VOLUME

1 m^3 = 10^3 liters = 10^6 cm^3 = 35.32 ft^3
1 liter = 10^3 cm^3 = 10^{-3} m^3 = 0.2642 gal
= 1.056 quart
1 ft^3 = 1728 in.^{3} = 2.832×10^{-2} m^3 = 28.32 liters
= 7.481 gal
1 U.S. gallon (gal) = 4 quarts = 0.1337 ft^3
= 3.785 liters

SPEED

1 m/s = 3.281 ft/s = 2.237 mi/h = 3.600 km/h
1 ft/s = 0.3048 m/s = 0.6818 mi/h = 1.097 km/h
1 km/h = 0.2778 m/s = 0.9113 ft/s
= 0.6214 mi/h
1 mi/h = 1.467 ft/s = 0.4470 m/s = 1.609 km/h
1 knot = 1 nmi/h = 1.152 mi/h = 1.852 km/h
= 1.688 ft/s = 0.5144 m/s

MASS

1 kilogram (kg) = 1000 grams (g) = 0.0685 slug
(*Note:* 1 kg corresponds to 2.21 lb in the sense that the *weight* of 1 kg is 2.21 lb at sea level. Similarly 1 lb corresponds to 453.6 g and 1 oz to 28.35 g.)
1 slug = 14.59 kg
1 atomic mass unit (u) = 1.660×10^{-27} kg
= 1.492×10^{-10} J = 931.5 MeV

FORCE

1 newton (N) = 0.2248 lb
1 pound (lb) = 4.448 N

PRESSURE

1 pascal (Pa) = 1 N/m^2 = 1.450×10^{-4} lb/in.^{2}
1 bar = 10^5 Pa = 14.50 lb/in.^{2}
1 lb/in.^{2} = 144 lb/ft^2 = 6.895×10^3 Pa
1 atm = 1.013×10^5 Pa = 1.013 bar
= 14.70 lb/in.^{2}
1 torr = 133.3 Pa

ENERGY

1 joule (J) = 0.7376 ft · lb = 2.390×10^{-4} kcal
= 9.484×10^{-4} Btu
= 2.778×10^{-7} kWh
1 foot-pound (ft · lb) = 1.356 J = 1.29×10^{-3} Btu
= 3.25×10^{-4} kcal
1 kilocalorie (kcal) = 4185 J = 3.968 Btu
= 3077 ft · lb
1 Btu = 0.252 kcal = 778 ft · lb = 1054 J
1 electron volt (eV) = 10^{-6} MeV = 10^{-9} GeV
= 1.602×10^{-19} J
1 kilowatt-hour (kWh) = 3.600×10^6 J
= 2.655×10^6 ft · lb
= 860.4 kcal